Chapter 4 Graphing Linear Equations

Linear equation in two variables: $ax + by = c$

A **graph** is an illustration of the set of points whose coordinates satisfy the equation.

Every **linear equation** of the form $ax + by = c$ will be a straight line when graphed.

To find the *y*-intercept (where the graph crosses the *y*-axis) set $x = 0$ and solve for y.

To find the *x*-intercept (where the graph crosses the *x*-axis) set $y = 0$ and solve for x.

$$\text{slope } (m) = \frac{\text{change in } y}{\text{change in } x} = \frac{y_2 - y_1}{x_2 - x_1}$$

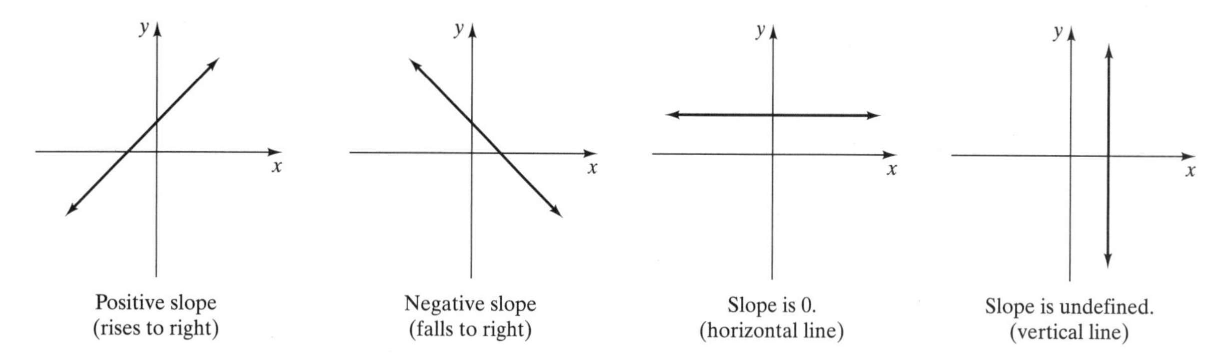

Linear Equations

Standard form of a linear equation: $ax + by = c$

Slope–intercept form of a linear equation: $y = mx + b$, where m is the slope and $(0, b)$ is the y-intercept.

Point–slope form of a linear equation: $y - y_1 = m(x - x_1)$, where m is the slope and (x_1, y_1) is a point on the line.

A **relation** is any set of ordered pairs.

A **function** is a set of ordered pairs in which each first component corresponds to exactly one second component.

Chapter 5 Systems of Linear Equations

The **solution** to a system of linear equations is the ordered pair or pairs that satisfy all equations in the system. A system of linear equations may have no solution, exactly one solution, or an infinite number of solutions.

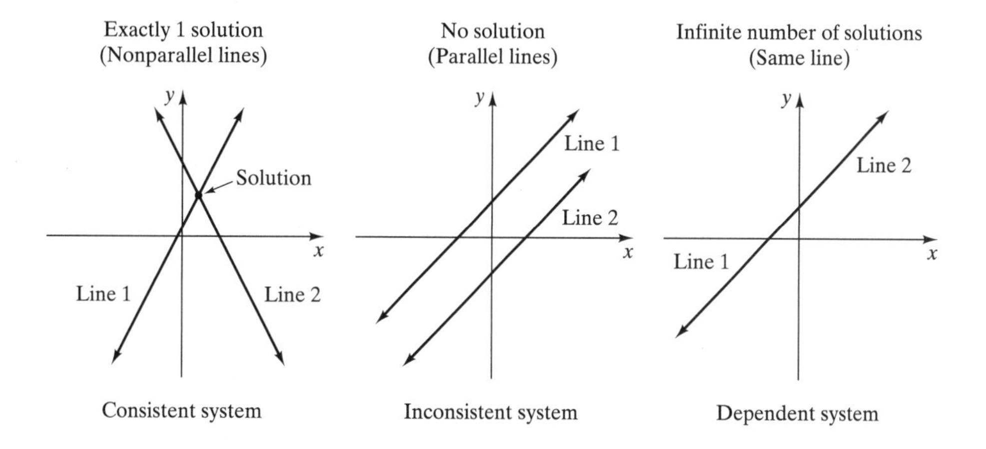

A system of linear equations may be solved graphically, or algebraically by the substitution method or by the addition (or elimination) method.

Chapter 6 Exponents and Polynomials

Rules of Exponents

1. $x^m \cdot x^n = x^{m+n}$	**product rule**
2. $\dfrac{x^m}{x^n} = x^{m-n}, x \neq 0$	**quotient rule**
3. $(x^m)^n = x^{m \cdot n}$	**power rule**
4. $x^0 = 1, x \neq 0$	**zero exponent rule**
5. $x^{-m} = \dfrac{1}{x^m}, x \neq 0$	**negative exponent rule**
6. $\left(\dfrac{ax}{by}\right)^m = \dfrac{a^m x^m}{b^m y^m}, b \neq 0, y \neq 0$	**expanded power rule**

FOIL method (*F*irst, *O*uter, *I*nner, *L*ast) of multiplying binomials:
$(a + b)(c + d) = ac + ad + bc + bd$

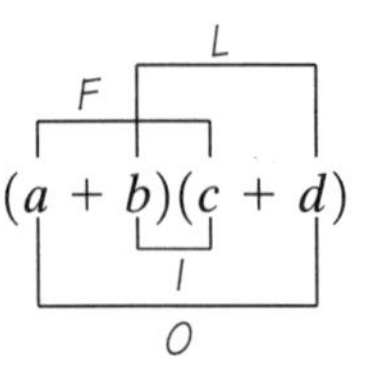

Product of the sum and difference of the same two terms:
$(a + b)(a - b) = a^2 - b^2$

Squares of binomials: $(a + b)^2 = a^2 + 2ab + b^2$
$(a - b)^2 = a^2 - 2ab + b^2$

Chapter 7 Factoring

If $a \cdot b = c$, then a and b are **factors** of c.
Difference of two squares: $a^2 - b^2 = (a + b)(a - b)$
Sum of two cubes: $a^3 + b^3 = (a + b)(a^2 - ab + b^2)$
Difference of two cubes: $a^3 - b^3 = (a - b)(a^2 + ab + b^2)$

To Factor a Polynomial

1. If all the terms of the polynomial have a greatest common factor other than 1, factor it out.
2. If the polynomial has two terms, determine if it is a difference of two squares or a sum or a difference of two cubes. If so, factor using the appropriate formula.
3. If the polynomial has three terms, factor the trinomial using one of the procedures discussed.
4. If the polynomial has more than three terms, try factoring by grouping.
5. As a final step, examine your factored polynomial to see if the terms in any factors listed have a common factor. If you find a common factor, factor it out at this point.

Quadratic equation: $ax^2 + bx + c = 0, a \neq 0$.

Zero-factor Property: If $ab = 0$, then $a = b$ or $b = 0$.

To Solve a Quadratic Equation by Factoring

1. Write the equation in standard form with the squared term positive. This will result in one side of the equation being 0.
2. Factor the side of the equation that is not 0.
3. Set each factor containing a variable equal zero and solve each equation.
4. Check the solution found in step 3 in the original equation.

Chapter 8 Rational Expressions and Equations

To Simplify Rational Expressions

1. Factor both the numerator and denominator as completely as possible.
2. Divide out any factors common to both the numerator and denominator.

To Multiply Rational Expressions

1. Factor all numerators and denominators completely.
2. Divide out common factors.
3. Multiply the numerators together and multiply the denominators together.

To Add or Subtract Two Rational Expressions

1. Determine the least common denominator (LCD).
2. Rewrite each fraction as an equivalent fraction with the LCD.
3. Add or subtract numerators while maintaining the LCD.
4. When possible, factor the remaining numerator and simplify the fraction.

To Solve Rational Expressions

1. Determine the LCD of all fractions in the equation.
2. Multiply both sides of the equation by the LCD. This will result in every term in the equation being multiplied by the LCD.
3. Remove any parentheses and combine like terms on each side of the equation.
4. Solve the equation.
5. Check your solution in the original equation.

Summary continues on back end sheets.

ELEMENTARY ALGEBRA
FOR COLLEGE STUDENTS: EARLY GRAPHING

ALLEN R. ANGEL
Monroe Community College

with assistance from

DONNA R. PETRIE
Monroe Community College

RICHARD SEMMLER
Northern Virginia Community College

PRENTICE HALL
Upper Saddle River, New Jersey 07458

Library of Congress Cataloging-in-Publication Data

Angel, Allen R.
Elementary algebra for college students: early graphing / Allen R. Angel.
p. cm.
Includes index.
ISBN 0-13-011645-9 (alk. paper)
1. Algebra. I. Title.
QA152.2.A54 2000
512.9—dc21 99-39265
CIP

Executive Editor: Karin E. Wagner
Editor-in-Chief: Jerome Grant
Editor-in-Chief, Development: Carol Trueheart
Senior Managing Editor: Linda Mihatov Behrens
Executive Managing Editor: Kathleen Schiaparelli
Assistant Vice President of Production and Manufacturing: David W. Riccardi
Marketing Manager: Eilish Main
Marketing Assistant: Amy Lysik/Vince Jansen
Manufacturing Buyer: Alan Fischer
Manufacturing Manager: Trudy Pisciotti
Editorial Assistant/Supplements Editor: Kate Marks
Associate Editor, Mathematics/Statistics Media: Audra J. Walsh
Art Director: Maureen Eide
Cover Designer: Joseph Sengotta
Interior Designer: Lorraine Castellano
Associate Creative Director: Amy Rosen
Director of Creative Services: Paula Maylahn/Paul Belfanti
Assistant to Art Director: John Christiana
Art Manager: Gus Vibal
Art Editor: Grace Hazeldine
Cover Image: James H. Carmichael, Jr. / The Image Bank/Grammatus Welk
"Ancistrolepis Grammatus," Yessno, Japan
Project Management: Elm Street Publishing Services, Inc.
Photo Researcher: Diana Gongora
Photo Research Administrator: Beth Boyd
Art Studio: Scientific Illustrators
Composition: Prepare, Inc./Emilcomp srl

Printed in the United States of America

10 9 8 7 6 5 4 3 2 1

ISBN 0-13-011645-9

Prentice-Hall International (UK) Limited, *London*
Prentice-Hall of Australia Pty. Limited, *Sydney*
Prentice-Hall Canada, Inc., *Toronto*
Prentice-Hall Hispanoamericana, S.A., *Mexico*
Prentice-Hall of India Private Limited, *New Delhi*
Prentice-Hall of Japan, Inc., *Tokyo*
Prentice-Hall Pte. Ltd., *Singapore*
Editora Prentice-Hall do Brasil, Ltda., *Rio de Janeiro*

To my wife, Kathy,

and my sons, Robert and Steven

Contents

3) Formulas and Applications of Algebra 161

4) Graphing Linear Equations 229

5) Systems of Linear Equations 293

6) Exponents and Polynomials 341

7) Factoring 395

8) Rational Expressions and Equations 453

Preface

This book was written for college students and other adults who have never been exposed to algebra or those who have been exposed but need a refresher course. My primary goal was to write a book that students can read, understand, and enjoy. To achieve this goal I have used short sentences, clear explanations, and many detailed worked-out examples. I have tried to make the book relevant to college students by using practical applications of algebra throughout the text.

Features of the Text

Four-Color Format Color is used pedagogically in the following ways:

- Important definitions and procedures are color screened.
- Color screening or color type is used to make other important items stand out.
- Artwork is enhanced and clarified with use of multiple colors.
- The four-color format allows for easy identification of important features by students.
- The four-color format makes the text more appealing and interesting to students.

Readability One of the most important features of the text is its readability. The book is very readable, even for those with weak reading skills. Short, clear sentences are used and more easily recognized, and easy-to-understand language is used whenever possible.

Accuracy Accuracy in a mathematics text is essential. To ensure accuracy in this book, mathematicians from around the country have read the pages carefully for typographical errors and have checked all the answers.

Connections Many of our students do not thoroughly grasp new concepts the first time they are presented. In this text we encourage students to make connections. That is, we introduce a concept, then later in the text briefly reintroduce it and build upon it. Often an important concept is used in many sections of the text. Students are reminded where the material was seen before, or where it will be used again. This also serves to emphasize the importance of the concept. Important concepts are also reinforced throughout the text in the Cumulative Review Exercises and Cumulative Review Tests.

Chapter Opening Application Each chapter begins with a real-life application related to the material covered in the chapter. By the time students complete the chapter, they should have the knowledge to work the problem.

Preview and Perspective This feature at the beginning of each chapter explains to the students why they are studying the material and where this material will be used again in other chapters of the book. This material helps students see the connections between various topics in the book and the connection to real-world situations.

Student's Solution Manual, Videotape, and Software Icons At the beginning of each section, *Student's Solution Manual*, videotape, and tutorial software icons are displayed. These icons tell the student that material in the section can be found in the Student's Solution Manual, on the videotapes, and in the tutorial software. Small videotape icons are also placed next to exercises that are worked out on the videotapes.

Keyed Section Objectives Each section opens with a list of skills that the student should learn in that section. The objectives are then keyed to the appropriate portions of the sections with symbols such as 1.

Problem Solving Polya's five-step problem-solving procedure is discussed in Section 1.2. Throughout the book problem solving and Polya's problem-solving procedure are emphasized.

Practical Applications Practical applications of algebra are stressed throughout the text. Students need to learn how to translate application problems into algebraic symbols. The problem-solving approach used throughout this text gives students ample practice in setting up and solving application problems. The use of practical applications motivates students.

Detailed Worked-Out Examples A wealth of examples have been worked out in a step-by-step, detailed manner. Important steps are highlighted in color, and no steps are omitted until after the student has seen a sufficient number of similar examples.

Now Try Exercise In each section, students are asked to work exercises that parallel the examples given in the text. These Now Try Exercises make the students *active*, rather than passive, learners and they reinforce the concepts as students work the exercises.

Study Skills Section Many students taking this course have poor study skills in mathematics. Section 1.1, the first section of this text, discusses the study skills needed to be successful in mathematics. This section should be very beneficial for your students and should help them to achieve success in mathematics.

Helpful Hints The helpful hint boxes offer useful suggestions for problem solving and other varied topics. They are set off in a special manner so that students will be sure to read them.

Avoiding Common Errors Errors that students often make are illustrated. The reasons why certain procedures are wrong are explained, and the correct procedure for working the problem is illustrated. These Avoiding Common Errors boxes will help prevent your students from making those errors we see so often.

Using Your Calculator The Using Your Calculator boxes, placed at appropriate intervals in the text, are written to reinforce the algebraic topics presented in the section and to give the student pertinent information on using the calculator to solve algebraic problems.

Using Your Graphing Calculator Using Your Graphing Calculator boxes are placed at appropriate locations throughout the text. They reinforce the algebraic topics taught and sometimes offer alternate methods of working problems. This book is designed to give the instructor the option of using or not using a graphing calculator in their course. Many Using Your Graphing Calculator boxes contain graphing calculator exercises, whose answers appear in the answer section of the book. The illustrations shown in the Using Your Graphing Calculator boxes are from a Texas Instrument 83 calculator. The Using Your Graphing Calculator boxes are written assuming that the student has no prior graphing calculator experience.

Exercise Sets

The exercise sets are broken into three main categories: Concept/Writing Exercises, Practice the Skills, and Problem Solving. Many exercise sets also contain Challenge Problems and/or Group Activities. Each exercise set is graded in difficulty. The early problems help develop the student's confidence, and then students are eased gradually into the more difficult problems. A sufficient number and variety of examples are given in each section for the student to successfully complete even the more difficult exercises. The number of exercises in each section is more than ample for student assignments and practice.

Concept/Writing Exercises Most exercise sets include exercises that require students to write out the answers in words. These exercises improve students' understanding and comprehension of the material. Many of these exercises involve problem solving and conceptualization and help develop better reasoning and critical thinking skills. Writing exercises are indicated by the symbol ✎.

Challenge Problems These exercises, which are part of many exercise sets, provide a variety of problems. Many were written to stimulate student thinking. Others provide additional applications of algebra or present material from future sections of the book so that students can see and learn the material on their own before it is covered in class. Others are more challenging than those in the regular exercise set.

Problem Solving Exercises These exercises have been added to help students become better thinkers and problem solvers. Many of these exercises involve real life applications of algebra.

Cumulative Review Exercises All exercise sets (after the first two) contain questions from previous sections in the chapter and from previous chapters. These cumulative review exercises will reinforce topics that were previously covered and help students retain the earlier material, while they are learning the new material. For the students' benefit the Cumulative Review Exercises are keyed to the section where the material is covered.

Group Activities Many exercise sets have group activity exercises that lead to interesting group discussions. Many students learn well in a cooperative learning atmosphere, and these exercises will get students talking mathematics to one another.

Chapter Summary At the end of each chapter is a chapter summary which includes a glossary and important chapter facts.

Review Exercises At the end of each chapter are review exercises that cover all types of exercises presented in the chapter. The review exercises are keyed to the sections where the material was first introduced.

Practice Tests The comprehensive end-of-chapter practice test will enable the students to see how well they are prepared for the actual class test. The Test Item File includes several forms of each chapter test that are similar to the student's practice test. Multiple choice tests are also included in the Test Item File.

Cumulative Review Test These tests, which appear at the end of each chapter, test the students' knowledge of material from the beginning of the book to the end of that chapter. Students can use these tests for review, as well as for preparation for the final exam. These exams, like the cumulative review exercises, will serve to reinforce topics taught earlier.

Answers The *odd answers* are provided for the exercise sets. *All answers* are provided for the Using Your Graphing Calculator Exercises, Cumulative Review Exercises, the Review Exercises, Practice Tests, and the Cumulative Practice Test. *Answers* are not provided for the Group Activity exercises since we want students to reach agreement by themselves on the answers to these exercises.

National Standards

Recommendations of the *Curriculum and Evaluation Standards for School Mathematics*, prepared by the National Council of Teachers of Mathematics, (NCTM) and *Crossroads in Mathematics: Standards for Introductory College Mathematics Before Calculus*, prepared by the American Mathematical Association of Two Year Colleges (AMATYC) are incorporated into this edition.

Prerequisite

This text assumes no prior knowledge of algebra. However, a working knowledge of arithmetic skills is important. Fractions are reviewed early in the text, and decimals are reviewed in Appendix A.

Modes of Instruction

The format and readability of this book lends itself to many different modes of instruction. The constant reinforcement of concepts will result in greater understanding and retention of the material by your students.

The features of the text and the large variety of supplements available make this text suitable for many types of instructional modes including:

- lecture
- self-paced instruction
- modified lecture
- cooperative or group study
- learning laboratory

Changes Between this Text and the Fourth Edition of Elementary Algebra for College Students

When I wrote this book I considered the many letters and reviews I got from students and faculty alike. I would like to thank the many instructors and students who wrote to inform me of how much they enjoyed and appreciated the text.

Some of the changes made in this text include:

- The Using Your Graphing Calculator boxes are designed so that instructors have the opportunity of using, or not using, a graphing calculator with this book. The Using Your Graphing Calculator boxes are written with the assumption that students have no prior knowledge of how to use a graphing calculator. No new algebraic information is introduced in these calculator boxes.
- Functions, now covered in the last section of Chapter 4, are treated more intuitively.
- Real-life chapter-opening applications have been added to each chapter.

- The Exercise sets have been rewritten and reorganized. They now start with Concept/Writing Exercises, followed by Practice the Skills Exercises, followed by Problem Solving Exercises.
- Problem Solving and George Polya's problem-solving procedure are stressed throughout the book. Problem-solving examples are worked using the following steps: Understand, Translate, Carry out, Check, and Answer. Problem solving is introduced in Section 1.2.
- Cumulative Review Tests are now at the end of every chapter.
- The Challenge Problems/Group Activity exercises have been broken up into separate categories.
- The Exercise Sets have a great variety of exercises. The Exercise Sets are graded in level of difficulty.
- Although the emphasis of this book remains on students mastering the basic skills, many problem-solving and thought-provoking exercises appear in the exercise sets for those instructors who wish to assign them.
- This text introduces graphing early (Chapter 4) for those instructors that wish to teach this topic earlier in the course.
- Systems of Equations are now covered in Chapter 5.
- The statistical topics of mean, median, and mode are discussed in Chapter 1 and students learn the meaning of, and how to find, these measures of central tendency.
- Circle (or pie), line, and bar graphs are now introduced in Section 1.2. They are used throughout the book. This provides ample opportunity for your students to learn to read and understand these graphs that they see daily.
- The sections on Exponents and Order of Operations have been combined into one section for clarity and continuity. Exponents are discussed in detail, when the rules of exponents are given, in the chapter on Polynomials.
- Motion and mixture problems have been moved to Chapter 3 so that these problems are discussed along with the other applications of algebra.
- The Using Your Calculator and Using Your Graphing Calculator boxes are colored differently for easy identification.
- The Common Student Error boxes have been renamed as Avoiding Common Errors boxes.
- The Exercise Sets have many more real-life applications.
- A more colorful and appealing design results in distinct features being more recognizable. The exciting design also results in students being more willing to read the text.
- Graphing AIE answers were moved to an appendix in the back of the text. This results in the students' text not having large blocks of empty space.
- Definitions are given in Definition Boxes and Procedures are given in Procedure Boxes.
- In the AIE, Teaching Tips provide ideas for exploration.
- The Practice Tests and Cumulative Review Tests have been made uniform. All Practice Tests now have 25 problems, and all Cumulative Review Tests now have 20 problems.
- Cumulative Review Tests are included at the end of each chapter.
- Now Try Exercises have been added in each section after many examples. Students are asked to work specific exercises after they read specific examples. Working these exercises reinforces what the student has just learned, and also serves to make students active, rather than passive, learners. The Now Try Exercises are marked in green in the Exercise Sets for easy identification by the student.
- Exercises that are worked on the videotapes are indicated by an icon next to the exercises.
- There are more writing exercises, that is, exercises that require a written answer. Writing exercises are indicated with a pencil icon.

Supplements

For this edition of the book the author has personally coordinated the development of the *Student's Solution Manual* and the *Instructor's Solution Manual*. Experienced mathematics professors who have prior experience in writing supplements, and whose works have been of superior quality, have been carefully selected for authoring the supplements.

For Instructors

Printed Supplements

Annotated Instructor's Edition

- Contains all of the content found in the student edition.
- Answers to all exercises are printed on the same

text page (graphed answers are in a special graphing answer section at the back of the text).
- Teaching Tips throughout the text are placed at key points in the margin.

Instructor's Solutions Manual

- Solutions to even-numbered section exercises.
- Solutions to every (even and odd) exercise found in the Chapter Reviews, Chapter Tests, and Cumulative Review Tests.

Instructor's Test Manual

- Two free-response Pretests per chapter.
- Eight Chapter Tests per chapter (3 multiple choice, 5 free response).
- Two Cumulative Review Tests (one multiple choice, one free response) every two chapters.
- Eight Final Exams (3 multiple choice, 5 free response).
- Twenty additional exercises per section for added test exercises if needed.

Media Supplements

TestPro4 Computerized Testing

- Algorithmically driven, text-specific testing program.
- Networkable for administering tests and capturing grades on-line.
- Edit and add your own questions—create nearly unlimited number of tests and drill worksheets.

Companion Web site

- www.prenhall.com/angel
- Links related to the chapter openers at the beginning of each chapter allow students to explore related topics and collect data needed in order to complete application problems.
- Additional links to helpful, generic sites include Fun Math and For Additional Help.
- Syllabus builder management program allows instructor to post course syllabus information and schedule on the Web site.

For Students

Printed Supplements

Student Solutions Manual

- Solutions to odd-numbered section exercises.
- Solutions to every (even and odd) exercise found in the Chapter Reviews, Chapter Tests, and Cumulative Review Tests.

Student Study Guide

- Includes additional worked-out examples, additional exercises, practice tests and answers.
- Includes information to help students study and succeed in mathematics class.
- Emphasizes important concepts.

New York Times *Themes of the Times*

- Contact your local Prentice Hall sales representative.

How to Study Mathematics

- Contact your local Prentice Hall sales representative.

Internet Guide

- Contact your local Prentice Hall sales representative.

Media Supplements

MathPro4 Computerized Tutorial

- Keyed to each section of the text for text specific tutorial exercises and instruction.
- Includes Warm up exercises and graded Practice Problems.
- Includes video Watch screens.
- Take chapter quizzes.
- Send and receive e-mail from and to your instructor.
- Algorithmically driven and fully networkable.

Videotape Series

- Keyed to each section of the text.
- Step by step solutions to exercises from each section of the text. Exercises from the text that are worked in the videos are marked with a video icon.

Companion Web site

- www.prenhall.com/angel
- Links related to the chapter openers at the beginning of each chapter allow students to explore related topics and collect data needed in order to complete application problems.
- Additional links to helpful, generic sites include Fun Math and For Additional Help.
- Syllabus builder management program allows instructor to post course syllabus information and schedule on the Web site.

Acknowledgments

Writing a textbook is a long and time-consuming project. Many people deserve thanks for encouraging and assisting me with this project. Most importantly I would like to thank my wife Kathy, and sons, Robert and Steven. Without their constant encouragement and understanding, this project would not have become a reality.

I would like to thank Richard Semmler of Northern Virginia Community College, Larry Clar and Donna Petrie of Monroe Community College, and Cindy Trimble and Teri Lovelace of Laurel Technical Services for their conscientiousness and the attention to details they provided in checking pages, artwork, and answers.

I would like to thank my editor at Prentice Hall, Karin Wagner, my developmental editor Don Gecewicz, and my production editor Ingrid Mount for their many valuable suggestions and conscientiousness with this project.

I would like to thank the following reviewers and proofreaders for their thoughtful comments and suggestions.

Helen Banes, *Kirkwood Community College*
Jon Becker, *Indiana University Northwest*
Paul Boisvert, *Robert Morris College*
Holly Broesamle, *Oakland Community College*
Charlotte Buffington, *New Hampshire Community Technical College—Stratham*
Connie Buller, *Metropolitan Community College—Omaha*
Gerald Busald, *San Antonio College*
Joan Capps, *Raritan Valley Community College*
Larry Clar, *Monroe Community College*
Elizabeth Condon, *Owens Community College*
Ann Corbeil, *Massasoit Community College*
John DeCoursey, *Vincennes University*
Abdollah Hajikandi, *Buffalo State College*
Cheryl Hobneck, *Illinois Valley Community College*
Bruce Hoelter, *Raritan Valley Community College*
Gisele Icore, *Baltimore City Community College*
Patricia Lanz, *Erie Community College—South*
David Lunsford, *Grossmont College*
Christopher McNally, *Tallahassee Community College*
Chuck Miller, *Albuquerque TVI*
Linda Mudge, *Illinois Valley Community College*
Elsie Newman, *Owens Community College*
Katherine Nickell, *College of DuPage*
Dorothy Pennington, *Tallahassee Community College*
Donna Petrie, *Monroe Community College*
Thomas Pomykalski, *Metropolitan Community College—Omaha*
Shawn Robinson, *Valencia Community College*
Robert Secrist, *Kellogg Community College*
Richard Semmler, *Northern Virginia Community College*

EMPHASIS ON *Problem Solving*

The Angel series places a stronger emphasis on problem solving than ever before. Problem solving is now introduced early and incorporated as a theme throughout the texts.

Five-Step Problem-Solving Procedure

The in-text examples demonstrate how to solve each exercise based on Polya's five-step problem-solving procedure: **Understand, Translate, Carry Out, Check,** and **State Answer.**

1.2 PROBLEM SOLVING

1) Learn the problem-solving procedure.
2) Solve problems involving bar, line, and circle graphs.
3) Solve problems involving statistics.

1) Learn the Problem-Solving Procedure

George Polya

One of the main reasons we study mathematics is that we can use it to solve many real-life problems. To solve most real-life problems mathematically, we need to be able to express the problem in mathematical symbols. This is an important part of the problem-solving procedure that we will present shortly. Throughout the book, we will be problem solving. In Chapter 3, we will also spend a great deal of time explaining how to express real-life applications mathematically.

We will now give the general five-step **problem-solving procedure** that was developed by George Polya and presented in his book *How to Solve It*. You can approach any problem by following this general procedure.

Guidelines for Problem Solving

1. **Understand the problem.**
   - Read the problem *carefully* at least twice. In the first reading, get a general overview of the problem. In the second reading, determine **(a)** exactly what you are being asked to find and **(b)** what information the problem provides.
   - Make a list of the given facts. Determine which are pertinent to solving the problem.
   - Determine whether you can substitute smaller or simpler numbers to make the problem more understandable.
   - If it will help you organize the information, list the information in a table.
   - If possible, make a sketch to illustrate the problem...abel the inf...

Problem-Solving Exercises

These exercises are designed to help students become better thinkers.

Problem Solving

69. What is the value of a if the graph of $ax + 4y = 8$ is to have an x-intercept of $(2, 0)$?

70. What is the value of a if the graph of $ax + 8y = 12$ is to have an x-intercept of $(3, 0)$?

71. What is the value of b if the graph of $3x + by = 10$ is to have a y-intercept of $(0, 5)$?

72. What is the value of b if the graph of $4x + by = 12$ is to have a y-intercept of $(0, -3)$?

The bar graphs in Exercises 73 and 74 display information. State whether the graph displays a linear relationship. Explain your answer.

73. **Calories Burned by Average 150–Pound Person Walking at 4.5 mph**

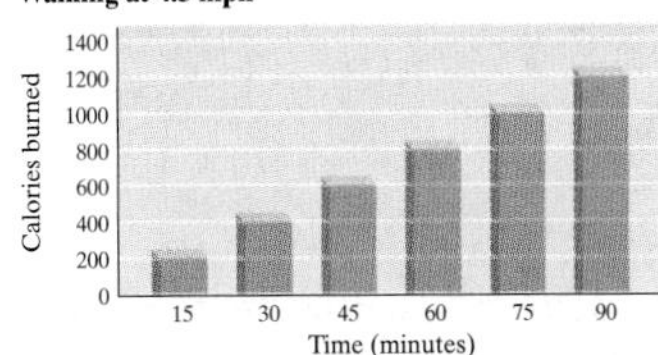

74. **Major League Baseball Attendance**

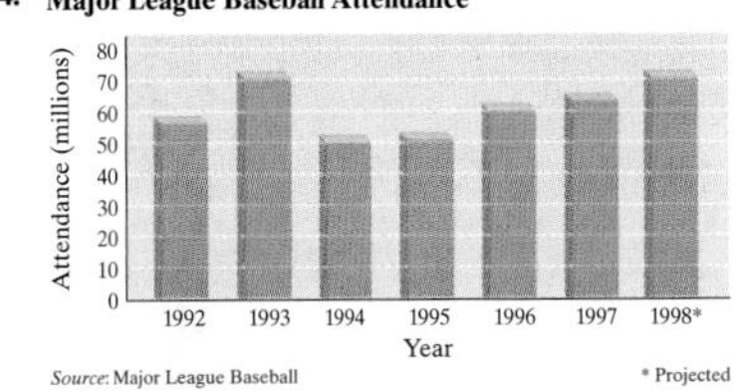

EMPHASIS ON *Applications*

Each chapter begins with an illustrated, real-world application to motivate students and encourage them to see algebra as an important part of their daily lives. Problems based on real data from a broad range of subjects appear throughout the text, in the end-of-chapter material, and in the exercise sets.

SOLVING LINEAR EQUATIONS AND INEQUALITIES

CHAPTER

Use the Angel Web site at www.prenhall.com/angel to be linked to an internet resource that will help you further explore the following application.

When trying to predict whether an athlete will set a new record, journalists and coaches often compare the athlete's current pace and performance with the pace set by the record holder at different times during a record-breaking season, On page 140, we use proportions to determine how many home runs during the first 48 games of a baseball season a player would need to hit to be on schedule to break Mark McGwire's record of 70 home runs from the 1998 baseball season.

Chapter-Opening Applications

New **chapter-opening applications** emphasize the use of mathematics in everyday life, and in the workplace giving students an applied, real-world introduction to the chapter material. The applications are often tied to examples presented in the section. The chapter openers direct students to the Angel Web site where additional information related to the chapter opening application may be found.

Real-World Applications

An abundance of wonderfully updated, **real-world applications** gives students needed practice with practical applications of algebra. Real data is used, and real-world situations emphasize the relevance of the material being covered to students' everyday lives.

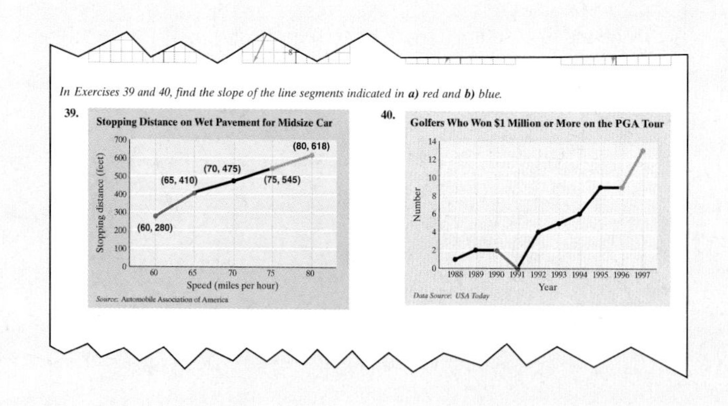

EMPHASIS ON *Exercises*

End-of-section exercise sets provide a thorough review of the section material. Each exercise set progresses in difficulty to help students gain confidence and succeed with more difficult exercises.

Practice the Skills

Determine which of the relations are also functions. Give the domain and range of each relation or function.

9. $\{(5,4),(2,2),(3,5),(1,3),(4,1)\}$

10. $\{(2,1),(4,0),(3,5),(2,2),(5,1)\}$

11. $\{(6,-2),(3,0),(1,2),(1,4),(2,4),(7,5)\}$

12. $\{(-2,1),(1,-3),(3,4),(4,5),(-2,0)\}$

13. $\{(5,0),(3,-4),(0,-1),(3,2),(1,1)\}$

14. $\{(-2,3),(-3,4),(0,3),(5,2),(3,5),(2,5)\}$

15. $\{(3,0),(0,-3),(1,5),(1,0),(1,2)\}$

16. $\{(4,-3),(3,-7),(4,-9),(3,5)\}$

17. $\{(0,3),(1,3),(2,3),(3,3),(4,3)\}$

18. $\{(3,5),(2,4),(1,0),(0,1),(-1,5)\}$

Practice the Skills Exercises

Practice the Skills exercises cover all types of exercises presented in the chapter.

Problem-Solving Exercises

These exercises are designed to help students become better thinkers.

Problem Solving

51. If a relation consists of six ordered pairs and the domain of the relation consists of five values of x, can the relation be a function? Explain.

52. If a relation consists of six ordered pairs and the range of the relation consists of five values of y, can the relation be a function? Explain.

Are the following graphs functions? Explain.

53.

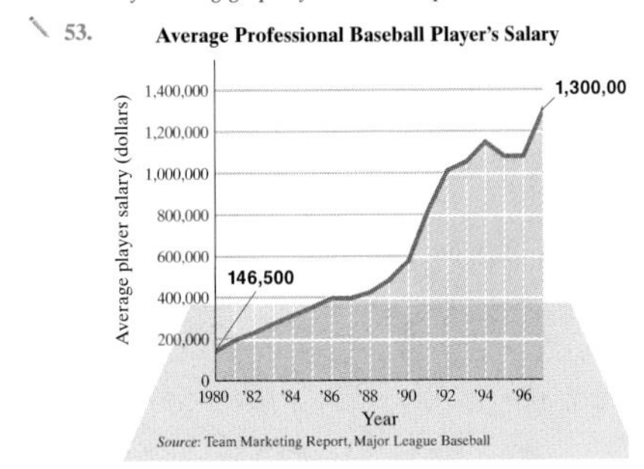

57. The cost, c, in dollars, of repairing a highway can be estimated by the function $c = 2000 + 6000m$, where m is the number of miles to be repaired.

a) Draw a graph of the function for up to and including 6 miles.

b) Estimate the cost of repairing 2 miles of road.

58. The cost, c, in dollars, of a cross-country train trip can be estimated by the function $c = 50 + 0.15m$, where m is the distance traveled in miles.

a) Draw a graph of the function for up to and including 3000 miles traveled.

b) Estimate the cost of a 1000-mile trip.

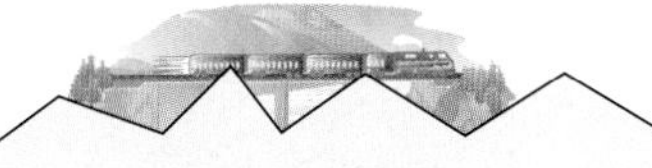

Concept/Writing Exercises

New Concept/Writing Exercises encourage students to analyze and write about the concepts they are learning, improving their understanding and comprehension of the material.

Exercise Set 4.5

Concept/Writing Exercises

1. When graphing inequalities that contain either $\leq$ or $\geq$, explain why the points on the line will be solutions to the inequality.

2. When graphing inequalities that contain either $<$ or $>$, explain why the points on the line will not be solutions to the inequality.

3. How do the graphs of $2x + 3y > 6$ and $2x + 3y < 6$ differ?

4. How do the graphs of $4x - 3y < 6$ and $4x - 3y \leq 6$ differ?

EMPHASIS ON *Exercises*

62. A monthly electric bill, m, in dollars, consists of a \$20 monthly fee plus \$0.07 per kilowatt-hour, k, of electricity used. The amount of the bill is a function of the kilowatt-hours used, $m = 20 + 0.07k$.
   a) Draw a graph for up to and including 3000 kilowatt-hours of electricity used in a month.
   b) Estimate the bill if 1500 kilowatt-hours of electricity are used.

Consider the following graphs. Recall from Section 2.5 that an open circle at the end of a line segment means that the endpoint is not included in the answer. A solid circle at the end of a line segment indicates that the endpoint is included in the answer. Determine whether the following graphs are functions. Explain your answer.

63.
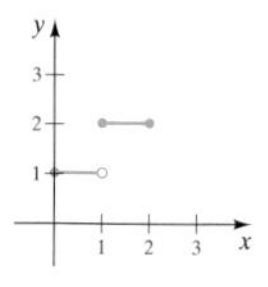

64.
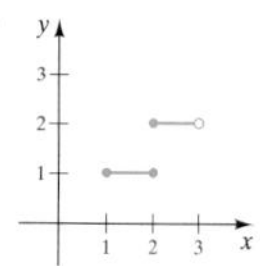

65.
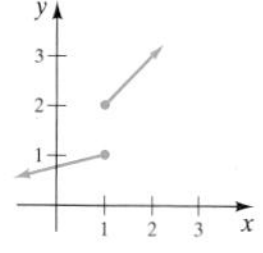

66.
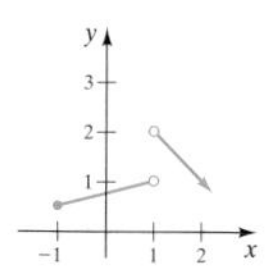

Challenge Problems

Challenge Problems with problems that may be conceptually and computationally more demanding.

Challenge Problems

c).

67. $f(x) = \frac{1}{2}x^2 - 3x + 5$; find **a)** $f(\frac{1}{2})$, **b)** $f(\frac{2}{3})$, **c)** $f(0.2)$

68. $f(x) = x^2 + 2x - 3$; find **a)** $f(1)$, **b)** $f(2)$, **c)** $f(a)$. Explain how you determined your answer to part **c)**.

Group Activities

Group Activities provide students with opportunities for collaborative learning.

Group Activity

Discuss and answer Exercises 69 and 70 as a group.

69. Submit three real-life examples (different from those already given) of a quantity that is a function of another. Write each as a function, and indicate what each variable represents.

70. In April, 1999 the cost of mailing a first class letter was 33 cents for the first ounce and 22 cents for each additional ounce. A graph showing the cost of mailing a letter first class is pictured below.

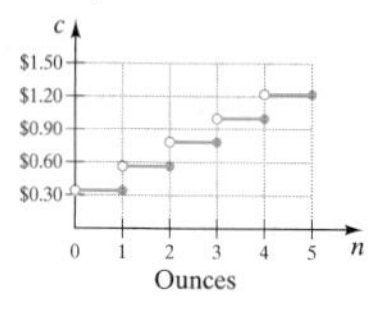

   a) Does this graph represent a function? Explain your answer.
   b) From the graph, estimate the cost of mailing a 4-ounce package first class.
   c) Determine the exact cost of mailing a 4-ounce package first class.
   d) From the graph, estimate the cost of mailing a 3.6-ounce package first class.
   e) Determine the exact cost of mailing a 3.6-ounce package first class.

Cumulative Review Exercises

Cumulative Review Exercises reinforce previously covered topics. These exercises are keyed to sections where the material is explained.

Cumulative Review Exercises

[1.3] 71. Evaluate $\frac{5}{9} - \frac{3}{7}$.

[1.10] 72. Name each illustrated property.
   a) $2 \cdot 5 = 5 \cdot 2$
   b) $(x + 2) + 3 = x + (2 + 3)$
   c) $2(x + 5) = 2x + 2 \cdot 5$

[2.5] 73. Solve the equation $2x - 3(x + 2) = 8$.

[3.3] 74. The cost of a taxi ride is \$2.00 for the first mile and \$1.50 for each additional mile or part thereof. Find the maximum distance Andrew Collins can ride in the taxi if he has only \$20.

EMPHASIS ON *Pedagogy*

Preview and Perspective

Every chapter begins with a **Preview and Perspective** to give students an overview of the chapter. The **Preview and Perspective** shows students the connections between the concepts presented in the text and the real world.

230 · Chapter 4 · Graphing Linear Equations

Preview and Perspective

In this chapter we explain how to graph linear equations. The graphs of linear equations are straight lines. Graphing is one of the most important topics in mathematics, and each year its importance increases. If you take additional mathematics courses you will graph many different types of equations. The material presented in this chapter should give you a good background for graphing in later courses. Graphs are also used in many professions and industries. They are used to display information and to make projections about future trends.

In Section 4.1 we introduce the Cartesian coordinate system and explain how to plot points. In Section 4.2 we discuss two methods for graphing linear equations: by plotting points and by using the x- and y-intercepts. The slope of a line is discussed in Section 4.3. In Section 4.4 we discuss a third procedure, using slope, for graphing a linear equation.

We solved inequalities in one variable in Section 2.6. In Section 4.5 we will solve and graph linear inequalities in two variables. Graphing linear inequalities is an extension of graphing linear equations.

Functions are a unifying concept in mathematics. In Section 4.6, we give a brief and somewhat informal introduction to functions. Functions will be discussed in much more depth in later mathematics courses.

This is an important chapter. If you plan on taking another mathematics course, graphs and functions will probably be a significant part of that course.

4.1 THE CARTESIAN COORDINATE SYSTEM AND LINEAR EQUATIONS IN TWO VARIABLES

SSM

VIDEO 4.1

CD Rom

1) Plot points in the Cartesian coordinate system.
2) Determine whether an ordered pair is a solution to a linear equation.

1) **Plot Points in the Cartesian Coordinate System**

Numbered Section Objectives

Each section begins with a **list of objectives.** Numbered icons connect the objectives to the appropriate sections of the text.

In-Text Examples

An abundance of **in-text examples** illustrate the concept being presented and provide a step-by-step annotated solution.

EXAMPLE 7 The weekly profit, p, of an ice skating rink is a function of the number of skaters per week, n. The function approximating the profit is $p = f(n) = 8n - 600$, where $0 \leq n \leq 400$.

a) Construct a graph showing the relationship between the number of skaters and the weekly profit.

b) Estimate the profit if there are 200 skaters in a given week.

EMPHASIS ON *Pedagogy*

Calculator

Using Your Calculator and **Using Your Graphing Calculator** boxes provide more optional exercises for use with technology than in the previous edition as well as keystroke instructions.

Using Your Graphing Calculator

To graph an equation on a graphing calculator, use the following steps.

1. Solve the equation for y, if necessary.
2. Press the Y= key and enter the equation.
3. Press the GRAPH key (to see the graph). You may need to adjust the window, as explained in the Using Your Graphing Calculator box on page 237.

In Example 2 when ...

If you do not get the graph in Figure 4.24 press WINDOW and determine whether you have the standard window −10, 10, 1, −10, 10, 1, as discussed in the Using Your Graphing Calculator box on page 237. If not, change to the standard window and press the GRAPH key again.

It is possible to graph two or more equations on your graphing calculator. If, for example, you wanted to graph both $y = 2x - 6$ and $y = -3x + 4$ on the same screen, you would begin

Avoiding Common Errors

Avoiding Common Errors boxes illustrate common mistakes, explain why certain procedures are wrong, and show the correct methods for working the problems.

Avoiding Common Errors

In Example 4**a)** we asked you to represent a cost, c, increased by 6%. Note, the answer is $c + 0.06c$. Often, students write the answer to this question as $c + 0.06$. It is important to realize that a percent of a quantity must always be a percent multiplied by some number or letter. Some phrases involving the word percent and the correct and incorrect interpretations follow.

PHRASE	CORRECT	INCORRECT
A $7\frac{1}{2}$% sales tax on c dollars	$0.075c$	~~0.075~~
The cost, c, increased by a $7\frac{1}{2}$% sales tax	$c + 0.075c$	~~$c + 0.075$~~
The cost, c, reduced by 25%	$c - 0.25c$	~~$c - 0.25$~~

Helpful Hints

Helpful Hints offer useful suggestions for problem solving and various other topics.

HELPFUL HINT

Since only two points are needed to determine a straight line, it is not absolutely necessary to determine and plot the check point in step 3. However, if you use only the x- and y-intercepts to draw your graph and one of those points is wrong, your graph will be incorrect and you will not know it. It is always a good idea to use three points when graphing a linear equation.

Procedures, Important Facts, and Definitions

Procedures and **Important Facts** are presented in boxes throughout the text to make it easy for students to focus on this material and find it when preparing for quizzes and tests. Definitions are set off in **Definition Boxes** for easy reference and review.

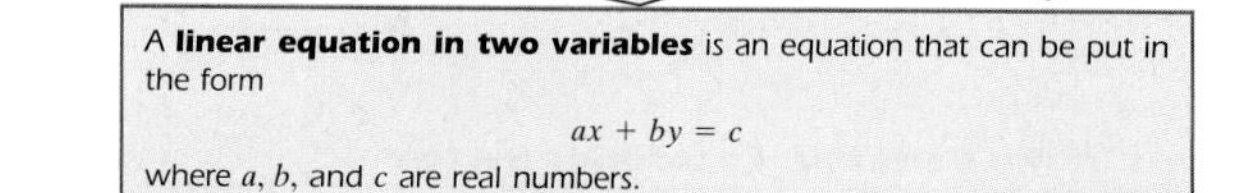

A **linear equation in two variables** is an equation that can be put in the form

$$ax + by = c$$

where a, b, and c are real numbers.

change ... found by subtra...

Slope of a Line Through the Points (x_1, y_1) and (x_2, y_2)

$$\text{slope} = \frac{\text{change in } y \text{ (vertical change)}}{\text{change in } x \text{ (horizontal change)}} = \frac{y_2 - y_1}{x_2 - x_1}$$

Now Try Exercises

Now Try Exercises appear after selected examples to reinforce important concepts. **Now Try Exercises** also provide students with immediate practice and make the student an active, rather than passive, learner.

EXAMPLE 3 Graph the equation $5x + 3y = 12$ by using the slope and y-intercept.

Solution Solve the equation for y.

$$5x + 3y = 12$$
$$3y = -5x + 12$$
$$y = \frac{-5x + 12}{3}$$
$$= -\frac{5}{3}x + 4$$

Thus, the slope is $-\frac{5}{3}$ and the y-intercept is $(0, 4)$. Begin by marking a point at 4 on the y-axis (Fig. 4.40). Then move 5 units down and 3 units to the right to determine the next point. Move down and to the right because the slope is negative and a line with a negative slope must fall as it goes from left to right. Finally, draw the straight line between the plotted points.

FIGURE 4.40

NOW TRY EXERCISE 21

To the Student

Algebra is a course that cannot be learned by observation. To learn algebra you must become an active participant. You must read the text, pay attention in class, and, most importantly, you must work the exercises. The more exercises you work, the better.

The text was written with you in mind. Short, clear sentences are used, and many examples are given to illustrate specific points. The text stresses useful applications of algebra. Hopefully, as you progress through the course, you will come to realize that algebra is not just another math course that you are required to take, but a course that offers a wealth of useful information and applications.

This text makes full use of color. The different colors are used to highlight important information. Important procedures, definitions, and formulas are placed within colored boxes.

The boxes marked **Helpful Hints** should be studied carefully, for they stress important information. The boxes marked **Avoiding Common Errors** should also be studied carefully. These boxes point out errors that students commonly make, and provide the correct procedures for doing these problems.

Ask your professor early in the course to explain the policy on when the calculator may be used. Pay particular attention to the **Using Your Calculator** boxes. You should also read the **Using Your Graphing Calculator** boxes even if you are not using a graphing calculator in class. You may find the information presented here helps you better understand the algebraic concepts.

Other questions you should ask your professor early in the course include: What supplements are available for use? Where can help be obtained when the professor is not available? Supplements that may be available include: Student's Study Guide, Student's Solutions Manual, tutorial software, and videotapes, including a tape on the study skills needed for success in mathematics.

You may wish to form a study group with other students in your class. Many students find that working in small groups provides an excellent way to learn the material. By discussing and explaining the concepts and exercises to one another you reinforce your own understanding. Once guidelines and procedures are determined by your group, make sure to follow them.

One of the first things you should do is to read Section 1.1, Study Skills Needed for Success in Mathematics. Read this section slowly and carefully, and pay particular attention to the advice and information given. Occasionally, refer back to this section. This could be the most important section of the book. Carefully read the material on doing your homework and on attending class.

At the end of all Exercise Sets (after the first two) are **Cumulative Review Exercises**. You should work these problems on a regular basis, even if they are not assigned. These problems are from earlier sections and chapters of the text, and they will refresh your memory and reinforce those topics. If you have a problem when working these exercises, read the appropriate section of the text or study your notes that correspond to that material. The section of the text where the Cumulative Review Exercise was introduced is indicated in brackets, [ ], to the left of the exercise. After reviewing the material, if you still have a problem, make an appointment to see your professor. Working the Cumulative Review Exercises throughout the semester will also help prepare you to take your final exam.

At the end of each chapter are a **Summary**, a set of **Review Exercises**, and a **Practice Test**. Before each examination you should review this material carefully and take the Practice Test. If you do well on the Practice Test, you should do well on the class test. The questions in the Review Exercises are marked to indicate the section in which that material was first introduced. If you have a problem with a Review Exercise question, reread the section indicated. You may also wish to take the **Cumulative Review Test** that appears at the end of every chapter.

In the back of the text there is an **answer section** which contains the answers to the *odd-numbered* exercises, including the Challenge Problems. Answers to *all* Using Your Graphing Calculator

Exercises, Cumulative Review Exercises, Review Exercises, Practice Tests, and Cumulative Review Tests are provided. Answers to the group exercises are not provided, for we wish students to reach agreement by themselves on answers to these exercises. The answers should be used only to check your work.

I have tried to make this text as clear and error free as possible. No text is perfect, however. If you find an error in the text, or an example or section that you believe can be improved, I would greatly appreciate hearing from you. If you enjoy the text, I would also appreciate hearing from you. You can contact me at *www.prenhall.com/angel.*

Allen R. Angel

REAL NUMBERS

CHAPTER 1

Use the Angel Web site at www.prenhall.com/angel to be linked to an internet resource that will help you further explore the following application.

Coverage provided by medical insurance policies differs according to the type of policy held by an individual. In addition to yearly premiums, certain health maintenance plans require a co-payment for each visit made to a doctor's office. Other types of policies require that the individual pay a certain dollar amount in medical expenses each year, after which the insurance company pays a large percentage of the remaining costs. On pages 11 and 12 we use problem solving techniques developed by a famous mathematician, George Polya, to determine the portion of a medical bill a person is responsible for and how much of that bill the insurance company will pay.

Preview and Perspective

In this chapter we provide the building blocks for this course and all other mathematics courses you may take. A review of addition, subtraction, multiplication, and division of numbers containing decimal points is provided in Appendix A. Percents are also discussed in Appendix A. You may wish to review this material now.

For many students, Section 1.1, Study Skills for Success in Mathematics, may be the most important section in this book. *Read it carefully and follow the advice given.* Following the study skills presented will greatly increase your chance of success in this course and in all other mathematics courses.

In Section 1.2 we introduce you to a procedure you can use in problem-solving situations. Operations on fractions are discussed in Section 1.3. *It is essential that you understand fractions* because we will work with fractions throughout the course.

In Section 1.4 we introduce the structure of the real number system. In Section 1.5, we present inequalities and informally discuss absolute value. Both topics are important, especially if you plan to take more mathematics courses. Operations on the real numbers are discussed in Sections 1.6 through 1.8. *Addition, subtraction, multiplication, and division of real numbers must be clearly understood before you go on to the next chapter.*

You must also master the order of operations to follow when evaluating expressions and formulas. We explain how to do this in Section 1.9, where we will also introduce exponents. In Section 1.10, we discuss the properties of the real number system, which we shall use throughout the book.

1.1 STUDY SKILLS FOR SUCCESS IN MATHEMATICS

1) **Recognize the goals of this text.**
2) **Learn proper study skills.**
3) **Prepare for and take exams.**
4) **Learn to manage time.**
5) **Purchase a calculator.**

This section is extremely important. Take the time to read it carefully and follow the advice given. For many of you this section may be the most important section of the text.

Most of you taking this course fall into one of three categories: (1) those who did not take algebra in high school, (2) those who took algebra in high school but did not understand the material, or (3) those who took algebra in high school and were successful but have been out of school for some time and need a refresher course. Whichever the case, you will need to acquire study skills for mathematics courses.

Before we discuss study skills, we will present the goals of this text. These goals may help you realize why certain topics are covered in the text and why they are covered as they are.

1) Recognize the Goals of This Text

The goals of this text include:

1. Presenting traditional algebra topics
2. Preparing you for more advanced mathematics courses
3. Building your confidence in, and your enjoyment of, mathematics

4. Improving your reasoning and critical thinking skills
5. Increasing your understanding of how important mathematics is in solving real-life problems
6. Encouraging you to think mathematically, so that you will feel comfortable translating real-life problems into mathematical equations, and then solving the problems.

It is important to realize that this course is the foundation for more advanced mathematics courses. A thorough understanding of algebra will make it easier for you to succeed in later mathematics courses and in life.

2) Learn Proper Study Skills

Now we will consider study skills and other items of importance.

Have a Positive Attitude

You may be thinking to yourself, "I hate math," or "I wish I did not have to take this class." You may have heard of "math anxiety" and feel you fit this category. The first thing to do to be successful in this course is to change your attitude to a more positive one. You must be willing to give this course, and yourself, a fair chance.

Based on past experiences in mathematics, you may feel that this is difficult. However, mathematics is something you need to work at. Many of you are more mature now than when you took previous mathematics courses. Your maturity and desire to learn are extremely important and can make a tremendous difference in your ability to succeed in mathematics. I believe you can be successful in this course, but you also need to believe it.

Prepare for and Attend Class

To be prepared for class, you need to do your homework. If you have difficulty with the homework, or some of the concepts, write down questions to ask your instructor. If you were given a reading assignment, read the appropriate material carefully before class. If you were not given a reading assignment, spend a few minutes previewing any new material in the textbook before class. At this point you don't have to understand everything you read. Just get a feeling for the definitions and concepts that will be discussed. This quick preview will help you understand what your instructor is explaining during class.

After the material is explained in class, read the corresponding sections of the text slowly and carefully, word by word.

You should plan to attend every class. Most instructors agree that there is an inverse relationship between absences and grades. That is, the more absences you have, the lower your grade will be. Every time you miss a class, you miss important information. If you must miss a class, contact your instructor ahead of time, and get the reading assignment and homework. If possible, before the next class, try to borrow and copy a friend's notes to help you understand the material you missed.

To be successful in this course, you must thoroughly understand the material in this chapter, especially fractions and adding and subtracting real numbers. If you are having difficulty with these topics, see your instructor for help.

In algebra and other mathematics courses, the material you learn is cumulative. That is, the new material is built on material that was presented previously. You must understand each section before moving on to the next section, and each chapter before moving on to the next chapter. Therefore, do not let yourself fall behind. Seek help as soon as you need it—do not wait! Make sure that you do all your homework assignments completely and study the text carefully. You will greatly increase your chance of success in this course by following the study skills presented in this section.

While in class, pay attention to what your instructor is saying. If you don't understand something, ask your instructor to repeat the material. If you have read the assigned material before class and have questions that have not been answered, ask your instructor. If you don't ask questions, your instructor will not know that you have a problem understanding the material.

In class, take careful notes. Write numbers and letters clearly, so that you can read them later. Make sure your x's do not look like y's and vice versa. It is not necessary to write down every word your instructor says. Copy the major points and the examples that do not appear in the text. You should not be taking notes so frantically that you lose track of what your instructor is saying. It is a mistake to believe that you can copy material in class without understanding it and then figure it out later.

Read the Text

A mathematics text is not a novel. Mathematics textbooks should be read slowly and carefully, word by word. If you don't understand what you are reading, reread the material. When you come across a new concept or definition, you may wish to underline it, so that it stands out. Then it will be easier to find later. When you come across an example, read and follow it line by line. Don't just skim it. Then work out the example on another sheet of paper. Also, work the **Now Try Exercises** that appear in the text next to many Examples. Make notes of anything you don't understand to ask your instructor.

This textbook has special features to help you. I suggest that you pay particular attention to these highlighted features, including the Avoiding Common Errors boxes, the Helpful Hint boxes, and important procedures and definitions identified by color. The **Avoiding Common Errors** boxes point out the most common errors made by students. Read and study this material very carefully and make sure that you understand what is explained. If you avoid making these common errors, your chances of success in this and other mathematics classes will be increased greatly. The **Helpful Hints** offer many valuable techniques for working certain problems. They may also present some very useful information or show an alternative way to work a problem.

Do the Homework

Two very important commitments that you must make to be successful in this course are attending class and doing your homework regularly. Your assignments must be worked conscientiously and completely. Do your homework as soon as possible, so the material presented in class will be fresh in your mind. Research has shown that for mathematics courses, studying and doing homework shortly after learning the material improves retention and performance. Mathematics cannot be learned by observation. You need to practice what was presented in class. It is through doing homework that you truly learn the material. While working homework you will become aware of the types of problems that you need further help with. If you do not work the assigned exercises, you will not know what questions to ask in class.

When you do your homework, make sure that you write it neatly and carefully. List the exercise number next to each problem and work each problem step by step. Then you can refer to it later and understand what is written. Pay particular attention to copying signs and exponents correctly.

Don't forget to check the answers to your homework assignments. This book contains the answers to the odd-numbered exercises in the back of the book. In addition, the answers to all the cumulative review and end-of-chapter review exercises, practice tests, and cumulative review tests are in the back of the book. Answers to the Group Activity Exercises are not provided for we want you to arrive at the answers as a group.

Ask questions in class about homework problems you don't understand. You should not feel comfortable until you understand all the concepts needed to work every assigned problem successfully.

Study for Class

Study in the proper atmosphere, in an area where you will not be constantly disturbed, so that your attention can be devoted to what you are reading. The area where you study should be well ventilated and well lit. You should have sufficient desk space to spread out all your materials. Your chair should be comfortable. There should be no loud music to distract you from studying.

Before you begin studying, make sure that you have all the materials you need (pencils, markers, calculator, etc.). You may wish to highlight the important points covered in class or in the book.

It is recommended that students study and do homework for at least two hours for each hour of class time. Some students require more time than others. It is important to spread your studying time out over the entire week rather than studying during one large block of time.

When studying, you should not only understand how to work a problem but also know *why* you follow the specific steps you do to work the problem. If you do not have an understanding of why you follow the specific process, you will not be able to transfer the process to solve similar problems.

This book has **Cumulative Review Exercises** at the end of every section after Section 1.2. Even if these exercises are not assigned for homework, I urge you to work them as part of your studying process. These exercises reinforce material presented earlier in the course, and you will be less likely to forget the material if you review it repeatedly throughout the course. They will also help prepare you for the final exam. If you forget how to work one of the Cumulative Review Exercises, turn to the section indicated in blue next to the problem and review that section. Then try the problem again.

Prepare for and Take Exams

If you study a little bit each day you should not need to cram the night before an exam. Begin your studying early. If you wait until the last minute, you may not have time to seek the help you may need if you find you cannot work a problem.

To prepare for an exam:

1. Read your class notes.
2. Review your homework assignments.
3. Study formulas, definitions, and procedures given in the text.

4. Read the Avoiding Common Errors boxes and Helpful Hint boxes carefully.
5. Read the summary at the end of each chapter.
6. Work the review exercises at the end of each chapter. If you have difficulties, restudy those sections. If you still have trouble, seek help.
7. Work the chapter practice test.
8. If your exam is a cumulative exam, work the Cumulative Review Test.

Prepare for Midterm and Final Exam

When studying for a comprehensive midterm or final exam follow the procedures discussed for preparing for an exam. However, also:

1. Study all your previous tests carefully. Make sure that you have learned to work the problems that you may have previously missed.
2. Work the cumulative review tests at the end of each chapter. These tests cover the material from the beginning of the book to the end of that chapter.
3. If your instructor has given you a worksheet or practice exam, make sure that you complete it. Ask questions about any problems you do not understand.
4. Begin your studying process early so that you can seek all the help you need in a timely manner.

Take an Exam

Make sure you get sufficient sleep the night before the test. If you studied properly, you should not have to stay up late preparing for a test. Arrive at the exam site early so that you have a few minutes to relax before the exam. If you rush into the exam, you will start out nervous and anxious. After you are given the exam, you should do the following:

1. Carefully write down any formulas or ideas that you need to remember.
2. Look over the entire exam quickly to get an idea of its length. Also make sure that no pages are missing.
3. Read the test directions carefully.
4. Read each question carefully. Answer each question completely, and make sure that you have answered the specific question asked.
5. Work the questions you understand best first; then go back and work those you are not sure of. Do not spend too much time on any one problem or you may not be able to complete the exam. Be prepared to spend more time on problems worth more points.
6. Attempt each problem. You may get at least partial credit even if you do not obtain the correct answer. If you make no attempt at answering the question, you will lose full credit.
7. Work carefully step by step. Copy all signs and exponents correctly when working from step to step, and make sure to copy the original question from the test correctly.
8. Write clearly so that your instructor can read your work. If your instructor cannot read your work, you may lose credit. Also, if your writing is not clear, it is easy to make a mistake when working from one step to the next. When appropriate, make sure that your final answer stands out by placing a box around it.

9. If you have time, check your work and your answers.
10. Do not be concerned if others finish the test before you or if you are the last to finish. Use any extra time to check your work.

Stay calm when taking your test. Do not get upset if you come across a problem you can't figure out right away. Go on to something else and come back to that problem later.

4) Learn to Manage Time

As mentioned earlier, it is recommended that students study and do homework for at least two hours for each hour of class time. Finding the necessary time to study is not always easy. Below are some suggestions that you may find helpful.

1. Plan ahead. Determine when you will study and do your homework. Do not schedule other activities for these periods. Try to space these periods evenly over the week.
2. Be organized, so that you will not have to waste time looking for your books, your pen, your calculator, or your notes.
3. If you are allowed to use a calculator, use it for tedious calculations.
4. When you stop studying, clearly mark where you stopped in the text.
5. Try not to take on added responsibilities. You must set your priorities. If your education is a top priority, as it should be, you may have to reduce time spent on other activities.
6. If time is a problem, do not overburden yourself with too many courses. Consider taking fewer credits. If you do not have sufficient time to study, your understanding and all your grades may suffer.

Use Supplements

This text comes with a large variety of supplements. Find out from your instructor early in the semester which supplements are available and might be beneficial for you to use. Supplements should not replace reading the text, but should be used to enhance your understanding of the material.

Seek Help

Be sure to get help as soon as you need it! Do not wait! In mathematics, one day's material is often based on the previous day's material. So, if you don't understand the material today, you may not be able to understand the material tomorrow.

Where should you seek help? There are often a number of resources on campus. Try to make a friend in the class with whom you can study. Often, you can help one another. You may wish to form a study group with other students in your class. Discussing the concepts and homework with your peers will reinforce your own understanding of the material.

You should know your instructor's office hours, and you should not hesitate to seek help from your instructor when you need it. Make sure that you have read the assigned material and attempted the homework before meeting with your instructor. Come prepared with specific questions to ask.

There are often other sources of help available. Many colleges have a mathematics lab or a mathematics learning center, where tutors are available. Ask your instructor early in the semester where and when tutoring is available. Arrange for a tutor as soon as you need one.

5) Purchase a Calculator

I strongly urge you to purchase a scientific or graphing calculator as soon as possible. Ask your instructor if he or she recommends a particular calculator for this or a future mathematics class. Also ask your instructor if you may use a calculator in class, on homework, and on tests. If so, you should use your calculator whenever possible to save time.

If a calculator contains a LOG key or SIN key, it is a scientific calculator. You *cannot* use the square root key $\sqrt{\ }$ to identify scientific calculators since both scientific calculators and nonscientific calculators may have this key. You should pay particular attention to the Using Your Calculator boxes in this book. The boxes explain how to use your calculator to solve problems. If you are using a graphing calculator, pay particular attention to the Using Your Graphing Calculator boxes. You may also need to use the reference manual that comes with your calculator at various times.

A Final Word

You can be successful at mathematics if you attend class regularly, pay attention in class, study your text carefully, do your homework daily, review regularly, and seek help as soon as you need it. Good luck in your course.

Exercise Set 1.1

Do you know:

1. Your professor's name and office hours?
2. Your professor's office location and telephone number?
3. Where and when you can obtain help if your professor is not available?
4. The name and phone number of a friend in your class?
5. What supplements are available to assist you in learning?
6. Is your instructor recommending the use of a particular calculator?
7. When can you use your calculator in this course?

If you do not know the answer to any of the questions just asked, you should find out as soon as possible.

8. What are your reasons for taking this course?
9. What are your goals for this course?
10. Are you beginning this course with a positive attitude? It is important that you do!
11. List the things you need to do to prepare properly for class.
12. Explain how a mathematics text should be read.
13. For each hour of class time, how many hours outside of class are recommended for studying and doing homework?
14. When studying, you should not only understand how to work a problem, but also why you follow the specific steps you do. Why is this important?
15. Two very important commitments that you must make to be successful in this course are **a)** doing homework regularly and completely and **b)** attending class regularly. Explain why these commitments are necessary.
16. Write a summary of the steps you should follow when taking an exam.
17. Have you given any thought to studying with a friend or a group of friends? Can you see any advantages in doing so? Can you see any disadvantages in doing so?

1.2 PROBLEM SOLVING

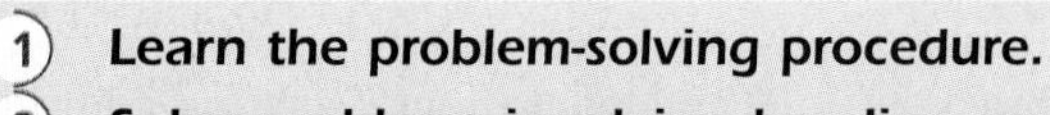

SSM VIDEO 1.2 CD Rom

1) **Learn the problem-solving procedure.**
2) **Solve problems involving bar, line, and circle graphs.**
3) **Solve problems involving statistics.**

1) Learn the Problem-Solving Procedure

George Polya

One of the main reasons we study mathematics is that we can use it to solve many real-life problems. To solve most real-life problems mathematically, we need to be able to express the problem in mathematical symbols. This is an important part of the problem-solving procedure that we will present shortly. Throughout the book, we will be problem solving. In Chapter 3, we will also spend a great deal of time explaining how to express real-life applications mathematically.

We will now give the general five-step **problem-solving procedure** that was developed by George Polya and presented in his book *How to Solve It*. You can approach any problem by following this general procedure.

Guidelines for Problem Solving

1. **Understand the problem.**
   - Read the problem *carefully* at least twice. In the first reading, get a general overview of the problem. In the second reading, determine **(a)** exactly what you are being asked to find and **(b)** what information the problem provides.
   - Make a list of the given facts. Determine which are pertinent to solving the problem.
   - Determine whether you can substitute smaller or simpler numbers to make the problem more understandable.
   - If it will help you organize the information, list the information in a table.
   - If possible, make a sketch to illustrate the problem. Label the information given.
2. **Translate the problem to mathematical language.**
   - This will generally involve expressing the problem in terms of an algebraic expression or equation. (We will explain how to express application problems as equations in Chapter 3.)
   - Determine whether there is a formula that can be used to solve the problem.
3. **Carry out the mathematical calculations necessary to solve the problem.**
4. **Check the answer obtained in step 3.**
   - Ask yourself, "Does the answer make sense?" "Is the answer reasonable?" If the answer is not reasonable, recheck your method for solving the problem and your calculations.
   - Check the solution in the original problem if possible.
5. **Make sure you have answered the question.**
   - State the answer clearly.

In step 2 we use the words *algebraic expression*. An **algebraic expression**, sometimes just referred to as an **expression**, is a general term for any collection of numbers, letters (called variables), grouping symbols (such as parentheses ( ) or brackets [ ]), and **operations** (such as addition, subtraction, multiplication, and division). In this section we will not be using variables, so we will discuss their use later.

Examples of Expressions

$3 + 4, \qquad 6(12 \div 3), \qquad (2)(7)$

The following examples show how to apply the guidelines for problem solving. We will sometimes provide the steps in the examples to illustrate the five-step procedure. However, in some problems it may not be possible or necessary to list every step in the procedure. In some of the examples, we use decimal numbers and percents. If you need to review procedures for adding, subtracting, multiplying, or dividing decimal numbers, or if you need a review of percents, read Appendix A before going on.

EXAMPLE 1 Chicago's O'Hare airport is the busiest in the world with about 65 million passengers arriving and departing annually. The airport express bus operates between the airport and downtown, a distance of 19 miles. A particular airport express bus makes 8 round trips daily between the airport and downtown and carries an average of 12 passengers per trip (each way). The fare each way is $15.50.

a) What are the bus's receipts from one day's operation?

b) If the one-way fare is increased by 10%, determine the new fare.

Solution **a) Understand the problem** A careful reading of the problem shows that the task is to find the bus's total receipts from one day's operation. Make a list of all the information given and determine which information is needed to find the total receipts.

Information Given	Pertinent to Solving the Problem
65 million passengers arrive/depart annually	no
19 miles from airport to downtown	no
8 round trips daily	yes
12 passengers per trip (each way)	yes
$15.50 fare (each way)	yes

To find the total receipts, it is not necessary to know the number of passengers who use the airport or the distance between the airport and downtown. Solving this problem involves realizing that the total receipts depend on the number of one-way trips per day, the average number of passengers per trip, and the one-way cost per passenger. The product of these three numbers will yield the total daily receipts. For the 8 round trips daily, there are 2×8 or 16 one-way trips daily.

Translate the problem into mathematical language

$$\begin{pmatrix}\text{receipts}\\ \text{for one}\\ \text{day}\end{pmatrix} = \begin{pmatrix}\text{number of}\\ \text{one-way}\\ \text{trips per day}\end{pmatrix} \times \begin{pmatrix}\text{number of}\\ \text{passengers}\\ \text{per trip}\end{pmatrix} \times \begin{pmatrix}\text{cost per}\\ \text{passenger}\\ \text{each way}\end{pmatrix}$$

Carry out the calculations

$$= 16 \times 12 \times \$15.50 = \$2976.00$$

We could also have used 8 round trips and a fare of $31.00 per person to obtain the answer. Can you explain why?

Check the answer The answer $2976.00 is a reasonable answer based on the information given.

Answer the question asked The receipts for one day's operation are $2976.00.

b) Understand If the fare is increased by 10%, the new fare becomes 10% greater than $15.50. Thus you need to add 10% of $15.50 to $15.50 to obtain the answer. When performing calculations, numbers given in percent are usually changed to decimal numbers.

Translate new fare = original fare + 10% of original fare

Carry Out new fare = $15.50 + 0.10($15.50)

= $15.50 + $1.55 = $17.05

Check The answer $17.05, which is a little larger than $15.50, seems reasonable.

Answer The fare when increased by 10% is $17.05.

EXAMPLE 2 In 1997, the fastest Intel processor, the Pentium II, could perform about 300 million operations per second (300 megahertz, symbolized 300 MHz). How many operations could it perform in 0.3 seconds?

Solution **Understand** We are given the name of the processor, a speed of about 300 million (300,000,000) operations per second, and 0.3 second. To determine the answer to this problem, the name of the processor, Pentium II, is not needed.

To obtain the answer, will we need to multiply or divide? Often a fairly simple problem seems more difficult because of the numbers involved. When very large or very small numbers make the problem seem confusing, try substituting commonly used numbers in the problem to determine how to solve the problem. Suppose the problem said that the processor can perform 6 operations per second. How many operations can it perform in 2 seconds? The answer to this question should be more obvious. It is 6×2 or 12. Since we multiplied to obtain this answer, we also will need to multiply to obtain the answer to the given problem.

Translate

number of operations in 0.3 seconds = 0.3(number of operations per second)

Carry Out = 0.3(300,000,000)

= 90,000,000 *From a calculator*

Check The answer, 90,000,000 operations, is less than the 300,000,000 operations per second, which makes sense because the processor is operating for less than a second.

Answer In 0.3 second, the processor can perform about 90,000,0000 operations.

EXAMPLE 3 Beth Rechsteiner's medical insurance policy is similar to that of many workers. Her policy requires that she pay the first $100 of medical expenses each calendar year (called a deductible). After the deductible is paid, she pays 20%

of the medical expenses (called a co-payment) and the insurance company pays 80%. (There is a maximum co-payment of $600 that she must pay each year. After that, the insurance company pays 100% of the fee schedule.) On January 1, Beth sprained her ankle playing tennis. She went to the doctor's office for an examination and X rays. The total bill of $325 was sent to the insurance company.

a) How much of the bill will Beth be responsible for?

b) How much will the insurance company be responsible for?

Solution **a) Understand** First we list all the *relevant* given information.

Given Information

$100 deductible

20% co-payment after deductible

80% paid by insurance company after deductible

$325 doctor bill

All the other information is not needed to solve the problem. Beth will be responsible for the first $100 and 20% of the remaining balance. The insurance company will be responsible for 80% of the balance after the deductible. Before we can find what Beth owes, we need to first find the balance of the bill after the deductible. The balance of the bill after the deductible is $\$325 - \$100 = \$225$.

Translate Beth's responsibility = deductible + 20% of balance of bill after deductible

Carry Out
$$\begin{aligned}\text{Beth's responsibility} &= 100 + 20\%(225)\\ &= 100 + 0.20(225)\\ &= 100 + 45\\ &= 145\end{aligned}$$

Check and Answer The answer appears reasonable. Beth will owe the doctor $145.

b) The insurance company will be responsible for 80% of the balance after the deductible.

$$\begin{aligned}\text{insurance company's responsibility} &= 80\% \text{ of balance after deductible}\\ &= 0.80(225)\\ &= 180\end{aligned}$$

Thus the insurance company is responsible for $180. This checks because the sum of Beth's responsibility and the insurance company's responsibility is equal to the doctor's bill.

$$\$145 + \$180 = \$325$$

We could have also found the answer to part **b)** by subtracting Beth's responsibility from the total amount of the bill, but to give you more practice with percents we decided to show the solution as we did.

NOW TRY EXERCISE 33

2) Solve Problems Involving Bar, Line, and Circle Graphs

Problem solving often involves understanding and reading graphs and sets of data (or numbers). We will be using bar, line, and circle (or pie) graphs and sets of data, throughout the book. We will illustrate a number of such graphs in this section and explain how to interpret them. The data needed to solve problems involving the graphs are generally provided on the graphs. To work Example 4, you must interpret bar and circle graphs and work with data.

EXAMPLE 4 The following information was provided by the American Association of Engineering Societies, Inc.

Engineering Degrees

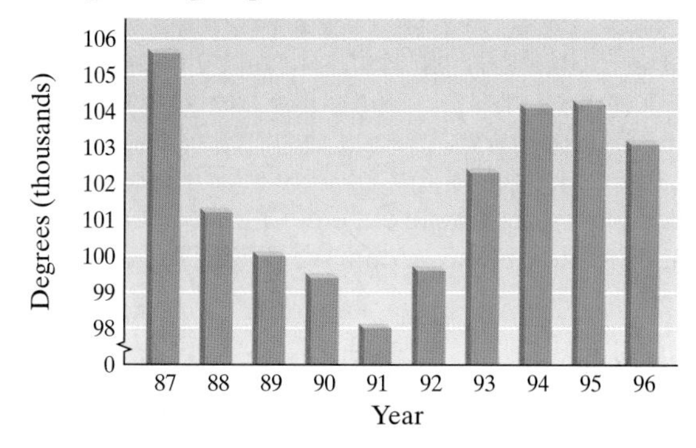

FIGURE 1.1

Engineering Degree Classification
(Based on 103,100 degrees in 1996)

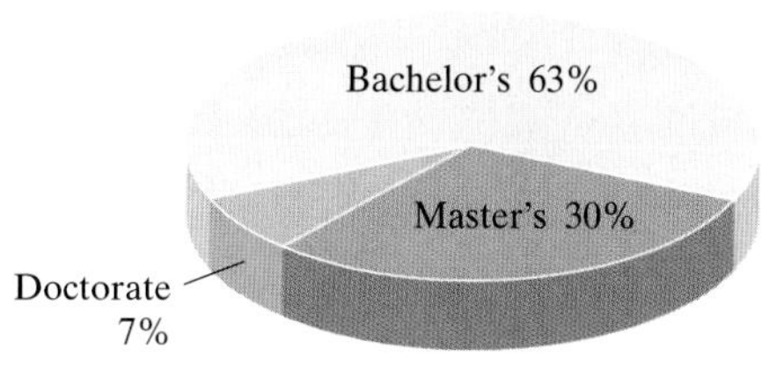

Source: American Association of Engineering Societies, Inc.

FIGURE 1.2

1996 Bachelor's Degrees in Engineering by Curriculum			
Mechanical	23.0%	Engineering science	2.0%
Electrical/electronic	21.0	General engineering	1.6
Civil	17.0	Biomedical	1.5
Chemical	10.0	Environmental	1.3
Computer	9.0	Materials & metallurgical	1.3
Industrial	5.0	Agricultural	1.0
Aerospace	3.0	Other	3.3

Notice that the total number of engineering degrees from 1987 through 1996 is presented by a **bar graph** (Fig. 1.1) while the degree classification in 1996 is given by a **circle**, or **pie, graph** (Fig. 1.2).

a) In which year were the fewest engineering degrees awarded? Estimate the number of engineering degrees awarded that year.

b) In 1996, about 103,100 engineering degrees were awarded. Determine the approximate number that were bachelor's degrees, master's degrees, and doctorate degrees.

c) Determine the approximate number of bachelor's degrees in electrical/electronic engineering that were awarded in 1996.

Solution **a)** The *vertical scale* (the up-and-down line on the left side of the graph) shows the number of engineering degrees awarded. The *horizontal scale* (the left-to-right line along the bottom of the graph) shows the years from 1987 through 1996. Notice the break ⦚ above the zero in the vertical scale. If all the values from 0 to 106 were shown, the graph would either be very large or it would be difficult to determine the values at the top of the bars. When we wish to omit the numbers at the bottom of a vertical scale, as we have done here, we use a symbol like ⦚ to break the axis.

By examining the bar graph, we see that the top of the bar labeled 1991 is lower than any of the other bars. Therefore, the fewest engineering degrees were awarded in 1991. To determine the approximate number of engineering degrees awarded in 1991, we estimate the number on the vertical scale that corresponds to the top of the bar labeled 1991. For a 3-dimensional bar graph, such as Figure 1.1, we will use the *front face* of the bar to determine the value. The number of engineering degrees awarded in 1991 was approximately 98,000.

b) Understand The degree classification circle graph indicates that 63% of the engineering degrees were bachelor's, 30% were master's, and 7% were doctorates. (Notice that 63% *of the total area of the circle* is the bachelor's sector, 30% is the master's sector, and 7% is the doctorate sector.) To determine the approximate number of bachelor's degrees in engineering, we need to find 63% of the total number of engineering degrees awarded. We do this by multiplying as follows.

Translate

$$\left(\begin{array}{c}\text{number of bachelor's}\\\text{degrees in 1996}\end{array}\right) = \left(\begin{array}{c}\text{percent of degrees}\\\text{that were bachelor's}\end{array}\right)(\text{total number of degrees})$$

Carry Out

$$\text{number of bachelor's degrees in 1996} = 0.63(103{,}100) = 64{,}953$$

Thus about 64,953 bachelor's degrees in engineering were awarded in 1996. To find the number of master's and doctorate degrees, we do similar calculations.

$$\text{number of master's degrees} = 0.30(103{,}100) = 30{,}930$$

$$\text{number of doctorate degrees} = 0.07(103{,}100) = 7217$$

Check If we add the number of bachelor's, master's, and doctorate degrees, we obtain the total number of degrees awarded, 103,100.

$$64{,}953 + 30{,}930 + 7217 = 103{,}100$$

Answer In 1996, approximately 64,953 bachelor's, 30,930 master's, and 7217 doctorate degrees were awarded in engineering.

c) Part **c)** is worked similarly to part **b)**. From the table we see that 21% of the bachelor's degrees in engineering were awarded for electrical/electronic engineering. In part **b)** we found that there were 64,953 bachelor's degrees awarded in engineering.

$$\text{number of electrical/electronic degrees awarded} = 0.21(64{,}953) = 13{,}640.13$$

Thus, about 13,640 bachelor's degrees in electrical/electronic engineering were awarded in 1996.

NOW TRY EXERCISE 35

In Example 5 we will use the symbol $\approx$, which is read "**is approximately equal to**." If, for example, the answer to a problem is 34.12432, we may write the answer as ≈ 34.1.

EXAMPLE 5 An article in the September 1996 issue of *Consumer Reports* titled "Turning up the Heat" included the **line graph** shown in Figure 1.3.

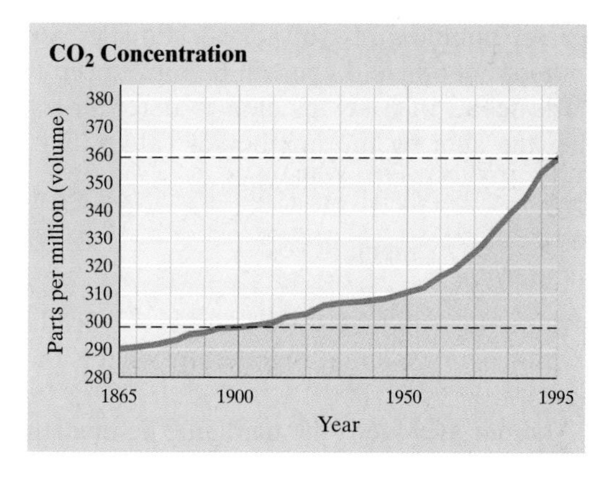

FIGURE 1.3

a) Estimate the CO_2 level in the atmosphere in 1900 and 1995.

b) How many parts per million (ppm) greater was the CO_2 level in 1995 than in 1900?

c) How many times greater was the CO_2 level in 1995 than in 1900?

Solution **a)** When reading a line graph where the line has some thickness, as in Figure 1.3, we will use the center of the line to make our estimates. By observing the dashed lines on the graph, we can estimate that the CO_2 level was about 298 ppm in 1900 and about 359 ppm in 1995.

b) We use the problem-solving procedure to answer the question.

Understand To determine how many parts per million greater the CO_2 level was in 1995 than in 1900 we need to subtract.

Translate difference in parts per million = ppm in 1995 − ppm in 1900

Carry Out difference in parts per million = 359 − 298 = 61

Check and Answer The answer appears reasonable. There were 61 ppm more CO_2 in 1995 than in 1900.

c) Understand If you examine parts **b)** and **c)**, they may appear to ask the same question, but they do not. In Section 1.1 we indicated that it is important to read a mathematics book carefully, word by word. The two parts are different in that part **b)** asks "how many parts per million greater" whereas part **c)** asks "how many *times* greater." To determine the number of times greater the CO_2 level was in 1995 than in 1900 we need to divide the amount of CO_2 in 1995 by the amount of CO_2 in 1900.

Translate $$\text{number of times greater} = \frac{CO_2 \text{ in ppm in 1995}}{CO_2 \text{ in ppm in 1990}}$$

Carry Out $$\text{number of times greater} = \frac{359}{298} \approx 1.20$$

Check and Answer The amount of CO_2 in 1995 was approximately 1.20 times greater than the amount of CO_2 in 1900.

3) Solve Problems Involving Statistics

Because understanding statistics is so important in our society, we will now discuss certain statistical topics and use them in solving problems.

The *mean* and *median* are two **measures of central tendency**, which are also referred to as *averages*. An average is a value that is representative of a set of data (or numbers). If you take a statistics course you will study these averages in more depth, and you will be introduced to other averages.

The **mean** of a set of data is determined by adding all the values and dividing the sum by the number of values. For example, to find the mean of 6, 9, 3, 12, 12, we do the following.

$$\text{mean} = \frac{6 + 9 + 3 + 12 + 12}{5} = \frac{42}{5} = 8.4$$

We divided the sum by 5 since there are five values. The mean is the most commonly used average and it is generally what is thought of when we use the word *average*.

Another average is the median. The **median** is the value in the middle of a set of **ranked data**. The data may be ranked from smallest to largest or largest to smallest. To find the median of 6, 9, 3, 12, 12, we can rank the data from smallest to largest as follows.

3, 6, 9, 12, 12

↑

Middle value

The value in the middle of the ranked set of data is 9. Therefore, the median is 9. Note that half the values will be above the median and half will be below the median.

If there is an even number of pieces of data, the median is halfway between the two middle pieces. For example, to find the median of 3, 12, 5, 12, 17, 9, we can rank the data as follows.

3, 5, 9, 12, 12, 17

↑

Middle values

Since there are six pieces of data (an even number), we find the value halfway between the two middle pieces, the 9 and the 12. To find the median, we add these values and divide the sum by 2.

$$\text{median} = \frac{9 + 12}{2} = \frac{21}{2} = 10.5$$

Thus, the median is 10.5. Note that half the values are above and half are below 10.5.

EXAMPLE 6 Alfonso Ramirez's first six exam grades are 90, 87, 76, 84, 78, and 62.

a) Find the mean for Alfonso's six grades.

b) If one more exam is to be given, what is the minimum grade that Alfonso can receive to obtain a B average (a mean average of 80 or better, but less than 90)?

c) Is it possible for Alfonso to obtain an A average (90 or better)? Explain.

Solution **a)** To obtain the mean, we add the six grades and divide by 6.

$$\text{mean} = \frac{90 + 87 + 76 + 84 + 78 + 62}{6} = \frac{477}{6} = 79.5$$

b) We will show the problem-solving steps for this part of the example.

Understand The answer to this part may be found in a number of ways. For the mean average of seven exams to be 80, the total points for the seven exams must be 7(80) or 560. Can you explain why? The minimum grade needed can be found by subtracting the sum of the first six grades from 560.

Translate

$$\begin{aligned}\text{minimum grade needed on the seventh exam} &= 560 - \text{sum of first six exam grades}\end{aligned}$$

Carry Out

$$\begin{aligned}&= 560 - (90 + 87 + 76 + 84 + 78 + 62)\\ &= 560 - 477\\ &= 83\end{aligned}$$

Check We can check to see that a seventh grade of 83 gives a mean of 80 as follows.

$$\text{mean} = \frac{90 + 87 + 76 + 84 + 78 + 62 + 83}{7} = \frac{560}{70} = 80$$

Answer A seventh grade of 83 or higher will result in a B average.

c) We can use the same reasoning as in part **b)**. For a 90 average, the total points that Alfonso will need to attain is $90 \cdot 7 = 630$. Since his total points are 477, he will need $630 - 477$ or 153 points to obtain an A average. Since the maximum number of points available on most exams is 100, Alfonso would not be able to obtain an A in the course.

NOW TRY EXERCISE 41

Exercise Set 1.2

Concept/Writing Exercises

1. Outline the five-step problem-solving procedure.
2. What is an expression?
3. If a problem is difficult to solve because the numbers in the problem are very large or very small, what can you do that may help make the problem easier to solve?
4. Explain how to find the mean of a set of data.
5. Explain how to find the median of a set of data.
6. What measure of central tendency do we generally think of as "the average"?
7. Consider the set of data 2, 3, 5, 6, 30. Without doing any calculations, can you determine whether the mean or the median is greater? Explain your answer.
8. Consider the set of data 3, 100, 102, 103. Without doing any calculations, determine which is greater, the mean or the median. Explain your answer.
9. To get a grade of B, a student must have a mean average of 80. Jerome Krasner has a mean average of 79 for 10 quizzes. He approaches his teacher and asks for a B, reasoning that he missed a B by only one point. What is wrong with his reasoning?
10. Consider the set of data 3, 3, 3, 4, 4, 4. If one 4 is changed to 5, which of the following will change, the mean and/or the median? Explain.

Practice the Skills

In this exercise set, use a calculator as needed to save time.

11. Jenna Webber's test grades are 78, 97, 59, 74, and 74. For Jenna's grades, determine the **a)** mean and **b)** median.

12. Eric Flemming's bowling scores for five games were 161, 131, 187, 163, and 145. For Eric's games, determine the **a)** mean and **b)** median.

13. Camilla Joyner's telephone bills for January through May, 1999, were \$109.62, \$62.73, \$83.79, \$74.74, and \$121.63. For Camilla's telephone bills, determine the **a)** mean and **b)** median.

14. The Foxes' electric bills for January through June, 1999, were \$96.56, \$108.78, \$87.23, \$85.90, \$79.55 and \$65.88. For these bills, determine the **a)** mean and **b)** median.

15. The top 6 business colleges in the United States, as of March 29, 1999, according to a ranking by *U.S. News and World Report*, are given in the table. Also provided is the cost for out of state tuition and fees. Determine **a)** the mean and **b)** the median out of state tuition and fees for the 6 colleges.

College	Out of State Tuition and Fees
1. Stanford University	\$24,990
2. Harvard University	\$26,260
2. Northwestern University	\$25,872
2. University of Pennsylvania	\$26,290
5. Massachusetts Institute of Technology	\$27,100
6. University of Chicago	\$26,284

Source: U.S. News & World Report, March 29, 1999.

16. The following table shows the total number of vehicles recalled for safety defects by the U.S. National Highway Traffic Safety Administration for the years 1992 through 1995. For the recalls shown from 1992 through 1995, determine the **a)** mean and **b)** median.

Year	Total Vehicles Recalled (in 1000s)
1992	10,122
1993	10,922
1994	6,063
1995	18,295

Source: U.S. National Highway Traffic Safety Administration

Problem Solving

17. Barbara Riedell earns a 5% commission on appliances she sells. Her sales last week totaled \$9400. Find her week's earnings.

18. a) The sales tax in Orange County is 7%. What was the sales tax that Bill Leonard paid on a used car that cost \$12,500 before tax?

b) What is the total cost of the car including tax?

19. The balance in Lois Heater's checking account is \$312.60. She purchased five compact disks at \$17.11 each including tax. If she pays by check, what is the new balance in her checking account?

20. In 1980, the fastest desktop personal computer could perform about 0.3 million operations per second. In 1997, the fastest desktop personal computer could perform about 300 million operations per second. How many times faster is the 1997 personal computer than the 1980 personal computer?

21. The Midtown Parking Lot charges \$1.50 for each hour of parking or part thereof. Bonita Hunsinger parks her car in the garage from 9:00 A.M. to 5:00 P.M., 5 days a week.

a) What is her weekly cost for parking?

b) How much money would she save by paying a weekly parking rate of \$35.00?

22. Al Friedberg wants to purchase a fax machine that sells for \$450. He can either pay the total amount at the time of purchase or agree to pay the store \$150 down and \$20 a month for 16 months.

a) If he pays the down payment and monthly charge, how much will he pay for the fax machine?

b) How much money can he save by paying the total amount at the time of purchase?

23. The table on the top of the next page gives the approximate energy values of some foods and the approximate energy consumption of some activities, in kilojoules (kJ).

Energy Value, Food	(kJ)	Energy Consumption, Activity	(kJ/min)
Chocolate milkshake	2200	Walking	25
Fried egg	460	Cycling	35
Hamburger	1550	Swimming	50
Strawberry shortcake	1440	Running	80
Glass of skim milk	350		

Determine how long it would take for you use up the energy from the following.

a) a hamburger by running

b) a chocolate milkshake by walking

c) a glass of skim milk by cycling

24. A person-to-person call made from a phone booth in Houston, Texas, to a friend on Cape Cod, Massachusetts, costs $3.50 for the first 3 minutes and $0.50 for each additional minute or part thereof. How much would an $18\frac{1}{2}$ minute phone call cost?

25. When the odometer in Tribet LaPierre's car reads 16,741.3, he fills his gas tank. The next time he fills his tank it takes 10.5 gallons, and his odometer reads 16,935.4. Determine the number of miles per gallon that his car gets.

26. The rental cost of a jet ski from Don's Ski Rental is $10.00 per 15 minutes, and the rental cost from Carol's Ski Rental is $25 per half hour. Suppose you plan to rent a jet ski for 3 hours.

a) Which is the better deal?

b) How much will you save?

27. Karin Sirk purchased four tires by mail order. She paid $62.30 plus $6.20 shipping and handling per tire. There was no sales tax on this purchase because it was an out-of-state purchase. When she received the tires, Karin had to pay $8.00 per tire for mounting and balancing. At a local tire store, her total cost for the four tires with mounting and balancing would have been $425 plus 8% sales tax. How much did Karin save by purchasing the tires through the mail?

28. At a local grocery store a six-pack of diet soda costs $3.45 and a carton of four six-packs costs $12.60. How much will Rachel Wicklund save by purchasing the carton instead of four individual six-packs?

29. The federal income tax rate schedule for a joint return in 1997 is illustrated in the following table.

Adjusted Gross Income	Taxes
$0–$38,000	15% of income
$38,001–$91,850	$5700 + 28% in excess of $38,000
$91,851–$140,000	$20,778.00 + 31% in excess of $91,850
$140,001–$250,000	$35,704.50 + 36% in excess of $140,000
$250,001 and up	$75,304.50 + 39.6% in excess of $250,000

a) If the Antonellis' adjusted gross income in 1997 was $26,420, determine their taxes.

b) If the Bachs' 1997 adiusted gross income was $47,835, determine their taxes.

30. a) What is 1 mile per hour equal to in feet per hour? One mile contains 5280 feet.

b) What is 1 mile per hour equal to in feet per second?

c) What is 60 miles per hour equal to in feet per second?

31. A faucet that leaks 1 ounce of water per minute wastes 11.25 gallons in a day.

a) How many gallons of water are wasted in a (non-leap) year?

b) If water costs $5.20 per 1000 gallons, how much additional money per year is being spent on the water bill?

32. When Eva Closson's car tire pressure is 28 pounds per square inch (psi), her car averages 17.3 miles per gallon (mpg) of gasoline. If her tire pressure is increased to 32 psi, it averages 18.0 mpg.

a) How much farther will she travel on a gallon of gas if her tires are inflated to the higher pressure?

b) If she drives an average of 12,000 miles per year, how many gallons of gasoline will she save in a year by increasing her tire pressure from 28 to 32 psi?

c) If gasoline costs $1.20 per gallon, how much money will she save in a year

33. Drivers under the age of 25 who pass a driver education course generally have their auto insurance premium decreased by 10%. Most insurers will offer this 10% deduction until the driver reaches 25. A particular driver education course costs $70. Andre DePue, who just turned 18, has auto insurance that costs $630 per year.

 a) How much would Andre save in auto insurance premiums, from the age of 18 until the age of 25, by taking the driver education course?

 b) What would be his net savings after considering the cost of the course?

34. The following graph illustrates the sources of retirement income for the typical retiree with at least $20,000 in annual income in 1998.

Retirement Income

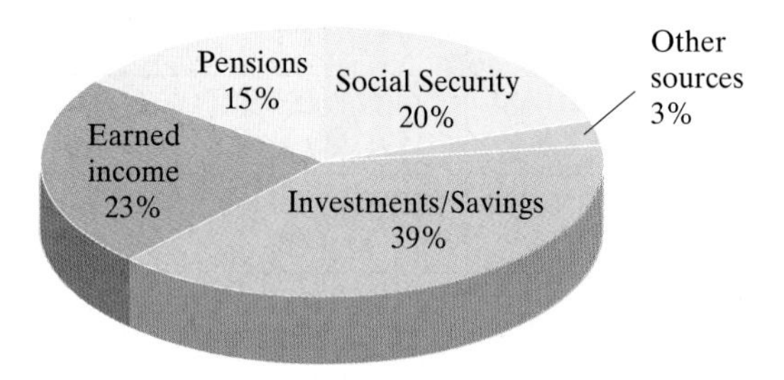

Source: U.S. Department of the Treasury

Mr. Johnson, a retiree, has an annual income of $30,000. Assuming his sources of retirement income are typical, determine Mr. Johnson's income from **a)** pension and **b)** investments/savings.

35. The following graph shows the cost to mail a letter in certain countries. The cost was determined using foreign exchange rates as of January 9, 1999.

Cost to Mail a Letter

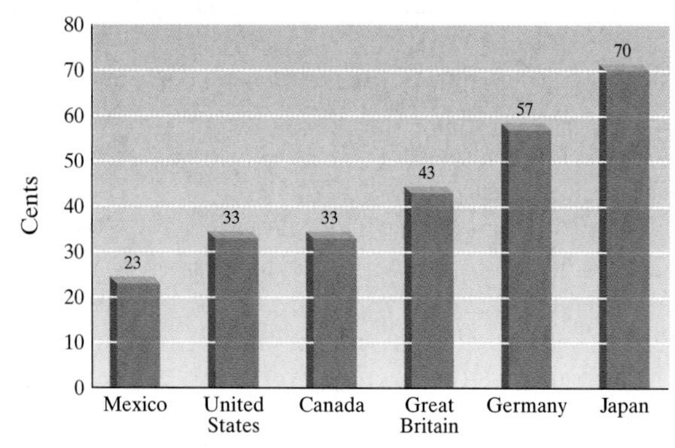

Source: U.S. Postal Service

 a) How much more did it cost to mail a letter in Japan than in Mexico?

 b) How many times greater was the cost of mailing a letter in Japan than in Mexico?

36. The following graph shows how the Sudden Infant Death Syndrome (SIDS) rate has declined as more and more parents place their babies to sleep on their backs instead of their stomachs.

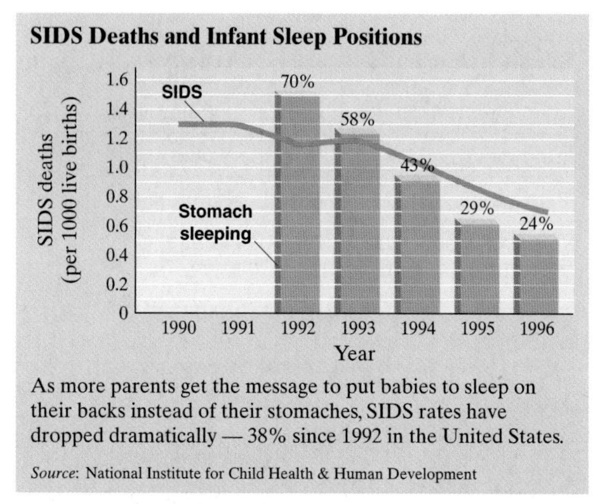

As more parents get the message to put babies to sleep on their backs instead of their stomaches, SIDS rates have dropped dramatically — 38% since 1992 in the United States.

Source: National Institute for Child Health & Human Development

 a) Estimate the SIDS rate (number of SIDS deaths per 1000 live births) in 1990 and 1996.

 b) Estimate the difference in the SIDS rate from 1990 to 1996.

 c) Determine the difference in the percent of babies put to sleep on their stomachs from 1992 to 1996.

37. The following chart provides some information about Hong Kong, China, and the United States.

Comparing Vital Statistics

	Hong Kong	China	United States
Population, in millions	5.5	1,203	263.4
Population per sq. mi.	15,158	327	75
Literacy	77%	78%	97%
Life expectancy	80	68	76
People per telephone	1.5	36.4	1.3
Per capita GNP	$21,650	$530	$25,860

 a) The table provides the number of *people per telephone*. Determine the *number* of telephones in Hong Kong and the United States. Explain how you determined your answer.

 b) Even though the *literacy* rate in the United States is greater than in Hong Kong, there are more illiterate people in the United States than in Hong Kong. Using this table, explain why.

38. Wanda Fadel's mean average on six exams is 78. Find the sum of her scores.

39. In 1997, the total number of points scored by the Notre Dame football team in eight games was 184. Determine the mean number of points scored per game by Notre Dame.

40. Construct a set of five pieces of data with a mean of 70 and no two values the same.

41. A mean of 60 on all exams is needed to pass a course. On his first five exams Lamond Paine's grades are 50, 59, 67, 80, and 56.

a) What is the minimum grade that Lamond can receive on the sixth exam to pass the course?

b) An average of 70 is needed to get a C in the course. Is it possible for Lamond to get a C? If so, what is the minimum grade that Lamond can receive on the sixth exam?

42. Consider the following data provided by the U.S. Bureau of the Census in 1995. The data shows the average lifetime earnings of individuals with different educational backgrounds.

Assume the average person works 40 years, for 40 hours per week, for 48 weeks a year (disregard paid holidays and vacations).

Average Lifetime Earnings

Education	Earnings
Professional degree	$3 million
Doctorate degree	$2.1 million
Master's degree	$1.7 million
Bachelor's degree	$1.5 million
Associate degree	$1.0 million
High school diploma	$821,000

a) Determine the number of hours worked in a lifetime.

b) Determine the average hourly wage of a person who has received a high school diploma.

c) Determine the average hourly wage for a person who has received a professional degree.

Group Activity

Discuss and answer Exercise 43 as a group.

43. a) Each member of your group should ask eight people in the class to select a number from 1 to 5. Each group member should ask a different set of people and record the eight results.

b) Each member should now compute the mean of his or her eight pieces of data.

c) Group member 1: Find the mean of the three means determined in part **b)**.

d) Group member 2: Compute the mean of all the individual pieces of data collected from each group member in part **a)**.

e) Group member 3: Determine the difference in the means obtained in parts **c)** and **d)**. Were the values fairly close?

f) Now as a group, construct a bar graph, with the numbers 1 through 5 on the horizontal scale, which shows how many times each number from 1 to 5 was selected.

1.3 FRACTIONS

1. **Learn multiplication symbols and recognize factors.**
2. **Simplify fractions.**
3. **Multiply fractions.**
4. **Divide fractions.**
5. **Add and subtract fractions.**
6. **Change mixed numbers to fractions and vice versa.**

Students taking algebra for the first time often ask, "What is the difference between arithmetic and algebra?" When doing arithmetic, all the quantities used in the calculations are known. In algebra, however, one or more of the quantities are unknown and must be found.

EXAMPLE 1 Mrs. Clark has 2 cups of flour. A recipe calls for 3 cups of flour. How many additional cups does she need?

Solution The answer is 1 cup.

Although very elementary, this is an example of an algebraic problem. The unknown quantity is the number of additional cups of flour needed.

An understanding of decimal numbers (see Appendix A) and fractions is essential to success in algebra. You will need to know how to simplify a fraction and how to add, subtract, multiply, and divide fractions. We will review these topics in this section. We will also explain the meaning of factors.

1) Learn Multiplication Symbols and Recognize Factors

In algebra we often use letters called **variables** to represent numbers. Letters commonly used as variables are x, y, and z. Variables are usually shown in italics. So that we do not confuse the variable x with the times sign, we often use different notation to indicate multiplication.

Multiplication Symbols

If a and b represent any two mathematical quantities, then each of the following may be used to indicate the product of a and b ("a times b").

$$ab \quad a \cdot b \quad a(b) \quad (a)b \quad (a)(b)$$

Examples

3 times 4 may be written:	3 times x may be written:	x times y may be written:
	$3x$	xy
$3(4)$	$3(x)$	$x(y)$
$(3)\,4$	$(3)x$	$(x)y$
$(3)(4)$	$(3)(x)$	$(x)(y)$
$3 \cdot 4$	$3 \cdot x$	$x \cdot y$

Throughout algebra we will be discussing factors.

Definition

The numbers or variables that are multiplied in a multiplication problem are called **factors**.

If $a \cdot b = c$, then a and b are *factors* of c.

For example, in $3 \cdot 5 = 15$, the numbers 3 and 5 are factors of the product 15. In $2 \cdot 15 = 30$, the numbers 2 and 15 are factors of the product 30. Note that 30 has many other factors. Since $5 \cdot 6 = 30$, the numbers 5 and 6 are also factors of 30. Since $3x$ means 3 times x, both the 3 and the x are factors of $3x$.

2) Simplify Fractions

Now we have the necessary information to discuss **fractions**. The top number of a fraction is called the **numerator**, and the bottom number is called the **denominator**. In the fraction $\frac{3}{5}$, the 3 is the numerator and the 5 is the denominator.

A fraction is **simplified** (or **reduced to its lowest terms**) when the numerator and denominator have no common factors other than 1. To simplify a fraction, follow these steps.

To Simplify a Fraction

1. Find the largest number that will divide (without remainder) both the numerator and the denominator. This number is called the **greatest common factor** (GCF).
2. Then divide both the numerator and the denominator by the greatest common factor.

If you do not remember how to find the greatest common factor of two or more numbers, read Appendix B.

EXAMPLE 2 Simplify **a)** $\frac{10}{25}$ **b)** $\frac{6}{18}$.

Solution **a)** The largest number that divides both 10 and 25 is 5. Therefore, 5 is the greatest common factor. Divide both the numerator and the denominator by 5 to simplify the fraction.

$$\frac{10}{25} = \frac{10 \div 5}{25 \div 5} = \frac{2}{5}$$

b) Both 6 and 18 can be divided by 1, 2, 3, and 6. The largest of these numbers, 6, is the greatest common factor. Divide both the numerator and the denominator by 6.

$$\frac{6}{18} = \frac{6 \div 6}{18 \div 6} = \frac{1}{3}$$

Note in Example **2b)** that both the numerator and denominator could have been written with a factor of 6. Then the common factor 6 could be divided out.

$$\frac{6}{18} = \frac{1 \cdot \cancel{6}}{3 \cdot \cancel{6}} = \frac{1}{3}$$

NOW TRY EXERCISE 17

When you work with fractions you should simplify your answers.

3 Multiply Fractions

To multiply two or more fractions, multiply their numerators together and multiply their denominators together.

To Multiply Fractions

$$\frac{a}{b} \cdot \frac{c}{d} = \frac{ac}{bd}$$

EXAMPLE 3 Multiply $\frac{7}{13}$ by $\frac{5}{12}$.

Solution $\frac{7}{13} \cdot \frac{5}{12} = \frac{7 \cdot 5}{13 \cdot 12} = \frac{35}{156}$

Before multiplying fractions, to help avoid having to simplify an answer, we often divide both a numerator and a denominator by a common factor.

EXAMPLE 4 Divide a numerator and a denominator by a common factor and then multiply.

$$\frac{6}{17} \cdot \frac{5}{12}$$

Solution Since the numerator 6 and the denominator 12 can both be divided by the common factor 6, we divide out the 6 first. Then we multiply.

$$\frac{6}{17} \cdot \frac{5}{12} = \frac{\overset{1}{\cancel{6}}}{17} \cdot \frac{5}{\underset{2}{\cancel{12}}} = \frac{1 \cdot 5}{17 \cdot 2} = \frac{5}{34}$$

EXAMPLE 5 Multiply $\frac{27}{40} \cdot \frac{16}{9}$.

Solution

$$\frac{27}{40} \cdot \frac{16}{9} = \frac{\overset{3}{\cancel{27}}}{40} \cdot \frac{16}{\underset{1}{\cancel{9}}} \qquad \text{Divide both 27 and 9 by 9.}$$

$$= \frac{\overset{3}{\cancel{27}}}{\underset{5}{\cancel{40}}} \cdot \frac{\overset{2}{\cancel{16}}}{\underset{1}{\cancel{9}}} \qquad \text{Divide both 40 and 16 by 8.}$$

$$= \frac{3 \cdot 2}{5 \cdot 1} = \frac{6}{5}$$

NOW TRY EXERCISE 45

The numbers 0, 1, 2, 3, 4, ... are called **whole numbers**. The three dots after the 4 indicate that the whole numbers continue indefinitely in the same manner. Thus the numbers 468 and 5043 are also whole numbers. Whole numbers will be discussed further in Section 1.4. To multiply a whole number by a fraction, write the whole number with a denominator of 1 and then multiply.

EXAMPLE 6 Some engines run on a mixture of gas and oil. A particular lawn mower engine requires a mixture of $\frac{5}{64}$ gallon of oil for each gallon of gasoline used. A lawn care company wishes to make a mixture for this engine using 12 gallons of gasoline. How much oil must be used?

Solution We must multiply 12 by $\frac{5}{64}$ to determine the amount of oil that must be used. First we write 12 as $\frac{12}{1}$, then we divide both 12 and 64 by their greatest common factor, 4, as follows.

$$12 \cdot \frac{5}{64} = \frac{12}{1} \cdot \frac{5}{64} = \frac{\overset{3}{\cancel{12}}}{1} \cdot \frac{5}{\underset{16}{\cancel{64}}} = \frac{3 \cdot 5}{1 \cdot 16} = \frac{15}{16}$$

Thus, $\frac{15}{16}$ gallon of oil must be added to the 12 gallons of gasoline to make the proper mixture.

4) Divide Fractions

To divide one fraction by another, invert the divisor (the second fraction if written with ÷) and proceed as in multiplication.

To Divide Fractions

$$\frac{a}{b} \div \frac{c}{d} = \frac{a}{b} \cdot \frac{d}{c} = \frac{ad}{bc}$$

EXAMPLE 7 Divide $\frac{3}{5} \div \frac{5}{6}$.

Solution $\frac{3}{5} \div \frac{5}{6} = \frac{3}{5} \cdot \frac{6}{5} = \frac{3 \cdot 6}{5 \cdot 5} = \frac{18}{25}$

Sometimes, rather than being asked to obtain the answer to a problem by adding, subtracting, multiplying, or dividing, you may be asked to evaluate an expression. To **evaluate** an expression means to obtain the answer to the problem using the operations given.

EXAMPLE 8 Evaluate $\frac{3}{8} \div 9$.

Solution Write 9 as $\frac{9}{1}$. Then invert the divisor and multiply.

$$\frac{3}{8} \div 9 = \frac{3}{8} \div \frac{9}{1} = \frac{\overset{1}{\cancel{3}}}{8} \cdot \frac{1}{\underset{3}{\cancel{9}}} = \frac{1}{24}$$

NOW TRY EXERCISE 53

5) Add and Subtract Fractions

Fractions that have the same (or a common) *denominator can be added or subtracted.* To add (or subtract) fractions with the same denominator, add (or subtract) the numerators and keep the common denominator.

To Add and Subtract Fractions

$$\frac{a}{c} + \frac{b}{c} = \frac{a+b}{c} \quad \text{or} \quad \frac{a}{c} - \frac{b}{c} = \frac{a-b}{c}$$

EXAMPLE 9 Add **a)** $\frac{9}{15} + \frac{2}{15}$ **b)** $\frac{8}{13} - \frac{5}{13}$

Solution **a)** $\frac{9}{15} + \frac{2}{15} = \frac{9+2}{15} = \frac{11}{15}$ **b)** $\frac{8}{13} - \frac{5}{13} = \frac{8-5}{13} = \frac{3}{13}$

To add (or subtract) fractions with unlike denominators, we must first rewrite each fraction with the same, or a common, denominator. The smallest number that is divisible by two or more denominators is called the **least common denominator** or **LCD**. *If you have forgotten how to find the least common denominator review Appendix B now.*

EXAMPLE 10 Add $\frac{1}{2} + \frac{1}{5}$.

Solution We cannot add these fractions until we rewrite them with a common denominator. Since the lowest number that both 2 and 5 divide into (without remainder) is 10, we will first rewrite both fractions with the least common denominator of 10.

$$\frac{1}{2} = \frac{1}{2} \cdot \frac{5}{5} = \frac{5}{10} \quad \text{and} \quad \frac{1}{5} = \frac{1}{5} \cdot \frac{2}{2} = \frac{2}{10}$$

Now we add.

$$\frac{1}{2} + \frac{1}{5} = \frac{5}{10} + \frac{2}{10} = \frac{7}{10}$$

Note that multiplying both the numerator and denominator by the same number is the same as multiplying by 1. Thus the value of the fraction does not change.

EXAMPLE 11 How much larger is $\frac{3}{4}$ inch than $\frac{2}{3}$ inch?

Solution To find how much larger, we need to subtract $\frac{2}{3}$ inch from $\frac{3}{4}$ inch.

$$\frac{3}{4} - \frac{2}{3}$$

The least common denominator is 12. Therefore, we rewrite both fractions with a denominator of 12.

$$\frac{3}{4} = \frac{3}{4} \cdot \frac{3}{3} = \frac{9}{12} \quad \text{and} \quad \frac{2}{3} = \frac{2}{3} \cdot \frac{4}{4} = \frac{8}{12}$$

Now we subtract.

$$\frac{3}{4} - \frac{2}{3} = \frac{9}{12} - \frac{8}{12} = \frac{1}{12}$$

NOW TRY EXERCISE 69

Therefore, $\frac{3}{4}$ inch is $\frac{1}{12}$ inch greater than $\frac{2}{3}$ inch.

Avoiding Common Errors

It is important that you realize that dividing out a common factor in the numerator of one fraction and the denominator of a different fraction can be performed only when multiplying fractions. *This process cannot be performed when adding or subtracting fractions.*

CORRECT MULTIPLICATION PROBLEMS	INCORRECT ADDITION PROBLEMS
$\frac{\overset{1}{\cancel{3}}}{5} \cdot \frac{1}{\underset{1}{\cancel{3}}}$	$\frac{\overset{1}{\cancel{3}}}{5} + \frac{1}{\underset{1}{\cancel{3}}}$ (crossed out)
$\frac{\overset{2}{\cancel{8}} \cdot 3}{\underset{1}{\cancel{4}}}$	$\frac{\overset{2}{\cancel{8}} + 3}{\underset{1}{\cancel{4}}}$ (crossed out)

6) Change Mixed Numbers to Fractions and Vice Versa

Consider the number $5\frac{2}{3}$. This is an example of a **mixed number**. A mixed number consists of a whole number followed by a fraction. The mixed number $5\frac{2}{3}$ means $5 + \frac{2}{3}$.

To Change a Mixed Number to a Fraction

1. Multiply the denominator of the fraction in the mixed number by the whole number preceding it.
2. Add the numerator of the fraction in the mixed number to the product obtained in step 1. This sum represents the numerator of the fraction we are seeking. The denominator of the fraction we are seeking is the same as the denominator of the fraction in the mixed number.

EXAMPLE 12 Change the mixed number $5\frac{2}{3}$ to a fraction.

Solution Multiply the denominator, 3, by the whole number, 5, to get a product of 15. To this product add the numerator, 2. This sum, 17, represents the numerator of the fraction. The denominator of the fraction we are seeking is the same as the denominator of the fraction in the mixed number, 3. Thus, $5\frac{2}{3} = \frac{17}{3}$.

$$5\frac{2}{3} = \frac{15 + 2}{3} = \frac{17}{3}$$

To Change a Fraction Greater Than 1 to a Mixed Number

1. Divide the numerator by the denominator. Note the quotient and remainder.
2. Write the mixed number. The quotient found in step 1 is the whole number part of the mixed number. The remainder is the numerator of the fraction in the mixed number. The denominator in the fraction of the mixed number will be the same as the denominator in the original fraction.

EXAMPLE 13 Change $\frac{17}{3}$ to a mixed number.

Solution

$$\begin{array}{r} 5 \\ 3\overline{)17} \\ \underline{15} \\ 2 \end{array}$$

5 ← *Whole number*

Denominator (or divisor) → 3

2 ← *Remainder*

$$\frac{17}{3} = 5\frac{2}{3}$$

2 ← *Remainder*

3 ← *Denominator (or divisor)*

5 ← *Whole number*

Thus, $\frac{17}{3}$ changed to a mixed number is $5\frac{2}{3}$.

To add, subtract, multiply, or divide mixed numbers, we often first change the mixed numbers to fractions.

EXAMPLE 14 To repair a plumbing leak, a coupling $\frac{1}{2}$ inch long is glued to a piece of plastic pipe that is $2\frac{3}{16}$ inches long. How long is the combined length? See Figure 1.4.

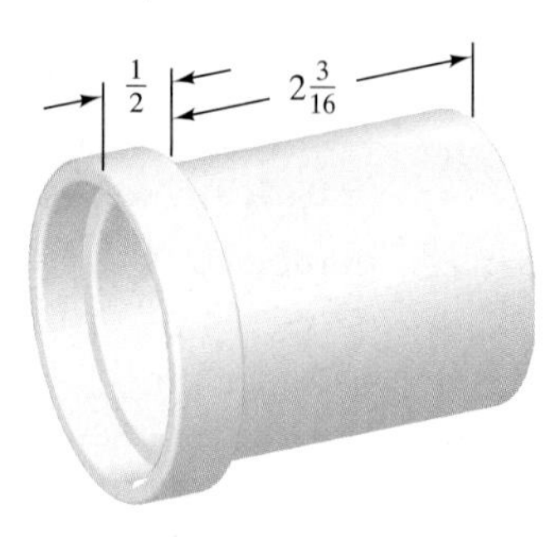

FIGURE 1.4

Solution **Understand and Translate** We need to add $2\frac{3}{16}$ inches and $\frac{1}{2}$ inch to obtain the combined lengths. We will change $2\frac{3}{16}$ to a fraction, then add the fractions.

Carry Out $$2\frac{3}{16} + \frac{1}{2} = \frac{35}{16} + \frac{1}{2} = \frac{35}{16} + \frac{8}{16} = \frac{35+8}{16} = \frac{43}{16} = 2\frac{11}{16}$$

Check and Answer The answer appears reasonable. Thus the total length is $2\frac{11}{16}$ inches.

EXAMPLE 15 The graph in Figure 1.5 shows the value of a share of Grove Tech stock on October 8 and October 9, 1999. How much had the value of a share increased from October 8 to October 9?

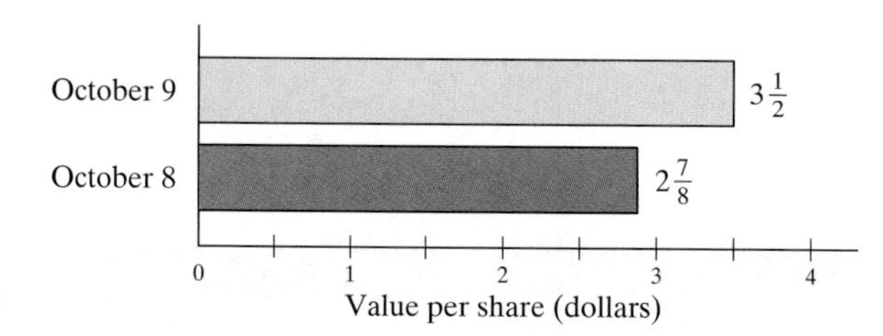

FIGURE 1.5

Solution **Understand and Translate** To find the increase, we need to subtract the value on October 8, $2\frac{7}{8}$, from the value on October 9, $3\frac{1}{2}$. We will change both mixed numbers to fractions, then subtract.

Carry Out $$3\frac{1}{2} - 2\frac{7}{8} = \frac{7}{2} - \frac{23}{8} = \frac{28}{8} - \frac{23}{8} = \frac{28-23}{8} = \frac{5}{8}$$

Check and Answer By examining the graph, we see that the answer is reasonable. Thus, the value of a share of Grove Tech stock increased by $\frac{5}{8}$ dollar per share.

EXAMPLE 16 A recipe calls for $1\frac{1}{4}$ teaspoon of teriyaki seasoning for each pound of beef. To cook $4\frac{1}{2}$ pounds of beef, how many teaspoons of teriyaki seasoning are needed?

Solution **Understand and Translate** We must multiply $1\frac{1}{4}$ by $4\frac{1}{2}$. To perform the multiplication, we change each mixed number to a fraction, and then multiply.

Carry Out $$1\frac{1}{4} \cdot 4\frac{1}{2} = \frac{5}{4} \cdot \frac{9}{2} = \frac{5 \cdot 9}{4 \cdot 2} = \frac{45}{8} = 5\frac{5}{8}$$

Check and Answer Because $1\frac{1}{4}$ teaspoons are used for 1 pound of beef, we would expect that $4\frac{1}{2}$ pounds of beef would use more than 4 teaspoons of seasoning. So the answer is reasonable. Therefore, $5\frac{5}{8}$ teaspoons of teriyaki seasoning are needed.

NOW TRY EXERCISE 91

EXAMPLE 17 A rectangular piece of material 3 feet wide by $12\frac{1}{2}$ feet long is cut into five equal strips, as shown in Figure 1.6. Find the dimensions of each strip.

3 ft

$12\frac{1}{2}$ ft

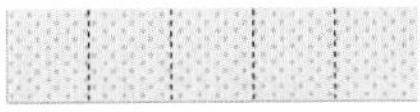

FIGURE 1.6

Solution **Understand and Translate** We know from the diagram that one side will have a width of 3 feet. To find the length of the strips, we need to divide $12\frac{1}{2}$ by 5.

Carry Out

$$12\frac{1}{2} \div 5 = \frac{25}{2} \div \frac{5}{1} = \frac{\overset{5}{\cancel{25}}}{2} \cdot \frac{1}{\underset{1}{\cancel{5}}} = \frac{5}{2} = 2\frac{1}{2}$$

Check and Answer If you multiply $2\frac{1}{2}$ times 5 you obtain the original length, $12\frac{1}{2}$. Thus the calculation was correct. The dimensions of each strip will be 3 feet by $2\frac{1}{2}$ feet.

Exercise Set 1.3

Concept/Writing Exercises

1. **a)** What are variables?
   b) What letters are often used to represent variables?
2. What are factors?
3. Show five different ways that "5 times x" may be written.
4. In a fraction, what is the name of the **a)** top number and **b)** bottom number?
5. Explain how to simplify a fraction.
6. Consider parts **a)** and **b)** below.

   a) $\frac{\overset{1}{\cancel{3}}}{5} \cdot \frac{1}{\underset{2}{\cancel{6}}}$ **b)** $\frac{\overset{1}{\cancel{3}}}{\underset{2}{\cancel{6}}}$

   Which part shows simplifying a fraction? Explain.
7. One of the following procedures is correct and one is incorrect. Indicate which one is correct and which one is incorrect. Explain the error in the incorrect procedure.

   a) $\frac{2}{\underset{1}{\cancel{3}}} + \frac{\overset{1}{\cancel{3}}}{5}$ **b)** $\frac{2}{\underset{1}{\cancel{3}}} \cdot \frac{\overset{1}{\cancel{3}}}{5}$
8. Explain how to multiply fractions.

In Exercises 9 and 10 indicate any parts where a common factor can be divided out as a first step in evaluating each expression. Explain your answer.

9. **a)** $\frac{4}{5} + \frac{1}{4}$ **b)** $\frac{4}{5} - \frac{1}{4}$ **c)** $\frac{4}{5} \cdot \frac{1}{4}$ **d)** $\frac{4}{5} \div \frac{1}{4}$
10. **a)** $6 + \frac{5}{12}$ **b)** $6 \cdot \frac{5}{12}$ **c)** $6 - \frac{5}{12}$ **d)** $6 \div \frac{5}{12}$
11. Explain how to divide fractions.
12. Explain how to add or subtract fractions.
13. Explain how to convert a mixed number to a fraction.
14. Explain how to convert a fraction whose numerator is greater than its denominator into a mixed number.

Practice the Skills

Simplify each fraction. If a fraction is already simplified, so state.

15. $\frac{3}{9}$
16. $\frac{30}{6}$
17. $\frac{10}{15}$
18. $\frac{19}{25}$
19. $\frac{13}{13}$
20. $\frac{9}{21}$
21. $\frac{36}{76}$
22. $\frac{16}{72}$
23. $\frac{40}{264}$
24. $\frac{60}{105}$
25. $\frac{12}{25}$
26. $\frac{80}{124}$

Convert each mixed number to a fraction.

27. $1\frac{7}{8}$
28. $5\frac{1}{3}$
29. $2\frac{13}{15}$
30. $6\frac{5}{12}$
31. $3\frac{3}{4}$
32. $8\frac{1}{2}$
33. $4\frac{13}{19}$
34. $3\frac{3}{32}$

Write each fraction as a mixed number.

35. $\frac{4}{3}$ **36.** $\frac{13}{3}$ **37.** $\frac{15}{4}$ **38.** $\frac{9}{2}$

39. $\frac{150}{20}$ **40.** $\frac{67}{13}$ **41.** $\frac{32}{7}$ **42.** $\frac{72}{14}$

Find each product or quotient. Simplify the answer.

43. $\frac{1}{2} \cdot \frac{3}{4}$ **44.** $\frac{3}{4} \cdot \frac{4}{5}$ **45.** $\frac{5}{12} \cdot \frac{4}{15}$ **46.** $\frac{1}{2} \cdot \frac{12}{15}$

47. $\frac{3}{4} \div \frac{1}{2}$ **48.** $\frac{15}{16} \cdot \frac{4}{3}$ **49.** $\frac{2}{5} \div \frac{1}{8}$ **50.** $\frac{3}{8} \cdot \frac{10}{11}$

51. $\frac{5}{12} \div \frac{4}{3}$ **52.** $\frac{15}{4} \cdot \frac{2}{3}$ **53.** $\frac{10}{3} \div \frac{5}{9}$ **54.** $\frac{12}{5} \div \frac{3}{7}$

55. $\left(2\frac{1}{5}\right)\left(\frac{7}{8}\right)$ **56.** $\frac{28}{13} \cdot \frac{2}{7}$ **57.** $5\frac{3}{8} \div 1\frac{1}{4}$ **58.** $4\frac{4}{5} \div \frac{8}{15}$

Add or subtract. Simplify each answer.

59. $\frac{1}{3} + \frac{2}{3}$ **60.** $\frac{3}{7} + \frac{1}{7}$ **61.** $\frac{5}{12} - \frac{1}{12}$ **62.** $\frac{18}{36} - \frac{1}{36}$

63. $\frac{8}{17} + \frac{2}{34}$ **64.** $\frac{3}{7} + \frac{17}{35}$ **65.** $\frac{4}{5} + \frac{6}{15}$ **66.** $\frac{5}{6} - \frac{3}{4}$

67. $\frac{1}{9} - \frac{1}{18}$ **68.** $\frac{11}{28} + \frac{1}{7}$ **69.** $\frac{5}{12} - \frac{1}{8}$ **70.** $\frac{5}{8} - \frac{4}{7}$

71. $\frac{7}{12} - \frac{2}{9}$ **72.** $\frac{3}{7} + \frac{5}{12}$ **73.** $\frac{7}{8} - \frac{1}{15}$ **74.** $\frac{1}{32} + \frac{5}{12}$

75. $4\frac{1}{3} - 3\frac{2}{9}$ **76.** $2\frac{3}{8} + \frac{1}{4}$ **77.** $5\frac{3}{4} - \frac{1}{3}$ **78.** $2\frac{1}{3} + 1\frac{1}{8}$

Problem Solving

79. The following graph shows Jamie Winston's height, in inches, on her 10th and 15th birthdays. How much had Jamie grown in the 5 years?

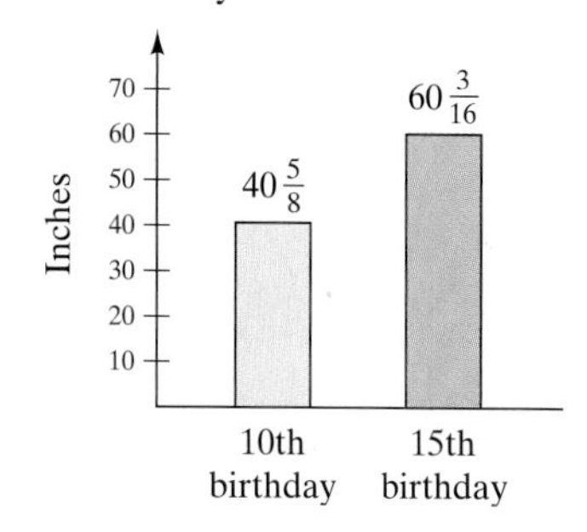

80. The following graph shows the progress of the Davenport Paving Company in paving the Memorial Highway. How much of the highway was paved from June through August?

Highway Paved in Selected Months

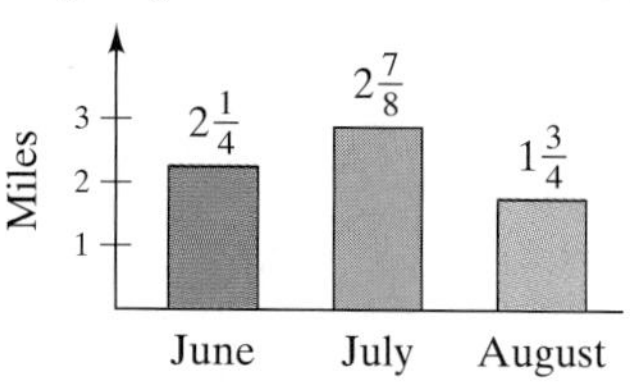

In many problems you will need to subtract a fraction from 1, where 1 represents "the whole" or the "total amount." Exercises 81–84 are answered by subtracting the given fraction from 1.

81. In the United States, $\frac{9}{10}$ of all natural disasters involve flooding. What fraction of natural disasters in the United States do not involve flooding?

82. The probability that an event does not occur may be found by subtracting the probability that the event does occur from 1. If the probability that global warming is occurring is $\frac{7}{9}$, find the probability that global warming is not occurring.

83. The following graph provides information about Earth's surface. What fraction of Earth's surface is covered by water?

Earth's Surface

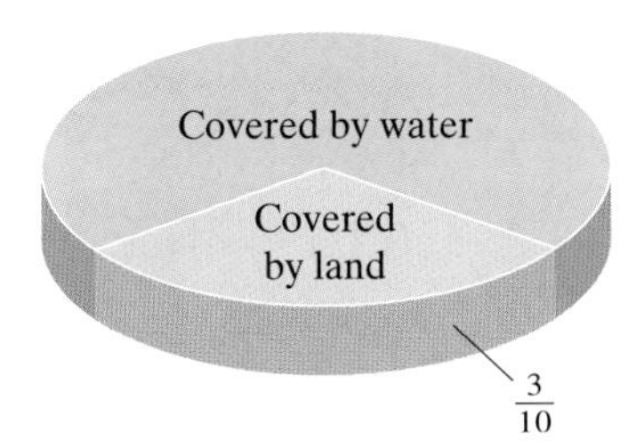

84. The following graph provides information about oats grown in the United States. What fraction of all the oats grown in the United States is not fed to animals?

Oats Grown in the United Stat

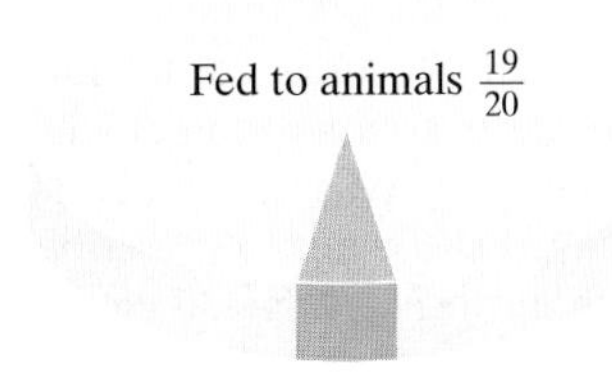

85. A flatbed tow truck weighing $4\frac{1}{2}$ tons is carrying two cars. One car weighs $1\frac{1}{6}$ tons, the other weighs $1\frac{3}{4}$ tons. What is the total weight of the tow truck and the two cars?

86. A length of $3\frac{1}{16}$ inches is cut from a piece of wood $16\frac{3}{4}$ inches long. What is the length of the remaining piece of wood?

87. At the beginning of the day a stock was selling for $11\frac{7}{8}$ dollars. At the close of the session it was selling for $13\frac{3}{4}$ dollars. How much did the stock gain that day?

88. Dawn Foster cuts a piece of wood measuring $3\frac{1}{8}$ inches into two equal pieces. How long is each piece?

89. The inseam on a new pair of pants is 30 inches. If Sean Leland's inseam is $28\frac{3}{8}$ inches by how much will the pants need to be shortened?

90. The instructions on a turkey indicate that a 12- to 16-pound turkey should bake at 325°F for about 22 minutes per pound. Donna Draus is planning to bake a $13\frac{1}{2}$-pound turkey. Approximately how long should the turkey be baked?

91. A recipe for pot roast calls for $\frac{1}{4}$ cup chopped onions for each pound of beef. For $5\frac{1}{2}$ pounds of beef, how many cups of chopped onions are needed?

92. Rick O'Shea wants to fence in his backyard as shown. The three sides to be fenced measure $16\frac{2}{3}$ yards, $22\frac{2}{3}$ yards, and $14\frac{1}{8}$ yards.

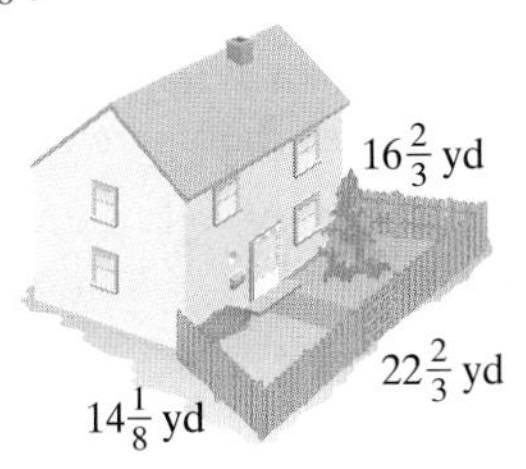

a) How much fence will Rick need?

b) If Rick buys 60 yards of fence, how much will be left over?

93. A bottle of shampoo contains 15 fluid ounces. If Tierra Bentley uses $\frac{3}{8}$ of an ounce each time she washes her hair, how many times can Tierra wash her hair using this bottle?

94. A nurse must give $\frac{1}{16}$ milligram of a drug for each kilogram of patient weight. If Mr. Duncan weighs (or has a mass of) 80 kilograms, find the amount of the drug Mr. Duncan should be given.

95. An insulated window for a house is made up of two pieces of glass, each $\frac{1}{4}$-inch thick, with a 1-inch space between them. What is the total thickness of this window?

96. Hal DeWitt normally weighs 160 pounds. He loses $\frac{1}{5}$ of his weight due to dieting.

a) How much weight did Hal lose?

b) What was his weight after dieting?

97. Rod Hamilton, a chemist, has $32\frac{1}{2}$ milliliters of a solution. He needs to pour this solution into vials that hold $2\frac{1}{2}$ milliliters each. How many vials can Rod fill with this solution?

98. Marcinda James is considering purchasing a mail order computer. The catalog describes the computer as $7\frac{1}{2}$ inches high and the monitor as $14\frac{3}{8}$ inches high. Marcinda is hoping to place the monitor on top of the computer and to place the computer and monitor together in the opening where the computer is shown.

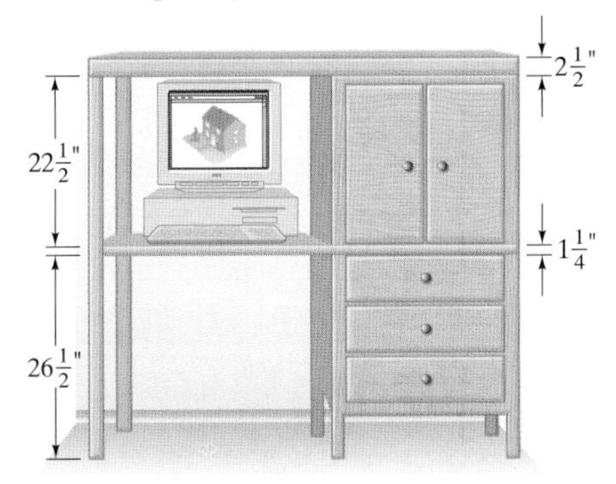

a) Will there be sufficient room to do this?

b) If so, how much extra space will she have?

c) Find the total height of the computer desk

99. A mechanic wishes to use a bolt to fasten a piece of wood $4\frac{1}{2}$ inches thick to a metal tube $2\frac{1}{3}$ inches thick. If the thickness of the nut is $\frac{1}{8}$ inch, find the length of the shaft of the bolt so that the nut fits flush with the end of the bolt (see the figure on the right).

100. If five 2-liter bottles of soda are split evenly among 30 people, how many liters of soda will each person get?

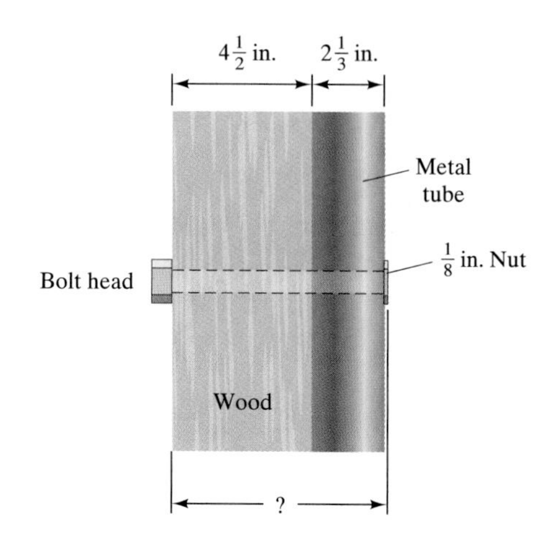

Challenge Problems

101. Add or subtract the following fractions using the rule discussed in this section. Your answer should be a single fraction, and it should contain the symbols given in the exercise.

a) $\frac{*}{a} + \frac{?}{a}$ **b)** $\frac{\odot}{?} - \frac{\square}{?}$

c) $\frac{\triangle}{\square} + \frac{4}{\square}$ **d)** $\frac{x}{3} - \frac{2}{3}$ **e)** $\frac{12}{x} - \frac{4}{x}$

102. Multiply the following fractions using the rule discussed in this section. Your answer should be a single fraction and it should contain the symbols given in the exercise.

a) $\frac{\triangle}{a} \cdot \frac{\square}{b}$ **b)** $\frac{6}{3} \cdot \frac{\triangle}{\square}$

c) $\frac{x}{a} \cdot \frac{y}{b}$ **d)** $\frac{3}{8} \cdot \frac{4}{y}$ **e)** $\frac{3}{x} \cdot \frac{x}{y}$

103. An allopurinal pill comes in 300-milligram doses. Dr. Highland wants a patient to get 450 milligrams each day by cutting the pills in half and taking $\frac{1}{2}$ pill three times a day. If she wants to prescribe enough pills for a 6-month period (assume 30 days per month), how many pills should she prescribe?

Group Activity

Discuss and answer Exercise 104 as a group.

104. The following table gives the amount of each ingredient recommended to make 2, 4, and 8 servings of Betty Crocker Potato Buds®.

Servings	**2**	**4**	**8**
Water	$\frac{2}{3}$ cup	$1\frac{1}{3}$ cups	$2\frac{2}{3}$ cups
Milk	2 tbsp	$\frac{1}{3}$ cup	$\frac{2}{3}$ cup
Butter*	1 tbsp	2 tbsp	4 tbsp
Salt†	$\frac{1}{4}$ tsp	$\frac{1}{2}$ tsp	1 tsp
Potato Buds®	$\frac{2}{3}$ cup	$1\frac{1}{3}$ cup	$2\frac{2}{3}$ cups

*or margarine
†Less salt can be used if desired.

Determine the amount of Potato Buds and milk needed to make 6 servings by the different methods described. When working with milk, 16 tbsp = 1 cup.

a) Group member 1: Determine the amount of Potato Buds and milk needed to make 6 servings by multiplying the amount for 2 servings by 3.

b) Group member 2: Determine the amounts by adding the amounts for 2 servings to the amounts for 4 servings.

c) Group member 3: Determine the amounts by finding the average (mean) of 4 and 8 servings.

d) As a group, determine the amount by subtracting the amount for 2 servings from the amount for 8 servings.

e) As a group, compare your answers from parts **a)** through **d)**. Are they all the same? If not, can you explain why? (This might be a little tricky.)

Cumulative Review Exercises

[1.1] **105.** What is your instructor's name and office hours?

[1.2] **106.** What is the mean of 4, 9, 17, 32, 16?

107. What is the median of 4, 9, 17, 32, 16?

[1.3] **108.** What are variables?

1.4 THE REAL NUMBER SYSTEM

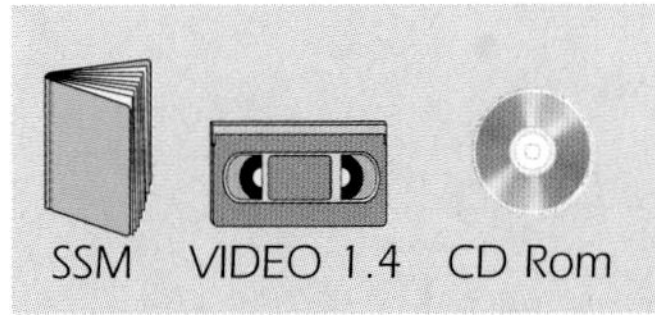

1) ***Identify sets of numbers.***
2) ***Know the structure of the real numbers.***

We will be talking about and using various types of numbers throughout the text. This section introduces you to some of those numbers and to the structure of the real number system. This section is a quick overview. Some of the sets of numbers we mention in this section, such as rational and irrational numbers, are discussed in greater depth later in the text.

1) Identify Sets of Numbers

A **set** is a collection of **elements** listed within braces. The set $\{a, b, c, d, e\}$ consists of five elements, namely a, b, c, d, and e. A set that contains no elements is called an **empty set** (or **null set**). The symbols $\{\ \}$ or $\varnothing$ are used to represent the empty set.

There are many different sets of numbers. Two important sets are the natural numbers and the whole numbers. The whole numbers were introduced earlier.

Natural numbers: $\{1, 2, 3, 4, 5, \dots\}$

Whole numbers: $\{0, 1, 2, 3, 4, 5, \dots\}$

An aid in understanding sets of numbers is the real number line (Fig. 1.7).

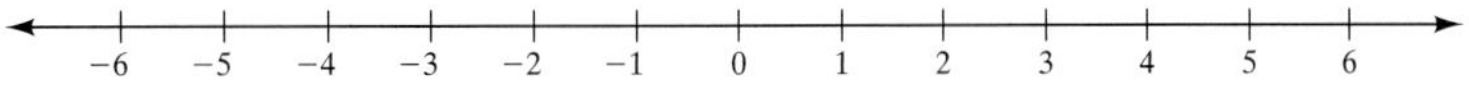

FIGURE 1.7

−6 −5 −4 −3 −2 −1 0 1 2 3 4 5 6

The real number line continues indefinitely in both directions. The numbers to the right of 0 are positive and those to the left of 0 are negative. Zero is neither positive nor negative (Fig. 1.8).

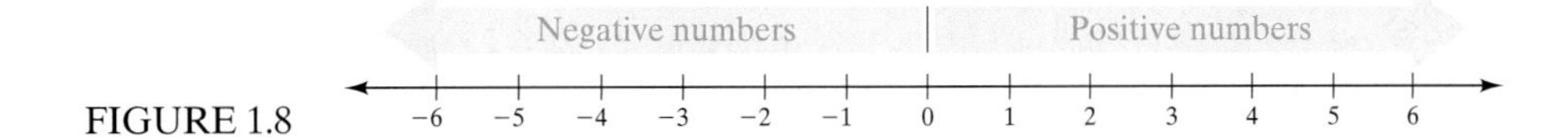

FIGURE 1.8

Figure 1.9 illustrates the natural numbers marked on a number line. The natural numbers are also called the **positive integers** or the **counting numbers**.

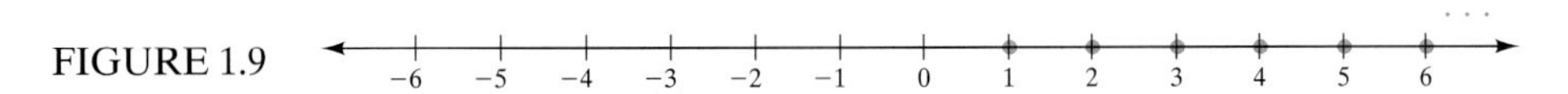

FIGURE 1.9

Another important set of numbers is the integers.

$$\textbf{Integers: } \{\underbrace{\dots, -5, -4, -3, -2, -1}_{\text{Negative integers}}, 0, \underbrace{1, 2, 3, 4, 5, \dots}_{\text{Positive integers}}\}$$

The integers consist of the negative integers, 0, and the positive integers. The integers are marked on the number line in Figure 1.10.

. . . −6 −5 −4 −3 −2 −1 0 1 2 3 4 5 6 . . .

FIGURE 1.10

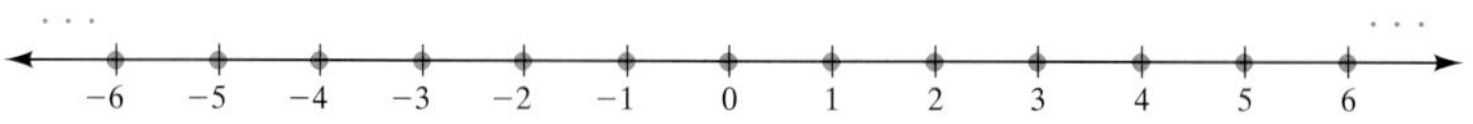

Can you think of any numbers that are not integers? You probably thought of "fractions" or "decimal numbers." Fractions and certain decimal numbers belong to the set of rational numbers. The set of **rational numbers** consists of all the numbers that can be expressed as a quotient of two integers, with the denominator not 0.

Rational numbers: {quotient of two integers, denominator not 0}

The fraction $\frac{1}{2}$ is a quotient of two integers (with the denominator not 0). Thus, $\frac{1}{2}$ is a rational number. The decimal number 0.4 can be written $\frac{4}{10}$ and is therefore a rational number. All integers are also rational numbers since they can be written with a denominator of 1: for example, $3 = \frac{3}{1}, -12 = \frac{-12}{1}$, and $0 = \frac{0}{1}$. Some rational numbers are illustrated on the number line in Figure 1.11.

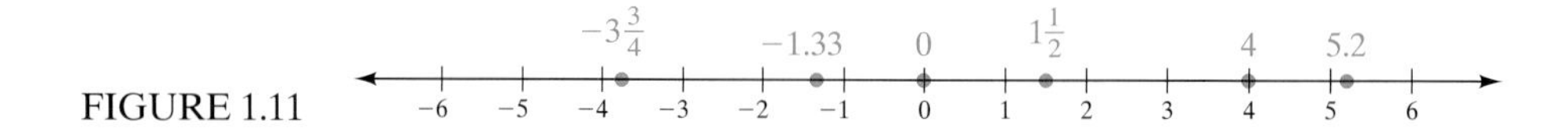

FIGURE 1.11

Most of the numbers that we use are rational numbers; however, some numbers are not rational. Numbers such as the square root of 2, written $\sqrt{2}$, are not rational numbers. Any number that can be represented on the number line that is not a rational number is called an **irrational number**. $\sqrt{2}$ is *approximately* 1.41. Some irrational numbers are illustrated on the number line in Figure 1.12. Rational and irrational numbers will be discussed further in later chapters.

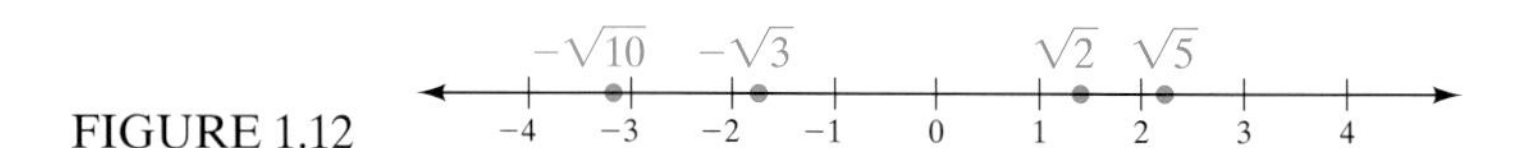

FIGURE 1.12

2) Know the Structure of the Real Numbers

Notice that many different types of numbers can be illustrated on a number line. Any number that can be represented on the number line is a **real number**.

Real numbers: {all numbers that can be represented on a real number line}

The symbol $\mathbb{R}$ is used to represent the set of real numbers. All the numbers mentioned thus far are real numbers. The natural numbers, the whole numbers, the integers, the rational numbers, and the irrational numbers are all real numbers. There are some types of numbers that are not real numbers, but these numbers are beyond the scope of this book. Figure 1.13 illustrates the relationships between the various sets of numbers within the set of real numbers.

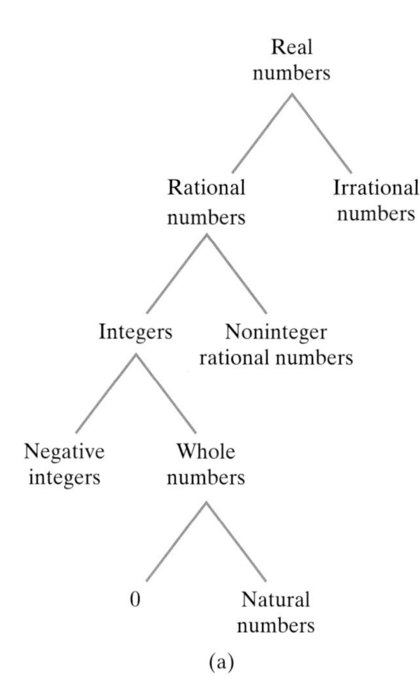

(a)

Real Numbers

Rational numbers (Integers and noninteger rational numbers)	Irrational numbers (Certain* square roots)
-12 4 0	$\sqrt{2}$ $\sqrt{5}$
$\frac{3}{8}$ -1.24	π $\sqrt{12}$
$-1\frac{3}{5}$ -2.463	

*Other higher roots like $\sqrt[3]{2}$ and $\sqrt[4]{5}$ are also irrational numbers.

(b)

FIGURE 1.13

In Figure 1.13a, we can see that when we combine the rational numbers and the irrational numbers we get the real numbers. When we combine the integers with the noninteger rational numbers (such as $\frac{1}{2}$ and 0.42), we get the rational numbers. When we combine the whole numbers and the negative integers, we get the integers.

Consider the natural number 5. If we follow the branches in Figure 1.13a upward, we see that the number 5 is also a whole number, an integer, a rational number, and a real number. Now consider the number $\frac{1}{2}$. It belongs to the noninteger rational numbers. If we follow the branches upward, we can see that $\frac{1}{2}$ is also a rational number and a real number.

EXAMPLE 1 Consider the following set of numbers.

$$\left\{-6, -0.5, 4\frac{1}{2}, -96, \sqrt{3}, 0, 9, -\frac{4}{7}, -2.9, \sqrt{7}, -\sqrt{5}\right\}$$

List the elements of the set that are

a) natural numbers. **b)** whole numbers. **c)** integers.
d) rational numbers. **e)** irrational numbers. **f)** real numbers.

Solution We will list the elements from left to right as they appear in the set. However, the elements may be listed in any order.

a) 9 **b)** 0, 9 **c)** $-6, -96, 0, 9$

d) $-6, -0.5, 4\frac{1}{2}, -96, 0, 9, -\frac{4}{7}, -2.9$

e) $\sqrt{3}, \sqrt{7}, -\sqrt{5}$

f) $-6, -0.5, 4\frac{1}{2}, -96, \sqrt{3}, 0, 9, -\frac{4}{7}, -2.9, \sqrt{7}, -\sqrt{5}$

NOW TRY EXERCISE 47

Exercise Set 1.4

Concept/Writing Exercises

1. What is a set?
2. What is a set that contains no elements called?
3. Describe a set that is an empty set.
4. How do the sets of natural numbers and whole numbers differ?
5. What are two other names for the set of natural numbers?
6. **a)** What is a rational number?
   b) Explain why every integer is a rational number.
7. Explain why the natural number 7 is also a
   a) whole number.
   b) rational number.
   c) real number.
8. Write a paragraph or two describing the structure of the real number system. Explain how whole numbers, counting numbers, integers, rational numbers, irrational numbers, and real numbers are related.

Practice the Skills

List each set of numbers.

9. Integers.
10. Counting numbers.
11. Natural numbers.
12. Positive integers.
13. Whole numbers.
14. Negative integers.

In Exercises 15–46, indicate whether each statement is true or false.

15. -1 is a negative integer.
16. 0 is a whole number.
17. 0 is an integer.
18. -4.25 is a real number.
19. $\frac{3}{5}$ is an integer.
20. 0.6 is an integer.
21. $\sqrt{2}$ is a rational number.
22. $\sqrt{5}$ is a real number.
23. $-\frac{1}{5}$ is a rational number.
24. 0 is a rational number.
25. $-2\frac{1}{3}$ is a rational number.
26. -8 is a rational number.
27. $-\frac{5}{3}$ is an irrational number.
28. $4\frac{5}{8}$ is an irrational number.
29. 0 is not a positive number.
30. Every integer is a rational number.
31. The symbol $\varnothing$ is used to represent the empty set.
32. Every integer is positive.
33. Every negative integer is a real number.
34. Every real number is a rational number.
35. Every rational number is a real number.
36. Every negative number is a negative integer.
37. Some real numbers are not rational numbers.
38. Some rational numbers are not real numbers.
39. When zero is added to the set of counting numbers, the set of whole numbers is formed.
40. All real numbers can be represented on a number line.
41. The symbol $\mathbb{R}$ is used to represent the set of real numbers.
42. Every number greater than zero is a positive integer.
43. Any number to the left of zero on a number line is a negative number.
44. Irrational numbers cannot be represented on a number line.
45. When the negative integers, the positive integers, and 0 are combined, the integers are formed.
46. The natural numbers, counting numbers, and positive integers are different names for the same set of numbers.
47. Consider the following set of numbers.

$$\left\{-\frac{4}{3}, 0, -2, 3, 5\frac{1}{2}, \sqrt{8}, -\sqrt{3}, 1.63, 77\right\}$$

List the numbers that are
a) positive integers.
b) whole numbers.
c) integers.
d) rational numbers.
e) irrational numbers.
f) real numbers.

48. Consider the following set of numbers.

$$\left\{-6, 7, 12.4, -\frac{9}{5}, -2\frac{1}{4}, \sqrt{3}, 0, 9, \sqrt{7}, 0.35\right\}$$

List the numbers that are
a) positive integers.
b) whole numbers.
c) integers.
d) rational numbers.
e) irrational numbers.
f) real numbers.

Problem Solving

Give three examples of numbers that satisfy the given conditions.

49. A real number but not an integer.

50. An integer but not a negative integer.

51. A rational number but not an integer.

52. An irrational number and a positive number.

53. A real number but not a rational number.

54. An integer and a rational number.

55. A negative integer and a rational number.

56. A negative integer and a real number.

57. A real number but not a positive rational number.

58. A rational number but not a negative number.

59. A real number but not an irrational number.

60. An integer but not a positive integer.

Three dots inside a set indicate that the set continues in the same manner. For example, $\{1, 2, 3, \ldots, 84\}$ is the set of natural numbers from 1 up to and including 84. In Exercises 61 and 62, determine the number of elements *in each set.*

61. $\{5, 6, 7, 8, \ldots, 87\}$

62. $\{-4, -3, -2, -1, 0, 1, \ldots, 64\}$

Challenge Problems

The diagrams in Exercises 63 and 64 are called Venn diagrams *(named after the English mathematician John Venn). Venn diagrams are used to illustrate sets. For example, in the diagrams, circle A contains all the elements in set A, and circle B contains all the elements in set B. For each diagram, determine **a)** set A, **b)** set B, **c)** the set of elements that belong to both set A* and *set B, and **d)** the set of elements that belong to either set A* or *set B.*

63.

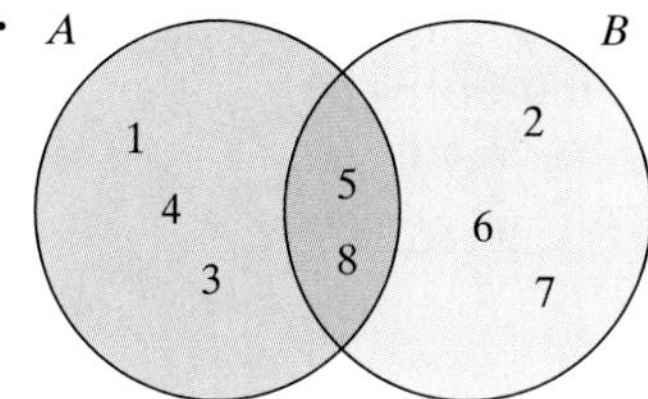

64.

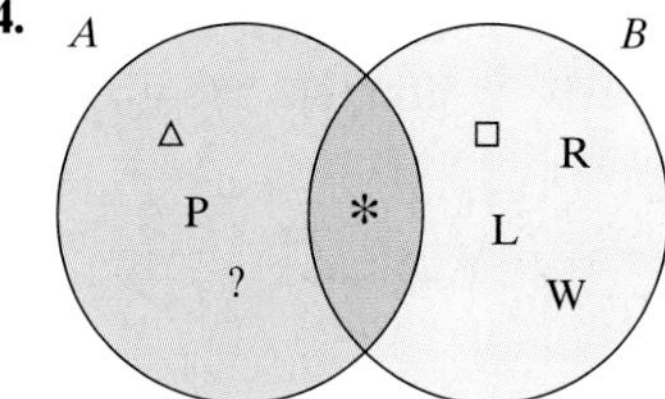

65. Consider the sets $A = \{1, 2, 3, 4\}$ and $B = \{1, 2, 3, 4, \ldots\}$.
a) Explain the difference between set A and set B.
b) How many elements are in set A?
c) How many elements are in set B?
d) Set A is an example of a *finite set*. Can you guess the name given to a set like set B?

66. How many decimal numbers are there
a) between 1.0 and 2.0;
b) between 1.4 and 1.5? Explain your answer.

67. How many fractions are there
a) between 1 and 2;
b) between $\frac{1}{3}$ and $\frac{1}{5}$? Explain your answer.

Group Activity

Discuss and answer Exercise 68 as a group.

68. Set A **union** set B, symbolized $A \cup B$, consists of the set of elements that belong to set A or set B (or both sets). Set A **intersection** set B, symbolized $A \cap B$, consists of the set of elements that both set A and set B have in common. Note that the elements that belong to both sets are listed only once in the union of the sets.

Consider the pairs of sets below.

Group member 1: $A = \{2, 3, 4, 6, 8, 9\}$ $B = \{1, 2, 3, 5, 7, 8\}$
Group member 2: $A = \{a, b, c, d, g, i, j\}$ $B = \{b, c, d, h, m, p\}$
Group member 3: $A = \{\text{red, blue, green, yellow}\}$ $B = \{\text{pink, orange, purple}\}$

a) Group member 1: Find the union and intersection of the sets marked Group member 1.
b) Group member 2: Find the union and intersection of the sets marked Group member 2.
c) Group member 3: Find the union and intersection of the sets marked Group member 3.
d) Now as a group, check each other's work. Correct any mistakes.
e) As a group, using group member 1's sets, construct a Venn diagram like those shown in Exercises 63 and 64.

Cumulative Review Exercises

[1.3] **69.** Convert $4\frac{2}{3}$ to a fraction.

70. Write $\frac{16}{3}$ as a mixed number.

71. Add $\frac{3}{5} + \frac{5}{8}$.

72. Multiply $\left(\frac{5}{9}\right)\left(4\frac{2}{3}\right)$.

1.5 INEQUALITIES

1) Determine which is the greater of two numbers.
2) Find the absolute value of a number.

1) Determine Which Is the Greater of Two Numbers

The number line, which shows numbers increasing from left to right, can be used to explain inequalities (see Fig. 1.14). When comparing two numbers, **the number to the right on the number line is the greater number, and the number to the left is the lesser number**. The symbol $>$ is used to represent the words "is greater than." The symbol $<$ is used to represent the words "is less than."

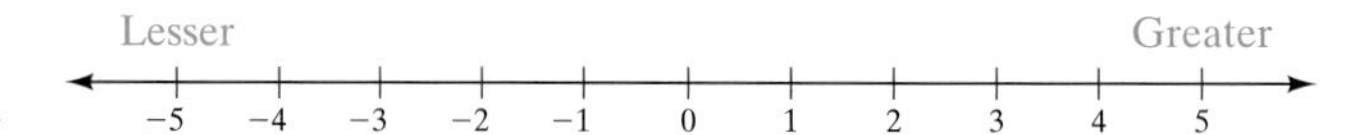

FIGURE 1.14

The statement that the number 3 is greater than the number 2 is written $3 > 2$. Notice that 3 is to the right of 2 on the number line. The statement that the number 0 is greater than the number -1 is written $0 > -1$. Notice that 0 is to the right of -1 on the number line.

Instead of stating that 3 is greater than 2, we could state that 2 is less than 3, written $2 < 3$. Notice that 2 is to the left of 3 on the number line. The statement that the number -1 is less than the number 0 is written $-1 < 0$. Notice that -1 is to the left of 0 on the number line.

EXAMPLE 1 Insert either $>$ or $<$ in the shaded area between each pair of numbers to make a true statement.

a) $-4 \;\square\; -2$ **b)** $-\frac{3}{2} \;\square\; 2.5$ **c)** $\frac{1}{2} \;\square\; \frac{1}{4}$ **d)** $-2 \;\square\; 4$

Solution The points given are shown on the number line (Fig. 1.15).

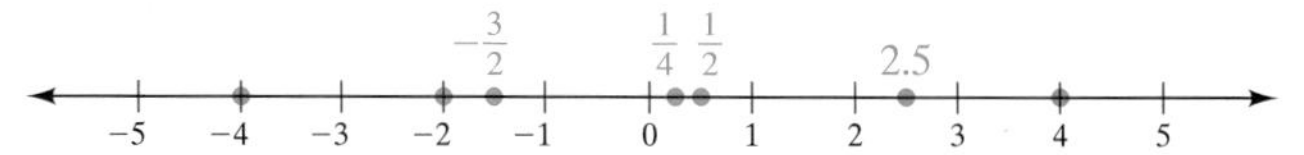

FIGURE 1.15

a) $-4 < -2$; notice that -4 is to the left of -2.

b) $-\frac{3}{2} < 2.5$; notice that $-\frac{3}{2}$ is to the left of 2.5.

c) $\frac{1}{2} > \frac{1}{4}$; notice that $\frac{1}{2}$ is to the right of $\frac{1}{4}$.

d) $-2 < 4$; notice that -2 is to the left of 4.

EXAMPLE 2 Insert either $>$ or $<$ in the shaded area between each pair of numbers to make a true statement.

a) $-1 \;\square\; -2$ **b)** $-1 \;\square\; 0$ **c)** $-2 \;\square\; 2$ **d)** $-4.09 \;\square\; -4.9$

Solution The numbers given are shown on the number line (Fig. 1.16).

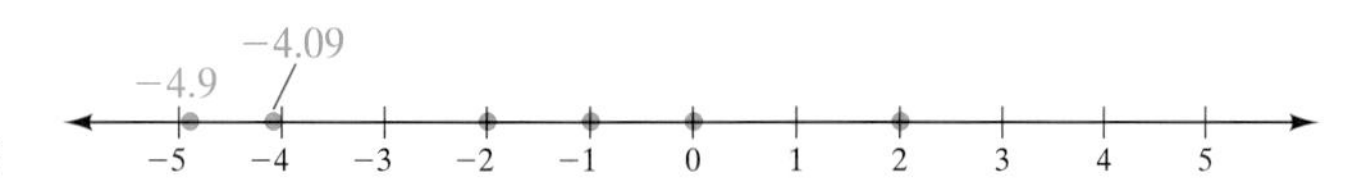

FIGURE 1.16

a) $-1 > -2$; notice that -1 is to the right of -2.

b) $-1 < 0$; notice that -1 is to the left of 0.

c) $-2 < 2$; notice that -2 is to the left of 2.

NOW TRY EXERCISE 25

d) $-4.09 > -4.9$; notice that -4.09 is to the right of -4.9.

2) Find the Absolute Value of a Number

The concept of absolute value can be explained with the help of the number line shown in Figure 1.17. The **absolute value** of a number can be considered the distance between the number and 0 on a number line. Thus, the absolute value of 3, written $|3|$, is 3 since it is 3 units from 0 on a number line. Similarly, the absolute value of -3, written $|-3|$, is also 3 since -3 is 3 units from 0.

$$|3| = 3 \qquad \text{and} \qquad |-3| = 3$$

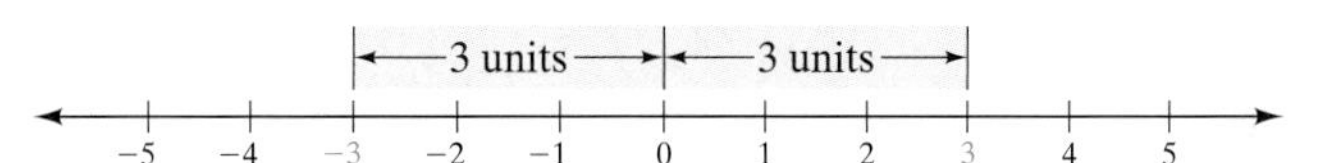

FIGURE 1.17

Since the absolute value of a number measures the distance (without regard to direction) of a number from 0 on the number line, *the absolute value of every number will be either positive or zero.*

Number	Absolute Value of Number
6	$\|6\| = 6$
-6	$\|-6\| = 6$
0	$\|0\| = 0$
$-\frac{1}{2}$	$\left\|-\frac{1}{2}\right\| = \frac{1}{2}$

The negative of the absolute value of a nonzero number will always be a negative number. For example,

$$-|2| = -(2) = -2 \qquad \text{and} \qquad -|-3| = -(3) = -3$$

EXAMPLE 3 Insert either $>$, $<$, or $=$ in each shaded area to make a true statement.

a) $|3| \;\square\; 3$ **b)** $|-2| \;\square\; |2|$ **c)** $-2 \;\square\; |-4|$ **d)** $|-5| \;\square\; 0$ **e)** $|12| \;\square\; |-18|$

Solution **a)** $|3| = 3$.

b) $|-2| = |2|$, since both $|-2|$ and $|2|$ equal 2.

c) $-2 < |-4|$, since $|-4| = 4$.

d) $|-5| > 0$, since $|-5| = 5$.

e) $|12| < |-18|$, since $|12| = 12$ and $|-18| = 18$.

NOW TRY EXERCISE 47

The concept of absolute value is very important in higher-level mathematics courses. If you take a course in intermediate algebra, you will learn a more formal definition of absolute value. We will use absolute value in Sections 1.6 and 1.7 to add and subtract real numbers.

Exercise Set 1.5

Concept/Writing Exercises

1. **a)** Draw a number line.
   b) Mark the numbers -2 and -4 on your number line.
   c) Is -2 less than -4 or is -2 greater than -4? Explain.
   In parts **d)** and **e)** write a correct statement using -2 and -4 and the symbol given.
   d) $<$
   e) $>$
2. What is the absolute value of a number?
3. **a)** Explain why the absolute value of 4, $|4|$, is 4.
   b) Explain why the absolute value of -4, $|-4|$, is 4.
   c) Explain why the absolute value of 0, $|0|$, is 0.
4. Are there any real numbers whose absolute value is not a positive number? Explain your answer.
5. Suppose a and b represent any two real numbers. If $a > b$ is true, will $b < a$ also be true? Explain and give some examples using specific numbers for a and b.
6. Will $|a| - |a| = 0$ always be true for any real number a? Explain

Practice the Skills

Evaluate.

7. $|7|$
8. $|-6|$
9. $|-15|$
10. $|-10|$
11. $|0|$
12. $|54|$
13. $-|-3|$
14. $-|92|$
15. $-|15|$
16. $-|-34|$

Insert either $<$ or $>$ in each shaded area to make a true statement.

17. $5 \;\square\; 8$
18. $4 \;\square\; -2$
19. $-4 \;\square\; 0$
20. $-6 \;\square\; -4$
21. $\frac{1}{2} \;\square\; -\frac{2}{3}$
22. $\frac{3}{5} \;\square\; \frac{4}{5}$
23. $0.9 \;\square\; 0.8$
24. $-0.2 \;\square\; -0.4$
25. $-\frac{1}{2} \;\square\; -1$
26. $0 \;\square\; -0.9$
27. $3 \;\square\; -3$
28. $-\frac{3}{4} \;\square\; -1$
29. $-2.1 \;\square\; -2$
30. $-1.83 \;\square\; -1.82$
31. $\frac{5}{9} \;\square\; -\frac{5}{9}$
32. $-9 \;\square\; -12$
33. $-\frac{3}{2} \;\square\; \frac{3}{2}$
34. $-4.09 \;\square\; -5.3$
35. $0.49 \;\square\; 0.43$
36. $-1.0 \;\square\; -0.7$
37. $5 \;\square\; -7$
38. $0.001 \;\square\; 0.002$
39. $-0.006 \;\square\; -0.007$
40. $\frac{1}{2} \;\square\; -\frac{1}{2}$

41. $\frac{5}{8}$ ▢ 0.6 42. 2.7 ▢ $\frac{10}{3}$ 43. $-\frac{2}{3}$ ▢ $-\frac{1}{3}$ 44. $\frac{9}{2}$ ▢ $\frac{7}{2}$

45. $-\frac{1}{2}$ ▢ $-\frac{3}{2}$ 46. -0.4 ▢ -0.5

Insert either <, >, or = in each shaded area to make a true statement.

47. 2 ▢ $|-1|$ 48. $|-8|$ ▢ $|-7|$ 49. $|-4|$ ▢ $\frac{2}{3}$ 50. $|-4|$ ▢ -3

51. $|0|$ ▢ $|-4|$ 52. $|-1.9|$ ▢ -1.8 53. 4 ▢ $\left|-\frac{9}{2}\right|$ 54. -5 ▢ $|5|$

55. $\left|-\frac{6}{2}\right|$ ▢ $\left|-\frac{2}{6}\right|$ 56. $\left|\frac{2}{5}\right|$ ▢ $|-0.40|$ 57. $|-4.6|$ ▢ $\left|-\frac{23}{5}\right|$ 58. $\left|-\frac{8}{3}\right|$ ▢ $|-3.5|$

Insert either >, <, or = in each shaded area to make a true statement.

59. $\frac{2}{3}+\frac{2}{3}+\frac{2}{3}+\frac{2}{3}$ ▢ $4\cdot\frac{2}{3}$ 60. $\frac{2}{3}\cdot\frac{2}{3}$ ▢ $\frac{2}{3}+\frac{2}{3}$ 61. $\frac{1}{2}\cdot\frac{1}{2}$ ▢ $\frac{1}{2}\div\frac{1}{2}$

62. $5\div\frac{2}{3}$ ▢ $\frac{2}{3}\div 5$ 63. $\frac{5}{8}-\frac{1}{2}$ ▢ $\frac{5}{8}\div\frac{1}{2}$ 64. $2\frac{1}{3}\cdot\frac{1}{2}$ ▢ $2\frac{1}{3}+\frac{1}{2}$

Problem Solving

65. What numbers are 4 units from 0 on a number line?
66. What numbers are 5 units from 0 on the number line?

In Exercises 67–74, give three real numbers that satisfy all the stated criteria. If no real numbers satisfy the criteria, so state and explain why.

67. greater than 4 and less than 6
68. less than -2
69. less than -2 and greater than -6
70. less than 4 and greater than 6
71. greater than -3 and greater than 3
72. less than -3 and less than 3
73. greater than $|-2|$ and less than $|-6|$
74. greater than $|-3|$ and less than $|3|$
75. **a)** Consider the word *between*. What does this word mean?
 b) List three real numbers between 4 and 6
 c) Is the number 4 between the numbers 4 and 6? Explain.
 d) Is the number 5 between the numbers 4 and 6? Explain.
 e) Is it true or false that the real numbers between 4 and 6 are the real numbers that are both greater than 4 and less than 6? Explain.
76. Many areas of the country are affected by tornadoes. The following graph shows the average number of tornadoes and tornado days (days when tornadoes occurred) each month in the United States for a 27-year period.

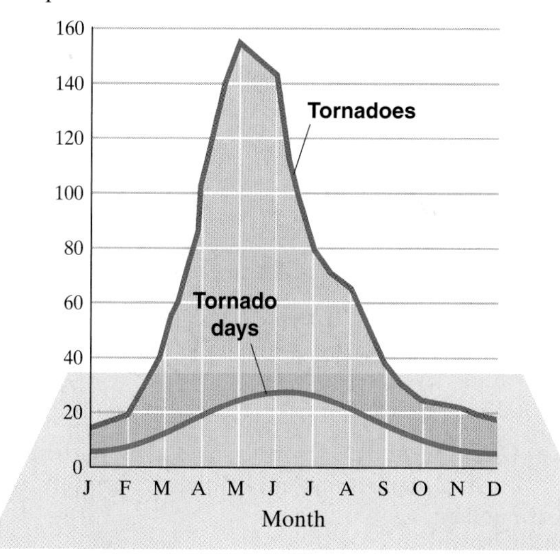

 a) In which months was the average number of tornado days greater than 20 (at any time during the month)?
 b) In which months was the average number of tornadoes less than 20?
 c) In which months was the average number of tornadoes greater than 100?
 d) In which month was the average number of tornadoes the greatest? Estimate the average number of tornadoes in that month.

77. At one time or another, many of you will purchase a home. An important consideration at the time of purchase is the interest rate because that influences your monthly mortgage payments. The following chart shows a history of mortgage rates from 1995 through 1997.

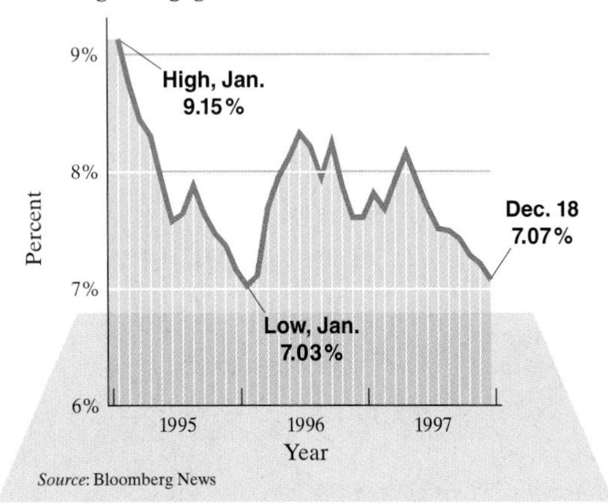

a) What is the highest interest rate shown on the graph? Indicate the month and year that it occurred.
b) What is the lowest interest rate shown on the graph? Indicate the month and year that it occurred.
c) In 1996, in which months was the interest rate less than 8%?
d) In 1996, in which months was the interest rate greater than 8%?
e) If you purchase a home with a 7%, 30-year conventional mortgage, your monthly mortgage payment is \$6.39 per \$1000 of mortgage. Determine your monthly mortgage payment on a \$40,000, 7%, 30-year conventional mortgage.
f) If you purchase a home with a 10%, 30-year conventional mortgage, your monthly mortgage payment is \$8.70 per \$1000 of mortgage. Determine your monthly mortgage payment on a \$40,000, 10%, 30-year conventional mortgage.

Challenge Problems

78. A number greater than 0 and less than 1 (or between 0 and 1) is multiplied by itself. Will the product be less than, equal to, or greater than the original number selected? Explain why this is always true.

79. A number between 0 and 1 is divided by itself. Will the quotient be less than, equal to, or greater than the original number selected? Explain why this is always true.

80. What two numbers can be substituted for x to make $|x| = 3$ a true statement?

81. Are there any values for x that would make $|x| = -|x|$ a true statement?

82. a) To what is $|x|$ equal if x represents a real number greater than or equal to 0? **b)** To what is $|x|$ equal if x represents a real number less than 0? **c)** Fill in the following shaded areas to make a true statement.

c) $|x| = \begin{cases} \blacksquare, & x \geq 0 \\ \blacksquare, & x < 0 \end{cases}$

Group Activity

Discuss and answer Exercise 83 as a group.

83. a) Group member 1: Draw a number line and mark points on the line to represent the following numbers.

$$|-2|, \quad -|3|, \quad -\left|\frac{1}{3}\right|$$

b) Group member 2: Do the same as in part **a)**, but on your number line mark points for the following numbers.

$$|-4|, \quad -|2|, \quad \left|-\frac{3}{5}\right|$$

c) Group member 3: Do the same as in parts **a)** and **b)**, but mark points for the following numbers.

$$|0|, \quad \left|\frac{16}{5}\right|, \quad -|-3|$$

d) As a group, construct one number line that contains all the points listed in parts **a)**, **b)**, and **c)**.

Cumulative Review Exercises

[1.3] 84. Subtract $1\frac{2}{3} - \frac{3}{8}$.

[1.4] 85. List the set of whole numbers.

86. List the set of counting numbers.

87. Consider the following set of numbers.

$$\left\{5, -2, 0, \frac{1}{3}, \sqrt{3}, -\frac{5}{9}, 2.3\right\}$$

List the numbers that are
a) natural numbers.
b) whole numbers.
c) integers.
d) rational numbers.
e) irrational numbers.
f) real numbers.

1.6 ADDITION OF REAL NUMBERS

1) **Add real numbers using a number line.**
2) **Identify opposites.**
3) **Add using absolute values.**
4) **Add using calculators.**

There are many practical uses for negative numbers. A submarine diving below sea level, a bank account that has been overdrawn, a business spending more than it earns, and a temperature below zero are some examples. In some European hotels, the floors below the registration lobby are given negative numbers.

The four basic **operations** of arithmetic are addition, subtraction, multiplication, and division. In the next few sections we will explain how to add, subtract, multiply, and divide numbers. We will consider both positive and negative numbers. In this section we discuss the operation of addition.

1) Add Real Numbers Using a Number Line

To add numbers, we make use of a number line. Represent the first number to be added (first *addend*) by an arrow starting at 0. The arrow is drawn to the right if the number is positive. If the number is negative, the arrow is drawn to the left. From the tip of the first arrow, draw a second arrow to represent the second addend. The second arrow is drawn to the right or left, as just explained. The sum of the two numbers is found at the tip of the second arrow. Note that with the exception of 0, *any number without a sign in front of it is positive*. For example, 3 means +3 and 5 means +5.

EXAMPLE 1 Evaluate $3 + (-4)$ using a number line.

Solution *Always begin at 0*. Since the first addend, the 3, is positive, the first arrow starts at 0 and is drawn 3 units to the right (Fig. 1.18).

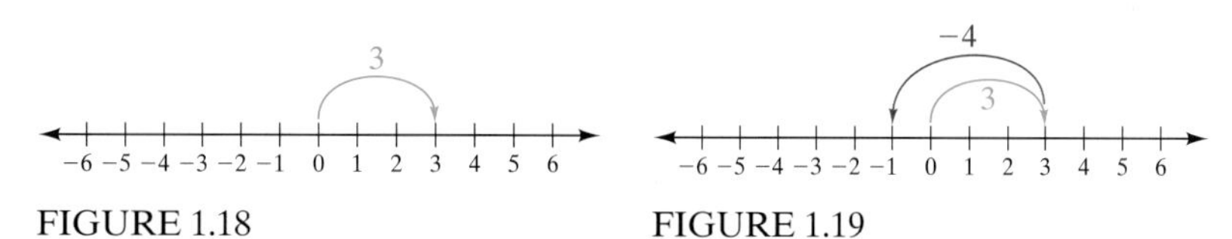

FIGURE 1.18

FIGURE 1.19

The second arrow starts at 3 and is drawn 4 units to the left, since the second addend is negative (Fig. 1.19). The tip of the second arrow is at -1. Thus

$$3 + (-4) = -1$$

EXAMPLE 2 Evaluate $-4 + 2$ using a number line.

Solution Begin at 0. Since the first addend is negative, -4, the first arrow is drawn 4 units to the left. From there, since 2 is positive, the second arrow is drawn 2 units to the right. The second arrow ends at -2 (Fig. 1.20).

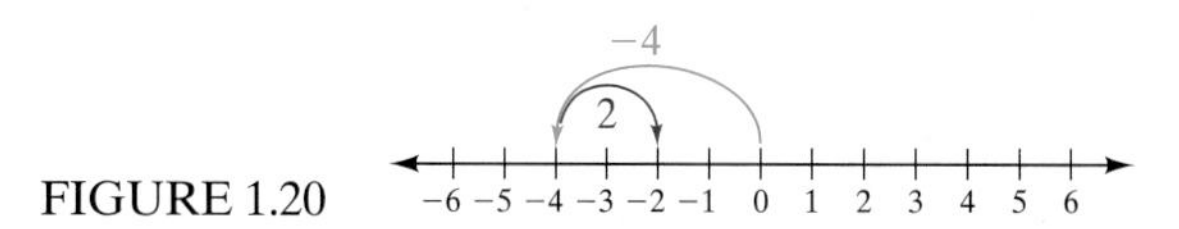

FIGURE 1.20

$$-4 + 2 = -2$$

EXAMPLE 3 Evaluate $-3 + (-2)$ using a number line.

Solution Start at 0. Since both numbers being added are negative, both arrows will be drawn to the left (Fig. 1.21).

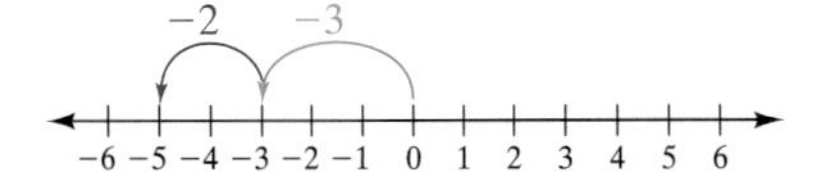

FIGURE 1.21

$$-3 + (-2) = -5$$

In Example 3, we can think of the expression $-3 + (-2)$ as combining a *loss* of 3 and a *loss* of 2 for a total *loss* of 5, or -5.

EXAMPLE 4 Add $5 + (-5)$ using a number line.

Solution The first arrow starts at 0 and is drawn 5 units to the right. The second arrow starts at 5 and is drawn 5 units to the left. The tip of the second arrow is at 0. Thus, $5 + (-5) = 0$ (Fig. 1.22).

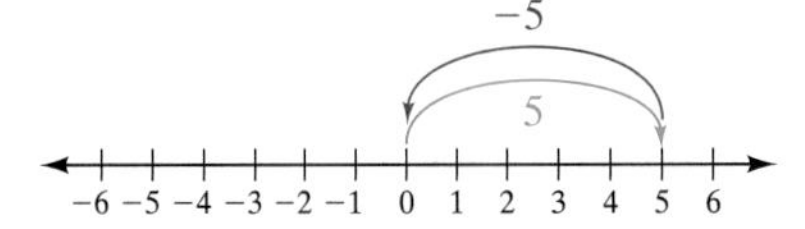

FIGURE 1.22

$$5 + (-5) = 0$$

EXAMPLE 5 A submarine dives 250 feet. Later it dives an additional 190 feet. Find the depth of the submarine (assume that depths below sea level are indicated by negative numbers).

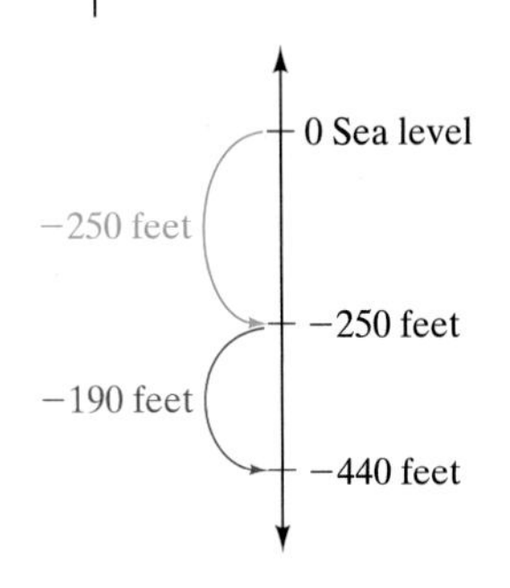

Solution A vertical number line (Fig. 1.23) may help you visualize this problem.

$$-250 + (-190) = -440 \text{ feet}$$

FIGURE 1.23

NOW TRY EXERCISE 87

2) Identify Opposites

Now let's consider **opposites**, or **additive inverses**.

Definition

Any two numbers whose sum is zero are said to be **opposites** (or **additive inverses**) of each other. In general, if we let a represent any real number, then its opposite is $-a$ and $a + (-a) = 0$.

In Example 4 the sum of 5 and -5 is 0. Thus -5 is the opposite of 5 and 5 is the opposite of -5.

EXAMPLE 6 Find the opposite of each number. **a)** 3 **b)** −4

Solution **a)** The opposite of 3 is −3, since $3 + (-3) = 0$.

NOW TRY EXERCISE 13 **b)** The opposite of −4 is 4, since $-4 + 4 = 0$.

 Add Using Absolute Values

Now that we have had some practice adding signed numbers on a number line, we give a rule (in two parts) for using absolute value to add signed numbers. Remember that the absolute value of a nonzero number will always be positive. The first part of the rule follows.

> **To add real numbers with the same sign** (either both positive or both negative), add their absolute values. The sum has the same sign as the numbers being added.

EXAMPLE 7 Add $4 + 8$.

Solution Since both numbers have the same sign, both positive, we add their absolute values: $|4| + |8| = 4 + 8 = 12$. Since both numbers being added are positive, the sum is positive. Thus $4 + 8 = 12$.

EXAMPLE 8 Add $-6 + (-9)$.

Solution Since both numbers have the same sign, both negative, we add their absolute values: $|-6| + |-9| = 6 + 9 = 15$. Since both numbers being added are negative, their sum is negative. Thus $-6 + (-9) = -15$.

The sum of two positive numbers will always be positive and the sum of two negative numbers will always be negative.

> **To add two signed numbers with different signs** (one positive and the other negative), subtract the smaller absolute value from the larger absolute value. The answer has the sign of the number with the larger absolute value.

EXAMPLE 9 Add $10 + (-6)$.

Solution The two numbers being added have different signs, so we subtract the smaller absolute value from the larger: $|10| - |-6| = 10 - 6 = 4$. Since $|10|$ is greater than $|-6|$ and the sign of 10 is positive, the sum is positive. Thus, $10 + (-6) = 4$.

EXAMPLE 10 Add $12 + (-18)$.

Solution The numbers being added have different signs, so we subtract the smaller absolute value from the larger: $|-18| - |12| = 18 - 12 = 6$. Since $|-18|$ is greater than $|12|$ and the sign of −18 is negative, the sum is negative. Thus, $12 + (-18) = -6$.

EXAMPLE 11 Add $-24 + 19$.

Solution The two numbers being added have different signs, so we subtract the smaller absolute value from the larger: $|-24| - |19| = 24 - 19 = 5$. Since $|-24|$ is greater than $|19|$, the sum is negative. Therefore, $-24 + 19 = -5$.

The sum of two signed numbers with different signs may be either positive or negative. The sign of the sum will be the same as the sign of the number with the larger absolute value.

NOW TRY EXERCISE 43

HELPFUL HINT

Architects often make a scale model of a building before starting construction of the building. This "model" helps them visualize the project and often helps them avoid problems.

Mathematicians also construct models. A mathematical *model* may be a physical representation of a mathematical concept. It may be as simple as using tiles or chips to represent specific numbers. For example, below we use a model to help explain addition of real numbers. This may help some of you understand the concepts better.

We let a red chip represent +1 and a green chip represent −1.

● = +1   ● = −1

If we add +1 and −1, or a red and a green chip, we get 0.
Now consider the addition problem $3 + (-5)$. We can represent this as

$$\underbrace{\bullet\bullet\bullet}_{3} + \underbrace{\bullet\bullet\bullet\bullet\bullet}_{-5}$$

If we remove 3 red chips and 3 green chips, or three zeros, we are left with 2 green chips, which represents a sum of −2. Thus, $3 + (-5) = -2$,

$$\not\bullet\not\bullet\not\bullet + \not\bullet\not\bullet\not\bullet\bullet\bullet$$

Now consider the problem $-4 + (-2)$. We can represent this as

$$\underbrace{\bullet\bullet\bullet\bullet}_{-4} + \underbrace{\bullet\bullet}_{-2}$$

Since we end up with 6 green chips, and each green chip represents −1, the sum is −6. Therefore, $-4 + (-2) = -6$.

EXAMPLE 12 The J.C. Ramos Company had a loss of $6000 for the first 6 months of the year and a profit of $19,500 for the second 6 months of the year. Find the net profit or loss for the year.

Solution **Understand and Translate** This problem can be represented as $-6000 + 19{,}500$. Since the two numbers being added have different signs, subtract the smaller absolute value from the larger.

Carry Out $|19{,}500| - |-6000| = 19{,}500 - 6000 = 13{,}500$

Check and Answer The answer is reasonable. Thus, the net profit for the year was $13,500.

NOW TRY EXERCISE 93

4) Add Using Calculators

Throughout the book we will provide information about calculators in the Using Your Calculator boxes. Some will be for scientific calculators; some will be for graphing calculators, also called graphers; and some will be for both. Below are pictures of a scientific calculator (on left) and a graphing calculator (on right). Graphing calculators can do everything that a scientific calculator can do and more. Ask your instructor if he or she is recommending a particular calculator for this course. If you plan on taking additional mathematics courses, you may want to consider purchasing the calculator that will be used in those courses.

Scientific Calculator

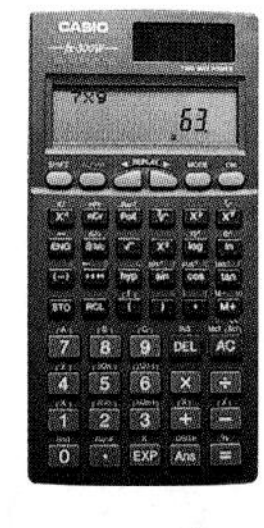

Graphing Calculator

It is important that you understand the procedures for adding, subtracting, multiplying, and dividing real numbers *without* using a calculator. *You should not need to rely on a calculator to work problems.* If you are permitted to use a calculator, you can, however, use the calculator to help save time on difficult calculations. If you have an understanding of the basic concepts, you should be able to tell if you have made an error entering information on the calculator if the answer shown does not seem reasonable.

Following is the first of many Using Your Calculator and Using Your Graphing Calculator boxes. Note that *no new material* will be presented in the boxes. The boxes are provided to help you use your calculator, if you are using one in this course. Do not be concerned if you do not know what all the calculator keys do. Your instructor will tell you which calculator keys you will need to know how to use.

NOW TRY EXERCISE 67

Using Your Calculator

Scientific Calculator

Entering Negative Numbers

Most scientific calculators contain a [+/–] key, which is used to enter a negative number. To enter the number −5, press 5 [+/–] and a −5 will be displayed. Now we show how to evaluate some addition problems on a scientific calculator

Addition of Real Numbers

EVALUATE	KEYSTROKES*	ANSWER DISPLAYED
$-9 + 24$	9 [+/–] [+] 24 [=]	15
$15 + (-22)$	15 [+] 22 [+/–] [=]	−7
$-30 + (-16)$	30 [+/–] [+] 16 [+/–] [=]	−46

*With some scientific calculators the negative sign is entered before the number, as is done with graphing calculators.

continued on next page

In Using Your Graphing Calculator boxes, when we show keystrokes or graphing screens (called windows), they will be for a Texas Instruments-83 calculator (TI-83). You should read your graphing calculator manual for more detailed instructions.

Using Your Graphing Calculator

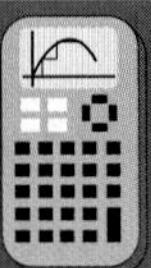

Graphing Calculator

Entering Negative Numbers

Graphing calculators contain two keys that look similar, as shown below.

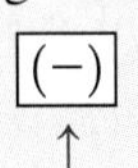

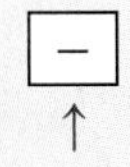

[(−)] ↑ *Used to make a number negative*   [−] ↑ *Used to subtract*

Addition of Real Numbers

EVALUATE	KEYSTROKES	ANSWER DISPLAYED
$-9+24$	[(−)] 9 + 24 [ENTER]	15
$15+(-22)$	15 + [(−)] 22 [ENTER]	−7
$-30+(-16)$	[(−)] 30 + [(−)] 16 [ENTER]	−46

Notice that to make a number negative on a scientific calculator you first enter the number, then press [+/−] key. To make a number negative on a graphing calculator, you first press the [(−)] key, then enter the number.

Exercise Set 1.6

Concept/Writing Exercises

1. What are the four basic operations of arithmetic?
2. **a)** What are opposites or additive inverses?
   b) Give an example of two numbers that are opposites.
3. **a)** Are the numbers $-\frac{2}{3}$ and $\frac{3}{2}$ opposites? Explain.
   b) If the numbers in part **a)** are not opposites, what number is the opposite of $-\frac{2}{3}$?
4. If we add two negative numbers, will the sum be a positive number or a negative number? Explain.
5. If we add a positive number and a negative number, will the sum be positive, negative, or can it be either? Explain.
6. **a)** If we add −24,692 and 30,519, will the sum be positive or negative? Without performing any calculations, explain how you determined your answer.
   b) Repeat part **a)** for the numbers 24,692 and −30,519.
   c) Repeat part **a)** for the numbers −24,692 and −30,519.
7. Explain in your own words how to add two numbers with like signs.
8. Explain in your own words how to add two numbers with unlike signs.
9. Mr. Dabskic charged $175 worth of goods on his charge card. He later made a payment of $93.
   a) Explain why the balance remaining on his card may be found by the addition $-175 + 93$.
   b) Find the sum of $-175 + 93$.
   c) In part **b)**, you should have obtained a sum of −82. Explain why this −82 indicates that Mr. Dabskic owes $82 on his credit card.
10. Mrs. Goldstein owed $163 on her credit card. She charged another item costing $56.
    a) Explain why the new balance on her credit card may be found by the addition $-163 + (-56)$.
    b) Find the sum of $-163 + (-56)$.
    c) In part **b)**, you should have obtained a sum of −219. Explain why this −219 indicates that Mrs. Goldstein owes $219 on her credit card.

Practice the Skills

Write the opposite of each number.

11. 5 **12.** −7 13. −28 **14.** 3

15. 0 **16.** 8 **17.** $\frac{5}{3}$ **18.** $-\frac{1}{4}$

19. $2\frac{3}{5}$ **20.** -1 **21.** 0.47 **22.** -0.721

Add.

23. $3 + 8$ **24.** $-7 + 2$ **25.** $4 + (-3)$ **26.** $8 + (-7)$

27. $-4 + (-2)$ **28.** $-3 + (-5)$ **29.** $6 + (-6)$ **30.** $-8 + 8$

31. $-4 + 4$ **32.** $-6 + 9$ **33.** $-8 + (-2)$ **34.** $6 + (-5)$

35. $-6 + 6$ **36.** $-5 + 3$ **37.** $-4 + (-5)$ **38.** $0 + (-3)$

39. $0 + 0$ **40.** $0 + (-0)$ **41.** $-6 + 0$ **42.** $-9 + 13$

43. $18 + (-9)$ **44.** $-7 + 7$ **45.** $-23 + (-31)$ **46.** $-27 + (-9)$

47. $-35 + (-9)$ **48.** $56 + (-14)$ **49.** $4 + (-30)$ **50.** $-16 + 9$

51. $-35 + 40$ **52.** $-12 + 17$ **53.** $180 + (-200)$ **54.** $-33 + (-92)$

55. $-67 + 28$ **56.** $183 + (-183)$ **57.** $184 + (-93)$ **58.** $-19 + 176$

59. $-452 + 312$ **60.** $-94 + (-98)$ **61.** $-26 + (-79)$ **62.** $49 + (-63)$

In Exercises 63–78, ***a)*** *determine by observation whether the sum will be a positive number, zero, or a negative number;* ***b)*** *find the sum using your calculator; and* ***c)*** *examine your answer to part* ***b)*** *to see whether it is reasonable and makes sense.*

63. $463 + (-197)$ **64.** $-140 + (-629)$ **65.** $-84 + (-289)$ **66.** $-647 + 352$

67. $-947 + 495$ **68.** $762 + (-762)$ **69.** $-496 + (-804)$ **70.** $-354 + 1090$

71. $-285 + 263$ **72.** $1127 + (-84)$ **73.** $-1833 + (-2047)$ **74.** $-426 + 572$

75. $3124 + (-2013)$ **76.** $-9095 + (-647)$ **77.** $-1025 + (-1025)$ **78.** $7513 + (-4361)$

Indicate whether each statement is true or false.

79. The sum of two negative numbers is always a negative number.

80. The sum of a negative number and a positive number is sometimes a negative number.

81. The sum of two positive numbers is never a negative number.

82. The sum of a positive number and a negative number is always a negative number.

83. The sum of a positive number and a negative number is always a positive number.

84. The sum of a number and its opposite is always equal to zero.

Problem Solving

Write an expression that can be used to solve each problem and then solve.

85. Mr. Peter owed \$94 on his bank credit card. He charged another item costing \$183. Find the amount that Mr. Peter owed the bank.

86. Mrs. Chu charged \$142 worth of goods on her charge card. Find her balance after she made a payment of \$87.

87. A football team lost 18 yards on one play and then lost 3 yards on the following play. What was the total loss in yardage?

88. Mrs. Poweski paid \$1823 in federal income tax. When she was audited, Mrs. Poweski had to pay an additional \$471. What was her total tax?

89. A company is drilling a well. During the first week they drilled 27 feet, and during the second week they drilled another 34 feet before they struck water. How deep is the well?

90. Mr. Vela hiked 847 meters down the Grand Canyon. He climbed back up 385 meters and then rested. Find his distance from the rim of the canyon when he rested.

91. *The Guinness Book of World Records,* 1997 edition, lists Mauna Kea in Hawaii as the tallest mountain in the world when measured from its base to its peak. The base of Mauna Kea is 19,684 feet below sea level. The total height of the mountain from its base to its peak is 33,480 feet. How high is the peak of Mauna Kea above sea level?

92. The U.S. Bureau of the Census projects that in 2000 there will be about 3,899,000 births and about 2,425,000 deaths in the United States. Determine the projected change in the population in 2000.

93. The Frenches opened a coffee bar. Their income and expenses for their first three months of operation are shown in the following graph.

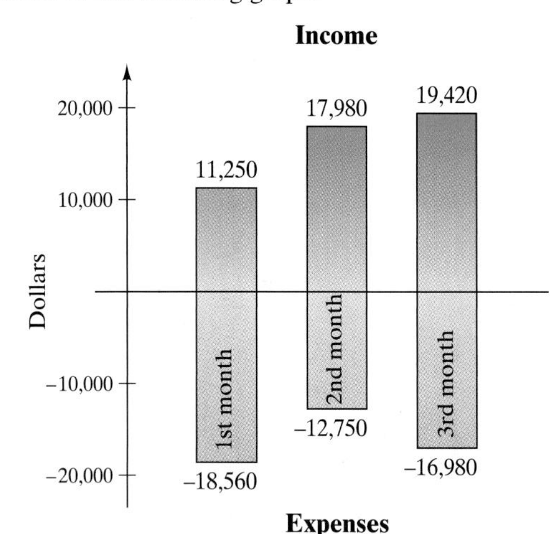

a) Find the net profit or loss (the sum of income and expenses) for the first month.

b) Find the net profit or loss for the second month.

c) Find the net profit or loss for the third month.

94. The following graph shows the percent change in corporate earnings in each of the four quarters for the Dow Jones Global–United States Index for the years 1995 and 1996.

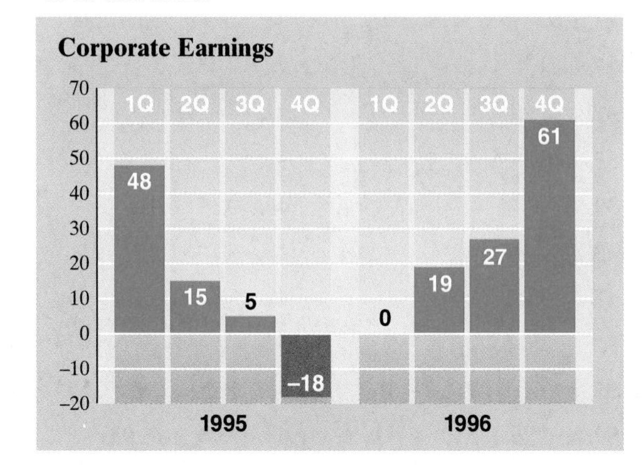

a) Determine the total percent gain (or loss) for the four quarters of 1995.

b) Determine the total percent gain (or loss) for the four quarters of 1996.

c) Which year had a greater percent change in the second quarter, 1995 or 1996?

d) In which year, 1995 or 1996, was the annual percent change greater? To find the annual percent change, add the four quarters.

Challenge Problems

Evaluate each exercise by adding the numbers from left to right. We will discuss problems like this shortly.

95. $(-4) + (-6) + (-12)$

96. $5 + (-7) + (-8)$

97. $29 + (-46) + 37$

98. $4 + (-5) + 6 + (-8)$

99. $(-12) + (-10) + 25 + (-3)$

100. $(-4) + (-2) + (-15) + (-27)$

Find the following sums. Explain how you determined your answer. (Hint: Pair small numbers with large numbers from the ends inward.)

101. $1 + 2 + 3 + \cdots + 10$

102. $1 + 2 + 3 + \cdots + 20$

Cumulative Review Exercises

[1.3] **103.** Multiply $\left(\frac{3}{5}\right)\left(1\frac{2}{3}\right)$.

104. Subtract $3 - \frac{5}{16}$.

[1.5] *Insert either <, >, or = in each shaded area to make a true statement.*

105. $|-3| \;\square\; 2$

106. $8 \;\square\; |-7|$

1.7 SUBTRACTION OF REAL NUMBERS

1. **Subtract numbers.**
2. **Subtract numbers mentally.**
3. **Evaluate expressions containing more than two numbers.**

1) Subtract Numbers

Any subtraction problem can be rewritten as an addition problem using the additive inverse.

To Subtract Real Numbers

In general, if a and b represent any two real numbers, then

$$a - b = a + (-b)$$

This rule says that to subtract b from a, add the opposite or additive inverse of b to a.

EXAMPLE 1 Evaluate $9 - (+4)$.

Solution We are subtracting a positive 4 from 9. To accomplish this, we add the opposite of +4, which is −4, to 9.

$$9 - (+4) = 9 + (-4) = 5$$

Subtract (under −), Positive 4 (under +4), Add (under +), Negative 4 (under −4)

We evaluated $9 + (-4)$ using the procedures for *adding* real numbers presented in Section 1.6.

Often in a subtraction problem, when the number being subtracted is a positive number, the + sign preceding the number being subtracted is not illustrated. For example, in the subtraction $9 - 4$,

$$9 - 4 \text{ means } 9 - (+4)$$

Thus, to evaluate $9 - 4$, we must add the opposite of 4, which is −4, to 9.

$$9 - 4 = 9 + (-4) = 5$$

Subtract (under −), Positive 4 (under 4), Add (under +), Negative 4 (under −4)

This procedure is illustrated in Example 2.

EXAMPLE 2 Evaluate $5 - 3$.

Solution We must subtract a positive 3 from 5. To change this problem to an addition problem, we add the opposite of 3, which is −3, to 5.

$$\underbrace{5 - 3}_{\text{Subtraction problem}} = \underbrace{5 + (-3)}_{\text{Addition problem}} = 2$$

Subtract (under −), Positive 3 (under 3), Add (under +), Negative 3 (under −3)

EXAMPLE 3 Evaluate $4 - 9$.

Solution Add the opposite of 9, which is -9, to 4.

NOW TRY EXERCISE 13

$$4 - 9 = 4 + (-9) = -5$$

EXAMPLE 4 Evaluate $-4 - 2$.

Solution Add the opposite of 2, which is -2, to -4.

$$-4 - 2 = -4 + (-2) = -6$$

EXAMPLE 5 Evaluate $4 - (-2)$.

Solution We are asked to subtract a negative 2 from 4. To do this, we add the opposite of -2, which is 2, to 4.

$$4 - (-2) = 4 + 2 = 6$$

Subtract (the first minus sign), *Negative 2* (the -2), *Add* (the plus sign), *Positive 2* (the 2)

HELPFUL HINT

By examining Example 5, we see that

$$4 - (-2) = 4 + 2$$

Two negative signs together (the $-$ and $-$ in $4 - (-2)$); *Plus* (the $+$ in $4 + 2$)

Whenever we subtract a negative number, we can replace the two negative signs with a plus sign.

EXAMPLE 6 Evaluate $6 - (-3)$.

Solution Since we are subtracting a negative number, adding the opposite of -3, which is 3, to 6 will result in the two negative signs being replaced by a plus sign.

$$6 - (-3) = 6 + 3 = 9$$

EXAMPLE 7 Evaluate $-15 - (-12)$.

Solution

NOW TRY EXERCISE 17

$$-15 - (-12) = -15 + 12 = -3$$

HELPFUL HINT

We will now indicate how we may illustrate subtraction using colored chips. Remember from the preceding section that a red chip represents +1 and a green chip −1.

● = +1   ● = −1

continued on next page

Consider the subtraction problem $2 - 5$. If we change this to an addition problem, we get $2 + (-5)$. We can then add, as was done in the preceding section. The figure below shows that $2 + (-5) = -3$.

Now consider $-2 - 5$. This means $-2 + (-5)$, which can be represented as follows:

Thus, $-2 - 5 = -7$.
Now consider the problem $-3 - (-5)$. This can be rewritten as $-3 + 5$, which can be represented as follows:

Thus, $-3 - (-5) = 2$.
Some students still have difficulty understanding why when you subtract a negative number you obtain a positive number. Let us look at the problem $3 - (-2)$. This time we will look at it from a slightly different point of view. Let's start with 3:

From this we wish to subtract a negative two. To the +3 shown above we will add two zeros by adding two +1 −1 combinations. Remember, +1 and −1 sum to 0.

+3 0 0

Now we can subtract or "take away" the two −1's as shown:

From this we see that we are left with $3 + 2$ or 5. Thus, $3 - (-2) = 5$.

EXAMPLE 8 Subtract 12 from 3.

Solution

$$3 - 12 = 3 + (-12) = -9$$

HELPFUL HINT

Example 8 asked us to "subtract 12 from 3." Some of you may have expected this to be written as $12 - 3$ since you may be accustomed to getting a positive answer. However, the correct method of writing this is $3 - 12$. Notice that the number following the word "from" is our starting point. That is where the calculation begins. For example:

Subtract 5 from −1 means $-1 - 5$.	From −1, subtract 5 means $-1 - 5$.
Subtract −4 from −2 means $-2 - (-4)$.	From −2, subtract −4 means $-2 - (-4)$.
Subtract −3 from 6 means $6 - (-3)$.	From 6, subtract −3 means $6 - (-3)$.
Subtract a from b means $b - a$.	From a, subtract b means $a - b$.

EXAMPLE 9 Subtract 5 from 5.

Solution

$$5 - 5 = 5 + (-5) = 0$$

EXAMPLE 10 Subtract -6 from 4.

Solution

NOW TRY EXERCISE 63

$$4 - (-6) = 4 + 6 = 10$$

EXAMPLE 11 Mary Jo Morin's checkbook indicated a balance of \$125 before she wrote a check for \$183. Find the balance in her checkbook.

Solution **Understand and Translate** We can obtain the balance by subtracting 183 from 125.

Carry Out $125 - 183 = 125 + (-183) = -58$

Check and Answer The negative indicates a deficit, which is what we expect. Therefore, Mary Jo is overdrawn by \$58.

EXAMPLE 12 Janet Weatherspoon made \$4200 in the stock market, while Mateo Farez lost \$3000. How much farther ahead is Janet than Mateo financially?

Solution **Understand and Translate** Janet's gain is represented as a positive number. Mateo's loss is represented as a negative number. We can obtain the difference in their financial positions by subtraction.

Carry Out $4200 - (-3000) = 4200 + 3000 = 7200$

NOW TRY EXERCISE 113

Check and Answer Janet is therefore \$7200 ahead of Mateo financially.

EXAMPLE 13 Evaluate.

a) $12 + (-4)$ **b)** $-16 - 3$ **c)** $19 + (-14)$
d) $6 - (-5)$ **e)** $-9 - (-3)$ **f)** $8 - 13$

Solution Parts **a)** and **b)** are addition problems, whereas the other parts are subtraction problems. We can rewrite each subtraction problem as an addition problem to evaluate.

a) $12 + (-4) = 8$ **b)** $-16 - 3 = -16 + (-3) = -19$
c) $19 + (-14) = 5$ **d)** $6 - (-5) = 6 + 5 = 11$
e) $-9 - (-3) = -9 + 3 = -6$ **f)** $8 - 13 = 8 + (-13) = -5$

2) Subtract Numbers Mentally

In the previous examples, we changed subtraction problems to addition problems. We did this because we know how to add real numbers. After this chapter, when we work out a subtraction problem, we will not show this step. *You need to practice and thoroughly understand how to add and subtract real numbers. You should understand this material so well that, when asked to evaluate an expression like* $-4 - 6$, *you will be able to compute the answer mentally. You should understand that* $-4 - 6$ *means the same as* $-4 + (-6)$, *but you should not need to write the addition to find the value of the expression,* -10.

Let us evaluate a few subtraction problems without showing the process of changing the subtraction to addition.

EXAMPLE 14 Evaluate. **a)** $-7 - 5$ **b)** $4 - 12$ **c)** $18 - 25$ **d)** $-20 - 12$

Solution **a)** $-7 - 5 = -12$ **b)** $4 - 12 = -8$
c) $18 - 25 = -7$ **d)** $-20 - 12 = -32$

In Example 14 **a)**, we may have reasoned that $-7 - 5$ meant $-7 + (-5)$, which is -12, but we did not need to show it.

3) Evaluate Expressions Containing More Than Two Numbers

In evaluating expressions involving more than one addition and subtraction, work from left to right unless parentheses or other grouping symbols appear.

EXAMPLE 15 Evaluate. **a)** $-6 - 12 - 4$ **b)** $-3 + 1 - 7$ **c)** $8 - 10 + 2$

Solution We work from left to right.

a) $\underbrace{-6 - 12} - 4$ **b)** $\underbrace{-3 + 1} - 7$ **c)** $\underbrace{8 - 10} + 2$

$= -18 - 4$ $= -2 - 7$ $= -2 + 2$

NOW TRY EXERCISE 97 $= -22$ $= -9$ $= 0$

After this section you will generally not see an expression like $3 + (-4)$. Instead, the expression will be written as $3 - 4$. Recall that $3 - 4$ means $3 + (-4)$ by our definition of subtraction. **Whenever we see an expression of the form $a + (-b)$, we can write the expression as $a - b$.** For example, $12 + (-15)$ can be written as $12 - 15$ and $-6 + (-9)$ can be written as $-6 - 9$.

As discussed earlier, **whenever we see an expression of the form $a - (-b)$, we can rewrite it as $a + b$.** For example $6 - (-13)$ can be rewritten as $6 + 13$ and $-12 - (-9)$ can be rewritten as $-12 + 9$. Using both of these concepts, the expression $9 + (-12) - (-8)$ may be simplified to $9 - 12 + 8$.

EXAMPLE 16 **a)** Evaluate $-5 - (-9) + (-12) + (-3)$.
b) Simplify the expression in part **a)**.
c) Evaluate the simplified expression in part **b)**.

Solution **a)** We work from left to right. The shading indicates the additions being performed to get to the next step.

$$\begin{aligned} -5 - (-9) + (-12) + (-3) &= -5 + 9 + (-12) + (-3) \\ &= 4 + (-12) + (-3) \\ &= -8 + (-3) \\ &= -11 \end{aligned}$$

b) The expression simplifies as follows:

$$-5 - (-9) + (-12) + (-3) = -5 + 9 - 12 - 3$$

c) Evaluate the simplified expression from left to right. Begin by adding $-5 + 9$ to obtain 4.

$$\begin{aligned} -5 + 9 - 12 - 3 &= 4 - 12 - 3 \\ &= -8 - 3 \\ &= -11 \end{aligned}$$

NOW TRY EXERCISE 103

When you come across an expression like the one in Example 16 **a)**, you should simplify it as we did in part **b)** and then evaluate the simplified expression.

Using Your Calculator

Subtraction

Scientific Calculator

In the Using Your Calculator box on page 47, we indicated that the [+/−] key is pressed after a number is entered to make the number negative. Following are some examples of subtraction on a scientific calculator.

EVALUATE	KEYSTROKES	ANSWER DISPLAYED
$-5 - 8$	5 [+/−] [−] 8 [=]	−13
$2 - (-7)$	2 [−] 7 [+/−] [=]	9

Graphing Calculator

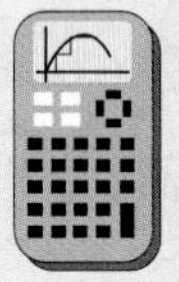

In the Using Your Calculator box on page 48 we mentioned that on a graphing calculator we press the [(−)] key before the number is entered to make the number negative. The [−] key on a graphing calculator is used to perform subtraction. Following are some examples of subtraction on a graphing calculator.

EVALUATE	KEYSTROKE	DISPLAY
$-5 - 8$	[(−)] 5 [−] 8 [ENTER]	−13
$2 - (-7)$	2 [−] [(−)] 7 [ENTER]	9

Exercise Set 1.7

Concept/Writing Exercises

1. Write an expression that illustrates 8 subtracted from 5.
2. Write an expression that illustrates −6 subtracted from −4.
3. Write an expression that illustrates ∗ subtracted from □.
4. Write an expression that illustrates ? subtracted from ☺.
5. **a)** Explain in your own words how to subtract a number b from a number a.
   b) Write an expression using addition that can be used to subtract 12 from 9.
   c) Evaluate the expression you determined in part **b)**.
6. **a)** Express the subtraction $a - (-b)$ in a simplified form.
   b) Write a simplified expression that can be used to evaluate $-9 - (-15)$.
   c) Evaluate the simplified expression obtained in part **b)**.
7. **a)** Express the subtraction $a - (+b)$ in a simplified form.
   b) Simplify the expression $7 - (+9)$.
   c) Evaluate the simplified expression obtained in part **b)**.
8. When we add three or more numbers without parentheses, how do we evaluate the expression?
9. **a)** Simplify $3 - (-6) + (-5)$ by eliminating two signs next to one another and replacing them with a single sign. (See Example 16.) Explain how you determined your answer.
   b) Evaluate the simplified expression obtained in part **a)**.
10. **a)** Simplify $-12 + (-5) - (-4)$. Explain how you determined your answer.
    b) Evaluate the simplified expression obtained in part **a)**.

Practice the Skills

Evaluate.

11. $9 - 5$
12. $-1 - 6$
13. $8 - 9$
14. $3 - 3$
15. $-4 - 2$
16. $-7 - (-4)$
17. $-4 - (-3)$
18. $-3 - 3$
19. $-4 - 4$
20. $7 - (-7)$
21. $0 - 6$
22. $6 - 6$
23. $0 - (-6)$
24. $9 - (-3)$
25. $-3 - 1$
26. $-5 - (-3)$
27. $5 - 3$
28. $9 - 3$
29. $6 - (-3)$
30. $6 - 10$
31. $4 - 4$
32. $-8 - 8$
33. $-9 - 11$
34. $-4 - (-2)$
35. $-4 - (-4)$
36. $14 - 7$
37. $-8 - (-12)$
38. $9 - 9$
39. $-6 - (-2)$
40. $8 - (-4)$
41. $-9 - 2$
42. $-25 - 16$
43. $-24 - (-8)$
44. $37 - 40$
45. $-100 - 80$
46. $-20 - 90$
47. $-45 - 37$
48. $-50 - (-40)$
49. $70 - (-70)$
50. $130 - (-90)$
51. $40 - 62$
52. $110 - (-16)$
53. $-61 - (-9)$
54. $-75 - (-16)$
55. Subtract 3 from −15.
56. Subtract −3 from −10.
57. Subtract 8 from −8.
58. Subtract 5 from −20.
59. Subtract −3 from −5.
60. Subtract 10 from −3.
61. Subtract −4 from 9.
62. Subtract 18 from −18.
63. Subtract 24 from 13.
64. Subtract −11 from −5.
65. Subtract −15 from −4.
66. Subtract 17 from −12.

*In Exercises 67–84, **a)** determine by observation whether the difference will be a positive number, zero, or a negative number; **b)** find the difference using your calculator; and **c)** examine your answer to part **b)** to see whether it is reasonable and makes sense.*

67. $296 - 197$
68. $483 - 569$
69. $-372 - 195$
70. $178 - (-377)$
71. $843 - (-745)$
72. $864 - (-762)$
73. $-408 - (-604)$
74. $-623 - 111$
75. $-1024 - (-576)$
76. $-104.7 - 27.6$
77. $165.7 - 49.6$
78. $-40.2 - (-12.6)$
79. Subtract 364 from 295.
80. Subtract −433 from −932.
81. Subtract 647 from −1023.
82. Subtract 2432 from −4120.
83. Subtract −7.62 from −7.62.
84. Subtract 21.5 from −69.1.

Evaluate.

85. $6 + 5 - (+4)$
86. $9 - (+6) - (+5)$
87. $-3 + (-4) + 5$
88. $9 - 4 + (-2)$
89. $-13 - (+5) + 3$
90. $7 - (+4) - (-3)$
91. $-9 - (-3) + 4$
92. $15 + (-7) - (-3)$
93. $5 - (-9) + (-1)$
94. $12 + (-5) - (-4)$
95. $25 + (-13) - (+5)$
96. $-7 + 6 - 3$
97. $-36 - 5 + 9$
98. $45 - 3 - 7$
99. $-2 + 7 - 9$
100. $-3 - 4 - 13$
101. $25 - 19 + 3$
102. $-4 - 1 + 5$
103. $-4 + (-3) + 5 - 7$
104. $-9 - 3 - (-4) + 5$
105. $17 + (-3) - 9 - (-7)$
106. $32 + 5 - 7 - 12$
107. $-9 + (-7) + (-5) - (-3)$
108. $6 - 9 - (-3) + 12$

Problem Solving

109. The Lands' End catalog department had 300 ladies' blue cardigan sweaters in stock on December 1. By December 9, they had taken orders for 343 of the sweaters.

a) How many sweaters were on back order?

b) If they wanted 100 sweaters in addition to those already ordered, how many sweaters would they need to back order?

110. According to the *Guinness Book of World Records*, the city with the greatest elevation in the United States is Leadville, Colorado at 10,152 feet. The city with the lowest elevation in the United States, at 184 feet below sea level, is Calipatria, California. What is the difference in the elevation of these cities?

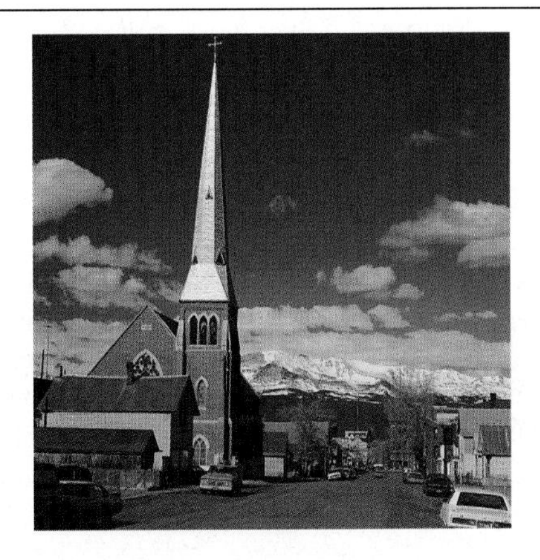

111. The Jacksons, who live near Myrtle Beach, South Carolina, have a house at an elevation of 42 feet above sea level. They hire the RL Schlicter Drilling Company to dig a well. After hitting water, the drilling company inform the Jacksons that they had to dig 58 feet to hit water. How deep is the well with regards to sea level?

112. An airplane is 2000 feet above sea level. A submarine is 1500 feet below sea level. How far above the submarine is the airplane?

113. The greatest change in temperature ever recorded within a 24-hour period occurred at Browning, Montana, on January 23, 1916. The temperature fell from 44°F to −56°F. How much did the temperature drop?

114. Two trains start at the same station at the same time. The Amtrak travels 68 miles in 1 hour. The Pacific Express travels 80 miles in 1 hour.

a) If the two trains travel in opposite directions, how far apart will they be in 1 hour? Explain.

b) If the two trains travel in the same direction, how far apart will they be in 1 hour?

115. The following information was found in the 1998 *Wall Street Journal Almanac*. It shows projections for the five occupations with the greatest increase and greatest decrease from 1984–2005.

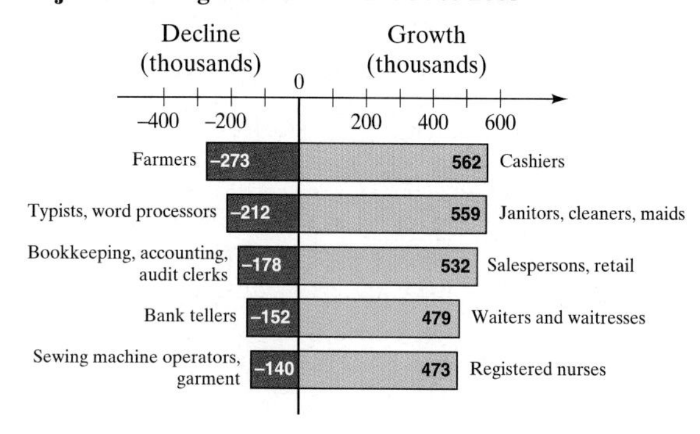

a) Estimate the difference in growth of cashiers and farmers (consider decline as negative growth).

b) Estimate the difference in growth of registered nurses and typists and word processors.

116. As our population ages, there is a greater strain on Social Security and other government programs. The following graph shows the Hospital Insurance Trust Fund total (also known as Medicare Part A), 1990–2006. The totals for 1997–2006 are projections.

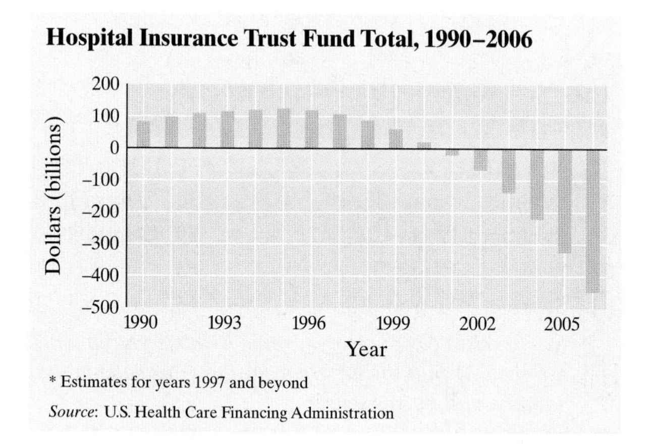

a) The HITF had the greatest balance in 1995. Estimate that balance.

b) In what year will the HITF show its first loss? Estimate that loss.

c) Estimate the change in the HITF from 1995 through 2006.

Challenge Problems

Find each sum.

117. $1 - 2 + 3 - 4 + 5 - 6 + 7 - 8 + 9 - 10$

118. $1 - 2 + 3 - 4 + 5 - 6 + \cdots + 99 - 100$

119. Consider a number line.

a) What is the distance, in units, between −2 and 5?

b) Write a subtraction problem to represent this distance (the distance is to be positive).

120. A model rocket is on a hill near the ocean. The hill's height is 62 feet above sea level. When ignited, the rocket climbs upward to 128 feet above sea level, then it falls and lands in the ocean and settles 59 feet below sea level. Find the total distance traveled by the rocket.

121. A ball rolls off a table and follows the path indicated in the figure. The maximum height reached by the ball on each bounce is 1 foot less than on the previous bounce.

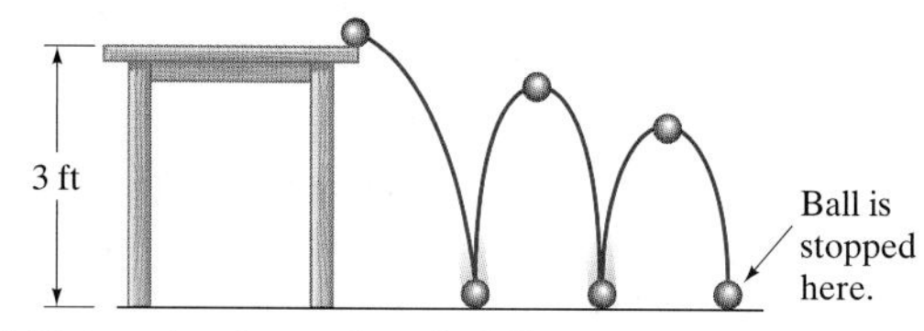

a) Determine the total vertical distance traveled by the ball.

b) If we consider the ball moving in a downward direction as negative, and the ball moving in an upward direction as positive, what was the net vertical distance traveled (from its starting point) by the ball?

Cumulative Review Exercises

[1.4] **122.** List the set of integers.

123. Explain the relationship between the set of rational numbers, the set of irrational numbers, and the set of real numbers.

[1.5] *Insert either* $>$, $<$, *or* $=$ *in each shaded area to make the statement true.*

124. $|-3| \;\square\; -5$

125. $|-6| \;\square\; |-7|$

1.8 MULTIPLICATION AND DIVISION OF REAL NUMBERS

1. **Multiply numbers.**
2. **Divide numbers.**
3. **Remove negative signs from denominators.**
4. **Evaluate divisions involving 0.**

1) Multiply Numbers

The following rules are used in determining the sign of the product when two numbers are multiplied.

The Sign of the Product of Two Real Numbers

1. The product of two numbers with **like** signs is a **positive** number.
2. The product of two numbers with **unlike** signs is a **negative** number.

By this rule, the product of two positive numbers or two negative numbers will be a positive number. The product of a positive number and a negative number will be a negative number.

EXAMPLE 1 Evaluate. **a)** $3(-5)$ **b)** $(-6)(7)$ **c)** $(-9)(-3)$

Solution **a)** Since the numbers have unlike signs, the product is negative.

$$3(-5) = -15$$

b) Since the numbers have unlike signs, the product is negative.

$$(-6)(7) = -42$$

c) Since the numbers have like signs, both negative, the product is positive.

$$(-9)(-3) = 27$$

EXAMPLE 2 Evaluate.

a) $(-8)(5)$ **b)** $(-4)(-8)$ **c)** $4(-9)$ **d)** $0(6)$ **e)** $0(-2)$ **f)** $-3(-6)$

Solution **a)** $(-8)(5) = -40$ **b)** $(-4)(-8) = 32$ **c)** $4(-9) = -36$
d) $0(6) = 0$ **e)** $0(-2) = 0$ **f)** $-3(-6) = 18$

NOW TRY EXERCISE 29

Note that zero multiplied by any real number equals zero.

HELPFUL HINT

At this point some students begin confusing problems like $-2 - 3$ with $(-2)(-3)$ and problems like $2 - 3$ with problems like $2(-3)$. If you do not understand the difference between problems like $-2 - 3$ and $(-2)(-3)$, make an appointment to see your instructor as soon as possible.

SUBTRACTION PROBLEMS	MULTIPLICATION PROBLEMS
$-2 - 3 = -5$	$(-2)(-3) = 6$
$2 - 3 = -1$	$(2)(-3) = -6$

EXAMPLE 3 Evaluate **a)** $\left(\frac{-1}{8}\right)\left(\frac{-3}{5}\right)$ **b)** $\left(\frac{3}{20}\right)\left(\frac{-3}{10}\right)$.

Solution **a)** $\left(\frac{-1}{8}\right)\left(\frac{-3}{5}\right) = \frac{(-1)(-3)}{8(5)} = \frac{3}{40}$ **b)** $\left(\frac{3}{20}\right)\left(\frac{-3}{10}\right) = \frac{3(-3)}{20(10)} = \frac{-9}{200}$

NOW TRY EXERCISE 41

Sometimes you may be asked to perform more than one multiplication in a given problem. When this happens, the sign of the final product can be determined by counting the number of *negative* numbers being multiplied. *The product of an even number of negative numbers will always be positive. The product of an odd number of negative numbers will always be negative.* Can you explain why?

EXAMPLE 4 Evaluate **a)** $(-2)(3)(-2)(-1)$ **b)** $(-3)(2)(-1)(-2)(-4)$.

Solution **a)** Since there are three negative numbers (an odd number of negatives), the product will be negative, as illustrated.

$$\begin{aligned}(-2)(3)(-2)(-1) &= (-6)(-2)(-1)\\ &= (12)(-1)\\ &= -12\end{aligned}$$

b) Since there are four negative numbers (an even number), the product will be positive.

$$\begin{aligned}(-3)(2)(-1)(-2)(-4) &= (-6)(-1)(-2)(-4)\\ &= (6)(-2)(-4)\\ &= (-12)(-4)\\ &= 48\end{aligned}$$

NOW TRY EXERCISE 35

2) Divide Numbers

The rules for dividing numbers are very similar to those used in multiplying numbers.

The Sign of the Quotient of Two Real Numbers

1. The quotient of two numbers with **like** signs is a **positive** number.
2. The quotient of two numbers with **unlike** signs is a **negative** number.

Therefore, the quotient of two positive numbers or two negative numbers will be a positive number. The quotient of a positive number and a negative number will be a negative number.

EXAMPLE 5 Evaluate. **a)** $\dfrac{20}{-5}$ **b)** $\dfrac{-45}{5}$ **c)** $\dfrac{-36}{-6}$

Solution **a)** Since the numbers have unlike signs, the quotient is negative.

$$\frac{20}{-5} = -4$$

b) Since the numbers have unlike signs, the quotient is negative.

$$\frac{-45}{5} = -9$$

c) Since the numbers have like signs, both negative, the quotient is positive.

$$\frac{-36}{-6} = 6$$

EXAMPLE 6 Evaluate $-16 \div (-2)$.

Solution Since the numbers have like signs, both negative, the quotient is positive.

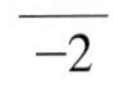

$$\frac{-16}{-2} = 8$$

EXAMPLE 7 Evaluate $\dfrac{-2}{3} \div \dfrac{-5}{7}$.

Solution Invert the *divisor*, $\dfrac{-5}{7}$, and then multiply.

$$\frac{-2}{3} \div \frac{-5}{7} = \left(\frac{-2}{3}\right)\left(\frac{7}{-5}\right) = \frac{-14}{-15} = \frac{14}{15}$$

HELPFUL HINT

For multiplication and division of two real numbers:

$$\left.\begin{array}{ll} (+)(+) = + & \dfrac{(+)}{(+)} = + \\ (-)(-) = + & \dfrac{(-)}{(-)} = + \end{array}\right\} \text{Like signs give positive products and quotients.}$$

$$\left.\begin{array}{ll} (+)(-) = - & \dfrac{(+)}{(-)} = - \\ (-)(+) = - & \dfrac{(-)}{(+)} = - \end{array}\right\} \text{Unlike signs give negative products and quotients.}$$

3) Remove Negative Signs from Denominators

We now know that the quotient of a positive and a negative number is a negative number. The fractions $-\frac{3}{4}$, $\frac{-3}{4}$, and $\frac{3}{-4}$ all represent the same negative number, negative three-fourths.

> If a and b represent any real numbers, $b \neq 0$, then
> $$\frac{a}{-b} = \frac{-a}{b} = -\frac{a}{b}$$

In mathematics we generally do not write a fraction with a negative sign in the denominator. When a negative sign appears in a denominator, we can move it to the numerator or place it in front of the fraction. For example, the fraction $\frac{5}{-7}$ should be written as either $-\frac{5}{7}$ or $\frac{-5}{7}$. Fractions also can be written using a slash, /. For example, the fraction $-\frac{5}{7}$ may be written $-5/7$ or $-(5/7)$.

EXAMPLE 8 Evaluate $\frac{2}{5} \div \left(\frac{-8}{15}\right)$.

Solution

$$\frac{2}{5} \div \left(\frac{-8}{15}\right) = \frac{\overset{1}{\cancel{2}}}{\underset{1}{\cancel{5}}} \cdot \left(\frac{\overset{3}{\cancel{15}}}{-\underset{4}{\cancel{8}}}\right) = \frac{1(3)}{1(-4)} = \frac{3}{-4} = -\frac{3}{4}$$

NOW TRY EXERCISE 73

The operations on real numbers are summarized in Table 1.1.

TABLE 1.1 Summary of Operations on Real Numbers

Signs of Numbers	Addition	Subtraction	Multiplication	Division
Both Numbers Are Positive	Sum Is Always Positive	Difference May Be Either Positive or Negative	Product Is Always Positive	Quotient Is Always Positive
Examples 6 and 2 2 and 6	 $6 + 2 = 8$ $2 + 6 = 8$	 $6 - 2 = 4$ $2 - 6 = -4$	 $6 \cdot 2 = 12$ $2 \cdot 6 = 12$	 $6 \div 2 = 3$ $2 \div 6 = \frac{1}{3}$
One Number Is Positive and the Other Number Is Negative	Sum May Be Either Positive or Negative	Difference May Be Either Positive or Negative	Product Is Always Negative	Quotient Is Always Negative
Examples 6 and −2 −6 and 2	 $6 + (-2) = 4$ $-6 + 2 = -4$	 $6 - (-2) = 8$ $-6 - 2 = -8$	 $6(-2) = -12$ $-6(2) = -12$	 $6 \div (-2) = -3$ $-6 \div 2 = -3$
Both Numbers Are Negative	Sum Is Always Negative	Difference May Be Either Positive or Negative	Product Is Always Positive	Quotient Is Always Positive
Examples −6 and −2 −2 and −6	 $-6 + (-2) = -8$ $-2 + (-6) = -8$	 $-6 - (-2) = -4$ $-2 - (-6) = 4$	 $-6(-2) = 12$ $-2(-6) = 12$	 $-6 \div (-2) = 3$ $-2 \div (-6) = \frac{1}{3}$

4 Evaluate Divisions Involving 0

Now let's look at divisions involving the number 0. What is $\frac{0}{1}$ equal to? Note that $\frac{6}{3} = 2$ because $3 \cdot 2 = 6$. We can follow the same procedure to determine the value of $\frac{0}{1}$. Suppose that $\frac{0}{1}$ is equal to some number, which we will designate by ? .

$$\text{If } \frac{0}{1} = \boxed{?} \quad \text{then} \quad 1 \cdot \boxed{?} = 0$$

Since only $1 \cdot 0 = 0$, the $\boxed{?}$ must be 0. Thus, $\frac{0}{1} = 0$. Using the same technique, we can show that zero divided by any nonzero number is zero.

> If a represents any real number except 0, then
>
> $$\frac{0}{a} = 0$$

Now what is $\frac{1}{0}$ equal to?

$$\text{If } \frac{1}{0} = \boxed{?} \quad \text{then} \quad 0 \cdot \boxed{?} = 1$$

But since 0 multiplied by any number will be 0, there is no value that can replace $\boxed{?}$. We say that $\frac{1}{0}$ is **undefined**. Using the same technique, we can show that any real number, except 0, divided by 0 is undefined.

> If a represents any real number except 0, then
>
> $\frac{a}{0}$ is **undefined**

What is $\frac{0}{0}$ equal to?

$$\text{If } \frac{0}{0} = \boxed{?} \quad \text{then} \quad 0 \cdot \boxed{?} = 0$$

But since the product of any number and 0 is 0, the $\boxed{?}$ can be replaced by any real number. Therefore the quotient $\frac{0}{0}$ cannot be determined, and so we will not use it in this course.*

> **Summary of Division Involving 0**
>
> If a represents any real number except 0, then
>
> $\frac{0}{a} = 0 \qquad \frac{a}{0}$ is undefined

EXAMPLE 9 Indicate whether each quotient is 0 or undefined.

a) $\frac{0}{2}$ **b)** $\frac{5}{0}$ **c)** $\frac{0}{-4}$ **d)** $\frac{-2}{0}$

Solution The answer to parts **a)** and **c)** is 0. The answer to parts **b)** and **d)** is undefined.

NOW TRY EXERCISE 95

---

*At this level, some professors prefer to call $\frac{0}{0}$ *indeterminate* while others prefer to call $\frac{0}{0}$ *undefined*. In higher-level mathematics courses, $\frac{0}{0}$ is sometimes referred to as the *indeterminate form*.

Using Your Calculator

Multiplication and Division

Below we show how numbers may be multiplied and divided on a calculator.

Scientific Calculator

EVALUATE	KEYSTROKES	ANSWER DISPLAYED
$6(-23)$	6 [×] 23 [+/−] [=]	−138
$\frac{-240}{-16}$	240 [+/−] [÷] 16 [+/−] [=]	15

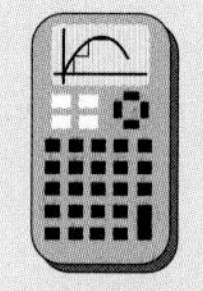

Graphing Calculator

EVALUATE	KEYSTROKES	ANSWER DISPLAYED
$6(-23)$	6 [×] [(−)] 23 [ENTER]	−138
$\frac{-240}{-16}$	[(−)] 240 [÷] [(−)] 16 [ENTER]	15

Since a positive number multiplied by a negative number will be negative, to obtain the product of $6(-23)$, you can multiply, $(6)(23)$ and write a negative sign before the answer. Since a negative number divided by a negative number is positive $\frac{-240}{-16}$ could have been found by dividing $\frac{240}{16}$.

Exercise Set 1.8

Concept/Writing Exercises

1. State the rules used to determine the sign of the product of two real numbers.
2. State the rules used to determine the sign of the quotient of two real numbers.
3. What is the product of 0 and any real number?
4. When multiplying three or more real numbers, explain how to determine the sign of the product of the numbers.

Determine the sign of each product. Explain how you determined the sign.

5. $(-3)(4)(-5)$
6. $(-9)(-12)(-15)$
7. $(-102)(-16)(24)(19)$
8. $(1054)(-92)(-16)(-37)$
9. $(-40)(-16)(30)(50)(-13)$
10. $(-1)(3)(-462)(-196)(-312)$
11. How do we generally rewrite a fraction of the form $\frac{a}{-b}$, where a and b represent any positive real numbers?
12. **a)** What is $\frac{0}{a}$ equal to where a is any nonzero real number?
    b) What is $\frac{a}{0}$ equal to where a is any nonzero real number?
13. **a)** Explain the difference between $3 - 5$ and $3(-5)$.
    b) Evaluate $3 - 5$ and $3(-5)$.
14. **a)** Explain the difference between $-4 - 2$ and $(-4)(-2)$.
    b) Evaluate $-4 - 2$ and $(-4)(-2)$.
15. **a)** Explain the difference between $x - y$ and $x(-y)$ where x and y represent any real numbers. If x is 5 and y is -2, find the value of **b)** $x - y$, **c)** $x(-y)$, and **d)** $-x - y$.
16. If x is -8 and y is 3, find the value of **a)** xy, **b)** $x(-y)$, **c)** $x - y$, and **d)** $-x - y$.

Practice the Skills

Find each product.

17. $(-4)(-3)$
18. $-4(2)$
19. $3(-3)$
20. $6(-2)$
21. $(-2)(4)$
22. $(-3)(2)$
23. $0(-8)$
24. $-1(8)$

25. $6(7)$ **26.** $-9(-4)$ **27.** $(-5)(-8)$ **28.** $7(-7)$

29. $(-5)(-6)$ **30.** $-2(5)$ **31.** $0(3)(8)$ **32.** $5(-4)(2)$

33. $(21)(-1)(4)$ **34.** $2(8)(-1)(-3)$ 35. $-1(-3)(3)(-8)$ **36.** $(-3)(-4)(-5)(-1)$

37. $(-4)(3)(-7)(1)$ **38.** $(-3)(2)(5)(3)$ **39.** $(-1)(3)(0)(-7)$ **40.** $(-6)(6)(4)(-4)$

Find each product.

41. $\left(\frac{-1}{2}\right)\left(\frac{3}{5}\right)$ **42.** $\left(\frac{1}{3}\right)\left(\frac{-3}{5}\right)$ **43.** $\left(\frac{-8}{9}\right)\left(\frac{-7}{12}\right)$ **44.** $\left(\frac{4}{5}\right)\left(\frac{-3}{10}\right)$

45. $\left(\frac{6}{-3}\right)\left(\frac{4}{-2}\right)$ **46.** $\left(\frac{8}{-11}\right)\left(\frac{6}{-5}\right)$ **47.** $\left(\frac{-3}{8}\right)\left(\frac{5}{6}\right)$ **48.** $\left(\frac{9}{10}\right)\left(\frac{7}{-8}\right)$

Find each quotient.

49. $\frac{6}{2}$ **50.** $25 \div (5)$ **51.** $-16 \div (-4)$ **52.** $\frac{-18}{9}$

53. $\frac{-36}{-9}$ **54.** $\frac{28}{-7}$ **55.** $\frac{36}{-2}$ **56.** $\frac{-15}{-1}$

57. $\frac{-12}{-1}$ **58.** $-15/(-3)$ **59.** $20/(-4)$ **60.** $\frac{-6}{-1}$

61. $\frac{-42}{7}$ **62.** $\frac{-25}{-5}$ **63.** $\frac{36}{-4}$ **64.** $\frac{-10}{10}$

65. $-40 \div (-8)$ **66.** $-64/(-4)$ **67.** Divide 0 by 4. **68.** Divide 26 by -13.

69. Divide 30 by -10. **70.** Divide -30 by -5. **71.** Divide -180 by 30. **72.** Divide -25 by -5.

Find each quotient.

73. $\frac{3}{12} \div \left(\frac{-5}{8}\right)$ **74.** $(-3) \div \frac{5}{19}$ **75.** $\frac{-7}{12} \div (-2)$ **76.** $\frac{-4}{9} \div \left(\frac{-6}{7}\right)$

77. $\frac{-15}{21} \div \left(\frac{-15}{21}\right)$ **78.** $\frac{6}{15} \div \left(\frac{7}{30}\right)$ **79.** $(-12) \div \frac{5}{12}$ **80.** $\frac{-16}{3} \div \left(\frac{5}{-9}\right)$

Evaluate.

81. $-4(8)$ **82.** $\frac{-18}{-2}$ **83.** $\frac{100}{-5}$ **84.** $-50 \div (-10)$

85. $-7(2)$ **86.** $-5(-12)$ **87.** $27 \div (-3)$ **88.** Divide -120 by -10.

89. $-100 \div 5$ **90.** $4(-2)(-1)(-5)$ **91.** Divide 60 by -60. **92.** $(6)(1)(-3)(4)$

Indicate whether each quotient is 0 or undefined.

93. $0 \div 6$ **94.** $0 \div (-7)$ 95. $\frac{5}{0}$ **96.** $\frac{-2}{0}$

97. $\frac{0}{1}$ **98.** $\frac{6}{0}$ **99.** 8 divided by 0 **100.** 0 divided by 12

Exercises 101–116, ***a)*** *determine by observation whether the product or quotient will be a positive number, zero, a negative number; or undefined;* ***b)*** *find the product or quotient on your calculator (an error message indicates that the quotient is undefined);* ***c)*** *examine your answer in part* ***b)*** *to see whether it is reasonable and makes sense.*

101. $(-212)(-87)$ **102.** $-168 \div 42$ **103.** $-240/15$ **104.** $190/10$

105. $243 \div (-27)$ **106.** $(323)(-115)$ **107.** $(171)(-89)$ **108.** $(1530)(0)$

109. $0 \div 5335$ **110.** $-86.4 \div (-36)$ **111.** $7.2 \div 0$ **112.** $-37.74 \div 37$

113. $8 \div (2.5)$ **114.** $(1.1)(9.72)(6.3)$ **115.** $(-3.0)(4.2)(-18)$ **116.** $-288.86/1.43$

Indicate whether each statement is true or false.

117. The product of a positive number and a negative number is a negative number.

118. The product of two negative numbers is a negative number.

119. The quotient of two negative numbers is a positive number.

120. The quotient of two numbers with unlike signs is a positive number.

121. The product of an even number of negative numbers is a positive number.

122. The product of an odd number of negative numbers is a negative number.

123. Zero divided by 1 is 1.

124. Six divided by 0 is 0.

125. One divided by 0 is 0.

126. Zero divided by 1 is undefined.

127. Five divided by 0 is undefined.

128. The product of 0 and any real number is 0.

129. Division by 0 does not result in a real number.

Problem Solving

130. A submarine is at a depth of −160 feet (160 feet below sea level). It dives to 3 times that depth. Find its new depth.

131. Leona DeVito's balance on her credit card is −\$450 (she owes \$450). She pays back $\frac{1}{3}$ of this balance.

a) How much did she pay back?

b) What is her new balance?

132. Dominike Jason owes a friend \$300. After he makes four payments of \$30 each, how much will he still owe?

133. If a stock loses $1\frac{1}{2}$ points each day for three successive days, how much has it lost in total?

134. On Monday in Minneapolis the wind chill was −30°F. On Tuesday the wind chill was only $\frac{1}{3}$ of what it was on Monday. What was the wind chill on Tuesday?

135. The following graph shows the federal debt in 1960, 1980, and a projection for 2000 (from the Bureau of Public Debt and Office of Management and Budget).

The Federal Debt

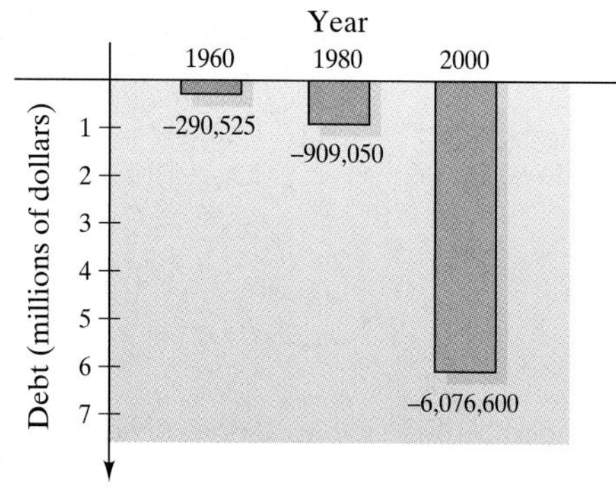

a) How many times larger was the debt in 1980 than in 1960?

b) How many times larger is the projected debt in 2000 than in 1980?

136. Most calculators have a *reciprocal key*, $\boxed{1/x}$.

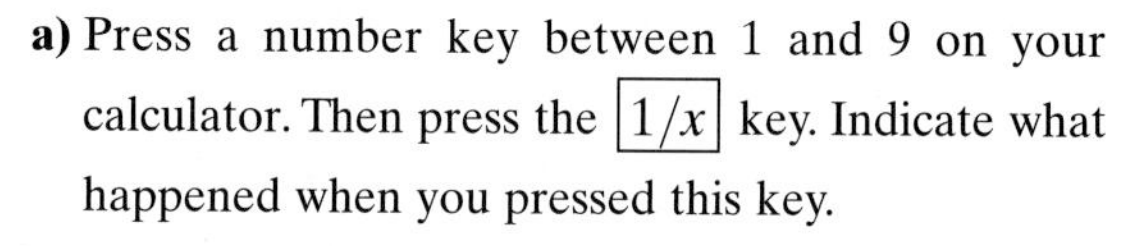

a) Press a number key between 1 and 9 on your calculator. Then press the $\boxed{1/x}$ key. Indicate what happened when you pressed this key.

b) Press the $\boxed{1/x}$ key a second time. What happens now?

c) What do you think will happen if you enter 0 and then press the $\boxed{1/x}$ key? Explain your answer.

d) Enter 0 and then press the $\boxed{1/x}$ key to see whether your answer to part **d)** was correct. If not, explain why.

137. The Johns Hopkins Medical Letter states that to find a person's *target heart rate* in beats per minute, follow this procedure. Subtract the person's age from 220, then multiply this difference by 60% and 75%. The difference multiplied by 60% gives the lower limit and the difference multiplied by 75% gives the upper limit.

a) Find the target heart rate rate range of a 50-year-old.

b) Find you own target heart rate.

Challenge Problems

We will learn in the next section that $2^3 = 2 \cdot 2 \cdot 2$ *and* $x^n = \underbrace{x \cdot x \cdot x \cdot \cdots \cdot x}_{n \text{ factors of } x}$.

Use this information to evaluate each expression.

138. 3^4

139. $(-2)^3$

140. $\left(\frac{2}{3}\right)^3$

141. 1^{100}

142. $(-1)^{81}$

143. Will the product of $(-1)(-2)(-3)(-4)\cdots(-10)$ be a positive number or a negative number? Explain how you determined your answer.

144. Will the product of $(1)(-2)(3)(-4)(5)(-6)\cdots(33)(-34)$ be a positive number or a negative number? Explain how you determined your answer.

Group Activity

Discuss and answer Exercise 145 as a group, according to the instructions.

145. a) Each member of the group is to do this procedure separately. At this time do not share your number with the other members of your group.

1. Choose a number between 2 and 10.
2. Multiply your number by 9.
3. Add the two digits in the product together.
4. Subtract 5 from the sum.
5. Now choose the corresponding letter of the alphabet that corresponds with the difference found. For example, 1 is a, 2 is b, 3 is c, and so on.
6. Choose a *one-word* country that starts with that letter.
7. Now choose a *one-word* animal that starts with the last letter of the country selected.
8. Finally, choose a color that starts with the last letter of the animal chosen.

b) Now share your final answer with the other members of your group. Did you all get the same answer?

c) Most people will obtain the answer *orange*. As a group, write a paragraph or two explaining why.

Cumulative Review Exercises

[1.3] **146.** Find the quotient $\frac{5}{7} \div \frac{1}{5}$.

[1.7] **147.** Subtract −18 from −20.

Evaluate.

148. $6 - 3 - 4 - 2$

149. $5 - (-2) + 3 - 7$

1.9 EXPONENTS, PARENTHESES, AND THE ORDER OF OPERATIONS

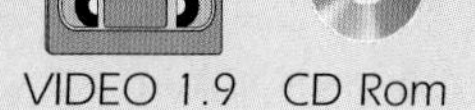

1) **Learn the meaning of exponents.**
2) **Evaluate expressions containing exponents.**
3) **Learn the difference between $-x^2$ and $(-x)^2$.**
4) **Learn the order of operations.**
5) **Learn the use of parentheses.**
6) **Evaluate expressions containing variables.**

Learn the Meaning of Exponents

To understand certain topics in algebra, you must understand exponents. Exponents are introduced in this section and are discussed in more detail in Chapter 6.

In the expression 4^2, the 4 is called the **base**, and the 2 is called the **exponent**. The number 4^2 is read "4 squared" or "4 to the second power" and means

$$\underbrace{4 \cdot 4}_{\text{2 factors of 4}} = 4^2$$

The number 4^3 is read "4 cubed" or "4 to the third power" and means

$$\underbrace{4 \cdot 4 \cdot 4}_{3 \text{ factors of } 4} = 4^3$$

In general, the number b to the nth power, written b^n, means

$$\underbrace{b \cdot b \cdot b \cdot \cdots \cdot b}_{n \text{ factors of } b} = b^n$$

Thus, $b^4 = b \cdot b \cdot b \cdot b$ or $bbbb$ and $x^3 = x \cdot x \cdot x$ or xxx.

2) Evaluate Expressions Containing Exponents

Now let's evaluate some expressions that contain exponents.

EXAMPLE 1 Evaluate. **a)** 3^2 **b)** 2^5 **c)** 1^5 **d)** $(-3)^2$ **e)** $(-2)^3$ **f)** $\left(\frac{2}{3}\right)^2$

Solution

a) $3^2 = 3 \cdot 3 = 9$

b) $2^5 = 2 \cdot 2 \cdot 2 \cdot 2 \cdot 2 = 32$

c) $1^5 = 1 \cdot 1 \cdot 1 \cdot 1 \cdot 1 = 1$ (1 raised to any power equals 1; why?)

d) $(-3)^2 = (-3)(-3) = 9$

e) $(-2)^3 = (-2)(-2)(-2) = -8$

f) $\left(\frac{2}{3}\right)^2 = \left(\frac{2}{3}\right)\left(\frac{2}{3}\right) = \frac{4}{9}$

NOW TRY EXERCISE 29

It is not necessary to write exponents of 1. Thus, when writing xxy, we write x^2y and not x^2y^1. **Whenever we see a letter or number without an exponent, we always assume that the letter or number has an exponent of 1.**

Examples of Exponential Notation

a) $xyxx = x^3y$ **b)** $xyzzy = xy^2z^2$

c) $3aabbb = 3a^2b^3$ **d)** $5xyyyy = 5xy^4$

e) $4 \cdot 4rrs = 4^2r^2s$ **f)** $5 \cdot 5 \cdot 5mmn = 5^3m^2n$

Notice in parts **a)** and **b)** that the order of the factors does not matter.

HELPFUL HINT

Note that $x + x + x + x + x + x = 6x$ and $x \cdot x \cdot x \cdot x \cdot x \cdot x = x^6$. Be careful that you do not get addition and multiplication confused.

3) Learn the Difference between $-x^2$ and $(-x)^2$

An exponent refers only to the number or letter that directly precedes it unless parentheses are used to indicate otherwise. For example, in the expression $3x^2$, only the x is squared. In the expression $-x^2$ only the x is squared. We can write $-x^2$ as $-1x^2$ because any real number may be multiplied by 1 without affecting its value.

$$-x^2 = -1x^2$$

By looking at $-1x^2$ we can see that only the x is squared, not the -1. If the entire expression $-x$ were to be squared, we would need to use parentheses and write $(-x)^2$. Note the difference in the following two examples:

$$-x^2 = -(x)(x)$$
$$(-x)^2 = (-x)(-x)$$

Consider the expressions -3^2 and $(-3)^2$. How do they differ?

$$-3^2 = -(3)(3) = -9$$
$$(-3)^2 = (-3)(-3) = 9$$

HELPFUL HINT

The expression $-x^2$ is read "negative x squared," or "the opposite of x squared." The expression $(-x)^2$ is read "negative x, quantity squared."

EXAMPLE 2 Evaluate. **a)** -5^2 **b)** $(-5)^2$ **c)** -2^3 **d)** $(-2)^3$

Solution **a)** $-5^2 = -(5)(5) = -25$ **b)** $(-5)^2 = (-5)(-5) = 25$
c) $-2^3 = -(2)(2)(2) = -8$ **d)** $(-2)^3 = (-2)(-2)(-2) = -8$

EXAMPLE 3 Evaluate. **a)** -2^4 **b)** $(-2)^4$

Solution **a)** $-2^4 = -(2)(2)(2)(2) = -16$ **b)** $(-2)^4 = (-2)(-2)(-2)(-2) = 16$

NOW TRY EXERCISE 87

Using Your Calculator

Use of $\boxed{x^2}$, $\boxed{y^x}$, and $\boxed{\wedge}$ Keys

The $\boxed{x^2}$ key is used to square a value. For example, to evaluate 5^2, we would do the following.

	Keystrokes	Answer Displayed
Scientific Calculator	5 $\boxed{x^2}$	25
Graphing Calculator	5 $\boxed{x^2}$ $\boxed{\text{ENTER}}$	25

To evaluate $(-5)^2$ on a calculator, we would do the following.

	Keystrokes	Answer Displayed
Scientific Calculator	5 $\boxed{+/-}$ $\boxed{x^2}$	25
Graphing Calculator	$\boxed{(}$ $\boxed{(-)}$ 5 $\boxed{)}$ $\boxed{x^2}$ $\boxed{\text{ENTER}}$	25

To raise a value to a power greater than 2, we use the $\boxed{y^x}$* or $\boxed{\wedge}$ key. To use these keys you enter the number, then press either the $\boxed{y^x}$ or $\boxed{\wedge}$ key, then enter the exponent. Following we show how to evaluate 2^5 and $(-2)^5$.

*Some calculators use a $\boxed{x^y}$ key instead of a $\boxed{y^x}$ key.

continued on next page

	Evaluate	Keystrokes	Answer Displayed
Scientific Calculator	2^5	2 [y^x] 5 [=]	32
Scientific Calculator	$(-2)^5$	2 [+/−] [y^x] 5 [=]**	−32
Graphing Calculator	2^5	2 [∧] 5 [ENTER]	32
Graphing Calculator	$(-2)^5$	[(] [(−)] 2 [)] [∧] 5 [ENTER]	−32

Possibly the easiest way to raise negative numbers to a power may to be raise the positive number to the power and then write a negative sign before the final answer if needed. *A negative number raised to an odd power will be negative, and a negative number raised to an even power will be positive.* Can you explain why this is true?

*Some calculators use a [x^y] key instead of a [y^x] key.

**Some scientific calculators cannot directly evaluate a negative number to an exponent greater than 2. On such calculators you will get an *error* message.

4) Learn the Order of Operations

Now that we have introduced exponents we can present the **order of operations**. Can you evaluate $2 + 3 \cdot 4$? Is it 20? Or is it 14? To answer this, you must know the order of operations to follow when evaluating a mathematical expression. You will often have to evaluate expressions containing multiple operations.

Order of Operations
To Evaluate Mathematical Expressions, Use the Following Order

1. First, evaluate the information within **parentheses** ( ), brakets [ ], or braces { }. These are **grouping symbols**, for they group information together. A fraction bar, —, also serves as a grouping symbol. If the expression contains nested parentheses (one pair of parentheses within another pair), evaluate the information in the innermost parentheses first.
2. Next, evaluate all **exponents.**
3. Next, evaluate all **multiplications** or **divisions** in the order in which they occur, working from left to right.
4. Finally, evaluate all **additions** or **subtractions** in the order in which they occur, working from left to right.

Some students remember the word PEMDAS or the phrase "Please Excuse My Dear Aunt Sally" to help them remember the order of operations. PEMDAS helps them remember the order: **P**arentheses, **E**xponents, **M**ultiplication, **D**ivision, **A**ddition, **S**ubtraction. Remember, this does not imply multiplication before division or addition before subtraction.

We can now evaluate $2 + 3 \cdot 4$. Since multiplications are performed before additions,

$$2 + 3 \cdot 4 \quad \text{means} \quad 2 + (3 \cdot 4) = 2 + 12 = 14$$

We now know that $2 + 3 \cdot 4$ means $2 + (3 \cdot 4)$ and has a value of 14. What will a calculator display if you key in the following?

2 [+] 3 [×] 4 [=]

The answer depends on your calculator. *Scientific and graphing calculators* will evaluate an expression following the rules just stated.

	Keystrokes	Answer Displayed
Scientific Calculator	2 [+] 3 [×] 4 [=]	14
Graphing Calculator	2 [+] 3 [×] 4 [ENTER]	14

Nonscientific calculators will perform operations in the order they are entered.

		Answer Displayed
Nonscientific Calculator	2 [+] 3 [×] 4 [=]	20

Remember that in algebra, unless otherwise instructed by parentheses, we always perform multiplications and divisions before additions and subtractions. In this course you should be using either a scientific or graphing calculator.

5) Learn the Use of Parentheses

Parentheses or brackets may be used (1) to change the order of operations to be followed in evaluating an algebraic expression or (2) to help clarify the understanding of an expression.

To evaluate the expression $2 + 3 \cdot 4$, we would normally perform the multiplication, $3 \cdot 4$, first. If we wished to have the addition performed before the multiplication, we could indicate this by placing parentheses around $2 + 3$:

$$(2 + 3) \cdot 4 = 5 \cdot 4 = 20$$

Consider the expression $1 \cdot 3 + 2 \cdot 4$. According to the order of operations, multiplications are to be performed before additions. We can rewrite this expression as $(1 \cdot 3) + (2 \cdot 4)$. Note that the order of operations was not changed. The parentheses were used only to help clarify the order to be followed.

HELPFUL HINT

If parentheses are not used to change the order of operations, multiplications and divisions are always performed before additions and subtractions. When a problem has only multiplications and divisions, work from left to right. Similarly, when a problem has only additions and subtractions, work from left to right.

EXAMPLE 4 Evaluate $2 + 3 \cdot 5^2 - 7$.

Solution Colored shading is used to indicate the order in which the expression is to be evaluated.

$$\begin{aligned} & 2 + 3 \cdot 5^2 - 7 && \textit{Exponent} \\ &= 2 + 3 \cdot 25 - 7 && \textit{Multiply} \\ &= 2 + 75 - 7 && \textit{Add/subtract, left to right} \\ &= 77 - 7 \\ &= 70 \end{aligned}$$

EXAMPLE 5 Evaluate $6 + 3[(12 \div 4) + 5]$.

Solution

$$\begin{aligned} & 6 + 3[(12 \div 4) + 5] && \textit{Innermost parentheses} \\ &= 6 + 3[3 + 5] && \textit{Brackets} \\ &= 6 + 3(8) && \textit{Multiply} \\ &= 6 + 24 \\ &= 30 \end{aligned}$$

EXAMPLE 6 Evaluate $(4 \div 2) + 4(5 - 2)^2$.

Solution

$$\begin{aligned} & (4 \div 2) + 4(5 - 2)^2 && \textit{Parentheses} \\ &= 2 + 4(3)^2 && \textit{Exponent} \\ &= 2 + 4 \cdot 9 && \textit{Multiply} \\ &= 2 + 36 \\ &= 38 \end{aligned}$$

NOW TRY EXERCISE 71

EXAMPLE 7 Evaluate $-8 - 81 \div 9 \cdot 2^2 + 7$.

Solution

$$\begin{aligned} &= -8 - 81 \div 9 \cdot 2^2 + 7 && \textit{Exponent} \\ &= -8 - 81 \div 9 \cdot 4 + 7 && \textit{Multiply/divide, left to right} \\ &= -8 - 9 \cdot 4 + 7 && \textit{Multiply} \\ &= -8 - 36 + 7 && \textit{Add/subtract, left to right} \\ &= -44 + 7 \\ &= -37 \end{aligned}$$

EXAMPLE 8 Evaluate. **a)** $-4^2 + 6 \div 3$ **b)** $(-4)^2 + 6 \div 3$

Solution

a)
$$\begin{aligned} & -4^2 + 6 \div 3 && \textit{Exponent} \\ &= -16 + 6 \div 3 && \textit{Divide} \\ &= -16 + 2 \\ &= -14 \end{aligned}$$

b)
$$\begin{aligned} & (-4)^2 + 6 \div 3 && \textit{Exponent} \\ &= 16 + 6 \div 3 && \textit{Divide} \\ &= 16 + 2 \\ &= 18 \end{aligned}$$

EXAMPLE 9 Evaluate $\frac{3}{8} - \frac{2}{5} \cdot \frac{1}{12}$.

Solution First perform the multiplication.

$$\begin{aligned} & \frac{3}{8} - \left(\frac{\overset{1}{\cancel{2}}}{5} \cdot \frac{1}{\underset{6}{\cancel{12}}}\right) && \textit{Multiply} \\ &= \frac{3}{8} - \frac{1}{30} && \textit{Subtract} \\ &= \frac{45}{120} - \frac{4}{120} \\ &= \frac{41}{120} \end{aligned}$$

NOW TRY EXERCISE 81

EXAMPLE 10 Write the following statements as mathematical expressions using parentheses and brackets and then evaluate: Subtract 3 from 15. Divide this difference by 2. Multiply this quotient by 4.

Solution

$15 - 3$ *Subtract 3 from 15.*

$(15 - 3) \div 2$ *Divide by 2.*

$4[(15 - 3) \div 2]$ *Multiply the quotient by 4.*

Now evaluate.

$$4[(15 - 3) \div 2]$$
$$= 4[12 \div 2]$$
$$= 4(6)$$
$$= 24$$

As shown in Example 10, sometimes brackets are used in place of parentheses to help avoid confusion. If only parentheses had been used, the preceding expression would appear as $4((15 - 3) \div 2)$.

Using Your Calculator

Using Parentheses

When evaluating an expression on a calculator where the order of operations is to be changed, you will need to use parentheses. If you are not sure whether they are needed, it will not hurt to add them. Consider $\frac{8}{4-2}$. Since we wish to divide 8 by the difference $4 - 2$, we neeed to use parentheses.

EVALUATE	KEYSTROKES	ANSWER DISPLAYED
$\frac{8}{4-2}$	8 [÷] [(] 4 [−] 2 [)] [=]*	4

What would you obtain if you evaluated 8 [÷] 4 [−] 2 [=] on a scientific calculator? Why would you get that result?

To evaluate $\left(\frac{2}{5}\right)^2$ on a scientific calculator, we press the following keys.

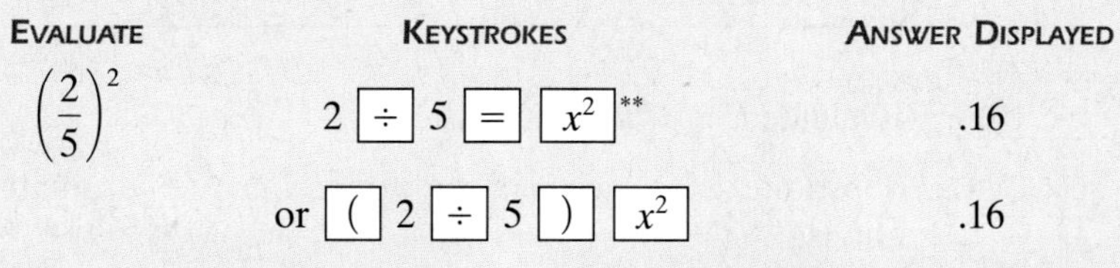

EVALUATE	KEYSTROKES	ANSWER DISPLAYED
$\left(\frac{2}{5}\right)^2$	2 [÷] 5 [=] [x^2]**	.16
	or [(] 2 [÷] 5 [)] [x^2]	.16

What would you obtain if you evaluated 2 [÷] 5 [x^2] on a scientific calculator? Why?

*If using a graphing calculator, replace [=] with [ENTER]. Everything else remains the same.

**On a graphing calculator, replace [=] with [ENTER] and press [ENTER] after [x^2].

6) Evaluate Expressions Containing Variables

Now we will evaluate some expressions for given values of the variables.

EXAMPLE 11 Evaluate $5x - 4$ when $x = 3$.

Solution Substitute 3 for x in the expression.

$$\begin{aligned} 5x - 4 &= 5(3) - 4 \\ &= 15 - 4 \\ &= 11 \end{aligned}$$

EXAMPLE 12 Evaluate **a)** x^2 and **b)** $-x^2$ when $x = 3$.

Solution Substitute 3 for x.

a) $x^2 = 3^2 = 3(3) = 9$

b) $-x^2 = -3^2 = -(3)(3) = -9$

EXAMPLE 13 Evaluate **a)** y^2 and **b)** $-y^2$ when $y = -4$.

Solution Substitute -4 for y.

a) $y^2 = (-4)^2 = (-4)(-4) = 16$

b) $-y^2 = -(-4)^2 = -(-4)(-4) = -16$

Note that $-x^2$ will always be a negative number for any nonzero value of x, and $(-x)^2$ will always be a positive number for any nonzero value of x. Can you explain why? See Exercise 6 on page 76.

Avoiding Common Errors

The expression $-x^2$ means $-(x^2)$. When asked to evaluate $-x^2$ for any real number x, many students will incorrectly treat $-x^2$ as $(-x)^2$. For example, to evaluate $-x^2$ when $x = 5$,

CORRECT	INCORRECT
$-5^2 = -(5^2) = -(5)(5) = -25$	~~$-5^2 = (-5)(-5) = 25$~~

EXAMPLE 14 Evaluate $(3x + 1) + 2x^2$ when $x = 4$.

Solution Substitute 4 for each x in the expression, then evaluate using the order of operations.

$$\begin{aligned} (3x + 1) + 2x^2 &= [3(4) + 1] + 2(4)^2 && \text{Substitute} \\ &= [12 + 1] + 2(4)^2 && \text{Multiply} \\ &= 13 + 2(16) && \text{Parentheses, exponent} \\ &= 13 + 32 \\ &= 45 \end{aligned}$$

EXAMPLE 15 Evaluate $-y^2 + 3(x + 2) - 5$ when $x = -3$ and $y = -2$.

Solution Substitute -3 for each x and -2 for each y, then evaluate using the order of operations.

$$
\begin{aligned}
-y^2 + 3(x + 2) - 5 &= -(-2)^2 + 3(-3 + 2) - 5 && \textit{Substitute} \\
&= -(-2)^2 + 3(-1) - 5 && \textit{Parentheses} \\
&= -(4) + 3(-1) - 5 && \textit{Exponent} \\
&= -4 - 3 - 5 && \textit{Multiply} \\
&= -7 - 5 && \textit{Subtract, left to right} \\
&= -12
\end{aligned}
$$

NOW TRY EXERCISE 113

Using Your Calculator

Evaluating Expressions on a Scientific Calculator

Later in this course you will need to evaluate an expression like $3x^2 - 2x + 5$ for various values of x. Below we show how to evaluate such expressions.

Evaluate	Keystrokes
a) $3x^2 - 2x + 5$, for $x = 4$	
$3(4)^2 - 2(4) + 5$	3 [×] 4 [x^2] [−] 2 [×] 4 [+] 5 [=] 45
b) $3x^2 - 2x + 5$, for $x = -6$	
$3(-6)^2 - 2(-6) + 5$	3 [×] 6 [+/−] [x^2] [−] 2 [×] 6 [+/−] [+] 5 [=] 125
c) $-x^2 - 3x - 5$, for $x = -2$	
$-(-2)^2 - 3(-2) - 5$	1 [+/−] [×] 2 [+/−] [x^2] [−] 3 [×] 2 [+/−] [−] 5 [=] −3

Remember in part **c)** that $-x^2 = -1x^2$.

Evaluating Expressions on a Graphing Calculator

All graphing calculators have a procedure for evaluating expressions. To do so you will generally need to enter the value to be used for the variable and the expression to be evaluated. After the [ENTER] key is pressed, the graphing calculator displays the answer. The procedure varies from calculator to calculator. Below we show how an expression is evaluated on a TI-82 or TI-83. On the TI-82 and TI-83, the store key—[STO▶]—is used to store a value. Stored values and expressions are separated using a colon, which is obtained by pressing [2nd] followed by [·] on the TI-82 and [ALPHA] followed by [·] on the TI-83.

Evaluate	Keystrokes on TI-83
	[(−)] 6 [STO▶] [X,T,θ,n] [ALPHA] [·] 3 [x] [x^2] [−] 2 [x] [+] 5 [ENTER]
$3x^2 - 2x + 5$ for $x = -6$	↓
Display shown:	−6 → x : 3 x^2 − 2 x + 5 125

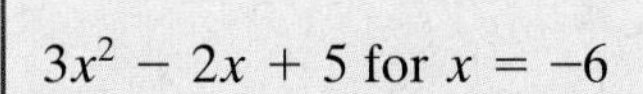

continued on next page

Notice that to obtain an x^2 on the display, we press the [X, T, θ, n] key* to select the variable x, then we press the [x^2] key, which is used to square the variable or number selected.

A nice feature of graphing calculators is that to evaluate an expression for different values of a variable you do not have to re-enter the expression each time. For example, on the TI-82 and TI-83 if you press [2nd] followed by [ENTER] it displays the expression again. Then you just need to go back and change the value stored for x to the new value and press [ENTER] to evaluate the expression with the new value of the variable.

Each brand of calculator uses different keys and procedures. We have given just a quick overview. Please read the manual that comes with your graphing calculator for a complete explanation of how to evaluate expressions.

*This key can be used to generate any of these letters (θ is a Greek letter). From this point on, in displays of keystrokes, to generate an x we will just show [x] rather than [X, T, θ, n].

Exercise Set 1.9

Concept/Writing Exercises

1. In the expression a^b, what is the a called and what is the b called?
2. Explain the meaning of **a)** 3^2, **b)** 5^4, and **c)** x^n
3. **a)** What is the exponent on a number or letter that has no exponent illustrated?
   b) In the expression $5x^3y^2z$, what is the exponent on the 5, the x, the y, and the z?
4. Write a simplified expression for the following.
   a) $y + y + y + y$
   b) $y \cdot y \cdot y \cdot y$
5. Write a simplified expression for the following.
   a) $x + x + x + x + x$
   b) $x \cdot x \cdot x \cdot x \cdot x$
6. **a)** Explain why $-x^2$ will always be a negative number for any nonzero real number selected for x.
   b) Explain why $(-x)^2$ will always be a positive number for any nonzero real number selected for x.
7. List the order of operations to be followed when evaluating a mathematical expression.
8. When an expression has only additions and subtractions or only multiplications or divisions, how is it evaluated?
9. If you evaluate $4 + 5 \times 2$ on a calculator and obtain an answer of 18, is the calculator a scientific calculator? Explain.
10. List two reasons why parentheses are used in an expression.
11. Determine the results obtained on a scientific calculator if the following keys are pressed.
    a) 15 [÷] 5 [−] 2 [=]
    b) 15 [÷] [(] 5 [−] 2 [)] [=]
    c) Which keystrokes, **a)** or **b)**, are used to evaluate $\frac{15}{5-2}$? Explain.
12. Determine the results obtained on a scientific calculator if the following keys are pressed.
    a) 15 [−] 10 [÷] 5 [=]
    b) [(] 15 [−] 10 [)] [÷] 5 [=]
    c) Which keystrokes, **a)** or **b)**, are used to evaluate $\frac{15-10}{5}$? Explain.

In Exercises 13 and 14, **a)** *write in your own words the step-by-step procedure you would use to evaluate the expression, and* **b)** *evaluate the expression.*

13. $[9 - (8 \div 2)]^2 - 6^3$
14. $[(8 \cdot 3) - 4^2]^2 - 5$

In Exercises 15 and 16, **a)** *write in your own words the step-by-step procedure you would use to evaluate the expression for the given value of the variable, and* **b)** *evaluate the expression for the given value of the variable.*

15. $-4x^2 + 3x - 6$ when $x = 5$
16. $-5x^2 - 2x + 8$ when $x = -2$

Practice the Skills

Evaluate.

17. 4^2
18. 2^3
19. 1^5
20. 3^3
21. -5^2
22. 6^3
23. $(-3)^2$
24. -6^3
25. $(-1)^3$
26. 2^5
27. $(-9)^2$
28. 5^3
29. $(-6)^2$
30. $(-3)^3$
31. 4^1
32. -7^2
33. $(-4)^4$
34. -1^4
35. -2^4
36. $3^2(4)^2$
37. $5^2 \cdot 3^2$
38. $(-1)^4(3)^3$
39. $2^3 \cdot 3^2$
40. $(-2)^3(-1)^3$

*In Exercises 41–52, **a)** determine by observation whether the answer should be positive or negative and explain your answer; **b)** evaluate the expression on your calculator; and **c)** determine whether your answer in part **b)** is reasonable and makes sense.*

41. 7^3
42. 4^6
43. 5^4
44. -2^5
45. $(-3)^5$
46. 10^3
47. $(-6)^4$
48. $(1.3)^3$
49. $(5.3)^4$
50. $(-3.3)^3$
51. $-\left(\frac{7}{8}\right)^2$
52. $\left(-\frac{3}{4}\right)^3$

Evaluate.

53. $2 + 5 \cdot 6$
54. $7 - 5^2 + 2$
55. $6 - 6 + 8$
56. $(6^2 \div 3) - (6 - 4)$
57. $1 + 3 \cdot 2^2$
58. $8 \cdot 3^2 + 1 \cdot 5$
59. $-3^3 + 27$
60. $(-2)^3 + 8 \div 4$
61. $(4 - 3) \cdot (5 - 1)^2$
62. $20 - 6 - 3 - 2$
63. $3 \cdot 7 + 4 \cdot 2$
64. $[1 - (4 \cdot 5)] + 6$
65. $4^2 - 3 \cdot 4 - 6$
66. $-2[-5 + (7 + 4)]$
67. $(6 \div 3)^3 + 4^2 \div 8$
68. $4 + (4^2 - 13)^4 - 3$
69. $[6 - (-2 - 3)]^2$
70. $(-2)^2 + 4^2 \div 2^2 + 3$
71. $(3^2 - 1) \div (3 + 1)^2$
72. $2[3(8 - 2^2) - 6]$
73. $[4 + ((5 - 2)^2 \div 3)^2]^2$
74. $(10 \div 5 \cdot 5 \div 5 - 5)^2$
75. $2.5 + 7.56 \div 2.1 + (9.2)^2$
76. $(8.4 + 3.1)^2 - (3.64 - 1.2)$
77. $2[1.55 + 5(3.7)] - 3.35$
78. $\frac{1}{2} + \frac{3}{4} \cdot \frac{5}{6}$
79. $\left(\frac{2}{7} + \frac{3}{8}\right) - \frac{3}{112}$
80. $\left(\frac{5}{6} \cdot \frac{4}{5}\right) + \left(\frac{2}{3} \cdot \frac{5}{8}\right)$
81. $\frac{3}{4} - 4 \cdot \frac{5}{40}$
82. $\frac{12 - (4 - 6)^2}{6 + 4^2 \div 2^2}$
83. $\frac{5 - [3(6 \div 3) - 2]}{5^2 - 4^2 \div 2}$
84. $\frac{[(7 - 3)^2 - 4]^2}{9 - 16 \div 8 - 4}$

*Evaluate **a)** x^2, **b)** $-x^2$, and **c)** $(-x)^2$ for the following values of x.*

85. 3
86. 8
87. -4
88. -5
89. 6
90. 7
91. $-\frac{1}{2}$
92. $\frac{3}{4}$

Evaluate each expression for the given value of the variable or variables.

93. $x + 4;\ x = -2$
94. $3x - 4x + 5;\ x = 1$
95. $5x - 1;\ x = 4$
96. $3(x - 2);\ x = 5$
97. $a^2 - 6;\ a = -3$
98. $b^2 - 8;\ b = 5$
99. $-4x^2 - 2x + 1;\ x = -1$
100. $2r^2 - 5r + 3;\ r = 1$
101. $3p^2 - 6p - 4;\ p = 2$
102. $-y^2 - 6y - 5;\ y = 4$
103. $-x^2 - 2x + 5;\ x = -4$
104. $2x^2 - 4x - 10;\ x = 5$
105. $4(x + 1)^2 - 6x;\ x = 5$
106. $3n^2(n - 1) + 5;\ n = -4$
107. $-6x + 3y;\ x = 2,\ y = 4$
108. $6x + 3y^2 - 5;\ x = 1,\ y = -3$
109. $r^2 - s^2;\ r = -2,\ s = -3$
110. $r^2 - s^2;\ r = 2,\ s = -4$
111. $2(x + y) + 4x - 3y;\ x = 2;\ y = -3$
112. $4(x + y)^2 + 2(x + y) + 3;\ x = 2,\ y = 4$
113. $6x^2 + 3xy - y^2;\ x = 2,\ y = -3$
114. $3(x - 4)^2 - (3y - 4)^2;\ x = -1,\ y = -2$

Problem Solving

Write the following statements as mathematical expressions using parentheses and brackets, and then evaluate.

115. Multiply 6 by 3. From this product, subtract 4. From this difference, subtract 2.

116. Add 4 to 9. Divide this sum by 2. Add 10 to this quotient.

117. Divide 20 by 5. Add 12 to this quotient. Subtract 8 from this sum. Multiply this difference by 9.

118. Multiply 6 by 3. To this product, add 27. Divide this sum by 8. Multiply this quotient by 10.

119. Add $\frac{4}{5}$ to $\frac{3}{7}$ Multiply this sum by $\frac{2}{3}$.

120. Multiply $\frac{3}{8}$ by $\frac{4}{5}$. To this product, add $\frac{7}{120}$. From this sum, subtract $\frac{1}{60}$.

121. For what value or values of x does $-(x^2) = -x^2$?

122. For what value or values of x does $x = x^2$?

123. If the sales tax on an item is 7%, the sales tax on an item costing d dollars can be found by the expression $0.07d$. Determine the sales tax on a compact disk that costs \$15.99.

124. If a car travels at 60 miles per hour, the distance it travels in t hours is $60t$. Determine how far a car traveling at 60 miles per hour travels in 2.5 hours.

125. If the sales tax on an item is 7%, then the total cost of an item d, including sales tax, can be found by the expression $d + 0.07d$. Find the total cost of a car that costs \$15,000.

126. The profit or loss of a business, in dollars, can be found by the expression $-x^2 + 60x - 100$ where x is the number of bicycles (from 0 to 30) sold in a week. Determine the profit or loss if 20 bicycles are sold in a week.

127. In the Using Your Calculator box on page 73 we showed that to evaluate $\left(\frac{2}{5}\right)^2$ we press the following keys:

2 [÷] 5 [=] [x^2]

We obtained an answer of .16. Indicate what a scientific (or graphing) calculator would display if the following keys are pressed. (On a graphing calculator, press [ENTER] instead of [=] and end part **b)** with [ENTER]). Explain your reason for each answer. Check your answer on your calculator.

a) 2 [÷] 5 [x^2] [=]

b) [(] 2 [÷] 5 [)] [x^2] .

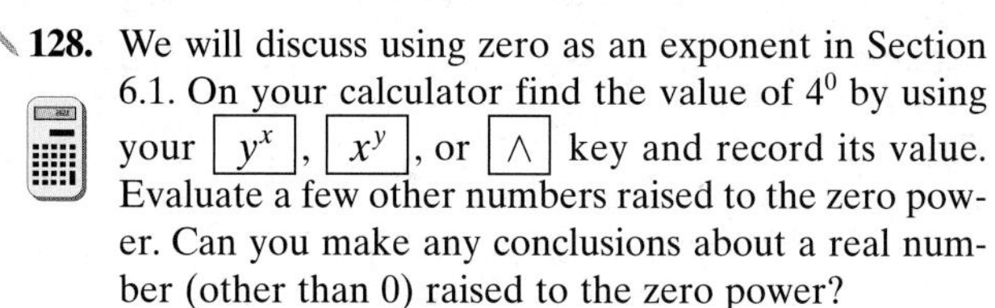

128. We will discuss using zero as an exponent in Section 6.1. On your calculator find the value of 4^0 by using your [y^x], [x^y], or [∧] key and record its value. Evaluate a few other numbers raised to the zero power. Can you make any conclusions about a real number (other than 0) raised to the zero power?

Challenge Problems

129. The rate of growth of grass in inches per week depends on a number of factors, including rainfall and temperature. For a certain region of the country, the growth per week can be approximated by the expression $0.2R^2 + 0.003RT + 0.0001T^2$, where R is the weekly rainfall, in inches, and T is the average weekly temperature, in degrees Fahrenheit. Find the amount of growth of grass for a week in which the rainfall is 2 inches and the average temperature is 70°F.

Insert one pair of parentheses to make each statement true.

130. $14 + 6 \div 2 \times 4 = 40$

131. $12 - 4 - 6 + 10 = 24$

132. $24 \div 6 \div 2 + 2 = 1$

Group Activity

Discuss and answer Exercises 133–136 as a group, according to the instructions. Each question has four parts. For parts **a)**, **b)**, *and* **c)**, *simplify the expression and write the answer in exponential form. Use the knowledge gained in parts* **a)–c)** *to answer part* **d)**. *(General rules that may be used to solve exercises like these will be discussed in Chapter 6.)*

a) *Group member 1:* Do part **a)** of each exercise.

b) *Group member 2:* Do part **b)** of each exercise.

c) *Group member 3:* Do part **c)** of each exercise.

d) *As a group,* answer part **d)** of each exercise. You may need to make up other examples like parts **a)–c)** to help you answer part **d)**.

133. **a)** $2^2 \cdot 2^3$ **b)** $3^2 \cdot 3^3$ **c)** $2^3 \cdot 2^4$ **d)** $x^m \cdot x^n$

134. **a)** $\frac{2^3}{2^2}$ **b)** $\frac{3^4}{3^2}$ **c)** $\frac{4^5}{4^3}$ **d)** $\frac{x^m}{x^n}$

135. **a)** $(2^3)^2$ **b)** $(3^3)^2$ **c)** $(4^2)^2$ **d)** $(x^m)^n$

136. **a)** $(2x)^2$ **b)** $(3x)^2$ **c)** $(4x)^3$ **d)** $(ax)^m$

Cumulative Review Exercises

[1.2] **137.** The graph shows the number of occupants in various houses selected at random in a neighborhood.

Occupants in Selected Houses

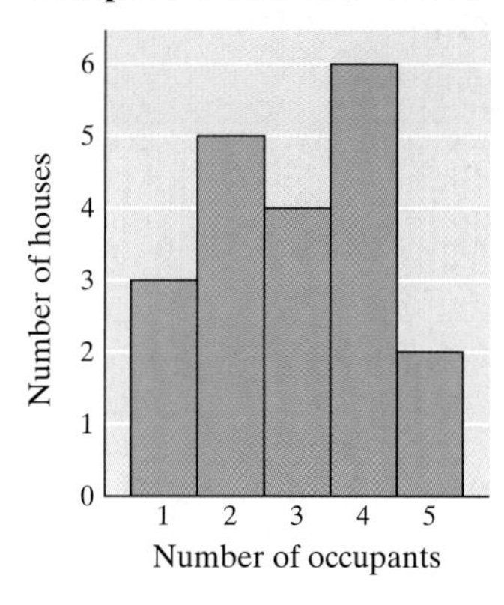

a) How many houses have three occupants?

b) Make a chart showing the number of houses that have one occupant, two occupants, three occupants, and so on.

c) How many occupants in total are there in all the houses?

d) Determine the mean number of occupants in all the houses surveyed.

138. Yellow Cab charges \$2.40 for the first $\frac{1}{2}$ mile plus 20 cents for each additional $\frac{1}{8}$ mile or part thereof. Find the cost of a 3-mile trip.

[1.8] *Evaluate.*

139. $(-2)(-4)(6)(-1)(-3)$

140. $\left(\frac{-5}{7}\right) \div \left(\frac{-3}{14}\right)$

1.10 PROPERTIES OF THE REAL NUMBER SYSTEM

1) Learn the commutative property.
2) Learn the associative property.
3) Learn the distributive property.

Here, we introduce various properties of the real number system. We will use these properties throughout the text.

1) Learn the Commutative Property

The **commutative property of addition** states that the order in which any two real numbers are added does not matter.

Commutative Property of Addition

If a and b represent any two real numbers, then

$$a + b = b + a$$

Notice that the commutative property involves a change in *order*. For example,

$$4 + 3 = 3 + 4$$
$$7 = 7$$

The **commutative property of multiplication** states that the order in which any two real numbers are multiplied does not matter.

Commutative Property of Multiplication

If a and b represent any two real numbers, then

$$a \cdot b = b \cdot a$$

For example,

$$6 \cdot 3 = 3 \cdot 6$$
$$18 = 18$$

The commutative property **does not hold** *for subtraction or division*. For example, $4 - 6 \neq 6 - 4$ and $6 \div 3 \neq 3 \div 6$.

2) Learn the Associative Property

The **associative property of addition** states that, in the addition of three or more numbers, parentheses may be placed around any two adjacent numbers without changing the results.

Associative Property of Addition

If a, b, and c represent any three real numbers, then

$$(a + b) + c = a + (b + c)$$

Notice that the associative property involves a change of *grouping*. For example,

$$(3 + 4) + 5 = 3 + (4 + 5)$$
$$7 + 5 = 3 + 9$$
$$12 = 12$$

In this example the 3 and 4 are grouped together on the left, and the 4 and 5 are grouped together on the right.

The **associative property of multiplication**, states that, in the multiplication of three or more numbers, parentheses may be placed around any two adjacent numbers without changing the results.

Associative Property of Multiplication

If a, b, and c represent any three real numbers, then

$$(a \cdot b) \cdot c = a \cdot (b \cdot c)$$

For example,

$$(6 \cdot 2) \cdot 4 = 6 \cdot (2 \cdot 4)$$
$$12 \cdot 4 = 6 \cdot 8$$
$$48 = 48$$

Since the associative property involves a change of grouping, when the associative property is used, the content within the parentheses changes.

The associative property **does not hold** *for subtraction or division.* For example, $(4 - 1) - 3 \neq 4 - (1 - 3)$ and $(8 \div 4) \div 2 \neq 8 \div (4 \div 2)$.

3) Learn the Distributive Property

A very important property of the real numbers is the **distributive property of multiplication over addition.** We often shorten the name to the **distributive property**.

Distributive Property

If a, b, and c represent any three real numbers, then

$$a(b + c) = ab + ac$$

For example, if we let $a = 2$, $b = 3$, and $c = 4$, then

$$2(3 + 4) = (2 \cdot 3) + (2 \cdot 4)$$
$$2 \cdot 7 = 6 + 8$$
$$14 = 14$$

Therefore, we may either add first and then multiply, or multiply first and then add. The distributive property will be discussed in more detail in Chapter 2.

HELPFUL HINT

The *commutative property* changes *order*.

The *associative property* changes *grouping*.

The *distributive property* involves *two operations*, usually multiplication and addition.

The following are additional illustrations of the commutative, associative, and distributive properties. If we assume that x represents any real number, then:

$x + 4 = 4 + x$ by the commutative property of addition.

$x \cdot 4 = 4 \cdot x$ by the commutative property of multiplication.

$(x + 4) + 7 = x + (4 + 7)$ by the associative property of addition.

$(x \cdot 4) \cdot 6 = x \cdot (4 \cdot 6)$ by the associative property of multiplication.

$3(x + 4) = (3 \cdot x) + (3 \cdot 4)$ or $3x + 12$ by the distributive property.

EXAMPLE 1 Name each property illustrated.

a) $4 + (-2) = -2 + 4$ **b)** $x + y = y + x$

c) $x \cdot y = y \cdot x$ **d)** $(-12 + 3) + 4 = -12 + (3 + 4)$

Solution

a) Commutative property of addition

b) Commutative property of addition

c) Commutative property of multiplication

d) Associative property of addition

NOW TRY EXERCISE 15

EXAMPLE 2 Name each property illustrated.

a) $2(x + 2) = (2 \cdot x) + (2 \cdot 2) = 2x + 4$

b) $4(x + y) = (4 \cdot x) + (4 \cdot y) = 4x + 4y$

c) $(3 \cdot 6) \cdot 5 = 3 \cdot (6 \cdot 5)$

Solution

a) Distributive property

b) Distributive property

c) Associative property of multiplication

HELPFUL HINT

Do not confuse the distributive property with the associative property of multiplication. Make sure you understand the difference.

Distributive Property	Associative Property of Multiplication
$3(4 + x) = 3 \cdot 4 + 3 \cdot x$	$3(4 \cdot x) = (3 \cdot 4)x$
$= 12 + 3x$	$= 12x$

For the distributive property to be used, within the parentheses, there must be two *terms*, separated by a plus or minus sign as in $3(4 + x)$.

Often when we add numbers we group the numbers so that we can add them easily. For example, when we add $70 + 50 + 30$ we may first add the $70 + 30$ to get 100. We are able to do this because of the commutative and associate properties. Notice that

$$
\begin{aligned}
(70 + 50) + 30 &= 70 + (50 + 30) && \textit{Associative property of addition} \\
&= 70 + (30 + 50) && \textit{Commutative property of addition} \\
&= (70 + 30) + 50 && \textit{Associative property of addition} \\
&= 100 + 50 && \textit{Addition facts} \\
&= 150
\end{aligned}
$$

Notice in the second step that the same numbers remained in parentheses but the order of the numbers changed, $50 + 30$ to $30 + 50$. Since this step involved a change in order (and not grouping), this is the commutative property of addition.

EXAMPLE 3 Name the property used to go from one step to the next.

a) $9 + 4(x + 5)$

b) $= 9 + 4x + 20$

c) $= 9 + 20 + 4x$

d) $= 29 + 4x$ *Addition facts*

e) $= 4x + 29$

Solution **(a to b)** Distributive property

(b to c) Commutative property of addition; $4x + 20 = 20 + 4x$

NOW TRY EXERCISE 39

(d to e) Commutative property of addition; $29 + 4x = 4x + 29$

The distributive property can be expanded in the following manner:

$$a(b + c + d + \cdots + n) = ab + ac + ad + \cdots + an$$

For example, $3(x + y + 5) = 3x + 3y + 15$.

Exercise Set 1.10

Concept/Writing Exercises

1. Explain the commutative property of addition and give an example of it.
2. Explain the commutative property of multiplication and give an example of it.
3. Explain the associative property of addition and give an example of it.
4. Explain the associative property of multiplication and give an example of it.
5. Explain the distributive property and give an example of it.
6. **a)** Explain the difference between $x + (y + z)$ and $x(y + z)$.

   b) Find the value of $x + (y + z)$ when $x = 4$, $y = 5$, and $z = 6$.

   c) Find the value of $x(y + z)$ when $x = 4$, $y = 5$, and $z = 6$.
7. Explain how you can tell the difference between the associative property of multiplication and the distributive property.
8. **a)** Write the associative property of addition using $x + (y + z)$.

   b) Write the distributive property using $x(y + z)$.

Practice the Skills

Name each property illustrated.

9. $4(3 + 5) = 4(3) + 4(5)$

10. $3 + y = y + 3$

11. $5 \cdot y = y \cdot 5$

12. $1(x + 3) = (1)(x) + (1)(3) = x + 3$

13. $2(x + 4) = 2x + 8$

14. $3(4 + x) = 12 + 3x$

15. $x \cdot (y \cdot z) = (x \cdot y) \cdot z$

16. $1(x + 4) = x + 4$

17. $1(x + 3) = x + 3$

18. $3 + (4 + x) = (3 + 4) + x$

Complete each equation to illustrate the given property. (Do not perform the operation shown.)

19. $3 + 4 =$
commutative property of addition

20. $-3 + 4 =$
commutative property of addition

21. $-6 \cdot (4 \cdot 2) =$
associative property of multiplication

22. $-4 + (5 + 3) =$
associative property of addition

23. $(6)(y) =$
commutative property of multiplication

24. $4(x + 3) =$
distributive property

25. $1(x + y) =$
distributive property

26. $6(x + y) =$
distributive property

27. $4x + 3y =$
commutative property of addition

28. $3(x + y) =$
distributive property

29. $(3 + x) + y =$
associative property of addition

30. $(x + 2)3 =$
commutative property of multiplication

31. $(3x + 4) + 6 =$
associative property of addition

32. $3(x + y) =$
commutative property of addition

33. $3(x + y) =$
commutative property of multiplication

34. $(3x)y =$
associative property of multiplication

35. $4(x + y + 3) =$
distributive property

36. $3(x + y + 2) =$
distributive property

Name the property used to go from one step to the next (see Example 3).

37. $(3 + x) + 4 = (x + 3) + 4$
38. $\qquad = x + (3 + 4)$
$\qquad = x + 7$

39. $6 + 5(x + 3) = 6 + 5x + 15$
40. $\qquad = 6 + 15 + 5x$
$\qquad = 21 + 5x$
41. $\qquad = 5x + 21$

42. $(x + 4)5 = 5(x + 4)$
43. $\qquad = 5x + 20$
44. $\qquad = 20 + 5x$

Problem Solving

Indicate whether the given processes are commutative. That is, does changing the order in which the actions are done result in the same final outcome? Explain each answer.

45. Putting sugar and then cream in coffee; putting cream and then sugar in coffee.

46. Brushing your teeth and then washing your face; washing your face and then brushing your teeth.

47. Applying suntan lotion and then sunning yourself; sunning yourself and then applying suntan lotion.

48. Putting on your socks and then your shoes; putting on your shoes and then your socks.

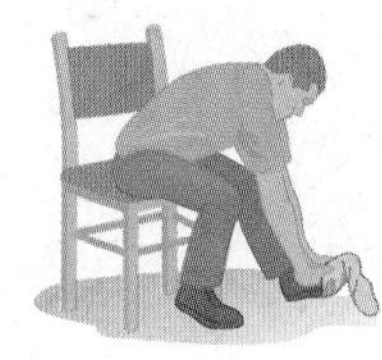

49. Writing on the blackboard and then erasing the blackboard; erasing the blackboard and then writing on the blackboard.

50. Turning on the lamp and then reading a book; reading a book and then turning on a lamp.

In Exercises 51–56, indicate whether the given processes are associative. For a process to be associative, the final outcome must be the same when the first two actions are performed first or when the last two actions are performed first. Explain each answer.

51. Brushing your teeth, washing your face, and combing your hair.

52. Cracking an egg, pouring out the egg, and cooking the egg.

53. Removing the gas cap, putting the nozzle in the tank, and turning on the gas.

54. A coffee machine dropping the cup, dispensing the coffee, and then adding the sugar.

55. Starting a car, moving the shift lever to drive, and then stepping on the gas.

56. Putting cereal, milk, and sugar in a bowl.

57. The commutative property of addition is $a + b = b + a$. Explain why $(3 + 4) + x = x + (3 + 4)$ also illustrates the commutative property of addition.

58. The commutative property of multiplication is $a \cdot b = b \cdot a$. Explain why $(3 + 4) \cdot x = x \cdot (3 + 4)$ also illustrates the commutative property of multiplication.

Challenge Problems

59. Consider $x + (3 + 5) = x + (5 + 3)$. Does this illustrate the commutative property of addition or the associative property of addition? Explain.

60. Consider $x + (3 + 5) = (3 + 5) + x$. Does this illustrate the commutative property of addition or the associative property of addition? Explain.

61. Consider $x + (3 + 5) = (x + 3) + 5$. Does this illustrate the commutative property of addition? Explain.

62. The commutative property of multiplication is $a \cdot b = b \cdot a$. Explain why $(3 + 4) \cdot (5 + 6) = (5 + 6) \cdot (3 + 4)$ also illustrates the commutative property of multiplication.

Cumulative Review Exercises

[1.3] 63. Add $2\frac{3}{5} + \frac{2}{3}$.

64. Subtract $3\frac{5}{8} - 2\frac{3}{16}$.

[1.9] *Evaluate.*

65. $12 - 24 \div 8 + 4 \cdot 3^2$

66. $-4x^2 + 6xy + 3y^2$; $x = 2$, $y = -3$

SUMMARY

Key Words and Phrases

1.2
Approximately equal to
Bar graph
Circle (or pie) graph
Expression
Line graph
Measures of central tendency
Mean
Median
Operations
Problem-solving procedure
Ranked data

1.3
Denominator
Evaluate
Factors
Fraction
Greatest common factor
Least common denominator
Mixed number
Numerator
Reduced to lowest terms
Variables
Whole numbers

1.4
Counting numbers
Elements of a set
Empty (or null) set
Integers
Irrational numbers
Natural numbers
Negative integers
Positive integers
Rational numbers
Real numbers
Set

1.5
Absolute value

1.6
Additive inverses
Operations
Opposites

1.8
Undefined

1.9
Base
Exponent
Grouping symbols
Order of operations
Parentheses

1.10
Associative properties
Commutative properties
Distributive property

IMPORTANT FACTS

Problem–Solving Procedure

1. Understand the question.
2. Translate into mathematical language.
3. Carry out the calculations.
4. Check the answer.
5. Answer the question asked.

Fractions:

$$\frac{a}{c} + \frac{b}{c} = \frac{a+b}{c} \qquad \frac{a}{c} - \frac{b}{c} = \frac{a-b}{c}$$

$$\frac{a}{b} \cdot \frac{c}{d} = \frac{ac}{bd} \qquad \frac{a}{b} \div \frac{c}{d} = \frac{a}{b} \cdot \frac{d}{c} = \frac{ad}{bc}$$

Sets of Numbers

Natural numbers: $\{1, 2, 3, 4, \ldots\}$
Whole numbers: $\{0, 1, 2, 3, 4, \ldots\}$
Integers: $\{\ldots, -3, -2, -1, 0, 1, 2, 3, \ldots\}$
Rational numbers: {quotient of two integers, denominator not 0}
Irrational numbers: {real numbers that are not rational numbers}
Real numbers: {all numbers that can be represented on a number line}

Operations on the Real Numbers

To *add real numbers with the same sign*, add their absolute values. The sum has the same sign as the numbers being added.

To *add real numbers with different signs*, subtract the smaller absolute value from the larger absolute value. The answer has the sign of the number with the larger absolute value.

To *subtract b from a*, add the opposite of b to a.

$$a - b = a + (-b)$$

The *products* and *quotients* of numbers with *like signs* will be *positive*. The *products* and *quotients* of numbers with *unlike signs* will be *negative*.

Division Involving 0

If a represents any real number except 0, then

$$\frac{0}{a} = 0$$

$$\frac{a}{0} \text{ is undefined}$$

Exponents

$$b^n = \underbrace{b \cdot b \cdot b \cdot \cdots \cdot b}_{n \text{ factors of } b}$$

Order of Operations

1. Evaluate expressions within parentheses.
2. Evaluate all expressions with exponents.
3. Perform multiplications or divisions working left to right.
4. Perform additions or subtractions working left to right.

Properties of the Real Number System

Property	Addition	Multiplication
Commutative	$a + b = b + a$	$ab = ba$
Associative	$(a + b) + c = a + (b + c)$	$(ab)c = a(bc)$
Distributive	$a(b + c) = ab + ac$	

Review Exercises

[1.2] *Solve*

1. Lynn Coates makes and sells pillows and mats. On August 4 he bought three large bags of foam rubber. Each bag contained 15 pounds of foam rubber. He used 8.2 pounds to make large pillows, 9.2 pounds to make small pillows, and 10.4 pounds to make mats. How many pounds of foam rubber did he have left?
2. Assume that the rate of inflation is 5% for the next two years. What will be the cost of goods two years from now, adjusted for inflation, if the goods cost $500.00 today?
3. Krystal Streeter wants to purchase a fax machine that sells for $300. She can either pay the total amount at the time of purchase, or she can agree to pay the store $30 down and $25 a month for 12 months. How much money can she save by paying the total amount at the time of purchase?
4. The cost of a car increases by 20% and then decreases by 20%. Is the resulting price of the car greater than, less than, or equal to the original price of the car? Explain.
5. On Kristen Reid's first five exams her grades were 75, 79, 86, 88, and 64. Find the **a)** mean and **b)** median of her grades.
6. The high temperatures, in degrees Fahrenheit, on July 1 in the last five years in Honolulu, Hawaii, were 76, 79, 84, 82, and 79. Find the **a)** mean and **b)** median of the temperatures.
7. The following graph shows the history and expected trends for CDs, cassettes, and LPs (long play records) from 1981–2001.

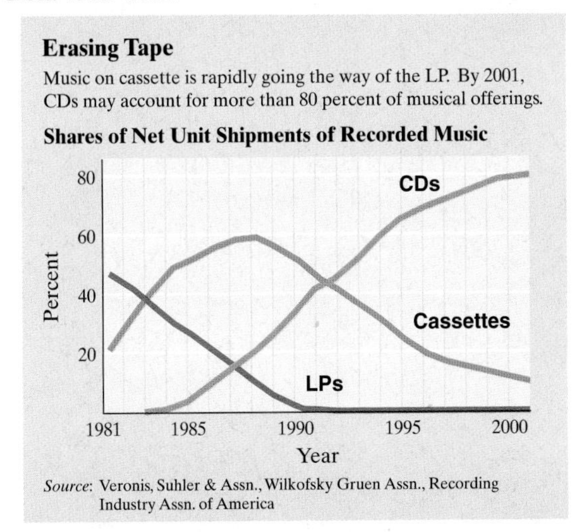

a) Estimate the year that the percent of shipments of CDs equaled the percent of shipments of cassettes.

b) Estimate the percent of shipments for both cassettes and CDs in 2000.

c) Estimate the percent difference in the shipments of CDs and cassettes in 2000.

d) Estimate the number of times greater the percent of shipments is for CDs is than for cassettes in 2000.

8. In July 1997, the Chairman of the Postal Service Board of Governors proposed to the Postal Rate Commission that first class stamps increase a penny, from 32¢ to 33¢. The new rate became effective in January, 1999. The graph shows the change in the cost of the first class stamp from March 1863 through January 1999.

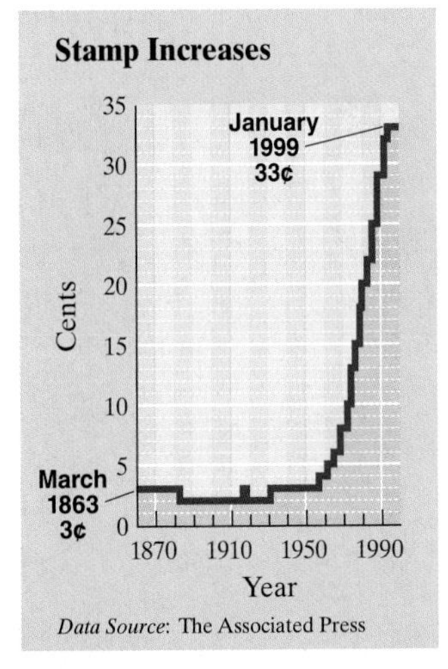

a) What was the cost of a first class stamp in November, 1863?

b) In which decade (10-year period) did the cost of a first class stamp reach 10 cents for the first time?

c) In the decade from 1900 to 1910, how many times was the price of a stamp increased?

d) How many times greater is the cost of the 33¢ stamp than a first class stamp in March 1863?

[1.3] *Perform each indicated operation. Simplify your answers.*

9. $\frac{3}{5} \cdot \frac{5}{6}$

10. $\frac{2}{5} \div \frac{10}{9}$

11. $\frac{5}{12} \div \frac{3}{5}$

12. $\frac{5}{6} + \frac{1}{3}$

13. $\frac{3}{8} - \frac{1}{9}$

14. $2\frac{1}{3} - 1\frac{1}{5}$

[1.4] **15.** List the set of natural numbers.

16. List the set of whole numbers.

17. List the set of integers.

18. Describe the set of rational numbers.

19. Describe the set of real numbers.

20. Consider the following set of numbers.

$$\left\{3, -5, -12, 0, \frac{1}{2}, -0.62, \sqrt{7}, 426, -3\frac{1}{4}\right\}$$

List the numbers that are

a) positive integers.
b) whole numbers.
c) integers.
d) rational numbers.
e) irrational numbers.
f) real numbers.

21. Consider the following set of numbers.

$$\left\{-2.3, -8, -9, 1\frac{1}{2}, \sqrt{2}, -\sqrt{2}, 1, -\frac{3}{17}\right\}$$

List the numbers that are

a) natural numbers.
b) whole numbers.
c) negative integers.
d) integers.
e) rational numbers.
f) real numbers.

[1.5] *Insert either <, >, or = in each shaded area to make a true statement.*

22. $-3 \;\square\; -5$

23. $-2.6 \;\square\; -3.6$

24. $0.50 \;\square\; 0.509$

25. $4.6 \;\square\; 4.06$

26. $-3.2 \;\square\; -3.02$

27. $5 \;\square\; |-3|$

28. $-3 \;\square\; |-7|$

29. $|-2.5| \;\square\; \left|\frac{5}{2}\right|$

[1.6–1.7] *Evaluate.*

30. $-4 + (-5)$

31. $-6 + 6$

32. $0 + (-3)$

33. $-10 + 4$

34. $-8 - (-2)$

35. $-9 - (-4)$

36. $4 - (-4)$

37. $2 - 12$

38. $7 - 2$

39. $2 - 7$

40. $0 - (-4)$

41. $-7 - 5$

Evaluate.

42. $6 - 4 + 3$

43. $-5 + 7 - 6$

44. $-5 - 4 - 3$

45. $-2 + (-3) - 2$

46. $7 - (+4) - (-3)$

47. $4 - (-2) + 3$

[1.8] *Evaluate.*

48. $-4(7)$

49. $(-9)(-3)$

50. $(-4)(-5)(-6)$

51. $\left(\frac{3}{5}\right)\left(\frac{-2}{7}\right)$

52. $\left(\frac{10}{11}\right)\left(\frac{3}{-5}\right)$

53. $\left(\frac{-5}{8}\right)\left(\frac{-3}{7}\right)$

54. $0\left(\frac{4}{9}\right)$

55. $4(-2)(-6)$

56. $(-4)(-6)(-2)(-3)$

Evaluate.

57. $15 \div (-3)$

58. $6 \div (-2)$

59. $-20 \div 5$

60. $0 \div 4$

61. $72 \div (-9)$

62. $-4 \div \left(\frac{-4}{9}\right)$

63. $\frac{28}{-3} \div \left(\frac{9}{-2}\right)$

64. $\frac{14}{3} \div \left(\frac{-6}{5}\right)$

65. $\left(\frac{-5}{12}\right) \div \left(\frac{-5}{12}\right)$

Indicate whether each quotient is 0 or undefined.

66. $0 \div 4$

67. $0 \div (-6)$

68. $8 \div 0$

69. $-4 \div 0$

70. $\frac{8}{0}$

71. $\frac{0}{-5}$

[1.6–1.8, 1.9] *Evaluate.*

72. $-4(2 - 8)$
73. $2(4 - 8)$
74. $(3 - 6) + 4$
75. $(-4 + 3) - (2 - 6)$
76. $[4 + 3(-2)] - 6$
77. $(-4 - 2)(-3)$
78. $[4 + (-4)] + (6 - 8)$
79. $9[3 + (-4)] + 5$
80. $-4(-3) + [4 \div (-2)]$
81. $(-3 \cdot 4) \div (-2 \cdot 6)$
82. $(-3)(-4) + 6 - 3$
83. $[-2(3) + 6] - 4$

[1.9] *Evaluate.*

84. 6^2
85. 9^3
86. 3^4
87. $(-3)^3$
88. $(-1)^9$
89. $(-2)^5$
90. $\left(\frac{-3}{5}\right)^2$
91. $\left(\frac{2}{5}\right)^3$

Express in exponential form.

92. xxy
93. $2 \cdot 2 \cdot 3 \cdot 3 \cdot 3xyy$
94. $5 \cdot 7 \cdot 7xxy$
95. $xyxyz$

Express as a product of factors.

96. $x^2 y$
97. xz^3
98. $y^3 z$
99. $2x^3 y^2$

Evaluate.

100. $3 + 5 \cdot 4$
101. $3 \cdot 5 + 4 \cdot 2$
102. $(3 - 7)^2 + 6$
103. $8 - 36 \div 4 \cdot 3$
104. $6 - 3^2 \cdot 5$
105. $[6 - (3 \cdot 5)] + 5$
106. $3[9 - (4^2 + 3)] \cdot 2$
107. $(-3^2 + 4^2) + (3^2 \div 3)$
108. $2^3 \div 4 + 6 \cdot 3$
109. $(4 \div 2)^4 + 4^2 \div 2^2$
110. $(15 - 2^2)^2 - 4 \cdot 3 + 10 \div 2$
111. $4^3 \div 4^2 - 5(2 - 7) \div 5$

Evaluate each expression for the given values.

112. $4x - 6; x = 5$
113. $6 - 4x; x = -5$
114. $x^2 - 5x + 3; x = 6$
115. $5y^2 + 3y - 2; y = -1$
116. $-x^2 + 2x - 3; x = 2$
117. $-x^2 + 2x - 3; x = -2$
118. $-3x^2 - 5x + 5; x = 1$
119. $3xy - 5x; x = 3, y = 4$
120. $-x^2 - 8x - 12y; x = -3, y = -2$

[1.6–1.9] ***a)*** *Use a calculator to evaluate each expression, and* ***b)*** *check to see whether your answer is reasonable.*

121. $158 + (-493)$
122. $324 - (-29.6)$
123. $\frac{-17.28}{6}$
124. $(-62)(-1.9)$
125. $(-3)^6$
126. $-(4.2)^3$

[1.10] *Name each indicated property.*

127. $(4 + 3) + 9 = 4 + (3 + 9)$
128. $6 \cdot x = x \cdot 6$
129. $4(x + 3) = 4x + 12$
130. $(x + 4)3 = 3(x + 4)$
131. $6x + 3x = 3x + 6x$
132. $(x + 7) + 4 = x + (7 + 4)$

Practice Test

1. While shopping, Ester Catching purchases two half-gallons of milk for $1.30 each, one Boston cream pie for $4.75, and three 2-liter bottles of soda for $1.10 each.

a) What is her total bill before tax?

b) If there is a 7% sales tax on the bottles of soda, how much is the sales tax?

c) How much is her total bill including tax?

d) How much change will she receive from a $50 bill?

2. The following graph shows the U.S. denim jeans market.

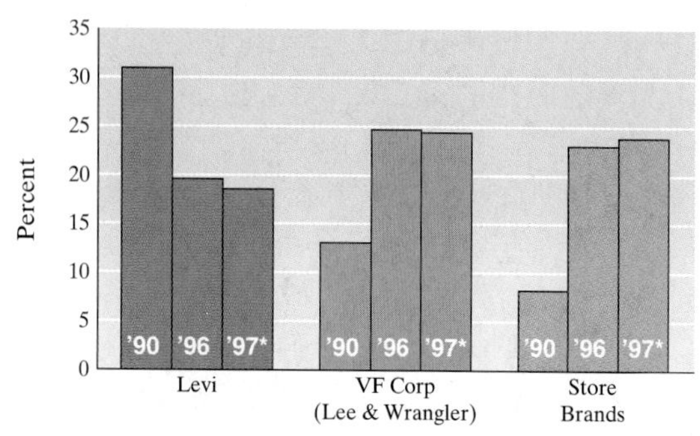

a) Estimate Levi's share in 1990 and 1997.

b) Estimate the VF Corporation's and the store brands' share in 1997.

c) Estimate the sum of the three categories in 1997.

3. The following graph shows the percent of all households with a TV on and tuned to the Kentucky Derby, Preakness, or Belmont Stakes, from 1960 through 1997.

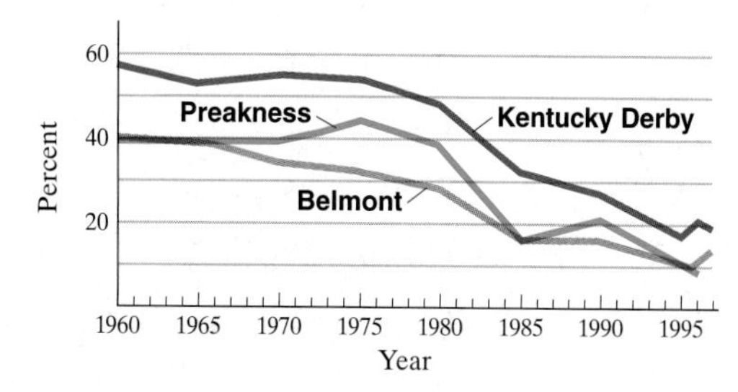

a) Estimate the percent of households who watched the Kentucky Derby in 1960 and in 1997.

b) For the Kentucky Derby, how much greater was the ratings share in 1960 than in 1997?

c) For the Kentucky Derby, how many times greater was the ratings share in 1960 than in 1997?

4. Consider the following set of numbers.

$$\left\{-6, 42, -3\frac{1}{2}, 0, 6.52, \sqrt{5}, \frac{5}{9}, -7, -1\right\}$$

List the numbers that are

a) natural numbers.

b) whole numbers.

c) integers.

d) rational numbers.

e) irrational numbers.

f) real numbers.

Insert either $<$, $>$, or $=$ in each shaded area to make a true statement.

5. $-6 \quad -3$

6. $|-3| \quad |-2|$

Evaluate.

7. $-4 + (-8)$

8. $-6 - 5$

9. $5 - 12 - 7$

10. $(-4 + 6) - 3(-2)$

11. $(-4)(-3)(2)(-1)$

12. $\left(\frac{-2}{9}\right) \div \left(\frac{-7}{8}\right)$

13. $\left(-12 \cdot \frac{1}{2}\right) \div 3$

14. $3 \cdot 5^2 - 4 \cdot 6^2$

15. $-6(-2 - 3) \div 5 \cdot 2$

16. $\left(\frac{3}{5}\right)^3$

17. Write $2 \cdot 2 \cdot 5 \cdot 5yyzzz$ in exponential form.

18. Write $2^2 3^3 x^4 y^2$ as a product of factors.

Evaluate for the given values.

19. $2x^2 - 6; x = -4$

20. $6x - 3y^2 + 4; x = 3, y = -2$

21. $-x^2 - 6x + 3; x = -2$

22. $-x^2 + xy + y^2; x = 1, y = -2$

Name each indicated property.

23. $x + 3 = 3 + x$

24. $4(x + 9) = 4x + 36$

25. $(2 + x) + 4 = 2 + (x + 4)$

SOLVING LINEAR EQUATIONS AND INEQUALITIES

CHAPTER

Use the Angel Web site at www.prenhall.com/angel to be linked to an internet resource that will help you further explore the following application.

When trying to predict whether an athlete will set a new record, journalists and coaches often compare the athlete's current pace and performance with the pace set by the record holder at different times during a record-breaking season, On page 140, we use proportions to determine how many home runs during the first 48 games of a baseball season a player would need to hit to be on schedule to break Mark McGwire's record of 70 home runs from the 1998 baseball season.

Preview and Perspective

When many students describe algebra they use the words "solving equations." Solving equations is an important part of algebra and of most other mathematics courses you may take. The major emphasis of this chapter is to teach you how to solve linear equations. We will be using principles learned in this chapter throughout the book.

To be successful in solving linear equations, you need to have a thorough understanding of adding, subtracting, multiplying, and dividing real numbers. This material was discussed in Chapter 1. The materials presented in the first four sections of this chapter are the building blocks for solving linear equations. In Section 2.5 we combine the material presented previously to solve a variety of linear equations.

In Section 2.6 we discuss ratios and proportions and how to set up and solve them. For many students, proportions may be the most common types of equations used to solve real-life applications. In Section 2.7 we discuss solving linear inequalities, which is an extension of solving linear equations.

2.1 COMBINING LIKE TERMS

1) **Identify terms.**
2) **Identify like terms.**
3) **Combine like terms.**
4) **Use the distributive property.**
5) **Remove parentheses when they are preceded by a plus or minus sign.**
6) **Simplify an expression.**

1) Identify Terms

In Section 1.3 and other sections of the text, we indicated that letters called **variables** are used to represent numbers. A variable can represent a variety of different numbers.

As was indicated in Chapter 1, an **expression** (sometimes referred to as an **algebraic expression**) is a collection of numbers, variables, grouping symbols, and operation symbols.

Examples of Expressions

$$5, \quad x^2 - 6, \quad 4x - 3, \quad 2(x + 5) + 6, \quad \frac{x + 3}{4}$$

When an algebraic expression consists of several parts, the parts that are *added* are called the **terms** of the expression. Consider the expression $2x - 3y - 5$. The expression can be written as $2x + (-3y) + (-5)$, and so the expression $2x - 3y - 5$ has three terms: $2x$, $-3y$, and -5. The expression $3x + 2xy + 5(x + y)$ also has three terms: $3x$, $2xy$, and $5(x + y)$.

When listing the terms of an expression, it is not necessary to list the + sign at the beginning of a term.

Expression	Terms
$-2x + 3y - 8$	$-2x,\ 3y,\ -8$
$3y - 2x + \frac{1}{2}$	$3y,\ -2x,\ \frac{1}{2}$
$7 + x + 4 - 5x$	$7,\ x,\ 4,\ -5x$
$3(x - 1) - 4x + 2$	$3(x - 1),\ -4x,\ 2$
$\frac{x + 4}{3} - 5x + 3$	$\frac{x + 4}{3},\ -5x,\ 3$

NOW TRY EXERCISE 1

The numerical part of a term is called its **numerical coefficient** or simply its **coefficient**. In the term $6x$, the 6 is the numerical coefficient. Note that $6x$ means the variable x is multiplied by 6.

Term	Numerical Coefficient
$3x$	3
$-\frac{1}{2}x$	$-\frac{1}{2}$
$4(x - 3)$	4
$\frac{2x}{3}$	$\frac{2}{3}$, since $\frac{2x}{3}$ means $\frac{2}{3}x$
$\frac{x + 4}{3}$	$\frac{1}{3}$, since $\frac{x + 4}{3}$ means $\frac{1}{3}(x + 4)$

Whenever a term appears without a numerical coefficient, we assume that the numerical coefficient is 1.

Examples

x means $1x$	$-x$ means $-1x$
x^2 means $1x^2$	$-x^2$ means $-1x^2$
xy means $1xy$	$-xy$ means $-1xy$
$(x + 2)$ means $1(x + 2)$	$-(x + 2)$ means $-1(x + 2)$

If an expression has a term that is a number (without a variable), we refer to that number as a **constant term**, or simply a **constant**. In the expression $x^2 + 3x - 4$, the -4 is a constant term, or a constant.

2) Identify Like Terms

Like terms are terms that have the same variables with the same exponents. Some examples of like terms and unlike terms follow. Note that if two terms are like terms, only their numerical coefficients may differ.

Like Terms	Unlike Terms	
$3x, \ -4x$	$3x, \ 2$	*(One term has a variable, the other is a constant.)*
$4y, \ 6y$	$3x, \ 4y$	*(Variables differ)*
$5, \ -6$	$x, \ 3$	*(One term has a variable, the other is a constant.)*
$3(x+1), \ -2(x+1)$	$2x, \ 3xy$	*(Variables differ)*
$3x^2, \ 4x^2$	$3x, \ 4x^2$	*(Exponents differ)*

EXAMPLE 1 Identify any like terms.

a) $2x + 3x + 4$ **b)** $2x + 3y + 2$ **c)** $x + 3 + y - \frac{1}{2}$

Solution **a)** $2x$ and $3x$ are like terms.

b) There are no like terms.

c) 3 and $-\frac{1}{2}$ are like terms.

EXAMPLE 2 Identify any like terms.

a) $5x - x + 6$ **b)** $3 - 2x + 4x - 6$ **c)** $12 + x + 7$

Solution **a)** $5x$ and $-x$ (or $-1x$) are like terms.

b) 3 and -6 are like terms; $-2x$ and $4x$ are like terms.

c) 12 and 7 are like terms.

3) Combine Like Terms

We often need to simplify expressions by combining like terms. **To combine like terms** means to add or subtract the like terms in an expression. To combine like terms, we can use the procedure that follows.

To Combine Like Terms

1. Determine which terms are like terms.
2. Add or subtract the coefficients of the like terms.
3. Multiply the number found in step 2 by the common variable(s).

Examples 3 through 8 illustrate this procedure.

EXAMPLE 3 Combine like terms: $4x + 3x$.

Solution $4x$ and $3x$ are like terms with the common variable x. Since $4 + 3 = 7$, then $4x + 3x = 7x$.

EXAMPLE 4 Combine like terms: $\frac{3}{5}x - \frac{2}{3}x$.

Solution Since $\frac{3}{5} - \frac{2}{3} = \frac{9}{15} - \frac{10}{15} = -\frac{1}{15}$, then $\frac{3}{5}x - \frac{2}{3}x = -\frac{1}{15}x$.

EXAMPLE 5 Combine like terms: $5.23a - 7.45a$.

Solution Since $5.23 - 7.45 = -2.22$, then $5.23a - 7.45a = -2.22a$.

EXAMPLE 6 Combine like terms: $3x + x + 5$.

Solution The $3x$ and x are like terms.

$$3x + x + 5 = 3x + 1x + 5 = 4x + 5$$

The order of the terms in the answer is not critical. Thus $5 + 4x$ is also an acceptable answer to Example 6. When writing answers, we generally list the terms containing variables in alphabetical order from left to right, and list the constant term on the right.

The commutative and associative properties of addition will be used to rearrange the terms in Examples 7 and 8.

EXAMPLE 7 Combine like terms: $3y + 4x - 3 - 2x$.

Solution The only like terms are $4x$ and $-2x$.

$4x - 2x + 3y - 3$ *Rearrange terms.*

$2x + 3y - 3$ *Combine like terms.*

EXAMPLE 8 Combine like terms: $-2x + 3y - 4x + 3 - y + 5$.

Solution

$-2x$ and $-4x$ are like terms.
$3y$ and $-y$ are like terms.
3 and 5 are like terms.

Grouping the like terms together gives

$$-2x - 4x + 3y - y + 3 + 5$$
$$-6x \quad + \quad 2y \quad + \quad 8$$

NOW TRY EXERCISE 29

Use the Distributive Property

We introduced the distributive property in Section 1.10. Because this property is so important, we will study it again. But before we do, let's go back briefly to the subtraction of real numbers. Recall from Section 1.7 that

$$6 - 3 = 6 + (-3)$$

In general,

> For any real numbers a and b,
> $$a - b = a + (-b)$$

We will use the fact that $a + (-b)$ means $a - b$ in discussing the distributive property.

> **Distributive Property**
>
> For any real numbers a, b, and c,
> $$a(b + c) = ab + ac$$

EXAMPLE 9 Use the distributive property to remove parentheses.

a) $2(x + 4)$ **b)** $-2(x + 4)$

Solution **a)** $2(x + 4) = 2x + 2(4) = 2x + 8$

b) $-2(x + 4) = -2x + (-2)(4) = -2x + (-8) = -2x - 8$

Note in part **b)** that, instead of leaving the answer $-2x + (-8)$, we wrote it as $-2x - 8$, which is the proper form of the answer.

EXAMPLE 10 Use the distributive property to remove parentheses.

a) $3(x - 2)$ **b)** $-2(4x - 3)$

Solution **a)** By the definition of subtraction, we may write $x - 2$ as $x + (-2)$.

$$3(x - 2) = 3[x + (-2)] = 3x + 3(-2)$$
$$= 3x + (-6)$$
$$= 3x - 6$$

b) $-2(4x - 3) = -2[4x + (-3)] = -2(4x) + (-2)(-3) = -8x + 6$

The distributive property is used often in algebra, so you need to understand it well. You should understand it so well that you will be able to simplify an expression using the distributive property without having to write down all the steps that we listed in working Examples 9 and 10. Study closely the Helpful Hint that follows.

HELPFUL HINT

With a little practice, you will be able to eliminate some of the intermediate steps when you use the distributive property. When using the distributive property, there are eight possibilities with regard to signs. Study and learn the eight possibilities that follow.

POSITIVE COEFFICIENT

a) $2(x) = 2x$
$$2(x + 3) = 2x + 6$$
$2(+3) = +6$

b) $2(x) = 2x$
$$2(x - 3) = 2x - 6$$
$2(-3) = -6$

c) $2(-x) = -2x$
$$2(-x + 3) = -2x + 6$$
$2(+3) = +6$

d) $2(-x) = -2x$
$$2(-x - 3) = -2x - 6$$
$2(-3) = -6$

NEGATIVE COEFFICIENT

e) $(-2)(x) = -2x$
$$-2(x + 3) = -2x - 6$$
$(-2)(+3) = -6$

f) $(-2)(x) = -2x$
$$-2(x - 3) = -2x + 6$$
$(-2)(-3) = +6$

g) $(-2)(-x) = 2x$
$$-2(-x + 3) = 2x - 6$$
$(-2)(+3) = -6$

h) $(-2)(-x) = 2x$
$$-2(-x - 3) = 2x + 6$$
$(-2)(-3) = +6$

The distributive property can be expanded as follows:

$$a(b + c + d + \cdots + n) = ab + ac + ad + \cdots + an$$

Examples of the Expanded Distributive Property

$$3(x + y + z) = 3x + 3y + 3z$$

$$2(x + y - 3) = 2x + 2y - 6$$

EXAMPLE 11 Use the distributive property to remove parentheses.

a) $4(x - 3)$ **b)** $-2(2x - 4)$ **c)** $-\frac{1}{2}(4x + 5)$ **d)** $-2(3x - 2y + 4z)$

Solution **a)** $4(x - 3) = 4x - 12$ **b)** $-2(2x - 4) = -4x + 8$

c) $-\frac{1}{2}(4x + 5) = -2x - \frac{5}{2}$ **d)** $-2(3x - 2y + 4z) = -6x + 4y - 8z$

The distributive property can also be used from the right, as in Example 12.

EXAMPLE 12 Use the distributive property to remove parentheses from the expression $(2x - 8y)4$.

Solution We distribute the 4 on the right side of the parentheses over the terms within the parentheses.

$$\begin{aligned}(2x - 8y)4 &= 2x(4) - 8y(4)\\ &= 8x - 32y\end{aligned}$$

NOW TRY EXERCISE 75

Example 12 could have been rewritten as $4(2x - 8y)$ by the commutative property of multiplication, and then the 4 could have been distributed from the left to obtain the same answer, $8x - 32y$.

5) Remove Parentheses When They Are Preceded by a Plus or Minus Sign

In the expression $(4x + 3)$, how do we remove parentheses? Recall that the coefficient of a term is assumed to be 1 if none is shown. Therefore, we may write

$$\begin{aligned}(4x + 3) &= 1(4x + 3)\\ &= 1(4x) + (1)(3)\\ &= 4x + 3\end{aligned}$$

Note that $(4x + 3) = 4x + 3$. **When no sign or a plus sign precedes parentheses, the parentheses may be removed without having to change the expression inside the parentheses.**

Examples

$$(x + 3) = x + 3$$

$$(2x - 3) = 2x - 3$$

$$+(2x - 5) = 2x - 5$$

$$+(x + 2y - 6) = x + 2y - 6$$

Now consider the expression $-(4x + 3)$. How do we remove parentheses in this expression? Here, the coefficient in front of the parentheses is -1, so each term within the parentheses is multiplied by -1.

$$\begin{aligned} -(4x + 3) &= -1(4x + 3) \\ &= -1(4x) + (-1)(3) \\ &= -4x + (-3) \\ &= -4x - 3 \end{aligned}$$

Thus, $-(4x + 3) = -4x - 3$. **When a negative sign precedes parentheses, the signs of all the terms within the parentheses are changed when the parentheses are removed.**

Examples

$$-(x + 4) = -x - 4$$

$$-(-2x + 3) = 2x - 3$$

$$-(5x - y + 3) = -5x + y - 3$$

$$-(-2x - 3y - 5) = 2x + 3y + 5$$

NOW TRY EXERCISE 69

6) Simplify an Expression

Combining what we learned in the preceding discussions, we have the following procedure for **simplifying an expression.**

To Simplify an Expression

1. Use the distributive property to remove any parentheses.
2. Combine like terms.

EXAMPLE 13 Simplify $6 - (2x + 3)$.

Solution

$$\begin{aligned} 6 - (2x + 3) &= 6 - 2x - 3 && \textit{Use the distributive property.} \\ &= -2x + 3 && \textit{Combine like terms.} \end{aligned}$$

Note: $3 - 2x$ is the same as $-2x + 3$; however, we generally write the term containing the variable first.

EXAMPLE 14 Simplify $6x + 4(2x + 3)$.

Solution

$$\begin{aligned} 6x + 4(2x + 3) &= 6x + 8x + 12 && \textit{Use the distributive property.} \\ &= 14x + 12 && \textit{Combine like terms.} \end{aligned}$$

EXAMPLE 15 Simplify $-(x - 5) + 8$.

Solution

$$\begin{aligned} -(x - 5) + 8 &= -x + 5 + 8 && \textit{Use the distributive property.} \\ &= -x + 13 && \textit{Combine like terms.} \end{aligned}$$

EXAMPLE 16 Simplify $2(x - 4) - 3(x - 2) - 4$.

Solution

$$2(x - 4) - 3(x - 2) - 4 = 2x - 8 - 3x + 6 - 4 \quad \textit{Use the distributive property.}$$
$$= 2x - 3x - 8 + 6 - 4 \quad \textit{Rearrange terms.}$$
$$= -x - 6 \quad \textit{Combine like terms.}$$

NOW TRY EXERCISE 95

HELPFUL HINT

It is important for you to have a clear understanding of the concepts of *term* and *factor*. When two or more expressions are **multiplied**, each expression is a **factor** of the product. For example, since $4 \cdot 3 = 12$, the 4 and the 3 are factors of 12. Since $3 \cdot x = 3x$, the 3 and the x are factors of $3x$. Similarly, in the expression $5xyz$, the 5, x, y, and z are all factors.

In an expression, the parts that are **added** are the **terms** of the expression. For example, the expression $2x^2 + 3x - 4$, has three terms, $2x^2$, $3x$, and -4. Note that the terms of an expression may have factors. For example, in the term $2x^2$, the 2 and the x^2 are factors because they are multiplied.

Exercise Set 2.1

Concept/Writing Exercises

1. **a)** What are the terms of an expression?
   b) What are the terms of $3x - 4y - 5$?
   c) What are the terms of $6xy + 3x - y - 9$?
2. **a)** What are like terms? Determine whether the following are like terms. If not, explain why.
   b) $3x, 4y$ **c)** $7, -2$
   d) $5x^2, 2x$ **e)** $4x, -5xy$
3. **a)** What are the factors of an expression?
   b) Explain why 3 and x are factors of $3x$.
   c) Explain why 5, x, and y are all factors of the expression $5xy$.
4. Consider the expression $2x - 5$.
   a) What is the x called?
   b) What is the -5 called?
   c) What is the 2 called?
5. **a)** What is the name given to the numerical part of a term? List the coefficient of the following terms.
   b) $4x$ **c)** x **d)** $-x$ **e)** $\frac{3x}{5}$ **f)** $\frac{2x + 3}{7}$
6. What does it mean to simplify an expression?
7. **a)** When no sign or a plus sign precedes an expression within parentheses, explain how to remove parentheses.
   b) Write $+(x - 8)$ without parentheses.
8. **a)** When a minus sign precedes an expression within parentheses, explain how to remove parentheses.
   b) Write $-(x - 8)$ without parentheses.

Practice the Skills

Combine like terms when possible. If not possible, rewrite the expression as is.

9. $5x + 3x$
10. $3x + 6$
11. $4x - 5x$
12. $4x + 3y$
13. $y + 3 + 4y$
14. $-2x - 3x$
15. $-2x + 5x$
16. $4x - 7x + 4$
17. $3 - 8x + 5$
18. $-4 + 5x + 12$
19. $-2x - 3x - 2 - 3$
20. $5x + 2y + 3 + y$
21. $-x + 2 - x - 2$
22. $8x - 2y - 1 - 3x$
23. $x - 4x + 3$
24. $y - 2y + 5$
25. $5 + 2x - 4x + 6$
26. $2y + 4 - x + 5x$
27. $x - 2 - 4 + 2x$
28. $2x + 4 - 3 + x$
29. $2 - 3x - 2x + y$

30. $7x - 3 - 2x$

31. $-2x + 4x - 3$

32. $4 - x + 4x - 8$

33. $x + 4 + \frac{3}{5}$

34. $\frac{3}{4}x + 2 + x$

35. $5.1x + 6.42 - 4.3x$

36. $13.4x + 1.2x + 8.3$

37. $\frac{1}{2}x + 3y + 1$

38. $x + \frac{1}{2}y - \frac{3}{8}y$

39. $2x + 3y + 4x + 5y$

40. $-4x - 3.1 - 5.2$

41. $-x + 2x + y$

42. $1 + x + 6 - 3x$

43. $2x - 7y - 5x + 2y$

44. $3x - 7 - 9 + 4x$

45. $6 - 3x + 9 - 2x$

46. $9x + y - 2 - 4x$

47. $-19.36 + 40.02x + 12.25 - 18.3x$

48. $52x - 52x - 63.5 - 63.5$

49. $\frac{3}{5}x - 3 - \frac{7}{4}x - 2$

50. $\frac{1}{2}y - 4 + \frac{3}{4}x - \frac{1}{5}y$

Use the distributive property to remove parentheses.

51. $2(x + 6)$

52. $4(x - 1)$

53. $5(x + 4)$

54. $-2(y + 8)$

55. $-2(x - 4)$

56. $2(-y + 5)$

57. $-\frac{1}{2}(2x - 4)$

58. $-4(x + 6)$

59. $1(-4 + x)$

60. $4(y + 3)$

61. $\frac{2}{5}(x - 5)$

62. $5(x - y + 5)$

63. $-0.3(3x + 5)$

64. $-(x - 3)$

65. $\frac{1}{2}(-2x + 6)$

66. $-2(x + y - z)$

67. $0.7(2x + 0.5)$

68. $-(x + 4y)$

69. $-(-x + y)$

70. $(3x + 4y - 6)$

71. $-(2x + 4y - 8)$

72. $-2(-2x + 6 + y)$

73. $1.1(3.1x - 5.2y + 2.8)$

74. $-2(-x + 3y + 5)$

75. $2\left(3x - 2y + \frac{1}{4}\right)$

76. $2\left(3 - \frac{1}{2}x + 4y\right)$

77. $(x + 3y - 9)$

78. $(-x + 5 - 2y)$

79. $-3(-x + 4 + 2y)$

80. $2.3(1.6x + 5.1y - 4.1)$

Simplify.

81. $4(x - 2) - x$

82. $2 + (x - 3)$

83. $-2(3 - x) + 7$

84. $-(3x - 3) + 5$

85. $6x + 2(4x + 9)$

86. $3(x + y) + 2y$

87. $2(x - y) + 2x + 3$

88. $6 + (x - 5) + 3x$

89. $(2x - y) + 2x + 5$

90. $4 + (2y + 2) + y$

91. $8x - (x - 3)$

92. $-(x - 5) - 3x + 4$

93. $2(x - 3) - (x + 3)$

94. $3y - (2x + 2y) - 6x$

95. $4(x - 1) + 2(3 - x) - 4$

96. $4(x + 3) - 2x$

97. $(x - 4) + 3x + 9$

98. $6 - 2(x + 3) + 5x$

99. $-3(x + 1) + 5x + 6$

100. $-(x + 2) + 3x - 6$

101. $4(x + 3) - 2x - 7$

102. $-3(x + 2y) + 3y + 4$

103. $0.4 + (y + 5) + 0.6 - 2$

104. $4 - (2 - x) + 3x$

105. $4 + (3x - 4) - 5$

106. $2y - 6(y - 2) + 3$

107. $4(x + 2) - 3(x - 4) - 5$

108. $4 - (y - 5) - 2x + 1$

109. $-0.2(6 - x) - 4(y + 0.4)$

110. $-5(2y - 8) - 3(1 + x) - 7$

111. $-6x + 7y - (3 + x) + (x + 3)$

112. $(x + 3) + (x - 4) - 6x$

113. $\frac{1}{2}(x + 3) + \frac{1}{3}(3x + 6)$

114. $\frac{2}{3}(x - 2) - \frac{1}{2}(x + 4)$

Problem Solving

If □ + □ + □ + ⊙ + ⊙ can be represented as 3□ + 2⊙, write an expression to represent each of the following.

115. □ + ⊖ + ⊖ + □ + ⊖

116. ⊗ + ☺ + ⊗ + ☺ + ☺ + ☺

117. $x + y + \triangle + \triangle + x + y + y$

118. $2 + x + 2 + x + x + 2 + y$

The positive factors of 6 are 1, 2, 3, and 6 since

$$1 \cdot 6 = 6$$
$$2 \cdot 3 = 6$$
$$\uparrow \uparrow$$
factors

119. List all the positive factors of 12.

120. List all the positive factors of 16.

Combine like terms.

121. 3△ + 5□ − △ − 3□

122. 8☺ − 4⊡ − 2⊡ − 3☺

Challenge Problems

Simplify.

123. $4x + 5y + 6(3x - 5y) - 4x + 3$

124. $2x^2 - 4x + 8x^2 - 3(x + 2) - x^2 - 2$

125. $x^2 + 2y - y^2 + 3x + 5x^2 + 6y^2 + 5y$

126. $2[3 + 4(x - 5)] - [2 - (x - 3)]$

Cumulative Review Exercises

[1.5] *Evaluate.*

127. $|-7|$

128. $-|-16|$

[1.9] **129.** Write a paragraph explaining the order of operations.

130. Evaluate $-x^2 + 5x - 6$ when $x = -1$.

2.2 THE ADDITION PROPERTY OF EQUALITY

SSM VIDEO 2.2 CD Rom

1. Identify linear equations.
2. Check solutions to equations.
3. Identify equivalent equations.
4. Use the addition property to solve equations.
5. Solve equations by doing some steps mentally.

1 Identify Linear Equations

A statement that shows two algebraic expressions are equal is called an **equation**. For example, $4x + 3 = 2x - 4$ is an equation. In this chapter we learn to solve **linear equations** in one variable.

Definition

A **linear equation** in one variable is an equation that can be written in the form

$$ax + b = c$$

where a, b, and c are real numbers and $a \neq 0$.

Examples of Linear Equations

$$x + 4 = 7$$
$$2x - 4 = 6$$

2) Check Solutions to Equations

The **solution to an equation** is the number or numbers that make the equation a true statement. For example, the solution to $x + 4 = 7$ is 3. We will shortly learn how to find the solution to an equation, or to **solve an equation**. But before we do this we will learn how to *check* the solution to an equation.

The solution to an equation may be **checked** by substituting the value that is believed to be the solution in the original equation. If the substitution results in a true statement, your solution is correct. If the substitution results in a false statement, then either your solution or your check is incorrect, and you need to go back and find your error. Try to check all your solutions.

When we show the check of a solution we shall use the $\stackrel{?}{=}$ notation. This notation is used when we are questioning whether a statement is true. For example, if we use

$$2 + 3 \stackrel{?}{=} 2(3) - 1$$

we are asking "Does $2 + 3 = 2(3) - 1$?"

To check whether 3 is the solution to $x + 4 = 7$, we substitute 3 for each x in the equation.

CHECK: $x = 3$

$$x + 4 = 7$$
$$3 + 4 \stackrel{?}{=} 7$$
$$7 = 7 \quad \text{True}$$

Since the check results in a true statement, 3 is a solution.

EXAMPLE 1 Consider the equation $2x - 4 = 6$. Determine whether

a) 3 is a solution.

b) 5 is a solution.

Solution **a)** To determine whether 3 is a solution to the equation, we substitute 3 for x.

CHECK: $x = 3$

$$2x - 4 = 6$$
$$2(3) - 4 \stackrel{?}{=} 6$$
$$6 - 4 \stackrel{?}{=} 6$$
$$2 = 6 \quad \text{False}$$

Since we obtained a false statement, 3 is not a solution.

b) Substitute 5 for x in the equation.

CHECK: $x = 5$

$$2x - 4 = 6$$
$$2(5) - 4 \stackrel{?}{=} 6$$
$$10 - 4 \stackrel{?}{=} 6$$
$$6 = 6 \quad \text{True}$$

Since the value 5 results in a true statement, 5 is a solution to the equation.

We can use the same procedures to check more complex equations, as shown in Examples 2 and 3.

EXAMPLE 2 Determine whether 18 is a solution to the following equation.

$$3x - 2(x + 3) = 12$$

Solution To determine whether 18 is a solution, we substitute 18 for each x in the equation. If the substitution results in a true statement, then 18 is a solution.

$$3x - 2(x + 3) = 12$$
$$3(18) - 2(18 + 3) \stackrel{?}{=} 12$$
$$54 - 2(21) \stackrel{?}{=} 12$$
$$54 - 42 \stackrel{?}{=} 12$$
$$12 = 12 \quad \text{True}$$

Since we obtain a true statement, 18 is a solution.

EXAMPLE 3 Determine whether $-\frac{3}{2}$ is a solution to the following equation.

$$3(x + 3) = 6 + x$$

Solution Substitute $-\frac{3}{2}$ for each x in the equation.

$$3(x + 3) = 6 + x$$
$$3\left(-\frac{3}{2} + 3\right) \stackrel{?}{=} 6 + \left(-\frac{3}{2}\right)$$
$$3\left(-\frac{3}{2} + \frac{6}{2}\right) \stackrel{?}{=} \frac{12}{2} - \frac{3}{2}$$
$$3\left(\frac{3}{2}\right) \stackrel{?}{=} \frac{9}{2}$$
$$\frac{9}{2} = \frac{9}{2} \quad \text{True}$$

Thus, $-\frac{3}{2}$ is a solution.

NOW TRY EXERCISE 21

Using Your Calculator

Checking Solutions

Calculators can be used to check solutions to equations. For example, to check whether $\frac{-10}{3}$ is a solution to the equation $2x + 3 = 5(x + 3) - 2$, we perform the following steps.

1. Substitute $\frac{-10}{3}$ for each x.

$$2x + 3 = 5(x + 3) - 2$$
$$2\left(\frac{-10}{3}\right) + 3 \stackrel{?}{=} 5\left(\frac{-10}{3} + 3\right) - 2$$

2. Evaluate each side of the equation separately using your calculator. If you obtain the same value on both sides, your solution checks.

continued on next page

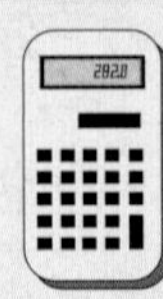

Scientific Calculator

To evaluate the left side of the equation, $2\left(\frac{-10}{3}\right) + 3$, press the following keys:

2 [×] [(] 10 [+/−] [÷] 3 [)] [+] 3 [=] −3.6666667

To evaluate the right side of the equation, $5\left(\frac{-10}{3} + 3\right) - 2$, press the following keys:

5 [×] [(] 10 [+/−] [÷] 3 [+] 3 [)] [−] 2 [=] −3.6666667

Since both sides give the same value, the solution checks. Note that because calculators differ in their electronics, sometimes the last digit of a calculation will differ.

Graphing Calculator

Left side of equation: 2 [(] [(−)] 10 [÷] 3 [)] [+] 3 [ENTER] −3.666666667

Right side of the equation: 5 [(] [(−)] 10 [÷] 3 [+] 3 [)] [−] 2 [ENTER] −3.666666667

Since both sides give the same solution, the solution checks.

3) Identify Equivalent Equations

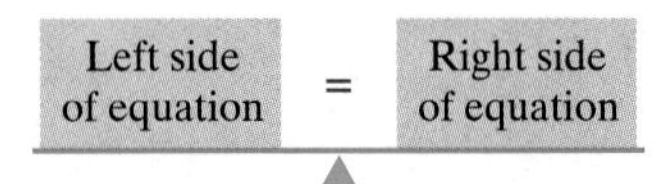

FIGURE 2.1

Now that we know how to check a solution to an equation we will discuss solving equations. Complete procedures for solving equations will be given shortly. For now, you need to understand that **to solve an equation, it is necessary to get the variable alone on one side of the equal sign. We say that we isolate the variable**. To isolate the variable, we make use of two properties: the addition and multiplication properties of equality. Look first at Figure 2.1.

Think of an equation as a balanced statement whose left side is balanced by its right side. When solving an equation, we must make sure that the equation remains balanced at all times. That is, both sides must always remain equal. **We ensure that an equation always remains equal by doing the same thing to both sides of the equation**. For example, if we add a number to the left side of the equation, we must add exactly the same number to the right side. If we multiply the right side of the equation by some number, we must multiply the left side by the same number.

When we add the same number to both sides of an equation or multiply both sides of an equation by the same nonzero number, we do not change the solution to the equation, just the form. Two or more equations with the same solution are called **equivalent equations**. The equations $2x - 4 = 2$, $2x = 6$, and $x = 3$ are equivalent, since the solution to each is 3.

CHECK: $x = 3$

$2x - 4 = 2$	$2x = 6$	$x = 3$
$2(3) - 4 \stackrel{?}{=} 2$	$2(3) \stackrel{?}{=} 6$	$3 = 3$ *True*
$6 - 4 \stackrel{?}{=} 2$	$6 = 6$ *True*	
$2 = 2$ *True*		

When solving an equation, we use the addition and multiplication properties to express a given equation as simpler equivalent equations until we obtain the solution.

Before stating the addition property of equality, we would like to give you an intuitive and visual introduction to solving equations using the addition property. The multiplication property of equality will be discussed in Section 2.3.

We stated that both sides of an equation must always stay balanced. That is, what you do to one side of the equation you must also do to the other side. Consider Figure 2.2. In this figure and in other figures involving a balance, we use a chocolate "kiss," to represent some number. We could have used a box, a tree, a letter, or any other symbol in place of the kiss. The symbol used is not important in understanding the addition property of equality. Whenever we use more than one kiss on a balance, the kisses all represent the same value. The value of the kiss may change with each example.

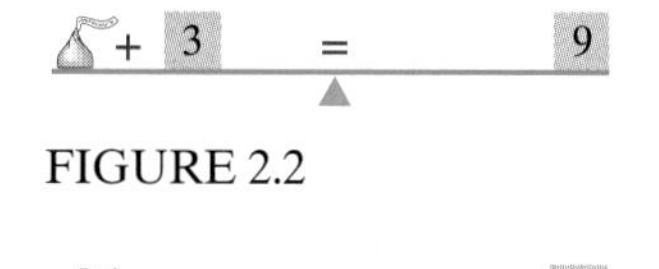

FIGURE 2.2

FIGURE 2.3

In Figure 2.2, can you determine the number the kiss represents if the left and right sides of the equation are to be balanced? If you answered 6, you answered correctly.

Note that one kiss plus 3 equals 9. If you subtract 3 from the left and right sides of the equation (or scale) you will be left with one kiss equals 6 (Fig. 2.3).

Now consider Figure 2.4. What is the value of the kiss if both sides are to be balanced? If you subtract 5 from both sides of the balance (Fig. 2.5), you see that one kiss equals 3.

FIGURE 2.4

FIGURE 2.5

In both of these problems we have actually used the addition property of equality to solve equations. Below we show Figures 2.2 and 2.4, the equations that can be determined from the figures, and the solutions to the equations. In the equations we have used the letter x to represent the value of a kiss. However, we could have used k, for kiss, or any other letter. To find the value of a kiss, we subtract the amount on the same side of the balance as the kiss from both sides of the balance. To solve each equation for the variable x, we subtract the constant on the same side of the equal sign as the variable from both sides of the equation.

Figure	Equation	Solution
 + 3 = 9	$x + 3 = 9$	$x = 6$
 8 = + 5	$8 = x + 5$	$x = 3$

4 Use the Addition Property to Solve Equations

Now that we have provided an informal introduction, let us define the **addition property of equality**.

Addition Property of Equality

If $a = b$, then $a + c = b + c$ for any real numbers a, b, and c.

This property means that the same number can be added to both sides of an equation without changing the solution. **The addition property is used to solve equations of the form $x + a = b$.** To isolate the variable x in equations of this form, add the opposite or additive inverse of a, $-a$, to both sides of the equation.

To isolate the variable when solving equations of the form $x + a = b$, **we use the addition property to eliminate the number on the same side of the equal sign as the variable**. (This is like isolating the kiss on the balance.) Study the following examples carefully.

Equation	To Solve, Use the Addition Property to Eliminate the Number
$x + 8 = 10$	8
$x - 7 = 12$	−7
$5 = x - 12$	−12
$-4 = x + 9$	9

Now let's work some examples.

EXAMPLE 4 Solve the equation $x - 4 = 3$.

Solution To isolate the variable, x, we must eliminate the −4 from the left side of the equation. To do this we add 4, the opposite of −4, to *both sides* of the equation.

$$x - 4 = 3$$
$$x - 4 + 4 = 3 + 4 \quad \text{Add 4 to both sides.}$$
$$x + 0 = 7$$
$$x = 7$$

Note how the process helps to isolate x.

CHECK:

$$x - 4 = 3$$
$$7 - 4 \stackrel{?}{=} 3$$
$$3 = 3 \quad \text{True}$$

EXAMPLE 5 Solve the equation $y - 3 = -5$.

Solution To solve this equation, we must isolate the variable, y. To eliminate the −3 from the left side of the equation, we add its opposite, 3, to *both sides* of the equation.

$$y - 3 = -5$$
$$y - 3 + 3 = -5 + 3 \quad \text{Add 3 to both sides.}$$
$$y + 0 = -2$$
$$y = -2$$

Note that we did not check the solution to Example 5. Space limitations prevent us from showing all checks. However, *you should check all of your answers*.

EXAMPLE 6 Solve the equation $x + 5 = 9$.

Solution To solve this equation, we must isolate the variable, x. Therefore, we must eliminate the 5 from the left side of the equation. To do this, we add the opposite of 5, −5, to *both sides* of the equation.

$$x + 5 = 9$$
$$x + 5 + (-5) = 9 + (-5) \quad \text{Add } -5 \text{ to both sides.}$$
$$x + 0 = 4$$
$$x = 4$$

In Example 6 we added −5 to both sides of the equation. From Section 1.7 we know that $5 + (-5) = 5 - 5$. Thus, we can see that adding a negative 5 to both sides of the equation is equivalent to subtracting a 5 from both sides of the equation. According to the addition property, the same number may be *added* to both sides of an equation. **Since subtraction is defined in terms of addition, the addition property also allows us to *subtract* the same number from both sides of the equation.** Thus, Example 6 could have also been worked as follows:

$$x + 5 = 9$$
$$x + 5 - 5 = 9 - 5 \quad \text{Subtract 5 from both sides.}$$
$$x + 0 = 4$$
$$x = 4$$

In this text, unless there is a specific reason to do otherwise, rather than adding a negative number to both sides of the equation, we will subtract a number from both sides of the equation.

EXAMPLE 7 Solve the equation $x + 7 = -3$.

Solution

$$x + 7 = -3$$
$$x + 7 - 7 = -3 - 7 \quad \text{Subtract 7 from both sides.}$$
$$x + 0 = -10$$
$$x = -10$$

CHECK:

$$x + 7 = -3$$
$$-10 + 7 \stackrel{?}{=} -3$$
$$-3 = -3 \quad \text{True}$$

NOW TRY EXERCISE 59

HELPFUL HINT

Remember that our goal in solving an equation is to get the variable alone on one side of the equation. To do this, we add or subtract **the number on the same side of the equation as the variable** to both sides of the equation.

Equation	Must Eliminate	Number to Add (or Subtract) to (or from) Both Sides of the Equation	Correct Results	Solution
$x - 5 = 8$	-5	add 5	$x - 5 + 5 = 8 + 5$	$x = 13$
$x - 3 = -12$	-3	add 3	$x - 3 + 3 = -12 + 3$	$x = -9$
$2 = x - 7$	-7	add 7	$2 + 7 = x - 7 + 7$	$9 = x$
$x + 12 = -5$	$+12$	subtract 12	$x + 12 - 12 = -5 - 12$	$x = -17$
$6 = x + 4$	$+4$	subtract 4	$6 - 4 = x + 4 - 4$	$2 = x$
$13 = x + 9$	$+9$	subtract 9	$13 - 9 = x + 9 - 9$	$4 = x$

Notice that under the *Correct Results* column, when the equation is simplified by combining terms, the x will become isolated because the sum of a number and its opposite is 0, and $x + 0$ equals x.

EXAMPLE 8 Solve the equation $4 = x - 5$.

Solution The variable x is on the right side of the equation. To isolate the x, we must eliminate the -5 from the right side of the equation. This can be accomplished by adding 5 to both sides of the equation.

$$4 = x - 5$$
$$4 + 5 = x - 5 + 5 \quad \text{Add 5 to both sides.}$$
$$9 = x + 0$$
$$9 = x$$

Thus, the solution is 9.

EXAMPLE 9 Solve the equation $-6.25 = x + 12.78$.

Solution The variable is on the right side of the equation. Subtract 12.78 from both sides of the equation to isolate the variable.

$$-6.25 = x + 12.78$$
$$-6.25 - 12.78 = x + 12.78 - 12.78 \quad \text{Subtract 12.78 from both sides.}$$
$$-19.03 = x + 0$$
$$-19.03 = x$$

NOW TRY EXERCISE 75

The solution is -19.03.

Avoiding Common Errors

When solving an equation, our goal is to get the variable alone on one side of the equal sign. Consider the equation $x + 3 = -4$. How do we solve it?

CORRECT

Remove the 3 from the left side of the equation.

$$x + 3 = -4$$
$$x + 3 - 3 = -4 - 3$$
$$x = -7$$

Variable is now isolated.

WRONG

Remove the -4 from the right side of the equation.

$$x + 3 = -4$$
$$x + 3 + 4 = -4 + 4$$
$$x + 7 = 0$$

Variable is ***not*** *isolated.*

Remember, use the addition property to *remove the number that is on the same side of the equation as the variable.*

5) Solve Equations by Doing Some Steps Mentally

Consider the following two problems.

a)
$$x - 5 = 12$$
$$x - 5 + 5 = 12 + 5$$
$$x + 0 = 12 + 5$$
$$x = 17$$

b)
$$15 = x + 3$$
$$15 - 3 = x + 3 - 3$$
$$15 - 3 = x + 0$$
$$12 = x$$

Note how the number on the same side of the equal sign as the variable is transferred to the opposite side of the equal sign when the addition property is used.

Also note that the sign of the number changes when transferred from one side of the equal sign to the other.

When you feel comfortable using the addition property, you may wish to do some of the steps mentally to reduce some of the written work. For example, the preceding two problems may be shortened as follows:

a)

$$x - 5 = 12$$
$$x - 5 + 5 = 12 + 5 \longleftarrow \text{Do this step mentally.}$$
$$x = 12 + 5$$
$$x = 17$$

Shortened Form

$$x - 5 = 12$$
$$x = 12 + 5$$
$$x = 17$$

b)

$$15 = x + 3$$
$$15 - 3 = x + 3 - 3 \longleftarrow \text{Do this step mentally.}$$
$$15 - 3 = x$$
$$12 = x$$

Shortened Form

$$15 = x + 3$$
$$15 - 3 = x$$
$$12 = x$$

NOW TRY EXERCISE 63

Exercise Set 2.2

Concept/Writing Exercises

1. What is an equation?
2. **a)** What is meant by the "solution to an equation"?
   b) What does it mean to "solve an equation"?
3. Explain how the solution to an equation may be checked.
4. In your own words, explain the addition property of equality.
5. What are equivalent equations?
6. To solve an equation we "isolate the variable."
   a) Explain what this means.
   b) Explain how to isolate the variable in the equations discussed in this section.
7. When solving the equation $x - 4 = 6$, would you add 4 to both sides of the equation or subtract 6 from both sides of the equation? Explain.
8. When solving the equation $5 = x + 3$, would you subtract 5 from both sides of the equation or subtract 3 from both sides of the equation? Explain.
9. Give an example of a linear equation in one variable
10. Explain why the following three equations are equivalent.

    $$2x + 3 = 5, \qquad 2x = 2, \qquad x = 1$$

11. Explain why the addition property allows us to subtract the same quantity from both sides of an equation.
12. To solve the equation $x - \square = \triangle$ for x, do we add $\square$ to both sides of the equation or do we subtract $\triangle$ from both sides of the equation? Explain.

Practice the Skills

13. Is $x = 4$ a solution of $3x - 3 = 9$?
14. Is $x = -6$ a solution of $2x + 1 = x - 5$?
15. Is $x = -3$ a solution of $2x - 5 = 5(x + 2)$?
16. Is $x = 1$ a solution of $2(x - 3) = -3(x + 1)$?
17. Is $x = 0$ a solution of $3x - 5 = 2(x + 3) - 11$?
18. Is $x = -2$ a solution of $-2(x - 3) = -5x + 3 - x$?
19. Is $x = 3.4$ a solution of $3(x + 2) - 3(x - 1) = 9$?
20. Is $x = \frac{1}{2}$ a solution of $x + 3 = 3x + 2$?

21. Is $x = \frac{1}{2}$ a solution of $4x - 4 = 2x - 2$?

22. Is $x = \frac{1}{2}$ a solution of $3x + 4 = 2x + 9$?

23. Is $x = \frac{11}{2}$ a solution of $3(x + 2) = 5(x - 1)$?

24. Is $x = 3$ a solution of $-(x - 5) - (x - 6) = 3x - 4$?

In Exercises 25–32, ***a)*** *represent each figure as an equation with the variable x, and* ***b)*** *solve the equation. Refer to page 105 for examples.*

25. + 5 = 8

26. + 2 = 9

27. 12 = + 3

28. 15 = + 4

29. 10 = + 7

30. + 9 = 10

31. + 6 = 4 + 11

32. 12 + 4 = + 3

Solve each equation and check your solution.

33. $x + 2 = 6$
34. $x - 4 = 13$
35. $x + 1 = -6$
36. $x - 4 = -8$
37. $x + 4 = -5$
38. $x - 16 = 36$
39. $x + 9 = 52$
40. $6 + x = 9$
41. $-8 + x = 14$
42. $7 = 9 + x$
43. $27 = x + 16$
44. $50 = x - 25$
45. $-18 = -14 + x$
46. $7 + x = -50$
47. $9 + x = 4$
48. $x + 29 = -29$
49. $4 + x = -9$
50. $9 = x - 3$
51. $5 + x = -18$
52. $x + 7 = 4$
53. $6 = 4 + x$
54. $9 + x = 12$
55. $-4 = x - 3$
56. $-13 = 4 + x$
57. $12 = 16 + x$
58. $40 = x - 13$
59. $15 + x = -5$
60. $-20 = 4 + x$
61. $-10 = -10 + x$
62. $8 = 8 + x$
63. $5 = x - 12$
64. $2 = x + 9$
65. $-50 = x - 24$
66. $-29 + x = -15$
67. $16 + x = -20$
68. $-25 = 74 + x$
69. $40.2 + x = -5.9$
70. $-27.23 + x = 9.77$
71. $-37 + x = 9.5$
72. $7.2 + x = 7.2$
73. $x - 8.77 = -17$
74. $6.1 + x = 10.2$
75. $9.32 = x + 3.75$
76. $139 = x - 117$

Problem Solving

77. Do you think the equation $x + 1 = x + 2$ has a real number as a solution? Explain. (We will discuss equations like this in Section 2.5)

78. Do you think the equation $x + 4 = x + 4$ has more than one real number as a solution? If so, how many solutions does it have? Explain.

79. In the next section we introduce the multiplication property. When discussing the multiplication property, we will use a figure like the one that follows.

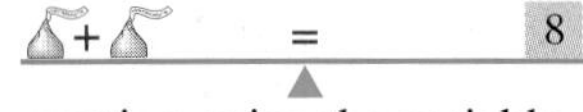

a) Write an equation, using the variable x, that can be used to represent this figure.
b) Solve the equation.

Follow the instructions in Exercise 79 for the following figures.

80. + + = 12

81. 20 = + + +

Challenge Problems

82. By checking, determine which of the following are solutions to $2x^2 - 7x + 3 = 0$.

a) 3 **b)** 0 **c)** $\frac{1}{2}$

We can solve equations that contain unknown symbols. Solve each equation for the symbol indicated by adding (or subtracting) a symbol to (or from) both sides of the equation. Explain each answer. (Remember that to solve the equation you want to isolate the symbol you are solving for on one side of the equation.)

83. $x - \triangle = \square$, for x

84. $\square + ☺ = \triangle$, for ☺

85. ☺ $= \square + \triangle$, for $\square$

86. $\sim\, = \triangle + ?$, for ?

Group Activity

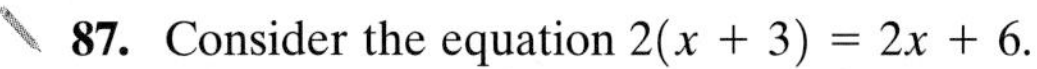

Discuss and answer Exercise 87 as a group.

87. Consider the equation $2(x + 3) = 2x + 6$.

a) Group member 1: Determine whether 4 is a solution to the equation.

b) Group member 2: Determine whether −2 is a solution to the equation.

c) Group member 3: Determine whether 0.3 is a solution to the equation.

d) Each group member: Select a number not used in parts **a)-c)** and determine whether that number is a solution to the equation.

e) As a group, write what you think is the solution to the equation $2(x + 3) = 2x + 6$ and write a paragraph explaining your answer.

Cumulative Review Exercises

[1.9] *Evaluate.*

88. $3x + 4(x - 3) + 2$ when $x = 4$

89. $6x - 2(2x + 1)$ when $x = -3$

[2.1] *Simplify.*

90. $4x + 3(x - 2) - 5x - 7$

91. $-(x - 3) + 7(2x - 5) - 3x$

2.3 THE MULTIPLICATION PROPERTY OF EQUALITY

SSM VIDEO 2.3 CD Rom

1) **Identify reciprocals.**
2) **Use the multiplication property to solve equations.**
3) **Solve equations of the form $-x = a$.**
4) **Do some steps mentally when solving equations.**

1) Identify Reciprocals

Before we discuss the multiplication property, let's discuss what is meant by the **reciprocal** of a number. Two numbers are reciprocals of each other when their product is 1. Some examples of numbers and their reciprocals follow.

Number	Reciprocal	Product
3	$\frac{1}{3}$	$(3)\left(\frac{1}{3}\right) = 1$
$-\frac{3}{5}$	$-\frac{5}{3}$	$\left(-\frac{3}{5}\right)\left(-\frac{5}{3}\right) = 1$
-1	-1	$(-1)(-1) = 1$

The reciprocal of a positive number is a positive number and the reciprocal of a negative number is a negative number. Note that 0 has no reciprocal. Why?

In general, if a represents any nonzero number, its reciprocal is $\frac{1}{a}$. For example, the reciprocal of 3 is $\frac{1}{3}$ and the reciprocal of −2 is $\frac{1}{-2}$ or $-\frac{1}{2}$. The reciprocal of $-\frac{3}{5}$ is $\frac{1}{-\frac{3}{5}}$, which can be written as $1 \div \left(-\frac{3}{5}\right)$. Simplifying, we get $\left(\frac{1}{1}\right)\left(-\frac{5}{3}\right) = -\frac{5}{3}$.

Thus, the reciprocal of $-\frac{3}{5}$ is $-\frac{5}{3}$.

2) Use the Multiplication Property to Solve Equations

In Section 2.2 we used the addition property of equality to solve equations of the form $x + a = b$, where a and b represent real numbers. In this section we use the multiplication property of equality to solve equations of the form $ax = b$, where a and b represent real numbers.

It is important that you recognize the difference between equations like $x + 2 = 8$ and $2x = 8$. In $x + 2 = 8$ the 2 is a *term* that is being added to x, so we use the addition property to solve the equation. In $2x = 8$ the 2 is a *factor* of $2x$. The 2 is the coefficient multiplying the x, so we use the multiplication property to solve the equation.

The multiplication property is used to solve linear equations where the coefficient of the x-term is a number other than 1. Below we give a visual interpretation of the difference between the equations. To write the equation, we have used x to represent the value of a kiss.

Figure	Equation	Property to Use to Solve Equation	Solution
	$x + 2 = 8$	Addition (the equation contains only one x)	6
	$x + x = 8$ or $2x = 8$	Multiplication (the left side of the equation contains more than one x)	4
	$15 = x + x + x$ or $15 = 3x$	Multiplication (the right side of the equation contains more than one x)	5

To help you understand the multiplication property of equality, we will give a visual interpretation of the property before stating it. Consider Figure 2.6. To find the value of one kiss, we need to redraw the balance with only one kiss on one side of the balance. We can eliminate one of the two kisses on the left side of the balance by either multiplying the two kisses by $\frac{1}{2}$ to get $\frac{1}{2}(2) = 1$, or by dividing the two kisses by 2 to get $\frac{2}{2} = 1$. The two processes are equivalent. We must remember that whatever we do to one side of the balance we must do to the other side. Thus, if we multiply the two kisses by $\frac{1}{2}$, we need to multiply the 8 by $\frac{1}{2}$ to get 4. If we divide the two kisses by 2, we need to divide the 8 by 2 to get 4. Either procedure results in the balance shown in Figure 2.7, where the value of a kiss is 4.

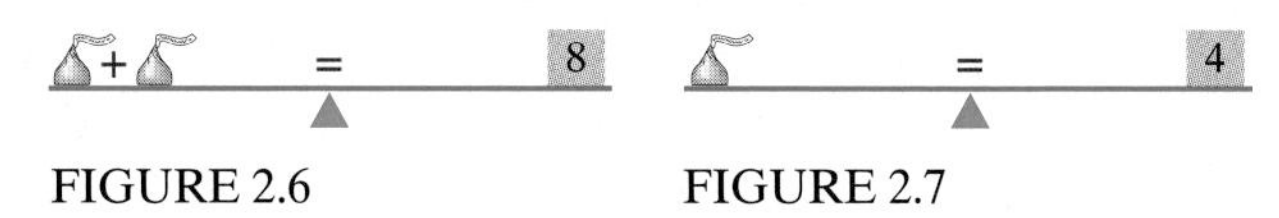

FIGURE 2.6 FIGURE 2.7

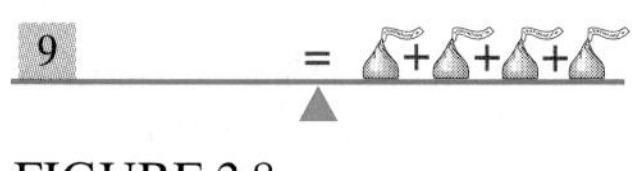

FIGURE 2.8

Now consider Figure 2.8. There are four kisses of equal value on the right side of the balance. To find the value of one kiss, we need to redraw the balance so that only one kiss appears on the right side. We can do this by multiplying the four kisses by $\frac{1}{4}$ to get $4\left(\frac{1}{4}\right) = 1$, or by dividing the four kisses by 4 to get $\frac{4}{4} = 1$. If we multiply the kisses on the right side of the balance by $\frac{1}{4}$, we need to multiply the 9 on the left side of the balance by $\frac{1}{4}$ to get $\frac{9}{4}$. If we divide the kisses on the right side of the balance by 4, we need to divide the 9 on the left side of the balance by 4 to get $\frac{9}{4}$. Either method results in the kiss on the right side of the balance having a value of $\frac{9}{4}$. (Fig. 2.9).

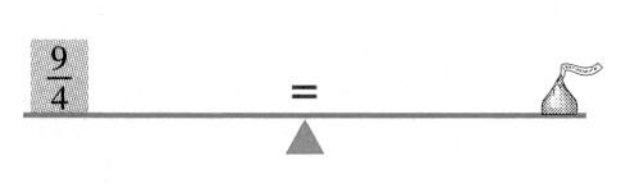

FIGURE 2.9

Next we illustrate how Figure 2.8 may be represented as an equation if we let x represent the value of a kiss.

Figure	Equation	Solution
9 = 4 kisses	$9 = x + x + x + x$ or $9 = 4x$	$\frac{9}{4}$

To solve the equation $9 = 4x$ we perform a process similar to that used in finding the value of one kiss on the balance. To find the value of x in the equation $9 = 4x$, we can *multiply* both sides of the equation *by the reciprocal of the number of x's that appear*. Since the right side of the equation contains 4 x's, we can multiply both sides of the equation by $\frac{1}{4}$. We can also solve the equation by *dividing* both sides of the equation *by the number of xs that appear*, 4. Using either method we find that $x = \frac{9}{4}$.

The information presented above may help you in solving equations using the multiplication property of equality. Now we present the *multiplication property of equality*.

Multiplication Property of Equality

If $a = b$, then $a \cdot c = b \cdot c$ for any real numbers a, b, and c.

The multiplication property means that both sides of an equation can be multiplied by the same nonzero number without changing the solution. **The multiplication property can be used to solve equations of the form $ax = b$.** We can isolate the variable in equations of this form by multiplying both sides of the equation by the reciprocal of a, which is $\frac{1}{a}$. Doing so makes the numerical coefficient of the variable, x, become 1, which can be omitted when we write the variable. By following this process, we say that we *eliminate* the coefficient from the variable.

Equation	To Solve, Use the Multiplication Property to Eliminate the Coefficient
$4x = 9$	4
$-5x = 20$	-5
$15 = \frac{1}{2}x$	$\frac{1}{2}$
$7 = -9x$	-9

Now let's work some examples.

EXAMPLE 1 Solve the equation $3x = 6$.

Solution To isolate the variable, x, we must eliminate the 3 from the left side of the equation. To do this, we multiply both sides of the equation by the reciprocal of 3, which is $\frac{1}{3}$.

$$3x = 6$$

$$\frac{1}{3} \cdot 3x = \frac{1}{3} \cdot 6 \qquad \text{Multiply both sides by } \frac{1}{3}.$$

$$\frac{1}{\cancel{3}_1} \cdot \cancel{3}^1 x = \frac{1}{\cancel{3}_1} \cdot \cancel{6}^2 \qquad \text{Divide out the common factors.}$$

$$1x = 2$$

$$x = 2$$

Notice in Example 1 that $1x$ is replaced by x in the next step. Usually we do this step mentally. How would you represent and solve the equation $3x = 6$ using a balance? Try this now.

EXAMPLE 2 Solve the equation $\frac{x}{2} = 4$.

Solution Since dividing by 2 is the same as multiplying by $\frac{1}{2}$, the equation $\frac{x}{2} = 4$ is the same as $\frac{1}{2}x = 4$. We will therefore multiply both sides of the equation by the reciprocal of $\frac{1}{2}$, which is 2.

$$\frac{x}{2} = 4$$

$$\overset{1}{\cancel{2}}\left(\frac{x}{\underset{1}{\cancel{2}}}\right) = 2 \cdot 4 \qquad \text{Multiply both sides by 2.}$$

$$x = 2 \cdot 4$$

$$x = 8$$

NOW TRY EXERCISE 23

EXAMPLE 3 Solve the equation $\frac{2}{3}x = 6$.

Solution The reciprocal of $\frac{2}{3}$ is $\frac{3}{2}$. We multiply both sides of the equation by $\frac{3}{2}$.

$$\frac{2}{3}x = 6$$

$$\frac{3}{2} \cdot \frac{2}{3}x = \frac{3}{2} \cdot 6 \qquad \text{Multiply both sides by } \frac{3}{2}.$$

$$1x = 9$$

$$x = 9$$

We will show a check of this solution.

CHECK:

$$\frac{2}{3}x = 6$$

$$\frac{2}{3}(9) \stackrel{?}{=} 6$$

$$6 = 6 \qquad \text{True}$$

NOW TRY EXERCISE 57

In Example 1, we multiplied both sides of the equation $3x = 6$ by $\frac{1}{3}$ to isolate the variable. We could have also isolated the variable by dividing both sides of the equation by 3, as follows:

$$3x = 6$$

$$\frac{\overset{1}{\cancel{3}}x}{\underset{1}{\cancel{3}}} = \frac{\overset{2}{\cancel{6}}}{\underset{1}{\cancel{3}}} \qquad \text{Divide both sides by 3.}$$

$$x = 2$$

We can do this because dividing by 3 is equivalent to multiplying by $\frac{1}{3}$. **Since division can be defined in terms of multiplication ($\frac{a}{b}$ means $a \cdot \frac{1}{b}$), the multipli-**

cation property also allows us to divide both sides of an equation by the same nonzero number. This process is illustrated in Examples 4 through 6.

EXAMPLE 4 Solve the equation $7p = 4$.

Solution

$$7p = 4$$

$$\frac{7p}{7} = \frac{4}{7} \quad \text{Divide both sides by 7.}$$

$$p = \frac{4}{7}$$

EXAMPLE 5 Solve the equation $-15 = -3x$.

Solution In this equation the variable, x, is on the right side of the equal sign. To isolate x, we divide both sides of the equation by -3.

$$-15 = -3x$$

$$\frac{-15}{-3} = \frac{-3x}{-3} \quad \text{Divide both sides by } -3.$$

$$5 = x$$

EXAMPLE 6 Solve the equation $0.24x = 1.20$.

Solution We begin by dividing both sides of the equation by 0.24 to isolate the variable x.

$$0.24x = 1.20$$

$$\frac{0.24x}{0.24} = \frac{1.20}{0.24} \quad \text{Divide both sides by 0.24.}$$

$$x = 5$$

NOW TRY EXERCISE 43

Working problems involving decimal numbers on a calculator will probably save you time.

HELPFUL HINT

When solving an equation of the form $ax = b$, we can isolate the variable by

1. multiplying both sides of the equation by the reciprocal of a, $\frac{1}{a}$, as was done in Examples 1, 2, and 3, or
2. dividing both sides of the equation by a, as was done in Examples 4, 5, and 6.

Either method may be used to isolate the variable. However, if the equation contains a fraction, or fractions, you will arrive at a solution more quickly by multiplying by the reciprocal of a. This is illustrated in Examples 7 and 8.

EXAMPLE 7 Solve the equation $-2x = \frac{3}{5}$.

Solution Since this equation contains a fraction, we will isolate the variable by multiplying both sides of the equation by $-\frac{1}{2}$, which is the reciprocal of -2.

$$-2x = \frac{3}{5}$$

$$\left(-\frac{1}{2}\right)(-2x) = \left(-\frac{1}{2}\right)\left(\frac{3}{5}\right) \quad \text{Multiply both sides by } -\frac{1}{2}.$$

$$1x = \left(-\frac{1}{2}\right)\left(\frac{3}{5}\right)$$

$$x = -\frac{3}{10}$$

In Example 7, if you wished to solve the equation by dividing both sides of the equation by -2, you would have to divide the fraction $\frac{3}{5}$ by -2.

EXAMPLE 8 Solve the equation $-6 = -\frac{3}{5}x$.

Solution Since this equation contains a fraction, we will isolate the variable by multiplying both sides of the equation by the reciprocal of $-\frac{3}{5}$, which is $-\frac{5}{3}$.

$$-6 = -\frac{3}{5}x$$

$$\left(-\frac{5}{3}\right)(-6) = \left(-\frac{5}{3}\right)\left(-\frac{3}{5}x\right) \quad \text{Multiply both sides by } -\frac{5}{3}.$$

NOW TRY EXERCISE 71

$$10 = x$$

In Example 8, the equation was written as $-6 = -\frac{3}{5}x$. This equation is equivalent to the equations $-6 = \frac{-3}{5}x$ and $-6 = \frac{3}{-5}x$. Can you explain why? All three equations have the same solution, 10.

3 Solve Equations of the Form $-x = a$

When solving an equation, we may obtain an equation like $-x = 7$. This is not a solution since $-x = 7$ means $-1x = 7$. The solution to an equation is of the form $x =$ some number. When an equation is of the form $-x = 7$, we can solve for x by multiplying both sides of the equation by -1, as illustrated in the following example.

EXAMPLE 9 Solve the equation $-x = 7$.

Solution $-x = 7$ means that $-1x = 7$. We are solving for x, not $-x$. We can multiply both sides of the equation by -1 to isolate x on the left side of the equation.

$$-x = 7$$

$$-1x = 7$$

$$(-1)(-1x) = (-1)(7) \quad \text{Multiply both sides by } -1.$$

$$1x = -7$$

$$x = -7$$

CHECK:

$$-x = 7$$
$$-(-7) \stackrel{?}{=} 7$$
$$7 = 7 \quad \text{True}$$

Thus, the solution is −7.

Example 9 may also be solved by dividing both sides of the equation by −1. Try this now and see that you get the same solution. Whenever we have the opposite (or negative) of a variable equal to a quantity, as in Example 9, we can solve for the variable by multiplying (or dividing) both sides of the equation by −1.

EXAMPLE 10 Solve the equation $-x = -5$.

Solution

$$-x = -5$$
$$-1x = -5$$
$$(-1)(-1x) = (-1)(-5) \quad \text{Multiply both sides by } -1.$$
$$1x = 5$$
$$x = 5$$

NOW TRY EXERCISE 31

HELPFUL HINT

For any real number a, if $-x = a$, then $x = -a$.

EXAMPLES

$$-x = 7 \qquad\qquad -x = -2$$
$$x = -7 \qquad\qquad x = -(-2)$$
$$x = 2$$

4) Do Some Steps Mentally When Solving Equations

When you feel comfortable using the multiplication property, you may wish to do some of the steps mentally to reduce some of the written work. Now we present two examples worked out in detail, along with their shortened form.

EXAMPLE 11 Solve the equation $-3x = -21$.

Solution

$$-3x = -21$$
$$\frac{-3x}{-3} = \frac{-21}{-3} \longleftarrow \text{Do this step mentally.}$$
$$x = \frac{-21}{-3}$$
$$x = 7$$

Shortened Form

$$-3x = -21$$
$$x = \frac{-21}{-3}$$
$$x = 7$$

EXAMPLE 12 Solve the equation $\frac{1}{3}x = 9$.

Solution

$$\frac{1}{3}x = 9$$
$$3\left(\frac{1}{3}x\right) = 3(9) \longleftarrow \text{Do this step mentally.}$$
$$x = 3(9)$$
$$x = 27$$

Shortened Form

$$\frac{1}{3}x = 9$$
$$x = 3(9)$$
$$x = 27$$

NOW TRY EXERCISE 55

In Section 2.2 we discussed the addition property and in this section we discussed the multiplication property. It is important that you understand the difference between the two. The following Helpful Hint should be studied carefully.

HELPFUL HINT

The **addition property** is used to solve equations of the form $x + a = b$. The *addition property* is used when a number is *added to or subtracted from* a variable.

$$x + 3 = -6 \qquad\qquad x - 5 = -2$$
$$x + 3 - 3 = -6 - 3 \qquad\qquad x - 5 + 5 = -2 + 5$$
$$x = -9 \qquad\qquad x = 3$$

The **multiplication property** is used to solve equations of the form $ax = b$. It is used when a variable is *multiplied* or *divided by* a number.

$$3x = 6 \qquad\qquad \frac{x}{2} = 4 \qquad\qquad \frac{2}{5}x = 12$$
$$\frac{3x}{3} = \frac{6}{3} \qquad\qquad 2\left(\frac{x}{2}\right) = 2(4) \qquad\qquad \left(\frac{5}{2}\right)\left(\frac{2}{5}x\right) = \left(\frac{5}{2}\right)(12)$$
$$x = 2 \qquad\qquad x = 8 \qquad\qquad x = 30$$

Exercise Set 2.3

Concept/Writing Exercises

1. In your own words, explain the multiplication property of equality.
2. Explain why the multiplication property allows us to divide both sides of an equation by a nonzero quantity.
3. **a)** If $-x = a$, where a represents any real number, what does x equal?
   b) If $-x = 5$, what is x?
   c) If $-x = -5$, what is x?
4. When solving the equation $3x = 5$, would you divide both sides of the equation by 3 or by 5? Explain.
5. When solving the equation $-2x = 5$, would you divide both sides of the equation by -2 or by 5? Explain.
6. When solving the equation $\frac{x}{2} = 3$, what would you do to isolate the variable? Explain.
7. When solving the equation $4 = \frac{x}{3}$, what would you do to isolate the variable? Explain.
8. When solving the equation $\triangle x = \square$ for x, you would divide both sides of the equation by $\triangle$. Explain why.

Practice the Skills

*In Exercises 9–16, **a)** represent the figure as an equation with the variable x, and **b)** solve the equation. Refer to page 113 for an example.*

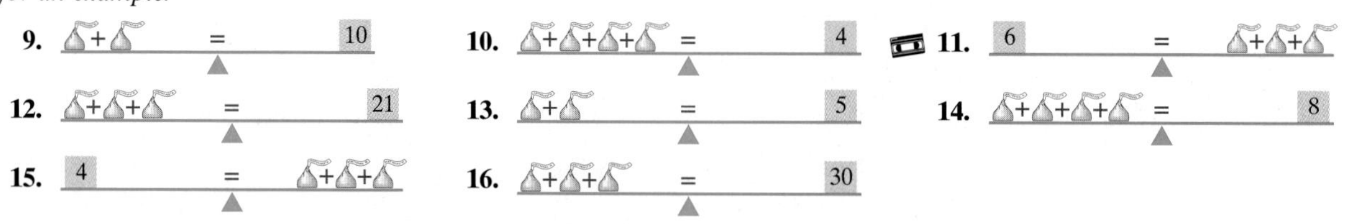

Solve each equation and check your solution.

17. $2x = 6$
18. $5x = 50$
19. $\frac{x}{2} = 4$
20. $\frac{x}{3} = 12$

21. $-4x = 12$

22. $8 = 16y$

23. $\frac{x}{4} = -2$

24. $\frac{x}{3} = -3$

25. $\frac{x}{5} = 1$

26. $-7x = 49$

27. $-32x = -96$

28. $16 = -4y$

29. $-6 = 4z$

30. $\frac{x}{8} = -3$

31. $-x = -11$

32. $-x = 9$

33. $10 = -y$

34. $-3 = \frac{x}{5}$

35. $-\frac{x}{7} = -7$

36. $2 = \frac{x}{7}$

37. $4 = -12x$

38. $12y = -15$

39. $-\frac{x}{3} = -2$

40. $-\frac{a}{8} = -7$

41. $13x = 10$

42. $-24x = -18$

43. $-4.2x = -8.4$

44. $-3.88 = 1.94y$

45. $7x = -7$

46. $3x = \frac{3}{5}$

47. $5x = -\frac{3}{8}$

48. $-2b = -\frac{4}{5}$

49. $15 = -\frac{x}{4}$

50. $\frac{x}{5} = 0$

51. $-\frac{x}{2} = -75$

52. $-x = -\frac{5}{9}$

53. $\frac{x}{5} = -7$

54. $-3r = -18$

55. $5 = \frac{x}{4}$

56. $-3 = \frac{x}{-5}$

57. $\frac{3}{5}d = -30$

58. $\frac{2}{7}x = 7$

59. $\frac{y}{-2} = -6$

60. $-6x = \frac{5}{2}$

61. $\frac{-7}{8}w = 0$

62. $-x = \frac{5}{8}$

63. $\frac{1}{5}x = 4.5$

64. $6 = \frac{3}{5}x$

65. $-4 = -\frac{2}{3}z$

66. $-8 = \frac{-4}{5}x$

67. $-1.4x = 28.28$

68. $-0.42x = -2.142$

69. $6x = \frac{5}{10}$

70. $6x = \frac{8}{3}$

71. $\frac{2}{3}x = 6$

72. $-\frac{1}{4}x = \frac{3}{4}$

Problem Solving

73. **a)** Explain the difference between $5 + x = 10$ and $5x = 10$.
b) Solve $5 + x = 10$.
c) Solve $5x = 10$.

74. **a)** Explain the difference between $3 + x = 6$ and $3x = 6$.
b) Solve $3 + x = 6$.
c) Solve $3x = 6$.

In the next section we will solve equations using both the addition and multiplication properties. We can use figures like those in Exercises 75–78 to illustrate such problems.

For each exercise,

a) Use the addition property first to get the kisses by themselves on one side of the balance. Then use the multiplication property to find the value of a kiss.

b) Write an equation with variable x that can be used to represent the figure.

c) Use the addition property and then the multiplication property to solve the equation and find the value of x. (The value of x should be the same as the value of a kiss.)

75. + + 6 = 14

76. 9 = + + + 9

77. 6 = + 4 +

78. 7 = + + 2 +

79. Consider the equation $\frac{2}{3}x = 4$. This equation could be solved by multiplying both sides of the equation by $\frac{3}{2}$, the reciprocal of $\frac{2}{3}$, or by dividing both sides of the equation by $\frac{2}{3}$. Which method do you feel would be easier? Explain your answer. Find the solution to the equation.

80. Consider the equation $4x = \frac{3}{5}$. Would it be easier to solve this equation by dividing both sides of the equation by 4 or by multiplying both sides of the equation by $\frac{1}{4}$, the reciprocal of 4? Explain your answer. Find the solution to the problem.

81. Consider the equation $\frac{3}{7}x = \frac{4}{5}$. Would it be easier to solve this equation by dividing both sides of the equation by $\frac{3}{7}$ or by multiplying both sides of the equation by $\frac{7}{3}$, the reciprocal of $\frac{3}{7}$? Explain your answer. Find the solution to the equation.

Challenge Problems

82. Consider the equation □ ⊙ = △.

a) To solve for ⊙, what symbol do we need to isolate?

b) How would you isolate the symbol you specified in part **a)**?

c) Solve the equation for ⊙.

83. Consider the equation ☺ = △ ⊡.

a) To solve for ⊡, what symbol do we need to isolate?

b) How would you isolate the symbol you specified in part **a)**?

c) Solve the equation for ⊡.

84. Consider the equation # = $\frac{☺}{△}$.

a) To solve for ☺, what symbol do we need to isolate?

b) How would you isolate the symbol you specified in part **a)**?

c) Solve the equation for ☺.

Cumulative Review Exercises

[1.7] **85.** Subtract -4 from -8.

86. Evaluate $6 - (-3) - 5 - 4$.

[2.1] **87.** Simplify $-(x + 3) - 5(2x - 7) + 6$.

[2.2] **88.** Solve the equation $-48 = x + 9$.

2.4 SOLVING LINEAR EQUATIONS WITH A VARIABLE ON ONLY ONE SIDE OF THE EQUATION

1) **Solve linear equations with a variable on only one side of the equal sign.**

SSM

VIDEO 2.4

CD Rom

1) Solve Linear Equations With a Variable on Only One Side of the Equal Sign

In this section we discuss how to solve linear equations using *both* the addition and multiplication properties of equality when a variable appears on only one side of the equal sign. In Section 2.5 we will discuss how to solve linear equations using both properties when a variable appears on both sides of the equal sign.

Below we show two balances, the equations that represent the balances, and the solutions to each equation. You probably cannot determine the solutions yet. Do not worry about this. The purpose of this section is to teach you the procedures for finding the solution to such problems.

Figure	Equation	Solution
(kiss) + (kiss) + (kiss) + 6 = 27	$3x + 6 = 27$	7
12 = (kiss) + (kiss) + 5	$12 = 2x + 5$	$\frac{7}{2}$

The general procedure we use to solve equations is to "isolate the variable." That is, get the variable, x, alone on one side of the equal sign. If we consider the balance, we need to eliminate all the numbers from the same side of the balance as the kisses.

No one method is the "best" to solve all linear equations. But the following general procedure can be used to solve linear equations when the variable appears on only one side of the equation and the equation does not contain fractions.

To Solve Linear Equations with a Variable on Only One Side of the Equal Sign

1. Use the distributive property to remove parentheses.
2. Combine like terms on the same side of the equal sign.
3. Use the addition property to obtain an equation with the term containing the variable on one side of the equal sign and a constant on the other side. This will result in an equation of the form $ax = b$.
4. Use the multiplication property to isolate the variable. This will give a solution of the form $x = \frac{b}{a}\left(\text{or } 1x = \frac{b}{a}\right)$.
5. Check the solution in the original equation.

When solving an equation you should always check your solution, as is indicated in step 5. To conserve space, we will not show all checks. We solved some equations containing fractions in Section 2.3. More complex equations containing fractions will be solved by a different procedure in Section 8.5.

When solving an equation remember that our goal is to isolate the variable on one side of the equation.

To help visualize the boxed procedure, consider the figure and corresponding equation below.

Figure	Equation
+ + 4 = 10	$2x + 4 = 10$

The equation $2x + 4 = 10$ contains no parentheses and no like terms on the same side of the equal sign. Therefore, we start with step 3, using the addition property. Remember that the addition property allows us to add (or subtract) the same quantity to (or from) both sides of an equation without changing its solution. Here we subtract 4 from both sides of the equation to isolate the term containing the variable.

Figure	Equation	
+ + 4 = 10	$2x + 4 = 10$	
	$2x + 4 - 4 = 10 - 4$	*Addition property*
+ = 6	or $2x = 6$	

Notice how the term containing the variable, $2x$, is now by itself on one side of the equal sign. Now we use the multiplication property, step 4, to isolate the variable x. Remember that the multiplication property allows us to multiply or divide both sides of the equation by the same nonzero number without changing its solution. Here we divide both sides of the equation by 2, the coefficient of the term containing the variable, to obtain the solution, 3.

$$2x = 6$$

$$\frac{\overset{1}{\cancel{2}}x}{\underset{1}{\cancel{2}}} = \frac{\overset{3}{\cancel{6}}}{\underset{1}{\cancel{2}}} \quad \textit{Multiplication property}$$

$$x = 3$$

EXAMPLE 1 Solve the equation $2x - 5 = 9$.

Solution We will follow the procedure outlined for solving equations. Since the equation contains no parentheses and since there are no like terms to be combined, we start with step 3.

Step 3

$$2x - 5 = 9$$

$$2x - 5 + 5 = 9 + 5 \quad \textit{Add 5 to both sides.}$$

$$2x = 14$$

Step 4

$$\frac{2x}{2} = \frac{14}{2} \quad \textit{Divide both sides by 2.}$$

$$x = 7$$

Step 5 Check:

$$2x - 5 = 9$$

$$2(7) - 5 \stackrel{?}{=} 9$$

$$14 - 5 \stackrel{?}{=} 9$$

$$9 = 9 \quad \textit{True}$$

Since the check is true, the solution is 7. Note that after completing step 3 we obtain $2x = 14$, which is an equation of the form $ax = b$. After completing step 4, we obtain the answer in the form $x =$ some real number.

HELPFUL HINT

When solving an equation that does not contain fractions, **the addition property (step 3) is to be used before the multiplication property (step 4)**. If you use the multiplication property before the addition property, it is still possible to obtain the correct answer. However, you will usually have to do more work, and you may end up working with fractions. What would happen if you tried to solve Example 1 using the multiplication property before the addition property?

EXAMPLE 2 Solve the equation $-2x - 6 = -3$.

Solution

$$-2x - 6 = -3$$

Step 3

$$-2x - 6 + 6 = -3 + 6 \quad \textit{Add 6 to both sides.}$$

$$-2x = 3$$

Step 4

$$\frac{-2x}{-2} = \frac{3}{-2} \quad \textit{Divide both sides by } -2.$$

$$x = -\frac{3}{2}$$

Step 5 Check:

$$-2x - 6 = -3$$
$$-2\left(-\frac{3}{2}\right) - 6 \stackrel{?}{=} -3$$
$$3 - 6 \stackrel{?}{=} -3$$
$$-3 = -3 \quad \text{True}$$

NOW TRY EXERCISE 27

The solution is $-\frac{3}{2}$.

Note that checks are always made with the *original* equation. In some of the following examples, the check will be omitted to save space.

EXAMPLE 3 Solve the equation $16 = 4x + 6 - 2x$.

Solution Again we must isolate the variable x. Since the right side of the equation has two like terms containing the variable x, we will first combine these like terms.

Step 2
$$16 = 4x + 6 - 2x$$
$$16 = 2x + 6 \quad \text{Like terms were combined.}$$

Step 3
$$16 - 6 = 2x + 6 - 6 \quad \text{Subtract 6 from both sides.}$$
$$10 = 2x$$

Step 4
$$\frac{10}{2} = \frac{2x}{2} \quad \text{Divide both sides by 2.}$$
$$5 = x$$

The preceding solution can be condensed as follows.

$$16 = 4x + 6 - 2x$$
$$16 = 2x + 6 \quad \text{Like terms were combined.}$$
$$10 = 2x \quad \text{6 was subtracted from both sides.}$$
$$5 = x \quad \text{Both sides were divided by 2.}$$

In Chapter 3 we will be solving many equations that contain decimal numbers. To solve such equations, we follow the same procedure as outlined earlier. Example 4 illustrates the solution to an equation that contains decimal numbers.

EXAMPLE 4 Solve the equation $x + 1.24 - 0.07x = 4.96$.

Solution

$$x + 1.24 - 0.07x = 4.96$$
$$0.93x + 1.24 = 4.96 \quad \text{Like terms were combined, } 1x - 0.07x = 0.93x.$$
$$0.93x + 1.24 - 1.24 = 4.96 - 1.24 \quad \text{Subtract 1.24 from both sides.}$$
$$0.93x = 3.72$$
$$\frac{0.93x}{0.93} = \frac{3.72}{0.93} \quad \text{Divide both sides by 0.93.}$$
$$x = 4$$

NOW TRY EXERCISE 49

EXAMPLE 5 Solve the equation $2(x + 4) - 5x = -3$.

Solution

$$2(x + 4) - 5x = -3$$

$$2x + 8 - 5x = -3 \qquad \textit{Distributive property was used.}$$

$$-3x + 8 = -3 \qquad \textit{Like terms were combined.}$$

$$-3x + 8 - 8 = -3 - 8 \qquad \textit{Subtract 8 from both sides.}$$

$$-3x = -11$$

$$\frac{-3x}{-3} = \frac{-11}{-3} \qquad \textit{Divide both sides by } -3.$$

$$x = \frac{11}{3}$$

The solution to Example 5 can be condensed as follows:

$$2(x + 4) - 5x = -3$$

$$2x + 8 - 5x = -3 \qquad \textit{The distributive property was used.}$$

$$-3x + 8 = -3 \qquad \textit{Like terms were combined.}$$

$$-3x = -11 \qquad \textit{8 was subtracted from both sides.}$$

$$x = \frac{11}{3} \qquad \textit{Both sides were divided by } -3.$$

EXAMPLE 6 Solve the equation $2x - (x + 2) = 6$.

Solution

$$2x - (x + 2) = 6$$

$$2x - x - 2 = 6 \qquad \textit{The distributive property was used.}$$

$$x - 2 = 6 \qquad \textit{Like terms were combined.}$$

$$x = 8 \qquad \textit{2 was added to both sides.}$$

NOW TRY EXERCISE 75

HELPFUL HINT

Some of the most commonly used terms in algebra are "evaluate," "simplify," "solve," and "check." Make sure you understand what each term means and when each term is used.

Evaluate: To *evaluate an expression* means to find its numerical value.

Evaluate $16 \div 2^2 + 36 \div 4$

$= 16 \div 4 + 36 \div 4$

$= 4 + 36 \div 4$

$= 4 + 9$

$= 13$

Evaluate $-x^2 + 3x - 2$ when $x = 4$

$= -4^2 + 3(4) - 2$

$= -16 + 12 - 2$

$= -4 - 2$

$= -6$

continued on next page

Simplify: To *simplify an expression* means to perform the operations and combine like terms.

Simplify $3(x - 2) - 4(2x + 3)$

$$3(x - 2) - 4(2x + 3) = 3x - 6 - 8x - 12$$
$$= -5x - 18$$

Note that when you simplify an expression containing variables you do not generally end up with just a numerical value unless all the variable terms happen to add to zero.

Solve: To *solve an equation* means to find the value or the values of the variable that make the equation a true statement.

Solve $2x + 3(x + 1) = 18$

$$2x + 3x + 3 = 18$$
$$5x + 3 = 18$$
$$5x = 15$$
$$x = 3$$

Check: To *check the proposed solution to an equation*, substitute the value in the original equation. If this substitution results in a true statement, then the answer checks. For example, to check the solution to the equation just solved, we substitute 3 for x in the equation.

Check $2x + 3(x + 1) = 18$

$$2(3) + 3(3 + 1) \stackrel{?}{=} 18$$
$$6 + 3(4) \stackrel{?}{=} 18$$
$$6 + 12 \stackrel{?}{=} 18$$
$$18 = 18 \quad \text{True}$$

Since we obtained a true statement, the 3 checks.

It is important to realize that expressions may be evaluated or simplified (depending on the type of problem) and equations are solved and then checked.

Exercise Set 2.4

Concept/Writing Exercises

1. Does the equation $x + 3 = 2x + 5$ contain a variable on only one side of the equation? Explain.
2. Does the equation $2x - 4 = 3$ contain a variable on only one side of the equation? Explain.
3. If $1x = \frac{5}{8}$, what does x equal?
4. If $1x = -\frac{3}{5}$, what does x equal?
5. If $-x = \frac{1}{2}$ what does x equal?
6. If $-x = \frac{5}{9}$, what does x equal?
7. If $-x = -\frac{3}{5}$, what does x equal?
8. If $-x = -\frac{4}{9}$, what does x equal?
9. Do you evaluate or solve an equation? Explain.
10. Do you evaluate or solve an expression? Explain.
11. **a)** In your own words, write the general procedure for solving an equation where the variable appears on only one side of the equal sign.
    b) Refer to page 121 to see whether you omitted any steps.
12. When solving equations that do not contain fractions, do we normally use the addition or multiplication property first in the process of isolating the variable? Explain your answer.
13. **a)** Explain, in a step-by-step manner, how to solve the equation $2(3x + 4) = -4$.
    b) Solve the equation by following the steps you listed in part **a)**.
14. **a)** Explain, in a step-by-step manner, how to solve the equation $4x - 2(x + 3) = 4$.
    b) Solve the equation by following the steps you listed in part **a)**.

Practice the Skills

*In Exercises 15–22, **a)** Represent the figure as an equation with the variable x and **b)** solve the equation. Refer to pages 121–122 for an example.*

15. + + 4 = 16

16. + + + 8 = 20

17. 

18. 27 = + + + 9

19. + + 10 + = 4

20. 3 = + 6 + +

21. 5 + + + = 12

22. 9 = + + 12

Solve each equation. You may wish to use a calculator to solve equations containing decimal numbers.

23. $2x + 4 = 10$

24. $2x - 4 = 8$

25. $-2x - 5 = 7$

26. $-4x + 6 = 20$

27. $5x - 6 = 19$

28. $6 - 3x = 18$

29. $5x - 2 = 10$

30. $-9x - 3 = 15$

31. $-x + 4 = 15$

32. $10 = 2x + 6$

33. $12 - x = 9$

34. $-3x - 3 = -12$

35. $8 + 3x = 19$

36. $-2x + 7 = -10$

37. $16x + 5 = -14$

38. $19 = 25 + 4x$

39. $-4.2 = 3x + 25.8$

40. $-24 + 16x = -24$

41. $6x - 29 = 7$

42. $-x + 4 = -8$

43. $56 = -6x + 2$

44. $15 = 7x + 1$

45. $-2x - 7 = -13$

46. $-2 - x = -12$

47. $2.3x - 9.34 = 6.3$

48. $x + 0.05x = 21$

49. $x + 0.07x = 16.05$

50. $-2.7 = -1.3 + 0.7x$

51. $28.8 = x + 1.40x$

52. $8.40 = 2.45x - 1.05x$

53. $3(x + 2) = 6$

54. $3(x - 2) = 12$

55. $4(3 - x) = 12$

56. $-2(x - 3) = 26$

57. $-4 = -(x + 5)$

58. $-3(2 - 3x) = 9$

59. $12 = 4(x - 3)$

60. $-2(x + 8) - 5 = 1$

61. $22 = -(3x - 4)$

62. $-2 = 5(3x + 1) - 12x$

63. $2x + 3(x + 2) = 11$

64. $4 = -2(x + 3)$

65. $x - 3(2x + 3) = 11$

66. $3(4 - x) + 5x = 9$

67. $5x + 3x - 4x - 7 = 9$

68. $4(x + 2) = 13$

69. $0.7(x - 3) = 1.4$

70. $21 + (x - 9) = 24$

71. $1.4(5x - 4) = -1.4$

72. $0.1(2.4x + 5) = 1.7$

73. $3 - 2(x + 3) + 2 = 1$

74. $2(3x - 4) - 4x = 12$

75. $1 + (x + 3) + 6x = 6$

76. $5x - 2x + 7x = -81$

77. $4.85 - 6.4x + 1.11 = 22.6$

78. $5.76 - 4.24x - 1.9x = 27.864$

Problem Solving

79. a) Explain why it is easier to solve the equation $3x + 2 = 11$ by first subtracting 2 from both sides of the equation rather than by first dividing both sides of the equation by 3.

b) Solve the equation.

80. a) Explain why it is easier to solve the equation $5x - 3 = 12$ by first adding 3 to both sides of the equation rather than by first dividing both sides of the equation by 5.

b) Solve the equation.

*In Section 2.5 we will give a step-by-step procedure to solve equations in which the variable appears on both sides of the equation. Let's see if you can solve such equations now. In Exercises 81–84, **a)** Express the figure as an equation. **b)** Solve the equation by first using the addition property to get all terms containing the variable on one side of the equation and all constant terms on the other side of the equation. Then use the multiplication property.*

81. + = + 3

82. + + 6 = + + + 4

83. + + 3 = + + + + 2

84. + + + 1 = + 4

Challenge Problems

Solve each equation.

85. $3(x - 2) - (x + 5) - 2(3 - 2x) = 18$

86. $-6 = -(x - 5) - 3(5 + 2x) - 4(2x - 4)$

87. $4[3 - 2(x + 4)] - (x + 3) = 13$

88. To solve the equation □☺ − △ = ∗ for ☺, we need to isolate ☺. To isolate the *term* containing ☺, we add △ to both sides of the equation. Then, to isolate ☺, we divide both sides of the equation by □. Solve this equation for ☺.

Group Activity

In Chapter 3 we will discuss procedures for writing application problems as equations. Let's look at an application now.

John Logan purchased 2 large chocolate kisses and a birthday card. The birthday card cost \$3. The total cost was \$9. What was the price of a single chocolate kiss?

This problem can be represented by the following balance.

△+△+ 3 = 9

Using this balance we can obtain the equation $2x + 3 = 9$, which can be used to solve the problem. Solving the equation we find that x, the price of a single kiss, is \$3.

For Exercises 89 and 90, each group member should do parts ***a)–c)****. Then do part* ***d)*** *as a group.*

a) *Represent the problem by a balance.*

b) *Obtain an equation that can be used to solve the problem.*

c) *Solve the equation and answer the question.*

d) *Compare and check each other's work.*

89. Eduardo Verner purchased three boxes of stationery. He also purchased wrapping paper and thank-you cards. If the wrapping paper and thank-you cards together cost \$6, and the total he paid was \$42, find the cost of a box of stationery.

90. Mahandi Ison purchased three rolls of peppermint candies and the local newspaper. The newspaper cost 50 cents. He paid \$2.75 in all. What did a roll of candies cost?

Cumulative Review Exercises

[1.3] **91.** Add $\frac{5}{8} + \frac{3}{5}$.

[1.9] **92.** Evaluate $[5(2 - 6) + 3(8 \div 4)^2]^2$.

[2.2] **93.** To solve an equation, what do you need to do to the variable?

[2.3] **94.** To solve the equation $7 = -4x$, would you add 4 to both sides of the equation or divide both sides of the equation by -4? Explain your answer.

2.5 SOLVING LINEAR EQUATIONS WITH THE VARIABLE ON BOTH SIDES OF THE EQUATION

SSM VIDEO 2.5 CD Rom

1) **Solve equations with the variable on both sides of the equal sign.**

2) **Identify identities and contradictions.**

1) Solve Equations With the Variable on Both Sides of the Equal Sign

Below we show two balances, the equations that represent the balances, and the solutions to the equations. In each case the variable appears on both sides of the equation. At this time you probably cannot determine the solutions. Do not worry about this. In this section we will teach you the procedure for solving equations of this type.

Figure	Equation	Solution
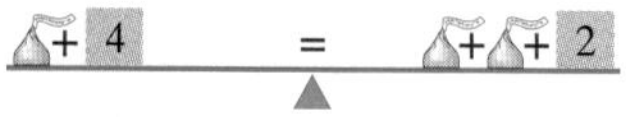 x + 4 = x + x + 2	$x + 4 = 2x + 2$	2
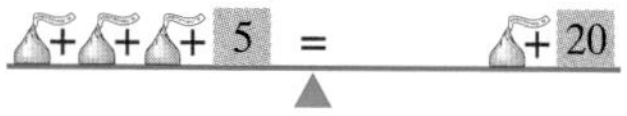 x + x + x + 5 = x + 20	$3x + 5 = x + 20$	$\frac{15}{2}$

Following is a general procedure, similar to the one outlined in Section 2.4, to solve linear equations with the variable on both sides of the equal sign.

To Solve Linear Equations with the Variable on Both Sides of the Equal Sign

1. Use the distributive property to remove parentheses.
2. Combine like terms on the same side of the equal sign.
3. Use the addition property to rewrite the equation with all terms containing the variable on one side of the equal sign and all terms not containing the variable on the other side of the equal sign. It may be necessary to use the addition property twice to accomplish this goal. You will eventually get an equation of the form $ax = b$.
4. Use the multiplication property to isolate the variable. This will give a solution of the form $x =$ some number.
5. Check the solution in the original equation.

The steps listed here are basically the same as the steps listed in the boxed procedure on page 121, except that in step 3 you may need to use the addition property more than once to obtain an equation of the form $ax = b$.

Remember that our goal in solving an equation is to isolate the variable, that is, to get the variable alone on one side of the equation. To help visualize the boxed procedure, consider the figure and corresponding equation that follow.

Figure	Equation
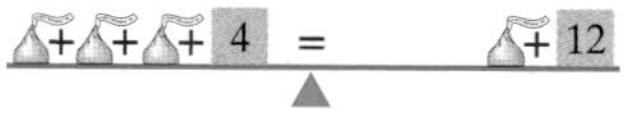 x + x + x + 4 = x + 12	$3x + 4 = x + 12$

The equation $3x + 4 = x + 12$ contains no parentheses and no like terms on the same side of the equal sign. Therefore, we start with step 3, the addition property. We will use the addition property twice in order to obtain an equation where the variable appears on only one side of the equal sign. We begin by subtracting x from both sides of the equation to get all the terms containing the variable on the left side of the equation. This will give the following:

Figure	Equation	
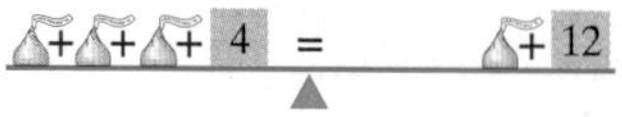 x + x + x + 4 = x + 12	$3x + 4 = x + 12$	
	$3x - x + 4 = x - x + 12$	*Addition property*
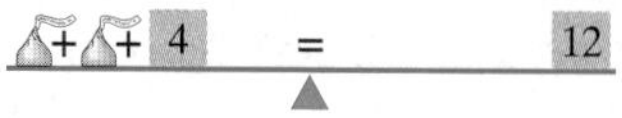 x + x + 4 = 12	or $2x + 4 = 12$	

Notice that the variable, x, now appears on only one side of the equation. However, +4 still appears on the same side of the equal sign as the $2x$. We use the addition property a second time to get the term containing the variable by itself on one side of the equation. Subtracting 4 from both sides of the equation gives $2x = 8$, which is an equation of the form $ax = b$.

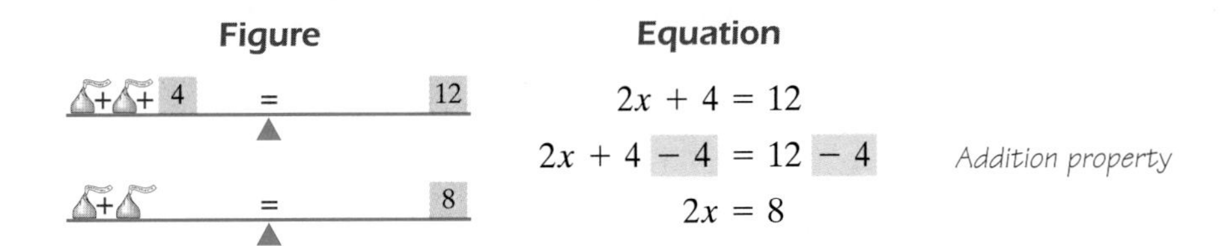

$$2x + 4 = 12$$
$$2x + 4 - 4 = 12 - 4 \quad \textit{Addition property}$$
$$2x = 8$$

Now that the $2x$ is by itself on one side of the equation, we can use the multiplication property to isolate the variable and solve the equation for x. We divide both sides of the equation by 2 to isolate the variable and solve the equation.

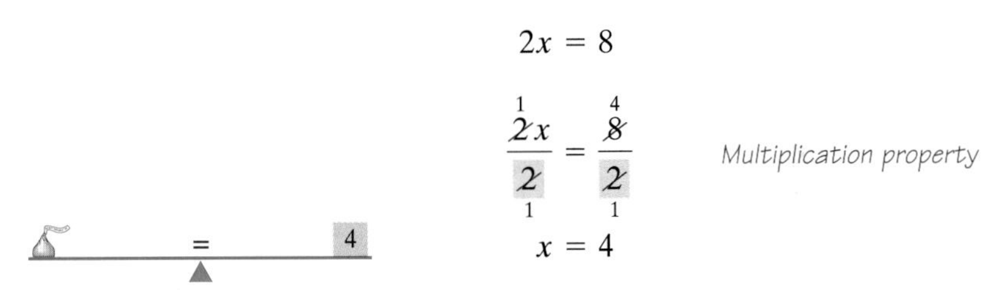

$$2x = 8$$
$$\frac{\overset{1}{\cancel{2}}x}{\underset{1}{\cancel{2}}} = \frac{\overset{4}{\cancel{8}}}{\underset{1}{\cancel{2}}} \quad \textit{Multiplication property}$$
$$x = 4$$

The solution to the equation is 4.

EXAMPLE 1 Solve the equation $4x + 6 = 2x + 4$.

Solution Our goal is to get all terms with the variable on one side of the equal sign and all terms without the variable on the other side. The terms with the variable may be collected on either side of the equal sign. Many methods can be used to isolate the variable. We will illustrate two. In method 1, we will isolate the variable on the left side of the equation. In method 2, we will isolate the variable on the right side of the equation. In both methods, we will follow the steps given in the box on page 128. Since this equation does not contain parentheses and there are no like terms on the same side of the equal sign, we begin with step 3.

Method 1: Isolating the variable on the left

$$4x + 6 = 2x + 4$$

Step 3 $\quad 4x - 2x + 6 = 2x - 2x + 4 \quad$ *Subtract 2x from both sides.*

$$2x + 6 = 4$$

Step 3 $\quad 2x + 6 - 6 = 4 - 6 \quad$ *Subtract 6 from both sides.*

$$2x = -2$$

Step 4 $\quad \dfrac{2x}{2} = \dfrac{-2}{2} \quad$ *Divide both sides by 2.*

$$x = -1$$

Method 2: Isolating the variable on the right

$$4x + 6 = 2x + 4$$

Step 3 $$4x - 4x + 6 = 2x - 4x + 4$$ *Subtract 4x from both sides.*

$$6 = -2x + 4$$

Step 3 $$6 - 4 = -2x + 4 - 4$$ *Subtract 4 from both sides.*

$$2 = -2x$$

Step 4 $$\frac{2}{-2} = \frac{-2x}{-2}$$ *Divide both sides by −2.*

$$-1 = x$$

The same answer is obtained whether we isolate the variable on the left or right. However, we need to divide both sides of the equation by a negative number in method 2.

Step 5 Check:

$$4x + 6 = 2x + 4$$

$$4(-1) + 6 \stackrel{?}{=} 2(-1) + 4$$

$$-4 + 6 \stackrel{?}{=} -2 + 4$$

$$2 = 2 \quad \text{True}$$

NOW TRY EXERCISE 27

EXAMPLE 2 Solve the equation $2x - 3 - 5x = 13 + 4x - 2$.

Solution We will choose to collect the terms containing the variable on the right side of the equation in order to create a positive coefficient of x. Since there are like terms *on the same side of the equal sign*, we will begin by combining these like terms.

Step 2 $$2x - 3 - 5x = 13 + 4x - 2$$

$$-3x - 3 = 4x + 11$$ *Like terms were combined.*

Step 3 $$-3x + 3x - 3 = 4x + 3x + 11$$ *Add 3x to both sides.*

$$-3 = 7x + 11$$

Step 3 $$-3 - 11 = 7x + 11 - 11$$ *Subtract 11 from both sides.*

$$-14 = 7x$$

Step 4 $$\frac{-14}{7} = \frac{7x}{7}$$ *Divide both sides by 7.*

$$-2 = x$$

Step 5 Check:

$$2x - 3 - 5x = 13 + 4x - 2$$

$$2(-2) - 3 - 5(-2) \stackrel{?}{=} 13 + 4(-2) - 2$$

$$-4 - 3 + 10 \stackrel{?}{=} 13 - 8 - 2$$

$$-7 + 10 \stackrel{?}{=} 5 - 2$$

$$3 = 3 \quad \text{True}$$

Since the check is true, the solution is −2.

The solution to Example 2 could be condensed as follows:

$$2x - 3 - 5x = 13 + 4x - 2$$

$$-3x - 3 = 4x + 11 \quad \textit{Like terms were combined.}$$

$$-3 = 7x + 11 \quad \textit{3x was added to both sides.}$$

$$-14 = 7x \quad \textit{11 was subtracted from both sides.}$$

$$-2 = x \quad \textit{Both sides were divided by 7.}$$

We solved Example 2 by moving the terms containing the variable to the right side of the equation. Now rework the problem by moving the terms containing the variable to the left side of the equation. You should obtain the same answer.

EXAMPLE 3 Solve the equation $5.74x + 5.42 = 2.24x - 9.28$.

Solution We first notice that there are no like terms on the same side of the equal sign that can be combined. We will elect to collect the terms containing the variable on the left side of the equation.

$$5.74x + 5.42 = 2.24x - 9.28$$

Step 3 $$5.74x - 2.24x + 5.42 = 2.24x - 2.24x - 9.28 \quad \textit{Subtract 2.24x from both sides.}$$

$$3.5x + 5.42 = -9.28$$

Step 3 $$3.5x + 5.42 - 5.42 = -9.28 - 5.42 \quad \textit{Subtract 5.42 from both sides.}$$

$$3.5x = -14.7$$

Step 4 $$\frac{3.5x}{3.5} = \frac{-14.7}{3.5} \quad \textit{Divide both sides by 3.5.}$$

$$x = -4.2$$

EXAMPLE 4 Solve the equation $2(p + 3) = -3p + 10$.

Solution

$$2(p + 3) = -3p + 10$$

Step 1 $$2p + 6 = -3p + 10 \quad \textit{Distributive property was used.}$$

Step 3 $$2p + 3p + 6 = -3p + 3p + 10 \quad \textit{Add 3p to both sides.}$$

$$5p + 6 = 10$$

Step 3 $$5p + 6 - 6 = 10 - 6 \quad \textit{Subtract 6 from both sides.}$$

$$5p = 4$$

Step 4 $$\frac{5p}{5} = \frac{4}{5} \quad \textit{Divide both sides by 5.}$$

$$p = \frac{4}{5}$$

The solution to Example 4 could be condensed as follows:

$$2(p + 3) = -3p + 10$$

$$2p + 6 = -3p + 10 \quad \textit{Distributive property was used.}$$

$$5p + 6 = 10 \quad \textit{3p was added to both sides.}$$

$$5p = 4 \quad \textit{6 was subtracted from both sides.}$$

$$p = \frac{4}{5} \quad \textit{Both sides were divided by 5.}$$

NOW TRY EXERCISE 33

HELPFUL HINT

After the distributive property was used in step 1 in Example 4, we obtained the equation $2p + 6 = -3p + 10$. Then we had to decide whether to collect terms with the variable on the left or the right side of the equal sign. If we wish the sum of the terms containing a variable to be positive, we use the addition property to eliminate the variable, with the *smaller* numerical coefficient from one side of the equation. Since -3 is smaller than 2, we added $3p$ to both sides of the equation. This eliminated $-3p$ from the right side of the equation and resulted in the sum of the variable terms on the left side of the equation, $5p$, being positive.

EXAMPLE 5 Solve the equation $2(x - 5) + 3 = 3x + 9$.

Solution

$$2(x - 5) + 3 = 3x + 9$$

Step 1 $2x - 10 + 3 = 3x + 9$ *Distributive property was used.*

Step 2 $2x - 7 = 3x + 9$ *Like terms were combined.*

Step 3 $-7 = x + 9$ *2x was subtracted from both sides.*

Step 3 $-16 = x$ *9 was subtracted from both sides.*

EXAMPLE 6 Solve the equation $7 - 2x + 5x = -2(-3x + 4)$.

Solution

$$7 - 2x + 5x = -2(-3x + 4)$$

Step 1 $7 - 2x + 5x = 6x - 8$ *Distributive property was used.*

Step 2 $7 + 3x = 6x - 8$ *Like terms were combined.*

Step 3 $7 = 3x - 8$ *3x was subtracted from both sides.*

Step 3 $15 = 3x$ *8 was added to both sides.*

Step 4 $5 = x$ *Both sides were divided by 3.*

NOW TRY EXERCISE 57 The solution is 5.

Identify Identities and Contradictions

Thus far all the equations we have solved have had a single value for a solution. Equations of this type are called **conditional equations**, for they are only true under specific conditions. Some equations, as in Example 7, are true for all values of x. Equations that are true for all values of x are called **identities**. A third type of equation, as in Example 8, has no solution and is called a **contradiction**.

EXAMPLE 7 Solve the equation $2x + 6 = 2(x + 3)$.

Solution

$$2x + 6 = 2(x + 3)$$
$$2x + 6 = 2x + 6$$

Since the same expression appears on both sides of the equal sign, the statement is true for all values of x. If we continue to solve this equation further, we might obtain

$2x = 2x$ *6 was subtracted from both sides.*

$0 = 0$ *2x was subtracted from both sides.*

Note: The solution process could have been stopped at $2x + 6 = 2x + 6$. Since one side is identical to the other side, the equation is true for all values of x. *Therefore, the solution to this equation is all real numbers.*

EXAMPLE 8 Solve the equation $-3x + 4 + 5x = 4x - 2x + 5$.

Solution

$$-3x + 4 + 5x = 4x - 2x + 5$$

$$2x + 4 = 2x + 5 \quad \textit{Like terms were combined.}$$

$$2x - 2x + 4 = 2x - 2x + 5 \quad \textit{Subtract 2x from both sides.}$$

$$4 = 5 \quad \textit{False}$$

When solving an equation, if you obtain an obviously false statement, as in this example, the equation has no solution. No value of x will make the equation a true statement. **Therefore, when giving the answer to this problem, you should use the words "*no solution.*"** An answer left blank may be marked wrong.

NOW TRY EXERCISE 37

HELPFUL HINT

Some students start solving equations correctly but do not complete the solution. Sometimes they are not sure that what they are doing is correct and they give up for lack of confidence. You must have confidence in yourself. As long as you follow the procedure on page 128, you should obtain the correct solution even if it takes quite a few steps. Remember two important things: (1) your goal is to isolate the variable, and (2) whatever you do to one side of the equation you must also do to the other side. That is, you must treat both sides of the equation equally.

Exercise Set 2.5

Concept/Writing Exercises

1. **a)** In your own words, write the general procedure for solving an equation that does not contain fractions where the variable appears on both sides of the equation.
   b) Refer to page 128 to see whether you omitted any steps.
2. What is a conditional equation?
3. **a)** What is an identity?
   b) What is the solution to an identity?
4. When solving an equation, how will you know if the equation is an identity?
5. Explain why the equation $x + 5 = x + 5$ must be an identity?
6. **a)** What is a contradiction?
   b) What is the solution to a contradiction?
7. When solving an equation, how will you know if the equation has no solution?
8. Explain why the equation $x + 5 = x + 4$ must be a contradiction.
9. **a)** Explain, in a step-by-step manner, how to solve the equation $4(x + 3) = 6(x - 5)$.
   b) Solve the equation by following the steps you listed in part **a)**.
10. **a)** Explain, in a step-by-step manner, how to solve the equation $4x + 3(x + 2) = 5x - 10$.
    b) Solve the equation by following the steps you listed in part **a)**.

Practice the Skills

*In exercises 11–18, **a)** represent the figure as an equation with the variable x, and **b)** solve the equation. Refer to pages 128–129 for an example.*

11.  + = + 6
12.  + 6 = + + 2
13.  5 + + = + 19
14. + + = + 14 
15. 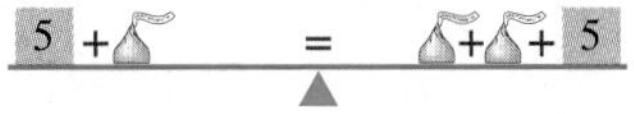 5 + = + + 5
16. 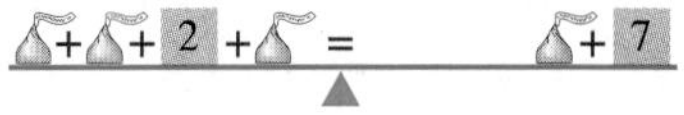 + + 2 + = + 7
17.  + + 8 = + 4
18. 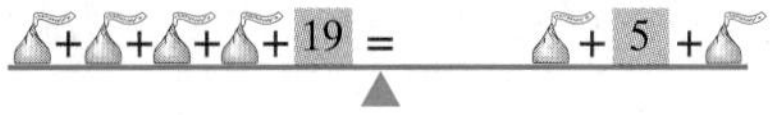 + + + + 19 = + 5 +

Solve each equation. You may wish to use a calculator to solve the equations containing decimal numbers.

19. $8x = 6x + 30$

20. $x + 4 = 2x - 7$

21. $-4x + 10 = 6x$

22. $6x = 4x + 8$

23. $5x + 3 = 6$

24. $-6x = 2x + 16$

25. $15 - 5x = 3x - 2x$

26. $8 - 3x = 4x + 50$

27. $2x - 4 = 3x - 6$

28. $-5x = -4x + 9$

29. $3 - 2y = 9 - 8y$

30. $124.8 - 9.4x = 4.8x + 32.5$

31. $9 - 0.5x = 4.5x + 8.50$

32. $8 + y = 2y - 6 + y$

33. $5x + 3 = 2(x + 6)$

34. $x - 14 = 3(x + 2)$

35. $x - 25 = 12x + 9 + 3x$

36. $5y + 6 = 2y + 3 - y$

37. $2(x - 2) = 4x - 6 - 2x$

38. $4r = 10 - 2(r - 4)$

39. $-(w + 2) = -6w + 32$

40. $12(4 + x) = 3(10 - 6x)$

41. $4 - (2x - 5) = 3x + 13$

42. $4(2x - 3) = -2(3x + 16)$

43. $0.1(x + 10) = 0.3x - 4$

44. $3(y - 1) + 9 = 8y + 6 - 5y$

45. $2(x + 4) = 4x + 3 - 2x + 5$

46. $5(3.2x - 3) = 2(x - 4)$

47. $9(-y - 3) = 6y - 15 + 3y + 6$

48. $-4(-y + 3) = 12y + 8 - 2y$

49. $-(3 - p) = -(2p + 3)$

50. $12 - 2x - 3(x + 2) = 4x + 6 - x$

51. $-(x + 4) + 5 = 4x + 1 - 5x$

52. $18x + 3(4x - 9) = -6x + 81$

53. $35(2x - 1) = 7(x + 4) + 3x$

54. $10(x - 10) + 5 = 5(2x - 20)$

55. $0.4(x + 0.7) = 0.6(x - 4.2)$

56. $3(x - 4) = 2(x - 8) + 5x$

57. $-(x - 5) + 2 = 3(4 - x) + 5x$

58. $0.5(6x - 8) = 1.4(x - 5) - 0.2$

59. $2(x - 6) + (x + 1) = x - 11$

60. $-2(-3x + 5) + 6 = 4(x - 2)$

61. $5 + 2x = 6(x + 1) - 5(x - 3)$

62. $4 - (6x + 6) = -(-2x + 10)$

63. $5 - (x - 5) = 2(x + 3) - 6(x + 1)$

64. $12 - 6x + 3(2x + 3) = 2x + 5$

Problem Solving

65. **a)** Construct a *conditional equation* containing three terms on the left side of the equal sign and two terms on the right side of the equal sign.
b) Explain how you know your answer to part **a)** is a conditional equation.
c) Solve the equation.

66. **a)** Construct a *conditional equation* containing two terms on the left side of the equal sign and three terms on the right side of the equal sign.
b) Explain how you know your answer to part **a)** is a conditional equation.
c) Solve the equation.

67. **a)** Construct an *identity* containing three terms on the left side of the equal sign and two terms on the right side of the equal sign.
b) Explain how you know your answer to part **a)** is an identity.
c) What is the solution to the equation?

68. **a)** Construct an *identity* containing two terms on the left side of the equal sign and three terms on the right side of the equal sign.
b) Explain how you know your answer to part **a)** is an identity.
c) What is the solution to the equation?

69. **a)** Construct a *contradiction* containing three terms on the left side of the equal sign and two terms on the right side of the equal sign.
b) Explain how you know your answer to part **a)** is a contradiction.
c) What is the solution to the equation?

70. **a)** Construct a *contradiction* containing three terms on the left side of the equal sign and four terms on the right side of the equal sign.
b) Explain how you know your answer to part **a)** is a contradiction.
c) What is the solution to the equation?

Challenge Problems

71. Solve the equation $5* - 1 = 4* + 5$ for $*$.

72. Solve the equation $2\triangle - 4 = 3\triangle + 5 - \triangle$ for $\triangle$.

73. Solve the equation 3☺ − 5 = 2☺ − 5 + ☺ for ☺.

74. Solve $-2(x + 3) + 5x = 3(4 - 2x) - (x + 2)$.

75. Solve $4 - [5 - 3(x + 2)] = x - 3$.

Group Activity

Discuss and answer Exercises 76–78 as a group. In the next chapter we will be discussing procedures for writing application problems as equations. Let's get some practice now.

76. Consider the following word problem.
Mary Kay purchased two large chocolate kisses. The total cost of the two kisses was equal to the cost of one kiss plus \$6. Find the cost of one chocolate kiss.

a) Each group member: Make a sketch using a balance like those in Exercises 11–18 to represent the problem.

b) Each group member: Represent your sketch as an equation with the variable x.

c) As a group, compare your answers to parts **a)** and **b)** and reach agreement on a sketch and equation to represent the problem.

d) Each group member: Solve the equation. Then compare your answers. If your answers disagree, determine the correct answer.

e) As a group, check your answer to make sure that it makes sense.

Repeat parts ***a)–e)*** *of Exercise 76 for Exercises 77 and 78.*

77. Three identical boxes are weighed. Their total weight is the same as (or equals) the weight of one of the boxes plus 20 pounds. Find the weight of a box.

78. Isaac Marcus purchased 4 gallons of skim milk. The price of the 4 gallons of milk is the same as the price of 2 gallons of milk plus some other groceries that cost \$5.20. What is the price of a gallon of skim milk?

Cumulative Review Exercises

[1.9] **79.** Evaluate $\left(\frac{2}{3}\right)^5$ on your calculator.

[2.1] **80.** Explain the difference between factors and terms.

[2.1] **81.** Simplify $2(x - 3) + 4x - (4 - x)$.

[2.4] **82.** Solve $2(x - 3) + 4x - (4 - x) = 0$.

83. Solve $(x + 4) - (4x - 3) = 16$.

2.6 RATIOS AND PROPORTIONS

SSM VIDEO 2.6 CD Rom

1) Understand ratios.
2) Solve proportions using cross-multiplication.
3) Solve applications.
4) Use proportions to change units.
5) Use proportions to solve problems involving similar figures.

1) Understand Ratios

A **ratio** is a quotient of two quantities. Ratios provide a way to compare two numbers or quantities. The ratio of the number a to the number b may be written

$$a \text{ to } b, \qquad a\!:\!b, \qquad \text{or} \qquad \frac{a}{b}$$

where a and b are called the **terms of the ratio**.

EXAMPLE 1 In a restaurant one evening, a survey was taken. It showed that 20 people ordered the corn beef and cabbage special and 12 people ordered the salmon special.

a) Find the ratio of those who ordered the corn beef and cabbage special to those who ordered the salmon special.

b) Find the ratio of those who ordered the salmon special to all those who ordered one of the specials.

Solution We will use our five-step problem-solving procedure.

a) Understand and Translate The ratio we are seeking is

number who ordered corn beef and cabbage : number who ordered salmon

Carry Out We substitute the appropriate values into the ratio.

$$20:12$$

Now we simplify by dividing each number in the ratio by 4, the greatest number that divides both terms in the ratio. This gives

$$5:3$$

Check and Answer Our division is correct. The ratio is 5:3.

b) We use the same procedure as in part **a)**. Twelve people ordered the salmon special. There were 20 + 12 or 32 people that ordered a special. Thus, the ratio is 12:32, which simplifies to 3:8.

In Example 1, part **a)** could also have been written $\frac{5}{3}$ or 5 to 3. Part **b)** could also have been written $\frac{3}{8}$ or 3 to 8.

NOW TRY EXERCISE 17

EXAMPLE 2 There are two types of cholesterol: low-density lipoprotein, (LDL—considered the harmful type of cholesterol) and high-density lipoprotein (HDL—considered the healthful type of cholesterol). Some doctors recommend that the ratio of low- to high-density cholesterol be less than or equal to 4:1. Mr. Kane's cholesterol test showed that his low-density cholesterol measured 167 milligrams per deciliter, and his high-density cholesterol measured 40 milligrams per deciliter. Is Mr. Kane's ratio of low- to high-density cholesterol less than or equal to the recommended 4:1 ratio?

Solution **Understand** We need to determine if Mr. Kane's low- to high-density cholesterol is less than or equal to 4:1.

Translate Mr. Kane's low- to high-density cholesterol is 167:40. To make the second term equal to 1, we divide both terms in the ratio by the second term, 40.

Carry Out

$$\frac{167}{40}:\frac{40}{40}$$

$$\text{or}\quad 4.175:1$$

Check and Answer Our division is correct. Therefore, Mr. Kane's ratio is not less than or equal to the desired 4:1 ratio.

EXAMPLE 3 Some power equipment, such as chainsaws and blowers, use a gas–oil mixture to run the engine. The instructions on a particular chainsaw indicate that 5 gallons of gasoline should be mixed with 40 ounces of special oil to obtain the proper gas–oil mixture. Find the ratio of gasoline to oil in the proper mixture.

Solution **Understand** To express these quantities in a ratio, both quantities must be in the same units. We can either convert 5 gallons to ounces or 40 ounces to gallons.

Translate Let's change 5 gallons to ounces. Since there are 128 ounces in 1 gallon, 5 gallons of gas equals 5(128) or 640 ounces. The ratio we are seeking is

ounces of gasoline : ounces of oil

Carry Out

$$640:40$$

$$\text{or}\quad 16:1$$ *Divide both terms by 40 to simplify.*

Check and Answer Our simplification is correct. The correct ratio of gas to oil for this chainsaw is 16:1.

EXAMPLE 4 The *gear ratio* of two gears is defined as

$$\text{gear ratio} = \frac{\text{number of teeth on the driving gear}}{\text{number of teeth on the driven gear}}$$

Find the gear ratio of the gears shown in Figure 2.10.

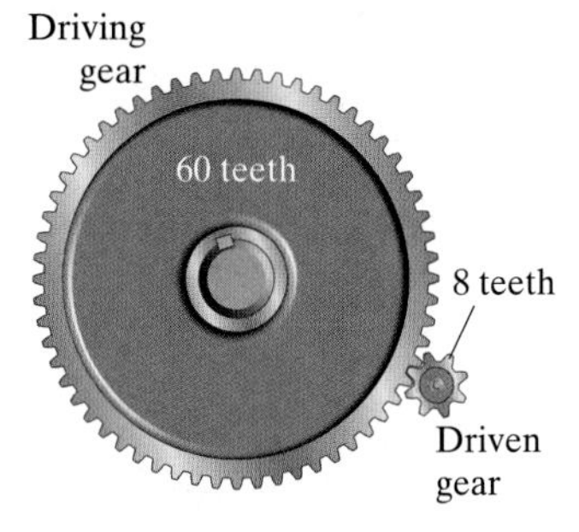

FIGURE 2.10

Solution Understand and Translate To find the gear ratio we need to substitute the appropriate values.

Carry Out $\text{gear ratio} = \frac{\text{number of teeth on driving gear}}{\text{number of teeth on driven gear}} = \frac{60}{8} = \frac{15}{2}$

Thus, the gear ratio is 15:2. Gear ratios are generally given as some quantity to 1. If we divide both terms of the ratio by the second term, we will obtain a ratio of some number to 1. Dividing both 15 and 2 by 2 gives a gear ratio of 7.5:1.

Check and Answer The gear ratio is 7.5:1. (A typical first gear ratio on a passenger car may be 3.545:1).

NOW TRY EXERCISE 27

2) Solve Proportions Using Cross-Multiplication

A **proportion** is a special type of equation. It is a statement of equality between two ratios. One way of denoting a proportion is $a:b = c:d$, which is read "a is to b as c is to d." In this text we write proportions as

$$\frac{a}{b} = \frac{c}{d}$$

The a and d are referred to as the **extremes**, and the b and c are referred to as the **means** of the proportion. One method that can be used in evaluating proportions is **cross-multiplication**.

Cross-Multiplication

If $\frac{a}{b} = \frac{c}{d}$, then $ad = bc$.

Note that *the product of the means is equal to the product of the extremes.*

If any three of the four quantities of a proportion are known, the fourth quantity can easily be found.

EXAMPLE 5 Solve $\frac{x}{3} = \frac{35}{15}$ for x by cross-multiplying.

Solution

$$\frac{x}{3} = \frac{35}{15}$$
$$x \cdot 15 = 3 \cdot 35$$
$$15x = 105$$
$$x = \frac{105}{15} = 7$$

Check:

$$\frac{x}{3} = \frac{35}{15}$$
$$\frac{7}{3} \stackrel{?}{=} \frac{35}{15}$$
$$\frac{7}{3} = \frac{7}{3} \quad \text{True}$$

EXAMPLE 6 Solve $\frac{-8}{3} = \frac{64}{x}$ for x by cross-multiplying.

Solution

$$\frac{-8}{3} = \frac{64}{x}$$
$$-8 \cdot x = 3 \cdot 64$$
$$-8x = 192$$
$$\frac{-8x}{-8} = \frac{192}{-8}$$
$$x = -24$$

Check:
$$\frac{-8}{3} = \frac{64}{x}$$
$$\frac{-8}{3} \stackrel{?}{=} \frac{64}{-24}$$
$$\frac{-8}{3} \stackrel{?}{=} \frac{8}{-3}$$
$$\frac{-8}{3} = \frac{-8}{3} \quad \textit{True}$$

NOW TRY EXERCISE 41

Solve Applications

Often, practical problems can be solved using proportions. To solve such problems, use the five-step problem-solving procedure we have been using throughout the book. Below we give that procedure with more specific directions for translating problems into proportions.

To Solve Problems Using Proportions

1. Understand the problem.
2. Translate the problem into mathematical language.
   a. First, represent the unknown quantity by a variable (a letter).
   b. Second, set up the proportion by listing the given ratio on the left side of the equal sign, and the unknown and the other given quantity on the right side of the equal sign. When setting up the right side of the proportion, the same respective quantities should occupy the same respective positions on the left and the right. For example, an acceptable proportion might be

$$\text{Given ratio} \left\{ \frac{\text{miles}}{\text{hour}} \right. = \frac{\text{miles}}{\text{hour}}$$

3. Carry out the mathematical calculations necessary to solve the problem.
   a. Once the proportion is correctly written, drop the units and cross-multiply.
   b. Solve the resulting equation.
4. Check the answer obtained in step 3.
5. Make sure you have answered the question.

Note that the two ratios* must have the same units. For example, if one ratio is given in miles/hour and the second ratio is given in feet/hour, one of the ratios must be changed before setting up the proportion.

---

*Strictly speaking, a quotient of two quantities with different units, such as $\frac{6\ miles}{1\ hour}$, is called a *rate*. However, few books make the distinction between ratios and rates when discussing proportions.

EXAMPLE 7 A 30-pound bag of fertilizer will cover an area of 2500 square feet.

a) How many pounds are needed to cover an area of 16,000 square feet?

b) How many bags of fertilizer are needed?

Solution **a) Understand** The given ratio is 30 pounds per 2500 square feet. The unknown quantity is the number of pounds necessary to cover 16,000 square feet.

Translate Let x = number of pounds.

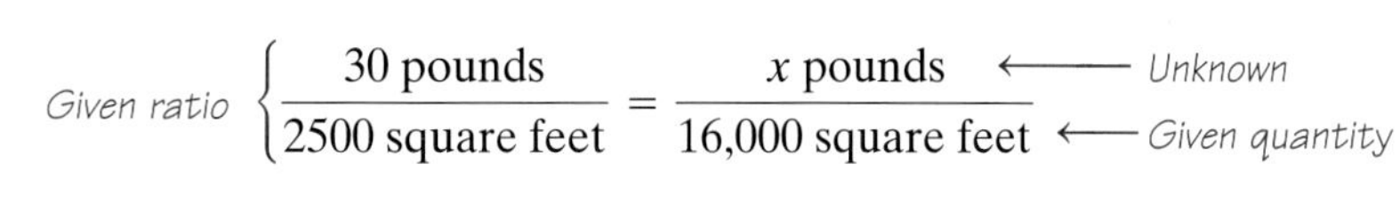

$$\text{Given ratio} \begin{cases} \dfrac{30 \text{ pounds}}{2500 \text{ square feet}} \end{cases} = \frac{x \text{ pounds}}{16{,}000 \text{ square feet}} \quad \begin{matrix} \longleftarrow \textit{Unknown} \\ \longleftarrow \textit{Given quantity} \end{matrix}$$

Note how the weight and the area are given in the same relative positions.

Carry Out

$$\frac{30}{2500} = \frac{x}{16{,}000}$$

$$30(16{,}000) = 2500x \qquad \textit{Cross-multiply.}$$

$$480{,}000 = 2500x \qquad \textit{Solve.}$$

$$\frac{480{,}000}{2500} = x$$

$$192 = x$$

Check Using a calculator, we determine that both ratios in the proportion, 30/2500 and 192/16,000, have a value of 0.012. Thus, the answer 192 pounds, checks.

Answer The amount of fertilizer needed to cover an area of 16,000 square feet is 192 pounds.

b) Since each bag weighs 30 pounds, the number of bags is found by division.

$$192 \div 30 = 6.4 \text{ bags}$$

NOW TRY EXERCISE 65 The number of bags needed is therefore 7, since one must purchase whole bags.

EXAMPLE 8 Many self-employed people pay estimated quarterly taxes. To do so, they need to estimate their annual income early in a given year. The Daniels own a bicycle sales and repair business. During the first 80 days of the year, their net income (or profit) was \$7280. Assuming that their income continues at the same rate throughout the year, estimate the Daniels' annual income.

Solution **Understand** The unknown quantity is the annual income. We know there are 365 days in a year. We will set up a proportion to estimate their annual income.

Translate We will let x represent the annual income.

$$\text{Given ratio} \begin{cases} \dfrac{\text{income}}{80 \text{ days}} \end{cases} = \frac{\text{annual income}}{365 \text{ days}}$$

Carry Out

$$\frac{7280}{80} = \frac{x}{365}$$

$$7280(365) = 80x \qquad \textit{Cross-multiply.}$$

$$2{,}657{,}200 = 80x \qquad \textit{Solve.}$$

$$33{,}215 = x$$

Check Both ratios 7280/80 and 33,215/365 equal 91 (use a calculator), so the answer is correct.

Answer Their annual income will be about \$33,215.

EXAMPLE 9 A doctor asks a nurse to give a patient 250 milligrams of the drug simethicone. The drug is available only in a solution whose concentration is 40 milligrams of simethicone per 0.6 milliliter of solution. How many milliliters of solution should the nurse give the patient?

Solution **Understand and Translate** We can set up the proportion using the medication on hand as the given ratio and the number of milliliters needed to be given as the unknown.

$$\text{Given ratio (medication on hand)} \left\{ \frac{40 \text{ milligrams}}{0.6 \text{ milliliter}} \right. = \frac{250 \text{ milligrams}}{x \text{ milliliters}} \begin{array}{l} \longleftarrow \text{Desired medication} \\ \longleftarrow \text{Unknown} \end{array}$$

Carry Out

$$\frac{40}{0.6} = \frac{250}{x}$$

$$40x = 0.6(250) \quad \textit{Cross-multiply.}$$

$$40x = 150 \quad \textit{Solve.}$$

$$x = \frac{150}{40} = 3.75$$

Check and Answer The nurse should administer 3.75 milliliters of the simethicone solution.

EXAMPLE 10 Mark McGwire, who plays for the St. Louis Cardinals baseball team, holds the record for most home runs, 70, in a 162-game season. In the first 48 games of a season, how many home runs would a baseball player need to hit to be on schedule to break McGwire's record?

Solution **Understand and Translate** We will let x represent the number of home runs needed during the first 48 games.

$$\text{Given ratio} \left\{ \frac{70 \text{ home runs}}{162 \text{ games}} \right. = \frac{x \text{ home runs}}{48 \text{ games}}$$

Carry Out

$$\frac{70}{162} = \frac{x}{48}$$

$$70(48) = 162x \quad \textit{Cross-multiply.}$$

$$3360 = 162x \quad \textit{Solve.}$$

$$20.7 \approx x$$

Check and Answer The calculations are correct. For a baseball player to be on schedule to break McGwire's record, he would need to hit 21 home runs by the time he had played 48 games. Notice that we rounded upward from 20.7. Can you explain why?

Avoiding Common Errors

When you set up a proportion, the same units should not be multiplied by themselves during cross-multiplication.

CORRECT	INCORRECT
$\frac{\text{miles}}{\text{hour}} = \frac{\text{miles}}{\text{hour}}$	~~$\frac{\text{miles}}{\text{hour}}$ ~~$\frac{\text{hour}}{\text{miles}}$~~

4) Use Proportions to Change Units

Proportions can also be used to convert from one quantity to another. For example, you can use a proportion to convert a measurement in feet to a measurement in meters, or to convert from pounds to kilograms. The following examples illustrate converting units.

EXAMPLE 11 Convert 18.36 inches to feet.

Solution **Understand and Translate** We know that 1 foot is 12 inches. We use this known fact in one ratio of our proportion. In the second ratio, we set the quantities with the same units in the same respective positions. The unknown quantity is the number of feet, which we will call x.

$$\text{Known ratio } \left\{ \frac{1 \text{ foot}}{12 \text{ inches}} \right. = \frac{x \text{ feet}}{18.36 \text{ inches}}$$

Note that both numerators contain the same units and both denominators contain the same units.

Carry Out Now drop the units and solve for x by cross-multiplying.

$$\frac{1}{12} = \frac{x}{18.36}$$

$$1(18.36) = 12x \quad \textit{Cross-multiply.}$$

$$18.36 = 12x \quad \textit{Solve.}$$

$$\frac{18.36}{12} = \frac{12x}{12}$$

$$1.53 = x$$

NOW TRY EXERCISE 75

Check and Answer Thus, 18.36 inches equals 1.53 feet.

EXAMPLE 12 Carmen Waide is going to send a box by the United Parcel Service (UPS). The mailing cost of the box increases significantly if the weight of the box exceeds 70 pounds. In the box Carmen will place individually wrapped packets of brochures to be given out to those attending a lecture. Each packet weighs 14 ounces.

a) If you disregard the weight of the box, how many packets can be sent in the box without exceeding the 70-pound weight limit?

b) One kilogram is equal to about 2.2 pounds. What is the weight, in kilograms, that is equal to 70 pounds?

Solution **a) Understand** Since the maximum weight of the box is given in pounds and the weight of the packets is given in ounces, we will need to convert one of these quantities so that the units are the same. After we do this, we can set up a proportion to solve the problem.

Translate Since 1 pound = 16 ounces, 70 pounds = 70(16) = 1120 ounces. We will let x represent the number of packets in our proportion.

$$\text{Given ratio } \left\{ \frac{1 \text{ packet}}{14 \text{ ounces}} \right. = \frac{x \text{ packets}}{1120 \text{ ounces}}$$

Carry Out

$$\frac{1}{14} = \frac{x}{1120}$$

$$1(1120) = 14x \quad \textit{Cross-multiply.}$$

$$1120 = 14x \quad \textit{Solve.}$$

$$80 = x$$

Check and Answer Thus 80 packets can be included in the box without exceeding the 70-pound weight limit.

b) Understand We are told that 1 kilogram = 2.2 pounds. We need to convert 70 pounds to kilograms. We will set up a proportion to do this.

Translate *Given ratio* $\left\{\frac{1 \text{ kilogram}}{2.2 \text{ pounds}}\right. = \frac{x \text{ kilograms}}{70 \text{ pounds}}$

Carry Out

$$\frac{1}{2.2} = \frac{x}{70}$$

$$1(70) = 2.2x \quad \textit{Cross-multiply.}$$

$$70 = 2.2x \quad \textit{Solve.}$$

$$31.8 \approx x$$

Check Since 31.8/70 and 1/2.2 both equal about 0.45 on a calculator, the answer is correct.

Answer Thus, 70 pounds is equal to about 31.8 kilograms.

HELPFUL HINT

Some of the problems we have just worked using proportions could have been done without using proportions. However, when working problems of this type, students often have difficulty in deciding whether to multiply or divide to obtain the correct answer. By setting up a proportion, you may be better able to understand the problem and have more success in obtaining the correct answer.

5 Use Proportions to Solve Problems Involving Similar Figures

Proportions can also be used to solve problems in geometry and trigonometry. The following examples illustrate how proportions may be used to solve problems involving **similar figures**. Two figures are said to be similar when their corresponding angles are equal and their corresponding sides are in proportion.

EXAMPLE 13 The figures to the left are similar. Find the length of the side indicated by the x.

Solution We set up a proportion of corresponding sides to find the length of side x.

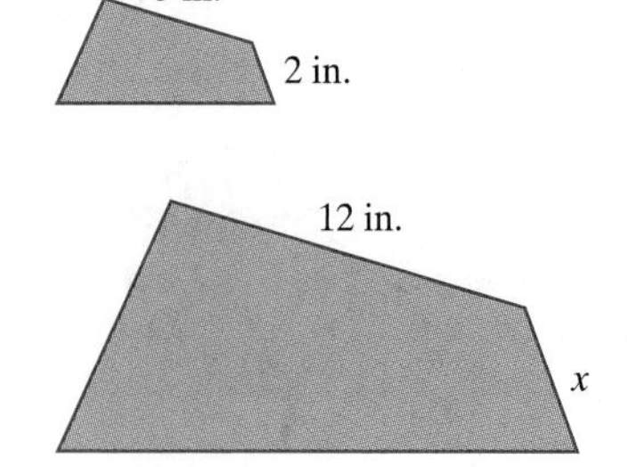

Lengths from smaller figure ↓ *Lengths from larger figure* ↓

5 inches and 12 inches are corresponding sides of similar figures. ⟶

2 inches and x are corresponding sides of similar figures. ⟶

$$\frac{5}{2} = \frac{12}{x}$$

$$5x = 24$$

$$x = \frac{24}{5} = 4.8$$

Thus, the side is 4.8 inches in length.

Note in Example 13 that the proportion could have also been set up as

$$\frac{5}{12} = \frac{2}{x}$$

because one pair of corresponding sides is in the numerators and another pair is in the denominators.

EXAMPLE 14 Triangles ABC and $AB'C'$ are similar triangles. Use a proportion to find the length of side AB'.

A, x, 15 in., 7.2 in., C', B', B, 9 in., C

Solution We set up a proportion of corresponding sides to find the length of side AB'. We will let x represent the length of side AB'. One proportion we can use is

$$\frac{\text{length of } AB}{\text{length of } BC} = \frac{\text{length of } AB'}{\text{length of } B'C'}$$

Now we insert the proper values and solve for the variable x.

$$\frac{15}{9} = \frac{x}{7.2}$$

$$(15)(7.2) = 9x$$

$$108 = 9x$$

$$12 = x$$

Thus, the length of side AB' is 12 inches.

NOW TRY EXERCISE 49

Exercise Set 2.6

Concept/Writing Exercises

1. What is a ratio?
2. In the ratio $a:b$, what are the a and b called?
3. List three ways to write the ratio of c to d.
4. What is a proportion?
5. As you have learned, proportions can be used to solve a wide variety of problems. What information is needed for a problem to be set up and solved using a proportion?

In Exercises 6–9, is the proportion set up correctly? Explain.

6. $\frac{\text{gal}}{\text{min}} = \frac{\text{gal}}{\text{min}}$
7. $\frac{\text{sq ft}}{\text{lb}} = \frac{\text{sq ft}}{\text{lb}}$
8. $\frac{\text{ft}}{\text{sec}} = \frac{\text{sec}}{\text{ft}}$
9. $\frac{\text{tax}}{\text{cost}} = \frac{\text{cost}}{\text{tax}}$
10. What are similar figures?
11. Must similar figures be the same size? Explain.
12. Must similar figures have the same shape? Explain.

Practice the Skills

The results of a mathematics examination are 6 As, 4 Bs, 9 Cs, 3 Ds, and 2 Fs. Write the following ratios in lowest terms.

13. A's to C's
14. A's to total grades
15. D's to F's
16. Grades better than C to total grades
17. Total grades to D's
18. Grades better than C to grades less than C

Determine the following ratios. Write each ratio in lowest terms.

19. 7 gallons to 4 gallons
20. 50 dollars to 60 dollars
21. 16 pounds to 24 pounds
22. 100 people to 80 people
23. 3 hours to 30 minutes
24. 6 feet to 4 yards
25. 26 ounces to 4 pounds
26. 7 dimes to 12 nickels

Find each gear ratio in lowest terms. (See Example 4.)

27. Driving gear, 40 teeth; driven gear, 5 teeth

28. Driven gear, 8 teeth; driving gear, 30 teeth

*In Exercises 29–32, **a)** Determine the indicated ratio, and **b)** write the ratio as some quantity to 1.*

29. A person running at 7.5 miles per hour will burn about 430 calories in 30 minutes. The same person running at 5.5 miles per hour for 30 minutes will burn about 320 calories. What is the ratio of calories burned in 30 minutes for a person running at 7.5 miles per hour to calories burned for a person running at 5.5 miles per hour?

30. According to the Social Security Administration, the average retiree in the United States with a retirement income of \$30,000 will receive about \$11,700 from personal savings, \$6900 from post-retirement employment, \$6000 from Social Security, \$4500 from pension, and \$900 from other sources. Find the ratio of the amount received from personal savings to the amount received from Social Security.

31. The estimated number of victims of child abuse and neglect in the United States was about 798,000 in 1990 and about 1,001,000 in 1995. Find the ratio of the child abuse cases in 1995 to the number of cases in 1990.

32. The number of cable television subscribers in the United States was 4,500,000 in 1970 and 58,000,000 in 1995. Find the ratio of cable subscribers in 1995 to 1970.

Exercises 33–36 show graphs that are self-explanatory. For each exercise, find each indicated ratio.

33. a) Estimate the ratio of the expected number of total on-line households in 2000 to those in 1995.

b) Estimate the ratio of households with PCs in 1995 to expected households with PCs in 2000.

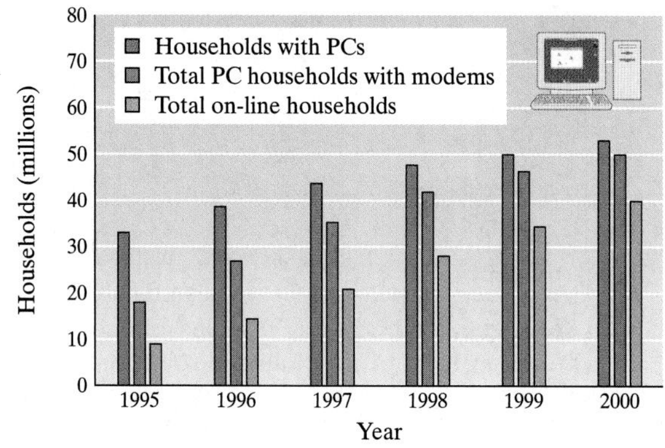

34. a) Estimate the ratio of the median sales price of existing homes in the United States in 1996 to the median sales price in 1970.

b) Estimate the ratio of the median sales price of existing homes in the Midwest in 1970 to the median sales price in 1996.

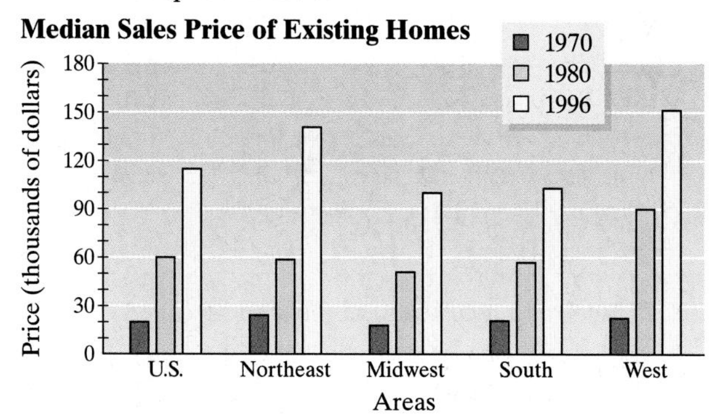

35. a) Determine the ratio of three-bedroom homes to homes with two or fewer bedrooms in 1970.

b) Repeat part **a)** for 1995.

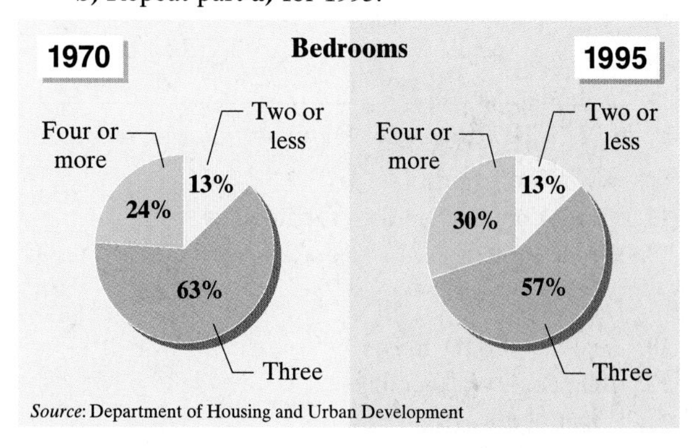

36. a) Determine the ratio of houses with $1\frac{1}{2}$ or fewer bathrooms to homes with more than $1\frac{1}{2}$ bathrooms in 1970.

b) Repeat part **a)** for 1995.

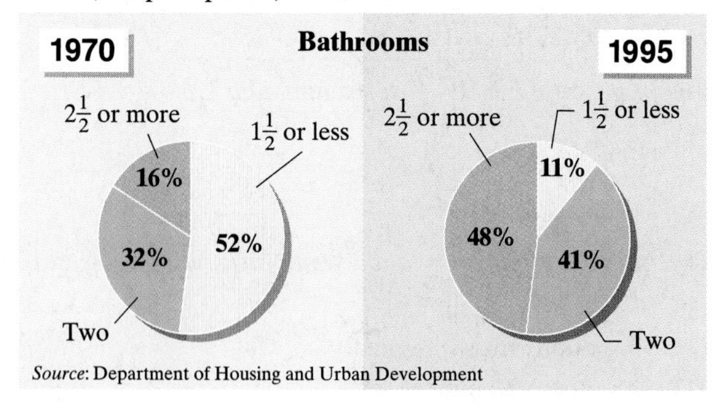

Solve each proportion for the variable by cross-multiplying.

37. $\dfrac{4}{x} = \dfrac{5}{20}$ **38.** $\dfrac{x}{8} = \dfrac{24}{48}$ **39.** $\dfrac{5}{3} = \dfrac{75}{x}$ **40.** $\dfrac{x}{3} = \dfrac{90}{30}$

41. $\dfrac{90}{x} = \dfrac{-9}{10}$ **42.** $\dfrac{3}{4} = \dfrac{x}{8}$ **43.** $\dfrac{15}{45} = \dfrac{x}{-6}$ **44.** $\dfrac{y}{6} = \dfrac{7}{42}$

45. $\dfrac{3}{z} = \dfrac{-1.5}{27}$ **46.** $\dfrac{3}{12} = \dfrac{-1.4}{z}$ **47.** $\dfrac{15}{20} = \dfrac{x}{8}$ **48.** $\dfrac{2}{20} = \dfrac{x}{200}$

The following figures are similar. For each pair, find the length of the side indicated by x.

49.

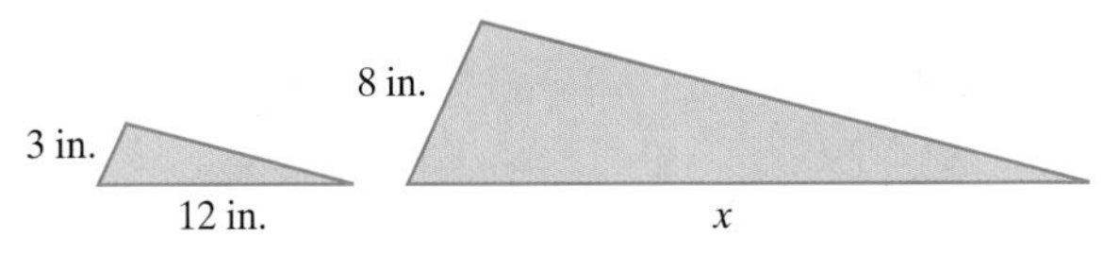

50.

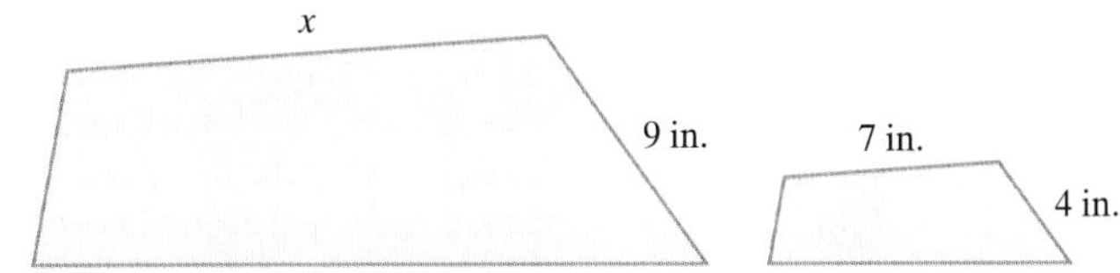

51.

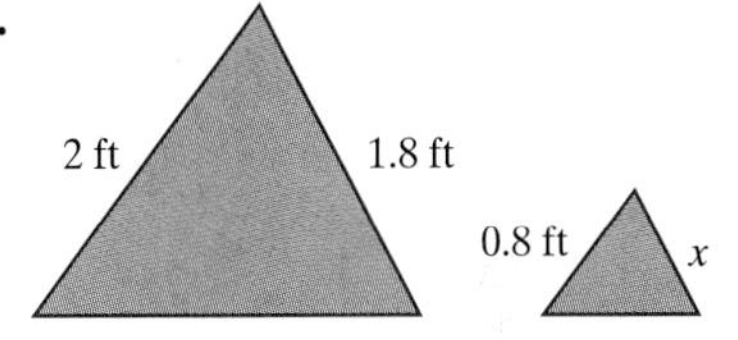

52.

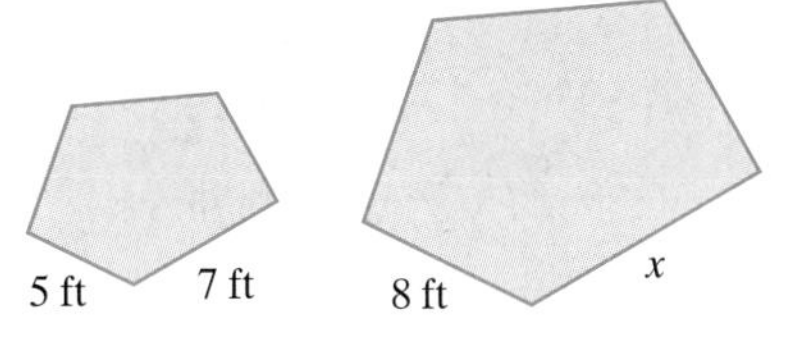

53.

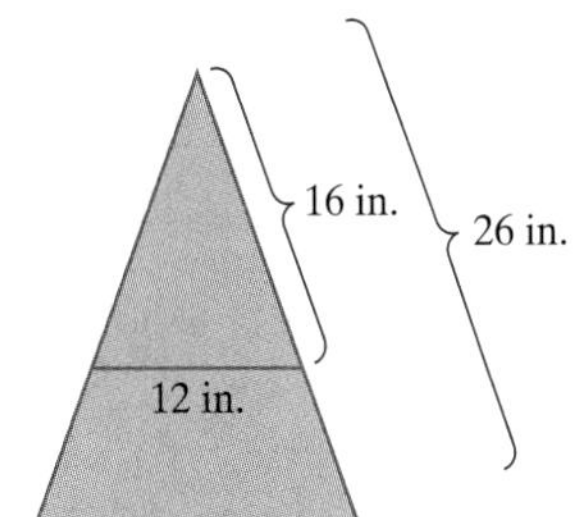

54.

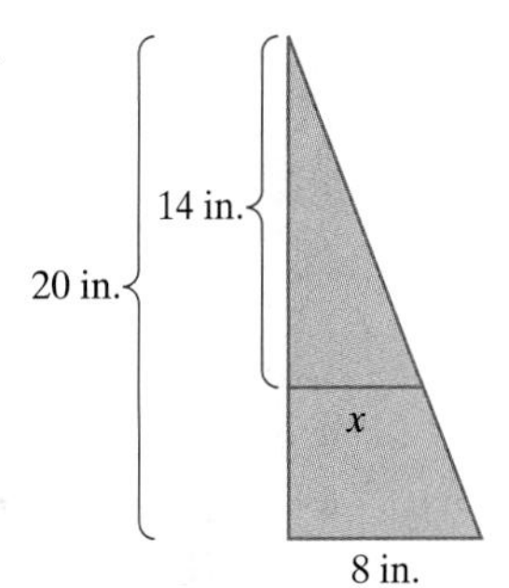

Problem Solving

55. A bottle of liquid Tide contains 100 fluid ounces. If one wash load requires 4 ounces of the detergent, how many washes can be done with one bottle of Tide?

56. Amy Mayfield, a waitress, earned $380 in a typical week. How much can she expect to earn in 50 weeks?

57. A 1999 Ford Mustang with the 4.6 liter engine is rated to get 23 miles per gallon (highway driving). How far can it travel on a full tank of gas, 15.7 gallons?

58. A gallon of paint covers 825 square feet. How much paint is needed to cover a house with a surface area of 5775 square feet?

59. Architects are planning a shopping mall. Their blueprints are in the scale of 1:120. That is, 1 foot on the blueprint represents 120 feet of actual length. Find the width of the corridor on the blue print if the corridor is to be 90 feet wide, as illustrated.

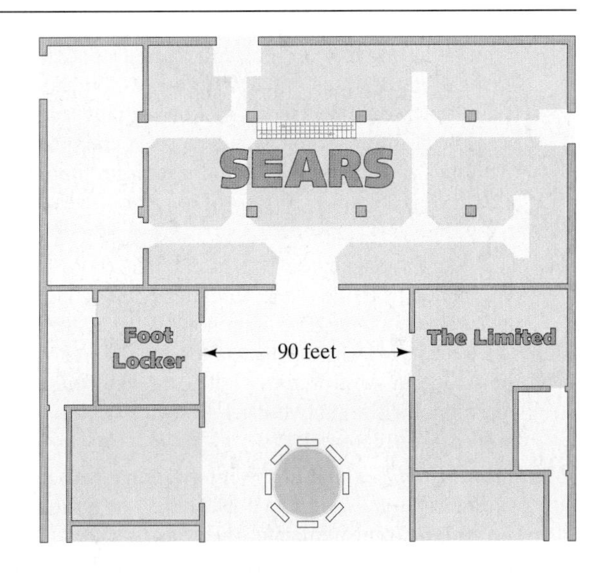

60. If a 40-pound bag of fertilizer covers 5000 square feet, how many pounds of fertilizer are needed to cover an area of 26,000 square feet?

61. The instructions on a bottle of liquid insecticide say "use 3 teaspoons of insecticide per gallon of water." If your sprayer has an 8-gallon capacity, how much insecticide should be used to fill the sprayer?

62. The property tax in the town of Plainview, Texas, is \$8.235 per \$1000 of assessed value. If the Littons' house is assessed at \$122,000, how much property tax will they owe?

63. A photograph shows a woman standing next to a monument. If the woman, who is actually 64-inches tall, measures 0.875 inch in the photograph, how tall is the monument that measures 3.117 inches in the photo?

64. A recipe for 6 servings of French onion soup requires $1\frac{1}{2}$ cups of thinly sliced onions. If the recipe were to be made for 15 servings, how many cups of onions would be needed?

65. Every ton of recycled paper saves approximately 17 trees. (It also saves dumping cost, landfill space, about 7000 gallons of water, and 4100 kilowatt-hours of electricity that would be used in making new paper products. Furthermore, to collect and recycle paper provides five times as many jobs as to harvest virgin timber.) If your college recycles 20 tons of paper in a year, how many trees has it saved?

66. The number of people who watch a television show is determined by the A.C. Nielsen ratings. One rating point means that about 994,000 households watched the show. The week of March 26, 1999, the top-rated TV show was the Academy Awards with a rating of 28.6 points. Approximately how many households watched the Academy Awards that week?

67. At the *Titanic* exhibit at the International Museum at St. Petersburg, Florida, in 1998 there was a model of the *Nautile*, the submarine that was used to explore the *Titanic*. The model was in the ratio of 1 to 3 to the actual submarine. If the length of the actual submarine was 26.24 feet, how long was the model?

68. When they returned home from vacation, the Duncans found that they had a foot of water in their basement. They contacted their fire department, which sent some equipment to pump out the water. After the pump had been on for 30 minutes, 3 inches of water had been removed. How long, from the time they started pumping, will it take to remove all the water from the basement?

69. A nurse must administer 220 micrograms of atropine sulfate. The drug is available in solution form. The concentration of the atropine sulfate solution is 400 micrograms per milliliter. How many milliliters should be given?

70. A doctor asks a nurse to administer 0.7 gram of meprobamate per square meter of body surface. The patient's body surface is 0.6 square meter. How much meprobamate should be given?

71. While on fast forward, the counter of your VCR goes from 0 to 250 in 30 seconds. Your videocassette tape contains two movies. The second movie starts at 800 on the VCR counter. If you are at the beginning of the tape, approximately how long will you keep the VCR on fast forward to reach the beginning of the second movie?

72. Mary read 40 pages of a novel in 30 minutes.

a) If she continues reading at the same rate, how long will it take her to read the entire 760-page book?

b) How long will it take her to finish the book from page 41?

73. It is estimated that each year in the United States 1 in every 15,000 (1:15,000) people is born with a genetic disorder called Prader–Willi syndrome. If there were approximately 3,895,000 births in the United States in 1997, approximately how many children were born with Prader–Willi syndrome?

74. In the United States in 1997, the birth rate was 14.6 per one thousand people. In the United States in 1997, there were approximately 3,895,000 births. What was the U.S. population in 1997?

In Exercises 75–86, use a proportion to make the conversion. Round your answers to two decimal places.

75. Convert 42 inches to feet.

76. Convert 22,704 feet to miles (5280 feet = 1 mile).

77. Convert 26.1 square feet to square yards (9 square feet = 1 square yard).

78. Convert 146.4 ounces to pounds.

79. One inch equals 2.54 centimeters. Find the length of a book in inches if it measures 26.67 centimeters.

80. One mile equals approximately 1.6 kilometers. Find the distance in kilometers from San Diego, California, to San Francisco, California—a distance of 520 miles.

81. Most countries of the world use the metric system. The following photo was taken at a farmers' market in Fiji.

a) Determine the price of 3.5 kilograms of Mung Dhau if Mung Dhau costs $3.00 per kilogram.

b) If 1 kilogram = 2.2 pounds, determine the cost of 1 pound of Mung Dhau.

82. The following photo was taken in Queenstown, New Zealand.

a) Determine the price of 2.5 kilograms of red peppers if red peppers cost $3.75 per kilogram.

b) If 1 kilogram = 2.2 pounds, determine the cost of 1 pound of red peppers.

83. If gold is selling for $408 per 480 grains (a troy ounce), what is the cost per grain?

84. In chemistry, we learn that 100 torr (a unit of measurement) equals 0.13 atmosphere. Find the number of torr in 0.39 atmosphere.

85. In a statistics course, we find that for one particular set of scores 15 points equals 3.75 standard deviations. How many points equals 1 standard deviation?

86. When Antonio Juarez visited the United States from Mexico, he exchanged 85 pesos for 10 U.S. dollars. If he exchanges his remaining 20,000 pesos for U.S. dollars, how much more in dollars will he receive?

87. Mrs. Ruff's low-density cholesterol level is 127 milligrams per deciliter (mg/dL). Her high-density cholesterol level is 60 mg/dL. Is Mrs. Ruff's ratio of low- to high-density cholesterol level less than or equal to the 4:1 recommended level? (See Example 2.)

88. **a)** Another ratio used by some doctors when measuring cholesterol level is the ratio of total cholesterol to high-density cholesterol.* Is this ratio increased or decreased if the total cholesterol remains the same but the high-density level is increased? Explain.

b) Doctors recommend that the ratio of total cholesterol to high-density cholesterol be less than or equal to 4.5:1. If Mike's total cholesterol is 220 mg/dL and his high-density cholesterol is 50 mg/dL, is his ratio less than or equal to 4.5:1? Explain.

89. For the proportion $\frac{a}{b} = \frac{c}{d}$, if a increases while b and d stay the same, what must happen to c? Explain.

90. For the proportion $\frac{a}{b} = \frac{c}{d}$, if a and c remain the same while d decreases, what must happen to b? Explain.

*Total cholesterol includes both low- and high-density cholesterol, plus other types of cholesterol.

Challenge Problems

91. A new Goodyear tire has a tread of about 0.34 inches. After 5000 miles the tread is about 0.31 inches. If the legal minimum amount of tread for a tire is 0.06 inches, how many more miles will the tires last? (Assume no problems with the car or tires and that the tires wear at an even rate.)

92. The recipe for the filling for an apple pie calls for

12 cups sliced apples	$\frac{1}{4}$ teaspoon salt
$\frac{1}{2}$ cup flour	2 tablespoons butter or margarine
1 teaspoon nutmeg	$1\frac{1}{2}$ cups sugar
1 teaspoon cinnamon	

Determine the amount of each of the other ingredients that should be used if only 8 cups of apples are available.

93. The size of a car tire may be denoted something like P205/60R15. The first number is the width of the tire in millimeters (see Fig. 2.11a). The number following the slash is the height of the tire (also known as the aspect ratio), which is the distance from the rim to the ground, in terms of the percent of the tire width (see Fig. 2.11b). The third number, following the R, is the diameter of the rim, in inches. Luxury car tires will generally have a higher aspect ratio, which indicates a tire that is taller in proportion to its width. This height leads to a more flexible tire and a softer ride. Sports car tires like a Corvette, usually have a smaller aspect ratio. This leads to a stiffer tire and improved handling (see Fig. 2.11c).

The rear tires that came standard on a 1998 Dodge Viper were size P335/35R17. The tires that came on a 1998 Cadillac Seville STS were size P235/60R16. Determine, in millimeters, the height of each tire (the distance from the rim to the ground).

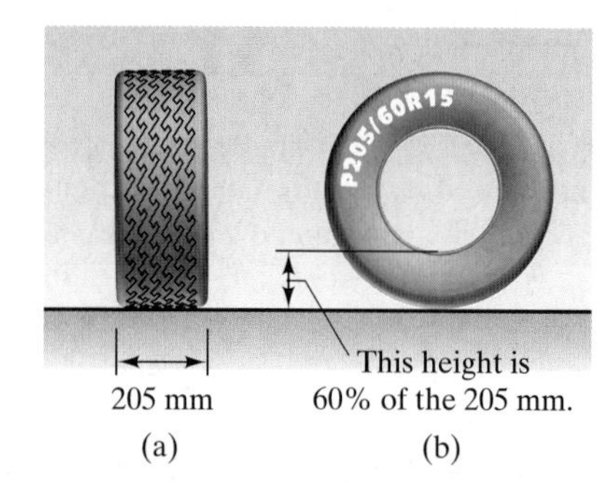

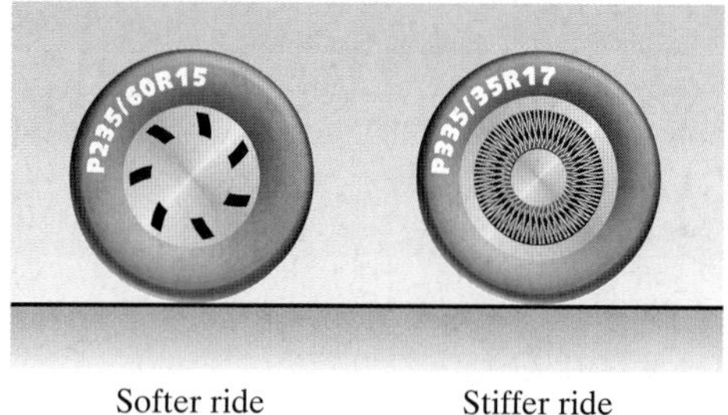

FIGURE 2.11

94. Insulin comes in 10-cubic-centimeter (cc) vials labeled in the number of units of insulin per cubic centimeter. Thus a vial labeled U40 means there are 40 units of insulin per cubic centimeter of fluid. If a patient needs 25 units of insulin, how many cubic centimeters of fluid should be drawn up into a syringe from the U40 vial?

Group Activity

Discuss and answer Exercises 95–97 as a group.

95. **a)** Each group member: Find the ratio of your height to your arm span (finger tips to finger tips) when your arms are extended horizontally outward. You will need help from your group in getting these measurements.

b) If a box were to be drawn about your body with your arms extended, would the box be a square or a rectangle? If a rectangle, would the longer length be your arm span or your height measurement? Explain.

c) Compare these results with other members of your group.

d) What one ratio would you use to report the height to arm span for your group as a whole? Explain.

96. A special ratio in mathematics is called the *golden ratio*. Do research in a history of mathematics book or another book recommended by your professor, and as a group write a paper that explains what the golden ratio is and why it is important.

97. An important concept, which we will discuss in Chapter 4, is *slope*. The slope of a line may be defined as a *ratio* of the vertical change to the horizontal change between any two points on a line. In the following figures, the vertical change is found using the red dashed lines, and the horizontal change is found using the green dashed lines.

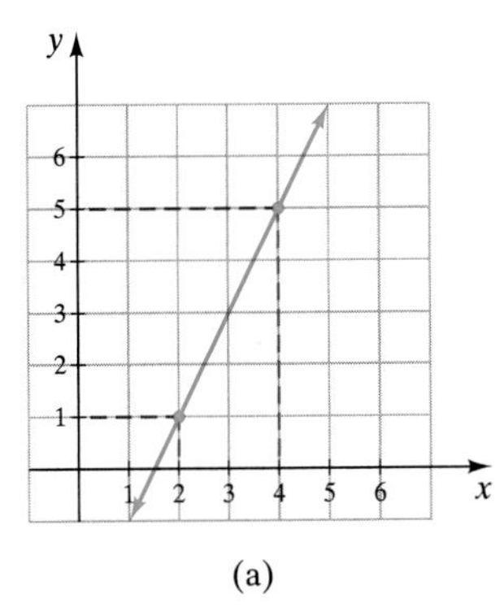

(a)

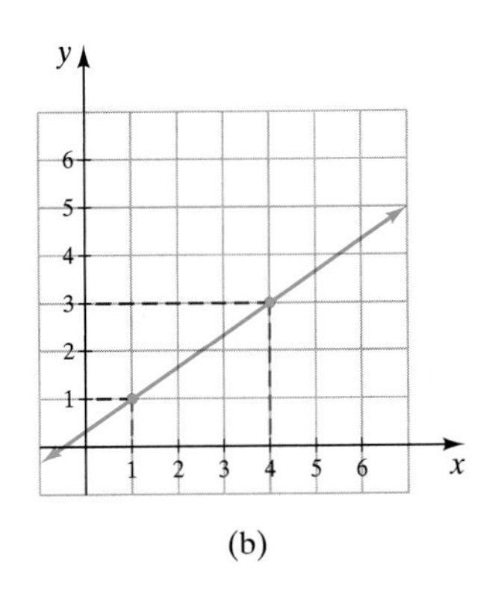

(b)

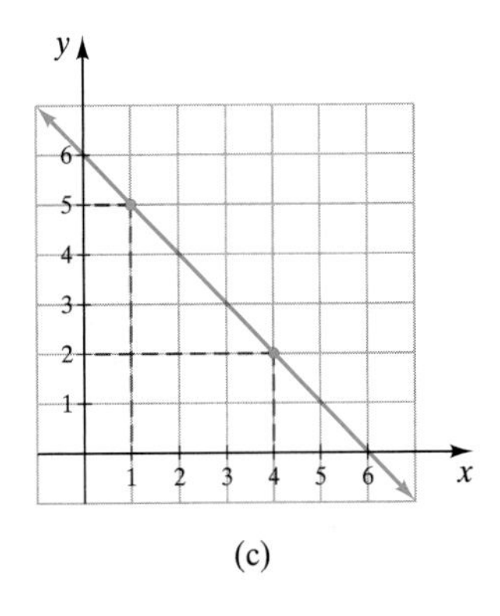

(c)

FIGURE 2.12

a) Group member 1: Determine the slope of Figure 2.12a.

Group member 2: Determine the slope of Figure 2.12b.

Group member 3: Determine the slope of Figure 2.12c.

b) Now compare answers and check each other's work.

c) As a group, list the slopes of the lines in Figure 2.12a, 2.12b, and 2.12c, respectively.

d) Consider the Figure 2.13. Notice the point at (1, 2). Through this point, draw a line that has a slope of $\frac{2}{3}$.

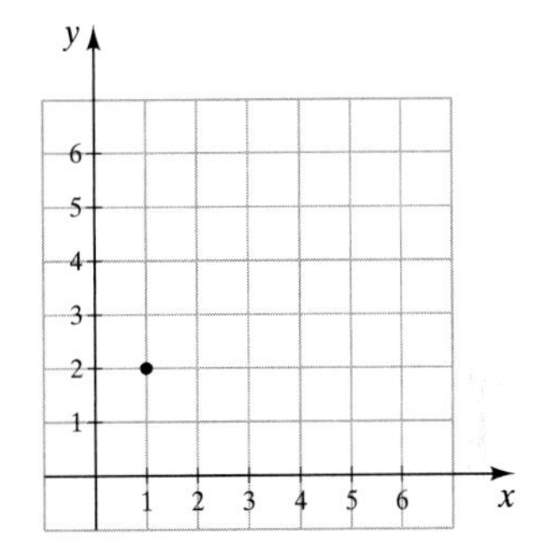

FIGURE 2.13

Cumulative Review Exercises

[1.11] *Name each illustrated property.*

98. $x + 3 = 3 + x$

99. $3(xy) = (3x)y$

100. $2(x - 3) = 2x - 6$

[2.5] **101.** Solve $-(2x + 6) = 2(3x - 6)$.

2.7 INEQUALITIES IN ONE VARIABLE

SSM VIDEO 2.7 CD Rom

1. Solve linear inequalities.
2. Solve linear inequalities that have all real numbers as their solution, or have no solution.

1 Solve Linear Inequalities

The is-greater-than symbol, >, and is-less-than symbol, <, were introduced in Section 1.5. The symbol $\geq$ means is greater than or equal to and $\leq$ means is less than or equal to. A mathematical statement containing one or more of these symbols is called an **inequality**. The direction of the symbol is sometimes called the **sense** or **order of the inequality**.

Examples of Inequalities in One Variable

$$x + 3 < 5 \qquad x + 4 \geq 2x - 6 \qquad 4 > -x + 3$$

To solve an inequality, we must get the variable by itself on one side of the inequality symbol. To do this, we make use of properties very similar to those used to solve equations. Here are four properties used to solve inequalities. Later in this section, we will introduce two additional properties.

Properties Used to Solve Inequalities

For real numbers, a, b, and c:

1. If $a > b$, then $a + c > b + c$.
2. If $a > b$, then $a - c > b - c$.
3. If $a > b$ **and $c > 0$**, then $ac > bc$.
4. If $a > b$ **and $c > 0$**, then $\frac{a}{c} > \frac{b}{c}$.

Property 1 says the same number may be added to both sides of an inequality. Property 2 says the same number may be subtracted from both sides of an inequality. Property 3 says the same *positive* number may be used to multiply both sides of an inequality. Property 4 says the same *positive* number may be used to divide both sides of an inequality. When any of these four properties is used, *the direction of the inequality symbol does not change.*

EXAMPLE 1 Solve the inequality $x - 4 > 7$, and graph the solution on a number line.

Solution To solve this inequality, we need to isolate the variable, x. Therefore, we must eliminate the −4 from the left side of the inequality. To do this, we add 4 to both sides of the inequality.

$$x - 4 > 7$$

$$x - 4 + 4 > 7 + 4 \qquad \text{Add 4 to both sides.}$$

$$x > 11$$

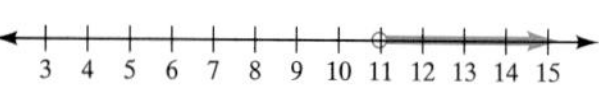

FIGURE 2.14

The solution is all real numbers greater than 11. We can illustrate the solution on a number line by placing an open circle at 11 on a number line and drawing an arrow to the right (Fig. 2.14).

The open circle at the 11 indicates that the 11 is *not* part of the solution. The arrow going to the right indicates that all the values greater than 11 are solutions to the inequality.

EXAMPLE 2 Solve the inequality $2x + 6 \le -2$, and graph the solution on a number line.

Solution To isolate the variable, we must eliminate the +6 from the left side of the inequality. We do this by subtracting 6 from both sides of the inequality.

$$2x + 6 \le -2$$

$$2x + 6 - 6 \le -2 - 6 \qquad \text{Subtract 6 from both sides.}$$

$$2x \le -8$$

$$\frac{2x}{2} \le \frac{-8}{2} \qquad \text{Divide both sides by 2.}$$

$$x \le -4$$

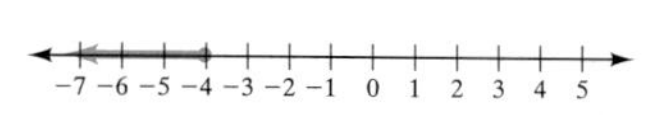

FIGURE 2.15

The solution is all real numbers less than or equal to −4. We can illustrate the solution on a number line by placing a closed, or darkened, circle at −4 and drawing an arrow to the left (Fig. 2.15).

The darkened circle at -4 indicates that -4 *is* a part of the solution. The arrow going to the left indicates that all the values less than -4 are also solutions to the inequality.

NOW TRY EXERCISE 23

Notice in properties 3 and 4 that we specified that $c > 0$. What happens when an inequality is multiplied or divided by a negative number? Examples 3 and 4 illustrate that **when an inequality is multiplied or divided by a negative number, the direction of the inequality symbol changes**.

EXAMPLE 3 Multiply both sides of the inequality $8 > -4$ by -2.

Solution

$$8 > -4$$

$$-2(8) < -2(-4) \quad \text{Change the direction of the inequality symbol.}$$

$$-16 < 8$$

EXAMPLE 4 Divide both sides of the inequality $8 > -4$ by -2.

Solution

$$8 > -4$$

$$\frac{8}{-2} < \frac{-4}{-2} \quad \text{Change the direction of the inequality symbol.}$$

$$-4 < 2$$

Now we state two additional properties, used when an inequality is multiplied or divided by a negative number.

Additional Properties Used to Solve Inequalities

5. If $a > b$ **and $c < 0$**, then $ac < bc$.
6. If $a > b$ **and $c < 0$**, then $\frac{a}{c} < \frac{b}{c}$.

EXAMPLE 5 Solve the inequality $-2x > 6$, and graph the solution on a number line.

Solution To isolate the variable, we must eliminate the -2 on the left side of the inequality. To do this, we can divide both sides of the inequality by -2. When we do this, however, we must remember to *change the direction* of the inequality symbol.

$$-2x > 6$$

$$\frac{-2x}{-2} < \frac{6}{-2} \quad \text{Divide both sides by } -2 \text{, and change the direction of the inequality symbol.}$$

$$x < -3$$

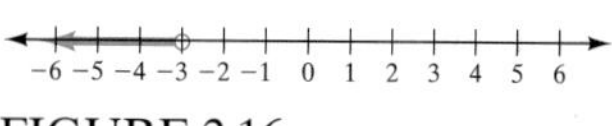

FIGURE 2.16

The solution is all real numbers less than −3. The solution is graphed on a number line in Figure 2.16.

EXAMPLE 6 Solve the inequality $4 \geq -5 - x$, and graph the solution on a number line. We will illustrate two methods that can be used to solve this inequality.

Solution **Method 1:**

$$4 \geq -5 - x$$

$$4 + 5 \geq -5 + 5 - x \qquad \textit{Add 5 to both sides.}$$

$$9 \geq -x$$

$$-1(9) \leq -1(-x) \qquad \textit{Multiply both sides by } -1\textit{, and change the direction of the inequality symbol.}$$

$$-9 \leq x$$

The inequality $-9 \leq x$ can also be written $x \geq -9$.

Method 2:

$$4 \geq -5 - x$$

$$4 + x \geq -5 - x + x \qquad \textit{Add } x \textit{ to both sides.}$$

$$4 + x \geq -5$$

$$4 - 4 + x \geq -5 - 4 \qquad \textit{Subtract 4 from both sides.}$$

$$x \geq -9$$

FIGURE 2.17

NOW TRY EXERCISE 27

The solution is graphed on a number line in Figure 2.17. Other methods could also be used to solve this problem.

Notice in Example 6, Method 1, we wrote $-9 \leq x$ as $x \geq -9$. Although the solution $-9 \leq x$ is correct, it is customary to write the solution to an inequality with the variable on the left. One reason we write the variable on the left is that it often makes it easier to graph the solution on the number line. How would you graph $-3 > x$? How would you graph $-5 \leq x$? If you rewrite these inequalities with the variable on the left side, the answer becomes clearer.

$$-3 > x \quad \text{means} \quad x < -3$$

$$-5 \leq x \quad \text{means} \quad x \geq -5$$

Notice that you can change an answer from an is-greater-than statement to an is-less-than statement or from an is-less-than statement to an is-greater-than statement. When you change the answer from one form to the other, remember that the inequality symbol must point to the letter or number to which it was pointing originally.

HELPFUL HINT

$a > x$ means $x < a$ *Note that both inequality symbols point to* x.

$a < x$ means $x > a$ *Note that both inequality symbols point to* a.

EXAMPLES

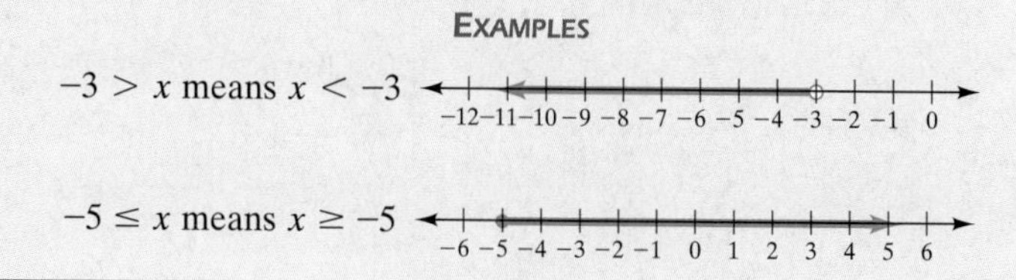

Now let's solve inequalities where the variable appears on both sides of the inequality symbol. To solve these inequalities, we use the same basic procedure that we used to solve equations. However, we must remember that whenever we multiply or divide both sides of an inequality by a negative number, we must change the direction of the inequality symbol.

EXAMPLE 7 Solve the inequality $2x + 4 < -x + 12$, and graph the solution on a number line.

Solution

$$2x + 4 < -x + 12$$
$$2x + x + 4 < -x + x + 12 \quad \text{Add } x \text{ to both sides.}$$
$$3x + 4 < 12$$
$$3x + 4 - 4 < 12 - 4 \quad \text{Subtract 4 from both sides.}$$
$$3x < 8$$
$$\frac{3x}{3} < \frac{8}{3} \quad \text{Divide both sides by 3.}$$
$$x < \frac{8}{3}$$

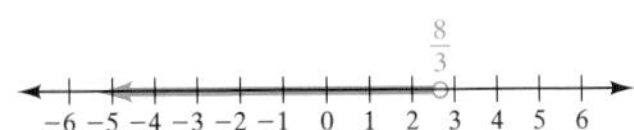

FIGURE 2.18

The solution is graphed on the number line in Figure 2.18.

EXAMPLE 8 Solve the inequality $-5x + 9 < -2x + 6$, and graph the solution on a number line.

Solution

$$-5x + 9 < -2x + 6$$
$$9 < 3x + 6 \quad 5x \text{ was added to both sides.}$$
$$3 < 3x \quad 6 \text{ was subtracted from both sides.}$$
$$1 < x \quad \text{Both sides were divided by 3.}$$
$$\text{or } x > 1$$

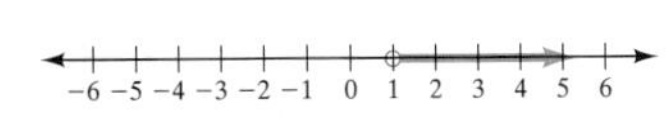

FIGURE 2.19

The solution is graphed in Figure 2.19. **NOW TRY EXERCISE 37**

2) Solve Linear Inequalities That Have All Real Numbers as Their Solution, or Have No Solution

In Examples 9 and 10 we illustrate two special types of inequalities. Example 9 is an inequality that is true for all real numbers, and Example 10 is an inequality that is never true for any real number.

EXAMPLE 9 Solve the inequality $2(x + 3) \le 5x - 3x + 8$, and graph the solution on a number line.

Solution

$$2(x + 3) \le 5x - 3x + 8$$
$$2x + 6 \le 5x - 3x + 8 \quad \text{Distributive property was used.}$$
$$2x + 6 \le 2x + 8 \quad \text{Like terms were combined.}$$
$$2x - 2x + 6 \le 2x - 2x + 8 \quad \text{Subtract } 2x \text{ from both sides.}$$
$$6 \le 8$$

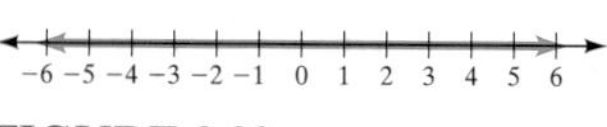

FIGURE 2.20

Since 6 is always less than or equal to 8, the solution is **all real numbers** (Fig. 2.20).

EXAMPLE 10 Solve the inequality $4(x + 1) > x + 5 + 3x$, and graph the solution on a number line.

Solution

$$4(x + 1) > x + 5 + 3x$$

$$4x + 4 > x + 5 + 3x \quad \text{Distributive property was used.}$$

$$4x + 4 > 4x + 5 \quad \text{Like terms were combined.}$$

$$4x - 4x + 4 > 4x - 4x + 5 \quad \text{Subtract } 4x \text{ from both sides.}$$

$$4 > 5$$

FIGURE 2.21

Since 4 is never greater than 5, the answer is **no solution** (Fig. 2.21). There is no real number that makes the statement true.

NOW TRY EXERCISE 47

Exercise Set 2.7

Concept/Writing Exercises

1. List the four inequality symbols given in this section and write how each is read.
2. Explain the difference between $>$ and $\geq$.
3. Are the following statements true or false? Explain.
   a) $3 > 3$
   b) $3 \geq 3$
4. If $a < b$ is a true statement, must $b > a$ also be a true statement? Explain.
5. When solving an inequality, under what conditions will it be necessary to change the direction of the inequality symbol?
6. List the six rules used to solve inequalities.
7. When solving an inequality, if you obtain the result $3 < 5$, what is the solution?
8. When solving an inequality, if you obtain the result $4 \geq 2$, what is the solution?
9. When solving an inequality, if you obtain the result $5 < 2$, what is the solution?
10. When solving an inequality, if you obtain the result $-4 \geq -2$, what is the solution?

Practice the Skills

Solve each inequality, and graph the solution on a number line.

11. $x + 2 > 6$
12. $x - 5 > -1$
13. $x + 7 \geq 4$
14. $4 - x \geq 3$
15. $-x + 3 < 8$
16. $4 < 3 + x$
17. $6 > x - 4$
18. $-6 \leq -x - 3$
19. $9 \leq 2 - x$
20. $2x < 4$
21. $-2x < 3$
22. $6 \geq -3x$
23. $2x + 3 \leq 5$
24. $-4x - 3 > 5$
25. $12x - 12 < -12$
26. $7x - 4 \leq 9$
27. $4 - 6x > -5$
28. $8 < 4 - 2x$

29. $15 > -9x + 50$

30. $3x - 4 < 5$

31. $4 < 3x + 12$

32. $-3x > 2x + 10$

33. $6x + 2 \leq 3x - 9$

34. $-2x - 4 \leq -5x + 12$

35. $x - 4 \leq 3x + 8$

36. $-3x - 5 \geq 4x - 29$

37. $-x + 4 < -3x + 6$

38. $2(x - 3) < 4x + 10$

39. $-3(2x - 4) > 2(6x - 12)$

40. $-(x + 3) \leq 4x + 5$

41. $x + 3 < x + 4$

42. $x + 5 \geq x - 2$

43. $6(3 - x) < 2x + 12$

44. $2(3 - x) + 4x < -6$

45. $-21(2 - x) + 3x > 4x + 4$

46. $-(x + 3) \geq 2x + 6$

47. $4x - 4 < 4(x - 5)$

48. $-2(-5 - x) > 3(x + 2) + 4 - x$

49. $5(2x + 3) \geq 6 + (x + 2) - 2x$

50. $-3(-2x + 12) < -4(x + 2) - 6$

Problem Solving

51. The following chart shows the average high and low monthly temperatures in Chicago over a 126-year period (*Source:* Richard Koeneman/ WGN-TV meteorologist). Notice that the months are not listed in order.

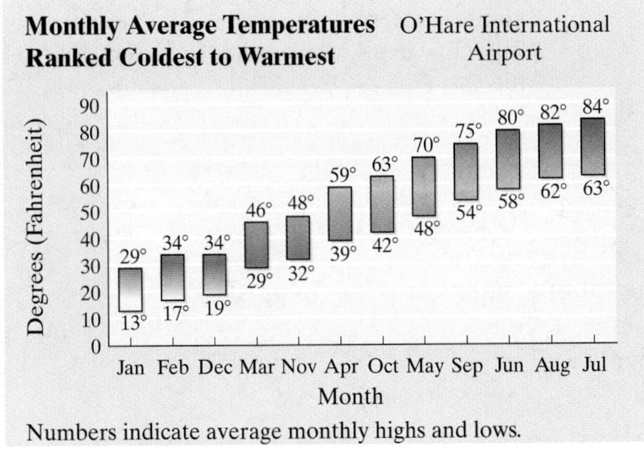

a) In what months was the average high temperature $> 65°F$?

b) In what months was the average high temperature $\leq 59°F$?

c) In what months was the average low temperature $< 29°F$?

d) In what months was the average low temperature $\geq 58°F$?

52. The following chart shows both international and U.S. quotas for Atlantic swordfish, in tons (*Source:* International Commission for the Conservation of Atlantic Tunas).

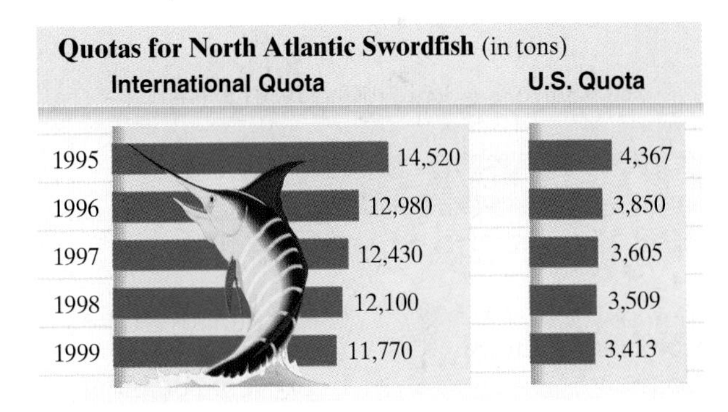

Source: International Commission for the Conservation of Atlantic Tunas

a) In which years was the international quota $\leq$ 12,000 tons?

b) In which years was the international quota $\geq$ 12,430 tons?

c) In which years was the U.S. quota $\leq$ 3605 tons?

d) In which years was the U.S. quota $>$ 3605 tons?

53. The inequality symbols discussed so far are $<, \leq, >,$ and $\geq$. Can you name an inequality symbol that we have not mentioned in this section?

54. Reproduced below is a portion of the Florida Individual and Joint Intangible Tax Return for 1998.

Tax Calculation Worksheet **(Complete only one column below)**

Instructions: Determine which column applies based on filing status and amount entered on Schedule A, Line 5.

Complete only the applicable column.	Individual		Joint	
	Column A If Schedule A, Line 5 is $100,000 or less, complete the following:	**Column B** If Schedule A, Line 5 is more than $100,000, complete the following:	**Column C** If Schedule A, Line 5 is $200,000 or less, complete the following:	**Column D** If Schedule A, Line 5 is more than $200,000, complete the following:
6A. Enter Total from Schedule A, Line 5.	$	$	$	$
6B. Multiply by Tax Rate	× 0.001	× 0.002	× 0.001	× 0.002
6C. Gross Tax	$	$	$	$
6D. Subtract Exemption	− $20.00	− $120.00	− $40.00	− $240.00
6E. Enter Total Tax Due Carry Amount to Schedule A, Line 6.	$	$	$	$

Use the Tax Calculation Worksheet to determine your total tax due (line 6e) if your total taxable assets from Schedule A line 5 are as follows.

a) $30,000 and your filing status is individual
b) $175,000 and your filing status is individual
c) $200,000 and your filing status is joint
d) $300,000 and your filing status is joint

55. Consider the inequality $xy > 6$, where x and y represent real numbers. Explain why we *cannot* do the following step:

$$\frac{xy}{y} > \frac{6}{y} \quad \text{Divide both sides by } y.$$

Challenge Problems

56. Solve the following inequality.

$$3(2 - x) - 4(2x - 3) \le 6 + 2x - 4x$$

57. Solve the following inequality.

$$6x - 6 > -4(x + 3) + 5(x + 6) - x$$

Cumulative Review Exercises

[1.9] **58.** Evaluate $-x^2$ for $x = 3$.

59. Evaluate $-x^2$ for $x = -5$.

[2.5] **60.** Solve $4 - 3(2x - 4) = 5 - (x + 3)$.

[2.6] **61.** The Milford Electric Company charges $0.174 per kilowatt-hour of electricity. The Cisneros's monthly electric bill was $87 for the month of July. How many kilowatt-hours of electricity did the Cisneros use in July?

SUMMARY

Key Words and Phrases

2.1
Combine like terms
Constant
Expression
Factor
Like terms
Numerical coefficient
Simplify an expression
Term
Variable
Check an equation
Equation
Equivalent equations
Isolate the variable
Linear equation
Solution to an equation
Solve an equation

2.2
Addition property of equality

2.3
Multiplication property of equality
Reciprocal

2.5
Conditional equations
Contradiction
Identity

2.6
Cross-multiplication
Extremes
Means
Proportion
Ratio
Similar figures
Slope
Terms of a ratio

2.7
Inequality
Sense (or order) of the inequality

IMPORTANT FACTS	
Distributive property	$a(b + c) = ab + ac$
Addition property	If $a = b$, then $a + c = b + c$.
Multiplication property	If $a = b$, then $a \cdot c = b \cdot c, c \neq 0$.
Cross-multiplication	If $\frac{a}{b} = \frac{c}{d}$, then $ad = bc$.

Properties used to solve inequalities

1. If $a > b$, then $a + c > b + c$.
2. If $a > b$, then $a - c > b - c$.
3. If $a > b$ and $c > 0$, then $ac > bc$.
4. If $a > b$ and $c > 0$, then $\frac{a}{c} > \frac{b}{c}$.
5. If $a > b$ and $c < 0$, then $ac < bc$.
6. If $a > b$ and $c < 0$, then $\frac{a}{c} < \frac{b}{c}$.

Review Exercises

[2.1] *Use the distributive property to simplify.*

1. $3(x + 4)$
2. $3(x - 2)$
3. $-2(x + 4)$
4. $-(x + 2)$
5. $-(x - 2)$
6. $-4(4 - x)$
7. $3(6 - 2x)$
8. $6(4x - 5)$
9. $-5(5x - 5)$
10. $4(-x + 3)$
11. $-2(3x - 2)$
12. $-(3 + 2y)$
13. $-(x + 2y - z)$
14. $-2(2x - 3y + 7)$

Simplify.

15. $5x + 3x$
16. $2y + 3y + 2$
17. $5 - 3y + 3$
18. $1 + 3x + 2x$
19. $-2x - x + 3y$
20. $2x + 3y + 4x + 5y$
21. $9x + 3y + 2$
22. $6x - 2x + 3y + 6$
23. $x + 8x - 9x + 3$
24. $-4x - 8x + 3$
25. $-2(x + 3) + 6$
26. $2x + 3(x + 4) - 5$
27. $4(3 - 2x) - 2x$
28. $6 - (-x + 6) - x$
29. $2(2x + 5) - 10 - 4$
30. $-6(4 - 3x) - 18 + 4x$
31. $6 - 3(x + y) + 6x$
32. $3x - 6y + 2(4y + 8)$
33. $3 - (x - y) + (x - y)$
34. $(x + y) - (2x + 3y) + 4$

[2.2–2.5] *Solve.*

35. $4x = 4$
36. $x + 6 = -7$
37. $x - 4 = 7$
38. $\frac{x}{3} = -9$
39. $2x + 4 = 8$
40. $14 = 3 + 2x$
41. $8x - 3 = -19$
42. $6 - x = 9$
43. $-x = -12$
44. $3(x - 2) = 6$
45. $-3(2x - 8) = -12$
46. $4(6 + 2x) = 0$
47. $3x + 2x + 6 = -15$
48. $4 = -2(x + 3)$
49. $27 = 46 + 2x - x$
50. $4x + 6 - 7x + 9 = 18$
51. $4 + 3(x + 2) = 10$
52. $-3 + 3x = -2(x + 1)$

53. $9x - 6 = -3x + 30$

54. $-(x + 2) = 2(3x - 6)$

55. $2x + 6 = 3x + 9 - 3$

56. $-5x + 3 = 2x + 10$

57. $3x - 12x = 24 - 9x$

58. $2(x + 4) = -3(x + 5)$

59. $4(2x - 3) + 4 = 8x - 8$

60. $6x + 11 = -(6x + 5)$

61. $2(x + 7) = 6x + 9 - 4x$

62. $-5(3 - 4x) = -6 + 20x - 9$

63. $4(x - 3) - (x + 5) = 0$

64. $-2(4 - x) = 6(x + 2) + 3x$

[2.6] *Determine the following ratios. Write each ratio in lowest terms.*

65. 12 feet to 20 feet

66. 80 ounces to 12 pounds

67. 32 ounces to 2 pounds

Solve each proportion.

68. $\frac{x}{4} = \frac{8}{16}$

69. $\frac{5}{20} = \frac{x}{80}$

70. $\frac{3}{x} = \frac{15}{45}$

71. $\frac{20}{45} = \frac{15}{x}$

72. $\frac{6}{5} = \frac{-12}{x}$

73. $\frac{x}{9} = \frac{8}{-3}$

74. $\frac{-4}{9} = \frac{-16}{x}$

75. $\frac{x}{-15} = \frac{30}{-5}$

The following pairs of figures are similar. For each pair, find the length of the side indicated by x.

76.

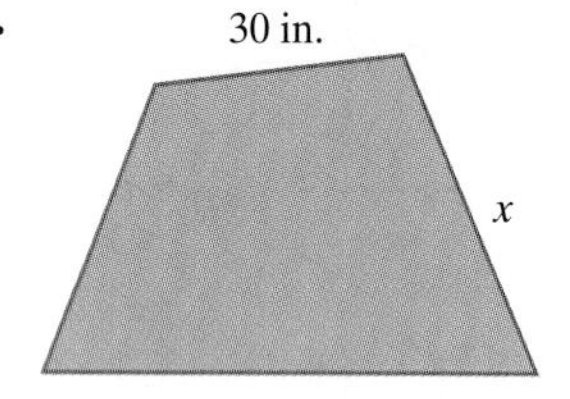

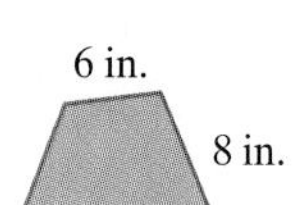

77.

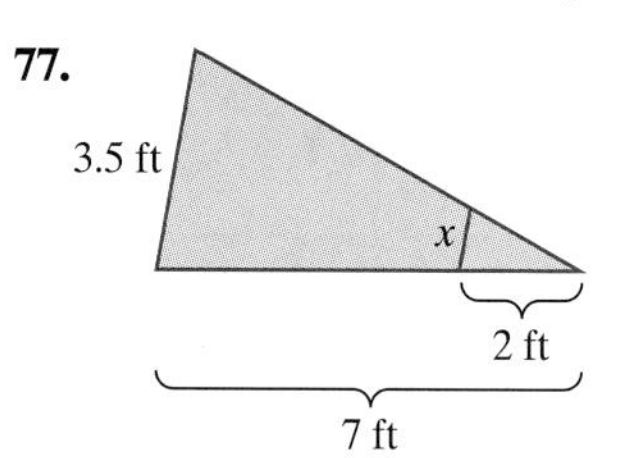

[2.7] *Solve each inequality, and graph the solution on a number line.*

78. $3x + 4 \geq 10$

79. $8 - 6x > 4x - 12$

80. $6 - 3x \leq 2x + 18$

81. $2(x + 4) \leq 2x - 5$

82. $2(x + 3) > 6x - 4x + 4$

83. $x + 6 > 9x + 30$

84. $x - 2 \leq -4x + 7$

85. $-(x + 2) < -2(-2x + 5)$

86. $-2(x - 4) \leq 3x + 6 - 5x$

87. $2(2x + 4) > 4(x + 2) - 6$

[2.6] *Set up a proportion and solve each problem.*

88. If a 4-ounce piece of cake has 160 calories, how many calories does a 6-ounce piece of that cake have?

89. If a copy machine can copy 20 pages per minute, how many pages can be copied in 22 minutes?

90. If the scale of a map is 1 inch to 60 miles, what distance on the map represents 380 miles?

91. Bryce Winston builds a model car to a scale of 1 inch to 0.9 feet. If the completed model is 10.5 inches, what is the size of the actual car?

92. If one U.S. dollar can be exchanged for 8.5410 Mexican pesos, find the value of 1 peso in terms of U.S. dollars.

93. If 3 radians equal 171.9 degrees, find the number of degrees in 1 radian.

94. If a machine can fill and cap 80 bottles of catsup in 50 seconds, how many bottles of catsup can it fill and cap in 2 minutes?

Practice Test

Use the distributive property to simplify.

1. $-6(4 - 2x)$

2. $-(x + 3y - 4)$

Simplify.

3. $5x - 8x + 4$

4. $4 + 2x - 3x + 6$

5. $-y - x - 4x - 6$

6. $x - 4y + 6x - y + 3$

7. $2x + 3 + 2(3x - 2)$

Solve.

8. $2x + 4 = 12$

9. $-x - 3x + 4 = 12$

10. $2x - 2 = 4x + 4$

11. $3(x - 2) = -(5 - 4x)$

12. $2x - 3(-2x + 4) = -13 + x$

13. $3x - 4 - x = 2(x + 5)$

14. $-3(2x + 3) = -2(3x + 1) - 7$

15. $\dfrac{9}{x} = \dfrac{3}{-15}$

16. $4(x - 3) - 2 = 2x - 14$

17. What do we call an equation that has **a)** exactly one solution, **b)** no solution, **c)** all real numbers as its solution?

Solve, and graph the solution on a number line.

18. $2x - 4 < 4x + 10$

19. $3(x + 4) \geq 5x - 12$

20. $4(x + 3) + 2x < 6x - 3$

21. $-(x - 2) - 3x = 4(1 - x) - 2$

22. The following figures are similar. Find the length of side x.

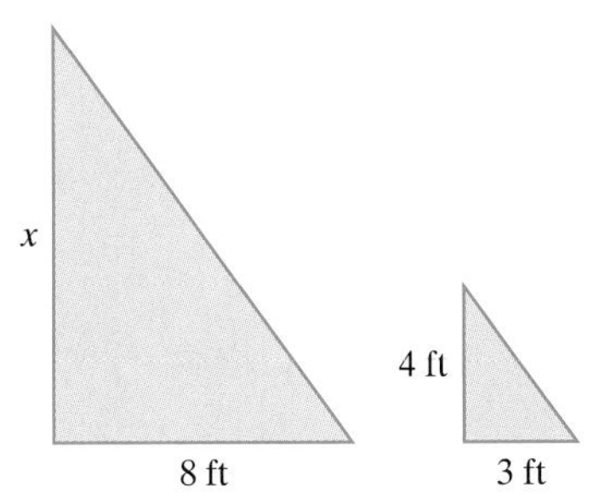

23. If 6 gallons of insecticide can treat 3 acres of land, how many gallons of insecticide are needed to treat 75 acres?

24. Assume a gas station owner makes a profit of 40 cents per gallon of gasoline sold. How many gallons of gasoline would he have to sell in a year to make a profit of $20,000 from gasoline sales?

25. While traveling, you notice that you traveled 25 miles in 35 minutes. If your speed does not change, how long will it take you to travel 125 miles?

Cumulative Review Test

1. Multiply $\dfrac{16}{20} \cdot \dfrac{4}{5}$.

2. Divide $\dfrac{8}{24} \div \dfrac{2}{3}$.

3. Insert $<$, $>$, or $=$ in the shaded area to make a true statement: $|-2|$ ☐ 1.

4. Evaluate $-7 - (-4) + 5 - 8$.

5. Subtract -6 from -7.

6. Evaluate $20 - 6 \div 3 \cdot 2$.

7. Evaluate $3[6 - (4 - 3^2)] - 30$.

8. Evaluate $-3x^2 - 4x + 5$ when $x = -2$.

9. Name the illustrated property.

$$(x + 4) + 6 = x + (4 + 6)$$

Simplify.

10. $8x + 2y + 4x - y$

11. $3x - 2x + 16 + 2x$

Solve.

12. $4x - 2 = 10$

13. $\dfrac{1}{4}x = -10$

14. $-6x - 5x + 6 = 28$

15. $4(x - 2) = 5(x - 1) + 3x + 2$

16. $\dfrac{15}{30} = \dfrac{3}{x}$

Solve, and graph the solution on a number line.

17. $x - 3 > 7$

18. $2x - 7 \leq 3x + 5$

19. A 36-pound bag of fertilizer can fertilize an area of 5000 square feet. How many pounds of fertilizer will Marisa Neilson need to fertilize her 22,000-square-foot lawn?

20. If Samuel earns $10.50 after working for 2 hours scrubbing boats at the marina, how much does he earn after 8 hours?

FORMULAS AND APPLICATIONS OF ALGEBRA

3

Use the Angel Web site at www.prenhall.com/angel to be linked to an internet resource that will help you further explore the following application.

Physical fitness has become an important part of people's daily lives. We find people using paths in local parks and nature trails to jog, bike, and rollerblade. Exercising with friends can be motivating and can add to the enjoyment. On page 212 we solve an equation based on the distance formula to determine how long it will take a person biking along the Pinellas Trail in Florida to meet a friend who had started rollerblading earlier in the day.

Preview and Perspective

The major goal of this chapter is to teach you the terminology and techniques to write real-life applications as equations. The equations are then solved using the techniques taught in Chapter 2. For mathematics to be relevant it must be useful. In this chapter we explain and illustrate many real-life applications of algebra. This is an important topic and we want you to learn it well and feel comfortable applying mathematics to real-life situations. Thus, we cover this chapter very slowly. You need to have confidence in your work, and you need to do all your homework. The more problems you attempt, the better you will become at setting up and solving application (or word) problems.

We begin this chapter with a discussion of formulas. We explain how to evaluate a formula and how to solve for a variable in a formula. You probably realize that most mathematics and science courses use a wide variety of formulas. Formulas are also used in many other disciplines, including the arts, business and economics, medicine, and technology.

In Section 3.2 we explain how to write real-life applications as equations. In Section 3.3 through 3.5 we solve a variety of application problems, including geometric, motion, and mixture problems.

3.1 FORMULAS

1) **Use the simple interest formula.**
2) **Use geometric formulas.**
3) **Solve for a variable in a formula.**

A **formula** is an equation commonly used to express a specific relationship mathematically. For example, the formula for the area of a rectangle is

$$\text{area} = \text{length} \cdot \text{width} \qquad \text{or} \qquad A = lw$$

To **evaluate a formula**, substitute the appropriate numerical values for the variables and perform the indicated operations.

1) Use the Simple Interest Formula

A formula used in banking is the **simple interest formula**

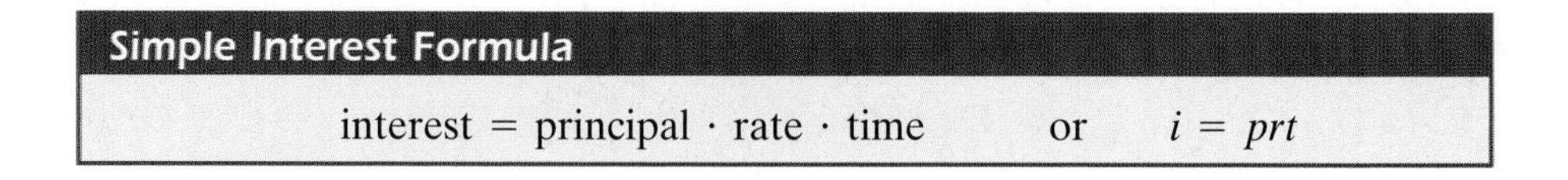

Simple Interest Formula

$$\text{interest} = \text{principal} \cdot \text{rate} \cdot \text{time} \qquad \text{or} \qquad i = prt$$

This formula is used to determine the simple interest, i, earned on some savings accounts, or the simple interest an individual must pay on certain loans. In the simple interest formula $i = prt$, p is the principal (the amount invested or borrowed), r is the interest rate in decimal form, and t is the amount of time of the investment or loan.

EXAMPLE 1 To buy a car, Lou McGuire borrowed \$6000 from a bank for 3 years. The bank charged 9% simple interest for the loan. How much interest will Lou owe the bank?

Solution **Understand and Translate** Since the bank charged simple interest, we use the simple interest formula to solve the problem. We are given that the rate, r, is 9%, or 0.09, in decimal form. The principal, p, is \$6000 and the time, t, is 3 years. We substitute these values in the simple interest formula and solve for the interest, i.

$$i = prt$$

Carry Out

$$i = 6000(0.09)(3)$$

$$i = 1620$$

Answer Lou will pay \$1620 interest. After 3 years, when he repays the loan, he will pay the principal, \$6000, plus the interest, \$1620, for a total of \$7620.

NOW TRY EXERCISE 81

EXAMPLE 2 Benito Moretti invests \$5000 in a savings account that earns simple interest for 2 years. If the interest earned from the account is \$800, find the rate.

Solution **Understand and Translate** We use the simple interest formula, $i = prt$. We are given the principal, p, the time, t, and the interest, i. We are asked to find the rate, r. We substitute the given values in the simple interest formula and solve the resulting equation for r.

$$i = prt$$

$$800 = 5000(r)(2)$$

Carry Out

$$800 = 10{,}000r$$

$$\frac{800}{10{,}000} = \frac{10{,}000r}{10{,}000}$$

$$0.08 = r$$

Answer The simple interest rate is 0.08, or 8% per year.

2) Use Geometric Formulas

The **perimeter**, P, is the sum of the lengths of the sides of a figure. Perimeters are measured in the same common unit as the sides. For example, perimeter may be measured in centimeters, inches, or feet. The **area**, A, is the total surface within the figure's boundaries. Areas are measured in square units. For example, area may be measured in square centimeters, square inches, or square feet. Table 3.1 gives the formulas for finding the areas and perimeters of triangles and quadrilaterals. **Quadrilateral** is a general name for a four-sided figure.

TABLE 3.1 Formulas for Areas and Perimeters of Quadrilaterals and Triangles

Figure	Sketch	Area	Perimeter
Square	s	$A = s^2$	$P = 4s$
Rectangle	w, l	$A = lw$	$P = 2l + 2w$
Parallelogram	h, w, l	$A = lh$	$P = 2l + 2w$
Trapezoid	b, a, h, c, d	$A = \frac{1}{2}h(b + d)$	$P = a + b + c + d$
Triangle	a, h, c, b	$A = \frac{1}{2}bh$	$P = a + b + c$

In Table 3.1, the letter h is used to represent the *height* of the figure. In the figure of the trapezoid, the sides b and d are called the *bases* of the trapezoid. In the triangle, the side labeled b is called the *base* of the triangle.

EXAMPLE 3 Sandra Ivey's rectangular vegetable garden is 12 feet long and 6 feet wide (Fig. 3.1).

a) If Sandra wants to put fencing around the garden to keep the animals out, how much fencing will she need?

b) What is the area of Sandra's garden?

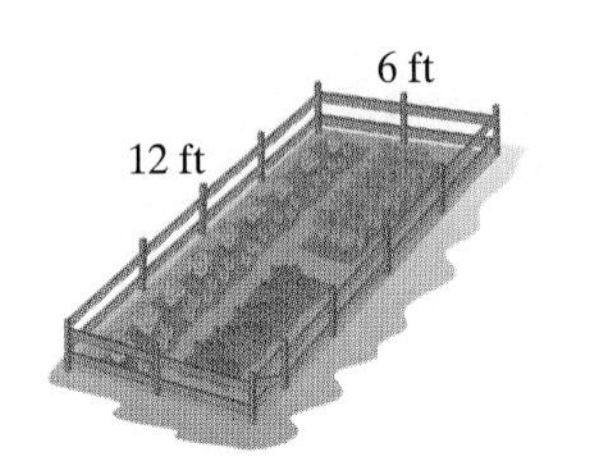

FIGURE 3.1

Solution a) Understand and Translate To find the amount of fencing required, we need to find the perimeter of the garden. Substitute 12 for l and 6 for w in the formula for the perimeter of a rectangle.

$$P = 2l + 2w$$

Carry Out $$P = 2(12) + 2(6) = 24 + 12 = 36$$

Answer Thus, Sandra will need 36 feet of fencing.

b) Substitute 12 for l and 6 for w in the formula for the area of a rectangle.

$$A = lw$$

$$A = (12)(6) = 72 \text{ square feet (or } 72 \text{ ft}^2)$$

Sandra's vegetable garden has an area of 72 square feet.

EXAMPLE 4 David Lopez recently purchased a new camera that can take panoramic photos, like the one shown, in addition to regular photos. A panoramic photo has a perimeter of 27 inches and a length of 10 inches. Find the width of a panoramic photo.

Solution **Understand and Translate** The perimeter, P, is 27 inches and the length, l, is 10 inches. Substitute these values into the formula for the perimeter of a rectangle and solve for the width, w.

$$P = 2l + 2w$$

$$27 = 2(10) + 2w$$

Carry Out

$$27 = 20 + 2w$$

$$27 - 20 = 20 - 20 + 2w \qquad \text{Subtract 20 from both sides.}$$

$$7 = 2w$$

$$\frac{7}{2} = \frac{2w}{2} \qquad \text{Divide both sides by 2.}$$

$$\frac{7}{2} = w$$

$$3.5 = w$$

NOW TRY EXERCISE 85

Answer The width of the photo is 3.5 inches.

EXAMPLE 5

Karin Wagner owns a small sailboat, and she will need to replace a triangular sail shortly because of wear. When ordering the sail, she needs to specify the base and height of the sail. She measures the base and finds that it is 5 feet (see Fig. 3.2). She also remembers that the sail has an area of 30 square feet. She does not want to have to take the sail down to find its height, so she uses algebra to find its height. Find the height of Karin's sail.

Solution **Understand and Translate** We use the formula for the area of a triangle given in Table 3.1.

$$A = \frac{1}{2}bh$$

$$30 = \frac{1}{2}(5)h$$

Carry Out

$$2 \cdot 30 = 2 \cdot \frac{1}{2}(5)h \qquad \text{Multiply both sides by 2.}$$

$$60 = 5h$$

$$\frac{60}{5} = \frac{5h}{5} \qquad \text{Divide both sides by 5.}$$

$$12 = h$$

FIGURE 3.2

Answer The height of the triangle is 12 feet, and thus the height of the sail is 12 feet.

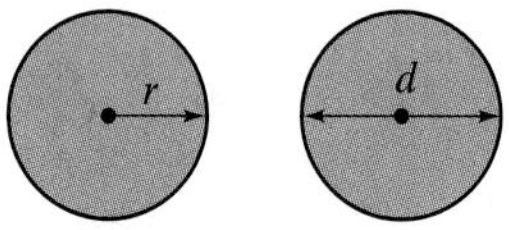

(a) (b)

FIGURE 3.3

Another figure that we see and use daily is the circle. The **circumference**, C, is the length (or perimeter) of the curve that forms a circle. The **radius**, r, is the line segment from the center of the circle to any point on the circle (Fig. 3.3a). The **diameter** of a circle is a line segment through the center whose endpoints both lie on the circle (Fig. 3.3b). *Note that the length of the diameter is twice the length of the radius.*

The formulas for both the area and the circumference of a circle are given in Table 3.2.

TABLE 3.2 Formulas for Circles

Circle	Area	Circumference
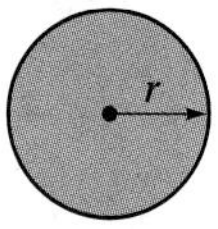	$A = \pi r^2$	$C = 2\pi r$

The value of **pi**, symbolized by the Greek lowercase letter π, is *approximately* 3.14.

Using Your Calculator

Scientific and graphing calculators have a key for finding the value of π. If you press the $\boxed{\pi}$ key, your calculator may display 3.1415927. This is only an approximation of π since π is an irrational number. If you own a scientific or graphing calculator, use the $\boxed{\pi}$ key when evaluating expressions containing π. If your calculator does not have a $\boxed{\pi}$ key, use 3.14 to approximate it. *When evaluating an expression containing π, we will use the $\boxed{\pi}$ key on a calculator to obtain the answer.* The final answer displayed in the text or answer section may therefore be slightly different (and more accurate) than yours if you use 3.14 for π.

EXAMPLE 6 The University Medical Center has a circular landing pad, for medical helicopters to land, near the entrance to the emergency room. Determine the area and circumference of the circular landing pad if its diameter is 40 feet.

Solution The radius is half its diameter, so $r = \dfrac{40}{2} = 20$ feet.

$$A = \pi r^2 \qquad\qquad C = 2\pi r$$
$$A = \pi(20)^2 \qquad\qquad C = 2\pi(20)$$
$$A = \pi(400) \qquad\qquad C \approx 125.66 \text{ feet}$$
$$A \approx 1256.64 \text{ square feet}$$

To obtain our answer of 1256.64, we used the $\boxed{\pi}$ key on a calculator and rounded our final answer to the nearest hundredth. If you do not have a calculator with a $\boxed{\pi}$ key and use 3.14 for π, your answer for the area would be 1256.

NOW TRY EXERCISE 89

Table 3.3 gives formulas for finding the volume of certain **three-dimensional figures**. **Volume** is measured in cubic units, such as cubic centimeters or cubic feet.

TABLE 3.3 Formulas for Volumes of Three-Dimensional Figures

Figure	Sketch	Volume
Rectangular solid	h, w, l	$V = lwh$
Right circular cylinder	h, r	$V = \pi r^2 h$
Right circular cone	h, r	$V = \frac{1}{3}\pi r^2 h$
Sphere	r	$V = \frac{4}{3}\pi r^3$

EXAMPLE 7 Paige Akins, who plays basketball regularly, is curious about the volume of air in a basketball. Using the Internet she finds that a basketball has a diameter of 9 inches (Fig. 3.4). Find the volume of air in the basketball.

FIGURE 3.4

Solution **Understand and Translate** Table 3.3 gives the formula for the volume of a sphere. The formula involves the radius, r. Since its diameter is 9 inches, its radius is $\frac{9}{2}$ or 4.5 inches.

$$V = \frac{4}{3}\pi r^3$$

Carry Out

$$V = \frac{4}{3}\pi(4.5)^3 = \frac{4}{3}\pi(91.125) \approx 381.70$$

Answer Therefore, a basketball contains about 381.70 cubic inches of air. If you used 3.14 for π, your answer would be 381.51.

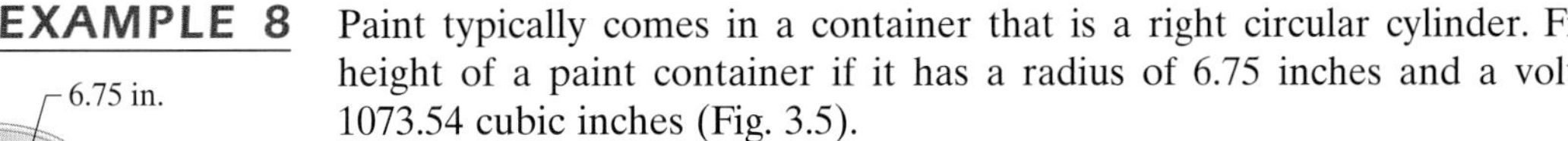

EXAMPLE 8 Paint typically comes in a container that is a right circular cylinder. Find the height of a paint container if it has a radius of 6.75 inches and a volume of 1073.54 cubic inches (Fig. 3.5).

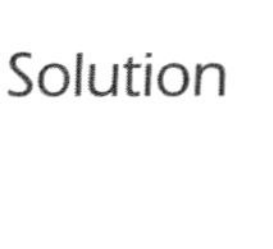

FIGURE 3.5

Solution **Understand and Translate** We are given that $V = 1073.54$ and $r = 6.75$. Substitute these values into the formula for the volume of a right circular cylinder and solve for the height, h.

$$V = \pi r^2 h$$

$$1073.54 = \pi(6.75)^2 h$$

Carry Out

$$1073.54 = \pi(45.5625)h$$

$$\frac{1073.54}{45.5625\pi} = \frac{\cancel{\pi(45.5625)}h}{\cancel{45.5625\pi}} \qquad \text{Divide both sides by } 45.5625\pi.$$

$$\frac{1073.54}{45.5625\pi} = h$$

$$7.5 \approx h$$

Answer Thus, the container is about 7.5 inches tall.

EXAMPLE 9 A quadrilateral is a polygon that has four sides (Fig. 3.6). Notice it has two diagonals. The number of diagonals, d, in a polygon of n sides is given by the formula $d = \frac{1}{2}n^2 - \frac{3}{2}n$.

a) How many diagonals does a pentagon (five sides) have?

b) How many diagonals does an octagon (eight sides) have?

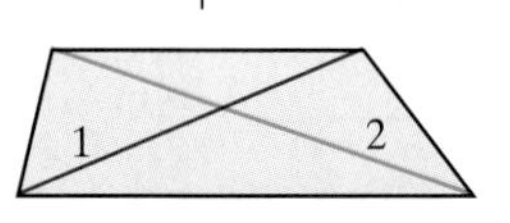

FIGURE 3.6

Solution

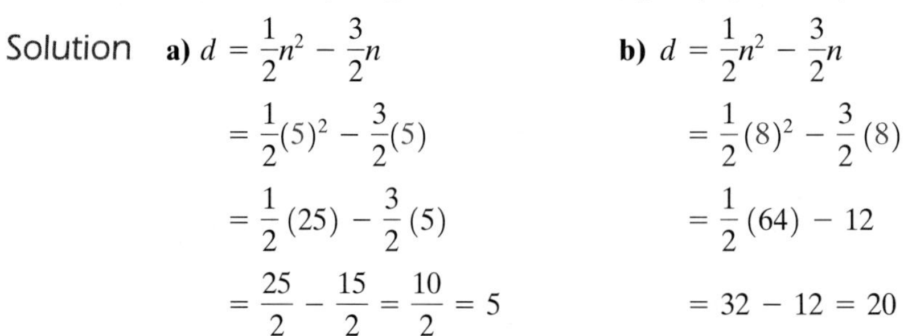

a) $d = \frac{1}{2}n^2 - \frac{3}{2}n$

$= \frac{1}{2}(5)^2 - \frac{3}{2}(5)$

$= \frac{1}{2}(25) - \frac{3}{2}(5)$

$= \frac{25}{2} - \frac{15}{2} = \frac{10}{2} = 5$

b) $d = \frac{1}{2}n^2 - \frac{3}{2}n$

$= \frac{1}{2}(8)^2 - \frac{3}{2}(8)$

$= \frac{1}{2}(64) - 12$

$= 32 - 12 = 20$

NOW TRY EXERCISE 67

A pentagon has 5 diagonals and an octagan has 20 diagonals.

3) Solve for a Variable in a Formula

Often in this course and in other mathematics and science courses, you will be given an equation or formula solved for one variable and have to solve it for a different variable. We will now learn how to do this. This material will reinforce what you learned about solving equations in Chapter 2. We will use the procedures learned here to solve problems in many other sections of the text.

To solve for a variable in a formula, treat each of the quantities, except the one for which you are solving, as if they were constants. Then solve for the desired variable by isolating it on one side of the equation, as you did in Chapter 2.

EXAMPLE 10 The formula for the area of a rectangle is $A = lw$. Solve this formula for the width, w.

Solution We must get w by itself on one side of the equation. We begin by removing the l from the right side of the equation to isolate the w.

$$A = lw$$

$$\frac{A}{l} = \frac{\cancel{l}w}{\cancel{l}} \qquad \text{Divide both sides by } l.$$

$$\frac{A}{l} = w$$

EXAMPLE 11 The formula for the perimeter of a rectangle is $P = 2l + 2w$. Solve this formula for the length, l.

Solution We must get l all by itself on one side of the equation. We begin by removing the $2w$ from the right side of the equation to isolate the term containing the l.

$$P = 2l + 2w$$

$$P - 2w = 2l + 2w - 2w \quad \text{Subtract } 2w \text{ from both sides.}$$

$$P - 2w = 2l$$

$$\frac{P - 2w}{2} = \frac{2l}{2} \quad \text{Divide both sides by 2.}$$

$$\frac{P - 2w}{2} = l \quad \left(\text{or} \quad l = \frac{P}{2} - w\right)$$

EXAMPLE 12 An important concept that we will discuss in Chapter 4 is slope. In the equation $y = mx + b$, the letter m represents the slope. Solve this equation for the slope, m.

Solution We must get the m by itself on one side of the equal sign.

$$y = mx + b$$

$$y - b = mx + b - b \quad \text{Subtract } b \text{ from both sides.}$$

$$y - b = mx$$

$$\frac{y - b}{x} = \frac{mx}{x} \quad \text{Divide both sides by } x.$$

$$\frac{y - b}{x} = m \quad \left(\text{or} \quad m = \frac{y}{x} - \frac{b}{x}\right)$$

NOW TRY EXERCISE 63

EXAMPLE 13 We used the simple interest formula, $i = prt$, in Example 2. Solve the simple interest formula for the principal, p.

Solution We must isolate the p. Since p is multiplied by both r and t, we divide both sides of the equation by rt.

$$i = prt$$

$$\frac{i}{rt} = \frac{prt}{rt}$$

$$\frac{i}{rt} = p$$

When we discuss graphing, in Chapter 4, we will need to solve many equations for the variable y. Also, when graphing an equation on a graphing calculator, you will need to solve the equation for y before you can graph it. The procedure to solve an equation for y is illustrated in Example 14.

EXAMPLE 14 **a)** Solve the equation $2x + 3y = 12$ for y.

b) Find the value of y when $x = 6$.

Solution **a)** Begin by isolating the term containing the variable y.

$$2x + 3y = 12$$
$$2x - 2x + 3y = 12 - 2x \qquad \text{Subtract } 2x \text{ from both sides.}$$
$$3y = 12 - 2x$$
$$\frac{3y}{3} = \frac{12 - 2x}{3} \qquad \text{Divide both sides by 3.}$$
$$y = \frac{12 - 2x}{3} \quad \left(\text{or} \quad y = \frac{12}{3} - \frac{2x}{3} = 4 - \frac{2}{3}x\right)$$

b) To find the value of y when x is 6, substitute 6 for x in the equation solved for y in part **a)**.

$$y = \frac{12 - 2x}{3}$$
$$y = \frac{12 - 2(6)}{3} = \frac{12 - 12}{3} = \frac{0}{3} = 0$$

NOW TRY EXERCISE 31

We see that when $x = 6$, $y = 0$.

Some formulas contain fractions. When a formula contains a fraction, we can eliminate the fraction by multiplying both sides of the equation by the denominator, as illustrated in Example 15.

EXAMPLE 15 In Section 1.2 we discussed the procedure to find the mean (average) of a set of data. The mean of two values, m and n, may be found by the formula $A = \frac{m + n}{2}$. Solve this formula for m.

Solution We begin by multiplying both sides of the equation by 2 to eliminate the fraction. Then we isolate the variable m.

$$A = \frac{m + n}{2}$$
$$2A = 2\left(\frac{m + n}{2}\right) \qquad \text{Multiply both sides by 2.}$$
$$2A = m + n$$
$$2A - n = m + n - n \qquad \text{Subtract } n \text{ from both sides.}$$
$$2A - n = m$$

Thus, $m = 2A - n$.

Exercise Set 3.1

Concept/Writing Exercises

1. What is a formula?
2. What does it mean to *evaluate a formula*?
3. Write the simple interest formula, then indicate what each letter in the formula represents.
4. What is a quadrilateral?
5. What is the relationship between the radius and the diameter of a circle?
6. Is π equal to 3.14? Explain your answer.
7. By using any formula for area, explain why area is measured in square units.
8. By using any formula for volume, explain why volume is measured in cubic units.

Practice the Skills

Use the formula to find the value of the variable indicated. Use a calculator to save time and where necessary, round your answer to the nearest hundredth.

9. $P = 4s$ (perimeter of a square); find P when $s = 4$.
10. $A = s^2$ (area of a square); find A when $s = 5$.
11. $A = lw$ (area of a rectangle); find A when $l = 12$ and $w = 8$.
12. $P = 2l + 2w$ (perimeter of a rectangle); find P when $l = 6$ and $w = 5$.
13. $c = 2.54i$ (to change inches to centimeters); find c when $i = 12$.
14. $f = 1.47m$ (to change speed from mph to ft/sec); find f when $m = 60$.
15. $A = \pi r^2$ (area of a circle); find A when $r = 4$.
16. $p = i^2 r$ (formula for finding electical power); find r when $p = 2000$ and $i = 4$.
17. $z = \frac{x - m}{s}$ (statistics formula for finding the z-score); find z when $x = 100$, $m = 80$, and $s = 10$.
18. $A = \frac{1}{2}bh$ (area of a triangle); find h when $A = 30$ and $b = 6$.
19. $V = \frac{1}{3}Bh$ (volume of a cone); find h when $V = 60$ and $B = 12$.
20. $P = 2l + 2w$ (perimeter of a rectangle); find l when $P = 28$ and $w = 6$.
21. $A = \frac{m + n}{2}$ (mean of two values); find n when $A = 36$ and $m = 16$.
22. $F = \frac{9}{5}C + 32$ (for converting Celsius temperature to Fahrenheit); find F when $C = 10$.
23. $A = P(1 + rt)$ (banking formula to find the amount in an account); find r when $A = 1050$, $t = 1$, and $P = 1000$.
24. $V = \pi r^2 h$ (volume of a cylinder); find h when $V = 678.24$ and $r = 6$.
25. $V = \frac{4}{3}\pi r^3$ (volume of a sphere); find V when $r = 6$.
26. $C = \frac{5}{9}(F - 32)$ (for converting Fahrenheit temperature to Celsius); find F when $C = 68$.
27. $B = \frac{703w}{h^2}$ (for finding body mass index); find w when $B = 24$ and $h = 61$.
28. $P = \frac{f}{1 + i}$ (investment banking formula); find i when $P = 3738.32$ and $f = 4000$.
29. $S = C + rC$ (for determining selling price when an item is marked up); find S when $C = 160$ and $r = 0.12$ (or 12%).
30. $S = R - rR$ (for determining sale price when an item is discounted); find R when $S = 92$ and $r = 0.08$ (or 8%).

*In Exercises 31–42, **a)** solve each equation for y, then **b)** find the value of y for the given value of x.*

31. $3x + y = 5, x = 2$
32. $6x + 2y = -12, x = -3$
33. $4x = 6y - 8, x = 10$
34. $-3x - 5y = -10, x = 0$
35. $2y = 6 - 3x, x = 2$
36. $15 = 3y - x, x = 3$
37. $-3x + 5y = -10, x = 4$
38. $3x - 2y = -18, x = -1$
39. $-3x = 18 - 6y, x = 0$
40. $-12 = -2x - 3y, x = -2$
41. $-8 = -x - 2y, x = -4$
42. $2x + 5y = 20, x = -5$

Solve for the indicated variable.

43. $P = 4s$, for s
44. $A = lw$, for w
45. $d = rt$, for r
46. $i = prt$, for p
47. $C = \pi d$, for d
48. $V = lwh$, for l
49. $A = \frac{1}{2}bh$, for h
50. $E = IR$, for I
51. $P = 2l + 2w$, for w
52. $PV = KT$, for T
53. $6n + 3 = m$, for n
54. $5t - 2r = 25$, for t
55. $y = mx + b$, for b
56. $y = mx + b$, for x
57. $A = P + Prt$, for r
58. $A = \frac{m + d}{2}$, for m
59. $A = \frac{m + 2d}{3}$, for d
60. $R = \frac{l + 3w}{2}$, for w
61. $d = a + b + c$, for b
62. $A = \frac{a + b + c}{3}$, for b
63. $ax + by = c$, for y
64. $ax + by + c = 0$, for y
65. $V = \pi r^2 h$, for h
66. $V = \frac{1}{3}\pi r^2 h$, for h

Use the formula in Example 9, $d = \frac{1}{2}n^2 - \frac{3}{2}n$, to find the number of diagonals in a figure with the given number of sides.

67. 10 sides

68. 6 sides

Use the formula $C = \frac{5}{9}(F - 32)$ to find the Celsius temperature (C) equivalent to the given Fahrenheit temperature (F).

69. $F = 50°$

70. $F = 86°$

Use the formula $F = \frac{9}{5}C + 32$, to find the Fahrenheit temperature (F) equivalent to the given Celsius tempertature (C).

71. $C = 35°$

72. $C = 10°$

In chemistry the ideal gas law is $P = KT/V$ where P is pressure, T is temperature, V is volume, and K is a constant. Find the missing quantity.

73. $T = 10, K = 1, V = 1$

74. $T = 30, P = 3, K = 0.5$

75. $P = 80, T = 100, V = 5$

76. $P = 100, K = 2, V = 6$

Problem Solving

77. Consider the formula for the area of a square, $A = s^2$. If the length of the side of a square, s, is doubled, what is the change in its area?

78. Consider the formula for the volume of a cube, $V = s^3$. If the length of the side of a cube, s, is doubled, what is the change in its volume?

The sum of the first n even numbers can be found by the formula $S = n^2 + n$. Find the sum of the numbers indicated.

79. First 5 even numbers

80. First 10 even numbers

In Exercises 81–84, use the simple interest formula.

81. Thang Tran decided to borrow \$6000 from Citibank to help pay for a car. His loan was for 3 years at a simple interest rate of 8%. How much interest will Thang pay?

82. Danielle Maderi lent her brother \$4000 for a period of 2 years. At the end of the 2 years, her brother repaid the \$4000 plus \$640 interest. What simple interest rate did her brother pay?

83. Kate Lynch invested a certain amount of money in a savings account paying 7% simple interest per year. When she withdrew her money at the end of 3 years, she received \$1050 in interest. How much money did Kate place in the savings account?

84. Peter Ostroushko put \$6000 in a savings account earning $7\frac{1}{2}\%$ simple interest per year. When he withdrew his money, he received \$1800 in interest. How long had he left his money in the account?

Use the formula given in Tables 3.1, 3.2, and 3.3 to work Exercises 85–98.

85. To display some items at a convention, Cynthia Lennon used a triangular table top whose sides were 12 feet, 8 feet, and 5 feet. Find the perimeter of the table top.

86. The screen of the Texas Instruments rectangular display screen (or window) is 2.5 inches by 1.5 inches. Find the area of the window.

87. A yield traffic sign is triangular with a base of 36 inches and a height of 31 inches. Find the area of the sign.

88. Milt McGowen has a rectangular lot that measures 100 feet by 60 feet. If Milt wants to fence in his lot, how much fencing will he need?

89. Katherine Butler purchases a set of five circular mats to place below potted plants. Each mat has a radius of 26 centimeters. Find the area of each mat.

90. David Hughes has a landscaping service. He is going to be placing a circular border of diameter 10 feet around each tree on Katherine Grassi's lot. Inside the border he will plant flowers. Find the circumference of each border.

91. Canter Martin made a sign to display at a baseball game. The sign was in the shape of a trapezoid. Its bases are 4 feet and 3 feet, and its height is 2 feet. Find the area of the sign.

92. The area of the smallest post office in America (in Ochopee, Florida) is 48 square feet. The length of the post office is 6 feet. Find the width of the post office.

93. On each end of a shuffleboard is a triangle (see the figure). Jagat Singh is building his own court and wishes to purchase green paint for the triangular shaded areas at each end (the "10 off" will be painted red). Find the triangular area shown in green.

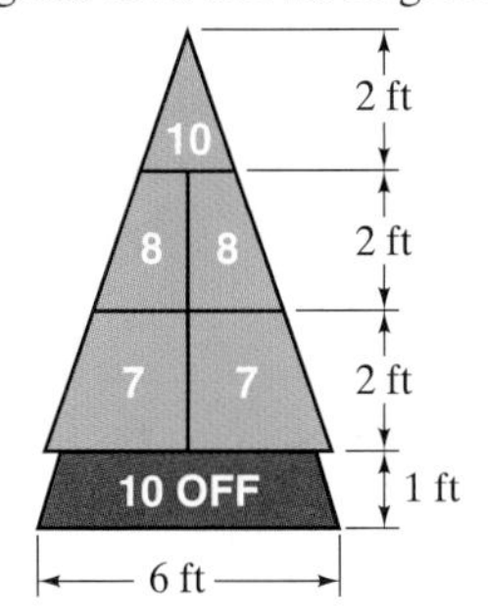

94. The largest banyan tree in the continental United States is at the Edison House in Fort Myers, Florida. The circumference of the aerial roots of the tree is 390 feet.

a) Find the radius of the aerial roots to the nearest tenth of a foot.

b) Find the diameter of the aerial roots to the nearest tenth of a foot.

95. The seats in an amphitheater are inside a trapezoidal area as shown in the figure.

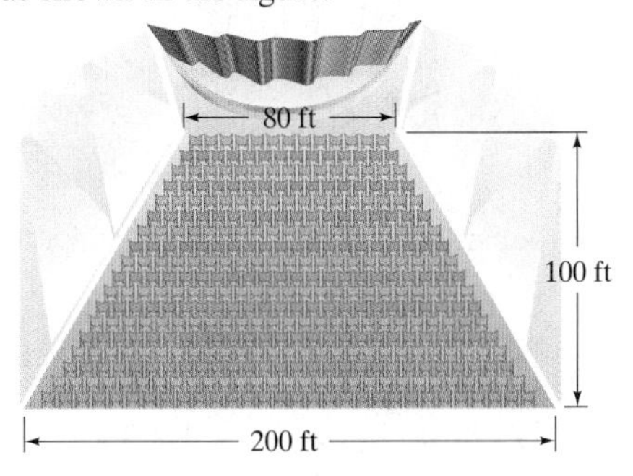

The bases of the trapezoidal area are 80 feet and 200 feet, and the height is 100 feet. Find the area of the floor occupied by seats.

96. Thomas Johnson has an empty oil drum that he uses for storage. The oil drum is 4 feet high and has a diameter of 24 inches. Find the volume of the drum in cubic feet.

97. Find the volume of an ice cream cone (cone only) if its diameter is 3 inches and its height is 5 inches.

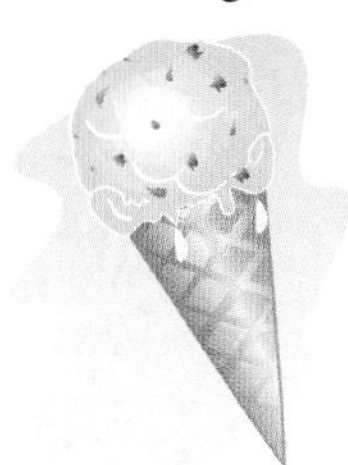

98. The top of the opening of a mailbox is semicircular and the bottom part is rectangular.

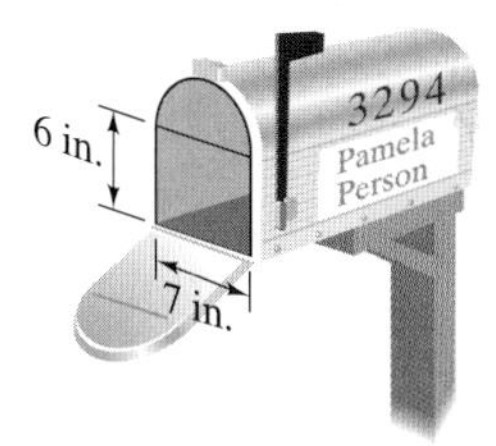

The rectangular section has the dimensions shown in the figure. Find the cross-sectional area of the mailbox opening.

99. A person's body mass index (BMI) is found by multiplying a person's weight, w, in pounds by 703, then dividing this product by the square of the person's height, h, in inches.

a) Write a formula to find the BMI.

b) Brandy Belmont is 5 feet 3 inches tall and weighs 135 pounds. Find her BMI.

100. Refer to Exercise 99. Mario Guzza's weight is 162 pounds, and he is 5 feet 7 inches tall. Find his BMI.

101. **a)** Consider the formula for the circumference of a circle, $C = 2\pi r$. If you solve this formula for π, what will you obtain?

b) If you take the ratio of the circumference of a circle to its diameter, about what numerical value will you obtain? Explain how you determined your answer.

c) Carefully draw a circle, at least 4 inches in diameter. Use a piece of string and a ruler to determine the circumference and diameter of the circle. Find the ratio of the circumference to the diameter. When you divide the circumference by the diameter, what value do you obtain?

Challenge Problems

102. **a)** Using the formulas presented in this section, write an equation in d that can be used to find the shaded area in the figure shown.

b) Find the shaded area when $d = 4$ feet.

c) Find the shaded area when $d = 6$ feet.

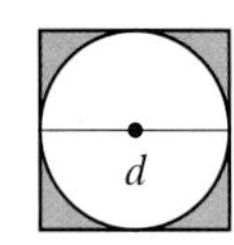

103. A cereal box is to be made by folding the cardboard along the dashed lines as shown in the figure on the right.

a) Using the formula

$$\text{volume} = \text{length} \cdot \text{width} \cdot \text{height}$$

write an equation for the volume of the box.

b) Find the volume of the box when $x = 7$ cm.

c) Write an equation for the surface area of the box.

d) Find the surface area when $x = 7$ cm.

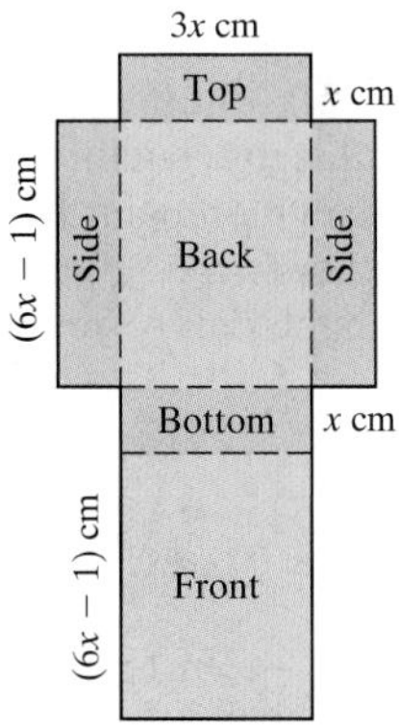

104. Linda Taneff owns a dairy farm near Columbus, Ohio. On the farm she has a silo whose bottom part is a right circular cylinder and whose top is half a sphere. She uses the silo to hold corn for her livestock. The silo has dimensions as shown in the figure. Determine the volume of corn the silo can hold.

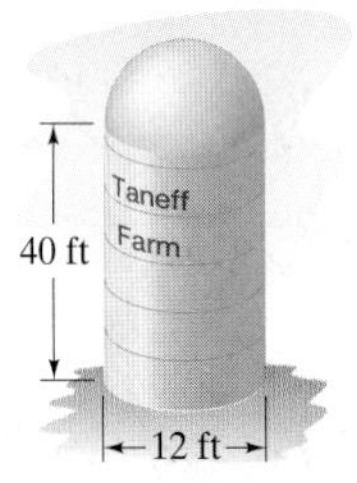

Group Activity

105. The end table shown has three levels. The top level is a right triangle. The middle level is a trapezoid and the bottom level is a rectangle. Assume that on each level the glass in the center is surrounded on all sides by a wood border $1\frac{1}{2}$ inches wide. The dimensions of each of the three levels (from the ends of the wood border) are shown in the figure below.

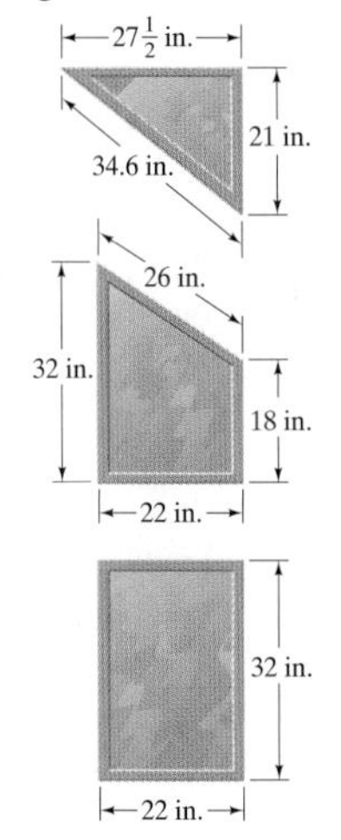

a) Group member 1: Find the area of the top level and the glass in the top level.

b) Group member 2: Repeat part **a)** for the middle level.

c) Group member 3: Repeat part **a)** for the lower level.

d) As a group, determine the total area of the table tops.

Cumulative Review Exercises

[1.9] **106.** Evaluate. $[4(12 \div 2^2 - 3)^2]^2$.

[2.6] **107.** A stable has four Morgan and six Arabian horses. Find the ratio of Arabians to Morgans.

108. It takes 3 minutes to siphon 25 gallons of water out of a swimming pool. How long will it take to empty a 13,500-gallon swimming pool by siphoning? Write a proportion that can be used to solve the problem, and then find the desired value.

[2.7] **109.** Solve $2(x - 4) \geq 3x + 9$.

3.2 CHANGING APPLICATION PROBLEMS INTO EQUATIONS

SSM VIDEO 3.2 CD Rom

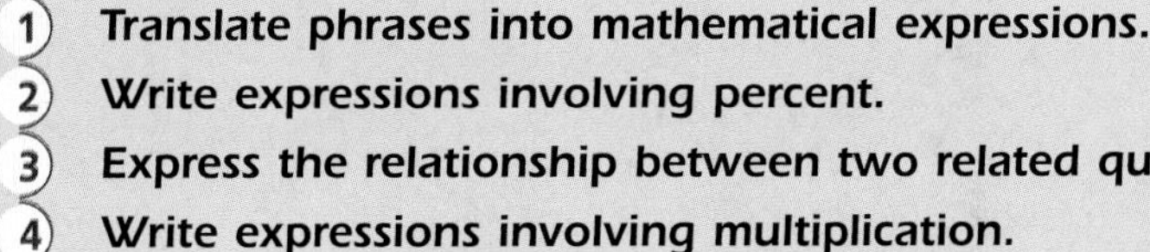

1) **Translate phrases into mathematical expressions.**
2) **Write expressions involving percent.**
3) **Express the relationship between two related quantities.**
4) **Write expressions involving multiplication.**
5) **Translate applications into equations.**

1) Translate Phrases into Mathematical Expressions

One practical advantage of knowing algebra is that you can use it to solve everyday problems involving mathematics. For algebra to be useful in solving everyday problems, you must first be able to *translate application problems into mathematical language*. The purpose of this section is to help you take an application problem, also referred to as a *word* or *verbal problem*, and write it as a mathematical equation.

Often the most difficult part of solving an application problem is translating it into an equation. Before you can translate a problem into an equation, you must understand the meaning of certain words and phrases and how they are expressed mathematically. Table 3.4 is a list of selected words and phrases and the operations they imply. We used the variable x. However, any variable could have been used.

TABLE 3.4

Word or Phrase	Operation	Statement	Algebraic Form
Added to More than Increased by The sum of	Addition	7 *added to* a number 5 *more than* a number A number *increased by* 3 *The sum of* a number and 4	$x + 7$ $x + 5$ $x + 3$ $x + 4$
Subtracted from Less than Decreased by The difference between	Subtraction	6 *subtracted from* a number 7 *less than* a number A number *decreased by* 5 The *difference between* a number and 9	$x - 6$ $x - 7$ $x - 5$ $x - 9$
Multiplied by The product of Twice a number, 3 times a number, etc. Of, when used with a percent or fraction	Multiplication	A number *multiplied by* 6 *The product of* 4 and a number *Twice a number* 20% *of* a number	$6x$ $4x$ $2x$ $0.20x$
Divided by The quotient of One-half of a number, one third, etc.	Division	A number *divided by* 8 *The quotient of* a number and 6 *One-seventh of a number*	$\frac{x}{8}$ $\frac{x}{6}$ $\frac{x}{7}$

Often a statement may contain more than one operation. The following chart provides some examples of this.

Statement	Algebraic Form
Four more than twice a number	$\underbrace{2x}_{\text{Twice a number}} + 4$
Five less than 3 times a number	$\underbrace{3x}_{\text{Three times a number}} - 5$
Three times the sum of a number and 8	$3\underbrace{(x + 8)}_{\text{The sum of a number and 8}}$
Twice the difference between a number and 4	$2\underbrace{(x - 4)}_{\text{The difference between a number and 4}}$

To give you more practice with the mathematical terms, we will also convert some algebraic expressions into statements. Often an algebraic expression can be written in several different ways. Following is a list of some of the possible statements that can be used to represent the given algebraic expression.

Algebraic	Statements
$2x + 3$	Three more than twice a number The sum of twice a number and 3 Twice a number, increased by 3 Three added to twice a number
$3x - 4$	Four less than 3 times a number Three times a number, decreased by 4 The difference between 3 times a number and 4 Four subtracted from 3 times a number

EXAMPLE 1 Express each statement as an algebraic expression.

a) The distance, d, increased by 10 miles

b) Six less than twice the area, a

c) Three pounds more than 4 times the weight, w

d) Twice the sum of the height, h, and 3 feet

Solution **a)** $d + 10$ **b)** $2a - 6$

c) $4w + 3$ **d)** $2(h + 3)$

EXAMPLE 2 Write three different statements to represent the following expressions.

a) $5x - 2$ **b)** $2x + 7$

Solution **a) 1.** Two less than 5 times a number

2. Five times a number, decreased by 2

3. The difference between 5 times a number and 2

b) **1.** Seven more than twice a number
2. Two times a number, increased by 7
3. The sum of twice a number and 7

EXAMPLE 3 Write a statement to represent each expression.

a) $3x - 4$ **b)** $3(x - 4)$

Solution **a)** One of many possible statements is 4 less than 3 times a number.

b) The expression within parentheses may be written "the difference between a number and 4." Therefore, the entire expression may be written as 3 times the difference between a number and 4.

NOW TRY EXERCISE 37

2) Write Expressions Involving Percent

Since percents are used so often, you must have a clear understanding of how to write expressions involving percent. Whenever we perform a calculation involving percent, we generally change it to a decimal or a fraction first.

EXAMPLE 4 Express each phrase as an algebraic expression.

a) The cost of a pair of boots, c, increased by 6%

b) The population in the town of Aikland, p, decreased by 12%

Solution **a)** When shopping we may see a "25% off" sales sign. We assume that this means 25% off *the original cost*, even though this is not stated. This question asks for the cost increased by 6%. We assume that this means the cost increased by 6% of the original cost, and write

$$c + 0.06c$$

Original cost → c; *Increased by* → $+$; *6% of the original cost* → $0.06c$

Thus, the answer is $c + 0.06c$.

b) Using the same reasoning as in part **a)** the answer is $p - 0.12p$.

Avoiding Common Errors

In Example 4**a)** we asked you to represent a cost, c, increased by 6%. Note, the answer is $c + 0.06c$. Often, students write the answer to this question as $c + 0.06$. It is important to realize that a percent of a quantity must always be a percent multiplied by some number or letter. Some phrases involving the word percent and the correct and incorrect interpretations follow.

PHRASE	CORRECT	INCORRECT
A $7\frac{1}{2}$% sales tax on c dollars	$0.075c$	~~0.075~~
The cost, c, increased by a $7\frac{1}{2}$% sales tax	$c + 0.075c$	~~$c + 0.075$~~
The cost, c, reduced by 25%	$c - 0.25c$	~~$c - 0.25$~~

3) Express the Relationship between Two Related Quantities

Sometimes in a problem, two numbers are related to each other in a certain way. We often represent the simplest, or most basic number that needs to be expressed, as a variable and the other as an expression containing that variable. Some examples follow.

Statement	One Number	Second Number
Two numbers differ by 3	x	$x + 3$
John's age now and John's age in 6 years	x	$x + 6$
One number is 6 times the other number	x	$6x$
One number is 12% less than the other	x	$x - 0.12x$

Note that often more than one pair of expressions can be used to represent the two numbers. For example, "two numbers differ by 3" can also be expressed as x and $x - 3$. Let's now look at two more statements.

Statement	One Number	Second Number
The sum of two numbers is 10.	x	$10 - x$
A 25-foot length of wood is cut in two pieces.	x	$25 - x$

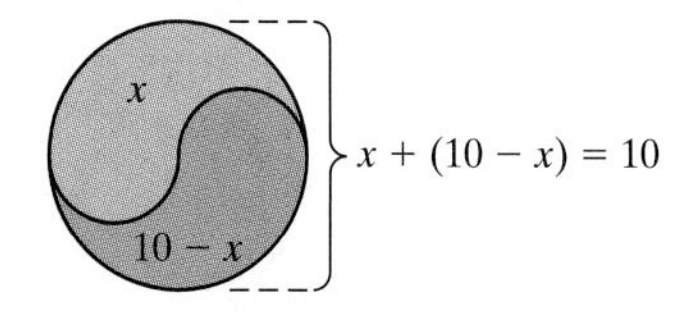

FIGURE 3.7

It may not be obvious why in "the sum of two numbers is 10" the two numbers are represented as x and $10 - x$. Suppose that one number is 2; what is the other number? Since the sum is 10, the second number must be $10 - 2$ or 8. Suppose that one number is 6; the second number must be $10 - 6$, or 4. In general, if the first number is x, the second number must be $10 - x$. Note that the sum of x and $10 - x$ is 10 (Fig. 3.7).

Consider the statement, "a 25-foot length of wood is cut in two pieces." If we call one length x, then the other length must be $25 - x$. For example, if one length is 6 feet, the other length must be $25 - 6$ or 19 feet (Fig. 3.8).

25 25

x $25 - x$ 6 $25 - 6 = 19$

FIGURE 3.8

EXAMPLE 5 For each relationship, select a variable to represent one quantity and state what that quantity represents. Then express the second quantity in terms of the variable.

a) A boy is 15 years older than his brother.

b) The speed of one car is 1.4 times the speed of another.

c) Two business partners share \$75.

d) John has \$5 more than 3 times the amount that Dee has.

e) The length of a rectangle is 3 feet less than 4 times its width.

f) A number is increased by 6%.

g) The profits of a business, in percent, are shared by two business partners.

Solution **a)** Let x be the age of the younger brother; then $x + 15$ is the age of the older brother.

b) Let s be the speed of the slower car; then $1.4s$ is the speed of the faster car.

c) Let d be the amount in dollars one partner receives; then $75 - d$ is the amount in dollars the other partner receives.

d) Let d be Dee's money in dollars; then $3d + 5$ is John's money in dollars.

e) Let w be the width of the rectangle; then $4w - 3$ is the length of the rectangle.

f) Let n be the number. Then that number increased by 6% is $n + 0.06n$.

g) Let p be the percent that one business partner gets; then $100 - p$ is the percent that the other partner gets. Notice that the total profit is 100% and that the sum of the two percents, $p + (100 - p)$, is 100.

NOW TRY EXERCISE 55

4) Write Expressions Involving Multiplication

Consider the statement "the cost of 3 items at \$5 each." How would you represent this quantity using mathematical symbols? You would probably reason that the cost would be 3 times \$5 and write $3 \cdot 5$ or $3(5)$.

Now consider the statement "the cost of x items at \$5 each." How would you represent this statement using mathematical symbols? If you use the same reasoning, you might write $x \cdot 5$ or $x(5)$. Another way to write this product is $5x$. Thus, the cost of x items at \$5 each could be represented as $5x$.

Finally, consider the statement "the cost of x items at y dollars each." Following the reasoning used in the previous two illustrations, you might write $x \cdot y$ or $x(y)$. Since these products can be written as xy, the cost of x items at y dollars each can be represented as xy.

EXAMPLE 6 Write each statement as an algebraic expression.

a) The cost of purchasing x pens at \$2 each

b) A 5% commission on x dollars in sales

c) The number of calories in x potato chips, if each potato chip has 8 calories

d) The increase in population in n years for a city growing by 300 persons per year

e) The distance traveled in t hours when 55 miles are traveled each hour

Solution **a)** We can reason like this: one pen would cost $1(2)$ dollars, two pens would cost $2(2)$ dollars, three pens $3(2)$, four pens $4(2)$, and so on. Continuing this reasoning process, we can see that x pens would cost $x(2)$ or $2x$ dollars.

b) A 5% commission on \$1 sales would be $0.05(1)$, on \$2 sales $0.05(2)$, on \$3 sales $0.05(3)$, on \$4 sales $0.05(4)$, and so on. Therefore, the commission on sales of x dollars would be $0.05(x)$ or $0.05x$.

c) $8x$

d) $300n$

e) $55t$

NOW TRY EXERCISE 9

EXAMPLE 7 The cost for seeing a movie at the Tinseltown Movie Theater is \$6.50 for adults and \$4.25 for children. Write an algebraic expression to represent the total income received by the theater if x adults and y children pay for admission.

Solution For x adults, the theater receives $6.50x$ dollars.

For y children, the theater receives $4.25y$ dollars.

NOW TRY EXERCISE 33

For x adults and y children, the theater receives $6.50x + 4.25y$ dollars.

EXAMPLE 8 Write an algebraic expression for each statement.

a) The number of ounces in x pounds

b) The number of cents in a dimes and b nickels

c) The number of seconds in x hours, y minutes, and z seconds (3600 seconds = 1 hour)

Solution **a)** Since each pound contains 16 ounces, x pounds is $16 \cdot x$ or $16x$ ounces.

b) Since a dimes is $10a$ cents and b nickels is $5b$ cents, the answer is $10a + 5b$.

c) $3600x + 60y + z$

Some terms that we will be using are consecutive integers, consecutive even integers, and consecutive odd integers. **Consecutive integers** are integers that differ by 1 unit. For example, the integers 6 and 7 are consecutive integers. Two consecutive integers may be represented as x and $x + 1$. **Consecutive even integers** are even integers that differ by 2 units. For example, 6 and 8 are consecutive even integers. **Consecutive odd integers** also differ by 2 units. For example, 7 and 9 are consecutive odd integers. Two consecutive even integers, or two consecutive odd integers, may be represented as x and $x + 2$.

5) Translate Applications into Equations

The word *is* in an application problem often means *is equal to* and is represented by an equal sign. Some examples of statements written as equations follow.

Statement	Equation
Six more than twice a number *is* 4.	$2x + 6 = 4$
A number decreased by 4 *is* 3 more than twice the number.	$x - 4 = 2x + 3$
The product of two consecutive integers *is* 56.	$x(x + 1) = 56$
One number is 4 more than 3 times the other number; their sum *is* 60.	$x + (3x + 4) = 60$
A number increased by 15% *is* 120.	$x + 0.15x = 120$
The sum of two consecutive odd integers *is* 24.	$x + (x + 2) = 24$

Now let's translate some equations into statements. Some examples of equations written as statements follow. We will write only two statements for each equation, but remember there are other ways these equations can be written.

Equation	Statements
$3x - 4 = 4x + 3$	Four less than 3 times a number *is* 3 more than 4 times the number.
	Three times a number, decreased by 4 is 4 times the number, increased by 3.
$3(x - 2) = 6x - 4$	Three times the difference between a number and 2 *is* 4 less than 6 times the number.
	The product of 3 and the difference between a number and 2 *is* 6 times the number, decreased by 4.

EXAMPLE 9 Write two statements to represent the equation $x - 2 = 3x - 5$.

Solution

1. A number decreased by 2 *is* 5 less than 3 times the number.
2. The difference between a number and 2 *is* the difference between 3 times the number and 5.

EXAMPLE 10 Write a statement to represent the equation $x + 2(x - 4) = 6$.

Solution The sum of a number and twice the difference between the number and 4 *is* 6.

EXAMPLE 11 Write each problem as an equation.

a) One number is 4 less than twice the other. Their sum is 14.

b) For two consecutive integers, the sum of the smaller and 3 times the larger is 23.

Solution **a)** First, we express the two numbers in terms of the variable.

$$\text{Let } x = \text{one number}$$

$$\text{then } 2x - 4 = \text{second number}$$

Now we write the equation using the information given.

$$\text{first number} + \text{second number} = 14$$

$$x + (2x - 4) = 14$$

b) First, we express the two consecutive integers in terms of the variable.

$$\text{Let } x = \text{smaller consecutive integer}$$

$$\text{then } x + 1 = \text{larger consecutive integer}$$

Now we write the equation using the information given.

$$\text{smaller} + 3 \text{ times the larger} = 23$$

NOW TRY EXERCISE 73

$$x + 3(x + 1) = 23$$

EXAMPLE 12 Write the following problem as an equation. One train travels 3 miles more than twice the distance another train travels. The total distance traveled by both trains is 800 miles.

Solution First express the distance traveled by each train in terms of the variable.

$$\text{Let } x = \text{distance traveled by one train}$$

$$\text{then } 2x + 3 = \text{distance traveled by second train}$$

Now write the equation using the information given.

$$\text{distance of train 1} + \text{distance of train 2} = \text{total distance}$$

$$x + (2x + 3) = 800$$

EXAMPLE 13 Write the following problem as an equation. Lori Soushon is 4 years older than 3 times the age of her son Ron. The difference in Lori's age and Ron's age is 26 years.

Solution Since Lori's age is given in terms of Ron's age, we will let the variable represent Ron's age.

$$\text{Let } x = \text{Ron's age}$$

$$\text{then } 3x + 4 = \text{Lori's age}$$

We are told that the difference in Lori's age and Ron's age is 26 years. The word *difference* indicates subtraction.

$$\text{Lori's age} - \text{Ron's age} = 26$$
$$(3x + 4) - x = 26$$

The word *was* is the past tense of the word *is*. Therefore, the word *was* often means *was equal to* and is represented by an equal sign.

EXAMPLE 14 Express each problem as an equation.

a) George Devenney rented a tiller for x days at a cost of \$22 per day. The cost of renting the tiller *was* \$88.

b) The distance Scott Borden traveled for x days at 600 miles per day *was* 1500 miles.

c) The population in the town of Rush is increasing by 500 people per year. The increase in population in t years *is* 2500.

d) The number of cents in d dimes *is* 120.

Solution **a)** The cost of renting the tiller for x days is $22x$. Therefore, the equation is $22x = 88$.

b) The distance traveled at 600 miles per day for x days is $600x$. Therefore, the equation is $600x = 1500$.

c) The increase in the population in t years is $500t$. Therefore, the equation is $500t = 2500$.

NOW TRY EXERCISE 81

d) The number of cents in d dimes is $10d$. Therefore, the equation is $10d = 120$.

Exercise Set 3.2

Concept/Writing Exercises

1. Give four phrases that indicate the operation of addition.
2. Give four phrases that indicate the operation of subtraction.
3. Give four phrases that indicate the operation of multiplication.
4. Give four phrases that indicate the operation of division.
5. Explain why $c + 0.25$ *does not* represent the cost of an item increased by 25 percent.
6. Explain why $c - 0.10$ *does not* represent the cost of an item decreased by 10 percent.

Practice the Skills

7. Wendy Bowes is n years old now. Write an expression that represents her age in 5 years.
8. Ajit Silva is t years old. Write an expression that represents David Alevy's age if he is 4 times as old as Ajit.
9. At a sale, a Dr. Grip pen costs \$4. Write an expression that represents the cost of purchasing x Dr. Grip pens.
10. An item that costs r dollars is increased by \$6. Write an expression that represents the new price.
11. Melissa Blum is selling her motorcycle. She was asking x dollars for the motorcycle but has cut the price in half. Write an expression that represents the new price.
12. Michael Capan is buying some suits. The store is having a two-for-one sale, where for each suit you purchase you are given a second suit free. Write an expression that represents the number of new suits he will get if he purchases y suits.
13. Julie Burgmeier pitches in a girl's softball league. Last year her fastest pitching speed was s miles per hour. This year her speed increased by 1.2 miles per hour. Write an expression that represents her new pitching speed.

14. John Debruzzi used to read p words per minute. After taking a speed-reading course, his speed increased by 60 words per minute. Write an expression that represents his new reading speed.

15. The entire population of the United States is P, and 16% of the population does not receive adequate nourishment. Write an expression that represents the number of people who do not receive adequate nourishment.

16. Each year t tires are disposed of in the United States. Only 7% of all tires disposed of are recycled. Write an expression that represents the number of tires that are recycled.

17. The United Kingdom consumes P British thermal units (Btus) of power annually. The United States consumes 9 less than 10 times the Btus used by the United Kingdom. Write an expression that represents the number of Btus the United States uses.

18. The second largest emitter of carbon dioxide (which causes global warming) is China. The United States, the largest producer of carbon dioxide, emits 0.4 less than twice the amount of carbon dioxide emitted by China. If the amount of carbon dioxide emitted by China is A, write an expression that represents the amount of carbon dioxide emitted by the United States.

19. Assume that last year Jose Rivera had a salary of m dollars. This year he received a promotion and his new salary is $16,000 plus eight-ninths of his previous salary. Write an expression that represents his new salary.

20. According to *Amusement Business*, in 1996, Disneyland in California was the theme park with the greatest attendance. (If you consider the combined attendance at Disney World's three theme parks in Florida, Disney World had the greatest attendance.) The park with the second greatest attendance was the Magic Kingdom at Walt Disney World. If n represents attendance at the Magic Kingdom in millions, and the attendance at Disneyland was 12.6 million less than twice that of the Magic Kingdom, write an expression that represents the attendance at Disneyland.

21. John Bartizal rented a truck for a trip. He paid a daily fee of $45 and a mileage fee of 40 cents a mile. Write an expression that represents his total cost when he travels x miles in one day.

22. The city of Clarkville has a population of 4000. If the population increases by 300 people per year, write an expression that represents the population in n years.

23. Rebecca Feist-Miller found that she had x quarters in her handbag. Write an expression that represents this quantity of money in cents.

24. Dennis DeValeriz's height is x feet and y inches. Write an expression that represents his height in inches.

25. Lisa Davis's weight is x pounds and y ounces. Write an expression that represents her weight in ounces.

26. Andrew Przewuzman's newborn baby is m minutes and s seconds old. Write an expression that represents the baby's age in seconds.

27. The attendance at Sea World in San Diego in 1996 increased by 4% over the attendance in 1995. If n represents the attendance in 1995, write an expression that represents the attendance in 1996.

28. As of May 31, 1997, the mutual fund that increased the most over the previous 12-month period was Fidelity Select Electronics. The mutual fund's price (net asset value) increased by 52.63% from its previous year's price. If its previous year's price was p dollars, write an expression that represents the price on May 31, 1997.

29. Through some careful tax planning, Allyson Williams was able to reduce her federal income tax from 1999 to 2000 by 18%. If her 1999 tax was t dollars, write an expression that represents her 2000 tax.

30. In 1996, the number of Burger King restaurants was about 42% less than the number of McDonald's restaurants. If m represents the number of McDonald's restaurants, write an expression that represents the number of Burger King restaurants.

31. Each slice of white bread has 110 calories and each teaspoon of strawberry preserves has 80 calories. If Donna Contoy makes a strawberry preserve sandwich (2 pieces of bread) and uses x teaspoons of preserves, write an expression that represents the number of calories the sandwich will contain.

32. According to Smith Travel Research, in 1998, the U.S. city with the greatest number of hotel rooms was Las Vegas, Nevada. The U.S. city with the second largest number was Orlando, Florida. The number of rooms in Orlando is n and the number of rooms in Las Vegas is 155,900 less than 3 times the number hotel rooms in Orlando. Write an expression that represents the number of hotel rooms in Las Vegas.

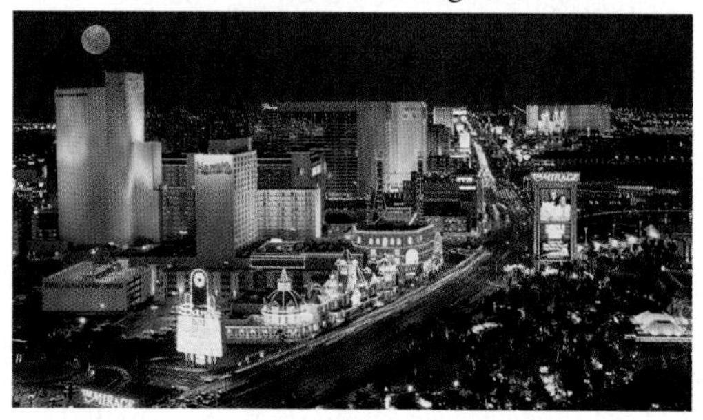

33. An average chicken egg contains about 275 milligrams (mg) of cholesterol and an ounce of chicken contains about 25 mg of cholesterol. Write an expression that represents the amount of cholesterol in x chicken eggs and y ounces of chicken.

34. According to U.S. guidelines, each gram of carbohydrates contains 4 calories, each gram of protein contains 4 calories, and each gram of fat contains 9 calories. Write an expression that represents the number of calories in a serving of a product that contains x grams of carbohydrates, y grams of protein, and z grams of fat.

Write each mathematical expression as a statement. (There are many acceptable answers.)

35. $x - 6$

36. $x + 3$

37. $4x + 1$

38. $3x - 4$

39. $5x - 7$

40. $2x - 3$

41. $4x - 2$

42. $5 - x$

43. $2 - 3x$

44. $4 + 6x$

45. $2(x - 1)$

46. $3(x + 2)$

In Exercises 47–62, ***a)*** *select a letter to represent one quantity and state what that variable represents.* ***b)*** *Express the second quantity in terms of the variable selected. (For example, if Marty's weekly salary is \$20 more than Don's salary, you might let d = Don's salary, then Marty's salary could be represented as $d + 20$.)*

47. Dana Meltzer is 3 years older than Chuck Greystone.

48. Professor Sandra Hakanson has been teaching for 4 years less than Professor Roger Howell.

49. Lois Heather's son is one-third as old as Lois.

50. Two consecutive integers.

51. Two consecutive even integers.

52. One hundred dollars divided between Sharon Koch and James Lloyd.

53. A Pontiac Firebird costs 1.1 times as much as a Chevrolet Camaro.

54. An 80-foot tree is cut into two pieces.

55. Shari Meffert and Tabitha McCaun share in the profits, in percent, of a cookie business.

56. The calories in a serving of mixed nuts is 280 calories less than twice the number of calories in a serving of cashew nuts.

57. According to MIPF Research, Inc., the average monthly rental rate for apartments in San Jose, California (first quarter of 1997) was \$177 less than twice that in Los Angeles, California.

58. The number of electoral college votes in New Jersey is 3 less than 6 times the number of electoral votes in North Dakota.

59. The number of coupons redeemed decreased from 1997 to 1999. The number of coupons redeemed in 1999 was 4.4 billion less than twice the number redeemed in 1997.

60. Total software sales in the United States, in billions of dollars, in 2000 is estimated to be \$98.5 billion greater than half the 1999 total sales.

61. According to the Telecommunity Industry Association, the number of cellular subscribers in 1996 was 1,500,000 more than 125 times the number of subscribers in 1985.

62. The projected gross income from the movie Armageddon was \$250,000,000 less than 3 times the estimated budget for making the movie.

*In Exercises 63–70, **a)** select a letter to represent the variable and indicate what the variable represents. **b)** Write an expression to represent the given statement. (For example, to represent the statement "The number of people who shop at Wegmans increased by 25%," you might answer **a)** let n = number of people who shop at Wegmans **b)** $n + 0.25n$.)*

63. Dianne Wooten is a sales representative for a book company. Her 1999 sales increased by 60% over her 1998 sales.

64. George Young compared his electricity use, in kilowatt hours, in 1999 with his use in 1998, and found that it decreased by 16%.

65. Kevin Mueller, an engineer, had a salary increase of 15% over last year's salary.

66. The number of cases of flu in Archville decreased by 12% from the previous year.

67. The cost of a new car purchased in Collier County included a 7% sales tax.

68. At a 25% off everything sale, William Winchief's cost of purchasing a new shirt.

69. The pollution level in Detroit decreased by 50%.

70. The number of students earning a grade of A in this course increased by 100%.

Problem Solving

*In Exercises 71–88, **a)** select a letter to represent the variable and state exactly what the variable represents. **b)** Write an equation to represent the problem.*

71. One number is 5 times another. The sum of the two numbers is 18.

72. Marie is 6 years older than Denise. The sum of their ages is 48.

73. The sum of two consecutive integers is 47.

74. The product of two consecutive even integers is 48.

75. Twice a number, decreased by 8 is 12.

76. For two consecutive integers, the sum of the smaller and twice the larger is 29.

77. One-fifth of the sum of a number and 10 is 150.

78. David Ostrow jogs 5 times as far as Patricia Einstein. The total distance traveled by both people is 8 miles.

79. An Amtrak train travels 4 miles less than twice the distance traveled by a Southern Pacific train. The total distance traveled by both trains is 890 miles.

80. According to the U.S. Bureau of the Census, the number of U.S. apartment households in 1995 was 2.5 million less than twice the number in 1970. The sum of the 1970 and 1995 households was 23 million.

81. Barbara Rose bought a new car. The cost of the car plus a 7% sales tax was $26,200.

82. James Porter Hamann purchased a sports coat at a 25% off sale. He paid $195 for the sport coat.

83. Beth Reschsteiner ate at a steakhouse. The cost of the meal plus a 15% tip was $32.50.

84. At the Better Buy Warehouse, Christas Giakoumopoulos purchased a video cassette recorder that was reduced by 10% for $208.

85. In 1950, New York Yankee baseball player Joe DiMaggio received a salary about 7.69 times that of the average professional baseball player. The difference between DiMaggio's salary and the average player's salary was $87,000.

86. In 1998, there were 122 less than twice the number of roller coasters in North America than there were in 1994. The difference in the number of roller coasters in 1998 and 1994 was 112.

87. According to the U.S. Department of Education, in 1995–1996, the average salary of Connecticut public school teachers was the highest in the nation whereas the average salary of public school teachers in South Dakota was the lowest. The average public school teacher's salary in Connecticut was $28,784 less than 3 times the average school teacher's salary in South Dakota. The sum of the average teacher's salaries in Connecticut and South Dakota was $76,600.

88. According to the Federal Communications Commission, the average basic local service charge for telephone bills in 1995 was $2.01 more than twice the average amount in 1980. The difference between the basic rates in 1995 and 1980 was $10.75.

In Exercises 89–100, express each equation as a statement. (There are many acceptable answers.)

89. $x + 3 = 6$

90. $x - 5 = 2x$

91. $3x - 1 = 2x + 4$

92. $x - 3 = 2x + 3$

93. $4(x - 1) = 6$

94. $3x + 2 = 2(x - 3)$

95. $5x + 6 = 6x - 1$

96. $x - 3 = 2(x + 1)$

97. $x + (x + 4) = 8$

98. $x + (2x + 1) = 5$

99. $2x + (x + 3) = 5$

100. $2x - (x + 3) = 6$

101. Explain why the cost of purchasing x items at 6 dollars each is represented as $6x$.

102. Explain why the cost of purchasing x items at y dollars each is represented as xy.

Challenge Problems

103. a) Write an algebraic expression for the number of seconds in d days, h hours, m minutes, and s seconds.

b) Use the expression found in part **a)** to determine the number of seconds in 4 days, 6 hours, 15 minutes, and 25 seconds.

104. At the time of this writing, the toll for southbound traffic on the Golden Gate Bridge is \$1.50 per vehicle *axle* (there is no toll for northbound traffic).

a) If the number of 2-, 3-, 4-, 5-, and 6-axle vehicles are represented with the letters r, s, t, u, and v, respectively, write an *expression* that represents the daily revenue of the Golden Gate Bridge Authority.

b) Write an *equation* that can be used to determine the daily revenue, d.

Group Activity

Exercises 105–108 will help prepare you for the next section, where we set up and solve application problems. Discuss and work each exercise as a group. For each exercise, **a)** *write down the quantity you are being asked to find and represent this quantity with a variable.* **b)** *Write an equation containing your variable that can be used to solve the problem. Do not solve the equation.*

105. An average bath uses 30 gallons of water and an average shower uses 6 gallons of water per minute. How long a shower would result in the same water usage as a bath?

106. The average American produces 40,000 pounds of carbon dioxide each year by driving a car, running air conditioners, lighting, and using appliances and other items that require the burning of fossil fuels. How long would it take for the average American to produce 1,000,000 pounds of carbon dioxide?

107. An employee has a choice of two salary plans. Plan A provides a weekly salary of \$200 plus a 5% commission on the employee's sales. Plan B provides a weekly salary of \$100 plus an 8% commission on the employee's sales. What must be the weekly sales for the two plans to give the same weekly salary?

108. The cost of renting an 18-foot truck from Mertz is \$20 a day plus 60 cents a mile. The cost of renting a similar truck from U-Hail is \$30 a day plus 45 cents a mile. How far would you have to drive the rental truck in 1 day for the total cost to be the same with both companies?

Cumulative Review Exercises

[2.6] *Write a proportion that can be used to solve each problem. Solve each problem and find the desired values.*

109. A recipe for chicken stew calls for $\frac{1}{2}$ teaspoon of thyme for each pound of poultry. If the poultry for the stew weighs 6.7 pounds, how much thyme should be used?

110. Cindy Trimble mixes water with dry cat food for her cat Max. If the directions say to mix 1 cup of water with every 3 cups of dry cat food, how much water will Cindy add to $\frac{1}{2}$ cup of dry cat food?

[3.1] **111.** $P = 2l + 2w$; find l when $P = 40$ and $w = 5$.

112. Solve $3x - 2y = 6$ for y. Then find the value of y when x has a value of 6.

3.3 SOLVING APPLICATION PROBLEMS

1) Set up and solve application problems.
2) Use the problem-solving procedure.
3) Select a mortgage.
4) Solve applications containing large numbers.

There are many types of application problems that can be solved using algebra. In this section we introduce several types. In Section 3.4 we introduce additional types of applications. They are also presented in many other sections and exercise sets throughout the book. Your instructor may not have time to cover all the applications given in this book. If not, you may still wish to spend some time on your own reading those problems just to get a feel for the types of applications presented.

To be prepared for this section, you must understand the material presented in Section 3.2. The best way to learn how to set up an application or word problem is to practice. The more problems you study and attempt, the easier it will become to solve them.

1) Set Up and Solve Application Problems

We often translate problems into mathematical terms without realizing it. For example, if you need 3 cups of milk for a recipe and the measuring cup holds only 2 cups, you reason that you need 1 additional cup of milk after the initial 2 cups. You may not realize it, but when you do this simple operation, you are using algebra.

Let x = number of additional cups of milk needed

Thought process: (initial 2 cups) + $\left(\begin{array}{c}\text{number of}\\ \text{additional cups}\end{array}\right)$ = total milk needed

Equation to represent problem: $2 + x = 3$

When you solve for x, you get 1 cup of milk.

You probably said to yourself: Why do I have to go through all this when I know that the answer is $3 - 2$ or 1 cup? When you perform this subtraction, you have mentally solved the equation $2 + x = 3$.

$$2 + x = 3$$
$$2 - 2 + x = 3 - 2 \qquad \textit{Subtract 2 from both sides.}$$
$$x = 3 - 2$$
$$x = 1$$

Let's look at some other examples.

EXAMPLE 1 Tricia Liscio, an accountant, is reviewing a client's expenses for the year. She notices that for 1 year (12 payments) her client paid \$8100 in rent. How much was the client's monthly rent?

Solution Many of you will be able to determine the answer without using algebra. However, the object of this section is for you to learn to set up and solve verbal problems algebraically. So let's do it.

Understand We are told that the total of 12 monthly payments is \$8100. We are asked to find the client's monthly rent. To solve the problem, we will use the fact that the client's rent for 12 months is \$8100. In Section 3.2 we learned that if x represents the rent for one month, then $12x$ represents the rent for 12 months.

Translate

Let x = the client's monthly rent

then $12x$ = the client's rent for 12 months

client's rent for 12 months = 8100

$$12x = 8100$$

Carry Out

$$\frac{12x}{12} = \frac{8100}{12} \quad \text{Divide both sides by 12.}$$

$$x = 675$$

Check If we multiply the monthly rent, \$675, by 12, we get \$8100. So our answer is correct.

Answer The client's monthly rent is \$675.

EXAMPLE 2 Suppose that you are at a supermarket, and your purchases so far total \$13.20. In addition to groceries, you wish to purchase as many packages of gum as possible, but you have a total of only \$18. If a package of gum costs \$1.15, how many can you purchase?

Solution **Understand** How can we represent this problem as an equation? We might reason as follows. We need to find the number of packages of gum. Let's call this unknown quantity x.

Let x = number of packages of gum

Thought process: cost of groceries + cost of gum = total cost

We substitute \$13.20 for the cost of groceries and \$18 for the total cost to get

$$13.20 + \text{cost of gum} = 18$$

At this point you might be tempted to replace the cost of gum with the letter x. But x represents the *number* of packages of gum, *not the cost of the gum*. The cost of x packages of gum at \$1.15 per package is $1.15x$. Now we substitute the cost of the x packages of gum, $1.15x$, into the equation.

Translate cost of groceries + cost of gum = total cost

$$13.20 + 1.15x = 18$$

Carry Out

$$13.20 - 13.20 + 1.15x = 18 - 13.20 \quad \text{Subtract 13.20 from both sides.}$$

$$1.15x = 4.80$$

$$\frac{1.15x}{1.15} = \frac{4.80}{1.15} \quad \text{Divide both sides by 1.15.}$$

$$x \approx 4.2$$

Check When we solve this equation, we obtain $x = 4.2$ packages (to the nearest tenth). Since you cannot purchase a part of a pack of gum, only four packages of gum can be purchased. Four packages of gum would cost $4 \times \$1.15 = \$4.60$. Adding this amount to the \$13.20 for groceries gives a total of \$17.80. This would leave change of only 20 cents from the \$18. This is not enough to purchase another pack of gum. Therefore, this answer checks.

NOW TRY EXERCISE 21

Answer Four packages of gum could be purchased.

2) Use the Problem-Solving Procedure

There are many types of application problems. The general problem-solving procedure given in Section 1.2 and used in Examples 1 and 2 can be used to solve all types of verbal problems. Below, we present the **five-step problem-solving procedure** again so you can easily refer to it. We have included some additional information under steps 1 and 2, since in this section we are going to emphasize translating application problems into equations.

Problem-Solving Procedure for Solving Applications

1. **Understand the problem.**
   Identify the quantity or quantities you are being asked to find.
2. **Translate the problem into mathematical language (express the problem as an equation).**
   a) Choose a variable to represent one quantity, *and write down exactly what it represents*. Represent any other quantity to be found in terms of this variable.
   b) Using the information from step a), write an equation that represents the application.
3. **Carry out the mathematical calculations (solve the equation).**
4. **Check the answer (using the *original* application).**
5. **Answer the question asked.**

Sometimes we will combine two steps in the problem-solving procedure when it helps to clarify the explanation. We may not show the check of a problem to save space. Even if we do not show a check, you should check the problem yourself and make sure your answer is reasonable and makes sense.

Let's now set up and solve some additional problems using this procedure.

EXAMPLE 3 Two subtracted from 4 times a number is 10. Find the number.

Solution **Understand** To solve this problem, we need to express the statement given as an equation. We are asked to find the unknown number.

Translate Let x = the unknown number. Now write the equation.

$$\underbrace{4x - 2}_{\text{2 subtracted from 4 times a number}} \overset{\text{is}}{=} \overset{10}{10}$$

Carry Out

$$4x = 12$$
$$x = 3$$

Check Substitute 3 for the number in the original problem, two subtracted from 4 times a number is 10.

$$4(3) - 2 \stackrel{?}{=} 10$$
$$10 = 10 \quad \text{True}$$

Answer Since the solution checks, the unknown number is 3.

EXAMPLE 4 The sum of two numbers is 17. Find the two numbers if the larger number is 5 more than twice the smaller number.

Solution **Understand** This problem involves finding two numbers. When finding two numbers, if a second number is expressed in terms of a first number, we generally let the variable represent the first number. Then we represent the second number as an expression containing the variable used for the first number. In this example, we are given that "the larger number is 5 more than twice the smaller number." Notice that the larger number is expressed in terms of the smaller number. Therefore, we will let the variable represent the smaller number.

Translate Let x = smaller number

then $2x + 5$ = larger number

The sum of the two numbers is 17. Therefore, we write the equation

$$\text{smaller number} + \text{larger number} = 17$$
$$x + (2x + 5) = 17$$

Carry Out Now we solve the equation.

$$3x + 5 = 17$$
$$3x = 12$$
$$x = 4$$

The smaller number is 4. Now we find the larger number.

$$\text{larger number} = 2x + 5$$
$$= 2(4) + 5 \quad \text{Substitute 4 for } x.$$
$$= 13$$

The larger number is 13.

Check The sum of the two numbers is 17.

$$4 + 13 \stackrel{?}{=} 17$$
$$17 = 17 \quad \text{True}$$

NOW TRY EXERCISE 7

Answer The two numbers are 4 and 17.

EXAMPLE 5 With the backing of friends and a loan from a bank, Candice *Co*tton and Thomas *Jo*hnson formed a corporation that manufactures walking sneaker-shoes that they have patented. Their corporation, called CoJo, has been producing 1200 pairs of sneaker-shoes per year. This year they plan to manufacture 550 pairs more than in previous years, and they plan to increase production by 550 pairs each year until their annual production is 10,000 pairs. How long will it take them to reach their production goal?

Solution **Understand** We are asked to find the *number of years* that it will take for their production to reach 10,000 pairs a year. This year their production will increase by 550 pairs. In the second year, their production will increase by 2(550) over the present year's production. In n years, their production will increase by $n(550)$ or $550n$. We will use this information when we write the equation to solve the problem.

Translate Let n = number of years

then $550n$ = increase in production over n years

$$\text{(present production)} + \left(\begin{array}{c}\text{increased production}\\ \text{over } n \text{ years}\end{array}\right) = \text{future production}$$

$$1200 + 550n = 10{,}000$$

Carry Out

$$550n = 8800$$

$$n = \frac{8800}{550}$$

$$n = 16 \text{ years}$$

Check and Answer A check will show that 16 years is the correct answer. In 16 years, CoJo's production will be 10,000 pairs of sneaker-shoes a year.

EXAMPLE 6 When reading the label on a bottle of a Tangy Taste fruit punch, Karl Abolafia found that the 32-ounce bottle contains 3.84 ounces of pure fruit juice. Find the percent of pure fruit juice in the punch.

Solution **Understand** We need to find the *percent* of juice in the punch. The formula

volume of punch · *percent* of pure juice in punch = *amount* of pure juice in punch

can be used to find the percent of pure juice in the punch. You will see variations of this formula when we study mixture problems in Section 3.5.

In this problem we are given the volume of punch, 32 fluid ounces, and the amount of pure juice, 3.84 fluid ounces. We use this information to find the percent of pure juice in the punch.

Translate Let x = percent of pure juice. Now substitute the appropriate values in the formula.

volume of punch · percent of pure juice in punch = amount of pure juice in punch

$$32 \cdot x = 3.84$$

Carry Out

$$\frac{32x}{32} = \frac{3.84}{32}$$

$$x = 0.12 \text{ or } 12\%$$

Check Twelve percent of 32 is $0.12(32) = 3.84$. Thus, the answer checks.

NOW TRY EXERCISE 19

Answer The punch is 12% pure fruit juice.

EXAMPLE 7 Alfredo Irizarry is moving. He plans to rent a truck for one day to make the local move. The cost of renting the truck is \$60 a day plus 40 cents a mile. Find the maximum distance Alfredo can drive if he has only \$92.

Solution **Understand** The total cost of renting the truck consists of two parts, a fixed cost of \$60 per day, and a variable cost of 40 cents per mile. We need to determine the *number of miles* that Alfredo can drive so that the total rental cost is \$92. Since the fixed cost is given in dollars, we will write the variable cost, or mileage cost, in dollars also.

Translate

$$\text{Let } x = \text{number of miles}$$

$$\text{then } 0.40x = \text{cost of driving } x \text{ miles}$$

$$\text{daily cost} + \text{mileage cost} = \text{total cost}$$

$$60 + 0.40x = 92$$

Carry Out

$$0.40x = 32$$

$$\frac{0.40x}{0.40} = \frac{32}{0.40}$$

$$x = 80$$

Check The cost of driving 80 miles at 40 cents a mile is $80(0.40) = \$32$. Adding the \$32 to the daily cost of \$60 gives \$92, so the answer checks.

Answer Alfredo can drive a maximum of 80 miles.

EXAMPLE 8 Jacqueline Johnson recently graduated from college and has accepted a position selling medical supplies and equipment. During her first year, she is given a choice of salary plans. Plan 1 is a \$450 base salary plus a 3% commission of sales. Plan 2 is a straight 10% commission of sales.

a) Jacqueline must select one of the plans but is not sure of the sales needed for her weekly salary to be the same under the two plans. Can you determine it?

b) If Jacqueline is certain that she can make \$8000 in sales per week, which plan should she select?

Solution **a) Understand** We are asked to find the *dollar sales* that will result in Jacqueline receiving the same total salary from both plans. To solve this problem, we write expressions to represent the salary from each of the plans. We then obtain the desired equation by setting the salaries from the two plans equal to one another.

Translate

$$\text{Let } x = \text{dollar sales}$$

$$\text{then } 0.03x = \text{commission from plan 1 sales}$$

$$\text{and } 0.10x = \text{commission from plan 2 sales}$$

$$\text{salary from plan 1} = \text{salary from plan 2}$$

$$\text{base salary} + 3\% \text{ commission} = 10\% \text{ commission}$$

$$450 + 0.03x = 0.10x$$

Carry Out

$$450 = 0.07x$$

$$\text{or} \quad 0.07x = 450$$

$$\frac{0.07x}{0.07} = \frac{450}{0.07}$$

$$x \approx 6428.57$$

Check We will leave it up to you to show that sales of \$6428.57 result in Jacqueline receiving the same weekly salary from both plans.

Answer Jacqueline's weekly salary will be the same from both plans if she sells \$6428.57 worth of medical supplies and equipment.

b) If Jacqueline's sales are \$8000, she would earn more weekly by working on straight commission, Plan 2. Check this out yourself by computing Jacqueline's salary under both plans and comparing them.

NOW TRY EXERCISE 37

EXAMPLE 9 National Airlines wishes to keep its supersaver airfare, including a 7% tax, between Denver, Colorado, and Seattle, Washington, at exactly \$320. Find the cost of the ticket before tax.

Solution **Understand** We are asked to find the cost of the ticket before tax. The cost of the ticket before tax plus the tax on the ticket must equal \$320.

Translate Let x = cost of the ticket before tax

then $0.07x$ = tax on the ticket

$$\begin{pmatrix}\text{cost of ticket}\\ \text{before tax}\end{pmatrix} + \begin{pmatrix}\text{tax on}\\ \text{the ticket}\end{pmatrix} = 320$$

$$x + 0.07x = 320$$

Carry Out

$$1.07x = 320$$

$$x = \frac{320}{1.07}$$

$$x \approx 299.07$$

Check and Answer A check will show that if the cost of the ticket before tax is \$299.07, the cost of a ticket including a 7% tax is \$320.

EXAMPLE 10 According to a will, an estate is to be divided among two grandchildren and two charities. The two grandchildren, Rayanna and Alisa Owens, are each to receive twice as much as each of the two charities, the Red Cross and the Salvation Army. If the estate is valued at \$240,000, how much will each grandchild and each charity receive?

Solution **Understand** The total estate is to be divided among two grandchildren and two charities. Since the amount each grandchild receives is twice the amount each charity receives, we will let the variable represent the amount each charity receives. If we select x to represent the amount each charity receives, then the amount each grandchild receives is $2x$, or twice the amount.

Translate Let x = amount each charity receives

then $2x$ = amount each grandchild receives

The total received by the two grandchildren and two charities is \$240,000. Thus, the equation we use is

$$\overbrace{x + x}^{\text{Each charity receives } x.} + \overbrace{2x + 2x}^{\text{Each grandchild receives } 2x.} = 240{,}000$$

Carry Out

$$6x = 240{,}000$$

$$x = 40{,}000$$

Check If each charity receives \$40,000, then each grandchild receives 2(40,000) or \$80,000. If we add \$40,000 + \$40,000 + \$80,000 + \$80,000 we get a total of \$240,000, so the answer checks.

NOW TRY EXERCISE 39

Answer Each charity will receive \$40,000 and each grandchild will receive \$80,000.

EXAMPLE 11

This house is
protected by
Moneywell
Security Systems

Randi Rosen plans to install a security system in her house. She has narrowed down her choices to two security system dealers: Moneywell and Doile. Moneywell's system costs \$3360 to install and their monitoring fee is \$17 per month. Doile's equivalent system costs only \$2260 to install, but their monitoring fee is \$28 per month.

a) Assuming that their monthly monitoring fees do not change, in how many months would the total cost of Moneywell's and Doile's system be the same?

b) If both dealers guarantee not to raise monthly fees for 10 years, and if Randi plans to use the system for 10 years, which system would be the least expensive?

Solution

Understand **a)** Doile's system has a smaller initial cost (\$2260 vs. \$3360); however, their monthly monitoring fees are greater (\$28 vs. \$17). We are asked to find the number of months after which the total cost of the two systems will be the same.

Translate

Let n = number of months

then $17n$ = monthly monitoring cost for Moneywell's system for n months

and $28n$ = monthly monitoring cost for Doile's system for n months

$$\text{total cost of Moneywell} = \text{total cost of Doile}$$

$$\begin{pmatrix}\text{initial}\\ \text{cost}\end{pmatrix} + \begin{pmatrix}\text{monthly cost}\\ \text{for } n \text{ months}\end{pmatrix} = \begin{pmatrix}\text{initial}\\ \text{cost}\end{pmatrix} + \begin{pmatrix}\text{monthly cost}\\ \text{for } n \text{ months}\end{pmatrix}$$

$$3360 + 17n = 2260 + 28n$$

Carry Out

$$1100 + 17n = 28n$$

$$1100 = 11n$$

$$100 = n$$

Check and Answer The total cost would be the same in 100 months or about 8.3 years. We will leave the check of this answer for you.

b) Over a 10-year period Moneywell's system would be less expensive. After 8.3 years Moneywell will be less expensive because of their lower monthly cost. Check this by determining the cost for both Moneywell and Doile for 10 years of use now.

3) Select a Mortgage

Many of you will purchase a house. Choosing the wrong mortgage can cost you thousands of extra dollars. Table 3.5 is used to determine monthly **mortgage** payments of principal and interest. The table gives the monthly mortgage payment per \$1000 of mortgage at different mortgage rates for various terms of the loan. For example, for a mortgage for 30 years at 7.5%, the monthly payment of principal and interest is \$7.00 per \$1000 borrowed (shaded in table). Thus, for a

$50,000 mortgage for 30 years at 7.5%, the monthly mortgage payment would be 50 times $7.00 or $350.

$$50(7.00) = \$350$$

TABLE 3.5 Any Bank, USA: Equal Monthly Payment to Amortize a Loan of $1,000

Rate	Payment for a Mortgage Period (years) of:				Rate	Payment for a Mortgage Period (years) of:			
(%)	15	20	25	30	(%)	15	20	25	30
4.500	7.65	6.33	5.56	5.07	8.625	9.93	8.76	8.14	7.78
4.625	7.71	6.39	5.63	5.14	8.750	10.00	8.84	8.23	7.87
4.750	7.78	6.46	5.70	5.22	8.875	10.07	8.92	8.31	7.96
4.875	7.84	6.53	5.77	5.29	9.000	10.15	9.00	8.40	8.05
5.000	7.91	6.60	5.85	5.37	9.125	10.22	9.08	8.48	8.14
5.125	7.97	6.67	5.92	5.44	9.250	10.30	9.16	8.57	8.23
5.250	8.04	6.73	6.00	5.52	9.375	10.37	9.24	8.66	8.32
5.375	8.10	6.81	6.07	5.60	9.500	10.45	9.33	8.74	8.41
5.500	8.17	6.88	6.14	5.68	9.625	10.52	9.41	8.83	8.50
5.625	8.24	6.95	6.22	5.76	9.750	10.60	9.49	8.92	8.60
5.750	8.30	7.02	6.29	5.84	9.875	10.67	9.57	9.00	8.69
5.875	8.37	7.09	6.37	5.92	10.000	10.75	9.66	9.09	8.78
6.000	8.44	7.16	6.44	6.00	10.125	10.83	9.74	9.18	8.87
6.125	8.51	7.24	6.52	6.08	10.250	10.90	9.82	9.27	8.97
6.250	8.57	7.31	6.60	6.16	10.375	10.98	9.90	9.36	9.06
6.375	8.64	7.38	6.67	6.24	10.500	11.06	9.99	9.45	9.15
6.500	8.71	7.46	6.75	6.32	10.625	11.14	10.07	9.54	9.25
6.625	8.78	7.53	6.83	6.40	10.750	11.21	10.16	9.63	9.34
6.750	8.85	7.60	6.91	6.49	10.875	11.29	10.24	9.72	9.43
6.875	8.92	7.68	6.99	6.57	11.000	11.37	10.33	9.81	9.53
7.000	8.99	7.76	7.07	6.66	11.125	11.45	10.41	9.90	9.62
7.125	9.06	7.83	7.15	6.74	11.250	11.53	10.50	9.99	9.72
7.250	9.13	7.91	7.23	6.83	11.375	11.61	10.58	10.08	9.81
7.375	9.20	7.98	7.31	6.91	11.500	11.69	10.67	10.17	9.91
7.500	9.28	8.06	7.39	7.00	11.625	11.77	10.76	10.26	10.00
7.625	9.35	8.14	7.48	7.08	11.750	11.85	10.84	10.35	10.10
7.750	9.42	8.21	7.56	7.17	11.875	11.93	10.93	10.44	10.20
7.875	9.49	8.29	7.64	7.26	12.000	12.01	11.02	10.54	10.29
8.000	9.56	8.37	7.72	7.34	12.125	12.09	11.10	10.63	10.39
8.125	9.63	8.45	7.81	7.43	12.250	12.17	11.19	10.72	10.48
8.250	9.71	8.53	7.89	7.52	12.375	12.25	11.28	10.82	10.58
8.375	9.78	8.60	7.97	7.61	12.500	12.33	11.37	10.91	10.68
8.500	9.85	8.68	8.06	7.69					

This payment does not include taxes or insurance, which are sometimes added to the mortgage payment and sometimes paid separately. Also, these figures may be slightly inaccurate because of round-off error.

Sometimes banks charge "points" when they give a loan. One point is 1% of the mortgage. Thus for a $50,000 mortgage, one point is $0.01(50{,}000) = \$500$ and 3 points is $0.03(50{,}000) = \$1500$.

EXAMPLE 12 Kristen Schwartz is buying her first home and needs to obtain a 30-year $50,000 mortgage. She has narrowed down her choices to Bank One and Collier's Bank. Bank One is charging 7.5% with no points and Collier's Bank is charging 7.125% with 3 points.

a) How long would it take for the total cost of both mortgages to be the same?

b) If Kristen is planning to sell the house in 5 years, which mortgage should she select?

c) If Kristen is planning on paying off the loan in 30 years, how much will she save by selecting the 7.125% mortgage?

Solution **a) Understand** With the 7.5% mortgage, Kristen's monthly payment is $50(7) = \$350$. With the 7.125% mortgage, Kristen's monthly payment is $50(6.74) = \$337$. In addition, Kristen must pay $0.03(50{,}000) = \$1500$ to cover the 3 points which was charged on the 7.125% mortgage.

Translate Let x = number of months when total payments from both mortgages are equal

then $350x$ = monthly payments for x months with 7.5% loan

and $337x$ = monthly payments for x months with 7.125% loan

Are you looking for a low interest mortgage to finance your new house? Call us at BANK ONE
Great rates available right now!
for example:
7.5% for 30 years with no points

Now set up an equation and solve.

$$\underset{\text{Bank One}}{\begin{pmatrix}\text{monthly payments}\\ \text{for 7.5\% mortgage}\end{pmatrix}} = \underset{\text{Collier's Bank}}{\begin{pmatrix}\text{monthly payments}\\ \text{for 7.125\% mortgage}\end{pmatrix}} + \begin{pmatrix}\text{points}\end{pmatrix}$$

$$350x = 337x + 1500$$

Carry Out

$$13x = 1500$$

$$x \approx 115.4 \text{ months}$$

Answer Thus, the two mortgages would be the same after about 115 months or about 9 years 7 months.

b) Because of the lower initial cost, Bank One's total cost will be lower until about 9 years 7 months. After this, Collier's Bank will have the lower total cost because of their lower monthly payment. If she plans to sell the house in 5 years, Kristen should select the one with the lower initial cost, that is, the 7.5% Bank One mortgage.

c) Over 30 years (360 months) the total cost of each plan is as follows:

Bank One	Collier's Bank
$350(360) = \$126{,}000$	$337(360) + 1500 = 121{,}320 + 1500$
	$= 122{,}820$

Thus, over 30 years Kristen would save $126{,}000 - 122{,}820 = \$3180$ with the 7.125% Collier's Bank mortgage.

NOW TRY EXERCISE 53

4 Solve Applications Containing Large Numbers

Now we will work an application that contains large numbers.

EXAMPLE 13 Many of you have read articles about the potential problems facing Social Security. An article in the July 27, 1998, issue of *USA Today* stated that the Social Security Trust Fund will have a maximum of about \$3777 trillion dollars in 2020. Shortly after that, the surplus is expected to decrease dramatically. Some project that the Social Security Trust Fund will decrease by about \$314 trillion dollars per year after 2020. If this projection holds true, in how many years after 2020 will the Social Security Trust Fund run out of money?

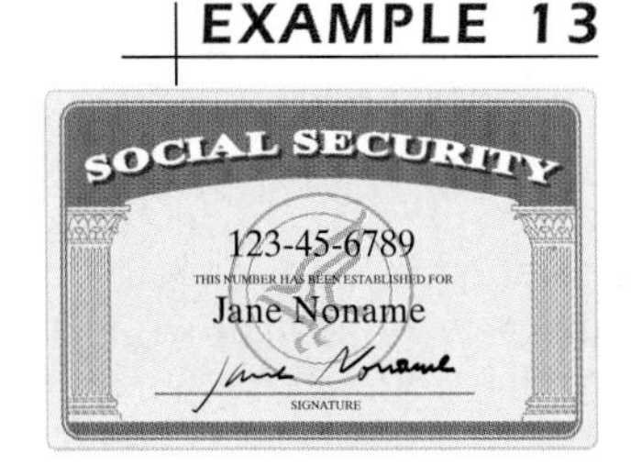

Solution **Understand** We are asked to determine the number of years after 2020 when the Social Security Trust Fund will run out of money, or have a balance of \$0.

Translate Let n = number of years

then $314n$ = decrease in surplus, in trillions of dollars, in n years

total surplus in 2020 − decrease in surplus in n years = remaining balance

$$3777 - 314n = 0$$

Carry Out

$$3777 - 314n + 314n = 0 + 314n$$

$$3777 = 314n$$

$$\frac{3777}{314} = \frac{314n}{314}$$

$$12.03 \approx n$$

Answer Twelve years after 2020, or in 2032, the Social Security Trust Fund will have a balance of \$0. (Of course, our government is considering this problem and may take steps to keep the system working longer.)

Notice in Example 13 that the numbers given were 3777 trillion and 314 trillion. Since both numbers were given in trillions, it was not necessary to write the numbers as 3,777,000,000,000,000 and 314,000,000,000,000, respectively, to set up and solve the equation. Had we written the equation as

$$3{,}777{,}000{,}000{,}000{,}000 - 314{,}000{,}000{,}000{,}000x = 0$$

NOW TRY EXERCISE 47

and solved this equation we would have obtained the same answer.

Exercise Set 3.3

Concept/Writing Exercises

1. Outline the five-step problem-solving procedure we use.
2. Explain the procedure we used to translate an application problem into mathematical language (step 2).
3. When you have a solution such as fruit punch that contains some pure fruit juice, explain how to find the amount of pure juice in the punch.
4. When translating a problem into an equation, suppose you obtain the equation $24{,}000 + 63{,}200x = 436{,}000$. Will the solution to this equation be the same or different from the solution of $2.4 + 6.32x = 43.6$? Explain.

Practice the Skills/Problem Solving

*For Exercises 5–48, **a)** set up an equation that can be used to solve the problem. **b)** Solve the equation and answer the question asked. Use a calculator where you feel it is appropriate.*

5. The sum of two consecutive integers is 85. Find the numbers.
6. The sum of two consecutive odd integers is 104. Find the numbers.
7. One number is 3 more than twice a second number. Their sum is 27. Find the numbers.
8. One number is 5 less than 3 times a second number. Their sum is 43. Find the numbers.
9. The sum of three consecutive integers is 39. Find the three integers.
10. The sum of three consecutive odd integers is 33. Find the three integers.
11. The larger of two integers is 8 less than twice the smaller. When the smaller number is subtracted from the larger, the difference is 17. Find the two numbers.
12. The sum of the two facing page numbers in an open book is 145. What are the page numbers?

13. The noise level of a rock concert is 40 decibels greater than the noise level of a vacuum cleaner. The sum of the noise level of a rock concert and a vacuum cleaner is 200 decibels. Find the noise level of a vacuum cleaner and of a rock concert.

14. In 1997, the *Corporate Travel Magazine* ranked 100 cities by the average cost of a one night hotel stay, three meals, and a rental car. The most expensive city was New York City, which was about twice the cost of the 42nd most expensive city, Santa Barbara, California. The sum of the costs for New York City and Santa Barbara was $609. Find the costs for Santa Barbara and New York City.

15. The amount of caffeine in a cup of regular brewed (automatic percolated) coffee is 26 times as much as in a cup of brewed decaffeinated coffee. If one cup of regular and one cup of decaffeinated have a total of 121.5 milligrams of caffeine, how much caffeine is in the decaffeinated coffee and in the regular coffee?

16. The calories burned by the average 150-pound person jogging at 5 mph for one hour is 115 less than twice the amount burned by walking at 4 mph for one hour. If the average 150-pound person burns 875 calories by walking for 1 hour and jogging for 1 hour, how many calories will be burned when walking 1 hour at 4 mph?

17. The price of a first class stamp in the United States is 33¢. This is a bargain compared to the cost of mailing a letter in many other countries. The cost of mailing a letter in Japan is 44 cents less than twice the cost of mailing a letter in Germany. The sum of the cost of mailing a letter in Germany and Japan is $1.27. Find the cost of mailing a letter in Germany and in Japan.

18. There are various classes of U.S. submarines. On one, the Los Angeles class, a crew of 141 is standard. If the number of enlisted men on the submarine is 1 greater than 9 times the number of officers, how many officers and how many enlisted men are in the crew?

19. An 18-karat gold bracelet weighing 20 grams contains 15 grams of pure gold. What is the percent of pure gold by weight contained in 18-karat gold?

20. Greg Middleton, a financial planner, charges an annual fee of 1% of the client's assets that he is managing. If his annual fee for managing Judy Mooney's retirement portfolio is $620.00, how much money is Greg managing for Judy?

21. Over the years Stefanie Rufenoarger has donated $6000 to her favorite charity. Beginning this year, if she donates $450 each year, how long will it take before she has donated a total of $10,050?

22. It cost Teshanna Ross $6.75 a week to wash and dry her clothing at the corner laundry. If a washer and dryer cost a total of $877.50, how many weeks would it take for Teshanna's laundry cost to equal the cost of purchasing a washer and dryer? (Disregard energy costs.)

23. Professor Laura Mann was teaching two sections of a mathematics course. Each section contained the same number of students. In the second section she tried a new teaching technique that requires the students to assume more responsibility. The number of students who were successful in the second section (a grade of C or greater) was 60 less than twice the number of students who were successful in the first section. Together, 153 students were successful. How many students were successful in the first section? In the second section?

24. At the Buyrite Warehouse, for a yearly fee of $60 you save 8% of the price of all items purchased in the store. How much would Mary need to spend during the year so that her savings equal the yearly fee?

25. A tennis star was hired to sign autographs at a convention. She was paid $3000 plus 3% of all admission fees collected at the door. The total amount she received for the day was $3750. Find the total amount collected at the door.

26. Jason O'Connor is considering buying the building where he presently has his office. His monthly rent for his office is \$890. If he makes a \$31,350 down payment on the building, his monthly mortgage payments would be \$560. How many months would it take for his down payment plus his monthly mortgage payments to equal the amount he spends on rent?

27. At a 1-day 20% off sale, Jane Demsky purchased a hat for \$25.99. What is the regular price of the hat?

28. A manufacturing plant is running at a deficit. To avoid layoffs, the workers agree on a temporary wage cut of 2%. If the average salary in the plant after the wage cut is \$28,600, what was the average salary before the wage cut?

29. During the 1999 contract negotiations, the city school board approved a 4% pay increase for its teachers effective in 2000. If Dana Frick, a first-grade teacher, projects his 2000 annual salary to be \$36,400, what is his present salary?

30. Bruce Gregory receives a weekly salary of \$350. He also receives a 6% commission on the total dollar volume of all sales he makes. What must his dollar volume be in a week, if he is to make a total of \$710?

31. Polk County has a 7% sales tax. How much does Jim Misenti's car cost before tax if the total cost of the car plus its sales tax is \$22,800?

32. Mary Beth Kodger, an electrician, wishes to charge her customers an hourly rate of \$50 per hour, which includes a 7% tax. What will be her hourly rate before tax?

33. The projected gross earnings of the movie *The Truman Show* were \$5 million more than twice its budget. The difference between its projected gross earnings and its budget was \$65 million. Find the budget and projected gross earnings of the movie.

34. In 1997, 96 billion pounds of food in the United States spoiled and was thrown out. (About 27% of the total U.S. food supply). Of the 96 billion pounds lost, the majority is lost in people's homes. If 7 billion pounds more than 14 times the amount of food is lost in homes than in all other places (like stores and restaurants), how much food is lost in places other than homes? In homes?

35. According to a test performed by *Consumer Report*, the time in seconds that it takes the 1998 Chevrolet Corvette to go from 0 to 60 miles per hour is 8.6 seconds less than twice the time it takes for the 1998 Porsche Boxster (or the BMW Z3) to go from 0 to 60 miles per hour. The difference in times between the Boxster and the Corvette is 1.5 seconds. Find the 0 to 60 mph time for both the Boxster and the Corvette.

36. Scott Montgomery is going to buy a washing machine. He is considering two machines, a Kenmore and a Neptune. The Neptune costs \$454 while the Kenmore costs \$362. Based on U.S. government tests, the energy guides indicate that the Kenmore will cost an estimated \$84 per year to operate and the Neptune will cost an estimated \$38 per year to operate. How long will it be before the total cost (purchase price plus energy cost) is the same for both washing machines?

37. Brooke Mills recently graduated from college with a two-year degree in computer information systems. She is being recruited by a number of high-tech companies. Data Technology Corporation has offered her an annual salary of \$40,000 per year plus a \$2400 increase per year. Nuteck has offered her an annual salary of \$49,600 per year plus a \$800 increase per year. In how many years will the salaries from the companies be the same?

38. Edward Wawrzewski left his estate of \$210,000 to his two children and his favorite charity. Each child received 3 times the amount left to his favorite charity. How much did each child and the charity receive?

39. Ninety-one hours of overtime must be split among four workers. The two younger workers are to be assigned the same number of hours. The third worker is to be assigned twice as much as each of the younger workers. The fourth worker is to be assigned three times as much as each of the younger workers. How much overtime should be assigned to each worker?

40. Irene Doo worked a 55-hour week last week. She is not sure of her hourly rate, but knows that she is paid $1\frac{1}{2}$ times her regular hourly rate for all hours over a 40-hour week. Her pay last week was \$600. What is her regular hourly rate?

41. Installing a water-saving showerhead saves about 60% of the water when you shower. If a 10-minute shower with a water-saving showerhead uses 24 gallons of water, how much water would a 10-minute shower without the special showerhead use?

42. The Coastline Racquet Club has two payment plans for its members. Plan 1 has a monthly fee of $20 plus $8 per hour court rental time. Plan 2 has no monthly fee, but court time is $16.25 per hour. If court time is rented in 1-hour intervals, how many hours would you have to play per month so that plan 1 becomes a better buy?

43. Betsy Beasley is on vacation and stops at a store to buy some film and other items. Her total bill before tax was $22. After tax the bill came to $23.76. Find the local sales tax rate.

44. After Linda Kodama is seated in a restaurant, she realizes that she has only $20. From this $20 she must pay a 7% tax and she wishes to leave a 15% tip. What is the maximum price for a meal that she can afford to pay?

45. The Holiday Health Club has reduced its annual membership fee by 10%. In addition, if you sign up on a Monday, they will take an additional $20 off the already reduced price. If Jorge Sanchez purchases a year's membership on a Monday and pays $250, what is the regular membership fee?

46. During the first week of a going-out-of-business sale, the Alpine Ski Shop reduced the price of all items by 20%. During the second week of the sale, they reduced the price of all items over $100 by an additional $25. Jane Manning Hyatt purchases a pair of Head skis during the second week for $231, what is the regular price of the skis?

47. According to the U.S. Environmental Protection Agency, the Magnesium Corporation of America in Rowley, Utah, was the facility in 1995 with the largest total release of toxic chemicals. The company with the second highest total release was Asarco Inc. in East Helena, Montana. The total toxic chemicals released at the Magnesium Corporation of America was about 63% greater than the total toxic chemicals released at Asarco. If about 64.3 million pounds of toxic chemicals were released at the Magnesium Corporation of America plant, how much was released at the Asarco plant? (See Example 13 for working with large numbers.)

48. A chain saw uses a mixture of gasoline and oil. For each part oil, you need 15 parts gasoline. If a total of 4 gallons of the oil–gas mixture is to be made, how much oil and how much gas will need to be mixed?

In Exercises 49 and 50, you will need to find the sum of three quantities to determine the answer.

49. According to Arthur D. Little, Inc., the average American household as of July, 1997, received 24 pieces of mail each week. The mail received is categorized as personal mail, bills and statements, or advertisements. The number of bills and statements is 1 greater than the number of pieces of personal mail, and the number of advertisements is 2 greater than 5 times the number of pieces of personal mail. Find the number of pieces of personal mail, bills and statements, and advertisements.

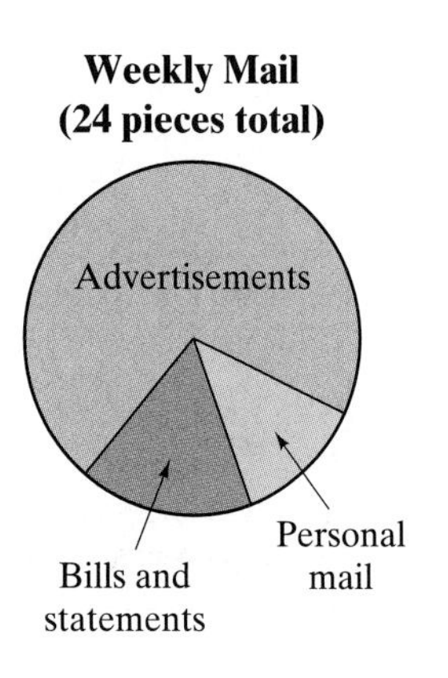

50. According to a Pennsylvania State University report, even though a large percent of women work outside the home they still spend an average of 35.1 hours weekly taking care of family business (childcare, cleaning, cooking, paying bills, home repairs, yardwork, shopping, etc.—men spend an average of 17.4 hours weekly on these chores). The amount of time women spend on cleaning and cooking is 1.1 hours more than the number of hours they spend on childcare. The time spent on all other family chores is 0.4 hours more than the number of hours they spend on childcare. Find the number of hours the average woman spends on childcare, cleaning and cooking, and all other family chores.

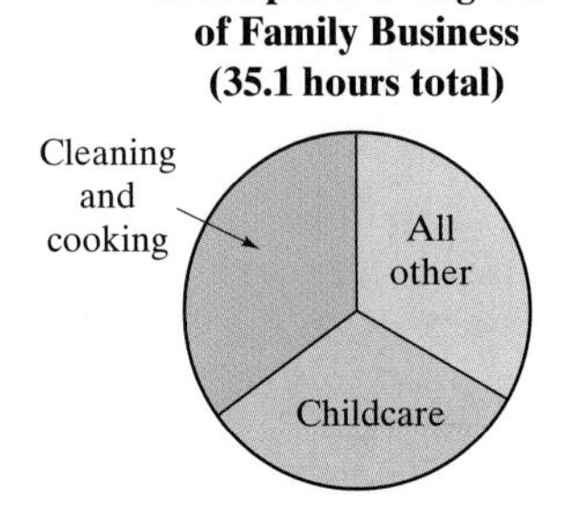

Solve the following problems.

51. In 1999, AT&T had a number of different rate plans. With one plan, called the *One Rate Plan*, each call is 15 cents per minute and there is no monthly charge. With another plan, called the *One Rate Plus Plan*, each call is 10 cents per minute and there is a $4.95 per month fee.

a) How many minutes of long distance calling in a month would result in both plans having the same monthly cost?

b) If you plan to speak long distance for an average of 200 minutes a month, what plan would result in the lowest cost? How much would you save by selecting that plan?

52. The SavUmor mail order prescription drug suppliers provide two membership plans. Under plan A you pay an annual $200 membership plus 50% of the manufacturer's list price of each drug. Under plan B you pay a $50 annual membership fee plus 75% of the manufacturer's list price of any drug.

a) How much per year must a family's drug bill total for the two plans to result in the same cost?

b) Mr. Renaud's allopurinol treatment has a manufacturer's list price of $7.50 for a month's supply. Mrs. Renaud's monthly treatment of Premarin and Medroxyprogesterone together have a list price of $48. They expect to average an additional $10 per month for other prescriptions. Which plan would be the least expensive?

53. Quincy McDonald is considering two banks for a 20-year $50,000 mortgage. M&T is charging 9.50% interest with no points, and Citibank is charging 8.00% interest with 4 points.

a) What is the monthly mortgage payment using M&T Bank?

b) What is the monthly mortgage payment using Citibank?

c) What is the cost of the 4 points with Citibank?

d) How long would it take for the total cost of the two mortgages to be the same?

e) If he plans to live in the house for the 20 years, which mortgage would be the least expensive?

54. Song Tran is considering two banks for a 30-year $60,000 mortgage. Marine Midland is charging 8.25% interest with no points and Chase Bank is charging 8.00% interest with 3 points.

a) What is the monthly mortgage payment using Marine Midland Bank?

b) What is the monthly mortgage payment using Chase Bank?

c) What is the cost of the 3 points with Chase Bank?

d) How long would it be for the total cost of the two mortgages to be the same?

e) If she plans to sell her house in 10 years, which mortgage would be the less expensive?

55. Jean-Pierre LaForce is considering two banks for a 20-year $100,000 mortgage. Key Mortgage Corp. is charging 9.00% interest with no points, and Countrywide Mortgage Corp. is charging 8.875% interest, also with no points. However, the credit check and application fee at Countrywide is $150 greater than that at Key Mortgage Corp.

a) How long would it take for the total cost of the two mortgages to be the same?

b) If Jean-Pierre plans to live in the house for 10 years, which mortgage would be less expensive?

56. Leeanne Fisher is considering two banks for a 30-year $75,000 mortgage. NationsBank is charging 9.5% interest with 1 point and Bank America is charging 9.25% interest with 2 points. The Bank America application fee is $150 greater than at NationsBank.

a) How long would it take for the total cost of the two mortgages to be the same?

b) If Lisa plans to sell her house in 8 years, which mortgage would be less expensive?

57. Because interest rates are low, Scott and Alice Barr are considering refinancing their house. They presently have $50,000 of their mortgage remaining and they are making monthly mortgage payments of principal and interest of $740. The bank they are considering will refinance their $50,000 mortgage for 20 years at 7.875% interest with no points. However, the closing cost for refinancing the house is $3000 (the closing costs are paid by the borrower when refinancing).

a) If the Barrs refinance, how long will it take for the money they save from the lower payment to equal the closing cost?

b) How much lower will their monthly payments be?

58. Dennis Williams is considering refinancing his house. He presently has $40,000 of his mortgage remaining and he is making monthly mortgage payments of $450. The bank he is considering will refinance his house for a 30-year period at 9.25% interest with 2 points. He would have an additional closing cost of $2500.

a) If he refinances, how long will it take for the money he saves from the lower monthly payments to equal the closing cost and points?

b) If he plans to live in the house for only 6 more years, does it pay for him to refinance?

Challenge Problems

59. To find the *average* of a set of values, you find the sum of the values and divide the sum by the number of values. **a)** If Paul Lavenski's first three test grades are 74, 88, and 76, write an equation that can be used to find the grade that Paul must get on his fourth exam to have an 80 average. **b)** Solve the equation from part **a)** and determine the grade Paul must receive.

For Exercises 60 and 61, set up an equation that can be used to solve the problem, then solve the equation and answer the question asked.

60. At a basketball game Emory University scored 78 points. Emory made 12 free throws (1 point each). Emory also made 4 times as many 2-point field goals as 3-point field goals (field goals made from more than 18 feet from the basket). How many 2-point field goals and how many 3-point field goals did Emory make?

61. A driver education course costs \$45 but saves those under age 25 10% of their annual insurance premiums until they reach age 25. Scott Day has just turned 18, and his insurance costs \$600 per year.

a) How long will it take for the amount saved from insurance to equal the price of the course?

b) When Scott turns 25, how much will he have saved?

Group Activity

Remodeling Costs and Return

National average cost of five of the most popular home remodeling projects and the average percent of cost returned when the home is sold.

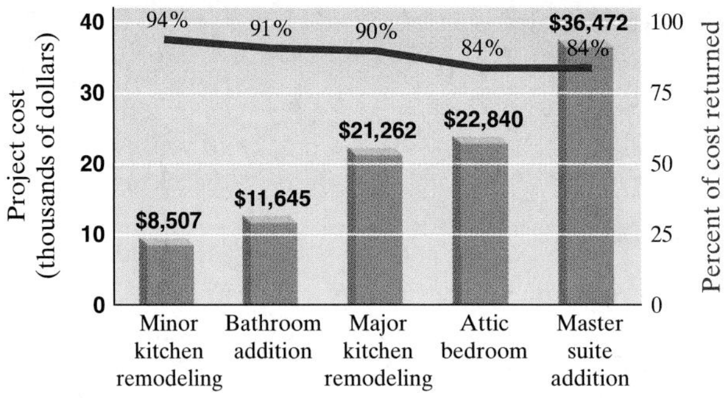

Data Source: National Association of Home Builders

Discuss and answer Exercise 62 as a group.

62. a) Each member of the group use the figure to make up your own verbal problem that can be solved algebraically. There are many different types of problems that can be created. Solve your problem algebraically.

b) Share the problem you made up with all the other members of your group.

c) For each problem given to you, set up an equation that may be used to solve the problem. Solve the equation and answer the question asked.

d) Compare your answers to each problem with all members of your group. If any member's answers do not match, work together to determine the correct answer.

Cumulative Review Exercises

[1.9] **63.** Evaluate $\frac{1}{4} + \frac{3}{4} \div \frac{1}{2} - \frac{1}{3}$.

Name each indicated property.

[1.10] **64.** $(x + y) + 5 = x + (y + 5)$

65. $xy = yx$

66. $x(x + y) = x^2 + xy$

[2.6] **67.** At the firefighter's annual chicken barbecue, the chef estimates that he will need $\frac{1}{2}$ pound of coleslaw for each 5 people. If he expects 560 residents to attend, how many pounds of coleslaw will he need?

[3.1] **68.** Solve the formula $M = \frac{a + b}{2}$ for b.

3.4 GEOMETRIC PROBLEMS

1) **Solve geometric problems.**

SSM

VIDEO 3.4

CD Rom

1) Solve Geometric Problems

This section serves two purposes. One is to reinforce the geometric formulas introduced in Section 3.1. The second is to reinforce procedures for setting up and solving verbal problems discussed in Sections 3.2 and 3.3. The more practice you have at setting up and solving the verbal problems, the better you will become at solving them.

EXAMPLE 1 Kimberly Morse, a carpenter, is planning on designing and building her own dining room table. Around the border of the table she will add a thin, slightly raised strip containing a design, as shown in Figure 3.9.

Kimberly wants the perimeter of the table (when expanded with the leaf) to be 240 inches. She also wants the length of the table to be 44 inches longer than the width. Find the dimensions of the table that Kimberly wants to build.

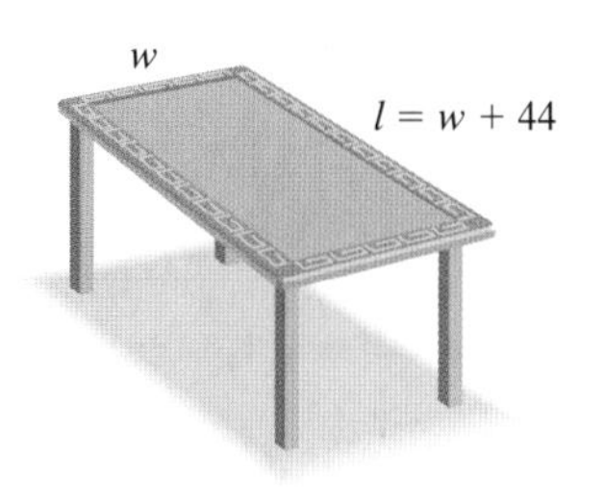

FIGURE 3.9

Solution **Understand** We are asked to find the dimensions of the table. We need to find both the length and the width. Since the length is given in terms of the width, we will let the variable represent the width. Then we can express the length in terms of the variable selected for the width. To solve this problem, we use the formula for the perimeter of a rectangle, $P = 2l + 2w$, where $P = 240$ inches.

Translate Let w = width of the table

then $w + 44$ = length of the table

$$P = 2l + 2w$$

$$240 = 2(w + 44) + 2w$$

Carry Out

$$240 = 2w + 88 + 2w$$

$$240 = 4w + 88$$

$$152 = 4w$$

$$38 = w$$

The width is 38 inches. Since the length is 44 inches greater than the width, the length is $38 + 44 = 82$ inches.

Check We will check the solution by substituting the appropriate values in the perimeter formula.

$$P = 2l + 2w$$
$$240 \stackrel{?}{=} 2(82) + 2(38)$$
$$240 = 240 \quad \text{True}$$

NOW TRY EXERCISE 17

Answer The width of the table is 38 inches and the length is 82 inches.

EXAMPLE 2

A triangle that contains two sides of equal length is called an **isosceles triangle**. In isosceles triangles, the angles opposite the two sides of equal length have equal measures. Mr. and Mrs. Harmon Katz have a corner lot that is an isosceles triangle. Two angles of their triangular lot are the same and the third angle is 30° greater than the other two. Find the measure of all three angles (see Fig. 3.10).

Solution

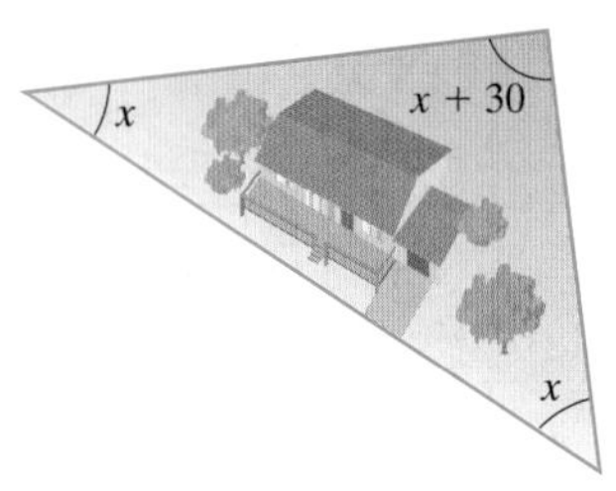

FIGURE 3.10

Understand To solve this problem, you must know that the sum of the angles of any triangle measures 180°. We are asked to find the measure of each of the three angles, where the two smaller angles have the same measure. We will let the variable represent the smaller angles, and then we will express the larger angle in terms of the variable selected for the smaller angles.

Translate Let x = each smaller angle

then $x + 30$ = larger angle

$$\text{sum of the 3 angles} = 180$$
$$x + x + (x + 30) = 180$$

Carry Out

$$3x + 30 = 180$$
$$3x = 150$$
$$x = \frac{150}{3} = 50°$$

The two smaller angles are each 50°. The larger angle is $x + 30°$ or $50° + 30° = 80°$.

Check and Answer Since $50° + 50° + 80° = 180°$, the answer checks. The two smaller angles are each 50° and the larger angle is 80°.

NOW TRY EXERCISE 15

Recall from Section 3.1 that a quadrilateral is a four-sided figure. Quadrilaterals include squares, rectangles, parallelograms, and trapezoids. The sum of the measures of the angles of any quadrilateral is 360°. We will use this information in Example 3.

EXAMPLE 3

Sarah Fuqua owns horses and uses a water trough whose ends are trapezoids. The measure of the two bottom angles of the trapezoid are the same, and the measure of the two top angles are the same. The bottom angles measure 15° less than twice the measure of the top angles. Find the measure of each angle.

FIGURE 3.11

Solution **Understand** To help visualize the problem, we draw a picture of the trapezoid, as in Figure 3.11. We use the fact that the sum of the measures of the four angles of a quadrilateral is 360°.

Translate Let x = the measure of each of the two smaller angles

then $2x - 15$ = the measure of each of the two larger angles

$$\begin{pmatrix}\text{measure of the}\\\text{two smaller angles}\end{pmatrix} + \begin{pmatrix}\text{measure of the}\\\text{two larger angles}\end{pmatrix} = 360$$

$$x + x + (2x - 15) + (2x - 15) = 360$$

Carry Out

$$x + x + 2x - 15 + 2x - 15 = 360$$
$$6x - 30 = 360$$
$$6x = 390$$
$$x = 65$$

Each smaller angle is 65°. Each larger angle is $2x - 15 = 2(65) - 15 = 115°$.

Check and Answer Since $65° + 65° + 115° + 115° = 360°$, the answer checks. Each smaller angle is 65° and each larger angle is 115°.

NOW TRY EXERCISE 19

EXAMPLE 4 Richard Jeffries recently started an ostrich farm He is separating the ostriches by fencing in three equal areas, as shown in Figure 3.12. The length of the fenced-in area, l, is to be 30 feet greater than the width and the total amount of fencing available is 660 feet. Find the length and width of the fenced-in area.

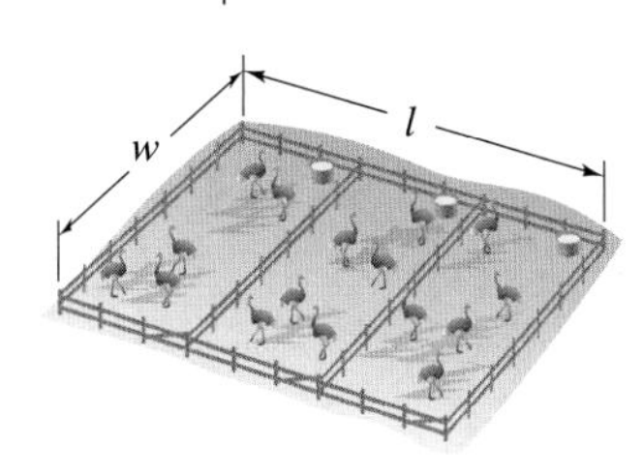

FIGURE 3.12

Solution **Understand** The fencing consists of four pieces of fence of length w, and two pieces of fence of length l.

Translate Let w = width of fenced-in area

then $w + 30$ = length of fenced-in area

$$\begin{pmatrix}\text{4 pieces of fence}\\\text{of length } w\end{pmatrix} + \begin{pmatrix}\text{two pieces of fence}\\\text{of length } w + 30\end{pmatrix} = 660$$

$$4w + 2(w + 30) = 660$$

Carry Out

$$4w + 2w + 60 = 660$$
$$6w + 60 = 660$$
$$6w = 600$$
$$w = 100$$

Since the width is 100 feet, the length is $w + 30$ or $100 + 30$ or 130 feet.

Check and Answer Since $4(100) + 2(130) = 660$, the answer checks. The width of the ostrich farm is 100 feet and the length is 130 feet.

NOW TRY EXERCISE 27

Exercise Set 3.4

Concept/Writing Exercises

1. In the equation $A = l \cdot w$, what happens to the area if the length is doubled and the width is halved? Explain your answer.
2. In the equation $A = s^2$, what happens to the area if the length of a side, s, is tripled? Explain your answer.
3. In the equation $V = l \cdot w \cdot h$, what happens to the volume if the length, width, and height are all doubled? Explain your answer.
4. In the equation $V = \frac{4}{3}\pi r^3$, what happens to the volume if the radius is tripled? Explain your answer.
5. What is an isosceles triangle?
6. What is a quadrilateral?
7. What is the sum of the measures of the angles of a triangle?
8. What is the sum of the measures of the angles of a quadrilateral?

Practice the Skills/Problem Solving

Solve the following geometric problems.

9. An **equilateral triangle** is a triangle that has three sides of the same length. The perimeter of an equilateral triangle is 28.5 inches. Find the length of each side. Equilateral triangles are discussed in Appendix C.

10. Two angles are **complementary angles** if the sum of their measures is 90°. Angle A and angle B are complementary angles, and angle A is 21° more than twice angle B. Find the measures of angle A and angle B.

Complementary Angles

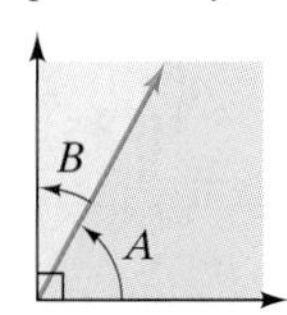

11. Two angles are **supplementary angles** if the sum of their measures is 180°. Angle A and angle B are supplementary angles, and angle B is 8° less than three times angle A. Find the measures of angle A and angle B.

Supplementary Angles

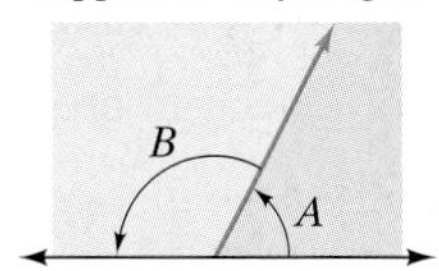

12. One angle of a triangle is 20° larger than the smallest angle, and the third angle is 6 times as large as the smallest angle. Find the measures of the three angles.

13. One angle of a triangle is 10° greater than the smallest angle, and the third angle is 30° less than twice the smallest angle. Find the measures of the three angles.

14. The length of a rectangle is 8 feet more than its width. What are the dimensions of the rectangle if the perimeter is 48 feet?

15. In an isosceles triangle, the third side is 2 meters less than each of the other sides. Find the length of each side if the perimeter is 10 meters.

16. The perimeter of a rectangle is 120 feet. Find the length and width of the rectangle if the length is twice the width.

17. The length of a regulation tennis court is 6 feet greater than twice its width. The perimeter of the court is 228 feet. Find the dimensions of the court.

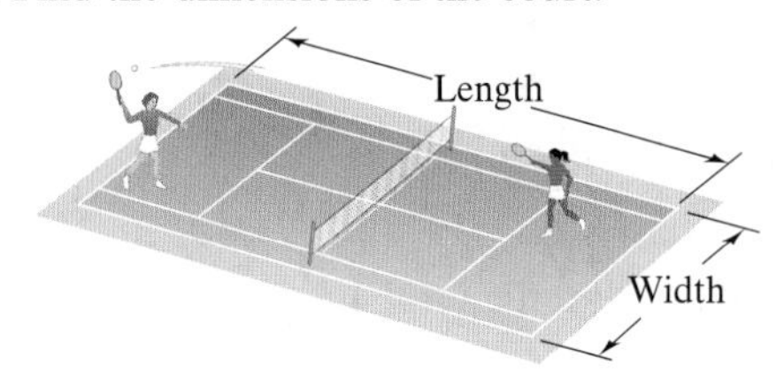

18. Mrs. Christine O'Connor is planning to build a sandbox for her daughter. She has 26 feet of lumber with which to build the perimeter. What should be the dimensions of the rectangular sandbox if the length is to be 3 feet longer than the width?

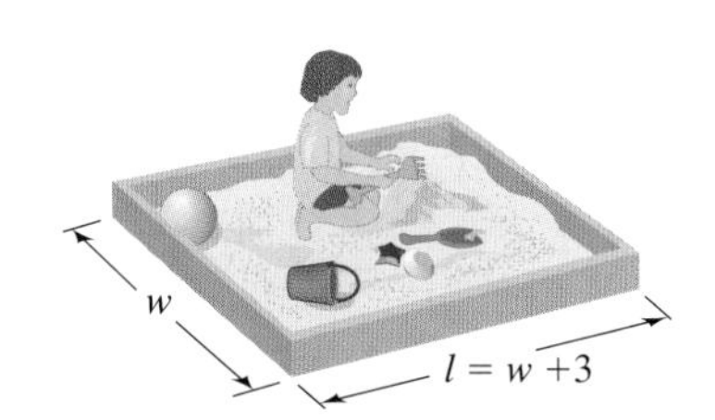

19. In a parallelogram the opposite angles have the same measures. Each of the two larger angles in a parallelogram is 20° less than 3 times the smaller angles. Find the measure of each angle.

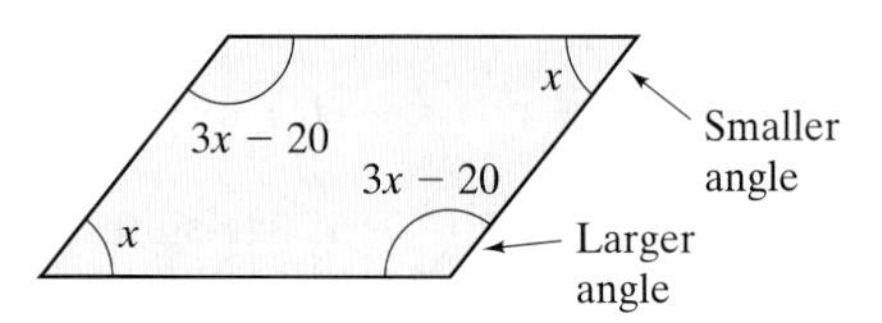

20. The two smaller angles of a parallelogram have equal measures, and the two larger angles each measure 27° less than twice each smaller angle. Find the measure of each angle.

21. The measure of one angle of a quadrilateral is 10° greater than the smallest angle; the third angle is 14° greater than twice the smallest angle; and the fourth angle is 21° greater than the smallest angle. Find the measures of the four angles of the quadrilateral.

22. The measure of one angle of a quadrilateral is twice the smallest angle; the third angle is 20° greater than the smallest angle; and the fourth angle is 20° less than twice the smallest angle. Find the measures of the four angles of the quadrilateral.

23. A bookcase is to have four shelves, including the top, as shown. The height of the bookcase is to be 3 feet more than the width. Find the width and height of the bookcase if only 30 feet of lumber is available.

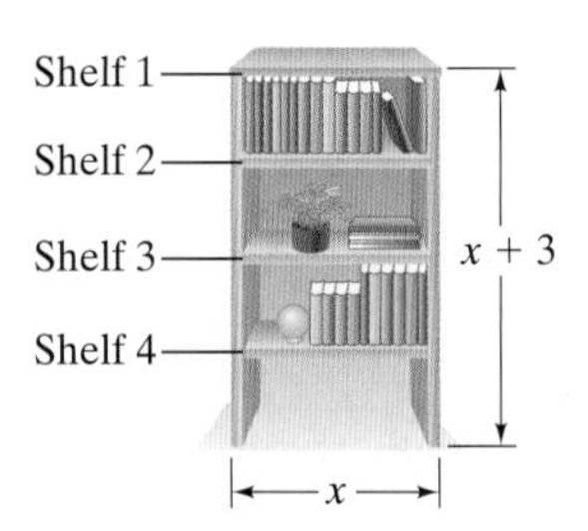

24. A bookcase is to have four shelves as shown. The height of the bookcase is to be 2 feet more than the width, and only 20 feet of lumber is available. What should be the width and height of the bookcase?

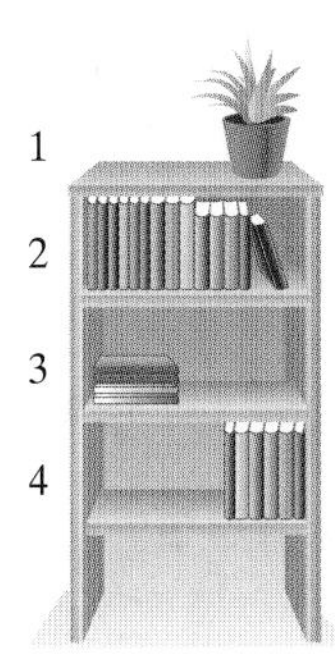

25. What should be the width and height of the bookcase in Exercise 24 if the height is to be twice the width?

26. Marty McKane plans to build storage shelves as shown. He has only 45 feet of lumber for the entire unit and wishes the width to be 3 times the height. Find the width and height of the unit.

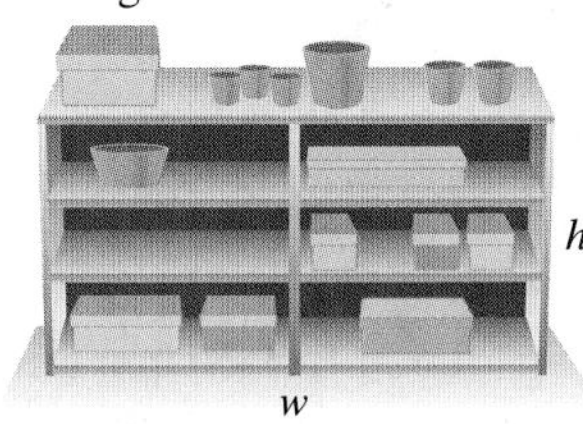

27. An area is to be fenced in along a straight river bank as illustrated. The length of the fenced-in area is to be 4 feet greater than the width, and the total amount of fencing to be used is 64 feet. Find the width and length of the fenced-in area.

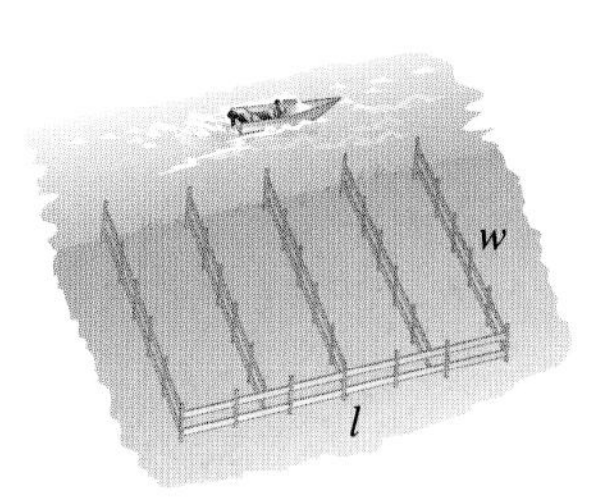

28. Trina Zimmerman is placing a border around and within a garden where she intends to plant flowers (see the figure). She has 60 feet of bordering, and the length of the garden is to be 2 feet greater than the width. Find the length and width of the area. The red shows the location of all the bordering in the figure.

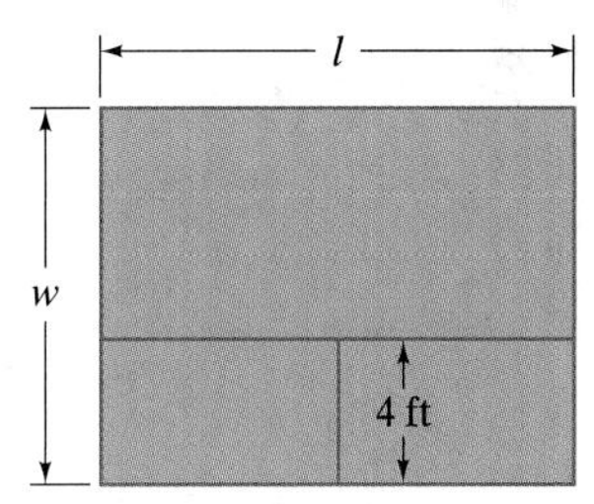

Challenge Problems

29. Consider the square in the figure.

a) Write a formula for determining the area of the shaded region.

b) Find the area of the shaded region when $S = 9$ inches and $s = 6$ inches.

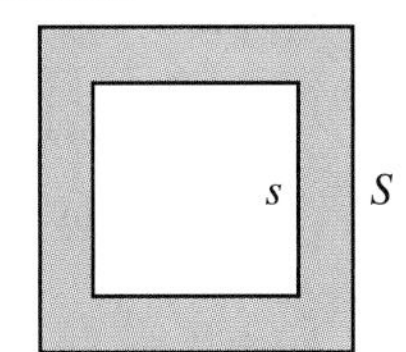

30. One way to express the area of the figure below is $(a + b)(c + d)$. Can you determine another expression, using the area of the four rectangles, to represent the area of the figure?

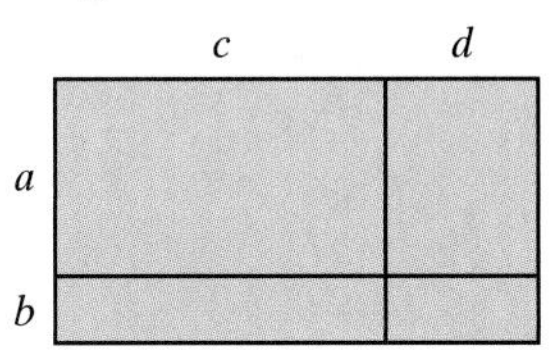

Group Activity

Discuss and answer Exercise 31 as a group.

31. Consider the four pieces shown. Two are squares and two are rectangles.

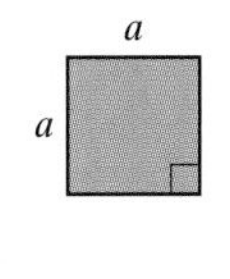

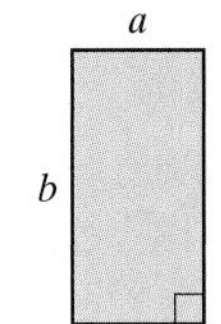

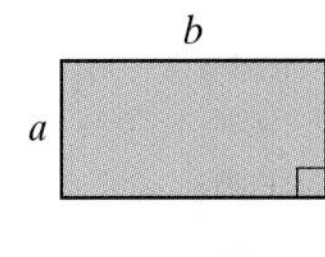

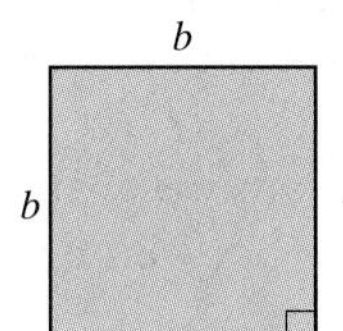

a) Individually, rearrange and place the four pieces together to form one square.

b) The area of the square you constructed is $(a + b)^2$. Write another expression for the area of the square by adding the four individual areas.

c) Compare your answers. If each member of the group did not get the same answers to parts **a)** and **b)**, work together to determine the correct answer.

d) Answer the following question as a group. If b is twice the length of a, and the perimeter of the square you created is 54 inches, find the length of a and b.

e) Use the values of a and b found in part **d)** to find the area of the square you created.

f) Use the values of a and b found in part **d)** to find the areas of the four individual pieces that make up the large square.

g) Does the sum of the areas of the four pieces found in part **f)** equal the area of the large square found in part **e)**? Is this what you expected? Explain.

Cumulative Review Exercises

Insert either >, <, or = in each shaded area to make the statement true.

[1.5] **32.** $-|-6| \;\square\; |-4|$

33. $|-3| \;\square\; -|3|$

[1.7] **34.** Evaluate $-6 - (-2) + (-4)$.

[2.1] **35.** Simplify $-6y + x - 3(x - 2) + 2y$.

[3.1] **36.** Solve $2x + 3y = 9$ for y; then find the value of y when $x = 3$.

3.5 MOTION AND MIXTURE PROBLEMS

1) Solve motion problems involving only one rate.
2) Solve motion problems involving two rates.
3) Solve mixture problems.

We now discuss two additional types of applications, motion and mixture problems. Motion and mixture problems are grouped in the same section because, as you will learn shortly, you use the same general multiplication procedure to solve them. We begin by discussing motion problems.

1) Solve Motion Problems Involving Only One Rate

A **motion problem** is one in which an object is moving at a specified rate for a specified period of time. A car traveling at a constant speed, a swimming pool being filled or drained (the water is added or removed at a specified rate), and spaghetti being cut on a conveyor belt (conveyor belt moving at a specified speed) are all motion problems.

The formula often used to solve motion problems follows.

Motion Formula

$$\text{amount} = \text{rate} \cdot \text{time}$$

The amount can be a measure of many different quantities, depending on the rate. For example, if the rate is measuring *distance* per unit time, the amount will be *distance*. If the rate is measuring *volume* per unit time, the amount will be *volume*; and so on.

EXAMPLE 1 Clarissa Skocy is filling her above-ground swimming pool. If her pool is being filled at a rate of 10 gallons per minute, how many gallons will be added in 25 minutes?

Solution **Understand and Translate** Since we are discussing gallons, which measure volume, the formula we will use is volume = rate · time. We are given the rate, 10 gallons per minute, and the time, 25 minutes. We are asked to find the volume.

$$\text{volume} = \text{rate} \cdot \text{time}$$

Carry out

$$= 10 \cdot 25 = 250$$

Answer Thus, the volume after 25 minutes is 250 gallons.

Let's look at the units of measurement in Example 1. The rate is given in gallons per minute and the time is given in minutes. If we analyze the units (a process called *dimensional analysis*), we see that the volume is measured in gallons.

$$\text{volume} = \text{rate} \cdot \text{time}$$

$$= \frac{\text{gallons}}{\cancel{\text{minute}}} \cdot \cancel{\text{minutes}}$$

$$= \text{gallons}$$

When the *amount* in the rate formula is *distance*, we often refer to the formula as the **distance formula**.

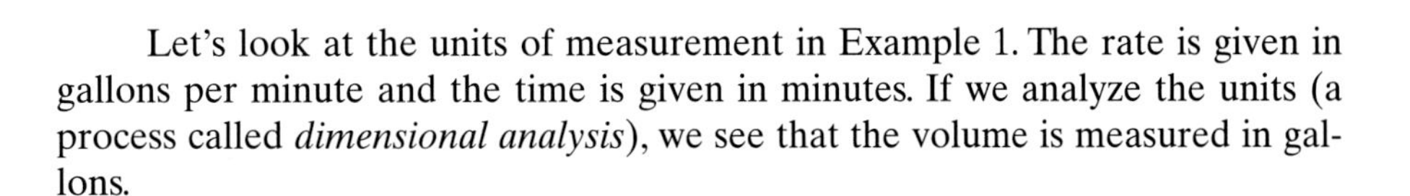

Distance Formula

distance = rate · time or $d = r \cdot t$

Example 2 illustrates the use of the distance formula.

EXAMPLE 2 Dave Salem, a geologist for Everon Oil Corporation, believes that his crew has found a large deposit of oil at a depth of about 1870 feet. Because they will need to dig through rock, they estimate their drilling speed will be 3 feet per hour. How long will it take for Everon Oil to drill to a depth of 1870 feet?

Solution **Understand and Translate** Since we are given a distance of 1870 feet, we will use the distance formula. We are given the distance and the rate and need to solve for the time, t.

$$\text{distance} = \text{rate} \cdot \text{time}$$

$$1870 = 3t$$

Carry out

$$\frac{1870}{3} = t$$

$$623.33 \approx t$$

$$\text{or} \quad t \approx 623.33 \text{ hours}$$

Answer It will take approximately 623 hours and 20 minutes to drill to a depth of 1870 feet.

NOW TRY EXERCISE 3

HELPFUL HINT

When working motion problems, the units must be consistent with each other. If you are given a problem where the units are not consistent, you will need to change one of the quantities so that the units will agree before you substitute the values into the formula. For example, if the rate is in feet per second and the distance is given in inches, you will either need to change the distance to feet or the rate to inches per second.

2) Solve Motion Problems Involving Two Rates

Now we will look at some motion problems that involve *two rates*, such as two trains traveling at different speeds. In these problems, we generally begin by letting the variable represent one of the unknown quantities, and then we represent the second unknown quantity in terms of the first unknown quantity. For example, suppose that one train travels 20 miles per hour faster than another train. We might let r represent the rate of the slower train and $r + 20$ represent the rate of the faster train.

To solve problems of this type that use the distance formula, we generally add the two distances, or subtract the smaller distance from the larger, or set the two distances equal to each other, depending on the information given in the problem.

Often, when working problems involving two different rates, we construct a table to organize the information. Examples 3 through 5 illustrate the procedures used.

EXAMPLE 3 The DiVito family decides to go on a camping trip. They will travel by canoes on the Erie Canal in New York state. They start in Tonawanda (near Buffalo) heading toward Rochester. Mike and Danny, the teenage boys, are in one canoe while Pete and Rita, the mother and father, are in a second canoe. About noon Mike and Danny decide to canoe faster than their parents to get to their campsite before their parents so they can set up camp. While the parents canoe at a leisurely speed of 2 miles per hour, the boys canoe at 6 miles per hour. In how many hours will the boys and their parents be 10 miles apart?

Solution **Understand and Translate** We are asked to find the time it takes for the canoes to become separated by 10 miles. We will construct a table to aid us in setting up the problem.

$$\text{Let } t = \text{time when canoes are 10 miles apart}$$

We draw a sketch to help visualize the problem (Fig. 3.13). When the two canoes are 10 miles apart, each has traveled for the same number of hours, t.

FIGURE 3.13

Canoe	Rate	Time	Distance
Parents'	2	t	$2t$
Sons'	6	t	$6t$

Since the canoes are traveling in the same direction, the distance between them is found by subtracting the distance traveled by the slower canoe from the distance traveled by the faster canoe.

$$\begin{pmatrix}\text{distance traveled}\\ \text{by faster canoe}\end{pmatrix} - \begin{pmatrix}\text{distance traveled}\\ \text{by slower canoe}\end{pmatrix} = 10 \text{ miles}$$

$$6t - 2t = 10$$

Carry out

$$4t = 10$$

$$t = 2.5$$

Answer After 2.5 hours the two canoes will be 10 miles apart.

EXAMPLE 4 Two construction crews are 20 miles apart working toward each other. Both are laying sewer pipe in a straight line that will eventually be connected together. Both crews will work the same hours. One crew has better equipment and more workers and can lay a greater length of pipe per day. The faster crew lays 0.4 mile of pipe per day more than the slower crew, and the two pipes are connected after 10 days. Find the rate at which each crew lays pipe.

Solution **Understand and Translate** We are asked to find the two rates. We are told that both crews work for 10 days.

$$\text{Let } r = \text{rate of slower crew}$$

$$\text{then } r + 0.4 = \text{rate of faster crew}$$

We make a sketch (Fig. 3.14) and set up a table of values.

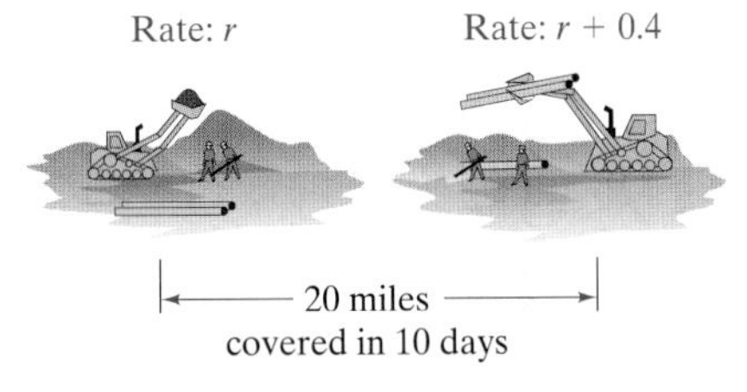

FIGURE 3.14

Crew	Rate	Time	Distance
Slower	r	10	$10r$
Faster	$r + 0.4$	10	$10(r + 0.4)$

The total distance covered by both crews is 20 miles.

$$\begin{pmatrix}\text{distance covered}\\ \text{by slower crew}\end{pmatrix} + \begin{pmatrix}\text{distance covered}\\ \text{by faster crew}\end{pmatrix} = 20 \text{ miles}$$

$$10r + 10(r + 0.4) = 20$$

Carry Out

$$10r + 10r + 4 = 20$$

$$20r + 4 = 20$$

$$20r = 16$$

$$\frac{20r}{20} = \frac{16}{20}$$

$$r = 0.8$$

Answer The slower crew lays 0.8 mile of pipe per day and the faster crew lays $r + 0.4$ or $0.8 + 0.4 = 1.2$ miles of pipe per day.

NOW TRY EXERCISE 19

EXAMPLE 5 The Pinellas Trail, used for recreational purposes, goes from St. Petersburg, Florida to Tarpon Springs, Florida, a distance of 26 miles. Al Ryckman gets on the trail in the town of Dunedin and starts rollerblading at 6 miles per hour. His friend Lois Stiller plans to meet him on the trail. Lois gets on the trail at the same point $\frac{1}{2}$ hour after Al and starts bike riding in the same direction. Lois rides at 10 miles per hour.

a) How long after Lois starts riding will they meet?

b) How far from their starting point will they be when they meet?

Solution **a) Understand and Translate** Since Lois will travel faster, she will cover the same distance in less time. When they meet, each has traveled the same distance, but Al will have been on the trail for $\frac{1}{2}$ hour more than Lois. The question asks, how long after Lois starts riding will they meet? Since the rate is given in miles per hour, the time will be in hours.

$$\text{Let } t = \text{time Lois is biking}$$

$$\text{then } t + \frac{1}{2} = \text{time Al is rollerblading}$$

Make a sketch (Fig. 3.15) and set up a table.

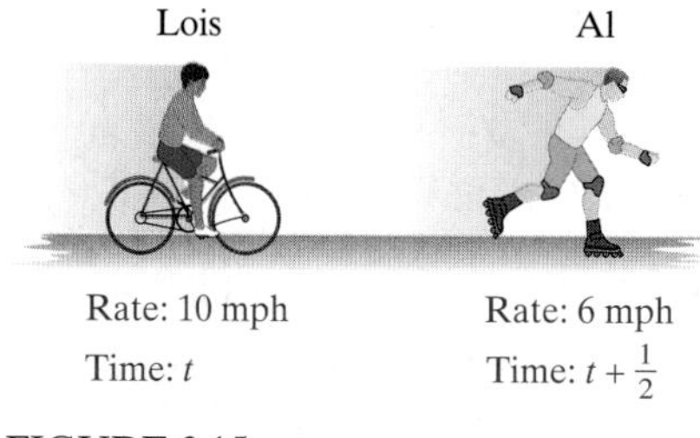

FIGURE 3.15

Traveler	Rate	Time	Distance
Lois	10	t	$10t$
Al	6	$t + \frac{1}{2}$	$6\left(t + \frac{1}{2}\right)$

$$\text{Lois's distance} = \text{Al's distance}$$

$$10t = 6\left(t + \frac{1}{2}\right)$$

Carry Out

$$10t = 6t + 3$$

$$10t - 6t = 6t - 6t + 3$$

$$4t = 3$$

$$t = \frac{3}{4}$$

Answer Thus, they will meet 3/4 hour after Lois starts biking.

b) To find the distance traveled, we will use Lois's rate and time.

$$d = r \cdot t$$

$$= 10 \cdot \frac{3}{4} = \frac{15}{2} = 7.5 \text{ miles}$$

NOW TRY EXERCISE 25 Al and Lois will meet 7.5 miles from their starting point.

3) Solve Mixture Problems

Now we will work some mixture problems. Any problem in which two or more quantities are combined to produce a different quantity or a single quantity is separated into two or more different quantities may be considered a **mixture problem**. Mixture problems are familiar to everyone, as we can see in the everyday examples that follow.

When solving mixture problems, we often let the variable represent one unknown quantity, and then we represent a second unknown quantity in terms of the first unknown quantity. For example, if we know that when two solutions are mixed they make a total of 80 liters, we may represent the number of liters of one of the solutions as x and the number of liters of the second solution as $80 - x$. Note that when we add x and $80 - x$ we get 80, the total amount. This is the same relationship we discussed on page 178.

We generally solve mixture problems by using the fact that the amount (or value) of one part of the mixture plus the amount (or value) of the second part of the mixture is equal to the total amount (or value) of the total mixture.

As we did with motion problems involving two rates, we will use a table to help analyze the problem.

EXAMPLE 6 Becky Bugos owns a coffee shop in Santa Fe, New Mexico. In her shop are many varieties of coffee. One, an orange-flavored coffee, sells for \$7 per pound, and a second, a hazelnut coffee, sells for \$4 per pound. One day, by mistake, she mixed some orange-flavored beans with some hazelnut beans, and she found that some of her customers liked the blend when they sampled the coffee. So Becky decided to make and sell a blend of the two coffees.

a) How much of the orange-flavored coffee should she mix with 12 pounds of the hazelnut coffee to get a mixture that sells for \$6 per pound?

b) How much of the mixture will be produced?

Solution **a) Understand and Translate** We are asked to find the number of pounds of orange flavored coffee.

Let x = number of pounds of orange-flavored coffee

We make a sketch of the situation (Fig. 3.16), then we construct a table.

FIGURE 3.16

The value of the coffee is found by multiplying the number of pounds by the price per pound.

Coffee	Price per Pound	Number of Pounds	Value of Coffee
Orange-flavored	7	x	$7x$
Hazelnut	4	12	$4(12)$
Mixture	6	$x + 12$	$6(x + 12)$

$$\begin{pmatrix}\text{value of}\\ \text{orange-flavored coffee}\end{pmatrix} + \begin{pmatrix}\text{value of}\\ \text{hazelnut coffee}\end{pmatrix} = \text{value of mixture}$$

$$7x + 4(12) = 6(x + 12)$$

Carry Out

$$7x + 48 = 6x + 72$$
$$x + 48 = 72$$
$$x = 24 \text{ pounds}$$

Answer Thus, 24 pounds of the orange-flavored coffee must be mixed with 12 pounds of the hazelnut coffee to make a mixture worth $6 per pound.

b) The number of pounds of the mixture is

NOW TRY EXERCISE 43

$$x + 12 = 24 + 12 = 36 \text{ pounds}$$

EXAMPLE 7 Luis Diaz just had a certificate of deposit mature and he now has $15,000 to invest. He is considering two investments. One is a loan he can make to another party through the Gibralta Mortgage Company. This investment pays him 11% simple interest for a year. A second investment, which is more secure, is a 1 year certificate of deposit that pays 5%. Luis decides that he wants to place some money in each investment, but he needs to earn a total of $1500 interest in 1 year from the two investments. How much money should Luis put in each investment?

Solution **Understand and Translate** We use the simple interest formula that was introduced in Section 3.1 to solve this problem: interest = principal · rate · time.

Let x = amount to be invested at 5%

then $15{,}000 - x$ = amount to be invested at 11%

Account	Principal	Rate	Time	Interest
CD	x	0.05	1	$0.05x$
Loan	$15{,}000 - x$	0.11	1	$0.11(15{,}000 - x)$

Since the sum of the interest from the two investments is $1500, we write the equation

$$\begin{pmatrix}\text{interest from}\\ \text{5\% CD}\end{pmatrix} + \begin{pmatrix}\text{interest from}\\ \text{11\% investment}\end{pmatrix} = \text{total interest}$$

$$0.05x + 0.11(15{,}000 - x) = 1500$$

Carry Out

$$0.05x + 0.11(15{,}000) - 0.11(x) = 1500$$
$$0.05x + 1650 - 0.11x = 1500$$
$$-0.06x + 1650 = 1500$$
$$-0.06x = -150$$
$$x = \frac{-150}{-0.06} = 2500$$

Check and Answer Thus, \$2500 should be invested at 5% interest. The amount to be invested at 11% is

$$15{,}000 - x = 15{,}000 - 2500 = 12{,}500$$

The total amount invested is \$15,000, which checks with the information given.

In Example 7 we let x represent the amount invested at 5%. If we had let x represent the amount invested at 11%, the answer would not have changed. Rework Example 7 now, letting x represent the amount invested at 11%.

NOW TRY EXERCISE 35

EXAMPLE 8 Mary Gallagher sells both small and large paintings at an art fair. The small paintings sell for \$50 each and large paintings sell for \$175 each. By the end of the day, Mary lost track of the number of paintings of each size she sold. However, by looking at her receipts she knows that she sold a total of 14 paintings for a total of \$1200. Determine the number of small and the number of large paintings she sold that day.

Solution **Understand and Translate** We are asked to find the number of paintings of each size sold.

$$\text{Let } x = \text{number of small paintings sold}$$
$$\text{then } 14 - x = \text{number of large paintings sold}$$

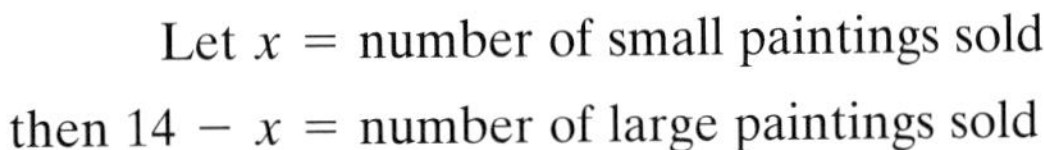

The income received from the sale of the small paintings is found by multiplying the number of small paintings sold by the cost of a small painting. The income received from the sale of the large paintings is found by multiplying the number of large paintings sold by the cost of a large painting. The total income received for the day is the sum of the income from the small paintings and the large paintings.

Painting	Cost	Number of Paintings	Income from Paintings
Small	50	x	$50x$
Large	175	$14 - x$	$175(14 - x)$

$$\begin{pmatrix}\text{income from}\\ \text{small paintings}\end{pmatrix} + \begin{pmatrix}\text{income from}\\ \text{large paintings}\end{pmatrix} = \text{total income}$$

$$50x + 175(14 - x) = 1200$$

Carry Out

$$\begin{aligned} 50x + 2450 - 175x &= 1200 \\ -125x + 2450 &= 1200 \\ -125x &= -1250 \\ x &= \frac{-1250}{-125} = 10 \end{aligned}$$

Check and Answer Ten small paintings and $14 - 10$ or 4 large paintings were sold.

Check:

$$\begin{aligned} \text{income from 10 small paintings} &= 500 \\ \text{income from 4 large paintings} &= \underline{700} \\ \text{total} &= 1200 \quad \textit{True} \end{aligned}$$

EXAMPLE 9 At Countryside High School, Mr. Paul Peterson, a chemistry teacher, needs a 10% acetic acid solution for a chemistry experiment. After checking the store room, he finds that there are only 5% and 20% acetic acid solutions available. Since there is not sufficient time to order the 10% solution, Mr. Peterson decides to make the 10% solution by combining the 5% and 20% solutions. How many liters of the 5% solution must he add to 8 liters of the 20% solution to get a solution that is 10% acetic acid?

Solution **Understand and Translate** We are asked to find the number of liters of the 5% acetic acid solution to mix with 8 liters of the 20% acetic acid solution.

Let x = number of liters of 5% acetic acid solution

Let's draw a sketch of the solution (Fig. 3.17)

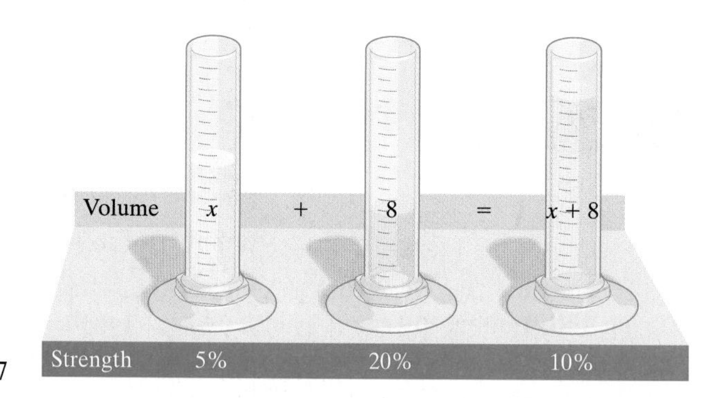

FIGURE 3.17

The amount of acid in a given solution is found by multiplying the percent strength by the number of liters.

Solution	Strength	Liters	Amount of Acetic Acid
5%	0.05	x	$0.05x$
20%	0.20	8	$0.20(8)$
Mixture	0.10	$x + 8$	$0.10(x + 8)$

$$\begin{pmatrix}\text{amount of acid}\\ \text{in 5\% solution}\end{pmatrix} + \begin{pmatrix}\text{amount of acid}\\ \text{in 20\% solution}\end{pmatrix} = \begin{pmatrix}\text{amount of acid}\\ \text{in 10\% mixture}\end{pmatrix}$$

$$0.05x + 0.20(8) = 0.10(x + 8)$$

Carry Out

$$0.05x + 1.6 = 0.10x + 0.8$$
$$0.05x + 0.8 = 0.10x$$
$$0.8 = 0.05x$$
$$\frac{0.8}{0.05} = x$$
$$16 = x$$

Answer Sixteen liters of 5% acetic acid solution must be added to the 8 liters of 20% acetic acid solution to get a 10% acetic acid solution. The total number of liters that will be obtained is 16 + 8 or 24.

NOW TRY EXERCISE 47

Exercise Set 3.5

Practice the Skills/Problem Solving

Set up an equation that can be used to solve each problem. Solve the equation, and answer the question. Use a calculator when you feel it is appropriate.

1. On his way from Providence, Rhode Island, to Boston, Massachusetts, on Route 95, Richard Chechile traveled 150 miles in 3 hours. What was his average speed?

2. An article found on the Internet states that a typical shower uses 30 gallons of water and lasts for 6 minutes. How much water is typically used per minute?

3. Lasers have many uses, from eye surgery to cutting through steel doors. One manufacturer of lasers produces a laser that is capable of cutting through steel at a rate of 0.2 centimeters per minute. During one of their tests, they went through a steel door in 12 minutes. How thick was the door?

4. Jerome Grant is at Mailboxes, Etc. making copies of an advertisement. While the copies are being run, he counts the copies made in 2.5 minutes and finds that 100 copies were made. At what rate is the copy machine running?

5. *Apollo 11* took approximately 87 hours to reach the moon, a distance of about 238,000 miles. Find the average speed of *Apollo 11*. Give the answer rounded to the nearest mile per hour.

6. Roy Shaw is laying tile. He can lay 30 square feet of tile per hour. How long will it take him to tile a room that is 420 square feet?

7. Janette Rider is in a hospital recovering from minor surgery. The nurse must administer 1500 cubic centimeters of an intravenous fluid that contains an antibiotic over a 6 hour period. What is the flow rate of the fluid?

8. A conveyer belt at a cement plant is transporting 600 pounds of crushed stone per minute. Find the time it will take for the conveyer belt to transport 20,000 pounds of crushed stone.

9. At present the fastest moving glacier is the Columbia Glacier, between Valdez and Anchorage, Alaska. The glacier's average movement is 82 feet per day. How long will it be before the glacier moves a distance of 5280 feet?

10. A production line at a factory checks and packages 620 rolls of film in an hour. How long will it take to check and pack 2170 rolls of film?

11. The record speed for a 500-mile auto race was obtained at the Michigan 500 on August 9, 1990. Al Unser Jr. completed the race in about 2.635 hours. Find his average speed during the race.

12. A standard T120 VHS videocassette tape contains 246 meters of tape. When played on standard play speed (SP), the tape will play for 2 hours. The tape will play for 4 hours on long play speed (LP) and will play for 6 hours on extended play speed (EP). Determine the rate of play of the tape at **a)** SP, **b)** LP, and **c)** EP.

Solve the following motion problems.

13. Willie and Shanna Johnston have hand-held walkie talkies. The walkie talkies, on an open flat plain, have a range of 16.8 miles. Willie and Shanna start at the same point and walk in opposite directions. If Willie walks at 3 miles per hour and Shanna walks at 4 miles per hour, how long will it take before they are out of range?

14. At a Navy Blue Angel air show assume that two F/A-18 Hornet jets travel toward each other at speeds of about 1000 miles per hour (the F/A-18 can travel up to 1200 mph or mach 1.7+). After they pass each other, if they were to keep flying in the same direction and speed, how long would it take for them to be 500 miles apart?

15. The Goodyear Tire Company is testing a new tire. The tires are placed on a machine that can simulate the tires riding on a road. The machine is first set to 60 miles per hour and the tires run at this speed for 7.2 hours. The machine is then set to a second speed and runs at this speed for 6.8 hours. After this 14-hour period, the machine indicates the tires have traveled the equivalent of 908 miles. Find the second speed to which the machine was set.

16. Tim Bozek is racing his sailboat. Heading north, with the wind at his back, he travels for 1.2 hours at 6 miles per hour. He then turns around a buoy and heads south, against the wind. From the buoy it took Tim 1.5 hours to return to the starting point. If the total distance he traveled was 14.4 miles, find the average speed that he maintained after he rounded the buoy and headed back to the start/finish line.

17. It is a beautiful Sunday in Manhattan, and Reba and Kantrell Wilsen decide to jog in Central Park. They enter the park on 59th street, across from the Mickey Mantle Restaurant, and start jogging on the trail together. Reba jogs at 5 miles per hour while Kantrell jogs at 7 miles per hour. By the time Reba reaches Strawberry Fields, Kantrell has passed Strawberry Fields, has circled the Jacqueline Kennedy Onassis Reservoir, and is by the Metropolitan Museum of Art. If Kantrell has jogged 4 miles farther than Reba, how long have they been jogging?

18. When an earthquake occurs, it generates two waves in circular paths that travel within the Earth from the earthquake's epicenter. The waves are called p-waves and s-waves (see the figure). The p-waves travel faster, but generally the s-waves do the most damage. The speed at which they travel depends on the material they are traveling through. For example, when traveling through granite, p-waves may have a velocity of 3.6 miles per second while s-waves may have a velocity of 1.8 miles per second. Assuming these speeds, how long after the earthquake will p-waves and s-waves be 80 miles apart?

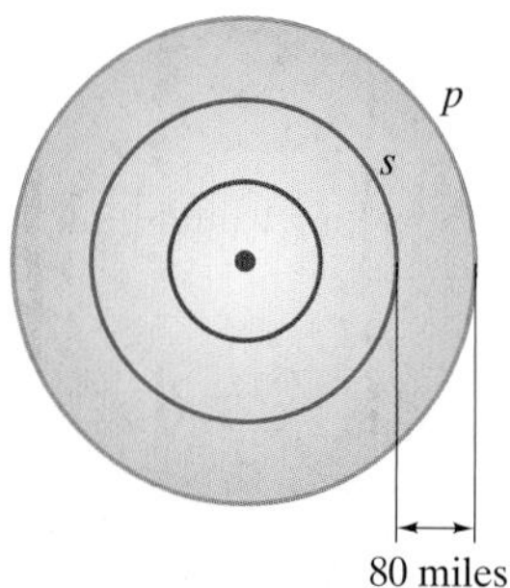

19. Two Coast Guard cutters are 225 miles apart traveling toward each other, one from the east and the other from the west, searching for a disabled boat. Because of the current, the eastbound cutter travels 5 miles per hour faster than the westbound cutter.

a) If the two cutters pass each other after 3 hours, find the average speed of each cutter.

b) Speeds at sea are generally measured in knots. One knot is approximately equal to 1.15 miles per hour. Find the speed of each boat in knots.

20. A T120 VHS videocassette tape plays at a rate of 6.72 feet per minute for 2 hours (120 minutes) on standard play speed (SP). What is the rate of the tape on extended play speed (EP) if the same tape plays for 6 hours on this speed?

21. A triathalon consists of three parts: swimming, cycling, and running. One of the more famous triathalons is the Hawaiian Ironman Triathalon held in Kailua-Kona, Hawaii (the island of Hawaii). Participants in the Ironman must swim a certain distance, cycle a certain distance, and then run for a certain distance. The 1997 women's winner was Heather Fuhr, age 29, from Canada. She swam at an average of about 2.35 miles per hour for about 1.03 hours, then she cycled at an average of 20.82 miles per hour for about 5.38 hours. Finally, she ran at an average of 8.45 miles per hour for about 3.1 hours.

a) Estimate the distance that Heather swam.

b) Estimate the distance that Heather cycled.

c) Estimate the distance that Heather ran.

d) Estimate the total distance covered during the triathalon.

e) Estimate the winning time of the triathalon.*

22. Two crews are laying blacktop on a road. They start at the same time at opposite ends of a 12-mile road and work toward one another. One crew lays blacktop at an average rate of 0.75 mile a day faster than the other crew. If the two crews meet after 3.2 days, find the rate of each crew.

23. Two sailboats are 9.8 miles apart and sailing toward each other. The larger boat, the *Pythagoras*, sails 4 miles per hour faster than the smaller boat, the *Apollo*. The two boats pass each other after 0.7 hour. Find the speed of each boat.

24. On Earth Day, two groups of people walk a 7-mile stretch of Myrtle Beach, cleaning up the beach. One group, headed by Auturo Perez, starts at one end, and the other group, headed by Jane Ivanov, starts at the

*The record winning time for the triathalon by a woman was by Californian Paula Newby-Frasier, an eight-time winner who in 1992 completed the course in 8:55:28. The actual distances are: swim, 2.4 mi; cycle, 112 mi; run 26 mi, 385 yds.

other end of the beach. They start at the same time and walk toward each other. Auturo's group is traveling at a rate of 0.5 miles per hour faster than Jane's group, and they meet in 2 hours. Find the speed of each group.

25. A motorboat carrying a bank robber leaves Galveston, Texas, and goes through the Gulf of Mexico heading toward Cozumel, Mexico. One-half hour after the boat leaves Galveston, the Coast Guard gets news of the bank robber's route and sends out a Coast Guard cutter to capture the bank robber. The motorboat is traveling at 25 miles per hour, and the Coast Guard cutter is traveling at 35 miles per hour.

a) How long will it take for the Coast Guard to catch the bank robber?

b) How far from shore will the boats be when they meet?

26. Serge and Francine Saville go mountain climbing together. Francine begins climbing the mountain 30 minutes before Serge and averages 18 feet per minute. When Serge begins climbing, he averages 20 feet per minute. How far up the mountain will they meet?

27. Ray Maloney is the quarterback for his high school football team and Pete Jovanovich is an end. From the start of a play, Pete runs down the field at about 25 feet per second. Two seconds after the play starts, Ray throws the ball, at about 50 feet per second, to Pete who is still running.

a) How long after Ray throws the ball will Pete catch it?

b) How far will Pete be from where the ball was thrown?

28. Dien and Phuong Vu belong to a health club and exercise together regularly. They start running on two treadmills at the same time. Dien's machine is set for 6 miles per hour and Phuong's machine is set for 4 miles per hour. When they finish, they compare the distances and find that together they have run a total of 11 miles. How long had they run?

29. The moving walkway in the United Airlines terminal at Chicago's O'Hare International Airport (also called a "travelator" by United Airlines) is like a flat escalator moving along the floor. When Derek Mpinga walked the floor along side the moving walkway at his normal speed of 100 feet per minute, it took him 2.75 minutes to walk the length of the moving walkway. When he returned the same distance, walking at his normal speed on the moving walkway, it took him only 1.25 minutes.

a) Find the length of the moving walkway.

b) Find the speed of the moving walkway.

30. Two rockets are launched in the same direction 1 hour apart. The first rocket is launched at noon and travels at 12,000 miles per hour. If the second rocket travels at 14,400 miles per hour, at what time will the rockets be the same distance from Earth?

31. An airplane is scheduled to leave San Diego, California, at 9 A.M. and arrive in Cleveland, Ohio, at 1 P.M. Because of mechanical problems on the plane, the plane is delayed by 0.2 hours. To arrive at Cleveland at its originally scheduled time, the plane will need to increase its planned speed by 30 miles per hour. Find the plane's planned speed and its increased speed.

32. Robin Temple started driving to the shopping mall at an average speed of 30 miles per hour. A short while later, she realized she had left her credit card on the kitchen counter. She turned around and headed back to the house, driving at 20 miles per hour (more traffic returning). If it took her a total of 0.6 hours to leave and then return home, how far had Robin driven before she turned around?

33. Because of cracks in their foundations, both a road leading to a bridge and the bridge must be replaced. The engineer estimates that it will take 20 days for the road crew to tear up and clear the road and an additional 60 days for the same road crew to dismantle and clear the bridge. The rate of clearing of the road is 1.2 feet per day faster than the rate of clearing of the bridge, and the total distance cleared is 124 feet. Find the rate for clearing the road and the rate for clearing the bridge.

34. Rich Poorman delivers the mail to residents along a 10.5-mile route. It normally takes him 5 hours to do the whole route. One Friday he needs to leave work early, so he gets his friend Keri Goldberg to help him. Rich will start delivering at one end and Keri will start at the other end 1 hour later, and they will meet somewhere along his route. If Keri covers 1.6 miles per hour, how long after Keri begins will the two meet?

Solve the following problems.

35. Butch Porter invested \$9400, part at 5% simple interest and the rest at 7% simple interest for a period of 1 year. How much did he invest at each rate if his total annual interest from both investments was \$610? (Use interest = principal · rate · time.)

36. Paul and Donna Petrie invested \$7000, part at 8% simple interest and the rest at 5% simple interest for a period of 1 year. If they received a total annual interest of \$476 from both investments, how much did they invest at each rate?

37. Aleksandra Tomich invested \$6000, part at 6% simple interest and part at 4% simple interest for a period of 1 year. How much did she invest at each rate if each account earned the same interest?

38. Larry and Carol Clar invested \$12,500, part at 7% simple interest and part at 6% simple interest for a period of 1 year. How much was invested at each rate if each account earned the same interest?

39. During the year the Public Service Commission approved a rate increase for the General Telephone Company. For a one-line household, the basic service rate increased from \$15.10 to \$16.40. While preparing for her income tax, Patricia Burgess found that she paid a total of \$183.80 for basic telephone service for the year. In what month did the rate increase take effect?

40. Violet Kokolakis knows that her subscription rate for the basic tier of cable television increased from \$13.20 to \$14.50 at some point during the calendar year. She knew that during the calendar year she paid a total of \$170.10 to the cable company. Determine the month of the rate increase.

41. Mihály Sarett holds two part-time jobs. One job, at Home Depot, pays \$6.50 an hour and the second job, at a veterinary clinic, pays \$7.00 per hour. Last week Mihály worked a total of 18 hours and earned \$122.00. How many hours did Mihály work at each job?

42. Jean Valjean owns a nut shop. In the shop he has many varieties of nuts, including walnuts that cost \$6.80 per pound and almonds that cost \$6.40 per pound. Jean gets a call from the Winchester School District regarding their end-of-the-year party. They specifically request a 30-pound mixture of walnuts and almonds that will cost \$6.65 per pound (they have budgeted only \$200 for the nuts). How many pounds of each type of nut should Jean mix to get the desired mixture?

43. Thaddeus Tarpey is the manager of Lupi's Coffee Shop in Sandusky, Ohio. His shop carries Bavarian chocolate cherry coffee beans that sell for \$6.50 per pound and Colombian coffee beans that sell for \$5.90 per pound. One customer likes to experiment with different coffees. She asks Thaddeus to mix 2 pounds of the Bavarian chocolate cherry coffee beans with 5 pounds of the Colombian coffee beans. What should be the cost per pound of the mixture?

44. At Downtown Disney World in Kissimmee, Florida, there is a 24-screen AMC theater. During October, 1998, the cost for an adult at an evening show was \$7.50 and the cost at a matinee was \$4.75. On one day there was one matinee and one evening showing of *There's Something About Mary*. On that day a total of 310 adult tickets were sold, which resulted in ticket sales of \$2022.50. How many adults went to the matinee and how many went to the evening show?

45. Chef Ramon marinates his beef, which he uses in making beef Wellington, overnight in a wine that is a blend of two red wines. To make his blend, he mixes 5 liters of a wine that is 12% alcohol by volume with 2 liters of a wine that is 9% alcohol by volume. Determine the alcohol content of the mixture.

46. A cup of Sunkist orange soda contains 15 calories per ounce. A cup of Sunkist diet orange soda contains 2.5 calories per ounce. Kevin Witt usually drinks regular soda but has only a small amount left. He therefore mixes 6 ounces of regular soda with 4 ounces of diet soda.

a) Find the calories per ounce in the mixture.

b) Find the total calories in his glass of soda.

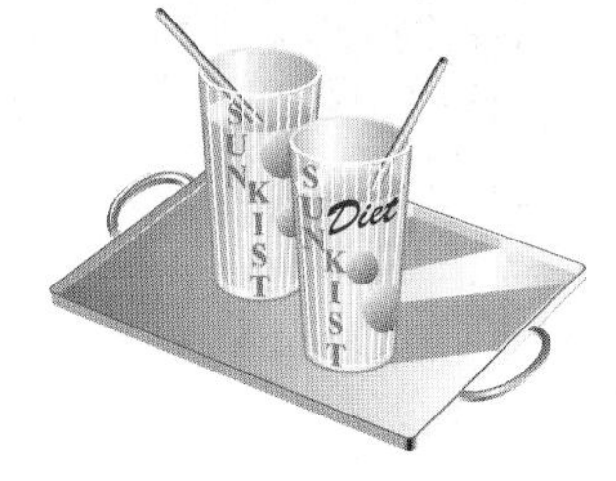

47. In chemistry class, Todd Corbin has 1 liter of a 20% sulfuric acid solution. How much of a 12% sulfuric acid solution must he mix with the 1 liter of 20% solution to make a 15% sulfuric acid solution?

48. Susan Staples, a pharmacist, has a 60% solution of the drug sodium iodite. She also has a 25% solution of the same drug. She gets a prescription calling for a 40% solution of the drug. How much of each solution should she mix to make 0.5 liter of the 40% solution?

49. Clorox bleach, used in washing machines, is 5.25% sodium hypochlorite by weight. Swimming pool shock treatments are 10.5% sodium hypochlorite by weight. Neither has any other *active* ingredient. (Clorox also contains a small amount of sodium chloride.) The instructions on the Clorox bottle say to add 8 ounces (1 cup) of Clorox to a quart of water (before placing it in

the wash). How much shock treatment should Xu Pingya very carefully add to a quart of water to get the same amount of sodium hypochlorite in the mixture as when 1 cup of Clorox is added to a quart of water?

50. Read Exercise 49 for the percents of sodium hypochlorite in Clorox and swimming pool shock treatment. Allen Angel is planning to shock his pool by very carefully adding 3.5 gallons of pool shock treatment. After he starts shocking his pool, he realizes that he has only 3.25 gallons of shock treatment. He has lots of Clorox on hand. How much Clorox should he carefully add to his pool to get the same effect as adding 3.5 gallons of pool shock treatment?

51. Read Exercise 49 to find out about Clorox and pool shock treatment. Often in warm, humid climates a black fungus forms on roofs and sidewalks. Kirsten Hale calls Home Depot to find out how to get rid of the fungus. Home Depot says they sell an 8% solution of sodium hypochlorite that she can brush on the fungus and let sit for 30 minutes, then wash off. However, they are out of the solution and do not expect another shipment for about a week. Kirsten has both Clorox and pool shock treatment at her house. How much of each should she very carefully mix to get 4 gallons of an 8% sodium hypochlorite solution?

52. Hans Kappus notices that in his medicine cabinet he has two different bottles of mouthwash, each containing a small amount. To save space, he decides to mix the mouthwashes together in one bottle. The label on the Listerine Cool Mint Antiseptic mouthwash says that it is 21.6% alcohol by volume. The label on the Scope Original Mint mouthwash says that it is 15.0% alcohol by volume. If Hans mixes 6 ounces of the Listerine with 4 ounces of the Scope, what is the percent alcohol content of the mixture?

53. Chuck Levy knows that reduced fat milk has 2% milkfat by weight and low fat milk has 1% milkfat by weight, but he does not know the milkfat content of whole milk. Chuck also knows a piece of trivia; that is, if you mix 4 gallons of whole milk with 5 gallons of low fat milk you obtain 9 gallons of reduced fat milk. Use this information to find the milkfat content of whole milk.

54. Mary Ann Terwilliger has made 6 quarts of an orange juice punch for a party. The punch contains 12% orange juice. She feels that she may need more punch, but she has no more orange juice so she adds $\frac{1}{2}$ quart of water to the punch. Find the percent of orange juice in the new mixture.

55. The label on a 12-ounce can of frozen concentrate Hawaiian Punch indicates that when the can of concentrate is mixed with 3 cans of cold water the resulting mixture is 10% juice. Find the percent of pure juice in the concentrate.

56. Scott's Family grass seed sells for \$2.45 per pound and Scott's Spot Filler grass seed sells for \$2.10 per pound. How many pounds of each should be mixed to get a 10-pound mixture that sells for \$2.20 per pound?

57. Suppose General Electric stock is selling at \$74 a share and PepsiCo stock is selling at \$35 a share. Mr. Gilbert has a maximum of \$8000 to invest. He wishes to purchase five times as many shares of PepsiCo as of General Electric. Only whole shares of stock can be purchased.

a) How many shares of each will he purchase?

b) How much money will be left over?

58. Suppose Wal Mart stock is selling at \$59 a share and Mattel stock is selling at \$28 a share. Amy Waller has a maximum of \$6000 to invest. She wishes to purchase four times as many shares of Wal Mart as of Mattel. Only whole shares of stock can be purchased.

a) How many shares of each will she purchase?

b) How much money will be left over?

Challenge Problems

59. The Navy's Blue Angels perform about 68 times a year. Their home base is at the Naval Air Station in Pensacola, Florida. They spend their winters at the Naval Air Facility (NAF) in El Centro, California. The jets used by the Blue Angels are F/A-18 Hornets. When they fly from Pensacola to El Centro assume they fly at about 900 miles per hour. On every trip they make, their C-130 transport (affectionately called Fat Albert) leaves before them carrying supplies as well as maintenance and support personnel. The C-130 generally travels at about 370 miles per hour. If the Blue Angels are making their trip from Pensacola to El Centro, how long before the Hornets leave should Fat Albert leave if it is to arrive 3 hours before the Hornets? The flying distance between Pensacola and El Centro is 1720 miles.

60. The radiator of Mark Jillian's 1999 Chevrolet holds 16 quarts. It is presently filled with a 20% antifreeze solution. How many quarts must Mark drain and replace with pure antifreeze for the radiator to contain a 50% antifreeze solution?

Group Activity

Discuss and answer Exercises 61 and 62 as a group.

61. According to the *Guinness Book of World Records*, the fastest race horse speed recorded was by a horse called Big Racket on February 5, 1945, in Mexico City, Mexico. Big Racket ran a $\frac{3}{4}$-mile race in 62.41 seconds. Find Big Racket's speed in miles per hour. Round your answer to the nearest hundredth.

62. An automatic garage door opener is designed to begin to open when a car is 100 feet from the garage. At what rate will the garage door have to open if it is to raise 6 feet by the time a car traveling at 4 miles per hour reaches it? (1 mile per hour $\approx$ 1.47 feet per second.)

Cumulative Review Exercises

[1.3] **63.** **a)** Divide $2\frac{3}{4} \div 1\frac{5}{8}$.

b) Add $2\frac{3}{4} + 1\frac{5}{8}$.

[2.5] **64.** Solve the equation $6(x - 3) = 4x - 18 + 2x$.

[2.6] **65.** Solve the proportion $\frac{6}{x} = \frac{72}{9}$.

[2.7] **66.** Solve the inequality $3x - 4 \leq -4x + 3(x - 1)$.

SUMMARY

Key Words and Phrases

3.1
Area
Circle
Circumference
Diameter
Evaluate a formula
Formula
Perimeter
Pi
Quadrilateral
Radius
Three-dimensional figure
Triangle
Volume

3.2
Consecutive even integer
Consecutive integer
Consecutive odd integer
Translate application problems into mathematical language

3.3
Mortgage points in a mortgage
Problem-solving procedure

3.4
Complementary angles
Equilateral triangle
Isosceles triangle
Supplementary angle

3.5
Mixture problems
Motion problems

IMPORTANT FACTS

Simple interest formula: $i = prt$

Distance formula: $d = rt$

The sum of the measures of the angles in any triangle is 180°.

The sum of the measures of the angles of a quadrilateral is 360°.

continued on next page

Problem-Solving Procedure for Solving Application Problems

1. Understand the problem.
   Identify the quantity or quantities you are being asked to find.
2. Translate the problem into mathematical language (express the problem as an equation).
   a) Choose a variable to represent one quantity, *and write down exactly what it represents.* Represent any other quantity to be found in terms of this variable.
   b) Using the information from part **a)**, write an equation that represents the application.
3. Carry out the mathematical calculations (solve the equation).
4. Check the answer (using the original application).
5. Answer the question asked.

Review Exercises

[3.1] *Use the formula to find the value of the indicated variable. Use a calculator to save time and round your answer to the nearest hundredth when necessary.*

1. $C = \pi d$ (circumference of a circle); find C when $d = 8$.

2. $P = 2l + 2w$ (perimeter of a rectangle); find P when $l = 4$ and $w = 5$.

3. $A = \frac{1}{2}bh$ (area of a triangle); find A when $b = 8$ and $h = 12$.

4. $K = \frac{1}{2}mv^2$ (energy formula); find m when $K = 200$ and $v = 4$.

5. $y = mx + b$ (slope–intercept form); find b when $y = 15$, $m = 3$, and $x = -2$.

6. $P = \dfrac{f}{1 + i}$ (investment formula); find f when $P = 4716.98$ and $i = 0.06$.

In Exercises 7–10 ***a)*** *solve each equation for y; then* ***b)*** *find the value of y for the given value of x.*

7. $2x = 2y + 4$, $x = 10$

8. $6x + 3y = -9$, $x = 12$

9. $4x - 3y = 15$, $x = 3$

10. $2x = 3y + 12$, $x = -6$

Solve for the indicated variable.

11. $F = ma$, for m

12. $A = \frac{1}{2}bh$, for h

13. $i = prt$, for t

14. $P = 2l + 2w$, for w

15. $V = \pi r^2 h$, for h

16. $A = \dfrac{B + C}{2}$, for B

Solve.

17. How much interest will Tom Proietti pay if he borrows \$600 for 2 years at 9% simple interest? (Use $i = prt$.)

18. The perimeter of a rectangle is 16 inches. Find the length of the rectangle if the width is 2 inches.

19. Express $x + (x + 5) = 9$ as a statement.

20. Express $x + (2x - 1) = 10$ as a statement.

[3.2,3.3] *Solve each problem.*

21. One number is 8 more than the other. Find the two numbers if their sum is 74.

22. The sum of two consecutive integers is 237. Find the two integers.

23. The larger of two integers is 3 more than 5 times the smaller integer. Find the two numbers if the smaller subtracted from the larger is 31.

24. Dan Sullivan recently purchased a new car. What was the cost of the car before tax if the total cost including a 7% tax was \$19,260?

25. In Ron Gigliotti's present position as a salesman he receives a base salary of \$500 per week plus a 3% commission on all sales he makes. He is considering moving to another company where he would sell the same goods. His base salary would be only \$400 per week, but his commission would be 8% on all sales he makes. What weekly dollar sales would he have to make for the total salaries from each company to be the same?

26. During the first week of a going-out-of-business sale, all prices were reduced by 20%. During the second week of the sale, all prices that still cost more than \$100 were reduced by an additional \$25. During the second week of the sale, Kathy Golladay purchased a camcorder for \$495. What was the original price of the camcorder?

27. The Johnsons are considering two banks for a 30-year \$60,000 mortgage. Comerica Bank is offering 8.875% interest with no points and Mellon Bank is offering 8.625% interest with 3 points.

a) How long would it take for the total payments from each bank to be the same?

b) If the Johnsons plan to keep the house for 20 years, to which bank should they apply?

28. Jennifer Rehklau is considering refinancing her house. The present balance on her mortgage is \$70,000 and her monthly payment of principal and interest is \$750. First Chicago Corporation is offering her a 20-year \$70,000 mortgage with 8.50% interest and 1 point. The closing cost, in addition to the 1 point, is \$3200.

a) How long would it take for the money she saves in monthly payments to equal the cost of the point and the closing cost?

b) If she plans to live in the house for only 10 more years, does it pay for her to refinance?

[3.4] *Solve each problem.*

29. One angle of a triangle measures 10° greater than the smallest angle, and the third angle measures 10° less than twice the smallest angle. Find the measures of the three angles.

30. One angle of a trapezoid measures 10° greater than the smallest angle; a third angle measures five times the smallest angle; and the fourth angle measures 20° greater than four times the smallest angle. Find the measure of the four angles.

31. Phaedra Gwyn has a rectangular garden whose length is 4 feet longer than its width. The perimeter of the garden is to be 70 feet. What are the dimensions of the garden?

32. Iram Hafeez is designing a house he plans to build. The basement will be rectangular with two areas. He has placed poles in the ground and attached string to the poles to mark off the two rooms (see the figure). The length of the basement is to be 30 feet greater than the width, and a total of 310 feet of string was used to mark off the rooms. Find the length and width of the basement.

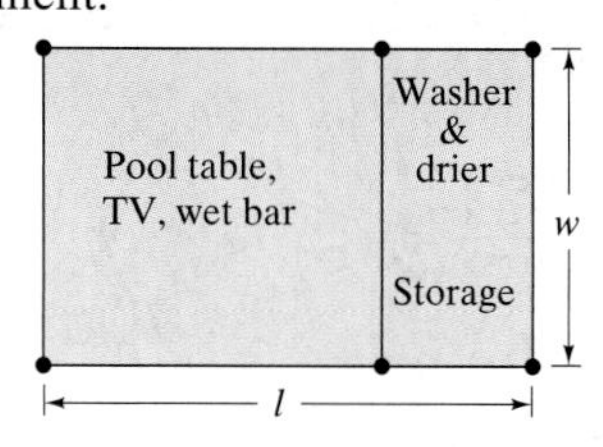

[3.5] *Solve each problem.*

33. Corey Christensen is filling a swimming pool using a hose. After 3.5 hours, the pool has a volume of 105 gallons. Find the rate of flow of the water.

34. Dave Morris completed the 26-mile Boston Marathon in 4 hours. Find his average speed.

35. Two joggers follow the same route. Harold Lowe jogs at 8 kilometers per hour and Susan Karney Fackert at 6 kilometers per hour. If they leave at the same time, how long will it take for them to be 4 kilometers apart?

36. Two trains going in opposite directions leave from the same station on parallel tracks. One train travels at 50 miles per hour and the other at 60 miles per hour. How long will it take for the trains to be 440 miles apart?

37. Kelly Scandalios wishes to place part of \$12,000 into a savings account earning 8% simple interest and part into a savings account earning $7\frac{1}{4}$% simple interest. How much should she invest in each if she wishes to earn \$900 in interest for the year?

38. Bruce Kennan, a chemist, wishes to make 2 liters of an 8% acid solution by mixing a 10% acid solution and a 5% acid solution. How many liters of each should he use?

[3.1–3.5] *Solve each problem.*

39. The sum of two consecutive odd integers is 208. Find the two integers.

40. What is the cost of a television set before tax if the total cost, including a 6% tax, is $477?

41. Mr. Vinny McAdams sells medical supplies. He receives a weekly salary of $300 plus a 5% commission on the sales he makes. If Mr. McAdams earned $900 last week, what were his sales in dollars?

42. One angle of a triangle is 8° greater than the smallest angle. The third angle is 4° greater than twice the smallest angle. Find the measure of the three angles of the triangle.

43. The Darchelle Leggett Company plans to increase its number of employees by 25 per year. If the company presently has 427 employees, how long will it take before they have 627 employees?

44. The two larger angles of a parallelogram each measure 40° greater than the two smaller angles. Find the measure of the four angles.

45. Two copy centers across the street from one another are competing for business and both have made special offers. Under Copy King's plan, for a monthly fee of $20 each copy made in that month costs only 4 cents. King Kopie charges a monthly fee of $25 plus 3 cents a copy.

a) How many copies made in a month would result in both copy centers charging the same amount?

b) If you need to make 1000 copies of an advertisement for your band, which center would cost less, and by how much?

46. Sisters Kathy and Chris Walter decide to swim to shore from a rowboat. Kathy starts swimming two minutes before Chris and averages 50 feet per minute. When Chris starts swimming, she averages 60 feet per minute.

a) How long after Chris begins swimming will they meet?

b) How far will they be from the rowboat when they meet?

47. A butcher combined hamburger that cost $3.50 per pound with hamburger that cost $4.10 per pound. How many pounds of each were used to make 80 pounds of a mixture that sells for $3.65 per pound?

48. Two brothers who are 230 miles apart start driving toward each other at the same time. The younger brother travels 5 miles per hour faster than the older brother, and the brothers meet after 2 hours. Find the speed traveled by each brotber.

49. How many liters of a 30% acid solution must be mixed with 2 liters of a 12% acid solution to obtain a 15% acid solution?

Practice Test

1. Liz Wood took out a $12,000 3-year simple interest loan. If the interest she paid on the loan was $3240, find the simple interest rate (use $i = prt$).

2. Use $P = 2l + 2w$ to find P when $l = 6$ feet and $w = 3$ feet.

3. Use $A = P + Prt$ to find A when $P = 100$, $r = 0.15$, and $t = 3$.

4. Use $A = \dfrac{m + n}{2}$ to find n when $A = 79$ and $m = 73$.

5. Use $C = 2\pi r$ to find r when $C = 50$.

6. a) Solve $4x = 3y + 9$ for y.

b) Find y when $x = 12$.

Solve for the indicated variable.

7. $P = IR$, for R

8. $A = \dfrac{a + b}{3}$, for a

9. $D = R(c + a)$, for c

10. Find the area of the skating rink shown. The ends of the rink are semicircular.

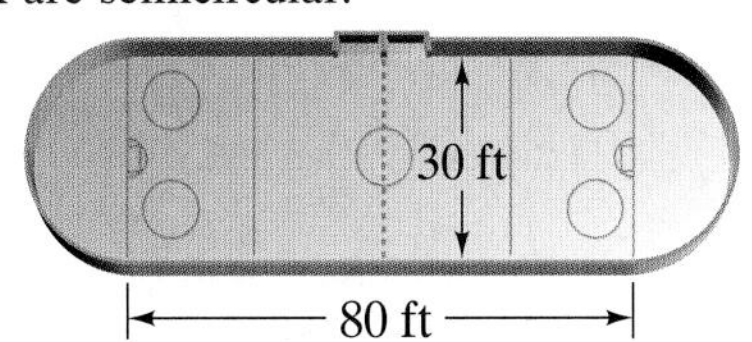

*In Exercises 11 and 12, **a)** select a letter to represent one quantity and state what the variable represents. **b)** Express the second quantity in terms of the variable.*

11. Three hundred dollars was divided between Matthew Maniscolco and Karen Sandberg.

12. The projected gross income from the movie *Deep Impact* was 45 million less than twice the projected gross income from the movie *Mask of Zorro*.

13. Express $x + (x + 3) = 7$ as a verbal statement.

*In Exercises 14–25, **a)** select a letter to represent the variable and state exactly what the variable represents. **b)** Write an equation to represent the problem. **c)** Solve the problem. Make sure to answer the question clearly.*

14. The sum of two integers is 158. Find the two integers if the larger is 10 less than twice the smaller.

15. The sum of three consecutive integers is 42. Find the three integers.

16. Tim Kent purchased a set of lawn furniture. The cost of the furniture, including a 6% tax, was \$2650. Find the cost before tax of the furniture.

17. Mark Sullivan has only \$40. He wishes to leave a 15% tip and must pay 7% tax. Find the price of the most expensive meal that he can order.

18. Three women formed a very successful business. Kathleen Luliucci and Corrina Schultz receive the same profit, and Kristen Hodge receives twice as much as Kathleen and Corrina do. If the profit for the year was \$120,000, how much will each receive?

19. Eric Gilmore is going to hire a service to plow his driveway whenever the snow totals 3 inches or more. Antoinette Payne offers a service where she charges an annual fee of \$80, plus \$5 each time she plows. For the same service, Ray Mesing charges an annual fee of \$50, plus \$10 each time he plows. How many times would the snow need to be plowed for the cost of both plans to be the same?

20. Marissa Feliberty is considering two banks, Bank One and NationsBank, for a 30-year \$60,000 mortgage. Bank One is charging 8.25% with 1 point and NationsBank is charging 7.75% with 4 points. The monthly mortgage payment with Bank One would be \$451.20, and the monthly mortgage payment with NationsBank would be \$430.20. How long would it take for the total cost of both mortgages to be the same?

21. A triangle has a perimeter of 75 inches. Find the three sides if one side is 15 inches larger than the smallest side, and the third side is twice the smallest side.

22. The sum of the angles of a parallelogram is 360°. The two smaller angles are equal, and the two larger angles are each 30° greater than twice the smaller angles. Find the measure of each angle.

23. The perimeter of Matthew Crispin's rectangular driveway is 144 feet. Find the dimensions of his driveway if the length is 12 feet more than 4 times the width.

24. Donald and Carrie Mattaini are digging a shallow 67.2-foot-long trench to lay electrical cable to a new outdoor light fixture they just installed. They start digging at the same time at opposite ends of where the trench is to go, and dig toward each other. Don digs at a rate of 0.2 feet per minute faster than Carrie, and they meet after 84 minutes. Find the speed that each digs.

25. How many liters of 20% salt solution must be added to 60 liters of 40% salt solution to get a solution that is 35% salt?

Cumulative Review Test

1. In the United States in 1998, about 222 million tons of trash were generated (about 4.40 pounds per person per day). The largest component of the trash generated was paper/paperboard, which accounted for about 88.4 million tons. The percents of certain types of waste material recycled in 1998 are shown in the following graph.

Recycled Trash

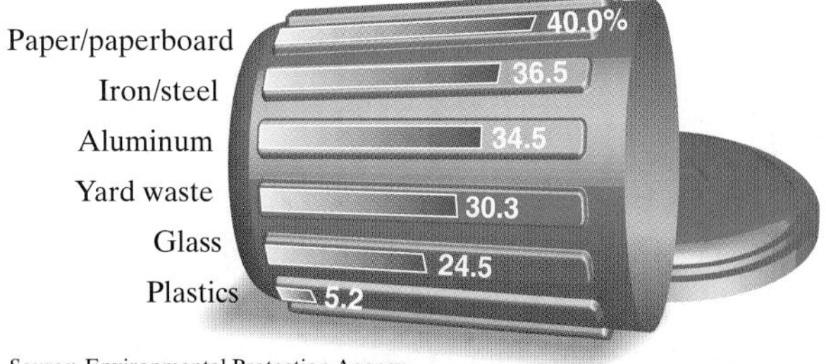

Source: Environmental Protection Agency

Determine the number of tons of paper/paperboard recycled in 1998.

2. The drug Viagra was launched on March 27, 1998, and quickly became one of the most successful prescription drugs in history. Its sales in 1998 beat the combined sale of the top five drugs in 1997. The following graph shows the number of new prescriptions and number of refills of Viagra from April 3 through July 31, 1998.

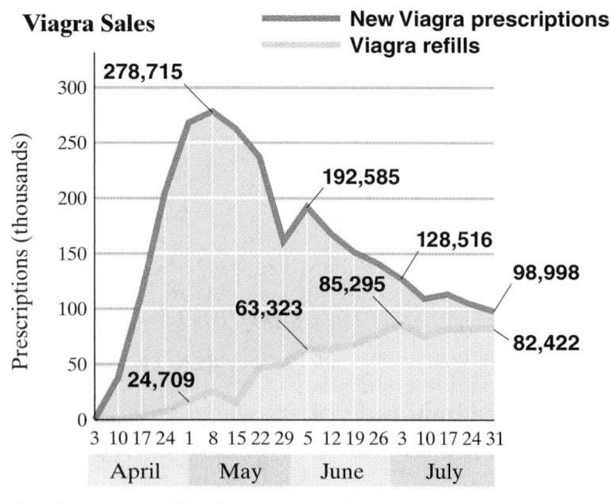

Data Source: Gannett News Service

a) On June 5, 1998, what was the difference in the number of new prescriptions of Viagra and the number of refills?

b) During the week of June 5, 1998, how many prescriptions of Viagra, including new prescriptions and refills, were filled?

c) During the week of June 5, 1998, how many times greater was the number of new prescriptions of Viagra than the number of refills? Use a calculator and round your answer to the nearest hundredth.

3. David Warner, an environmentalist, was checking the level of carbon dioxide in the air. On five readings, he got the following results.

Test	Carbon Dioxide (parts per million)
1	5
2	6
3	8
4	12
5	5

a) Find the mean level of carbon dioxide detected.

b) Find the median level of carbon dioxide detected.

4. Evaluate $\frac{5}{12} \div \frac{3}{4}$.

5. How much larger is $\frac{2}{3}$ inch than $\frac{3}{8}$ inch?

6. a) List the set of natural numbers.

b) List the set of whole numbers.

c) What is a rational number?

7. a) Evaluate $|-3|$.

b) Which is greater, $|-5|$ or $|-3|$? Explain.

8. Evaluate $12 - 6 \div 2 \cdot 2$.

9. Simplify $4(2x - 3) - 2(3x + 5) - 6$.

In Exercises 10–12, solve the equation.

10. $3x - 6 = x + 12$

11. $6r = 2(r + 3) - (r + 5)$

12. $2(x + 5) = 3(2x - 4) - 4x$

13. If Lisa Shough's car can travel 50 miles on 2 gallons of gasoline, how many gallons of gas will it need to travel 225 miles?

14. Solve the inequality $3x - 4 \le -1$ and graph the solution on the number line.

15. If $A = \pi r^2$, find A when $r = 6$.

16. Consider the equation $3x + 6y = 12$.

a) Solve the equation for y.

b) Find y when $x = -4$.

17. Solve the formula $P = 2l + 2w$ for l.

18. The sum of two numbers is 29. Find the two numbers if the larger is 11 greater than twice the smaller.

19. Lori Sypher is considering two cellular telephone plans. Plan A has a monthly charge of $19.95 plus 35 cents per minute. Plan B has a monthly charge of $29.95 plus 10 cents per minute. How long would Lori need to talk in a month for the two plans to have the same total cost?

20. One angle of a quadrilateral measures 5° larger than the smallest angle; the third angle measures 50° larger than the smallest angle; and the fourth angle measures 25° greater than 4 times the smallest angle. Find the measure of each angle of the quadrilateral.

GRAPHING LINEAR EQUATIONS

CHAPTER

4

Use the Angel Web site at www.prenhall.com/angel to be linked to an internet resource that will help you further explore the following application.

Have you ever wondered how salespeople are paid? In some cases, weekly salaries are determined by an hourly wage. Other salespeople are commissioned. Their income is dependent on the dollar amount of goods they sell each week. A common scenario is for a salesperson to receive a small weekly salary plus a commission on all sales. We can use linear equations to calculate a salary, and graphs to show potential earnings. On page 247 we estimate the salary of a salesperson from the graph of a linear equation.

Preview and Perspective

In this chapter we explain how to graph linear equations. The graphs of linear equations are straight lines. Graphing is one of the most important topics in mathematics, and each year its importance increases. If you take additional mathematics courses you will graph many different types of equations. The material presented in this chapter should give you a good background for graphing in later courses. Graphs are also used in many professions and industries. They are used to display information and to make projections about future trends.

In Section 4.1 we introduce the Cartesian coordinate system and explain how to plot points. In Section 4.2 we discuss two methods for graphing linear equations: by plotting points and by using the x- and y-intercepts. The slope of a line is discussed in Section 4.3. In Section 4.4 we discuss a third procedure, using slope, for graphing a linear equation.

We solved inequalities in one variable in Section 2.6. In Section 4.5 we will solve and graph linear inequalities in two variables. Graphing linear inequalities is an extension of graphing linear equations.

Functions are a unifying concept in mathematics. In Section 4.6, we give a brief and somewhat informal introduction to functions. Functions will be discussed in much more depth in later mathematics courses.

This is an important chapter. If you plan on taking another mathematics course, graphs and functions will probably be a significant part of that course.

4.1 THE CARTESIAN COORDINATE SYSTEM AND LINEAR EQUATIONS IN TWO VARIABLES

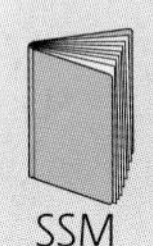
SSM

VIDEO 4.1

CD Rom

1) Plot points in the Cartesian coordinate system.
2) Determine whether an ordered pair is a solution to a linear equation.

1) Plot Points in the Cartesian Coordinate System

René Descartes

In this chapter we discuss several procedures that can be used to draw graphs. A **graph** shows the relationship between two variables in an equation. Many algebraic relationships are easier to understand if we can see a picture of them. We draw graphs using the **Cartesian (or rectangular) coordinate system.** The Cartesian coordinate system is named for its developer, the French mathematician and philosopher René Descartes (1596–1650).

The Cartesian coordinate system provides a means of locating and identifying points just as the coordinates on a map help us find cities and other locations. Consider the map of the Great Smoky Mountains (see Fig. 4.1). Can you find Cades Cove on the map? If we tell you that it is in grid A3, you can probably find it much more quickly and easily.

The Cartesian coordinate system is a grid system, like that of a map, except that it is formed by two axes (or number lines) drawn perpendicular to

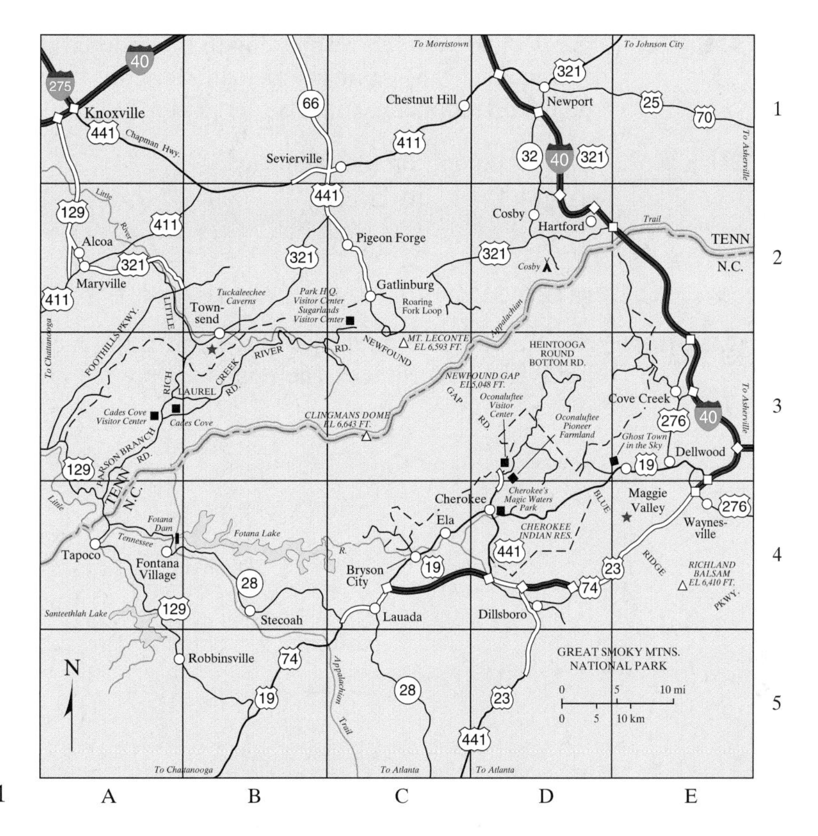

FIGURE 4.1 A B C D E

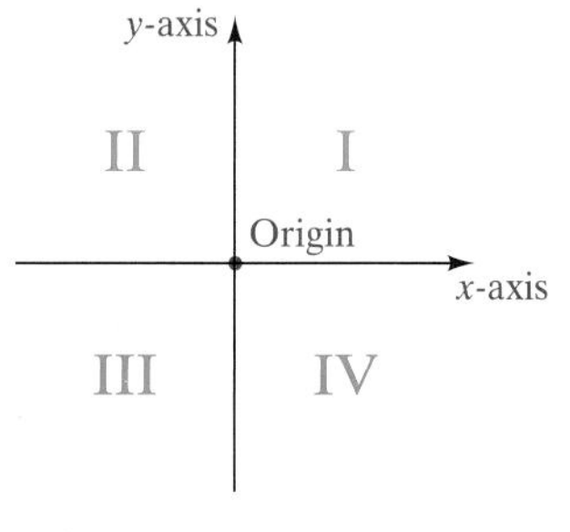

FIGURE 4.2

each other. The two intersecting axes form four **quadrants**, numbered I through IV in Fig. 4.2.

The horizontal axis is called the ***x*-axis.** The vertical axis is called the ***y*-axis.** The point of intersection of the two axes is called the **origin.** At the origin the value of x is 0 and the value of y is 0. Starting from the origin and moving to the right along the x-axis, the numbers increase. Starting from the origin and moving to the left, the numbers decrease (Fig. 4.3). Starting from the origin and moving up the y-axis, the numbers increase. Starting from the origin and moving down, the numbers decrease.

To locate a point, it is necessary to know both the value of x and the value of y, or the **coordinates,** of the point. When the x- and y-coordinates of a point are placed in parentheses, *with the x-coordinate listed first*, we have an **ordered pair.** In the ordered pair (3, 5) the x-coordinate is 3 and the y-coordinate is 5. The point corresponding to the ordered pair (3, 5) is plotted in Figure 4.4. The

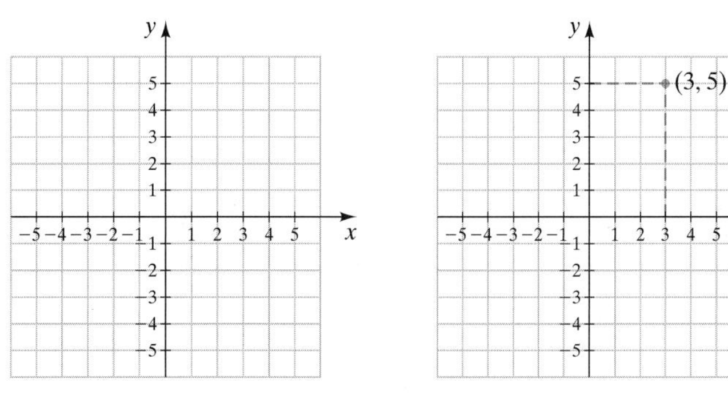

FIGURE 4.3 FIGURE 4.4

phrase "the point corresponding to the ordered pair (3, 5)" is often abbreviated "the point (3, 5)." For example, if we write "the point (−1, 2)," it means "the point corresponding to the ordered pair (−1, 2)."

EXAMPLE 1 Plot each point on the same axes.

a) $A(4, 2)$ **b)** $B(2, 4)$ **c)** $C(-3, 1)$

d) $D(4, 0)$ **e)** $E(-2, -5)$ **f)** $F(0, -3)$

g) $G(0, 3)$ **h)** $H\left(6, -\frac{7}{2}\right)$ **i)** $I\left(-\frac{3}{2}, -\frac{5}{2}\right)$

Solution The first number in each ordered pair is the x-coordinate and the second number is the y-coordinate. The points are plotted in Figure 4.5.

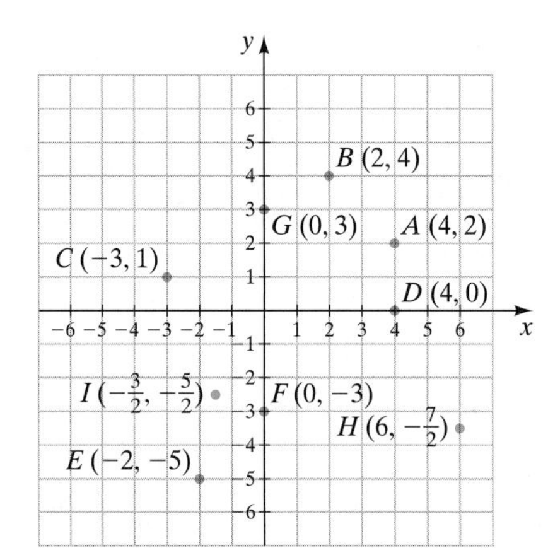

FIGURE 4.5

Note that when the x-coordinate is 0, as in Example 1 **f)** and 1 **g)**, the point is on the y-axis. When the y-coordinate is 0, as in Example 1 **d)**, the point is on the x-axis.

NOW TRY EXERCISE 29

EXAMPLE 2 List the ordered pairs for each point shown in Figure 4.6.

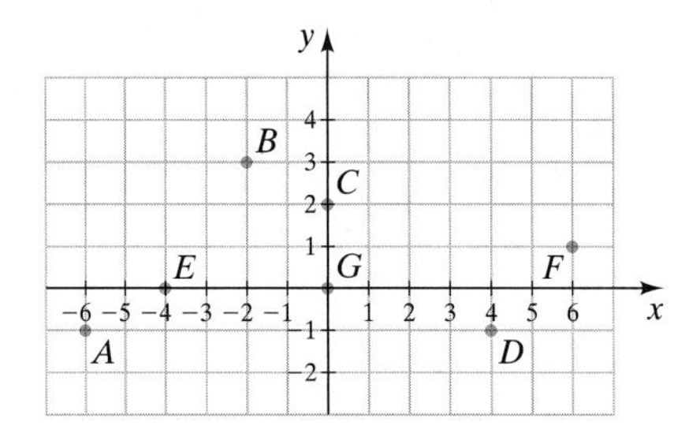

FIGURE 4.6

Solution Remember to give the x-value first in the ordered pair.

Point	Ordered Pair
A	$(-6, -1)$
B	$(-2, 3)$
C	$(0, 2)$
D	$(4, -1)$
E	$(-4, 0)$
F	$(6, 1)$
G	$(0, 0)$

NOW TRY EXERCISE 27

2) Determine Whether an Ordered Pair Is a Solution to a Linear Equation

In Section 4.2 we will learn to graph linear equations in two variables. Below we explain how to identify a **linear equation in two variables.**

> A **linear equation in two variables** is an equation that can be put in the form
>
> $$ax + by = c$$
>
> where a, b, and c are real numbers.

The graphs of equations of the form $ax + by = c$ are straight lines. For this reason such equations are called linear. Linear equations may be written in various forms, as we will show later. A linear equation in the form $ax + by = c$ is said to be in **standard form.**

Examples of Linear Equations

$$3x - 2y = 4$$

$$y = 5x + 3$$

$$x - 3y + 4 = 0$$

Note in the examples that only the equation $3x - 2y = 4$ is in standard form. However, the bottom two equations can be written in standard form, as follows:

$$y = 5x + 3 \qquad x - 3y + 4 = 0$$

$$-5x + y = 3 \qquad x - 3y = -4$$

Most of the equations we have discussed thus far have contained only one variable. Exceptions to this include formulas used in application sections. Consider the linear equation in *one* variable, $2x + 3 = 5$. What is its solution?

$$2x + 3 = 5$$

$$2x = 2$$

$$x = 1$$

This equation has only one solution, 1.

CHECK:

$$2x + 3 = 5$$

$$2(1) + 3 \stackrel{?}{=} 5$$

$$5 = 5 \quad \textit{True}$$

Now consider the linear equation in *two* variables, $y = x + 1$. What is the solution? Since the equation contains two variables, its solutions must contain two numbers, one for each variable. One pair of numbers that satisfies this equation is $x = 1$ and $y = 2$. To see that this is true, we substitute both values into the equation and see that the equation checks.

CHECK:

$$y = x + 1$$

$$2 \stackrel{?}{=} 1 + 1$$

$$2 = 2 \quad \textit{True}$$

We write this answer as an ordered pair by writing the x- and y-values within parentheses separated by a comma. Remember the x-value is always listed first since the form of an ordered pair is (x, y). Therefore, one possible solution to this equation is the ordered pair $(1, 2)$. The equation $y = x + 1$ has other possible solutions, as follows.

	Solution		Solution		Solution	
	$x = 2, y = 3$		$x = -3, y = -2$		$x = -\frac{1}{3}, y = \frac{2}{3}$	
CHECK:	$y = x + 1$		$y = x + 1$		$y = x + 1$	
	$3 \stackrel{?}{=} 2 + 1$		$-2 \stackrel{?}{=} -3 + 1$		$\frac{2}{3} \stackrel{?}{=} -\frac{1}{3} + 1$	
	$3 = 3$	*True*	$-2 = -2$	*True*	$\frac{2}{3} = \frac{2}{3}$	*True*

Solution Written as an Ordered Pair

$(2, 3)$	$(-3, -2)$	$\left(-\frac{1}{3}, \frac{2}{3}\right)$

How many possible solutions does the equation $y = x + 1$ have? The equation $y = x + 1$ has an unlimited or *infinite number* of possible solutions. Since it is not possible to list all the specific solutions, the solutions are illustrated with a graph.

Definition

A **graph** of an equation is an illustration of a set of points whose coordinates satisfy the equation.

Figure 4.7a shows the points $(2, 3)$, $(-3, -2)$, and $\left(-\frac{1}{3}, \frac{2}{3}\right)$ plotted in the Cartesian coordinate system. Figure 4.7b shows a straight line drawn through the three points. Arrowheads are placed at the ends of the line to show the line continues in both directions. Every point on this line will satisfy the equation $y = x + 1$, so this graph illustrates all the solutions of $y = x + 1$. The ordered pair $(1, 2)$, which is on the line, also satisfies the equation.

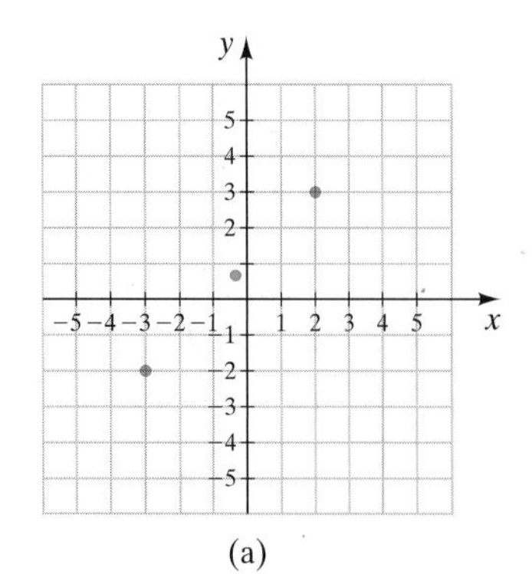

(a)

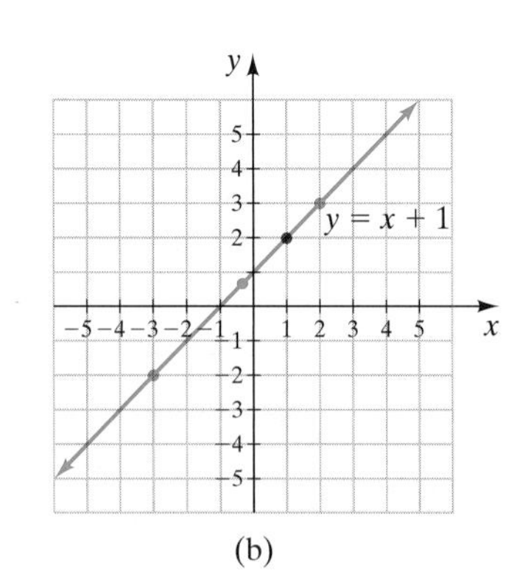

(b)

FIGURE 4.7

EXAMPLE 3 Determine which of the following ordered pairs satisfy the equation $2x + 3y = 12$.

a) $(2, 3)$ **b)** $(3, 2)$ **c)** $\left(8, -\frac{4}{3}\right)$

Solution To determine whether the ordered pairs are solutions, we substitute them into the equation.

Check: $(2, 3)$

a)
$$2x + 3y = 12$$
$$2(2) + 3(3) \stackrel{?}{=} 12$$
$$4 + 9 \stackrel{?}{=} 12$$
$$13 = 12 \quad \textit{False}$$

$(2, 3)$ is *not* a solution.

Check: $(3, 2)$

b)
$$2x + 3y = 12$$
$$2(3) + 3(2) \stackrel{?}{=} 12$$
$$6 + 6 \stackrel{?}{=} 12$$
$$12 = 12 \quad \textit{True}$$

$(3, 2)$ is a solution.

Check: $\left(8, -\frac{4}{3}\right)$

c)
$$2x + 3y = 12$$
$$2(8) + 3\left(-\tfrac{4}{3}\right) \stackrel{?}{=} 12$$
$$16 - 4 \stackrel{?}{=} 12$$
$$12 = 12 \quad \textit{True}$$

$\left(8, -\frac{4}{3}\right)$ is a solution.

In Example 3, if we plotted the two solutions $(3, 2)$ and $\left(8, -\frac{4}{3}\right)$ and connected the two points with a straight line, we would get the graph of the equation $2x + 3y = 12$. The coordinates of every point on this line would satisfy the equation.

In Figure 4.7b, what do you notice about the points $(2, 3)$, $(1, 2)$, $\left(-\frac{1}{3}, \frac{2}{3}\right)$, and $(-3, -2)$? You probably noticed that they are in a straight line. A set of points that are in a straight line are said to be **collinear.** *In Section 4.2 when you graph linear equations by plotting points, the points you plot should all be collinear.*

EXAMPLE 4 Determine whether the three points given appear to be collinear.

a) $(2, 7)$, $(0, 3)$, and $(-2, -1)$

b) $(0, 5)$, $\left(\frac{5}{2}, 0\right)$, and $(5, -5)$

c) $(-2, -5)$, $(0, 1)$, and $(5, 8)$

Solution We plot the points to determine whether they appear to be collinear. The solution is shown in Figure 4.8.

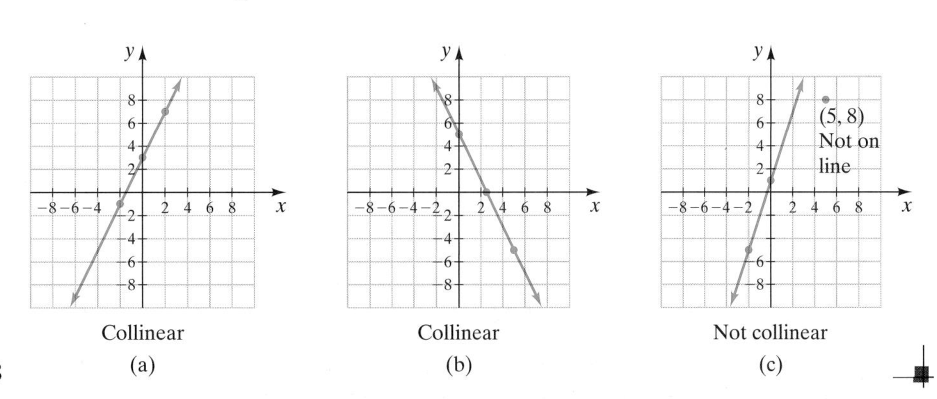

FIGURE 4.8

NOW TRY EXERCISE 33

How many points do you need to graph a linear equation? As mentioned earlier, *the graph of every linear equation of the form $ax + by = c$ will be a*

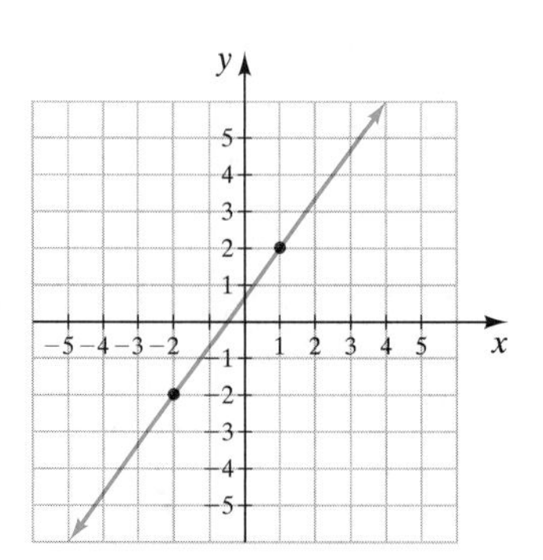

FIGURE 4.9

straight line. Since only two points are needed to draw a straight line (Fig. 4.9), only two points are needed to graph a linear equation. However, if you graph a linear equation using only two points and you have made an error in determining or plotting one of those points, your graph will be wrong and you will not know it. If you use at least three points to plot your graph, as in Figure 4.7b, and they appear to be collinear, you probably have not made a mistake.

EXAMPLE 5 **a)** Determine which of the following ordered pairs satisfy the equation $2x + y = 4$.

$$(2, 0), (0, 4), (3, 3), (-1, 6)$$

b) Plot all the points that satisfy the equation on the same axes and draw a straight line through the points.

c) What does this straight line represent?

Solution **a)** We substitute values for x and y into the equation $2x + y = 4$ and determine whether they check.

CHECK:

(2, 0)		(0, 4)	
$2x + y = 4$		$2x + y = 4$	
$2(2) + 0 \stackrel{?}{=} 4$		$2(0) + 4 \stackrel{?}{=} 4$	
$4 = 4$	*True*	$4 = 4$	*True*

(3, 3)		(−1, 6)	
$2x + y = 4$		$2x + y = 4$	
$2(3) + 3 \stackrel{?}{=} 4$		$2(-1) + 6 \stackrel{?}{=} 4$	
$9 = 4$	*False*	$4 = 4$	*True*

All the points except (3, 3) check and are solutions to the equation.

b) Figure 4.10 shows the three points that satisfy the equation. A straight line drawn through the three points shows that they appear to be collinear.

c) The straight line represents all solutions of $2x + y = 4$. The coordinates of every point on this line satisfy the equation $2x + y = 4$.

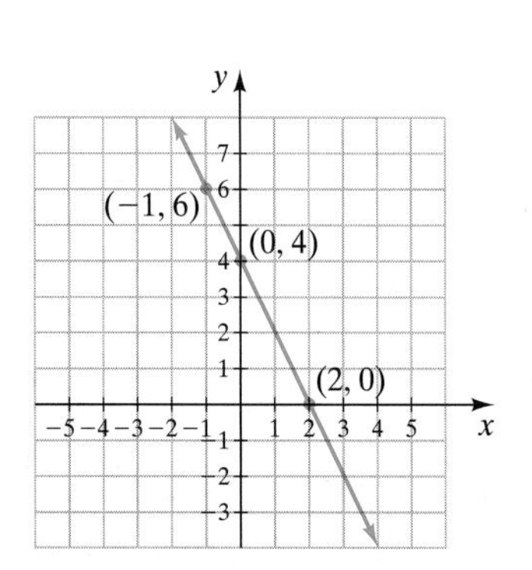

FIGURE 4.10

Using Your Graphing Calculator

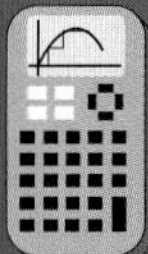

Some of you may have graphing calculators. In this chapter we will give a number of Using Your Graphing Calculator boxes that will give you some information on using your calculator. Because the instructions will be general, you may need to refer to your calculator manual for more specific instructions. The keystrokes we will show as examples in this book are for use on the Texas Instruments-83 calculator. The keystrokes you will use will depend on the brand and model of your calculator. All graphing calculator screens we show in this book are from a TI-83 graphing calculator.

A primary use of a graphing calculator is to graph equations. A graphing calculator *window* is the rectangular screen in which a graph is displayed. Figure 4.11 shows a TI-83 calculator window with some information illustrated. Figure 4.12 shows the meaning of the information given in Figure 4.11. These are the *standard window settings* for a graphing calculator screen.

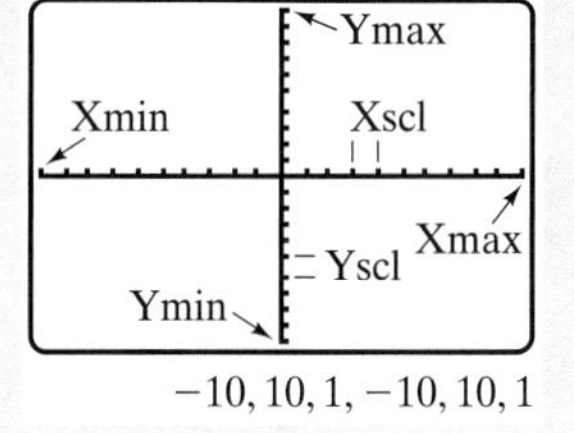

FIGURE 4.11

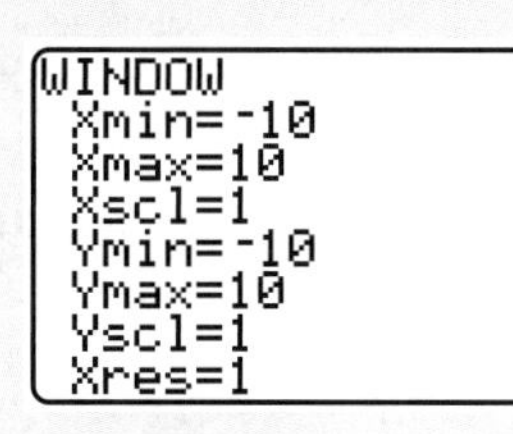

FIGURE 4.12

The x-axis on the standard window goes from −10 (the minimum value of x, Xmin) to 10 (the maximum value of x, Xmax) with a scale of 1. Therefore, each tick mark represents 1 unit (Xscl = 1). The y-axis goes from −10 (the minimum value of y, Ymin) to 10 (the maximum value of y, Ymax) with a scale of 1(Yscl = 1). The numbers below the graph in Figure 4.11 indicate, in order, the window settings: Xmin, Xmax, Xscl, Ymin, Ymax, Yscl. *When no settings are shown below a graph always assume the standard window setting is used.* Since the window is rectangular, the distance between tick marks on the standard window are greater on the x-axis than on the y-axis.

When graphing, you will often need to change the window settings. Read your graphing calculator manual to learn how to change the window settings. On the TI-82 and TI-83, you press the WINDOW key and then change the settings.

Now, turn on your calculator and press the WINDOW key. If necessary, adjust the window so it looks like the window in Figure 4.12. Use the (−) key, if necessary, to make negative numbers. (You can also obtain the standard window settings on a TI-82 or TI-83 by pressing the ZOOM key and then pressing option 6, ZStandard.) Next press the GRAPH key. Your screen should resemble the screen in Figure 4.11 (without the words that were added). Now press the WINDOW key again. Then use the appropriate keys to change the window setting so it is the same as that shown in Figure 4.13.

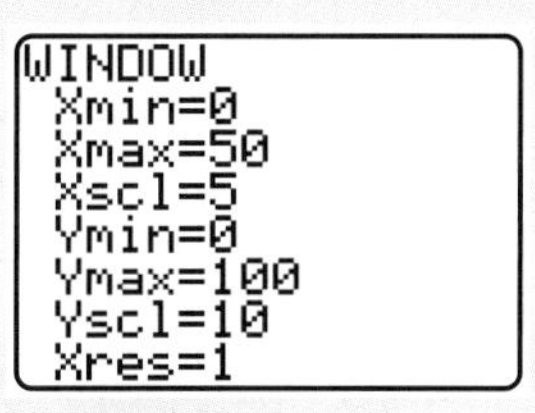

FIGURE 4.13

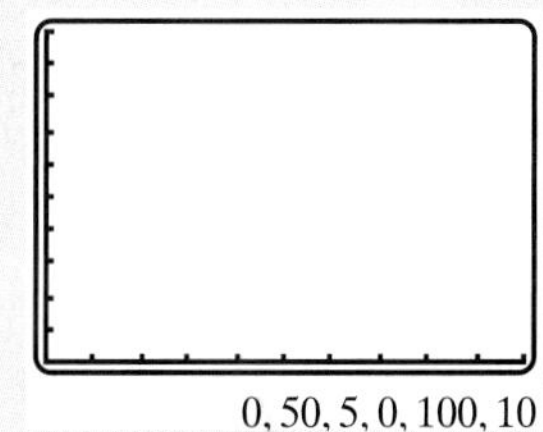

FIGURE 4.14

Now press the GRAPH key again. You should get the screen shown in Figure 4.14. In Figure 4.14, the x-axis starts at 0 and goes to 50, and each tick mark represents 5 units (represented by the first 3 numbers under the window). The y-axis starts at 0 and goes to 100, and each tick mark represents 10 units (represented by the last 3 numbers under the window).

Exercises

For Exercises 1 and 2, set your window to the values shown. Then use the GRAPH *key to show the axes formed.*

1. Xmin = −20, Xmax = 40, Xscl = 5,
   Ymin = −10, Ymax = 60, Yscl = 10
2. Xmin = −200, Xmax = 400, Xscl = 100,
   Ymin = −500, Ymax = 1000, Yscl = 200

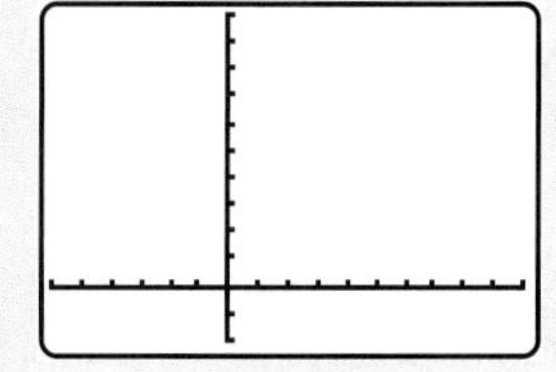

FIGURE 4.15

3. Consider the screen in Figure 4.15. If Xmin = −300 and Xmax = 500, find Xscl. Explain how you determined your answer.
4. In Figure 4.15, if Ymin = −200 and Ymax = 1000, find Yscl. Explain how you determined your answer.

Exercise Set 4.1

Concept/Writing Exercises

1. In an ordered pair, which coordinate is always listed first?
2. What is another name for the Cartesian coordinate system?
3. a) Is the *horizontal axis* the x- or y-axis in the Cartesian coordinate system?
   b) Is the *vertical axis* the x- or y-axis?
4. What is the *origin* in the Cartesian coordinate system?
5. We can refer to the *x-axis* and we can refer to the *y-axis*. We can also refer to the x- and *y-axes*. Explain when we use the word *axis* and when we use the word *axes*.
6. Explain how to plot the point $(-3, 5)$ in the Cartesian coordinate system.
7. What does the graph of a linear equation illustrate?
8. Why are arrowheads added to the ends of graphs of linear equations?
9. What will the graph of a linear equation look like?
10. a) How many points are needed to graph a linear equation?
    b) Why is it always a good idea to use three or more points when graphing a linear equation?
11. What is the standard form of a linear equation?
12. When graphing linear equations, the points that are plotted should all be *collinear.* Explain what this means.
13. In the Cartesian coordinate system there are four quadrants. Draw the x- and y-axes and mark the four quadrants, I through IV, on your axes.
14. How many solutions does a linear equation in two variables have?

Practice the Skills

Indicate the quadrant in which each of the points belongs.

15. $(3, 2)$ **16.** $(-3, 6)$ **17.** $(4, -2)$ **18.** $(5, -3)$

19. $(-8, 5)$ **20.** $(-6, 30)$ **21.** $(-16, -87)$ **22.** $(63, -47)$

23. $(-124, -132)$ **24.** $(75, -200)$ **25.** $(-8, 42)$ **26.** $(-46, -192)$

27. List the ordered pairs corresponding to each point.

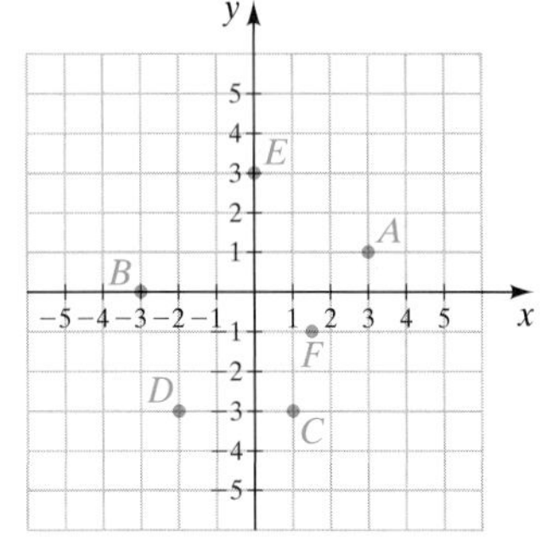

28. List the ordered pairs corresponding to each point.

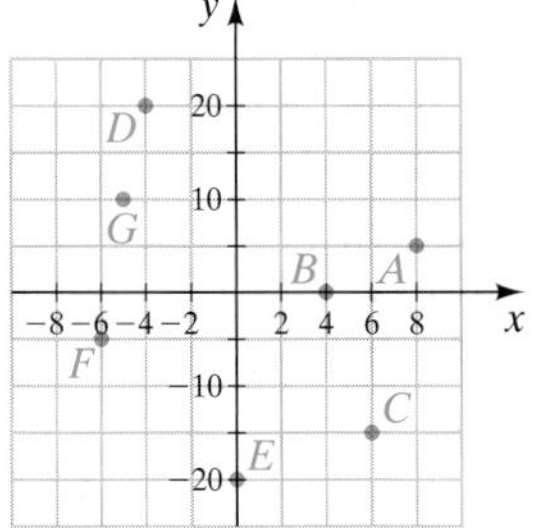

Plot each point on the same axes.

29. $A(4, 2), B(-3, 2), C(0, -3), D(-2, 0), E(-3, -4), F\left(-4, -\frac{5}{2}\right)$

30. $A(-3, -1), B(2, 0), C(3, 2), D\left(\frac{1}{2}, -4\right), E(-4, 2), F(0, 4)$

31. $A(4, 0), B(-1, 3), C(2, 4), D(0, -2), E(-3, -3), F(2, -3)$

32. $A(-3, 4), B(2, 3), C(0, 3), D(-1, 0), E(-2, -2), F(2, -4)$

Plot the following points. Then determine whether they appear to be collinear.

33. $A(1, -1), B(5, 3), C(-3, -5), D(0, -2), E(2, 0)$

34. $A(1, -2), B(0, -5), C(3, 1), D(-1, -8), E\left(\frac{1}{2}, -\frac{7}{2}\right)$

35. $A(0, 2), B(-1, 3), C(-2, 4), D(3, 0), E(4, -2)$

36. $A(1, -1), B(3, 5), C(0, -3), D(-2, -7), E(2, 1)$

In Exercises 37–42, **a)** *determine which of the four ordered pairs does not satisfy the given equation.* **b)** *Plot all the points that satisfy the equation on the same axes and draw a straight line through the points.*

37. $y = x + 1$, **a)** $(0, 1)$ **b)** $(-1, 0)$ **c)** $(2, 3)$ **d)** $(1, 1)$

38. $2x + y = -4$, **a)** $(-2, 0)$ **b)** $(-2, 1)$ **c)** $(0 -4)$ **d)** $(-1, -2)$

39. $3x - 2y = 6$, **a)** $(4, 0)$ **b)** $(2, 0)$ **c)** $\left(\frac{2}{3}, -2\right)$ **d)** $\left(\frac{4}{3}, -1\right)$

40. $2x - 4y = 0$, **a)** $\left(3, \frac{3}{2}\right)$ **b)** $(2, 1)$ **c)** $(0, 0)$ **d)** $(1, -1)$

41. $\frac{1}{2}x + 4y = 4$, **a)** $(2, -1)$ **b)** $\left(2, \frac{3}{4}\right)$ **c)** $(0, 1)$ **d)** $\left(-4, \frac{3}{2}\right)$

42. $y = \frac{1}{2}x + 2$, **a)** $(0, 2)$ **b)** $(2, 0)$ **c)** $(-2, 1)$ **d)** $(4, 4)$

Problem Solving

Consider the linear equation $y = 3x - 2$. In Exercises 43–46, find the value of y that makes the given ordered pair a solution to the equation.

43. $(2, y)$ **44.** $(-1, y)$ **45.** $(0, y)$ **46.** $(3, y)$

Consider the linear equation $2x + 3y = 12$. In Exercises 47–50, find the value of y that makes the given ordered pair a solution to the equation.

47. $(3, y)$ **48.** $(0, y)$ **49.** $\left(\frac{1}{2}, y\right)$ **50.** $(-5, y)$

51. What is the value of y at the point where a straight line crosses the x-axis? Explain.

52. What is the value of x at the point where a straight line crosses the y-axis? Explain.

Group Activity

In Section 4.2 we discuss how to find ordered pairs to plot when graphing linear equations. Let's see if you can draw some graphs now. Individually work parts **a)** *through* **c)** *in Exercises 53–56.*

a) *Select any three values for x and find the corresponding values of y.*

b) *Plot the points (they should appear to be collinear).*

c) *Draw the graph.*

d) *As a group, compare your answers. You should all have the same lines.*

53. $y = x$ **54.** $y = 2x$ **55.** $y = x + 1$ **56.** $2x + y = 4$

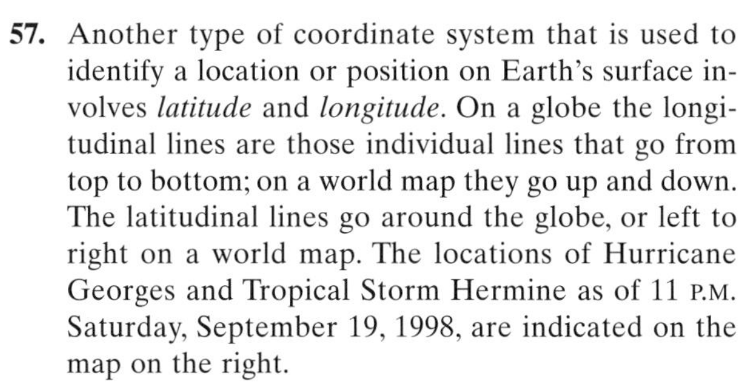

57. Another type of coordinate system that is used to identify a location or position on Earth's surface involves *latitude* and *longitude*. On a globe the longitudinal lines are those individual lines that go from top to bottom; on a world map they go up and down. The latitudinal lines go around the globe, or left to right on a world map. The locations of Hurricane Georges and Tropical Storm Hermine as of 11 P.M. Saturday, September 19, 1998, are indicated on the map on the right.

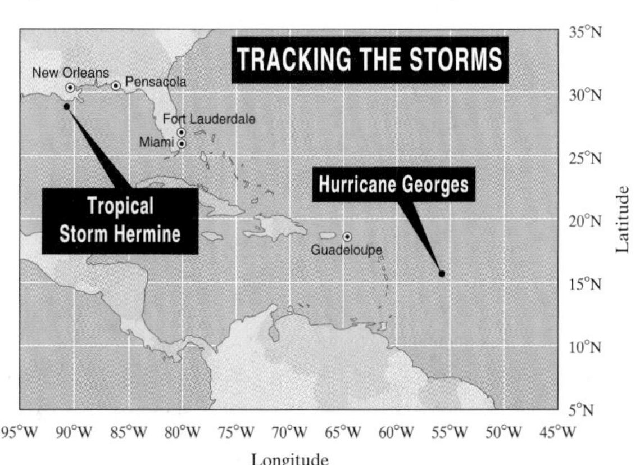

Source: National Weather Service

a) Group member 1: Estimate the latitude and longitude of Hurricane Georges.

b) Group member 2: Estimate the latitude and longitude of Tropical Storm Hermine.

c) Group member 3: Estimate the latitude and longitude of the city of Miami.

d) As a group, use either a map or a globe to estimate the latitude and longitude of your college.

Cumulative Review Exercises

Answer each question.

[2.2] **58.** What is a linear equation in one variable?

59. What is a conditional linear equation in one variable?

60. What is an identity?

[3.1] **61.** Find the circumference and area of the circle shown on the right.

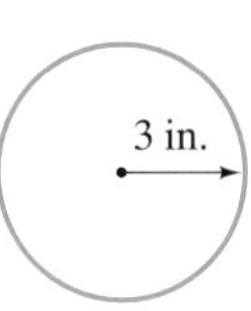

62. Solve the equation $2x - 5y = 6$ for y.

4.2 GRAPHING LINEAR EQUATIONS

1) **Graph linear equations by plotting points.**
2) **Graph linear equations of the form $ax + by = 0$.**
3) **Graph linear equations using the x- and y-intercepts.**
4) **Graph horizontal and vertical lines.**
5) **Study applications of graphs.**

In Section 4.1 we explained the Cartesian coordinate system, how to plot points, and how to recognize linear equations in two variables. Now we are ready to graph linear equations. *In this section we discuss two methods that can be used to graph linear equations: (1) graphing by plotting points and (2) graphing using the x- and y-intercepts.* In Section 4.4 we discuss graphing using the slope and the y-intercept.

1) Graph Linear Equations by Plotting Points

Graphing by plotting points is the most versatile method of graphing because we can also use it to graph second- and higher-degree equations. In Chapter 10, we will graph quadratic equations, which are second-degree equations, by plotting points.

To Graph Linear Equations by Plotting Points

1. Solve the linear equation for the variable y. That is, get the variable y by itself on the left side of the equal sign.
2. Select a value for the variable x. Substitute this value in the equation for x and find the corresponding value of y. Record the ordered pair (x, y).
3. Repeat step 2 with two different values of x. This will give you two additional ordered pairs.
4. Plot the three ordered pairs. The three points should appear to be collinear. If they do not, recheck your work for mistakes.
5. With a straightedge, draw a straight line through the three points. Draw arrowheads on each end of the line to show that the line continues indefinitely in both directions.

In step 1, you need to solve the equation for y. If you have forgotten how to do this, review Section 3.1. In steps 2 and 3, you need to select values for x. The values you choose to select are up to you. However, you should choose values small enough so that the ordered pairs obtained can be plotted on the axes. Since y is often easy to find when $x = 0$, 0 is always a good value to select for x.

EXAMPLE 1 Graph the equation $y = 3x + 6$.

Solution First we determine that this is a linear equation. Its graph must therefore be a straight line. The equation is already solved for y. We select three values for x, substitute them in the equation, and find the corresponding values for y. We will arbitrarily select the values 0, 2, and -3 for x. The calculations that follow show that when $x = 0$, $y = 6$, when $x = 2$, $y = 12$, and when $x = -3$, $y = -3$.

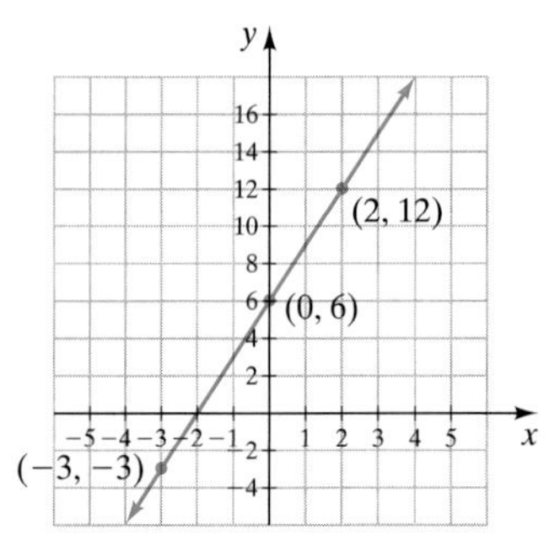

FIGURE 4.16

x	$y = 3x + 6$	Ordered Pair
0	$y = 3(0) + 6 = 6$	(0, 6)
2	$y = 3(2) + 6 = 12$	(2, 12)
-3	$y = 3(-3) + 6 = -3$	$(-3, -3)$

x	y
0	6
2	12
-3	-3

It is sometimes convenient to list the x- and y-values in a table. Then we plot the three ordered pairs on the same axes (Fig. 4.16).

Since the three points appear to be collinear, the graph appcars correct. Connect the three points with a straight line and place arrowheads at the ends of the line to show that the line continues infinitely in both directions.

To graph the equation $y = 3x + 6$, we arbitrarily used the three values $x = 0$, $x = 2$, and $x = -3$. We could have selected three entirely different values and obtained exactly the same graph. When selecting values to substitute for x, use values that make the equation easy to evaluate.

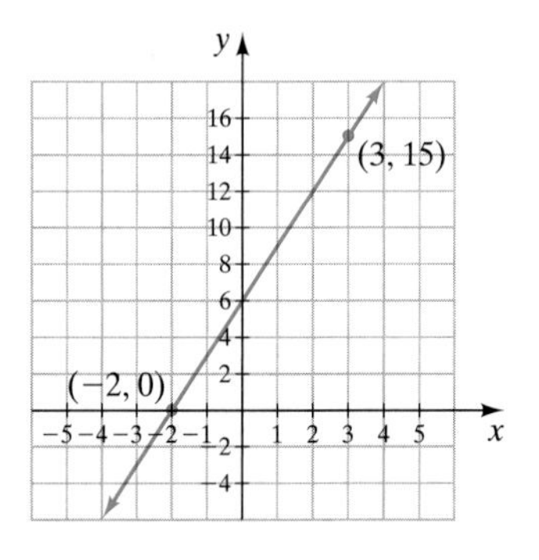

FIGURE 4.17

The graph drawn in Example 1 represents the set of *all* ordered pairs that satisfy the equation $y = 3x + 6$. If we select any point on this line, the ordered pair represented by that point will be a solution to the equation $y = 3x + 6$. Similarly, any solution to the equation will be represented by a point on the line. Let us select some points on the line, say, (3, 15) and $(-2, 0)$, and verify that they are solutions to the equation (Fig. 4.17).

Check: (3, 15)

$$y = 3x + 6$$
$$15 \stackrel{?}{=} 3(3) + 6$$
$$15 \stackrel{?}{=} 9 + 6$$
$$15 = 15 \quad \textit{True}$$

Check: $(-2, 0)$

$$y = 3x + 6$$
$$0 \stackrel{?}{=} 3(-2) + 6$$
$$0 \stackrel{?}{=} -6 + 6$$
$$0 = 0 \quad \textit{True}$$

Remember, a graph of an equation is an illustration of the set of points whose coordinates satisfy the equation.

EXAMPLE 2 Graph the equation $2y = 4x - 12$.

Solution We begin by solving the equation for y. This will make it easier to determine ordered pairs that satisfy the equation. To solve the equation for y, we divide both sides of the equation by 2.

$$2y = 4x - 12$$
$$y = \frac{4x - 12}{2}$$
$$= \frac{4x}{2} - \frac{12}{2}$$
$$= 2x - 6$$

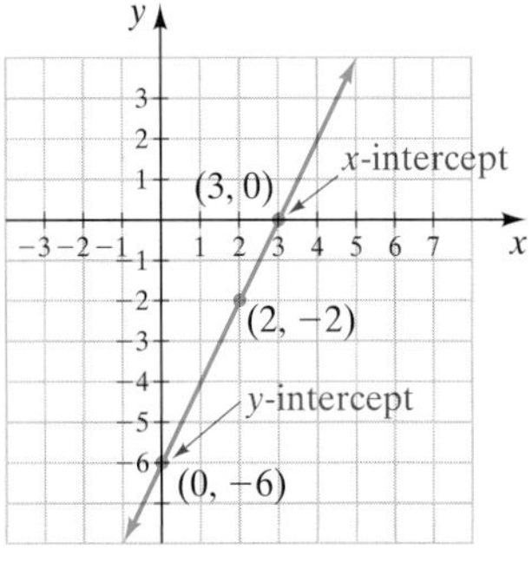

FIGURE 4.18

Now we select three values for x and find the corresponding values for y using the equation $y = 2x - 6$.

	$y = 2x - 6$
Let $x = 0$	$y = 2(0) - 6 = -6$
Let $x = 2$	$y = 2(2) - 6 = -2$
Let $x = 3$	$y = 2(3) - 6 = 0$

x	y
0	−6
2	−2
3	0

NOW TRY EXERCISE 25 Finally, we plot the points and draw the straight line (Fig. 4.18).

2) Graph Linear Equations of the Form $ax + by = 0$

In Example 3 we graph an equation of the form $ax + by = 0$, which is a linear equation whose constant is 0.

EXAMPLE 3 Graph the equation $2x + 5y = 0$.

Solution We begin by solving the equation for y.

$$2x + 5y = 0$$
$$5y = -2x$$
$$y = -\frac{2x}{5} \quad \text{or} \quad y = -\frac{2}{5}x$$

Now we select values for x and find the corresponding values of y. Which values shall we select for x? Notice that the coefficient of the x-term is a fraction, with the denominator 5. If we select values for x that are multiples of the denominator, such as ..., −15, −10, −5, 0, 5, 10, 15, ..., the 5 in the denominator will divide out. This will give us integer values for y. We will arbitrarily select the values $x = -5$, $x = 0$, and $x = 5$.

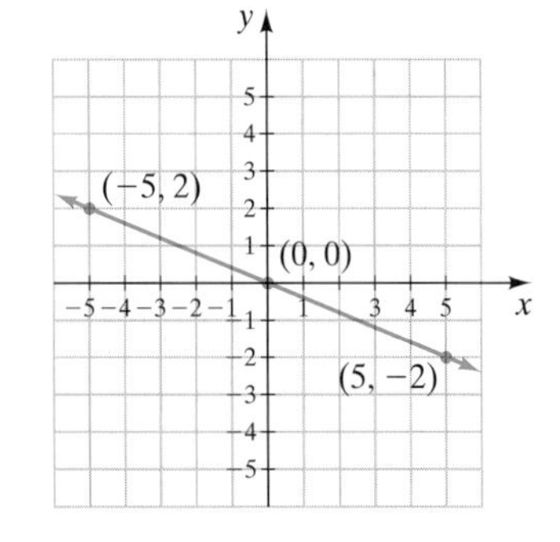

FIGURE 4.19

	$y = -\frac{2}{5}x$
Let $x = -5$	$y = \left(-\frac{2}{5}\right)(-5) = 2$
Let $x = 0$	$y = \left(-\frac{2}{5}\right)(0) = 0$
Let $x = 5$	$y = -\frac{2}{5}(5) = -2$

x	y
−5	2
0	0
5	−2

NOW TRY EXERCISE 37 Now we plot the points and draw the graph (Fig. 4.19).

The graph in Example 3 passes through the origin. The graph of every linear equation with a constant of 0 (equations of the form $ax + by = 0$) will pass through the origin.

3) Graph Linear Equations Using the x- and y-Intercepts

Now we discuss graphing linear equations using the x- and y-intercepts. The **x-intercept** is the point at which a graph crosses the x-axis and the **y-intercept** is the point where the graph crosses the y-axis. Let's examine two points on the graph in Figure 4.18. Note that the graph crosses the x-axis at 3. Therefore $(3, 0)$ is the x-intercept. Since the graph crosses the x-axis at 3, we might say the x-intercept is *at* 3 (on the x-axis). In general, the x-intercept is $(x, 0)$, and the x-intercept is *at* x (on the x-axis).

Note that the graph in Figure 4.18 crosses the y-axis at -6. Therefore $(0, -6)$ is the y-intercept. Since the graph crosses the y-axis at -6, we might say the y-intercept is *at* -6 (on the y-axis). In general, the y-intercept is $(0, y)$, and the y-intercept is *at* y (on the y-axis).

Note that the graph in Figure 4.19 crosses both the x- and y-axes at the origin. Thus, both the x- and y-intercepts of this graph are $(0, 0)$.

It is often convenient to graph linear equations by finding their x- and y-intercepts. To graph an equation using the x- and y-intercepts, use the following procedure.

To Graph Linear Equations Using the x- and y-Intercepts

1. Find the y-intercept by setting x in the given equation equal to 0 and finding the corresponding value of y.
2. Find the x-intercept by setting y in the given equation equal to 0 and finding the corresponding value of x.
3. Determine a check point by selecting a nonzero value for x and finding the corresponding value of y.
4. Plot the y-intercept (where the graph crosses the y-axis), the x-intercept (where the graph crosses the x-axis), and the check point. The three points should appear to be collinear. If not, recheck your work.
5. Using a straightedge, draw a straight line through the three points. Draw an arrowhead at both ends of the line to show that the line continues indefinitely in both directions.

HELPFUL HINT

Since only two points are needed to determine a straight line, it is not absolutely necessary to determine and plot the check point in step 3. However, if you use only the x- and y-intercepts to draw your graph and one of those points is wrong, your graph will be incorrect and you will not know it. It is always a good idea to use three points when graphing a linear equation.

EXAMPLE 4 Graph the equation $3y = 6x + 12$ by plotting the x- and y-intercepts.

Solution To find the y-intercept (where the graph crosses the y-axis), set $x = 0$ and find the corresponding value of y.

$$3y = 6x + 12$$
$$3y = 6(0) + 12$$
$$3y = 0 + 12$$
$$3y = 12$$
$$y = \frac{12}{3} = 4$$

The graph crosses the y-axis at 4. The ordered pair representing the y-intercept is $(0, 4)$. To find the x-intercept (where the graph crosses the x-axis), set $y = 0$ and find the corresponding value of x.

$$3y = 6x + 12$$
$$3(0) = 6x + 12$$
$$0 = 6x + 12$$
$$-12 = 6x$$
$$\frac{-12}{6} = x$$
$$-2 = x$$

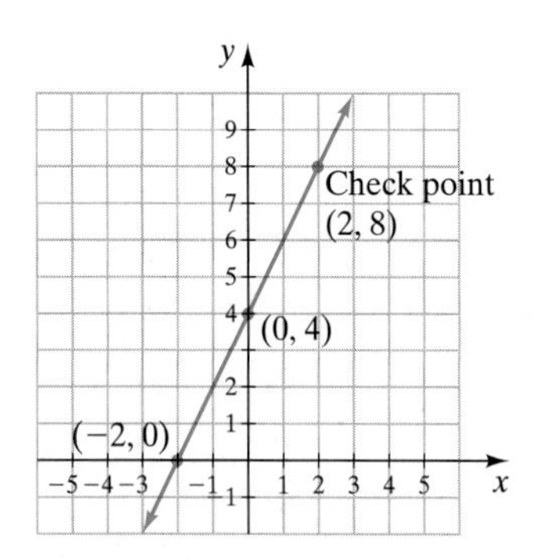

FIGURE 4.20

The graph crosses the x-axis at -2. The ordered pair representing the x-intercept is $(-2, 0)$. Now plot the intercepts (Fig. 4.20).

Before graphing the equation, select a nonzero value for x, find the corresponding value of y, and make sure that it is collinear with the x- and y-intercepts. This third point is the check point.

$$\text{Let } x = 2$$
$$3y = 6x + 12$$
$$3y = 6(2) + 12$$
$$3y = 12 + 12$$
$$3y = 24$$
$$y = \frac{24}{3} = 8$$

Plot the check point $(2, 8)$. Since the three points appear to be collinear, draw the straight line through all three points.

EXAMPLE 5 Graph the equation $2x + 3y = 9$ by finding the x- and y-intercepts.

Solution

Find y-Intercept	Find x-Intercept	Check Point
Let $x = 0$	Let $y = 0$	Let $x = 2$
$2x + 3y = 9$	$2x + 3y = 9$	$2x + 3y = 9$
$2(0) + 3y = 9$	$2x + 3(0) = 9$	$2(2) + 3y = 9$
$0 + 3y = 9$	$2x + 0 = 9$	$4 + 3y = 9$
$3y = 9$	$2x = 9$	$3y = 5$
$y = 3$	$x = \frac{9}{2}$	$y = \frac{5}{3}$

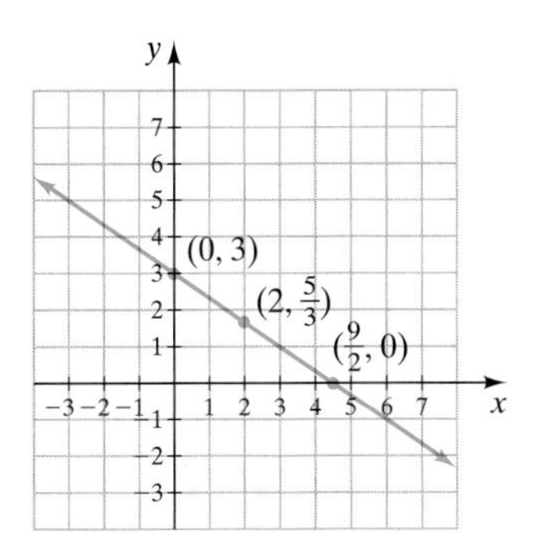

FIGURE 4.21

The three ordered pairs are $(0, 3)$, $\left(\frac{9}{2}, 0\right)$, and $\left(2, \frac{5}{3}\right)$.

The three points appear to be collinear. Draw a straight line through all three points (Fig. 4.21).

NOW TRY EXERCISE 45

EXAMPLE 6 Graph the equation $y = 20x + 60$.

Solution

Find y-Intercept	Find x-Intercept	Check Point
Let $x = 0$	Let $y = 0$	Let $x = 3$
$y = 20x + 60$	$y = 20x + 60$	$y = 20x + 60$
$y = 20(0) + 60$	$0 = 20x + 60$	$y = 20(3) + 60$
$y = 60$	$-60 = 20x$	$y = 60 + 60$
	$-3 = x$	$y = 120$

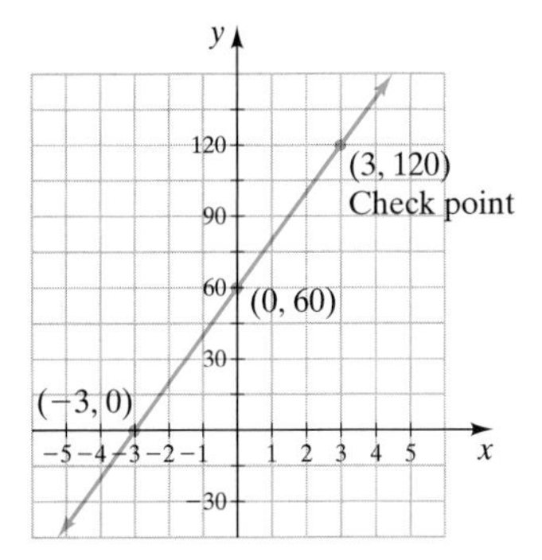

FIGURE 4.22

The three ordered pairs are $(0, 60)$, $(-3, 0)$, and $(3, 120)$. Since the values of y are large, we let each interval on the y-axis be 15 units rather than 1 (Fig. 4.22). Sometimes you will have to use different scales on the x- and y-axes, as illustrated, to accommodate the graph. Now we plot the points and draw the graph.

When selecting the scales for your axes, you should realize that different scales will result in the same equation having a different appearance. Consider the graphs shown in Figure 4.23. Both graphs represent the same equation, $y = x$. In Figure 4.23 both the x- and y-axes have the same scale. In Figure 4.23b, the x- and y-axes do not have the same scale. Both graphs are correct in that each represents the graph of $y = x$. The difference in appearance is due to the difference in scales on the x-axis. When possible, keep the scales on the x- and y-axes the same, as in Figure 4.23a.

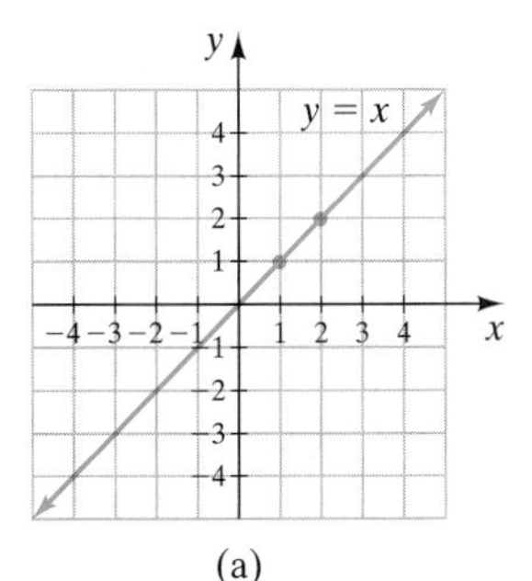

(a)

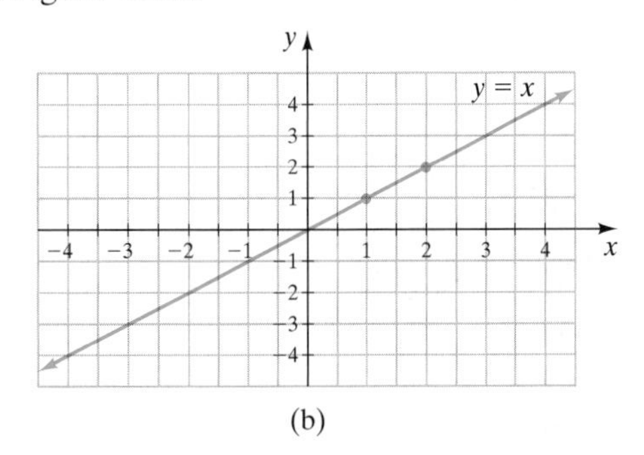

(b)

FIGURE 4.23

Using Your Graphing Calculator

To graph an equation on a graphing calculator, use the following steps.

1. Solve the equation for y, if necessary.
2. Press the [Y =] key and enter the equation.
3. Press the [GRAPH] key (to see the graph). You may need to adjust the window, as explained in the Using Your Graphing Calculator box on page 237.

In Example 2, when we solved the equation $2y = 4x - 12$ for y we obtained $y = 2x - 6$. If you press [Y =], enter $2x - 6$ as Y_1, and then press [GRAPH], you should get the graph shown in Figure 4.24.

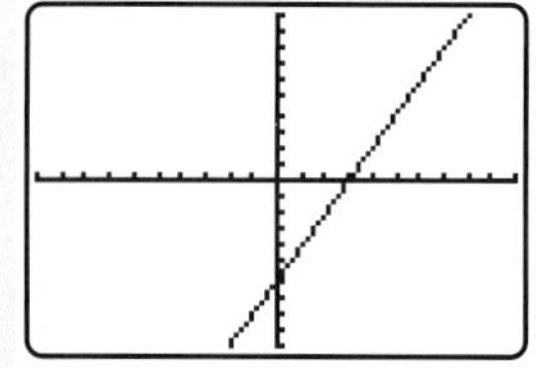

FIGURE 4.24

If you do not get the graph in Figure 4.24 press [WINDOW] and determine whether you have the standard window $-10, 10, 1, -10, 10, 1$, as discussed in the Using Your Graphing Calculator box on page 237. If not, change to the standard window and press the [GRAPH] key again.

It is possible to graph two or more equations on your graphing calculator. If, for example, you wanted to graph both $y = 2x - 6$ and $y = -3x + 4$ on the same screen, you would begin by pressing the [Y =] key. Then you would let $Y_1 = 2x - 6$ and $Y_2 = -3x + 4$. After you enter both equations and press the [GRAPH] key, both equations will be graphed. Try graphing both equations on your graphing calculator now.

Exercises

Graph each equation on your graphing calculator.

1. $y = 3x - 5$ **2.** $y = -2x + 6$

3. $2x - 3y = 6$ **4.** $5x + 10y = 20$

4 Graph Horizontal and Vertical Lines

When a linear equation contains only one variable, its graph will be either a horizontal or a vertical line, as is explained in Examples 7 and 8.

EXAMPLE 7 Graph the equation $y = 3$.

Solution This equation can be written as $y = 3 + 0x$. Thus, for any value of x selected, y will be 3. The graph of $y = 3$ is illustrated in Figure 4.25.

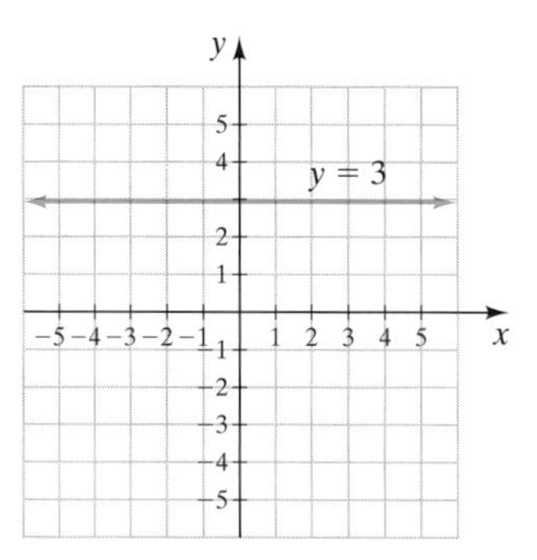

FIGURE 4.25

The graph of an equation of the form $y = b$ is a **horizontal line** *whose y-intercept is $(0, b)$.*

EXAMPLE 8 Graph the equation $x = -2$.

Solution This equation can be written as $x = -2 + 0y$. Thus, for any value of y selected, x will have a value of -2. The graph of $x = -2$ is illustrated in Figure 4.26.

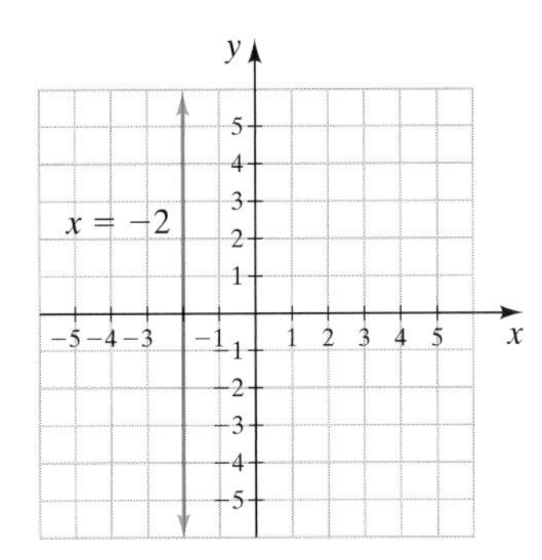

FIGURE 4.26

NOW TRY EXERCISE 21

The graph of an equation of the form $x = a$ is a **vertical line** *whose x-intercept is $(a, 0)$.*

5) Study Applications of Graphs

Before we leave this section, let's look at an application of graphing. We will see additional applications of graphing linear equations in Sections 4.4 and 4.5.

EXAMPLE 9 Brooke Mills recently graduated from college. She accepted a position selling and installing medical software in doctors' offices. She will receive a salary of \$200 per week plus a 7% commission on all sales, s.

a) Write an equation for the salary Brooke will receive, R, in terms of the sales, s.

b) Graph the salary for sales of \$0 up to and including \$20,000.

c) From the graph, estimate the salary if Brooke's weekly sales are \$15,000.

d) From the graph, estimate the sales needed for Brooke to earn a salary of \$800.

Solution **a)** Since s is the amount of sales, a 7% commission on s dollars in sales is $0.07s$.

$$\text{salary received} = \$200 + \text{commission}$$
$$R = 200 + 0.07s$$

b) We select three values for s and find the corresponding values of R.

$$R = 200 + 0.07s$$

Let $s = 0$ $\quad R = 200 + 0.07(0) = 200$

Let $s = 10{,}000$ $\quad R = 200 + 0.07(10{,}000) = 900$

Let $s = 20{,}000$ $\quad R = 200 + 0.07(20{,}000) = 1600$

s	R
0	200
10,000	900
20,000	1600

The graph is illustrated in Figure 4.27. Notice that since we only graph the equation for values of s from 0 to 20,000, we do not place arrowheads on the ends of the graph.

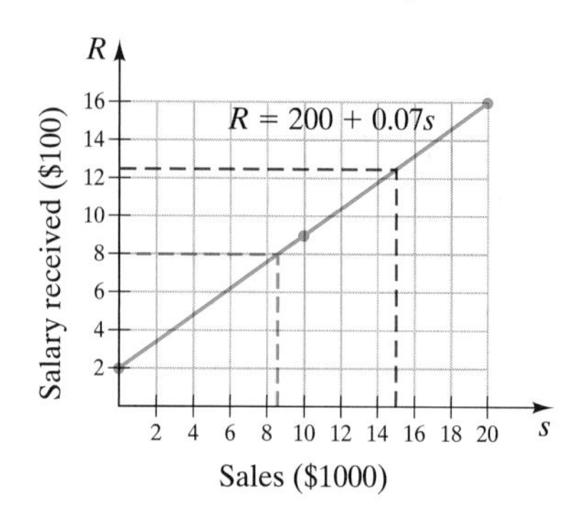

FIGURE 4.27

c) To determine Brooke's weekly salary on sales of $15,000, locate $15,000 on the sales axis. Then draw a vertical line up to where it intersects the graph, the *red* line in Figure 4.27. Now draw a horizontal line across to the salary axis. Since the horizontal line crosses the salary axis at about $1250, weekly sales of $15,000 would result in a weekly salary of about $1250. We can find the exact salary by substituting 15,000 for s in the equation $R = 200 + 0.07s$ and finding the value of R. Do this now.

d) To find the sales needed for Brooke to earn a weekly salary of $800, we find $800 on the salary axis. We then draw a horizontal line from the point to the graph, as shown with the *green* line in Figure 4.27. We then draw a vertical line from the point of intersection of the graph to the sales axis. This value on the sales axis represents the sales needed for Brooke to earn $800. Thus, sales of about $8600 per week would result in a salary of $800. We can find an exact answer by substituting 800 for R in the equation $R = 200 + 0.07s$ and solving the equation for s. Do this now.

NOW TRY EXERCISE 75

Using Your Graphing Calculator

Before we leave this section, let's spend a little more time discussing the graphing calculator. In the previous Using Your Graphing Calculator, we graphed $y = 2x - 6$. The graph of $y = 2x - 6$ is shown again using the standard window in Figure 4.28. After obtaining the graph, if you press the TRACE key you may (depending on your calculator) obtain the graph in Figure 4.29.

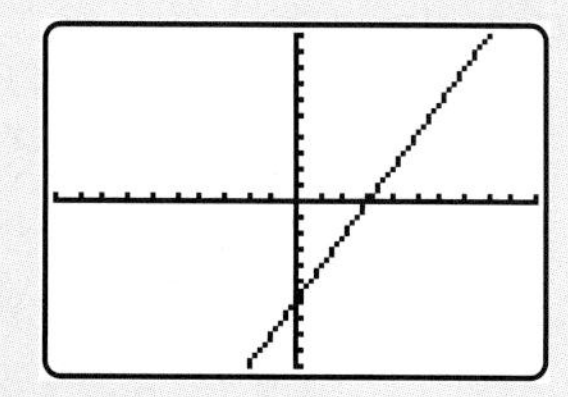

FIGURE 4.28

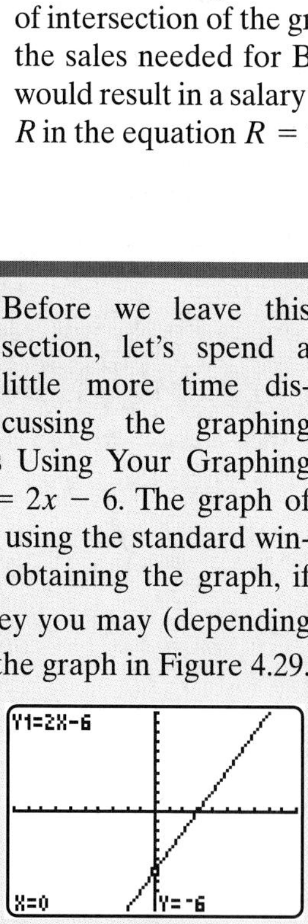

Y1=2X-6
X=0 Y=-6

FIGURE 4.29

Notice that the cursor, the blinking box in Figure 4.29, is on the y-intercept, that is, where $x = 0$ and $y = -6$. By pressing the left or the right arrow keys, you can move the cursor. The corresponding values of x and y change with the position of the cursor. You can magnify the part of the graph by the cursor by using the ZOOM key. Read your calculator manual to learn how to use the ZOOM feature and the various ZOOM options available.

Another important key on many graphing calculators is the TABLE feature. On the TI-83 when you press 2nd GRAPH to get the TABLE feature, you may get the display shown in Figure 4.30.

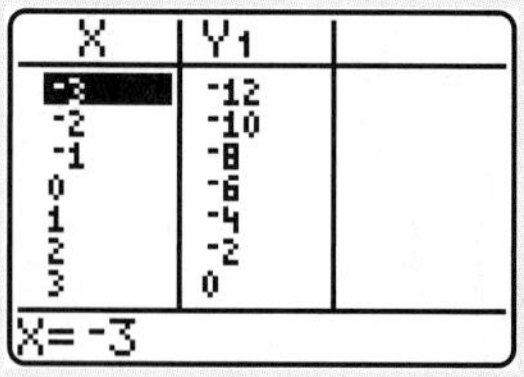

X	Y1
-3	-12
-2	-10
-1	-8
0	-6
1	-4
2	-2
3	0

X=-3

FIGURE 4.30

continued on next page

If your display does not show the screen in Figure 4.30, use your up and down arrows until you get the screen shown. The table gives x-values and corresponding y-values for the graph. Notice from the table that when $x = 3$, $y = 0$ (the x-intercept) and when $x = 0$, $y = -6$ (the y-intercept).

You can change the table features by pressing TBLSET, (press 2nd WINDOW on the TI-83). For example, if you want the table to give values of x in tenths, you could change ΔTbl to 0.1 instead of the standard 1 unit.

Exercises

Use the TABLE feature of your calculator to find the x- and y-intercepts of the graphs of the following equations. In Exercises 3 and 4, you will need to set ΔTbl to tenths.

1. $y = 4x - 8$
2. $y = -2x - 6$
3. $4x - 5y = 10$
4. $2y = 6x - 9$

Exercise Set 4.2

Concept/Writing Exercises

1. Explain how to find the x- and y-intercepts of a line.
2. How many points are needed to graph a straight line? How many points should be used? Why?
3. What will the graph of $y = b$ look like for any real number b?
4. What will the graph of $x = a$ look like for any real number a?
5. In Example 9(c) and 9(d), we made an estimate. Why is it sometimes not possible to obtain an exact answer from a graph?
6. In Example 9 does the salary, R, depend on the sales, s, or do the sales depend on the salary? Explain.
7. Will the equation $2x - 4y = 0$ go through the origin? Explain.
8. Write an equation, other than the ones given in this section, whose graph will go through the origin. Explain how you determined your answer.

Practice the Skills

Find the missing coordinate in the given solutions for $2x + y = 6$.

9. $(3, ?)$
10. $(-1, ?)$
11. $(?, -5)$
12. $(?, -3)$
13. $(?, 0)$
14. $\left(\frac{1}{2}, ?\right)$

Find the missing coordinate in the given solutions for $3x - 2y = 8$.

15. $(4, ?)$
16. $(0, ?)$
17. $(?, 0)$
18. $\left(?, -\frac{1}{2}\right)$
19. $(-3, ?)$
20. $(?, -5)$

Graph each equation.

21. $x = -3$
22. $x = \frac{3}{2}$
23. $y = 4$
24. $y = -\frac{5}{3}$

Graph by plotting points. Plot at least three points for each graph.

25. $y = 3x - 1$
26. $y = -x + 3$
27. $y = 6x + 2$
28. $y = x - 4$
29. $y = -\frac{1}{2}x + 3$
30. $2y = 2x + 4$
31. $2x - 4y = 4$
32. $4x - y = 5$
33. $5x - 2y = 8$
34. $3x + 2y = 0$
35. $6x + 5y = 30$
36. $-2x - 3y = 6$
37. $-4x + 5y = 0$
38. $12y - 24x = 36$
39. $y = -20x + 60$
40. $2y - 50 = 100x$
41. $y = \frac{2}{3}x$
42. $y = -\frac{3}{5}x$
43. $y = \frac{1}{2}x + 4$
44. $y = -\frac{2}{5}x + 2$

Graph using the x- and y-intercepts.

45. $y = 3x + 3$
46. $y = -2x + 6$
47. $y = 2x - 3$
48. $y = -3x + 8$
49. $y = -6x + 5$
50. $y = 4x + 16$
51. $4y + 6x = 24$
52. $4x = 3y - 9$
53. $\frac{1}{2}x + y = 4$
54. $30x + 25y = 50$
55. $6x - 12y = 24$
56. $25x + 50y = 100$
57. $8y = 6x - 12$
58. $-3y - 2x = -6$
59. $30y + 10x = 45$
60. $20x - 240 = -60y$
61. $\frac{1}{3}x + \frac{1}{4}y = 12$
62. $\frac{1}{5}x - \frac{2}{3}y = 60$
63. $\frac{1}{2}x = \frac{2}{5}y - 80$
64. $\frac{2}{3}y = \frac{5}{4}x + 120$

Write the equation represented by the given graph.

65.

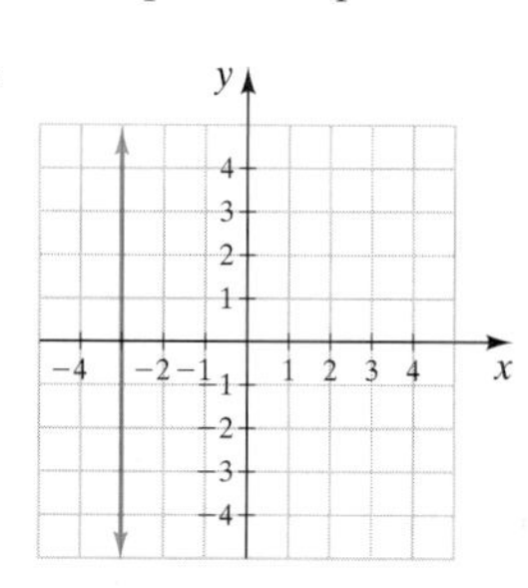

66.

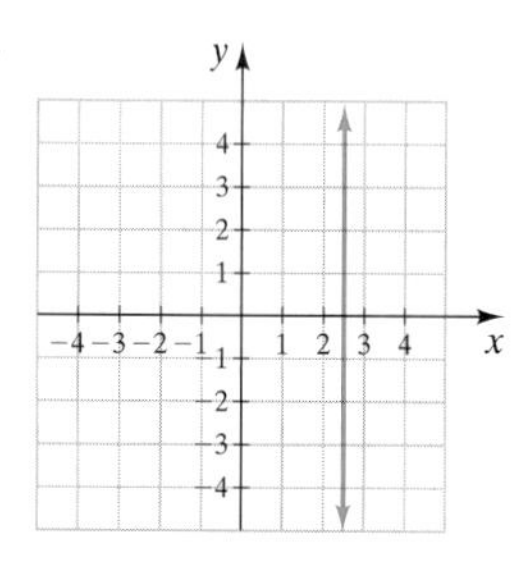

67.

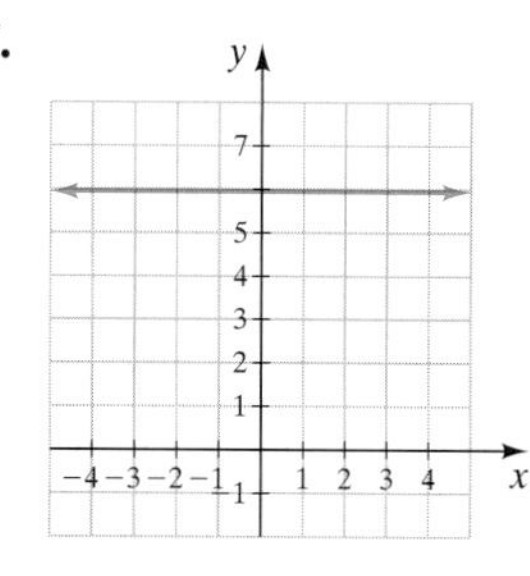

68.

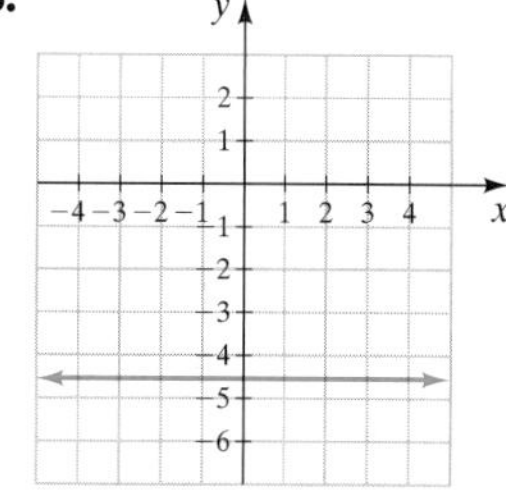

Problem Solving

69. What is the value of a if the graph of $ax + 4y = 8$ is to have an x-intercept of $(2, 0)$?
70. What is the value of a if the graph of $ax + 8y = 12$ is to have an x-intercept of $(3, 0)$?
71. What is the value of b if the graph of $3x + by = 10$ is to have a y-intercept of $(0, 5)$?
72. What is the value of b if the graph of $4x + by = 12$ is to have a y-intercept of $(0, -3)$?

The bar graphs in Exercises 73 and 74 display information. State whether the graph displays a linear relationship. Explain your answer.

73. **Calories Burned by Average 150–Pound Person Walking at 4.5 mph**

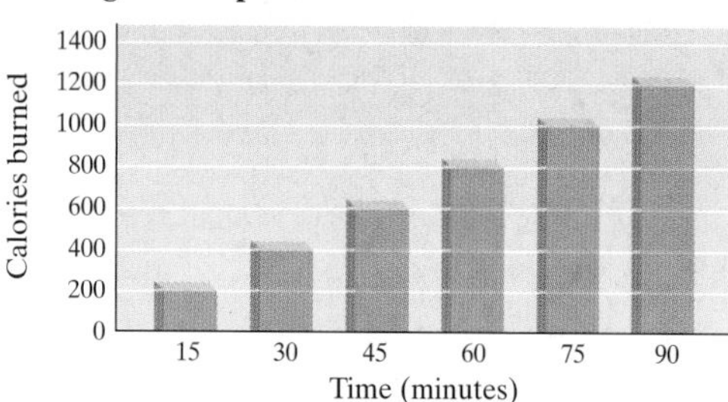

74. **Major League Baseball Attendance**

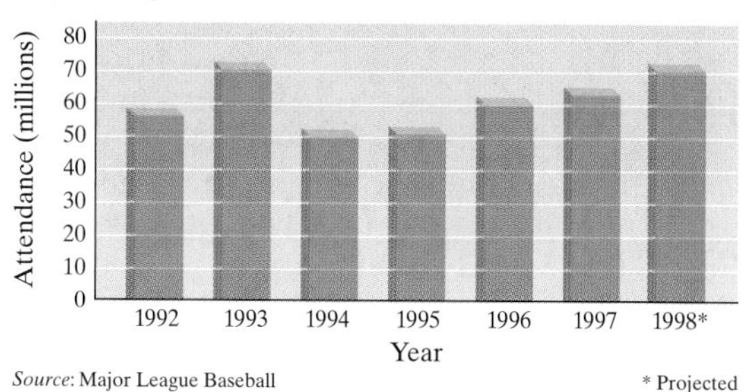

Source: Major League Baseball

* Projected

Review Example 9 before working Exercises 75–80.

75. Jerry Correa is a handyman. For laying tile, he charges $50 plus $1 per square foot of tile laid.
   a) Write an equation for the cost, C, he charges in terms of the square feet, s, of tile he lays.
   b) Graph the equation for values up to and including 150 square feet of area.
   c) Estimate the cost for Jerry's laying 100 square feet of tile.
   d) If his bill for laying the tile comes to $125, how many square feet of tile did he lay?

76. Distance traveled is calculated using the formula distance = rate · time or $d = rt$. Assume the rate of a car is a constant 60 miles per hour.
   a) Write an equation for the distance, d, in terms of time, t.
   b) Graph the equation for times of 0 to 10 hours inclusive.
   c) Estimate the distance traveled in 6 hours.
   d) If the distance traveled is 300 miles, estimate the time traveled.

77. Jayne Demsky needs a large truck to move some furniture. She found that the cost C, of renting a truck is \$40 per day plus \$1 per mile, m.

a) Write an equation for the cost in terms of the miles driven.

b) Graph the equation for values up to and including 100 miles.

c) Estimate the cost of driving 50 miles in one day.

d) Estimate the miles driven if the cost for one day is \$60.

78. Simple interest is calculated by the simple interest formula, interest = principal · rate · time or $I = prt$. Suppose the principal is \$10,000 and the rate is 5%.

a) Write an equation for simple interest in terms of time.

b) Graph the equation for times of 0 to 20 years inclusive.

c) What is the simple interest for 10 years?

d) If the simple interest is \$500, find the length of time.

79. The weekly profit, P of a video rental store can be approximated by the formula $P = 1.5n - 200$, where n is the number of tapes rented weekly.

a) Draw a graph of profit in terms of tape rentals for up to and including 1000 tapes.

b) Estimate the weekly profit if 500 tapes are rented.

c) Estimate the number of tapes rented if the week's profit is \$1000.

80. The cost, C, of playing tennis in the Downtown Tennis Club includes an annual \$200 membership fee plus \$10 per hour, h, of court time.

a) Write an equation for the annual cost of playing tennis at the Downtown Tennis Club in terms of hours played.

b) Graph the equation for up to and including 300 hours.

c) Estimate the cost for playing 200 hours in a year.

d) If the annual cost for playing tennis was \$1200, estimate how many hours of tennis were played.

Determine the coefficients to be placed in the shaded areas so that the graph of the equation will be a line with the x- and y-intercepts specified. Explain how you determined your answer.

81. $\square x + \square y = 20$; x-intercept at 4, y-intercept at 5

82. $\square x + \square y = 18$; x-intercept at -3, y-intercept at 6

83. $\square x - \square y = -12$; x-intercept at -2, y-intercept at 3

84. $\square x - \square y = 30$; x-intercept at -5, y-intercept at -15

Challenge Problems

85. In Chapter 10 we will be graphing quadratic equations. The graphs of quadratic equations are *not* straight lines. Graph the quadratic equation $y = x^2 - 4$ by selecting values for x and find the corresponding values of y, then plot the points. Make sure you plot a sufficient number of points to get an accurate graph.

Group Activity

Discuss and answer Exercises 86 and 87 as a group.

86. Let's study the graphs of the equations $y = 2x + 4$, $y = 2x + 2$, and $y = 2x - 2$ to see how they are similar and how they differ. Each group member should start with the same axes.

a) Group member 1: Graph $y = 2x + 4$.

b) Group member 2: Graph $y = 2x + 2$.

c) Group member 3: Graph $y = 2x - 2$.

d) Now transfer all three graphs onto the same axes. (You can use one of the group members' graphs or you can construct new axes.)

e) Explain what you notice about the three graphs.

f) Explain what you notice about the y-intercepts.

87. Consider the following equations: $y = 2x - 1$, $y = -x + 5$.

a) Carefully graph both equations on the same axes.

b) Determine the point of intersection of the two graphs.

c) Substitute the values for x and y at the point of intersection into each of the two equations and determine whether the point of intersection satisfies each equation.

d) Do you believe there are any other ordered pairs that satisfy both equations? Explain your answer. (We will study equations like these, called systems of equations, in Chapter 5.)

Cumulative Review Exercises

[1.9] **88.** Evaluate $2[6 - (4 - 5)] \div 2 - 5^2$.

89. Evaluate $\dfrac{-3^2 \cdot 4 \div 2}{\sqrt{9} - 2^2}$.

[2.6] **90.** According to the instructions on a bottle of concentrated household cleaner, 8 ounces of the cleaner should be mixed with 3 gallons of water. If your bucket has 2.5 gallons of water, how much cleaner should you use?

[3.3] **91.** The larger of two integers is 1 more than 3 times the smaller. If the sum of the two integers is 37, find the two integers.

4.3 SLOPE OF A LINE

SSM VIDEO 4.3 CD Rom

1) Find the slope of a line.
2) Recognize positive and negative slopes.
3) Examine the slopes of horizontal and vertical lines.

1) Find the Slope of a Line

The slope of a line is an important concept in many areas of mathematics. A knowledge of slope is helpful in understanding linear equations.

The **slope of a line** is a measure of the *steepness* of the line.

Definition

The **slope of a line** is a ratio of the vertical change to the horizontal change between any two selected points on the line.

As an example, consider the line that goes through the two points (3, 6) and (1, 2). (see Fig. 4.31a).

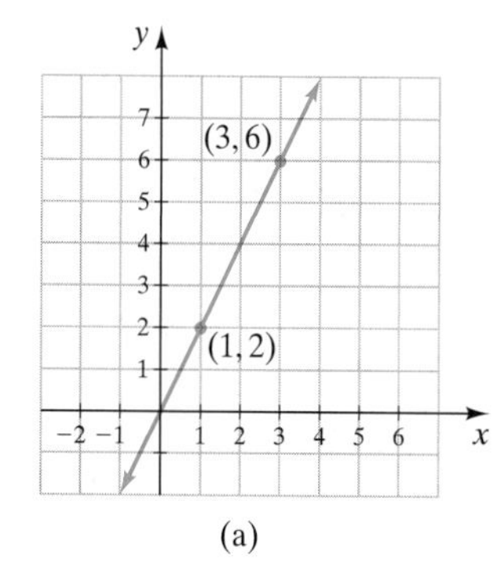

(a)

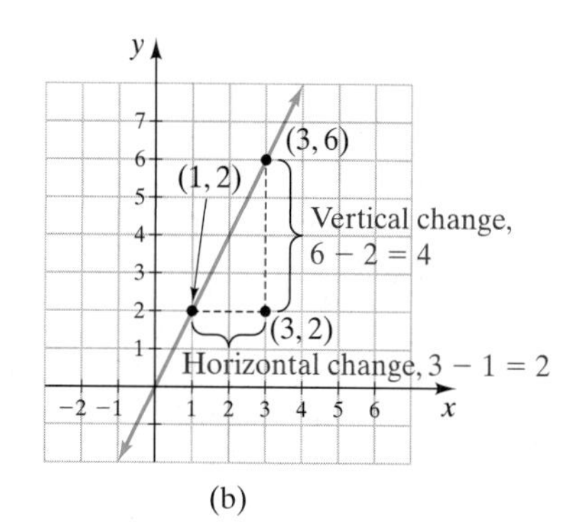

(b)

FIGURE 4.31

If we draw a line parallel to the x-axis through the point (1, 2) and a line parallel to the y-axis through the point (3, 6), the two lines intersect at (3, 2), see Figure 4.31b. From the figure, we can determine the slope of the line. The vertical change (along the y-axis) is $6 - 2$, or 4 units. The horizontal change (along the x-axis) is $3 - 1$, or 2 units.

$$\text{slope} = \frac{\text{vertical change}}{\text{horizontal change}} = \frac{4}{2} = 2$$

FIGURE 4.32

Thus, the slope of the line through these two points is 2. By examining the line connecting these two points, we can see that as the graph moves up 2 units on the y-axis it moves to the right 1 unit on the x-axis (Fig. 4.32).

Now we present the procedure to find the slope of a line between any two points (x_1, y_1) and (x_2, y_2). Consider Figure 4.33.

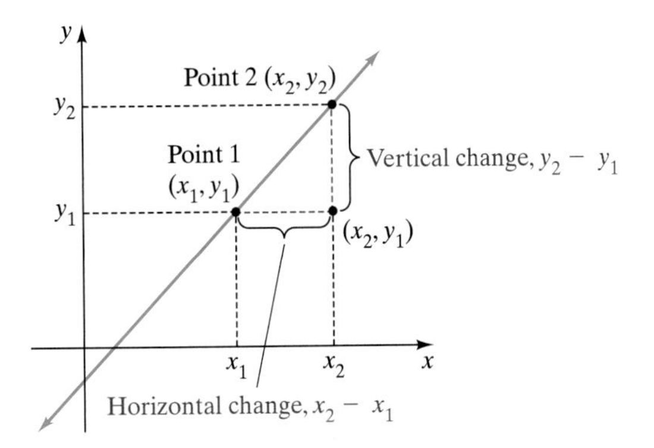

FIGURE 4.33

The vertical change can be found by subtracting y_1 from y_2. The horizontal change can be found by subtracting x_1 from x_2.

Slope of a Line Through the Points (x_1, y_1) and (x_2, y_2)

$$\text{slope} = \frac{\text{change in } y \text{ (vertical change)}}{\text{change in } x \text{ (horizontal change)}} = \frac{y_2 - y_1}{x_2 - x_1}$$

It makes no difference which two points are selected when finding the slope of a line. It also makes no difference which point you label (x_1, y_1) or (x_2, y_2). The Greek capital letter delta, Δ, is often used to represent the words "the change in." Thus, the slope, which is symbolized by the letter m, is indicated as

$$m = \frac{\Delta y}{\Delta x} = \frac{y_2 - y_1}{x_2 - x_1}$$

EXAMPLE 1 Find the slope of the line through the points $(-6, -1)$ and $(3, 5)$.

Solution We will designate $(-6, -1)$ as (x_1, y_1) and $(3, 5)$ as (x_2, y_2).

$$\begin{aligned} m &= \frac{y_2 - y_1}{x_2 - x_1} \\ &= \frac{5 - (-1)}{3 - (-6)} \\ &= \frac{5 + 1}{3 + 6} = \frac{6}{9} = \frac{2}{3} \end{aligned}$$

Thus, the slope is $\frac{2}{3}$.

If we had designated $(3, 5)$ as (x_1, y_1) and $(-6, -1)$ as (x_2, y_2), we would have obtained the same results.

$$\begin{aligned} m &= \frac{y_2 - y_1}{x_2 - x_1} \\ &= \frac{-1 - 5}{-6 - 3} = \frac{-6}{-9} = \frac{2}{3} \end{aligned}$$

NOW TRY EXERCISE 11

Avoiding Common Errors

Students sometimes subtract the x's and y's in the slope formula in the wrong order. For instance, using the problem in Example 1:

$$m = \frac{y_2 - y_1}{x_1 - x_2} = \frac{5 - (-1)}{-6 - 3} = \frac{5 + 1}{-6 - 3} = \frac{6}{-9} = -\frac{2}{3}$$

Notice that subtracting in this incorrect order results in a negative slope, when the actual slope of the line is positive. The same sign error will occur each time subtraction is done incorrectly in this manner.

2) Recognize Positive and Negative Slopes

A straight line for which the value of y increases as x increases has a **positive slope**, see Figure 4.34a. A line with a positive slope rises as it moves from left to right. A straight line for which the value of y decreases as x increases has a **negative slope**, see Figure 4.34b. A line with a negative slope falls as it moves from left to right.

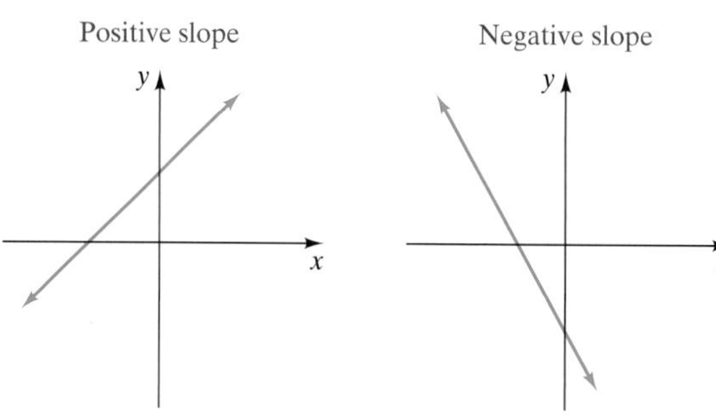

FIGURE 4.34

EXAMPLE 2 Consider the line in Figure 4.35.

a) Determine the slope of the line by observing the vertical change and horizontal change between the points (1, 5) and (0, 2).

b) Calculate the slope of the line using the two given points.

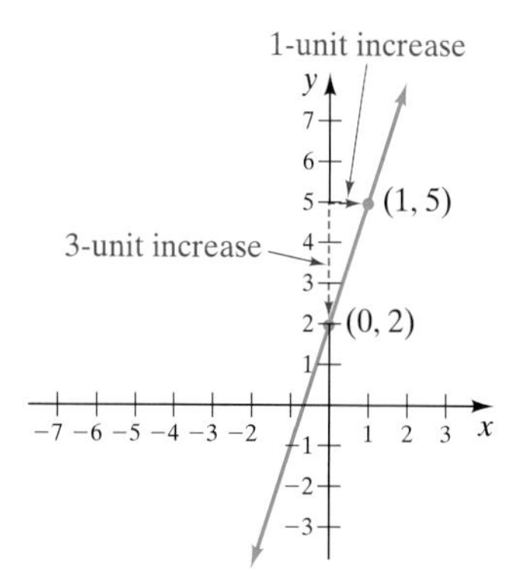

FIGURE 4.35

Solution

a) The first thing you should notice is that the slope is positive since the line rises from left to right. Now determine the vertical change between the two points. The vertical change is +3 units. Next determine the horizontal change between the two points. The horizontal change is +1 unit. Since the slope is the ratio of the vertical change to the horizontal change between any two points, and since the slope is positive, the slope of the line is $\frac{3}{1}$ or 3.

b) We can use any two points on the line to determine its slope. Since we are given the ordered pairs (1, 5) and (0, 2), we will use them.

Let (x_2, y_2) be $(1, 5)$. Let (x_1, y_1) be $(0, 2)$.

$$m = \frac{y_2 - y_1}{x_2 - x_1} = \frac{5 - 2}{1 - 0} = \frac{3}{1} = 3$$

Note that the slope obtained in part **b)** agrees with the slope obtained in part **a)**. If we had designated $(1, 5)$ as (x_1, y_1) and $(0, 2)$ as (x_2, y_2), the slope would not have changed. Try it and see that you will still obtain a slope of 3.

NOW TRY EXERCISE 23

EXAMPLE 3 Find the slope of the line in Figure 4.36 by observing the vertical change and horizontal change between the two points shown.

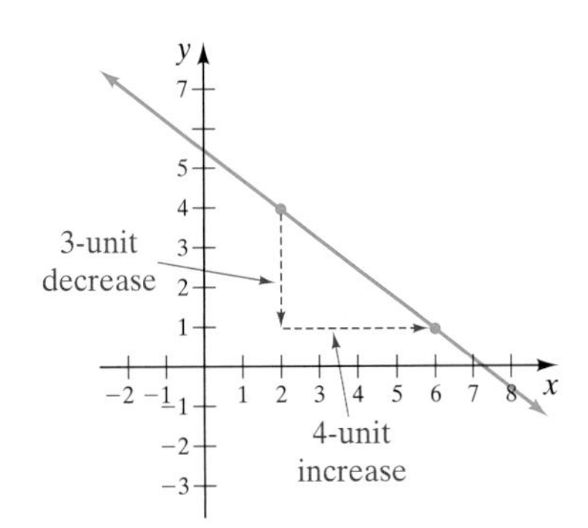

FIGURE 4.36

Solution Since the graph falls from left to right, you should realize that the line has a negative slope. The vertical change between the two given points is -3 units since it is decreasing. The horizontal change between the two given points is 4 units since it is increasing. Since the ratio of the vertical change to the horizontal change is -3 units to 4 units, the slope of this line is $\frac{-3}{4}$ or $-\frac{3}{4}$.

NOW TRY EXERCISE 27

Using the two points shown in Figure 4.36 and the definition of slope, calculate the slope of the line in Example 3. You should obtain the same answer.

3) Examine the Slopes of Horizontal and Vertical Lines

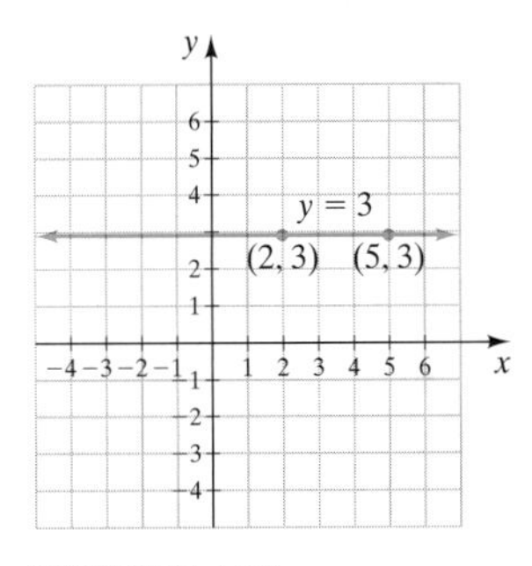

FIGURE 4.37

Now we consider the slope of horizontal and vertical lines. Consider the graph of $y = 3$ (Fig. 4.37). What is its slope?

The graph is parallel to the x-axis and goes through the points $(2, 3)$ and $(5, 3)$. Arbitrarily select $(5, 3)$ as (x_2, y_2) and $(2, 3)$ as (x_1, y_1). Then the slope of the line is

$$m = \frac{y_2 - y_1}{x_2 - x_1} = \frac{3 - 3}{5 - 2} = \frac{0}{3} = 0$$

Since there is no change in y, this line has a slope of 0.

NOW TRY EXERCISE 35

Every horizontal line has a slope of 0.

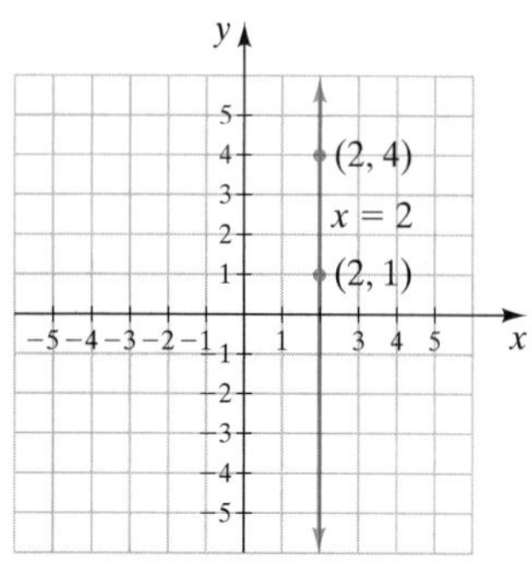

FIGURE 4.38

Now we discuss vertical lines. Consider the graph of $x = 2$ (Fig. 4.38). What is its slope?

The graph is parallel to the y-axis and goes through the points $(2, 1)$ and $(2, 4)$. Arbitrarily select $(2, 4)$ as (x_2, y_2) and $(2, 1)$ as (x_1, y_1). Then the slope of the line is

$$m = \frac{y_2 - y_1}{x_2 - x_1} = \frac{4 - 1}{2 - 2} = \frac{3}{0}$$

We learned in Section 1.8 that $\frac{3}{0}$ is undefined. Thus, we say that the slope of this line is undefined.

> The slope of any vertical line is undefined.

Exercise Set 4.3

Concept/Writing Exercises

1. Explain what is meant by the slope of a line.
2. Explain how to find the slope of a line.
3. Describe the appearance of a line that has a positive slope.
4. Describe the appearance of a line that has a negative slope.
5. Explain how to tell by observation whether a line has a positive slope or negative slope.
6. What is the slope of any horizontal line? Explain your answer.
7. Do vertical lines have a slope? Explain.
8. What letter is used to represent the slope?

Practice the Skills

Find the slope of the line through the given points.

9. $(4, 1)$ and $(5, 6)$
10. $(8, -2)$ and $(6, -4)$
11. $(9, 0)$ and $(5, -2)$
12. $(5, -6)$ and $(6, -5)$
13. $\left(3, \frac{1}{2}\right)$ and $\left(-3, \frac{1}{2}\right)$
14. $(-4, 2)$ and $(6, 5)$
15. $(-4, 6)$ and $(-2, 6)$
16. $(9, 3)$ and $(5, -6)$
17. $(3, 4)$ and $(3, -2)$
18. $(-7, 5)$ and $(3, -4)$
19. $(6, 0)$ and $(-2, 3)$
20. $(-2, 3)$ and $(-2, -1)$
21. $\left(0, \frac{3}{2}\right)$ and $\left(-\frac{3}{4}, 1\right)$
22. $(-1, 7)$ and $\left(\frac{1}{3}, -2\right)$

By observing the vertical and horizontal change of the line between the two points indicated, determine the slope of each line.

23.
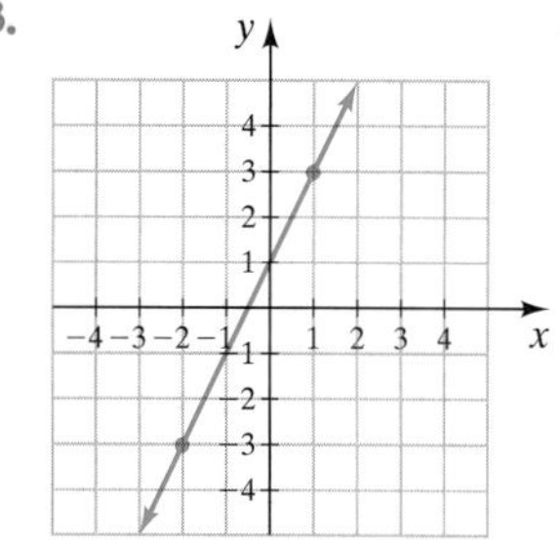

24.
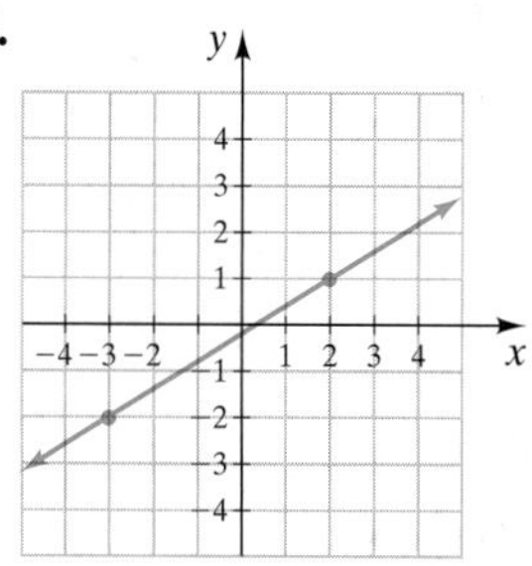

25.
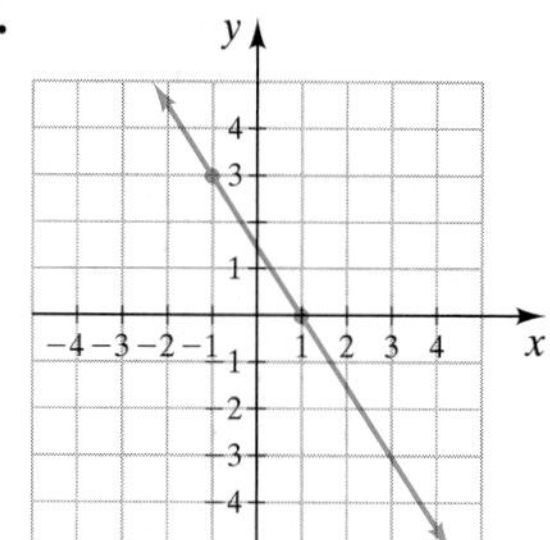

26.
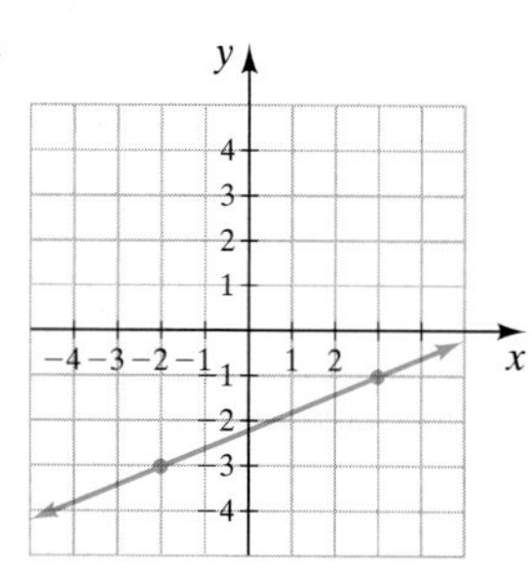

27.
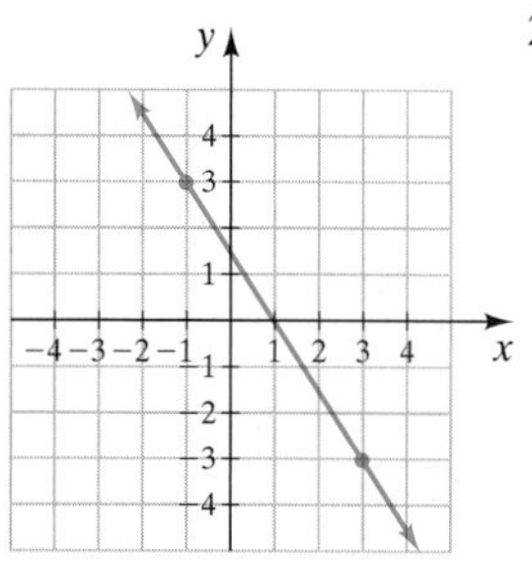

28.
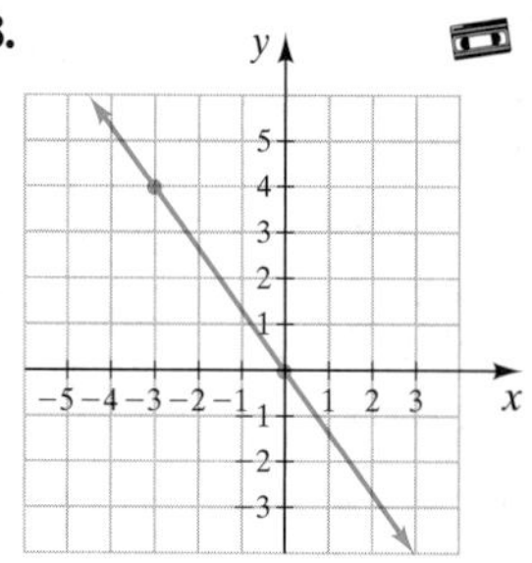

29.
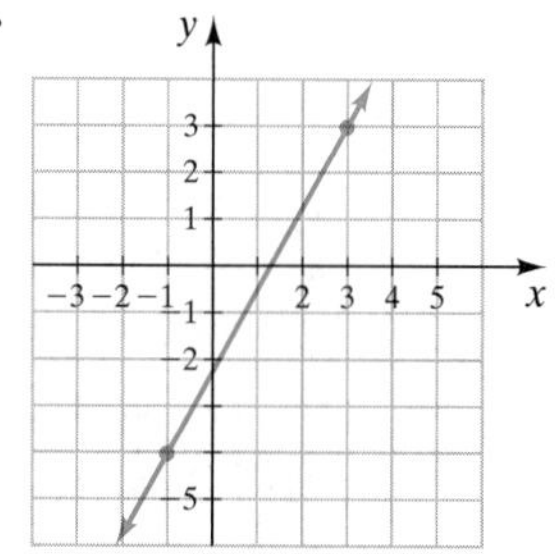

30.
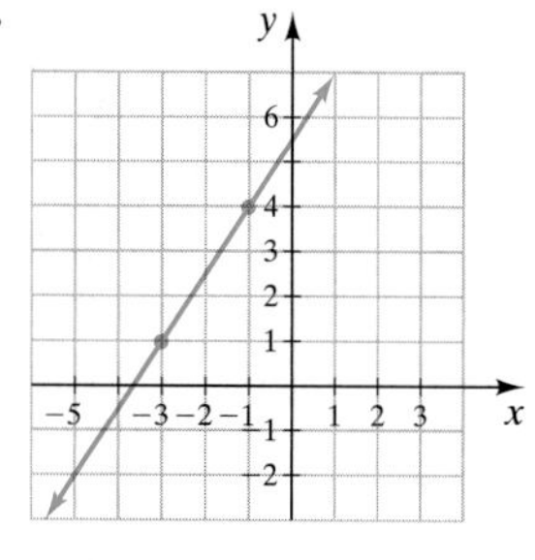

31.

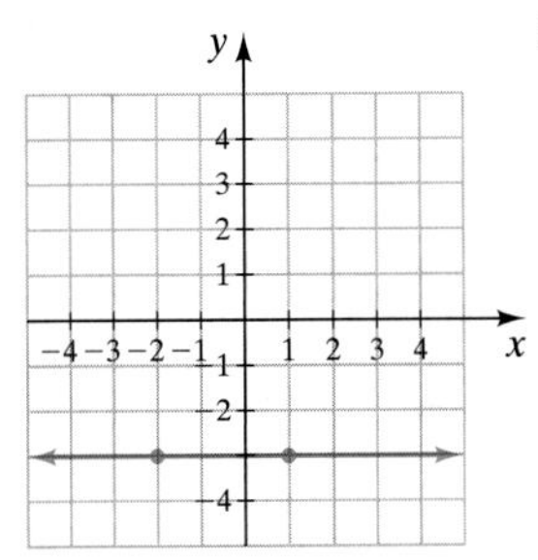

32.

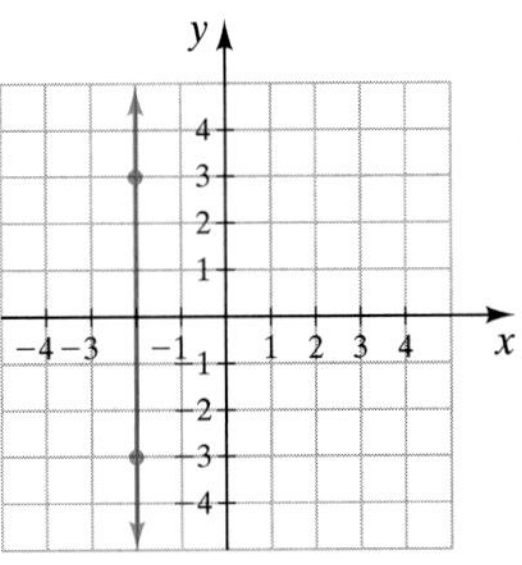

33.

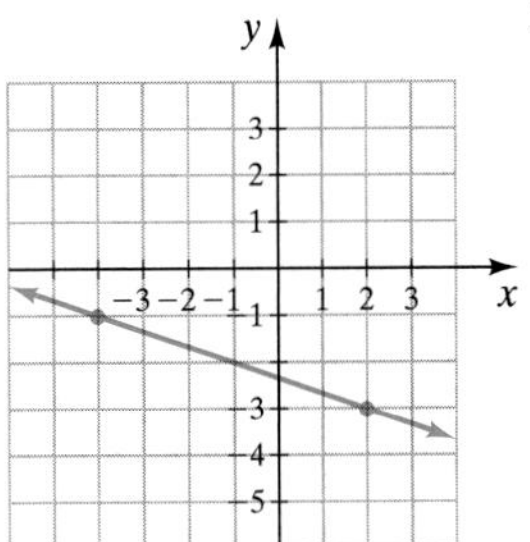

34.

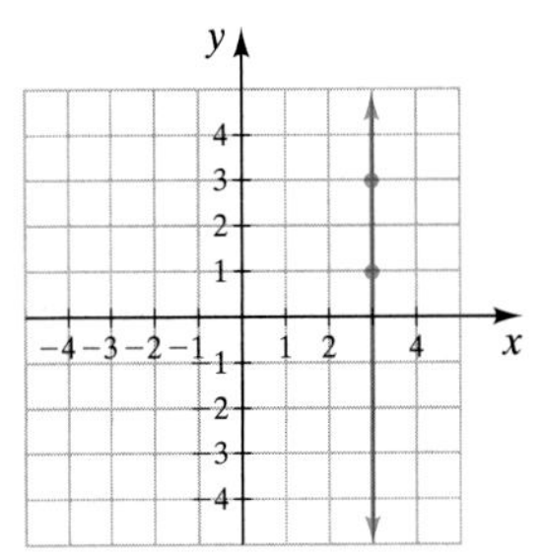

35.

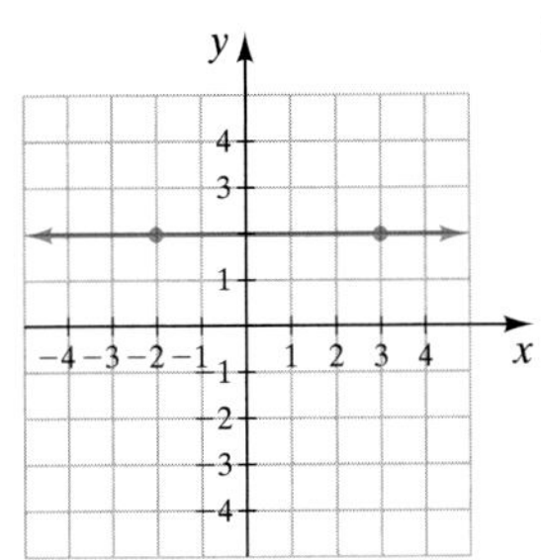

36.

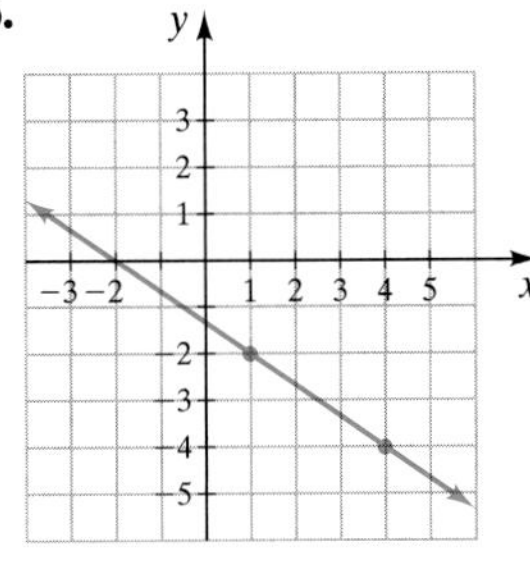

Problem Solving

In Exercises 37 and 38, determine which line (the first or second) has the greater slope. Explain your answer. Notice that the scales on the x- and y-axes are different.

37.

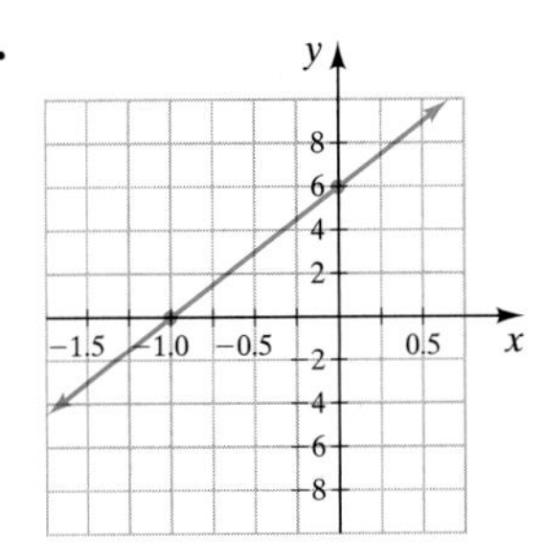

38.

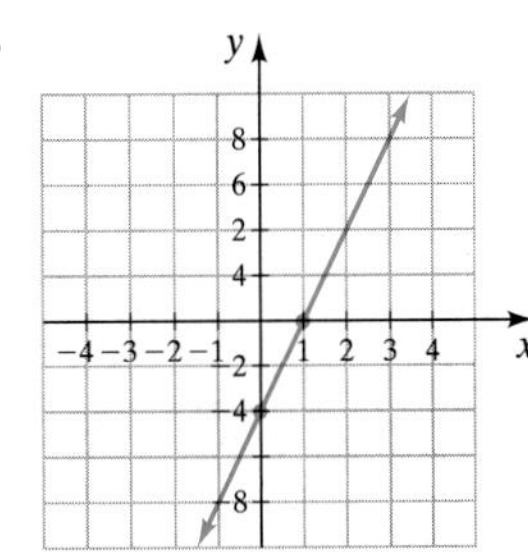

*In Exercises 39 and 40, find the slope of the line segments indicated in **a)** red and **b)** blue.*

39.

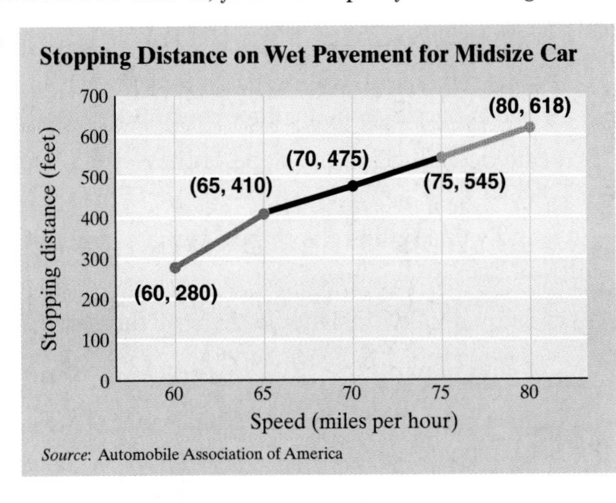

40.

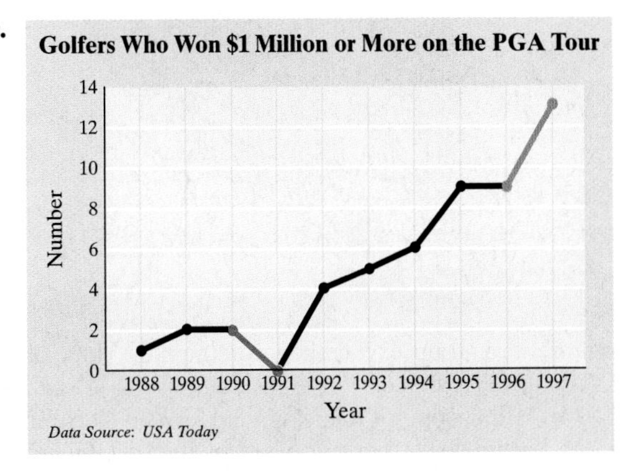

In the next section of the book we will plot a point on a line and then find a second point on the line using the slope, m. In Exercises 41–44, one point on a line and the slope of the line are given. Plot the given point, and then use the given slope to find a second point on the line. (Hint: For a positive slope, move up and to the right; for a negative slope, move down and to the right.)

41. $(3, 2)$, $m = 4$

42. $(-3, 5)$, $m = -2$

43. $(2, 3)$, $m = 0$

44. $(-2, -5)$, m is undefined

Challenge Problems

45. Find the slope of the line through the points $\left(\frac{1}{2}, -\frac{3}{8}\right)$ and $\left(-\frac{4}{9}, -\frac{7}{2}\right)$.

46. If one point on a line is $(6, -4)$ and the slope of the line is $-\frac{5}{3}$, identify another point on the line.

47. A quadrilateral (a four-sided figure) has four vertices (the points where the sides meet). Vertex A is at $(0, 1)$, vertex B is at $(6, 2)$, vertex C is at $(5, 4)$, and vertex D is at $(1, -1)$.

a) Graph the quadrilateral in the Cartesian coordinate system.

b) Find the slopes of sides AC, CB, DB, and AD.

c) Do you think this figure is a parallelogram? Explain.

48. The following graph shows the world's population estimated to the year 2016.

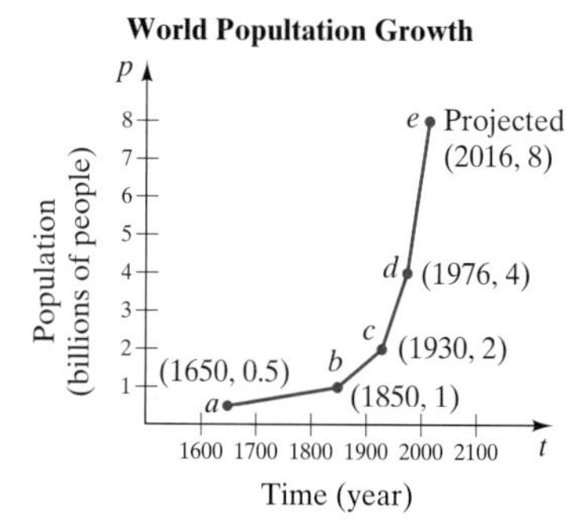

a) Find the slope of the line segment between each pair of points, that is, ab, bc, and so on. Remember, the second coordinate is in billions. Thus, for example, 0.5 billion is actually 500,000,000.

b) Would you say that this graph represents a linear equation? Explain.

Group Activity

Discuss and answer Exercises 49 and 50 as a group, according to the instructions.

49. The slope of a hill and the slope of a line both measure steepness. However, there are several important differences.

a) As a group, explain how you think the slope of a hill is determined.

b) Is the slope of a line, graphed in the Cartesian coordinate system, measured in any specific unit?

c) Is the slope of a hill measured in any specific unit?

50. Consider the graph below.

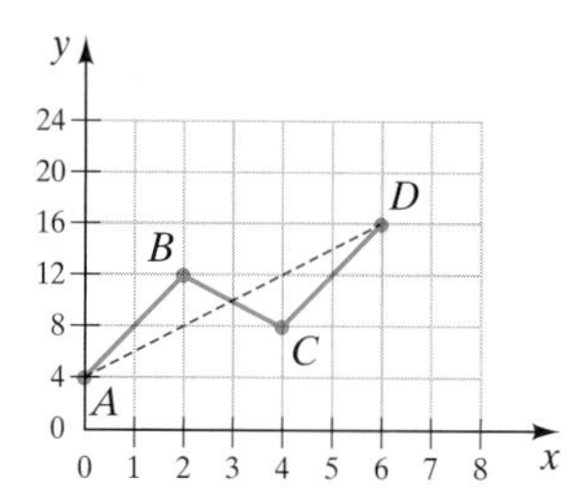

a) As a group, determine whether the slope of the dashed line from A to D will be the same as the (mean) average of the slopes of the three solid lines?

b) Individually, find the slope of the dashed line from A to D.

c) Individually, find the slope of each of the three solid lines.

d) Individually, find the average of the three slopes found in part **c)**.

e) As a group, determine whether your answer in part **a)** appears correct.

f) As a group, indicate what you think this means.

Cumulative Review Exercises

[1.9] **51.** Evaluate $4x^2 + 3x + \frac{x}{2}$ when $x = 0$.

[2.3] **52.** **a)** If $-x = -\frac{3}{2}$, what is the value of x?

b) If $5x = 0$, what is the value of x?

[2.5] **53.** Solve the equation $2x - 3(x - 2) = x + 2$.

[4.2] **54.** Find the x- and y-intercepts for the line whose equation is $5x - 3y = 15$.

4.4 SLOPE–INTERCEPT AND POINT–SLOPE FORMS OF A LINEAR EQUATION

1. **Write a linear equation in slope–intercept form.**
2. **Graph a linear equation using the slope and y-intercept.**
3. **Use the slope–intercept form to determine the equation of a line.**
4. **Determine whether two lines are parallel.**
5. **Use the point–slope form to determine the equation of a line.**
6. **Compare the three methods of graphing linear equations.**

In Section 4.1 we introduced the *standard form* of a linear equation, $ax + by = c$. In this section we introduce two more forms, the slope–intercept form and the point–slope form. We begin our discussion with the slope–intercept form.

1) Write a Linear Equation in Slope–Intercept Form

A very important form of a linear equation is the **slope–intercept form**, $\boldsymbol{y = mx + b}$. The graph of an equation of the form $y = mx + b$ will always be a straight line with a **slope of $\boldsymbol{m}$** and a **$\boldsymbol{y}$-intercept $\boldsymbol{(0, b)}$**. For example, the graph of the equation $y = 3x - 4$ will be a straight line with a slope of 3 and a y-intercept $(0, -4)$. The graph of $y = -2x + 5$ will be a straight line with a slope of -2 and a y-intercept $(0, 5)$.

Slope–Intercept Form of a Linear Equation

$$y = mx + b$$

where m is the slope, and $(0, b)$ is the y-intercept of the line.

slope y-intercept

$$y = mx + b$$

Equations in Slope–Intercept Form	Slope	y-Intercept
$y = 3x - 6$	3	$(0, -6)$
$y = \frac{1}{2}x + \frac{3}{2}$	$\frac{1}{2}$	$\left(0, \frac{3}{2}\right)$
$y = -5x + 3$	-5	$(0, 3)$
$y = -\frac{2}{3}x - \frac{3}{5}$	$-\frac{2}{3}$	$\left(0, -\frac{3}{5}\right)$

To write a linear equation in slope–intercept form, solve the equation for y.

Once the equation is solved for y, the numerical coefficient of the x-term will be the slope, and the constant term will give the y-intercept.

EXAMPLE 1 Write the equation $-3x + 4y = 8$ in slope–intercept form. State the slope and y-intercept.

Solution To write this equation in slope–intercept form, we solve the equation for y.

$$\begin{aligned} -3x + 4y &= 8 \\ 4y &= 3x + 8 \\ y &= \frac{3x + 8}{4} \\ y &= \frac{3}{4}x + \frac{8}{4} \\ y &= \frac{3}{4}x + 2 \end{aligned}$$

The slope is $\frac{3}{4}$, and the y-intercept is $(0, 2)$.

2) Graph a Linear Equation Using the Slope and y-Intercept

In Section 4.2 we discussed two methods of graphing a linear equation. They were (1) by plotting points and (2) using the x- and y-intercepts. Now we present a third method. This method makes use of the slope and the y-intercept. Remember that when we solve an equation for y we put the equation in slope–intercept form. Once it is in this form we can determine the slope and y-intercept of the graph from the equation. The procedure to use to graph by this method follows.

To Graph Linear Equations Using the Slope and y-Intercept

1. Solve the linear equation for y. That is, get the equation in slope–intercept form, $y = mx + b$.
2. Note the slope, m, and y-intercept, $(0, b)$.
3. Plot the y-intercept on the y-axis.
4. Use the slope to find a second point on the graph.
   a) If the slope is positive, a second point can be found by moving up and to the right. Thus, when the slope is in fraction form $\frac{p}{q}$ where $p > 0$ and $q > 0$, we can find a second point by moving p units up and q units to the right.
   b) If the slope is negative, a second point can be found by moving down and to the right. Thus, when the slope is in fraction form, $-\frac{p}{q}$ where $p > 0$ and $q > 0$, we can find a second point by moving p units down and q units to the right.
5. With a ruler, draw a straight line through the two points. Draw arrowheads at the ends of the line to show that the line continues indefinitely in both directions.

EXAMPLE 2 Write the equation $-3x + 4y = 8$ in slope–intercept form; then use the slope and y-intercept to graph $-3x + 4y = 8$.

Solution In Example 1 we solved $-3x + 4y = 8$ for y. We found that

$$y = \frac{3}{4}x + 2$$

The slope of the line is $\frac{3}{4}$ and the y-intercept is $(0, 2)$. We mark the first point, the y-intercept at 2 on the y-axis (Fig. 4.39). Now we use the slope $\frac{3}{4}$, to find a second point. Since the slope is positive, we move 3 units up and 4 units to the right to find the second point. A second point will be at $(4, 5)$. We can continue this process to obtain a third point at $(8, 8)$. Now we draw a straight line through the three points. Notice that the line has a positive slope, which is what we expected.

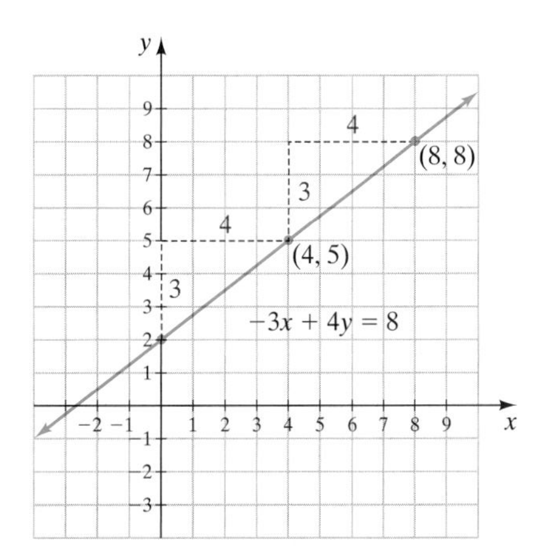

FIGURE 4.39

EXAMPLE 3 Graph the equation $5x + 3y = 12$ by using the slope and y-intercept.

Solution Solve the equation for y.

$$5x + 3y = 12$$
$$3y = -5x + 12$$
$$y = \frac{-5x + 12}{3}$$
$$= -\frac{5}{3}x + 4$$

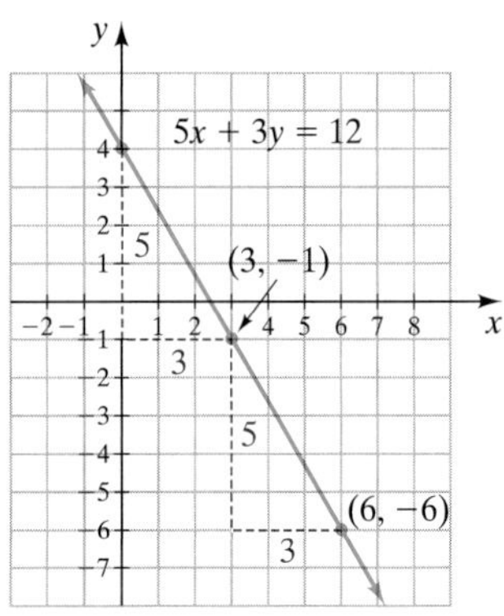

FIGURE 4.40

NOW TRY EXERCISE 21

Thus, the slope is $-\frac{5}{3}$ and the y-intercept is $(0, 4)$. Begin by marking a point at 4 on the y-axis (Fig. 4.40). Then move 5 units down and 3 units to the right to determine the next point. Move down and to the right because the slope is negative and a line with a negative slope must fall as it goes from left to right. Finally, draw the straight line between the plotted points.

3) Use the Slope–Intercept Form to Determine the Equation of a Line

Now that we know how to use the slope–intercept form of a line, we can use it to write the equation of a given line. To do so, we need to determine the slope, m, and y-intercept of the line. Once we determine these values we can write the equation in slope–intercept form, $y = mx + b$. For example, if we determine the slope of a line is -4 and the y-intercept is at 6, the equation of the line is $y = -4x + 6$.

EXAMPLE 4 Determine the equation of the line shown in Figure 4.41.

Solution The graph shows that the y-intercept is at -3. Now we need to determine the slope of the line. Since the graph falls from left to right, it has a negative slope. We can see that the vertical change is 2 units for each horizontal change of 1 unit. Thus, the slope of the line is -2. The slope can also be determined by selecting any two points on the line and calculating the slope. Let's use the point $(-2, 1)$ to represent (x_2, y_2) and the point $(0, -3)$ to represent (x_1, y_1).

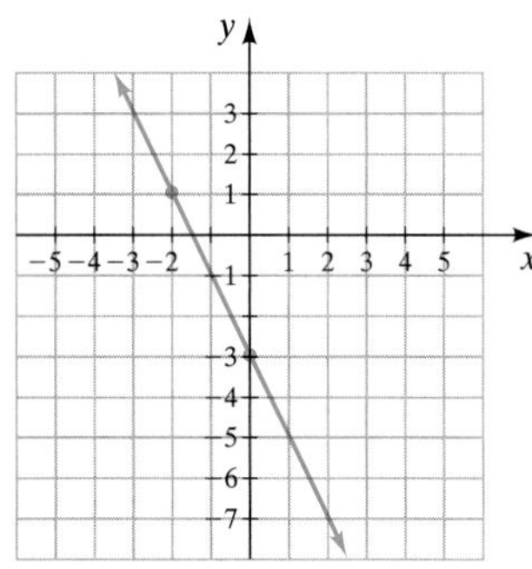

FIGURE 4.41

$$m = \frac{\Delta y}{\Delta x} = \frac{y_2 - y_1}{x_2 - x_1}$$

$$= \frac{1 - (-3)}{-2 - 0}$$

$$= \frac{1 + 3}{-2} = \frac{4}{-2} = -2$$

Again we obtain a slope of -2. Substituting -2 for m and -3 for b into the slope–intercept form of a line gives us the equation of the line in Figure 4.41, which is $y = -2x - 3$.

NOW TRY EXERCISE 29

Now let's look at an application of graphing.

EXAMPLE 5 Ross Miller owns a small business where he manufactures wooden clocks. His business has a fixed monthly cost (office rent, heat, etc.) and a variable cost per clock manufactured (cost of materials, cost of labor, etc.) The total monthly cost for x clocks manufactured is illustrated by the graph in Figure 4.42.

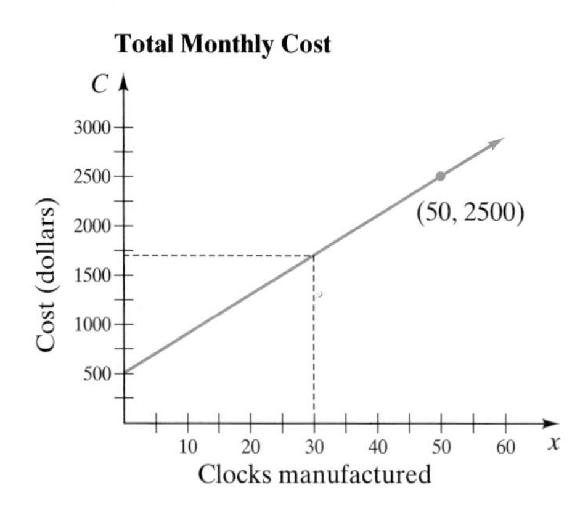

FIGURE 4.42

a) Find the equation of the total monthly cost when x units are produced.

b) Use the equation found in part **a)** to find the total monthly cost if 30 clocks are produced.

c) Use the graph in Figure 4.42 to see whether your answer in part **b)** appears correct.

Solution **a) Understand and Translate** Notice that the vertical axis is cost, C, and not y. The names used on the axes do not change the way we solve the problem. We will use slope–intercept form to write the equation of the line. However, since y is replaced by C, we will use $C = mx + b$, where b is where the graph crosses the vertical or C-axis. We first note that the graph crosses the vertical axis at 500. Thus, b is 500. Now we need to find the slope of the line. Let's use the point (0, 500) as (x_1, y_1) and (50, 2500) as (x_2, y_2).

Carry Out

$$m = \frac{y_2 - y_1}{x_2 - x_1}$$

$$= \frac{2500 - 500}{50 - 0} = \frac{2000}{50} = 40$$

Answer The slope is 40. The equation in slope–intercept form is

$$C = mx + b$$

$$= 40x + 500$$

b) To find the monthly cost, we substitute 30 for x.

$$C = 40x + 500$$

$$= 40(30) + 500$$

$$= 1200 + 500 = 1700$$

c) If we draw a vertical line up from 30 on the x-axis (the red line), we see that the corresponding cost is about \$1700. Thus, our answer in part **b)** appears correct.

4) Determine Whether Two Lines Are Parallel

We will discuss the meaning of parallel lines shortly, but before we do we will work Example 6.

EXAMPLE 6 Determine whether both equations represent lines that have the same slope.

$$6x + 3y = 8$$

$$-4x - 2y = -3$$

Solution Solve each equation for y to get the equations in slope–intercept form.

$$6x + 3y = 8 \qquad\qquad -4x - 2y = -3$$

$$3y = -6x + 8 \qquad\qquad -2y = 4x - 3$$

$$y = \frac{-6x + 8}{3} \qquad\qquad y = \frac{4x - 3}{-2}$$

$$y = -2x + \frac{8}{3} \qquad\qquad y = -2x + \frac{3}{2}$$

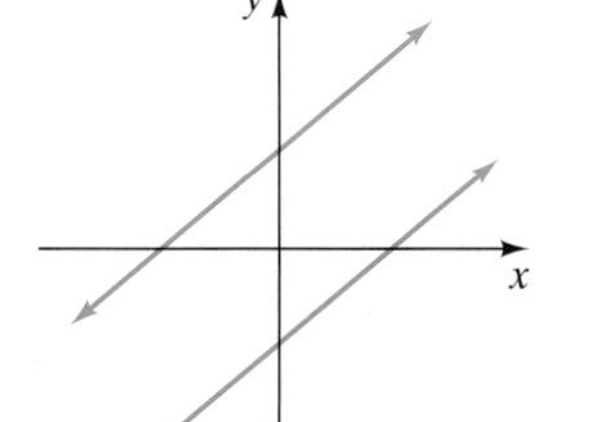

FIGURE 4.43

Both lines have the same slope of −2. Notice, however, that their y-intercepts are different.

Two lines are **parallel** when they do not intersect no matter how far they are extended. Figure 4.43 illustrates two parallel lines. *Parallel lines have the* *same slope*. The graphs of the equations in Example 6 are parallel lines since they both have the same slope, −2. Note that the two equations represent different lines since their y-intercepts are different.

To Determine Whether Two Lines Are Parallel

Write both equations in slope–intercept form and compare the slopes of the two lines. If both lines have the same slope but different y-intercepts, then the lines are parallel. If the slopes are not the same, the lines are not parallel. If both equations have the same slope and the same y-intercept then both equations represent the same line.

EXAMPLE 7 **a)** Determine whether the following equations represent parallel lines.

b) Graph both equations on the same axes.

$$y = 2x + 4$$
$$-4x + 2y = -2$$

Solution **a)** Write each equation in slope–intercept form and compare their slopes. The equation $y = 2x + 4$ is already in slope–intercept form.

$$-4x + 2y = -2$$
$$2y = 4x - 2$$
$$y = \frac{4x - 2}{2} = 2x - 1$$

The two equations are now

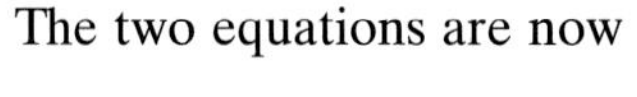

$$y = 2x + 4$$
$$y = 2x - 1$$

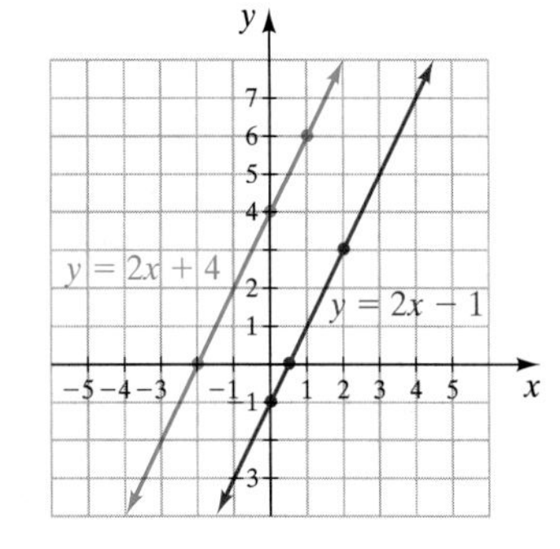

FIGURE 4.44

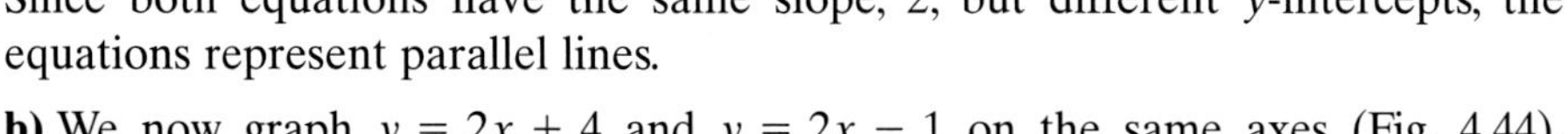

Since both equations have the same slope, 2, but different y-intercepts, the equations represent parallel lines.

b) We now graph $y = 2x + 4$ and $y = 2x - 1$ on the same axes (Fig. 4.44). Remember that $y = 2x - 1$ is the equation $-4x + 2y = -2$ in slope–intercept form.

NOW TRY EXERCISE 37

5) Use the Point–Slope Form to Determine the Equation of a Line

Thus far we have discussed the standard form of a linear equation, $ax + by = c$, and the slope–intercept form of a linear equation, $y = mx + b$. Now we will discuss another form, called the *point–slope form*.

When the slope of a line and a point on the line are known, we can use the point–slope form to determine the equation of the line. The **point–slope form** can be obtained by beginning with the slope between any selected point (x, y) and a fixed point (x_1, y_1) on a line.

$$m = \frac{y - y_1}{x - x_1} \quad \text{or} \quad \frac{m}{1} = \frac{y - y_1}{x - x_1}$$

Now cross-multiply to obtain

$$m(x - x_1) = y - y_1 \qquad \text{or} \qquad y - y_1 = m(x - x_1)$$

Point–Slope Form of a Linear Equation

$$y - y_1 = m(x - x_1)$$

where m is the slope of the line and (x_1, y_1) is a point on the line.

EXAMPLE 8 Write an equation, in slope–intercept form, of the line that goes through the point (2, 3) and has a slope of 4.

Solution Since we are given a point on the line and the slope of the line, we begin by writing the equation in point–slope form. The slope m is 4. The point on the line is (2, 3); we will use this point for (x_1, y_1) in the formula. We substitute 4 for m, 2 for x_1, and 3 for y_1 in the point–slope form of a linear equation.

$$y - y_1 = m(x - x_1)$$

$$y - 3 = 4(x - 2) \qquad \textit{Equation in point–slope form}$$

$$y - 3 = 4x - 8 \qquad \textit{Distributive Property}$$

$$y = 4x - 5 \qquad \textit{Equation in slope–intercept form}$$

The graph of $y = 4x - 5$ has a slope of 4 and passes through the point (2, 3).

The answer to Example 8 was given in slope–intercept form. If we were asked to give the answer in standard form, two acceptable answers would be $-4x + y = -5$ and $4x - y = 5$. Your instructor may specify the form in which the equation is to be given.

In Example 9, we will work an example very similar to Example 8. However, in the solution to Example 9, the equation will contain a fraction. We solved some equations that contained fractions in Section 2.2. Recall that to simplify an equation that contains a fraction, we multiply both sides of the equation by the denominator of the fraction.

EXAMPLE 9 Write an equation, in slope–intercept form, of the line that goes through the point (6, −2) and has a slope $\frac{2}{3}$.

Solution We will begin with the point–slope form of a line, where m is $\frac{2}{3}$, 6 is x_1, and −2 is y_1.

$$y - y_1 = m(x - x_1)$$

$$y - (-2) = \frac{2}{3}(x - 6) \qquad \textit{Equation in point–slope form}$$

$$y + 2 = \frac{2}{3}(x - 6)$$

$$3(y + 2) = \cancel{3} \cdot \frac{2}{\cancel{3}}(x - 6) \qquad \textit{Multiply both sides by 3.}$$

$$3y + 6 = 2(x - 6) \quad \textit{Distributive Property}$$
$$3y + 6 = 2x - 12 \quad \textit{Distributive Property}$$
$$3y = 2x - 18 \quad \textit{Subtract 6 from both sides}$$
$$y = \frac{2x - 18}{3} \quad \textit{Divide both sides by 3.}$$
$$y = \frac{2}{3}x - 6 \quad \textit{Equation in slope–intercept form}$$

HELPFUL HINT

We have discussed three forms of a linear equation. We summarize the three forms below.

STANDARD FORM	EXAMPLES
$ax + by = c$	$2x - 3y = 8$
	$-5x + y = -2$

SLOPE–INTERCEPT FORM	EXAMPLES
$y = mx + b$	$y = 2x - 5$
m is the slope, $(0, b)$ is the y-intercept	$y = -\frac{3}{2}x + 2$

POINT–SLOPE FORM	EXAMPLES
$y - y_1 = m(x - x_1)$	$y - 3 = 2(x + 4)$
m is the slope, (x_1, y_1) is a point on the line	$y + 5 = -4(x - 1)$

We now discuss how to use the point–slope form to determine the equation of a line when two points on the line are known.

EXAMPLE 10 Find an equation of the line through the points $(-1, -3)$ and $(4, 2)$. Write the equation in slope–intercept form.

Solution To use the point–slope form, we must first find the slope of the line through the two points. To determine the slope, let's designate $(-1, -3)$ as (x_1, y_1) and $(4, 2)$ as (x_2, y_2).

$$m = \frac{y_2 - y_1}{x_2 - x_1} = \frac{2 - (-3)}{4 - (-1)} = \frac{2 + 3}{4 + 1} = \frac{5}{5} = 1$$

We can use either point (one at a time) in determining the equation of the line. This example will be worked out using both points to show that the solutions obtained are identical.

Using the point $(-1, -3)$ as (x_1, y_1),

$$y - y_1 = m(x - x_1)$$
$$y - (-3) = 1[x - (-1)]$$
$$y + 3 = x + 1$$
$$y = x - 2$$

Using the point $(4, 2)$ as (x_1, y_1),

$$y - y_1 = m(x - x_1)$$
$$y - 2 = 1(x - 4)$$
$$y - 2 = x - 4$$
$$y = x - 2$$

NOW TRY EXERCISE 53 Note that the equations for the line are identical.

HELPFUL HINT

In the exercise set at the end of this section, you will be asked to write a linear equation in slope–intercept form. Even though you will eventually write the equation in slope–intercept form, you may need to start your work with the point–slope form. Below we indicate the initial form to use to solve the problem.

Begin with the **slope–intercept form** if you know:

The slope of the line and the y-intercept

Begin with the **point–slope form** if you know:

a) The slope of the line and a point on the line, or

b) Two points on the line (first find the slope, then use the point–slope form)

6 Compare the Three Methods of Graphing Linear Equations

We have discussed three methods to graph a linear equation: (1) plotting points, (2) using the x- and y-intercepts, and (3) using the slope and y-intercept. In Example 11 we graph an equation using all three methods. No single method is always the easiest to use. If the equation is given in slope–intercept form, $y = mx + b$, then graphing by plotting points or by using the slope and y-intercept might be easier. If the equation is given in standard form, $ax + by = c$, then graphing using the intercepts might be easier. Unless your teacher specifies that you should graph by a specific method, you may use the method with which you feel most comfortable. Graphing by plotting points is the most versatile method since it can also be used to graph equations that are not straight lines.

EXAMPLE 11 Graph $3x - 2y = 8$ **a)** by plotting points; **b)** using the x- and y-intercepts; and **c)** using the slope and y-intercept.

Solution For parts **a)** and **c)** we must write the equation in slope–intercept form.

$$3x - 2y = 8$$
$$-2y = -3x + 8$$
$$y = \frac{-3x + 8}{-2} = \frac{3}{2}x - 4$$

a) Plotting points We substitute values for x and find the corresponding values of y. Then we plot the ordered pairs and draw the graph (Fig. 4.45).

$$y = \frac{3}{2}x - 4$$

x	y
0	−4
2	−1
4	2

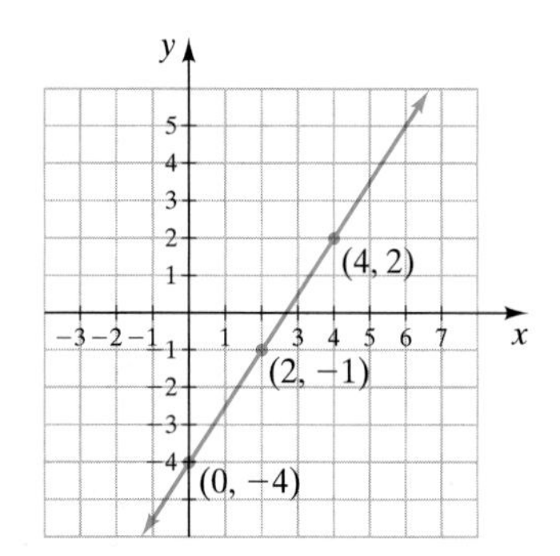

FIGURE 4.45

b) Intercepts We find the x- and y- intercepts and a check point. Then we plot the points and draw the graph (Fig. 4.46).

$$3x - 2y = 8$$

x-Intercept	y-Intercept	Check Point
Let $y = 0$	Let $x = 0$	Let $x = 2$
$3x - 2y = 8$	$3x - 2y = 8$	$3x - 2y = 8$
$3x - 2(0) = 8$	$3(0) - 2y = 8$	$3(2) - 2y = 8$
$3x = 8$	$-2y = 8$	$6 - 2y = 8$
$x = \frac{8}{3}$	$y = -4$	$-2y = 2$
		$y = -1$

The three ordered pairs are $\left(\frac{8}{3}, 0\right)$, $(0, -4)$ and $(2, 1)$.

c) Slope and y-intercept The y-intercept is $(0, -4)$, therefore we place a point at -4 on the y-axis. Since the slope is $\frac{3}{2}$, we obtain a second point by moving 3 units up and 2 units to the right. The graph is illustrated in Figure 4.47.

Notice that we get the same line by all three methods.

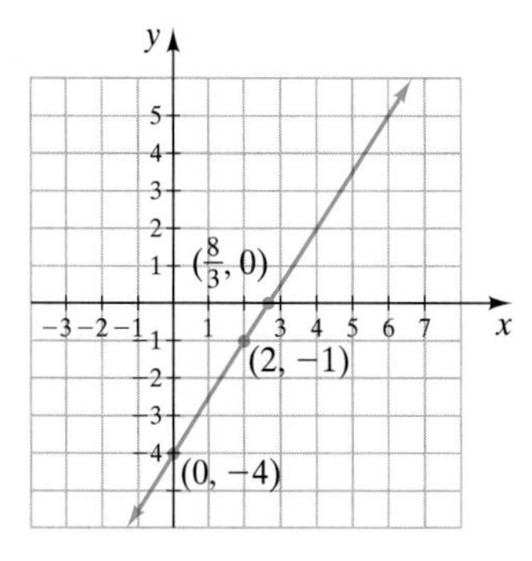

FIGURE 4.46

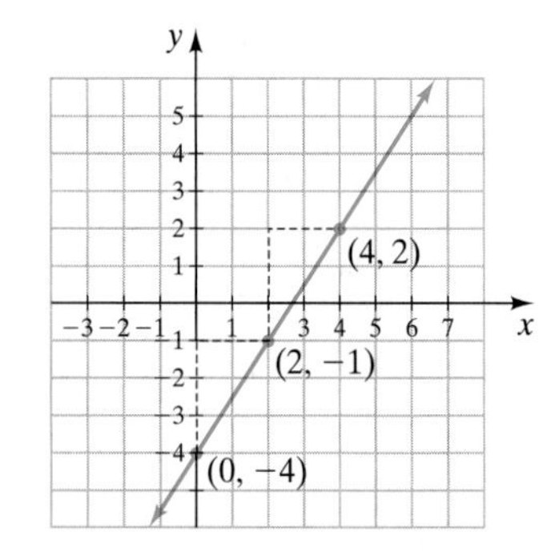

FIGURE 4.47

Using Your Graphing Calculator

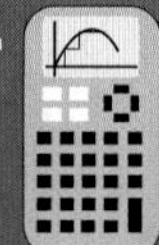

In Example 11 we graphed the equation $y = \frac{3}{2}x - 4$. Let's see what the graph of $y = \frac{3}{2}x - 4$ looks like on a graphing calculator using three different window settings. Figure 4.48 shows the graph of $y = \frac{3}{2}x - 4$ using the *standard window setting*. To obtain the standard window, press the ZOOM key and then choose option 6, ZStandard. Notice that in the standard window setting, the units on the y-axis are not as long as the units on the x-axis.

In Figure 4.49, we illustrate the graph of $y = \frac{3}{2}x - 4$ using the *square window setting*. To obtain the square window setting, press the ZOOM key and then choose option 5, ZSquare. This setting makes the units on both axes the same size. This allows the line to be displayed at the correct orientation to the axes.

In Figure 4.50, we illustrate the graph using the *decimal window setting*. To obtain the decimal window setting, press the ZOOM key and then choose option 4, ZDecimal. The decimal window setting also makes the units the same on both axes, but sets the increment from one pixel (dot) to the next at 0.1 on both axes.

STANDARD WINDOW SETTING (ZOOM: OPTION 6)

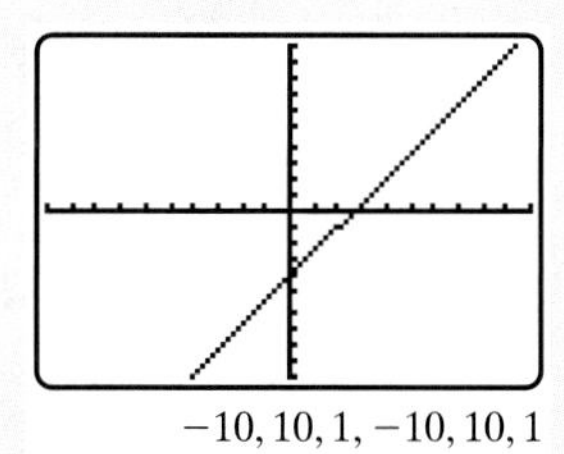

$-10, 10, 1, -10, 10, 1$

FIGURE 4.48

SQUARE WINDOW SETTING (ZOOM: OPTION 5)

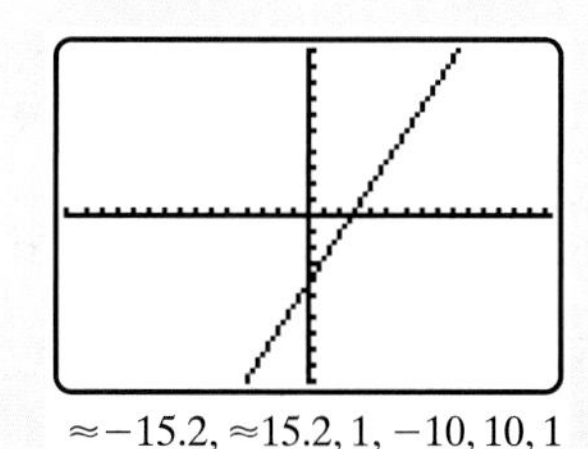

$\approx -15.2, \approx 15.2, 1, -10, 10, 1$

FIGURE 4.49

DECIMAL WINDOW SETTING (ZOOM: OPTION 4)

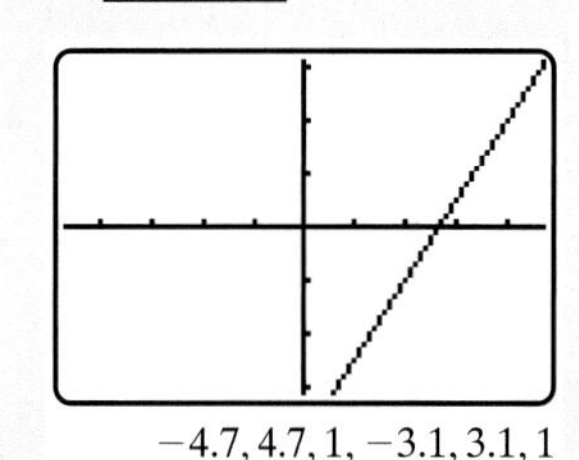

$-4.7, 4.7, 1, -3.1, 3.1, 1$

FIGURE 4.50

Exercises

*Graph each equation using the **a)** standard window, **b)** square window, and **c)** decimal window settings.*

1. $y = 4x - 6$ **2.** $y = -\frac{1}{5}x + 4$

Exercise Set 4.4

Concept/Writing Exercises

1. Give the slope–intercept form of a linear equation.
2. When you are given an equation in a form other than slope–intercept form, how can you change it to slope–intercept form?
3. What is the equation of a line, in slope–intercept form, if the slope is 4 and the y-intercept is at -2?
4. What is the equation of a line, in slope–intercept form, if the slope is -3 and the y-intercept is at 5?
5. Explain how you can determine whether two lines are parallel without actually graphing them.
6. Explain how you can determine whether two equations represent the same line without graphing the equations?
7. Give the point–slope form of a linear equation.
8. Assume the slope of a line is 2 and the line goes through the origin. Write the equation of the line in point–slope form.

Practice the Skills

Determine the slope and y-intercept of the line represented by the given equation.

9. $y = 3x - 7$ **10.** $y = -6x + 19$ **11.** $4x - 3y = 15$ **12.** $7x = 5y + 20$

Determine the slope and y-intercept of the line represented by each equation. Graph the line using the slope and y-intercept.

13. $y = x - 1$

14. $y = -x + 5$

15. $y = 3x + 2$

16. $y = 2x$

17. $y = -4x$

18. $2x + y = 5$

19. $-2x + y = -3$

20. $3x + 3y = 9$

21. $5x - 2y = 10$

22. $-x + 2y = 8$

23. $5x + 10y = 15$

24. $4x = 6y + 9$

25. $-6x = -2y + 8$

26. $16y = 8x + 32$

27. $3x = 2y - 4$

28. $20x = 80y + 40$

Determine the equation of each line.

29.

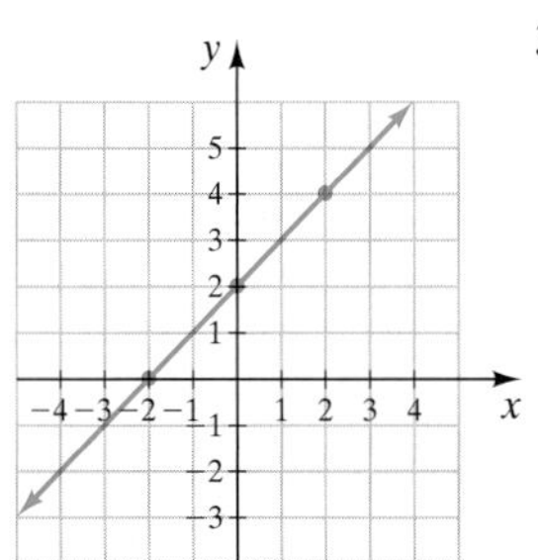

30.

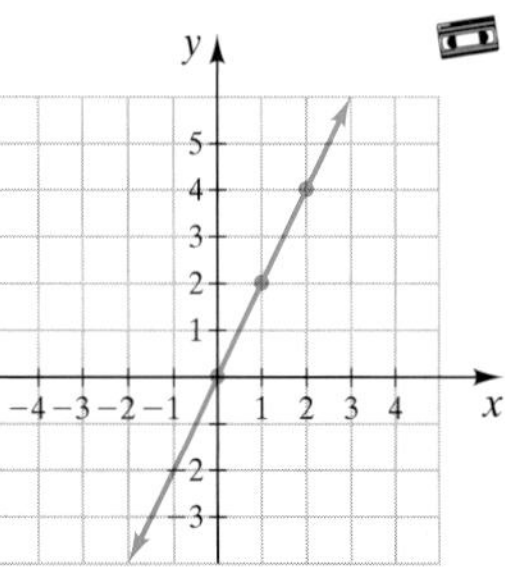

31.

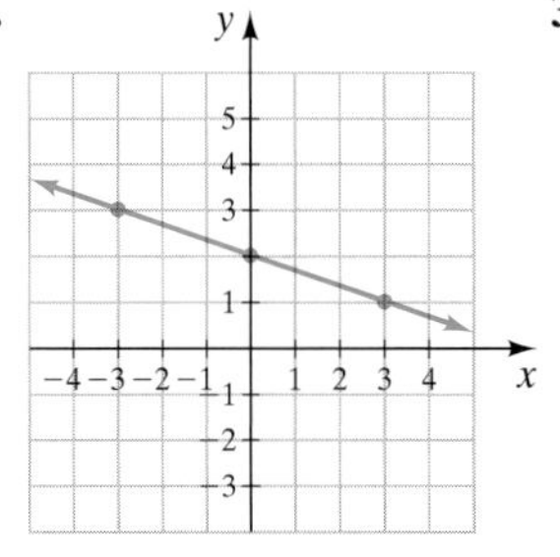

32.

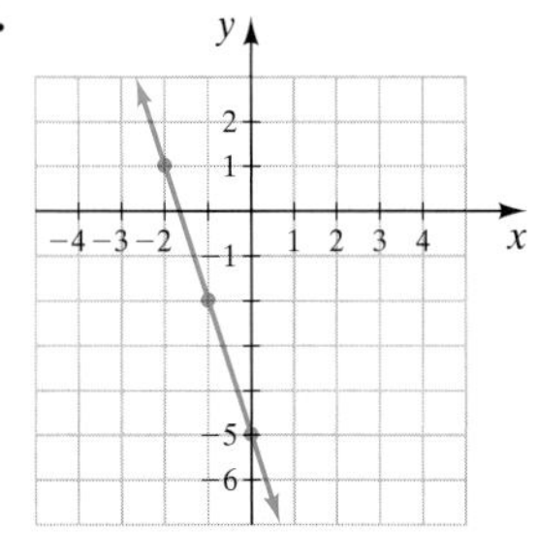

33.

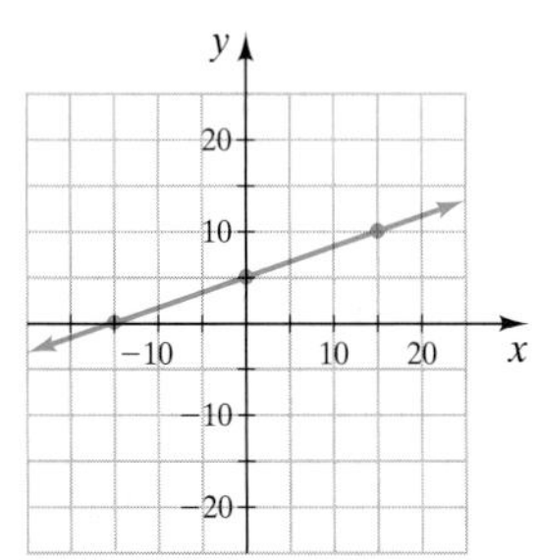

34.

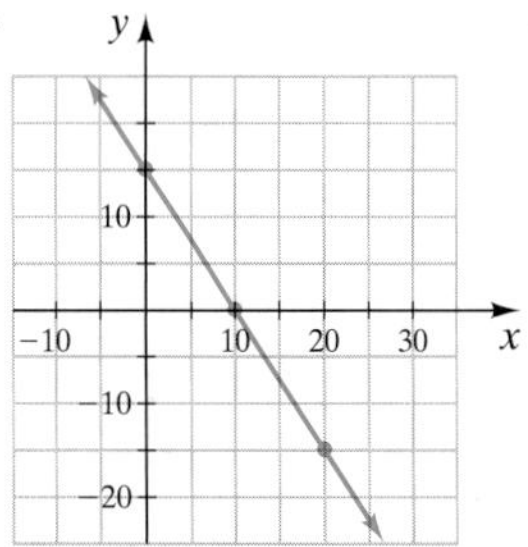

35.

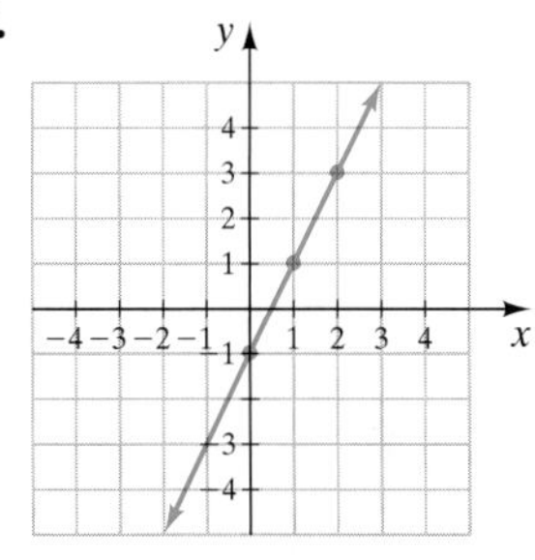

36.

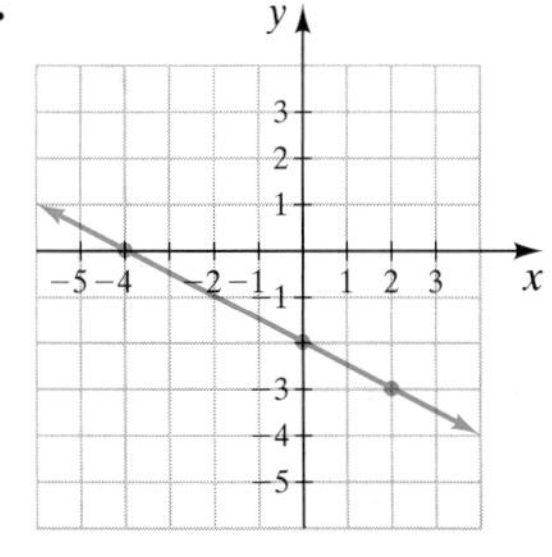

Determine whether each pair of lines are parallel.

37. $y = 5x - 2$
$y = 5x + 3$

38. $2x + 3y = 8$
$y = -\frac{2}{3}x + 5$

39. $4x + 2y = 9$
$8x = 4y + 4$

40. $3x - 5y = 7$
$5y + 3x = 2$

41. $3x + 5y = 9$
$6x = -10y + 9$

42. $6x + 2y = 8$
$4x - 9 = -y$

43. $y = \frac{1}{2}x - 6$
$3y = 6x + 9$

44. $2y - 6 = -5x$
$y = -\frac{5}{2}x - 2$

Problem Solving

Write the equation of each line, with the given properties, in slope–intercept form.

45. Slope = 3, through (0, 2)

46. Slope = 4, through (2, 3)

47. Slope = −2, through (−4, 5)

48. Slope = −2, through (4, 0)

49. Slope = $\frac{1}{2}$, through (−1, −5)

50. Slope = $-\frac{2}{3}$, through (−1, −2)

51. Slope = $\frac{2}{5}$, y-intercept is (0, 6)

52. Slope = $\frac{4}{9}$, y-intercept is $\left(0, -\frac{2}{3}\right)$

53. Through (−4, −2) and (−2, 4)

54. Through (6, 3) and (5, 2)

55. Through (−6, 9) and (6, −9)

56. Through (1, 0) and (−2, 4)

57. Through $(10, 3)$ and $(0, -2)$

58. Through $(-6, -2)$ and $(5, -3)$

59. Slope $= 5.2$, y-intercept $(0, -1.6)$

60. Slope $= -\frac{5}{8}$, y-intercept $\left(0, -\frac{7}{10}\right)$

61. Stacy Best owns a weight loss clinic. She charges her clients a one-time membership fee. She also charges per pound of weight lost. Therefore, the more successful she is at helping clients lose weight, the more income she will receive. The following graph shows a client's cost for losing weight.

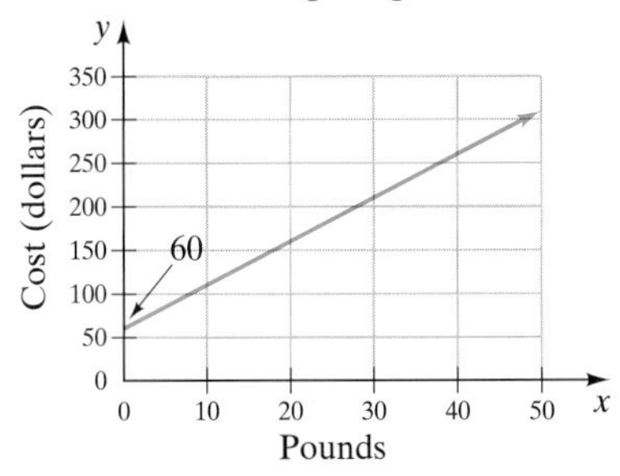

a) Find the equation that represents the cost for a client who loses x pounds.

b) Use the equation found in part **a)** to determine the cost for a client who loses 30 pounds.

62. A submarine is submerged below sea level. Tom Johnson, the captain, orders the ship to dive slowly. The following graph illustrates the submarine's depth at a time t minutes after the submarine begins to dive.

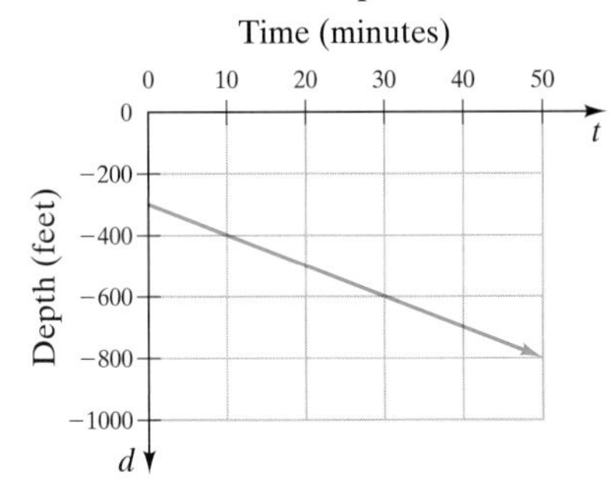

a) Find the equation that represents the depth at time t.

b) Use the equation found in part **a)** to find the submarine's depth after 20 minutes.

63. Suppose that you were asked to write the equation of a line with the properties given below. Which form of a linear equation—standard form, slope–intercept form, or point–slope form—would you start with? Explain your answer.

a) The slope of the line and the y-intercept of the line

b) The slope and a point on the line

c) Two points on the line

64. Consider the two equations $40x - 60y = 100$ and $-40x + 60y = 80$.

a) When these equations are graphed, will the two lines have the same slope? Explain how you determined your answer.

b) When these two equations are graphed, will they be parallel lines?

65. Assume the slope of a line is 2 and two points on the line are $(-5, -4)$ and $(2, 10)$.

a) If you use $(-5, -4)$ as (x_1, y_1) and then $(2, 10)$ as (x_1, y_1) will the appearance of the two equations be the same in point–slope form? Explain.

b) Find the equation, in point–slope form, using $(-5, -4)$ as (x_1, y_1).

c) Find the equation, in point–slope form, using $(2, 10)$ as (x_1, y_1).

d) Write the equation obtained in part **b)** in slope–intercept form.

e) Write the equation obtained in part **c)** in slope–intercept form.

f) Are the equations obtained in parts **d)** and **e)** the same? If not, explain why.

66. Assume the slope of a line is -3 and two points on the line are $(-1, 8)$ and $(2, -1)$.

a) If you use $(-1, 8)$ as (x_1, y_1) and then $(2, -1)$ as (x_1, y_1) will the appearance of the two equations be the same in point–slope form? Explain.

b) Find the equation, in point–slope form, using $(-1, 8)$ as (x_1, y_1).

c) Find the equation, in point–slope form, using $(2, -1)$ as (x_1, y_1).

d) Write the equation obtained in part **b)** in slope–intercept form.

e) Write the equation obtained in part **c)** in slope–intercept form.

f) Are the equations obtained in parts **d)** and **e)** the same? If not, explain why.

Challenge Problems

67. Determine the equation of the line with y-intercept at 4 that is parallel to the line whose equation is $2x + y = 6$. Explain how you determined your answer.

68. Will a line through the points $(60, 30)$ and $(20, 90)$ be parallel to the line with x-intercept at 2 and y-intercept at 3? Explain how you determined your answer.

69. Write an equation of the line parallel to the graph of $3x - 4y = 6$ that passes through the point $(-4, -1)$.

70. Two lines are **perpendicular** and cross at right angles (see the figure) when their slopes are negative reciprocals of each other. The negative reciprocal of any nonzero number a is $-\frac{1}{a}$. For example, the negative reciprocal of 2 is $-\frac{1}{2}$ and the negative reciprocal of $-\frac{3}{4}$ is $\frac{4}{3}$. Write an equation of the line perpendicular to the graph of $y = 2x + 4$ that passes through the point $(2, 0)$.

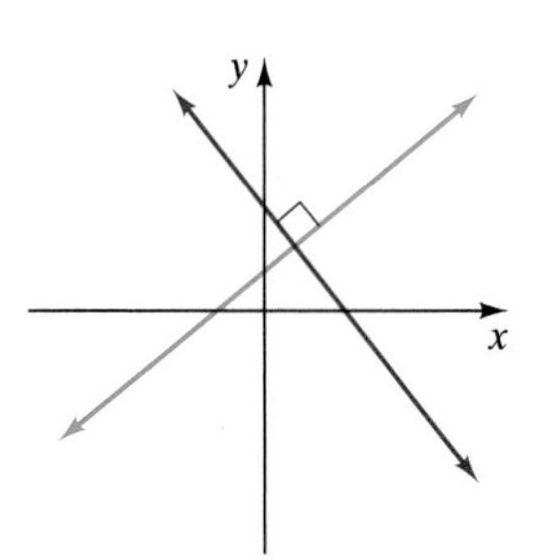

Group Activity

Discuss and answer Exercises 71–73 as a group, according to the instructions.

71. Consider the equation $-3x + 2y = 4$.

a) Group member 1: Explain how to graph this equation by plotting points. Then graph the equation by plotting points.

b) Group member 2: Explain how to graph this equation using the intercepts. Then graph the equation using the intercepts.

c) Group member 3: Explain how to graph this equation using the slope and y-intercept. Then graph the equation using the slope and y-intercept.

d) As a group, compare your graphs. Did you all obtain the same graph? If not, determine why.

72. The following graph shows the relationship between Fahrenheit temperature and Celsius temperature. Work parts **a)–e)** individually.

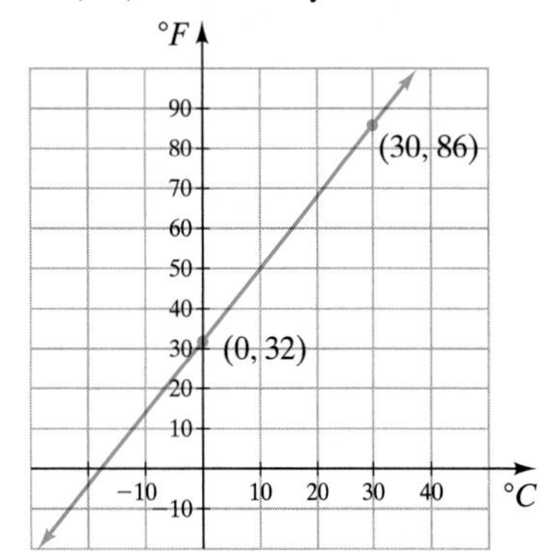

a) Determine the slope of the line.

b) Determine the equation of the line in slope–intercept form.

c) Use the equation (or formula) you obtained in part **b)** to find the Fahrenheit temperature when the Celsius temperature is 20°.

d) Use the graph to estimate the Celsius temperature when the Fahrenheit temperature is 100°.

e) Estimate the Celsius temperature that corresponds to a Fahrenheit temperature of 0°.

f) Compare your results with the other members of your group. You should all have the same results. If not, determine why and correct any wrong answers.

73. Determine the equation of the straight line that intersects the greatest number of shaded points on the following graph.

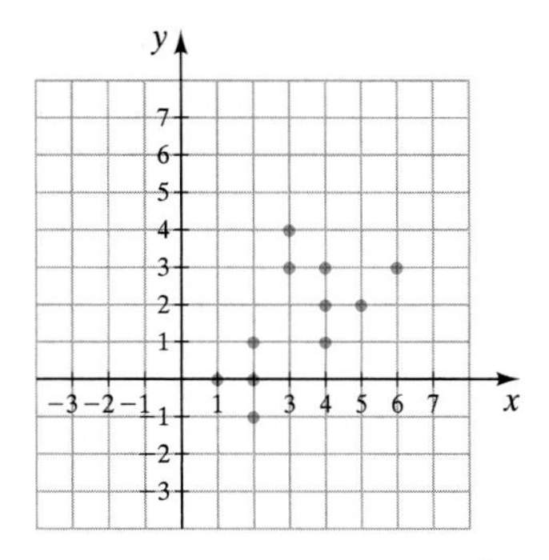

Cumulative Review Exercises

[1.5] **74.** Insert either >, <, or = in the shaded area to make the statement true: $|-4|$ ▭ $|-6|$.

[1.8] *Indicate whether each statement is true or false.*

75. The product of two negative numbers is always a negative number.

76. The sum of two negative numbers is always a negative number.

77. The difference of two negative numbers is always a negative number.

78. The quotient of two negative numbers is always a negative number.

[1.9] **79.** Evaluate 5^3.

4.5 GRAPHING LINEAR INEQUALITIES

1) **Graph linear inequalities in two variables.**

SSM VIDEO 4.5 CD Rom

1) Graph Linear Inequalities in Two Variables

A **linear inequality** results when the equal sign in a linear equation is replaced with an inequality sign.

Examples of Linear Inequalities in Two Variables

$3x + 2y > 4$	$-x + 3y < -2$
$-x + 4y \geq 3$	$4x - y \leq 4$

To Graph a Linear Inequality in Two Variables

1. Replace the inequality symbol with an equal sign.
2. Draw the graph of the equation in step 1. If the original inequality contained the symbol $\geq$ or $\leq$, draw the graph using a solid line. If the original inequality contained the symbol $>$ or $<$, draw the graph using a dashed line.
3. Select any point not on the line and determine whether this point is a solution to the original inequality. If the selected point is a solution, shade the region on the side of the line containing this point. If the selected point does not satisfy the inequality, shade the region on the side of the line not containing this point.

EXAMPLE 1 Graph the inequality $y < 2x - 4$.

Solution First we graph the equation $y = 2x - 4$ (Fig. 4.51). Since the original inequality contains an is less-than sign, $<$, we use a dashed line when drawing the graph. The dashed line indicates that the points on this line are not solutions to the inequality $y < 2x - 4$.

Next we select a point not on the line and determine whether this point satisfies the inequality. Often the easiest point to use is the origin, $(0, 0)$. In the check we will use the symbol $\overset{?}{<}$ until we determine whether the statement is true or false.

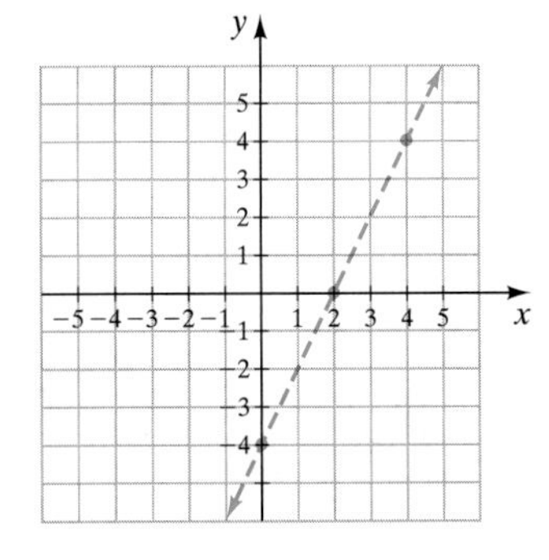

FIGURE 4.51

Check:

$$y < 2x - 4$$

$$0 \overset{?}{<} 2(0) - 4$$

$$0 \overset{?}{<} 0 - 4$$

$$0 < -4 \qquad \textit{False}$$

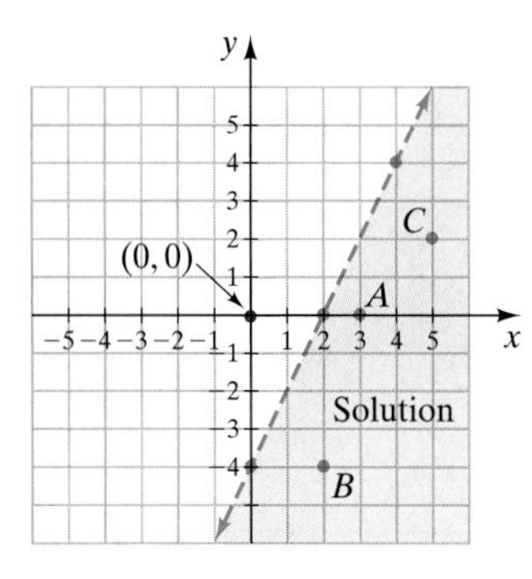

FIGURE 4.52

Since 0 is not less than −4, the point (0, 0) does not satisfy the inequality. The solution will therefore be all the points on the opposite side of the line from the point (0, 0). We shade this region (Fig. 4.52).

Every point in the shaded region satisfies the given inequality. Let's check a few selected points A, B, and C.

Point A	Point B	Point C
(3, 0)	(2, −4)	(5, 2)
$y < 2x - 4$	$y < 2x - 4$	$y < 2x - 4$
$0 \overset{?}{<} 2(3) - 4$	$-4 \overset{?}{<} 2(2) - 4$	$2 \overset{?}{<} 2(5) - 4$
$0 < 2$ *True*	$-4 < 0$ *True*	$2 < 6$ *True*

All points in the shaded region in Figure 4.52 satisfy the inequality $y < 2x - 4$. The points in the unshaded region to the left of the dashed line would satisfy the inequality $y > 2x - 4$.

NOW TRY EXERCISE 11

EXAMPLE 2 Graph the inequality $y \geq -\frac{1}{2}x$.

Solution Graph the equation $y = -\frac{1}{2}x$. Since the inequality symbol is ≥, we use a solid line to indicate that the points on the line are solutions to the inequality (Fig. 4.53). Since the point (0, 0) is on the line, we cannot select it as our test point. Let's select the point (3, 1).

$$y \geq -\frac{1}{2}x$$

$$1 \overset{?}{\geq} -\frac{1}{2}(3)$$

$$1 \geq -\frac{3}{2} \quad \textit{True}$$

Since the ordered pair (3, 1) satisfies the inequality, every point on the same side of the line as (3, 1) will also satisfy the inequality $y \geq -\frac{1}{2}x$. We shade this region (Fig. 4.54). Every point in the shaded region as well as every point on the line satisfies the inequality.

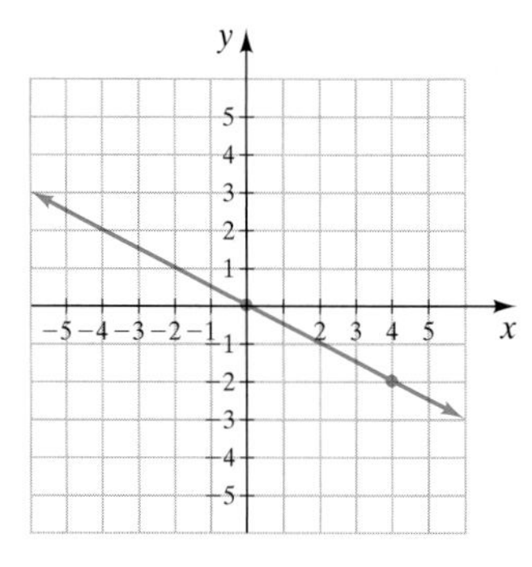

FIGURE 4.53

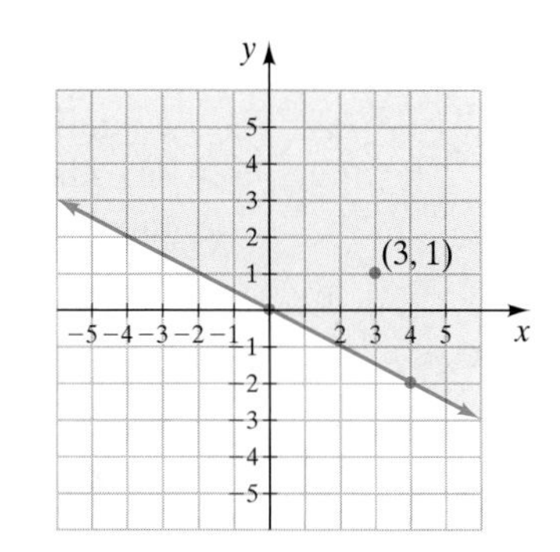

FIGURE 4.54

NOW TRY EXERCISE 9

In some of the exercises you may need to solve the inequality for y before graphing. For example, to graph $-2x + y < -4$ you would solve the inequality for y to obtain $y < 2x - 4$. Then you would graph the inequality $y < 2x - 4$. Note that $y < 2x - 4$ was graphed in Figure 4.52.

Exercise Set 4.5

Concept/Writing Exercises

1. When graphing inequalities that contain either $\le$ or $\ge$, explain why the points on the line will be solutions to the inequality.

2. When graphing inequalities that contain either $<$ or $>$, explain why the points on the line will not be solutions to the inequality.

3. How do the graphs of $2x + 3y > 6$ and $2x + 3y < 6$ differ?

4. How do the graphs of $4x - 3y < 6$ and $4x - 3y \le 6$ differ?

Practice the Skills

Graph each inequality.

5. $y > -3$
6. $x > 3$
7. $x \ge \frac{5}{2}$
8. $y < x$
9. $y \ge 2x$
10. $y > -2x$
11. $y < x - 4$
12. $y < 2x + 1$
13. $y < -3x + 4$
14. $y \ge 2x + 4$
15. $y \ge \frac{1}{2}x - 4$
16. $y > -\frac{x}{2} + 2$
17. $y \le \frac{1}{3}x + 3$
18. $y > \frac{1}{2}x - 2$
19. $3x + y \le 5$
20. $3x - 2 < y$
21. $2x + y \le 3$
22. $3y > 2x - 3$
23. $y - 4 \le -x$
24. $4x - 2y \le 6$

Problem Solving

25. Determine whether (4, 2) is a solution to each inequality.

 a) $2x + 4y < 16$
 b) $2x + 4y > 16$
 c) $2x + 4y \ge 16$
 d) $2x + 4y \le 16$

26. Determine whether $(-3, 5)$ is a solution to each inequality.

 a) $-2x + 3y < 9$
 b) $-2x + 3y > 9$
 c) $-2x + 3y \ge 9$
 d) $-2x + 3y \le 9$

27. If an ordered pair is not a solution to the inequality $ax + by < c$, must the ordered pair be a solution to $ax + by > c$? Explain.

28. If an ordered pair is not a solution to the inequality $ax + by \le c$, must the ordered pair be a solution to $ax + by > c$? Explain.

29. If an ordered pair is a solution to $ax + by > c$, is it possible for the ordered pair to be a solution to $ax + by \le c$? Explain.

30. Is it possible for an ordered pair to be a solution to both $ax + by < c$ and $ax + by > c$? Explain.

Challenge Problems

A toy company must ship x toy cars to one outlet and y toy cars to a second outlet. The maximum number of toy cars that the manufacturer can produce and ship is 200. We can represent this situation with the inequality $x + y \le 200$. Use this illustration as an aid to work Exercise 31.

31. An auto dealer wishes to sell x cars and y trucks this year, and he needs to sell a total of at least 100 vehicles.

 a) Represent this situation as an inequality.
 b) Graph the inequality in the first quadrant only, that is, where $x \ge 0$ and $y \ge 0$.

32. Which of the following inequalities have the same graphs? Explain how you determined your answer.

 a) $2x - y > 4$ **b)** $-2x + y < -4$
 c) $y < 2x - 4$ **d)** $-2y + 4x < -8$

Group Activity

Discuss and answer Exercises 33 and 34 as a group.

33. Determine whether the given phrase means: less than, less than or equal to, greater than, or greater than or equal to.

a) no more than **b)** no less than

c) at most **d)** at least

34. Consider the two inequalities $2x + 1 > 5$ and $2x + y > 5$.

a) How many variables does the inequality $2x + 1 > 5$ contain?

b) How many variables does the inequality $2x + y > 5$ contain?

c) What is the solution to $2x + 1 > 5$? Indicate the solution on a number line.

d) Graph $2x + y > 5$.

Cumulative Review Exercises

[1.4] **35.** Consider the set of numbers

$$\left\{2, -5, 0, \sqrt{7}, \frac{2}{5}, -6.3, \sqrt{3}, -\frac{23}{34}\right\}.$$

List those that are **a)** natural numbers; **b)** whole numbers; **c)** rational numbers; **d)** irrational numbers; **e)** real numbers.

[1.8] **36. a)** To what is $\frac{0}{1}$ equal?

b) How do we refer to an expression like $\frac{1}{0}$?

[1.9] **37.** Give the order of operations to be followed when evaluating a mathematical expression.

[2.5] **38.** Solve the equation $2(x + 3) + 2x = x + 4$.

4.6 FUNCTIONS

1. Find the domain and range of a relation.
2. Recognize functions.
3. Evaluate functions.
4. Graph linear functions.

In this section we introduce relations and functions. As you will learn shortly, a function is a special type of relation. Functions are a common thread in mathematics courses from algebra through calculus. In this section we give an informal introduction to relations and functions.

1 Find the Domain and Range of a Relation

First we will discuss **relations.**

Definition A **relation** is any set of ordered pairs.

Since a relation is *any* set of points, *every graph will represent a relation.*

Examples of Relations

$$\{(3, 5), (4, 6), (5, 9), (7, 12)\}$$

$$\{(1, 2), (2, 2), (3, 2), (4, 2)\}$$

$$\{(3, 2), (3, 3), (3, 4), (3, 5), (3, 6)\}$$

In the ordered pair (x, y), the x and y are called the **components of the ordered pair**. The **domain** of a relation is the set of *first components* in the set of ordered pairs. For example,

Relation	Domain
$\{(3, 5), (4, 6), (5, 9), (7, 12)\}$	$\{3, 4, 5, 7\}$
$\{(1, 2), (2, 2), (3, 2), (4, 2)\}$	$\{1, 2, 3, 4\}$
$\{(3, 2), (3, 3), (3, 4), (3, 5), (3, 6)\}$	$\{3\}$

The **range** of a relation is the set of *second components* in the set of ordered pairs. For example,

Relation	Range
$\{(3, 5), (4, 6), (5, 9), (7, 12)\}$	$\{5, 6, 9, 12\}$
$\{(1, 2), (2, 2), (3, 2), (4, 2)\}$	$\{2\}$
$\{(3, 2), (3, 3), (3, 4), (3, 5), (3, 6)\}$	$\{2, 3, 4, 5, 6\}$

In relations, the sets can contain elements other than numbers. For example,

Relation

{(Carol, Seat 1), (Mary, Seat 2), (John, Seat 3), (Olonso, Seat 4)}

Domain

{Carol, Mary, John, Olonso}

Range

{Seat 1, Seat 2, Seat 3, Seat 4}

Figure 4.55 illustrates the relation between the person and the seat number.

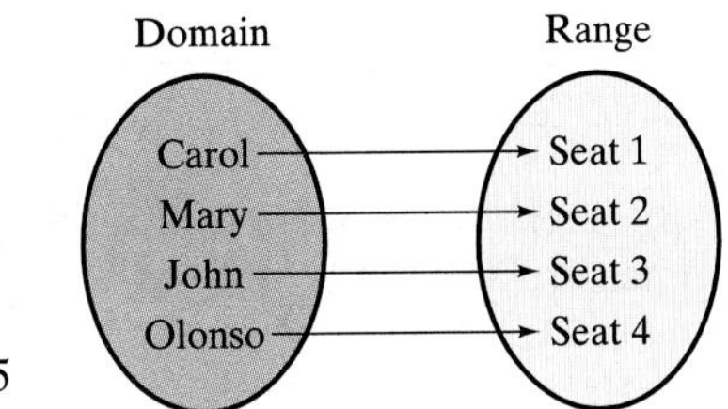

FIGURE 4.55

2) Recognize Functions

Now we are ready to discuss functions. Consider the relation shown in Figure 4.55. Notice that each member in the domain corresponds with exactly one member of the range. That is, each person is assigned to exactly one seat. This is an example of a **function**.

Definition

A **function** is a set of ordered pairs in which each first component corresponds to exactly one second component.

Since a function is a special type of relation, our discussion of domain and range applies to functions. In the definition of a function, the set of first components represents the domain of the function and the set of second components represents the range of the function.

EXAMPLE 1 Consider the relations in Figures 4.56a, 4.56b, and 4.56c. Which relations are functions?

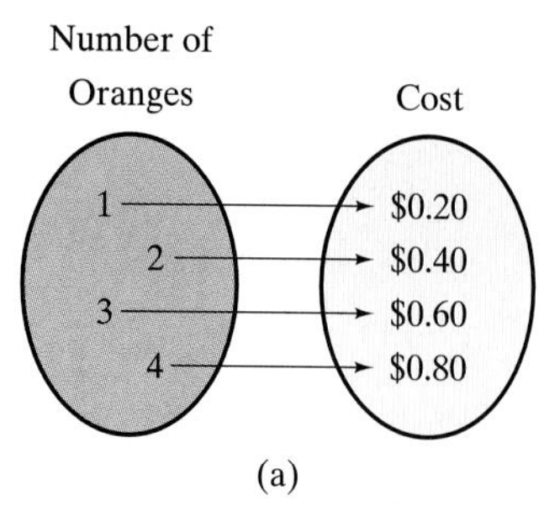

(a)

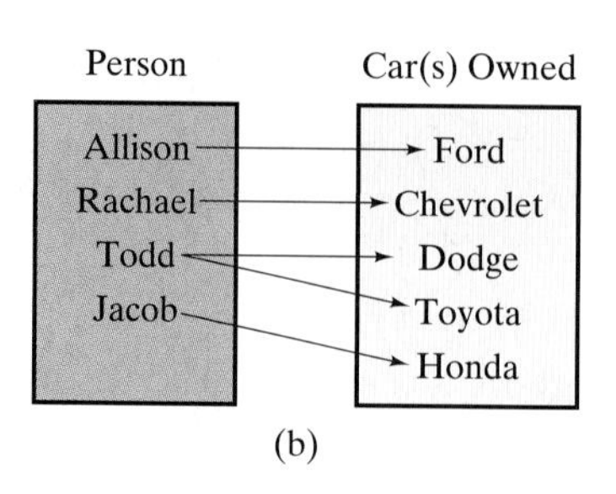

(b)

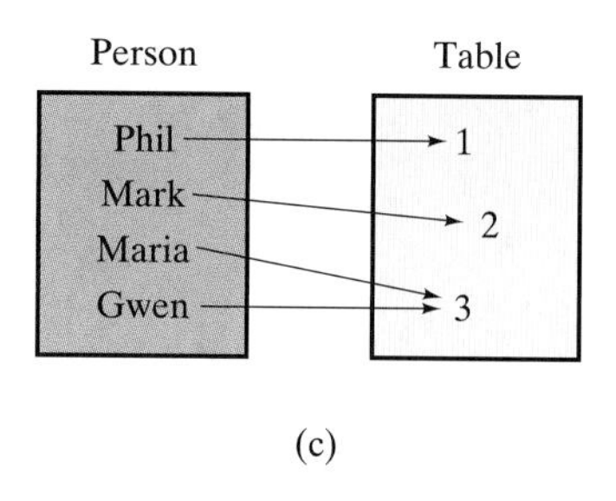

(c)

FIGURE 4.56

Solution **a)** If we wished, we could represent the information given in the figure as the following set of ordered pairs: $\{(1, \$0.20), (2, \$0.40), (3, \$0.60), (4, \$0.80)\}$. Notice that each first component corresponds to exactly one second component. Therefore, this relation is a function.

b) If we look at Figure 4.56b, we can see that Todd does *not* correspond to exactly one car. If we were to list the set of ordered pairs to represent this relation, the set would contain the ordered pairs (Todd, Dodge) and (Todd, Toyota). Therefore, each first component does not correspond to exactly one second component and this relation is not a function.

c) Although both Maria and Gwen share a table, each person corresponds to exactly one table. Therefore, this relation is a function.

NOW TRY EXERCISE 19

The functions given in Example 1**a)** and 1**c)** were determined by looking at correspondences in figures. Most functions have an infinite number of ordered pairs and are usually defined with an equation (or rule) that tells how to obtain the second component when you are given the first component. In Example 2, we determine a function from the information provided.

EXAMPLE 2 Assume each apple costs \$0.25. Write a function to determine the cost, c, when n apples are purchased.

Solution When one apple is purchased, the cost is \$0.25. When two apples are purchased, the cost is 2(\$0.25), and when n apples are purchased, the cost is $n(\$0.25)$ or $\$0.25n$. The function $c = 0.25n$ will give the cost, c, in dollars, when n apples are purchased. Note that for any value of n, there is exactly one value of c.

EXAMPLE 3 Determine whether the following sets of ordered pairs are functions.

a) $\{(4, 5), (3, 2), (-2, -3), (2, 5), (1, 6)\}$

b) $\{(4, 5), (3, 2), (-2, -3), (4, 1), (5, -2)\}$

Solution **a)** Since each first component corresponds with exactly one second component, this set of ordered pairs is a function.

b) The ordered pairs $(4, 5)$ and $(4, 1)$ contain the same first component. Therefore, each first component does not correspond to exactly one second component, and this set of ordered pairs is not a function.

In Figures 4.57a and 4.57b we plot the ordered pairs from Example 3**a)** and 3**b)**, respectively.

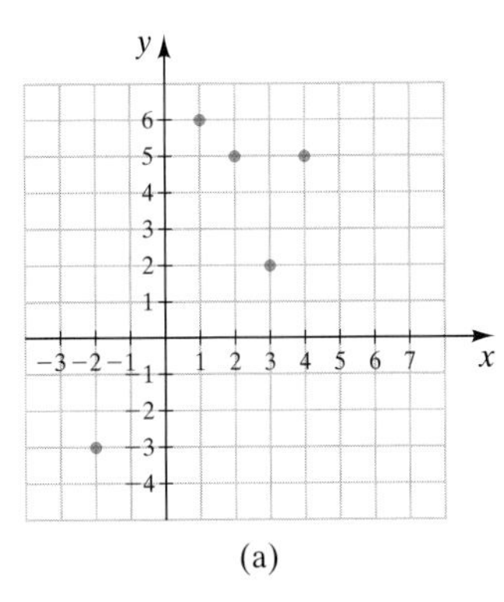

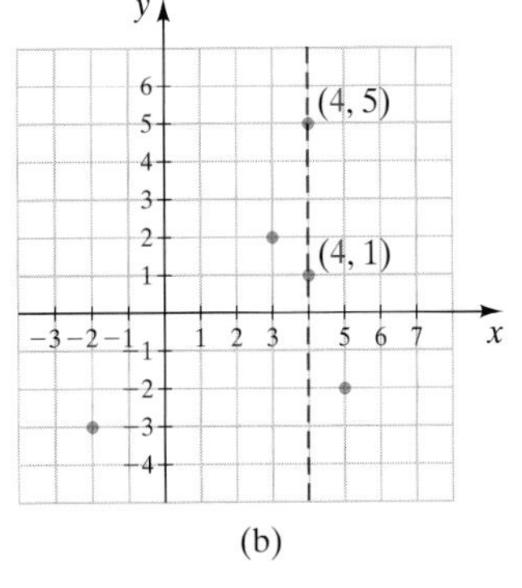

(a) First set of ordered pairs, Function

(b) Second set of ordered pairs, Not a function

FIGURE 4.57

Consider Figure 4.57a. If a vertical line is drawn through each point, no vertical line intersects more than one point. This indicates that no two ordered pairs have the same first (or x) coordinate and that each value of x in the domain corresponds to exactly one value of y in the range. Therefore, this set of points represents a function.

Now look at Figure 4.57b. If a vertical line is drawn through each point, one vertical line passes through two points. The dashed red vertical line intersects both $(4, 5)$ and $(4, 1)$. Each element in the domain *does not* correspond to exactly one element in the range. The number 4 in the domain corresponds to two numbers, 5 and 1, in the range. Therefore, this set of ordered pairs does *not* represent a function.

To determine whether a graph represents a function, we can use the **vertical line test** just described.

Vertical Line Test

If a vertical line can be drawn through any part of a graph and the vertical line intersects another part of the graph, then each value of x does not correspond to exactly one value of y and the graph does not represent a function.

If a vertical line cannot be drawn to intersect the graph at more than one point, each value of x corresponds to exactly one value of y and the graph represents a function.

EXAMPLE 4 Using the vertical line test, determine whether the following graphs represent functions.

a)

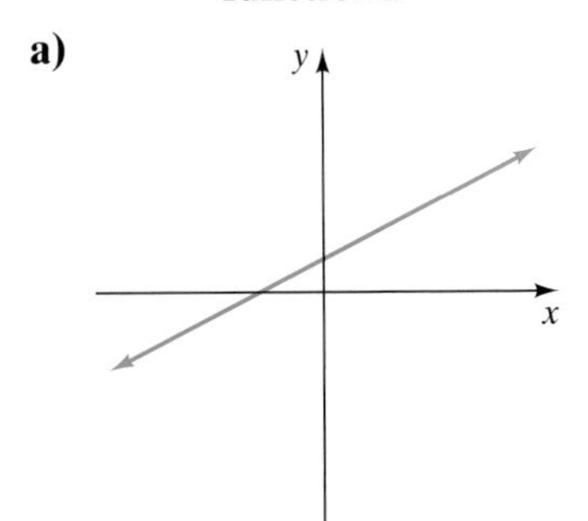

b)

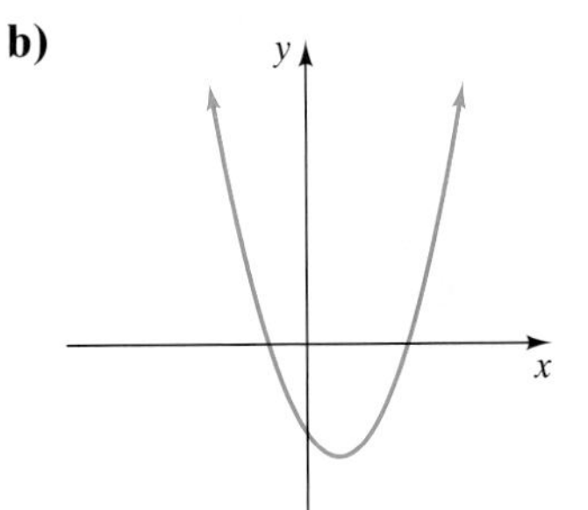

c)

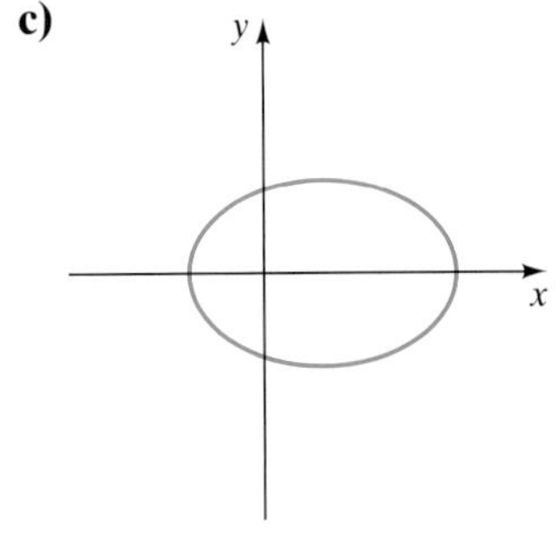

Solution

a)

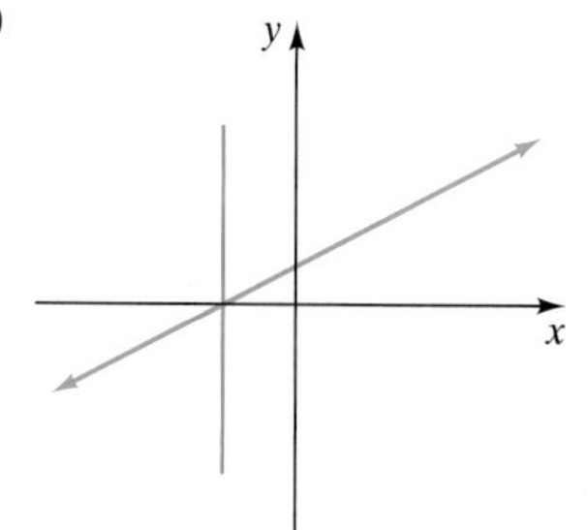

b)

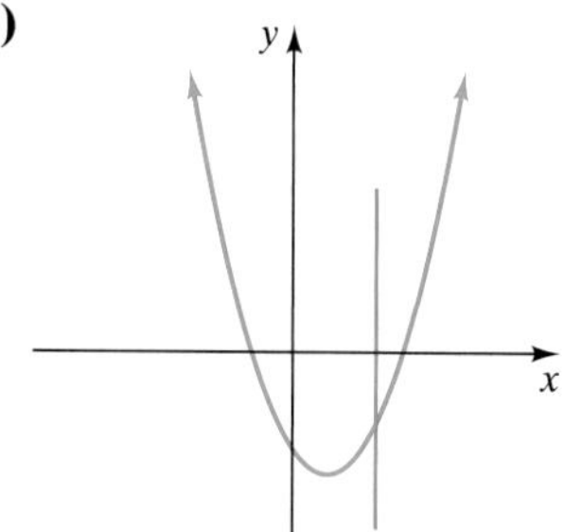

c)

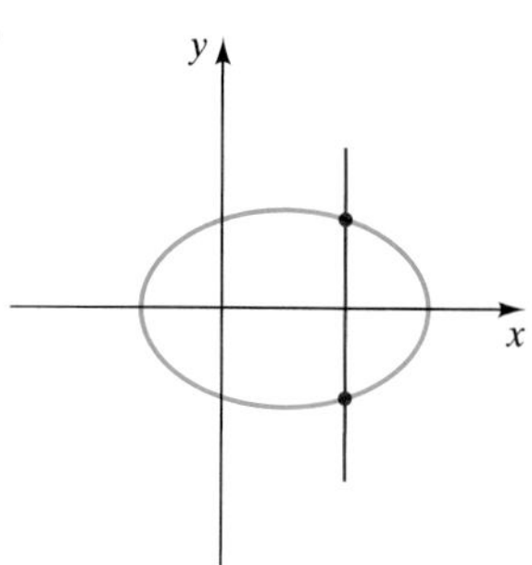

The graphs in parts **a)** and **b)** represent functions since it is not possible to draw a vertical line that intersects the graph at more than one point. The graph in part **c)** does not represent a function since a vertical line can be drawn to intersect the graph at more than one point.

NOW TRY EXERCISE 31

We see functions all around us. Consider the information provided in Table 4.1, which may apply to some salespeople. This table of values is a function because each amount of sales corresponds to exactly one income.

TABLE 4.1 Monthly Income

Sales (dollars)	Income (dollars)
0	$1200
$5000	$1500
$10,000	$1800
$15,000	$2100
$20,000	$2400
$25,000	$2700

Consider the graph shown in Figure 4.58. This graph represents a function. Notice that each year corresponds with a unique number of donors. Both graphs in Figure 4.59 represent functions. Each graph passes the vertical line test.

NOW TRY EXERCISE 53

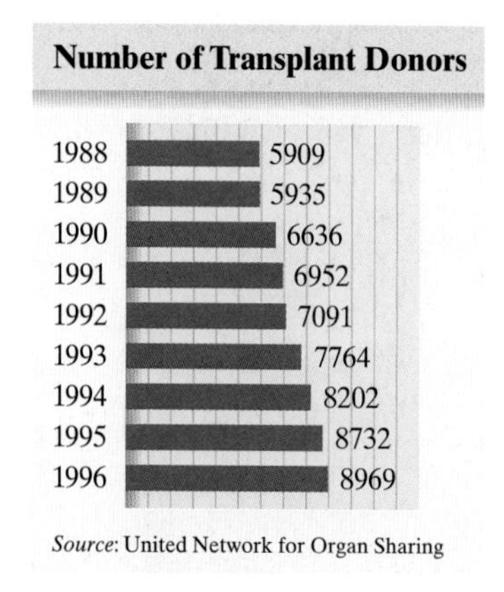

FIGURE 4.58

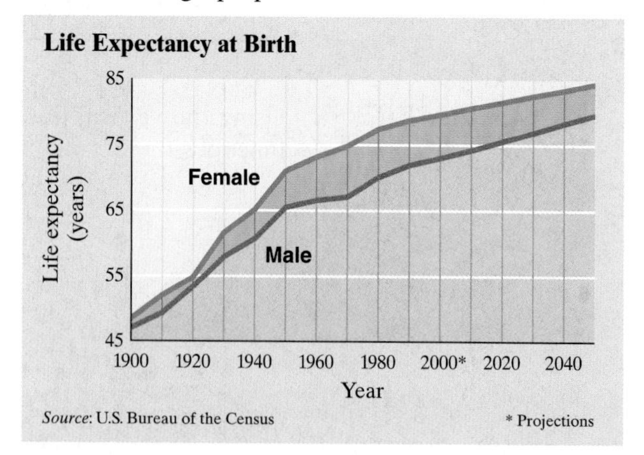

FIGURE 4.59

3) Evaluate Functions

When a function is represented by an equation, it is often convenient to use **function notation**, $f(x)$. If we were to graph $y = x + 2$, we would see it is a function since its graph passes the vertical line test. The value of y in the equation or function depends on the value of x. Therefore, we say that y is a function of x, and we can write $y = f(x)$. The notation $y = f(x)$ is used to show that y is a function of the variable x. If we wish, we can write

$$y = f(x) = x + 2 \qquad \text{or simply} \qquad f(x) = x + 2$$

The notation $f(x)$ is read "f of x" *and does not mean* f times x.

To evaluate a function for a specific value of x, we substitute that value for x everywhere the x appears in the function. For example, to evaluate the function $f(x) = x + 2$ at $x = 1$, we do the following:

$$f(x) = x + 2$$

$$f(1) = 1 + 2 = 3$$

Thus, when x is 1, $f(x)$ or y is 3.

When $x = 4$, $f(x)$ or $y = 6$, as illustrated below.

$$y = f(x) = x + 2$$

$$y = f(4) = 4 + 2 = 6$$

The notation $f(1)$ is read "f of 1" and $f(4)$ is read "f of 4."

EXAMPLE 5 For the function $f(x) = x^2 + 2x - 3$, find **a)** $f(4)$ and **b)** $f(-5)$. **c)** If $x = -1$, determine the value of y.

Solution **a)** Substitute 4 for each x in the function, and then evaluate.

$$f(x) = x^2 + 2x - 3$$

$$f(4) = 4^2 + 2(4) - 3$$

$$= 16 + 8 - 3 = 21$$

b)

$$f(x) = x^2 + 2x - 3$$

$$f(-5) = (-5)^2 + 2(-5) - 3$$

$$= 25 - 10 - 3 = 12$$

c) Since $y = f(x)$, we evaluate $f(x)$ at -1.

$$f(x) = x^2 + 2x - 3$$
$$f(-1) = (-1)^2 + 2(-1) - 3$$
$$= 1 - 2 - 3 = -4$$

NOW TRY EXERCISE 39 Thus, when $x = -1$, $y = -4$.

4) Graph Linear Functions

The graphs of all equations of the form $y = ax + b$ will be straight lines that are functions. Therefore, we may refer to equations of the form $y = ax + b$ as **linear functions**. Equations of the form $f(x) = ax + b$ are also linear functions since $f(x)$ is the same as y. We may graph linear functions as shown in Example 6.

EXAMPLE 6 Graph $f(x) = 2x + 4$.

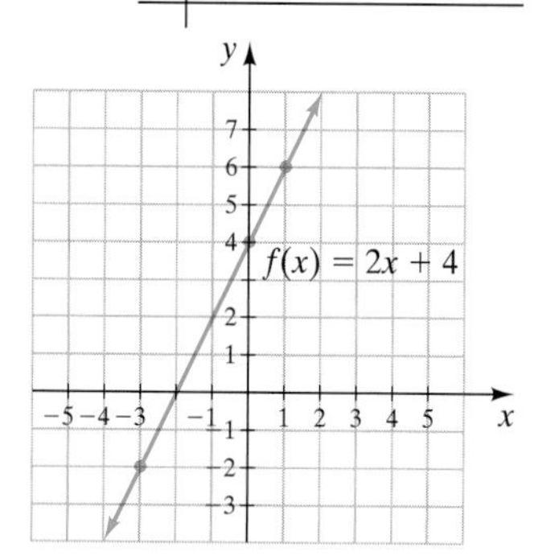

FIGURE 4.60

Solution Since $f(x)$ is the same as y, write $y = f(x) = 2x + 4$. Select values for x and find the corresponding values for y or $f(x)$.

	$y = f(x) = 2x + 4$
Let $x = -3$	$y = f(-3) = 2(-3) + 4 = -2$
Let $x = 0$	$y = f(0) = 2(0) + 4 = 4$
Let $x = 1$	$y = f(1) = 2(1) + 4 = 6$

x	y
−3	−2
0	4
1	6

NOW TRY EXERCISE 45

Now plot the points and draw the graph of the function (Fig. 4.60).

EXAMPLE 7 The weekly profit, p, of an ice skating rink is a function of the number of skaters per week, n. The function approximating the profit is $p = f(n) = 8n - 600$, where $0 \le n \le 400$.

a) Construct a graph showing the relationship between the number of skaters and the weekly profit.

b) Estimate the profit if there are 200 skaters in a given week.

Solution **a)** Select values for n, and find the corresponding values for p. Then draw the graph (Fig. 4.61). Notice there are no arrowheads on the line because the function is defined only for values of n between 0 and 400 inclusive.

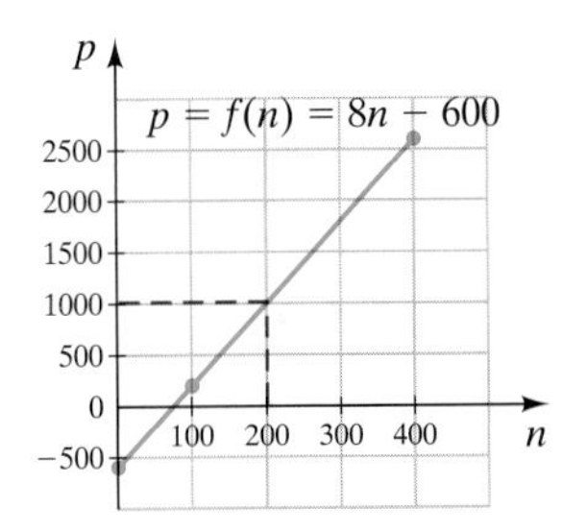

FIGURE 4.61

$$p = f(n) = 8n - 600$$

Let $n = 0$ $\quad p = f(0) = 8(0) - 600 = -600$

Let $n = 100$ $\quad p = f(100) = 8(100) - 600 = 200$

Let $n = 400$ $\quad p = f(400) = 8(400) - 600 = 2600$

n	p
0	−600
100	200
400	2600

NOW TRY EXERCISE 57

b) Using the red dashed line on the graph, we can see that if there are 200 skaters the weekly profit is \$1000.

Exercise Set 4.6

Concept/Writing Exercises

1. What is a relation?
2. Is every graph in the Cartesian coordinate system a relation? Explain.
3. What is a function?
4. Is every graph in the Cartesian coordinate system a function? Explain.
5. **a)** What is the domain of a relation or function?
   b) What is the range of a relation or function?
6. **a)** Is every relation a function?
   b) Is every function a relation? Explain your answer.
7. If two distinct ordered pairs in a relation have the same first coordinate, can the relation be a function? Explain.
8. In a function is it necessary for each value of y in the range to correspond to exactly one value of x in the domain? Explain.

Practice the Skills

Determine which of the relations are also functions. Give the domain and range of each relation or function.

9. $\{(5, 4), (2, 2), (3, 5), (1, 3), (4, 1)\}$
10. $\{(2, 1), (4, 0), (3, 5), (2, 2), (5, 1)\}$
11. $\{(6, -2), (3, 0), (1, 2), (1, 4), (2, 4), (7, 5)\}$
12. $\{(-2, 1), (1, -3), (3, 4), (4, 5), (-2, 0)\}$
13. $\{(5, 0), (3, -4), (0, -1), (3, 2), (1, 1)\}$
14. $\{(-2, 3), (-3, 4), (0, 3), (5, 2), (3, 5), (2, 5)\}$
15. $\{(3, 0), (0, -3), (1, 5), (1, 0), (1, 2)\}$
16. $\{(4, -3), (3, -7), (4, -9), (3, 5)\}$
17. $\{(0, 3), (1, 3), (2, 3), (3, 3), (4, 3)\}$
18. $\{(3, 5), (2, 4), (1, 0), (0, 1), (-1, 5)\}$

In the figures in Exercises 19–22, the domain and range of a relation are illustrated. ***a)*** *Construct a set of ordered pairs that represent the relation.* ***b)*** *Determine whether the relation is a function. Explain your answer.*

19.
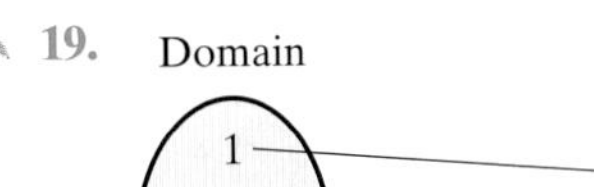

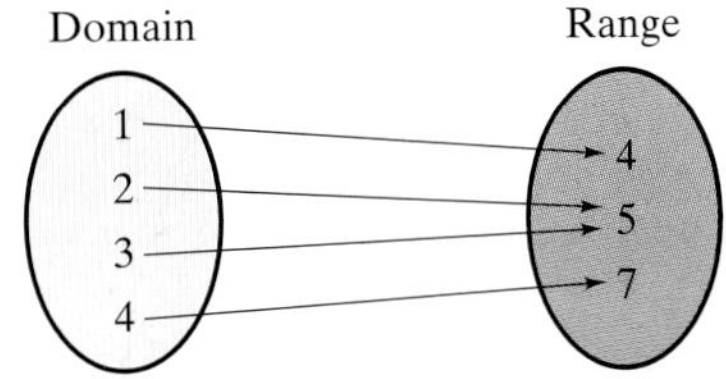

20.
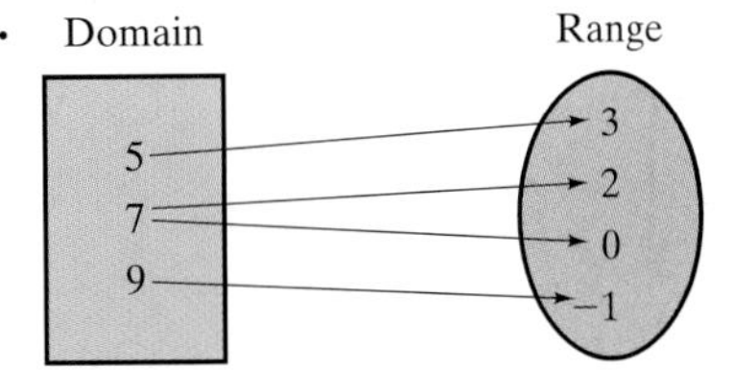

21.

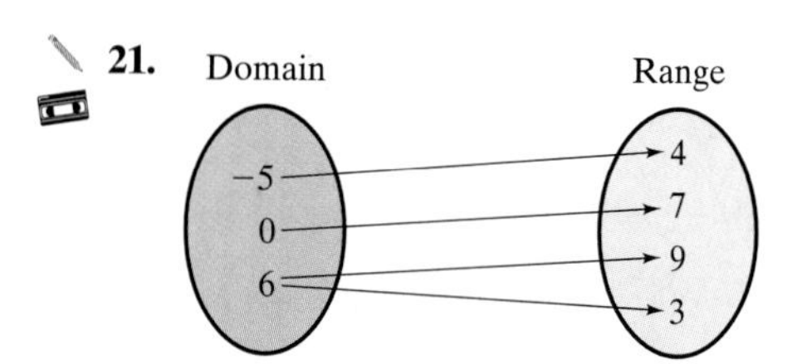

22.

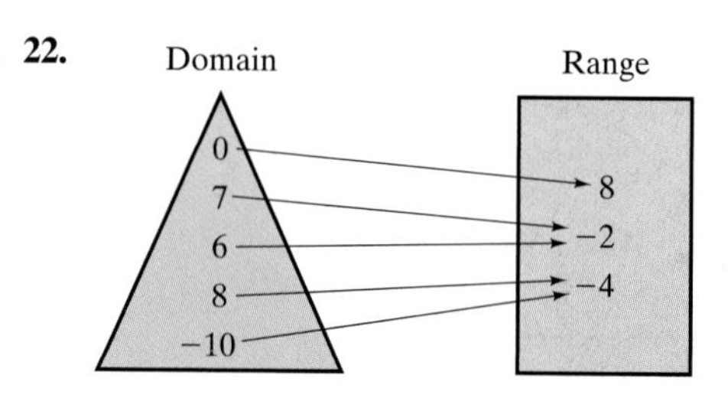

Use the vertical line test to determine whether each relation is also a function.

23.

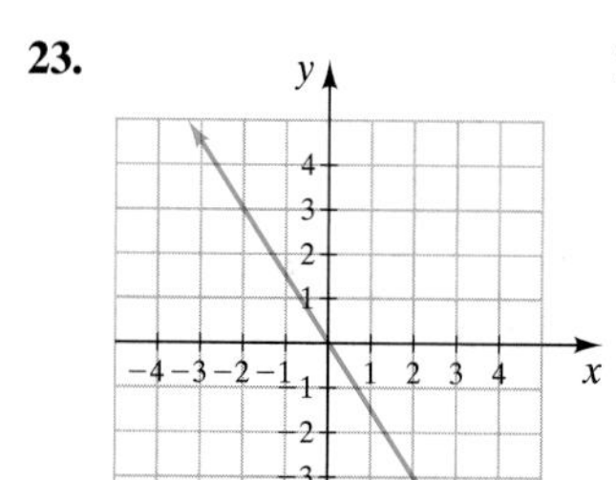

24.

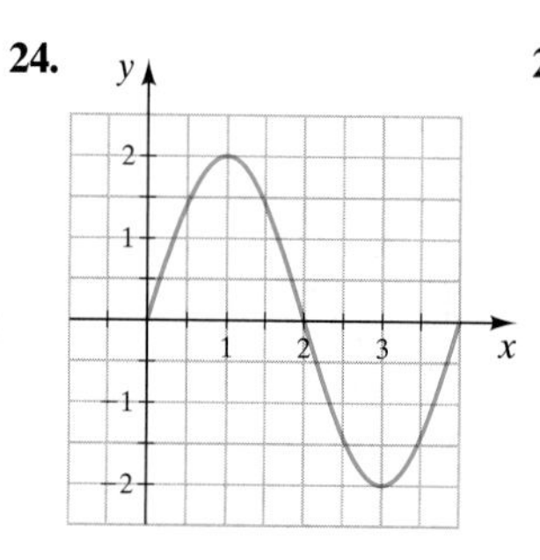

25.

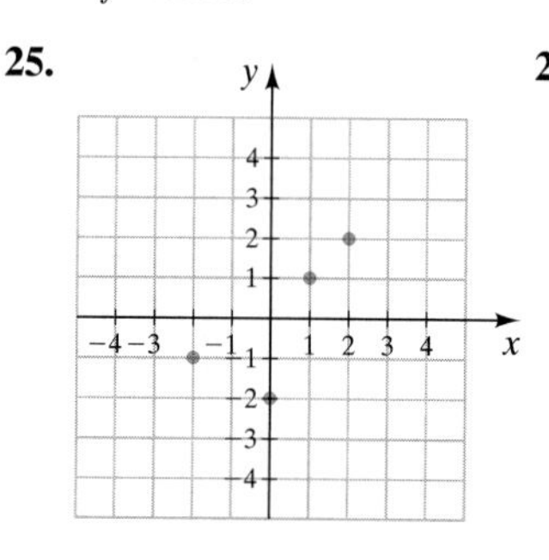

26.

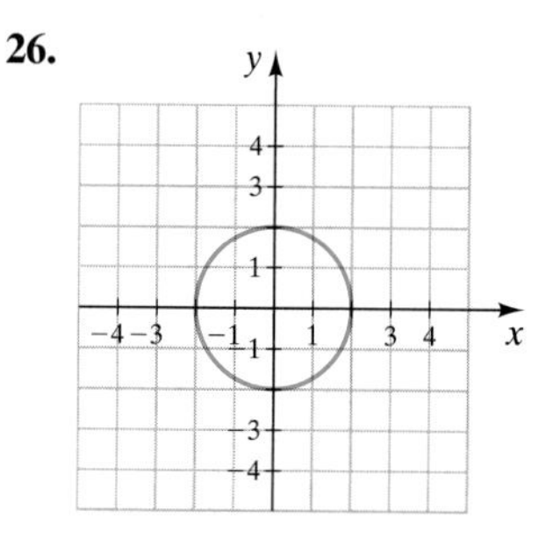

27.

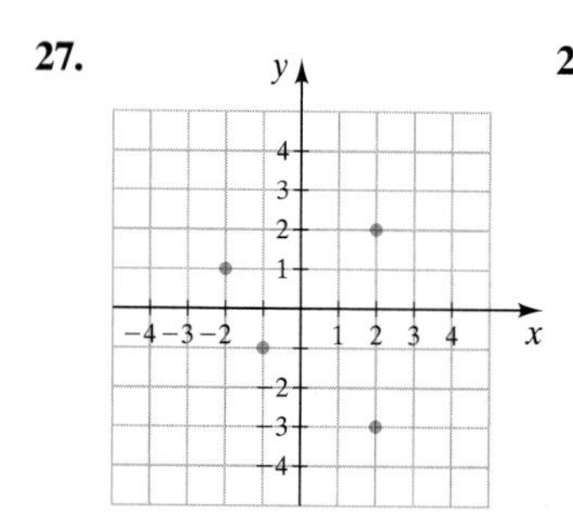

28.

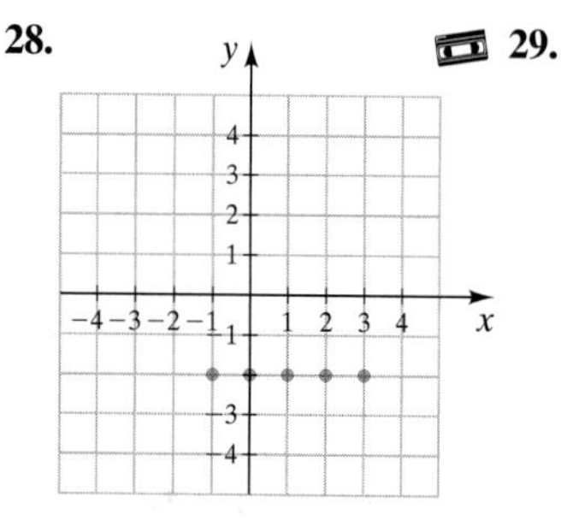

29.

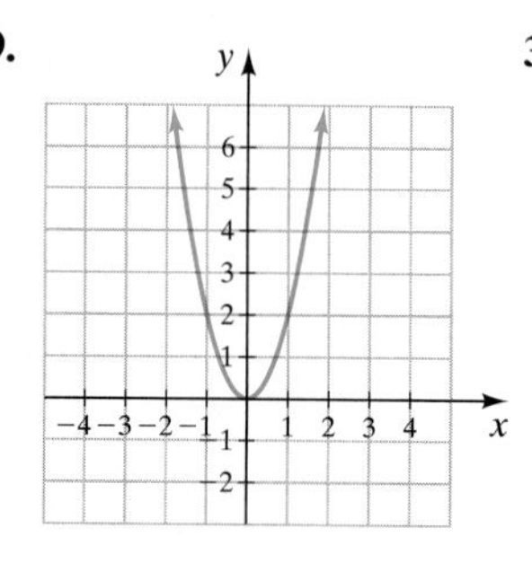

30.

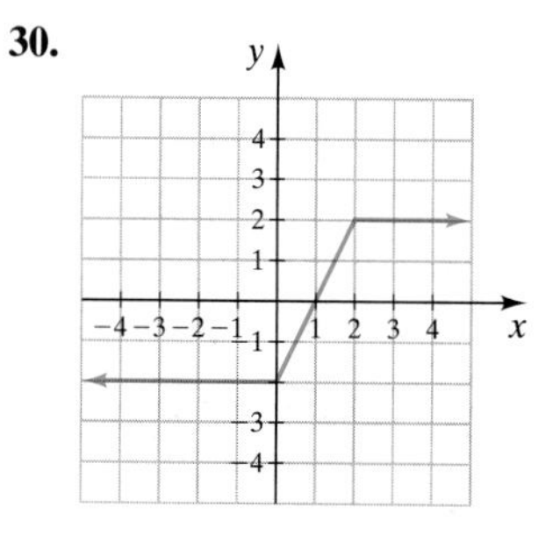

31.

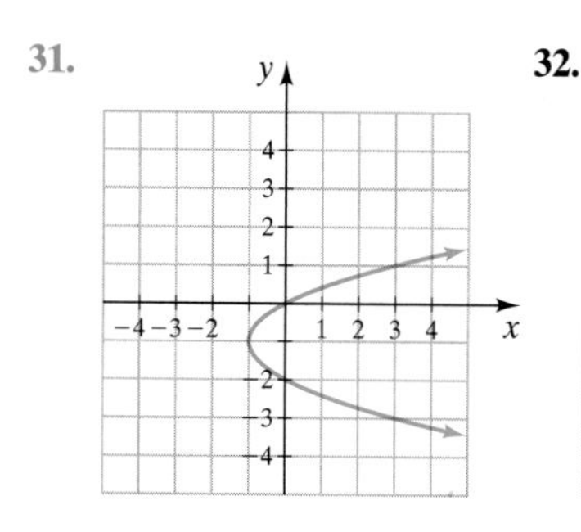

32.

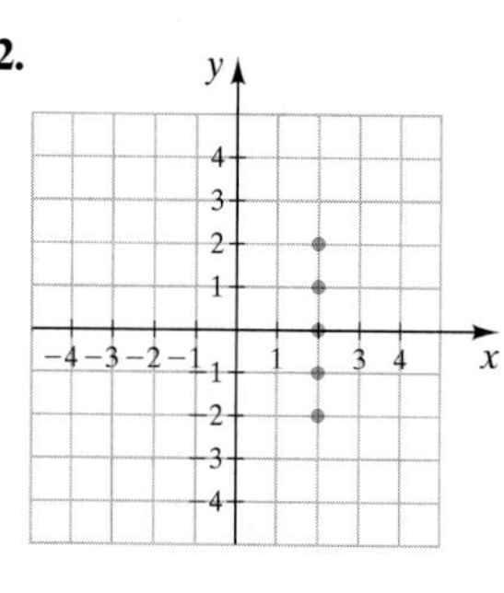

33.

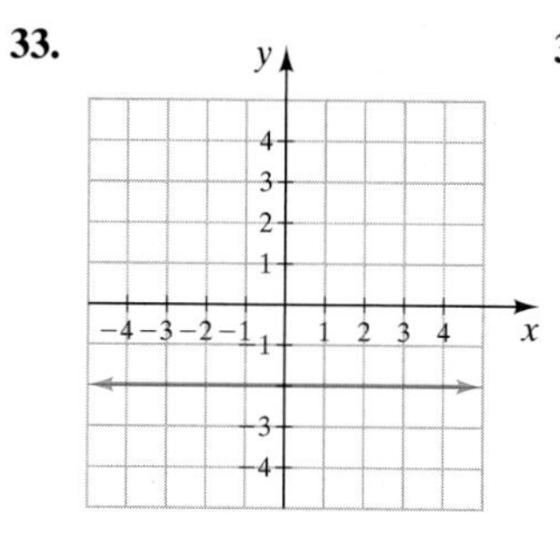

34.

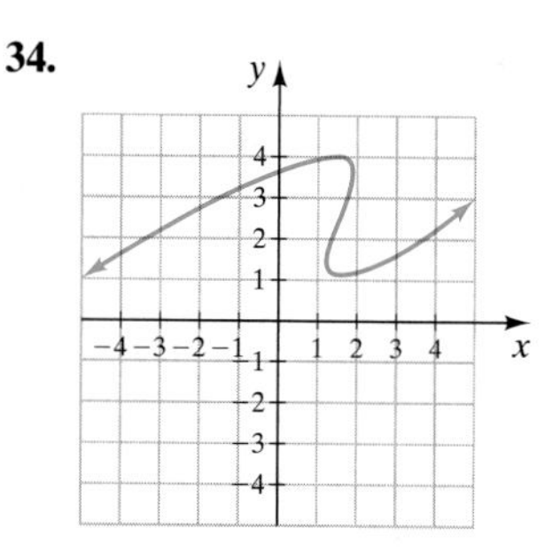

Evaluate each function at the indicated values.

35. $f(x) = 4x + 2$; find **a)** $f(3)$, **b)** $f(-1)$

36. $f(x) = -4x + 7$; find **a)** $f(0)$, **b)** $f(1)$

37. $f(x) = x^2 - 3$; find **a)** $f(3)$, **b)** $f(-2)$

38. $f(x) = 2x^2 + 3x - 4$; find **a)** $f(1)$, **b)** $f(-3)$

39. $f(x) = 2x^2 - x + 5$; find **a)** $f(0)$, **b)** $f(2)$

40. $f(x) = \frac{1}{2}x - 4$; find **a)** $f(10)$, **b)** $f(-4)$

41. $f(x) = \frac{x + 4}{2}$; find **a)** $f(2)$, **b)** $f(6)$

42. $f(x) = \frac{1}{2}x^2 + 6$; find **a)** $f(2)$, **b)** $f(-2)$

Graph each function.

43. $f(x) = x + 1$

44. $f(x) = -x + 4$

45. $f(x) = 2x - 1$

46. $f(x) = 4x + 2$

47. $f(x) = -2x + 4$

48. $f(x) = -x + 5$

49. $f(x) = -3x - 3$

50. $f(x) = -4x$

Problem Solving

51. If a relation consists of six ordered pairs and the domain of the relation consists of five values of x, can the relation be a function? Explain.

52. If a relation consists of six ordered pairs and the range of the relation consists of five values of y, can the relation be a function? Explain.

Are the following graphs functions? Explain.

53.

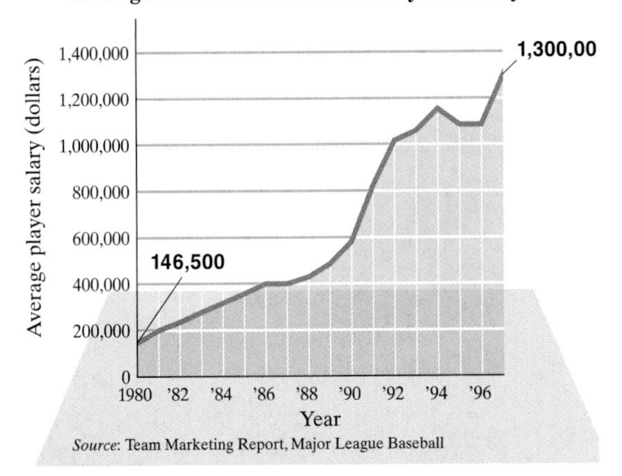

54.

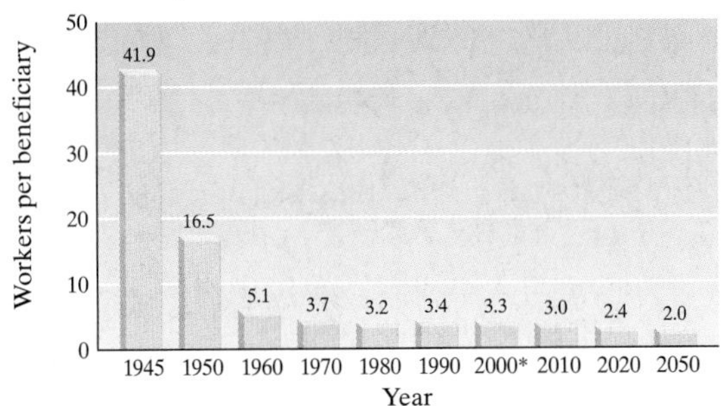

55. Shelia Becker is using a hose to fill her pool. The amount of water going into a swimming pool, w, from a hose delivering 2 gallons of water per minute can be found by using the function $w = 2t$, where t is the number of minutes.

a) Draw a graph of the function for times up to and including 60 minutes.

b) Estimate the number of gallons of water added to the pool if water is flowing from the hose for 20 minutes.

56. Rachel Falk and Tony Mathias go bike riding together. If they ride at a constant rate of 8 miles per hour, the distance they travel, d, can be found by the function $d = 8t$, where t is time in hours.

a) Draw a graph of the function for times up to and including 6 hours.

b) Estimate the distance they will travel if they ride for 2 hours.

57. The cost, c, in dollars, of repairing a highway can be estimated by the function $c = 2000 + 6000m$, where m is the number of miles to be repaired.

a) Draw a graph of the function for up to and including 6 miles.

b) Estimate the cost of repairing 2 miles of road.

58. The cost, c, in dollars, of a cross-country train trip can be estimated by the function $c = 50 + 0.15m$, where m is the distance traveled in miles.

a) Draw a graph of the function for up to and including 3000 miles traveled.

b) Estimate the cost of a 1000-mile trip.

59. A stock broker's commission, c, on stock trades is \$25 plus 2% of the sales value, s. Therefore, the broker's commission is a function of the sales, $c = 25 + 0.02s$.

a) Draw a graph illustrating the broker's commission on sales up to and including \$10,000.

b) If the sales value of a trade is \$8000, estimate the broker's commission.

60. A state's auto registration fee, f, is \$20 plus \$15 per 1000 pounds of the vehicle's gross weight. The registration fee is a function of the vehicle's weight, $f = 20 + 0.015w$, where w is the weight of the vehicle in pounds.

a) Draw a graph of the function for vehicle weights up to and including 10,000 pounds.

b) Estimate the registration fee of a vehicle whose gross weight is 4000 pounds.

61. A new singing group, Three Forks and a Spoon, sign a recording contract with the Smash Record label. Their contract provides them with a signing bonus of \$10,000, plus an 8% royalty on the sales, s, of their new record, *There's Mud in Your Eye!* Their income, i, is a function of their sales, $i = 10{,}000 + 0.08s$.

a) Draw a graph of the function for sales of up to and including \$100,000.

b) Estimate their income if their sales are \$20,000.

62. A monthly electric bill, m, in dollars, consists of a \$20 monthly fee plus \$0.07 per kilowatt-hour, k, of electricity used. The amount of the bill is a function of the kilowatt-hours used, $m = 20 + 0.07k$.

a) Draw a graph for up to and including 3000 kilowatt-hours of electricity used in a month.

b) Estimate the bill if 1500 kilowatt-hours of electricity are used.

Consider the following graphs. Recall from Section 2.5 that an open circle at the end of a line segment means that the endpoint is not included in the answer. A solid circle at the end of a line segment indicates that the endpoint is included in the answer. Determine whether the following graphs are functions. Explain your answer.

63.

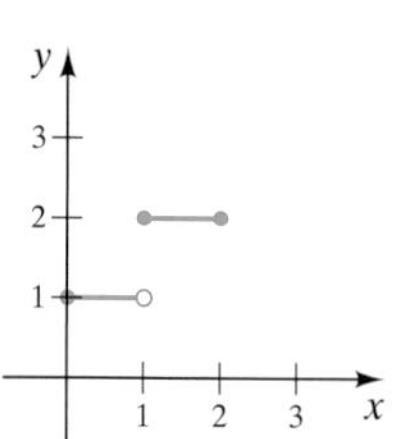

64.

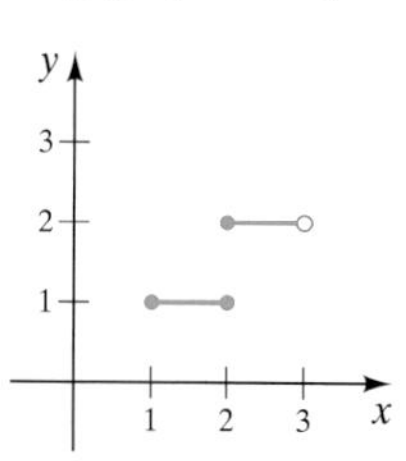

65.

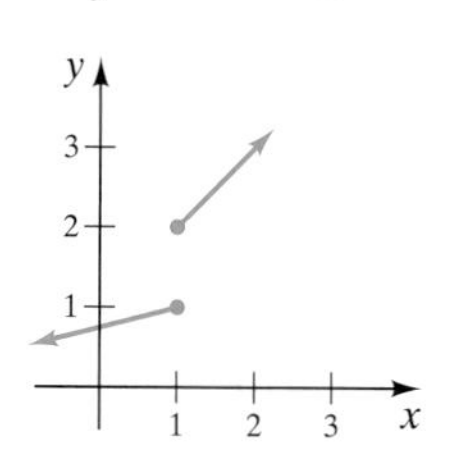

66.

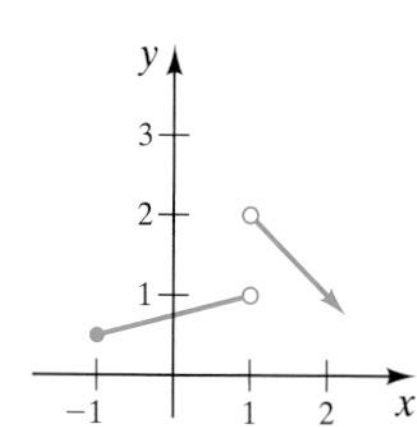

Challenge Problems

67. $f(x) = \frac{1}{2}x^2 - 3x + 5$; find **a)** $f\left(\frac{1}{2}\right)$, **b)** $f\left(\frac{2}{3}\right)$, **c)** $f(0.2)$

68. $f(x) = x^2 + 2x - 3$; find **a)** $f(1)$, **b)** $f(2)$, **c)** $f(a)$. Explain how you determined your answer to part **c)**.

Group Activity

Discuss and answer Exercises 69 and 70 as a group.

69. Submit three real-life examples (different from those already given) of a quantity that is a function of another. Write each as a function, and indicate what each variable represents.

70. In April, 1999 the cost of mailing a first class letter was 33 cents for the first ounce and 22 cents for each additional ounce. A graph showing the cost of mailing a letter first class is pictured below.

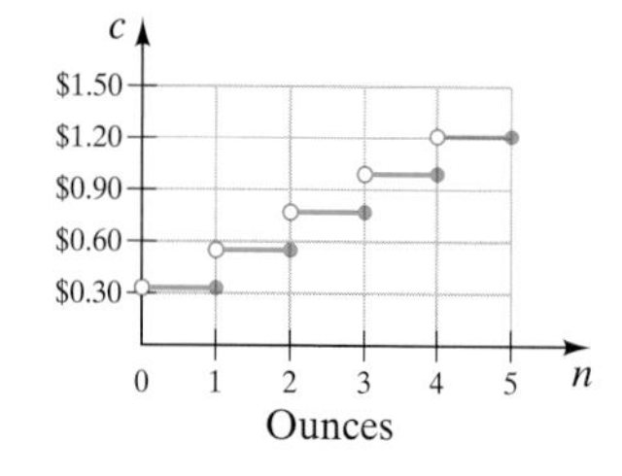

a) Does this graph represent a function? Explain your answer.

b) From the graph, estimate the cost of mailing a 4-ounce package first class.

c) Determine the exact cost of mailing a 4-ounce package first class.

d) From the graph, estimate the cost of mailing a 3.6-ounce package first class.

e) Determine the exact cost of mailing a 3.6-ounce package first class.

Cumulative Review Exercises

[1.3] **71.** Evaluate $\frac{5}{9} - \frac{3}{7}$.

[1.10] **72.** Name each illustrated property.

a) $2 \cdot 5 = 5 \cdot 2$

b) $(x + 2) + 3 = x + (2 + 3)$

c) $2(x + 5) = 2x + 2 \cdot 5$

[2.5] **73.** Solve the equation $2x - 3(x + 2) = 8$.

[3.3] **74.** The cost of a taxi ride is \$2.00 for the first mile and \$1.50 for each additional mile or part thereof. Find the maximum distance Andrew Collins can ride in the taxi if he has only \$20.

SUMMARY

Key Words and Phrases

4.1
Cartesian coordinate system
Collinear
Coordinates
Graph
Linear equation in two variables
Ordered pairs
Origin
Quadrants
Standard form of a linear equation
x-axis
y-axis

4.2
Horizontal line
Vertical line
x-intercept
y-intercept

4.3
Negative slope
Positive slope
Slope of a line

4.4
Parallel lines
Perpendicular lines
Point–slope form
Slope–intercept form

4.5
Linear inequality

4.6
Components of an ordered pair
Domain
Function
Function notation
Linear function
Range
Relation
Vertical line test

IMPORTANT FACTS

To Find the *x*-Intercept: Set $y = 0$ and find the corresponding value of x.

To Find the *y*-Intercept: Set $x = 0$ and find the corresponding value of y.

Slope of Line, *m*, through points (x_1, y_1) and (x_2, y_2):

$$m = \frac{y_2 - y_1}{x_2 - x_1}$$

Methods of Graphing
$y = 3x - 4$

By Plotting Points

Using Intercepts

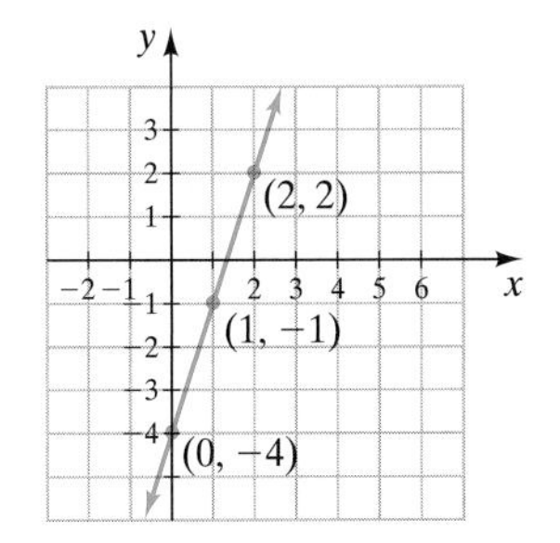

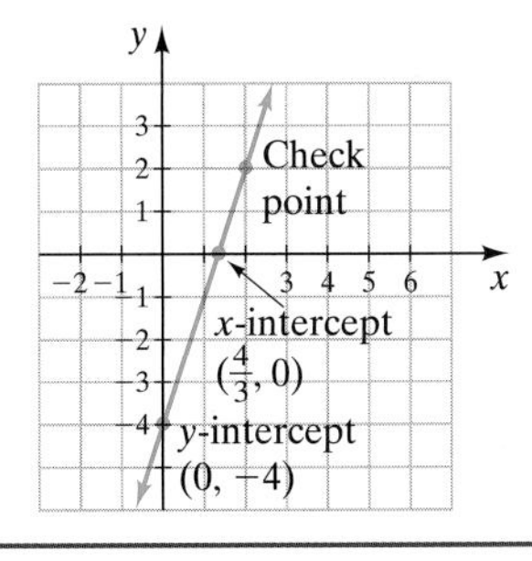

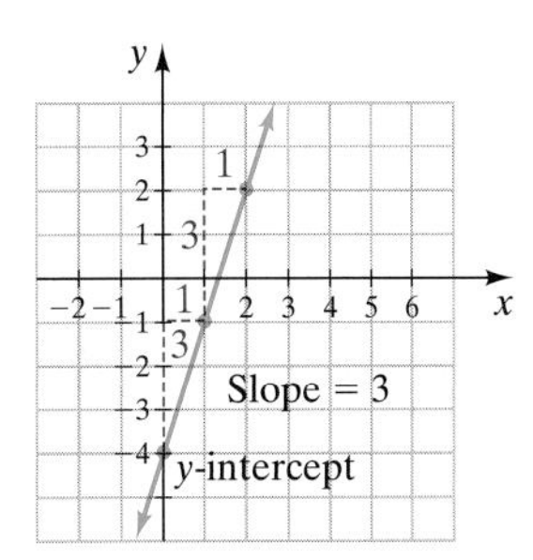

Standard Form of a Linear Equation: $ax + by = c$.

Slope–Intercept Form of a Linear Equation: $y = mx + b$.

Point–Slope Form of a Linear Equation: $y - y_1 = m(x - x_1)$.

continued on next page

Review of Slope

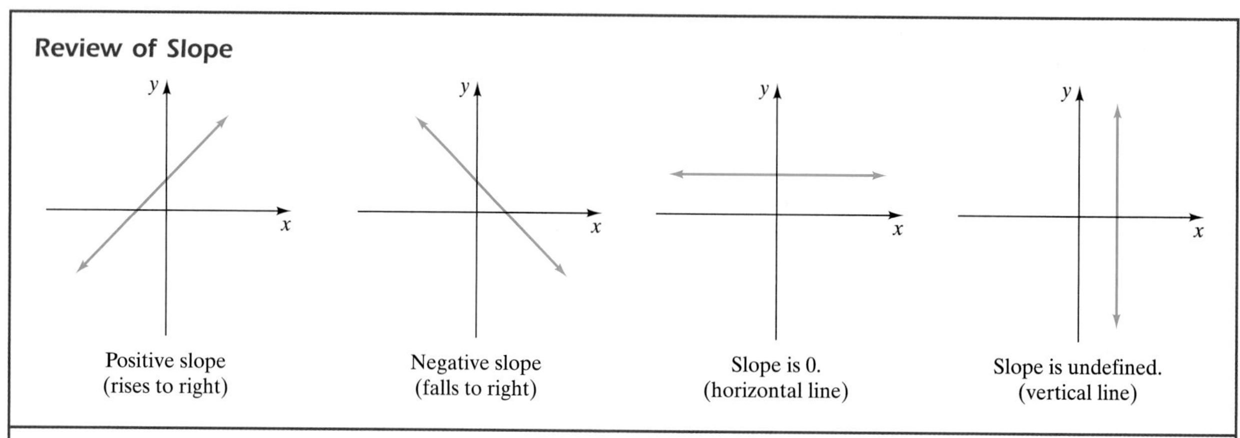

Vertical Line Test: If a vertical line can be drawn through any part of a graph and the vertical line intersects another part of the graph, the graph is not a function.

Review Exercises

[4.1] **1.** Plot each ordered pair on the same axes.

a) $A(5, 3)$ **b)** $B(0, 6)$ **c)** $C\left(5, \frac{1}{2}\right)$

d) $D(-4, 3)$ **e)** $E(-6, -1)$ **f)** $F(-2, 0)$

2. Determine whether the following points are collinear.

$(0, -4), (6, 8), (-2, 0), (4, 4)$

3. Which of the following ordered pairs satisfy the equation $2x + 3y = 9$?

a) $(4, 3)$ **b)** $(0, 3)$ **c)** $(-1, 4)$ **d)** $\left(2, \frac{5}{3}\right)$

[4.2] **4.** Find the missing coordinate in the following solutions to $3x - 2y = 8$.

a) $(4, ?)$ **b)** $(0, ?)$ **c)** $(?, 4)$ **d)** $(?, 0)$

Graph each equation using the method of your choice.

5. $y = -3$

6. $x = 2$

7. $y = 3x$

8. $y = 2x - 1$

9. $y = -3x + 4$

10. $y = -\frac{1}{2}x + 4$

11. $2x + 3y = 6$

12. $3x - 2y = 12$

13. $2y = 3x - 6$

14. $-5x - 2y = 10$

15. $25x + 50y = 100$

16. $\frac{2}{3}x = \frac{1}{4}y + 20$

[4.3] *Find the slope of the line through the given points.*

17. $(8, -7)$ and $(-2, 5)$

18. $(-4, -2)$ and $(8, -3)$

19. $(-2, -1)$ and $(-4, 3)$

20. What is the slope of a horizontal line?

21. What is the slope of a vertical line?

22. Define the slope of a straight line.

Find the slope of each line.

23.

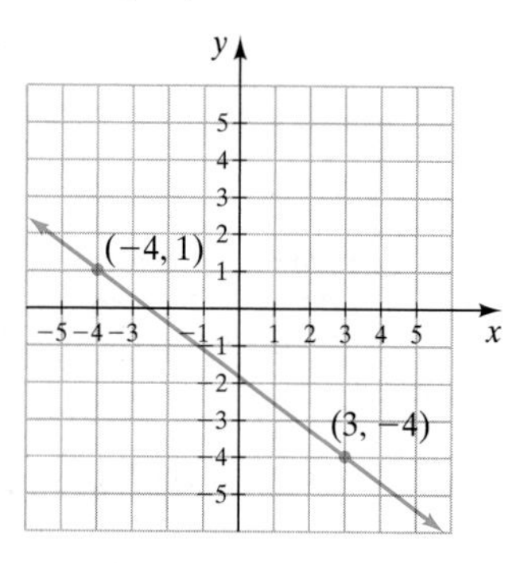

24.

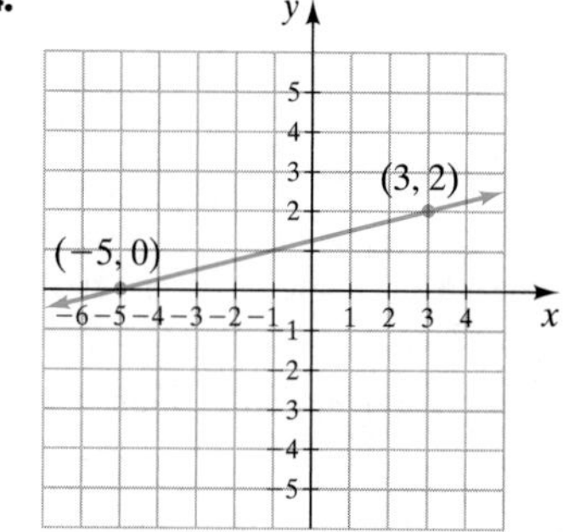

25. Find the slope of the line segment in **a)** red and **b)** blue.

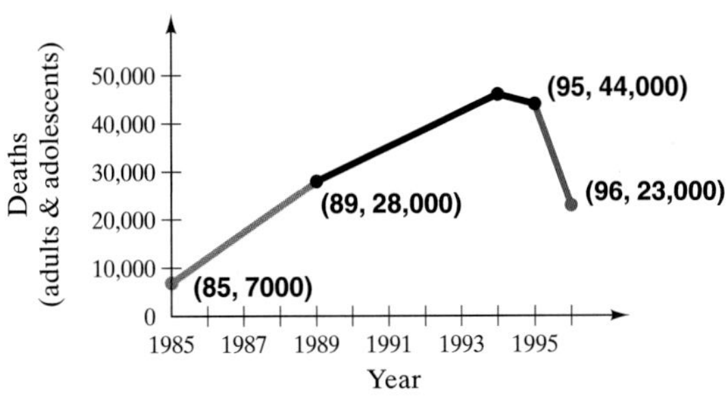

[4.4] *Determine the slope and y-intercept of the graph of each equation.*

26. $9x + 7y = 15$

27. $2x + 5 = 0$

28. $3y + 9 = 0$

Write the equation of each line.

29.

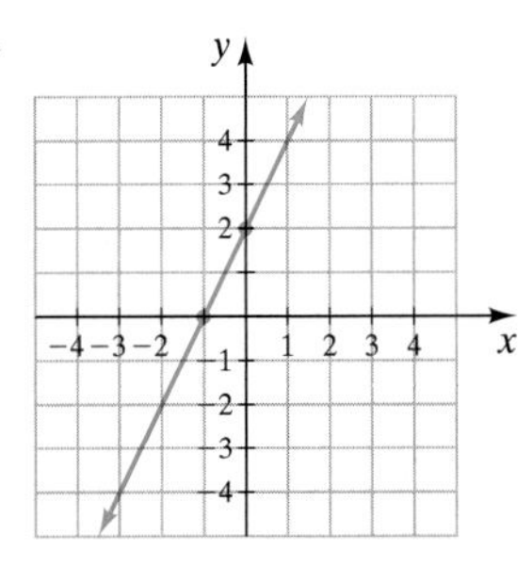

30.

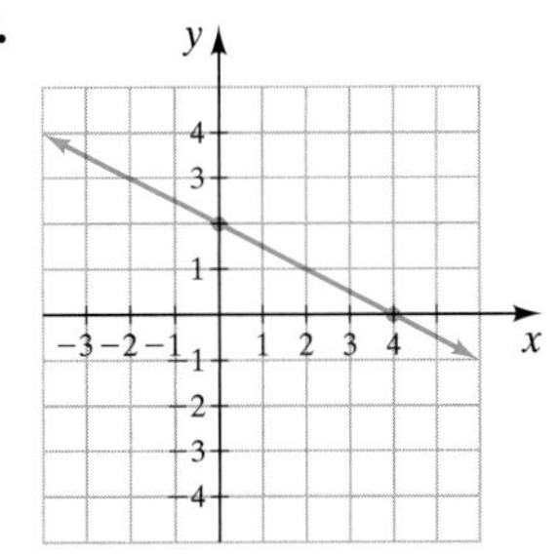

Determine whether each pair of lines are parallel.

31. $y = 2x - 6$
$6y = 12x + 6$

32. $2x - 3y = 9$
$3x - 2y = 6$

Find the equation of each line with the given properties.

33. Slope $= 2$, through $(3, 4)$

34. Slope $= -\frac{2}{3}$, through $(3, 2)$

35. Slope $= 0$, through $(4, 2)$

36. Slope is undefined, through $(4, 2)$

37. Through $(-2, 3)$ and $(0, -4)$

38. Through $(-4, -2)$ and $(-4, 3)$

[4.5] *Graph each inequality.*

39. $y \geq -3$

40. $x < 4$

41. $y < 3x$

42. $y > 2x + 1$

43. $-6x + y \geq 5$

44. $3y + 6 \leq x$

[4.6] *Determine which of the following relations are also functions. Give the domain and range of each.*

45. $\{(0, 2), (4, -3), (1, 5), (2, -1), (6, 4)\}$

46. $\{(3, 1), (4, 2), (4, 5), (6, 1), (7, 0)\}$

47. $\{(3, 1), (4, 1), (5, 1), (6, 2), (3, -3)\}$

48. $\{(5, -2), (3, -2), (4, -2), (9, -2), (-2, -2)\}$

In Exercises 49 and 50, the domain and range of a relation are illustrated. ***a)*** *Construct a set of ordered pairs that represent the relation.* ***b)*** *Determine whether the relation is a function. Explain your answer.*

49.

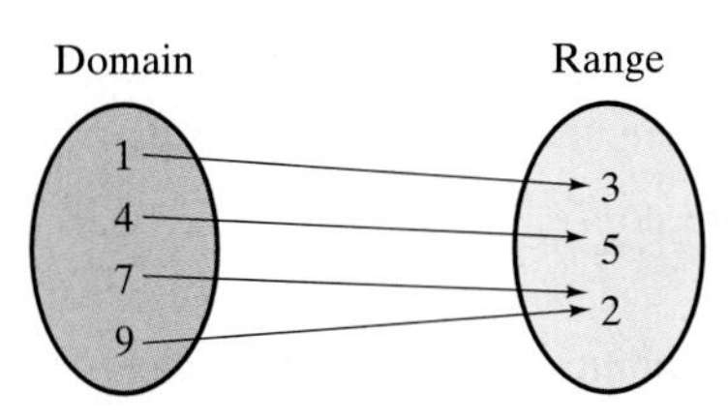

50.

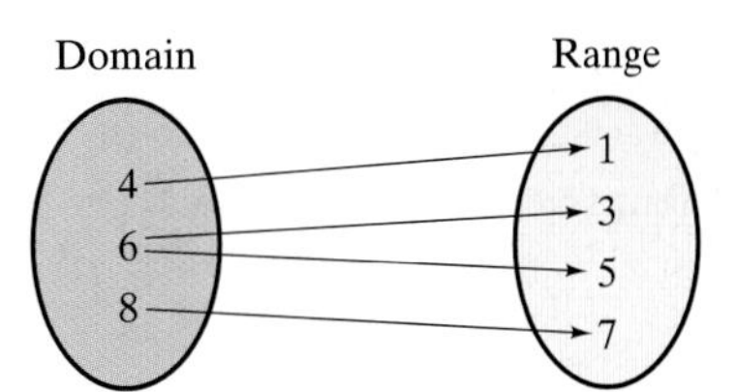

Use the vertical line test to determine whether the relation is also a function.

51. **52.**

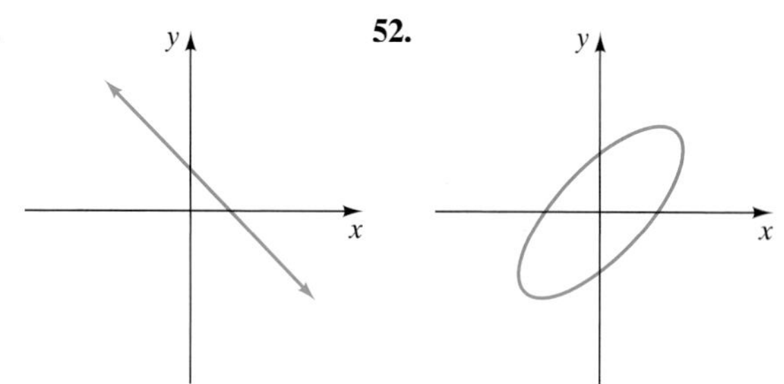

53.

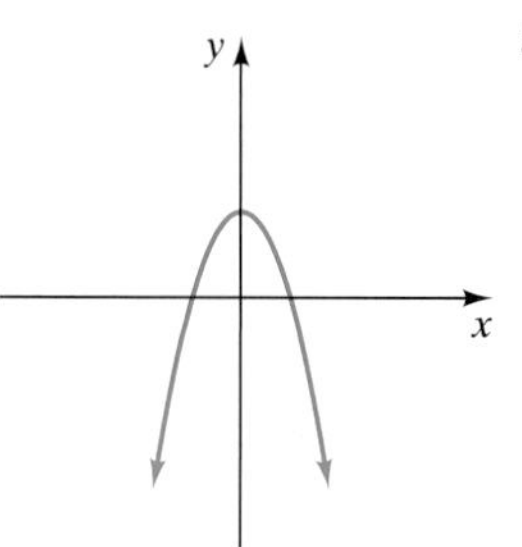

54.

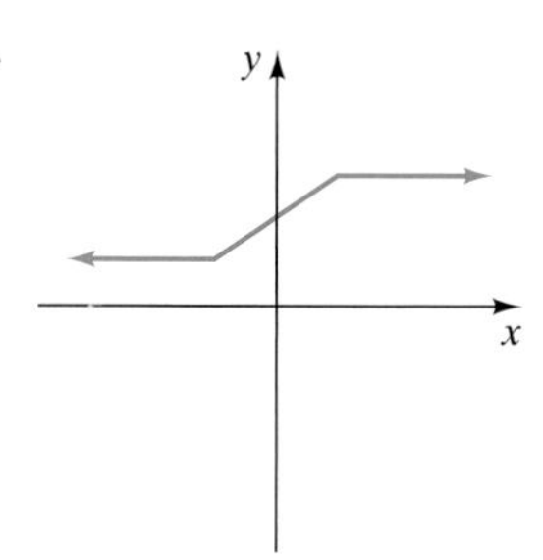

Evaluate each function at the indicated values.

55. $f(x) = 5x - 3$; find **a)** $f(2)$, **b)** $f(-5)$

56. $f(x) = -4x - 5$; find **a)** $f(-4)$, **b)** $f(8)$

57. $f(x) = \frac{1}{3}x - 5$; find **a)** $f(3)$, **b)** $f(-9)$

58. $f(x) = 2x^2 - 4x + 6$; find **a)** $f(3)$, **b)** $f(-5)$

Determine whether the following graphs are functions. Explain your answer.

59. **McGwire Home Runs**

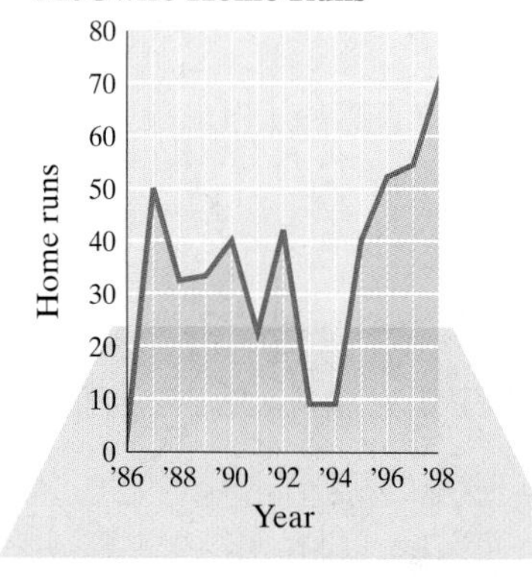

60. **Personal Bankruptcies**

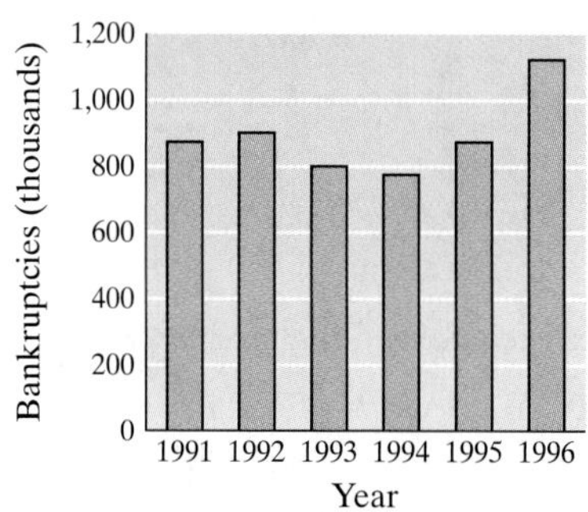

Source: Federal Reserve Bank

Graph the following functions.

61. $f(x) = 3x - 5$

62. $f(x) = -2x + 3$

63. A discount stock broker charges $25 plus 3 cents per share of stock bought or sold. A customer's cost, c, in dollars, is a function of the number of shares, n, bought or sold, $c = 25 + 0.03n$.

a) Draw a graph illustrating a customer's cost for up to and including 10,000 shares of stock.

b) Estimate the cost if 1000 shares of a stock are purchased.

64. The monthly profit, p, of an Everything for a Dollar store can be estimated by the function $p = 4x - 1600$, where x represents the number of items sold.

a) Draw a graph of the function for up to and including 1000 items sold.

b) Estimate the profit if 400 items are sold.

Practice Test

1. What is a graph?
2. In which quadrants do the following points lie?

   a) $(-4, -5)$

   b) $\left(2, -\frac{1}{2}\right)$

3. **a)** What is the standard form of a linear equation?

   b) What is the slope–intercept form of a linear equation?

   c) What is the point–slope form of a linear equation?

4. Which of the following ordered pairs satisfy the equation $3y = 5x - 9$?

a) $(3, 2)$ b) $\left(\frac{9}{5}, 0\right)$

c) $(-2, -6)$ d) $(0, 3)$

5. Find the slope of the line through the points $(-4, 3)$ and $(2, -5)$.
6. Find the slope and y-intercept of $4x - 9y = 15$.
7. Write an equation of the graph in the accompanying figure.

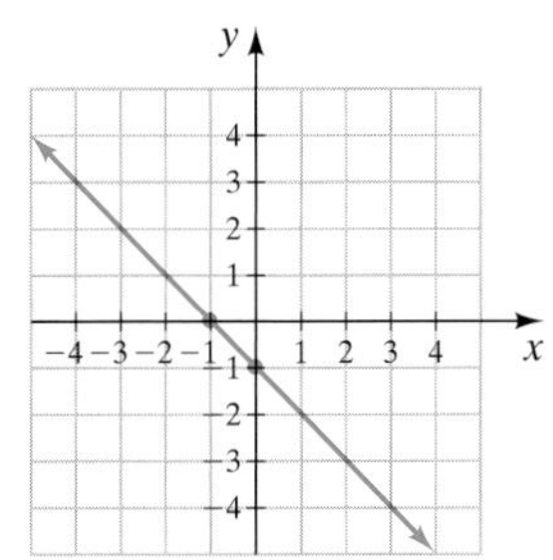

8. Graph $x = -5$.
9. Graph $y = 2$.
10. Graph $y = 3x - 2$ by plotting points.
11. a) Solve the equation $2x - 4y = 8$ for y.

 b) Graph the equation by plotting points.
12. Graph $3x + 5y = 15$ using the intercepts.
13. Write, in slope–intercept form, an equation of the line with a slope of 3 passing through the point $(1, 3)$.
14. Write, in slope–intercept form, an equation of the line passing through the points $(3, -1)$ and $(-4, 2)$.
15. Determine whether the following equations represent parallel lines. Explain how you determined your answer.

$$2y = 3x - 6 \quad \text{and} \quad y - \frac{3}{2}x = -5$$

16. Graph $y = 3x - 4$ using the slope and y-intercept.
17. Graph $3x - 2y = 8$ using the slope and y-intercept.
18. Define a function.
19. a) Determine whether the following relation is a function. Explain your answer.

$$\{(1, 2), (3, -4), (5, 3), (1, 0), (6, 5)\}$$

 b) Give the domain and range of the relation or function.
20. Determine whether the following graphs are functions. Explain how you determined your answer.

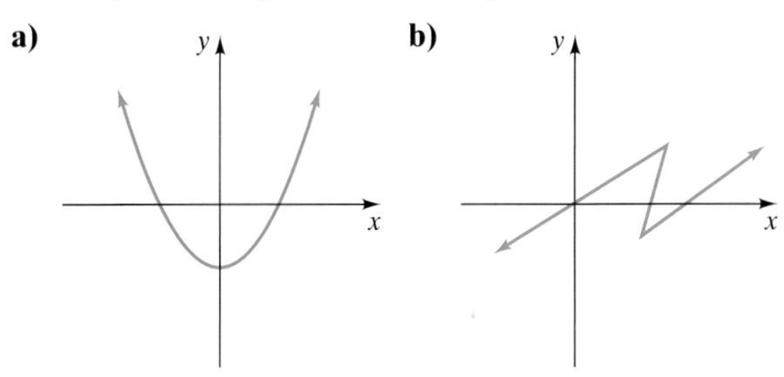

21. If $f(x) = 2x^2 + 3x$, find **a)** $f(2)$ and **b)** $f(-3)$.
22. Graph the function $f(x) = 2x - 4$.
23. Graph $y \geq -3x + 5$.
24. Graph $y < 4x - 2$.
25. Kate Moore, a salesperson, has a weekly income, i, that can be determined by the function $i = 200 + 0.05s$, where s is her weekly sales.

 a) Draw a graph of her weekly income for sales from \$0 to \$10,000.

 b) Estimate her weekly income if her sales are \$3000.

Cumulative Review Test

1. Write the set of

 a) natural numbers.

 b) whole numbers.
2. Name each indicated property.

 a) $3(x + 2) = 3x + 3 \cdot 2$

 b) $a + b = b + a$
3. Evaluate $-3 - (-2)$.
4. Evaluate $4^2 + 8 \div (-2)$.
5. Evaluate $16 \div (4 - 6) \cdot 5$.
6. Solve the equation $2x + 5 = 3(x - 5)$.
7. Solve the equation $3(x - 2) - (x + 4) = 2x - 10$.
8. Solve the inequality $2x - 14 > 5x + 1$. Graph the solution on a number line.
9. Solve $v = lwh$ for w.
10. At Tsong Hsu's Grocery Store, 3 cans of chicken soup sell for \$1.25. Find the cost of 8 cans.
11. Eleven increased by twice a number is 19. Find the number.
12. The length of a rectangle is 4 more than twice the width. Find the dimensions of the rectangle if its perimeter is 26 feet.

13. Two runners start at the same point and run in opposite directions. One runs at 6 mph and the other runs at 8 mph. In how many hours will they be 28 miles apart?

14. Give three ordered pairs that satisfy the equation $2x + 4y = 8$.

15. Graph $y = 3x - 5$ by plotting points.

16. Graph $2x + 6y = 12$ using the intercepts.

17. Find the slope and y-intercept of $3x + 5y = 12$.

18. Graph $y = \frac{2}{3}x - 3$ using the slope and y-intercept.

19. Write the equation, in point–slope form, of the line with a slope of 3 passing through the point $(5, 2)$.

20. Determine whether the following relations are functions? Explain your answer.

a)

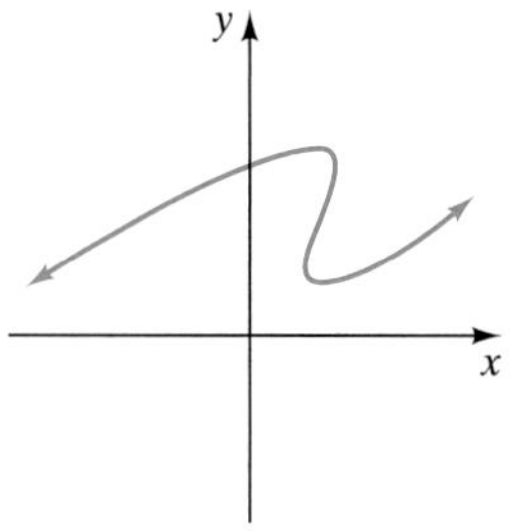

y

x

b) $\{(1, 2), (5, 3), (7, 3), (-2, 0)\}$

SYSTEMS OF LINEAR EQUATIONS

CHAPTER

5

Use the Angel Web site at www.prenhall.com/angel to be linked to an internet resource that will help you further explore the following application.

The unpredictable and destructive nature of hurricanes make them difficult to research. Hurricane hunters fly airplanes into the eye of a hurricane to collect data which can be analyzed. As they fly with the wind, the speed of their plane increases. When they fly against the wind, the speed of their plane decreases. On page 320, we determine the speed of an airplane and the speed of the hurricane wind in which it is flying by solving a system of equations.

Preview and Perspective

In this chapter we learn how to express applications as systems of linear equations and how to solve systems of linear equations. People in business work with many variables and unknown quantities. For example, a company's owners consider overhead costs, costs of material, labor costs, selling price of the item, and a host of other items when seeking to maximize profit. The owners may express the relationships between the variables in several equations or inequalities. These equations or inequalities form a system of equations or inequalities. The solution to the system of equations or inequalities gives the value or values of the variables for which the company can maximize profits.

In this chapter we explain three procedures for solving systems of equations. In Section 5.1 we solve systems using graphs. In Section 5.2 we solve systems using substitution. In Section 5.3 we solve systems using the addition (or elimination) method. In Section 5.4 we explain how to express real-life applications as systems of linear equations. Lastly, in Section 5.5 we build on the graphical solution presented in Section 5.1, and solve systems of linear *inequalities* graphically.

To be successful with Section 5.1, solving systems of equations graphically, you need to understand the procedure for graphing straight lines presented in Sections 4.2 through 4.4. To be successful with Section 5.5, systems of inequalities, you need to understand how to graph linear inequalities, which was presented in Section 4.5.

5.1 SOLVING SYSTEMS OF EQUATIONS GRAPHICALLY

1) **Determine if an ordered pair is a solution to a system of equations.**
2) **Determine if a system of equations is consistent, inconsistent, or dependent.**
3) **Solve a system of equations graphically.**

1) Determine if an Ordered Pair is a Solution to a System of Equations

When we seek a common solution to two or more linear equations, the equations are called a **system of linear equations**. An example of a system of linear equations follows:

$$\left.\begin{aligned}(1)\ y &= x + 5\\ (2)\ y &= 2x + 4\end{aligned}\right\}\quad \textit{System of linear equations}$$

The **solution to a system of equations** is the ordered pair or pairs that satisfy all equations in the system. The solution to the system above is (1, 6).

CHECK:

In Equation (1)	In Equation (2)
$(1, 6)$	$(1, 6)$
$y = x + 5$	$y = 2x + 4$
$6 \stackrel{?}{=} 1 + 5$	$6 \stackrel{?}{=} 2(1) + 4$
$6 = 6$ *True*	$6 = 6$ *True*

Because the ordered pair (1, 6) satisfies *both* equations, it is a solution to the system of equations. Notice that the ordered pair (2, 7) satisfies the first equation but does not satisfy the second equation.

CHECK:	In Equation (1)		In Equation (2)	
	$(2, 7)$		$(2, 7)$	
	$y = x + 5$		$y = 2x + 4$	
	$7 \stackrel{?}{=} 2 + 5$		$7 \stackrel{?}{=} 2(2) + 4$	
	$7 = 7$	*True*	$7 = 8$	*False*

Since the ordered pair (2, 7) does not satisfy *both* equations, it is *not* a solution to the system of equations.

EXAMPLE 1 Determine which of the following ordered pairs satisfy the system of equations.

$$y = 2x - 8$$
$$2x + y = 4$$

a) $(2, -4)$ **b)** $(3, -2)$

Solution **a)** Substitute 2 for x and -4 for y in each equation.

$y = 2x - 8$		$2x + y = 4$	
$-4 \stackrel{?}{=} 2(2) - 8$		$2(2) + (-4) \stackrel{?}{=} 4$	
$-4 \stackrel{?}{=} 4 - 8$		$4 - 4 \stackrel{?}{=} 4$	
$-4 = -4$	*True*	$0 = 4$	*False*

Since $(2, -4)$ does not satisfy both equations, it is not a solution to the system of equations.

b) Substitute 3 for x and -2 for y in each equation.

$y = 2x - 8$		$2x + y = 4$	
$-2 \stackrel{?}{=} 2(3) - 8$		$2(3) + (-2) \stackrel{?}{=} 4$	
$-2 \stackrel{?}{=} 6 - 8$		$6 - 2 \stackrel{?}{=} 4$	
$-2 = -2$	*True*	$4 = 4$	*True*

Since $(3, -2)$ satisfies both equations, it is a solution to the system of linear equations.

NOW TRY EXERCISE 13

In this chapter we discuss three methods for finding the solution to a system of equations: the *graphical method*, the *substitution method*, and the *addition method*. In this section we discuss the graphical method.

2) Determine if a System of Equations is Consistent, Inconsistent, or Dependent

The **solution to a system of linear equations** is the ordered pair (or pairs) common to all lines in the system when the lines are graphed. When two lines are graphed, three situations are possible, as illustrated in Figure 5.1.

In Figure 5.1a, lines 1 and 2 are not parallel lines. They intersect at exactly one point. This system of equations has *exactly one solution*. This is an example of a **consistent system of equations**. A consistent system of equations is a system of equations that has a solution.

In Figure 5.1b, lines 1 and 2 are two different parallel lines. The lines do not intersect, and this system of equations has *no solution*. This is an example of

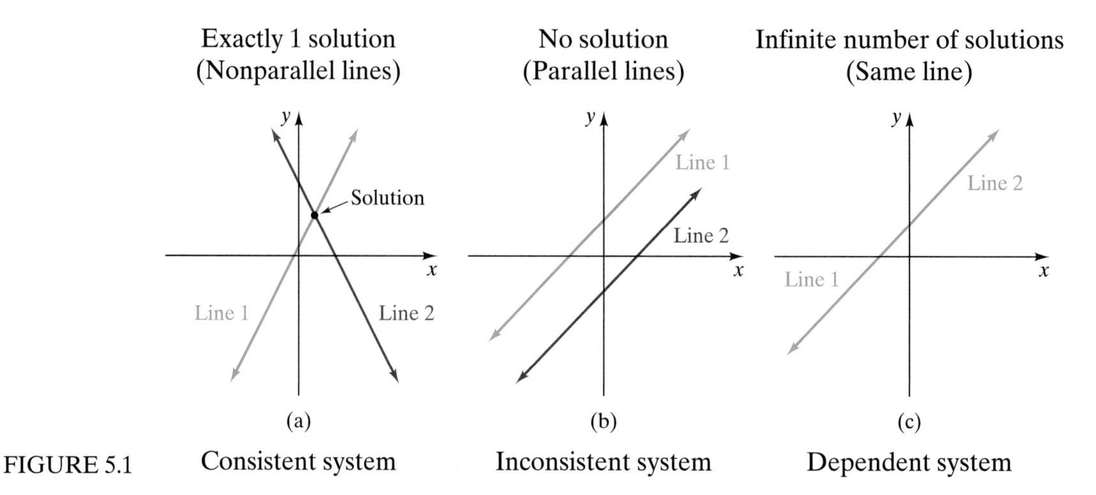

FIGURE 5.1 Consistent system | Inconsistent system | Dependent system

an **inconsistent system of equations**. An inconsistent system of equations is a system of equations that has no solution.

In Figure 5.1c, lines 1 and 2 are actually the same line. In this case, every point on the line satisfies both equations and is a solution to the system of equations. This system has *an infinite number of solutions*. This is an example of a **dependent system of equations**. A dependent system of linear equations is a system of equations that has an infinite number of solutions. If a system of two linear equations is dependent, then both equations represent the same line. *Note that a dependent system is also a consistent system since it has a solution.*

We can determine if a system of linear equations is consistent, inconsistent, or dependent by writing each equation in slope–intercept form and comparing the slopes and y-intercepts. If the slopes of the lines are different (Fig. 5.1a), the system is consistent. If the slopes are the same but the y-intercepts are different (Fig. 5.1b), the system is inconsistent. If both the slopes and the y-intercepts are the same (Fig. 5.1c), the system is dependent.

EXAMPLE 2 Determine whether the following system has exactly one solution, no solution, or an infinite number of solutions.

$$3x + 4y = 8$$
$$6x + 8y = 4$$

Solution Write each equation in slope–intercept form and then compare the slopes and the y-intercepts.

$$\begin{aligned} 3x + 4y &= 8 \\ 4y &= -3x + 8 \\ y &= \frac{-3x + 8}{4} \\ y &= -\frac{3}{4}x + \frac{8}{4} \\ y &= -\frac{3}{4}x + 2 \end{aligned} \qquad \begin{aligned} 6x + 8y &= 4 \\ 8y &= -6x + 4 \\ y &= \frac{-6x + 4}{8} \\ y &= -\frac{6}{8}x + \frac{4}{8} \\ y &= -\frac{3}{4}x + \frac{1}{2} \end{aligned}$$

Since the lines have the same slope, $-\frac{3}{4}$, and different y-intercepts, the lines are parallel. This system of equations is therefore inconsistent and has no solution.

NOW TRY EXERCISE 29

3) Solve a System of Equations Graphically

Now we will see how to solve systems of equations graphically.

To Obtain the Solution to a System of Equations Graphically

Graph each equation and determine the point or points of intersection.

EXAMPLE 3 Solve the following system of equations graphically.

$$2x + y = 11$$
$$x + 3y = 18$$

Solution Find the x- and y-intercepts of each graph; then draw the graphs.

$2x + y = 11$	Ordered Pair	$x + 3y = 18$	Ordered Pair
Let $x = 0$; then $y = 11$	$(0, 11)$	Let $x = 0$; then $y = 6$	$(0, 6)$
Let $y = 0$; then $x = \frac{11}{2}$	$\left(\frac{11}{2}, 0\right)$	Let $y = 0$; then $x = 18$	$(18, 0)$

The two graphs (Fig. 5.2) appear to intersect at the point (3, 5). The point (3, 5) may be the solution to the system of equations. To be sure, however, we must check to see that (3, 5) satisfies *both* equations.

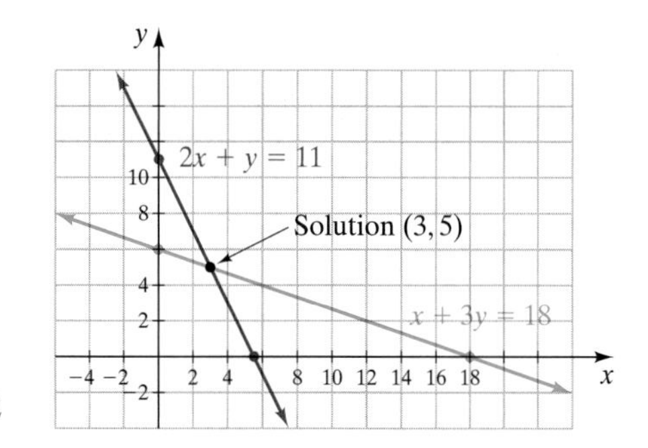

FIGURE 5.2

CHECK:

$$2x + y = 11 \qquad\qquad x + 3y = 18$$
$$2(3) + 5 \stackrel{?}{=} 11 \qquad\qquad 3 + 3(5) \stackrel{?}{=} 18$$
$$11 = 11 \quad \text{True} \qquad\qquad 18 = 18 \quad \text{True}$$

Since the ordered pair (3, 5) checks in both equations, it is the solution to the system of equations. This system of equations is consistent.

EXAMPLE 4 Solve the following system of equations graphically.

$$2x + y = 3$$
$$4x + 2y = 12$$

Solution Find the x- and y-intercepts of each graph; then draw the graphs.

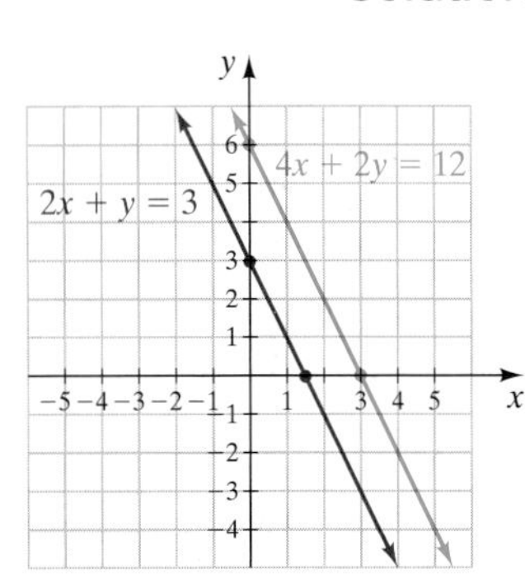

FIGURE 5.3

$2x + y = 3$	Ordered Pair	$4x + 2y = 12$	Ordered Pair
Let $x = 0$; then $y = 3$	$(0, 3)$	Let $x = 0$; then $y = 6$	$(0, 6)$
Let $y = 0$; then $x = \frac{3}{2}$	$\left(\frac{3}{2}, 0\right)$	Let $y = 0$; then $x = 3$	$(3, 0)$

The two lines (Fig. 5.3) appear to be parallel.

To show that the two lines are indeed parallel, write each equation in slope–intercept form.

$$2x + y = 3 \qquad\qquad 4x + 2y = 12$$

$$y = -2x + 3 \qquad\qquad 2y = -4x + 12$$

$$y = -2x + 6$$

Both equations have the same slope, -2, and different y-intercepts; thus the lines must be parallel. Since parallel lines do not intersect, this system of equations has no solution. This system of equations is inconsistent.

NOW TRY EXERCISE 55

EXAMPLE 5 Solve the following system of equations graphically.

$$x - \frac{1}{2}y = 2$$

$$y = 2x - 4$$

Solution Find the x- and y-intercepts of each graph; then draw the graphs.

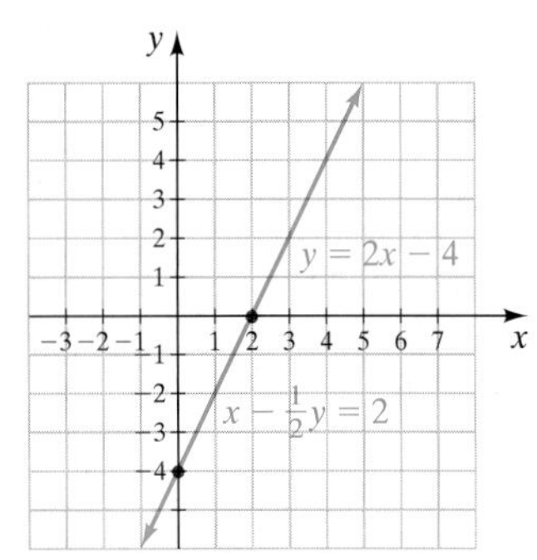

FIGURE 5.4

$x - \frac{1}{2}y = 2$	Ordered Pair	$y = 2x - 4$	Ordered Pair
Let $x = 0$; then $y = -4$	$(0, -4)$	Let $x = 0$; then $y = -4$	$(0, -4)$
Let $y = 0$; then $x = 2$	$(2, 0)$	Let $y = 0$; then $x = 2$	$(2, 0)$

Because the lines have the same x- and y-intercepts, both equations represent the same line (Fig. 5.4). When the equations are written in slope–intercept form, it becomes clear that the equations are identical and the system is dependent.

$$x - \frac{1}{2}y = 2 \qquad\qquad y = 2x - 4$$

$$2\left(x - \frac{1}{2}y\right) = 2(2)$$

$$2x - y = 4$$

$$-y = -2x + 4$$

$$y = 2x - 4$$

The solution to this system of equations is all the points on the line.

NOW TRY EXERCISE 53

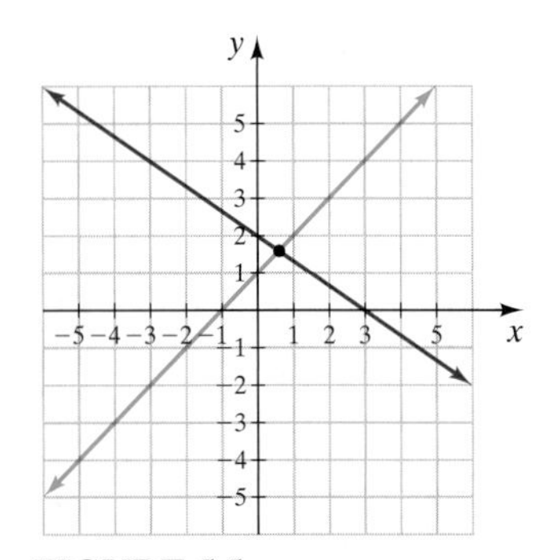

FIGURE 5.5

When graphing a system of equations, the intersection of the lines is not always easy to read on the graph. For example, can you determine the solution to the system of equations shown in Figure 5.5? You may estimate the solution to be $\left(\frac{7}{10}, \frac{3}{2}\right)$ when it may actually be $\left(\frac{4}{5}, \frac{8}{5}\right)$. The accuracy of your answer will depend on how carefully you draw the graphs and on the scale of the graph paper used. In Section 5.2 we present algebraic methods that give exact solutions to systems of equations.

Using Your Graphing Calculator

A graphing calculator can be used to solve or check systems of equations. To solve a system of equations graphically, graph both equations as was explained in Chapter 4. The point or points of intersection of the graphs is the solution to the system of equations.

The first step in solving the system graphically is to solve each equation for y. For example, consider the system of equations given in Example 3.

System of Equations	Equations Solved for y
$2x + y = 11$	$y = -2x + 11$
$x + 3y = 18$	$y = -\frac{1}{3}x + 6$

$$\text{Let } y_1 = -2x + 11$$

$$y_2 = -\frac{1}{3}x + 6$$

Figure 5.6 shows the screen of a TI-83 with these two equations graphed.

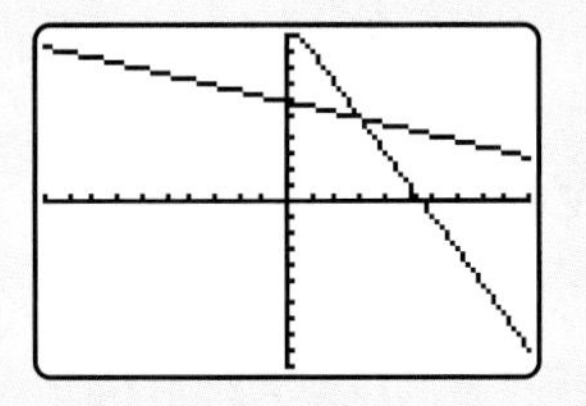

FIGURE 5.6

X	Y_1	Y_2
0	11	6
1	9	5.6667
2	7	5.3333
3	5	5
4	3	4.6667
5	1	4.3333
6	-1	4

X=3

FIGURE 5.7

To find the intersection of the two graphs, you can use the TRACE and ZOOM keys or the TABLE feature as was explained earlier. Figure 5.7 shows a table of values for equations y_1 and y_2. Notice when $x = 3$, y_1 and y_2 both have the same value, 5. Therefore both graphs intersect at (3, 5), and (3, 5) is the solution.

Some graphing calculators have a feature that displays the intersection of graphs by pressing a sequence of keys. For example, on the TI-83 if you go to CALC (which stands for calculate) by pressing 2nd TRACE, you get the screen shown in Figure 5.8. Now press 5, intersect. Once the *intersect* feature has been selected, the calculator will display the graph and the question

FIRST CURVE?

At this time, move the cursor along the first curve until it is close to the point of intersection. Then press ENTER. The calculator now shows

SECOND CURVE?

and has the cursor on the second curve. If the cursor is not close to the point of intersection, move it along this curve until this happens. Then press ENTER. The calculator now displays

GUESS?

Now press ENTER and the point of intersection is displayed, see Figure 5.9. The point of intersection is (3, 5).

FIGURE 5.8

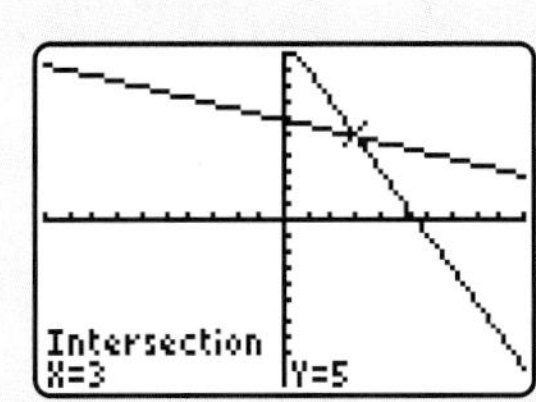

FIGURE 5.9

Exercises

Use your graphing calculator to find the solution to each system of equations. Round noninteger answers to the nearest tenth.

1. $x + 2y = -11$
$2x - y = -2$

2. $x - 3y = -13$
$-2x - 2y = 2$

3. $2x - y = 7.7$
$-x - 3y = 1.4$

4. $3x + 2y = 7.8$
$-x + 3y = 15.0$

In Section 3.3, Example 11, we solved a problem involving security systems using only one variable. In the example that follows, we will work that same problem using two variables and illustrate the solution in the form of a graph. Although an answer may sometimes be easier to obtain using only one variable, a graph of the situation may help you to visualize the total picture better.

EXAMPLE 6 Vanessa Juenger plans to install a security system in her house. She has narrowed her choices to two security dealers: Moneywell and Doile. Moneywell's system costs \$3360 to install and their monitoring fee is \$17 per month. Doile's equivalent system costs only \$2260 to install, but their monitoring fee is \$28 per month.

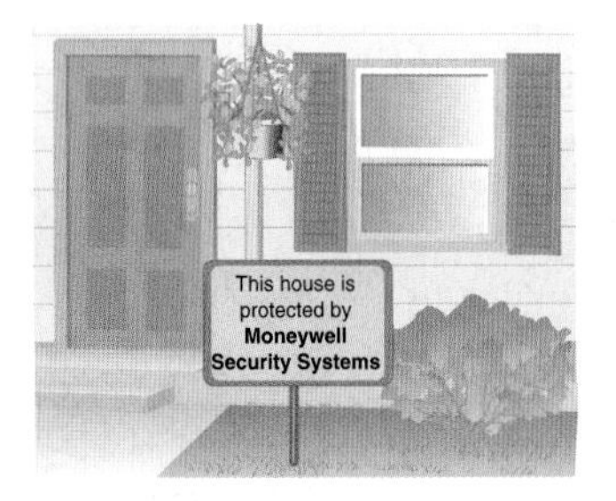
This house is
protected by
Moneywell
Security Systems

a) Assuming that their monthly monitoring fees do not change, in how many months would the total cost of Moneywell's and Doile's systems be the same?

b) If both dealers guarantee not to raise monthly fees for 10 years, and if Vanessa plans to use the system for 10 years, which system would be the least expensive?

Solution **a) Understand and Translate** We need to determine the number of months for which both systems will have the same total cost.

Let n = number of months

c = total cost of the security system over n months

Now we can write an equation to represent the cost of each system using the two variables c and n.

Moneywell

$$\text{Total cost} = \begin{pmatrix}\text{initial}\\ \text{cost}\end{pmatrix} + \begin{pmatrix}\text{fees over}\\ n\text{ months}\end{pmatrix}$$

$$c = 3360 + 17n$$

Doile

$$\text{Total cost} = \begin{pmatrix}\text{initial}\\ \text{cost}\end{pmatrix} + \begin{pmatrix}\text{fees over}\\ n\text{ months}\end{pmatrix}$$

$$c = 2260 + 28n$$

Thus, our system of equations is

$$c = 3360 + 17n$$
$$c = 2260 + 28n$$

Carry out Now let's graph each equation.

$c = \mathbf{3360 + 17}n$

Let $n = 0$ $\quad c = 3360 + 17(0) = 3360$

Let $n = 100$ $\quad c = 3360 + 17(100) = 5060$

Let $n = 150$ $\quad c = 3360 + 17(150) = 5910$

n	c
0	3360
100	5060
150	5910

$c = \mathbf{2260 + 28}n$

Let $n = 0$ $\quad c = 2260 + 28(0) = 2260$

Let $n = 100$ $\quad c = 2260 + 28(100) = 5060$

Let $n = 150$ $\quad c = 2260 + 28(150) = 6460$

n	c
0	2260
100	5060
150	6460

Check and Answer The graph (Fig. 5.10) shows that the total cost of the two security systems would be the same in 100 months. This is the same answer we obtained in Example 11 in Section 3.3.

b) Since 10 years is 120 months, we draw a dashed vertical line at $n = 120$ months and see where it intersects the two lines. Since at 120 months the Doile

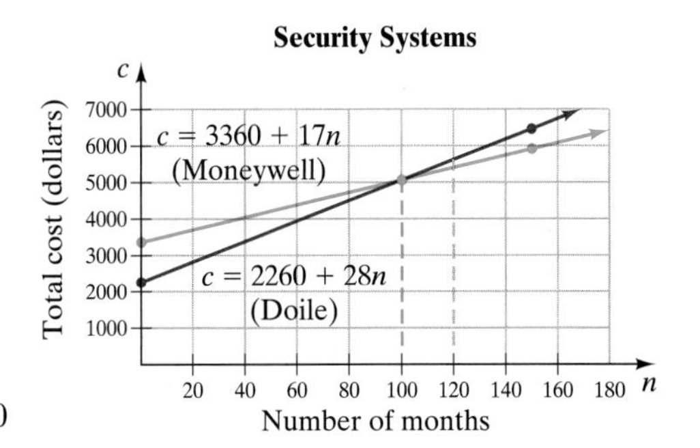

FIGURE 5.10

line is higher than the Moneywell line, the cost for the Doile system for 120 months is more than the cost of the Moneywell system. Therefore, the cost of the Moneywell system would be less expensive for 10 years.

NOW TRY EXERCISE 71

We will discuss applications of systems of equations further in Section 5.4.

HELPFUL HINT

An equation in one variable may be solved using a system of linear equations. Consider the equation $3x - 1 = x + 1$. Its solution is 1, as illustrated below.

$$3x - 1 = x + 1$$
$$2x - 1 = 1$$
$$2x = 2$$
$$x = 1$$

Let us set each side of the equation $3x - 1 = x + 1$ equal to y to obtain the following system of equations:

$$y = 3x - 1$$
$$y = x + 1$$

If we determine the ordered pair (x, y) that satisfies *both* equations, then for the value of x in the ordered pair, $3x - 1 = x + 1$.

The graphical solution of this system of equations is illustrated in Figure 5.11.

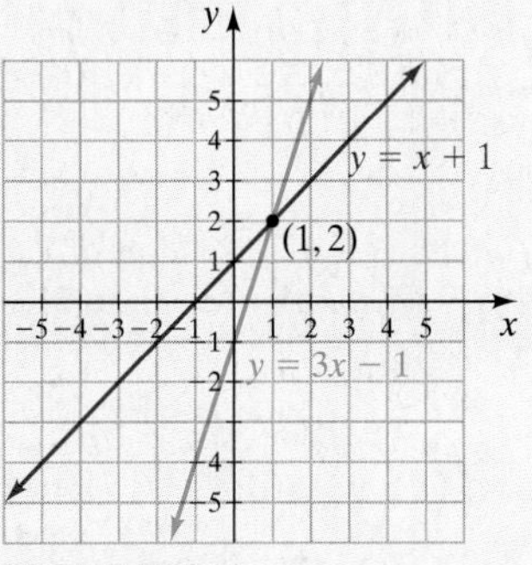

FIGURE 5.11

The solution to the system is (1, 2). Notice that the x-coordinate of the solution of the system, 1, is the solution to the linear equation in one variable, $3x - 1 = x + 1$. If you have a difficult equation to solve and you have a graphing calculator, you can solve the equation on your calculator. The x-coordinate of the solution to the system will be the solution to the linear equation in one variable.

Exercise Set 5.1

Concept/Writing Exercises

1. What does the solution to a system of equations represent?
2. **a)** What is a consistent system of equations? **b)** What is an inconsistent system of equations? **c)** What is a dependent system of equations?
3. Explain how to determine without graphing if a system of linear equations has exactly one solution, no solution, or an infinite number of solutions.
4. When a dependent system of two linear equations is graphed, what will be the result?
5. Explain why it may be difficult to obtain an exact answer to a system of equations graphically.
6. Is a dependent system of equations a consistent system or an inconsistent system? Explain.

Practice the Skills

Determine which, if any, of the following ordered pairs satisfy each system of linear equations.

7. $y = 2x - 6$
   $y = -x + 3$
   a) $(-1, 2)$ **b)** $(3, 0)$ **c)** $(2, 2)$
8. $y = -4x$
   $y = -2x + 8$
   a) $(0, 0)$ **b)** $(-4, 16)$ **c)** $(2, -8)$
9. $y = 2x - 3$
   $y = x + 5$
   a) $(8, 13)$ **b)** $(4, 5)$ **c)** $(4, 9)$
10. $x + 2y = 4$
    $y = 3x + 3$
    a) $(0, 2)$ **b)** $(-2, 3)$ **c)** $(4, 15)$
11. $3x - y = 6$
    $4x + y = 10$
    a) $(3, 3)$ **b)** $(2, -2)$ **c)** $(4, 6)$
12. $y = 2x + 4$
    $y = 2x - 1$
    a) $(0, 4)$ **b)** $(3, 10)$ **c)** $(-2, 0)$
13. $2x - 3y = 6$
    $y = \frac{2}{3}x - 2$
    a) $(3, 0)$ **b)** $(3, -2)$ **c)** $(6, 2)$
14. $y = -x + 4$
    $2y = -2x + 8$
    a) $(2, 5)$ **b)** $(0, 4)$ **c)** $(5, -1)$
15. $3x - 4y = 8$
    $2y = \frac{2}{3}x - 4$
    a) $(0, -2)$ **b)** $(1, -6)$ **c)** $\left(-\frac{1}{3}, -\frac{9}{4}\right)$
16. $2x + 3y = 6$
    $-x + \frac{5}{2} = \frac{1}{2}y$
    a) $\left(\frac{1}{2}, \frac{5}{3}\right)$ **b)** $(2, 1)$ **c)** $\left(\frac{9}{4}, \frac{1}{2}\right)$

Identify each system of linear equations (lines are labeled 1 and 2) as consistent, inconsistent, or dependent. State whether the system has exactly one solution, no solution, or an infinite number of solutions.

17.
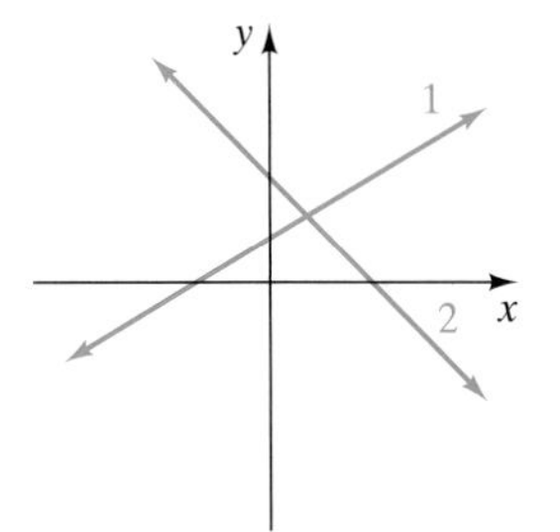

18.
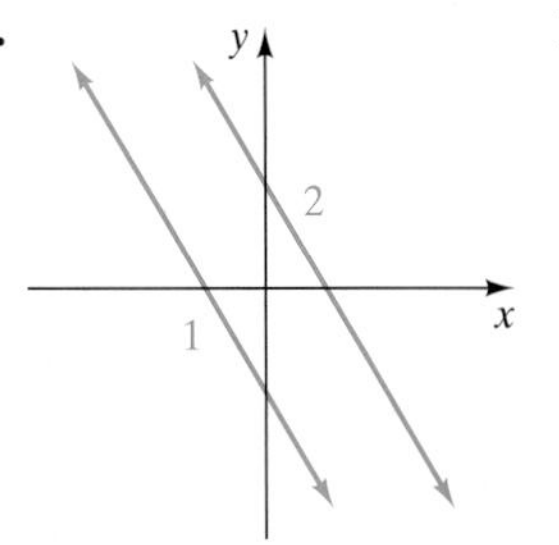

19.
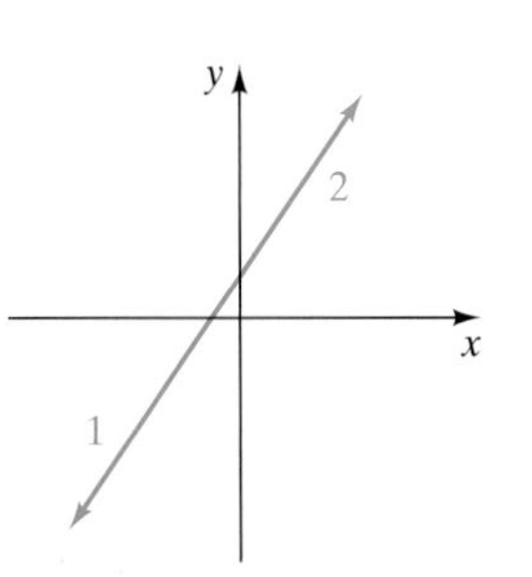

20.
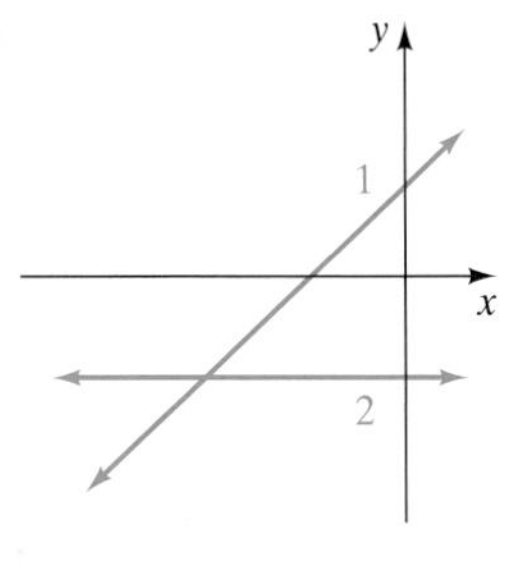

21.

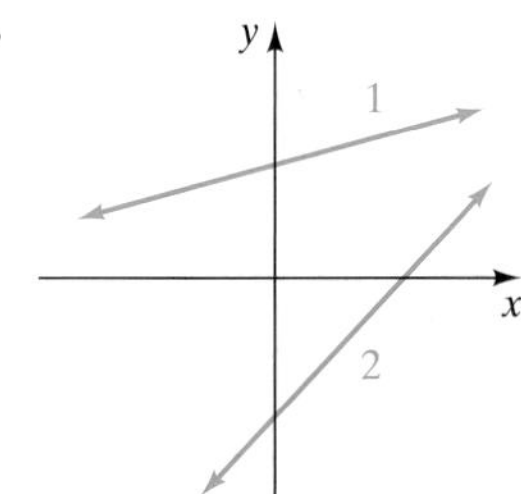

22.

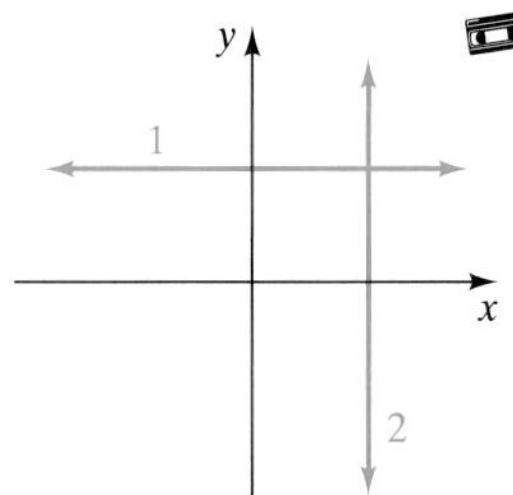

23.

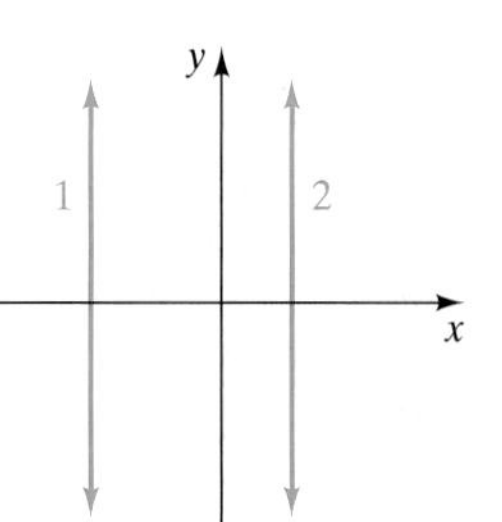

24.

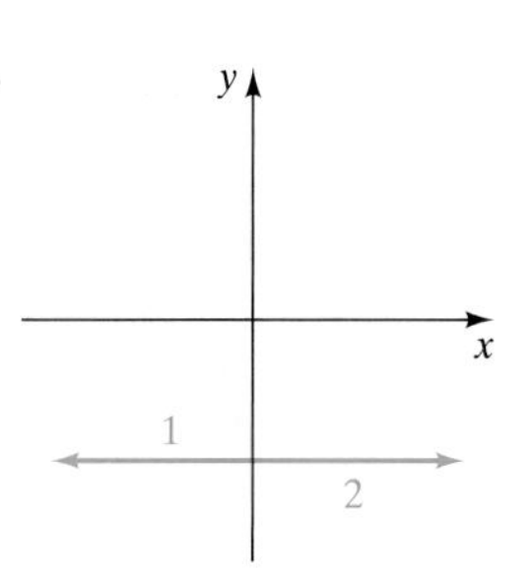

Express each equation in slope–intercept form. Without graphing the equations, state whether the system of equations has exactly one solution, no solution, or an infinite number of solutions.

25. $y = 4x - 1$
$2y = 4x - 6$

26. $x + y = 6$
$x - y = 6$

27. $2y = 3x + 3$
$y = \frac{3}{2}x - 2$

28. $y = \frac{1}{2}x + 4$
$2y = x + 8$

29. $5x = y - 6$
$3x = 4y + 5$

30. $x + 2y = 6$
$2x + y = 4$

31. $2x = 3y + 4$
$6x - 9y = 12$

32. $x - y = 2$
$2x - 2y = -2$

33. $3x + 5y = -7$
$-3x - 5y = -7$

34. $x - y = 3$
$\frac{1}{2}x - 2y = -6$

35. $y = \frac{3}{2}x + \frac{1}{2}$
$3x - 2y = -\frac{1}{2}$

36. $y = \frac{7}{3}x - 9$
$3y = 6x + 9$

Determine the solution to each system of equations graphically. If the system is dependent or inconsistent, so state.

37. $y = x + 2$
$y = -x + 2$

38. $y = 2x + 4$
$y = -3x - 6$

39. $y = 3x - 6$
$y = -x + 6$

40. $y = 2x - 1$
$2y = 4x + 6$

41. $2x = 4$
$y = -3$

42. $x + y = 5$
$2y = x - 2$

43. $y = x + 2$
$x + y = 4$

44. $2x + y = 6$
$2x - y = -2$

45. $y = -\frac{1}{2}x + 4$
$x + 2y = 6$

46. $x + 2y = -4$
$2x - y = -3$

47. $x + 2y = 8$
$2x - 3y = 2$

48. $4x - y = 5$
$2y = 8x - 10$

49. $2x + 3y = 6$
$4x = -6y + 12$

50. $2x + 3y = 6$
$2x + y = -2$

51. $y = 3$
$y = 2x - 3$

52. $x = 3$
$y = 2x - 2$

53. $x - 2y = 4$
$2x - 4y = 8$

54. $3x + y = -6$
$2x = 1 + y$

55. $2x + y = -2$
$6x + 3y = 6$

56. $y = 2x - 3$
$y = -x$

57. $4x - 3y = 6$
$2x + 4y = 14$

58. $2x + 6y = 6$
$y = -\frac{1}{3}x + 1$

59. $2x - 3y = 0$
$x + 2y = 0$

60. $2x = 4y - 12$
$-4x + 8y = 8$

Problem Solving

61. Given the system of equations $5x - 4y = 10$ and $12y = 15x - 20$, determine without graphing whether the graphs of the two equations will be parallel lines. Explain how you determined your answer.

62. Given the system of equations $4x - 8y = 12$ and $2x - 8 = 4y$, determine without graphing whether the graphs of the two equations will be parallel lines. Explain how you determined your answer.

63. If a system of linear equations has solutions (2, 3) and (4, 5), how many solutions does the system have? Explain.

64. If the slope of one line in a system of linear equations is 2 and the slope of the second line in the system is 3, how many solutions does the system have? Explain.

65. If two lines are parallel and have different y-intercepts, how many solutions does the system have? Explain.

66. If two different lines in a linear system of equations pass through the origin, must the solution to the system be (0, 0)? Explain.

67. Consider the system $x = 4$, $y = 3$. How many solutions does the system have? What is the solution?

68. A system of linear equations has (2, 1) as its solution. If one line in the system is vertical and the other line is horizontal, determine the equations in the system.

In Exercises 69–72, find each solution by graphing the system of equations.

69. Edith Hall's furnace is 10 years old and has a problem. The furnace repair man indicates that it will cost Edith $600 to repair her furnace. She can purchase a new, more efficient furnace for $1800. Her present furnace averages about $650 per year for energy cost and the new furnace would average about $450 per year.

We can represent the total cost, c, of repair or replacement, plus energy cost over n years by the following system of equations.

(repair) $c = 600 + 650n$

(replacement) $c = 1800 + 450n$

Find the number of years for which the total cost of repair would equal the total cost of replacement.

70. Juan Varges is considering the two security systems discussed in Example 6. If Moneywell's system costs $4400 plus $15 per month and Doile's system costs $3400 plus $25 per month, after how many months would the total cost of the two systems be the same?

71. One brokerage firm, the Patricia Morris Agency, charges a base charge, c, of $50 plus 10 cents per share of stock, s, purchased or sold. A second brokerage firm, Scott Sambucci and Associates, charges 30 cents per share of stock purchased or sold. The equations that represent the cost of purchasing or selling stock with these firms are

$$c = 50 + 0.10s$$
$$c = 0.30s$$

Determine the number of shares of stock that must be purchased (or sold) for the total cost to be the same.

72. The Evergreen Landscape Service charges a consultation fee of $200 plus $60 per hour for labor. The Out of Sight Landscape Service charges a consultation fee of $300 plus $40 per hour for labor. We can represent this situation with the system of equations

$$c = 200 + 60h$$
$$c = 300 + 40h$$

where c is the total cost and h is the number of hours of labor. Find the number of hours of labor for the two services to have the same total cost.

Group Activity

Discuss and answer Exercises 73–78 as a group. Suppose that a system of three linear equations in two variables is graphed on the same axes. Find the maximum number of points where two or more of the lines can intersect if:

73. the three lines have the same slope but different y-intercepts.

74. the three lines have the same slope and the same y-intercept.

75. two lines have the same slope but different y-intercepts and the third line has a different slope.

76. the three lines have different slopes but the same y-intercept.

77. the three lines have different slopes but two have the same y-intercept.

78. the three lines have different slopes and different y-intercepts.

Cumulative Review Exercises

[2.1] **79.** Simplify $3x - (x - 6) + 4(3 - x)$.

[3.1] **81.** If $A = p(1 + rt)$, find r when $A = 1000$, $t = 2$, and $p = 500$.

[2.5] **80.** Solve the equation $2(x + 3) - x = 5x + 2$.

82. Solve the formula $A = \frac{1}{2}bh$ for h.

5.2 SOLVING SYSTEMS OF EQUATIONS BY SUBSTITUTION

1) **Solve systems of equations by substitution.**

As we stated in Section 5.1, a graphic solution to a system of equations may be inaccurate since you must estimate the coordinates of the point of intersection. When an exact solution is necessary, the system should be solved algebraically, either by substitution or by addition of equations.

1) Solve Systems of Equations by Substitution

The procedure for solving a system of equations by **substitution** is illustrated in Example 1. The procedure for solving by addition is presented in Section 5.3. Regardless of which of the two algebraic techniques is used to solve a system of equations, our immediate goal remains the same, that is, *to obtain one equation containing only one unknown.*

EXAMPLE 1 Solve the following system of equations by substitution.

$$2x + y = 11$$
$$x + 3y = 18$$

Solution Begin by solving for one of the variables in either of the equations. You may solve for any of the variables; however, if you solve for a variable with a numerical coefficient of 1, you may avoid working with fractions. In this system the y-term in $2x + y = 11$ and the x-term in $x + 3y = 18$ both have a numerical coefficient of 1.

Solve for y in $2x + y = 11$.

$$2x + y = 11$$
$$y = -2x + 11$$

Next, substitute $-2x + 11$ for y in the *other equation*, $x + 3y = 18$, and solve for the remaining variable, x.

$$x + 3y = 18$$
$$x + 3(-2x + 11) = 18 \quad \text{Substitution step}$$
$$x - 6x + 33 = 18$$
$$-5x + 33 = 18$$
$$-5x = -15$$
$$x = 3$$

Finally, substitute $x = 3$ in the equation that is solved for y and find the value of y.

$$y = -2x + 11$$
$$y = -2(3) + 11$$
$$y = -6 + 11$$
$$y = 5$$

A check using $x = 3$ and $y = 5$ in both equations will show that the solution is the ordered pair $(3, 5)$.

Note that the solution in Example 1 is identical to the graphical solution obtained in Example 3 of Section 5.1.

To Solve a System of Equations by Substitution

1. Solve for a variable in either equation. (If possible, solve for a variable with a numerical coefficient of 1 to avoid working with fractions.)
2. Substitute the expression found for the variable in step 1 into the other equation.
3. Solve the equation determined in step 2 to find the value of one variable.
4. Substitute the value found in step 3 into the equation obtained in step 1 to find the value of the other variable.
5. Check by substituting both values in each original equation.

EXAMPLE 2 Solve the following system of equations by substitution.

$$2x + y = 3$$
$$4x + 2y = 12$$

Solution Solve for y in $2x + y = 3$.

$$2x + y = 3$$
$$y = -2x + 3$$

Now substitute the expression $-2x + 3$ for y in the *other equation*, $4x + 2y = 12$, and solve for x.

$$4x + 2y = 12$$
$$4x + 2(-2x + 3) = 12$$
$$4x - 4x + 6 = 12$$
$$6 = 12 \quad \textit{False}$$

Since the statement $6 = 12$ is false, the system has no solution. (Therefore, the graphs of the equations will be parallel lines and the system is inconsistent because it has no solution.)

Note that the solution in Example 2 is identical to the graphical solution obtained in Example 4 of Section 5.1. Figure 5.3 on page 298 shows the parallel lines.

NOW TRY EXERCISE 21

EXAMPLE 3 Solve the following system of equations by substitution.

$$x - \frac{1}{2}y = 2$$

$$y = 2x - 4$$

Solution The equation $y = 2x - 4$ is already solved for y. Substitute $2x - 4$ for y in the other equation, $x - \frac{1}{2}y = 2$, and solve for x.

$$x - \frac{1}{2}y = 2$$

$$x - \frac{1}{2}(2x - 4) = 2$$

$$x - x + 2 = 2$$

$$2 = 2 \quad \text{True}$$

Notice that the sum of the x terms is 0, and when simplified x is no longer part of the equation. *Since the statement $2 = 2$ is true, this system has an infinite number of solutions. Therefore, the graphs of the equations represent the same line and the system is dependent.*

Note that the solution in Example 3 is identical to the solution obtained graphically in Example 5 of Section 5.1. Figure 5.4 on page 298 shows that the graphs of both equations are the same line.

NOW TRY EXERCISE 13

EXAMPLE 4 Solve the following system of equations by substitution.

$$2x + 4y = 6$$

$$4x - 2y = -8$$

Solution None of the variables in either equation has a numerical coefficient of 1. However, since the numbers 4 and 6 are both divisible by 2, if you solve the first equation for x, you will avoid having to work with fractions.

$$2x + 4y = 6$$

$$2x = -4y + 6$$

$$\frac{2x}{2} = \frac{-4y + 6}{2}$$

$$x = -\frac{4}{2}y + \frac{6}{2}$$

$$x = -2y + 3$$

Now substitute $-2y + 3$ for x in the other equation, $4x - 2y = -8$, and solve for the remaining variable, y.

$$4x - 2y = -8$$

$$4(-2y + 3) - 2y = -8$$

$$-8y + 12 - 2y = -8$$

$$-10y + 12 = -8$$

$$-10y = -20$$

$$y = 2$$

Finally, solve for x by substituting $y = 2$ in the equation previously solved for x.

$$x = -2y + 3$$
$$x = -2(2) + 3 = -4 + 3 = -1$$

The solution is $(-1, 2)$.

HELPFUL HINT

Remember that a solution to a system of linear equations must contain both an x- and a y-value. Don't solve the system for one of the variables and forget to solve for the other. Write the solution as an ordered pair.

EXAMPLE 5 Solve the following system of equations by substitution.

$$4x + 4y = 3$$
$$2x = 2y + 5$$

Solution We will elect to solve for x in the second equation.

$$2x = 2y + 5$$
$$x = \frac{2y + 5}{2}$$
$$x = y + \frac{5}{2}$$

Now substitute $y + \frac{5}{2}$ for x in the other equation.

$$4x + 4y = 3$$
$$4\left(y + \frac{5}{2}\right) + 4y = 3$$
$$4y + 10 + 4y = 3$$
$$8y + 10 = 3$$
$$8y = -7$$
$$y = -\frac{7}{8}$$

Finally, find the value of x.

$$x = y + \frac{5}{2}$$
$$x = -\frac{7}{8} + \frac{5}{2} = -\frac{7}{8} + \frac{20}{8} = \frac{13}{8}$$

NOW TRY EXERCISE 19

The solution is the ordered pair $\left(\frac{13}{8}, -\frac{7}{8}\right)$.

Exercise Set 5.2

Concept/Writing Exercises

1. When solving the system of equations
$$3x + 6y = 9$$
$$4x + 3y = 5$$
by substitution, which variable, in which equation, would you choose to solve for to make the solution easier? Explain your answer.

2. When solving the system of equations
$$4x + 2y = 8$$
$$3x - 9y = 8$$
by substitution, which variable, in which equation, would you choose to solve for to make the solution easier? Explain your answer.

3. When solving a system of linear equations by substitution, how will you know if the system is inconsistent?

4. When solving a system of linear equations by substitution, how will you know if the system is dependent?

Practice the Skills

Find the solution to each system of equations by substitution.

5. $x + 2y = 5$
$2x - 3y = 3$

6. $y = x + 3$
$y = -x - 5$

7. $x + y = -2$
$x - y = 0$

8. $2x + y = 3$
$3y = 9 - 6x$

9. $3x + y = 3$
$3x + y + 5 = 0$

10. $y = 2x + 4$
$y = -2$

11. $x = 3$
$x + y + 5 = 0$

12. $y = 2x - 13$
$-4x - 7 = 9y$

13. $x - \frac{1}{2}y = 2$
$y = 2x - 4$

14. $2x + 3y = 7$
$6x - y = 1$

15. $2x + y = 9$
$y = 4x - 3$

16. $y = -2x + 5$
$x + 3y = 0$

17. $y = \frac{1}{3}x - 2$
$x - 3y = 6$

18. $x = y + 4$
$3x + 7y = -18$

19. $2x + 3y = 7$
$6x - 2y = 10$

20. $4x - 3y = 6$
$2x + 4y = 5$

21. $3x - y = 14$
$6x - 2y = 10$

22. $5x - 2y = -7$
$5 = y - 3x$

23. $4x - 5y = -4$
$3x = 2y - 3$

24. $3x + 4y = 10$
$4x + 5y = 14$

25. $5x + 4y = -7$
$x - \frac{5}{3}y = -2$

26. $\frac{1}{2}x + y = 4$
$3x + \frac{1}{4}y = 6$

Problem Solving

27. In September 1998, the mortgage rates were very low, so Sybil Geraud was considering refinancing her house. She found that the cost of refinancing her house would be a one time charge of \$1200. With her reduced mortgage rate her monthly interest and principal payments would be about \$752. Her total cost, c, for n months could be determined by the equation $c = 1200 + 752n$. At the higher interest rate that she presently has, her mortgage payments are \$832 per month and the total cost for n months can be determined by the equation $c = 832n$. **a)** Determine the number of months for which both mortgage plans would have the same total cost. **b)** If Sybil plans to remain in her house for 12 years, should she refinance?

28. In Seattle, Washington, the temperature is 82°F, but it is decreasing by 2 degrees per hour. The temperature, T, at time, t, in hours, is represented by $T = 82 - 2t$. In Spokane, Washington, the temperature is 64°F, but it is increasing by 2.5 degrees per hour. The temperature, T, can be represented by $T = 64 + 2.5t$. **a)** If the temperature continues decreasing and increasing at the same rate in these cities, how long will it be before

both cities have the same temperature? **b)** When both cities have the same temperature, what will that temperature be?

29. Jim Ruppel, a salesperson for a video tape distribution company, earns a weekly salary of \$300 plus \$3 for each video tape he sells. His weekly salary can be represented by $S = 300 + 3n$. He is being offered a chance to change salary plans. Under the new plan he will earn a weekly salary of \$400 plus \$2 for each video tape he sells. His weekly salary under this new plan can be represented by $S = 400 + 2n$. How many tapes would Jim need to sell in a week for his salary to be the same under both plans?

30. John Tweedt's car is at the 100 mile marker on the interstate highway 15 miles behind Mary Shapiro's car. John's car is traveling at 65 miles per hour and Mary's car is traveling at 60 miles per hour. The mile marker that John will be at in t hours can be found by $m = 100 + 65t$ and the mile marker that Mary will be at in t hours can be found by $m = 115 + 60t$. **a)** Determine the time it will take for John to catch Mary. **b)** At what mile marker will they be when they meet?

Group Activity

Answer parts **a)** *through* **d)** *on your own.*

31. In a laboratory, during an experiment on heat transfer, a large metal ball is heated to a temperature of 180°F. This metal ball is then placed in a gallon of oil at a temperature of 20°F. Assume that when the ball is placed in the oil it loses temperature at the rate of 10 degrees per minute while the oil's temperature rises at a rate of 6 degrees per minute.

a) Write an equation that can be used to determine the ball's temperature t minutes after being placed in the oil.

b) Write an equation that can be used to determine the oil's temperature t minutes after the ball is placed in it.

c) Determine how long it will take for the ball and oil to reach the same temperature.

d) When the ball and oil reach the same temperature, what will the temperature be?

e) Check your results with the other members of your group.

Discuss and work Exercise 32 as a group.

32. In intermediate algebra you may solve systems containing three equations with three variables. As a group, solve the system of equations on the right. Your answer will be in the form of an **ordered triple** (x, y, z).

$$x = 4$$
$$2x - y = 6$$
$$-x + y + z = -3$$

Cumulative Review Exercises

[3.1] **33.** The diameter of a willow tree grows about 3.5 inches per year. What is the approximate age of a willow tree whose diameter is 25 inches?

[3.3] **34.** Steve Salone takes the Transit Authority bus to and from work each day. The one-way fare is \$1.60. The Transit Authority offers a monthly bus pass for \$ 33, which provides unlimited rides. How many days would Steve have to travel to and from work in a month to make it worthwhile for him to purchase the monthly bus pass?

[4.2] **35.** Graph $4x - 8y = 16$ using the intercepts.

[4.4] **36.** Find the slope and y-intercept of the graph of the equation $3x - 5y = 8$.

5.3 SOLVING SYSTEMS OF EQUATIONS BY THE ADDITION METHOD

1) **Solve systems of equations by the addition method.**

SSM VIDEO 5.3 CD Rom

1) Solve Systems of Equations by the Addition Method

A third, and often the easiest, method of solving a system of equations is by the *addition* (*or elimination*) *method. The object of this process is to obtain two equations whose sum will be an equation containing only one variable.* Always keep in mind that our immediate goal is to obtain one equation containing only one unknown.

EXAMPLE 1 Solve the following system of equations using the addition method.

$$x + y = 6$$
$$2x - y = 3$$

Solution Note that one equation contains $+y$ and the other contains $-y$. By adding the equations, we can eliminate the variable y and obtain one equation containing only one variable, x. When added, $+y$ and $-y$ sum to 0, and so the variable y is eliminated.

$$\begin{array}{r} x + y = 6 \\ \underline{2x - y = 3} \\ 3x \qquad = 9 \end{array}$$

Now we solve for the remaining variable, x.

$$\frac{3x}{3} = \frac{9}{3}$$
$$x = 3$$

Finally, we solve for y by substituting $x = 3$ in either of the original equations.

$$x + y = 6$$
$$3 + y = 6$$
$$y = 3$$

The solution is (3, 3).

We check the answer in *both* equations.

CHECK:

$$x + y = 6 \qquad\qquad 2x - y = 3$$
$$3 + 3 \stackrel{?}{=} 6 \qquad\qquad 2(3) - 3 \stackrel{?}{=} 3$$
$$6 = 6 \quad \text{True} \qquad\qquad 6 - 3 \stackrel{?}{=} 3$$
$$3 = 3 \quad \text{True}$$

To Solve a System of Equations by the Addition (or Elimination) Method

1. If necessary, rewrite each equation so that the terms containing variables appear on the left side of the equal sign and any constants appear on the right side of the equal sign.
2. If necessary, multiply one or both equations by a constant(s) so that when the equations are added the resulting sum will contain only one variable.
3. Add the equations. This will result in a single equation containing only one variable.
4. Solve for the variable in the equation from step 3.
5. Substitute the value found in step 4 into either of the original equations. Solve that equation to find the value of the remaining variable.
6. Check the values obtained in all original equations.

In step 2 we indicate it may be necessary to multiply one or both equations by a constant. In this text we will use brackets [ ], to indicate that both sides of the equation within the brackets are to be multiplied by some constant. Thus, for example, $2[x + y = 1]$ means that both sides of the equation $x + y = 1$ are to be multiplied by 2. We write

$$2[x + y = 1] \quad \text{gives} \quad 2x + 2y = 2$$

Similarly, $-3[4x - 2y = 5]$ means both sides of the equation $4x - 2y = 5$ are to be multiplied by -3. We write

$$-3[4x - 2y = 5] \quad \text{gives} \quad -12x + 6y = -15$$

The use of this notation may make it easier for you to follow the procedure used to solve the problem.

EXAMPLE 2 Solve the following system of equations using the addition method.

$$x + 2y = 6$$
$$x + 3y = 5$$

Solution The object of the addition process is to obtain two equations whose sum will be an equation containing only one variable. If we add these two equations, none of the variables will be eliminated. However, if we multiply either equation by -1 and then add, the terms containing x will sum to 0, and we will accomplish our goal. We will multiply the top equation by -1.

$$-1[x + 2y = 6] \quad \text{gives} \quad -x - 2y = -6$$
$$x + 3y = 5 \qquad\qquad x + 3y = 5$$

Remember that both sides of the equation must be multiplied by −1. This process changes the sign of each term in the equation being multiplied without changing the solution to the system of equations. Now add the two equations on the right.

$$\begin{array}{r} -x - 2y = -6 \\ \underline{x + 3y = 5} \\ y = -1 \end{array}$$

Now we solve for x in either of the original equations.

$$\begin{aligned} x + 2y &= 6 \\ x + 2(-1) &= 6 \\ x - 2 &= 6 \\ x &= 8 \end{aligned}$$

A check will show that the solution is (8, −1).

EXAMPLE 3 Solve the following system of equations using the addition method.

$$\begin{aligned} 2x + y &= 11 \\ x + 3y &= 18 \end{aligned}$$

Solution To eliminate the variable x, multiply the second equation by −2 and add the two equations.

$$\begin{aligned} 2x + y &= 11 & & & 2x + y &= 11 \\ -2[x + 3y &= 18] & &\text{gives} & -2x - 6y &= -36 \end{aligned}$$

Now add:

$$\begin{array}{r} 2x + y = 11 \\ \underline{-2x - 6y = -36} \\ -5y = -25 \\ y = 5 \end{array}$$

Solve for x.

$$\begin{aligned} 2x + y &= 11 \\ 2x + 5 &= 11 \\ 2x &= 6 \\ x &= 3 \end{aligned}$$

The solution is (3, 5).

Note that the solution in Example 3 is the same as the solutions obtained graphically in Example 3 of Section 5.1 and by substitution in Example 1 of Section 5.2.

In Example 3, we could have multiplied the first equation by −3 to eliminate the variable y. At this time, rework Example 3 by eliminating the variable y to see that you get the same answer.

EXAMPLE 4 Solve the following system of equations using the addition method.

$$\begin{aligned} 4x + 2y &= -18 \\ -2x - 5y &= 10 \end{aligned}$$

Solution To eliminate the variable x, we can multiply the second equation by 2 and then add.

$$\begin{aligned} 4x + 2y &= -18 & & & 4x + 2y &= -18 \\ 2[-2x - 5y &= 10] & &\text{gives} & -4x - 10y &= 20 \end{aligned}$$

$$\begin{array}{r} 4x + 2y = -18 \\ \underline{-4x - 10y = 20} \\ -8y = 2 \\ y = -\dfrac{1}{4} \end{array}$$

Solve for x.

$$4x + 2y = -18$$
$$4x + 2\left(-\frac{1}{4}\right) = -18$$
$$4x - \frac{1}{2} = -18$$
$$2\left(4x - \frac{1}{2}\right) = 2(-18) \quad \text{Multiply both sides by 2 to remove fractions.}$$
$$8x - 1 = -36$$
$$8x = -35$$
$$x = -\frac{35}{8}$$

The solution is $\left(-\frac{35}{8}, -\frac{1}{4}\right)$.

Check the solution $\left(-\frac{35}{8}, -\frac{1}{4}\right)$ in both equations.

CHECK:

$$4x + 2y = -18 \qquad\qquad -2x - 5y = 10$$
$$4\left(-\frac{35}{8}\right) + 2\left(-\frac{1}{4}\right) \stackrel{?}{=} -18 \qquad\qquad -2\left(-\frac{35}{8}\right) - 5\left(-\frac{1}{4}\right) \stackrel{?}{=} 10$$
$$-\frac{35}{2} - \frac{1}{2} \stackrel{?}{=} -18 \qquad\qquad \frac{35}{4} + \frac{5}{4} \stackrel{?}{=} 10$$
$$-\frac{36}{2} \stackrel{?}{=} -18 \qquad\qquad \frac{40}{4} \stackrel{?}{=} 10$$
$$-18 = -18 \quad \text{True} \qquad\qquad 10 = 10 \quad \text{True}$$

Note that the solution to Example 4 contains fractions. You should not always expect to get integers as answers.

EXAMPLE 5 Solve the following system of equations using the addition method.

$$2x + 3y = 6$$
$$5x - 4y = -8$$

Solution The variable x can be eliminated by multiplying the first equation by -5 and the second by 2 and then adding the equations.

$$-5[2x + 3y = 6] \quad \text{gives} \quad -10x - 15y = -30$$
$$2[5x - 4y = -8] \quad \text{gives} \quad 10x - 8y = -16$$

$$\begin{aligned} -10x - 15y &= -30 \\ 10x - 8y &= -16 \\ \hline -23y &= -46 \\ y &= 2 \end{aligned}$$

Solve for x.

$$\begin{aligned} 2x + 3y &= 6 \\ 2x + 3(2) &= 6 \\ 2x + 6 &= 6 \\ 2x &= 0 \\ x &= 0 \end{aligned}$$

The solution is $(0, 2)$.

In Example 5, the same value could have been obtained for y by multiplying the first equation by 5 and the second by -2 and then adding. Try it now and see.

NOW TRY EXERCISE 19

EXAMPLE 6 Solve the following system of equations using the addition method.

$$\begin{aligned} 2x + y &= 3 \\ 4x + 2y &= 12 \end{aligned}$$

Solution The variable y can be eliminated by multiplying the first equation by -2 and then adding the two equations.

$$\begin{aligned} -2[2x + y = 3] \quad &\text{gives} \quad -4x - 2y = -6 \\ 4x + 2y = 12 \quad & \quad\quad\quad\; 4x + 2y = 12 \end{aligned}$$

$$\begin{aligned} -4x - 2y &= -6 \\ 4x + 2y &= 12 \\ \hline 0 &= 6 \quad \textit{False} \end{aligned}$$

Since $0 = 6$ is a false statement, this system has no solution. The system is inconsistent. The graphs of the equations will be parallel lines.

Note that the solution in Example 6 is identical to the solutions obtained by graphing in Example 4 of Section 5.1 and by substitution in Example 2 of Section 5.2.

NOW TRY EXERCISE 25

EXAMPLE 7 Solve the following system of equations using the addition method.

$$\begin{aligned} x - \frac{1}{2}y &= 2 \\ y &= 2x - 4 \end{aligned}$$

Solution First align the x- and y-terms on the left side of the second equation by subtracting $2x$ from both sides of the second equation.

$$\begin{aligned} x - \frac{1}{2}y &= 2 \\ -2x + y &= -4 \end{aligned}$$

Now proceed as in the previous examples.

$$\begin{aligned} 2\left[x - \frac{1}{2}y = 2\right] \quad &\text{gives} \quad 2x - y = 4 \\ -2x + y = -4 \quad & \quad\quad\quad -2x + y = -4 \end{aligned}$$

$$\begin{aligned} 2x - y &= 4 \\ -2x + y &= -4 \\ \hline 0 &= 0 \quad \textit{True} \end{aligned}$$

Since $0 = 0$ is a true statement, the system is dependent and has an infinite number of solutions. When graphed, both equations will be the same line.

NOW TRY EXERCISE 21

The solution in Example 7 is the same as the solutions obtained by graphing in Example 5 of Section 5.1 and by substitution in Example 3 of Section 5.2.

EXAMPLE 8 Solve the following system of equations using the addition method.

$$2x + 3y = 7$$
$$5x - 7y = -3$$

Solution We can eliminate the variable x by multiplying the first equation by -5 and the second by 2.

$$-5[2x + 3y = 7] \quad \text{gives} \quad -10x - 15y = -35$$
$$2[5x - 7y = -3] \quad \text{gives} \quad 10x - 14y = -6$$

$$\begin{aligned} -10x - 15y &= -35 \\ 10x - 14y &= -6 \\ \hline -29y &= -41 \\ y &= \frac{41}{29} \end{aligned}$$

We can now find x by substituting $y = \frac{41}{29}$ into one of the original equations and solving for x. If you try this, you will see that although it can be done, the calculations are messy. An easier method of solving for x is to go back to the original equations and eliminate the variable y.

$$7[2x + 3y = 7] \quad \text{gives} \quad 14x + 21y = 49$$
$$3[5x - 7y = -3] \quad \text{gives} \quad 15x - 21y = -9$$

$$\begin{aligned} 14x + 21y &= 49 \\ 15x - 21y &= -9 \\ \hline 29x \qquad &= 40 \\ x &= \frac{40}{29} \end{aligned}$$

NOW TRY EXERCISE 35

The solution is $\left(\frac{40}{29}, \frac{41}{29}\right)$.

HELPFUL HINT

We have illustrated three methods for solving a system of linear equations: graphing, substitution, and addition. When you are given a system of equations, which method should you use to solve the system? When you need an exact solution, graphing should not be used. Of the two algebraic methods, the addition method may be easier to use if there are no numerical coefficients of 1 in the system. If one or more of the variables has a coefficient of 1, you can use either substitution or addition.

Exercise Set 5.3

Concept/Writing Exercises

1. When solving the following system of equations by the addition method, what will your first step be in solving the system? Explain your answer. Do not solve the system.

$$-x + 3y = 4$$
$$2x + 5y = 2$$

2. When solving the following system of equations by the addition method, what will your first step be in solving the system? Explain your answer. Do not solve the system.

$$2x + 4y = -8$$
$$3x - 2y = 10$$

3. When solving a system of linear equations by the addition method, how will you know if the system is inconsistent?

4. When solving a system of linear equations by the addition method, how will you know if the system is dependent?

Practice the Skills

Solve each system of equations using the addition method.

5. $x + y = 6$
   $x - y = 4$

6. $x - y = 6$
   $x + y = 8$

7. $-x + y = 5$
   $x + y = 1$

8. $x + y = 10$
   $-x + y = -2$

9. $x + 2y = 15$
   $x - 2y = -7$

10. $3x + y = 10$
    $4x - y = 4$

11. $4x + y = 6$
    $-8x - 2y = 20$

12. $6x + 3y = 30$
    $4x + 3y = 18$

13. $-5x + y = 14$
    $-3x + y = -2$

14. $x - y = 2$
    $3x - 3y = 0$

15. $3x + y = 10$
    $3x - 2y = 16$

16. $-4x + 3y = 0$
    $5x - 6y = 9$

17. $4x - 3y = 8$
    $2x + y = 14$

18. $2x - 3y = 4$
    $2x + y = -4$

19. $5x + 3y = 12$
    $2x - 4y = 10$

20. $6x - 4y = 9$
    $2x - 8y = 3$

21. $4x - 2y = 6$
    $y = 2x - 3$

22. $4x - 2y = -4$
    $-3x - 4y = -30$

23. $3x - 2y = -2$
    $3y = 2x + 4$

24. $5x + 4y = 10$
    $-3x - 5y = 7$

25. $5x - 4y = 1$
    $-10x + 8y = -4$

26. $2x - 3y = 11$
    $-3x = -5y - 17$

27. $3x - 5y = 0$
    $2x + 3y = 0$

28. $4x - 2y = 8$
    $4y = 8x - 16$

29. $-5x + 4y = -20$
    $3x - 2y = 15$

30. $5x = 2y - 4$
    $3x - 5y = 6$

31. $7x - 3y = 4$
    $2y = 4x - 6$

32. $4x - 3y = -4$
    $3x - 5y = 10$

33. $4x + 5y = 0$
    $3x = 6y + 4$

34. $4x - 3y = 8$
    $-3x + 4y = 9$

35. $x - \frac{1}{2}y = 4$
    $3x + y = 6$

36. $2x - \frac{1}{3}y = 6$
    $5x - y = 4$

Problem Solving

37. Construct a system of two equations that has no solution. Explain how you know the system has no solution.

38. Construct a system of two equations that has an infinite number of solutions. Explain how you know the system has an infinite number of solutions.

39. a) Solve the system of equations

$$4x + 2y = 1000$$
$$2x + 4y = 800$$

b) If we divide all the terms in the top equation by 2 we get the following system:

$$2x + y = 500$$
$$2x + 4y = 800$$

How will the solution to this system compare to the solution in part **a)**? Explain and then check your explanation by solving this system.

40. Suppose we divided all the terms in both equations given in Exercise 39 **a)** by 2, and then solved the system. How will the solution to this system compare to the solution in part **a)**? Explain and then check your explanation by solving each system.

Challenge Problems

Solve each system of equations using the addition method. (Hint: First remove all fractions by multiplying both sides of the equation by the LCD.)

41. $\frac{x+2}{2} - \frac{y+4}{3} = 4$

$\frac{x+y}{2} = \frac{1}{2} + \frac{x-y}{3}$

42. $\frac{5}{2}x + 3y = \frac{9}{2} + y$

$\frac{1}{4}x - \frac{1}{2}y = 6x + 12$

Group Activity

*Work parts **a)** and **b)** of Exercise 43 on your own. Then discuss and work parts **c)** and **d)** as a group.*

43. How difficult is it to construct a system of linear equations that has a specific solution? It is really not too difficult to do. Consider:

$$2(3) + 4(5) = 26$$
$$4(3) - 7(5) = -23$$

The system of equations

$$2x + 4y = 26$$
$$4x - 7y = -23$$

has solution (3, 5).

a) Using the information provided, determine another system of equations that has (3, 5) as a solution.

b) Determine a system of linear equations that has (2, 3) as a solution.

c) Compare your answer with the answers of the other members of your group.

d) As a group, determine the number of systems of equations that have (2, 3) as a solution.

Discuss and work Exercise 44 as a group.

44. In intermediate algebra you may solve systems of three equations with three unknowns. Solve the following system.

$$x + 2y - z = 2$$
$$2x - y + z = 3$$
$$2x + y + z = 7$$

(*Hint:* Work with *one pair* of equations to get one equation in two unknowns. Then work with *a different pair* of the original equations to get another equation in the same two unknowns. Then solve the system of two equations in two unknowns.)

Cumulative Review Exercises

[1.9] **45.** Evaluate 5^3.

[2.5] **46.** Solve the equation $2(2x - 3) = 2x + 8$.

[2.7] **47.** Solve the inequality $2x - 4 < 4x - 2$ and graph the solution on a number line.

[4.6] **48.** If $f(x) = 2x^2 - 4$, find $f(-3)$.

5.4 SYSTEMS OF EQUATIONS: APPLICATIONS AND PROBLEM SOLVING

1) **Use systems of equations to solve application problems.**

SSM VIDEO 5.4 CD Rom

1) Use Systems of Equations to Solve Application Problems

The method you use to solve a system of equations may depend on whether you wish to see "the entire picture" or are interested in finding the exact solution. If you are interested in the trend as the variable changes, you might decide to graph the equations. If you want only the solution—that is, the ordered pair common to both equations—you might use one of the two algebraic methods to find the common solution.

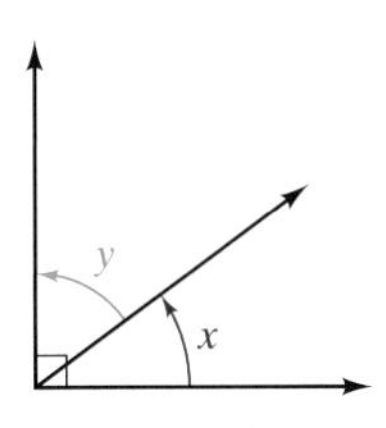

FIGURE 5.12

Many of the application problems solved in earlier chapters using only one variable can also be solved using two variables. In Example 1 we use **complementary angles**, which are two angles whose sum measures 90°. Figure 5.12 illustrates complementary angles x and y.

EXAMPLE 1 Hal Balmer is building a rectangular patio out of cement (Fig. 5.13). When he measures the two angles formed by a diagonal he finds that angle x is 24° greater than angle y. Find the two angles.

Solution **Understand and Translate** A rectangle has 4 right (or 90°) angles. Angles x and y are therefore complementary, and the sum of angles x and y is 90°. Thus one equation in the system of equations is $x + y = 90$. Since angle x is 24° greater than angle y, the second equation is $x = y + 24$.

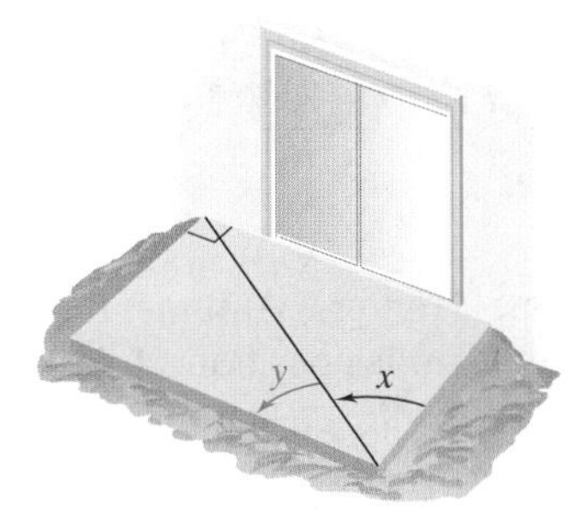

FIGURE 5.13

System of equations $\begin{cases} x + y = 90 \\ x = y + 24 \end{cases}$

Carry Out Subtract y from each side of the second equation. Then use the addition method to solve.

$$\begin{aligned} x + y &= 90 \\ x - y &= 24 \\ \hline 2x &= 114 \\ x &= 57 \end{aligned}$$

Now substitute 57 for x in the first equation and solve for y.

$$\begin{aligned} x + y &= 90 \\ 57 + y &= 90 \\ y &= 33 \end{aligned}$$

Check and Answer Angle x is 57° and angle y is 33°. Note that their sum is 90° and angle x is 24° greater than angle y.

EXAMPLE 2 Nancy Moritz brings a rectangular picture she bought at a flea market into an art supply and framing store to have the picture framed. The clerk measures the picture and tells Nancy that the perimeter of the picture is 144 inches and the length of the picture is twice its width. Later, Nancy wonders what the dimensions of the picture are and figures it out using algebra. Determine the dimensions of the picture.

Solution **Understand and Translate** The formula for the perimeter of a rectangle is $P = 2l + 2w$.

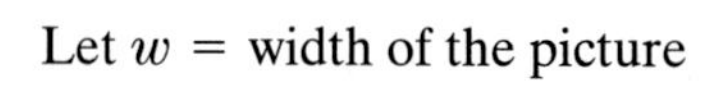

Let w = width of the picture

l = length of the picture

Since the perimeter is 144 inches, one equation in the system is $144 = 2l + 2w$. Since the length is twice the width, the other equation is $l = 2w$.

System of equations
$$\begin{cases} 144 = 2l + 2w \\ l = 2w \end{cases}$$

Carry Out Solve this system by substitution. Since $l = 2w$, substitute $2w$ for l in the equation $144 = 2l + 2w$ to obtain

$$144 = 2l + 2w$$
$$144 = 2(2w) + 2w$$
$$144 = 4w + 2w$$
$$144 = 6w$$
$$24 = w$$

Check and Answer The width is 24 inches. Since the length is twice the width, the length is 2(24) or 48 inches. Notice that the perimeter is $2(48) + 2(24) = 144$ inches.

NOW TRY EXERCISE 7

EXAMPLE 3 Major Peter Ostroushko is a pilot with the Hurricane Hunters. Assume that during one storm he flies with the wind (a tailwind) at 600 miles per hour. Leaving the storm, he flies against the wind (a headwind) at 450 miles per hour. If he did not adjust his plane's speed, find the speed of the plane in still air and the speed of the wind.

Solution **Understand and Translate**

Let p = speed of the plane in still air

w = speed of the wind

If p equals the speed of the plane in still air and w equals the speed of the wind, then $p + w$ equals the speed of the plane flying with the wind, and $p - w$ equals the speed of the plane flying against the wind. Make use of this information when writing the system of equations.

Speed of plane flying with wind: $p + w = 600$
Speed of plane flying against wind: $p - w = 450$
System of equations

Carry Out Now use the addition method because the sum of the terms containing w is zero.

$$\begin{aligned} p + w &= 600 \\ p - w &= 450 \\ \hline 2p \quad &= 1050 \\ p &= 525 \end{aligned}$$

Now substitute 525 for p in the first equation and solve for w.

$$\begin{aligned} p + w &= 600 \\ 525 + w &= 600 \\ w &= 75 \end{aligned}$$

Answer The plane's speed is 525 miles per hour in still air, and the wind's speed is 75 miles per hour.

NOW TRY EXERCISE 13

EXAMPLE 4 Only two-axle vehicles are permitted to cross a bridge that leads to Honeymoon Island State Park. The toll for the bridge is 50 cents for motorcycles and $1.00 for cars and trucks. On Saturday, the toll booth attendant collected a total of $150, and the vehicle counter recorded 170 vehicles crossing the bridge. How many motorcycles and how many cars and trucks crossed the bridge that day?

Solution **Understand and Translate**

$$\begin{aligned} \text{Let } x &= \text{number of motorcycles} \\ y &= \text{number of cars and trucks} \end{aligned}$$

Since a total of 170 vehicles crossed the bridge, one equation is $x + y = 170$. The second equation comes from the tolls collected.

$$\begin{array}{ccccc} \text{Tolls from motorcycles} & + & \text{tolls from cars and trucks} & = 150 \\ 0.50x & + & 1.00y & = 150 \end{array}$$

System of equations $\begin{cases} x + y = 170 \\ 0.50x + 1.00y = 150 \end{cases}$

Carry Out Since the first equation can be easily solved for y, solve this system by substitution. Solving for y in $x + y = 170$ gives $y = 170 - x$. Substitute $170 - x$ for y in the second equation and solve for x.

$$\begin{aligned} 0.50x + 1.00y &= 150 \\ 0.50x + 1.00(170 - x) &= 150 \\ 0.50x + 170 - 1.00x &= 150 \\ 170 - 0.5x &= 150 \\ -0.5x &= -20 \\ \frac{-0.5x}{-0.5} &= \frac{-20}{-0.5} \\ x &= 40 \end{aligned}$$

Answer Forty motorcycles crossed the bridge. The total number of vehicles that crossed the bridge is 170. Therefore, $170 - 40$ or 130 cars and trucks crossed the bridge that Saturday. A check will show that the total collected for these 170 vehicles is $150.

EXAMPLE 5 Kenna Ose needs to purchase a new engine for her car and have it installed by a mechanic. She is considering two garages: Sally's garage and Scotty's garage. At Sally's garage, the parts cost $800 and the labor cost is $25 per hour. At Scotty's garage, the parts cost $575 and labor cost is $50 per hour.

a) How many hours would the repairs need to take for the total cost at each garage to be the same?

b) If both garages estimate that the repair will take 8 hours, which garage would be the least expensive?

Solution **a) Understand and Translate** We will write equations for the total cost for both Sally's garage and Scotty's garage.

Let n = number of hours of labor

c = total cost of repairs

Sally's	**Scotty's**
total cost = parts + labor	total cost = parts + labor
$c = 800 + 25n$	$c = 575 + 50n$

Now we have our system of equations.

System of equations $$\begin{cases} c = 800 + 25n \\ c = 575 + 50n \end{cases}$$

Carry Out total cost at Sally's = total cost at Scotty's

$$800 + 25n = 575 + 50n$$
$$225 + 25n = 50n$$
$$225 = 25n$$
$$9 = n$$

Answer If 9 hours of labor is required, the cost at both garages would be equal.

b) If repairs take 8 hours, Scotty's would be less expensive, as shown below.

Sally's	**Scotty's**
$c = 800 + 25n$	$c = 575 + 50n$
$c = 800 + 25(8)$	$c = 575 + 50(8)$
$c = 1000$	$c = 975$

Simple Interest Problems

We introduced the simple interest formula, interest = principal × rate × time or $i = prt$ in Section 3.1. Now we will use this formula to solve a simple interest problem using a system of equations. In Example 6 we will organize the given information in a table.

EXAMPLE 6 Angel Martinez has invested a total of $12,000 in two savings accounts. One account earns 5% simple interest and the other earns 8% simple interest. Find the amount invested in each account if he receives a total of $840 interest after 1 year.

Solution **Understand and Translate** We use the information provided to determine our system of equations.

Let x = principal invested at 5%

y = principal invested at 8%

Account Type	Principal	Rate	Time	Interest
5% account	x	0.05	1	$0.05x$
8% account	y	0.08	1	$0.08y$

Since the total interest is $840, one of our equations is

$$0.05x + 0.08y = 840$$

Because the total principal invested is $12,000, our second equation is

$$x + y = 12{,}000$$

Thus,

$$\left.\begin{aligned} 0.05x + 0.08y &= 840 \\ x + y &= 12{,}000 \end{aligned}\right\} \quad \text{System of equations}$$

Carry Out We will multiply our first equation by 100 to eliminate the decimal numbers. This gives the system

$$5x + 8y = 84{,}000$$

$$x + y = 12{,}000$$

To eliminate the x, we will multiply the second equation by -5 and then add the result to the first equation.

$$\begin{aligned} 5x + 8y &= 84{,}000 \\ -5[x + y &= 12{,}000] \end{aligned} \qquad \text{gives} \qquad \begin{aligned} 5x + 8y &= 84{,}000 \\ -5x - 5y &= -60{,}000 \\ \hline 3y &= 24{,}000 \\ y &= 8000 \end{aligned}$$

Now we solve for x.

$$x + y = 12{,}000$$

$$x + 8000 = 12{,}000$$

$$x = 4000$$

NOW TRY EXERCISE 21

Answer Thus, $4000 is invested at 5% and $8000 is invested at 8%.

Motion Problems with Two Rates

We introduced the distance formula, distance = rate · time or $d = rt$, in Section 3.5. Now we introduce a method using two variables and a system of equations to solve motion problems that involve two rates. Often when working motion problems with two different rates it is helpful to construct a table to summarize the information given. We will do this in Examples 7 and 8.

EXAMPLE 7 It is a beautiful day, so Martel Perman decides to take his motorboat out for the day. His trip begins at the Tennessee River near Savannah, Tennessee. He travels with the current down the river to Florence, Alabama, a distance of 48 miles. The trip from Savannah to Florence takes him 3 hours. The return trip, against the current, takes him 4 hours.

a) Find the speed of the motorboat in still water.

b) Find the current.

Solution **a) Understand and Translate** Let us make a sketch of the situation (Fig. 5.14).

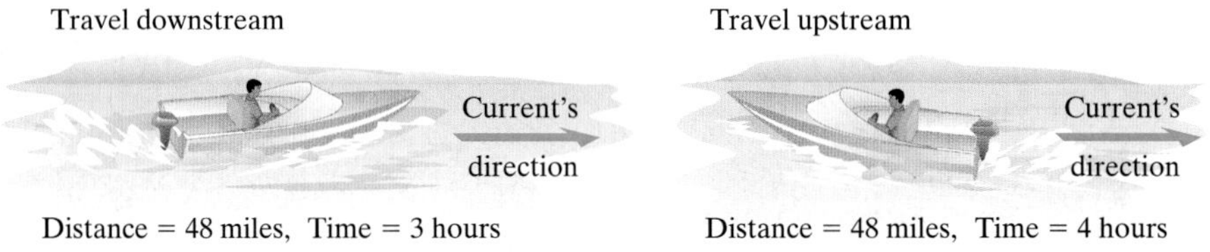

Travel downstream

Current's direction

Distance = 48 miles, Time = 3 hours

Travel upstream

Current's direction

Distance = 48 miles, Time = 4 hours

FIGURE 5.14

Let m = speed of the motorboat in still water

c = the current

Boat's Direction	Rate	Time	Distance
With current	$m + c$	3	$3(m + c)$
Against current	$m - c$	4	$4(m - c)$

Since the distances traveled downstream and upstream are both 48 miles, our system of equations is

$$3(m + c) = 48$$
$$4(m - c) = 48$$

Carry Out If we divide both sides of the top equation by 3 and both sides of the bottom equation by 4, we obtain a simplified system of equations.

$$\frac{3(m + c)}{3} = \frac{48}{3} \qquad \frac{4(m - c)}{4} = \frac{48}{4}$$
$$m + c = 16 \qquad m - c = 12$$

Now we solve the simplified system of equations.

$$\begin{aligned} m + c &= 16 \\ m - c &= 12 \\ \hline 2m \quad &= 28 \\ m &= 14 \end{aligned}$$

Answer Therefore, the speed of the boat in still water is 14 miles per hour.

b) The current may be found by substituting 14 for m in either of the simplified equations. We will use $m + c = 16$.

$$\begin{aligned} m + c &= 16 \\ 14 + c &= 16 \\ c &= 2 \end{aligned}$$

Thus, the current is 2 miles per hour.

EXAMPLE 8 Marlene Olsavsky has been rollerblading at 5 miles per hour for 0.75 hour along a trail near Pebble Beach, California. She receives a phone call on her cellular phone from her friend Richard Rowe. Richard tells her that he is where she started, at the beginning of the trail. He tells her to keep rollerblading and says that he will ride his bike to catch up with her. Richard plans to ride at 10.5 miles per hour.

a) How long after Richard starts will Marlene be 1 mile ahead of him?

b) How long after Richard starts will they meet?

c) When they meet, how far from the trail's beginning will they be?

Solution **a)** **Understand and Translate** We need to find the time it takes for Marlene to be 1 mile ahead of Richard.

$$\text{Let } x = \text{time Marlene is rollerblading}$$

$$y = \text{time Richard is riding}$$

Traveler	Rate	Time	Distance
Marlene	5	x	$5x$
Richard	10.5	y	$10.5y$

Since Marlene began 0.75 hour before Richard, our first equation is

$$x = y + 0.75$$

Note that if we add 0.75 hour to Richard's time we get the time Marlene has been rollerblading.

Our second equation is obtained from the fact that the distance between Marlene and Richard must be 1 mile. Since Marlene is ahead of Richard, we subtract his distance from hers.

$$\text{Marlene's distance} - \text{Richard's distance} = 1 \text{ mile}$$
$$5x - 10.5y = 1$$

Our system of equations is

$$x = y + 0.75$$
$$5x - 10.5y = 1$$

Carry Out The first equation, $x = y + 0.75$, is already solved for x. Substituting $y + 0.75$ for x in the second equation, we get

$$5x - 10.5y = 1$$
$$5(y + 0.75) - 10.5y = 1$$
$$5y + 3.75 - 10.5y = 1$$
$$-5.5y + 3.75 = 1$$
$$-5.5y = -2.75$$
$$y = \frac{-2.75}{-5.5} = 0.5$$

Check and Answer Richard will have been biking for 0.5 hours when the distance between them is 1 mile. When Richard has been biking for 0.5 hours, Marlene has been rollerblading for $y + 0.75$ or $0.5 + 0.75$ or 1.25 hours. Check to verify that this answer is correct.

b) Understand and Translate We will still use the equation relating their times, $x = y + 0.75$, as the first equation in our system of equations. Our second equation comes from the fact that when they meet they both will have traveled the same distance. Therefore, our second equation is $5x = 10.5y$.

System of equations

$$\begin{cases} x = y + 0.75 \\ 5x = 10.5y \end{cases}$$

Carry Out We substitute $y + 0.75$ for x in the second equation.

$$\begin{aligned} 5x &= 10.5y \\ 5(y + 0.75) &= 10.5y \\ 5y + 3.75 &= 10.5y \\ 3.75 &= 5.5y \\ \frac{3.75}{5.5} &= y \\ 0.68 &\approx y \end{aligned}$$

Answer Thus, they will meet about 0.68 hours after Richard begins biking.

c) The distance from their starting point can be found using the distance formula. We will find the distance Richard traveled.

$$\begin{aligned} d &= rt \\ d &\approx 10.5(0.68) \approx 7.14 \end{aligned}$$

Thus, they will be approximately 7.14 miles from the trail's beginning when they meet.

NOW TRY EXERCISE 29

Mixture Problems

Mixture problems were solved with one variable in Section 3.5. Now we will solve mixture problems using two variables and systems of equations. Recall that any problem in which two or more quantities are combined to produce a different quantity, or a single quantity is separated into two or more quantities, may be considered a mixture problem. Again, we will use a table to summarize the information given.

EXAMPLE 9 Colleen Odell owns a candy and nut shop. In the shop she has both bulk and packaged candy and nuts. A customer comes into the shop and explains that he needs a large amount of a mixture of chocolate covered cherries and amaretto cordials. Each chocolate covered cherry and each amaretto cordial is individually wrapped in the bulk food bins. The customer explains that he plans to make individual packages of the mixture and give a package to each guest at a dinner party. The chocolate covered cherries sell for \$7.50 per pound and the amaretto cordials sell for \$6.00 per pound.

a) How many pounds of the amaretto cordials must be mixed with 12 pounds of chocolate covered cherries to obtain a mixture that sells for \$6.50 per pound?

b) How many pounds of the mixture will there be?

Solution **a) Understand and Translate** We are asked to find the number of pounds of amaretto cordials to be mixed.

Let x = number of pounds of amaretto cordials

y = number of pounds of mixture

Often it is helpful to make a sketch of the situation. After we draw a sketch, we will construct a table. In our sketch we will use bins to mix the candy (Fig. 5.15).

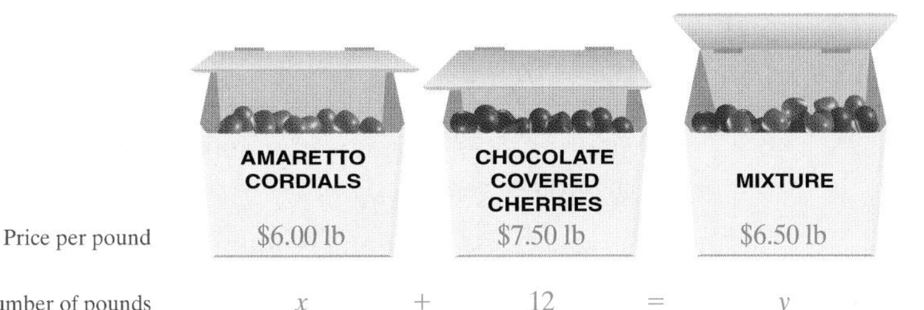

FIGURE 5.15

The value of the candy is found by multiplying the number of pounds by the price per pound.

Candy	Price	Number of Pounds	Value of Candy
Cordials	6	x	$6x$
Cherries	7.50	12	$7.50(12)$
Mixture	6.50	y	$6.50y$

Our two equations come from the following information:

$$\begin{pmatrix}\text{number of pounds}\\ \text{of cordials}\end{pmatrix} + \begin{pmatrix}\text{number of pounds}\\ \text{of cherries}\end{pmatrix} = \begin{pmatrix}\text{number of pounds}\\ \text{of mixture}\end{pmatrix}$$

$$x \quad + \quad 12 \quad = \quad y$$

$$\text{value of cordials} + \text{value of cherries} = \text{value of mixture}$$

$$6x \quad + \quad 7.50(12) \quad = \quad 6.50y$$

System of equations
$$\begin{cases} x + 12 = y \\ 6x + 7.50(12) = 6.50y \end{cases}$$

Carry Out Since $y = x + 12$, we substitute $x + 12$ for y in the second equation and solve for x.

$$\begin{aligned} 6x + 7.50(12) &= 6.50y \\ 6x + 7.50(12) &= 6.50(x + 12) \\ 6x + 90 &= 6.50x + 78 \\ 90 &= 0.50x + 78 \\ 12 &= 0.50x \\ 24 &= x \end{aligned}$$

Answer Thus, 24 pounds of the amaretto cordials must be mixed with 12 pounds of the chocolate covered cherries.

b) The total mixture will weigh 24 + 12 or 36 pounds.

Now we will work an example similar to Example 9 in Section 3.5, but this time we will use a system of equations to solve the problem.

EXAMPLE 10 Eilish Main is a chemistry professor. She has a number of labs coming up shortly where her students will need to use a 15% solution of sulfuric acid. Her assistant informs her that they are out of the 15% solution and do not have sufficient time to reorder. Eilish checks the supply room and finds they have a large supply of both 5% and 30% sulfuric acid solutions. She decides to use the 5% and 30% solutions to make 60 liters of a 15% solution. How many liters of the 5% solution and the 30% solution should she mix?

Solution **Understand and Translate** Eilish will combine the 5% and 30% solutions to get 60 liters of a 15% solution of sulfuric acid. We need to determine how much of each should be mixed.

$$\text{Let } x = \text{number of liters of 5\% solution}$$
$$y = \text{number of liters of 30\% solution}$$

The problem is displayed in Figure 5.16 and the information is summarized in the following table.

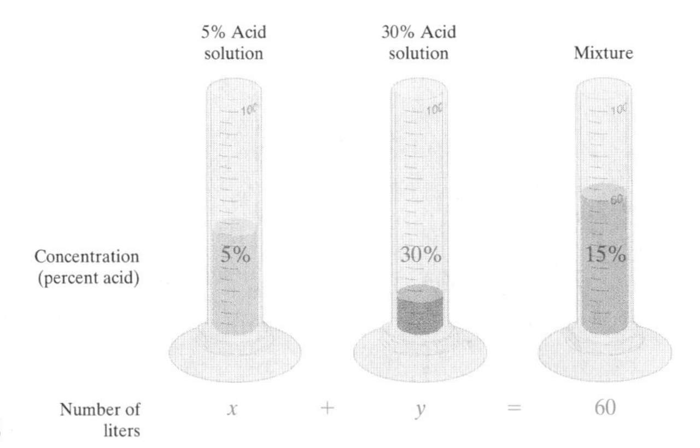

FIGURE 5.16

Solution	Number of Liters	Concentration	Acid Content
5% Solution	x	0.05	$0.05x$
30% Solution	y	0.30	$0.30y$
Mixture	60	0.15	0.15(60)

Because the total volume of the combination is 60 liters, we have

$$x + y = 60$$

From the table we see that

$$\begin{pmatrix}\text{acid content of}\\ \text{5\% solution}\end{pmatrix} + \begin{pmatrix}\text{acid content of}\\ \text{30\% solution}\end{pmatrix} = \begin{pmatrix}\text{acid content}\\ \text{of mixture}\end{pmatrix}$$

$$0.05x \quad + \quad 0.30y \quad = \quad 0.15(60)$$

System of equations
$$\begin{cases} x + y = 60 \\ 0.05x + 0.30y = 0.15(60) \end{cases}$$

Carry Out We will solve this system by substitution. First we solve for y in the first equation.

$$x + y = 60$$
$$y = 60 - x$$

Then we substitute $60 - x$ for y in the second equation.

$$0.05x + 0.30y = 0.15(60)$$
$$0.05x + 0.30(60 - x) = 9$$
$$0.05x + 18 - 0.30x = 9$$
$$-0.25x + 18 = 9$$
$$-0.25x = -9$$
$$x = \frac{-9}{-0.25} = 36$$

Now we solve for y.

$$y = 60 - x$$
$$y = 60 - 36$$
$$y = 24$$

Answer Thus, 36 liters of the 5% acid solution should be mixed with 24 liters of the 30% acid solution to obtain 60 liters of a 15% acid solution.

NOW TRY EXERCISE 31

Exercise Set 5.4

Practice the Skills and Problem Solving

Express each exercise as a system of linear equations, then find the solution. Use a calculator where appropriate.

1. The sum of two integers is 29. Find the numbers if one number is 3 greater than the other.
2. The difference of two integers is 25. Find the two numbers if the larger is 1 less than three times the smaller.
3. Angles A and B are complementary angles. If angle B is 18° greater than angle A, find the measure of each angle. (See Example 1.)
4. Two angles are **supplementary angles** when the sum of their measures is 180°. If angles A and B are supplementary angles, and angle A is four times as large as angle B, find the measure of each angle.
5. If angles A and B are supplementary angles (see Exercise 4) and angle A is 48° greater than angle B, find the measure of each angle.
6. A rectangular picture frame will be made from a piece of molding 60 inches long. What dimensions will the frame have if the length is to be 6 inches greater than the width? (See Example 2.)
7. Stacy Prock just purchased a new high resolution television. She noticed that the perimeter of the screen is 124 inches and the width of the screen is 8 inches greater than its height. Find the dimensions of the screen.
8. Lois Heater, an insurance agent, sent out a total of 230 envelopes by first class mail. Some weighed under 1 ounce and required a 33-cent stamp. The rest weighed 1 ounce or more but less than 2 ounces and required 55-cents postage. If her total postage cost was $106.70, determine how many envelopes were sent at each postage rate.

9. Celeste Nossiter plants corn and wheat on her 100-acre farm near Albuquerque, New Mexico. She estimates that her income after deducting expenses is \$450 per acre of corn and \$430 per acre of wheat. Find the number of acres of corn and wheat planted if her total income after expenses is \$44,400.

10. Countryside Hand Car Wash sells coupon books for \$21 which contain 20 coupons. Each coupon allows the customer to have his or her car washed for \$8.00 instead of the regular price of \$11.50.

a) After how many car washes, using a coupon with each wash, would the amount the customer saves equal the cost of the coupon book?

b) If Boris purchased and used 2 coupon books in a year, how much would he have saved?

11. On Oct 1, 1998, Barbara Reidell bought five times as many shares of Disney stock as she did of Microsoft stock. The Disney stock cost \$26 a share and the Microsoft stock cost \$85 a share. If her total cost for all the stock was \$10,750, how many shares of each did she purchase?

12. Two commercial airplanes are flying in the same vicinity but in opposite directions. The pilot of an American Airlines jet flying with the wind reports that his plane's speed is 560 miles per hour. The pilot of a US Airways jet flying against the wind indicates that his plane's speed is 510 miles per hour. Jim Spivey, in the control tower at a nearby airport, reports to both pilots that if it were not for the wind, they would be flying at identical speeds. Find the speed of the planes in still air and the speed of the wind (see Example 3).

13. Shane Stagg is kayaking in the St. Lawrence River. He can paddle 4.7 miles per hour with the current and 3.4 miles per hour against the current. Find the speed of the kayak in still water and the current.

See Exercise 13.

14. The population of Green Mountain is 40,000 and it is growing by 800 per year. The population of Pleasant View Valley is 66,000 and it is decreasing by 500 per year. How long will it take for both areas to have the same population?

15. Sol's Club Discount Warehouse has two membership plans. Under plan A the customer pays a \$50 annual membership and 85% of the manufacturer's recommended list price. Under plan B the annual membership fee is \$100 and the customer pays 80% of the manufacturer's recommended list price. How much merchandise, in dollars, would one have to purchase to pay the same amount under both plans?

16. Susan Summerlin, a salesperson, is considering two job offers. She would be selling the same product at each company. At the Medtec Company, Susan's salary would be \$300 per week plus a 5% commission of sales. At the Genzone Company, her salary would be \$200 per week plus an 8% commission of sales.

a) What weekly dollar volume of sales would Susan need to make for the total income from both companies to be the same?

b) If she expects to make sales of \$4000, which company would give the greater salary?

17. Lisa Nodora just purchased a high speed copier for her office and wants to purchase a service contract on the copier. She is considering two sources for the contract. The Kate Spence Copier Sales and Service Company charges \$18 a month plus 2 cents a copy. Office Copier Depot charges \$25 a month but only 1.5 cents a copy.

a) Assuming the prices do not change, how many copies would Lisa need to make for the monthly cost of both plans to be the same?

b) If Lisa plans to make 2500 copies a month, which plan would be the least expensive?

18. Chris Suit is a Financial Planner for Walsh and Associates. His salary is a flat 40% commission of sales. As an employee he has no overhead. He is considering starting his own company. Then 100% of sales would be income to him. However, he estimates his monthly overhead for office rent, secretary, utilities, and so on, would be about \$1500 per month.

a) How much in sales would Chris's own company need to make in a month for him to make the same income he does as an employee of Walsh and Associates?

b) Suppose when Chris opens his own office that in addition to the $1500 per month overhead, he has a one-time cost of $6000 for the purchase of office equipment. If he estimates his monthly sales at $3000, how long would it take to recover the initial $6000 cost?

19. The following bar graph shows the cost per square yard of the same carpet at two different carpet stores along with the installation cost (including matting) at each store.

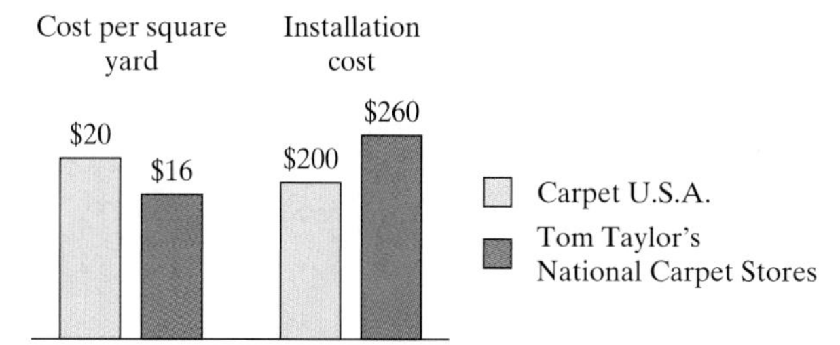

a) Find the number of square yards of carpet that Dorothy Rosene must purchase for the total cost of the carpet plus installation to be the same from both stores.

b) If she needs to purchase and have installed 25 square yards of carpet, which store will be less expensive?

20. Jorge Perez has written a book that he plans to self-publish. He has consulted two printers to determine the cost of printing his book. Each company has a one-time setup charge and a charge for each book printed. The cost information is shown on the following bar graph.

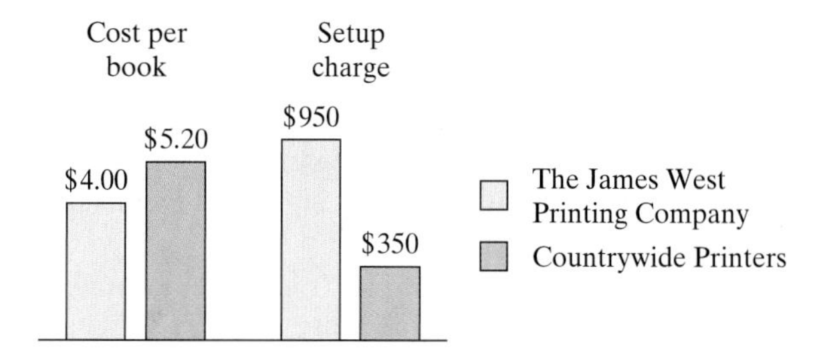

a) How many books would Jorge need to have printed for the total cost of his books to be the same from both printers?

b) If he plans to print 1000 books, which printer would be the least expensive?

Review Example 6 before working Exercises 21–24.

21. Mr. and Mrs. Vinny McAdams invest a total of $8000 in two savings accounts. One account yields 10% simple interest and the other 8% simple interest. Find the amount placed in each account if they receive a total of $750 in interest after 1 year.

22. Carol Horton invested a total of $10,000. Part of the money was placed in a savings account paying 5% simple interest. The rest was placed in a fixed annuity paying 6% simple interest. If the total interest received for the year was $540, how much had been invested in each account?

23. John Bragg invested $10,000 in a CD paying a specific simple interest rate and $6000 in a money market account paying a different simple interest rate. The total interest received from both accounts after 1 year was $740. If the interest rate of the CD is 1% (or 0.01) greater than the interest rate of the money market account, find the interest rates of the CD and the money market account.

24. Christine Culman invested $12,000 in a CD paying a specific simple interest rate and $10,000 in a money market account paying a different simple interest rate. The total interest received from both accounts after 1 year is $1120. If the interest rate of the CD is 2% (or 0.02) greater than the interest rate of the money market account, find both simple interest rates.

Review Examples 7 and 8 before working Exercises 25–30.

25. During a race, Elizabeth Kell's speed boat travels 4 miles per hour faster than Melissa Suarez's boat. If Elizabeth's boat finishes the race in 3 hours and Melissa's finishes the race in 3.2 hours, find the speed of each boat.

26. Dave Visser started driving from Columbus, Ohio, toward Lincoln, Nebraska—a distance of 903 miles. At the same time Alice Harra started driving to Columbus, Ohio, from Lincoln, Nebraska. If the two meet after 7 hours and Alice's speed averages 15 miles per hour greater than Dave's speed, find the speed of each car.

27. Two Lucent Technology cable crews are digging a ditch to lay fiber-optic cable. One crew, headed by John Mayleben, and a second crew, headed by Leigh Sumeral, start at opposite ends of a 30-mile stretch of land. John's crew digs at an average speed that is 0.1 miles per hour faster than Leigh's crew. If the two crews meet after 50 hours, find the average speed of each crew.

28. Sometimes two trains need to use the same tracks. When this happens, a switch on the tracks is thrown before the trains reach one another to move one of the trains to different tracks. Two trains are 560 miles apart using the same tracks traveling toward each other. One train is traveling 6 miles per hour faster than the other. If the tracks were not switched, the trains would meet in 4 hours. Find the speed of the trains.

29. Amanda Rodriguez and Delores Melendez go jogging along the River Walk Trail in San Antonio, Texas. They start at the same point, but Amanda starts 0.3 hours before Delores does. If Amanda jogs at a rate of 5 miles per hour and Delores jogs at a rate of 8 miles per hour, how long after Delores starts will Delores catch up to Amanda?

30. Bill Leonard trots his horse Trixie east at 8 miles per hour. One half-hour later, Mary Mullaley starts at the same point and canters her horse Pegarno west at 16 miles per hour. How long after Mary starts riding will Mary and Bill be separated by 10 miles?

Review Examples 9 and 10 before working Exercises 31–38.

31. Theresa Morgan, a chemist, has a 25% hydrochloric acid solution and a 50% hydrochloric acid solution. How many liters of each should she mix to get 10 liters of a hydrochloric acid solution with a 40% acid concentration?

32. Tamuka Williams, a pharmacist, needs 1000 milliliters of a 10% phenobarbital solution. She has only 5% and 25% phenobarbital solutions available. How many milliliters of each solution should she mix to obtain the desired solution?

33. Julie Hildebrand is selecting tile for her foyer and living room. She wants to make a pattern using two different colors and types of tile. One type costs \$3 per tile (per square foot) and the other type costs \$5 per tile (per square foot). She needs a total of 380 tiles but does not want to spend more than \$1500 on the tile. What is the maximum number of \$5 tiles she can purchase?

34. Fred Whittingham offers Donna Stansell a glass of chardonnay wine, with some ice in it. The wine originally had an alcohol content of 13% by volume. Donna receives a lengthy phone call during which the ice melts. The alcohol content of the 2.6 ounce mixture after the ice melts is 10%. Find the volume of wine and the volume of the ice that was added.

35. Wayne Froelich, a dairy farmer, has milk that is 5% butterfat and skim milk without butterfat. How much 5% milk and how much skim milk should he mix to make 100 gallons of milk that is 3.5% butterfat?

36. Lynn Hicks wishes to mix soybean meal that is 16% protein and cornmeal that is 7% protein to get a 300-pound mixture that is 10% protein. How much of each should be used?

37. The All Natural Juice Company sells apple juice for 12 cents an ounce and apple drink for 6 cents an ounce. They wish to market and sell for 10 cents an ounce cans of juice drink that are part juice and part drink. How many ounces of each will be used if the juice drink is to be sold in 8-ounce cans?

38. Pierre LaRue's recipe for quiche lorraine calls for 16 ounces (or 2 cups) of light cream, which is 20% buttermilk. It is often difficult to find light cream with 20% buttermilk at the supermarket. What is commonly found is heavy cream, which is 36% buttermilk, and half-and-half, which is 10.5% buttermilk. How many ounces of the heavy cream and how much of the half-and-half should be mixed to obtain 16 ounces of light cream that is 20% buttermilk?

---

39. A brother and sister, Sean and Meghan O'Donnell, jog to school daily. Sean, who is older, jogs at 9 miles per hour. Meghan jogs at 5 miles per hour. When Sean reached the school, Meghan is $\frac{1}{2}$ mile away. How far is the school from their house?

40. By weight, an alloy of brass is 70% copper and 30% zinc. Another alloy of brass is 40% copper and 60% zinc. How many grams of each of these alloys must be melted and combined to obtain 300 grams of a brass alloy that is 60% copper and 40% zinc?

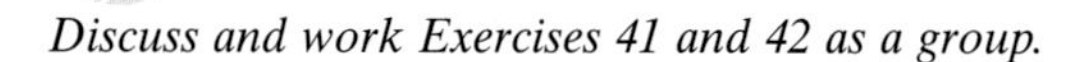

Discuss and work Exercises 41 and 42 as a group.

41. Two pressurized tanks are connected by a controlled pressure valve, as shown in the figure.

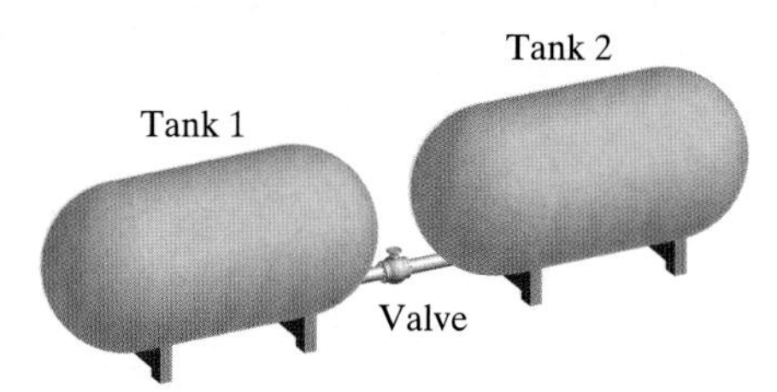

Initially, the internal pressure in tank 1 is 200 pounds per square inch, and the internal pressure in tank 2 is 20 pounds per square inch. The pressure valve is opened slightly to reduce the pressure in tank 1 by 2 pounds per square inch per minute. This increases the pressure in tank 2 by 2 pounds per square inch per minute. At this rate, how long will it take for the pressure to be equal in both tanks?

42. Debby Patterson is considering cars for purchase. Car A has a list price of $16,500 and gets an average of 40 miles per gallon. Car B has a list price of $15,500 and gets an average of 20 miles per gallon. Being a conservationist, Debby wishes to purchase car A but is concerned about its greater initial cost. She plans to keep the car for many years. If she purchases car A, how many miles would she need to drive for the total cost of car A to equal the total cost of car B? Assume gasoline costs of $1.25 per gallon.

[1.10] **43.** Name the properties illustrated.

a) $x + 4 = 4 + x$

b) $(3x)y = 3(xy)$

c) $4(x + 2) = 4x + 8$

[2.5] **44.** Solve the equation $3x + 4 = -(x - 6)$.

[3.4] **45.** The perimeter of a rectangle is 22 feet. Find the dimensions of the rectangle if the length is two more than twice the width.

[4.1] **46.** What is a graph?

5.5 SOLVING SYSTEMS OF LINEAR INEQUALITIES

1) Solve systems of linear inequalities graphically.

SSM VIDEO 5.5 CD Rom

1) Solve Systems of Linear Inequalities Graphically

In Section 4.5, we learned how to graph linear inequalities in two variables. In Section 5.1, we learned how to solve systems of equations graphically. In this section, we discuss how to solve systems of linear inequalities graphically. The **solution to a system of linear inequalities** is the set of points that satisfies all inequalities in the system. Although a system of linear inequalities may contain more than two inequalities, in this book, except in the Group Activity Exercises, we will consider systems with only two inequalities.

To Solve a System of Linear Inequalities Graphically

Graph each inequality on the same axes. The solution is the set of points that satisfies all the inequalities in the system.

EXAMPLE 1 Determine the solution to the following system of inequalities.

$$x + 2y \le 6$$
$$y > 2x - 4$$

Solution First graph the inequality $x + 2y \le 6$ (Fig. 5.17).

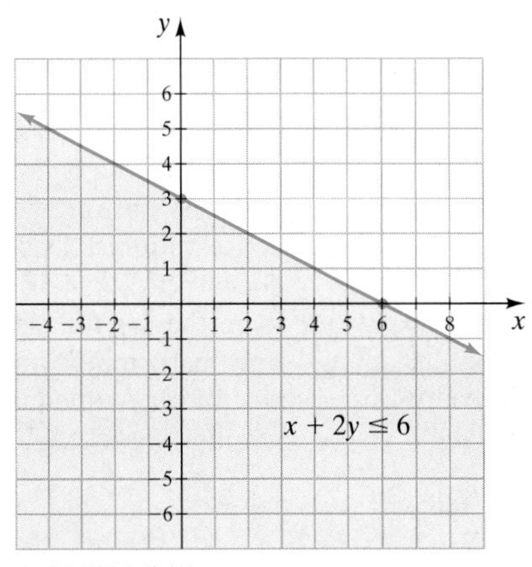

FIGURE 5.17

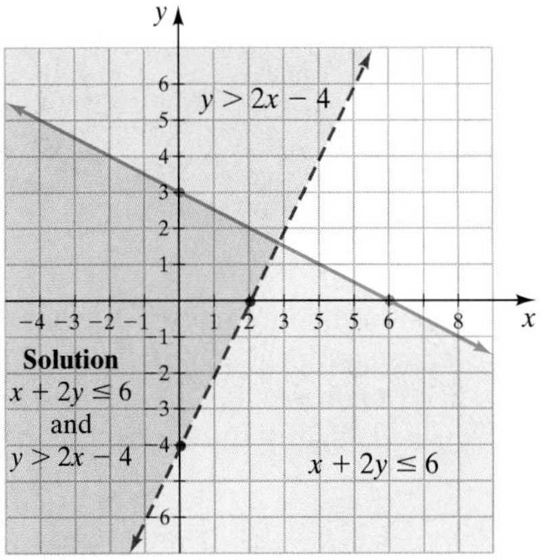

FIGURE 5.18

Now, on the same axes, graph the inequality $y > 2x - 4$ (Fig. 5.18). Note that the line is dashed. Why?

The solution is the set of points common to both inequalities—the part of the graph that contains both shadings (the purple color). The dashed line is not part of the solution. However, the part of the solid line that satisfies both inequalities is part of the solution.

EXAMPLE 2 Determine the solution to the system of inequalities.

$$2x + 3y \ge 4$$
$$2x - y > -6$$

Solution Graph $2x + 3y \ge 4$ (Fig. 5.19). Graph $2x - y > -6$ on the same axes (Fig. 5.20). The solution is the part of the graph with both shadings and the part of the solid line that satisfies both inequalities.

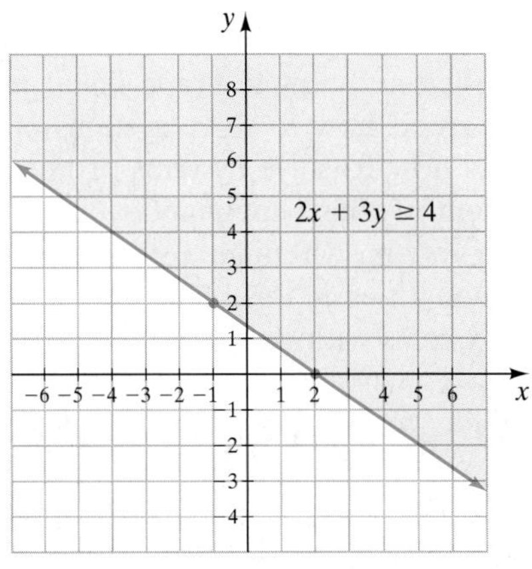

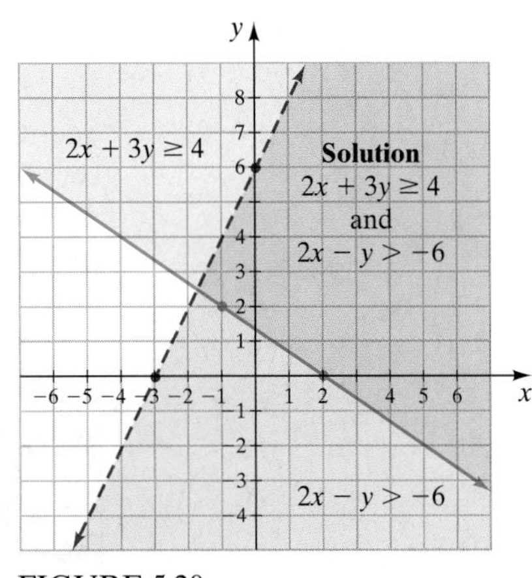

NOW TRY EXERCISE 11

FIGURE 5.19

FIGURE 5.20

EXAMPLE 3 Determine the solution to the following system of inequalities.

$$y < 2$$
$$x > -3$$

Solution Graph both inequalities on the same axes (Fig. 5.21). The solution is the part of the graph that contains both shadings.

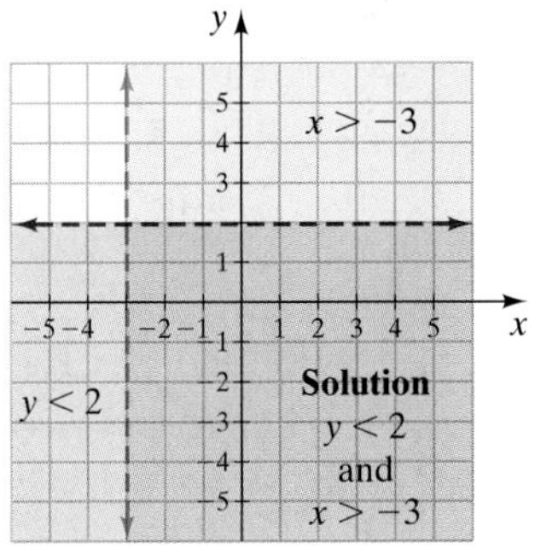

NOW TRY EXERCISE 15

FIGURE 5.21

Exercise Set 5.5

1. If an ordered pair satisfies both inequalities in a system of linear inequalities, must that ordered pair be in the solution to the system? Explain.
2. If an ordered pair satisfies only one inequality in a system of linear inequalities, is it possible for that ordered pair to be in the solution to the system? Explain.
3. Can a system of linear inequalities have no solution? Explain your answer with the use of your own example.
4. Is it possible to construct a system of two nonparallel linear inequalities that has no solution? Explain.

Determine the solution to each system of inequalities.

5. $x + y > 2$, $x - y < 2$
6. $y \leq 3x - 2$, $y > -4x$
7. $y \leq x$, $y < -2x + 4$
8. $2x + 3y < 6$, $4x - 2y \geq 8$
9. $y > x + 1$, $y \geq 3x + 2$
10. $x + 3y \geq 6$, $2x - y > 4$
11. $x - 2y < 6$, $y \leq -x + 4$
12. $y \leq 3x + 4$, $y < 2$
13. $4x + 5y < 20$, $x \geq -3$
14. $3x - 4y \leq 12$, $y > -x + 4$
15. $x \leq 4$, $y \geq -2$
16. $x \geq 0$, $y \leq 0$
17. $x > -3$, $y > 1$
18. $4x + 2y > 8$, $y \leq 2$
19. $-2x + 3y \geq 6$, $x + 4y \geq 4$

20. Construct a system of two linear inequalities that has no solution. Explain how you determined your answer.
21. Is it possible for a system of two linear inequalities to have only one solution? Explain.

Group Activity

In more advanced mathematics courses and when working in many industries, you may need to graph more than two linear inequalities. When a system has more than two inequalities, the solution is the point or points that satisfy all inequalities in the system. As a group, determine the solutions to the systems of inequalities in Exercises 22 and 23.

22. $x + 2y \leq 6$
$2x - y < 2$
$y > 2$

23. $x \geq 0$
$y \geq 0$
$y \leq 2x + 4$
$y \leq -x + 6$

Cumulative Review Exercises

[2.7] **24.** Solve the inequality and graph the solution on a number line: $6(x - 2) < 4x - 3 + 2x$.

[3.1] **25.** Solve the equation $2x - 5y = 6$ for y.

[4.2] **26.** Graph $2x + y = 4$ by plotting points.

[4.3] **27.** A line passes through the points $(-4, 6)$ and $(3, -2)$. Find the slope of the line.

SUMMARY

Key Words and Phrases

5.1
Consistent system of equations
Dependent system of equations
Graphical method
Inconsistent system of equations
Solution to a system of linear equations
System of linear equations

5.2
Ordered triple
Substitution method

5.3
Addition (or elimination) method

5.4
Complementary angles
Supplementary angles

5.5
Solution to a system of linear inequalities
System of linear inequalities

IMPORTANT FACTS

Three methods that can be used to solve a system of linear equations are the (1) graphical method, (2) substitution method, and (3) addition (or elimination) method. When solving a system of linear equations by either substitution or the addition method, if you get a false statement, such as $6 = 0$, the system is inconsistent and has no solution. If you get a true statement, such as $0 = 0$, the system is dependent and has an infinite number of solutions.

Review Exercises

[5.1] *Determine which, if any, of the ordered pairs satisfy each system of equations.*

1. $y = 3x - 2$
$2x + 3y = 5$
a) $(0, -2)$ **b)** $(2, 4)$ **c)** $(1, 1)$

2. $y = -x + 4$
$3x + 5y = 15$
a) $\left(\frac{5}{2}, \frac{3}{2}\right)$ **b)** $(0, 4)$ **c)** $\left(\frac{1}{2}, \frac{3}{5}\right)$

Identify each system of linear equations as consistent, inconsistent, or dependent. State whether the system has exactly one solution, no solution, or an infinite number of solutions.

3.

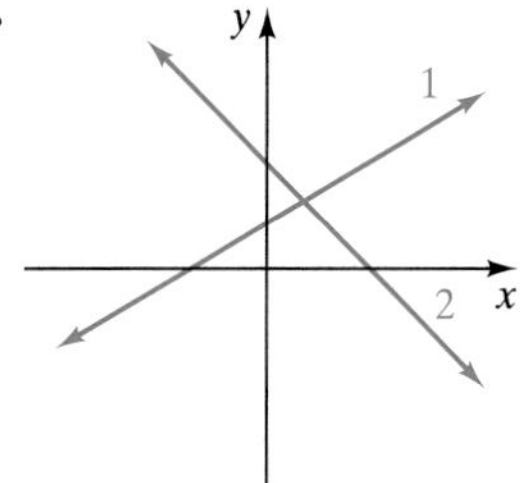

4.

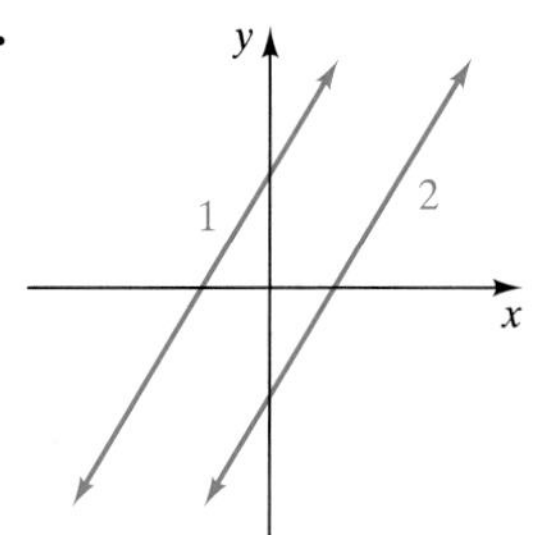

5.

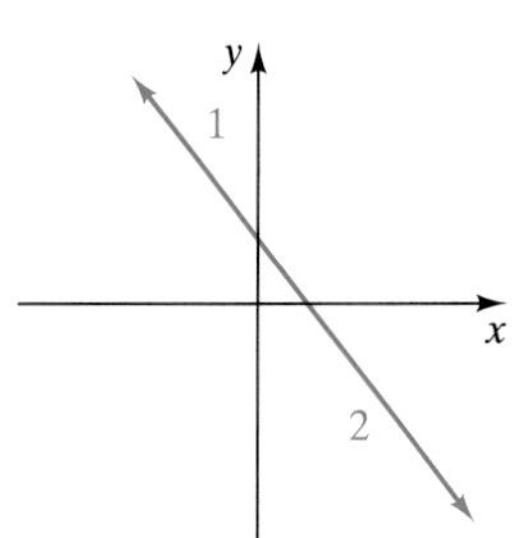

6.

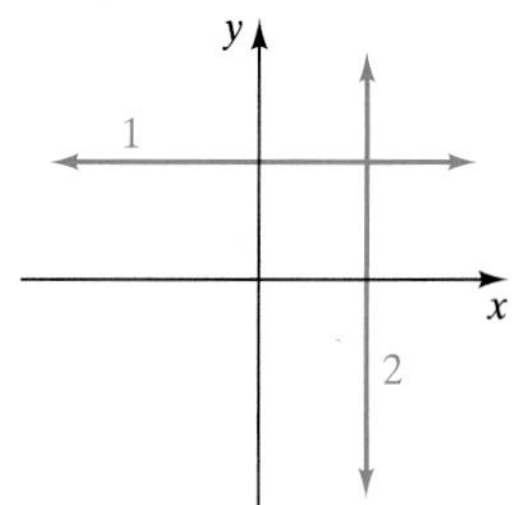

Write each equation in slope–intercept form. Without graphing or solving the system of equations, state whether the system of linear equations has exactly one solution, no solution, or an infinite number of solutions.

7. $x + 2y = 8$
$3x + 6y = 12$

8. $y = -3x - 6$
$2x + 5y = 8$

9. $y = \frac{1}{2}x - 4$
$x - 2y = 8$

10. $6x = 4y - 8$
$4x = 6y + 8$

Determine the solution to each system of equations graphically.

11. $y = x - 5$
$y = 2x - 8$

12. $x = -2$
$y = 3$

13. $y = 3$
$y = -2x + 5$

14. $x + 3y = 6$
$y = 2$

15. $x + 2y = 8$
$2x - y = -4$

16. $y = x - 3$
$2x - 2y = 6$

17. $2x + y = 0$
$4x - 3y = 10$

18. $x + 2y = 4$
$\frac{1}{2}x + y = -2$

[5.2] *Find the solution to each system of equations by substitution.*

19. $y = 2x - 8$
$2x - 5y = 0$

20. $x = 3y - 9$
$x + 2y = 1$

21. $2x - y = 6$
$x + 2y = 13$

22. $x = -3y$
$x + 4y = 6$

23. $4x - 2y = 10$
$y = 2x + 3$

24. $2x + 4y = 8$
$4x + 8y = 16$

25. $2x - 3y = 8$
$6x + 5y = 10$

26. $4x - y = 6$
$x + 2y = 8$

[5.3] *Find the solution to each system of equations using the addition method.*

27. $x + y = 6$
$x - y = 10$

28. $x + 2y = -3$
$2x - 2y = 6$

29. $x + y = 12$
$2x + y = 5$

30. $4x - 3y = 8$
$2x + 5y = 8$

31. $-2x + 3y = 15$
$3x + 3y = 10$

32. $2x + y = 9$
$-4x - 2y = 4$

33. $3x + 4y = 10$
$-6x - 8y = -20$

34. $2x - 5y = 12$
$3x - 4y = -6$

[5.4] *Express each exercise as a system of linear equations, then find the solution.*

35. The sum of two integers is 48. Find the two numbers if the larger is 3 less than twice the smaller.

36. A plane flies 600 miles per hour with the wind and 530 miles per hour against the wind. Find the speed of the wind and the speed of the plane in still air.

37. ABC Truck Rental charges $20 per day plus 50 cents per mile, while Murtz Truck Rental charges $35 per day plus 40 cents per mile. How far would you have to travel in one day for the total cost from both rental companies to be the same?

38. Moura Hakala invested a total of $16,000. Part of the money was placed in a savings account paying 4% simple interest. The rest was placed in a savings account paying 6% simple interest. If the total interest received for the year was $760, how much had she invested in each account?

39. Liz Wood drives from Charleston, South Carolina to Louisville, Kentucky—a distance of 600 miles. At the same time, Mary Mayer starts driving from Louisville to Charleston along the same route. If the two meet after driving 5 hours and Mary's average speed was 6 miles per hour greater than Liz's, find the average speed of each car.

40. Green Turf's grass seed costs 60 cents a pound and Agway's grass seed costs 45 cents a pound. How many pounds of each were used to make a 40-pound mixture that cost $20.25?

41. A chemist has a 30% acid solution and a 50% acid solution. How much of each must be mixed to get 6 liters of a 40% acid solution?

[5.5] *Determine the solution to each system of inequalities.*

42. $x + y > 2$
$2x - y \le 4$

43. $2x - 3y \le 6$
$x + 4y > 4$

44. $2x - 6y > 6$
$x > -2$

45. $x < 2$
$y \ge -3$

Practice Test

1. Determine which, if any, of the ordered pairs satisfy the system of equations.

$$x + 2y = -6$$
$$3x + 2y = -12$$

a) $(0, -6)$ **b)** $\left(-3, -\frac{3}{2}\right)$ **c)** $(2, -4)$

Identify each system as consistent, inconsistent, or dependent. State whether the system has exactly one solution, no solution, or an infinite number of solutions.

2.

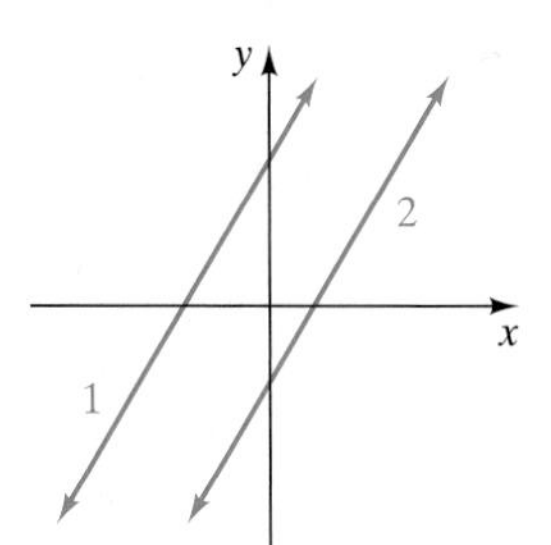

3.

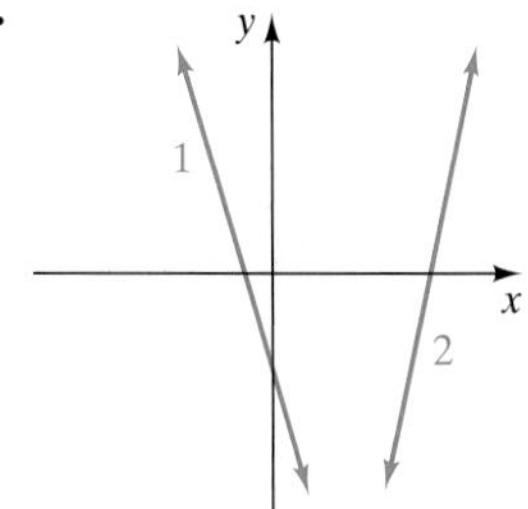

4.

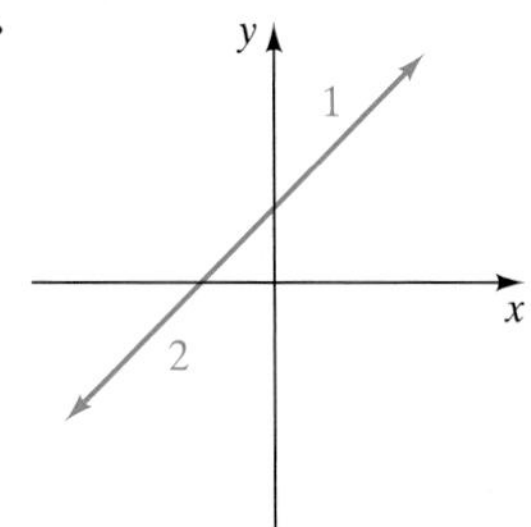

Write each equation in slope–intercept form. Then determine, without graphing or solving the system, whether the system of equations has exactly one solution, no solution, or an infinite number of solutions.

5. $3y = 6x - 9$
$2x - y = 6$

6. $3x + 2y = 10$
$3x - 2y = 10$

7. $4x = 6y - 12$
$2x - 3y = -6$

8. When solving a system of linear equations by the substitution or the addition methods, how will you know if the system is **a)** inconsistent, **b)** dependent?

Solve each system of equations graphically.

9. $y = 3x - 2$
$y = -2x + 8$

10. $3x - 2y = -3$
$3x + y = 6$

11. $y = 2x + 4$
$4x - 2y = 6$

Solve each system of equations by substitution.

12. $3x + y = 8$
$x - y = 6$

13. $4x - 3y = 9$
$2x + 4y = 10$

14. $y = 5x - 7$
$y = 3x + 5$

Solve each system of equations using the addition method.

15. $2x + y = 5$
$x + 3y = -10$

16. $3x + 2y = 12$
$-2x + 5y = 8$

17. $5x - 10y = 20$
$x = 2y + 4$

Solve each system of equations using the method of your choice.

18. $y = 3x - 4$
$2x + y = 6$

19. $3x + 5y = 20$
$6x + 3y = -12$

20. $4x - 6y = 8$
$3x + 5y = 10$

Express each exercise as a system of linear equations, and then find the solution.

21. Budget Rent a Car Agency charges $40 per day plus 8 cents per mile to rent a certain model car. Hertz charges $45 per day plus 3 cents per mile to rent the same car. How many miles will have to be driven in one day for the cost of Budget's car to equal the cost of Hertz's car?

22. Albert's Grocery sells individually wrapped lemon candies for $6.00 a pound and individually wrapped butterscotch candies for $4.50 a pound. How much of each must Albert mix to get 20 pounds of a mixture that he can sell for $5.00 per pound?

23. During a race, Dante Hull's speed boat travels 4 miles per hour faster than Deja Rocket's speed boat. If Dante's boat finishes the race in 3 hours and Deja's finishes the race in 3.2 hours, find the speed of each boat.

Determine the solution to each system of inequalities.

24. $2x + 4y < 8$
$x - 3y \geq 6$

25. $x + 3y \geq 6$
$y < 3$

Cumulative Review Test

1. The following bar graph shows deaths and damages caused by hurricanes from 1900 through 1997.

Hurricane Deaths and Damage Costs

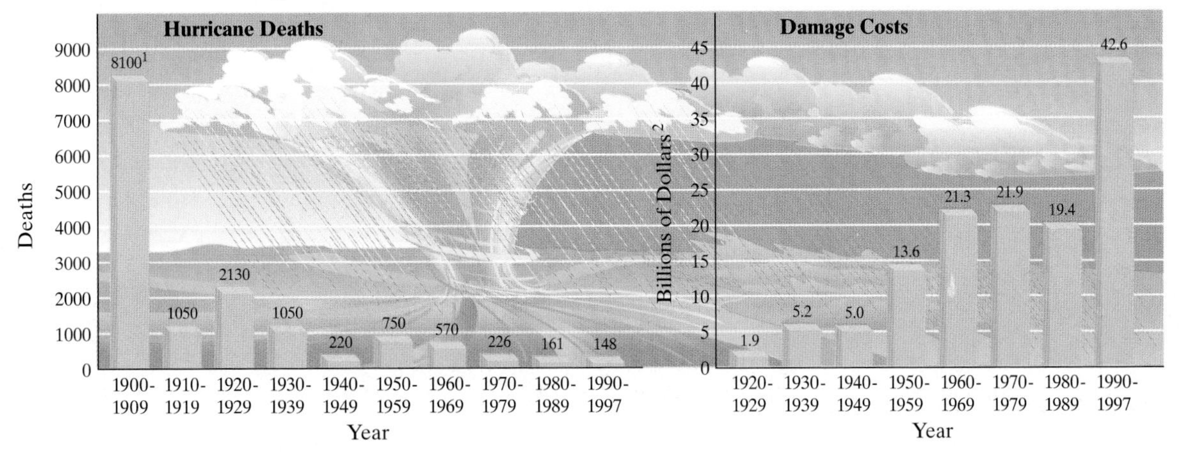

Data Source: The USA TODAY Weather Book, revised edition, by Jack Williams.
1 – Includes the Sept. 8, 1900 Galveston, Texas, hurricane that killed more than 6000 people
2 – Adjusted to 1997 values.

a) In which time period were there the most deaths?

b) In which time period were there the fewest deaths?

c) In which time period was there the most property damage?

2. Janalee Morse's first five test scores were 83, 78, 92, 70, and 99. Find her median test score.

3. Find the mean of the five test scores in Exercise 2.

4. Add $\frac{3}{8} + \frac{9}{10}$.

5. Consider the set of numbers $\{-6, -0.2, \frac{3}{5}, \sqrt{7}, -\sqrt{2}, 7, 0, -\frac{5}{9}, 1.34\}$. List the elements that are

a. natural numbers.

b. rational numbers.

c. irrational numbers.

d. real numbers.

6. Which is greater, $|-4|$ or $-|2|$? Explain your answer.

7. Simplify $-64 + 74 + (-192)$.

8. Evaluate $-(2x + 1) - 3x^2$ when $x = 3$.

9. Simplify $4 - (3x - 2) + 2(x + 3)$.

10. Solve the equation $2(x - 4) + 2 = 3x - 4$.

11. A 20-pound bag of fertilizer covers 8000 square feet of lawn. How many pounds of fertilizer are needed to cover 25,000 square feet of lawn?

12. Solve the inequality $3x - 4 \leq x + 6$. Graph the solution on a number line.

13. Solve the formula $P = 2l + 2w$ for w.

14. Maria Gentile recently graduated from college and has accepted a position selling software. She is given a choice of two salary plans. Plan A is a straight 12% commission on sales. Plan B is $350 per week plus 6% commission on sales. How much must Maria sell in a week for both plans to have the same weekly salary?

15. One angle of a triangle measures 20° greater than the smallest angle. The third angle measures 6 times the smallest angle. Find the measure of all three angles.

16. Graph $2x - 4y = 8$.

17. Graph $\frac{1}{3}x + \frac{1}{2}y = 12$.

18. Determine if the system of equations

$$3x - y = 6$$
$$\frac{3}{2}x - 3 = \frac{1}{2}y$$

has one solution, no solution, or an infinite number of solutions. Explain how you determined your answer.

19. Solve the following system of equations graphically.

$$2x + y = 5$$
$$x - 2y = 0$$

20. Solve the following system of equations using the addition method.

$$3x - 2y = 8$$
$$-6x + 3y = 2$$

EXPONENTS AND POLYNOMIALS

CHAPTER 6

Use the Angel Web site at www.prenhall.com/angel to be linked to an internet resource that will help you further explore the following application.

Advancements in technology are making it possible for computers to perform calculations more quickly with each passing year. In October 1998, the White House announced the unveiling of *Blue Pacific*, a computer capable of performing calculations 15,000 times faster than the average desktop computer. To describe its speed another way, it would take you about 63,000 years using a hand calculator to perform the same number of calculations *Blue Pacific* can complete in one second! In Example 5, page 362, we will use scientific notation to determine how long it would take *Blue Pacific* to complete 7 billion calculations.

Preview and Perspective

In this chapter we discuss exponents and polynomials. In Sections 6.1 and 6.2 we discuss the rules of exponents. The rules of exponents are used and expanded upon in Section 9.7 when we discuss fractional exponents. When discussing scientific notation in Section 6.3, we use the rules of exponents to solve real-life application problems that involve very large or very small numbers. A knowledge of scientific notation may also help you in science and other courses.

In Sections 6.4 through 6.6 we explain how to add, subtract, multiply, and divide polynomials. To be successful with this material you must understand the rules of exponents presented in the first two sections of this chapter. *To understand factoring, which is covered in Chapter 7, you need to understand polynomials, especially multiplication of polynomials.* As you will learn, factoring polynomials is the reverse of multiplying polynomials. We will be working with polynomials throughout the book.

6.1 EXPONENTS

1) **Review exponents.**
2) **Learn the rules of exponents.**
3) **Simplify an expression before using the expanded power rule.**

1) Review Exponents

To use polynomials, we need to expand our knowledge of exponents. Exponents were introduced in Section 1.9. Let's review the fundamental concepts. In the expression x^n, x is referred to as the **base** and n is called the **exponent**. x^n is read "x to the nth power."

$$x^2 = \underbrace{x \cdot x}_{\text{2 factors of } x}$$

$$x^4 = \underbrace{x \cdot x \cdot x \cdot x}_{\text{4 factors of } x}$$

$$x^m = \underbrace{x \cdot x \cdot x \cdot \cdots \cdot x}_{m \text{ factors of } x}$$

EXAMPLE 1 Write $xxxxyy$ using exponents.

Solution

$$\underbrace{x\,x\,x\,x}_{\text{4 factors of } x} \quad \underbrace{y\,y}_{\text{2 factors of } y} = x^4y^2$$

Remember, when a term containing a variable is given without a numerical coefficient, the numerical coefficient of the term is assumed to be 1. For example $x = 1x$ and $x^2y = 1x^2y$.

Also recall that when a variable or numerical value is given without an exponent, the exponent of that variable or numerical value is assumed to be 1. For example, $x = x^1$, $xy = x^1y^1$, $x^2y = x^2y^1$, and $2xy^2 = 2^1x^1y^2$.

2) Learn the Rules of Exponents

Now we will learn the rules of exponents.

EXAMPLE 2 Multiply $x^4 \cdot x^3$.

Solution

$$\underbrace{x \cdot x \cdot x \cdot x}_{x^4} \cdot \underbrace{x \cdot x \cdot x}_{x^3} = x^7$$

Example 2 illustrates that when multiplying expressions with the same base we keep the base and *add* the exponents. This is the **product rule for exponents**.

Product Rule for Exponents

$$x^m \cdot x^n = x^{m+n}$$

In Example 2 we showed that $x^4 \cdot x^3 = x^7$. This problem could also be done using the product rule: $x^4 \cdot x^3 = x^{4+3} = x^7$.

EXAMPLE 3 Multiply each expression using the product rule.

a) $3^2 \cdot 3$ **b)** $2^4 \cdot 2^2$ **c)** $x \cdot x^4$ **d)** $x^3 \cdot x^6$ **e)** $y^4 \cdot y^7$

Solution

a) $3^2 \cdot 3 = 3^2 \cdot 3^1 = 3^{2+1} = 3^3$ or 27

b) $2^4 \cdot 2^2 = 2^{4+2} = 2^6$ or 64

c) $x \cdot x^4 = x^1 \cdot x^4 = x^{1+4} = x^5$

d) $x^3 \cdot x^6 = x^{3+6} = x^9$

e) $y^4 \cdot y^7 = y^{4+7} = y^{11}$

NOW TRY EXERCISE 17

Avoiding Common Errors

Note in Example 3**a)** that $3^2 \cdot 3^1$ is 3^3 and not 9^3. When multiplying powers of the same base, *do not multiply the bases.*

CORRECT	INCORRECT
$3^2 \cdot 3^1 = 3^3$	~~$3^2 \cdot 3^1 = 9^3$~~

Example 4 will help you understand the **quotient rule for exponents**.

EXAMPLE 4 Divide $x^5 \div x^3$.

Solution

$$\frac{x^5}{x^3} = \frac{\cancel{x} \cdot \cancel{x} \cdot \cancel{x} \cdot x \cdot x}{\cancel{x} \cdot \cancel{x} \cdot \cancel{x}} = \frac{1x^2}{1} = x^2$$

When dividing expressions with the same base, keep the base and *subtract* the exponent in the denominator from the exponent in the numerator.

Quotient Rule for Exponents

$$\frac{x^m}{x^n} = x^{m-n}, \qquad x \neq 0$$

In Example 4 we showed that $x^5/x^3 = x^2$. This problem could also be done using the quotient rule: $x^5/x^3 = x^{5-3} = x^2$.

EXAMPLE 5 Divide each expression using the quotient rule.

a) $\dfrac{3^5}{3^2}$ **b)** $\dfrac{5^4}{5}$ **c)** $\dfrac{x^{12}}{x^5}$ **d)** $\dfrac{x^9}{x^5}$ **e)** $\dfrac{y^7}{y}$

Solution **a)** $\dfrac{3^5}{3^2} = 3^{5-2} = 3^3$ or 27 **b)** $\dfrac{5^4}{5} = \dfrac{5^4}{5^1} = 5^{4-1} = 5^3$ or 125

c) $\dfrac{x^{12}}{x^5} = x^{12-5} = x^7$ **d)** $\dfrac{x^9}{x^5} = x^{9-5} = x^4$

e) $\dfrac{y^7}{y} = \dfrac{y^7}{y^1} = y^{7-1} = y^6$

NOW TRY EXERCISE 31

Avoiding Common Errors

Note in Example 5**a)** that $3^5/3^2$ is 3^3 and not 1^3. When dividing powers of the same base, *do not divide out the bases.*

CORRECT	INCORRECT
$\dfrac{3^3}{3^1} = 3^2$ or 9	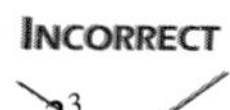

The answer to Example 5**c)**, x^{12}/x^5, is x^7. We obtained this answer using the quotient rule. This answer could also be obtained by dividing out the common factors in both the numerator and denominator as follows.

$$\frac{x^{12}}{x^5} = \frac{(\cancel{x} \cdot \cancel{x} \cdot \cancel{x} \cdot \cancel{x} \cdot \cancel{x}) \cdot x \cdot x \cdot x \cdot x \cdot x \cdot x \cdot x}{(\cancel{x} \cdot \cancel{x} \cdot \cancel{x} \cdot \cancel{x} \cdot \cancel{x})} = x^7$$

We divided out the product of five x's, which is x^5. We can indicate this process in shortened form as follows.

$$\frac{x^{12}}{x^5} = \frac{\cancel{x^5} \cdot x^7}{\cancel{x^5}} = x^7$$

In this section, to simplify an expression when the numerator and denominator have the same base and the exponent in the denominator is greater than the exponent in the numerator, we divide out common factors. For example, x^5/x^{12} can be simplified by dividing out the common factor, x^5, as follows.

$$\frac{x^5}{x^{12}} = \frac{\cancel{x^5}}{\cancel{x^5} \cdot x^7} = \frac{1}{x^7}$$

We will now simplify some expressions by dividing out common factors.

EXAMPLE 6 Simplify by dividing out a common factor in both the numerator and denominator.

a) $\dfrac{x^9}{x^{12}}$ **b)** $\dfrac{y^4}{y^9}$

Solution **a)** Since the numerator is x^9, we write the denominator with a factor of x^9. Since $x^9 \cdot x^3 = x^{12}$, we rewrite x^{12} as $x^9 \cdot x^3$.

$$\frac{x^9}{x^{12}} = \frac{\cancel{x^9}}{\cancel{x^9} \cdot x^3} = \frac{1}{x^3}$$

b) $\frac{y^4}{y^9} = \frac{\cancel{y^4}}{\cancel{y^4} \cdot y^5} = \frac{1}{y^5}$

In the next section, we will show another way to evaluate expressions like $\frac{x^9}{x^{12}}$ by using the negative exponent rule.

Example 7 leads us to our next rule, the **zero exponent rule**.

EXAMPLE 7 Divide $\frac{x^3}{x^3}$.

Solution By the quotient rule,

$$\frac{x^3}{x^3} = x^{3-3} = x^0$$

However,

$$\frac{x^3}{x^3} = \frac{1x^3}{1x^3} = \frac{1 \cdot \cancel{x} \cdot \cancel{x} \cdot \cancel{x}}{1 \cdot \cancel{x} \cdot \cancel{x} \cdot \cancel{x}} = \frac{1}{1} = 1$$

Since $x^3/x^3 = x^0$ and $x^3/x^3 = 1$, then x^0 must equal 1.

Zero Exponent Rule

$$x^0 = 1, \quad x \neq 0$$

By the zero exponent rule, any real number, except 0, raised to the zero power equals 1. Note that 0^0 is undefined.

EXAMPLE 8 Simplify each expression. Assume $x \neq 0$.

a) 3^0 **b)** x^0 **c)** $3x^0$ **d)** $(3x)^0$

Solution **a)** $3^0 = 1$

b) $x^0 = 1$

c) $3x^0 = 3(x^0)$
$= 3 \cdot 1 = 3$

Remember, the exponent refers only to the immediately preceding symbol unless parentheses are used.

d) $(3x)^0 = 1$

Avoiding Common Errors

An expression raised to the zero power is not equal to 0; it is equal to 1.

CORRECT	INCORRECT
$x^0 = 1$	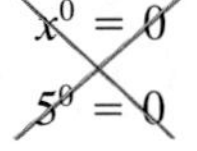
$5^0 = 1$	

The **power rule** will be explained with the aid of Example 9.

EXAMPLE 9 Simplify $(x^3)^2$.

Solution

$$(x^3)^2 = \underbrace{x^3 \cdot x^3}_{\text{2 factors of } x^3} = x^{3+3} = x^6$$

Power Rule for Exponents

$$(x^m)^n = x^{m \cdot n}$$

The power rule indicates that when we raise an exponential expression to a power, we keep the base and *multiply* the exponents. Example 9 could also be simplified using the power rule: $(x^3)^2 = x^{3 \cdot 2} = x^6$.

EXAMPLE 10 Simplify. **a)** $(x^3)^5$ **b)** $(3^4)^2$ **c)** $(y^3)^8$

Solution **a)** $(x^3)^5 = x^{3 \cdot 5} = x^{15}$ **b)** $(3^4)^2 = 3^{4 \cdot 2} = 3^8$ **c)** $(y^3)^8 = y^{3 \cdot 8} = y^{24}$

HELPFUL HINT

Students often confuse the product and power rules. Note the difference carefully.

PRODUCT RULE	POWER RULE
$x^m \cdot x^n = x^{m+n}$	$(x^m)^n = x^{m \cdot n}$
$2^3 \cdot 2^5 = 2^{3+5} = 2^8$	$(2^3)^5 = 2^{3 \cdot 5} = 2^{15}$

NOW TRY EXERCISE 47

Example 11 will help us in explaining the **expanded power rule**. As the name suggests, this rule is an expansion of the power rule.

EXAMPLE 11 Simplify $\left(\frac{ax}{by}\right)^4$.

Solution

$$\left(\frac{ax}{by}\right)^4 = \frac{ax}{by} \cdot \frac{ax}{by} \cdot \frac{ax}{by} \cdot \frac{ax}{by}$$

$$= \frac{a \cdot a \cdot a \cdot a \cdot x \cdot x \cdot x \cdot x}{b \cdot b \cdot b \cdot b \cdot y \cdot y \cdot y \cdot y} = \frac{a^4 \cdot x^4}{b^4 \cdot y^4} = \frac{a^4 x^4}{b^4 y^4}$$

Expanded Power Rule for Exponents

$$\left(\frac{ax}{by}\right)^m = \frac{a^m x^m}{b^m y^m}, \quad b \neq 0, y \neq 0$$

The expanded power rule illustrates that every factor within parentheses is raised to the power outside the parentheses when the expression is simplified.

EXAMPLE 12 Simplify each expression.

a) $(4x)^2$ **b)** $(-x)^3$ **c)** $(2xy)^3$ **d)** $\left(\frac{-3x}{2y}\right)^2$

Solution **a)** $(4x)^2 = 4^2x^2 = 16x^2$

b) $(-x)^3 = (-1x)^3 = (-1)^3x^3 = -1x^3 = -x^3$

c) $(2xy)^3 = 2^3x^3y^3 = 8x^3y^3$

d) $\left(\frac{-3x}{2y}\right)^2 = \frac{(-3)^2x^2}{2^2y^2} = \frac{9x^2}{4y^2}$

3) Simplify an Expression before Using the Expanded Power Rule

Whenever we have an expression raised to a power, it helps to simplify the expression in parentheses before using the expanded power rule. This procedure is illustrated in Examples 13 and 14.

EXAMPLE 13 Simplify $\left(\dfrac{9x^3y^2}{3xy^2}\right)^3$.

Solution We first simplify the expression within parentheses by dividing out common factors.

$$\left(\frac{9x^3y^2}{3xy^2}\right)^3 = \left(\frac{9}{3}\cdot\frac{x^3}{x}\cdot\frac{y^2}{y^2}\right)^3 = (3x^2)^3$$

Now we use the expanded power rule to simplify further.

$$(3x^2)^3 = 3^3(x^2)^3 = 27x^6$$

Thus, $\left(\dfrac{9x^3y^2}{3xy^2}\right)^3 = 27x^6$.

EXAMPLE 14 Simplify $\left(\dfrac{25x^4y^3}{5x^2y^7}\right)^4$.

Solution Begin by simplifying the expression within parentheses.

$$\left(\frac{25x^4y^3}{5x^2y^7}\right)^4 = \left(\frac{25}{5}\cdot\frac{x^4}{x^2}\cdot\frac{y^3}{y^7}\right)^4 = \left(\frac{5x^2}{y^4}\right)^4$$

Now use the expanded power rule to simplify further.

$$\left(\frac{5x^2}{y^4}\right)^4 = \frac{5^4x^8}{y^{16}} = \frac{625x^8}{y^{16}}$$

Thus, $\left(\dfrac{25x^4y^3}{5x^2y^7}\right)^4 = \dfrac{625x^8}{y^{16}}$.

NOW TRY EXERCISE 87

Avoiding Common Errors

Students sometimes make errors in simplifying expressions containing exponents. It is very important that you have a thorough understanding of exponents. One of the most common errors follows. Study this error carefully to make sure you do not make this error.

CORRECT	INCORRECT
$\dfrac{4}{2x} = \dfrac{\overset{2}{\cancel{4}}}{\underset{1}{\cancel{2}}x} = \dfrac{2}{x}$	$\dfrac{4}{x+2} = \dfrac{\overset{2}{\cancel{4}}}{x+\underset{1}{\cancel{2}}} = \dfrac{2}{x+1}$
$\dfrac{x}{xy} = \dfrac{\overset{1}{\cancel{x}}}{\underset{1}{\cancel{x}}y} = \dfrac{1}{y}$	$\dfrac{x}{x+y} = \dfrac{\overset{1}{\cancel{x}}}{\underset{1}{\cancel{x}}+y} = \dfrac{1}{1+y}$
$\dfrac{5x^3y^2}{y^2} = \dfrac{5x^3\overset{1}{\cancel{y^2}}}{\underset{1}{\cancel{y^2}}} = 5x^3$	$\dfrac{5x^3+y^2}{y^2} = \dfrac{5x^3+\overset{1}{\cancel{y^2}}}{\underset{1}{\cancel{y^2}}} = 5x^3+1$

The simplifications on the right side are not correct because only common *factors* can be divided out (remember, factors are multiplied together). In the first denominator on the right, $x + 2$, the x and 2 are terms, not factors, since they are being added. Similarly, in the second denominator, $x + y$, the x and the y are terms, not factors, since they are being added. Also, in the numerator $5x^3 + y^2$, the $5x^3$ and y^2 are terms, not factors. No common factors can be divided out in the fractions on the right.

EXAMPLE 15 Simplify $(2x^2y^3)^4(xy^2)$.

Solution First simplify $(2x^2y^3)^4$ by using the expanded power rule.

$$(2x^2y^3)^4 = 2^4x^{2\cdot4}y^{3\cdot4} = 16x^8y^{12}$$

Now use the product rule to simplify further.

$$\begin{aligned}(2x^2y^3)^4(xy^2) &= (16x^8y^{12})(x^1y^2)\\ &= 16 \cdot x^8 \cdot x^1 \cdot y^{12} \cdot y^2\\ &= 16x^{8+1}y^{12+2}\\ &= 16x^9y^{14}\end{aligned}$$

NOW TRY EXERCISE 119

Thus, $(2x^2y^3)^4(xy^2) = 16x^9y^{14}$.

Summary of the Rules of Exponents Presented in This Section

1. $x^m \cdot x^n = x^{m+n}$	**product rule**
2. $\dfrac{x^m}{x^n} = x^{m-n}, \quad x \neq 0$	**quotient rule**
3. $x^0 = 1, \quad x \neq 0$	**zero exponent rule**
4. $(x^m)^n = x^{m\cdot n}$	**power rule**
5. $\left(\dfrac{ax}{by}\right)^m = \dfrac{a^m x^m}{b^m y^m}, \quad b \neq 0, \quad y \neq 0$	**expanded power rule**

Exercise Set 6.1

Concept/Writing Exercises

1. In the exponential expression c^r, what is the c called? What is the r called?
2. **a)** Write the product rule for exponents.
   b) In your own words, explain the product rule.
3. **a)** Write the quotient rule for exponents.
   b) In your own words, explain the quotient rule.
4. **a)** Write the zero exponent rule.
   b) In your own words, explain the zero exponent rule.
5. **a)** Write the power rule for exponents.
   b) In your own words, explain the power rule.
6. **a)** Write the expanded power rule for exponents.
   b) In your own words, explain the expanded power rule.
7. For what value of x is $x^0 \neq 1$?
8. Explain the difference between the product rule and the power rule. Give an example of each.

Practice the Skills

Simplify.

9. $x^2 \cdot x^4$
10. $x^5 \cdot x^4$
11. $y \cdot y^2$
12. $4^2 \cdot 4$
13. $3^2 \cdot 3^3$
14. $x^4 \cdot x^2$
15. $y^3 \cdot y^2$
16. $x^3 \cdot x^4$
17. $z^3 \cdot z^5$
18. $2^2 \cdot 2^2$
19. $y^6 \cdot y$
20. $x^4 \cdot x^4$

Simplify.

21. $\dfrac{2^2}{2}$
22. $\dfrac{x^4}{x^3}$
23. $\dfrac{x^{10}}{x^3}$
24. $\dfrac{y^5}{y}$
25. $\dfrac{5^4}{5^2}$
26. $\dfrac{4^5}{4^3}$
27. $\dfrac{y^2}{y}$
28. $\dfrac{x^3}{x^5}$

29. $\frac{z^3}{z^3}$ 30. $\frac{x^{13}}{x^4}$ 31. $\frac{y^{12}}{y^9}$ 32. $\frac{3^4}{3^4}$

Simplify.

33. x^0 34. 5^0 35. $3x^0$ 36. $-2x^0$
37. $(3x)^0$ 38. $-(4x)^0$ 39. $(-4x)^0$ 40. $-(-x)^0$

Simplify.

41. $(x^5)^2$ 42. $(x^2)^3$ 43. $(x^5)^5$ 44. $(y^5)^2$
45. $(x^3)^1$ 46. $(x^3)^2$ 47. $(x^4)^3$ 48. $(x^5)^4$
49. $(x^5)^3$ 50. $(3x)^2$ 51. $(1.3x)^2$ 52. $(-3x)^2$
53. $(-3x^3)^3$ 54. $(xy)^4$ 55. $(2x^2y)^3$ 56. $(4x^3y^2)^3$

Simplify.

57. $\left(\frac{x}{4}\right)^3$ 58. $\left(\frac{2}{x}\right)^3$ 59. $\left(\frac{y}{x}\right)^4$ 60. $\left(\frac{3}{y}\right)^4$
61. $\left(\frac{6}{x}\right)^3$ 62. $\left(\frac{2x}{y}\right)^3$ 63. $\left(\frac{3x}{y}\right)^3$ 64. $\left(\frac{2x^2}{y}\right)^2$
65. $\left(\frac{3x}{5}\right)^2$ 66. $\left(\frac{3x^4}{y}\right)^3$ 67. $\left(\frac{2y^3}{x}\right)^4$ 68. $\left(\frac{-4x^2}{5}\right)^2$

Simplify.

69. $\frac{x^5y}{xy^3}$ 70. $\frac{x^3y^5}{x^7y}$ 71. $\frac{10x^3y^8}{2xy^{10}}$ 72. $\frac{5x^{12}y^2}{10xy^9}$
73. $\frac{2xy}{16x^3y^3}$ 74. $\frac{20x^4y^6}{5xy^9}$ 75. $\frac{35x^4y^9}{15x^9y^{12}}$ 76. $\frac{20x^8y^{12}}{5x^8y^7}$
77. $-\frac{36xy^7z}{12x^4y^5z}$ 78. $\frac{4x^4y^7z^3}{32x^5y^4z^9}$ 79. $-\frac{6x^2y^7z}{3x^5y^9z^6}$ 80. $-\frac{25x^4y^{10}}{30x^3y^7z}$

Simplify.

81. $\left(\frac{4x^4}{2x^6}\right)^3$ 82. $\left(\frac{4x^4}{8x^8}\right)^3$ 83. $\left(\frac{6y^6}{2y^3}\right)^3$ 84. $\left(\frac{125y^4}{25y^{10}}\right)^3$
85. $\left(\frac{13x^9}{17x^5}\right)^0$ 86. $\left(\frac{8y^6}{24y^{10}}\right)^3$ 87. $\left(\frac{x^4y^3}{x^2y^5}\right)^2$ 88. $\left(\frac{2x^7y^2}{4xy}\right)^3$
89. $\left(\frac{9y^2z^7}{18y^9z}\right)^4$ 90. $\left(\frac{y^7z^5}{y^8z^4}\right)^{10}$ 91. $\left(\frac{4xy^5}{y}\right)^3$ 92. $\left(\frac{-64xy^6}{32xy^9}\right)^4$

Simplify.

93. $(3xy^4)^2$ 94. $(2x^4y)(-y^5)$ 95. $(6xy^5)(3x^2y^4)$ 96. $(-2xy)(3xy)$
97. $(2x^4y^2)(5x^2y)$ 98. $(5x^2y)(3xy^5)$ 99. $(5xy)(2xy^6)$ 100. $(3x^2y)^2(xy)$
101. $(2xy)^3(9x^4y^5)^0$ 102. $(3x^2)^4(2xy^5)$

Simplify.

103. $\left(-\frac{x^3y^5}{xy^7}\right)^3$ 104. $(2xy^4)^3$ 105. $(-x)^2$ 106. $(2x^2y^5)(3x^5y^4)^3$
107. $(2.5x^3)^2$ 108. $(-x^4y^5z^6)^3$ 109. $\frac{x^7y^2}{xy^6}$ 110. $(xy^4)(xy^4)^3$
111. $\left(-\frac{x^5}{y^2}\right)^3$ 112. $\left(-\frac{12x}{16x^7y^2}\right)^2$ 113. $(-6x^3y^2)^3$ 114. $(3x^6y)^2(4xy^8)$
115. $(-2x^4y^2z)^3$ 116. $\left(\frac{x}{3}\right)^2$ 117. $(7x^2y^4)^2$ 118. $(3x^4z^{10})^2(2x^2z^8)$

119. $(4x^2y)(3xy^2)^3$

120. $\dfrac{x^2y^6}{x^4y}$

121. $(8.6x^2y^5)^2$

122. $\left(\dfrac{-3x^3}{4}\right)^3$

123. $(x^7y^5)(xy^2)^4$

124. $(5x^4y^7)(2x^3y)^3$

125. $\left(\dfrac{-x^4z^7}{x^2z^5}\right)^4$

126. $(x^4y^6)^3(3x^2y^5)$

Study the Avoiding Common Errors box on page 347. Simplify the following expressions by dividing out common factors. If the expression cannot be simplified by dividing out common factors, so state.

127. $\dfrac{x+y}{x}$

128. $\dfrac{xy}{x}$

129. $\dfrac{x^2+2}{x}$

130. $\dfrac{x+4}{2}$

131. $\dfrac{5yz^4}{yz^2}$

132. $\dfrac{x^2+y^2}{x^2}$

133. $\dfrac{x}{x+1}$

134. $\dfrac{x^4}{x^2y}$

Problem Solving

135. What is the value of x^2y if $x = 3$ and $y = 2$?

136. What is the value of xy^2 if $x = -2$ and $y = -4$?

137. What is the value of $(xy)^0$ if $x = 2$ and $y = 4$?

138. What is the value of $(xy)^0$ if $x = -2$ and $y = 3$?

139. Consider the expression $(-x^5y^7)^9$. When the expanded power rule is used to simplify the expression, will the *sign* of the simplified expression be positive or negative? Explain how you determined your answer.

140. Consider the expression $(-9x^4y^6)^8$. When the expanded power rule is used to simplify the expression, will the *sign* of the simplified expression be positive or negative? Explain how you determined your answer.

Write an expression for the total area of the figure or figures shown.

141.

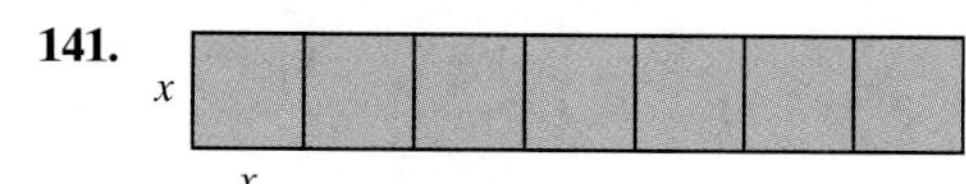

142.

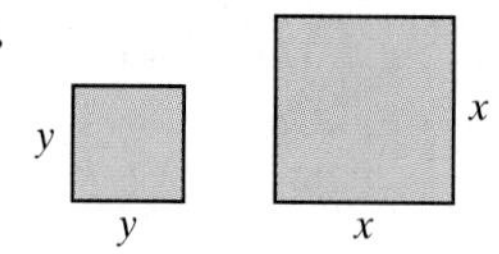

143.

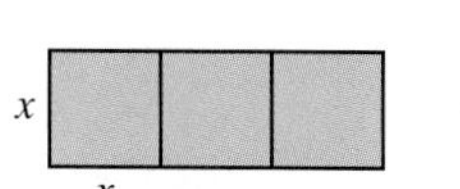

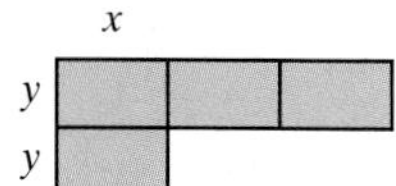

144.

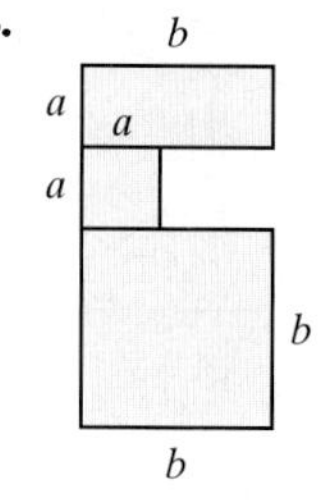

Challenge Problems

Simplify.

145. $\left(\dfrac{3x^4y^5}{6x^6y^8}\right)^3\left(\dfrac{9x^7y^8}{3x^3y^5}\right)^2$

146. $(3yz^2)^2\left(\dfrac{2y^3z^5}{10y^6z^4}\right)^0(4y^2z^3)^3$

Group Activity

Discuss and answer Exercise 147 as a group, according to the instructions.

147. In the next section we will be working with negative exponents. To prepare for that work, use the expression $\dfrac{3^2}{3^3}$ to work parts **a)** through **c)**. Work parts **a)** through **d)** individually, then part **e)** as a group.

a) Divide out common factors in the numerator and denominator and determine the value of the expression.

b) Use the quotient rule on the given expression and write down your results.

c) Write a statement of equality using the results of part **a)** and **b)** above.

d) Repeat parts **a)** through **c)** for the expression $\frac{2^3}{2^4}$.

e) As a group, compare your answers to parts **a)** through **d)**, then write an exponential expression for $\frac{1}{x^m}$.

Cumulative Review Exercises

[2.5] **148.** Solve the equation
$2(x + 4) - 3 = 5x + 4 - 3x + 1$.

[3.1] **149.** **a)** Use the formula $P = 2l + 2w$ to find the length of the sides of the rectangle shown if the perimeter of the rectangle is 26 inches.

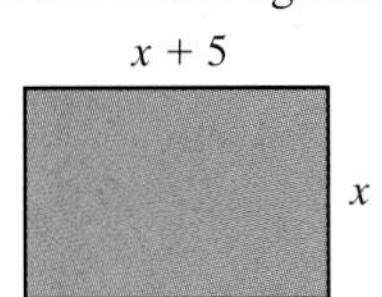

b) Solve the formula $P = 2l + 2w$ for w.

[4.2] **150.** Graph the equation $y = -2x - 4$.

[4.3] **151.** Find the slope of the line that goes through the points $(-6, 2)$ and $(-3, 4)$.

6.2 NEGATIVE EXPONENTS

SSM VIDEO 6.2 CD Rom

1) Understand the negative exponent rule.
2) Simplify expressions containing negative exponents.

1) Understand the Negative Exponent Rule

One additional rule that involves exponents is the negative exponent rule. You will need to understand negative exponents to be successful with scientific notation in the next section.

The negative exponent rule will be developed using the quotient rule illustrated in Example 1.

EXAMPLE 1 Simplify x^3/x^5 by **a)** using the quotient rule and **b)** dividing out common factors.

Solution **a)** By the quotient rule,

$$\frac{x^3}{x^5} = x^{3-5} = x^{-2}$$

b) By dividing out common factors,

$$\frac{x^3}{x^5} = \frac{\not{x} \cdot \not{x} \cdot \not{x}}{\not{x} \cdot \not{x} \cdot \not{x} \cdot x \cdot x} = \frac{1}{x^2}$$

In Example 1 we see that x^3/x^5 is equal to both x^{-2} and $1/x^2$. Therefore, x^{-2} must equal $1/x^2$. That is, $x^{-2} = 1/x^2$. This is an example of the **negative exponent rule.**

Negative Exponent Rule

$$x^{-m} = \frac{1}{x^m}, \quad x \neq 0$$

When a variable or number is raised to a negative exponent, the expression may be rewritten as 1 divided by the variable or number to that positive exponent.

Examples

$$x^{-6} = \frac{1}{x^6} \qquad 4^{-2} = \frac{1}{4^2} = \frac{1}{16}$$

$$y^{-7} = \frac{1}{y^7} \qquad 5^{-3} = \frac{1}{5^3} = \frac{1}{125}$$

Avoiding Common Errors

Students sometimes believe that a negative exponent automatically makes the value of the expression negative. This is not true.

EXPRESSION	CORRECT	INCORRECT	ALSO INCORRECT
3^{-2}	$\frac{1}{3^2} = \frac{1}{9}$	~~-3^2~~	~~$-\frac{1}{3^2}$~~
x^{-3}	$\frac{1}{x^3}$	~~$-x^3$~~	~~$-\frac{1}{x^3}$~~

To help you see that the negative exponent rule makes sense, consider the following sequence of exponential expressions and their corresponding values.

$2^4 = 16$

$2^3 = 8$ *(One-half of 16 is 8.)*

$2^2 = 4$ *(One-half of 8 is 4.)*

$2^1 = 2$ *(One-half of 4 is 2.)*

$2^0 = 1$ *(One-half of 2 is 1.)*

$2^{-1} = \frac{1}{2^1}$ or $\frac{1}{2}$ *(One-half of 1 is $\frac{1}{2}$.)*

$2^{-2} = \frac{1}{2^2}$ or $\frac{1}{4}$ *(One-half of $\frac{1}{2}$ is $\frac{1}{4}$.)*

$2^{-3} = \frac{1}{2^3}$ or $\frac{1}{8}$ *(One-half of $\frac{1}{4}$ is $\frac{1}{8}$.)*

$2^{-4} = \frac{1}{2^4}$ or $\frac{1}{16}$ *(One-half of $\frac{1}{8}$ is $\frac{1}{16}$.)*

Note that each time the exponent decreases by 1 the value of the expression is halved. For example, when we go from 2^4 to 2^3, the value of the expression goes from 16 to 8. If we continue decreasing the exponents beyond $2^0 = 1$, the next exponent in the pattern is -1. And if we take half of 1 we get $\frac{1}{2}$. This pattern illustrates that $x^{-m} = \frac{1}{x^m}$.

2) Simplify Expressions Containing Negative Exponents

Generally, when you are asked to simplify an exponential expression **your final answer should contain no negative exponents.** You may simplify exponential expressions using the negative exponent rule and the rules of exponents presented in the previous section. The following examples indicate how exponential expressions containing negative exponents may be simplified.

EXAMPLE 2 Use the negative exponent rule to write each expression with a positive exponent.

a) x^{-3} **b)** y^{-4} **c)** 3^{-2} **d)** 5^{-1}

Solution **a)** $x^{-3} = \dfrac{1}{x^3}$ **b)** $y^{-4} = \dfrac{1}{y^4}$

c) $3^{-2} = \dfrac{1}{3^2} = \dfrac{1}{9}$ **d)** $5^{-1} = \dfrac{1}{5}$

NOW TRY EXERCISE 11

EXAMPLE 3 Use the negative exponent rule to write each expression with a positive exponent.

a) $\dfrac{1}{x^{-2}}$ **b)** $\dfrac{1}{4^{-1}}$

Solution First use the negative exponent rule on the denominator. Then simplify further.

a) $\dfrac{1}{x^{-2}} = \dfrac{1}{1/x^2} = \dfrac{1}{1} \cdot \dfrac{x^2}{1} = x^2$ **b)** $\dfrac{1}{4^{-1}} = \dfrac{1}{1/4} = \dfrac{1}{1} \cdot \dfrac{4}{1} = 4$

HELPFUL HINT

From Examples 2 and 3, we can see that when a factor is moved from the denominator to the numerator or from the numerator to the denominator, the sign of the *exponent* changes.

$$x^{-4} = \frac{1}{x^4} \qquad \frac{1}{x^{-4}} = x^4$$

$$3^{-5} = \frac{1}{3^5} \qquad \frac{1}{3^{-5}} = 3^5$$

Now let's look at additional examples that combine two or more of the rules presented so far.

EXAMPLE 4 Simplify. **a)** $(y^{-3})^8$ **b)** $(4^2)^{-3}$

Solution **a)**
$$\begin{aligned}(y^{-3})^8 &= y^{(-3)(8)} && \textit{By the power rule}\\ &= y^{-24}\\ &= \frac{1}{y^{24}} && \textit{By the negative exponent rule}\end{aligned}$$

b)
$$\begin{aligned}(4^2)^{-3} &= 4^{(2)(-3)} && \textit{By the power rule}\\ &= 4^{-6}\\ &= \frac{1}{4^6} && \textit{By the negative exponent rule}\end{aligned}$$

NOW TRY EXERCISE 25

EXAMPLE 5 Simplify. **a)** $x^3 \cdot x^{-5}$ **b)** $3^{-4} \cdot 3^{-7}$

Solution **a)** $x^3 \cdot x^{-5} = x^{3+(-5)}$ *By the product rule*

$= x^{-2}$

$= \dfrac{1}{x^2}$ *By the negative exponent rule*

b) $3^{-4} \cdot 3^{-7} = 3^{-4+(-7)}$ *By the product rule*

$= 3^{-11}$

$= \dfrac{1}{3^{11}}$ *By the negative exponent rule*

NOW TRY EXERCISE 51

Avoiding Common Errors

What is the sum of $3^2 + 3^{-2}$? Look carefully at the correct solution.

CORRECT	INCORRECT
$3^2 + 3^{-2} = 9 + \dfrac{1}{9}$ $= 9\dfrac{1}{9}$	~~$3^2 + 3^{-2} = 0$~~

Note that $3^2 \cdot 3^{-2} = 3^{2+(-2)} = 3^0 = 1$.

EXAMPLE 6 Simplify. **a)** $\dfrac{x^{-4}}{x^{10}}$ **b)** $\dfrac{5^{-7}}{5^{-4}}$

Solution **a)** $\dfrac{x^{-4}}{x^{10}} = x^{-4-10}$ *By the quotient rule*

$= x^{-14}$

$= \dfrac{1}{x^{14}}$ *By the negative exponent rule*

b) $\dfrac{5^{-7}}{5^{-4}} = 5^{-7-(-4)}$ *By the quotient rule*

$= 5^{-7+4}$

$= 5^{-3}$

$= \dfrac{1}{5^3}$ or $\dfrac{1}{125}$ *By the negative exponent rule*

NOW TRY EXERCISE 73

HELPFUL HINT

Consider a division problem where a variable has a negative exponent in either its numerator or its denominator, such as in Example 6**a)**. Another way to simplify such an expression is to move the variable with the negative exponent from the numerator to the denominator, or from the denominator to the numerator, and change the sign of the exponent. For example,

$$\frac{x^{-4}}{x^5} = \frac{1}{x^5 \cdot x^4} = \frac{1}{x^{5+4}} = \frac{1}{x^9}$$

$$\frac{y^3}{y^{-7}} = y^3 \cdot y^7 = y^{3+7} = y^{10}$$

continued on next page

Now consider a division problem where a number or variable has a negative exponent in both its numerator and its denominator, such as in Example 6**b)**. Another way to simplify such an expression is to move the variable with the more negative exponent from the numerator to the denominator, or from the denominator to the numerator, and change the sign of the exponent from negative to positive. For example,

$$\frac{x^{-8}}{x^{-3}} = \frac{1}{x^{8} \cdot x^{-3}} = \frac{1}{x^{8-3}} = \frac{1}{x^{5}} \qquad \text{Note that } -8 < -3.$$

$$\frac{y^{-4}}{y^{-7}} = y^{7} \cdot y^{-4} = y^{7-4} = y^{3} \qquad \text{Note that } -7 < -4.$$

EXAMPLE 7 Simplify. **a)** $4x^2(5x^{-5})$ **b)** $\frac{8x^3y^{-2}}{4xy^2}$ **c)** $\frac{2x^2y^5}{8x^7y^{-3}}$

Solution **a)** $4x^2(5x^{-5}) = 4 \cdot 5 \cdot x^2 \cdot x^{-5} = 20x^{-3} = \frac{20}{x^3}$

b)
$$\frac{8x^3y^{-2}}{4xy^2} = \frac{8}{4} \cdot \frac{x^3}{x} \cdot \frac{y^{-2}}{y^2} = 2 \cdot x^2 \cdot \frac{1}{y^4} = \frac{2x^2}{y^4}$$

c)
$$\frac{2x^2y^5}{8x^7y^{-3}} = \frac{2}{8} \cdot \frac{x^2}{x^7} \cdot \frac{y^5}{y^{-3}} = \frac{1}{4} \cdot \frac{1}{x^5} \cdot y^8 = \frac{y^8}{4x^5}$$

NOW TRY EXERCISE 103

In Example 7**b)**, the variable with the negative exponent, y^{-2}, was moved from the numerator to the denominator. In Example 7**c)**, the variable with the negative exponent, y^{-3}, was moved from the denominator to the numerator. In each case, the sign of the exponent was changed from negative to positive when the variable factor was moved.

EXAMPLE 8 Simplify $(4x^{-3})^{-2}$.

Solution Begin by using the expanded power rule.

$$\begin{aligned}(4x^{-3})^{-2} &= 4^{-2}x^{(-3)(-2)} \\ &= 4^{-2}x^{6} \\ &= \frac{1}{4^2}x^6 \\ &= \frac{x^6}{16}\end{aligned}$$

Avoiding Common Errors

Can you explain why the simplification on the right is incorrect?

CORRECT	INCORRECT
$\frac{x^3y^{-2}}{w} = \frac{x^3}{wy^2}$	~~$\frac{x^3 + y^{-2}}{w} = \frac{x^3}{w + y^2}$~~

The simplification on the right is incorrect because in the numerator $x^3 + y^{-2}$ the y^{-2} *is not a factor*; it is a term. We will learn how to simplify expressions like this when we study complex fractions in Section 8.5.

Summary of Rules of Exponents	
1. $x^m \cdot x^n = x^{m+n}$	**product rule**
2. $\dfrac{x^m}{x^n} = x^{m-n}, \quad x \neq 0$	**quotient rule**
3. $x^0 = 1, \quad x \neq 0$	**zero exponent rule**
4. $(x^m)^n = x^{m \cdot n}$	**power rule**
5. $\left(\dfrac{ax}{by}\right)^m = \dfrac{a^m x^m}{b^m y^m}, \quad b \neq 0, y \neq 0$	**expanded power rule**
6. $x^{-m} = \dfrac{1}{x^m}, \quad x \neq 0$	**negative exponent rule**

Exercise Set 6.2

Concept/Writing Exercises

1. In your own words, describe the negative exponent rule.
2. Is the expression x^{-2} simplified? Explain. If not simplified, then simplify.
3. Is the expression x^5y^{-3} simplified? Explain. If not simplified, then simplify.
4. Can the expression a^6b^{-2} be simplified to $1/a^6b^{-2}$? If not, what is the correct simplification? Explain.
5. Can the expression $(y^4)^{-3}$ be simplified to $1/y^7$? If not, what is the correct simplification? Explain.
6. Are the following expressions simplified? If an expression is not simplified, explain why and then simplify.

   a) $\dfrac{5}{x^3}$ **b)** x^{-7}

   c) $\dfrac{x^{-3}}{2}$ **d)** $\dfrac{x^{-4}}{x^4}$

7. **a)** Identify the terms in the numerator of the expression $\dfrac{x^5y^2}{z^3}$.

   b) Identify the factors in the numerator of the expression.
8. **a)** Identify the terms in the numerator of the expression $\dfrac{x^{-4}y^3}{z^5}$.

   b) Identify the factors in the numerator of the expression.
9. Describe what happens to the exponent of a factor when the factor is moved from the numerator to the denominator of a fraction.
10. Describe what happens to the exponent of a factor when the factor is moved from the denominator to the numerator of a fraction.

Practice the Skills

Simplify.

11. x^{-4} **12.** y^{-5} **13.** 3^{-1} **14.** 5^{-2}

15. $\dfrac{1}{x^{-3}}$ **16.** $\dfrac{1}{x^{-2}}$ **17.** $\dfrac{1}{x^{-1}}$ **18.** $\dfrac{1}{y^{-4}}$

19. $\dfrac{1}{4^{-2}}$ **20.** $\dfrac{1}{5^{-3}}$ **21.** $(x^{-2})^3$ **22.** $(x^{-4})^2$

23. $(y^{-5})^4$ **24.** $(y^3)^{-8}$ **25.** $(x^4)^{-2}$ **26.** $(x^{-9})^{-2}$

27. $(2^{-3})^{-2}$ **28.** $(2^{-3})^2$ **29.** $y^4 \cdot y^{-2}$ **30.** $x^{-3} \cdot x^1$

31. $x^7 \cdot x^{-5}$ **32.** $x^{-3} \cdot x^{-2}$ **33.** $3^{-2} \cdot 3^4$ **34.** $6^{-3} \cdot 6^6$

35. $\dfrac{r^4}{r^6}$ **36.** $\dfrac{x^2}{x^{-1}}$ **37.** $\dfrac{p^0}{p^{-3}}$ **38.** $\dfrac{x^{-2}}{x^5}$

39. $\frac{x^{-7}}{x^{-3}}$ 40. $\frac{z^{-11}}{z^{-10}}$ 41. $\frac{3^2}{3^{-1}}$ 42. $\frac{4^2}{4^{-1}}$

43. 3^{-3} 44. x^{-7} 45. $\frac{1}{z^{-9}}$ 46. $\frac{1}{4^{-3}}$

47. $(x^2)^{-5}$ 48. $(x^{-3})^{-4}$ 49. $(y^{-2})^{-3}$ 50. $z^9 \cdot z^{-12}$

51. $x^3 \cdot x^{-7}$ 52. $x^{-3} \cdot x^{-5}$ 53. $x^{-8} \cdot x^{-7}$ 54. $4^{-3} \cdot 4^3$

55. $\frac{x^{-3}}{x^5}$ 56. $\frac{y^6}{y^{-8}}$ 57. $\frac{y^9}{y^{-1}}$ 58. $\frac{3^{-4}}{3}$

59. $\frac{2^{-3}}{2^{-3}}$ 60. $(6^3 y^{-4})^0$ 61. $(2^{-1} + 3^{-1})^0$ 62. $5^0 y^3$

63. $\frac{1}{1^{-5}}$ 64. $(z^{-7})^{-3}$ 65. $(x^{-4})^{-2}$ 66. $(x^{-3})^0$

67. $(x^0)^{-3}$ 68. $(2^{-2})^{-1}$ 69. $2^{-3} \cdot 2$ 70. $6^4 \cdot 6^{-2}$

71. $6^{-4} \cdot 6^2$ 72. $\frac{z^{-3}}{z^{-7}}$ 73. $\frac{x^{-1}}{x^{-4}}$ 74. $\frac{x^{-4}}{x^{-1}}$

75. $(3^2)^{-1}$ 76. $(5^{-2})^{-2}$ 77. $\frac{5}{5^{-2}}$ 78. $\frac{x^6}{x^7}$

79. $\frac{3^{-4}}{3^{-2}}$ 80. $x^{-12} \cdot x^8$ 81. $\frac{7^{-1}}{7^{-1}}$ 82. $2x^{-1}y$

83. $(6x^2)^{-2}$ 84. $(3z^3)^{-1}$ 85. $3x^{-2}y^2$ 86. $5x^4y^{-1}$

87. $5x^{-5}y^{-2}$ 88. $(3x^2y^3)^{-2}$ 89. $(x^5y^{-3})^{-3}$ 90. $3x(5x^{-4})$

91. $(4y^{-2})(5y^{-3})$ 92. $2x^5(3x^{-6})$ 93. $6x^4(-2x^{-2})$ 94. $(9x^5)(-3x^{-7})$

95. $(4x^2y)(3x^3y^{-1})$ 96. $(2x^{-3}y^{-2})(x^4y^0)$ 97. $(2y^2)(4y^{-3}z^5)$ 98. $(3y^{-2})(5x^{-1}y^3)$

99. $\frac{8x^4}{4x^{-1}}$ 100. $\frac{8z^{-4}}{32z^{-2}}$ 101. $\frac{36x^{-4}}{9x^{-2}}$ 102. $\frac{12x^{-2}y^0}{2x^3y^2}$

103. $\frac{3x^4y^{-2}}{6y^3}$ 104. $\frac{16x^{-7}y^{-2}}{4x^5y^2}$ 105. $\frac{32x^4y^{-2}}{4x^{-2}y^0}$ 106. $\frac{21x^{-3}z^2}{7xz^{-3}}$

Problem Solving

107. a) Does $a^{-1}b^{-1} = \frac{1}{ab}$? Explain your answer.

b) Does $a^{-1} + b^{-1} = \frac{1}{a+b}$? Explain your answer.

108. a) Does $\frac{x^{-1}y^2}{z} = \frac{y^2}{xz}$? Explain your answer.

b) Does $\frac{x^{-1} + y^2}{z} = \frac{y^2}{x+z}$? Explain your answer.

Evaluate.

109. $4^2 + 4^{-2}$ 110. $3^2 + 3^{-2}$ 111. $5^2 + 5^{-2}$ 112. $6^{-3} + 6^3$

Determine the number that when placed in the shaded area makes the statement true.

113. $5^{\square} = \frac{1}{25}$ 114. $\frac{1}{3^{\square}} = 9$

Evaluate.

115. $3^0 - 3^{-1}$ 116. $4^{-1} - 3^{-1}$ 117. $2 \cdot 4^{-1} + 4 \cdot 3^{-1}$

Challenge Problems

In Exercises 118–120, determine the number (or numbers) that when placed in the shaded area (or areas) make the statement true.

118. $(x^{\square} y^3)^{-2} = \dfrac{x^4}{y^6}$

119. $(\square x^{\square} y^{-2})^3 = \dfrac{8}{x^9 y^6}$

120. $(x^4 y^{-3})^{\square} = \dfrac{y^9}{x^{12}}$

121. Evaluate $-\left(-\dfrac{2}{5}\right)^{-1}$.

122. For any nonzero real number a, if $a^{-1} = x$, describe the following in terms of x. **a)** $-a^{-1}$ **b)** $\dfrac{1}{a^{-1}}$

Group Activity

Discuss and answer Exercises 123 and 124 as a group.

123. Often problems involving exponents can be done in more than one way. Consider

$$\left(\frac{3x^2y^3}{x}\right)^{-2}$$

a) Group member 1: Simplify this expression by first simplifying the expression within parentheses.

b) Group member 2: Simplify this expression by first using the expanded power rule.

c) Group member 3: Simplify this expression by first using the negative exponent rule.

d) Compare your answers. If you did not all get the same answers, determine why.

e) As a group, decide which method—**a)**, **b)**, or **c)**—was the easiest way to simplify this expression.

124. Consider $(3^{-1} + 2^{-1})^0$. We know this is equal to 1 by the zero exponent rule. As a group, determine the error in the following calculation. Explain your answer.

$$\begin{aligned}(3^{-1} + 2^{-1})^0 &= (3^{-1})^0 + (2^{-1})^0\\ &= 3^{-1(0)} + 2^{-1(0)}\\ &= 3^0 + 2^0\\ &= 1 + 1 = 2\end{aligned}$$

Cumulative Review Exercises

[1.9] **125.** Evaluate $2[6 - (4 - 5)] \div 2 - 5^2$.

126. Evaluate $\dfrac{-3^2 \cdot 4 \div 2}{\sqrt{9} - 2^2}$.

[2.6] **127.** According to the instructions on a bottle of concentrated household cleaner, 8 ounces of the cleaner should be mixed with 3 gallons of water. If your bucket holds only 2.5 gallons of water, how much cleaner should you use?

[3.3] **128.** The larger of two integers is 1 more than 3 times the smaller. If the sum of the two integers is 37, find the two integers.

6.3 SCIENTIFIC NOTATION

SSM VIDEO 6.3 CD Rom

1) **Convert numbers to and from scientific notation.**
2) **Recognize numbers in scientific notation with a coefficient of 1.**
3) **Do calculations using scientific notation.**

1) Convert Numbers to and from Scientific Notation

We often see, and sometimes use, very large or very small numbers. For example, in January 1999, the world population was about 5,920,000,000 people. You may have read that an influenza virus is about 0.0000001 meters in diameter. Because it is difficult to work with many zeros, we can express such numbers using exponents. For example, the number 5,920,000,000 could be written 5.92×10^9 and the number 0.0000001 could be written 1.0×10^{-7}.

Numbers such as 5.92×10^9 and 1.0×10^{-7} are in a form called **scientific notation**. Each number written in scientific notation is written as a number greater than or equal to 1 and less than 10 ($1 \le a < 10$) multiplied by some power of 10.

Examples of Numbers in Scientific Notation

$$1.2 \times 10^6$$
$$3.762 \times 10^3$$
$$8.07 \times 10^{-2}$$
$$1 \times 10^{-5}$$

Below we change the number 68,400 to scientific notation.

$$68{,}400 = 6.84 \times 10{,}000$$
$$= 6.84 \times 10^4 \qquad \text{Note that } 10{,}000 = 10 \cdot 10 \cdot 10 \cdot 10 = 10^4.$$

Therefore, $68{,}400 = 6.84 \times 10^4$. To go from 68,400 to 6.84 the decimal point was moved four places to the left. Note that the exponent on the 10, the 4, is the same as the number of places the decimal point was moved to the left.

Following is a simplified procedure for writing a number in scientific notation.

To Write a Number in Scientific Notation

1. Move the decimal point in the original number to the right of the first nonzero digit. This will give a number greater than or equal to 1 and less than 10.
2. Count the number of places you moved the decimal point to obtain the number in step 1. If the original number was 10 or greater, the count is to be considered positive. If the original number was less than 1, the count is to be considered negative.
3. Multiply the number obtained in step 1 by 10 raised to the count (power) found in step 2.

EXAMPLE 1 Write the following numbers in scientific notation.

a) 10,700 **b)** 0.000386 **c)** 972,000 **d)** 0.0083

Solution

a) The original number is greater than 10; therefore, the exponent is positive. The decimal point in 10,700 belongs after the last zero.

$10{,}700. = 1.07 \times 10^4$ (4 places)

b) The original number is less than 1; therefore, the exponent is negative.

$0.000386 = 3.86 \times 10^{-4}$ (4 places)

c) $972{,}000. = 9.72 \times 10^5$ (5 places)

d) $0.0083 = 8.3 \times 10^{-3}$ (3 places)

NOW TRY EXERCISE 17

When we write a number in scientific notation, we are allowed to leave our answer with a negative exponent, as in Example 1**b)** and 1**d)**.

Now we explain how to write a number in scientific notation as a number without exponents, or in decimal form.

To Convert a Number from Scientific Notation to Decimal Form

1. Observe the exponent of the power of 10.
2. a) If the exponent is positive, move the decimal point in the number (greater than or equal to 1 and less than 10) to the right the same number of places as the exponent. It may be necessary to add zeros to the number. This will result in a number greater than or equal to 10.
   b) If the exponent is 0, do not move the decimal point. Drop the factor 10^0 since it equals 1. This will result in a number greater than or equal to 1 but less than 10.
   c) If the exponent is negative, move the decimal point in the number to the left the same number of places as the exponent (dropping the negative sign). It may be necessary to add zeros. This will result in a number less than 1.

EXAMPLE 2 Write each number without exponents.

a) 3.2×10^4 **b)** 6.28×10^{-3} **c)** 7.95×10^8

Solution **a)** Move the decimal point four places to the right.

$$3.2 \times 10^4 = 3.2 \times 10{,}000 = 32{,}000$$

b) Move the decimal point three places to the left.

$$6.28 \times 10^{-3} = 0.00628$$

c) Move the decimal point eight places to the right.

$$7.95 \times 10^8 = 795{,}000{,}000$$

NOW TRY EXERCISE 39

2) Recognize Numbers in Scientific Notation with a Coefficient of 1

When reading scientific magazines, you will sometimes see numbers written as powers of 10, but without a numerical coefficient. For example, in Figure 6.1 we see that the

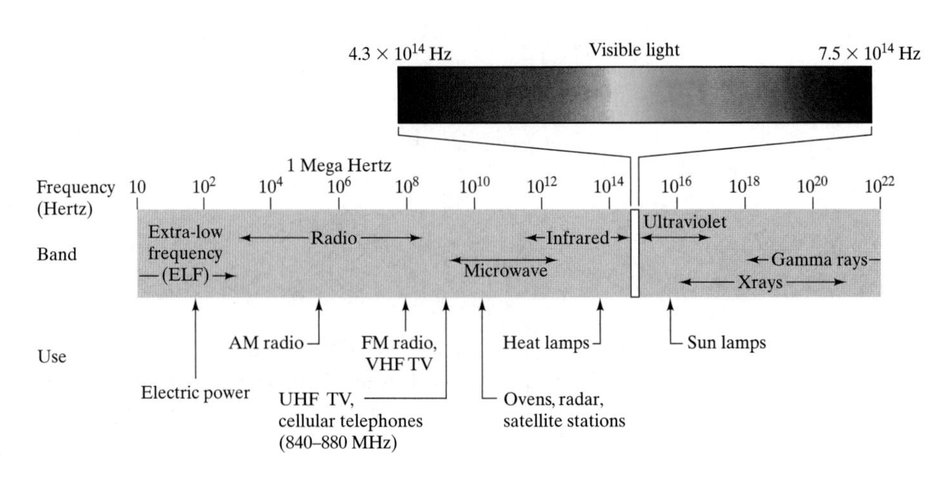

FIGURE 6.1

frequency of FM radio and VHF TV is about 10^8 hertz (or cycles per second). If no numerical coefficient is indicated, the numerical coefficient is always assumed to be 1. Thus, the frequency of FM radio is $10^8 = 1.0 \times 10^8 = 100{,}000{,}000$ hertz (or 100 megahertz since 1 megahertz $= 10^6$ hertz).

Now write the frequency of microwave ovens, 10^{10}, as a frequency in hertz. Your answer should be 10,000,000,000 (which equals 10,000 megahertz).

3) Do Calculations Using Scientific Notation

We can use the rules of exponents presented in Sections 6.1 and 6.2 when working with numbers written in scientific notation.

EXAMPLE 3 Multiply $(4.2 \times 10^6)(2 \times 10^{-4})$.

Solution By the commutative and associative properties of multiplication we can rearrange the expression as follows.

$$\begin{aligned}(4.2 \times 10^6)(2 \times 10^{-4}) &= (4.2 \times 2)(10^6 \times 10^{-4}) \\ &= 8.4 \times 10^{6+(-4)} \quad \text{By the product rule} \\ &= 8.4 \times 10^2 \text{ or } 840\end{aligned}$$

NOW TRY EXERCISE 47

EXAMPLE 4 Divide $\dfrac{6.2 \times 10^{-5}}{2 \times 10^{-3}}$.

Solution

$$\begin{aligned}\frac{6.2 \times 10^{-5}}{2 \times 10^{-3}} &= \left(\frac{6.2}{2}\right)\left(\frac{10^{-5}}{10^{-3}}\right) \\ &= 3.1 \times 10^{-5-(-3)} \quad \text{By the quotient rule} \\ &= 3.1 \times 10^{-5+3} \\ &= 3.1 \times 10^{-2} \text{ or } 0.031\end{aligned}$$

NOW TRY EXERCISE 53

Using Your Calculator

What will your calculator show when you multiply very large or very small numbers? The answer depends on whether your calculator has the ability to display an answer in scientific notation. On calculators without the ability to express numbers in scientific notation, you will probably get an error message because the answer will be too large or too small for the display. For example, on a calculator without scientific notation:

8000000 [×] 600000 [=] Error

On scientific calculators and graphing calculators the answer to the example above might be displayed in the following ways.

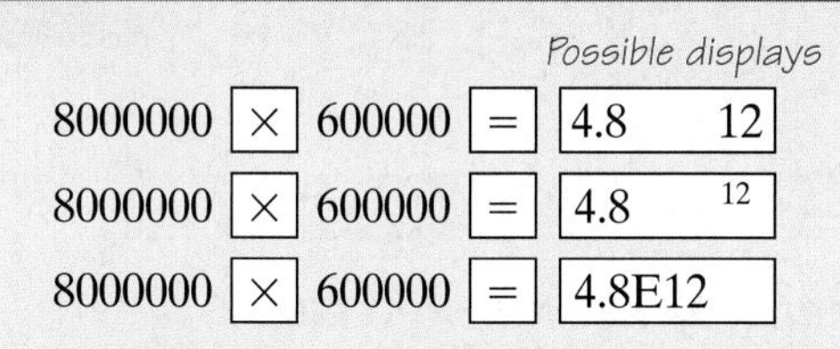

Each answer means 4.8×10^{12}. Let's look at one more example.

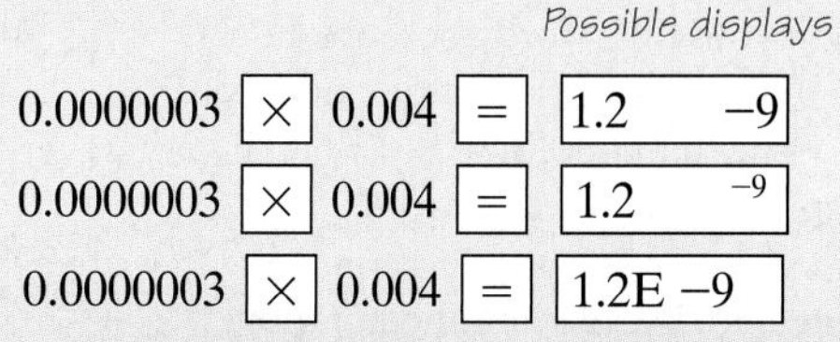

Each display means 1.2×10^{-9}. On some calculators you will press the [ENTER] key instead of the [=] key. The TI-82 and TI-83 graphing calculators display answers using E, such as 4.8E12.

EXAMPLE 5 As of January 1999, the fastest computer in the world, called *Blue Pacific*, located at the Lawrence Livermore National Laboratory in California, could perform a single calculation in about 0.00000000000026 second (twenty six hundred trillionths of a second). How long would it take this computer to do 7 billion (7,000,000,000) calculations?

Solution **Understand** The computer could do 1 calculation in 1(0.00000000000026) second, 2 calculations in 2(0.00000000000026) second, 3 calculations in 3(0.00000000000026) second, and 7 billion operations in 7,000,000,000(0.00000000000026) second.

Translate We will multiply by converting each number to scientific notation.

$$7{,}000{,}000{,}000(0.00000000000026) = (7 \times 10^9)(2.6 \times 10^{-13})$$

Carry Out

$$= (7 \times 2.6)(10^9 \times 10^{-13})$$

$$= 18.2 \times 10^{-4}$$

$$= 0.00182$$

Answer The *Blue Pacific* would take about 0.00182 second to perform 7 billion calculations.

NOW TRY EXERCISE 67

Exercise Set 6.3

Concept/Writing Exercises

1. Describe the form of a number given in scientific notation.
2. a) In your own words, describe how to write a number 10 or greater in scientific notation.
   b) Using the procedure described in part **a)**, write 2370 in scientific notation.
3. a) In your own words, describe how to write a number less than 1 in scientific notation.
   b) Using the procedure described in part **a)**, write 0.00469 in scientific notation.
4. How many places, and in what direction, will you move the decimal point when you convert a number from scientific notation to decimal form when the exponent on the base 10 is 4?
5. How many places, and in what direction, will you move the decimal point when you convert a number from scientific notation to decimal form when the exponent on the base 10 is −6?
6. When changing a number to scientific notation, under what conditions will the exponent on the base 10 be positive?
7. When changing a number to scientific notation, under what conditions will the exponent on the base 10 be negative?
8. In writing the number 71,129 in scientific notation, will the exponent on the base 10 be positive or negative? Explain.

9. In writing the number 0.00734 in scientific notation, will the exponent on the base 10 be positive or negative? Explain.
10. Write the number 1,000,000 in scientific notation.
11. Write the number 0.000001 in scientific notation.
12. a) Is 82.39×10^4 written in scientific notation? If not, how should it be written?
b) Is 0.083×10^{-5} written in scientific notation? If not, how should it be written?

Practice the Skills

Express each number in scientific notation.

13. 42,000
14. 3,610,000
15. 450
16. 0.00062
17. 0.053
18. 0.0000462
19. 19,000
20. 5,260,000,000
21. 0.00000186
22. 0.0075
23. 0.00000914
24. 94,000
25. 110,100
26. 0.02
27. 0.887
28. 416,000

Express each number without exponents.

29. 7.4×10^3
30. 1.63×10^{-4}
31. 4×10^7
32. 6.15×10^5
33. 2.13×10^{-5}
34. 9.64×10^{-7}
35. 6.25×10^5
36. 4.6×10^1
37. 9×10^6
38. 6.475×10^1
39. 5.35×10^2
40. 1.04×10^{-2}
41. 2.991×10^3
42. 2.17×10^{-6}
43. 1×10^4
44. 7.13×10^{-4}

Perform each indicated operation and express each number without exponents.

45. $(4 \times 10^2)(3 \times 10^5)$
46. $(2 \times 10^{-3})(3 \times 10^2)$
47. $(2.7 \times 10^{-1})(9 \times 10^5)$
48. $(1.6 \times 10^{-2})(4 \times 10^{-3})$
49. $(1.3 \times 10^{-8})(1.74 \times 10^6)$
50. $(9 \times 10^8)(1.2 \times 10^{-4})$
51. $\dfrac{6.4 \times 10^5}{2 \times 10^3}$
52. $\dfrac{8 \times 10^{-3}}{2 \times 10^1}$
53. $\dfrac{7.5 \times 10^6}{3 \times 10^3}$
54. $\dfrac{14 \times 10^6}{4 \times 10^8}$
55. $\dfrac{4 \times 10^5}{2 \times 10^4}$
56. $\dfrac{16 \times 10^3}{8 \times 10^{-3}}$

Perform each indicated operation by first converting each number to scientific notation. Write the answer in scientific notation.

57. (700,000)(6,000,000)
58. (500,000)(25,000)
59. (0.003)(0.00015)
60. (230,000)(3000)
61. $\dfrac{1{,}400{,}000}{700}$
62. $\dfrac{0.00004}{200}$
63. $\dfrac{0.00035}{0.000002}$
64. $\dfrac{150{,}000}{0.0005}$
65. List the following numbers from smallest to largest: 4.8×10^5, 3.2×10^{-1}, 4.6, 8.3×10^{-4}.
66. List the following numbers from smallest to largest: 7.3×10^2, 3.3×10^{-4}, 1.75×10^6, 5.3.

Problem Solving

In Exercises 67–81, write the answer without exponents unless asked to do otherwise.

67. In 1997, The Boeing Corporation sold five 777s to China for a cost of $685 million dollars. **a)** Write 685 million dollars as a number in scientific notation. **b)** What was the cost of a single Boeing 777 airplane? Write your answer in scientific notation.
68. Avogadro's number, named after the nineteenth century Italian chemist Amedeo Avogadro, is roughly 6.02×10^{23}. It represents the number of atoms in 12 grams of pure carbon.
a) If this number were written out in decimal form, how many digits would it contain?
b) What is the number of atoms in 1 gram of pure carbon? Write your answer in scientific notation.
69. The gross ticket sales of the top five movies in the United States as of 1998 are listed below.

Movie	Year Released	Approximate U.S. Gross Ticket Sales
1. Titanic	1997	$601,000,000
2. Star Wars	1977	$461,000,000
3. E.T.	1982	$400,000,000
4. Jurassic Park	1993	$357,000,000
5. Forrest Gump	1994	$330,000,000

a) What is the total gross ticket sales from all five movies? Write your answer in scientific notation.

b) How much more did *Titanic* gross than did *Forrest Gump*? Write your answer in scientific notation.

70. The half-life of a radioactive isotope is the time required for half the quantity of the isotope to decompose. The half-life of uranium 238 is 4.5×10^9 years, and the half-life of uranium 234 is 2.5×10^5 years. How many times greater is the half-life of uranium 238 than uranium 234?

71. A treaty between the United States and Canada requires that during the tourist season a minimum of 100,000 cubic feet of water per second flows over Niagara Falls (another 130,000 to 160,000 cubic feet/sec is diverted for power generation). Find the minimum volume of water that will flow over the falls in a 24-hour period during the tourist season.

72. If a computer can do a calculation in 0.000002 second, how long, in seconds, would it take the computer to do 8 trillion (8,000,000,000,000) calculations? Write your answer in scientific notation.

73. Many airlines have magazines available to passengers that include information on the types of planes the airline uses. In a Quantas magazine they mention that one plane they use, the Boeing 747-438, can hold up to 396 passengers and carry a load of up to 14,300 kilograms. Another plane they use is a Boeing 767-238 ER, which can carry up to 201 passengers and carry a load of 8300 kilograms.

a) Write the load of the 747 in kilograms, in scientific notation.

b) Write the load of the 767, in kilograms, in scientific notation.

c) How much larger a load can the 747 carry than can the 767?

74. Laid end to end the 18 billion disposable diapers thrown away in the United States each year would reach the moon and back seven times (seven round trips).

a) Write 18 billion in scientific notation.

b) If the distance from Earth to the moon is 2.38×10^5 miles, what is the length of all these diapers placed end to end? Write your answer in scientific notation and as a number without exponents.

75. In the realm of microelectronics, smaller means faster. During the last two decades, the number of transistors crammed onto an integrated circuit (IC) chip has doubled roughly every two years. Nowadays, the channels defining each transistor on a silicon surface are typically 3.5×10^{-5} (or 35 micrometers wide), small enough that about 1.6×10^7 transistors can occupy a single chip. Looking a decade into the future, the semiconductor industry is expecting to manufacture chips with 5.0×10^8 transistors.

a) Write the number of transistors presently on a chip as a number without exponents.

b) Write the number of transistors that are expected to be on a chip a decade from now as a number without exponents.

c) How many million transistors are on a chip now?

d) How many million transistors are expected to be on a chip a decade from now?

76. The mass of Earth, Earth's moon, and the planet Jupiter are listed below.

Earth: 5,794,000,000,000,000,000,000,000 metric tons

Moon: 73,400,000,000,000,000,000 metric tons

Jupiter: 1,899,000,000,000,000,000,000,000,000 metric tons

a) Write the mass of Earth, the moon, and Jupiter in scientific notation.

b) How many times greater is the mass of Earth than the mass of the moon?

c) How many times greater is the mass of Jupiter than the mass of Earth?

77. The following chart appeared in the November 1997 issue of *Scientific American*.

Average Weights of U.S. Coins

Denomination	Mass (grams)	Weight (newtons)
Dime	2.264	2.220×10^{-2}
Penny	3.110	3.049×10^{-2}
Nickel	4.999	4.902×10^{-2}
Quarter	5.669	5.559×10^{-2}
Half-Dollar	11.50	1.128×10^{-1}
Dollar	26.73	2.621×10^{-1}

a) Divide the mass of a quarter, in grams, by the mass of a dime, in grams. What quotient do you obtain? Round your answer to the nearest thousandth.

b) Divide the weight of a quarter, in newtons, by the weight of a dime, in newtons. What quotient do you obtain? Round your answer to the nearest thousandth.

c) Did you get the same quotient in parts **a)** and **b)**?

d) How many newtons greater is the weight of a half-dollar than the weight of a dime?

78. According to the U.S. Bureau of Labor Statistics, the occupation with the greatest anticipated decline in the number of jobs from 1994 to 2005 is farming. The bureau estimates a loss of 273 thousand farmers during this time period. It also estimates that there will be a decline of 26 thousand personnel clerks from 1994 to 2005. For this time period, how many times greater is the loss of farming jobs than the loss of personnel clerk jobs?

79. The following bar graph shows the estimated insured loss from Hurricane Andrew in 1992 and from Hurricane Bob in 1991. From the graph, determine the sum of the total estimated insured loss from both hurricanes. Write your answer in scientific notation.

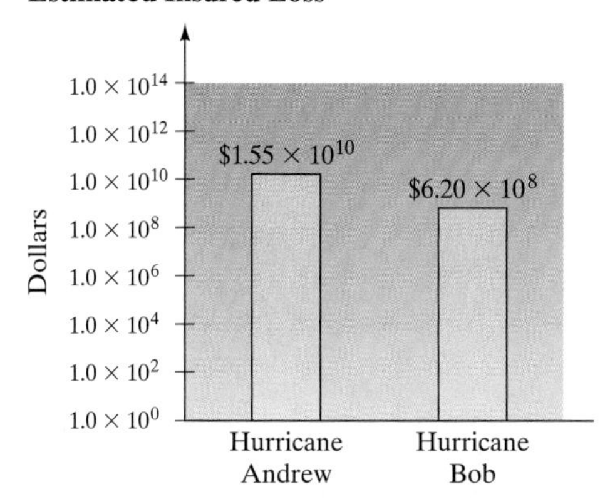

80. The following line graph shows the approximate attendance at Disneyland in Anaheim, California, and Busch Gardens in Tampa, Florida, for the years 1994, 1995, and 1996.

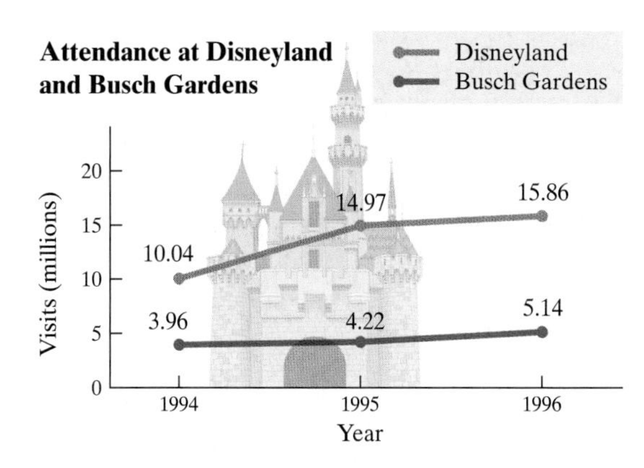

a) By how much did attendance increase at Disneyland from 1994 to 1996? Write your answer in scientific notation.

b) In 1995, how many times greater was the number of people who attended Disneyland than the number who attended Busch Gardens?

81. It took all of human history for the world's population to reach 5.92×10^9 in 1999. At current rates, the world population will double in about 53 years.

a) Estimate the world's population in 2052. Write your answer in scientific notation.

b) Assuming 365 days in a year, estimate the average number of people added to the world's population each day between 1999 and 2052. Write your answer in scientific notation and without using exponents.

82. Consider the problem $\dfrac{(4 \cdot 10^3)(6 \cdot 10^{■})}{24 \cdot 10^{-5}} = 1$. Find the exponent that when placed in the shaded area makes this statement true. Explain how you found your answer.

83. Consider the problem $\dfrac{25 \cdot 10^8}{(5 \cdot 10^{-3})(5 \cdot 10^{■})} = 1$. Find the exponent that when placed in the shaded area makes this statement true. Explain how you found your answer.

Challenge Problems

84. a) Light travels at a speed of 1.86×10^5 miles per second. A *light year* is the distance the light travels in one year. Determine the number of miles in a light year.

b) Earth is approximately 93,000,000 miles from the sun. How long does it take light from the sun to reach Earth?

Group Activity

Discuss and answer Exercise 85 as a group.

85. Many students have no idea of the difference in size between a million (1,000,000), a billion (1,000,000,000), and a trillion (1,000,000,000,000).

a) Write a million, a billion, and a trillion in scientific notation.

b) Group member 1: Determine how long it would take to spend a million dollars if you spent $1000 a day.

c) Group member 2: Repeat part **b)** for a billion dollars.

d) Group member 3: Repeat part **b)** for a trillion dollars.

e) As a group, determine how many times greater a billion dollars is than a million dollars.

Cumulative Review Exercises

[1.9] **86.** Evaluate $4x^2 + 3x + \frac{x}{2}$ when $x = 0$.

[2.3] **87. a)** If $-x = -\frac{3}{2}$, what is the value of x?

b) If $5x = 0$, what is the value of x?

[2.5] **88.** Solve the equation $2x - 3(x - 2) = x + 2$.

[6.1] **89.** Simplify $\left(\frac{-2x^5y^7}{8x^8y^3}\right)^3$.

6.4 ADDITION AND SUBTRACTION OF POLYNOMIALS

1) Identify polynomials.
2) Add polynomials.
3) Subtract polynomials.

1) Identify Polynomials

A **polynomial in *x*** is an expression containing the sum of a finite number of terms of the form ax^n, for any real number a and any *whole number n*.

Examples of Polynomials	Not Polynomials	
$2x$	$4x^{1/2}$	*(Fractional exponent)*
$\frac{1}{3}x - 4$	$3x^2 + 4x^{-1} + 5$	*(Negative exponent)*
$x^2 - 2x + 1$	$4 + \frac{1}{x}$	$\left(\frac{1}{x} = x^{-1}\text{, negative exponent}\right)$

A polynomial is written in **descending order** (or **descending powers**) **of the variable** when the exponents on the variable decrease from left to right.

Example of Polynomial in Descending Order

$$2x^4 + 4x^2 - 6x + 3$$

Note in the example that the constant term 3 is last because it can be written as $3x^0$. Remember that $x^0 = 1$.

A polynomial with one term is called a **monomial**. A **binomial** is a two-termed polynomial. A **trinomial** is a three-termed polynomial. Polynomials containing more than three terms are not given special names. The prefix "poly" means "many." The chart that follows summarizes this information.

Type of Polynomial	Number of Terms	Examples
Monomial	One	$5, \quad 4x, \quad -6x^2$
Binomial	Two	$x + 4, \quad x^2 - 6, \quad 2x^2 - 5x$
Trinomial	Three	$x^2 - 2x + 3, \quad 5x^2 - 6x + 7$

The **degree of a term** of a polynomial in one variable is the exponent on the variable in that term.

Term	Degree of Term	
$4x^2$	Second	
$2y^5$	Fifth	
$-5x$	First	*($-5x$ can be written $-5x^1$.)*
3	Zero	*(3 can be written $3x^0$.)*

The **degree of a polynomial** in one variable is the same as that of its highest-degree term.

Polynomial	Degree of Polynomial	
$8x^3 + 2x^2 - 3x + 4$	Third	*($8x^3$ is highest-degree term.)*
$x^2 - 4$	Second	*(x^2 is highest-degree term.)*
$2x - 1$	First	*($2x$ or $2x^1$ is highest-degree term.)*
4	Zero	*(4 or $4x^0$ is highest-degree term.)*

NOW TRY EXERCISE 37

2) Add Polynomials

In Section 2.1 we stated that like terms are terms having the same variables and the same exponents. That is, like terms differ only in their numerical coefficients.

Examples of Like Terms	
$3,$	-5
$2x$	x
$-2x^2,$	$4x^2$
$3y^2,$	$5y^2$
$3xy^2,$	$5xy^2$

To Add Polynomials

To add polynomials, combine the like terms of the polynomials.

EXAMPLE 1 Simplify $(4x^2 + 6x + 3) + (2x^2 + 5x - 1)$.

Solution Remember that $(4x^2 + 6x + 3) = 1(4x^2 + 6x + 3)$ and $(2x^2 + 5x - 1) = 1(2x^2 + 5x - 1)$. We can use the distributive property to remove the parentheses, as shown below.

$$\begin{aligned} &(4x^2 + 6x + 3) + (2x^2 + 5x - 1) \\ &= 1(4x^2 + 6x + 3) + 1(2x^2 + 5x - 1) \\ &= 4x^2 + 6x + 3 + 2x^2 + 5x - 1 && \textit{Use Distributive Property.} \\ &= \underbrace{4x^2 + 2x^2} + \underbrace{6x + 5x} + \underbrace{3 - 1} && \textit{Rearrange terms.} \\ &= \quad 6x^2 \quad + \quad 11x \quad + \quad 2 && \textit{Combine like terms.} \end{aligned}$$

NOW TRY EXERCISE 51

In the following examples we will not show the multiplication by 1 as was shown in Example 1.

EXAMPLE 2 Simplify $(4x^2 + 3x + y) + (x^2 - 6x + 3)$.

Solution

$$\begin{aligned} &(4x^2 + 3x + y) + (x^2 - 6x + 3) \\ &= 4x^2 + 3x + y + x^2 - 6x + 3 && \textit{Remove parentheses.} \\ &= \underbrace{4x^2 + x^2} + \underbrace{3x - 6x} + y + 3 && \textit{Rearrange terms.} \\ &= \quad 5x^2 \quad - \quad 3x \quad + y + 3 && \textit{Combine like terms.} \end{aligned}$$

EXAMPLE 3 Simplify $(3x^2y - 4xy + y) + (x^2y + 2xy + 3y)$.

Solution

$$\begin{aligned} &(3x^2y - 4xy + y) + (x^2y + 2xy + 3y) \\ &= 3x^2y - 4xy + y + x^2y + 2xy + 3y && \textit{Remove parentheses.} \\ &= \underbrace{3x^2y + x^2y} \underbrace{- 4xy + 2xy} \underbrace{+ y + 3y} && \textit{Rearrange terms.} \\ &= \quad 4x^2y \quad - \quad 2xy \quad + 4y && \textit{Combine like terms.} \end{aligned}$$

NOW TRY EXERCISE 61

Usually when we add polynomials we will do so as in Examples 1 through 3. That is, we will list the polynomials horizontally. However, in Section 6.6, when we divide polynomials, there will be steps where we add polynomials in columns.

To Add Polynomials in Columns

1. Arrange polynomials in descending order one under the other with like terms in the same columns.
2. Add the terms in each column.

EXAMPLE 4 Add $4x^2 - 2x + 2$ and $-2x^2 - x + 4$ using columns.

Solution

$$\begin{array}{r} 4x^2 - 2x + 2 \\ \underline{-2x^2 - x + 4} \\ 2x^2 - 3x + 6 \end{array}$$

EXAMPLE 5 Add $(3x^3 + 2x - 4)$ and $(2x^2 - 6x - 3)$ using columns.

Solution Since the polynomial $3x^3 + 2x - 4$ does not have an x^2 term, we will add the term $0x^2$ to the polynomial. This procedure sometimes helps in aligning like terms.

$$\begin{array}{r} 3x^3 + 0x^2 + 2x - 4 \\ 2x^2 - 6x - 3 \\ \hline 3x^3 + 2x^2 - 4x - 7 \end{array}$$

3) Subtract Polynomials

Now let's subtract polynomials.

To Subtract Polynomials

1. Use the distributive property to remove parentheses. (This will have the effect of changing the sign of *every* term within the parentheses of the polynomial being subtracted.)
2. Combine like terms.

EXAMPLE 6 Simplify $(3x^2 - 2x + 5) - (x^2 - 3x + 4)$.

Solution $(3x^2 - 2x + 5)$ means $1(3x^2 - 2x + 5)$ and $(x^2 - 3x + 4)$ means $1(x^2 - 3x + 4)$. We use this information in the solution, as shown below.

$$\begin{aligned} (3x^2 - 2x + 5) - (x^2 - 3x + 4) &= 1(3x^2 - 2x + 5) - 1(x^2 - 3x + 4) \\ &= 3x^2 - 2x + 5 - x^2 + 3x - 4 && \text{Remove parentheses.} \\ &= \underbrace{3x^2 - x^2}\underbrace{- 2x + 3x}\underbrace{+ 5 - 4} && \text{Rearrange terms.} \\ &= 2x^2 + x + 1 && \text{Combine like terms.} \end{aligned}$$

NOW TRY EXERCISE 85

Remember from Section 2.1 that when a negative sign precedes the parentheses, the sign of every term within the parentheses is changed when the parentheses are removed. This was shown in Example 6. In Example 7, we will not show the multiplication by -1, as was done in Example 6.

EXAMPLE 7 Subtract $(-x^2 - 2x + 3)$ from $(x^3 + 4x + 6)$.

Solution

$$\begin{aligned} &(x^3 + 4x + 6) - (-x^2 - 2x + 3) \\ &= x^3 + 4x + 6 + x^2 + 2x - 3 && \text{Remove parentheses.} \\ &= x^3 + x^2 \underbrace{+ 4x + 2x}\underbrace{+ 6 - 3} && \text{Rearrange terms.} \\ &= x^3 + x^2 + 6x + 3 && \text{Combine like terms.} \end{aligned}$$

NOW TRY EXERCISE 91

Avoiding Common Errors

One of the most common mistakes occurs when subtracting polynomials. When subtracting one polynomial from another, **the sign of each term in the polynomial being subtracted must be changed, not just the sign of the first term.**

CORRECT	INCORRECT
$6x^2 - 4x + 3 - (2x^2 - 3x + 4)$	~~$6x^2 - 4x + 3 - (2x^2 - 3x + 4)$~~
$= 6x^2 - 4x + 3 - 2x^2 + 3x - 4$	~~$= 6x^2 - 4x + 3 - 2x^2 - 3x + 4$~~
$= 4x^2 - x - 1$	~~$= 4x^2 - 7x + 7$~~
	Do not make this mistake!

4) Subtract Polynomials in Columns

Polynomials can be subtracted as well as added using columns.

To Subtract Polynomials in Columns

1. Write *the polynomial being subtracted* below the polynomial from which it is being subtracted. List like terms in the same column.
2. *Change the sign of each term* in the polynomial being subtracted. (This step can be done mentally, if you like.)
3. Add the terms in each column.

EXAMPLE 8 Subtract $(x^2 - 6x + 5)$ from $(3x^2 + 5x + 7)$ using columns.

Solution Align like terms in columns (step 1).

$$\begin{array}{r} 3x^2 + 5x + 7 \\ \underline{-(x^2 - 6x + 5)} \end{array} \quad \text{Align like terms.}$$

Change *all* signs in the second row (step 2); then add (step 3).

$$\begin{array}{r} 3x^2 + 5x + 7 \\ -x^2 + 6x - 5 \\ \hline 2x^2 + 11x + 2 \end{array} \quad \begin{array}{l} \\ \text{Change all signs.} \\ \text{Add.} \end{array}$$

EXAMPLE 9 Subtract $(2x^2 - 6)$ from $(-3x^3 + 4x - 3)$ using columns.

Solution To help align like terms, write each expression in descending order. If any power of x is missing, write that term with a numerical coefficient of 0.

$$-3x^3 + 4x - 3 = -3x^3 + 0x^2 + 4x - 3$$
$$2x^2 - 6 = 2x^2 + 0x - 6$$

Align like terms.

$$\begin{array}{r} -3x^3 + 0x^2 + 4x - 3 \\ \underline{-(2x^2 + 0x - 6)} \end{array}$$

Change all signs in the second row; then add the terms in each column.

$$\begin{array}{r} -3x^3 + 0x^2 + 4x - 3 \\ -2x^2 - 0x + 6 \\ \hline -3x^3 - 2x^2 + 4x + 3 \end{array}$$

NOW TRY EXERCISE 99

Note: Many of you will find that you can change the signs mentally and can therefore align and change the signs in one step.

Exercise Set 6.4

Concept/Writing Exercises

1. What is a polynomial?
2. a) What is a monomial? Make up three examples.
   b) What is a binomial? Make up three examples.
   c) What is a trinomial? Make up three examples.
3. a) Explain how to find the degree of a term in one variable.
   b) Explain how to find the degree of a polynomial in one variable.

4. Make up your own fifth-degree polynomial with three terms. Explain why it is a fifth-degree polynomial with three terms.
5. Explain why $(3x + 2) - (4x - 6) \neq 3x + 2 - 4x - 6$.
6. Explain how to write a polynomial in descending order of the variable.
7. Why is the constant term always written last when writing a polynomial in descending order?
8. Explain how to add polynomials.
9. a) In your own words, describe how to add polynomials in columns.
   b) How will you rewrite $4x^3 + 5x - 7$ in order to add it to $3x^3 + x^2 - 4x + 8$ using columns? Explain.
10. Is $4x^{-3} + 9$ a polynomial? Explain.
11. Is $4x^2 - 6x^{1/2}$ a polynomial? Explain.
12. Is $5x + \frac{2}{x}$ a polynomial? Explain.

Practice the Skills

Indicate, which expressions are polynomials. If the polynomial has a specific name—monomial, binomial, or trinomial—give that name.

13. $x^2 + 5$
14. $2x^2 - 6x + 7$
15. 13
16. $4x^{-2}$
17. $4x^3 - 8$
18. $6x^2 - 2x + 8$
19. $7x^3$
20. $3x^{1/2} + 2x$
21. $2x + 5$
22. $x^3 - 8x^2 + 8$
23. $3x^3 - 6x^2 + 4x - 5$
24. $10x^2$
25. $5 - x^2 - 6x$
26. $2x^{-2}$

Express each polynomial in descending order. If the polynomial is already in descending order, so state. Give the degree of each polynomial.

27. $3 + 2x$
28. 5
29. $-4 + x^2 - 2x$
30. $6x - 5$
31. $x + 3x^2 - 8$
32. $x^3 - 6$
33. $-x - 1$
34. $2x^2 + 5x - 8$
35. $3x^3 - x + 4$
36. 15
37. $-4 + x - 3x^2 + 4x^3$
38. $1 - x^3 + 3x$
39. $5x + 3x^2 - 6 - 2x^3$
40. $5 - 2x^4 + 8x^2 - 7x$

Add.

41. $(2x + 3) + (x - 5)$
42. $(5x - 6) + (2x - 3)$
43. $(-4x + 8) + (2x + 3)$
44. $(-5x - 3) + (-2x + 3)$
45. $(x + 7) + (-6x - 8)$
46. $(4x - 3) + (3x - 3)$
47. $(x^2 + 2.6x - 3) + (4x + 3.8)$
48. $(-2x^2 + 3x - 9) + (-2x - 3)$
49. $(5x - 7) + (2x^2 + 3x + 12)$
50. $(-x^2 - 2x - 4) + (4x^2 + 3)$
51. $(2x^2 - 3x + 5) + (-x^2 + 6x - 8)$
52. $(x^2 - 6x + 7) + (-x^2 + 3x + 5)$
53. $(-x^2 - 4x + 8) + \left(5x - 2x^2 + \frac{1}{2}\right)$
54. $(8x^2 + 3x - 5) + \left(x^2 + \frac{1}{2}x + 2\right)$
55. $(8x^2 + 4) + (-2.6x^2 - 5x)$
56. $(8x^3 + 4x^2 + 6) + (0.2x^2 + 5x)$
57. $(-7x^3 - 3x^2 + 4) + (4x + 5x^3 - 7)$
58. $(6x^3 - 4x^2 - 7) + (3x^2 + 3x - 3)$
59. $(3x^2 + 2xy + 8) + (-x^2 - 5xy - 3)$
60. $(x^2y + 6x^2 - 3xy^2) + (-x^2y - 12x^2 + 4xy^2)$
61. $(2x^2y + 2x - 3) + (3x^2y - 5x + 5)$
62. $(x^2y + x - y) + (2x^2y + 2x - 6y + 3)$

Add using columns.

63. Add $3x - 6$ and $4x + 5$.
64. Add $-2x + 5$ and $-3x - 5$.
65. Add $5x^2 - 2x + 4$ and $3x + 1$.
66. Add $7x^2 - 3x - 1$ and $-3x^2 + 6$.
67. Add $-x^2 - 3x + 3$ and $5x^2 + 5x - 7$.
68. Add $-5x^2 - 3$ and $x^2 + 2x - 9$.
69. Add $2x^3 + 3x^2 + 6x - 9$ and $7 - 4x^2$
70. Add $-3x^3 + 3x + 9$ and $2x^2 - 4$.
71. Add $6x^3 - 4x^2 + x - 9$ and $-x^3 - 3x^2 - x + 7$
72. Add $7x^3 + 5x - 6$ and $3x^3 - 4x^2 - x + 8$.

Subtract.

73. $(3x - 4) - (2x + 2)$

74. $(3x - 2) - (4x + 3)$

75. $(-2x - 3) - (-5x - 7)$

76. $(12x - 3) - (-2x + 7)$

77. $(-x + 8) - (3x + 5)$

78. $(4x + 8) - (3x + 9)$

79. $(9x^2 + 7x - 5) - (3x^2 + 3.5)$

80. $(-2x^2 + 4x - 5.2) - (5x^2 + 3x + 7.5)$

81. $(5x^2 - x - 1) - (-3x^2 - 2x - 5)$

82. $(-x^2 + 3x + 12) - (-6x^2 - 3)$

83. $(-5x^2 - 2x) - (2x^2 - 7x + 9)$

84. $(7x - 0.6) - (-2x^2 + 4x - 8)$

85. $(8x^3 - 2x^2 - 4x + 5) - (5x^2 + 8)$

86. $\left(9x^3 - \frac{1}{5}\right) - (x^2 + 5x)$

87. $(2x^3 - 4x^2 + 5x - 7) - \left(3x + \frac{3}{5}x^2 - 5\right)$

88. $(-3x^2 + 4x - 7) - \left(x^3 + 4x^2 - \frac{3}{4}x\right)$

89. Subtract $(5x + 4)$ from $(8x + 2)$

90. Subtract $(-4x + 7)$ from $(-3x - 9)$

91. Subtract $(5x - 6)$ from $(2x^2 - 4x + 8)$

92. Subtract $(3x^2 - 5x - 3)$ from $(-x^2 + 3x + 10)$

93. Subtract $(4x^3 - 6x^2)$ from $(3x^3 + 5x^2 + 9x - 7)$

94. Subtract $(-4x^2 + 8x - 7)$ from $(-5x^3 - 6x^2 + 7)$

Perform each subtraction using columns.

95. Subtract $(3x - 3)$ from $(6x + 5)$.

96. Subtract $(6x + 8)$ from $(2x - 5)$.

97. Subtract $(-9x - 4)$ from $(-5x + 3)$.

98. Subtract $(-3x + 8)$ from $(6x^2 - 5x + 3)$.

99. Subtract $(6x^2 - 1)$ from $(7x^2 - 3x - 4)$.

100. Subtract $(4x^2 + 7x - 9)$ from $(x^2 - 6x + 3)$.

101. Subtract $(-4x^2 + 6x)$ from $(x - 6)$.

102. $(5x^2 + 4)$ from $(x^2 + 4)$.

103. Subtract $(x^2 + 6x - 7)$ from $(4x^3 - 6x^2 + 7x - 9)$.

104. Subtract $(2x^3 + 4x^2 - 9x)$ from $(-5x^3 + 4x - 12)$.

Problem Solving

105. Make up your own addition problem where the sum of two binomials is $-2x + 4$.

106. Make up your own addition problem where the sum of two trinomials is $2x^2 + 5x - 6$.

107. Make up your own subtraction problem where the difference of two trinomials is $x - 2$.

108. Make up your own subtraction problem where the difference of two trinomials is $-x^2 + 4x - 5$.

109. When two binomials are added, will the sum **a)** always, **b)** sometimes, or **c)** never be a binomial? Explain your answer and give examples to support your answer.

110. When one binomial is subtracted from another, will the difference **a)** always, **b)** sometimes, or **c)** never be a binomial? Explain your answer and give examples to support your answer.

111. When two trinomials are added, will the sum **a)** always, **b)** sometimes, or **c)** never be a trinomial? Explain your answer and give examples to support your answer.

112. When one trinomial is subtracted from another, will the difference **a)** always, **b)** sometimes, or **c)** never be a trinomial? Explain your answer and give examples to support your answer.

113. Write a fifth-degree trinomial in the variable x that has neither a third- nor second-degree term.

114. Write a sixth-degree trinomial in the variable x that has no fifth-, fourth-, or zero-degree term.

115. Is it possible to have a fifth-degree trinomial in x that has no fourth-, third-, second-, or first-degree term and contains no like terms? Explain.

116. Is it possible to have a fourth-degree trinomial in x that has no third-, second-, or zero-degree term and contains no like terms? Explain.

Write a polynomial that represents the area of each figure shown.

117.

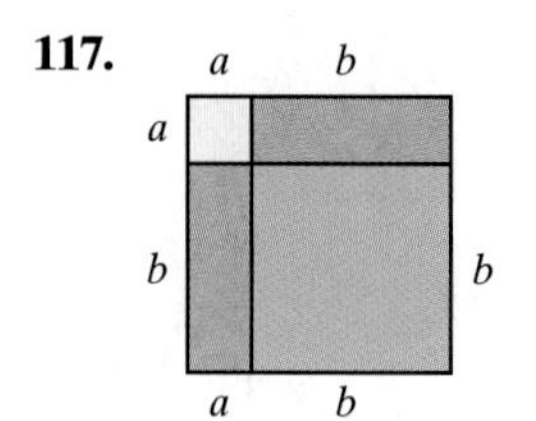

118.

x x x
x
y

119.

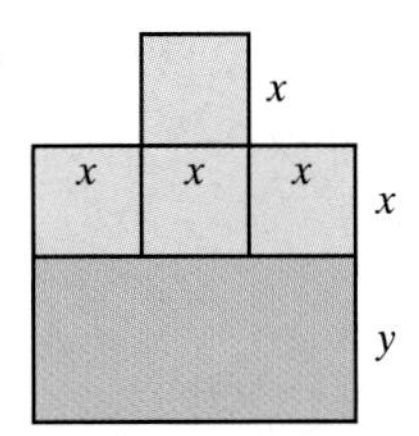

120.

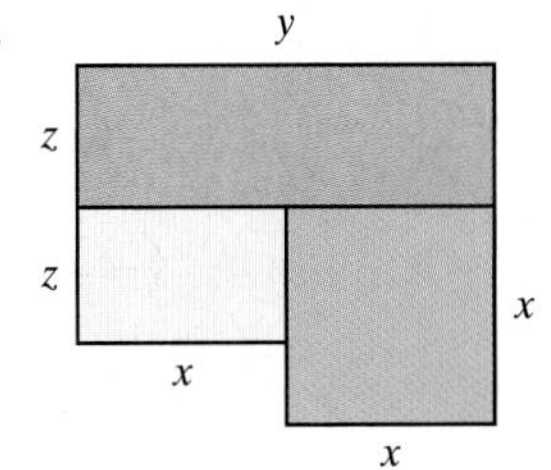

Challenge Problems

Simplify.

121. $(3x^2 - 6x + 3) - (2x^2 - x - 6) - (x^2 + 7x - 9)$

122. $3x^2y - 6xy - 2xy + 9xy^2 - 5xy + 3x$

123. $4(x^2 + 2x - 3) - 6(2 - 4x - x^2) - 2x(x + 2)$

Group Activity

Discuss and answer Exercise 124 as a group.

124. Make up an expression in x, containing a trinomial, a binomial, and a different trinomial such that (first trinomial) + (binomial) − (second trinomial) = 0.

Cumulative Review Exercises

[1.5] **125.** Insert either >, <, or = in the shaded area to make the statement true: $|-4| \;\square\; |-6|$.

[1.6–1.8] *Indicate whether each statement is true or false.*

126. The product of two negative numbers is always a negative number.

127. The sum of two negative numbers is always a negative number.

128. The difference of two negative numbers is always a negative number.

129. The quotient of two negative numbers is always a negative number.

[6.1] **130.** Simplify $\left(\frac{3x^4y^5}{6x^7y^4}\right)^3$.

6.5 MULTIPLICATION OF POLYNOMIALS

1. **Multiply a monomial by a monomial.**
2. **Multiply a polynomial by a monomial.**
3. **Multiply binomials using the distributive property.**
4. **Multiply binomials using the FOIL method.**
5. **Multiply binomials using formulas for special products.**
6. **Multiply any two polynomials.**

1) Multiply a Monomial by a Monomial

We begin our discussion of multiplication of polynomials by multiplying a monomial by a monomial. To multiply two monomials, multiply their coefficients and use the product rule of exponents to determine the exponents on the variables. Problems of this type were done in Section 6.1.

EXAMPLE 1 Multiply $(4x^2)(5x^5)$.

Solution $(4x^2)(5x^5) = 4 \cdot 5 \cdot x^2 \cdot x^5 = 20x^{2+5} = 20x^7$

EXAMPLE 2 Multiply $(-2x^6)(3x^4)$.

Solution $(-2x^6)(3x^4) = (-2)(3) \cdot x^6 \cdot x^4 = -6x^{6+4} = -6x^{10}$

EXAMPLE 3 Multiply $(6x^2y)(7x^5y^4)$.

Solution Remember that when a variable is given without an exponent we assume that the exponent on the variable is 1.

$$(6x^2y)(7x^5y^4) = 42x^{2+5}y^{1+4} = 42x^7y^5$$

EXAMPLE 4 Multiply $6xy^2z^5(-3x^4y^7z)$.

Solution $6xy^2z^5(-3x^4y^7z) = -18x^5y^9z^6$

EXAMPLE 5 Multiply $(-4x^4z^9)(-3xy^7z^3)$.

Solution $(-4x^4z^9)(-3xy^7z^3) = 12x^5y^7z^{12}$

NOW TRY EXERCISE 21

2) Multiply a Polynomial by a Monomial

To multiply a polynomial by a monomial, we use the distributive property presented earlier.

$$a(b + c) = ab + ac$$

The distributive property can be expanded to

$$a(b + c + d + \cdots + n) = ab + ac + ad + \cdots + an$$

EXAMPLE 6 Multiply $2x(3x^2 + 4)$.

Solution

$$\begin{aligned} 2x(3x^2 + 4) &= (2x)(3x^2) + (2x)(4) \\ &= 6x^3 + 8x \end{aligned}$$

Notice that the use of the distributive property results in monomials being multiplied by monomials. If we study Example 6, we see that the $2x$ and $3x^2$ are both monomials, as are the $2x$ and 4.

EXAMPLE 7 Multiply $-3x(4x^2 - 2x - 1)$.

Solution

$$\begin{aligned} -3x(4x^2 - 2x - 1) &= (-3x)(4x^2) + (-3x)(-2x) + (-3x)(-1) \\ &= -12x^3 + 6x^2 + 3x \end{aligned}$$

EXAMPLE 8 Multiply $5x^2(4x^3 - 2x + 7)$.

Solution

$$\begin{aligned} 5x^2(4x^3 - 2x + 7) &= (5x^2)(4x^3) + (5x^2)(-2x) + (5x^2)(7) \\ &= 20x^5 - 10x^3 + 35x^2 \end{aligned}$$

EXAMPLE 9 Multiply $2x(3x^2y - 6xy + 5)$.

Solution
$$2x(3x^2y - 6xy + 5) = (2x)(3x^2y) + (2x)(-6xy) + (2x)(5)$$
$$= 6x^3y - 12x^2y + 10x$$

EXAMPLE 10 Multiply $(3x^2 - 2xy + 3)4x$.

Solution
$$(3x^2 - 2xy + 3)4x = (3x^2)(4x) + (-2xy)(4x) + (3)(4x)$$
$$= 12x^3 - 8x^2y + 12x$$

NOW TRY EXERCISE 41

The problem in Example 10 could be written as $4x(3x^2 - 2xy + 3)$ by the commutative property of multiplication, and then simplified as in Examples 6 through 9.

3) Multiply Binomials Using the Distributive Property

Now we will discuss multiplying a binomial by a binomial. Before we explain how to do this, consider the multiplication problem $43 \cdot 12$.

	43 ⟵	*Multiplicand*
	12 ⟵	*Multiplier*
2(4) ⟶	86 ⟵	2(3)
1(4) ⟶	43 ⟵	1(3)
	516 ⟵	*Product*

Note how the 2 multiplies both the 3 and the 4 and the 1 also multiplies both the 3 and the 4. That is, every digit in the multiplier multiplies every digit in the multiplicand. We can also illustrate the multiplication process as follows.

$$\begin{aligned}(43)(12) &= (40 + 3)(10 + 2)\\ &= (40 + 3)(10) + (40 + 3)(2)\\ &= (40)(10) + (3)(10) + (40)(2) + (3)(2)\\ &= 400 + 30 + 80 + 6\\ &= 516\end{aligned}$$

Whenever any two polynomials are multiplied, the same process must be followed. That is, **every term in one polynomial must multiply every term in the other polynomial.**

Consider multiplying $(a + b)(c + d)$. Treating $(a + b)$ as a single term and using the distributive property, we get

$$(a + b)(c + d) = (a + b)c + (a + b)d$$

Using the distributive property a second time gives

$$= ac + bc + ad + bd$$

Notice how each term of the first polynomial was multiplied by each term of the second polynomial, and all the products were added to obtain the answer.

EXAMPLE 11 Multiply $(3x + 2)(x - 5)$.

Solution

$$\begin{aligned}(3x + 2)(x - 5) &= (3x + 2)x + (3x + 2)(-5) \\ &= 3x(x) + 2(x) + 3x(-5) + 2(-5) \\ &= 3x^2 + \underbrace{2x - 15x}_{} - 10 \\ &= 3x^2 - 13x - 10\end{aligned}$$

Note that after performing the multiplication like terms must be combined.

EXAMPLE 12 Multiply $(x - 4)(y + 3)$.

Solution

$$\begin{aligned}(x - 4)(y + 3) &= (x - 4)y + (x - 4)3 \\ &= xy - 4y + 3x - 12\end{aligned}$$

4) Multiply Binomials Using the FOIL Method

A common method used to multiply two binomials is the **FOIL method.** This procedure also results in each term of one binomial being multiplied by each term in the other binomial. Students often prefer to use this method when multiplying two binomials.

The FOIL Method

Consider $(a + b)(c + d)$.

F stands for **first**—multiply the first terms of each binomial together:

$$\overbrace{(a + b)(c} ^{F}+ d) \qquad \text{product } ac$$

O stands for **outer**—multiply the two outer terms together:

$$\overbrace{(a + b)(c + d)}^{O} \qquad \text{product } ad$$

I stands for **inner**—multiply the two inner terms together:

$$(a + \overbrace{b)(c}^{I} + d) \qquad \text{product } bc$$

L stands for **last**—multiply the last terms together:

$$(a + \overbrace{b)(c + d}^{L}) \qquad \text{product } bd$$

The product of the two binomials is the sum of these four products.

$$(a + b)(c + d) = ac + ad + bc + bd$$

The FOIL method is not actually a different method used to multiply binomials, but rather an acronym to help students remember to correctly apply the distributive property. We could have used IFOL or any other arrangement of the four letters. However, FOIL is easier to remember than the other arrangements.

EXAMPLE 13 Using the FOIL method, multiply $(2x - 3)(x + 4)$.

Solution

$$(2x - 3)(x + 4)$$

F O I L

$$(2x)(x) + (2x)(4) + (-3)(x) + (-3)(4)$$
$$= 2x^2 + 8x - 3x - 12$$
$$= 2x^2 + 5x - 12$$

Thus, $(2x - 3)(x + 4) = 2x^2 + 5x - 12$.

EXAMPLE 14 Multiply $(4 - 2x)(6 - 5x)$.

Solution

$$(4 - 2x)(6 - 5x)$$

F O I L

$$4(6) + 4(-5x) + (-2x)(6) + (-2x)(-5x)$$
$$= 24 - 20x - 12x + 10x^2$$
$$= 10x^2 - 32x + 24$$

NOW TRY EXERCISE 63

Thus, $(4 - 2x)(6 - 5x) = 10x^2 - 32x + 24$.

EXAMPLE 15 Multiply $(x - 6)(x + 6)$.

Solution

F O I L

$$(x)(x) + (x)(6) + (-6)(x) + (-6)(6)$$
$$= x^2 + 6x - 6x - 36$$
$$= x^2 - 36$$

Thus, $(x - 6)(x + 6) = x^2 - 36$.

EXAMPLE 16 Multiply $(2x + 3)(2x - 3)$.

Solution

F O I L

$$(2x)(2x) + (2x)(-3) + (3)(2x) + (3)(-3)$$
$$= 4x^2 - 6x + 6x - 9$$
$$= 4x^2 - 9$$

Thus, $(2x + 3)(2x - 3) = 4x^2 - 9$.

5) Multiply Binomials Using Formulas for Special Products

Examples 15 and 16 are examples of a special product, the product of the sum and difference of the same two terms.

Product of the Sum and Difference of the Same Two Terms

$$(a + b)(a - b) = a^2 - b^2$$

In this special product, a represents one term and b the other term. Then $(a + b)$ is the sum of the terms and $(a - b)$ is the difference of the terms. This special product is also called the **difference of two squares formula** because the expression on the right side of the equal sign is the difference of two squares.

EXAMPLE 17 Use the rule for finding the product of the sum and difference of two quantities to multiply each expression.

a) $(x + 3)(x - 3)$ **b)** $(2x + 4)(2x - 4)$ **c)** $(3x + 2y)(3x - 2y)$

Solution **a)** If we let $x = a$ and $3 = b$, then

$$\begin{aligned} (a + b)(a - b) &= a^2 - b^2 \\ \downarrow\ \ \downarrow\ \ \downarrow\ \ \downarrow\quad &\quad\ \downarrow\quad\ \ \downarrow \\ (x + 3)(x - 3) &= (x)^2 - (3)^2 \\ &= x^2 - 9 \end{aligned}$$

b)

$$\begin{aligned} (a + b)(a - b) &= a^2 - b^2 \\ \downarrow\ \ \downarrow\ \ \downarrow\ \ \downarrow\quad &\quad\ \downarrow\quad\ \ \downarrow \\ (2x + 4)(2x - 4) &= (2x)^2 - (4)^2 \\ &= 4x^2 - 16 \end{aligned}$$

c)

$$\begin{aligned} (a + b)\ (a - b) &= a^2 - b^2 \\ \downarrow\ \ \downarrow\ \ \downarrow\ \ \downarrow\quad &\quad\ \downarrow\quad\ \ \downarrow \\ (3x + 2y)(3x - 2y) &= (3x)^2 - (2y)^2 \\ &= 9x^2 - 4y^2 \end{aligned}$$

NOW TRY EXERCISE 77

Example 17 could also have been done using the FOIL method.

EXAMPLE 18 Using the FOIL method, find $(x + 3)^2$.

Solution $(x + 3)^2 = (x + 3)(x + 3)$

$$\begin{aligned} &\quad\ \ \text{F}\qquad\ \ \text{O}\qquad\ \ \text{I}\qquad\quad \text{L} \\ &= x(x) + x(3) + 3(x) + (3)(3) \\ &= x^2 + 3x + 3x + 9 \\ &= x^2 + 6x + 9 \end{aligned}$$

Example 18 is an example of the **square of a binomial**, another special product.

Square of Binomial Formulas

$$(a + b)^2 = (a + b)(a + b) = a^2 + 2ab + b^2$$
$$(a - b)^2 = (a - b)(a - b) = a^2 - 2ab + b^2$$

To square a binomial, add the square of the first term, twice the product of the terms, and the square of the second term.

EXAMPLE 19 Use the square of a binomial formula to multiply each expression.

a) $(x + 5)^2$ **b)** $(2x - 4)^2$

c) $(3x + 2y)(3x + 2y)$ **d)** $(x - 3)(x - 3)$

Solution **a)** If we let $x = a$ and $5 = b$, then

$$(a + b)(a + b) = a^2 + 2ab + b^2$$
$$\downarrow\ \downarrow\ \downarrow\ \downarrow \qquad \downarrow \quad \downarrow\ \downarrow \quad \downarrow$$
$$(x + 5)^2 = (x + 5)(x + 5) = (x)^2 + 2(x)(5) + (5)^2$$
$$= x^2 + 10x + 25$$

b)

$$(a - b)\ (a - b) = a^2 - 2ab + b^2$$
$$\downarrow\ \downarrow\ \downarrow\ \downarrow \qquad \downarrow \quad \downarrow\ \downarrow \quad \downarrow$$
$$(2x - 4)^2 = (2x - 4)(2x - 4) = (2x)^2 - 2(2x)(4) + (4)^2$$
$$= 4x^2 - 16x + 16$$

c)

$$(a + b)\ (a + b) = a^2 + 2ab + b^2$$
$$\downarrow\ \downarrow\ \downarrow\ \downarrow \qquad \downarrow \quad \downarrow\ \downarrow \quad \downarrow$$
$$(3 + 2y)^2 = (3x + 2y)(3x + 2y) = (3x)^2 + 2(3x)(2y) + (2y)^2$$
$$= 9x^2 + 12xy + 4y^2$$

d)

$$(a - b)(a - b) = a^2 - 2ab + b^2$$
$$\downarrow\ \downarrow\ \downarrow\ \downarrow \qquad \downarrow \quad \downarrow\ \downarrow \quad \downarrow$$
$$(x - 3)(x - 3) = (x - 3)(x - 3) = (x)^2 - 2(x)(3) + (3)^2$$
$$= x^2 - 6x + 9$$

NOW TRY EXERCISE 83

Example 19 could also have been done using the FOIL method.

Avoiding Common Errors

CORRECT	INCORRECT
$(a + b)^2 = a^2 + 2ab + b^2$	~~$(a + b)^2 = a^2 + b^2$~~
$(a - b)^2 = a^2 - 2ab + b^2$	~~$(a - b)^2 = a^2 - b^2$~~

Do not forget the middle term when you square a binomial.

$$(x + 2)^2 \neq x^2 + 4$$
$$(x + 2)^2 = (x + 2)(x + 2)$$
$$= x^2 + 4x + 4$$

6) Multiply Any Two Polynomials

When multiplying a binomial by a binomial, we saw that every term in the first binomial was multiplied by every term in the second binomial. When multiplying any two polynomials, each term of one polynomial must be multiplied by each term of the other polynomial. In the multiplication $(3x + 2)(4x^2 - 5x - 3)$ we use the distributive property as follows:

$$\begin{aligned} &(3x + 2)(4x^2 - 5x - 3) \\ &= 3x(4x^2 - 5x - 3) + 2(4x^2 - 5x - 3) \\ &= 12x^3 - 15x^2 - 9x + 8x^2 - 10x - 6 \\ &= 12x^3 - 7x^2 - 19x - 6 \end{aligned}$$

Thus, $(3x + 2)(4x^2 - 5x - 3) = 12x^3 - 7x^2 - 19x - 6$.

Multiplication problems can be performed by using the distributive property, as we just illustrated. However, many students prefer to multiply a polynomial by a polynomial using a vertical procedure. On page 375, we showed that when multiplying the number 43 by the number 12, we multiply each digit in the number 43 by each digit in the number 12. Review that example now. We can follow a similar procedure when multiplying a polynomial by a polynomial, as illustrated in the following examples. We must be careful, however, to align like terms in the same columns when performing the individual multiplications.

EXAMPLE 20 Multiply $(3x + 4)(2x + 5)$.

Solution First write the polynomials one beneath the other.

$$\begin{array}{r} 3x + 4 \\ \underline{2x + 5} \end{array}$$

Next, multiply each term in $(3x + 4)$ by 5.

$$\begin{array}{rr} & 3x + 4 \\ & \underline{2x + 5} \\ 5(3x + 4) \longrightarrow & 15x + 20 \end{array}$$

Next, multiply each term in $(3x + 4)$ by $2x$ and align like terms.

$$\begin{array}{rrl} & 3x + 4 & \\ & \underline{2x + 5} & \\ & 15x + 20 & \\ 2x(3x + 4) \longrightarrow & \underline{6x^2 + 8x } & \\ & 6x^2 + 23x + 20 & \text{Add like terms in columns} \end{array}$$

The same answer for Example 20 would be obtained using the FOIL method.

EXAMPLE 21 Multiply $(5x - 2)(2x^2 + 3x - 4)$.

Solution For convenience, we place the shorter expression on the bottom, as illustrated.

$$\begin{array}{rl} 2x^2 + 3x - 4 & \\ \underline{5x - 2} & \\ -4x^2 - 6x + 8 & \text{Multiply the top polynomial by } -2. \\ \underline{10x^3 + 15x^2 - 20x } & \text{Multiply the top polynomial by } 5x\text{; align like terms.} \\ 10x^3 + 11x^2 - 26x + 8 & \text{Add like terms in columns.} \end{array}$$

EXAMPLE 22 Multiply $x^2 - 3x + 2$ by $2x^2 - 3$.

Solution

$$\begin{array}{rrrrr} & & x^2 & -3x & +2 \\ & & 2x^2 & & -3 \\ \hline & & -3x^2 & +9x & -6 \\ 2x^4 & -6x^3 & +4x^2 & & \\ \hline 2x^4 & -6x^3 & +x^2 & +9x & -6 \end{array}$$

Multiply the top polynomial by -3.
Multiply the top polynomial by $2x^2$; align like terms.
Add like terms in columns.

NOW TRY EXERCISE 89

EXAMPLE 23 Multiply $(3x^3 - 2x^2 + 4x + 6)(x^2 - 5x)$.

Solution

$$\begin{array}{rrrrrr} & & 3x^3 & -2x^2 & +4x & +6 \\ & & & & x^2 & -5x \\ \hline & -15x^4 & +10x^3 & -20x^2 & -30x & \\ 3x^5 & -2x^4 & +4x^3 & +6x^2 & & \\ \hline 3x^5 & -17x^4 & +14x^3 & -14x^2 & -30x & \end{array}$$

Multiply the top polynomial by $-5x$.
Multiply the top polynomial by x^2.
Add like terms in columns.

Exercise Set 6.5

Concept/Writing Exercises

1. What is the name of the property used when multiplying a monomial by a polynomial?
2. Explain how to multiply a monomial by a monomial.
3. What do the letters in the acronym FOIL represent?
4. How does the FOIL method work when multiplying two binomials?
5. When multiplying two binomials, will you get the same answer if you multiply using the order LOIF instead of FOIL? Explain your answer.
6. Why is the special product $(a + b)(a - b) = a^2 - b^2$ also called the difference of two squares formula?
7. Write the square of binomial formulas.
8. Use your own words to describe how to square a binomial.
9. Does $(x + 5)^2 = x^2 + 5^2$? Explain. If not, what is the correct result?
10. Does $(x + 3)^2 = x^2 + 3^2$? Explain. If not, what is the correct result?
11. Make up a multiplication problem where a monomial in x is multiplied by a binomial in x. Determine the product.
12. Make up a multiplication problem where a monomial in y is multiplied by a trinomial in y. Determine the product.
13. Make up a multiplication problem where two binomials in x are multiplied. Determine the product.
14. When multiplying two polynomials, is it necessary for each term in one polynomial to multiply each term in the other polynomial?

Practice the Skills

Multiply.

15. $x^2 \cdot 3xy$
16. $6xy^2 \cdot 3xy^4$
17. $5x^3y^5(4x^2y)$
18. $-5x^2y^4(2x^3y^2)$
19. $4x^4y^6(-7x^2y^9)$
20. $12x^6y^2(2x^9y^7)$
21. $9xy^6 \cdot 6x^5y^8$
22. $(7x^3y^3)(2x^4)$
23. $(6x^2y)\left(\frac{1}{2}x^4\right)$
24. $\frac{2}{3}x(6x^2y^3)$
25. $(3.3x^4)(1.8x^4y^3)$
26. $(2.3x^5)(4.1x^2y^4)$

Multiply.

27. $3(x + 4)$
28. $3(x - 4)$
29. $-3x(2x - 2)$
30. $-4x(-2x + 6)$
31. $-2(8y + 5)$
32. $2x(x^2 + 3x - 1)$
33. $-2x(x^2 - 2x + 5)$
34. $-3x(-2x^2 + 5x - 6)$
35. $5x(-4x^2 + 6x - 4)$

36. $(x^2 - x + 1)x$

37. $0.5x^2(x^3 - 6x^2 - 1)$

38. $1.4(3x^2 + 4)$

39. $0.3x(2xy + 5x - 6y)$

40. $-\frac{1}{2}x(2x^2 + 4x - 6y^2)$

41. $(x - y - 3)y$

42. $\frac{1}{3}y(y^2 - 6y + 3x)$

Multiply.

43. $(x + 3)(x + 4)$

44. $(2x - 3)(x + 5)$

45. $(2x + 5)(3x - 6)$

46. $(3x - 1)(x + 4)$

47. $(2x - 4)(2x + 4)$

48. $(4 + 3x)(2 + x)$

49. $(5 - 3x)(6 + 2x)$

50. $(-x + 3)(2x + 5)$

51. $(6x - 1)(-2x + 5)$

52. $(3x + 4)(3x - 2)$

53. $(x - 2)(4x - 2)$

54. $(2x + 3)(x + 5)$

55. $(3x - 6)(4x - 2)$

56. $(2x - 5)(4x - 1)$

57. $(x - 1)(x + 1)$

58. $(3x - 8)(2x + 3)$

59. $(2x - 3)(2x - 3)$

60. $(7x + 3)(2x + 4)$

61. $(4x - 4)(7 - x)$

62. $(6 - 2x)(5x - 3)$

63. $(2x + 3)(4 - 2x)$

64. $(5 - 6x)(2x - 7)$

65. $(x + y)(x - y)$

66. $(x + 2y)(2x - 3)$

67. $(2x - 3y)(3x + 2y)$

68. $(x + 3)(2y - 5)$

69. $(3x + y)(2 + 2x)$

70. $(2x - 0.1)(x + 2.4)$

71. $(x + 0.6)(x + 0.3)$

72. $(3x - 6)\left(x + \frac{1}{3}\right)$

73. $(2y - 4)\left(\frac{1}{2}x - 1\right)$

74. $(x + 4)\left(x - \frac{1}{2}\right)$

Multiply using a special product formula.

75. $(x + 5)(x - 5)$

76. $(x + 3)^2$

77. $(3x - 3)(3x + 3)$

78. $(x - 3)(x - 3)$

79. $(x + y)^2$

80. $(2x - 3)(2x + 3)$

81. $(x - 0.2)^2$

82. $(x + 4y)(x - 4y)$

83. $(4x + 5)(4x + 5)$

84. $(5x + 4)(5x - 4)$

85. $(0.4x + y)^2$

86. $\left(x - \frac{1}{2}y\right)^2$

Multiply.

87. $(x + 3)(2x^2 + 4x - 1)$

88. $(2x + 3)(4x^2 - 5x + 6)$

89. $(3x + 2)(4x^2 - x + 5)$

90. $(x - 1)(3x^2 + 3x + 2)$

91. $(-2x^2 - 4x + 1)(7x - 3)$

92. $(4x^2 + 9x - 2)(x - 2)$

93. $(-2x + 6)(2x^2 + 3x - 4)$

94. $(4x^2 + 1)(2x + 1)$

95. $(3x^2 - 2x + 4)(2x^2 + 3x + 1)$

96. $(x^2 - 2x + 3)(x^2 - 4)$

97. $(x^2 - x + 3)(x^2 - 2x)$

98. $(6x + 4)(2x^2 + 2x - 4)$

99. $(a + b)(a^2 - ab + b^2)$

100. $(a - b)(a^2 + ab + b^2)$

Problem Solving

101. Will the product of a monomial and a monomial always be a monomial? Explain your answer.

102. Will the product of a monomial and a binomial ever be a trinomial? Explain your answer.

103. Will the product of two binomials after like terms are combined always be a trinomial? Explain your answer.

104. Will the product of any polynomial and a binomial always be a polynomial? Explain.

Consider the multiplications in Exercises 105 and 106. Determine the exponents to be placed in the shaded areas.

105. $3x^2(2x^{\square} - 5x^{\square} + 3x^{\square}) = 6x^8 - 15x^5 + 9x^3$.

106. $4x^3(x^{\square} + 2x^{\square} - 5x^{\square}) = 4x^7 + 8x^5 - 20x^4$.

107. Suppose that one side of a rectangle is represented as $x + 2$ and a second side is represented as $2x + 1$.

a) Express the area of the rectangle in terms of x.

b) Find the area if $x = 4$ feet.

c) What value of x, in feet, would result in the rectangle being a square? Explain how you determined your answer.

108. Suppose that a rectangular solid has length $x + 5$, width $3x + 4$, and height $2x - 2$ (see the figure).

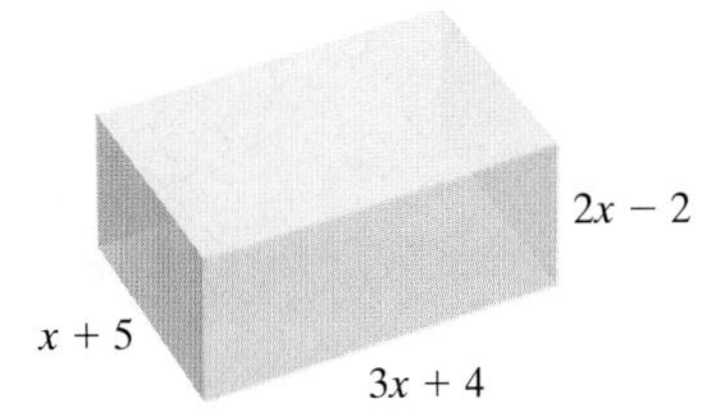

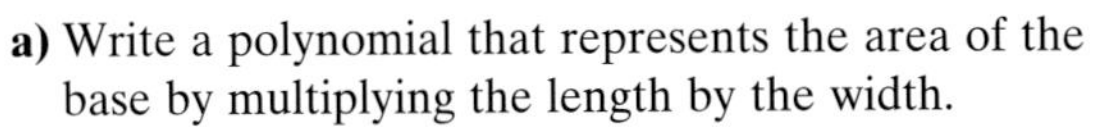

a) Write a polynomial that represents the area of the base by multiplying the length by the width.

b) The volume of the figure can be found by multiplying the area of the base by the height. Write a polynomial that represents the volume of the figure.

c) Using the polynomial in part **b)**, find the volume of the figure if x is 4 feet.

d) Using the binomials given for the length, width, and height, find the volume if x is 4 feet.

e) Are your answers to parts **c)** and **d)** the same? If not, explain why.

109. Consider the figure below.

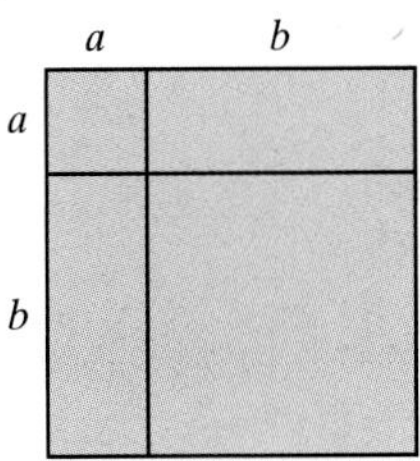

a) Write an expression for the length of the top.

b) Write an expression for the length of the left side.

c) Is this figure a square? Explain.

d) Express the area of this square as the square of a binomial.

e) Determine the area of the square by summing the areas of the four individual pieces.

f) Using the figure and your answer to part **e)**, complete the following.

$$(a + b)^2 = ?$$

Challenge Problems

Multiply.

110. $\left(\frac{1}{2}x + \frac{2}{3}\right)\left(\frac{2}{3}x - \frac{2}{5}\right)$

111. $(2x^3 - 6x^2 + 5x - 3)(3x^3 - 6x + 4)$

Group Activity

112. Consider the trinomial $2x^2 + 7x + 3$.

a) As a group, determine whether there is a maximum number of pairs of binomials whose product is $2x^2 + 7x + 3$. That is, how many different pairs of binomials can go in the shaded areas?

$$2x^2 + 7x + 3 = (\quad)(\quad)$$

b) Individually, find a pair of binomials whose product is $2x^2 + 7x + 3$.

c) Compare your answer to part **b)** with the other members of your group. If you did not all arrive at the same answer, explain why.

Cumulative Review Exercises

[3.3] **113.** The cost of a taxi ride is \$2.00 for the first mile and \$1.50 for each additional mile or part thereof. Find the maximum distance Ingrid Mount can ride in the taxi if she has only \$20.

[6.1] **114.** Simplify $\left(\frac{4x^8y^5}{8x^8y^6}\right)^4$.

[6.1–6.2] **115.** Evaluate the following.

a) -6^3 **b)** 6^{-3}

[6.4] **116.** Subtract $5x^2 - 4x - 3$ from $-x^2 - 6x + 5$.

6.6 DIVISION OF POLYNOMIALS

SSM VIDEO 6.6 CD Rom

1) **Divide a polynomial by a monomial.**
2) **Divide a polynomial by a binomial.**
3) **Check division of polynomial problems.**
4) **Write polynomials in descending order when dividing.**

1) Divide a Polynomial by a Monomial

Now let's see how to divide polynomials. We begin by dividing a polynomial by a monomial.

To divide a polynomial by a monomial

Divide each term of the polynomial by the monomial.

EXAMPLE 1 Divide: $\dfrac{2x + 16}{2}$.

Solution $\dfrac{2x + 16}{2} = \dfrac{2x}{2} + \dfrac{16}{2} = x + 8$

EXAMPLE 2 Divide: $\dfrac{4x^2 - 8x}{2x}$.

Solution $\dfrac{4x^2 - 8x}{2x} = \dfrac{4x^2}{2x} - \dfrac{8x}{2x} = 2x - 4$

Avoiding Common Errors

CORRECT	INCORRECT
$\dfrac{x+2}{2} = \dfrac{x}{2} + \dfrac{2}{2} = \dfrac{x}{2} + 1$	$\dfrac{x+\overset{1}{\cancel{2}}}{\underset{1}{\cancel{2}}} = \dfrac{x+1}{1} = x + 1$ (crossed out)
$\dfrac{x+2}{x} = \dfrac{x}{x} + \dfrac{2}{x} = 1 + \dfrac{2}{x}$	$\dfrac{\overset{1}{\cancel{x}}+2}{\underset{1}{\cancel{x}}} = \dfrac{1+2}{1} = 3$ (crossed out)

Can you explain why the procedures on the right are not correct?

EXAMPLE 3 Divide: $\dfrac{4x^5 - 6x^4 + 8x - 3}{2x^2}$.

Solution
$$\frac{4x^5 - 6x^4 + 8x - 3}{2x^2} = \frac{4x^5}{2x^2} - \frac{6x^4}{2x^2} + \frac{8x}{2x^2} - \frac{3}{2x^2}$$
$$= 2x^3 - 3x^2 + \frac{4}{x} - \frac{3}{2x^2}$$

EXAMPLE 4 Divide: $\dfrac{3x^3 - 6x^2 + 4x - 1}{-3x}$.

Solution A negative sign appears in the denominator. Usually it is easier to divide if the divisor is positive. We can multiply both numerator and denominator by -1 to get a positive denominator.

$$\begin{aligned}\frac{(-1)(3x^3 - 6x^2 + 4x - 1)}{(-1)(-3x)} &= \frac{-3x^3 + 6x^2 - 4x + 1}{3x}\\ &= \frac{-3x^3}{3x} + \frac{6x^2}{3x} - \frac{4x}{3x} + \frac{1}{3x}\\ &= -x^2 + 2x - \frac{4}{3} + \frac{1}{3x}\end{aligned}$$

NOW TRY EXERCISE 37

2) Divide a Polynomial by a Binomial

We divide a polynomial by a binomial in much the same way as we perform long division. This procedure will be explained in Example 5.

EXAMPLE 5 Divide: $\dfrac{x^2 + 6x + 8}{x + 2}$. $\begin{array}{l}\longleftarrow \textit{dividend}\\ \longleftarrow \textit{divisor}\end{array}$

Solution Rewrite the division problem in the following form:

$$x + 2\overline{)x^2 + 6x + 8}$$

Divide x^2 (the first term in the dividend) by x (the first term in the divisor).

$$\frac{x^2}{x} = x$$

Place the quotient, x, above the like term containing x in the dividend.

$$\begin{array}{r} x\phantom{{}+6x+8} \\ x + 2\overline{)x^2 + 6x + 8} \end{array}$$

Next, multiply the x by $x + 2$ as you would do in long division and place the terms of the product under their like terms.

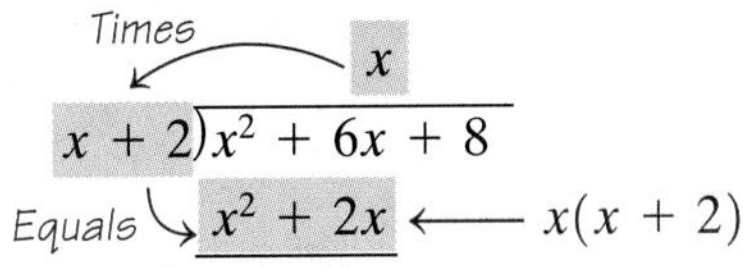
Times
Equals
x(x + 2)

Now subtract $x^2 + 2x$ from $x^2 + 6x$. When subtracting, remember to change the sign of the terms being subtracted and then add the like terms.

$$\begin{array}{r} x\phantom{{}+6x+8} \\ x + 2\overline{)\ x^2 + 6x + 8} \\ \underline{\overset{-}{x^2} \overset{-}{\cancel{+}} 2x}\phantom{{}+8} \\ 4x\phantom{{}+8} \end{array}$$

Next, bring down the 8, the next term in the dividend.

$$\begin{array}{r} x\phantom{{}+6x+8} \\ x + 2\overline{)x^2 + 6x + 8} \\ \underline{x^2 + 2x}\phantom{{}+8} \\ 4x + 8 \end{array}$$

Now divide $4x$, the first term at the bottom, by x, the first term in the divisor.

$$\frac{4x}{x} = +4$$

Write the +4 in the quotient above the constant in the dividend.

$$\begin{array}{r} x + 4 \\ x + 2\overline{)x^2 + 6x + 8} \\ \underline{x^2 + 2x \phantom{{}+8}} \\ 4x + 8 \end{array}$$

Multiply the $x + 2$ by 4 and place the terms of the product under their like terms.

Times

$$\begin{array}{r} x + 4 \\ x + 2\overline{)x^2 + 6x + 8} \\ \underline{x^2 + 2x \phantom{{}+8}} \\ 4x + 8 \\ 4x + 8 \longleftarrow 4(x+2) \end{array}$$

Equals

Now subtract.

$$\begin{array}{r} x + 4 \longleftarrow \text{Quotient} \\ x + 2\overline{)x^2 + 6x + 8} \\ \underline{x^2 + 2x \phantom{{}+8}} \\ 4x + 8 \\ \underline{{}^{-}4x \mathbin{\not+} {}^{-}8} \\ 0 \longleftarrow \text{Remainder} \end{array}$$

Thus,

$$\frac{x^2 + 6x + 8}{x + 2} = x + 4$$

There is no remainder.

EXAMPLE 6 Divide: $\dfrac{6x^2 - 5x + 5}{2x + 3}$.

Solution

$$\frac{6x^2}{2x} \qquad \frac{-14x}{2x}$$

$$\begin{array}{r} 3x - 7 \\ 2x + 3\overline{)\ 6x^2 - 5x + 5} \\ \underline{{}^{-}6x^2 \mathbin{\not+} {}^{-}9x \phantom{{}+5}} \longleftarrow 3x(2x+3) \\ -14x + 5 \\ \underline{\overset{+}{\not-} 14x \overset{+}{\not-} 21} \longleftarrow -7(2x+3) \\ 26 \longleftarrow \text{Remainder} \end{array}$$

When there is a remainder, as in this example, list the quotient, plus the remainder above the divisor. Thus,

NOW TRY EXERCISE 57

$$\frac{6x^2 - 5x + 5}{2x + 3} = 3x - 7 + \frac{26}{2x + 3}$$

3) Check Division of Polynomial Problems

The answer to a division problem can be checked. Consider the division problem $13 \div 5$.

$$\begin{array}{r} 2 \\ 5\overline{)13} \\ \underline{10} \\ 3 \end{array}$$

Note that the divisor times the quotient, plus the remainder, equals the dividend:

$$\text{(divisor} \times \text{quotient)} + \text{remainder} = \text{dividend}$$

$$(5 \cdot 2) + 3 = 13$$

$$10 + 3 = 13$$

$$13 = 13 \quad \textit{True}$$

This same procedure can be used to check all division problems.

To Check Division of Polynomials

(divisor × quotient) + remainder = dividend

Let's check the answer to Example 6. The divisor is $2x + 3$, the quotient is $3x - 7$, the remainder is 26, and the dividend is $6x^2 - 5x + 5$.

CHECK: (divisor × quotient) + remainder = dividend

$$(2x + 3)(3x - 7) + 26 \stackrel{?}{=} 6x^2 - 5x + 5$$

$$(6x^2 - 5x - 21) + 26 \stackrel{?}{=} 6x^2 - 5x + 5$$

$$6x^2 - 5x + 5 = 6x^2 - 5x + 5 \quad \textit{True}$$

4) Write Polynomials in Descending Order When Dividing

When dividing a polynomial by a binomial, both the polynomial and binomial should be listed in descending order. If a given power term is missing, it is often helpful to include that term with a numerical coefficient of 0 as a placeholder. This will help in keeping like terms aligned. For example, to divide $(6x^2 + x^3 - 4)/(x - 2)$, we begin by writing $(x^3 + 6x^2 + 0x - 4)/(x - 2)$.

EXAMPLE 7 Divide $(-x + 9x^3 - 28)$ by $(3x - 4)$.

Solution First we rewrite the dividend in descending order to get $(9x^3 - x - 28) \div (3x - 4)$. Since there is no x^2 term in the dividend, we will add $0x^2$ to help align like terms.

$$\frac{9x^3}{3x}, \quad \frac{12x^2}{3x}, \quad \frac{15x}{3x}$$

$$\downarrow \qquad \downarrow \qquad \downarrow$$

$$\begin{array}{r l}
 & 3x^2 + 4x + 5 \\
3x - 4 \overline{) \; 9x^3 + 0x^2 - x - 28} & \\
9x^3 - 12x^2 & \leftarrow 3x^2(3x - 4) \\
\hline
12x^2 - x & \\
12x^2 - 16x & \leftarrow 4x(3x - 4) \\
\hline
15x - 28 & \\
15x - 20 & \leftarrow 5(3x - 4) \\
\hline
-8 & \leftarrow \text{Remainder}
\end{array}$$

Thus, $\frac{-x + 9x^3 - 28}{3x - 4} = 3x^2 + 4x + 5 - \frac{8}{3x - 4}$. Check this division yourself using the procedure just discussed.

Exercise Set 6.6

Concept/Writing Exercises

1. Explain how to divide a polynomial by a monomial.
2. Explain how to check a division problem.
3. Explain why $\frac{\cancel{x} + 5}{\cancel{x}} \neq \frac{1 + 5}{1}$. Then, correctly divide the binomial by the monomial.
4. Explain why $\frac{\cancel{2}x + 8}{\cancel{2}} \neq \frac{x + 8}{1}$. Then, correctly divide the binomial by the monomial.
5. How should the terms of a polynomial and binomial be listed when dividing a polynomial by a binomial?
6. How would you rewrite $\frac{x^2 - 5}{x - 1}$ so that it is easier to complete the division?
7. How would you rewrite $\frac{x^3 + 5x - 1}{x + 2}$ so that it is easier to complete the division?
8. Show that $\frac{x^2 - 3x + 7}{x + 2} = x - 5 + \frac{17}{x + 2}$ by checking the division.
9. Show that $\frac{x^2 + 2x - 17}{x - 3} = x + 5 - \frac{2}{x - 3}$ by checking the division.
10. Show that $\frac{x^3 + 2x - 3}{x + 1} = x^2 - x + 3 - \frac{6}{x + 1}$ by checking the division.

Rewrite each multiplication problem as a division problem. There is more than one correct answer.

11. $(x + 3)(x - 5) = x^2 - 2x - 15$
12. $(x + 3)(3x - 1) = 3x^2 + 8x - 3$
13. $(2x + 3)(x + 1) = 2x^2 + 5x + 3$
14. $(2x - 3)(x + 4) = 2x^2 + 5x - 12$
15. $(2x + 3)(2x - 3) = 4x^2 - 9$
16. $(2x + 2)(x - 3) = 2x^2 - 4x - 6$

Practice the Skills

Divide.

17. $\frac{2x + 4}{2}$

18. $\frac{4x - 6}{2}$

19. $\frac{2x + 6}{2}$

20. $(-3x - 8) \div 4$

21. $\frac{3x + 8}{2}$

22. $\frac{5x - 10}{5}$

23. $\frac{-6x + 4}{2}$

24. $\frac{-4x + 5}{-3}$

25. $\frac{-9x - 3}{-3}$

26. $\frac{5x - 4}{-5}$

27. $\frac{2x + 16}{4}$

28. $\frac{4x - 3}{2x}$

29. $\frac{4 - 12x}{-3}$

30. $\frac{6 - 5x}{-3x}$

31. $(3x^2 + 6x - 9) \div 3x^2$

32. $\frac{12x^2 - 6x + 3}{3}$

33. $\frac{-4x^5 + 6x + 8}{2x^2}$

34. $\frac{5x^2 + 4x - 8}{2}$

35. $(x^6 + 4x^4 - 3) \div x^3$

36. $(6x^2 - 7x + 9) \div 3x$

37. $\frac{6x^5 - 4x^4 + 12x^3 - 5x^2}{2x^3}$

38. $\frac{7x^2 + 14x - 5}{-7}$

39. $\frac{4x^3 + 6x^2 - 8}{-4x}$

40. $\frac{-12x^4 + 6x^2 - 15x + 4}{-3x}$

41. $\frac{9x^6 + 3x^4 - 10x^2 - 9}{3x^2}$

42. $\frac{-10x^3 - 6x^2 + 15}{-5x^3}$

Divide.

43. $\frac{x^2 + 4x + 3}{x + 1}$

44. $(2x^2 + 3x - 35) \div (x + 5)$

45. $\frac{2x^2 + 13x + 15}{x + 5}$

46. $\frac{2x^2 + x - 10}{x - 2}$

47. $\frac{6x^2 + 16x + 8}{3x + 2}$

48. $\frac{2x^2 + 13x + 15}{2x + 3}$

49. $\frac{x^2 - 9}{-3 + x}$

50. $\frac{8x^2 - 26x + 15}{4x - 3}$

51. $(2x^2 + 7x - 18) \div (2x - 3)$

52. $\frac{x^2 - 25}{x - 5}$

53. $(4x^2 - 9) \div (2x - 3)$

54. $\frac{9x^2 - 16}{3x - 4}$

55. $\frac{6x + 8x^2 - 25}{4x + 9}$

56. $\frac{10x + 3x^2 + 6}{x + 2}$

57. $\frac{6x + 8x^2 - 12}{2x + 3}$

58. $\frac{x^3 + 3x^2 + 5x + 3}{x + 1}$

59. $\frac{3x^3 + 18x^2 - 5x - 30}{x + 6}$

60. $\frac{2x^3 - 3x^2 - 3x + 6}{x - 1}$

61. $\frac{2x^3 - 4x^2 + 12}{x - 2}$

62. $\frac{2x^3 + 6x - 4}{x + 4}$

63. $(x^3 - 8) \div (x - 3)$

64. $\frac{x^3 + 8}{x + 2}$

65. $\frac{x^3 - 27}{x - 3}$

66. $\frac{x^3 + 27}{x + 3}$

67. $\frac{4x^3 - 5x}{2x - 1}$

68. $\frac{9x^3 - x + 3}{3x - 2}$

69. $\frac{-x^3 - 6x^2 + 2x - 3}{x - 1}$

70. $\frac{-x^3 + 3x^2 + 14x + 16}{x + 3}$

Problem Solving

71. When dividing a binomial by a monomial, must the quotient be a binomial? Explain and give an example to support your answer.

72. When dividing a trinomial by a monomial, must the quotient be a trinomial? Explain and give an example to support your answer.

73. If the divisor is $x + 4$, the quotient is $2x + 3$, and the remainder is 4, find the dividend (or the polynomial being divided).

74. If the divisor is $2x - 3$, the quotient is $3x - 1$, and the remainder is -2, find the dividend.

75. If a second-degree polynomial in x is divided by a first-degree polynomial in x, what will be the degree of the quotient? Explain.

76. If a third-degree polynomial in x is divided by a first-degree polynomial in x, what will be the degree of the quotient? Explain.

Determine the expression to be placed in the shaded area to make a true statement. Explain how you determined your answer

77. $\dfrac{16x^4 + 20x^3 - 4x^2 + 12x}{\blacksquare} = 4x^3 + 5x^2 - x + 3$

78. $\dfrac{9x^5 - 6x^4 + 3x^2 + 12}{\blacksquare} = 3x^3 - 2x^2 + 1 + \dfrac{4}{x^2}$

Determine the exponents to be placed in the shaded areas to make a true statement. Explain how you determined your answer.

79. $\dfrac{8x^{\blacksquare} + 4x^{\blacksquare} - 20x^{\blacksquare} - 5x^{\blacksquare}}{2x^2} = 4x^3 + 2x - 10 - \dfrac{5}{2x}$

80. $\dfrac{15x^{\blacksquare} + 25x^{\blacksquare} + 5x^{\blacksquare} + 10x^{\blacksquare}}{5x^2} = 3x^5 + 5x^4 + x^2 + 2$

Challenge Problems

Divide. The quotient in Exercises 81 and 83 will contain fractions.

81. $\dfrac{4x^3 - 4x + 6}{2x + 3}$

82. $\dfrac{3x^3 - 5}{3x - 2}$

83. $\dfrac{3x^2 + 6x - 10}{-x - 3}$

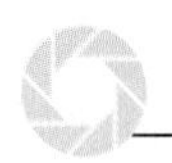

Group Activity

Discuss and answer Exercises 84 and 85 as a group. Determine the polynomial that when substituted in the shaded area results in a true statement. Explain how you determined your answer.

84. $\dfrac{\blacksquare}{x + 4} = x + 2 + \dfrac{2}{x + 4}$

85. $\dfrac{\blacksquare}{x + 3} = x + 1 - \dfrac{1}{x + 3}$

Cumulative Review Exercises

[1.4] **86.** Consider the set of numbers

$$\{2, -5, 0, \sqrt{7}, \tfrac{2}{5}, -6.3, \sqrt{3}, -23/34\}.$$

List those that are **a)** natural numbers; **b)** whole numbers; **c)** rational numbers; **d)** irrational numbers; and **e)** real numbers.

[1.8] **87. a)** To what is 0/1 equal?

b) How do we refer to an expression like 1/0?

[1.9] **88.** Give the order of operations to be followed when evaluating a mathematical expression.

[2.5] **89.** Solve the equation $2(x + 3) + 2x = x + 4$.

[4.2] *Graph using the x- and y-intercepts.*

90. $y = 2x + 3$

91. $y = -2x + 8$

SUMMARY

Key Words and Phrases

6.1
Base
Exponent

6.3
Scientific notation

6.4
Binomial
Degree of a polynomial
Degree of a term
Descending order
Monomial
Polynomial
Trinomial

IMPORTANT FACTS

Rules of Exponents

1.	$x^m x^n = x^{m+n}$	**product rule**
2.	$\dfrac{x^m}{x^n} = x^{m-n}, \quad x \neq 0$	**quotient rule**
3.	$x^0 = 1, \quad x \neq 0$	**zero exponent rule**
4.	$(x^m)^n = x^{m \cdot n}$	**power rule**
5.	$\left(\dfrac{ax}{by}\right)^m = \dfrac{a^m x^m}{b^m y^m}, \quad b \neq 0, y \neq 0$	**expanded power rule**
6.	$x^{-m} = \dfrac{1}{x^m}, \quad x \neq 0$	**negative exponent rule**

Product of sum and difference of the same two terms (also called the difference of two squares):

$$(a + b)(a - b) = a^2 - b^2$$

FOIL Method to Multiply Two Binomials (First, Outer, Inner, Last)

$$(a + b)(c + d)$$

Square of a binomial

$$(a + b)^2 = a^2 + 2ab + b^2$$

$$(a - b)^2 = a^2 - 2ab + b^2$$

Review Exercises

[6.1] *Simplify.*

1. $x^4 \cdot x^2$ **2.** $x^2 \cdot x^4$ **3.** $3^2 \cdot 3^3$ **4.** $2^4 \cdot 2$

5. $\dfrac{x^4}{x}$ **6.** $\dfrac{x^6}{x^6}$ **7.** $\dfrac{5^5}{5^3}$ **8.** $\dfrac{4^5}{4^3}$

9. $\dfrac{x^6}{x^8}$ **10.** $\dfrac{y^4}{y}$ **11.** x^0 **12.** $4x^0$

13. $(3x)^0$ **14.** 4^0 **15.** $(5x)^2$ **16.** $(3x)^3$

17. $(-4y)^2$ **18.** $(-3x)^3$ **19.** $(2x^2)^4$ **20.** $(-x^4)^3$

21. $(-x^3)^4$ **22.** $\left(\dfrac{2x^3}{y}\right)^2$ **23.** $\left(\dfrac{8y^2}{2x}\right)^2$ **24.** $6x^2 \cdot 4x^3$

25. $\dfrac{16x^2y}{4xy^2}$ **26.** $2x(3xy^3)^2$ **27.** $\left(\dfrac{9x^2y}{3xy}\right)^2$ **28.** $(2x^2y)^3(3xy^4)$

29. $4x^2y^3(2x^3y^4)^2$ **30.** $4y^2(2x^2y^3)^2$ **31.** $\left(\dfrac{8x^4y^3}{2xy^5}\right)^2$ **32.** $\left(\dfrac{21x^4y^3}{7y^2}\right)^3$

[6.2] *Simplify.*

33. x^{-3} **34.** 3^{-3} **35.** 5^{-2} **36.** $\dfrac{1}{x^{-3}}$

37. $\dfrac{1}{x^{-7}}$ **38.** $\dfrac{1}{3^{-2}}$ **39.** $y^5 \cdot y^{-8}$ **40.** $x^{-2} \cdot x^{-3}$

41. $x^4 \cdot x^{-7}$ **42.** $x^{-2} \cdot x^{-2}$ **43.** $\dfrac{x^2}{x^{-3}}$ **44.** $\dfrac{x^5}{x^{-2}}$

45. $\dfrac{x^{-3}}{x^3}$ **46.** $(3x^4)^{-2}$ **47.** $(4x^{-3}y)^{-3}$ **48.** $(-2x^{-2}y)^2$

49. $5y^{-2} \cdot 2y^4$ **50.** $(3x^{-2}y)^3$ **51.** $(4x^{-2}y^3)^{-2}$ **52.** $2x(3x^{-2})$

53. $(5x^{-2}y)(2x^4y)$ **54.** $4x^5(6x^{-7}y^2)$ **55.** $4y^{-2}(3x^2y)$ **56.** $\dfrac{6xy^4}{2xy^{-1}}$

57. $\dfrac{9x^{-2}y^3}{3xy^2}$ **58.** $\dfrac{49x^2y^{-3}}{7x^{-3}y}$ **59.** $\dfrac{36x^4y^7}{9x^5y^{-3}}$ **60.** $\dfrac{4x^5y^{-2}}{8x^7y^3}$

[6.3] *Express each number in scientific notation.*

61. 364,000 **62.** 0.153 **63.** 0.00763

64. 47,000 **65.** 1,370,000 **66.** 0.000314

Express each number without exponents.

67. 4.2×10^{-3} **68.** 6.52×10^{-4} **69.** 9.7×10^5

70. 4.38×10^{-6} **71.** 9.14×10^{-1} **72.** 1.103×10^7

Perform each indicated operation and write your answer without exponents.

73. $(1.7 \times 10^5)(3.4 \times 10^{-4})$ **74.** $(4.2 \times 10^{-3})(3 \times 10^5)$ **75.** $(3.5 \times 10^{-2})(7.0 \times 10^3)$

76. $\dfrac{6.8 \times 10^3}{2 \times 10^{-2}}$ **77.** $\dfrac{6.5 \times 10^4}{2.0 \times 10^6}$ **78.** $\dfrac{15 \times 10^{-3}}{5 \times 10^2}$

Convert each number to scientific notation. Then calculate. Express your answer in scientific notation.

79. (60,000)(20,000) **80.** (12,500)(400,000) **81.** (0.00023)(40,000)

82. $\dfrac{250}{500{,}000}$ **83.** $\dfrac{0.000068}{0.02}$ **84.** $\dfrac{850{,}000}{0.025}$

85. Many educators have their retirement accounts with the Teachers Insurance and Annuity Association, College Retirement Equity Fund (TIAA–CREF). As of November 1998, the CREF Stock Fund had net assets of \$97.2 billion and the CREF Growth Fund had net assets of \$5.1 billion.

a) Write \$97.2 billion as a number in scientific notation.

b) How much more is invested in the CREF Stock Fund than is invested in the CREF Growth Fund? Write your answer in scientific notation and also as a decimal number.

86. If one light year is 5.87×10^{12} miles, how far away in miles is the star Alpha Orion (the brightest star in Orion), which is 520 light years from Earth? Write your answer in scientific notation.

[6.4] *Indicate whether each expression is a polynomial. If the polynomial has a specific name, give that name. If the polynomial is not written in descending order, rewrite it in descending order. State the degree of each polynomial.*

87. $x^{-4} - 8$

88. -2

89. $x^2 - 4 + 3x$

90. $-3 - x + 4x^2$

91. $13x^3 - 4$

92. $4x^{1/2} - 6$

93. $x - 4x^2$

94. $x^3 + x^{-2} + 3$

95. $2x^3 - 7 + 4x^2 - 3x$

[6.4–6.6] *Perform each indicated operation.*

96. $(x + 3) + (2x + 4)$

97. $(4x - 6) + (3x + 15)$

98. $(-x - 10) + (-2x + 5)$

99. $(-x^2 + 6x - 7) + (-2x^2 + 4x - 8)$

100. $(12x^2 + 4x - 8) + (-x^2 - 6x + 5)$

101. $(2x - 4.3) - (x + 2.4)$

102. $(-4x + 8) - (-2x + 6)$

103. $(4x^2 - 9x) - (3x + 15)$

104. $(6x^2 - 6x + 1) - (12x + 5)$

105. $(-2x^2 + 8x - 7) - (3x^2 + 12)$

106. $(x^2 + 7x - 3) - (x^2 + 3x - 5)$

107. $\frac{1}{2}x(14x + 20)$

108. $-3x(5x + 4)$

109. $3x(2x^2 - 4x + 7)$

110. $-x(3x^2 - 6x - 1)$

111. $-4x(-6x^2 + 4x - 2)$

112. $(x + 4)(x + 5)$

113. $(3x + 6)(-4x + 1)$

114. $(-2x + 6)^2$

115. $(6 - 2x)(2 + 3x)$

116. $(x + 4)(x - 4)$

117. $(3x + 1)(x^2 + 2x + 4)$

118. $(x - 1)(3x^2 + 4x - 6)$

119. $(-4x + 2)(3x^2 - x + 7)$

120. $\dfrac{2x + 4}{2}$

121. $\dfrac{10x + 12}{2}$

122. $\dfrac{8x^2 + 4x}{x}$

123. $\dfrac{6x^2 + 9x - 4}{3}$

124. $\dfrac{4x^2 - 8x - 6}{2x}$

125. $\dfrac{8x^5 - 4x^4 + 3x^2 - 2}{2x}$

126. $\dfrac{16x - 4}{-2}$

127. $\dfrac{5x^2 - 6x + 15}{3x}$

128. $\dfrac{5x^3 + 10x + 2}{2x^2}$

129. $\dfrac{x^2 + x - 12}{x - 3}$

130. $\dfrac{8x^2 - 16x - 10}{2x - 5}$

131. $\dfrac{5x^2 + 28x - 10}{x + 6}$

132. $\dfrac{4x^3 + 12x^2 + x - 12}{2x + 3}$

133. $\dfrac{4x^2 - 12x + 9}{2x - 3}$

Practice Test

Simplify each expression.

1. $4x^3 \cdot 3x^2$

2. $(3xy^2)^3$

3. $\dfrac{8x^4}{2x}$

4. $\left(\dfrac{3x^2y}{6xy^3}\right)^3$

5. $(2x^3y^{-2})^{-2}$

6. $\dfrac{15x^4y^2}{45x^{-1}y}$

7. $(4x^0)(3x^2)^0$

Convert each number to scientific notation. Then calculate. Express your answer in scientific notation.

8. $(42{,}000)(30{,}000)$

9. $\dfrac{0.0008}{4000}$

Determine whether each expression is a polynomial. If the polynomial has a specific name, give that name.

10. $x^2 - 4 + 6x$

11. y

12. $x^{-2} + 4$

13. Write the polynomial $-5 + 6x^3 - 2x^2 + 5x$ in descending order, and give its degree.

In Exercises 14–24, perform each indicated operation.

14. $(2x + 4) + (3x^2 - 5x - 3)$

15. $(x^2 - 4x + 7) - (3x^2 - 8x + 7)$

16. $(4x^2 - 5) - (x^2 + x - 8)$

17. $-4x(2x + 4)$

18. $(4x + 7)(2x - 3)$

19. $(6 - 4x)(5 + 3x)$

20. $(3x - 5)(2x^2 + 4x - 5)$

21. $\dfrac{16x^2 + 8x - 4}{4}$

22. $\dfrac{3x^2 - 6x + 5}{-3x}$

23. $\dfrac{8x^2 - 2x - 15}{2x - 3}$

24. $\dfrac{12x^2 + 7x - 12}{4x + 5}$

25. The half-life of an element is the time it takes one half the amount of a radioactive element to decay. The half-life of carbon 14 (C^{14}) is 5730 years. The half-life of uranium 238 (U^{238}) is 4.46×10^9 years.

a) Write the half-life of C^{14} in scientific notation.

b) How many times longer is the half-life of U^{238} than

Cumulative Review Test

1. Evaluate $4 + 8 \div 2 + 3$.

2. Solve the equation $3x + 5 = 4(x - 2)$.

3. Solve the equation $2(x + 3) + 3x - 5 = 4x - 2$.

4. Solve the inequality $3x - 11 < 5x - 2$ and graph the solution on a number line.

5. Solve the equation $5x - 2 = y - 7$ for y.

6. Find the slope of the line through the points $(1, 3)$ and $(5, 1)$.

7. Determine whether the following relation is a function. Give the domain and range of the relation or function. $\{(-4, -5), (1, 5), (2, 7), (5, 13)\}$

8. Find the solution to the following system of equations by substitution.

$$3x + 2y = 10$$
$$x - y = 5$$

9. Find the solution to the following system of equations graphically.

$$2x - y = 7$$
$$x - 3y = -4$$

10. Simplify $(3x^2y^4)^3(5x^2y)$.

11. Write the polynomial $-5x + 2 - 7x^2$ in descending order and give the degree.

Perform each indicated operation.

12. $(x^2 + 4x - 3) + (2x^2 + 5x + 1)$

13. $(8x^2 - 3x + 2) - (4x^2 - 3x + 3)$

14. $(3y - 5)(2y + 3)$

15. $(2x - 1)(3x^2 - 5x + 2)$

16. $\dfrac{6x^2 + 12x - 9}{3x}$

17. $\dfrac{6x^2 + 11x - 10}{3x - 2}$

18. At LeAnn's Grocery Store, three cans of chicken soup sell for \$1.25. Find the cost of eight cans.

19. The length of a rectangle is 2 less than 3 times the width. Find the dimensions of the rectangle if its perimeter is 28 feet.

20. Bob Dolan drives from Jackson, Mississippi, to Tallulah, Louisiana, a distance of 60 miles. At the same time, Nick Reide starts driving from Tallulah to Jackson along the same route, Route 20. If Bob and Nick meet after 0.5 hour and Nick's average speed was 7 miles per hour greater than Bob's, find the average speed of each car.

FACTORING

CHAPTER

7

Use the Angel Web site at www.prenhall.com/angel to be linked to an internet resource that will help you further explore the following application.

Have you ever watched an object falling and wondered how long it would take for it to hit the ground? Whether it be a ski pole accidentally dropped from a chair lift or an apple dropped from a tree house, the time it takes an object to reach the ground below is governed by laws of physics—you may be surprised by the results. In Example 10, page 444, we factor to solve a quadratic equation to determine the time it takes an object to drop 64 feet.

Preview and Perspective

In this chapter, we introduce *factoring polynomials*. Factoring polynomials is the reverse process of multiplying polynomials. When we factor a polynomial, we rewrite it as a product of two or more factors. In Sections 7.1 through 7.4, we discuss factoring techniques for several types of polynomials. In Section 7.5, we present some special factoring formulas, which can simplify the factoring process for some polynomials. In Section 7.6, we introduce quadratic equations and use factoring as one method for solving this type of equation.

It is essential that you have a thorough understanding of factoring, especially Sections 7.3 through 7.5, to complete Chapter 8 successfully. The factoring techniques used in this chapter are used throughout Chapter 8.

7.1 FACTORING A MONOMIAL FROM A POLYNOMIAL

1) **Identify factors.**
2) **Determine the greatest common factor of two or more numbers.**
3) **Determine the greatest common factor of two or more terms.**
4) **Factor a monomial from a polynomial.**

1) Identify Factors

In Chapter 6 you learned how to multiply polynomials. In this chapter we focus on factoring, the reverse process of multiplication. In Section 6.5 we showed that $2x(3x^2 + 4) = 6x^3 + 8x$. In this chapter we start with an expression like $6x^3 + 8x$ and determine that its factors are $2x$ and $3x^2 + 4$, and write $6x^3 + 8x = 2x(3x^2 + 4)$. To **factor an expression** means to write the expression as a product of its factors. Factoring is important because it can be used to solve equations.

If $a \cdot b = c$, then a and b are said to be *factors* of c.

$3 \cdot 5 = 15$; thus 3 and 5 are factors of 15.

$x^3 \cdot x^4 = x^7$; thus x^3 and x^4 are factors of x^7.

$x(x + 2) = x^2 + 2x$; thus x and $x + 2$ are factors of $x^2 + 2x$.

$(x - 1)(x + 3) = x^2 + 2x - 3$; thus $x - 1$ and $x + 3$ are factors of $x^2 + 2x - 3$.

A given number or expression may have many factors. Consider the number 30.

$$1 \cdot 30 = 30, \quad 2 \cdot 15 = 30, \quad 3 \cdot 10 = 30, \quad 5 \cdot 6 = 30$$

Thus, the positive factors of 30 are 1, 2, 3, 5, 6, 10, 15, and 30. Factors can also be negative. Since $(-1)(-30) = 30$, -1 and -30 are also factors of 30. In fact, for each factor a of an expression, $-a$ must also be a factor. Other factors of 30 are therefore $-1, -2, -3, -5, -6, -10, -15$, and -30. When asked to list the factors of an expression that contains a positive numerical coefficient with a variable, we generally list only positive factors.

EXAMPLE 1 List the factors of $6x^3$.

Solution

$$\overbrace{1}^{\text{factors}} \cdot \overbrace{6x^3}^{} = 6x^3 \qquad \overbrace{x}^{\text{factors}} \cdot \overbrace{6x^2}^{} = 6x^3$$

$$2 \cdot 3x^3 = 6x^3 \qquad 2x \cdot 3x^2 = 6x^3$$

$$3 \cdot 2x^3 = 6x^3 \qquad 3x \cdot 2x^2 = 6x^3$$

$$6 \cdot x^3 = 6x^3 \qquad 6x \cdot x^2 = 6x^3$$

The factors of $6x^3$ are 1, 2, 3, 6, x, $2x$, $3x$, $6x$, x^2, $2x^2$, $3x^2$, $6x^2$, x^3, $2x^3$, $3x^3$, and $6x^3$. The opposite (or negative) of each of these factors is also a factor, but these opposites are generally not listed unless specifically asked for.

Here are examples of multiplying and factoring: Notice again that factoring is the reverse process of multiplying.

Multiplying	Factoring
$2(3x + 4) = 6x + 8$	$6x + 8 = 2(3x + 4)$
$5x(x + 4) = 5x^2 + 20x$	$5x^2 + 20x = 5x(x + 4)$
$(x + 1)(x + 3) = x^2 + 4x + 3$	$x^2 + 4x + 3 = (x + 1)(x + 3)$

2) Determine the Greatest Common Factor of Two or More Numbers

To factor a monomial from a polynomial, we make use of the *greatest common factor (GCF)*. If after studying the following material you wish to see additional material on obtaining the GCF, you may read Appendix B, where one of the topics discussed is finding the GCF.

Recall from Section 1.3 that the **greatest common factor** of two or more numbers is the greatest number that divides into all the numbers. The greatest common factor of the numbers 6 and 8 is 2. Two is the greatest number that divides into both 6 and 8. What is the GCF of 48 and 60? When the GCF of two or more numbers is not easily found, we can find it by writing each number as a product of prime numbers. A **prime number** is an integer greater than 1 that has exactly two factors, itself and one. The first 13 prime numbers are

$$2, 3, 5, 7, 11, 13, 17, 19, 23, 29, 31, 37, 41$$

A positive integer (other than 1) that is not prime is called **composite**. The number 1 is neither prime nor composite, it is called a **unit**.

To write a number as a product of prime numbers, follow the procedure illustrated in Examples 2 and 3.

EXAMPLE 2 Write 48 as a product of prime numbers.

Solution Select any two numbers whose product is 48. Two possibilities are $6 \cdot 8$ and $4 \cdot 12$, but there are other choices. Continue breaking down the factors until all the factors are prime, as illustrated in Figure 7.1. Note that no matter how you select your initial factors,

$$48 = 2 \cdot 2 \cdot 2 \cdot 2 \cdot 3 = 2^4 \cdot 3$$

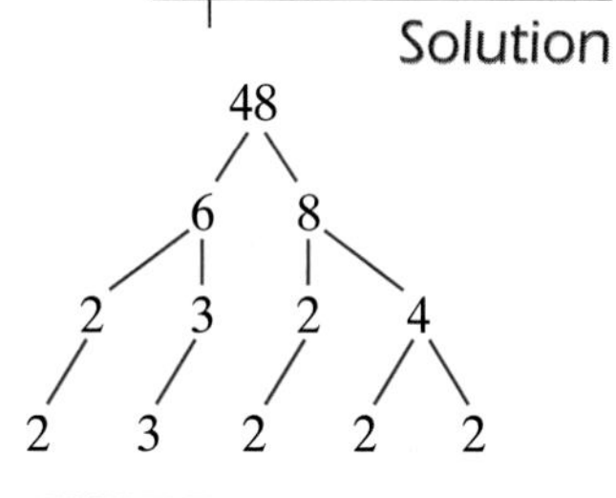

FIGURE 7.1

In Example 2, we found that $48 = 2 \cdot 2 \cdot 2 \cdot 2 \cdot 3 = 2^4 \cdot 3$. The $2 \cdot 2 \cdot 2 \cdot 2 \cdot 3$ or $2^4 \cdot 3$ may also be referred to as **prime factorizations** of 48.

EXAMPLE 3 Write 60 as a product of its prime factors

Solution One way to find the prime factors is shown in Figure 7.2. Therefore, $60 = 2 \cdot 2 \cdot 3 \cdot 5 = 2^2 \cdot 3 \cdot 5$.

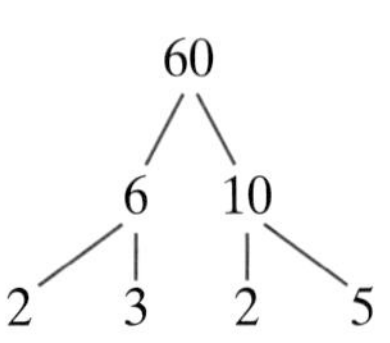

FIGURE 7.2

To Determine the GCF of Two or More Numbers

1. Write each number as a product of prime factors.
2. Determine the prime factors common to all the numbers.
3. Multiply the common factors found in step 2. The product of these factors is the GCF.

EXAMPLE 4 Determine the greatest common factor of 48 and 60.

Solution From Examples 2 and 3 we know that

$$48 = 2 \cdot 2 \cdot 2 \cdot 2 \cdot 3 = 2^4 \cdot 3$$
$$60 = 2 \cdot 2 \cdot 3 \cdot 5 = 2^2 \cdot 3 \cdot 5$$

Two factors of 2 and one factor of 3 are common to both numbers. The product of these factors is the GCF of 48 and 60:

$$\text{GCF} = 2 \cdot 2 \cdot 3 = 12$$

The GCF of 48 and 60 is 12. Twelve is the greatest number that divides into both 48 and 60.

EXAMPLE 5 Determine the GCF of 18 and 24.

Solution

$$18 = 2 \cdot 3 \cdot 3 = 2 \cdot 3^2$$
$$24 = 2 \cdot 2 \cdot 2 \cdot 3 = 2^3 \cdot 3$$

One factor of 2 and one factor of 3 are common to both 18 and 24.

NOW TRY EXERCISE 15

$$\text{GCF} = 2 \cdot 3 = 6$$

3) Determine the Greatest Common Factor of Two or More Terms

The GCF of several terms containing variables is easily found. Consider the terms x^3, x^4, x^5, and x^6. The GCF of these terms is x^3, since x^3 is the highest power of x that divides into all four terms. We can illustrate this by writing the terms in factored form, with x^3 as one factor.

$$x^3 = x^3 \cdot 1$$
$$x^4 = x^3 \cdot x$$
$$x^5 = x^3 \cdot x^2$$
$$x^6 = x^3 \cdot x^3$$

GCF of all four terms is x^3.

Notice that $\dfrac{x^3}{x^3} = 1$, that $\dfrac{x^4}{x^3} = x$, that $\dfrac{x^5}{x^3} = x^2$, and that $\dfrac{x^6}{x^3} = x^3$.

EXAMPLE 6 Determine the GCF of the terms n^5, n^3, n^8, and n^7.

Solution The GCF is n^3 because n^3 is the highest power of n that divides into each term.

EXAMPLE 7 Determine the GCF of the terms x^2y^3, x^3y^2 and xy^4.

Solution The highest power of x common to all three terms is x^1 or x. The highest power of y common to all three terms is y^2. So the GCF of the three terms is xy^2.

To Determine the Greatest Common Factor of Two or More Terms

To determine the GCF of two or more terms, take each factor the *fewest* number of times that it appears in any of the terms.

EXAMPLE 8 Determine the GCF of the terms xy, x^2y^2, and x^3.

Solution The GCF is x. The smallest power of x that appears in any of the terms is x. Since the term x^3 does not contain a power of y, the GCF does not contain y.

EXAMPLE 9 Determine the GCF of each set of terms.

a) $18y^2, 15y^3, 27y^5$ **b)** $-20x^2, 12x, 40x^3$ **c)** $x^2, 7x^2, x^3$

Solution **a)** The GCF of 18, 15, and 27 is 3. The GCF of y^2, y^3, and y^5 is y^2. Therefore, the GCF of the three terms is $3y^2$.

b) The GCF of −20, 12, and 40 is 4. The GCF of x^2, x, and x^3 is x. Therefore, the GCF of the three terms is $4x$.

NOW TRY EXERCISE 33

c) The GCF is x^2.

EXAMPLE 10 Determine the GCF of each pair of terms.

a) $x(x + 3)$ and $2(x + 3)$ **b)** $x(x - 2)$ and $x - 2$

c) $2(x + y)$ and $3x(x + y)$

Solution **a)** The GCF is $(x + 3)$.

b) $x - 2$ can be written as $1(x - 2)$. Therefore, the GCF of $x(x - 2)$ and $1(x - 2)$ is $x - 2$.

NOW TRY EXERCISE 39

c) The GCF is $(x + y)$.

4 Factor a Monomial from a Polynomial

To Factor a Monomial from a Polynomial

1. Determine the greatest common factor of all terms in the polynomial.
2. Write each term as the product of the GCF and its other factor.
3. Use the distributive property to factor out the GCF.

In step 3 of the process, we indicate that we use the distributive property. The distributive property is actually used in reverse. For example, if we have $4 \cdot x + 4 \cdot 2$, we use the distributive property in reverse to write $4(x + 2)$.

EXAMPLE 11 Factor $6x + 18$.

Solution The GCF is 6.

$$6x + 18 = 6 \cdot x + 6 \cdot 3$$

Write each term as a product of the GCF and some other factor.

$$= 6(x + 3)$$

Distributive property

To check the factoring process, multiply the factors using the distributive property. If the factoring is correct, the product will be the polynomial you started with. Following is a check of the factoring in Example 11.

CHECK: $6(x + 3) = 6x + 18$

EXAMPLE 12 Factor $10x - 15$.

Solution The GCF is 5.

$$10x - 15 = 5 \cdot 2x - 5 \cdot 3$$
$$= 5(2x - 3)$$

Check that the factoring is correct by multiplying.

EXAMPLE 13 Factor $6y^2 + 9y^5$.

Solution The GCF is $3y^2$.

$$6y^2 + 9y^5 = 3y^2 \cdot 2 + 3y^2 \cdot 3y^3$$
$$= 3y^2(2 + 3y^3)$$

Check to see that the factoring is correct.

EXAMPLE 14 Factor $8x^3 + 12x^2 - 16x$.

Solution The GCF is $4x$.

$$8x^3 + 12x^2 - 16x = 4x \cdot 2x^2 + 4x \cdot 3x - 4x \cdot 4$$
$$= 4x(2x^2 + 3x - 4)$$

CHECK: $4x(2x^2 + 3x - 4) = 8x^3 + 12x^2 - 16x$

EXAMPLE 15 Factor $45x^2 - 30x + 5$.

Solution The GCF is 5.

$$45x^2 - 30x + 5 = 5 \cdot 9x^2 - 5 \cdot 6x + 5 \cdot 1$$
$$= 5(9x^2 - 6x + 1)$$

NOW TRY EXERCISE 81

EXAMPLE 16 Factor $4x^3 + x^2 + 8x^2y$.

Solution The GCF is x^2.

$$4x^3 + x^2 + 8x^2y = x^2 \cdot 4x + x^2 \cdot 1 + x^2 \cdot 8y$$
$$= x^2(4x + 1 + 8y)$$

Notice in Examples 15 and 16 that when one of the terms is itself the GCF, we express it in factored form as the product of the term itself and 1.

EXAMPLE 17 Factor $x(5x - 2) + 7(5x - 2)$.

Solution The GCF of $x(5x - 2)$ and $7(5x - 2)$ is $(5x - 2)$. Factoring out the GCF gives

$$x(5x - 2) + 7(5x - 2) = (5x - 2)(x + 7)$$

Check that the factoring is correct by multiplying.

EXAMPLE 18 Factor $2x(3x + 1) + 5(3x + 1)$.

Solution The GCF of $2x(3x + 1)$ and $5(3x + 1)$ is $(3x + 1)$. Factoring out the GCF gives

$$2x(3x + 1) + 5(3x + 1) = (3x + 1)(2x + 5)$$

Recall from Section 1.10 that the commutative property of multiplication states that the order in which any two real numbers are multiplied does not matter. Therefore, $(3x + 1)(2x + 5)$ can also be written $(2x + 5)(3x + 1)$.

NOW TRY EXERCISE 87

EXAMPLE 19 Factor $8x(x - 1) - 5(x - 1)$.

Solution The GCF of $8x(x - 1)$ and $-5(x - 1)$ is $(x - 1)$. Factoring out the GCF gives

$$8x(x - 1) - 5(x - 1) = (8x - 5)(x - 1)$$

Note that we wrote the binomial factor $x - 1$ on the right in our answer.

EXAMPLE 20 Factor $3x^2 + 5xy + 7y^2$.

Solution The only factor common to all three terms is 1. Whenever the only factor common to all the terms in a polynomial is 1, the polynomial cannot be factored by the method presented in this section.

Whenever you are factoring a polynomial by any of the methods presented in this chapter, the first step will always be to see if there is a common factor (other than 1) to all the terms in the polynomial. If so, factor the greatest common factor from each term using the distributive property.

HELPFUL HINT

Checking a Factoring Problem

Every factoring problem may be checked by multiplying the factors. The product of the factors should be identical to the expression that was originally factored. You should check all factoring problems.

Exercise Set 7.1

Concept/Writing Exercises

1. What is a prime number?
2. What is a composite number?
3. Is the number 1 a prime number? If not, what is it called?
4. What does it mean to factor an expression?
5. What is the greatest common factor of two or more numbers?
6. In your own words, explain how to factor a monomial from a polynomial.
7. How may any factoring problem be checked?
8. One student factored $5x^2 + 15x + 5$ and wrote $5(x^2 + 3x + 1)$ as the answer. A second student factored $5x^2 + 15x + 5$ and wrote $5(x^2 + 3x)$ as the answer. Which student was correct? Why was the answer given by the other student incorrect?

Practice the Skills

Write each number as a product of prime numbers.

9. 48
10. 120
11. 90
12. 540
13. 200
14. 96

Determine the greatest common factor for each pair of numbers.

15. 20, 24
16. 45, 27
17. 70, 98
18. 120, 96
19. 72, 90
20. 76, 68

Determine the greatest common factor for each set of terms.

21. x^3, x, x^2
22. x^2, x^5, x^7
23. $3x, 6x^2, 9x^3$
24. $6p, 4p^2, 8p^3$
25. x, y, z
26. xy, x, x^2
27. xy, xy^2, xy^3
28. $x^2y, 2x^3y, 4x^2$
29. $x^3y^7, x^7y^{12}, x^5y^5$
30. $6x, 12y, 18x^2$
31. $-5, 20x, 30x^2$
32. $18r^4, 6r^2s, 9s^2$
33. $9x^3y^4, 8x^2y^4, 12x^4y^2$
34. $16x^9y^{12}, 8x^5y^3, 20x^4y^2$
35. $40x^3, 27x, 30x^4y^2$
36. $-8x^2y, -9x^3, 12xy^3$
37. $2(x + 3), 3(x + 3)$
38. $4(x - 5), 3x(x - 5)$
39. $x^2(2x - 3), 5(2x - 3)$
40. $x(7x + 5), 7x + 5$
41. $3x - 4, y(3x - 4)$
42. $x(x + 7), x + 7$
43. $x - 4, y(x - 4)$
44. $3y(x + 2), 3(x + 2)$

Factor the GCF from each term in the expression. If an expression cannot be factored, so state.

45. $5x + 10$
46. $4x + 2$
47. $15x - 5$
48. $12x + 15$
49. $13x + 5$
50. $14x^2 + 7x$
51. $9x^2 - 12x$
52. $24y - 6y^2$
53. $26p^2 - 8p$
54. $8x + 16x^2$
55. $4x^3 - 10x$
56. $7x^5 - 9x^4$
57. $36x^{12} + 24x^8$
58. $45y^{12} + 30y^{10}$
59. $24y^{15} - 9y^3$
60. $38x^4 - 16x^5$
61. $x + 3xy^2$
62. $4x^2y - 6x$
63. $6x + 5y$
64. $3x^2y + 6x^2y^2$
65. $16xy^2z + 4x^3y$
66. $80x^5y^3z^4 - 36x^2yz^3$
67. $34x^2y^2 + 16xy^4$
68. $56xy^5z^{13} - 24y^4z^2$
69. $25x^2yz^3 + 25x^3yz$
70. $19x^4y^{12}z^{13} - 8x^5y^3z^9$
71. $13y^5z^3 - 11xy^2z^5$
72. $7x^4y^9 - 21x^3y^7z^5$
73. $3x^2 + 6x + 9$
74. $x^3 + 6x^2 - 4x$
75. $9x^2 + 18x + 3$
76. $4x^2 + 8x + 24$
77. $4x^3 - 8x^2 + 12x$
78. $12x^2 - 9x + 9$
79. $35x^2 - 16x + 10$
80. $5x^3 - 7x^2 + x$
81. $15p^2 - 6p + 9$
82. $45y^3 - 63y^2 + 18y$
83. $24x^6 + 8x^4 - 4x^3$
84. $44x^5y + 11x^3y + 22x^2$
85. $8x^2y + 12xy^2 + 9xy$
86. $52x^2y^2 + 16xy^3 + 26z$

87. $x(x + 4) + 3(x + 4)$

88. $5x(2x - 5) - 8(2x - 5)$

89. $7x(4x - 3) - 4(4x + 3)$

90. $3x(7x + 1) - 2(7x + 1)$

91. $4x(2x + 1) + 1(2x + 1)$

92. $3x(4x - 5) + 1(4x - 5)$

93. $4x(2x + 1) + 2x + 1$

94. $3x(4x - 5) + 4x - 5$

Problem Solving

Factor each expression, if possible. Treat the unknown symbol as if it were a variable.

95. $3\bigstar + 6$

96. $12\nabla - 6\nabla^2$

97. $35\Delta^3 - 7\Delta^2 + 14\Delta$

98. $\copyright + 11\Delta$

Challenge Problems

99. Factor $4x^2(x - 3)^3 - 6x(x - 3)^2 + 4(x - 3)$.

100. Factor $6x^5(2x + 7) + 4x^3(2x + 7) - 2x^2(2x + 7)$.

101. Factor $x^{7/3} + 5x^{4/3} + 6x^{1/3}$. Begin by factoring $x^{1/3}$ from all three terms.

102. Factor $15x^{1/2} + 5x^{-1/2}$. Begin by factoring out $5x^{-1/2}$.

103. Factor $x^2 + 2x + 3x + 6$. (*Hint:* Factor the first two terms, then factor the last two terms, then factor the resulting two terms. We will discuss factoring problems of this type in Section 7.2.)

Group Activity

104. We are going to create **binomials** and determine their factors. Group members 1 and 2 will make cards containing the *terms* of the binomials, while group member 3 will make cards that will be used to illustrate the factoring of the binomial.

a) Group member 1: Write each of the following expressions on a 3×5 card. Each expression is to go on an individual card. Each of these expressions represents the *first term* of a binomial.

$$7x,\ 12x,\ 4x,\ 13x,\ 9x^2$$

b) Group member 2: Follow the instructions for part **a)** using the following expressions. Each of these expressions represents the *second term* of a binomial.

$$15,\ 5,\ 14,\ -12x,\ 2$$

c) Group member 3: Follow the instructions for part **a)** using the following. These expressions, and the phrase "cannot be factored," will be used to show the factoring of the binomial.

$$(3x - 4),\ (2x + 1),\ (x + 2),\ (4x + 5),$$
$$3x,\ 2,\ 7,\ 3,\ \text{cannot be factored}$$

d) As a group, use the cards from parts **a)**, **b)**, and **c)** to illustrate five different factoring problems. For each factoring problem, you will take one card from part **a)**, one card from part **b)**, and if the binomial can be factored, two cards from part **c)**. If the expression in part **b)** is positive, you will need to place a plus sign between the expressions in part **a)** and **b)**. Each of the expressions in **a)**, **b)**, and **c)** should be used, but only once.

Cumulative Review Exercises

[2.1] **105.** Simplify $2x - (x - 5) + 4(3 - x)$.

[2.5] **106.** Solve the equation $4 + 3(x - 8) = x - 4(x + 2)$.

[3.1] **107.** Solve the equation $4x - 5y = 20$ for y.

[4.2] **108.** **a)** Find the x- and y-intercepts of $2x + 3y = 12$.

b) Use the intercepts to graph the equation.

[6.1] **109.** Simplify $\left(\dfrac{3x^2y^3}{2x^5y^2}\right)^2$

7.2 FACTORING BY GROUPING

1) **Factor a polynomial containing four terms by grouping.**

SSM VIDEO 7.2 CD Rom

1) Factor a Polynomial Containing Four Terms by Grouping

It may be possible to factor a polynomial containing four or more terms by removing common factors from groups of terms. This process is called **factoring by grouping**. In Sections 7.3 and 7.4 we discuss factoring trinomials. One of the methods we will use in Section 7.4 requires a knowledge of factoring by grouping. Example 1 illustrates the procedure for factoring by grouping.

EXAMPLE 1 Factor $ax + ay + bx + by$.

Solution There is no factor (other than 1) common to all four terms. However, a is common to the first two terms and b is common to the last two terms. Factor a from the first two terms and b from the last two terms.

$$ax + ay + bx + by = a(x + y) + b(x + y)$$

This factoring gives two terms, and $(x + y)$ is common to both terms. Proceed to factor $(x + y)$ from each term, as shown below.

$$a(x + y) + b(x + y) = (a + b)(x + y)$$

Notice that when $(x + y)$ is factored out we are left with $a + b$, which becomes the other factor. Thus, $ax + ay + bx + by = (a + b)(x + y)$.

To Factor a Four-Term Polynomial Using Grouping

1. Determine whether there are any factors common to all four terms. If so, factor the greatest common factor from each of the four terms.
2. If necessary, arrange the four terms so that the first two terms have a common factor and the last two have a common factor.
3. Use the distributive property to factor each group of two terms.
4. Factor the greatest common factor from the results of step 3.

EXAMPLE 2 Factor $x^2 + 3x + 4x + 12$ by grouping.

Solution No factor is common to all four terms. However, you can factor x from the first two terms and 4 from the last two terms.

$$x^2 + 3x + 4x + 12 = x(x + 3) + 4(x + 3)$$

Notice that the expression on the right of the equal sign has two *terms* and that the factor $(x + 3)$ is common to both terms. Factor out the $(x + 3)$ using the distributive property.

$$x(x + 3) + 4(x + 3) = (x + 4)(x + 3)$$

NOW TRY EXERCISE 11

Thus, $x^2 + 3x + 4x + 12 = (x + 4)(x + 3)$.

In Example 2, the $3x$ and $4x$ are like terms and may be combined. However, since we are explaining how to factor four terms by grouping we will not combine them. Some four-term polynomials, such as in Example 9, have no like terms that can be combined. When we discuss factoring trinomials in Section 7.4, we will sometimes start with a trinomial and rewrite it using four terms. For example, we may start with a trinomial like $2x^2 + 11x + 12$ and rewrite it as $2x^2 + 8x + 3x + 12$. We then factor the resulting four terms by grouping. This is one method that can be used to factor trinomials, as will be explained later.

EXAMPLE 3 Factor $15x^2 + 10x + 12x + 8$ by grouping.

Solution

$$\begin{aligned} 15x^2 + 10x + 12x + 8 &= 5x(3x + 2) + 4(3x + 2) \\ &= (5x + 4)(3x + 2) \end{aligned}$$

Factor 5x from the first two terms and 4 from the last two terms.

A factoring by grouping problem can be checked by multiplying the factors using the FOIL method. If you have not made a mistake, your result will be the polynomial you began with. Here is a check of Example 3.

CHECK:

$$\begin{aligned} (5x + 4)(3x + 2) &= \overset{F}{(5x)(3x)} + \overset{O}{(5x)(2)} + \overset{I}{(4)(3x)} + \overset{L}{(4)(2)} \\ &= 15x^2 + 10x + 12x + 8 \end{aligned}$$

Because this is the polynomial we started with, the factoring is correct.

EXAMPLE 4 Factor $15x^2 + 12x + 10x + 8$ by grouping.

Solution

$$\begin{aligned} 15x^2 + 12x + 10x + 8 &= 3x(5x + 4) + 2(5x + 4) \\ &= (3x + 2)(5x + 4) \end{aligned}$$

Notice that Example 4 is the same as Example 3 with the two middle terms interchanged. The answers to Examples 3 and 4 are the same. When factoring by grouping, if the two middle terms are like terms, the two like terms may be interchanged and the answer will remain the same.

EXAMPLE 5 Factor $x^2 + 3x + x + 3$ by grouping.

Solution In the first two terms, x is the common factor. Is there a common factor in the last two terms? Yes; remember that 1 is a factor of every term. Factor 1 from the last two terms.

$$\begin{aligned} x^2 + 3x + x + 3 &= x^2 + 3x + 1 \cdot x + 1 \cdot 3 \\ &= x(x + 3) + 1(x + 3) \\ &= (x + 1)(x + 3) \end{aligned}$$

Note that $x + 3$ was expressed as $1(x + 3)$.

EXAMPLE 6 Factor $6x^2 - 3x - 2x + 1$ by grouping.

Solution When $3x$ is factored from the first two terms, we get

$$6x^2 - 3x - 2x + 1 = 3x(2x - 1) - 2x + 1$$

What should we factor from the last two terms? We wish to factor $-2x + 1$ in such a manner that we end up with an expression that is a multiple of $(2x - 1)$.

Whenever we wish to change the sign *of each term of an expression, we can factor out a negative number from each term.* In this case, we factor out −1.

$$-2x + 1 = -1(2x - 1)$$

Now, we rewrite $-2x + 1$ as $-1(2x - 1)$.

$$3x(2x - 1) \; -2x + 1 = 3x(2x - 1) \; -1(2x - 1)$$

Now we factor out the common factor $(2x - 1)$.

$$3x(2x - 1) - 1(2x - 1) = (3x - 1)(2x - 1)$$

EXAMPLE 7 Factor $x^2 + 3x - x - 3$ by grouping.

Solution

$$\begin{aligned} x^2 + 3x - x - 3 &= x(x + 3) - x - 3 && \text{Factor out } x. \\ &= x(x + 3) - 1(x + 3) && \text{Factor out } -1. \\ &= (x - 1)(x + 3) && \text{Factor out } (x + 3). \end{aligned}$$

NOW TRY EXERCISE 19

Note that we factored −1 from $-x - 3$ to get $-1(x + 3)$.

EXAMPLE 8 Factor $3x^2 - 6x - 4x + 8$ by grouping.

Solution

$$\begin{aligned} 3x^2 - 6x - 4x + 8 &= 3x(x - 2) - 4(x - 2) \\ &= (3x - 4)(x - 2) \end{aligned}$$

Note: $-4x + 8 = -4(x - 2)$.

HELPFUL HINT

When factoring four terms by grouping, if the coefficient of the third term is positive, as in Examples 2 through 5, you will generally factor out a positive coefficient from the last two terms. *If the coefficient of the third term is negative*, as in Examples 6 through 8, *you will generally factor out a negative coefficient from the last two terms*. The sign of the coefficient of the third term in the expression *must be included* so that the factoring results in two terms. For example,

$$2x^2 + 8x + 3x + 12 = 2x(x + 4) + 3(x + 4) = (2x + 3)(x + 4)$$

$$3x^2 - 15x - 2x + 10 = 3x(x - 5) - 2(x - 5) = (3x - 2)(x - 5)$$

In the examples illustrated so far, the two middle terms have been like terms. This need not be the case, as illustrated in Example 9.

EXAMPLE 9 Factor $xy + 3x - 2y - 6$ by grouping.

Solution

This problem contains two variables, x and y. The procedure to factor here is basically the same as before. Factor x from the first two terms and −2 from the last two terms.

$$\begin{aligned} xy + 3x - 2y - 6 &= x(y + 3) - 2(y + 3) \\ &= (x - 2)(y + 3) && \text{Factor out } (y + 3). \end{aligned}$$

NOW TRY EXERCISE 41

EXAMPLE 10 Factor $2x^2 + 4xy + 3xy + 6y^2$.

Solution

We will factor out $2x$ from the first two terms and $3y$ from the last two terms.

$$2x^2 + 4xy + 3xy + 6y^2 = 2x(x + 2y) + 3y(x + 2y)$$

Now we factor out the common factor $(x + 2y)$ from each term on the right.

$$2x(x + 2y) + 3y(x + 2y) = (2x + 3y)(x + 2y)$$

CHECK: $(2x + 3y)(x + 2y) = \overset{F}{(2x)(x)} + \overset{O}{(2x)(2y)} + \overset{I}{(3y)(x)} + \overset{L}{(3y)(2y)}$

$$= 2x^2 + 4xy + 3xy + 6y^2$$

If Example 10 were given as $2x^2 + 3xy + 4xy + 6y^2$, would the results be the same? Try it and see.

EXAMPLE 11 Factor $6r^2 - 9rs + 8rs - 12s^2$.

Solution Factor $3r$ from the first two terms and $4s$ from the last two terms.

$$6r^2 - 9rs + 8rs - 12s^2 = 3r(2r - 3s) + 4s(2r - 3s)$$

NOW TRY EXERCISE 31

$$= (3r + 4s)(2r - 3s)$$

EXAMPLE 12 Factor $3x^2 - 15x + 6x - 30$.

Solution *The first step in any factoring problem is to determine whether all the terms have a common factor. If so, we factor out that common factor.* In this polynomial, 3 is common to every term. Therefore, we begin by factoring out the 3.

$$3x^2 - 15x + 6x - 30 = 3(x^2 - 5x + 2x - 10)$$

Now we factor the expression in parentheses by grouping. We factor out x from the first two terms and 2 from the last two terms.

$$\begin{aligned} 3(x^2 - 5x + 2x - 10) &= 3[x(x - 5) + 2(x - 5)] \\ &= 3[(x + 2)(x - 5)] \\ &= 3(x + 2)(x - 5) \end{aligned}$$

Thus, $3x^2 - 15x + 6x - 30 = 3(x + 2)(x - 5)$.

Exercise Set 7.2

Concept/Writing Exercises

1. What is the first step in any factoring by grouping problem?
2. How can you check the solution to a factoring by grouping problem?
3. A polynomial of four terms is factored by grouping and the result is $(x - 2)(x + 4)$. Find the polynomial that was factored, and explain how you determined the answer.
4. A polynomial of four terms is factored by grouping and the result is $(x - 2y)(x - 3)$. Find the polynomial that was factored, and explain how you determined the answer.
5. In your own words, describe the steps you take to factor a polynomial of four terms by grouping.
6. What number when factored from each term in an expression changes the sign of each term in the original expression?

Practice the Skills

Factor by grouping.

7. $x^2 + 3x + 2x + 6$
8. $x^2 + 5x + 2x + 10$
9. $x^2 + 3x + 4x + 12$
10. $x^2 - x + 3x - 3$
11. $x^2 + 2x + 5x + 10$
12. $x^2 - 6x + 5x - 30$
13. $x^2 + 3x - 5x - 15$
14. $x^2 + 3x - 2x - 6$
15. $4x^2 - 14x + 14x - 49$
16. $4x^2 - 6x + 6x - 9$

17. $3x^2 + 9x + x + 3$
18. $x^2 + 4x + x + 4$
19. $6x^2 + 3x - 2x - 1$
20. $5x^2 + 20x - x - 4$
21. $8x^2 + 32x + x + 4$
22. $8x^2 - 4x - 2x + 1$
23. $3x^2 - 2x + 3x - 2$
24. $12x^2 + 42x - 10x - 35$
25. $2x^2 - 4x - 3x + 6$
26. $35x^2 - 40x + 21x - 24$
27. $3x^2 - 9x + 5x - 15$
28. $10x^2 - 15x - 8x + 12$
29. $x^2 + 2xy - 3xy - 6y^2$
30. $x^2 - 3xy + 2xy - 6y^2$
31. $3x^2 + 2xy - 9xy - 6y^2$
32. $3x^2 - 18xy + 4xy - 24y^2$
33. $10x^2 - 12xy - 25xy + 30y^2$
34. $12x^2 - 9xy + 4xy - 3y^2$
35. $x^2 + bx + ax + ab$
36. $x^2 - bx - ax + ab$
37. $xy + 4x - 2y - 8$
38. $x^2 - 2x + ax - 2a$
39. $a^2 + 3a + ab + 3b$
40. $2x^2 - 8x + 3xy - 12y$
41. $xy - x + 5y - 5$
42. $y^2 - yb + ya - ab$
43. $12 + 8y - 3x - 2xy$
44. $3y - 9 - xy + 3x$
45. $z^3 + 3z^2 + z + 3$
46. $x^3 - 3x^2 + 2x - 6$
47. $x^3 + 4x^2 - 3x - 12$
48. $y^3 + 2y^2 - 4y - 8$
49. $2x^2 - 12x + 8x - 48$
50. $3x^2 - 3x - 3x + 3$
51. $4x^2 + 8x + 8x + 16$
52. $2x^4 - 5x^3 - 6x^3 + 15x^2$
53. $6x^3 + 9x^2 - 2x^2 - 3x$
54. $9x^3 + 6x^2 - 45x^2 - 30x$
55. $x^3 - 3x^2y + 2x^2y - 6xy^2$
56. $18x^2 + 27xy + 12xy + 18y^2$

Rearrange the terms so that the first two terms have a common factor and the last two terms have a common factor (other than 1). Then factor by grouping. There may be more than one way to arrange the factors. However, the answer should be the same regardless of the arrangement selected.

57. $2x + 4y + 8 + xy$
58. $3a + 6y + ay + 18$
59. $6x + 5y + xy + 30$
60. $ax - 10 - 5x + 2a$
61. $ax + by + ay + bx$
62. $ax - 21 - 3a + 7x$
63. $cd - 12 - 4d + 3c$
64. $ca - 2b + 2a - cb$
65. $ac - bd - ad + bc$
66. $ax + 2by + 2bx + ay$

Problem Solving

67. If you know that a polynomial with four terms is factorable by a specific arrangement of the terms, then will *any* arrangement of the terms be factorable by grouping? Explain, and support your answer with an example.

Factor each expression, if possible. Treat the unknown symbol as if it were a variable.

68. $\Delta^2 + 3\Delta + 4\Delta + 12$
69. $\Delta^2 + 3\Delta - 5\Delta - 15$
70. $\odot^2 + 3\odot + 2\odot + 7$

Challenge Problems

In Section 7.4 we will factor trinomials of the form $ax^2 + bx + c$, $a \neq 1$ using grouping. To do this we rewrite the middle term of the trinomial, bx, as a sum or difference of two terms. Then we factor the resulting polynomial of four terms by grouping. For Exercises 71–76, **a)** *rewrite the trinomial as a polynomial of four terms by replacing the bx-term with the sum or difference given.* **b)** *Factor the polynomial of four terms. Note that the factors obtained are the factors of the trinomial.*

71. $3x^2 + 10x + 8,\ 10x = 6x + 4x$
72. $3x^2 + 10x + 8,\ 10x = 4x + 6x$
73. $2x^2 - 11x + 15,\ -11x = -6x - 5x$
74. $2x^2 - 11x + 15,\ -11x = -5x - 6x$
75. $4x^2 - 17x - 15,\ -17x = -20x + 3x$
76. $4x^2 - 17x - 15,\ -17x = 3x - 20x$

Factor each expression, if possible. Treat the unknown symbols as if they were variables.

77. $\bigstar\odot + 3\bigstar + 2\odot + 6$

78. $2\Delta^2 - 4\Delta\bigstar - 8\Delta\bigstar + 16\bigstar^2$

Cumulative Review Exercises

[2.5] **79.** Solve $5 - 3(2x - 7) = 4(x + 5) - 6$.

[3.5] **80.** Ed and Beatrice Petrie own a small grocery store near Leesport, Pennsylvania. The store carries a variety of bulk candy. To celebrate the 5th anniversary of the store's opening, the Petries decide to create a special candy mixture containing chocolate wafers and hard peppermint candies. The chocolate wafers sell for \$6.25 per pound and the peppermint candies sell for \$2.50 per pound. How many pounds of each type of candy will be needed to make a 50-pound mixture that will sell for \$4.75 per pound?

See Exercise 80

[6.6] **81.** Divide $\dfrac{15x^3 - 6x^2 - 9x + 5}{3x}$.

82. Divide $\dfrac{x^2 - 9}{x - 3}$.

7.3 FACTORING TRINOMIALS OF THE FORM $ax^2 + bx + c$, $a = 1$

SSM VIDEO 7.3 CD Rom

1) **Factor trinomials of the form $ax^2 + bx + c$, where $a = 1$.**

2) **Remove a common factor from a trinomial.**

An Important Note Regarding Factoring Trinomials

Factoring trinomials is important in algebra, higher-level mathematics, physics, and other science courses. Because it is important, and also to be successful in Chapter 8, you should study and learn Sections 7.3 and 7.4 well.

In this section we learn to factor trinomials of the form $ax^2 + bx + c$, where a, the numerical coefficient of the squared term, is 1. That is, we will be factoring trinomials of the form $x^2 + bx + c$. One example of this type of trinomial is $x^2 + 5x + 6$. Recall that x^2 means $1x^2$.

In Section 7.4 we will learn to factor trinomials of the form $ax^2 + bx + c$, where $a \neq 1$. One example of this type of trinomial is $2x^2 + 7x + 3$.

1) Factor Trinomials of the Form $ax^2 + bx + c$, where $a = 1$

Now we discuss how to factor trinomials of the form $ax^2 + bx + c$, where a, the numerical coefficient of the squared term, is 1. Examples of such trinomials are

$$x^2 + 7x + 12 \qquad\qquad x^2 - 2x - 24$$
$$a = 1, b = 7, c = 12 \qquad\qquad a = 1, b = -2, c = -24$$

Recall that factoring is the reverse process of multiplication. We can show with the FOIL method that

$$(x + 3)(x + 4) = x^2 + 7x + 12 \quad \text{and} \quad (x - 6)(x + 4) = x^2 - 2x - 24$$

Therefore, $x^2 + 7x + 12$ and $x^2 - 2x - 24$ factor as follows:

$$x^2 + 7x + 12 = (x + 3)(x + 4) \quad \text{and} \quad x^2 - 2x - 24 = (x - 6)(x + 4)$$

Notice that each of these trinomials when factored results in the product of two binomials in which the first term of each binomial is x and the second term is a number (including its sign). In general, when we factor a trinomial of the form $x^2 + bx + c$ we will get a pair of binomial factors as follows:

$$x^2 + bx + c = (x + \square)(x + \square)$$

Numbers go here

If, for example, we find that the numbers that go in the shaded areas of the factors are 4 and −6, the factors are written $(x + 4)$ and $(x - 6)$. Notice that instead of listing the second factor as $(x + (-6))$, we list it as $(x - 6)$.

To determine the numbers to place in the shaded areas when factoring a trinomial of the form $x^2 + bx + c$, write down factors of the form $(x + \square)\,(x + \square)$ and then try different sets of factors of the constant, c, in the shaded areas of the parentheses. We multiply each pair of factors using the FOIL method, and continue until we find the pair whose sum of the products of the outer and inner terms is the same as the x-term in the trinomial. For example, to factor the trinomial $x^2 + 7x + 12$ we determine the possible factors of 12. Then we try each pair of factors until we obtain a pair whose product from the FOIL method contains $7x$, the same x-term as in the trinomial. This method for factoring is called **trial and error**. In Example 1, we factor $x^2 + 7x + 12$ by trial and error.

EXAMPLE 1 Factor $x^2 + 7x + 12$ by trial and error.

Solution Begin by listing the factors of 12 (see the left-hand column of the chart below). Then list the possible factors of the trinomial, and the products of these factors. Finally, determine which, if any, of these products gives the correct middle term, $7x$.

Factors of 12	Possible Factors of Trinomial	Product of Factors
(1)(12)	$(x + 1)(x + 12)$	$x^2 + 13x + 12$
(2)(6)	$(x + 2)(x + 6)$	$x^2 + 8x + 12$
(3)(4)	$(x + 3)(x + 4)$	$x^2 + 7x + 12$
(−1)(−12)	$(x - 1)(x - 12)$	$x^2 - 13x + 12$
(−2)(−6)	$(x - 2)(x - 6)$	$x^2 - 8x + 12$
(−3)(−4)	$(x - 3)(x - 4)$	$x^2 - 7x + 12$

In the last column, we find the trinomial we are seeking in the third line. Thus,

NOW TRY EXERCISE 25

$$x^2 + 7x + 12 = (x + 3)(x + 4)$$

Now let's consider how we may more easily determine the correct factors of 12 to place in the shaded areas when factoring the trinomial in Example 1. In Section 6.5 we illustrated how the FOIL method is used to multiply two binomials. Let's multiply $(x + 3)(x + 4)$ using the FOIL method.

$$\begin{aligned}(x + 3)(x + 4) &= \overset{F}{x^2} + \overset{O}{4x} + \overset{I}{3x} + \overset{L}{12} \\ &= x^2 + 7x + 12\end{aligned}$$

We see that $(x + 3)(x + 4) = x^2 + 7x + 12$.

Note that the *sum of the outer and inner terms is 7x and the product of the last terms is 12.* To factor $x^2 + 7x + 12$, we look for two numbers whose product is 12 and whose sum is 7. We list the factors of 12 first and then list the sum of the factors.

Factors of 12	Sum of Factors
$(1)(12) = 12$	$1 + 12 = 13$
$(2)(6) = 12$	$2 + 6 = 8$
$(3)(4) = 12$	$3 + 4 = 7$
$(-1)(-12) = 12$	$-1 + (-12) = -13$
$(-2)(-6) = 12$	$-2 + (-6) = -8$
$(-3)(-4) = 12$	$-3 + (-4) = -7$

The only factors of 12 whose sum is a positive 7 are 3 and 4. The factors of $x^2 + 7x + 12$ will therefore be $(x + 3)$ and $(x + 4)$.

$$x^2 + 7x + 12 = (x + 3)(x + 4)$$

In the previous illustration all the possible factors of 12 were listed so that you could see them. However, when working a problem, once you find the specific factors you are seeking you need go no further.

To Factor Trinomials of the Form $ax^2 + bx + c$, where $a = 1$

1. Find two numbers whose product equals the constant, c, and whose sum equals the coefficient of the x-term, b.
2. Use the two numbers found in step 1, including their signs, to write the trinomial in factored form. The trinomial in factored form will be

$$(x + \text{one number})(x + \text{second number})$$

How do we find the two numbers mentioned in steps 1 and 2? The sign of the constant, c, is a key in finding the two numbers. *The Helpful Hint that follows is very important and useful. Study it carefully.*

HELPFUL HINT

When asked to factor a trinomial of the form $x^2 + bx + c$, first observe the sign of the constant.

a) If the constant, c, is positive, both numbers in the factors will have the same sign, either both positive or both negative. Furthermore, that common sign will be the same as b, the sign of the coefficient of the x-term of the trinomial being factored.

Example:

$$x^2 + 7x + 12 = (x + 3)(x + 4)$$

The coefficient, b, is positive — *The constant, c, is positive* — *positive* — *positive*

Both factors have positive numbers.

Example:

$$x^2 - 5x + 6 = (x - 2)(x - 3)$$

The coefficient, b, is negative — *The constant, c, is positive* — *negative* — *negative*

Both factors have negative numbers.

b) If the constant is negative, the two numbers in the factors will have opposite signs. That is, one number will be positive and the other number will be negative.

Example:

$$x^2 + x - 6 = (x + 3)(x - 2)$$

The coefficient, b, is positive — *The constant, c, is negative* — *positive* — *negative*

One factor has a positive number and the other factor has a negative number.

Example:

$$x^2 - 3x - 10 = (x + 2)(x - 5)$$

The coefficient, b, is negative — *The constant, c, is negative* — *positive* — *negative*

One factor has a positive number and the other factor has a negative number.

We will use this information as a starting point when factoring trinomials.

EXAMPLE 2 Consider a trinomial of the form $x^2 + bx + c$. Use the signs of b and c given below to determine the signs of the numbers in the factors.

a) b is negative and c is positive. **b)** b is negative and c is negative.

c) b is positive and c is negative. **d)** b is positive and c is positive.

Solution **a)** Since the constant, c, is positive, both numbers must have the same sign. Since the coefficient of the x-term, b, is negative, both factors will contain negative numbers.

b) Since the constant, c, is negative, one factor will contain a positive number and the other will contain a negative number.

c) Since the constant, c, is negative, one factor will contain a positive number and the other will contain a negative number.

d) Since the constant, c, is positive, both numbers must have the same sign. Since the coefficient of the x-term, b, is positive, both factors will contain positive numbers.

EXAMPLE 3 Factor $x^2 + x - 6$.

Solution We must find two numbers whose product is the constant, −6, and whose sum is the coefficient of the x-term, 1. Remember that x means $1x$. Since the constant is negative, one number must be positive and the other negative. Recall that the product of two numbers with unlike signs is a negative number. We now list the factors of −6 and look for the two factors whose sum is 1.

Factors of −6	Sum of Factors
$1(-6) = -6$	$1 + (-6) = -5$
$2(-3) = -6$	$2 + (-3) = -1$
$3(-2) = -6$	$3 + (-2) = 1$
$6(-1) = -6$	$6 + (-1) = 5$

Note that the factors 1 and −6 in the top row are different from the factors −1 and 6 in the bottom row, and their sums are different.

The numbers 3 and −2 have a product of −6 and a sum of 1. Thus, the factors are $(x + 3)$ and $(x - 2)$.

$$x^2 + x - 6 = (x + 3)(x - 2)$$

The order of the factors is not crucial. Therefore, $x^2 + x - 6 = (x - 2)(x + 3)$ is also an acceptable answer.

As mentioned earlier, **trinomial factoring problems can be checked by multiplying the factors using the FOIL method**. If the factoring is correct, the product obtained using the FOIL method will be identical to the original trinomial. Let's check the factors obtained in Example 3.

CHECK: $(x + 3)(x - 2) = x^2 - 2x + 3x - 6 = x^2 + x - 6$

Since the product of the factors is identical to the original trinomial, the factoring is correct.

EXAMPLE 4 Factor $x^2 - x - 6$.

Solution The factors of −6 are illustrated in Example 3. The factors whose product is −6 and whose sum is −1 are 2 and −3.

Factors of −6	Sum of Factors
$2(-3) = -6$	$2 + (-3) = -1$

Therefore, $x^2 - x - 6 = (x + 2)(x - 3)$

EXAMPLE 5 Factor $x^2 - 5x + 6$.

Solution We must find two numbers whose product is 6 and whose sum is −5. Since the constant, 6, is positive, both factors must have the same sign. Since the coefficient of the x-term, -5, is negative, both numbers must be negative. Recall that the product of a negative number and a negative number is positive. We now list the negative factors of 6 and look for the pair whose sum is -5.

Factors of 6	Sum of Factors
$(-1)(-6)$	$-1 + (-6) = -7$
$(-2)(-3)$	$-2 + (-3) = -5$

The factors of 6 whose sum is -5 are -2 and -3.

NOW TRY EXERCISE 27

$$x^2 - 5x + 6 = (x - 2)(x - 3)$$

EXAMPLE 6 Factor $x^2 + 3x - 10$.

Solution We must find the factors of -10 whose sum is 3. Since the constant is negative, one factor will be positive and the other factor will be negative.

Factors of −10	Sum of Factors
$(1)(-10)$	$1 + (-10) = -9$
$(2)(-5)$	$2 + (-5) = -3$
$(5)(-2)$	$5 + (-2) = 3$

Since we have found the two numbers, 5 and -2, whose product is -10 and whose sum is 3, we need go no further.

$$x^2 + 3x - 10 = (x + 5)(x - 2)$$

EXAMPLE 7 Factor $x^2 - 6x + 9$.

Solution We must find the factors of 9 whose sum is -6. Both factors must be negative. (Can you explain why?) The two factors whose product is 9 and whose sum is -6 are -3 and -3.

$$x^2 - 6x + 9 = (x - 3)(x - 3)$$
$$= (x - 3)^2$$

EXAMPLE 8 Factor $x^2 - 5x - 66$.

Solution We must find two numbers whose product is -66 and whose sum is -5. Since the constant is negative, one factor must be positive and the other negative. The desired factors are -11 and 6 because $(-11)(6) = -66$ and $-11 + 6 = -5$.

NOW TRY EXERCISE 49

$$x^2 - 5x - 66 = (x - 11)(x + 6)$$

EXAMPLE 9 Factor $x^2 + 7x + 18$.

Solution Let's first find the two numbers whose product is 18 and whose sum is 7. Since both the constant and the coefficient of the x-term are positive, the two numbers must also be positive.

Factors of 18	Sum of Factors
$(1)(18)$	$1 + 18 = 19$
$(2)(9)$	$2 + 9 = 11$
$(3)(6)$	$3 + 6 = 9$

Note that there are no two integers whose product is 18 and whose sum is 7. When two integers cannot be found to satisfy the given conditions, the trinomial cannot be factored by the method presented in this section. Therefore, we write *"cannot be factored"* as our answer.

When factoring a trinomial of the form $x^2 + bx + c$, there is at most one pair of numbers whose product is c and whose sum is b. For example, when factoring $x^2 - 2x - 24$, the two numbers whose product is -24 and whose sum is -2 are -6 and 4. No other pair of numbers will satisfy these specific conditions. Thus, the only factors of $x^2 - 2x - 24$ are $(x - 6)(x + 4)$.

A slightly different type of problem is illustrated in Example 10.

EXAMPLE 10 Factor $x^2 + 2xy + y^2$.

Solution In this problem the second term contains two variables, x and y, and the last term is not a constant. The procedure used to factor this trinomial is similar to that outlined previously. You should realize, however, that the product of the first terms of the factors we are looking for must be x^2, and the product of the last terms of the factors must be y^2.

We must find two numbers whose product is 1 (from $1y^2$) and whose sum is 2 (from $2xy$). The two numbers are 1 and 1. Thus

$$x^2 + 2xy + y^2 = (x + 1y)(x + 1y) = (x + y)(x + y) = (x + y)^2$$

EXAMPLE 11 Factor $x^2 - xy - 6y^2$.

Solution Find two numbers whose product is -6 and whose sum is -1. The numbers are -3 and 2. The last terms must be $-3y$ and $2y$ to obtain $-6y^2$.

NOW TRY EXERCISE 61

$$x^2 - xy - 6y^2 = (x - 3y)(x + 2y)$$

2) Remove a Common Factor from a Trinomial

Sometimes each term of a trinomial has a common factor. When this occurs, factor out the common factor first, as explained in Section 7.1. **Whenever the numerical coefficient of the highest-degree term is not 1, you should check for a common factor**. After factoring out any common factor, you should factor the remaining trinomial further, if possible.

EXAMPLE 12 Factor $2x^2 + 2x - 12$.

Solution Since the numerical coefficient of the squared term is not 1, we check for a common factor. Because 2 is common to each term of the polynomial, we factor it out.

$$2x^2 + 2x - 12 = 2(x^2 + x - 6) \qquad \textit{Factor out the common factor.}$$

Now we factor the remaining trinomial $x^2 + x - 6$ into $(x + 3)(x - 2)$. Thus,

$$2x^2 + 2x - 12 = 2(x + 3)(x - 2).$$

Note that the trinomial $2x^2 + 2x - 12$ is now completely factored into *three* factors: two binomial factors, $x + 3$ and $x - 2$, and a monomial factor, 2. After 2 has been factored out, it plays no part in the factoring of the remaining trinomial.

EXAMPLE 13 Factor $3x^3 + 24x^2 - 60x$.

Solution We see that $3x$ divides into each term of the polynomial and therefore is a common factor. After factoring out the $3x$, we factor the remaining trinomial.

$$3x^3 + 24x^2 - 60x = 3x(x^2 + 8x - 20) \quad \text{Factor out the common factor.}$$
$$= 3x(x + 10)(x - 2) \quad \text{Factor the remaining trinomial.}$$

Exercise Set 7.3

Concept/Writing Exercises

For each trinomial, determine the signs that will appear in the binomial factors. Explain how you determined your answer.

1. $x^2 + 180x + 8000$
2. $x^2 - 180x + 8000$
3. $x^2 + 20x - 8000$
4. $x^2 - 20x - 8000$
5. $x^2 - 240x + 8000$

Write the trinomial whose factors are listed. Explain how you determined your answer.

6. $(x - 3)(x - 8)$
7. $(x - 3y)(x + 6y)$
8. $2(x - 5y)(x + y)$
9. $3(x + y)(x - y)$

10. How can a trinomial factoring problem be checked?

11. Explain how to determine the factors when factoring a trinomial of the form $x^2 + bx + c$.

12. In your own words, describe the trial and error method of factoring.

Practice the Skills

Factor each expression. If an expression cannot be factored by the method presented in this section, so state.

13. $x^2 - 7x + 12$
14. $x^2 + 5x + 6$
15. $x^2 + 6x + 8$
16. $x^2 - 3x + 2$
17. $x^2 + 7x + 12$
18. $x^2 - x - 12$
19. $x^2 + 5x - 9$
20. $y^2 - 6y + 8$
21. $y^2 - 16y + 15$
22. $x^2 + 3x - 28$
23. $x^2 + x - 6$
24. $p^2 + 3p - 10$
25. $r^2 - 2r - 15$
26. $x^2 - 6x + 8$
27. $b^2 - 11b + 18$
28. $x^2 + 11x - 30$
29. $x^2 - 8x - 15$
30. $x^2 - 10x + 9$
31. $a^2 + 12a + 11$
32. $x^2 + 16x + 64$
33. $x^2 + 13x - 30$
34. $x^2 - 30x - 64$
35. $x^2 + 4x + 4$
36. $x^2 - 4x + 4$
37. $k^2 + 6k + 9$
38. $k^2 - 6k + 9$
39. $x^2 - 10x + 25$
40. $x^2 - 10x - 25$
41. $w^2 - 18w + 45$
42. $x^2 - 11x + 10$
43. $x^2 + 10x - 39$
44. $x^2 - 2x + 8$
45. $x^2 - x - 20$
46. $x^2 - 17x - 60$
47. $y^2 + 9y + 14$
48. $x^2 + 15x + 56$
49. $x^2 + 12x - 64$
50. $x^2 - 18x + 80$
51. $x^2 + 14x - 24$
52. $x^2 - 13x + 36$
53. $x^2 - 2x - 80$
54. $x^2 + 19x + 48$
55. $x^2 - 18x + 65$
56. $x^2 + 5x - 24$
57. $x^2 - 8xy + 15y^2$
58. $x^2 - 2xy + y^2$
59. $x^2 - 4xy + 4y^2$
60. $x^2 - 6xy + 8y^2$
61. $x^2 + 8xy + 15y^2$
62. $x^2 + 16xy - 17y^2$

Factor completely.

63. $7x^2 - 42x + 35$
64. $3x^2 + 6x - 24$
65. $5x^2 + 20x + 15$
66. $4x^2 + 12x - 16$
67. $2x^2 - 14x + 24$
68. $3y^2 - 33y + 54$
69. $x^3 - 3x^2 - 18x$
70. $x^3 + 11x^2 - 42x$
71. $2x^3 + 6x^2 - 56x$
72. $3x^3 - 36x^2 + 33x$
73. $x^3 + 8x^2 + 16x$
74. $2x^3y - 12x^2y + 10xy$

Problem Solving

75. The first two columns in the following table describe the signs of the x-term and constant term of a trinomial of the form $x^2 + bx + c$. Determine whether the third column should contain "both positive," "both negative," or "one positive and one negative." Explain how you determined your answer.

Sign of Coefficient of x-term	Sign of Constant of Trinomial	Signs of Constant Terms in the Binomial Factors
−	+	
−	−	
+	−	
+	+	

76. Assume that a trinomial of the form $x^2 + bx + c$ is factorable. Determine whether the constant terms in the factors are "both positive," "both negative," or "one positive and one negative" for the given signs of b and c. Explain your answer.

a) $b > 0, c > 0$

b) $b < 0, c > 0$

c) $b > 0, c < 0$

d) $b < 0, c < 0$

77. Write a trinomial whose binomial factors contain constant terms that sum to 5 and have a product of 4. Show the factoring of the trinomial.

78. Write a trinomial whose binomial factors contain constant terms that sum to −11 and have a product of 28. Show the factoring of the trinomial.

79. Write a trinomial whose binomial factors contain constant terms that sum to 11 and have a product of 28. Show the factoring of the trinomial.

80. Write a trinomial whose binomial factors contain constant terms that sum to 5 and have a product of −14. Show the factoring of the trinomial.

Challenge Problems

Factor.

81. $x^2 + 0.6x + 0.08$

82. $x^2 - 0.5x - 0.06$

83. $x^2 + \frac{2}{5}x + \frac{1}{25}$

84. $x^2 - \frac{2}{3}x + \frac{1}{9}$

85. $-x^2 - 6x - 8$

86. $-x^2 - 2x + 3$

87. $x^2 + 5x - 300$

88. $x^2 - 24x - 256$

Cumulative Review Exercises

[2.5] **89.** Solve the equation $3(3x - 4) = 5x + 12$.

[3.5] **90.** Karen Moreau, a chemist, mixes 4 liters of an 18% acid solution with 1 liter of a 26% acid solution. Find the strength of the mixture.

[4.2] **91.** Graph the equation $y = 3x - 2$.

[6.5] **92.** Multiply $(2x^2 + 5x - 6)(x - 2)$.

[6.6] **93.** Divide $3x^2 - 10x - 10$ by $x - 4$.

[7.2] **94.** Factor $3x^2 + 5x - 6x - 10$ by grouping.

See Exercise 90.

7.4 FACTORING TRINOMIALS OF THE FORM $ax^2 + bx + c$, $a \neq 1$

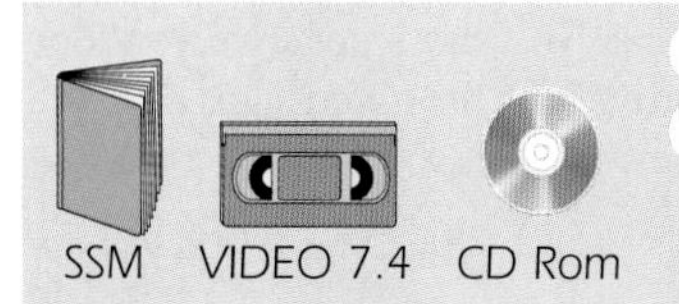

1) **Factor trinomials of the form $ax^2 + bx + c$, $a \neq 1$, by trial and error.**
2) **Factor trinomials of the form $ax^2 + bx + c$, $a \neq 1$, by grouping.**

An Important Note to Students

In this section we discuss two methods of factoring trinomials of the form $ax^2 + bx + c$, $a \neq 1$. That is, we will be factoring trinomials whose squared term has a numerical coefficient not equal to 1, after removing any common factors. Examples of trinomials with $a \neq 1$ are

$$2x^2 + 11x + 12 \ (a = 2) \qquad 4x^2 - 3x + 1 \ (a = 4)$$

The methods we discuss are (1) **factoring by trial and error** and (2) **factoring by grouping**. We present two different methods for factoring these trinomials because some students, and some instructors, prefer one method, while others prefer the second method. You may use either method unless your instructor asks you to use a specific method. We will use the same examples to illustrate both methods so that you can make a comparison. Each method is treated independently of the other. If your teacher asks you to use a specific method, either factoring by grouping or factoring by trial and error, you need only read the material related to that specific method. Factoring by trial and error was covered in Section 7.3 and factoring by grouping was covered in Section 7.2.

1) Factor Trinomials of the Form $ax^2 + bx + c$, $a \neq 1$, by Trial and Error

Let's now discuss factoring trinomials of the form $ax^2 + bx + c$, $a \neq 1$, by the trial and error method, introduced in Section 7.3. It may be helpful for you to reread that material before going any further.

Recall that factoring is the reverse of multiplying. Consider the product of the following two binomials:

$$\begin{aligned}(2x + 3)(x + 5) &= \overset{F}{2x(x)} + \overset{O}{(2x)(5)} + \overset{I}{3(x)} + \overset{L}{3(5)}\\ &= 2x^2 + 10x + 3x + 15\\ &= 2x^2 + 13x + 15\end{aligned}$$

Notice that the product of the first terms of the binomials gives the x-squared term of the trinomial, $2x^2$. Also notice that the product of the last terms of the binomials gives the last term, or constant, of the trinomial, $+ 15$. Finally, notice that the sum of the products of the outer terms and inner terms of the binomials gives the middle term of the trinomial, $+ 13x$. When we factor a trinomial using trial and error, we make use of these important facts. Note that $2x^2 + 13x + 15$ in factored form is $(2x + 3)(x + 5)$.

$$2x^2 + 13x + 15 = (2x + 3)(x + 5)$$

When factoring a trinomial of the form $ax^2 + bx + c$ by trial and error, the product of the first terms in the binomial factors must equal the first term of the trinomial, ax^2. Also, the product of the constants in the binomial factors, including their signs, must equal the constant, c, of the trinomial.

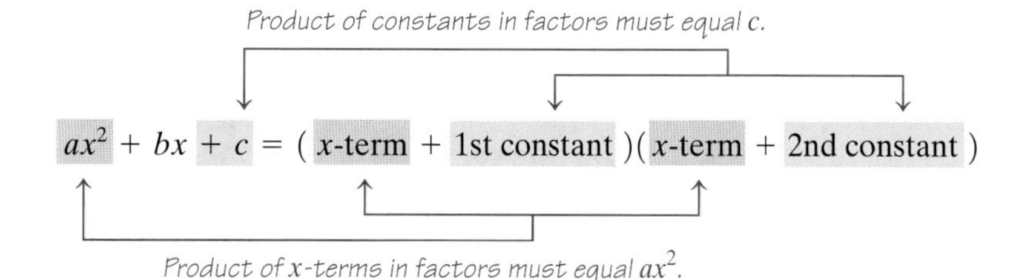

Product of constants in factors must equal c.
ax² + bx + c = (x-term + 1st constant)(x-term + 2nd constant)
Product of x-terms in factors must equal ax².

For example, when factoring the trinomial $2x^2 + 7x + 6$, each of the following pairs of factors has a product of the first terms equal to $2x^2$ and a product of the last terms equal to 6.

Trinomial	Possible Factors	Product of First Terms	Product of Last Terms
$2x^2 + 7x + 6$	$(2x + 1)(x + 6)$	$2x(x) = 2x^2$	$1(6) = 6$
	$(2x + 2)(x + 3)$	$2x(x) = 2x^2$	$2(3) = 6$
	$(2x + 3)(x + 2)$	$2x(x) = 2x^2$	$3(2) = 6$
	$(2x + 6)(x + 1)$	$2x(x) = 2x^2$	$6(1) = 6$

Each of these pairs of factors is a possible answer, but only one has the correct factors. How do we determine which is the correct factoring of the trinomial $2x^2 + 7x + 6$? The key lies in the x-term. We know that when we multiply two binomials using the FOIL method the sum of the products of the outer and inner terms gives us the x-term of the trinomial. We use this concept in reverse to determine the correct pair of factors. We need to find the pair of factors whose sum of the products of the outer and inner terms is equal to the x-term of the trinomial.

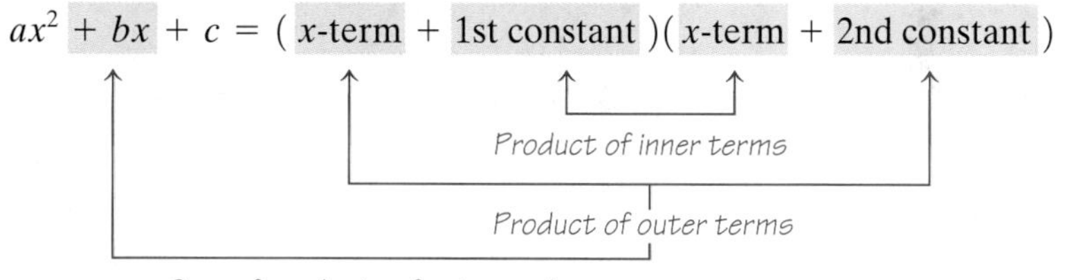

Now look at the possible pairs of factors we obtained for $2x^2 + 7x + 6$ to see if any yield the correct x-term, $7x$.

Trinomial	Possible Factors	Product of the First Terms	Product of the Last Terms	Sum of the Products of Outer and Inner Terms
$2x^2 + 7x + 6$	$(2x + 1)(x + 6)$	$2x^2$	6	$2x(6) + 1(x) = 13x$
	$(2x + 2)(x + 3)$	$2x^2$	6	$2x(3) + 2(x) = 8x$
	$(2x + 3)(x + 2)$	$2x^2$	6	$2x(2) + 3(x) = 7x$
	$(2x + 6)(x + 1)$	$2x^2$	6	$2x(1) + 6(x) = 8x$

Since $(2x + 3)(x + 2)$ yields the correct x-term, $7x$, the factors of the trinomial $2x^2 + 7x + 6$ are $(2x + 3)$ and $(x + 2)$.

$$2x^2 + 7x + 6 = (2x + 3)(x + 2)$$

We can check this factoring using the FOIL method.

CHECK:

$$\begin{aligned}(2x + 3)(x + 2) &= \overset{F}{2x(x)} + \overset{O}{2x(2)} + \overset{I}{3(x)} + \overset{L}{3(2)} \\ &= 2x^2 + 4x + 3x + 6 \\ &= 2x^2 + 7x + 6\end{aligned}$$

Since we obtained the original trinomial, our factoring is correct.

Note in the preceding illustration that $(2x + 1)(x + 6)$ are different factors than $(2x + 6)(x + 1)$, because in one case 1 is paired with $2x$ and in the second case 1 is paired with x. The factors $(2x + 1)(x + 6)$ and $(x + 6)(2x + 1)$ are, however, the same set of factors with their order reversed.

HELPFUL HINT

When factoring a trinomial of the form $ax^2 + bx + c$, remember that the sign of the constant, c, and the sign of the x-term, bx, offer valuable information. When factoring a trinomial by trial and error, first check the sign of the constant. If it is positive, the signs in both factors will be the same as the sign of the x-term. If the constant is negative, one factor will contain a plus sign and the other a negative sign.

Now we outline the procedure to factor trinomials of the form $ax^2 + bx + c$, $a \neq 1$, by trial and error. Keep in mind that the more you practice, the better you will become at factoring.

To Factor Trinomials of the Form $ax^2 + bx + c$, $a \neq 1$, by Trial and Error

1. Determine whether there is any factor common to all three terms. If so, factor it out.
2. Write all pairs of factors of the coefficient of the squared term, a.
3. Write all pairs of factors of the constant term, c.
4. Try various combinations of these factors until the correct middle term, bx, is found.

When factoring using this procedure, if there is more than one pair of numbers whose product is a, we generally begin with the middle-size pair. We will illustrate the procedure in Examples 1 through 8.

EXAMPLE 1 Factor $3x^2 + 20x + 12$.

Solution We first determine that all three terms have no common factors other than 1. Since the first term is $3x^2$, one factor must contain a $3x$ and the other an x. Therefore, the factors will be of the form $(3x + \square)(x + \square)$. Now we must find the numbers to place in the shaded areas. The product of the last terms in the factors must be 12. Since the constant and the coefficient of the x-term are both positive, only the positive factors of 12 need be considered. We will list the positive factors of 12, the possible factors of the trinomial, and the sum of the products of the outer and inner terms. Once we find the factors of 12 that yield the proper sum of the products of the outer and inner terms, $20x$, we can write the answer.

Factors of 12	Possible Factors of Trinomial	Sum of the Products of the Outer and Inner Terms
1(12)	$(3x + 1)(x + 12)$	$37x$
2(6)	$(3x + 2)(x + 6)$	$20x$
3(4)	$(3x + 3)(x + 4)$	$15x$
4(3)	$(3x + 4)(x + 3)$	$13x$
6(2)	$(3x + 6)(x + 2)$	$12x$
12(1)	$(3x + 12)(x + 1)$	$15x$

Since the product of $(3x + 2)$ and $(x + 6)$ yields the correct x-term, $20x$, they are the correct factors.

$$3x^2 + 20x + 12 = (3x + 2)(x + 6)$$

In Example 1, our first factor could have been written with an x and the second with a $3x$. Had we done this, we still would have obtained the correct answer: $(x + 6)(3x + 2)$. We also could have stopped once we found the pair of factors that yielded the $20x$. Instead, we listed all the factors so that you could study them.

EXAMPLE 2 Factor $5x^2 - 7x - 6$.

Solution One factor must contain a $5x$ and the other an x. We now list the factors of -6 and look for the pair of factors that yields $-7x$.

Factors of −6	Possible Factors	Sum of the Products of the Outer and Inner Terms
−1(6)	$(5x - 1)(x + 6)$	$29x$
−2(3)	$(5x - 2)(x + 3)$	$13x$
−3(2)	$(5x - 3)(x + 2)$	$7x$
−6(1)	$(5x - 6)(x + 1)$	$-x$

Since we did not obtain the desired quantity, $-7x$, by writing the negative factor with the $5x$, we will now try listing the negative factor with the x.

Factors of −6	Possible Factors	Sum of the Products of the Outer and Inner Terms
1(−6)	$(5x + 1)(x - 6)$	$-29x$
2(−3)	$(5x + 2)(x - 3)$	$-13x$
3(−2)	$(5x + 3)(x - 2)$	$-7x$
6(−1)	$(5x + 6)(x - 1)$	x

We see that $(5x + 3)(x - 2)$ gives the $-7x$ we are looking for. Thus,

$$5x^2 - 7x - 6 = (5x + 3)(x - 2)$$

Again we listed all the possible combinations for you to study.

HELPFUL HINT

In Example 2, we were asked to factor $5x^2 - 7x - 6$. When we considered the product of $-3(2)$ in the first set of possible factors, we obtained

Factors of -6	Possible Factors	Sum of the Products of the Outer and Inner Terms
$-3(2)$	$(5x - 3)(x + 2)$	$7x$

Later in the solution we tried the factors $3(-2)$ and obtained the correct answer.

$3(-2)$	$(5x + 3)(x - 2)$	$-7x$

When factoring a trinomial with a *negative constant,* if you obtain the x-term whose sign is the opposite of the one you are seeking, *reverse the signs on the constants* in the factors. This should give you the set of factors you are seeking.

EXAMPLE 3 Factor $8x^2 + 33x + 4$.

Solution There are no factors common to all three terms. Since the first term is $8x^2$, there are a number of possible combinations for the first terms in the factors. Since $8 = 8 \cdot 1$ and $8 = 4 \cdot 2$, the possible factors may be of the form $(8x \quad)(x \quad)$ or $(4x \quad)(2x \quad)$. When this situation occurs, we generally start with the middle-size pair of factors. Thus, we begin with $(4x \quad)(2x \quad)$. If this pair does not lead to the solution, we will then try $(8x \quad)(x \quad)$. We now list the factors of the constant, 4. Since all signs are positive, we list only the positive factors of 4.

Factors of 4	Possible Factors	Sum of the Products of the Outer and Inner Terms
1(4)	$(4x + 1)(2x + 4)$	$18x$
2(2)	$(4x + 2)(2x + 2)$	$12x$
4(1)	$(4x + 4)(2x + 1)$	$12x$

Since we did not obtain the factors with $(4x \quad)(2x \quad)$, we now try $(8x \quad)(x \quad)$.

Factors of 4	Possible Factors	Sum of the Products of the Outer and Inner Terms
1(4)	$(8x + 1)(x + 4)$	$33x$
2(2)	$(8x + 2)(x + 2)$	$18x$
4(1)	$(8x + 4)(x + 1)$	$12x$

Since the product of $(8x + 1)$ and $(x + 4)$ yields the correct x-term, $33x$, they are the correct factors.

NOW TRY EXERCISE 7

$$8x^2 + 33x + 4 = (8x + 1)(x + 4)$$

EXAMPLE 4 Factor $9x^2 - 6x + 1$.

Solution The factors must be of the form $(9x \quad)(x \quad)$ or $(3x \quad)(3x \quad)$. We will start with the middle-size factors $(3x \quad)(3x \quad)$. Since the constant is positive and the coefficient of the x-term is negative, both factors must be negative.

Factors of 1	Possible Factors	Sum of the Products of the Outer and Inner Terms
$(-1)(-1)$	$(3x - 1)(3x - 1)$	$-6x$

Since we found the correct factors, we can stop.

$$9x^2 - 6x + 1 = (3x - 1)(3x - 1) = (3x - 1)^2$$

EXAMPLE 5 Factor $2x^2 + 3x + 7$.

Solution The factors will be of the form $(2x \quad)(x \quad)$. We need only consider the positive factors of 7. Can you explain why?

Factors of 7	Possible Factors	Sum of the Products of the Outer and Inner Terms
1(7)	$(2x + 1)(x + 7)$	$15x$
7(1)	$(2x + 7)(x + 1)$	$9x$

Since we have tried all possible combinations and we have not obtained the x-term, $3x$, this trinomial cannot be factored.

If you come across a trinomial that cannot be factored, as in Example 5, do not leave the answer blank. Instead, write "cannot be factored." However, before you write the answer "cannot be factored," recheck your work and make sure you have tried every possible combination.

EXAMPLE 6 Factor $4x^2 + 7xy + 3y^2$.

Solution This trinomial is different from the other trinomials in that the last term is not a constant but contains y^2. Don't let this scare you. The factoring process is the same, except that the second term of both factors will contain y. We begin by considering factors of the form $(2x \quad)(2x \quad)$. If we cannot find the factors, then we try factors of the form $(4x \quad)(x \quad)$.

Factors of 3	Possible Factors	Sum of the Products of the Outer and Inner Terms
1(3)	$(2x + y)(2x + 3y)$	$8xy$
3(1)	$(2x + 3y)(2x + y)$	$8xy$
1(3)	$(4x + y)(x + 3y)$	$13xy$
3(1)	$(4x + 3y)(x + y)$	$7xy$

$$4x^2 + 7xy + 3y^2 = (4x + 3y)(x + y)$$

CHECK: $(4x + 3y)(x + y) = 4x^2 + 4xy + 3xy + 3y^2 = 4x^2 + 7xy + 3y^2$

EXAMPLE 7 Factor $6x^2 - 13xy - 8y^2$.

Solution We begin with factors of the form $(3x \quad)(2x \quad)$. If we cannot find the solution from these, we will try $(6x \quad)(x \quad)$. Since the last term, $-8y^2$, is negative, one factor will contain a plus sign and the other will contain a minus sign.

Factors of −8	Possible Factors	Sum of the Products of the Outer and Inner Terms
$1(-8)$	$(3x + y)(2x - 8y)$	$-22xy$
$2(-4)$	$(3x + 2y)(2x - 4y)$	$-8xy$
$4(-2)$	$(3x + 4y)(2x - 2y)$	$2xy$
$8(-1)$	$(3x + 8y)(2x - y)$	$13xy$ ←

We are looking for $-13xy$. When we considered $8(-1)$, we obtained $13xy$. As explained in the Helpful Hint on page 422, if we reverse the signs of the numbers in the factors, we will obtain the factors we are seeking.

$$(3x + 8y)(2x - y) \quad \text{Gives } 13xy$$

$$(3x - 8y)(2x + y) \quad \text{Gives } -13xy$$

NOW TRY EXERCISE 53

Therefore, $6x^2 - 13xy - 8y^2 = (3x - 8y)(2x + y)$.

Now we will look at an example in which all the terms of the trinomial have a common factor.

EXAMPLE 8 Factor $4x^3 + 10x^2 + 6x$.

Solution *The first step in any factoring problem is to determine whether all the terms contain a common factor. If so, factor out that common factor first.* In this example, $2x$ is common to all three terms. We begin by factoring out the $2x$. Then we continue factoring by trial and error.

$$\begin{aligned} 4x^3 + 10x^2 + 6x &= 2x(2x^2 + 5x + 3) \\ &= 2x(2x + 3)(x + 1) \end{aligned}$$

2) Factor Trinomials of the Form $ax^2 + bx + c$, $a \neq 1$, by Grouping

We will now discuss the use of grouping. The steps in the box that follow give the procedure for factoring trinomials by grouping.

To Factor Trinomials of the Form $ax^2 + bx + c$, $a \neq 1$, by Grouping

1. Determine whether there is a factor common to all three terms. If so, factor it out.
2. Find two numbers whose product is equal to the product of a times c, and whose sum is equal to b.
3. Rewrite the middle term, bx, as the sum or difference of two terms using the numbers found in step 2.
4. Factor by grouping as explained in Section 7.2.

This process will be made clear in Example 9. We will rework Examples 1 through 8 here using factoring by grouping. Example 9, which follows, is the same trinomial given in Example 1. After you study this method and try some exercises, you will gain a feel for which method you prefer using.

EXAMPLE 9 Factor $3x^2 + 20x + 12$.

Solution First determine whether there is a factor common to all the terms of the polynomial. There are no common factors (other than 1) to the three terms.

$$a = 3 \qquad b = 20 \qquad c = 12$$

1. We must find two numbers whose product is $a \cdot c$ and whose sum is b. We must therefore find two numbers whose product equals $3 \cdot 12 = 36$ and whose sum equals 20. Only the positive factors of 36 need be considered since all signs of the trinomial are positive.

Factors of 36	Sum of Factors
(1)(36)	$1 + 36 = 37$
(2)(18)	$2 + 18 = 20$
(3)(12)	$3 + 12 = 15$
(4)(9)	$4 + 9 = 13$
(6)(6)	$6 + 6 = 12$

The desired factors are 2 and 18.

2. Rewrite $20x$ as the sum or difference of two terms using the values found in step 1. Therefore, we rewrite $20x$ as $2x + 18x$.

$$\begin{aligned} &3x^2 + \overbrace{20x}\, + 12 \\ &= 3x^2 + 2x + 18x + 12 \end{aligned}$$

3. Now factor by grouping. Start by factoring out a common factor from the first two terms and a common factor from the last two terms. This procedure was discussed in Section 7.2.

$$\begin{aligned} &\overbrace{3x^2 + 2x}^{x \text{ is common factor}} + \overbrace{18x + 12}^{6 \text{ is common factor}} \\ &= x(3x + 2) + 6(3x + 2) \\ &= (x + 6)(3x + 2) \end{aligned}$$

Note that in step 2 of Example 9 we rewrote $20x$ as $2x + 18x$. Would it have made a difference if we had written $20x$ as $18x + 2x$? Let's work it out and see.

$$\begin{aligned} &3x^2 + \overbrace{20x}\, + 12 \\ &= 3x^2 + 18x + 2x + 12 \end{aligned}$$

$$\begin{aligned} &\overbrace{3x^2 + 18x}^{3x \text{ is common factor}} + \overbrace{2x + 12}^{2 \text{ is common factor}} \\ &= 3x(x + 6) + 2(x + 6) \\ &= (3x + 2)(x + 6) \end{aligned}$$

Since $(3x + 2)(x + 6) = (x + 6)(3x + 2)$, the factors are the same. We obtained the same answer by writing the $20x$ as either $2x + 18x$ or $18x + 2x$. *In general, when rewriting the middle term of the trinomial using the specific factors found, the terms may be listed in either order.* You should, however, check after you list the two terms to make sure that the sum of the terms you listed equals the middle term.

EXAMPLE 10 Factor $5x^2 - 7x - 6$.

Solution There are no common factors other than 1.

$$a = 5, \qquad b = -7, \qquad c = -6$$

The product of a times c is $5(-6) = -30$. We must find two numbers whose product is -30 and whose sum is -7.

Factors of −30	Sum of Factors
$(-1)(30)$	$-1 + 30 = 29$
$(-2)(15)$	$-2 + 15 = 13$
$(-3)(10)$	$-3 + 10 = 7$
$(-5)(6)$	$-5 + 6 = 1$
$(-6)(5)$	$-6 + 5 = -1$
$(-10)(3)$	$-10 + 3 = -7$
$(-15)(2)$	$-15 + 2 = -13$
$(-30)(1)$	$-30 + 1 = -29$

Rewrite the middle term of the trinomial, $-7x$, as $-10x + 3x$.

$$\begin{aligned} &5x^2 - 7x - 6 \\ &= 5x^2 - 10x + 3x - 6 \\ &= 5x(x - 2) + 3(x - 2) \\ &= (5x + 3)(x - 2) \end{aligned}$$

Now factor by grouping.

NOW TRY EXERCISE 31

In Example 10, we could have expressed the $-7x$ as $3x - 10x$ and obtained the same answer. Try working Example 10 by rewriting $-7x$ as $3x - 10x$.

HELPFUL HINT

Notice in Example 10 that we were looking for two factors of −30 whose sum was −7. When we considered the factors −3 and 10, we obtained a sum of 7. The factors we eventually obtained that gave a sum of −7 were 3 and −10. Note that when the *constant of the trinomial is negative*, if we switch the signs of the constants in the factors, the sign of the sum of the factors changes. Thus, when trying pairs of factors to obtain the middle term, if you obtain the opposite of the coefficient you are seeking, reverse the signs in the factors. This should give you the coefficient you are seeking.

Using Your Graphing Calculator

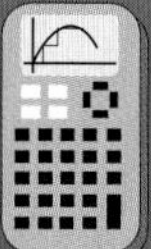

In example 10, we began factoring $5x^2 - 7x - 6$ by finding two numbers whose product equals -30 and whose sum equals -7. The graphing calculator can be used to help find those numbers.

We can represent two numbers whose product is -30 by x and $-\dfrac{30}{x}$. Notice $x\left(-\dfrac{30}{x}\right) = -30$. Let $Y_1 = -\dfrac{30}{x}$, the second factor in $x\left(-\dfrac{30}{x}\right)$. Let $Y_2 = -\dfrac{30}{x} + x$, the sum of the factors. Use the TABLE feature to create pairs of numbers whose product is -30 (the product of columns X and Y_1 is -30) and whose sum is in column Y_2, see Fig. 7.3. Look down the Y_2 column until you find the sum you are looking for, -7.

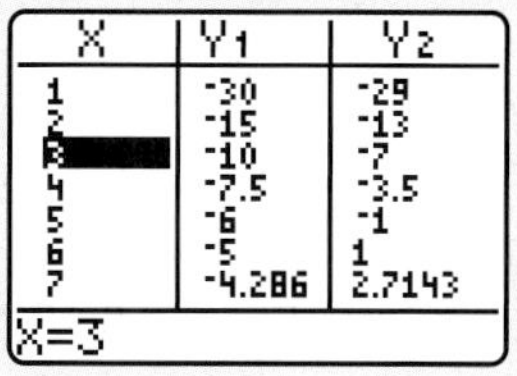

X	Y1	Y2
1	-30	-29
2	-15	-13
3	-10	-7
4	-7.5	-3.5
5	-6	-1
6	-5	1
7	-4.286	2.7143

X=3

Figure 7.3

Use the value for X, 3, and the value for Y_1, -10, to rewrite the middle term of the trinomial, $-7x$, as $3x - 10x$. Continue factoring by grouping.

Exercises

Use your graphing calculator as an aid in factoring each trinomial.

1. $6x^2 - 13x - 28$

2. $10x^2 - x - 24$

3. $12x^2 + 16x - 35$

4. $27x^2 + 57x + 20$

EXAMPLE 11 Factor $8x^2 + 33x + 4$.

Solution There are no common factors other than 1. We must find two numbers whose product is $8 \cdot 4$ or 32 and whose sum is 33. The numbers are 1 and 32.

Factors of 32	Sum of Factors
(1)(32)	$1 + 32 = 33$

Rewrite $33x$ as $32x + x$. Then factor by grouping.

$$
\begin{aligned}
&8x^2 + 33x + 4 \\
&= 8x^2 + 32x + x + 4 \\
&= 8x(x + 4) + 1(x + 4) \\
&= (8x + 1)(x + 4)
\end{aligned}
$$

Notice in Example 11 that we rewrote $33x$ as $32x + x$ rather than $x + 32x$. We did this to reinforce factoring out 1 from the last two terms of an expression. You should obtain the same answer if you rewrite $33x$ as $x + 32x$. Try this now.

EXAMPLE 12 Factor $9x^2 - 6x + 1$.

Solution There are no common factors other than 1. We must find two numbers whose product is $9 \cdot 1$ or 9 and whose sum is -6. Since the product of a times c is positive and the coefficient of the x-term is negative, both numerical factors must be negative.

Factors of 9	Sum of Factors
$(-1)(-9)$	$-1 + (-9) = -10$
$(-3)(-3)$	$-3 + (-3) = -6$

The desired factors are −3 and −3.

$$
\begin{aligned}
&9x^2 - 6x + 1 \\
&= 9x^2 - 3x - 3x + 1 && \text{Rewrite } -6x \text{ as } -3x - 3x. \\
&= 3x(3x - 1) - 3x + 1 \\
&= 3x(3x - 1) - 1(3x - 1) && \text{Rewrite } -3x + 1 \text{ as } -1(3x - 1). \\
&= (3x - 1)(3x - 1) \text{ or } (3x - 1)^2
\end{aligned}
$$

HELPFUL HINT

When attempting to factor a trinomial, if there are no two integers whose product equals $a \cdot c$ and whose sum equals b, the trinomial cannot be factored.

EXAMPLE 13 Factor $2x^2 + 3x + 7$.

Solution There are no common factors other than 1. We must find two numbers whose product is 14 and whose sum is 3. We need consider only positive factors of 14. Why?

Factors of 14	Sum of Factors
(1)(14)	1 + 14 = 15
(2)(7)	2 + 7 = 9

Since there are no factors of 14 whose sum is 3, we conclude that this trinomial cannot be factored.

EXAMPLE 14 Factor $4x^2 + 7xy + 3y^2$.

Solution There are no common factors other than 1. This trinomial contains two variables. It is factored in basically the same manner as the previous examples. Find two numbers whose product is $4 \cdot 3$ or 12 and whose sum is 7. The two numbers are 4 and 3.

$$
\begin{aligned}
&4x^2 + 7xy + 3y^2 \\
&= 4x^2 + 4xy + 3xy + 3y^2 \\
&= 4x(x + y) + 3y(x + y) \\
&= (4x + 3y)(x + y)
\end{aligned}
$$

EXAMPLE 15 Factor $6x^2 - 13xy - 8y^2$.

Solution There are no common factors other than 1. Find two numbers whose product is 6(−8) or −48 and whose sum is −13. Since the product is negative, one factor must be positive and the other negative. Some factors are given below.

Product of Factors	Sum of Factors
(1)(−48)	1 + (−48) = −47
(2)(−24)	2 + (−24) = −22
(3)(−16)	3 + (−16) = −13

There are many other factors, but we have found the pair we were looking for. The two numbers whose product is −48 and whose sum is −13 are 3 and −16.

$$\begin{aligned} &6x^2 - \overbrace{13xy}^{} - 8y^2 \\ &= 6x^2 + 3xy - 16xy - 8y^2 \\ &= 3x(2x + y) - 8y(2x + y) \\ &= (3x - 8y)(2x + y) \end{aligned}$$

CHECK: $(3x - 8y)(2x + y)$

$$\begin{aligned} &\quad\; F \qquad\quad O \qquad\quad\; I \qquad\qquad L \\ &= (3x)(2x) + (3x)(y) + (-8y)(2x) + (-8y)(y) \\ &= 6x^2 + 3xy - 16xy - 8y^2 \\ &= 6x^2 - 13xy - 8y^2 \end{aligned}$$

If you rework Example 15 by writing $-13xy$ as $-16xy + 3xy$, what answer would you obtain? Try it now and see.

Remember that in any factoring problem our first step is to determine whether all terms in the polynomial have a common factor other than 1. If so, we use the distributive property to factor the GCF from each term. We then continue to factor the trinomial, if possible.

EXAMPLE 16 Factor $4x^3 + 10x^2 + 6x$.

Solution The factor $2x$ is common to all three terms. Factor the $2x$ from each term of the polynomial.

$$4x^3 + 10x^2 + 6x = 2x(2x^2 + 5x + 3)$$

Now continue by factoring $2x^2 + 5x + 3$. The two numbers whose product is $2 \cdot 3$ or 6 and whose sum is 5 are 2 and 3.

$$\begin{aligned} &2x(2x^2 + \overbrace{5x}^{} + 3) \\ &= 2x(2x^2 + 2x + 3x + 3) \\ &= 2x[2x(x + 1) + 3(x + 1)] \\ &= 2x(2x + 3)(x + 1) \end{aligned}$$

NOW TRY EXERCISE 45

HELPFUL HINT

Which Method Should You Use to Factor a Trinomial?

If your instructor asks you to use a specific method, you should use that method. If your instructor does not require a specific method, you should use the method you feel most comfortable with. You may wish to start with the trial and error method if there are only a few possible factors to try. If you cannot find the factors by trial and error or if there are many possible factors to consider, you may wish to use the grouping procedure. With time and practice you will learn which method you feel most comfortable with and which method gives you greater success.

Exercise Set 7.4

Concept/Writing Exercises

1. What is the relationship between factoring trinomials and multiplying binomials?
2. When factoring a trinomial of the form $ax^2 + bx + c$, what must the product of the first terms of the binomial factors equal?
3. When factoring a trinomial of the form $ax^2 + bx + c$, what must the product of the constants in the binomial factors equal?
4. Explain in your own words the procedure used to factor a trinomial of the form $ax^2 + bx + c, a \neq 1$.

Practice the Skills

Factor completely. If an expression cannot be factored, so state.

5. $2x^2 + 11x + 5$
6. $3x^2 + 5x + 2$
7. $3x^2 + 14x + 8$
8. $5x^2 + 13x + 6$
9. $3x^2 + 4x + 1$
10. $3y^2 + 17y + 10$
11. $2x^2 + 13x + 20$
12. $3x^2 - 2x - 8$
13. $4x^2 + 4x - 3$
14. $4x^2 - 11x + 7$
15. $5y^2 - 8y + 3$
16. $5m^2 - 17m + 6$
17. $5a^2 - 12a + 6$
18. $2x^2 - x - 1$
19. $6x^2 + 19x + 3$
20. $6y^2 - 19y + 15$
21. $5x^2 + 11x + 4$
22. $3x^2 - 2x - 5$
23. $5y^2 - 16y + 3$
24. $5x^2 + 2x + 9$
25. $7x^2 + 43x + 6$
26. $4x^2 + 4x - 15$
27. $7x^2 - 8x + 1$
28. $15x^2 - 19x + 6$
29. $3x^2 - 10x + 7$
30. $3y^2 - 22y + 7$
31. $5z^2 - 6z - 8$
32. $3z^2 - 11z - 6$
33. $4y^2 + 5y - 6$
34. $8y^2 - 2y - 1$
35. $10x^2 - 27x + 5$
36. $6x^2 + 7x - 10$
37. $8x^2 - 2x - 15$
38. $8x^2 + 13x + 6$
39. $6x^2 + 33x + 15$
40. $12x^2 - 13x - 35$
41. $6x^2 + 16x + 10$
42. $12z^2 + 32z + 20$
43. $6x^3 - 5x^2 - 4x$
44. $8x^3 + 8x^2 - 6x$
45. $12x^3 + 28x^2 + 8x$
46. $18x^3 - 21x^2 - 9x$
47. $6x^3 + 4x^2 - 10x$
48. $300x^2 - 400x - 400$
49. $60x^2 + 40x + 5$
50. $28x^2 - 28x + 7$
51. $2x^2 + 5xy + 2y^2$
52. $8x^2 - 8xy - 6y^2$
53. $2x^2 - 7xy + 3y^2$
54. $15x^2 - xy - 6y^2$
55. $12x^2 + 10xy - 8y^2$
56. $12a^2 - 34ab + 24b^2$
57. $6x^2 - 15xy - 36y^2$
58. $60x^2 - 125xy + 60y^2$

Problem Solving

Write the polynomial whose factors are listed. Explain how you determined your answer.

59. $4x + 1, x - 7$
60. $4x - 2, 5x - 7$
61. $5, x + 3, 2x + 1$
62. $3, 2x + 3, x - 4$
63. $x^2, x + 1, 2x - 3$
64. $5x^2, 3x - 7, 2x + 3$
65. **a)** If you know one binomial factor of a trinomial, explain how you can use division to find the second binomial factor of the trinomial (see Section 6.6).
    b) One factor of $18x^2 + 93x + 110$ is $3x + 10$. Use division to find the second factor.
66. One factor of $30x^2 - 17x - 247$ is $6x - 19$. Find the other factor.
67. One factor of $105x^2 - 449x + 480$ is $7x - 15$. Find the other factor.

Challenge Problems

Factor each trinomial.

68. $18x^2 + 9x - 20$
69. $8x^2 - 99x + 36$
70. $15x^2 - 124x + 160$
71. $16x^2 - 62x - 45$
72. $72x^2 - 180x - 200$
73. $72x^2 + 417x - 420$
74. Two factors of $6x^3 + 235x^2 + 2250x$ are x and $3x + 50$, determine the other factor. Explain how you determined your answer.
75. Two factors of the polynomial $2x^3 + 11x^2 + 3x - 36$ are $x + 3$ and $2x - 3$. Determine the third factor. Explain how you determined your answer.

Cumulative Review Exercises

[1.9] **76.** Evaluate $-x^2 - 4(y + 3) + 2y^2$ when $x = -3$ and $y = -5$.

[3.5] **77.** Jeff Gordon won the 1999 Daytona 500 in a time of about 3.75 hours, including cautions and pit stops. If the 500 miles were covered by circling the 2.5 mile track 200 times, find the average speed of the race.

[4.6] **78.** $f(x) = x^2 + x + 5$; find **a)** $f(2)$, **b)** $f(-4)$

[7.1] **79.** Factor $36x^4y^3 - 12xy^2 + 24x^5y^6$.

[7.3] **80.** Factor $x^2 - 15x + 54$.

See Exercise 77.

7.5 SPECIAL FACTORING FORMULAS AND A GENERAL REVIEW OF FACTORING

SSM VIDEO 7.5 CD Rom

1) **Factor the difference of two squares.**
2) **Factor the sum and difference of two cubes.**
3) **Learn the general procedure for factoring a polynomial.**

There are special formulas for certain types of factoring problems that are often used. The special formulas we focus on in this section are the *difference of two squares, the sum of two cubes, and the difference of two cubes*. There is no special formula for the sum of two squares; this is because the sum of two squares cannot be factored using the set of real numbers. You will need to memorize the three formulas in this section so that you can use them whenever you need them.

1) Factor the Difference of Two Squares

Let's begin with the difference of two squares. Consider the binomial $x^2 - 9$. Note that each term of the binomial can be expressed as the square of some expression.

$$x^2 - 9 = x^2 - 3^2$$

This is an example of a **difference of two squares**. To factor the difference of two squares, it is convenient to use the difference of two squares formula (which was introduced in Section 6.5).

Difference of Two Squares

$$a^2 - b^2 = (a + b)(a - b)$$

EXAMPLE 1 Factor $x^2 - 9$.

Solution If we write $x^2 - 9$ as a difference of two squares, we have $x^2 - 3^2$. Using the difference of two squares formula, where a is replaced by x and b is replaced by 3, we obtain the following:

$$a^2 - b^2 = (a + b)(a - b)$$
$$x^2 - 3^2 = (x + 3)(x - 3)$$

EXAMPLE 2 Factor using the difference of two squares formula.

a) $x^2 - 16$ **b)** $4x^2 - 25$ **c)** $36x^2 - 49y^2$

Solution **a)** $x^2 - 16 = (x)^2 - (4)^2 = (x + 4)(x - 4)$ **b)** $4x^2 - 25 = (2x)^2 - (5)^2 = (2x + 5)(2x - 5)$

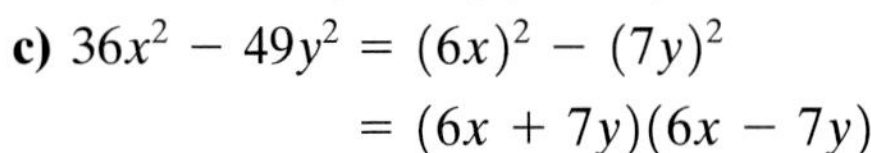

c) $36x^2 - 49y^2 = (6x)^2 - (7y)^2$
$= (6x + 7y)(6x - 7y)$

EXAMPLE 3 Factor each difference of two squares.

a) $16x^4 - 9y^4$ **b)** $x^6 - y^4$

Solution **a)** Rewrite $16x^4$ as $(4x^2)^2$ and $9y^4$ as $(3y^2)^2$, then use the difference of two squares formula.

$$16x^4 - 9y^4 = (4x^2)^2 - (3y^2)^2$$
$$= (4x^2 + 3y^2)(4x^2 - 3y^2)$$

b) Rewrite x^6 as $(x^3)^2$ and y^4 as $(y^2)^2$, then use the difference of two squares formula.

$$x^6 - y^4 = (x^3)^2 - (y^2)^2$$
$$= (x^3 + y^2)(x^3 - y^2)$$

NOW TRY EXERCISE 25

EXAMPLE 4 Factor $4x^2 - 16y^2$ using the difference of two squares formula.

Solution First remove the common factor, 4.

$$4x^2 - 16y^2 = 4(x^2 - 4y^2)$$

Now use the formula for the difference of two squares.

$$4(x^2 - 4y^2) = 4[(x)^2 - (2y)^2]$$
$$= 4(x + 2y)(x - 2y)$$

NOW TRY EXERCISE 73

Notice in Example 4 that $4x^2 - 16y^2$ is the difference of two squares, $(2x)^2 - (4y)^2$. If you factor this difference of squares without first removing the common factor 4, the factoring may be more difficult. After you factor this difference of squares you will need to factor out the common factor 2 from each binomial factor, as illustrated below.

$$4x^2 - 16y^2 = (2x)^2 - (4y)^2$$
$$= (2x + 4y)(2x - 4y)$$
$$= 2(x + 2y)2(x - 2y)$$
$$= 4(x + 2y)(x - 2y)$$

We obtain the same answer as we did in Example 4. However, since we did not factor out the common factor 4 first, we had to work a little harder to obtain the answer.

EXAMPLE 5 Factor $z^4 - 16$ using the difference of two squares formula.

Solution We rewrite z^4 as $(z^2)^2$ and 16 as 4^2, then use the difference of two squares formula.

$$z^4 - 16 = (z^2)^2 - 4^2$$
$$= (z^2 + 4)(z^2 - 4)$$

Notice that the second factor, $z^2 - 4$, is also the difference of two squares. To complete the factoring, we use the difference of two squares formula again to factor $z^2 - 4$.

$$= (z^2 + 4)(z^2 - 4)$$
$$= (z^2 + 4)(z + 2)(z - 2)$$

Avoiding Common Errors

The difference of two squares can be factored. However, it is not possible to factor the sum of two squares using real numbers.

CORRECT	INCORRECT
$a^2 - b^2 = (a + b)(a - b)$	~~$a^2 + b^2 = (a + b)(a + b)$~~

2) Factor the Sum and Difference of Two Cubes

We begin our discussion of the sum and difference of two cubes with a multiplication of polynomials problem. Consider the product of $(a + b)(a^2 - ab + b^2)$.

$$\begin{array}{rrrr} & a^2 & - ab & + b^2 \\ & & a & + b \\ \hline & a^2b & - ab^2 & + b^3 \\ a^3 & - a^2b & + ab^2 & \\ \hline a^3 & & & + b^3 \end{array}$$

Thus, $(a + b)(a^2 - ab + b^2) = a^3 + b^3$. Since factoring is the opposite of multiplying, we may factor $a^3 + b^3$ as follows:

$$a^3 + b^3 = (a + b)(a^2 - ab + b^2)$$

We see, using the same procedure, that $a^3 - b^3 = (a - b)(a^2 + ab + b^2)$. The expression $a^3 + b^3$ is a sum of two cubes and the expression $a^3 - b^3$ is a difference of two cubes. The formulas for factoring **the sum and the difference of two cubes** follow.

Sum of Two Cubes

$$a^3 + b^3 = (a + b)(a^2 - ab + b^2)$$

Difference of Two Cubes

$$a^3 - b^3 = (a - b)(a^2 + ab + b^2)$$

Note that the trinomials $a^2 - ab + b^2$ and $a^2 + ab + b^2$ cannot be factored further. Now let's do some factoring problems using the sum and the difference of two cubes.

EXAMPLE 6 Factor $x^3 + 8$.

Solution We rewrite $x^3 + 8$ as a sum of two cubes: $x^3 + 8 = (x)^3 + (2)^3$. Using the sum of two cubes formula, if we let a correspond to x and b correspond to 2, we get

$$\begin{array}{c} a^3 + b^3 = (a + b)(a^2 - a\,b + b^2) \\ \downarrow \quad \downarrow \quad \downarrow \quad \downarrow \quad \downarrow \quad \downarrow \quad \downarrow \quad \downarrow \\ x^3 + 2^3 = (x)^3 + (2)^3 = (x + 2)[x^2 - x(2) + 2^2] \\ = (x + 2)(x^2 - 2x + 4) \end{array}$$

You can check the factoring by multiplying $(x + 2)(x^2 - 2x + 4)$. If factored correctly, the product of the factors will equal the original expression, $x^3 + 8$. Try it and see.

EXAMPLE 7 Factor $y^3 - 125$.

Solution We rewrite $y^3 - 125$ as a difference of two cubes: $(y)^3 - (5)^3$. Using the difference of two cubes formula, if we let a correspond to y and b correspond to 5, we get

$$\begin{array}{c} a^3 - b^3 = (a - b)(a^2 + a\,b + b^2) \\ \downarrow \quad \downarrow \quad \downarrow \quad \downarrow \quad \downarrow \quad \downarrow \quad \downarrow \quad \downarrow \\ y^3 - 125 = (y)^3 - (5)^3 = (y - 5)[y^2 + y(5) + 5^2] \\ = (y - 5)(y^2 + 5y + 25) \end{array}$$

EXAMPLE 8 Factor $8p^3 - k^3$.

Solution We rewrite $8p^3 - k^3$ as a difference of two cubes. Because $(2p)^3 = 8p^3$, we can write

$$\begin{aligned} 8p^3 - k^3 &= (2p)^3 - (k)^3 \\ &= (2p - k)[(2p)^2 + (2p)(k) + k^2] \\ &= (2p - k)(4p^2 + 2pk + k^2) \end{aligned}$$

EXAMPLE 9 Factor $8r^3 + 27s^3$.

Solution We rewrite $8r^3 + 27s^3$ as a sum of two cubes. Since $8r^3 = (2r)^3$ and $27s^3 = (3s)^3$, we write

$$\begin{aligned} 8r^3 + 27s^3 &= (2r)^3 + (3s)^3 \\ &= (2r + 3s)[(2r)^2 - (2r)(3s) + (3s)^2] \\ &= (2r + 3s)(4r^2 - 6rs + 9s^2) \end{aligned}$$

NOW TRY EXERCISE 45

3) Learn the General Procedure for Factoring a Polynomial

In this chapter we have presented several methods of factoring. We now combine techniques from this and previous sections.

Here is a general procedure for factoring any polynomial:

General Procedure for Factoring a Polynomial

1. If all the terms of the polynomial have a greatest common factor other than 1, factor it out.
2. If the polynomial has two terms (or is a binomial), determine whether it is a difference of two squares or a sum or a difference of two cubes. If so, factor using the appropriate formula.
3. If the polynomial has three terms, factor the trinomial using the methods discussed in Sections 7.3 and 7.4.
4. If the polynomial has more than three terms, try factoring by grouping.
5. As a final step, examine your factored polynomial to determine whether the terms in any factors have a common factor. If you find a common factor, factor it out at this point.

EXAMPLE 10 Factor $3x^4 - 27x^2$.

Solution First determine whether the terms have a greatest common factor other than 1. Since $3x^2$ is common to both terms, factor it out.

$$3x^4 - 27x^2 = 3x^2(x^2 - 9)$$
$$= 3x^2(x + 3)(x - 3)$$

Note that $x^2 - 9$ is a difference of two squares.

EXAMPLE 11 Factor $3x^2y^2 - 6xy^2 - 24y^2$.

Solution Begin by factoring the GCF, $3y^2$, from each term. Then factor the remaining trinomial.

$$3x^2y^2 - 6xy^2 - 24y^2 = 3y^2(x^2 - 2x - 8)$$
$$= 3y^2(x - 4)(x + 2)$$

EXAMPLE 12 Factor $10a^2b - 15ab + 20b$.

Solution

$$10a^2b - 15ab + 20b = 5b(2a^2 - 3a + 4)$$

Since $2a^2 - 3a + 4$ cannot be factored, we stop here.

EXAMPLE 13 Factor $3xy + 6x + 3y + 6$

Solution Always begin by determining whether all the terms in the polynomial have a common factor. In this example, 3 is the GCF. Factor 3 from each term.

$$3xy + 6x + 3y + 6 = 3(xy + 2x + y + 2)$$

Now factor by grouping.

$$= 3[x(y + 2) + 1(y + 2)]$$
$$= 3(x + 1)(y + 2)$$

In Example 13, what would happen if we forgot to factor out the common factor 3. Let's rework the problem without first factoring out the 3, and see what happens. Factor $3x$ from the first two terms, and 3 from the last two terms.

$$\begin{aligned} 3xy + 6x + 3y + 6 &= 3x(y + 2) + 3(y + 2) \\ &= (3x + 3)(y + 2) \end{aligned}$$

In step 5 of the general factoring procedure on page 435 we are reminded to examine the factored polynomial to see whether the terms in any factor have a common factor. If we study the factors, we see that the factor $3x + 3$ has a common factor of 3. If we factor out the 3 from $3x + 3$ we will obtain the same answer obtained in Example 13.

$$(3x + 3)(y + 2) = 3(x + 1)(y + 2)$$

EXAMPLE 14 Factor $12x^2 + 12x - 9$.

Solution First factor out the common factor 3. Then factor the remaining trinomial by one of the methods discussed in Section 7.4 (either by grouping or trial and error).

$$\begin{aligned} 12x^2 + 12x - 9 &= 3(4x^2 + 4x - 3) \\ &= 3(2x + 3)(2x - 1) \end{aligned}$$

EXAMPLE 15 Factor $2x^4y + 54xy$.

Solution First factor out the common factor $2xy$.

$$\begin{aligned} 2x^4y + 54xy &= 2xy(x^3 + 27) \\ &= 2xy(x + 3)(x^2 - 3x + 9) \end{aligned}$$

Note that $x^3 + 27$ is a sum of two cubes.

Exercise Set 7.5

Concept/Writing Exercises

1. a) Write the formula for factoring the difference of two squares.
   b) In your own words, explain how to factor the difference of two squares.
2. a) Write the formula for factoring the sum of two cubes.
   b) In your own words, explain how to factor the sum of two cubes.
3. a) Write the formula for factoring the difference of two cubes.
   b) In your own words, explain how to factor the difference of two cubes.
4. Why is it important to memorize the special factoring formulas?
5. Is there a special formula for factoring the sum of two squares?
6. In your own words, describe the general procedure for factoring a polynomial.

Practice the Skills

Factor each difference of two squares.

7. $y^2 - 16$ **8.** $x^2 - 4$ **9.** $y^2 - 100$

10. $z^2 - 64$ **11.** $x^2 - 49$ **12.** $x^2 - a^2$

13. $x^2 - y^2$ **14.** $4x^2 - 9$ **15.** $9y^2 - 25z^2$

16. $49y^2 - 25$ **17.** $64a^2 - 36b^2$ **18.** $100x^2 - 81y^2$

19. $25x^2 - 36$
20. $y^4 - 121$
21. $z^4 - 81x^2$
22. $9x^4 - 16y^4$
23. $9x^4 - 81y^2$
24. $4x^4 - 25y^4$
25. $36m^4 - 49n^2$
26. $2x^4 - 50y^2$
27. $20x^2 - 180$
28. $4x^3 - xy^2$
29. $16x^2 - 100y^4$
30. $27x^4 - 3y^2$

Factor each sum or difference of two cubes.

31. $x^3 + y^3$
32. $x^3 - y^3$
33. $a^3 - b^3$
34. $a^3 + b^3$
35. $x^3 + 8$
36. $x^3 - 8$
37. $x^3 - 27$
38. $a^3 + 27$
39. $a^3 + 1$
40. $a^3 - 1$
41. $27x^3 - 1$
42. $27y^3 + 8$
43. $27a^3 - 125$
44. $64 + x^3$
45. $27 - 8y^3$
46. $1 + 27y^3$
47. $27x^3 + 64y^3$
48. $64x^3 - 27y^3$

Factor completely.

49. $8x^2 + 16x + 8$
50. $3x^2 - 9x - 12$
51. $a^2b - 9b$
52. $2x^2 - 32$
53. $3x^2 + 12x + 12$
54. $3x^2 + 9x + 12x + 36$
55. $5x^2 - 10x - 15$
56. $4x^2 - 36$
57. $3xy - 6x + 9y - 18$
58. $x^2y + 2xy - 6xy - 12y$
59. $2x^2 - 50$
60. $5a^2y - 80y^3$
61. $3x^2y - 27y$
62. $2x^3 - 72x$
63. $3x^3y^2 + 3y^2$
64. $x^4 - 125x$
65. $2x^3 - 16$
66. $x^3 - 27y^3$
67. $6x^2 - 4x + 24x - 16$
68. $4x^2y - 6xy - 20xy + 30y$
69. $3x^3 - 10x^2 - 8x$
70. $4x^4 - 26x^3 + 30x^2$
71. $4x^2 + 5x - 6$
72. $12a^2 + 36a + 27$
73. $25b^2 - 100$
74. $3b^2 - 75c^2$
75. $a^5b^2 - 4a^3b^4$
76. $12x^2 + 8x - 18x - 12$
77. $3x^4 - 18x^3 + 27x^2$
78. $2a^6 + 4a^4b^2$
79. $x^3 + 25x$
80. $8y^2 - 23y - 3$
81. $y^4 - 16$
82. $16m^3 + 250$
83. $60a^2 - 25ab - 10b^2$
84. $ac + 2a + bc + 2b$
85. $2ab + 4a - 3b - 6$
86. $x^3 - 100x$
87. $9 - 9y^4$

Problem Solving

88. Explain why the sum of two squares, $a^2 + b^2$, cannot be factored using real numbers.

89. Have you ever seen the proof that 1 is equal to 2? Here it is.
Let $a = b$, then square both sides of the equation:

$$a^2 = b^2$$

$$a^2 = b \cdot b$$

$$a^2 = ab \qquad \text{Substitute } a = b.$$

$$a^2 - b^2 = ab - b^2 \qquad \text{Subtract } b^2 \text{ from both sides of the equation.}$$

$$(a + b)(a - b) = b(a - b) \qquad \text{Factor both sides of the equation.}$$

$$\frac{(a + b)\cancel{(a - b)}}{\cancel{(a - b)}} = \frac{b\cancel{(a - b)}}{\cancel{(a - b)}} \qquad \text{Divide both sides of the equation by } (a - b) \text{ and divide out common factors.}$$

$$a + b = b$$

$$b + b = b \qquad \text{Substitute } a = b.$$

$$2b = b$$

$$\frac{2\overset{1}{\cancel{b}}}{\underset{1}{\cancel{b}}} = \frac{\overset{1}{\cancel{b}}}{\underset{1}{\cancel{b}}} \qquad \text{Divide both sides of the equation by } b.$$

$$2 = 1$$

Obviously, $2 \neq 1$. Therefore, we must have made an error somewhere. Can you find it?

Factor each expression. Treat the unknown symbols as if they were variables.

90. ◆✸ + 2◆ + ☺✸ + 2☺

91. 2◆^{6} + 4◆^{4}✸^{2}

92. 4◆^{2}✸ − 6◆✸ − 20✸◆ + 30✸

Challenge Problems

93. Factor $x^6 + 1$.

94. Factor $x^6 - 27y^9$.

95. Factor $x^2 - 6x + 9 - 4y^2$. (*Hint:* Write the first three terms as the square of a binomial.)

96. Factor $x^2 + 10x + 25 - y^2 + 4y - 4$. (*Hint:* Group the first three terms and the last three terms.)

97. Factor $x^6 - y^6$. (*Hint:* Factor initially as the difference of two squares.)

Cumulative Review Exercises

[2.7] **98.** Solve the inequality $3x - 2(x + 4) \geq 2x - 9$, and graph the solution on a number line.

[3.1] **99.** Solve the formula $A = \frac{1}{2}h(b + d)$ for d.

[4.3] **100.** Find the slope of the line through the points $(4, -6)$ and $(-2, 3)$.

[6.1] **101.** Simplify $\left(\frac{4x^4y}{6xy^5}\right)^3$.

[6.2] **102.** Simplify $x^{-2}x^{-3}$.

7.6 SOLVING QUADRATIC EQUATIONS USING FACTORING

1) Recognize quadratic equations.

2) Solve quadratic equations using factoring.

3) Solve applications by factoring quadratic equations.

1) Recognize Quadratic Equations

In this section we introduce **quadratic equations**, which are equations that contain a second-degree term and no term of a higher degree.

Quadratic Equation

Quadratic equations have the form

$$ax^2 + bx + c = 0$$

where a, b, and c are real numbers, $a \neq 0$.

Examples of Quadratic Equations

$$x^2 + 4x - 12 = 0$$
$$2x^2 - 5x = 0$$
$$3x^2 - 2 = 0$$

Quadratic equations like these, in which one side of the equation is written in descending order of the variable and the other side of the equation is 0, are said to be in **standard form**.

Some quadratic equations can be solved by factoring. Two methods for solving quadratic equations that cannot be solved by factoring are given in Chapter 10. To solve a quadratic equation by factoring, we use the **zero-factor property**.

You know that if you multiply by 0, the product is 0. That is, if $a = 0$ or $b = 0$, then $ab = 0$. The reverse is also true. If a product equals 0, at least one of its factors must be 0.

Zero-Factor Property

If $ab = 0$, then $a = 0$ or $b = 0$.

We now illustrate how the zero-factor property is used in solving equations.

EXAMPLE 1 Solve the equation $(x + 3)(x + 4) = 0$.

Solution Since the product of the factors equals 0, according to the zero-factor property, one or both factors must equal 0. Set each factor equal to 0, and solve each resulting equation.

$$x + 3 = 0 \quad \text{or} \quad x + 4 = 0$$
$$x + 3 - 3 = 0 - 3 \qquad x + 4 - 4 = 0 - 4$$
$$x = -3 \qquad x = -4$$

Thus, if x is either -3 or -4, the product of the factors is 0. The solutions to the equation are -3 and -4.

CHECK: $x = -3$

$$(x + 3)(x + 4) = 0$$
$$(-3 + 3)(-3 + 4) \stackrel{?}{=} 0$$
$$0(1) \stackrel{?}{=} 0$$
$$0 = 0 \quad \textit{True}$$

$x = -4$

$$(x + 3)(x + 4) = 0$$
$$(-4 + 3)(-4 + 4) \stackrel{?}{=} 0$$
$$-1(0) \stackrel{?}{=} 0$$
$$0 = 0 \quad \textit{True}$$

EXAMPLE 2 Solve the equation $(4x - 3)(2x + 4) = 0$.

Solution Set each factor equal to 0 and solve for x.

$$4x - 3 = 0 \quad \text{or} \quad 2x + 4 = 0$$
$$4x = 3 \qquad 2x = -4$$
$$x = \frac{3}{4} \qquad x = -2$$

NOW TRY EXERCISE 11

The solutions to the equation are $\frac{3}{4}$ and -2.

2) Solve Quadratic Equations Using Factoring

Now we give a general procedure for solving quadratic equations using factoring.

To Solve a Quadratic Equation Using Factoring

1. Write the equation in standard form with the squared term having a positive coefficient. This will result in one side of the equation being 0.
2. Factor the side of the equation that is not 0.
3. Set each factor *containing a variable* equal to 0 and solve each equation.
4. Check each solution found in step 3 in the *original* equation.

EXAMPLE 3 Solve the equation $2x^2 = 12x$.

Solution To make the right side of the equation equal to 0, we subtract $12x$ from both sides of the equation. Then we factor out $2x$ from both terms. Why did we make the right side of the equation equal to 0 instead of the left side?

$$2x^2 = 12x$$
$$2x^2 - 12x = 12x - 12x$$
$$2x^2 - 12x = 0$$
$$2x(x - 6) = 0$$

Now set each factor equal to 0.

$$2x = 0 \quad \text{or} \quad x - 6 = 0$$
$$x = \frac{0}{2} \qquad\qquad x = 6$$
$$x = 0$$

The solutions to the quadratic equation are 0 and 6. Check by substituting $x = 0$, then $x = 6$ in $2x^2 = 12x$.

EXAMPLE 4 Solve the equation $x^2 + 10x + 28 = 4$.

Solution To make the right side of the equation equal to 0, we subtract 4 from both sides of the equation. Then we factor and solve.

$$x^2 + 10x + 24 = 0$$
$$(x + 4)(x + 6) = 0$$
$$x + 4 = 0 \quad \text{or} \quad x + 6 = 0$$
$$x = -4 \qquad\qquad x = -6$$

NOW TRY EXERCISE 23

The solutions are -4 and -6. A check will show that both values satisfy the equation $x^2 + 10x + 28 = 4$.

EXAMPLE 5 Solve the equation $3x^2 + 2x - 12 = -7x$.

Solution Since all terms are not on the same side of the equation, add $7x$ to both sides of the equation.

$$3x^2 + 9x - 12 = 0$$

Factor out the common factor.

$$3(x^2 + 3x - 4) = 0$$

Factor the remaining trinomial.

$$3(x + 4)(x - 1) = 0$$

Now solve for x.

$$x + 4 = 0 \quad \text{or} \quad x - 1 = 0$$
$$x = -4 \qquad\qquad x = 1$$

Since 3 is a factor that does not contain a variable, we do not set it equal to zero. The solutions to the quadratic equation are -4 and 1.

EXAMPLE 6 Solve the equation $-x^2 + 5x + 6 = 0$.

Solution When the squared term is negative, we generally make it positive by multiplying both sides of the equation by -1.

$$-1(-x^2 + 5x + 6) = -1 \cdot 0$$
$$x^2 - 5x - 6 = 0$$

Note that the sign of each term on the left side of the equation changed and that the right side of the equation remained zero. Why? Now proceed as before.

$$x^2 - 5x - 6 = 0$$
$$(x - 6)(x + 1) = 0$$
$$x - 6 = 0 \quad \text{or} \quad x + 1 = 0$$
$$x = 6 \qquad\qquad x = -1$$

NOW TRY EXERCISE 33

A check using the original equation will show that the solutions are 6 and -1.

Avoiding Common Errors

Be careful not to confuse factoring a polynomial with using factoring as a method to solve an equation.

CORRECT

Factor: $x^2 + 3x + 2$

$(x + 2)(x + 1)$

INCORRECT

Factor: $x^2 + 3x + 2$

$(x + 2)(x + 1)$

~~$x + 2 = 0$ or $x + 1 = 0$~~

~~$x = -2$ $\quad$ $x = -1$~~

Do you know what is wrong with the example on the right? The expression $x^2 + 3x + 2$ is a polynomial (a trinomial), not an equation. Since it is not an equation, it cannot be solved. When you are given a polynomial, you cannot just include "= 0" to change it to an equation.

CORRECT

Solve: $x^2 + 3x + 2 = 0$

$$(x + 2)(x + 1) = 0$$
$$x + 2 = 0 \quad \text{or} \quad x + 1 = 0$$
$$x = -2 \qquad\qquad x = -1$$

EXAMPLE 7 Solve the equation $x^2 = 25$.

Solution Subtract 25 from both sides of the equation; then factor using the difference of two squares formula.

$$x^2 - 25 = 0$$
$$(x + 5)(x - 5) = 0$$
$$x + 5 = 0 \quad \text{or} \quad x - 5 = 0$$
$$x = -5 \qquad\qquad x = 5$$

The graph of a quadratic equation in two variables is called a **parabola**. In Chapter 10 we will learn how to graph parabolas. Although we have not yet studied parabolas, we do know how to make a table of values and how to determine x-intercepts. In the Using Your Graphing Calculator box that follows, we use a calculator to find the solutions to a quadratic equation using a table of values and using the x-intercepts of a parabola.

Using Your Graphing Calculator

If you solve the equation $x^2 - x - 6 = 0$ by factoring, you should obtain the solutions −2 and 3. Solve the equation now. The solutions to this equation can be found on a graphing calculator in a number of ways. If we replace the 0 with y we obtain $y = x^2 - x - 6$. Figure 7.4 shows the graph of $Y_1 = x^2 - x - 6$ using the standard window settings.

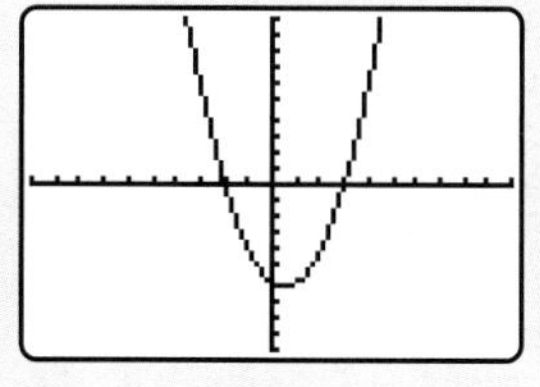

FIGURE 7.4

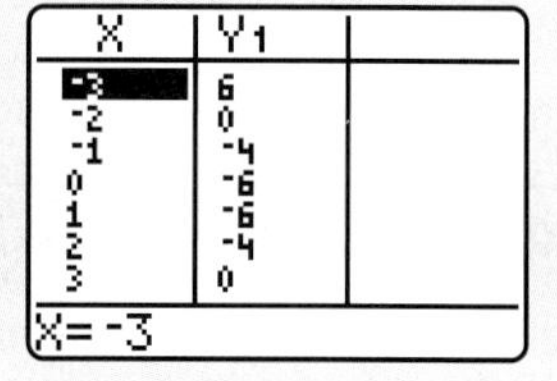

FIGURE 7.5

Some graphing calculators have a TABLE feature which shows a table of values On the TI-82 and TI-83 press [2nd] [GRAPH] to use the TABLE feature. In the Using Your Graphing Calculator Box on page 249, we explained how to use the TABLE feature. Please re-read that material now and notice how to change the table settings. From the graph, observe that both x-intercepts occur to the right of $x = -3$. We will therefore set TblStart = −3. We will let ΔTbl = 1. After you get the table, you can scroll up and down the table using the up and down arrows. Figure 7.5 shows a table of values for $Y_1 = x^2 - x - 6$. Note that $Y_1 = 0$ when X = −2 and X = 3. These values are the solution to the equation $x^2 - x - 6 = 0$. The values −2 and 3 are also called **zeros** (or **roots**), for when they are substituted for x in $y = x^2 - x - 6$, y has a value of zero. One drawback to using the TABLE feature to find the zeros is that if you do not use the appropriate table settings, you may not find the zeroes. Observe the TABLE when we use ΔTbl = 2, see Figure 7.6. Note that the value of Y_1 is zero only when X = 3. The other zero is not displayed because only odd values of x are shown in the table. If an x-intercept is not at an integer value, then it will not show up in a table if ΔTbl = 1. Can you explain why? If this happens you can change your table setting, or use the CALC menu, as explained on the right.

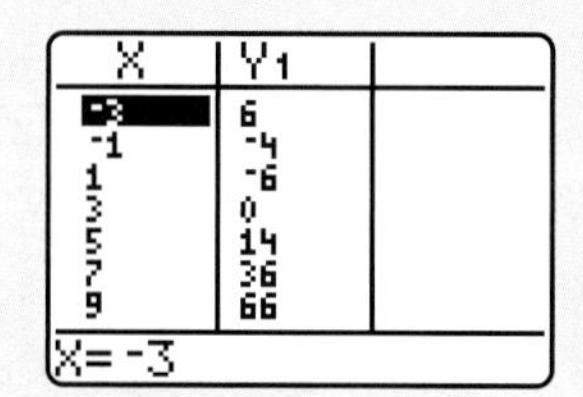

FIGURE 7.6

FIGURE 7.7

Another method that can be used to find the solution to $x^2 - x - 6 = 0$ is to graph $y = x^2 - x - 6$ and use the CALC (which stands for calculate) menu to find the x-intercepts. On the TI-82 and TI-83 you press [2nd] [TRACE] to get to the CALC menu. The CALC menu of a TI-83 is shown in Figure 7.7. The CALC menu for a TI-82 is identical, except in option 2: the word *root* is used instead of *zero*.

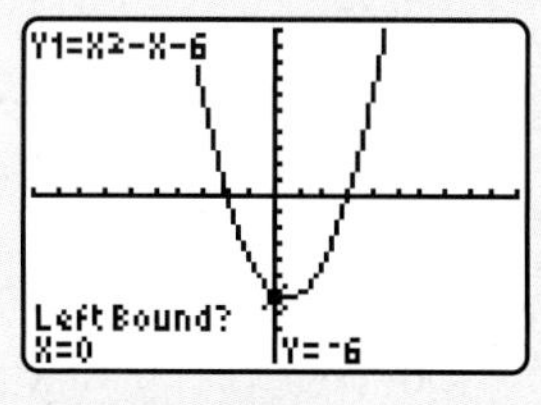

FIGURE 7.8

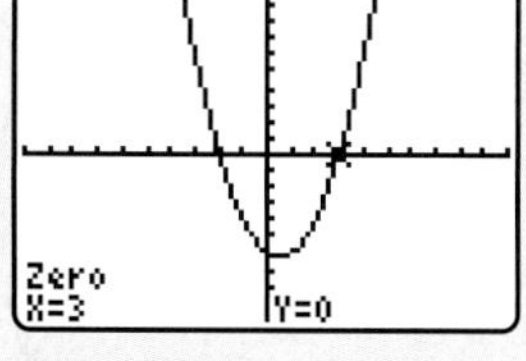

FIGURE 7.9

Scroll down to option 2: zero and press [ENTER] which then gives the screen in Figure 7.8. Under the words *Left Bound?** you see $x = 0$. Use the arrow key to move the cursor to the right until it is just to the left of the positive x-intercept (this puts the cursor slightly below the x-axis). Note that the value of Y is negative. Press [ENTER]. The screen now shows *Right Bound?* Move the cursor so it is slightly to the right of the x-intercept (this puts the cursor slightly above the x-axis), then press [ENTER]. Note that the sign of the value of Y has changed, and is positive. The screen now shows *Guess?* Press [ENTER] again and the screen shows that the zero is 3, see Figure 7.9. That is, when $x = 3$, $y = 0$. If you make a mistake, clear your calculator, re-enter $y_1 = x^2 - x - 6$, and start the procedure again.

continued on next page

You can also use this procedure to determine the other zero, or the other solution to $x^2 - x - 6 = 0$. To obtain the other zero, press [2nd] [TRACE] to get to the CALC menu again. Then using the left arrow, move the cursor from its current position to just left of the other x-intercept. In this case, it will be slightly *above* the x-axis, and the value of Y will be *positive*. Press [ENTER] to get *Left Bound*? Finish the procedure by moving the cursor right until it is slightly *below* the x-axis. Press [ENTER] to get the *Right Bound*? and once more to get *Guess*? The screen shows that the zero is −2.

Exercises

Use your graphing calculator to determine the solutions to each of the following quadratic equations. You may have to adjust the window settings to see the entire graph,

1. $x^2 - 2x - 8 = 0$
2. $x^2 - 7x + 10 = 0$
3. $2x^2 - 6x + 4 = 0$
4. $2x^2 - 8x - 10 = 0$

* On the TI-82, the calculator displays *Lower Bound* and *Upper Bound* rather than *Left Bound* and *Right Bound*.

3) Solve Applications by Factoring Quadratic Equations

Now let's solve some applied problems using quadratic equations.

EXAMPLE 8 The product of two numbers is 66. Find the two numbers if one number is 5 more than the other.

Solution **Understand and Translate** Our goal is to find the two numbers.

Let x = smaller number

$x + 5$ = larger number

$$x(x + 5) = 66$$

Carry Out

$$x^2 + 5x = 66$$
$$x^2 + 5x - 66 = 0$$
$$(x - 6)(x + 11) = 0$$
$$x - 6 = 0 \quad \text{or} \quad x + 11 = 0$$
$$x = 6 \qquad\qquad x = -11$$

Remember that x represents the smaller of the two numbers. This problem has two possible solutions.

	Solution 1	Solution 2
Smaller number	6	−11
Larger number	$x + 5 = 6 + 5 = 11$	$x + 5 = -11 + 5 = -6$

Thus the two possible solutions are 6 and 11, and −11 and −6.

CHECK:	6 AND 11	−11 AND −6
Product of the two numbers is 66.	$6 \cdot 11 = 66$	$(-11)(-6) = 66$
One number is 5 more than the other number.	11 is 5 more than 6.	−6 is 5 more than −11.

Answer One solution is: smaller number 6, larger number 11. A second solution is: smaller number −11, larger number −6. You must give both solutions. If the question had stated "the product of two *positive* numbers is 66," the only solution would be 6 and 11.

NOW TRY EXERCISE 65

EXAMPLE 9 The marketing department of a large publishing house is planning to make a large rectangular sign to advertise a new book at a convention. They want the length of the sign to be 3 feet longer than the width (Fig. 7.10). Signs at the convention may have a maximum area of 54 square feet. Find the length and width of the sign if the area is to be 54 square feet.

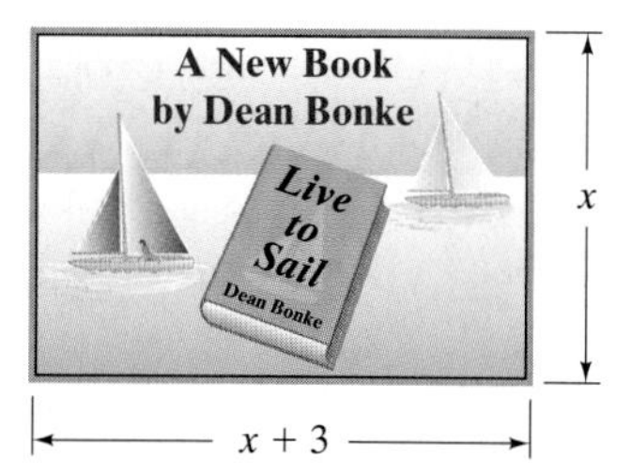

FIGURE 7.10

Solution **Understand and Translate** We need to find the length and width of the sign. We will use the formula for the area of a rectangle.

$$\text{Let } x = \text{width}$$
$$x + 3 = \text{length}$$
$$\text{area} = \text{length} \cdot \text{width}$$
$$54 = (x + 3)x$$

Carry Out

$$54 = x^2 + 3x$$
$$0 = x^2 + 3x - 54$$
$$\text{or} \quad x^2 + 3x - 54 = 0$$
$$(x - 6)(x + 9) = 0$$
$$x - 6 = 0 \quad \text{or} \quad x + 9 = 0$$
$$x = 6 \qquad\qquad x = -9$$

Check and Answer Since the width of the sign cannot be a negative number, the only solution is

$$\text{width} = x = 6 \text{ feet} \qquad \text{length} = x + 3 = 6 + 3 = 9 \text{ feet}$$

The area, length · width, is 54 square feet, and the length is 3 feet more than the width, so the answer checks.

NOW TRY EXERCISE 69

EXAMPLE 10 In Earth's gravitational field, the distance, d, in feet, that an object released at rest falls after t seconds can be found by the formula $d = 16t^2$. While waiting for the chairlift to resume its climb to the top of a mountain, a skier adjusting his gear drops his ski pole. How long does it take the ski pole to reach the ground, 64 feet below?

Solution **Understand and Translate** Substitute 64 for d in the formula and then solve for t.

$$d = 16t^2$$
$$64 = 16t^2$$

Carry Out

$$\frac{64}{16} = t^2$$
$$4 = t^2$$

Now subtract 4 from both sides of the equation and write the equation with 0 on the right side to put the quadratic equation in standard form.

$$4 - 4 = t^2 - 4$$
$$0 = t^2 - 4$$
$$\text{or} \quad t^2 - 4 = 0$$
$$(t + 2)(t - 2) = 0$$
$$t + 2 = 0 \quad \text{or} \quad t - 2 = 0$$
$$t = -2 \qquad\qquad t = 2$$

Check and Answer Since t represents the number of seconds, it must be a positive number. Thus, the only possible answer is 2 seconds. It takes 2 seconds for the ski pole (or any other object falling under the influence of gravity) to fall 64 feet.

Exercise Set 7.6

Concept/Writing Exercises

1. In your own words explain the zero-factor property.
2. What is a quadratic equation?
3. What is the standard form of a quadratic equation?
4. Create a quadratic equation whose solutions are 6 and 3. Explain how you determined your answer.
5. a) When solving the equation $(x + 1)(x - 2) = 4$, explain why we **cannot** solve the equation by first writing $x + 1 = 4$ or $x - 2 = 4$ and then solving each equation for x.
   b) Solve the equation $(x + 1)(x - 2) = 4$.
6. When solving an equation such as $3(x - 4)(x + 5) = 0$, we set the factors $x - 4$ and $x + 5$ equal to 0, but we do not set the 3 equal to 0. Can you explain why?

Practice the Skills

Solve.

7. $x(x + 1) = 0$
8. $3x(x + 4) = 0$
9. $7x(x - 8) = 0$
10. $(x + 3)(x + 5) = 0$
11. $(2x + 5)(x - 3) = 0$
12. $(2x + 3)(x - 5) = 0$
13. $x^2 - 16 = 0$
14. $x^2 - 25 = 0$
15. $x^2 - 12x = 0$
16. $x^2 + 7x = 0$
17. $9x^2 + 27x = 0$
18. $x^2 + 6x + 5 = 0$
19. $x^2 + x - 20 = 0$
20. $x^2 + 6x + 9 = 0$
21. $x^2 + 12x = -20$
22. $3y^2 - 2 = -y$
23. $z^2 + 3z = 18$
24. $3x^2 = -21x - 18$
25. $3x^2 + 6x - 24 = 0$
26. $x^2 = 4x + 21$
27. $x^2 + 19x = 42$
28. $3x^2 - 9x - 30 = 0$
29. $2y^2 + 22y + 60 = 0$
30. $w^2 + 45 + 18w = 0$
31. $-2x - 15 = -x^2$
32. $-9x + 20 = -x^2$
33. $-x^2 + 29x + 30 = 0$
34. $-y^2 + 12y - 11 = 0$
35. $x^2 - 3x - 18 = 0$
36. $z^2 - 8z = -16$
37. $3p^2 = 22p - 7$
38. $2x^2 - 5 = 3x$
39. $3r^2 + r = 2$
40. $3x^2 = 7x + 20$
41. $4x^2 + 4x - 48 = 0$
42. $6x^2 + 13x + 6 = 0$
43. $8x^2 + 2x = 3$
44. $2x^2 + 4x - 6 = 0$
45. $2x^2 - 10x = -12$
46. $x^2 - 49 = 0$
47. $2x^2 = 32x$
48. $4x^2 - 25 = 0$
49. $x^2 = 100$
50. $2x^2 - 50 = 0$
51. $x^2 = 9$
52. $x^2 = \frac{25}{4}$
53. $-t^2 = -81$
54. $-6r^2 = 5r - 6$

55. A video store owner finds that her daily profit, P, is approximated by the formula $P = x^2 - 15x - 50$, where x is the number of videos she sells. How many videos must she sell in a day for her profit to be $400?

56. The cost, C, for manufacturing x water sprinklers is given by the formula $C = x^2 - 27x - 20$. Determine the number of water sprinklers manufactured at a cost of $70.

57. The sum, s, of the first n even numbers is given by the formula $s = n^2 + n$. Determine n for the given sums:

a) $s = 20$ **b)** $s = 90$

58. For a switchboard that handles n telephone lines, the maximum number of telephone connections, C, that it can make simultaneously is given by the formula

$$C = \frac{n(n-1)}{2}.$$

a) How many telephone connections can a switchboard make simultaneously if it handles 15 lines?

b) How many lines does a switchboard have if it can make 55 telephone connections simultaneously?

Problem Solving

59. Create a quadratic equation whose only solution is 6. Explain how you determined your answer.

60. Create a quadratic equation whose solutions are $\frac{1}{2}$ and $\frac{2}{3}$. Explain how you determined your answer.

61. If each side of a square is increased by 4 meters, the area becomes 49 square meters. Determine the length of a side of the original square.

62. If the length of the sign in Example 9 is to be 2 feet longer than the width and the area is to be 35 square feet, determine the dimensions of the sign.

63. How long would it take for an egg dropped from a helicopter to fall 256 feet to the ground? See Example 10.

Express each problem as an equation, then solve.

64. The product of two consecutive positive odd integers is 63. Determine the two integers.

65. The product of two positive integers is 117. Determine the two numbers if one is 4 more than the other.

66. The product of two positive integers is 64. Determine the two integers if one number is 4 times the other.

67. The area of a rectangle is 36 square feet. Determine the length and width if the length is 4 times the width.

68. The area of a rectangle is 84 square inches. Determine the length and width if the length is 2 inches less than twice the width.

69. Corey Good wants to make a rectangular garden whose width is 2/3 its length. He has enough fertilizer to cover 150 square feet. What dimensions, should Corey's garden be?

70. Shameeya Wilson wishes to purchase a wallpaper border to go along the top of one wall in her living room. The length of the wall is 7 feet greater than its height.

a) Find the length and height of the wall if the area of the wall is 120 square feet.

b) What is the length of the border she will need?

c) If the border cost $4 per linear foot, how much will the border cost?

71. Historians found the following recorded in the Berlin Papyrus, one of the writings of the Middle Kingdom (c. 2160–1700 B.C.) of early Egypt. "Divide 100 square measures into two squares such that the side of one of the squares shall be three-fourths the side of the other." This problem can be stated mathematically as a quadratic equation and its solution represented the first known solution to a quadratic equation. Today, the problem presented by the early Egyptians might be written: The length of the side of one square is 3/4 the side of a second square. The sum of their areas is 100 square cm. Determine the length of the sides of each square.

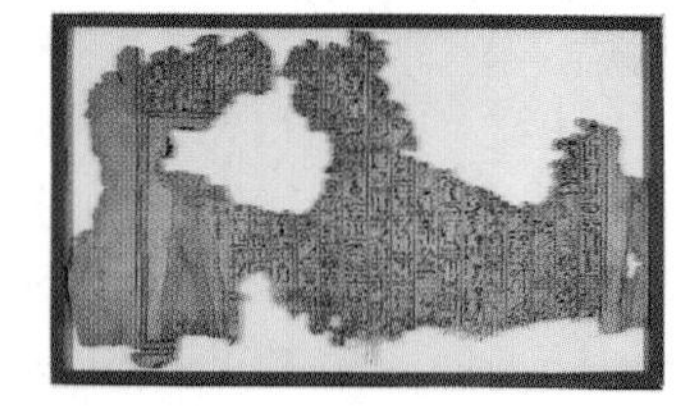

Challenge Problems

72. Solve the equation $x^3 + 3x^2 - 10x = 0$.

73. Create an equation whose solutions are 0, 3, and 5. Explain how you determined your answer.

74. The product of two numbers is -40. Determine the numbers if their sum is 3.

75. The sum of two numbers is 9. The sum of the squares of the two numbers is 45. Determine the two numbers.

Group Activity

Discuss and solve Exercises 76–78 in groups.

76. The break-even point for a manufacturer occurs when its cost of production, C, is equal to its revenue, R. The cost equation for a company is $C = 2x^2 - 20x + 600$ and its revenue equation is $R = x^2 + 50x - 400$, where x is the number of units produced and sold. How many units must be produced and sold for the manufacturer to break even? There are two values.

77. When a certain cannon is fired, the height, in feet, of the cannonball at time t can be found by using the formula $h = -16t^2 + 128t$.

a) Determine the height of the cannonball 3 seconds after being fired.

b) Determine the time it takes for the cannonball to hit the ground. (*Hint:* What is the value of h at impact?)

78. **a)** Write a word problem in a form that others can understand and that can be solved using a quadratic equation.

b) Complete the solution on a separate piece of paper. Use the problem-solving technique.

c) Exchange the word problem your group wrote in part **a)** with that of another group.

d) Solve the word problem that was given to you by the other group.

c) Compare your group's answer with the answer of the group who wrote the problem. If your group obtained a different answer, explain why.

Cumulative Review Exercises

[4.3] 79. Explain why the slope of a vertical line is undefined.

[6.4] 80. Subtract $x^2 - 4x + 6$ from $3x + 2$.

[6.5] 81. Multiply $(3x^2 + 2x - 4)(2x - 1)$.

[6.6] 82. Divide $\frac{6x^2 - 19x + 15}{3x - 5}$ by dividing the numerator by the denominator.

[7.4] 83. Divide $\frac{6x^2 - 19x + 15}{3x - 5}$ by factoring the numerator and dividing out common factors.

SUMMARY

Key Words and Phrases

7.1
Composite number
Factor an expression
Factors
Greatest common factor
Prime factors
Prime number
Unit

7.2
Factor by grouping

7.3
Factor out the common factor
Trial and error method

7.4
Factor by grouping
Factor by trial and error

7.5
Difference of two squares
Difference of two cubes
Sum of two cubes

7.6
Quadratic equation
Standard form of a quadratic equation
Zero-factor property

IMPORTANT FACTS

Difference of two squares:

$a^2 - b^2 = (a + b)(a - b)$

Note: The sum of two squares, $a^2 + b^2$, cannot be factored using real numbers.

Sum of two cubes:

$a^3 + b^3 = (a + b)(a^2 - ab + b^2)$

Difference of two cubes:

$a^3 - b^3 = (a - b)(a^2 + ab + b^2)$

Zero-factor property:

If $a \cdot b = 0$, then $a = 0$ or $b = 0$.

General Procedure to Factor a Polynomial

1. If all the terms of the polynomial have a greatest common factor other than 1, factor it out.
2. If the polynomial has two terms, determine whether it is a difference of two squares or a sum or difference of two cubes. If so, factor using the appropriate formula.
3. If the polynomial has three terms, factor the trinomial using the methods discussed in Sections 7.3 and 7.4.
4. If the polynomial has more than three terms, try factoring by grouping.
5. As a final step, examine your factored polynomial to determine whether the terms in any factor have a common factor. If you find a common factor, factor it out at this point.

Review Exercises

[7.1] *Find the greatest common factor for each set of terms.*

1. $2y^2, y^4, y^3$
2. $3p, 6p^2, 9p^3$
3. $20a^3, 15a^2, 35a^5$
4. $40x^2y^3, 36x^3y^4, 16x^5y^2z$
5. $9xyz, 12xz, 36, x^2y$
6. $4ab, 12a^2b, 16, a^2b^2$
7. $4(x - 5), x - 5$
8. $x(x + 5), x + 5$

Factor each expression. If an expression cannot be factored, so state.

9. $3x - 9$
10. $12x + 6$
11. $24y^2 - 4y$
12. $55p^3 - 20p^2$
13. $48a^2b - 36ab^2$
14. $6xy - 12x^2y$
15. $60x^4y^2 + 6x^9y^3 - 18x^5y^2$
16. $24x^2 - 13y^2 + 6xy$
17. $14a^2b - 7b - a^3$
18. $x(5x + 3) - 2(5x + 3)$
19. $3x(x - 1) - 2(x - 1)$
20. $2x(4x - 3) + 4x - 3$

[7.2] *Factor by grouping.*

21. $x^2 + 5x + 2x + 10$
22. $x^2 - 3x + 4x - 12$
23. $y^2 - 7y - 7y + 49$
24. $3a^2 - 3ab - a + b$
25. $3xy + 3x + 2y + 2$
26. $x^2 + 3x - 2xy - 6y$
27. $5x^2 + 20x - x - 4$
28. $5x^2 - xy + 20xy - 4y^2$
29. $4x^2 + 12xy - 5xy - 15y^2$
30. $6a^2 - 10ab - 3ab + 5b^2$
31. $ab - a + b - 1$
32. $3x^2 - 9xy + 2xy - 6y^2$
33. $7a^2 + 14ab - ab - 2b^2$
34. $6x^2 + 9x - 2x - 3$

[7.3] *Factor completely. If an expression cannot be factored, so state.*

35. $x^2 - 7x - 8$
36. $x^2 + 8x - 15$
37. $x^2 - 13x + 42$
38. $x^2 + x - 20$
39. $x^2 + 11x + 30$
40. $x^2 - 15x + 56$
41. $x^2 - 12x - 44$
42. $x^2 + 11x - 24$
43. $x^3 - 17x^2 + 72x$
44. $x^3 - 3x^2 - 40x$
45. $x^2 - 2xy - 15y^2$
46. $4x^3 + 32x^2y + 60xy^2$

[7.4] *Factor completely. If an expression cannot be factored, so state.*

47. $2x^2 + 11x + 12$
48. $3x^2 - 13x + 4$
49. $4x^2 - 9x + 5$
50. $5x^2 - 13x - 6$
51. $9x^2 + 3x - 2$
52. $5x^2 - 32x + 12$
53. $3x^2 + 13x + 8$
54. $6x^2 + 31x + 5$
55. $5x^2 + 37x - 24$
56. $6x^2 + 11x - 10$
57. $8x^2 - 18x - 35$
58. $9x^2 - 6x + 1$
59. $9x^3 - 12x^2 + 4x$
60. $18x^3 - 24x^2 - 10x$
61. $16a^2 - 22ab - 3b^2$
62. $4a^2 - 16ab + 15b^2$

[7.5] *Factor completely.*

63. $x^2 - 25$
64. $x^2 - 100$
65. $4x^2 - 16$
66. $81x^2 - 9y^2$
67. $49 - x^2$
68. $64 - x^2$
69. $16x^4 - 49y^2$
70. $100x^4 - 121y^4$
71. $x^3 - y^3$
72. $x^3 + y^3$
73. $x^3 - 1$
74. $x^3 + 8$
75. $a^3 + 27$
76. $x^3 - 8$
77. $125a^3 + b^3$
78. $27 - 8y^3$
79. $27x^4 - 75y^2$
80. $2x^3 - 128y^3$

[7.1–7.5] *Factor completely.*

81. $x^2 - 16x + 60$
82. $3x^2 - 18x + 27$
83. $4a^2 - 64$
84. $4y^2 - 36$
85. $8x^2 + 16x - 24$
86. $x^2 - 6x - 27$
87. $9x^2 + 3x - 2$
88. $4x^2 + 7x - 2$
89. $8x^3 - 8$
90. $x^3y - 27y$
91. $a^2b - 2ab - 15b$
92. $6x^3 + 30x^2 + 9x^2 + 45x$
93. $x^2 + 5xy + 6y^2$
94. $2x^2 - xy - 10y^2$
95. $4x^2 - 20xy + 25y^2$
96. $25a^2 - 49b^2$
97. $xy - 7x + 2y - 14$
98. $16y^5 - 25y^7$
99. $2x^3 + 12x^2y + 16xy^2$
100. $6x^2 + 5xy - 21y^2$
101. $16x^4 - 8x^3 - 3x^2$
102. $a^4 - 1$

[7.6] *Solve.*

103. $x(x + 6) = 0$
104. $(x - 2)(x + 8) = 0$
105. $(x + 5)(4x - 3) = 0$
106. $x^2 - 3x = 0$
107. $5x^2 + 20x = 0$
108. $7x^2 + 14x = 0$
109. $x^2 + 7x + 10 = 0$
110. $x^2 - 12 = -x$
111. $x^2 - 3x = -2$
112. $3x^2 + 15x + 12 = 0$
113. $x^2 - 6x + 8 = 0$
114. $6x^2 + 6x - 12 = 0$
115. $8x^2 - 3 = -10x$
116. $2x^2 + 15x = 8$
117. $4x^2 - 16 = 0$
118. $49x^2 - 100 = 0$
119. $8x^2 - 14x + 3 = 0$
120. $-48x = -12x^2 - 45$

[7.6] *Express each problem as an equation, then solve.*

121. The product of two consecutive positive integers is 156. Determine the two integers.

122. The product of two consecutive positive even integers is 48. Determine the two integers.

123. The product of two positive integers is 56. Determine the integers if the larger is 6 more than twice the smaller.

124. The area of a rectangle is 63 square feet. Determine the length and width of the rectangle if the length is 2 feet greater than the width.

125. The length of each side of a square is made smaller by 4 inches. If the area of the resulting square is 25 square inches, determine the length of a side of the original square.

126. The Pine Hills Neighborhood Association has determined that the cost, C, to make x dozen cookies can be estimated by the formula $C = x^2 - 79x + 20$. If they have \$100 to be used to make the cookies, how many dozen cookies can the association make to sell at a fund raiser?

Practice Test

1. Determine the greatest common factor of $3y^2$, $15y^3$, and $27y^4$.
2. Determine the greatest common factor of $6x^2y^3$, $9xy^2$, and $12xy^5$.

Factor completely.

3. $10x^3y^2 - 8xy$
4. $8a^3b - 12a^2b^2 + 28a^2b$
5. $x^2 - 8x + 2x - 16$
6. $5x^2 - 15x + 2x - 6$
7. $a^2 - 4ab - 5ab + 20b^2$
8. $x^2 + 12x + 32$
9. $x^2 + 5x - 24$
10. $25a^2 - 5ab - 6b^2$
11. $4x^2 - 16x - 48$
12. $2x^3 - 3x^2 + x$
13. $12x^2 - xy - 6y^2$
14. $x^2 - 9y^2$
15. $x^3 + 27$

Solve.

16. $(5x - 3)(x - 1) = 0$
17. $x^2 - 6x = 0$
18. $x^2 = 64$
19. $x^2 - 14x + 49 = 0$
20. $x^2 + 6 = -5x$
21. $x^2 - 7x + 12 = 0$

Solve.

22. The product of two positive integers is 36. Determine the two integers if the larger is 1 more than twice the smaller.
23. The product of two positive consecutive odd integers is 99. Determine the integers.
24. The area of a rectangle is 24 square meters. Determine the length and width of the rectangle if its length is 2 meters greater than its width.
25. How long would it take for an object dropped from a hot air balloon to fall 1600 feet to the ground?

Cumulative Review Test

1. Evaluate $3 - 7(x + 4x^2 - 21)$ when $x = -5$.
2. Evaluate $5x^2 - 3y + 7(2 + y^2 - 4x)$ when $x = 3$ and $y = -2$.
3. According to the U.S. Bureau of the Census, by the year 2010, 13.2% of the total population will be age 65 or older. Using the information provided in the table, estimate the U.S. population in 2010.

U.S. Population, Age 65 and Older (number in thousands)

Census Date	Number*	% of Total Population
1900	3,080	4.1%
1910	3,949	4.3
1920	4,933	4.7
1930	6,634	5.4
1940	9,019	6.8
1950	12,269	8.1
1960	16,560	9.2
1970	19,980	9.8
1980	25,550	11.3
1990	31,235	12.5
2000	34,709	12.6
2010	39,408	13.2
2020	53,220	16.5
2030	69,379	20.0
2040	75,233	20.3
2050	78,859	20.0

Source: U.S. Bureau of the Census

*Numbers for 2000 and later are projections.

4. Solve the equation $3x - 4 = 5(x - 7) + 2x$ for x.
5. Solve the inequality $3x - 5 \geq 10(6 - x)$, and graph the solution a number line.
6. Solve the equation $2x - 3y = 6$ for y.
7. Two cross-country skiers follow the same trail in a local park. Brooke Stoner skies at a rate of 8 kilometers per hour and Bob Thoresen skies at a rate of 4 kilometers per hour. How long will it take Brooke to catch Bob if she leaves 15 minutes after he does?

8. In determining where to send an employee for a multi-day training seminar, a San Francisco company considers airfare and per diem business travel costs.** Business-class round trip airfare is \$1370 to Orlando and \$919 to New York City. How many days would the conference have to last for the total cost of the conference to be less expensive in Orlando than New York

(Footnote appears on page 451.)

City? (Assume that the conference registration fees are equal.)

1997/1998 Per Diem Business Travel Costs

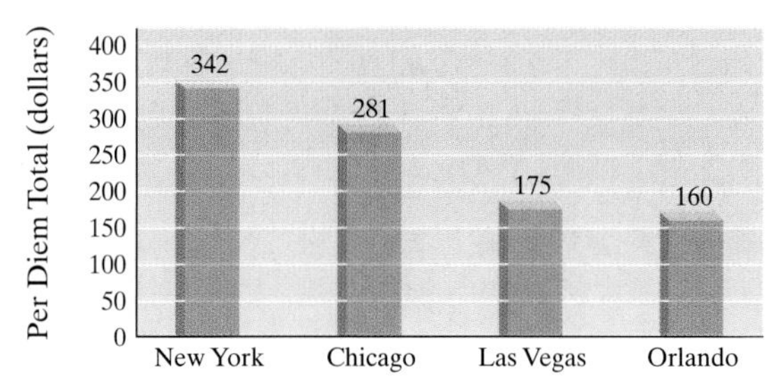

Data Source: The Wall Street Journal Almanac, 1999

9. Determine the x- and y-intercepts of $4x + 6y = 24$.

10. Determine the slope of the line through $(-3, 5)$ and $(6, -8)$.

11. Write an equation of the line, in slope–intercept form, whose graph goes through $(4, -3)$ and has a slope of $\frac{1}{2}$.

12. Graph $y < 2x - 3$.

13. Chris Gill is 6 years older than his sister, Mallory. One year ago, Chris was twice as old as Mallory. How old is Mallory now?

14. Determine the solution to the following system of equations graphically.

$$y = x$$
$$x = 5$$

15. Solve the following system of equations.

$$y = x + 1$$
$$y - 3x = -1$$

16. Simplify $\left(\frac{3x}{2y}\right)^3$.

17. Simplify $(2x^{-3})^{-2}(4x^{-3}y^2)^3$

18. Subtract $(2x^2 - 3x + 7)$ from $(x^3 - x^2 + 6x - 5)$.

19. Factor $x^2 - x - 110$.

20. Factor $5x^3 - 125x$.

** The per diem totals represent average costs for the typical business traveler, and include breakfast, lunch, and dinner in business-class restaurants and single-rate lodging in business-class hotels and motels.

RATIONAL EXPRESSIONS AND EQUATIONS

8

Use the Angel Web site at www.prenhall.com/angel to be linked to an internet resource that will help you further explore the following application.

Technological advances in transportation in the past 50 years have changed the way people travel and have significantly decreased traveling time. Rail travelers in Japan have benefited from the country's commitment to high-speed railroads. On page 503, we use a rational equation to determine the distance between two railroad stations by comparing a traveler's time riding one of Japan's new bullet trains to the time the traveler would have spent making the same trip on an older train.

Preview and Perspective

You worked with rational numbers when you worked with fractions in arithmetic. Now you will expand your knowledge to include fractions that contain variables. Fractions that contain variables are often referred to as rational expressions. The same basic procedures you used to simplify (or reduce), add, subtract, multiply, and divide arithmetic fractions will be used with rational expressions in Sections 8.1 through 8.5. You might wish to review Section 1.3 since the material presented in this chapter builds upon the procedures presented there.

Many real-life applications involve equations that contain rational expressions. Such equations are called rational equations. We will solve rational equations in Section 8.6 and give various real-life applications of rational equations, including the use of several familiar formulas, in Sections 8.6 and 8.7.

To be successful in this chapter you need to have a complete understanding of factoring, which was presented in Chapter 7. Sections 7.3 and 7.4 are especially important.

8.1 SIMPLIFYING RATIONAL EXPRESSIONS

SSM VIDEO 8.1 CD Rom

1) **Determine the values for which a rational expression is defined.**
2) **Understand the three signs of a fraction.**
3) **Simplify rational expressions.**
4) **Factor a negative 1 from a polynomial.**

1) Determine the Values for Which a Rational Expression Is Defined

A **rational expression** is an expression of the form p/q, where p and q are polynomials and $q \neq 0$.

Examples of Rational Expressions

$$\frac{4}{5}, \quad \frac{x-6}{x}, \quad \frac{x^2+2x}{x-3}, \quad \frac{x}{x^2-4}$$

The denominator of a rational expression cannot equal 0 since division by 0 is not defined. In the expression $(x+3)/x$, the value of x cannot be 0 because the denominator would then equal 0. We say that the expression $(x+3)/x$ is *defined* for all real numbers except 0. It is *undefined* when x is 0. In $(x^2+4x)/(x-3)$, the value of x cannot be 3 because the denominator would then be 0. What values of x cannot be used in the expression $x/(x^2-4)$? If you answered 2 and -2, you answered correctly. **Whenever we have a rational expression containing a variable in the denominator, we always assume that the value or values of the variable that make the denominator 0 are excluded.**

One method that can be used to determine the value or values of the variable that are excluded is to set the denominator equal to 0 and then solve the resulting equation for the variable.

EXAMPLE 1 Determine the value or values of the variable for which the rational expression is defined.

a) $\dfrac{x+3}{2x-5}$ **b)** $\dfrac{x+3}{x^2+6x-7}$

Solution **a)** We need to determine the value or values of x that make $2x - 5$ equal to 0 and exclude these. We can do this by setting $2x - 5$ equal to 0 and solving the equation for x.

$$2x - 5 = 0$$
$$2x = 5$$
$$x = \frac{5}{2}$$

Thus, we do not consider $x = \frac{5}{2}$ when we consider the rational expression $(x + 3)/(2x - 5)$. This expression is defined for all real numbers except $x = \frac{5}{2}$. We will sometimes shorten our answer and write $x \neq \frac{5}{2}$.

b) To determine the value or values that are excluded, we set the denominator equal to zero and solve the equation for the variable.

$$x^2 + 6x - 7$$
$$(x + 7)(x - 1) = 0$$
$$x + 7 = 0 \quad \text{or} \quad x - 1 = 0$$
$$x = -7 \qquad\qquad x = 1$$

Therefore, we do not consider the values $x = -7$ or $x = 1$ when we consider the rational expression $(x + 3)/(x^2 + 6x - 7)$. Both $x = -7$ and $x = 1$ make the denominator zero. This expression is defined for all real numbers except $x = -7$ and $x = 1$. Thus, $x \neq -7$ and $x \neq 1$.

NOW TRY EXERCISE 19

2) Understand the Three Signs of a Fraction

Three **signs** are associated with any fraction: the sign of the numerator, the sign of the denominator, and the sign of the fraction.

Sign of numerator

Sign of fraction → $+\frac{-a}{+b}$

Sign of denominator

Whenever any of the three signs is omitted, we assume it to be positive. For example,

$$\frac{a}{b} \quad \text{means} \quad +\frac{+a}{+b}$$
$$\frac{-a}{b} \quad \text{means} \quad +\frac{-a}{+b}$$
$$-\frac{a}{b} \quad \text{means} \quad -\frac{+a}{+b}$$

Changing any two of the three signs of a fraction does not change the value of a fraction. Thus,

$$\frac{-a}{b} = -\frac{a}{b} = \frac{a}{-b}$$

Generally, we do not write a fraction with a negative denominator. For example, the expression $\frac{2}{-5}$ would be written as either $\frac{-2}{5}$ or $-\frac{2}{5}$. The expression $\frac{x}{-(4-x)}$ can be written $\frac{x}{x-4}$ since $-(4-x) = -4 + x$ or $x - 4$.

3) Simplify Rational Expressions

A rational expression is **simplified** or **reduced to its lowest terms** when the numerator and denominator have no common factors other than 1. The fraction $\frac{9}{12}$ is not simplified because 9 and 12 both contain the common factor 3. When the 3 is factored out, the simplified fraction is $\frac{3}{4}$.

$$\frac{9}{12} = \frac{\overset{1}{\cancel{3}} \cdot 3}{\underset{1}{\cancel{3}} \cdot 4} = \frac{3}{4}$$

The rational expression $\frac{ab - b^2}{2b}$ is not simplified because both the numerator and denominator have a common factor, b. To simplify this expression, factor b from each term in the numerator, then divide it out.

$$\frac{ab - b^2}{2b} = \frac{\cancel{b}(a - b)}{2\cancel{b}} = \frac{a - b}{2}$$

Thus, $\frac{ab - b^2}{2b}$ becomes $\frac{a - b}{2}$ when simplified.

To Simplify Rational Expressions

1. Factor both the numerator and denominator as completely as possible.
2. Divide out any factors common to both the numerator and denominator.

EXAMPLE 2 Simplify $\frac{5x^3 + 10x^2 - 25x}{10x^2}$.

Solution Factor the greatest common factor, $5x$, from each term in the numerator. Since $5x$ is a factor common to both the numerator and denominator, divide it out.

$$\frac{5x^3 + 10x^2 - 25x}{10x^2} = \frac{\cancel{5x}(x^2 + 2x - 5)}{\cancel{5x} \cdot 2x}$$

$$= \frac{x^2 + 2x - 5}{2x}$$

NOW TRY EXERCISE 25

EXAMPLE 3 Simplify $\frac{x^2 + 2x - 3}{x + 3}$.

Solution Factor the numerator; then divide out the common factor.

$$\frac{x^2 + 2x - 3}{x + 3} = \frac{\cancel{(x + 3)}(x - 1)}{\cancel{x + 3}} = x - 1$$

EXAMPLE 4 Simplify $\dfrac{x^2 - 16}{x - 4}$.

Solution Factor the numerator; then divide out common factors.

$$\frac{x^2 - 16}{x - 4} = \frac{(x + 4)\cancel{(x - 4)}}{\cancel{x - 4}} = x + 4$$

EXAMPLE 5 Simplify $\dfrac{2x^2 + 7x + 6}{x^2 - x - 6}$.

Solution Factor both the numerator and denominator, then divide out common factors.

$$\frac{2x^2 + 7x + 6}{x^2 - x - 6} = \frac{(2x + 3)\cancel{(x + 2)}}{(x - 3)\cancel{(x + 2)}} = \frac{2x + 3}{x - 3}.$$

NOW TRY EXERCISE 33

Note that $\dfrac{2x + 3}{x - 3}$ cannot be simplified any further.

Avoiding Common Errors

Remember: Only common *factors* can be divided out from expressions.

CORRECT	INCORRECT
$\dfrac{\overset{5}{\cancel{20}}\overset{x}{\cancel{x^2}}}{\underset{1}{\cancel{4}}\underset{1}{\cancel{x}}} = 5x$	$\dfrac{\overset{x}{\cancel{x^2}} - \overset{5}{\cancel{20}}}{\underset{1}{\cancel{x}} - \underset{1}{\cancel{4}}}$ (crossed out)

In the denominator of the example on the left, $4x$, the 4 and x are factors since they are *multiplied* together. The 4 and the x are also both factors of the numerator $20x^2$, since $20x^2$ can be written $4 \cdot x \cdot 5x$.

Some students incorrectly divide out *terms*. In the expression $\dfrac{x^2 - 20}{x - 4}$, the x and -4 are *terms* of the denominator, not factors, and therefore cannot be divided out.

Using Your Graphing Calculator

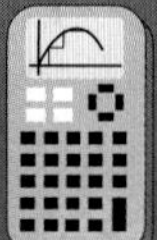

The graphing calculator can be used to determine whether a rational expression has been simplified correctly. To check the simplification below,

$$\frac{3x + 6}{6x^2 - 24} = \frac{1}{2(x - 2)}$$

we let $Y_1 = \dfrac{3x + 6}{6x^2 - 24}$ and $Y_2 = \dfrac{1}{2(x - 2)}$. Then we use the TABLE feature to compare results, as shown in Figure 8.1.

X	Y1	Y2
3	.5	.5
4	.25	.25
5	.16667	.16667
6	.125	.125
7	.1	.1
8	.08333	.08333
9	.07143	.07143

X=3

FIGURE 8.1

Since Y_1 and Y_2 have the same values for each value of x, we have not made a mistake. This procedure can only tell you if a mistake has been made. It cannot tell you if the fraction has been simplified completely. For example, $\dfrac{3x + 6}{6x^2 - 24}$ and $\dfrac{3}{6x - 12}$ will give you the same set of values, but $\dfrac{3}{6x - 12}$ is not simplified completely. Try this example again, using TblMin $= -3$. What do you notice about the values for Y_1 and Y_2? Do you know why the table is slightly different?

Exercises

Use the TABLE feature of your graphing calculator to determine whether the following rational expressions are equal for all values of x where the expression on the left is defined.

1. $\dfrac{x^2 - 4x + 4}{x^2 - 2x}$; $\dfrac{x - 2}{x}$

2. $\dfrac{x + 8}{x^2 - 64}$; $\dfrac{1}{x - 8}$

3. $\dfrac{x^2 - 4x + 4}{x^2 - 3x - 10}$; $\dfrac{x - 2}{x - 5}$

4. $\dfrac{5x^2 + 30x}{x + 6}$; $5x$

4) Factor a Negative 1 from a Polynomial

Recall that when −1 is factored from a polynomial, the sign of each term in the polynomial changes.

Examples

$$-3x + 5 = -1(3x - 5) = -(3x - 5)$$
$$6 - 2x = -1(-6 + 2x) = -(2x - 6)$$
$$-2x^2 + 3x - 4 = -1(2x^2 - 3x + 4) = -(2x^2 - 3x + 4)$$

Whenever the terms in a numerator and denominator differ only in their signs (one is the opposite or additive inverse of the other), we can factor out −1 from either the numerator or denominator and then divide out the common factor. This procedure is illustrated in Examples 6 and 7.

EXAMPLE 6 Simplify $\frac{3x - 7}{7 - 3x}$.

Solution Since each term in the numerator differs only in sign from its like term in the denominator, we will factor −1 from each term in the denominator.

$$\frac{3x - 7}{7 - 3x} = \frac{3x - 7}{-1(-7 + 3x)}$$
$$= \frac{\cancel{3x - 7}}{-\cancel{(3x - 7)}} = -1$$

EXAMPLE 7 Simplify $\frac{4x^2 - 23x - 6}{6 - x}$.

Solution

$$\frac{4x^2 - 23x - 6}{6 - x} = \frac{(4x + 1)(x - 6)}{6 - x}$$

The terms in $x - 6$ differ only in sign from the terms in $6 - x$.

$$= \frac{(4x + 1)\cancel{(x - 6)}}{-1\cancel{(x - 6)}}$$

Factor −1 from each term in denominator.

$$= \frac{4x + 1}{-1} = -(4x + 1)$$

NOW TRY EXERCISE 41 Note that $-4x - 1$ is also an acceptable answer.

Exercise Set 8.1

Concept/Writing Exercises

1. a) In your own words, define a rational expression.
   b) Give three examples of rational expressions.
2. Explain how to determine the value or values of the variable that makes a rational expression undefined.
3. In any rational expression with a variable in the denominator, what do we always assume about the variable?

Explain why the following expressions cannot be simplified.

4. $\frac{2 + 3x}{6}$
5. $\frac{3x + 4y}{12xy}$
6. In your own words, explain how to simplify a rational expression.

Explain why x can represent any real number in the following expressions.

7. $\dfrac{x - 5}{x^2 + 1}$

8. $\dfrac{x + 3}{x^2 + 4}$

In Exercises 9 and 10, determine what values, if any, x cannot represent. Explain.

9. $\dfrac{x + 3}{x - 3}$

10. $\dfrac{x}{(x - 4)^2}$

11. Is $-\dfrac{x + 4}{4 - x}$ equal to −1? Explain.

12. Is $-\dfrac{3x + 2}{-3x - 2}$ equal to 1? Explain.

Practice the Skills

Determine the value or values of the variable where each expression is defined.

13. $\dfrac{x - 1}{x}$

14. $\dfrac{3}{x + 5}$

15. $\dfrac{5}{2x - 6}$

16. $\dfrac{5}{2x - 3}$

17. $\dfrac{x + 4}{x^2 - 4}$

18. $\dfrac{7}{x^2 - 7x + 6}$

19. $\dfrac{x - 3}{x^2 + 6x - 16}$

20. $\dfrac{x^2 + 3}{2x^2 - 13x + 15}$

Simplify.

21. $\dfrac{x}{x + xy}$

22. $\dfrac{3x}{6x + 9}$

23. $\dfrac{5x + 15}{x + 3}$

24. $\dfrac{3x^2 + 6x}{3x^2 + 9x}$

25. $\dfrac{x^3 + 6x^2 + 3x}{2x}$

26. $\dfrac{x^2y^2 - xy + 3y}{y}$

27. $\dfrac{x^2 + 2x + 1}{x + 1}$

28. $\dfrac{x - 1}{x^2 + 4x - 5}$

29. $\dfrac{x^2 + 2x}{x^2 + 4x + 4}$

30. $\dfrac{x^2 + 3x - 18}{3x - 9}$

31. $\dfrac{x^2 - x - 6}{x^2 - 4}$

32. $\dfrac{x^2 - 6x + 9}{x^2 - 9}$

33. $\dfrac{x^2 - 2x - 3}{x^2 - x - 6}$

34. $\dfrac{4x^2 - 12x - 40}{2x^2 - 16x + 30}$

35. $\dfrac{2x - 3}{3 - 2x}$

36. $\dfrac{4x - 6}{3 - 2x}$

37. $\dfrac{x^2 - 2x - 8}{4 - x}$

38. $\dfrac{7 - x}{x^2 - 5x - 14}$

39. $\dfrac{x^2 + 3x - 18}{-2x^2 + 6x}$

40. $\dfrac{2x^2 + 5x - 12}{2x - 3}$

41. $\dfrac{2x^2 + 5x - 3}{1 - 2x}$

42. $\dfrac{x^2 - 16}{x^2 - 2x - 8}$

43. $\dfrac{6x^2 + x - 2}{2x - 1}$

44. $\dfrac{2x^2 - 11x + 15}{(x - 3)^2}$

45. $\dfrac{6x^2 + 7x - 20}{2x + 5}$

46. $\dfrac{16x^2 + 24x + 9}{4x + 3}$

47. $\dfrac{6x^2 - 13x + 6}{3x - 2}$

48. $\dfrac{x^2 - 25}{(x + 5)^2}$

49. $\dfrac{x^2 - 3x + 4x - 12}{x - 3}$

50. $\dfrac{x^2 - 2x + 4x - 8}{2x^2 + 3x + 8x + 12}$

51. $\dfrac{2x^2 - 8x + 3x - 12}{2x^2 + 8x + 3x + 12}$

52. $\dfrac{x^3 + 1}{x^2 - x + 1}$

53. $\dfrac{x^3 - 8}{x - 2}$

54. $\dfrac{x^3 - 125}{x^2 - 25}$

Problem Solving

Simplify the following expressions, if possible. Treat the unknown symbol as if it were a variable.

55. $\dfrac{5\Delta}{15}$

56. $\dfrac{\Delta}{\Delta + 7\Delta^2}$

57. $\dfrac{7\Delta}{14\Delta + 21}$

58. $\dfrac{\Delta^2 + 2\Delta}{\Delta^2 + 4\Delta + 4}$

59. $\dfrac{3\Delta - 2}{2 - 3\Delta}$

60. $\dfrac{(\Delta - 3)^2}{\Delta^2 - 6\Delta + 9}$

Determine the denominator that will make each statement true. Explain how you obtained your answer.

61. $\dfrac{x^2 - x - 6}{\square} = x - 3$

62. $\dfrac{2x^2 + 11x + 12}{\square} = x + 4$

Determine the numerator that will make each statement true. Explain how you obtained your answer.

63. $\dfrac{\square}{x + 4} = x + 3$

64. $\dfrac{\square}{x - 5} = 2x - 1$

Challenge Problems

In Exercises 65–67, ***a)*** *determine the value or values that x cannot represent.* ***b)*** *Simplify the expression.*

65. $\dfrac{x + 3}{x^2 - 2x + 3x - 6}$

66. $\dfrac{x - 4}{2x^2 - 5x - 8x + 20}$

67. $\dfrac{x + 5}{2x^3 + 7x^2 - 15x}$

Simplify. Explain how you determined your answer.

68. $\dfrac{\frac{1}{5}x^5 - \frac{2}{3}x^4}{x^4}$

69. $\dfrac{\frac{1}{5}x^5 - \frac{2}{3}x^4}{\frac{1}{5}x^5 - \frac{2}{3}x^4}$

70. $\dfrac{\frac{1}{5}x^5 - \frac{2}{3}x^4}{\frac{2}{3}x^4 - \frac{1}{5}x^5}$

Group Activity

Discuss and answer Exercise 71 as a group.

71. a) As a group, determine the values of the variable where the expression $\dfrac{x^2 - 25}{x^3 + 2x^2 - 15x}$ is undefined.

b) As a group, simplify the rational expression.

c) Group member 1: Substitute 6 in the *original expression* and evaluate.

d) Group member 2: Substitute 6 in the *simplified expression* from part **b)** and compare your result to that of Group member 1.

e) Group member 3: Substitute −2 in the original expression and in the simplified expression in part **b)**. Compare your answers.

f) As a group, discuss the results of your work in parts **c)–e)**.

g) Now, as a group, substitute −5 in the original expression and in the simplified expression. Discuss your results.

h) Is $\dfrac{x^2 - 25}{x^3 + 2x^2 - 15x}$ always equal to its simplified form for *any* value of x? Explain your answer.

Cumulative Review Exercises

[3.1] **72.** Solve the formula $z = \dfrac{x - y}{2}$ for y.

[3.4] **73.** Find the measures of the three angles of a triangle if one angle is 30° greater than the smallest angle, and the third angle is 10° greater than 3 times the smallest angle.

[5.3] **74.** Solve the following system of equations using the addition method.

$$3x - 4y = 1$$
$$2x - 2y = 2$$

[6.1] **75.** Simplify $\left(\dfrac{3x^6y^2}{9x^4y^3}\right)^2$.

[6.4] **76.** Subtract $6x^2 - 4x - 8 - (-3x^2 + 6x + 9)$.

8.2 MULTIPLICATION AND DIVISION OF RATIONAL EXPRESSIONS

1) **Multiply rational expressions.**
2) **Divide rational expressions.**

1) Multiply Rational Expressions

In Section 1.3 we reviewed multiplication of numerical fractions. Recall that to multiply two fractions we multiply their numerators together and multiply their denominators together.

To Multiply Two Fractions

$$\frac{a}{b} \cdot \frac{c}{d} = \frac{a \cdot c}{b \cdot d}, \qquad b \neq 0 \quad \text{and} \quad d \neq 0$$

EXAMPLE 1 Multiply $\left(\frac{3}{5}\right)\left(\frac{-2}{9}\right)$.

Solution First divide out common factors; then multiply.

$$\frac{\overset{1}{\cancel{3}}}{5} \cdot \frac{-2}{\underset{3}{\cancel{9}}} = \frac{1 \cdot (-2)}{5 \cdot 3} = -\frac{2}{15}$$

The same principles apply when multiplying rational expressions containing variables. Before multiplying, you should first divide out any factors common to both a numerator and a denominator.

To Multiply Rational Expressions

1. Factor all numerators and denominators completely.
2. Divide out common factors.
3. Multiply numerators together and multiply denominators together.

EXAMPLE 2 Multiply $\frac{3x^2}{2y} \cdot \frac{4y^3}{3x}$.

Solution This problem can be represented as

$$\frac{3xx}{2y} \cdot \frac{4yyy}{3x}$$

$$\frac{\overset{1}{\cancel{3}}\overset{1}{\cancel{x}}x}{2y} \cdot \frac{4yyy}{\underset{1}{\cancel{3}}\underset{1}{\cancel{x}}}$$ *Divide out the 3's and x's.*

$$\frac{\overset{1}{\cancel{3}}\overset{1}{\cancel{x}}x}{\underset{1}{\cancel{2}}\underset{1}{\cancel{y}}} \cdot \frac{\overset{2}{\cancel{4}}\overset{1}{\cancel{y}}yy}{\underset{1}{\cancel{3}}\underset{1}{\cancel{x}}}$$ *Divide both the 4 and the 2 by 2, and divide out the y's.*

Now we multiply the remaining numerators together and the remaining denominators together.

$$\frac{2xy^2}{1} \quad \text{or} \quad 2xy^2$$

Rather than illustrating this entire process when multiplying rational expressions, we will often proceed as follows:

$$\frac{3x^2}{2y} \cdot \frac{4y^3}{3x}$$

$$= \frac{\overset{1}{\cancel{3}}\overset{x}{\cancel{x^2}}}{\underset{1}{\cancel{2}}\underset{1}{\cancel{y}}} \cdot \frac{\overset{2}{\cancel{4}}\overset{y^2}{\cancel{y^3}}}{\underset{1}{\cancel{3}}\underset{1}{\cancel{x}}} = 2xy^2$$

EXAMPLE 3 Multiply $-\frac{3y^2}{2x^3} \cdot \frac{5x^2}{7y^2}$.

Solution

$$-\frac{3\cancel{y^2}}{2\underset{x}{\cancel{x^3}}} \cdot \frac{5\cancel{x^2}}{7\cancel{y^2}} = -\frac{15}{14x}$$

NOW TRY EXERCISE 39

In Example 3 when the y^2 was divided out from both the numerator and denominator we did not place a 1 above and below the y^2 factors. When a factor that appears in a numerator and denominator is factored out, we will generally not show the 1s.

EXAMPLE 4 Multiply $(x - 5) \cdot \frac{7}{x^3 - 5x^2}$.

Solution

$$(x - 5) \cdot \frac{7}{x^3 - 5x^2} = \frac{\cancel{x - 5}}{1} \cdot \frac{7}{x^2\cancel{(x - 5)}} = \frac{7}{x^2}$$

EXAMPLE 5 Multiply $\frac{(x + 2)^2}{6x^2} \cdot \frac{3x}{x^2 - 4}$.

Solution

$$\frac{(x + 2)^2}{6x^2} \cdot \frac{3x}{x^2 - 4} = \frac{(x + 2)(x + 2)}{6x^2} \cdot \frac{3x}{(x + 2)(x - 2)}$$

$$= \frac{\cancel{(x + 2)}(x + 2)}{\underset{2}{\cancel{6}}\underset{x}{\cancel{x^2}}} \cdot \frac{\overset{1}{\cancel{3}}\overset{1}{\cancel{x}}}{\cancel{(x + 2)}(x - 2)} = \frac{x + 2}{2x(x - 2)}$$

NOW TRY EXERCISE 15

In Example 5 we could have multiplied the factors in the denominator to get $(x + 2)/(2x^2 - 4x)$. This is also a correct answer. In this section we will leave rational answers with the numerator as a polynomial (in unfactored form) and the denominators in factored form, as was given in Example 5. This is consistent with how we will leave rational answers when we add and subtract rational expressions in later sections.

EXAMPLE 6 Multiply $\frac{x - 3}{2x} \cdot \frac{4x}{3 - x}$.

Solution

$$\frac{x - 3}{\underset{1}{\cancel{2}}\cancel{x}} \cdot \frac{\overset{2}{\cancel{4}}\cancel{x}}{3 - x} = \frac{2(x - 3)}{3 - x}.$$

This problem is still not complete. In Section 8.1 we showed that $3 - x$ is $-1(-3 + x)$ or $-1(x - 3)$. Thus,

$$\frac{2(x-3)}{3-x} = \frac{2\cancel{(x-3)}}{-1\cancel{(x-3)}} = -2$$

NOW TRY EXERCISE 13

HELPFUL HINT

When only the signs differ in a numerator and denominator in a multiplication problem, factor out -1 *from either the numerator or denominator*; then divide out the common factor.

$$\frac{a-b}{x} \cdot \frac{y}{b-a} = \frac{\cancel{a-b}}{x} \cdot \frac{y}{-1\cancel{(a-b)}} = -\frac{y}{x}$$

EXAMPLE 7 Multiply $\frac{3x+2}{2x-1} \cdot \frac{4-8x}{3x+2}$.

Solution

$$\frac{3x+2}{2x-1} \cdot \frac{4-8x}{3x+2} = \frac{3x+2}{2x-1} \cdot \frac{4(1-2x)}{3x+2} \quad \textit{Factor.}$$

$$= \frac{\cancel{3x+2}}{2x-1} \cdot \frac{4(1-2x)}{\cancel{3x+2}}. \quad \textit{Divide out common factors.}$$

Note that the factor $(1 - 2x)$ in the numerator of the second fraction differs only in sign from $2x - 1$, the denominator of the first fraction. We will therefore factor -1 from each term of the $(1 - 2x)$ in the numerator of the second fraction.

$$= \frac{\cancel{3x+2}}{2x-1} \cdot \frac{4(-1)(2x-1)}{\cancel{3x+2}} \quad \textit{Factor } -1 \textit{ from the second numerator.}$$

$$= \frac{\cancel{3x+2}}{\cancel{2x-1}} \cdot \frac{-4\cancel{(2x-1)}}{\cancel{3x+2}} \quad \textit{Divide out common factors.}$$

$$= \frac{-4}{1} = -4$$

EXAMPLE 8 Multiply $\frac{2x^2+7x-15}{4x^2-8x+3} \cdot \frac{2x^2+x-1}{x^2+6x+5}$.

Solution Factor all numerators and denominators, and then divide out common factors.

$$\frac{2x^2+7x-15}{4x^2-8x+3} \cdot \frac{2x^2+x-1}{x^2+6x+5} = \frac{(2x-3)(x+5)}{(2x-3)(2x-1)} \cdot \frac{(2x-1)(x+1)}{(x+1)(x+5)}$$

$$= \frac{\cancel{(2x-3)}\cancel{(x+5)}}{\cancel{(2x-3)}\cancel{(2x-1)}} \cdot \frac{\cancel{(2x-1)}\cancel{(x+1)}}{\cancel{(x+1)}\cancel{(x+5)}} = 1$$

EXAMPLE 9 Multiply $\frac{2x^3-14x^2+12x}{6y^2} \cdot \frac{-2y}{3x^2-3x}$.

Solution

$$\frac{2x^3 - 14x^2 + 12x}{6y^2} \cdot \frac{-2y}{3x^2 - 3x} = \frac{2x(x^2 - 7x + 6)}{6y^2} \cdot \frac{-2y}{3x(x-1)}$$

$$= \frac{2x(x-6)(x-1)}{6y^2} \cdot \frac{-2y}{3x(x-1)}$$

$$= \frac{\overset{}{\cancel{2}}\cancel{x}(x-6)\cancel{(x-1)}}{\underset{3\ \ y}{\cancel{6}\cancel{y^2}}} \cdot \frac{-2\cancel{y}}{3\cancel{x}\cancel{(x-1)}}$$

$$= \frac{-2(x-6)}{9y} = \frac{-2x+12}{9y}$$

EXAMPLE 10 Multiply $\frac{x^2 - y^2}{x + y} \cdot \frac{x + 2y}{2x^2 - xy - y^2}$.

Solution

$$\frac{x^2 - y^2}{x + y} \cdot \frac{x + 2y}{2x^2 - xy - y^2} = \frac{(x+y)(x-y)}{x+y} \cdot \frac{x+2y}{(2x+y)(x-y)}$$

$$= \frac{\cancel{(x+y)}\cancel{(x-y)}}{\cancel{x+y}} \cdot \frac{x+2y}{(2x+y)\cancel{(x-y)}}$$

$$= \frac{x+2y}{2x+y}$$

NOW TRY EXERCISE 19

2) Divide Rational Expressions

In Chapter 1 we learned that to divide one fraction by a second we invert the divisor and multiply.

To Divide Two Fractions

$$\frac{a}{b} \div \frac{c}{d} = \frac{a}{b} \cdot \frac{d}{c} = \frac{ad}{bc}, \quad b \neq 0, \quad d \neq 0, \quad \text{and} \quad c \neq 0$$

EXAMPLE 11 Divide.

a) $\frac{3}{5} \div \frac{4}{5}$ **b)** $\frac{2}{3} \div \frac{5}{6}$

Solution **a)** $\frac{3}{5} \div \frac{4}{5} = \frac{3}{\underset{1}{\cancel{5}}} \cdot \frac{\overset{1}{\cancel{5}}}{4} = \frac{3 \cdot 1}{1 \cdot 4} = \frac{3}{4}$ **b)** $\frac{2}{3} \div \frac{5}{6} = \frac{2}{\underset{1}{\cancel{3}}} \cdot \frac{\overset{2}{\cancel{6}}}{5} = \frac{2 \cdot 2}{1 \cdot 5} = \frac{4}{5}$

The same principles are used to **divide rational expressions**.

To Divide Rational Expressions

Invert the divisor (the second fraction) and multiply.

EXAMPLE 12 Divide $\frac{7x^3}{z} \div \frac{5z^3}{3}$.

Solution Invert the divisor (the second fraction), and then multiply.

$$\frac{7x^3}{z} \div \frac{5z^3}{3} = \frac{7x^3}{z} \cdot \frac{3}{5z^3} = \frac{21x^3}{5z^4}$$

NOW TRY EXERCISE 23

EXAMPLE 13 Divide $\dfrac{x^2 - 9}{x + 4} \div \dfrac{x - 3}{x + 4}$.

Solution

$$\frac{x^2 - 9}{x + 4} \div \frac{x - 3}{x + 4} = \frac{x^2 - 9}{x + 4} \cdot \frac{x + 4}{x - 3}$$

Invert the divisor and multiply.

$$= \frac{(x + 3)\cancel{(x - 3)}}{\cancel{x + 4}} \cdot \frac{\cancel{x + 4}}{\cancel{x - 3}} = x + 3$$

Factor, and divide out common factors.

EXAMPLE 14 Divide $\dfrac{-1}{2x - 3} \div \dfrac{3}{3 - 2x}$.

Solution

$$\frac{-1}{2x - 3} \div \frac{3}{3 - 2x} = \frac{-1}{2x - 3} \cdot \frac{3 - 2x}{3}$$

Invert the divisor and multiply.

$$= \frac{-1}{\cancel{2x - 3}} \cdot \frac{-1\cancel{(2x - 3)}}{3}$$

Factor out −1, then divide out common factors.

$$= \frac{(-1)(-1)}{(1)(3)} = \frac{1}{3}$$

EXAMPLE 15 Divide $\dfrac{x^2 + 8x + 15}{x^2} \div (x + 3)^2$.

Solution $(x + 3)^2$ means $\dfrac{(x + 3)^2}{1}$. Invert the divisor and multiply.

$$\frac{x^2 + 8x + 15}{x^2} \div (x + 3)^2 = \frac{x^2 + 8x + 15}{x^2} \cdot \frac{1}{(x + 3)^2}$$

$$= \frac{(x + 5)\cancel{(x + 3)}}{x^2} \cdot \frac{1}{\cancel{(x + 3)}(x + 3)}$$

$$= \frac{x + 5}{x^2(x + 3)}$$

EXAMPLE 16 Divide $\dfrac{12x^2 - 22x + 8}{3x} \div \dfrac{3x^2 + 2x - 8}{2x^2 + 4x}$.

Solution

$$\frac{12x^2 - 22x + 8}{3x} \div \frac{3x^2 + 2x - 8}{2x^2 + 4x} = \frac{12x^2 - 22x + 8}{3x} \cdot \frac{2x^2 + 4x}{3x^2 + 2x - 8}$$

$$= \frac{2(6x^2 - 11x + 4)}{3x} \cdot \frac{2x(x + 2)}{(3x - 4)(x + 2)}$$

$$= \frac{2\cancel{(3x - 4)}(2x - 1)}{3\cancel{x}} \cdot \frac{2\cancel{x}\cancel{(x + 2)}}{\cancel{(3x - 4)}\cancel{(x + 2)}}$$

$$= \frac{4(2x - 1)}{3} = \frac{8x - 4}{3}$$

NOW TRY EXERCISE 33

Exercise Set 8.2

Concept/Writing Exercises

1. In your own words, explain how to multiply rational expressions.

2. In your own words, explain how to divide rational expressions.

What polynomial should be in the shaded area of the second fraction to make each statement true? Explain how you determined your answer.

3. $\frac{x+3}{x-4} \cdot \frac{\square}{x+3} = x+2$

4. $\frac{x-5}{x+2} \cdot \frac{\square}{x-5} = 2x-3$

5. $\frac{x-4}{x+5} \cdot \frac{x+5}{\square} = \frac{1}{x+3}$

6. $\frac{2x-1}{x-3} \cdot \frac{x-3}{\square} = \frac{1}{x-6}$

Practice the Skills

Multiply.

7. $\frac{5x}{3y} \cdot \frac{y^2}{10}$

8. $\frac{15x^3y^2}{z} \cdot \frac{z}{5xy^3}$

9. $\frac{16x^2}{y^4} \cdot \frac{5x^2}{y^2}$

10. $\frac{5n^3}{32m} \cdot \frac{-4}{15m^2n^3}$

11. $\frac{6x^5y^3}{5z^3} \cdot \frac{6x^4}{5yz^4}$

12. $\frac{x^2-4}{x^2-9} \cdot \frac{x+3}{x-2}$

13. $\frac{3x-2}{3x+2} \cdot \frac{4x-1}{1-4x}$

14. $\frac{x-6}{2x+5} \cdot \frac{2x}{-x+6}$

15. $\frac{x^2+7x+12}{x+4} \cdot \frac{1}{x+3}$

16. $\frac{x^2+3x-10}{2x} \cdot \frac{x^2-3x}{x^2-5x+6}$

17. $\frac{a}{a^2-b^2} \cdot \frac{a+b}{a^2+ab}$

18. $\frac{x^2-25}{x^2-3x-10} \cdot \frac{x+2}{x}$

19. $\frac{6x^2-14x-12}{6x+4} \cdot \frac{x+3}{2x^2-2x-12}$

20. $\frac{2x^2-9x+9}{8x-12} \cdot \frac{2x}{x^2-3x}$

21. $\frac{x+3}{x-3} \cdot \frac{x^3-27}{x^2+3x+9}$

22. $\frac{x^3+8}{x^2-x-6} \cdot \frac{x+3}{x^2-2x+4}$

Divide.

23. $\frac{9x^3}{y^2} \div \frac{3x}{y^3}$

24. $\frac{9x^3}{4} \div \frac{1}{16y^2}$

25. $\frac{25xy^2}{7z} \div \frac{5x^2y^2}{14z^2}$

26. $\frac{36y}{7z^2} \div \frac{3xy}{2z}$

27. $\frac{xy}{7a^2b} \div \frac{6xy}{7}$

28. $2xz \div \frac{4xy}{z}$

29. $\frac{3x^2+6x}{x} \div \frac{2x+4}{x^2}$

30. $\frac{x-3}{4y^2} \div \frac{x^2-9}{2xy}$

31. $\frac{x^2+3x-18}{x} \div \frac{x-3}{1}$

32. $\frac{1}{x^2-17x+30} \div \frac{1}{x^2+7x-18}$

33. $\frac{x^2-12x+32}{x^2-6x-16} \div \frac{x^2-x-12}{x^2-5x-24}$

34. $\frac{a-b}{9a+9b} \div \frac{a^2-b^2}{a^2+2a+1}$

35. $\frac{2x^2+9x+4}{x^2+7x+12} \div \frac{2x^2-x-1}{(x+3)^2}$

36. $\frac{a^2-b^2}{9} \div \frac{3a-3b}{27x^2}$

37. $\frac{x^2-y^2}{x^2-2xy+y^2} \div \frac{x+y}{y-x}$

38. $\frac{9x^2-9y^2}{6x^2y^2} \div \frac{3x+3y}{12x^2y^5}$

Perform each indicated operation.

39. $\frac{9x}{6y^2} \cdot \frac{24x^2y^4}{9x}$

40. $\frac{y^3}{8} \cdot \frac{9x^2}{y^3}$

41. $\frac{45a^2b^3}{12c^3} \cdot \frac{4c}{9a^3b^5}$

42. $\frac{-2xw}{y^5} \div \frac{6x^2}{y^6}$

43. $\frac{-xy}{a} \div \frac{-2ax}{6y}$

44. $\frac{27x}{5y^2} \div 3x^2y^2$

45. $\dfrac{80m^4}{49x^5y^7} \cdot \dfrac{14x^{12}y^5}{25m^5}$

46. $\dfrac{-18x^2y}{11z^2} \cdot \dfrac{22z^3}{x^2y^5}$

47. $(3x + 5) \cdot \dfrac{1}{6x + 10}$

48. $\dfrac{1}{4x - 3} \cdot (20x - 15)$

49. $\dfrac{1}{7x^2y} \div \dfrac{1}{21x^3y}$

50. $\dfrac{x^2y^5}{3z} \div \dfrac{3z}{2x}$

Problem Solving

Perform each indicated operation. Treat Δ and ☺ as if they were variables.

51. $\dfrac{6\Delta^2}{12} \cdot \dfrac{12}{36\Delta^5}$

52. $\dfrac{\Delta - 6}{2\Delta + 5} \cdot \dfrac{2\Delta}{-\Delta + 6}$

53. $\dfrac{\Delta - \text{☺}}{9\Delta - 9\text{☺}} \div \dfrac{\Delta^2 - \text{☺}^2}{\Delta^2 + 2\Delta\text{☺} + \text{☺}^2}$

54. $\dfrac{\Delta^2 - \text{☺}^2}{\Delta^2 - 2\Delta\text{☺} + \text{☺}^2} \div \dfrac{\Delta + \text{☺}}{\text{☺} - \Delta}$

For each equation, fill in the shaded area with a binomial or trinomial to make the statement true. Explain how you determined your answer.

55. $\dfrac{\square}{x + 2} = x + 1$

56. $\dfrac{x + 3}{\square} = \dfrac{1}{x - 3}$

57. $\dfrac{\square}{x - 5} = x + 2$

58. $\dfrac{\square}{x^2 - 7x + 10} = \dfrac{1}{x - 5}$

59. $\dfrac{\square}{x^2 - 4} \cdot \dfrac{x + 2}{x - 1} = 1$

60. $\dfrac{x + 4}{x^2 + 9x + 20} \cdot \dfrac{\square}{x - 2} = 1$

Challenge Problems

Simplify.

61. $\left(\dfrac{x + 2}{x^2 - 4x - 12} \cdot \dfrac{x^2 - 9x + 18}{x - 2}\right) \div \dfrac{x^2 + 5x + 6}{x^2 - 4}$

62. $\left(\dfrac{x^2 - x - 6}{2x^2 - 9x + 9} \div \dfrac{x^2 + x - 12}{x^2 + 3x - 4}\right) \cdot \dfrac{2x^2 - 5x + 3}{x^2 + x - 2}$

63. $\left(\dfrac{x^2 + 4x + 3}{x^2 - 6x - 16}\right) \div \left(\dfrac{x^2 + 5x + 6}{x^2 - 9x + 8} \cdot \dfrac{x^2 - 1}{x^2 + 4x + 4}\right)$

64. $\left(\dfrac{x^2 + 4x + 3}{x^2 - 6x - 16} \div \dfrac{x^2 + 5x + 6}{x^2 - 9x + 8}\right) \cdot \left(\dfrac{x^2 - 1}{x^2 + 4x + 4}\right)$

Group Activity

Discuss and answer Exercises 65–67 as a group. For Exercises 65 and 66, determine the polynomials that when placed in the shaded areas make the statement true. Explain how you determined your answer.

65. $\dfrac{\square}{\square} \cdot \dfrac{x^2 + 3x - 4}{x^2 - 4x + 3} = \dfrac{x - 2}{x - 5}$

66. $\dfrac{\square}{x^2 + x - 2} \cdot \dfrac{x^2 + 6x + 8}{\square} = \dfrac{x + 3}{x + 5}$

67. Consider the three problems that follow:

(1) $\left(\dfrac{x + 2}{x - 3}\right) \div \left(\dfrac{x^2 - 5x + 6}{x - 2} \cdot \dfrac{x + 2}{x - 3}\right)$

(2) $\left(\dfrac{x + 2}{x - 3} \div \dfrac{x^2 - 5x + 6}{x - 2}\right) \cdot \left(\dfrac{x + 2}{x - 3}\right)$

(3) $\left(\dfrac{x + 2}{x - 3}\right) \div \left(\dfrac{x^2 - 5x + 6}{x - 2}\right) \cdot \left(\dfrac{x + 2}{x - 3}\right)$

a) Without working the problem, decide as a group which of the problems will have the same answer. Explain.

b) Individually, simplify each of the three problems.

c) Compare your answers to part **b)** with the other members of your group. If you did not get the same answers, determine why.

Cumulative Review Exercises

[6.5] **68.** Multiply $(4x^3y^2z^4)(5xy^3z^7)$.

[6.6] **69.** Divide $\dfrac{4x^3 - 5x}{2x - 1}$.

[7.4] **70.** Factor $3x^2 - 9x - 30$.

[7.6] **71.** Solve $3x^2 - 9x - 30 = 0$.

8.3 ADDITION AND SUBTRACTION OF RATIONAL EXPRESSIONS WITH A COMMON DENOMINATOR AND FINDING THE LEAST COMMON DENOMINATOR

1) **Add and subtract rational expressions with a common denominator.**
2) **Find the least common denominator.**

1) Add and Subtract Rational Expressions with a Common Denominator

Recall that when adding (or subtracting) two arithmetic fractions with a common denominator we add (or subtract) the numerators while keeping the common denominator.

To Add or Subtract Two Fractions

$$\frac{a}{c}+\frac{b}{c}=\frac{a+b}{c},\quad c\neq 0 \qquad \frac{a}{c}-\frac{b}{c}=\frac{a-b}{c},\quad c\neq 0$$

EXAMPLE 1 **a)** Add $\frac{7}{16}+\frac{8}{16}$. **b)** Subtract $\frac{5}{9}-\frac{1}{9}$.

Solution **a)** $\frac{7}{16}+\frac{8}{16}=\frac{7+8}{16}=\frac{15}{16}$ **b)** $\frac{5}{9}-\frac{1}{9}=\frac{5-1}{9}=\frac{4}{9}$

Note in Example **1a)** that we did not simplify $\frac{8}{16}$ to $\frac{1}{2}$. The fractions are given with a common denominator, 16. If $\frac{8}{16}$ was simplified to $\frac{1}{2}$, you would lose the common denominator that is needed to add or subtract fractions.

The same principles apply when **adding or subtracting rational expressions** containing variables.

To Add or Subtract Rational Expressions with a Common Denominator

1. Add or subtract the numerators.
2. Place the sum or difference of the numerators found in step 1 over the common denominator.
3. Simplify the fraction if possible.

EXAMPLE 2 Add $\frac{6}{x-5}+\frac{x+2}{x-5}$.

Solution

$$\frac{6}{x-5}+\frac{x+2}{x-5}=\frac{6+(x+2)}{x-5}=\frac{x+8}{x-5}$$

EXAMPLE 3 Add $\frac{2x^2+5}{x+3}+\frac{6x-5}{x+3}$.

Solution

$$\frac{2x^2+5}{x+3}+\frac{6x-5}{x+3}=\frac{(2x^2+5)+(6x-5)}{x+3}$$

$$=\frac{2x^2+5+6x-5}{x+3}$$

$$=\frac{2x^2+6x}{x+3}.$$

Now factor $2x$ from each term in the numerator and simplify.

$$=\frac{2x\cancel{(x+3)}}{\cancel{x+3}}=2x$$

NOW TRY EXERCISE 19

EXAMPLE 4 Add $\dfrac{x^2+3x-2}{(x+5)(x-2)}+\dfrac{4x+12}{(x+5)(x-2)}$.

Solution

$$\frac{x^2+3x-2}{(x+5)(x-2)}+\frac{4x+12}{(x+5)(x-2)}=\frac{(x^2+3x-2)+(4x+12)}{(x+5)(x-2)}$$ *Write as a single fraction.*

$$=\frac{x^2+3x-2+4x+12}{(x+5)(x-2)}$$ *Remove parentheses in the numerator.*

$$=\frac{x^2+7x+10}{(x+5)(x-2)}$$ *Combine like terms.*

$$=\frac{\cancel{(x+5)}(x+2)}{\cancel{(x+5)}(x-2)}$$ *Factor, divide out common factor.*

$$=\frac{x+2}{x-2}$$

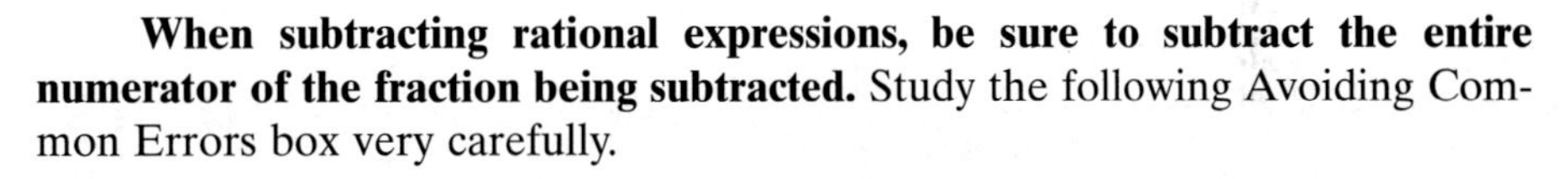

When subtracting rational expressions, be sure to subtract the entire numerator of the fraction being subtracted. Study the following Avoiding Common Errors box very carefully.

Avoiding Common Errors

Consider the subtraction

$$\frac{4x}{x-2}-\frac{2x+1}{x-2}$$

Many people begin problems of this type incorrectly. Here are the correct and incorrect ways of working this problem.

CORRECT

$$\frac{4x}{x-2}-\frac{2x+1}{x-2}=\frac{4x-(2x+1)}{x-2}$$

$$=\frac{4x-2x-1}{x-2}$$

$$=\frac{2x-1}{x-2}$$

INCORRECT

$$\cancel{\frac{4x}{x-2}-\frac{2x+1}{x-2}=\frac{4x-2x+1}{x-2}}$$

Note that the entire numerator of the second fraction (not just the first term) **must be subtracted.** Also note that the sign of *each* term of the numerator being subtracted will change when the parentheses are removed.

EXAMPLE 5 Subtract $\frac{x^2 - 2x + 3}{x^2 + 7x + 12} - \frac{x^2 - 4x - 5}{x^2 + 7x + 12}$.

Solution

$$\frac{x^2 - 2x + 3}{x^2 + 7x + 12} - \frac{x^2 - 4x - 5}{x^2 + 7x + 12} = \frac{(x^2 - 2x + 3) - (x^2 - 4x - 5)}{x^2 + 7x + 12}$$ *Write as a single fraction.*

$$= \frac{x^2 - 2x + 3 - x^2 + 4x + 5}{x^2 + 7x + 12}$$ *Remove parentheses*

$$= \frac{2x + 8}{x^2 + 7x + 12}$$ *Combine like terms.*

$$= \frac{2\cancel{(x + 4)}}{(x + 3)\cancel{(x + 4)}}$$ *Factor, divide out common factor.*

$$= \frac{2}{x + 3}$$

EXAMPLE 6 Subtract $\frac{5x}{x - 6} - \frac{4x^2 - 16x - 18}{x - 6}$

Solution

$$\frac{5x}{x - 6} - \frac{4x^2 - 16x - 18}{x - 6} = \frac{5x - (4x^2 - 16x - 18)}{x - 6}$$ *Write as a single fraction.*

$$= \frac{5x - 4x^2 + 16x + 18}{x - 6}$$ *Remove parentheses.*

$$= \frac{-4x^2 + 21x + 18}{x - 6}$$ *Combine like terms.*

$$= \frac{-(4x^2 - 21x - 18)}{x - 6}$$ *Factor out −1.*

$$= \frac{-(4x + 3)\cancel{(x - 6)}}{\cancel{x - 6}}$$ *Factor, divide out common factor.*

$$= -(4x + 3) \text{ or } -4x - 3$$

NOW TRY EXERCISE 41

2) Find the Least Common Denominator

To add two fractions with unlike denominators, we must first obtain a common denominator. Now we explain how to find the **least common denominator** for rational expressions. We will use this information in Section 8.4 when we add and subtract rational expressions.

EXAMPLE 7 Add $\frac{5}{7} + \frac{2}{3}$.

Solution The least common denominator (LCD) of the fractions $\frac{5}{7}$ and $\frac{2}{3}$ is 21. Twenty-one is the smallest number that is divisible by both denominators, 7 and 3. Rewrite each fraction so that its denominator is 21.

$$\frac{5}{7}+\frac{2}{3}=\frac{5}{7}\cdot\frac{3}{3}+\frac{2}{3}\cdot\frac{7}{7}$$

$$=\frac{15}{21}+\frac{14}{21}=\frac{29}{21}\text{ or }1\frac{8}{21}$$

To add or subtract rational expressions, we must write each expression with a common denominator.

To Find the Least Common Denominator of Rational Expressions

1. Factor each denominator completely. Any factors that occur more than once should be expressed as powers. For example, $(x-3)(x-3)$ should be expressed as $(x-3)^2$.
2. List all different factors (other than 1) that appear in any of the denominators. When the same factor appears in more than one denominator, write that factor with the highest power that appears.
3. The least common denominator is the product of all the factors listed in step 2.

EXAMPLE 8 Find the least common denominator.

$$\frac{1}{5}+\frac{1}{y}$$

Solution The only factor (other than 1) of the first denominator is 5. The only factor (other than 1) of the second denominator is y. The LCD is therefore $5\cdot y=5y$.

EXAMPLE 9 Find the LCD.

$$\frac{2}{x^2}-\frac{3}{7x}$$

Solution The factors that appear in the denominators are 7 and x. List each factor with its highest power. The LCD is the product of these factors.

Highest power of x

$$\text{LCD}=7\cdot x^2=7x^2$$

EXAMPLE 10 Find the LCD.

$$\frac{1}{18x^3y}+\frac{5}{27x^2y^3}$$

Solution Write both 18 and 27 as products of prime factors: $18=2\cdot 3^2$ and $27=3^3$. *If you have forgotten how to write a number as a product of prime factors, read Section 7.1 or Appendix B now.*

$$\frac{1}{18x^3y}+\frac{5}{27x^2y^3}=\frac{1}{2\cdot 3^2x^3y}+\frac{5}{3^3x^2y^3}$$

The factors that appear are 2, 3, x, and y. List the highest powers of each of these factors.

NOW TRY EXERCISE 59

$$\text{LCD}=2\cdot 3^3\cdot x^3\cdot y^3=54x^3y^3$$

EXAMPLE 11 Find the LCD.

$$\frac{5}{x} - \frac{7y}{x+3}$$

Solution The factors in the denominators are x and $(x + 3)$. *Note that the x in the second denominator, x + 3, is a term, not a factor.*

$$\text{LCD} = x(x+3)$$

EXAMPLE 12 Find the LCD.

$$\frac{7}{3x^2 - 6x} + \frac{x^2}{x^2 - 4x + 4}$$

Solution Factor both denominators.

$$\frac{7}{3x^2 - 6x} + \frac{x^2}{x^2 - 4x + 4} = \frac{7}{3x(x-2)} + \frac{x^2}{(x-2)(x-2)}$$

$$= \frac{7}{3x(x-2)} + \frac{x^2}{(x-2)^2}$$

The factors in the denominators are 3, x, and $x - 2$. List the highest powers of each of these factors.

NOW TRY EXERCISE 75

$$\text{LCD} = 3 \cdot x \cdot (x-2)^2 = 3x(x-2)^2.$$

EXAMPLE 13 Find the LCD.

$$\frac{5x}{x^2 - x - 12} - \frac{6x^2}{x^2 - 7x + 12}$$

Solution Factor both denominators.

$$\frac{5x}{x^2 - x - 12} - \frac{6x^2}{x^2 - 7x + 12} = \frac{5x}{(x+3)(x-4)} - \frac{6x^2}{(x-3)(x-4)}$$

The factors in the denominators are $x + 3$, $x - 4$, and $x - 3$.

$$\text{LCD} = (x+3)(x-4)(x-3)$$

Although $x - 4$ is a common factor of each denominator, the highest power of that factor that appears in each denominator is 1.

EXAMPLE 14 Find the LCD.

$$\frac{3x}{x^2 - 14x + 48} + x + 9$$

Solution Factor the denominator of the first term.

$$\frac{3x}{x^2 - 14x + 48} + x + 9 = \frac{3x}{(x-6)(x-8)} + x + 9$$

Since the denominator of $x + 9$ is 1, the expression can be rewritten as

$$\frac{3x}{(x-6)(x-8)} + \frac{x+9}{1}$$

NOW TRY EXERCISE 79

The LCD is therefore $1(x-6)(x-8)$ or simply $(x-6)(x-8)$.

Exercise Set 8.3

Concept/Writing Exercises

1. In your own words, explain how to add or subtract rational expressions with a common denominator.
2. When subtracting rational expressions, what must happen to the sign of each term of the numerator being subtracted?
3. In your own words, explain how to find the least common denominator of two rational expressions.

Determine the LCD to be used to perform each indicated operation. Explain how you determined the LCD. Do not perform the operations.

4. $\frac{3}{x+4} - \frac{2}{x}$
5. $\frac{2}{x+5} + \frac{3}{5}$
6. $\frac{2}{x+3} + \frac{1}{x} + \frac{1}{3}$
7. $\frac{6}{x-3} + \frac{1}{x} - \frac{1}{3}$

In Exercises 8–10 **a)** *Explain why the expression on the left side of the equal sign is not equal to the expression on the right side of the equal sign.* **b)** *Show what the expression on the right side should be for it to be equal to the one on the left.*

8. $\frac{4x-3}{5x+4} - \frac{2x-7}{5x+4} \neq \frac{4x-3-2x-7}{5x+4}$
9. $\frac{6x-2}{x^2-4x+3} - \frac{3x^2-4x+5}{x^2-4x+3} \neq \frac{6x-2-3x^2-4x+5}{x^2-4x+3}$
10. $\frac{4x+5}{x^2-6x} - \frac{-x^2+3x+6}{x^2-6x} \neq \frac{4x+5+x^2+3x+6}{x^2-6x}$

Practice the Skills

Add or subtract.

11. $\frac{x-1}{8} + \frac{2x}{8}$
12. $\frac{x-7}{3} - \frac{7}{3}$
13. $\frac{5x+3}{7} - \frac{x}{7}$
14. $\frac{3x+6}{2} - \frac{x}{2}$
15. $\frac{2}{x} + \frac{x+4}{x}$
16. $\frac{3x+4}{x+1} + \frac{6x+5}{x+1}$
17. $\frac{x-7}{x} + \frac{x+7}{x}$
18. $\frac{x-6}{x} - \frac{x+4}{x}$
19. $\frac{x}{x-1} + \frac{4x+7}{x-1}$
20. $\frac{4x-3}{x-7} - \frac{2x+8}{x-7}$
21. $\frac{9x+7}{6x^2} - \frac{3x+4}{6x^2}$
22. $\frac{-2x-4}{x^2+2x+1} + \frac{3x+5}{x^2+2x+1}$
23. $\frac{5x+4}{x^2-x-12} + \frac{-4x-1}{x^2-x-12}$
24. $\frac{-x-4}{x^2-16} + \frac{2(x+4)}{x^2-16}$
25. $\frac{x+4}{3x+2} - \frac{x+4}{3x+2}$
26. $\frac{2x+4}{(x+2)(x-3)} - \frac{x+7}{(x+2)(x-3)}$
27. $\frac{3x-5}{x+7} - \frac{6x+5}{x+7}$
28. $\frac{x^2-15}{4x} - \frac{x^2+4x-5}{4x}$
29. $\frac{x^2+4x+3}{x+2} - \frac{5x+9}{x+2}$
30. $\frac{-4x+2}{3x+6} + \frac{4(x-1)}{3x+6}$
31. $\frac{-2x+5}{5x-10} + \frac{2(x-5)}{5x-10}$
32. $\frac{x^2}{x+4} + \frac{16}{x+4}$
33. $\frac{x^2-2x-3}{x^2-x-6} + \frac{x-3}{x^2-x-6}$
34. $\frac{4x+12}{3-x} - \frac{3x+15}{3-x}$
35. $\frac{-x-7}{2x-9} - \frac{-3x-16}{2x-9}$
36. $\frac{x^2-2}{x^2+6x-7} - \frac{-4x+19}{x^2+6x-7}$
37. $\frac{x^2+2x}{(x+6)(x-3)} - \frac{15}{(x+6)(x-3)}$
38. $\frac{x^2-13}{x+5} - \frac{12}{x+5}$
39. $\frac{3x^2-7x}{4x^2-8x} + \frac{x}{4x^2-8x}$
40. $\frac{3x^2+15x}{x^3+2x^2-8x} + \frac{2x^2+5x}{x^3+2x^2-8x}$
41. $\frac{3x^2-4x+4}{3x^2+7x+2} - \frac{10x+9}{3x^2+7x+2}$
42. $\frac{x^3-10x^2+35x}{x(x-6)} - \frac{x^2+5x}{x(x-6)}$
43. $\frac{x^2+3x-6}{x^2-5x+4} - \frac{-2x^2+4x-4}{x^2-5x+4}$
44. $\frac{4x^2+5}{9x^2-64} - \frac{x^2-x+29}{9x^2-64}$
45. $\frac{5x^2+40x+8}{x^2-64} + \frac{x^2+9x}{x^2-64}$
46. $\frac{20x^2+5x+1}{6x^2+x-2} - \frac{8x^2-12x-5}{6x^2+x-2}$

Find the least common denominator for each expression.

47. $\frac{x}{7} + \frac{x+4}{7}$

48. $\frac{3+x}{5} - \frac{12}{5}$

49. $\frac{1}{3x} + \frac{1}{3}$

50. $\frac{1}{x+1} - \frac{4}{7}$

51. $\frac{3}{5x} + \frac{7}{4}$

52. $\frac{5}{2x} + 1$

53. $\frac{4}{x} + \frac{3}{x^3}$

54. $\frac{5x}{x+1} + \frac{6}{x+2}$

55. $\frac{x+3}{6x+5} + x$

56. $\frac{x+4}{2x} + \frac{3}{7x}$

57. $\frac{x}{x+1} + \frac{4}{x^2}$

58. $\frac{x}{3x^2} + \frac{9}{7x^3}$

59. $\frac{x+1}{12x^2y} - \frac{7}{9x^3}$

60. $\frac{-3}{8x^2y^2} + \frac{6}{5x^4y^5}$

61. $\frac{x^2-7}{24x} - \frac{x+3}{9(x+5)}$

62. $\frac{x-7}{3x+5} - \frac{6}{x+5}$

63. $\frac{5x-2}{x^2+x} - \frac{x^2}{x}$

64. $\frac{10}{(x+4)(x+2)} - \frac{5-x}{x+2}$

65. $\frac{21}{24x^2y} + \frac{x+4}{15xy^3}$

66. $\frac{x^2-4}{x^2-16} + \frac{3}{x+4}$

67. $\frac{3}{3x+12} + \frac{3x+6}{2x+4}$

68. $6x^2 + \frac{9x}{x-3}$

69. $\frac{9x+4}{x+6} - \frac{3x-6}{x+5}$

70. $\frac{x+1}{x^2+11x+18} - \frac{x^2-4}{x^2-3x-10}$

71. $\frac{x-2}{x^2-5x-24} + \frac{3}{x^2+11x+24}$

72. $\frac{6x+5}{x^2-4} - \frac{3x}{x^2-5x-14}$

73. $\frac{7}{x+4} - \frac{x+5}{x^2-3x-4}$

74. $\frac{3x+5}{x^2-1} + \frac{x^2-8}{(x+1)^2}$

75. $\frac{2x}{x^2+6x+5} - \frac{5x^2}{x^2+4x+3}$

76. $\frac{6x+5}{x+2} + \frac{4x}{(x+2)^2}$

77. $\frac{3x-5}{x^2-6x+9} + \frac{3}{x-3}$

78. $\frac{9x+7}{(x+3)(x+2)} - \frac{4x}{(x-3)(x+2)}$

79. $\frac{8x^2}{x^2-7x+6} + x - 3$

80. $\frac{x-1}{x^2-25} + x - 4$

81. $\frac{x}{3x^2+16x-12} + \frac{6}{3x^2+17x-6}$

82. $\frac{2x-7}{2x^2+5x+2} + \frac{x^2}{3x^2+4x-4}$

83. $\frac{2x-3}{4x^2+4x+1} + \frac{x^2-4}{8x^2+10x+3}$

84. $\frac{3x+1}{6x^2+5x-6} + \frac{x^2-5}{9x^2-12x+4}$

Problem Solving

List the polynomial to be placed in each shaded area to make a true statement. Explain how you determined your answer.

85. $\frac{x^2-6x+3}{x+3} + \frac{\blacksquare}{x+3} = \frac{2x^2-5x-6}{x+3}$

86. $\frac{4x^2-6x-7}{x^2-4} - \frac{\blacksquare}{x^2-4} = \frac{2x^2+x-3}{x^2-4}$

87. $\frac{-x^2-4x+3}{2x+5} + \frac{\blacksquare}{2x+5} = \frac{5x-7}{2x+5}$

88. $\frac{-3x^2-9}{(x+4)(x-2)} - \frac{\blacksquare}{(x+4)(x-2)} = \frac{x^2+3x}{(x+4)(x-2)}$

89. What is the least common denominator in the following subtraction? Explain how you determined your answer.

$$\frac{3x}{x^2-x-20} - \frac{2x}{-x^2+x+20}$$

Find the least common denominator of each expression.

90. $\frac{3}{\Delta} + \frac{4}{5\Delta}$

91. $\frac{5}{8\Delta^2 \text{☺}^2} + \frac{6}{5\Delta^4 \text{☺}^5}$

92. $\frac{8}{\Delta^2 - 9} - \frac{2}{\Delta + 3}$

93. $\frac{6}{\Delta + 3} - \frac{\Delta + 5}{\Delta^2 - 4\Delta + 3}$

Challenge Problems

Perform each indicated operation.

94. $\frac{4x - 1}{x^2 - 25} - \frac{3x^2 - 8}{x^2 - 25} + \frac{8x - 3}{x^2 - 25}$

95. $\frac{x^2 - 8x + 2}{x + 7} + \frac{2x^2 - 5x}{x + 7} - \frac{3x^2 + 7x + 6}{x + 7}$

Find the least common denominator for each expression.

96. $\frac{7}{6x^5y^9} - \frac{9}{2x^3y} + \frac{4}{5x^{12}y^2}$

97. $\frac{12}{x - 3} - \frac{5}{x^2 - 9} + \frac{7}{x + 3}$

98. $\frac{4}{x^2 - x - 12} + \frac{3}{x^2 - 6x + 8} + \frac{5}{x^2 + x - 6}$

99. $\frac{4}{x^2 - 4} - \frac{11}{3x^2 + 5x - 2} + \frac{5}{3x^2 - 7x + 2}$

100. $\frac{7}{2x^2 + x - 10} + \frac{5}{2x^2 + 7x + 5} - \frac{4}{-x^2 + x + 2}$

Cumulative Review Exercises

[1.3] **101.** Subtract $4\frac{3}{5} - 2\frac{5}{9}$.

[2.5] **102.** Solve $6x + 4 = -(x + 2) - 3x + 4$.

[2.6] **103.** The instructions on a bottle of concentrated hummingbird food indicate that 6 ounces of the concentrate should be mixed with 1 gallon (128 ounces) of water. If you wish to mix the concentrate with only 48 ounces of water, how much concentrate should be used?

[3.3] **104.** The Northern Fitness Club has two payment plans. Plan 1 is a yearly membership fee of \$125 plus \$2.50 per hour for use of the tennis court. Plan 2 is an annual membership fee of \$300 with no charge for court time.

a) How many hours would Malcom Wu have to play in a year to make the cost of Plan 1 equal to the cost of Plan 2?

b) If Malcom plans to play an average of 4 hours per week for the year, which plan should he use?

[4.2] **105.** Graph the equation $2x + 3y = 6$.

[4.3] **106.** What is the slope of the line in the graph below?

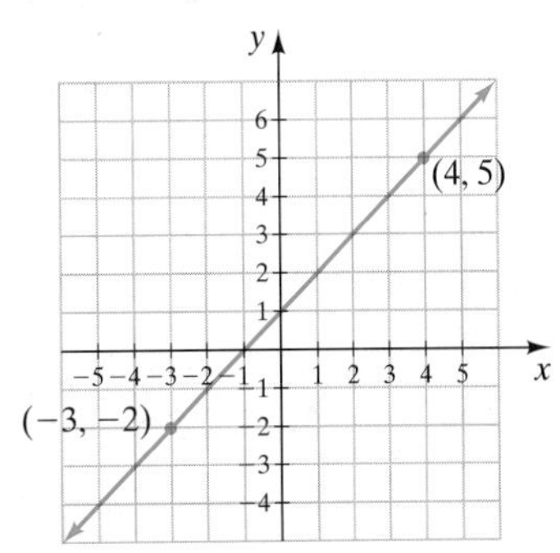

8.4 ADDITION AND SUBTRACTION OF RATIONAL EXPRESSIONS

1) **Add and subtract rational expressions.**

In Section 8.3 we discussed how to add and subtract rational expressions with a common denominator. Now we discuss adding and subtracting rational expressions that are not given with a common denominator.

1) Add and Subtract Rational Expressions

The method used to add and subtract rational expressions with unlike denominators is outlined in Example 1.

EXAMPLE 1 Add $\frac{7}{x} + \frac{3}{y}$.

Solution First we determine the LCD as outlined in Section 8.3.

$$\text{LCD} = xy$$

We write each fraction with the LCD. We do this by multiplying **both** the numerator and denominator of each fraction by any factors needed to obtain the LCD.

In this problem, the fraction on the left must be multiplied by y/y and the fraction on the right must be multiplied by x/x.

$$\frac{7}{x} + \frac{3}{y} = \frac{y}{y} \cdot \frac{7}{x} + \frac{3}{y} \cdot \frac{x}{x} = \frac{7y}{xy} + \frac{3x}{xy}$$

By multiplying both the numerator and denominator by the same factor, we are in effect multiplying by 1, which does not change the value of the fraction, only its appearance. Thus, the new fraction is equivalent to the original fraction.

Now we add the numerators, while leaving the LCD alone.

$$\frac{7y}{xy} + \frac{3x}{xy} = \frac{7y + 3x}{xy} \quad \text{or} \quad \frac{3x + 7y}{xy}$$

To Add or Subtract Two Rational Expressions with Unlike Denominators

1. Determine the LCD.
2. Rewrite each fraction as an equivalent fraction with the LCD. This is done by multiplying both the numerator and denominator of each fraction by any factors needed to obtain the LCD.
3. Add or subtract the numerators while maintaining the LCD.
4. When possible, factor the remaining numerator and simplify the fraction.

EXAMPLE 2 Add $\frac{5}{4x^2y} + \frac{3}{14xy^3}$.

Solution The LCD is $28x^2y^3$. We must write each fraction with the denominator $28x^2y^3$. To do this, we multiply the fraction on the left by $7y^2/7y^2$ and the fraction on the right by $2x/2x$.

$$\begin{aligned}\frac{5}{4x^2y} + \frac{3}{14xy^3} &= \frac{7y^2}{7y^2} \cdot \frac{5}{4x^2y} + \frac{3}{14xy^3} \cdot \frac{2x}{2x} \\ &= \frac{35y^2}{28x^2y^3} + \frac{6x}{28x^2y^3} \\ &= \frac{35y^2 + 6x}{28x^2y^3} \text{ or } \frac{6x + 35y^2}{28x^2y^3}\end{aligned}$$

NOW TRY EXERCISE 15

HELPFUL HINT

In Example 2 we multiplied the first fraction by $\frac{7y^2}{7y^2}$ and the second fraction by $\frac{2x}{2x}$ to get two fractions with a common denominator. How did we know what to multiply each fraction by? Many of you can determine this by observing the LCD and then determining what each denominator needs to be multiplied by to get the LCD. If this is not obvious, you can divide the LCD by the given denominator to determine what the numerator and denominator of each fraction should be multiplied by. In Example 2, the LCD is $28x^2y^3$. If we divide $28x^2y^3$ by each given denominator, $4x^2y$ and $14xy^3$, we can determine what the numerator and denominator of each respective fraction should be multiplied by.

$$\frac{28x^2y^3}{4x^2y} = 7y^2 \qquad \frac{28x^2y^3}{14xy^3} = 2x$$

Thus, $\frac{5}{4x^2y}$ should be multiplied by $\frac{7y^2}{7y^2}$ and $\frac{3}{14xy^3}$ should be multipied by $\frac{2x}{2x}$ to obtain the LCD $28x^2y^3$.

EXAMPLE 3 Add $\frac{3}{x+2} + \frac{5}{x}$.

Solution We must write each fraction with the LCD, which is $x(x + 2)$. To do this, we multiply the fraction on the left by x/x and the fraction on the right by $(x + 2)/(x + 2)$.

$$\begin{aligned}\frac{3}{x+2} + \frac{5}{x} &= \frac{x}{x} \cdot \frac{3}{x+2} + \frac{5}{x} \cdot \frac{x+2}{x+2} \\ &= \frac{3x}{x(x+2)} + \frac{5(x+2)}{x(x+2)} && \text{Rewrite each fraction as an equivalent fraction with the LCD.} \\ &= \frac{3x}{x(x+2)} + \frac{5x+10}{x(x+2)} && \text{Distributive property.}\end{aligned}$$

$$= \frac{3x + (5x + 10)}{x(x + 2)}$$ *Write as a single fraction.*

$$= \frac{3x + 5x + 10}{x(x + 2)}$$ *Remove parentheses in the numerator.*

$$= \frac{8x + 10}{x(x + 2)}$$ *Combine like terms in the numerator.*

NOW TRY EXERCISE 25

HELPFUL HINT

Look at the answer to Example 3, $\frac{8x + 10}{x(x + 2)}$. Notice that the numerator could have been factored to obtain $\frac{2(4x + 5)}{x(x + 2)}$. Also notice that the denominator could have been multiplied to get $\frac{8x + 10}{x^2 + 2x}$. All three of these answers are equivalent and each is correct. In this section, when writing answers, unless there is a common factor in the numerator and denominator we will leave the numerator in unfactored form and the denominator in factored form. If both the numerator and denominator have a common factor, we will factor the numerator and simplify the fraction.

EXAMPLE 4 Subtract $\frac{x}{x + 5} - \frac{2}{x - 3}$.

Solution The LCD is $(x + 5)(x - 3)$. The fraction on the left must be multiplied by $(x - 3)/(x - 3)$ to obtain the LCD. The fraction on the right must be multiplied by $(x + 5)/(x + 5)$ to obtain the LCD.

$$\frac{x}{x + 5} - \frac{2}{x - 3} = \frac{x - 3}{x - 3} \cdot \frac{x}{x + 5} - \frac{2}{x - 3} \cdot \frac{x + 5}{x + 5}$$

$$= \frac{x(x - 3)}{(x - 3)(x + 5)} - \frac{2(x + 5)}{(x - 3)(x + 5)}$$ *Rewrite each fraction as an equivalent fraction with the LCD.*

$$= \frac{x^2 - 3x}{(x - 3)(x + 5)} - \frac{2x + 10}{(x - 3)(x + 5)}$$

$$= \frac{(x^2 - 3x) - (2x + 10)}{(x - 3)(x + 5)}$$ *Write as a single fraction.*

$$= \frac{x^2 - 3x - 2x - 10}{(x - 3)(x + 5)}$$ *Remove parentheses in the numerator.*

$$= \frac{x^2 - 5x - 10}{(x - 3)(x + 5)}$$ *Combine like terms in the numerator.*

EXAMPLE 5 Subtract $\frac{x+2}{x-4} - \frac{x+3}{x+4}$.

Solution The LCD is $(x - 4)(x + 4)$.

$$\frac{x+2}{x-4} - \frac{x+3}{x+4} = \frac{x+4}{x+4} \cdot \frac{x+2}{x-4} - \frac{x+3}{x+4} \cdot \frac{x-4}{x-4}$$

$$= \frac{(x+4)(x+2)}{(x+4)(x-4)} - \frac{(x+3)(x-4)}{(x+4)(x-4)}$$ *Rewrite each fraction as an equivalent fraction with the LCD.*

Use the FOIL method to multiply each numerator.

$$= \frac{x^2+6x+8}{(x+4)(x-4)} - \frac{x^2-x-12}{(x+4)(x-4)}$$

$$= \frac{(x^2+6x+8)-(x^2-x-12)}{(x+4)(x-4)}$$ *Write as a single fraction.*

$$= \frac{x^2+6x+8-x^2+x+12}{(x+4)(x-4)}$$ *Remove parentheses in the numerator.*

$$= \frac{7x+20}{(x+4)(x-4)}$$ *Combine like terms in the numerator.*

NOW TRY EXERCISE 37

Consider the problem

$$\frac{6}{x-2} + \frac{x+3}{2-x}$$

How do we add these rational expressions? We could write each fraction with the denominator $(x - 2)(2 - x)$. However, there is an easier way. Study the following Helpful Hint.

HELPFUL HINT

When adding or subtracting fractions whose denominators are opposites (and therefore differ only in signs), multiply both the numerator *and* the denominator of *either* of the fractions by -1. Then both fractions will have the same denominator.

$$\frac{x}{a-b} + \frac{y}{b-a} = \frac{x}{a-b} + \frac{-1}{-1} \cdot \frac{y}{b-a}$$

$$= \frac{x}{a-b} + \frac{-y}{a-b}$$

$$= \frac{x-y}{a-b}$$

EXAMPLE 6 Add $\frac{6}{x-2} + \frac{x+3}{2-x}$.

Solution Since the denominators differ only in sign, we may multiply both the numerator and the denominator of either fraction by -1. Here we will multiply the numerator and denominator of the second fraction by -1 to obtain the common denominator $x - 2$.

$$\frac{6}{x-2} + \frac{x+3}{2-x} = \frac{6}{x-2} + \frac{-1}{-1} \cdot \frac{x+3}{2-x}$$

$$= \frac{6}{x-2} + \frac{-x-3}{x-2}$$

$$= \frac{6 + (-x-3)}{x-2} \quad \text{Write as a single fraction.}$$

$$= \frac{6 - x - 3}{x-2} \quad \text{Remove parentheses in the numerator.}$$

$$= \frac{-x+3}{x-2} \quad \text{Combine like terms in the numerator.}$$

NOW TRY EXERCISE 31

Recall from Section 8.1 that we can change any *two* signs of a fraction without changing its value. In Example 7 we change two signs of the fraction to simplify our work.

EXAMPLE 7 Subtract $\frac{x+2}{2x-5} - \frac{3x+5}{5-2x}$.

Solution The denominators of the two fractions differ only in sign. If we change the signs of one of the denominators, we will have a common denominator. Here we change *two* of the signs in the second fraction to obtain a common denominator.

$$\frac{x+2}{2x-5} - \frac{3x+5}{5-2x} = \frac{x+2}{2x-5} + \frac{3x+5}{-(5-2x)}$$

$$= \frac{x+2}{2x-5} + \frac{3x+5}{2x-5}$$

$$= \frac{(x+2) + (3x+5)}{2x-5} \quad \text{Write as a single fraction.}$$

$$= \frac{x + 2 + 3x + 5}{2x-5} \quad \text{Remove parentheses in the numerator.}$$

$$= \frac{4x+7}{2x-5} \quad \text{Combine like terms in the numerator.}$$

In Example 7 we elected to change two signs of the second fraction. The same results could be obtained by multiplying both the numerator and denominator of either the first or second fraction by -1, as was done in Example 6. Try this now.

EXAMPLE 8 Add $\dfrac{3}{x^2 + 5x + 6} + \dfrac{1}{3x^2 + 8x - 3}$.

Solution $$\frac{3}{x^2 + 5x + 6} + \frac{1}{3x^2 + 8x - 3} = \frac{3}{(x + 2)(x + 3)} + \frac{1}{(3x - 1)(x + 3)}$$

The LCD is $(x + 2)(x + 3)(3x - 1)$.

$$= \frac{3x - 1}{3x - 1} \cdot \frac{3}{(x + 2)(x + 3)} + \frac{1}{(3x - 1)(x + 3)} \cdot \frac{x + 2}{x + 2}$$

$$= \frac{9x - 3}{(3x - 1)(x + 2)(x + 3)} + \frac{x + 2}{(3x - 1)(x + 2)(x + 3)}$$

$$= \frac{(9x - 3) + (x + 2)}{(3x - 1)(x + 2)(x + 3)}$$

$$= \frac{9x - 3 + x + 2}{(3x - 1)(x + 2)(x + 3)}$$

$$= \frac{10x - 1}{(3x - 1)(x + 2)(x + 3)}$$

NOW TRY EXERCISE 51

EXAMPLE 9 Subtract $\dfrac{5}{x^2 - 5x} - \dfrac{x}{5x - 25}$.

Solution $$\frac{5}{x^2 - 5x} - \frac{x}{5x - 25} = \frac{5}{x(x - 5)} - \frac{x}{5(x - 5)}$$

The LCD is $5x(x - 5)$.

$$= \frac{5}{5} \cdot \frac{5}{x(x - 5)} - \frac{x}{5(x - 5)} \cdot \frac{x}{x}$$

$$= \frac{25}{5x(x - 5)} - \frac{x^2}{5x(x - 5)}$$

$$= \frac{25 - x^2}{5x(x - 5)}$$

$$= \frac{(5 - x)(5 + x)}{5x(x - 5)}$$ *Factor the numerator.*

$$= \frac{-1(x - 5)(x + 5)}{5x(x - 5)}$$ $5 - x = -1(x - 5)$

$$= \frac{-1\cancel{(x - 5)}(x + 5)}{5x\cancel{(x - 5)}}$$ *Simplify.*

$$= \frac{-1(x + 5)}{5x} \quad \text{or} \quad -\frac{x + 5}{5x}$$

Avoiding Common Errors

A common error in an addition or subtraction problem is to add or subtract the numerators and the denominators. Here is one such example.

CORRECT

$$\frac{1}{x} + \frac{x}{1} = \frac{1}{x} + \frac{x^2}{x}$$

$$= \frac{1 + x^2}{x} \quad \text{or} \quad \frac{x^2 + 1}{x}$$

INCORRECT

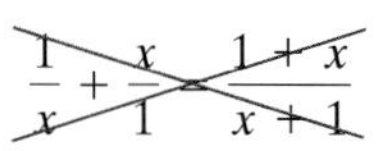

$$\frac{1}{x} - \frac{x}{1} = \frac{1 - x}{x - 1}$$

Remember that to add or subtract fractions you must first have a common denominator. Then you add or subtract the numerators while maintaining the common denominator.

Another common mistake is to treat an addition or subtraction problem as a multiplication problem. You can divide out common factors only when *multiplying* expressions, not when adding or subtracting them.

CORRECT

$$\frac{1}{x} \cdot \frac{x}{1} = \frac{1}{\cancel{x}_1} \cdot \frac{\cancel{x}^1}{1}$$

$$= 1 \cdot 1 = 1$$

INCORRECT

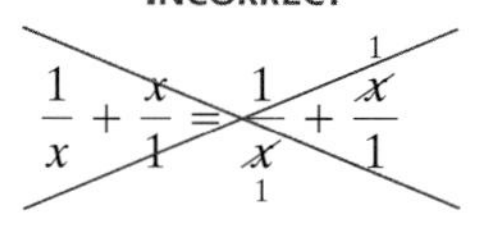

$$= 1 + 1 = 2$$

Exercise Set 8.4

Concept/Writing Exercises

1. When adding or subtracting fractions with unlike denominators, how can you determine what each denominator should be multiplied by to get the LCD?
2. When you multiply both the numerator and denominator of a fraction by the factors needed to obtain the LCD, why are you not changing the value of the fraction?
3. **a)** Explain in your own words a step-by-step procedure to add or subtract two rational expressions that have unlike denominators.

   b) Using the procedure outlined in part **a)**, add $\frac{x}{x^2 - x - 6} + \frac{3}{x^2 - 4}$.
4. Explain how to add or subtract fractions whose denominators are opposites. Give an example.
5. Consider $\frac{x}{3y} + \frac{5}{6y^2}$.

   a) What is the LCD?

   b) Perform the indicated operation.

   c) If you mistakenly used $18y^3$ for the LCD when adding, would you eventually obtain the correct answer? Explain.
6. Would you use the LCD to perform the following indicated operations? Explain.

   a) $\frac{3}{x + 2} - \frac{4}{x} + \frac{5}{2}$

   b) $\frac{1}{x + 7} \cdot \frac{5}{x}$

   c) $x + \frac{2}{3}$

   d) $\frac{5}{x^2 - 9} \div \frac{2}{x - 3}$

Practice the Skills

Add or subtract.

7. $\frac{1}{3x} + \frac{5}{x}$

8. $\frac{1}{4x} + \frac{1}{x}$

9. $\frac{4}{x^2} + \frac{3}{2x}$

10. $2 - \frac{1}{x^2}$

11. $3 + \frac{5}{x}$

12. $\frac{5}{6y} + \frac{3}{4y^2}$

13. $\frac{2}{x^2} + \frac{3}{5x}$

14. $\frac{3}{x} - \frac{5}{x^2}$

15. $\frac{7}{4x^2y} + \frac{3}{5xy^2}$

16. $\frac{5}{12x^4y} - \frac{1}{5x^2y^3}$

17. $y + \frac{x}{y}$

18. $x + \frac{x}{y}$

19. $\frac{3a - 1}{2a} + \frac{2}{3a}$

20. $\frac{3}{x} + 4$

21. $\frac{4x}{y} + \frac{2y}{xy}$

22. $\frac{3}{5p} - \frac{5}{2p^2}$

23. $\frac{6}{b} - \frac{4}{5a^2}$

24. $\frac{x - 3}{x} - \frac{1}{4x}$

25. $\frac{3}{x} + \frac{7}{x - 4}$

26. $6 - \frac{3}{x - 3}$

27. $\frac{9}{p + 3} + \frac{2}{p}$

28. $\frac{a}{a + b} + \frac{a - b}{a}$

29. $\frac{3}{4x} - \frac{x}{3x + 5}$

30. $\frac{2}{x - 3} - \frac{4}{x - 1}$

31. $\frac{4}{p - 3} + \frac{2}{3 - p}$

32. $\frac{3}{x - 2} - \frac{1}{2 - x}$

33. $\frac{9}{x + 7} - \frac{5}{-x - 7}$

34. $\frac{6}{7x - 1} - \frac{5}{1 - 7x}$

35. $\frac{3}{a - 2} + \frac{a}{2a - 4}$

36. $\frac{4}{y - 1} + \frac{3}{y + 1}$

37. $\frac{x + 5}{x - 5} - \frac{x - 5}{x + 5}$

38. $\frac{x + 7}{x + 3} - \frac{x - 3}{x + 7}$

39. $\frac{x}{x^2 - 9} + \frac{4}{x + 3}$

40. $\frac{5}{(x + 4)^2} + \frac{2}{x + 4}$

41. $\frac{x + 2}{x^2 - 4} - \frac{2}{x + 2}$

42. $\frac{3}{(x - 2)(x + 3)} + \frac{5}{(x + 2)(x + 3)}$

43. $\frac{2x + 3}{x^2 - 7x + 12} - \frac{2}{x - 3}$

44. $\frac{x + 3}{x^2 - 3x - 10} - \frac{2}{x - 5}$

45. $\frac{x^2}{x^2 + 2x - 8} - \frac{x - 4}{x + 4}$

46. $\frac{x + 1}{x^2 - 2x + 1} - \frac{x + 1}{x - 1}$

47. $\frac{x - 3}{x^2 + 10x + 25} + \frac{x - 3}{x + 5}$

48. $\frac{x}{x^2 - xy} - \frac{y}{xy - x^2}$

49. $\frac{5}{a^2 - 9a + 8} - \frac{3}{a^2 - 6a - 16}$

50. $\frac{3}{a^2 + a - 6} + \frac{1}{a^2 + 3a - 10}$

51. $\frac{1}{x^2 - 4} + \frac{3}{x^2 + 5x + 6}$

52. $\frac{x}{2x^2 + 7x - 4} + \frac{2}{x^2 - x - 20}$

53. $\frac{x}{2x^2 + 7x + 3} - \frac{3}{3x^2 + 7x - 6}$

54. $\frac{x}{6x^2 + 7x + 2} + \frac{5}{2x^2 - 3x - 2}$

55. $\frac{x}{4x^2 + 11x + 6} - \frac{2}{8x^2 + 2x - 3}$

56. $\frac{x}{5x^2 - 9x - 2} - \frac{2}{3x^2 - 7x + 2}$

57. $\frac{3x + 12}{x^2 + x - 12} - \frac{2}{x - 3}$

58. $\frac{5x + 10}{x^2 - 5x - 14} - \frac{4}{x - 7}$

Problem Solving

For what value(s) of x is each expression defined?

59. $\frac{3}{x} + 4$

60. $\frac{2}{x-1} - \frac{3}{x}$

61. $\frac{5}{x-3} + \frac{7}{x+6}$

62. $\frac{4}{x^2-9} - \frac{1}{x+3}$

Add or subtract. Treat the unknown symbols as if they were variables.

63. $\frac{3}{\Delta - 2} - \frac{1}{2 - \Delta}$

64. $\frac{\Delta}{2\Delta^2 + 7\Delta - 4} + \frac{2}{\Delta^2 - \Delta - 20}$

Challenge Problems

Under what conditions is each expression defined? Explain your answers.

65. $\frac{5}{a+b} + \frac{3}{a}$

66. $\frac{x+2}{x+5y} - \frac{y-3}{2x}$

Perform each indicated operation.

67. $\frac{x}{x^2-4} + \frac{3x}{x+2} + \frac{3x^2-5x}{4-x^2}$

68. $\frac{5x}{x^2+x-6} + \frac{x}{x+3} - \frac{2}{x-2}$

69. $\frac{x+6}{4-x^2} - \frac{x+3}{x+2} + \frac{x-3}{2-x}$

70. $\frac{3x-1}{x+2} + \frac{x}{x-3} - \frac{4}{2x+3}$

71. $\frac{2}{x^2-x-6} + \frac{3}{x^2-2x-3} + \frac{1}{x^2+3x+2}$

72. $\frac{3x}{x^2-4} + \frac{4}{x^3+8}$

Group Activity

Discuss and answer Exercise 73 as a group.

73. a) As a group, find the LCD of

$$\frac{x+3y}{x^2+3xy+2y^2} + \frac{y-x}{2x^2+3xy+y^2}$$

b) As a group, perform the indicated operation, but do not simplify your answer.

c) As a group, simplify your answer.

d) Group member 1: Substitute 2 for x and 1 for y in the fraction on the left in part **a)** and evaluate.

e) Group member 2: Substitute 2 for x and 1 for y in the fraction on the right in part **a)** and evaluate.

f) Group member 3: Add the numerical fractions found in parts **d)** and **e)**.

g) Individually, substitute 2 for x and 1 for y in the expression obtained in part **b)** and evaluate.

h) Individually, substitute 2 for x and 1 for y in the expression obtained in part **c)**, evaluate, and compare your answers.

i) As a group, discuss what you discovered from this activity.

j) Do you think your results would have been similar for any numbers substituted for x and y (for which the denominator is not 0)? Why?

Cumulative Review Exercises

[2.6] **74.** A videocassette recorder counter will go from 0 to 18 in 2 minutes. There are two movies on a VCR tape. If Gustav Schmidt wishes to watch the second movie and the first movie is $2\frac{1}{4}$ hours long, what will be the number on the counter at the end of the first movie, where the second movie starts?

[2.7] **75.** Solve the inequality $3(x-2)+2 < 4(x+1)$ and graph the solution on a number line.

[6.6] **76.** Divide $(8x^2+6x-13) \div (2x+3)$.

[8.2] **77.** Multiply $\frac{x^2+xy-6y^2}{x^2-xy-2y^2} \cdot \frac{y^2-x^2}{x^2+2xy-3y^2}$.

8.5 COMPLEX FRACTIONS

1) **Simplify complex fractions by combining terms.**
2) **Simplify complex fractions using multiplication first to clear fractions.**

1) Simplify Complex Fractions by Combining Terms

A **complex fraction** is one that has a fraction in its numerator or its denominator or in both its numerator and denominator.

Examples of Complex Fractions

$$\frac{\frac{3}{5}}{4} \qquad \frac{\frac{x+1}{x}}{2x} \qquad \frac{\frac{x}{y}}{x+1} \qquad \frac{\frac{a+b}{a}}{\frac{a-b}{b}}$$

Numerator of complex fraction $\left\{ \frac{a+b}{a} \right.$

———— ← *Main fraction line*

Denominator of complex fraction $\left\{ \frac{a-b}{b} \right.$

The expression above the main fraction line is the numerator, and the expression below the main fraction line is the denominator of the complex fraction.

There are two methods to simplify complex fractions. The first reinforces many of the concepts used in this chapter because we may need to add, subtract, multiply, and divide simpler fractions as we simplify the complex fraction. Many students prefer to use the second because the answer may be obtained more quickly. We will give three examples using the first method and then work the same three examples using the second method.

To Simplify a Complex Fraction by Combining Terms

1. Add or subtract the fractions in both the numerator and denominator of the complex fraction to obtain single fractions in both the numerator and the denominator.
2. Invert the denominator of the complex fraction and multiply the numerator by it.
3. Simplify further if possible.

EXAMPLE 1 Simplify $\dfrac{\frac{1}{3}+\frac{4}{5}}{\frac{4}{5}-\frac{1}{3}}$.

Solution First, add the terms in the numerator and subtract the terms in the denominator to obtain single fractions in both the numerator and denominator. The LCD of both the numerator and the denominator is 15.

$$\frac{\frac{1}{3}+\frac{4}{5}}{\frac{4}{5}-\frac{1}{3}} = \frac{\frac{5}{5}\cdot\frac{1}{3}+\frac{4}{5}\cdot\frac{3}{3}}{\frac{3}{3}\cdot\frac{4}{5}-\frac{1}{3}\cdot\frac{5}{5}} = \frac{\frac{5}{15}+\frac{12}{15}}{\frac{12}{15}-\frac{5}{15}} = \frac{\frac{17}{15}}{\frac{7}{15}}$$

Next, invert the denominator and multiply the numerator by it.

$$= \frac{\frac{17}{15}}{\frac{7}{15}} = \frac{17}{\cancel{15}}\cdot\frac{\cancel{15}}{7} = \frac{17}{7}$$

NOW TRY EXERCISE 9

EXAMPLE 2 Simplify $\dfrac{a+\frac{1}{x}}{x+\frac{1}{a}}$.

Solution Express both the numerator and denominator of the complex fraction as single fractions. The LCD of the numerator is x and the LCD of the denominator is a.

$$\frac{a+\frac{1}{x}}{x+\frac{1}{a}} = \frac{\frac{x}{x}\cdot a+\frac{1}{x}}{\frac{a}{a}\cdot x+\frac{1}{a}} = \frac{\frac{ax}{x}+\frac{1}{x}}{\frac{ax}{a}+\frac{1}{a}} = \frac{\frac{ax+1}{x}}{\frac{ax+1}{a}}$$

Now invert the denominator and multiply the numerator by it.

$$= \frac{\cancel{ax+1}}{x}\cdot\frac{a}{\cancel{ax+1}} = \frac{a}{x}$$

EXAMPLE 3 Simplify $\dfrac{x}{\frac{1}{x}+\frac{1}{y}}$.

Solution

$$\frac{x}{\frac{1}{x}+\frac{1}{y}} = \frac{x}{\frac{y}{y}\cdot\frac{1}{x}+\frac{1}{y}\cdot\frac{x}{x}} = \frac{x}{\frac{y}{xy}+\frac{x}{xy}} = \frac{x}{\frac{y+x}{xy}} = \frac{x}{1}\div\frac{y+x}{xy}$$

Now invert the divisor and multiply.

$$= \frac{x}{1}\cdot\frac{xy}{y+x} = \frac{x^2y}{y+x} \quad\text{or}\quad \frac{x^2y}{x+y}$$

NOW TRY EXERCISE 17

2) Simplify Complex Fractions Using Multiplication First to Clear Fractions

Here is the second method for simplifying complex fractions.

To Simplify a Complex Fraction Using Multiplication First

1. Find the least common denominator of *all* the denominators appearing in the complex fraction.
2. Multiply both the numerator and denominator of the complex fraction by the LCD found in step 1.
3. Simplify when possible.

We will now rework Examples 1, 2, and 3 using the given procedure.

EXAMPLE 4 Simplify $\dfrac{\frac{1}{3}+\frac{4}{5}}{\frac{4}{5}-\frac{1}{3}}$.

Solution The denominators in the complex fraction are 3 and 5. The LCD of 3 and 5 is 15. Thus 15 is the LCD of the complex fraction. Multiply both the numerator and denominator of the complex fraction by 15.

$$\frac{\frac{1}{3}+\frac{4}{5}}{\frac{4}{5}-\frac{1}{3}} = \frac{15}{15} \cdot \frac{\left(\frac{1}{3}+\frac{4}{5}\right)}{\left(\frac{4}{5}-\frac{1}{3}\right)} = \frac{15\left(\frac{1}{3}\right)+15\left(\frac{4}{5}\right)}{15\left(\frac{4}{5}\right)-15\left(\frac{1}{3}\right)}$$

Now simplify.

$$= \frac{5+12}{12-5} = \frac{17}{7}$$

NOW TRY EXERCISE 9

Note that the answers to Examples 1 and 4 are the same.

EXAMPLE 5 Simplify $\dfrac{a+\frac{1}{x}}{x+\frac{1}{a}}$.

Solution The denominators in the complex fraction are x and a. Therefore, the LCD of the complex fraction is ax. Multiply both the numerator and denominator of the complex fraction by ax.

$$\frac{a+\frac{1}{x}}{x+\frac{1}{a}} = \frac{ax}{ax} \cdot \frac{\left(a+\frac{1}{x}\right)}{\left(x+\frac{1}{a}\right)} = \frac{a^2x+a}{ax^2+x}$$

$$= \frac{a\cancel{(ax+1)}}{x\cancel{(ax+1)}} = \frac{a}{x}$$

Note that the answers to Examples 2 and 5 are the same.

EXAMPLE 6 Simplify $\dfrac{x}{\frac{1}{x}+\frac{1}{y}}$.

Solution The denominators in the complex fraction are x and y. Therefore, the LCD of the complex fraction is xy. Multiply both the numerator and denominator of the complex fraction by xy.

$$\frac{x}{\frac{1}{x}+\frac{1}{y}} = \frac{xy}{xy} \cdot \frac{x}{\left(\frac{1}{x}+\frac{1}{y}\right)}$$

$$= \frac{x^2y}{xy\left(\frac{1}{x}\right)+xy\left(\frac{1}{y}\right)}$$

$$= \frac{x^2y}{y+x}$$

NOW TRY EXERCISE 17

Note that the answers to Examples 3 and 6 are the same. When asked to simplify a complex fraction, you may use either method unless you are told by your instructor to use a specific method.

Exercise Set 8.5

Concept/Writing Exercises

1. What is a complex fraction?

2. What is the numerator and denominator of each complex fraction?

a) $\dfrac{5}{\dfrac{3}{x^2 + 5x + 6}}$

b) $\dfrac{\dfrac{5}{3}}{x^2 + 5x + 6}$

3. What is the numerator and denominator of each complex fraction?

a) $\dfrac{\dfrac{x - 1}{5}}{\dfrac{2}{x^2 + 5x + 6}}$

b) $\dfrac{\dfrac{1}{2y} + x}{\dfrac{3}{y} + x}$

4. a) Select the method you prefer to use to simplify complex fractions. Then write in your own words a step-by-step procedure for simplifying complex fractions using that method.

b) Using the procedure you wrote in part **a)**, simplify the following complex fraction.

$$\dfrac{\dfrac{2}{x} - \dfrac{3}{y}}{x + \dfrac{1}{y}}$$

Practice the Skills

Simplify.

5. $\dfrac{3 + \dfrac{2}{3}}{1 + \dfrac{1}{3}}$

6. $\dfrac{3 + \dfrac{4}{5}}{1 - \dfrac{9}{16}}$

7. $\dfrac{2 + \dfrac{3}{8}}{1 + \dfrac{1}{3}}$

8. $\dfrac{\dfrac{1}{5} + \dfrac{5}{6}}{\dfrac{2}{7} + \dfrac{3}{5}}$

9. $\dfrac{\dfrac{2}{7} - \dfrac{1}{4}}{6 - \dfrac{2}{3}}$

10. $\dfrac{1 - \dfrac{x}{y}}{x}$

11. $\dfrac{\dfrac{xy^2}{6}}{\dfrac{3}{y}}$

12. $\dfrac{\dfrac{12a}{b^3}}{\dfrac{b^2}{4}}$

13. $\dfrac{\dfrac{8x^2y}{3z^3}}{\dfrac{4xy}{9z^5}}$

14. $\dfrac{\dfrac{36x^4}{5y^4z^5}}{\dfrac{9xy^2}{15z^5}}$

15. $\dfrac{a - \dfrac{a}{b}}{\dfrac{1 + a}{b}}$

16. $\dfrac{a + \dfrac{1}{b}}{\dfrac{a}{b}}$

17. $\dfrac{\dfrac{9}{x} + \dfrac{3}{x^2}}{3 + \dfrac{1}{x}}$

18. $\dfrac{\dfrac{2}{a} + \dfrac{1}{2a}}{a + \dfrac{a}{2}}$

19. $\dfrac{5 - \dfrac{1}{x}}{4 - \dfrac{1}{x}}$

20. $\dfrac{\dfrac{x}{x - y}}{\dfrac{x^2}{y}}$

21. $\dfrac{\dfrac{y}{x} + \dfrac{x}{y}}{\dfrac{x - y}{x}}$

22. $\dfrac{1}{\dfrac{1}{x} + y}$

23. $\dfrac{\dfrac{a^2}{b} - b}{\dfrac{b^2}{a} - a}$

24. $\dfrac{\dfrac{1}{x^2} - \dfrac{3}{x}}{3 + \dfrac{1}{x^2}}$

25. $\dfrac{-\frac{a}{b}+3}{\frac{a}{b}-3}$ **26.** $\dfrac{\frac{x}{y}-2}{\frac{-x}{y}+2}$ **27.** $\dfrac{\frac{a^2-b^2}{a}}{\frac{a+b}{a^3}}$ **28.** $\dfrac{\frac{1}{a}+\frac{1}{b}}{ab}$

29. $\dfrac{\frac{1}{a}-\frac{1}{b}}{\frac{1}{ab}}$ **30.** $\dfrac{\frac{1}{ab}}{\frac{1}{a}+\frac{1}{b}}$ **31.** $\dfrac{\frac{a}{b}+\frac{1}{a}}{\frac{b}{a}+\frac{1}{a}}$ **32.** $\dfrac{\frac{1}{a}+\frac{1}{b}}{\frac{1}{a}}$

33. $\dfrac{\frac{1}{x}+\frac{1}{y}}{\frac{1}{x}-\frac{1}{y}}$ **34.** $\dfrac{\frac{4}{x^2}+\frac{4}{x}}{\frac{4}{x}+\frac{4}{x^2}}$ **35.** $\dfrac{\frac{3}{a}+\frac{3}{a^2}}{\frac{3}{b}+\frac{3}{b^2}}$ **36.** $\dfrac{\frac{x}{y}-\frac{1}{x}}{\frac{y}{x}+1}$

Problem Solving

For the complex fractions in Exercises 37–40,
a) *determine which of the two methods discussed in this section you would use to simplify the fraction. Explain why.*
b) *Simplify by the method you selected in part **a)**.*
c) *Simplify by the method you did not select in part **a)**. If your answers to parts **b)** and **c)** are not the same, explain why.*

37. $\dfrac{5+\frac{3}{7}}{\frac{1}{9}-4}$ **38.** $\dfrac{\frac{x+y}{x^3}-\frac{1}{x}}{\frac{x-y}{x^5}+5}$ **39.** $\dfrac{\frac{x-y}{x+y}+\frac{3}{x+y}}{2-\frac{7}{x+y}}$ **40.** $\dfrac{\frac{25}{x-y}+\frac{4}{x+y}}{\frac{5}{x-y}-\frac{3}{x+y}}$

*In Exercises 41 and 42, **a)** write a complex fraction to represent the statement, and **b)** simplify the complex fraction.*

41. The numerator of the complex fraction consists of one term: 5 is divided by $12x$. The denominator of the complex fraction consists of two terms: 4 divided by $3x$ is subtracted from 8 divided by x^2.

42. The numerator of the complex fraction consists of two terms: 3 divided by $2x$ is subtracted from 6 divided by x. The denominator of the complex fraction consists of two terms: the sum of x and the quantity 1 divided by x.

Simplify. (Hint: Refer to Section 6.2, which discusses negative exponents.)

43. $\dfrac{x^{-1}+y^{-1}}{2}$ **44.** $\dfrac{x^{-1}+y^{-1}}{x^{-1}}$ **45.** $\dfrac{x^{-1}+y^{-1}}{x^{-1}y^{-1}}$ **46.** $\dfrac{x^{-2}-y^{-2}}{y^{-1}-x^{-1}}$

Challenge Problems

47. The efficiency of a jack, E, is expressed by the formula $E=\dfrac{\frac{1}{2}h}{h+\frac{1}{2}}$, where h is determined by the pitch of the jack's thread. Determine the efficiency of a jack if h is **a)** $\frac{2}{3}$ **b)** $\frac{4}{5}$

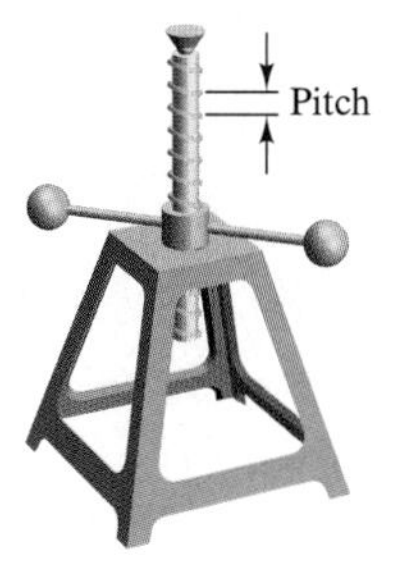

Simplify.

48. $\dfrac{\frac{x}{y}+\frac{y}{x}+\frac{1}{x}}{\frac{x}{y}+y}$ **49.** $\dfrac{\frac{a}{b}+b-\frac{1}{a}}{\frac{a}{b^2}-\frac{b}{a}+\frac{1}{a^2}}$ **50.** $\dfrac{x}{1+\frac{x}{1+x}}$

Cumulative Review Exercises

[2.5] **51.** Solve the equation
$2x - 8(5 - x) = 9x - 3(x + 2)$.

[5.3] **52.** Solve the following system of equations.

$$2x + y = 7$$
$$3x - 2y = 21$$

[6.4] **53.** What is a polynomial?

[8.4] **54.** Subtract $\frac{x}{3x^2 + 17x - 6} - \frac{2}{x^2 + 3x - 18}$.

8.6 SOLVING RATIONAL EQUATIONS

1) **Solve rational equations.**

SSM VIDEO 8.6 CD Rom

1) Solve Rational Equations

In Sections 8.1 through 8.5 we focused on how to add, subtract, multiply, and divide rational expressions. Now we are ready to solve rational equations. A **rational equation** is one that contains one or more rational expressions.

To Solve Rational Equations

1. Determine the LCD of all fractions in the equation.
2. Multiply **both** sides of the equation by the LCD. **This will result in every term in the equation being multiplied by the LCD.**
3. Remove any parentheses and combine like terms on each side of the equation.
4. Solve the equation using the properties discussed in earlier chapters.
5. Check your solution in the *original* equation.

The purpose of multiplying both sides of the equation by the LCD (step 2) is to eliminate all fractions from the equation. After both sides of the equation are multiplied by the LCD, the resulting equation should contain no fractions. We will omit some of the checks to save space.

EXAMPLE 1 Solve $\frac{x}{2} + 4x = 9$.

Solution Multiply both sides of the equation by the LCD, 2.

$$2\left(\frac{x}{2} + 4x\right) = 9 \cdot 2$$

$$\cancel{2}\left(\frac{x}{\cancel{2}}\right) + 2 \cdot 4x = 9 \cdot 2 \qquad \text{Distributive Property}$$

$$x + 8x = 18$$

$$9x = 18$$

$$x = 2$$

CHECK:

$$\frac{x}{2} + 4x = 9$$
$$\frac{2}{2} + 4(2) \stackrel{?}{=} 9$$
$$1 + 8 \stackrel{?}{=} 9$$
$$9 = 9 \quad \textit{True}$$

EXAMPLE 2 Solve the equation $\frac{2}{5} + \frac{3x}{4} = \frac{x}{2}$.

Solution Multiply both sides of the equation by the LCD, 20.

$$20\left(\frac{2}{5} + \frac{3x}{4}\right) = \frac{x}{2} \cdot 20$$
$$\overset{4}{\cancel{20}}\left(\frac{2}{\cancel{5}}\right) + \overset{5}{\cancel{20}}\left(\frac{3x}{\cancel{4}}\right) = \frac{x}{\cancel{2}} \cdot \overset{10}{\cancel{20}} \quad \textit{Distributive Property}$$
$$8 + 15x = 10x$$
$$8 = -5x$$
$$-\frac{8}{5} = x$$

NOW TRY EXERCISE 23

A check will show that $-\frac{8}{5}$ is the solution.

EXAMPLE 3 Solve the equation $\frac{x}{4} + 3 = 2(x - 2)$.

Solution Multiply both sides of the equation by the LCD, 4.

$$\frac{x}{4} + 3 = 2(x - 2)$$
$$4\left(\frac{x}{4} + 3\right) = 4[2(x - 2)]$$
$$4\left(\frac{x}{4}\right) + 4(3) = 4[2(x - 2)] \quad \textit{Distributive Property}$$
$$\cancel{4}\left(\frac{x}{\cancel{4}}\right) + 4(3) = 8(x - 2)$$
$$x + 12 = 8(x - 2) \quad \textit{Simplify.}$$
$$x + 12 = 8x - 16 \quad \textit{Distributive Property}$$
$$12 = 7x - 16 \quad \textit{Subtract } x \textit{ from both sides.}$$
$$28 = 7x \quad \textit{Add 16 to both sides.}$$
$$4 = x \quad \textit{Divide both sides by 7.}$$

WARNING: **Whenever a variable appears in any denominator of a rational equation, it is necessary to check your answer in the original equation. If the answer obtained makes any denominator equal to zero, that value is not a solution to the equation.** Such values are called **extraneous roots** or **extraneous solutions.**

EXAMPLE 4 Solve the equation $3 - \frac{4}{x} = \frac{5}{2}$.

Solution Multiply both sides of the equation by the LCD, $2x$.

$$2x\left(3 - \frac{4}{x}\right) = \left(\frac{5}{2}\right) \cdot 2x$$

$$2x(3) - 2x\left(\frac{4}{x}\right) = \left(\frac{5}{2}\right) \cdot 2x$$

$$6x - 8 = 5x$$

$$x - 8 = 0$$

$$x = 8$$

CHECK:

$$3 - \frac{4}{x} = \frac{5}{2}$$

$$3 - \frac{4}{8} \stackrel{?}{=} \frac{5}{2}$$

$$3 - \frac{1}{2} \stackrel{?}{=} \frac{5}{2}$$

$$\frac{5}{2} = \frac{5}{2} \quad \text{True}$$

Since 8 does check, it is the solution to the equation.

EXAMPLE 5 Solve the equation $\frac{x-7}{x+2} = \frac{1}{4}$.

Solution The LCD is $4(x + 2)$. Multiply both sides of the equation by the LCD.

$$4(x+2) \cdot \frac{x-7}{x+2} = \frac{1}{4} \cdot 4(x+2)$$

$$4(x - 7) = 1(x + 2)$$

$$4x - 28 = x + 2$$

$$3x - 28 = 2$$

$$3x = 30$$

$$x = 10$$

A check will show that 10 is the solution.

In Section 2.6 we illustrated that proportions of the form

$$\frac{a}{b} = \frac{c}{d}$$

can be cross-multiplied to obtain $a \cdot d = b \cdot c$. Example 5 is a proportion and can also be solved by cross-multiplying, as is done in Example 6.

EXAMPLE 6 Use cross-multiplication to solve the equation $\frac{6}{x+3} = \frac{5}{x-2}$.

Solution

$$\frac{6}{x+3} = \frac{5}{x-2}$$

$$6(x-2) = 5(x+3) \quad \textit{Cross multiply.}$$

$$6x - 12 = 5x + 15 \quad \textit{Distributive property.}$$

$$x - 12 = 15 \quad \textit{Subtract 5x from both sides.}$$

$$x = 27 \quad \textit{Add 12 to both sides.}$$

NOW TRY EXERCISE 31

A check will show that 27 is the solution to the equation.

Now let's examine some examples that involve quadratic equations. Recall from Section 7.6 that quadratic equations have the form $ax^2 + bx + c = 0$, where $a \neq 0$.

EXAMPLE 7 Solve the equation $x + \frac{12}{x} = -7$.

Solution

$$x \cdot \left(x + \frac{12}{x}\right) = -7 \cdot x \quad \textit{Multiply both sides by } x.$$

$$x(x) + \not{x}\left(\frac{12}{\not{x}}\right) = -7x \quad \textit{Distributive Property}$$

$$x^2 + 12 = -7x$$

$$x^2 + 7x + 12 = 0 \quad \textit{Add 7x to both sides.}$$

$$(x+3)(x+4) = 0 \quad \textit{Factor.}$$

$$x + 3 = 0 \quad \text{or} \quad x + 4 = 0 \quad \textit{Zero-factor property}$$

$$x = -3 \qquad\qquad x = -4$$

CHECK:

$x = -3$		$x = -4$	
$x + \frac{12}{x} = -7$		$x + \frac{12}{x} = -7$	
$-3 + \frac{12}{-3} \stackrel{?}{=} -7$		$-4 + \frac{12}{-4} \stackrel{?}{=} -7$	
$-3 + (-4) \stackrel{?}{=} -7$		$-4 + (-3) \stackrel{?}{=} -7$	
$-7 = -7$	*True*	$-7 = -7$	*True*

NOW TRY EXERCISE 41

The solutions are −3 and −4.

EXAMPLE 8 Solve the equation $\frac{2x-5}{x-6} = \frac{7}{x-6}$.

Solution Using cross-multiplication we get

$$\frac{2x-5}{x-6} = \frac{7}{x-6}$$
$$(2x-5)(x-6) = 7(x-6)$$
$$2x^2 - 17x + 30 = 7x - 42$$
$$2x^2 - 24x + 72 = 0$$
$$2(x^2 - 12x + 36) = 0$$
$$2(x-6)(x-6) = 0$$
$$x - 6 = 0 \quad \text{or} \quad x - 6 = 0$$
$$x = 6 \qquad\qquad x = 6$$

CHECK: $x = 6$

$$\frac{2x-5}{x-6} = \frac{7}{x-6}$$
$$\frac{2(6)-5}{6-6} \stackrel{?}{=} \frac{7}{6-6}$$
$$\frac{7}{0} = \frac{7}{0} \qquad \textit{Not a real number}$$

Since 7/0 is not a real number, 6 is an extraneous solution. Thus, this equation has *no solution*.

NOW TRY EXERCISE 35

EXAMPLE 9 Solve the equation $\frac{x^2}{x-3} = \frac{9}{x-3}$.

Solution If we try to solve this proportion using cross-multiplication we will get $x^2(x-3) = 9(x-3)$, which simplifies to $x^3 - 3x^2 = 9x - 27$. This is an example of a *cubic equation* since the greatest exponent on the variable x is 3. Since solving cubic equations is beyond the scope of this course, we will need to try another procedure to solve the original equation. If we multiply both sides of the original equation by the least common denominator, $x - 3$, we can solve the equation as follows.

$$\cancel{x-3} \cdot \frac{x^2}{\cancel{x-3}} = \frac{9}{\cancel{x-3}} \cdot \cancel{x-3}$$
$$x^2 = 9$$
$$x^2 - 9 = 0 \qquad \textit{This is a difference of two squares.}$$
$$(x+3)(x-3) = 0 \qquad \textit{Factor}$$
$$x + 3 = 0 \quad \text{or} \quad x - 3 = 0 \qquad \textit{Zero-factor property}$$
$$x = -3 \qquad\qquad x = 3$$

CHECK: $x = -3$ | $x = 3$

$$\frac{x^2}{x-3} = \frac{9}{x-3} \qquad \frac{x^2}{x-3} = \frac{9}{x-3}$$

$$\frac{(-3)^2}{-3-3} \stackrel{?}{=} \frac{9}{-3-3} \qquad \frac{(3)^2}{3-3} \stackrel{?}{=} \frac{9}{3-3}$$

$$\frac{9}{-6} \stackrel{?}{=} \frac{9}{-6} \qquad \frac{9}{0} = \frac{9}{0} \quad \text{Not a real number}$$

$$-\frac{3}{2} = -\frac{3}{2} \quad \text{True}$$

Since 3 results in a denominator of 0, $x = 3$ is *not* a solution to the equation. The 3 is an *extraneous root*. The only solution to the equation is $x = -3$.

EXAMPLE 10 Solve the equation $\dfrac{5x}{x^2-4} + \dfrac{1}{x-2} = \dfrac{4}{x+2}$.

Solution First factor $x^2 - 4$.

$$\frac{5x}{(x+2)(x-2)} + \frac{1}{x-2} = \frac{4}{x+2}$$

Multiply both sides of the equation by the LCD, $(x+2)(x-2)$.

$$(x+2)(x-2)\left[\frac{5x}{(x+2)(x-2)} + \frac{1}{x-2}\right] = \frac{4}{x+2} \cdot (x+2)(x-2)$$

$$(x+2)(x-2) \cdot \frac{5x}{(x+2)(x-2)} + (x+2)(x-2) \cdot \frac{1}{x-2} = \frac{4}{x+2} \cdot (x+2)(x-2)$$

$$\cancel{(x+2)}\cancel{(x-2)} \cdot \frac{5x}{\cancel{(x+2)}\cancel{(x-2)}} + (x+2)\cancel{(x-2)} \cdot \frac{1}{\cancel{x-2}} = \frac{4}{\cancel{x+2}} \cdot \cancel{(x+2)}(x-2)$$

$$5x + (x+2) = 4(x-2)$$

$$6x + 2 = 4x - 8$$

$$2x + 2 = -8$$

$$2x = -10$$

$$x = -5$$

NOW TRY EXERCISE 49 A check will show that -5 is the solution to the equation.

HELPFUL HINT

Some students confuse adding and subtracting rational expressions with solving rational equations. When adding or subtracting rational expressions, we must rewrite each expression with a common denominator. When solving a rational equation, we multiply both sides of the equation by the LCD to eliminate fractions from the equation. Consider the following two problems. Note that the one on the right is an equation because it contains an equal sign. We will work both problems. The LCD for both problems is $x(x + 4)$.

Adding Rational Expressions

$$\frac{x+2}{x+4} + \frac{3}{x}$$

We rewrite each fraction with the LCD, $x(x + 4)$.

$$= \frac{x}{x} \cdot \frac{x+2}{x+4} + \frac{3}{x} \cdot \frac{x+4}{x+4}$$

$$= \frac{x(x+2)}{x(x+4)} + \frac{3(x+4)}{x(x+4)}$$

$$= \frac{x^2+2x}{x(x+4)} + \frac{3x+12}{x(x+4)}$$

$$= \frac{x^2+2x+3x+12}{x(x+4)}$$

$$= \frac{x^2+5x+12}{x(x+4)}$$

Solving Rational Equations

$$\frac{x+2}{x+4} = \frac{3}{x}$$

We eliminate fractions by multiplying both sides of the equation by the LCD, $x(x + 4)$.

$$(x)(x+4)\left(\frac{x+2}{x+4}\right) = \frac{3}{x}(x)(x+4)$$

$$x(x+2) = 3(x+4)$$

$$x^2 + 2x = 3x + 12$$

$$x^2 - x - 12 = 0$$

$$(x-4)(x+3) = 0$$

$$x - 4 = 0 \quad \text{or} \quad x + 3 = 0$$

$$x = 4 \qquad\qquad x = -3$$

The numbers 4 and −3 on the right will both check and are thus solutions to the equation.

Note that when adding and subtracting rational expressions we usually end up with an algebraic expression. When solving rational equations, the solution will be a numerical value or values. The equation on the right could also be solved using cross-multiplication.

Exercise Set 8.6

Concept/Writing Exercises

1. **a)** Explain in your own words the steps to use to solve rational equations.

b) Using the procedure you wrote in part **a)**, solve the equation $\frac{1}{x-1} - \frac{1}{x+1} = \frac{3x}{x^2-1}$.

2. Consider the equation $\frac{x^2}{x+3} = \frac{25}{x+3}$.

a) Explain why you may have difficulty trying to solve this equation by cross-multiplication.

b) Find the solution to the equation.

3. Consider the following problems.

Simplify: $\frac{x}{3} - \frac{x}{4} + \frac{1}{x-1}$

Solve: $\frac{x}{3} - \frac{x}{4} = \frac{1}{x-1}$

a) Explain the difference between the two types of problems.

b) Explain how you would work each problem to obtain the correct answer.

c) Find the correct answer to each problem.

4. Consider the following problems.

Simplify: $\frac{x}{2} - \frac{x}{3} + \frac{5}{2x + 7}$

Solve: $\frac{x}{2} - \frac{x}{3} = \frac{5}{2x + 7}$

a) Explain the difference between the two types of problems.

b) Explain how you would work each problem to obtain the correct answer.

c) Find the correct answer to each problem then solve.

5. Under what conditions must you check rational equations for extraneous solutions?

In Exercises 6–8, indicate whether it is necessary to check the solution obtained to see if it is extraneous. Explain your answer.

6. $\frac{3}{x - 2} + \frac{1}{x} = 6$

7. $\frac{4}{7} + \frac{x - 2}{5} = 5$

8. $\frac{2}{x + 4} + 3 = 6$

Practice the Skills

Solve each equation, and check your solution.

9. $\frac{4}{7} = \frac{x}{21}$

10. $\frac{5}{y} = \frac{15}{33}$

11. $\frac{6}{13} = \frac{24}{x}$

12. $\frac{x}{4} = \frac{-20}{16}$

13. $\frac{12}{10} = \frac{z}{25}$

14. $\frac{9c}{10} = \frac{9}{5}$

15. $\frac{-2}{6} = \frac{3c}{9}$

16. $\frac{80}{2x} = \frac{8}{5}$

17. $\frac{x + 4}{7} = \frac{3}{7}$

18. $\frac{1}{4} = \frac{z + 1}{4}$

19. $\frac{4x + 5}{6} = \frac{7}{2}$

20. $\frac{a}{5} = \frac{a - 3}{2}$

21. $6 - \frac{x}{4} = \frac{x}{8}$

22. $\frac{n}{10} = 9 - \frac{n}{5}$

23. $\frac{x}{3} - \frac{3x}{4} = \frac{1}{12}$

24. $\frac{2}{8} + \frac{3}{4} = \frac{w}{5}$

25. $\frac{5}{2} - x = 3x$

26. $x - \frac{3}{4} = -2x$

27. $\frac{5}{3x} + \frac{3}{x} = 1$

28. $\frac{x}{4} - \frac{x}{6} = \frac{1}{4}$

29. $\frac{x - 1}{x - 5} = \frac{4}{x - 5}$

30. $\frac{2x + 3}{x + 2} = \frac{3}{2}$

31. $\frac{3 - 5y}{4} = \frac{2 - 4y}{3}$

32. $\frac{5}{x + 2} = \frac{1}{x - 4}$

33. $\frac{5}{-x - 6} = \frac{2}{x}$

34. $\frac{4}{y - 3} = \frac{6}{y + 3}$

35. $\frac{2x - 3}{x - 4} = \frac{5}{x - 4}$

36. $\frac{3}{x} + 4 = \frac{3}{x}$

37. $\frac{x + 4}{x - 3} = \frac{x + 10}{x + 2}$

38. $\frac{x - 3}{x + 1} = \frac{x - 6}{x + 5}$

39. $\frac{2x - 1}{3} - \frac{3x}{4} = \frac{5}{6}$

40. $x + \frac{3}{x} = \frac{12}{x}$

41. $x + \frac{20}{x} = -9$

42. $x - \frac{21}{x} = 4$

43. $\frac{3y - 2}{y + 1} = 4 - \frac{y + 2}{y - 1}$

44. $\frac{2b}{b + 1} = 2 - \frac{5}{2b}$

45. $\frac{1}{x + 3} + \frac{1}{x - 3} = \frac{-5}{x^2 - 9}$

46. $\frac{x}{x - 3} + \frac{3}{2} = \frac{3}{x - 3}$

47. $a - \frac{a}{3} + \frac{a}{5} = 26$

48. $\frac{3x}{x^2 - 9} + \frac{1}{x - 3} = \frac{3}{x + 3}$

49. $\frac{3}{x - 5} - \frac{4}{x + 5} = \frac{11}{x^2 - 25}$

50. $\frac{x + 1}{x + 3} + \frac{x - 3}{x - 2} = \frac{2x^2 - 15}{x^2 + x - 6}$

51. $\frac{y}{2y + 2} + \frac{2y - 16}{4y + 4} = \frac{y - 3}{y + 1}$

52. $\frac{3}{x + 3} + \frac{5}{x + 4} = \frac{12x + 19}{x^2 + 7x + 12}$

53. $\frac{1}{y - 1} + \frac{1}{2} = \frac{2}{y^2 - 1}$

54. $\frac{2y}{y + 2} = \frac{y}{y + 3} - \frac{3}{y^2 + 5y + 6}$

55. $\frac{x - 4}{x^2 - 2x} = \frac{-4}{x^2 - 4}$

56. $\frac{2}{x - 2} - \frac{1}{x + 1} = \frac{5}{x^2 - x - 2}$

Problem Solving

In Exercises 57–62, determine the solution by observation. Explain how you determined your answer.

57. $\dfrac{3}{x-2} = \dfrac{x-2}{x-2}$

58. $\dfrac{1}{2} + \dfrac{x}{2} = \dfrac{5}{2}$

59. $\dfrac{x}{x-3} + \dfrac{x}{x-3} = 0$

60. $\dfrac{x}{2} + \dfrac{x}{2} = x$

61. $\dfrac{x-2}{3} + \dfrac{x-2}{3} = \dfrac{2x-4}{3}$

62. $\dfrac{2}{x} - \dfrac{1}{x} = \dfrac{1}{x}$

63. A formula frequently used in optics is

$$\frac{1}{p} + \frac{1}{q} = \frac{1}{f}$$

where p represents the distance of the object from a mirror (or lens), q represents the distance of the image from the mirror (or lens), and f represents the focal length of the mirror (or lens). If a mirror has a focal length of 10 centimeters, how far from the mirror will the image appear when the object is 30 centimeters from the mirror?

Challenge Problems

64. In electronics the total resistance R_T, of resistors wired in a parallel circuit is determined by the formula

$$\frac{1}{R_T} = \frac{1}{R_1} + \frac{1}{R_2} + \frac{1}{R_3} + \cdots + \frac{1}{R_n}$$

where $R_1, R_2, R_3, \ldots, R_n$ are the resistances of the individual resistors (measured in ohms) in the circuit.

a) Find the total resistance if two resistors, one of 200 ohms and the other of 300 ohms, are wired in a parallel circuit.

b) If three identical resistors are to be wired in parallel, what should be the resistance of each resistor if the total resistance of the circuit is to be 300 ohms?

65. Can an equation of the form $\dfrac{a}{x} + 1 = \dfrac{a}{x}$ have a real number solution for any real number a? Explain your answer.

Group Activity

Discuss and answer Exercise 66 as a group.

66. a) As a group, discuss two different methods you can use to solve the equation $\dfrac{x+3}{5} = \dfrac{x}{4}$.

b) Group member 1: Solve the equation by obtaining a common denominator.
Group member 2: Solve the equation by cross-multiplying.
Group member 3: Check the results of group member 1 and group member 2.

c) Individually, create another equation by taking the reciprocal of each term in the equation in part **a)**. Compare your results. Do you think that the reciprocal of the answer you found in part **b)** will be the solution to this equation? Explain.

d) Individually, solve the equation you found in part **c)** and check your answer. Compare your work with the other group members. Was the conclusion you came to in part **c)** correct? Explain.

e) As a group, solve the equation $\dfrac{1}{x} + \dfrac{1}{3} = \dfrac{2}{x}$. Check your result.

f) As a group, create another equation by taking the reciprocal of each term of the equation in part **e)**. Do you think that the reciprocal of the answer you found in part **e)** will be the solution to this equation? Explain.

g) Individually, solve the equation you found in part **f)** and check your answer. Compare your work with the other group members. Did your group make the correct conclusion in part **f)**? Explain.

h) As a group, discuss the relationship between the solution to the equation $\dfrac{7}{x-9} = \dfrac{3}{x}$ and the solution to the equation $\dfrac{x-9}{7} = \dfrac{x}{3}$. Explain your answer.

Cumulative Review Exercises

[3.3] **67.** An Internet service offers two plans for its customers. One plan includes 5 hours of use and costs \$7.95 per month. Each additional minute after the 5 hours costs \$0.15. The second plan costs \$19.95 per month and provides unlimited Internet access. How many hours would Jake LaRue have to use the Internet monthly to make the second plan the less expensive?

[3.4] **68.** Two angles are supplementary angles if the sum of their measures is 180°. Find the two supplementary angles if the smaller angle is 30° less than half the larger angle.

69. How long will it take to fill a 600-gallon Jacuzzi if water is flowing into the Jacuzzi at a rate of 8 gallons a minute?

[7.6] **70.** Explain the difference between a linear equation and a quadratic equation, and give an example of each.

8.7 RATIONAL EQUATIONS: APPLICATIONS AND PROBLEM SOLVING

SSM VIDEO 8.7 CD Rom

1) **Set up and solve applications containing rational expressions.**
2) **Set up and solve motion problems.**
3) **Set up and solve work problems.**

1) Set Up and Solve Applications Containing Rational Expressions

Many applications of algebra involve rational equations. After we represent the application as an equation, we solve the rational equation as we did in Section 8.6.

The first type of application we will consider is a **geometry problem.**

EXAMPLE 1 Mary and Larry Armstrong are interested in purchasing a carpet whose area is 60 square feet. Determine the length and width if the width is 5 feet less than $\frac{3}{5}$ of the length, see Figure 8.2.

Solution **Understand and Translate**

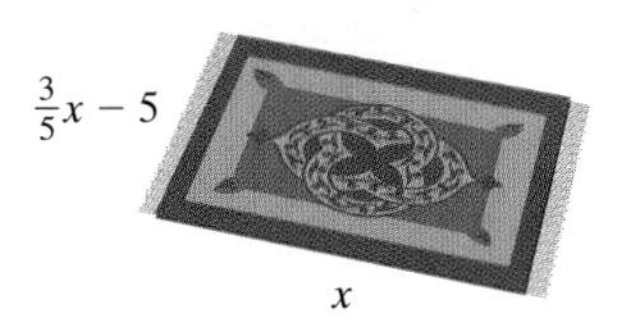

FIGURE 8.2

Let x = length

then $\frac{3}{5}x - 5$ = width

area = length · width

$$60 = x\left(\frac{3}{5}x - 5\right)$$

Carry Out

$$60 = \frac{3}{5}x^2 - 5x$$

$$5(60) = 5\left(\frac{3}{5}x^2 - 5x\right) \qquad \text{Multiply both sides by 5.}$$

$$300 = 3x^2 - 25x \qquad \text{Distributive Property}$$

$$0 = 3x^2 - 25x - 300 \quad \text{Subtract 300 from both sides.}$$

$$\text{or} \quad 3x^2 - 25x - 300 = 0$$

$$(3x + 20)(x - 15) = 0 \quad \text{Factor.}$$

$$3x + 20 = 0 \quad \text{or} \quad x - 15 = 0$$

$$3x = -20 \qquad x = 15$$

$$x = -\frac{20}{3}$$

Check and Answer Since the length of a rectangle cannot be negative, we can eliminate $-\frac{20}{3}$ as an answer to our problem.

$$\text{length} = x = 15 \text{ feet}$$

$$\text{width} = \frac{3}{5}(15) - 5 = 4 \text{ feet}$$

CHECK:

$$a = lw$$

$$60 \stackrel{?}{=} 15(4)$$

$$60 = 60 \quad \text{True}$$

NOW TRY EXERCISE 5

Therefore, the length is 15 feet and the width is 4 feet.

EXAMPLE 2 One number is 4 times another number. The sum of their reciprocals is 2. Determine the numbers.

Solution **Understand and Translate**

$$\text{Let } x = \text{first number}$$

$$\text{then } 4x = \text{second number}$$

The reciprocal of the first number is $\frac{1}{x}$ and the reciprocal of the second number is $\frac{1}{4x}$. The sum of their reciprocals is 2, thus

$$\frac{1}{x} + \frac{1}{4x} = 2$$

Carry Out

$$4x\left(\frac{1}{x} + \frac{1}{4x}\right) = 4x(2) \quad \text{Multiply both sides by } 4x.$$

$$4x\left(\frac{1}{x}\right) + 4x\left(\frac{1}{4x}\right) = 8x \quad \text{Distributive Property}$$

$$4 + 1 = 8x$$

$$5 = 8x$$

$$\frac{5}{8} = x$$

Check The first number is $\frac{5}{8}$. The second number is therefore $4x = 4\left(\frac{5}{8}\right) = \frac{5}{2}$. Let's now check if the sum of the reciprocals is 2.

The reciprocal of $\frac{5}{8}$ is $\frac{8}{5}$. The reciprocal of $\frac{5}{2}$ is $\frac{2}{5}$. The sum of the reciprocals is

$$\frac{8}{5} + \frac{2}{5} = \frac{10}{5} = 2$$

NOW TRY EXERCISE 9

Answer The two numbers are $\frac{5}{8}$ and $\frac{5}{2}$.

2) Set Up and Solve Motion Problems

In Chapter 3 we discussed **motion problems.** Recall that

$$\text{distance} = \text{rate} \cdot \text{time}$$

If we solve this equation for time, we obtain

$$\text{time} = \frac{\text{distance}}{\text{rate}} \quad \text{or} \quad t = \frac{d}{r}$$

This equation is useful in solving motion problems when the total time of travel for two objects or the time of travel between two points is known.

EXAMPLE 3 Cindy Kilborn lives near the Colorado River and loves to canoe in a section of the river where the current is not too great. One Saturday she went canoeing. She learned from a friend that the current in the river was 2 miles per hour. If it took Cindy the same amount of time to travel 10 miles downstream as 2 miles upstream, determine the speed at which Cindy's canoe would travel in still water.

Solution **Understand and Translate**

Let r = the canoe's speed in still water

then $r + 2$ = the canoe's speed traveling downstream (with current)

and $r - 2$ = the canoe's speed traveling upstream (against current)

Direction	Distance	Rate	Time
Downstream	10	$r + 2$	$\frac{10}{r+2}$
Upstream	2	$r - 2$	$\frac{2}{r-2}$

Since the time it takes to travel 10 miles downstream is the same as the time to travel 2 miles upstream, we set the times equal to each other and then solve the resulting equation.

$$\text{time downstream} = \text{time upstream}$$

$$\frac{10}{r+2} = \frac{2}{r-2}$$

Carry Out

$$10(r - 2) = 2(r + 2) \qquad \textit{Cross-multiply.}$$

$$10r - 20 = 2r + 4$$

$$8r = 24$$

$$r = 3$$

Check and Answer Since this rational equation contains a variable in a denominator, the solution must be checked. A check will show that 3 satisfies the equation. Thus, the canoe would travel at 3 miles per hour in still water.

NOW TRY EXERCISE 13

EXAMPLE 4 Officer DeWolf rides his bicycle every Saturday morning in the Kahana Valley State Park in Oahu, Hawaii, as part of his patrolling duties. During the first part of the ride he is pedaling mostly uphill and his average speed is 6 miles an hour. After a certain point, he is traveling mostly downhill and averages 9 miles per hour. If the total distance he travels is 30 miles and the total time he rides is 4 hours, how long did he ride at each speed?

Solution **Understand and Translate**

Let d = distance traveled at 6 miles per hour

then $30 - d$ = distance traveled at 9 miles per hour

Direction	Distance	Rate	Time
Uphill	d	6	$\frac{d}{6}$
Downhill	$30 - d$	9	$\frac{30 - d}{9}$

Since the total time spent riding is 4 hours, we write

time going uphill + time going downwill = 4 hours

$$\frac{d}{6} + \frac{30 - d}{9} = 4$$

Carry Out

$$18\left(\frac{d}{6} + \frac{30 - d}{9}\right) = 18 \cdot 4 \qquad \textit{Multiply both sides by the LCD, 18.}$$

$$\overset{3}{\cancel{18}}\left(\frac{d}{\cancel{6}}\right) + \overset{2}{\cancel{18}}\left(\frac{30 - d}{\cancel{9}}\right) = 72 \qquad \textit{Distributive Property}$$

$$3d + 2(30 - d) = 72$$

$$3d + 60 - 2d = 72$$

$$d + 60 = 72$$

$$d = 12$$

Answer The answer to the problem is not 12. Remember that the question asked us to *find the time spent* traveling at each speed. The variable d does not represent time, but represents the distance traveled at 6 miles per hour. To find the time traveled and to answer the question asked, we need to evaluate $\frac{d}{6}$ and $\frac{30 - d}{9}$ for $d = 12$

Time at 6 mph

$$\frac{d}{6} = \frac{12}{6} = 2$$

Time at 9 mph

$$\frac{30 - d}{9} = \frac{30 - 12}{9} = \frac{18}{9} = 2$$

Thus, 2 hours were spent traveling at each rate. We leave the check for you to do.

EXAMPLE 5 Bullet trains in Japan have been known to average 240 kilometers per hour. Prior to using bullet trains, trains in Japan traveled at an average speed of 120 kilometers per hour. If a bullet train traveling from the station in Shin-Osaka to Hakata can complete its trip in 2.3 hours less time than an older train, determine the distance from Shin-Osaka to Hakata.

Solution **Understand and Translate** Let d = the distance from Shin-Osaka to Hakata.

Train	Distance	Rate	Time
Bullet train	d	240	$\frac{d}{240}$
Older train	d	120	$\frac{d}{120}$

We are given that a trip on the bullet train takes 2.3 hours less than a trip on the older train. Therefore, to make the two travel times equal, we need to subtract 2.3 hours from the travel time of the older train. Using this information, we set up the following equation.

$$\text{time for bullet train} = \text{time for older train} - 2.3 \text{ hours}$$

$$\frac{d}{240} = \frac{d}{120} - 2.3$$

Carry Out

$$240\left(\frac{d}{240}\right) = 240\left(\frac{d}{120} - 2.3\right) \quad \text{Multiply both sides by the LCD, 240.}$$

$$d = 240\left(\frac{d}{120}\right) - 240(2.3) \quad \text{Distributive Property}$$

$$d = 2d - 552$$

$$-d = -552$$

$$d = 552$$

Check and Answer The distance appears to be 552 kilometers. Let's check the answer. The traveling time from Shin-Osaka to Hakata by the bullet train is 552/240 hours, or 2.3 hours. The traveling time from Shin-Osaka to Hakata by the older train is 552/120 hours, or 4.6 hours, which is 2.3 hours longer than the time of the bullet train. Therefore, the distance from the station in Shin-Osaka to the station in Hakata is 552 kilometers.

NOW TRY EXERCISE 23

In Example 5 we subtracted 2.3 hours from the time of the older train to obtain an equation. We could have added 2.3 hours to the time of the bullet train to obtain an equivalent equation. Rework Example 5 now by adding 2.3 hours to the bullet train's time.

3) Set Up and Solve Work Problems

Problems in which two or more machines or people work together to complete a certain task are sometimes referred to as **work problems**. Work problems often involve equations containing fractions. Generally, work problems are based on the fact that the fractional part of the work done by person 1 (or machine 1) plus the fractional part of the work done by person 2 (or machine 2) is equal to the total amount of work done by both people (or both machines). *We represent the total amount of work done by the number 1, which represents one whole job completed.*

part of task done by first person or machine	+	part of task done by second person or machine	=	1 (one whole task completed)

To determine the part of the task completed by each person or machine, we use the formula

part of task completed = rate · time

This formula is very similar to the formula

$$\text{amount} = \text{rate} \cdot \text{time}$$

that was discussed in Section 3.5. To determine the part of the task completed, we need to determine the rate. Suppose that Paul can do a particular task in 6 hours. Then he would complete $1/6$ of the task per hour. Thus, his rate is $1/6$ of the task per hour. If Audrey can do a particular task in 5 minutes, her rate is $1/5$ of the task per minute. In general, if a person or machine can complete a task in t units of time, the rate is $1/t$.

EXAMPLE 6 Joe Martinez can landscape the Antonelli's yard by himself in 20 hours. Mike Morley can landscape the same yard by himself in 30 hours. How long will it take them to landscape the yard if they work together?

Solution **Understand and Translate** Let t = the time, in hours, for both men to landscape the yard together. We will construct a table to help us in finding the part of the task completed by Joe and by Mike in t hours.

Landscaper	Rate of Work (part of the task completed per hour)	Time Worked	Part of Task
Joe	$\frac{1}{20}$	t	$\frac{t}{20}$
Mike	$\frac{1}{30}$	t	$\frac{t}{30}$

$$\left(\begin{array}{c}\text{part of the landscaping}\\ \text{completed by Joe in } t \text{ hours}\end{array}\right) + \left(\begin{array}{c}\text{part of the landscaping}\\ \text{completed by Mike in } t \text{ hours}\end{array}\right) = 1 \text{ (entire yard landscaped)}$$

$$\frac{t}{20} + \frac{t}{30} = 1$$

Carry Out Now multiply both sides of the equation by the LCD, 60.

$$60\left(\frac{t}{20} + \frac{t}{30}\right) = 60 \cdot 1$$

$$\overset{3}{\cancel{60}}\left(\frac{t}{\cancel{20}}\right) + \overset{2}{\cancel{60}}\left(\frac{t}{\cancel{30}}\right) = 60 \qquad \textit{Distributive Property}$$

$$3t + 2t = 60$$

$$5t = 60$$

$$t = 12$$

Answer Thus, the two men working together can landscape the Antonelli's yard in 12 hours. We leave the check for you to do.

NOW TRY EXERCISE 25

EXAMPLE 7

At the Spring Hill Water Treatment Plant, one pipe can fill a water tank in 3 hours and another pipe can empty it in 5 hours. If the valves to both pipes are open, how long will it take to fill the tank?

Solution **Understand and Translate** Let t = amount of time to fill the tank with both valves open.

Pipe	Rate of Work	Time	Part of Task
Pipe filling tank	$\frac{1}{3}$	t	$\frac{t}{3}$
Pipe emptyng tank	$\frac{1}{5}$	t	$\frac{t}{5}$

As one pipe is filling, the other is emptying the tank. Thus, the pipes are working against each other. Therefore, instead of adding the parts of the task, as was done in Example 6 where the people worked together, we will subtract the parts of the task.

$$\begin{pmatrix}\text{part of tank}\\ \text{filled in } t \text{ hours}\end{pmatrix} - \begin{pmatrix}\text{part of tank}\\ \text{emptied in } t \text{ hours}\end{pmatrix} = 1 \text{ (total tank filled)}$$

$$\frac{t}{3} - \frac{t}{5} = 1$$

Carry Out

$$15\left(\frac{t}{3} - \frac{t}{5}\right) = 15 \cdot 1 \quad \textit{Multiply both sides by the LCD, 15.}$$

$$\overset{5}{\cancel{15}}\left(\frac{t}{\cancel{3}}\right) - \overset{3}{\cancel{15}}\left(\frac{t}{\cancel{5}}\right) = 15 \quad \textit{Distributive Property}$$

$$5t - 3t = 15$$

$$2t = 15$$

$$t = 7\tfrac{1}{2}$$

Check and Answer The tank will be filled in $7\frac{1}{2}$ hours. This answer is reasonable because we expect it to take longer than 3 hours when the tank is being drained at the same time.

EXAMPLE 8

Brooke Cashion and Amy Alevy own a cleaning service. When Brooke cleans the Moose Club by herself, it takes her 7 hours. When Brooke and Amy work together to clean the same club, it takes them 4 hours. How long does it take Amy by herself to clean the Moose Club?

Solution **Understand and Translate** Let t = time for Amy to clean the club by herself. Then Amy's rate of work is $1/t$. Let's make a table to help analyze the problem. In the table, we use the fact that together they can clean the club in 4 hours.

Worker	Rate of Work	Time	Part of Task
Brooke	$\frac{1}{7}$	4	$\frac{4}{7}$
Amy	$\frac{1}{t}$	4	$\frac{4}{t}$

$$\left(\begin{matrix}\text{part of club cleaned}\\ \text{by Brooke}\end{matrix}\right) + \left(\begin{matrix}\text{part of club cleaned}\\ \text{by Amy}\end{matrix}\right) = 1$$

$$\frac{4}{7} + \frac{4}{t} = 1$$

Carry Out

$$7t\left(\frac{4}{7} + \frac{4}{t}\right) = 7t \cdot 1 \qquad \textit{Multiply both sides by the LCD, } 7t.$$

$$7t\left(\frac{4}{7}\right) + 7t\left(\frac{4}{t}\right) = 7t \qquad \textit{Distributive Property}$$

$$4t + 28 = 7t$$

$$28 = 3t$$

$$9\frac{1}{3} = t$$

Check and Answer Thus, it takes Amy $9\frac{1}{3}$ hours or 9 hours and 20 minutes to clean the Moose Club by herself. This answer is reasonable because we expect it to take longer for Amy to clean the Moose Club alone than it would take Amy and Brooke working together.

EXAMPLE 9 Peter and Kaitlyn Kewin are handwriting thank-you notes to guests who attended their 20th wedding anniversary party. Kaitlyn by herself could write all the notes in 8 hours and Peter could write all the notes by himself in 7 hours. After Kaitlyn has been writing thank-you notes for 5 hours by herself, she must leave town on business. Peter then continues the task of writing the thank-you notes. How long will it take Peter to finish writing the remaining notes?

Solution **Understand and Translate** Let t = time it will take Peter to finish writing the notes.

Person	Rate of Work	Time	Part of Task
Kaitlyn	$\frac{1}{8}$	5	$\frac{5}{8}$
Peter	$\frac{1}{7}$	t	$\frac{t}{7}$

$$\left(\begin{matrix}\text{part of cards written}\\ \text{by Kaitlyn}\end{matrix}\right) + \left(\begin{matrix}\text{part of cards written}\\ \text{by Peter}\end{matrix}\right) = 1$$

$$\frac{5}{8} + \frac{t}{7} = 1$$

Carry Out

$$56\left(\frac{5}{8} + \frac{t}{7}\right) = 56 \cdot 1 \qquad \text{Multiply both sides by the LCD, 56.}$$

$$56\left(\frac{5}{8}\right) + 56\left(\frac{t}{7}\right) = 56 \qquad \text{Distributive Property}$$

$$35 + 8t = 56$$

$$8t = 21$$

$$t = \frac{21}{8} \quad \text{or} \quad 2\frac{5}{8}$$

Answer Thus, it will take Peter $2\frac{5}{8}$ hours to complete the cards.

Exercise Set 8.7

Concept/Writing Exercises

1. Geometric formulas, which were discussed in Section 3.1, are often rational equations. Give three formulas that are rational equations.
2. Suppose that car 1 and car 2 are traveling to the same location 60 miles away and that car 1 travels 10 miles per hour faster than car 2.

   a) Let r represent the speed of car 2. Write an expression, using time $= \dfrac{\text{distance}}{\text{rate}}$, for the time it takes car 1 to reach its destination.

   b) Now let r represent the speed of car 1. Write an expression, using time $= \dfrac{\text{distance}}{\text{rate}}$, for the time it takes car 2 to reach its destination.
3. In an equation for a work problem, one side of the equation is set equal to 1. What does the 1 represent in the problem?
4. Suppose Tracy Augustine can complete a particular task in 3 hours, while the same task can be completed by John Bailey in 7 hours. How would you represent the part of the task completed by Tracy in 1 hour? By John in 1 hour?

Practice the Skills/Problem Solving

In Exercises 5–36, write an equation that can be used to solve each problem. Then solve the equation and answer the question.

Geometry Problems

5. The Phillips Paper Company has been commissioned by a large computer manufacturer to make rectangular pieces of cardboard for packing computers. The sheets of cardboard are to have an area of 90 square inches, and the length of a sheet is to be 4 inches more than $\frac{2}{3}$ its width. Determine the length and width of the cardboard to be manufactured.
6. Often books and other items used in the United States are printed in other countries and then imported to the United States. Since most countries use the metric system of measurement, a map of the state of Idaho being printed in Mexico will have specifications given in metric units. In the upper right-hand corner of the state map is a city map of Boise. If the map of Boise is to have an area of 192 square centimeters, and the width of the map is to be $\frac{3}{4}$ the length, determine the length and width of the map of Boise.
7. Pillsbury Crescent Rolls are packaged in tubes that contain perforated triangles of dough. To make a roll, you open the tube and break off a triangular piece of dough, then roll the dough into the crescent shape and place it in the oven. The base of the triangular piece of dough is about 5 centimeters more than its height. Determine the approximate base and height of a piece of dough if the area is about 42 square centimeters.
8. Yield right of way signs used in the United States are triangles. The area of the sign is about 558 square inches. The height of the sign is about 5 inches less than its base. Determine the approximate length of the base of a yield right of way sign.

Number Problems

9. One number is 10 times larger than another. The difference of their reciprocals is 3. Determine the two numbers.
10. One number is 3 times larger than another. The sum of their reciprocals is $\frac{4}{3}$. Determine the two numbers.
11. The numerator of the fraction $\frac{3}{4}$ is increased by an amount so that the value of the resulting fraction is $\frac{5}{2}$. Determine the amount by which the numerator was increased.
12. The denominator of the fraction $\frac{6}{19}$ is decreased by an amount so that the value of the resulting fraction is $\frac{3}{8}$. Determine the amount by which the denominator was decreased.

Motion Problems

13. In the Mississippi River near New Orleans, the Creole Queen paddleboat travels 4 miles upstream (against the current) in the same amount of time it travels 6 miles downstream (with the current). If the current of the river is 2 miles per hour, determine the speed of the Creole Queen in still water.

14. Kathy Boothby-Sestak can paddle her kayak 6 miles per hour in still water. It takes her as long to paddle 3 miles upstream as 4 miles downstream in the Wabash River near Lafayette, Indiana. Determine the current.
15. Ms. James took her two sons water skiing in still water at Lake Cochituate near Natick, Massachusetts. She drove the motor boat one way on the water pulling the younger son, Matthew, at 30 miles per hour. Then she turned around and pulled her older son, Jason, in the opposite direction the same distance at 30 miles per hour. If the total time spent skiing was $\frac{1}{2}$ hour, how far did each son travel?
16. Brandy Dawson and Jason Dodge start a motorcycle trip at the same point a little north of Fort Worth, Texas. Both are traveling to San Antonio, Texas, a distance of about 400 kilometers. Brandy rides 30 kilometers per hour faster than Jason does. When Brandy reaches her destination, Jason has only traveled to Austin, Texas, a distance of about 250 kilometers. Determine the approximate speed of each motorcycle.
17. Elenore Morales traveled 1600 miles by commercial jet from Kansas City, Missouri, to Spokane, Washington. She then traveled an additional 500 miles on a private propeller plane from Spokane to Billings, Montana. If the speed of the jet was 4 times the speed of the propeller plane and the total time in the air was 6 hours, determine the speed of each plane.
18. As part of his exercise regiment, Chris Barker walks a distance of 2 miles on an indoor track and then jogs at twice his walking speed for another mile. If the total time spent on the track was $\frac{3}{4}$ of an hour, determine the speeds at which he walks and jogs.
19. To go to work, Mary Fernandez drives her car a specific distance at 25 miles per hour and then takes a train, which travels at 50 miles per hour, the remainder of the distance. If the total distance traveled is 70 miles, and the total time spent commuting (in the car and on the train) to work is 1.6 hours, determine the distance she drives and the distance she travels by train.
20. Jayme Heffler jogs and then walks in alternating intervals. When she jogs she averages 5 miles per hour and when she walks she averages 2 miles per hour. If she walks and jogs a total of 3 miles in a total of 0.9 hour, how much time is spent jogging and how much time is spent walking?
21. A Boeing 747 flew from San Francisco to Honolulu, a distance of 2800 miles. Flying with the wind, it averaged 600 miles per hour. When the wind changed from a tailwind to a headwind, the plane's speed dropped to 500 miles per hour. If the total time of the trip was 5 hours, determine the length of time it flew at each speed.
22. A Boeing 747 and the Concorde travel the same route from New York City to Paris. The planes leave New York at the same time. The Boeing 747 averages 521 miles per hour and the Concorde averages 992 miles per hour. If the Concorde arrives in Paris 3.4 hours before the Boeing 747, determine the distance traveled, to the nearest mile, by air from New York City to Paris.
23. Bill Beville swims freestyle at an average speed of 40 meters per minute. He swims using the breaststroke at an average speed of 30 meters per minute. Bill decides to swim freestyle across Lake Alice in Waterport, New York. After resting, he swims back across the lake using the breaststroke. If his return trip took 20 minutes longer than his trip going, what is the width of the lake where he crossed?

24. Alana Bradley and her father Tim begin skiing the same cross-country ski trail in Elmwood Park in Sioux Falls, South Dakota, at the same time. If Alana, who averages 9 miles per hour, finishes the trail 0.25 hours sooner than her father, who averages 6 miles per hour, determine the length of the trail.

Work Problems

25. Reynaldo and Felicia Fernandez decide to wallpaper their family room. Felicia, who has wallpapering experience, can wallpaper the room in 6 hours. Reynaldo can wallpaper the same room in 8 hours. How long will it take them to wallpaper the family room if they work together?

26. Shortly after a tornado struck, a rural town was declared to be in a state of emergency. Because the town's drinking water had been polluted, the American Red Cross arranged for fresh water to be brought in on two trucks. Each family was to be given one 5-gallon jug of water per day. The crew on the smaller of the two trucks could dispense the jugs of water to all the families in the town by itself in 10 hours. The larger truck had a larger crew, and that crew could dispense the jugs of water to all the families in the town in only 5 hours. How long would it take both crews working together to dispense the jugs of water?

27. Pam and Loren Fornieri recently had a new hot tub installed on their deck. They planned to fill the tub, heat the water, and then get in. Their instruction booklet indicated that they could expect the hot tub to be filled in 2 hours. Pam turned on the water and returned to the house. After 2 hours, she went out to the tub and saw, to her disappointment, that the tub was only partly filled because the drain had been left open. The instruction booklet indicated that a full tub would completely drain (if no water was being added) in 3 hours. If the water was on and the drain was left open, how long would it take the tub to fill completely?

28. At a cement plant, a conveyer belt can fill a large holding tank with cement ore in 6 hours. At the bottom of the tank is a stone crusher that crushes and dispenses the ore. If the holding tank is full and no more ore is being added, the crusher can crush the contents and empty the entire tank in 12 hours. If the tank is empty and the conveyer belt and the stone crusher are both turned on at the same time, how long will it take for the tank to fill completely?

29. Jackie Fitzgerald, a mathematics instructor, can grade the final exams for her Elementary Algebra classes in 8 hours. If Jackie and her husband Tim, who is also a mathematics instructor, work together the grading can be completed in 3 hours. How long would it take Tim to grade the exams if he were working alone?

30. At the NCNB Savings Bank it takes a computer 4 hours to process and print payroll checks. When a second computer is used and the two computers work together, the checks can be processed and printed in 3 hours. How long would it take the second computer by itself to process and print the payroll checks?

31. A construction company with two backhoes has contracted to dig a long trench for drainage pipes. The larger backhoe can dig the entire trench by itself in 12 days. The smaller backhoe can dig the entire trench by itself in 15 days. The large backhoe begins working on the trench by itself, but after 5 days it is transferred to a different job and the smaller backhoe begins working on the trench. How long will it take for the smaller backhoe to complete the job?

32. Chadwick and Melissa Wicker deliver institutional size cans of condiments to various warehouses in New York State and Pennsylvania. They begin in Buffalo, NY, and drop off their supplies in Alfred, NY, Jamestown, NY, Erie, PA, and Newcastle, PA. If Chadwick drove the entire trip, the trip would take about 10 hours. Melissa is a faster driver. If Melissa drove the entire trip, the trip would take about 8 hours. After Melissa had been driving for 4 hours, Chadwick takes over the driving. About how much longer will Chadwick drive before they reach their final destination?

33. Ken and Bettina Wikendt live in Minneapolis, Minnesota. Following a severe snowstorm, Ken and Bettina must clear their driveway and sidewalk. Ken can clear the snow by himself in 4 hours, and Bettina can clear the snow by herself in 6 hours. After Bettina has been working for 3 hours, Ken is able to join her. How much longer will it take them working together to remove the rest of the snow?

34. A large farming cooperative near Hutchinson, Kansas, owns three hay balers. The oldest baler can pick up and bale an acre of hay in 3 hours and each of the two new balers can work an acre in 2 hours.

a) How long would it take the three balers working together to pick up and bale 1 acre?

b) How long would it take for the three balers working together to pick up and bale the farm's 375 acres of hay?

35. A boat designed to skim oil off the surface of the water has two skimmers. One skimmer can fill the boat's holding tank in 60 hours while the second skimmer can fill the boat's holding tank in 50 hours. There is also a valve in the holding tank that is used to transfer the oil to a larger vessel. If no new oil is coming into the holding tank, a full holding tank of skimmed oil can be transferred to a larger tank in 30 hours. If both skimmers begin skimming and the valve on the holding tank is opened, how long will it take for the empty holding tank on the boat to fill?

36. Assume that the residents of continent A can deplete all the world's fresh water supply in 300 years; the residents of continent B can deplete the world's fresh water supply in 100 years; and the residents of continent C can deplete all the world's fresh water supply in 200 years. How long would it take for the three populations together to deplete all the world's fresh water? (Some resources report that fresh water is still abundant. However, because the supply is limited and water pollution has increased, it can support at most one more doubling of demand, which is expected to occur in 20 to 30 years.)

Challenge Problems

37. If 5 times a number is added to 4 times the reciprocal of the number, the answer is 12. Determine the number(s).

38. The reciprocal of the difference of a certain number and 3 is twice the reciprocal of the difference of twice the number and 6. Determine the number(s).

39. Ed and Samantha Weisman, whose parents own a fruit farm, must each pick the same number of pints of blueberries each day during the season. Ed, who is older, picks an average of 8 pints per hour, while Samantha picks an average of 4 pints per hour. If Ed and Samantha begin picking blueberries at the same time, and Samantha finishes 1 hour after Ed, how many pints of blueberries must each pick?

40. A mail processing machine can sort a large bin of mail in 1 hour. A newer model can sort the same quantity of mail in 40 minutes. If they operate together, how long will it take them to sort the bin of mail?

Cumulative Review Exercises

[2.1] **41.** Simplify $\frac{1}{2}(x + 3) - (2x + 6)$.

[4.2] **42.** Graph the equation $x = 4$.

[8.2] **43.** Divide $\frac{x^2 - 14x + 48}{x^2 - 5x - 24} \div \frac{2x^2 - 13x + 6}{2x^2 + 5x - 3}$.

[8.4] **44.** Subtract $\frac{x}{6x^2 - x - 15} - \frac{5}{9x^2 - 12x - 5}$.

SUMMARY

Key Words and Phrases

8.1
Rational expression
Signs of a fraction
Simplify a rational expression

8.2
Divide rational expressions
Multiply rational expressions

8.3
Add rational expressions with a common denominator
Least common denominator
Subtract rational expressions with a common denominator

8.4
Add rational expressions with unlike denominators
Subtract rational expressions with unlike denominators

8.5
Complex fraction

8.6
Extraneous root or extraneous solution to a rational equation
Rational equation

8.7
Geometry problems
Motion problems
Work problems

IMPORTANT FACTS

For any fraction: $-\frac{a}{b} = \frac{-a}{b} = \frac{a}{-b}, b \neq 0$

To add fractions: $\frac{a}{c} + \frac{b}{c} = \frac{a+b}{c}, c \neq 0$

To subtract fractions: $\frac{a}{c} - \frac{b}{c} = \frac{a-b}{c}, c \neq 0$

To multiply fractions: $\frac{a}{b} \cdot \frac{c}{d} = \frac{ac}{bd}, b \neq 0, d \neq 0$

To divide fractions: $\frac{a}{b} \div \frac{c}{d} = \frac{a}{b} \cdot \frac{d}{c} = \frac{ad}{bc}, b \neq 0, c \neq 0, d \neq 0$

Time $= \frac{\textbf{distance}}{\textbf{rate}}$

Review Exercises

[8.1] *Determine the values of the variable for which the following expressions are defined.*

1. $\frac{4}{3x - 15}$ **2.** $\frac{2}{x^2 - 8x + 15}$ **3.** $\frac{2}{5x^2 + 4x - 1}$

Simplify.

4. $\frac{y}{xy - y}$ **5.** $\frac{x^3 + 4x^2 + 12x}{x}$ **6.** $\frac{9x^2 + 6xy}{3x}$

7. $\frac{x^2 + 2x - 8}{x - 2}$ **8.** $\frac{x^2 - 25}{x - 5}$ **9.** $\frac{2x^2 - 7x + 3}{3 - x}$

10. $\frac{x^2 - 2x - 24}{x^2 + 6x + 8}$ **11.** $\frac{4x^2 - 11x - 3}{4x^2 - 7x - 2}$ **12.** $\frac{2x^2 - 21x + 40}{4x^2 - 4x - 15}$

[8.2] *Multiply.*

13. $\frac{3a}{4b} \cdot \frac{2}{4a^2 b}$ **14.** $\frac{15x^2y^3}{3z} \cdot \frac{6z^3}{5xy^3}$ **15.** $\frac{40a^3b^4}{7c^3} \cdot \frac{14c^5}{5a^5b}$

16. $\frac{1}{x - 4} \cdot \frac{4 - x}{3}$ **17.** $\frac{-x + 3}{5} \cdot \frac{10x}{x - 3}$ **18.** $\frac{a - 2}{a + 3} \cdot \frac{a^2 + 4a + 3}{a^2 - a - 2}$

Divide.

19. $\frac{8x^4}{y} \div \frac{x^4}{8y}$ **20.** $\frac{8xy^2}{z} \div \frac{x^4y^2}{4z^2}$ **21.** $\frac{5a + 5b}{a^2} \div \frac{a^2 - b^2}{a^2}$

22. $\frac{1}{a^2 + 8a + 15} \div \frac{3}{a + 5}$ **23.** $(x + 7) \div \frac{x^2 + 3x - 28}{x - 4}$ **24.** $\frac{x^2 + xy - 2y^2}{4y} \div \frac{x + 2y}{12y^2}$

[8.3] *Add or subtract.*

25. $\frac{x}{x + 3} - \frac{3}{x + 3}$ **26.** $\frac{3x}{x + 7} + \frac{21}{x + 7}$ **27.** $\frac{9x - 4}{x + 8} + \frac{76}{x + 8}$

28. $\frac{7x - 3}{x^2 + 7x - 30} - \frac{3x + 9}{x^2 + 7x - 30}$ **29.** $\frac{6x^2 - 3x - 21}{x + 6} - \frac{2x^2 - 24x - 3}{x + 6}$ **30.** $\frac{6x^2 - 4x}{2x - 3} - \frac{-3x + 12}{2x - 3}$

Find the least common denominator for each expression.

31. $\frac{a}{7}+\frac{4a}{3}$

32. $\frac{8}{x+3}+\frac{4x}{x+3}$

33. $\frac{5}{4xy^3}-\frac{7}{10x^2y}$

34. $\frac{6}{x+1}-\frac{3x}{x}$

35. $\frac{4}{x+2}+\frac{6x-3}{x-3}$

36. $\frac{7x-12}{x^2+x}-\frac{4}{x+1}$

37. $\frac{12x-9}{x-y}-\frac{5x+7}{x^2-y^2}$

38. $\frac{4x^2}{x-7}+8x^2$

39. $\frac{19x-5}{x^2+2x-35}+\frac{3x-2}{x^2+9x+14}$

[8.4] *Add or subtract.*

40. $\frac{6}{3y}+\frac{y}{2y^2}$

41. $\frac{2x}{xy}+\frac{1}{5x}$

42. $\frac{5x}{3xy}-\frac{4}{x^2}$

43. $6-\frac{2}{x+2}$

44. $\frac{x-y}{y}-\frac{x+y}{x}$

45. $\frac{7}{x+4}+\frac{4}{x}$

46. $\frac{2}{3x}-\frac{3}{3x-6}$

47. $\frac{3}{x+2}+\frac{7}{(x+2)^2}$

48. $\frac{x+2}{x^2-x-6}+\frac{x-3}{x^2-8x+15}$

[8.2–8.4] *Perform each indicated operation.*

49. $\frac{x+2}{x+5}-\frac{x-5}{x+2}$

50. $3+\frac{x}{x-4}$

51. $\frac{a+2}{b}\div\frac{a-2}{4b^2}$

52. $\frac{x+3}{x^2-9}+\frac{2}{x+3}$

53. $\frac{4x+4y}{x^2y}\cdot\frac{y^3}{8x}$

54. $\frac{4}{(x+2)(x-3)}-\frac{4}{(x-2)(x+2)}$

55. $\frac{x+7}{x^2+9x+14}-\frac{x-10}{x^2-49}$

56. $\frac{x-y}{x+y}\cdot\frac{xy+x^2}{x^2-y^2}$

57. $\frac{2x^2-18y^2}{15}\div\frac{(x-3y)^2}{3}$

58. $\frac{a^2-9a+20}{a-4}\cdot\frac{a^2-8a+15}{a^2-10a+25}$

59. $\frac{a}{a^2-1}-\frac{2}{3a^2-2a-5}$

60. $\frac{2x^2+6x-20}{x^2-2x}\div\frac{x^2+7x+10}{4x^2-16}$

[8.5] *Simplify each complex fraction.*

61. $\dfrac{2+\frac{2}{3}}{\frac{6}{9}}$

62. $\dfrac{1+\frac{5}{8}}{4-\frac{9}{16}}$

63. $\dfrac{\frac{12ab}{9c}}{\frac{4a}{c^2}}$

64. $\dfrac{\frac{36x^4y^2}{9xy^5}}{4z^2}$

65. $\dfrac{a-\frac{a}{b}}{\frac{1+a}{b}}$

66. $\dfrac{x+\frac{1}{y}}{y^2}$

67. $\dfrac{\frac{4}{x}+\frac{2}{x^2}}{6-\frac{1}{x}}$

68. $\dfrac{\frac{x}{x+y}}{\frac{x^2}{2x+2y}}$

69. $\dfrac{\frac{3}{x}}{\frac{3}{x^2}}$

70. $\dfrac{\frac{1}{a}+2}{\frac{1}{a}+\frac{1}{a}}$

71. $\dfrac{\frac{1}{x^2}+\frac{1}{x}}{\frac{1}{x^2}-\frac{1}{x}}$

72. $\dfrac{\frac{3x}{y}-x}{\frac{y}{x}-1}$

[8.6] *Solve.*

73. $\frac{7}{35}=\frac{5}{x}$

74. $\frac{3}{x}=\frac{9}{3}$

75. $\frac{5}{9}=\frac{5}{x+3}$

76. $\dfrac{x}{6} = \dfrac{x-4}{2}$

77. $\dfrac{3x+4}{5} = \dfrac{2x-8}{3}$

78. $\dfrac{x}{5} + \dfrac{x}{2} = -14$

79. $\dfrac{x+3}{2} = \dfrac{x}{2}$

80. $\dfrac{3}{x} - \dfrac{1}{6} = \dfrac{1}{x}$

81. $\dfrac{1}{x-7} + \dfrac{1}{x+7} = \dfrac{1}{x^2-49}$

82. $\dfrac{x-3}{x-2} + \dfrac{x+1}{x+3} = \dfrac{2x^2+x+1}{x^2+x-6}$

83. $\dfrac{a}{a^2-64} + \dfrac{4}{a+8} = \dfrac{3}{a-8}$

[8.7] *Solve.*

84. It takes John and Amy Brogan 6 hours to build a sandcastle. It takes Paul and Cindy Carter 5 hours to make the same sandcastle. How long will it take all four children together to build the sandcastle?

85. A $\frac{3}{4}$-inch-diameter hose can fill a swimming pool in 7 hours. A $\frac{5}{16}$-inch-diameter hose can siphon all the water out of a full pool in 12 hours. How long will it take to fill the pool if while one hose is filling the pool the other hose is siphoning water from the pool?

86. One number is five times as large as another. The sum of their reciprocals is 6. Determine the numbers.

87. Robert Johnston can travel 3 miles on his roller blades in the same time Tran Lee can travel 8 miles on his mountain bike. If Tran's speed on his bike is 3.5 miles per hour faster than that of Robert on his roller blades, determine Robert's and Tran's speeds.

Practice Test

Simplify.

1. $\dfrac{x-5}{5-x}$

2. $\dfrac{x^3-1}{x^2-1}$

Perform each indicated operation.

3. $\dfrac{3x^2y}{4z^2} \cdot \dfrac{8xz^3}{9xy^4}$

4. $\dfrac{a^2-9a+14}{a-2} \cdot \dfrac{a^2-4a-21}{(a-7)^2}$

5. $\dfrac{x^2-x-6}{x^2-9} \cdot \dfrac{x^2-6x+9}{x^2+4x+4}$

6. $\dfrac{x^2-1}{x+2} \cdot \dfrac{2x+4}{2-2x^2}$

7. $\dfrac{x^2-9y^2}{3x+6y} \div \dfrac{x+3y}{x+2y}$

8. $\dfrac{15}{y^2+2y-15} \div \dfrac{3}{y-3}$

9. $\dfrac{x^2+x-20}{x-2} \div \dfrac{x^2-6x+8}{2-x}$

10. $\dfrac{4x+3}{2y} + \dfrac{2x-5}{2y}$

11. $\dfrac{7x^2-4}{x+3} - \dfrac{6x+7}{x+3}$

12. $\dfrac{5}{x} + \dfrac{3}{2x^2}$

13. $\dfrac{4}{xy} - \dfrac{3}{xy^3}$

14. $3 - \dfrac{6x}{x+4}$

15. $\dfrac{x-5}{x^2-16} - \dfrac{x-2}{x^2+2x-8}$

Simplify.

16. $\dfrac{4+\dfrac{1}{2}}{3-\dfrac{1}{5}}$

17. $\dfrac{x+\dfrac{x}{y}}{\dfrac{1}{x}}$

18. $\dfrac{2+\dfrac{3}{x}}{\dfrac{2}{x}-5}$

Solve.

19. $6 + \frac{2}{x} = 7$

20. $\frac{2x}{3} - \frac{x}{4} = x + 1$

21. $\frac{x}{x - 8} + \frac{6}{x - 2} = \frac{x^2}{x^2 - 10x + 16}$

Solve.

22. Mr. Johnson, on his tractor, can level a 1-acre field in 8 hours. Mr. Hackett, on his tractor, can level a 1-acre field in 5 hours. If they work together, how long will it take them to level a 1-acre field?

23. The sum of a positive number and its reciprocal is 2. Determine the number.

24. The area of a triangle is 27 square inches. If the height is 3 inches less than 2 times the base, determine the height and base of the triangle.

25. LaConya Bertrell exercises for $1\frac{1}{2}$ hours each day. During the first part of her routine, she rides a bicycle and averages 10 miles per hour. For the remainder of the time, she rollerblades and averages 4 miles per hour. If the total distance she travels is 12 miles, how far did she travel on the rollerblades?

Cumulative Review Test

1. Evaluate $4x^2 - 7xy^2 + 3$ when $x = -3$ and $y = -2$.

2. Solve the equation $5z + 4 = -3(z - 7)$.

3. Simplify $\left(\frac{6x^2y^3}{2x^5y}\right)^3$.

4. Solve the formula $P = 2E + 3R$ for R.

5. Simplify $(6x^2 - 3x - 5) - (-2x^2 - 8x - 9)$.

6. Multiply $(4x^2 - 6x + 3)(3x - 5)$.

7. Factor $6a^2 - 6a - 5a + 5$.

8. Factor $13x^2 + 26x - 39$.

9. Write the equation $4x - 3y = 12$ in slope–intercept form.

10. Solve the following system of equations by substitution.

$$x - y = 5$$
$$2x - 3y = 14$$

11. Determine the solution to the following system of inequalities graphically.

$$x + y > 5$$
$$y \leq x - 4$$

12. Solve $2x^2 = 11x - 12$.

13. Multiply $\frac{x^2 - 9}{x^2 - x - 6} \cdot \frac{x^2 - 2x - 8}{2x^2 - 7x - 4}$.

14. Subtract $\frac{x}{x + 4} - \frac{3}{x - 5}$.

15. Add $\frac{4}{x^2 - 3x - 10} + \frac{2}{x^2 + 5x + 6}$.

16. Solve the equation $\frac{x}{9} - \frac{x}{6} = \frac{1}{12}$.

17. Solve the equation $\frac{7}{x+3} + \frac{5}{x+2} = \frac{5}{x^2+5x+6}$.

18. A school district allows its employees to choose from two medical plans. With plan 1, the employee pays 10% of all medical bills (the school district pays the balance). With plan 2, the employee pays the school district a one-time payment of $100, then the employee pays 5% of all medical bills. What total medical bills would result in the employee paying the same amount with the two plans?

19. A feed store owner wishes to make his own store-brand mixture of bird seed by mixing sunflower seed that costs $0.50 per pound with a premixed assorted seed that costs $0.15 per pound. How many pounds of each will he have to use to make a 50-pound mixture that will cost $14.50?

20. During the first leg of a race, the 30-foot sailboat *Thumper* sailed at an average speed of 6.5 miles per hour. During the second leg of the race, the winds increased and *Thumper* sailed at an average speed of 9.5 miles per hour. If the total distance sailed by *Thumper* was 12.75 miles, and the total time spent racing was 1.5 hours, determine the distance traveled by *Thumper* on each leg of the race.

ROOTS AND RADICALS

CHAPTER

9

Use the Angel Web site at www.prenhall.com/angel to be linked to an internet resource that will help you further explore the following application.

Have you ever read a newspaper article or watched a news broadcast and wondered how law enforcement officials are able to determine the speed of a vehicle that has been involved in an accident? It may seem like educated guesswork, but the length of the skid marks left by the vehicle provides the answer. On page 553, we use a formula containing a radical expression to determine the original speed of a vehicle that skidded to a stop. The formula is based on the length of the longest skid mark left on the road surface.

Preview and Perspective

In this chapter we study roots, radical expressions, and radical equations, with an emphasis on square roots. Square roots are one type of radical expression. In Sections 9.1 through 9.4, we learn how to evaluate a square root, how to simplify square root expressions, and how to add, subtract, multiply, and divide expressions that contain square roots. In Section 9.5, we discuss solving equations that contain square roots. Section 9.6, Radicals: Applications and Problem Solving, is an extension of Section 9.5 and presents some real-life uses of square roots. In Section 9.7, we discuss cube roots and higher roots. We strongly suggest that you use a scientific calculator or graphing calculator for this and the next chapter.

Radical expressions and equations play an important part in mathematics and the sciences. Many mathematical and scientific formulas involve radicals. As you will see in Chapter 10, one of the most important formulas in mathematics, the quadratic formula, contains a square root.

9.1 EVALUATING SQUARE ROOTS

1) Evaluate square roots of real numbers.
2) Recognize that not all square roots represent real numbers.
3) Determine whether the square root of a real number is rational or irrational.
4) Write square roots as exponential expressions.

1) Evaluate Square Roots of Real Numbers

In this section we introduce a number of important concepts related to radicals. We first discuss some terminology. Let's begin with square roots. Square roots are one type of radical expression that you will use in both mathematics and science.

$\sqrt{x}$ is read "the square root of x."

The $\sqrt{}$ is called the **radical sign.** The number or expression inside the radical sign is called the **radicand.**

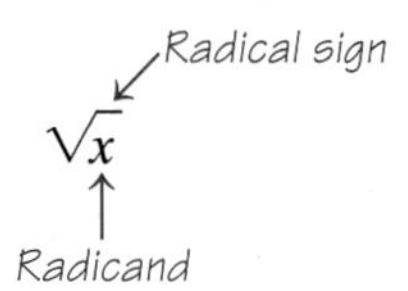

The entire expression, including the radical sign and radicand, is called the **radical expression.**

Another part of a radical expression is its **index**. The index tells the "root" of the expression. Square roots have an index of 2. The index of a square root is generally not written.

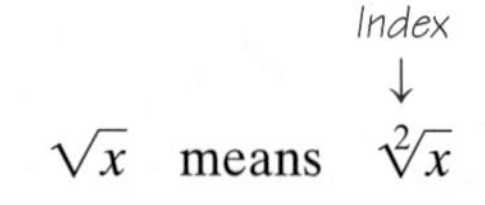

Other types of radical expressions have different indices. For example, $\sqrt[3]{x}$, which is read "the cube root of x," has an index of 3. Cube roots are discussed in Section 9.7.

Examples of Square Roots	How Read	Radicand
$\sqrt{8}$	the square root of 8	8
$\sqrt{5x}$	the square root of $5x$	$5x$
$\sqrt{\frac{x}{2y}}$	the square root of x over $2y$	$\frac{x}{2y}$

Every positive number has two square roots, a principal or positive square root and a negative square root.

Definition

The **principal or positive square root** of a positive real number x, written $\sqrt{x}$, is that *positive* number whose square equals x.

Examples

$\sqrt{25} = 5$ since $5^2 = 5 \cdot 5 = 25$

$\sqrt{49} = 7$ since $7^2 = 7 \cdot 7 = 49$

$\sqrt{\frac{1}{4}} = \frac{1}{2}$ since $\left(\frac{1}{2}\right)^2 = \left(\frac{1}{2}\right)\left(\frac{1}{2}\right) = \frac{1}{4}$

$\sqrt{\frac{4}{9}} = \frac{2}{3}$ since $\left(\frac{2}{3}\right)^2 = \left(\frac{2}{3}\right)\left(\frac{2}{3}\right) = \frac{4}{9}$

The **negative square root** of a positive real number x, written $-\sqrt{x}$, is the additive inverse or opposite of the principal square root. For example, $-\sqrt{25} = -5$ and $-\sqrt{36} = -6$. **Whenever we use the term square root in this book, we mean the principal or positive square root.**

EXAMPLE 1 Evaluate.

a) $\sqrt{81}$ **b)** $\sqrt{100}$

Solution **a)** $\sqrt{81} = 9$ since $9^2 = (9)(9) = 81$

b) $\sqrt{100} = 10$ since $(10)^2 = 100$

EXAMPLE 2 Evaluate.

a) $-\sqrt{81}$ **b)** $-\sqrt{100}$

Solution **a)** $\sqrt{81} = 9$. Now we take the opposite of both sides to get

$$-\sqrt{81} = -9$$

NOW TRY EXERCISE 21

b) Similarly, $-\sqrt{100} = -10$.

2) Recognize That Not All Square Roots Represent Real Numbers

You must understand that **square roots of negative numbers are not real numbers.** Consider $\sqrt{-4}$; to what is $\sqrt{-4}$ equal? To evaluate $\sqrt{-4}$, we must find some number whose square equals -4. But we know that the square of any nonzero real number must be a positive number. Therefore, no real number squared equals -4, so $\sqrt{-4}$ has no real value. Numbers like $\sqrt{-4}$, or square roots of any negative numbers, are called **imaginary numbers**. The study of imaginary numbers is beyond the scope of this book.

EXAMPLE 3 Indicate whether the radical expression is a real or an imaginary number.
a) $-\sqrt{25}$ **b)** $\sqrt{-25}$ **c)** $\sqrt{-43}$ **d)** $-\sqrt{43}$

Solution **a)** Real (equal to -5) **b)** Imaginary **c)** Imaginary **d)** Real

Suppose we have an expression like $\sqrt{x}$, where x represents some number. For the radical $\sqrt{x}$ to be a real number, and not imaginary, we must assume that x is a nonnegative number.

In this chapter, unless stated otherwise, we will assume that all expressions that are radicands represent nonnegative numbers.

3) Determine Whether the Square Root of a Real Number Is Rational or Irrational

To help in our discussion of rational and irrational numbers, we will define perfect squares. The numbers 1, 4, 9, 16, 25, 36, 49, ... are called **perfect squares** because each number is *the square of a natural number*. When a perfect square is a factor of a radicand, we may refer to it as a **perfect square factor.**

$1, \; 2, \; 3, \; 4, \; 5, \; 6, \; 7, \; \ldots$ *Natural numbers*

$1^2, \; 2^2, \; 3^2, \; 4^2, \; 5^2, \; 6^2, \; 7^2, \; \ldots$ *The squares of the natural numbers*

$1, \; 4, \; 9, \; 16, \; 25, \; 36, \; 49, \; \ldots$ *Perfect squares*

What are the next two perfect squares? Note that the square root of a perfect square is an integer. That is, $\sqrt{1} = 1$, $\sqrt{4} = 2$, $\sqrt{9} = 3$, $\sqrt{16} = 4$, and so on.

Table 9.1 illustrates the 20 smallest perfect squares. You may wish to refer to this table when simplifying radical expressions.

TABLE 9.1

Perfect Square	Square Root of Perfect Square		Value	Perfect Square	Square Root of Perfect Square		Value
1	$\sqrt{1}$	=	1	121	$\sqrt{121}$	=	11
4	$\sqrt{4}$	=	2	144	$\sqrt{144}$	=	12
9	$\sqrt{9}$	=	3	169	$\sqrt{169}$	=	13
16	$\sqrt{16}$	=	4	196	$\sqrt{196}$	=	14
25	$\sqrt{25}$	=	5	225	$\sqrt{225}$	=	15
36	$\sqrt{36}$	=	6	256	$\sqrt{256}$	=	16
49	$\sqrt{49}$	=	7	289	$\sqrt{289}$	=	17
64	$\sqrt{64}$	=	8	324	$\sqrt{324}$	=	18
81	$\sqrt{81}$	=	9	361	$\sqrt{361}$	=	19
100	$\sqrt{100}$	=	10	400	$\sqrt{400}$	=	20

A **rational number** is one that can be written in the form $\frac{a}{b}$, where a and b are integers, and $b \neq 0$. Examples of rational numbers are $\frac{1}{2}, \frac{3}{5}, -\frac{9}{2}$, 4, and 0.

All integers are rational numbers since they can be expressed with a denominator of 1. For example, $4 = \frac{4}{1}$ and $0 = \frac{0}{1}$. The square roots of perfect squares are also rational numbers since each is an integer. When a rational number is written as a decimal, it will be either a terminating or repeating decimal.

Terminating Decimals	**Repeating Decimals**
$\frac{1}{2} = 0.5$	$\frac{1}{3} = 0.333\ldots$
$\frac{5}{8} = 0.625$	$\frac{4}{9} = 0.444\ldots$
$\sqrt{4} = 2.0$	$\frac{1}{6} = 0.1666\ldots$

Real numbers that are not rational numbers are called **irrational numbers.** Irrational numbers when written as decimals are nonterminating, nonrepeating decimals. The square root of every positive integer that is not a perfect square is an irrational number. For example, $\sqrt{2}$ and $\sqrt{3}$ are irrational numbers. The 20 square roots listed in Table 9.1 are rational numbers. All other square roots of integers between 1 and 400 are irrational numbers. For example, since $\sqrt{230}$ is less than $\sqrt{400}$ and is not in Table 9.1, it is an irrational number. Furthermore, since $\sqrt{230}$ is between $\sqrt{225}$ and $\sqrt{256}$ in Table 9. 1, the value of $\sqrt{230}$ is between 15 and 16. Figure 9.1 shows the relationship between these different types of numbers.

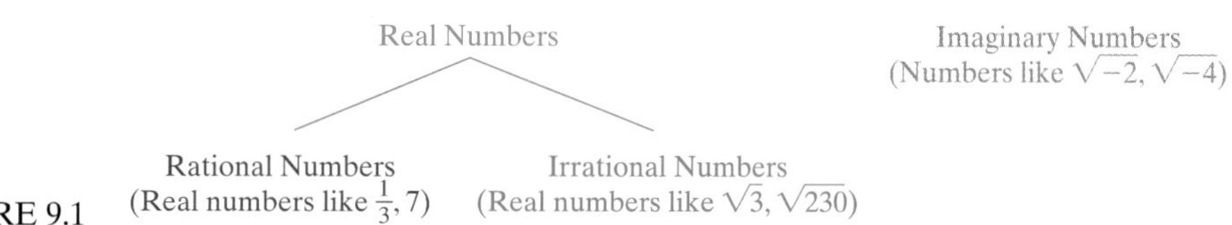

FIGURE 9.1

Using Your Calculator

Evaluating Square Roots on a Calculator

The square root key on calculators can be used to find square roots of nonnegative numbers.

On a scientific calculator, to find the square root of 4, we press

4 [$\sqrt{x}$] 2 ← Answer displayed

On the TI-82 or TI-83 graphing calculator, to find the square root of 4, we press

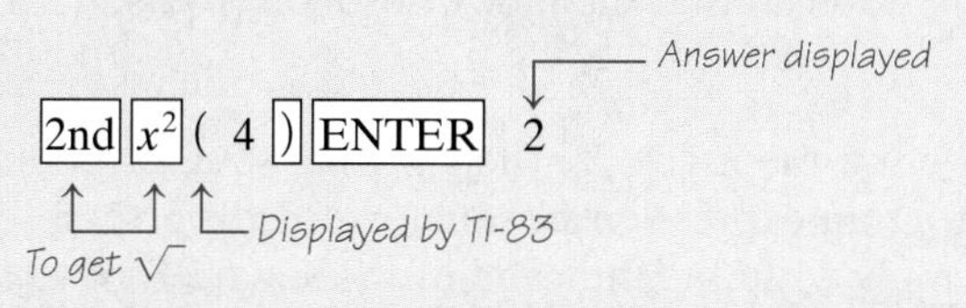

Both calculators display the answer 2. Since 2 is an integer, $\sqrt{4}$ is a rational number.

What would the calculator display if we evaluate $\sqrt{7}$? The display would show 2.6457513. Note that $\sqrt{7}$ is an irrational number, or a nonrepeating, nonterminating decimal. The decimal value of $\sqrt{7}$, or any other irrational number, can never be given exactly. The answers given on a calculator display are only close approximations of their value.

continued on next page

Suppose we tried to evaluate $\sqrt{-4}$ on a calculator. What would the calculator give as an answer? On a scientific calculator, we press

$$4 \boxed{+/-} \boxed{\sqrt{x}} \text{ Error}$$

← Answer displayed

On a TI-82 or TI-83 graphing calculator, we press

$$\boxed{2nd} \boxed{x^2} \; (\; \boxed{(-)} \; 4 \boxed{)} \boxed{ENTER} \text{ ERR: NONREAL ANS}$$

← Answer displayed

Both calculators would give an error message, because the square root of −4, or the square root of any other negative number, is not a real number.

Exercises

Use your calculator to evaluate each square root.

1. $\sqrt{11}$ **2.** $\sqrt{151}$ **3.** $\sqrt{-9}$ **4.** $\sqrt{27}$

When evaluating radicals, we may use the "is approximately equal to symbol." For example, we may write $\sqrt{2} \approx 1.414$. This is read "the square root of 2 is approximately equal to 1.414." Recall that $\sqrt{2}$ is not a perfect square, so its square root cannot be evaluated exactly.

EXAMPLE 4 Use your calculator or Table 9.1 to determine whether the following square roots are rational or irrational numbers.

a) $\sqrt{118}$ **b)** $\sqrt{121}$ **c)** $\sqrt{100}$ **d)** $\sqrt{300}$

Solution **a)** Irrational **b)** Rational, equal to 11

c) Rational, equal to 10 **d)** Irrational

NOW TRY EXERCISE 61

4) Write Square Roots as Exponential Expressions

Radical expressions can be written in exponential form. Since we are discussing square roots, we will show how to write square roots in exponential form. Writing other radicals in exponential form will be discussed in Section 9.7. We introduce this information here because your instructor may wish to use exponential form to help explain certain concepts.

Recall that the index of square roots is 2. For example,

$$\sqrt{x} \quad \text{means} \quad \sqrt[2]{x}$$

We use the index, 2, when writing square roots in exponential form. To change from an expression in square root form to an expression in exponential form, simply write the radicand of the square root to the $1/2$ power, as follows:

Writing a Square Root in Exponential Form

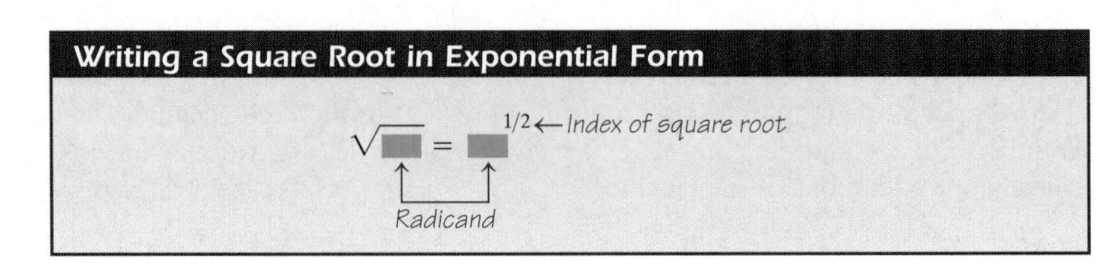

For example, $\sqrt{7}$ in exponential form is $7^{1/2}$, and $\sqrt{3ab} = (3ab)^{1/2}$. Other examples are

Square Root Form		Exponential Form
$\sqrt{16}$	=	$(16)^{1/2}$
$\sqrt{2x}$	=	$(2x)^{1/2}$
$\sqrt{15x^2y}$	=	$(15x^2y)^{1/2}$

EXAMPLE 5 Write each radical expression in exponential form.

a) $\sqrt{5}$ **b)** $\sqrt{12x}$

Solution **a)** $5^{1/2}$ **b)** $(12x)^{1/2}$

NOW TRY EXERCISE 69

We can also convert an expression from exponential form to radical form. To do so, we reverse the process. For example, $(6x)^{1/2}$ can be written $\sqrt{6x}$ and $(20x^4)^{1/2}$ can be written $\sqrt{20x^4}$.

The rules of exponents presented in Sections 6.1 and 6.2 apply to rational (or fractional) exponents. For example,

$$(x^2)^{1/2} = x^{2(1/2)} = x^1 = x$$

$$(xy)^{1/2} = x^{1/2}y^{1/2}$$

and $$x^{1/2} \cdot x^{3/2} = x^{(1/2)+(3/2)} = x^{4/2} = x^2$$

Exercise Set 9.1

Concept/Writing Exercises

1. What is the principal square root of a positive real number x?
2. **a)** What does the index indicate in a radical expression?
   b) What is the expression inside the radical sign called?
3. In your own words, explain the difference between a rational number and an irrational number.
4. Whenever we see an expression in a square root, what assumption do we make about the expression? Why do we make this assumption?
5. In your own words, explain how you would determine whether the square root of a positive integer less than 400 is a rational or irrational number **a)** by using a calculator, and **b)** without the use of a calculator.
6. In your own words, explain why the square root of a negative number is not a real number.
7. Is $\sqrt{16} = 4$? Explain.
8. Is $\sqrt{16} = -4$? Explain.
9. Is $\sqrt{-16} = -4$? Explain.
10. Is $-\sqrt{16} = -4$? Explain.
11. Is $\sqrt{\frac{9}{25}}$ a rational number? Explain.
12. Is $\sqrt{\frac{13}{15}}$ a rational number? Explain.

Practice the Skills

Evaluate each square root.

13. $\sqrt{0}$
14. $\sqrt{9}$
15. $\sqrt{1}$
16. $\sqrt{64}$
17. $-\sqrt{81}$
18. $\sqrt{4}$
19. $\sqrt{400}$
20. $\sqrt{100}$
21. $-\sqrt{25}$
22. $-\sqrt{36}$
23. $\sqrt{144}$
24. $\sqrt{49}$
25. $\sqrt{169}$
26. $\sqrt{225}$
27. $-\sqrt{1}$
28. $-\sqrt{100}$
29. $\sqrt{81}$
30. $-\sqrt{49}$
31. $-\sqrt{121}$
32. $-\sqrt{196}$
33. $\sqrt{\frac{1}{4}}$
34. $\sqrt{\frac{9}{4}}$
35. $\sqrt{\frac{16}{9}}$
36. $\sqrt{\frac{25}{64}}$

37. $-\sqrt{\frac{25}{36}}$ 38. $-\sqrt{\frac{100}{144}}$ 39. $\sqrt{\frac{36}{49}}$ 40. $\sqrt{\frac{121}{169}}$

Use your calculator to evaluate each square root.

41. $\sqrt{10}$ 42. $\sqrt{2}$ 43. $\sqrt{15}$ 44. $\sqrt{30}$
45. $\sqrt{80}$ 46. $\sqrt{79}$ 47. $\sqrt{81}$ 48. $\sqrt{121}$
49. $\sqrt{97}$ 50. $\sqrt{5}$ 51. $\sqrt{3}$ 52. $\sqrt{40}$

Indicate whether each statement is true or false.

53. $\sqrt{49}$ is a rational number.
54. $\sqrt{-25}$ is a real number.
55. $\sqrt{25}$ is a rational number.
56. $\sqrt{5}$ is an irrational number.
57. $\sqrt{9}$ is an irrational number.
58. $\sqrt{\frac{1}{4}}$ is a rational number.
59. $\sqrt{\frac{4}{9}}$ is a rational number.
60. $\sqrt{231}$ is a rational number.
61. $\sqrt{125}$ is an irrational number.
62. $\sqrt{27}$ is an irrational number.
63. $\sqrt{(18)^2}$ is an integer.
64. $\sqrt{(12)^2}$ is an integer.

Write in exponential form.

65. $\sqrt{3}$ 66. $\sqrt{31}$ 67. $\sqrt{17}$ 68. $\sqrt{16}$
69. $\sqrt{6y}$ 70. $\sqrt{5x}$ 71. $\sqrt{12x^2}$ 72. $\sqrt{25x^2y}$
73. $\sqrt{15ab^2}$ 74. $\sqrt{34x^3y}$ 75. $\sqrt{50a^3}$ 76. $\sqrt{36x^3y^3}$

Problem Solving

77. Classify each number as rational, irrational, or imaginary.

$$7.24, \sqrt{-9}, \frac{5}{7}, 0.666\ldots, 5, \sqrt{\frac{4}{49}}, \frac{3}{7}, \sqrt{\frac{5}{16}}, -\sqrt{9}, -\sqrt{-16}$$

78. Classify each number as rational, irrational, or imaginary.

$$\sqrt{5}, 8.23, \sqrt{-7}, 10, \frac{1}{3}, 0.33, \sqrt{\frac{25}{64}}, -\sqrt{90}$$

79. Between what two integers is the square root of 47? Do not use your calculator or Table 9.1. Explain how you determined your answer.

80. Between what two integers is the square root of 88? Do not use your calculator or Table 9.1. Explain how you determined your answer.

81. **a)** Explain how you can determine without using a calculator whether 4.6 or $\sqrt{20}$ is greater.
b) Without using a calculator, determine which is greater.

82. **a)** Explain how you can determine without using a calculator whether 7.2 or $\sqrt{58}$ is greater.
b) Without using a calculator, determine which is greater.

83. Arrange the following list from smallest to largest. Do not use a calculator or Table 9.1.

$$-\sqrt{4}, 3, -\sqrt{7}, 12, 2.5, -\frac{1}{2}, 4.01, \sqrt{16}$$

84. Arrange the following list from smallest to largest. Do not use a calculator or Table 9.1.

$$-\frac{1}{3}, -\sqrt{9}, 5, 0, \sqrt{9}, 8, -2, 3.25$$

85. Match each number in the column on the left with the corresponding answer in the column on the right.

$\sqrt{4}$	Imaginary number
$6^{1/2}$	2
$-\sqrt{9}$	≈ 2.45
$-(25)^{1/2}$	-5
$(30)^{1/2}$	-3
$(-4)^{1/2}$	≈ 5.48

86. Match each number in the column on the left with the corresponding answer in the column on the right.

$-\sqrt{36}$	10
$(40)^{1/2}$	-7
$\sqrt{100}$	imaginary number
$(10)^{1/2}$	≈ 6.32
$-(49^{1/2})$	≈ 3.16
$(-16)^{1/2}$	-6

87. Is $\sqrt{0}$ **a)** a real number? **b)** a positive number? **c)** a negative number? **d)** a rational number? **e)** an irrational number? Explain your answer.

Challenge Problems

We discuss the following concepts in Sections 9.2 and 9.3.

88. **a)** Is $\sqrt{4} \cdot \sqrt{9}$ equal to $\sqrt{4 \cdot 9}$?

b) Is $\sqrt{9} \cdot \sqrt{25}$ equal to $\sqrt{9 \cdot 25}$?

c) Using these two examples, can you guess what $\sqrt{a} \cdot \sqrt{b}$ is equal to (provided that $a \geq 0, b \geq 0$)?

d) Create your own problem like those given in parts **a)** and **b)** and see if the answer you gave in part **c)** works with your numbers.

89. **a)** Is $\sqrt{2^2}$ equal to 2?

b) Is $\sqrt{5^2}$ equal to 5?

c) Using these two examples, can you guess what $\sqrt{a^2}$, $a \geq 0$, is equal to?

d) Create your own problem like those given in parts **a)** and **b)** and see if the answer you gave in part **c)** works with your numbers.

90. **a)** Is $\dfrac{\sqrt{16}}{\sqrt{4}}$ equal to $\sqrt{\dfrac{16}{4}}$?

b) Is $\dfrac{\sqrt{36}}{\sqrt{9}}$ equal to $\sqrt{\dfrac{36}{9}}$?

c) Using these two examples, can you guess what $\dfrac{\sqrt{a}}{\sqrt{b}}$ is equal to (provided that $a \geq 0, b > 0$)?

d) Create your own problem like those given in parts **a)** and **b)** and see if the answer you gave in part **c)** works with your numbers.

The rules of exponents we discussed in Chapter 6 also apply with rational exponents. Use the rules of exponents to simplify the following expressions. We will discuss problems like this in Section 9.7.

91. $(x^3)^{1/2}$ **92.** $(x^4)^{1/2}$ **93.** $x^{1/2} \cdot x^{5/2}$ **94.** $x^{3/2} \cdot x^{1/2}$

Cumulative Review Exercises

[4.3] **95.** Determine the slope of the line through the points $(-5, 3)$ and $(6, 7)$.

[4.6] **96.** If $f(x) = x^2 - 4x - 5$, find $f(-3)$.

[8.7] *Solve.*

97. $\dfrac{2x}{x^2 - 4} + \dfrac{1}{x - 2} = \dfrac{2}{x + 2}$

98. $\dfrac{4x}{x^2 + 6x + 9} - \dfrac{2x}{x + 3} = \dfrac{x + 1}{x + 3}$

9.2 MULTIPLYING AND SIMPLIFYING SQUARE ROOTS

SSM VIDEO 9.2 CD Rom

1) Use the product rule to simplify square roots containing constants.

2) Use the product rule to simplify square roots containing variables.

1) Use the Product Rule to Simplify Square Roots Containing Constants

To simplify square roots in this section we will make use of the **product rule for square roots.**

Product Rule for Square Roots	
$\sqrt{a} \cdot \sqrt{b} = \sqrt{a \cdot b}$, provided $a \geq 0, b \geq 0$	Rule 1

The product rule states that the product of two square roots is equal to the square root of the product. The product rule applies only when both a and b are nonnegative, since the square roots of negative numbers are not real numbers.

Examples of the Product Rule

$$\left.\begin{aligned}\sqrt{1}\cdot\sqrt{60} &= \sqrt{1\cdot 60}\\ \sqrt{2}\cdot\sqrt{30} &= \sqrt{2\cdot 30}\\ \sqrt{3}\cdot\sqrt{20} &= \sqrt{3\cdot 20}\\ \sqrt{4}\cdot\sqrt{15} &= \sqrt{4\cdot 15}\\ \sqrt{6}\cdot\sqrt{10} &= \sqrt{6\cdot 10}\end{aligned}\right\} = \sqrt{60}$$

Note that $\sqrt{60}$ can be factored into any of these forms.

When two square roots are placed next to one another, the square roots are to be multiplied. Thus $\sqrt{a}\,\sqrt{b}$ means $\sqrt{a}\cdot\sqrt{b}$.

To Simplify the Square Root of a Constant

1. Write the constant as a product of the largest perfect square factor and another factor.
2. Use the product rule to write the expression as a product of square roots, with each square root containing one of the factors.
3. Find the square root of the perfect square factor.

EXAMPLE 1 Simplify $\sqrt{60}$.

Solution The only perfect square factor of 60 is 4.

$$\begin{aligned}\sqrt{60} &= \sqrt{4\cdot 15}\\ &= \sqrt{4}\cdot\sqrt{15}\\ &= 2\sqrt{15}\end{aligned}$$

Since 15 is not a perfect square and has no perfect square factors, this expression cannot be simplified further. The expression $2\sqrt{15}$ is read "two times the square root of fifteen."

EXAMPLE 2 Simplify $\sqrt{12}$.

Solution

$$\begin{aligned}\sqrt{12} = \sqrt{4\cdot 3} &= \sqrt{4}\cdot\sqrt{3}\\ &= 2\sqrt{3}\end{aligned}$$

EXAMPLE 3 Simplify $\sqrt{80}$.

Solution

$$\begin{aligned}\sqrt{80} = \sqrt{16\cdot 5} &= \sqrt{16}\cdot\sqrt{5}\\ &= 4\sqrt{5}\end{aligned}$$

EXAMPLE 4 Simplify $\sqrt{147}$.

Solution

$$\begin{aligned}\sqrt{147} = \sqrt{49\cdot 3} &= \sqrt{49}\cdot\sqrt{3}\\ &= 7\sqrt{3}\end{aligned}$$

HELPFUL HINT

When simplifying a square root, it is not uncommon for students to use a perfect square factor that is not the *largest* perfect square factor of the radicand. Let's consider Example 3 again. Four is also a perfect square factor of 80.

$$\sqrt{80} = \sqrt{4 \cdot 20} = \sqrt{4} \cdot \sqrt{20} = 2\sqrt{20}$$

Since 20 itself contains a perfect square factor of 4, the problem is not complete. Rather than starting the entire problem again, you can continue the simplification process as follows.

$$\sqrt{80} = 2\sqrt{20} = 2\sqrt{4 \cdot 5} = 2\sqrt{4} \cdot \sqrt{5} = 2 \cdot 2 \cdot \sqrt{5} = 4\sqrt{5}$$

Now the result checks with the answer in Example 3.

EXAMPLE 5 Simplify $\sqrt{132}$.

Solution

$$\sqrt{132} = \sqrt{4 \cdot 33} = \sqrt{4} \cdot \sqrt{33}$$
$$= 2\sqrt{33}$$

Although 33 can be factored into $3 \cdot 11$, neither of these factors is a perfect square. Thus, the answer cannot be simplified any further.

NOW TRY EXERCISE 21

Using Your Graphing Calculator

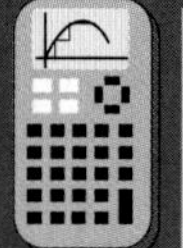

You can use the TABLE feature of a graphing calculator to help you simplify the square root of a constant. For example, suppose you wish to simplify $\sqrt{147}$. Note that the product $x\left(\frac{147}{x}\right) = 147$. If we designate $\frac{147}{x}$ as y_1, then $xy_1 = 147$. Now let $y_2 = \sqrt{y_1} = \sqrt{\frac{147}{x}}$. We are going to use the expression $\sqrt{\frac{147}{x}}$ to look for the largest perfect square factor of 147. Figure 9.2 shows the natural numbers in column X. Column Y_1 shows the value of $\frac{147}{x}$ for the corresponding value of x in the first column. Column Y_2 shows the square roots of the numbers in column Y_1. Reading down the table, determine the *first* natural number that appears in column Y_2. This number is the square root of the largest perfect square factor of the radicand. The largest perfect square factor of the radicand is found in column Y_1 adjacent to this natural number. In this example, the first natural number in column Y_2 is 7, which is the square root of 49, the largest perfect square factor of 147.

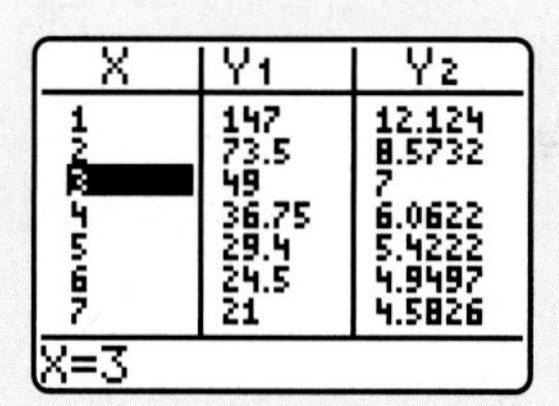

X	Y1	Y2
1	147	12.124
2	73.5	8.5732
3	49	7
4	36.75	6.0622
5	29.4	5.4222
6	24.5	4.9497
7	21	4.5826

X=3

FIGURE 9.2

Write

$$\sqrt{147} = \sqrt{49 \cdot 3} = \sqrt{49} \cdot \sqrt{3} = 7\sqrt{3}.$$

From Y_1 From X From Y_2

What do you think it means if you do not find a natural number, other than 1, in column Y_2?

Exercises

Use your graphing calculator to simplify the following expressions.

1. $\sqrt{180}$ **2.** $\sqrt{175}$

3. $\sqrt{192}$ **4.** $\sqrt{1083}$

2) Use the Product Rule to Simplify Square Roots Containing Variables

Now we will simplify square roots that contain variables in the radicand.

In Section 9.1 we noted that certain numbers were **perfect squares.** We will also refer to certain expressions that contain a variable as perfect squares. When a radical contains a variable (or number) raised to an **even exponent,** that variable (or number) and exponent together also form a perfect square. For example, in the expression $\sqrt{x^4}$, the x^4 is a perfect square since the exponent 4, is even. In the expression $\sqrt{x^5}$, the x^5 is not a perfect square since the exponent is odd. However, x^4 is a **perfect square factor** of x^5 because x^4 is a perfect square and x^4 is a factor of x^5. Note that $x^5 = x^4 \cdot x$.

To evaluate square roots when the radicand is a perfect square, we use the following rule.

$$\sqrt{a^{2\cdot n}} = a^n, \qquad a \geq 0 \qquad \text{Rule 2}$$

This rule states that **the square root of a variable raised to an even power equals the variable raised to one-half that power.** To explain this rule, we can write the square root expression $\sqrt{a^{2n}}$ in exponential form, and then simplify as follows.

$$\sqrt{a^{2n}} = \left(a^{2n}\right)^{1/2} = a^n$$

Examples of rule 2 follow.

Examples

$$\sqrt{x^2} = x$$
$$\sqrt{a^4} = a^2$$
$$\sqrt{x^{14}} = x^7$$
$$\sqrt{y^{20}} = y^{10}$$

A special case of rule 2 (when $n = 1$) is

$$\boldsymbol{\sqrt{a^2} = a, \qquad a \geq 0}$$

EXAMPLE 6 Simplify.

a) $\sqrt{x^{34}}$ **b)** $\sqrt{x^4y^6}$ **c)** $\sqrt{a^{10}b^2}$ **d)** $\sqrt{y^8z^{12}}$

Solution **a)** $\sqrt{x^{34}} = x^{17}$ **b)** $\sqrt{x^4y^6} = \sqrt{x^4}\sqrt{y^6} = x^2y^3$

c) $\sqrt{a^{10}b^2} = a^5b$ **d)** $\sqrt{y^8z^{12}} = y^4z^6$

NOW TRY EXERCISE 35

To Simplify the Square Root of a Radicand Containing a Variable Raised to an Odd Power

1. Express the variable as the product of two factors, one of which has an exponent of 1 (the other will therefore be a perfect square factor).
2. Use the product rule to simplify.

Examples 7 and 8 illustrate this procedure.

EXAMPLE 7 Simplify. **a)** $\sqrt{x^3}$ **b)** $\sqrt{y^7}$ **c)** $\sqrt{a^{79}}$

Solution **a)** $\sqrt{x^3} = \sqrt{x^2 \cdot x} = \sqrt{x^2} \cdot \sqrt{x}$ *(Remember that x means x^1.)*

$= x \cdot \sqrt{x}$ or $x\sqrt{x}$

b) $\sqrt{y^7} = \sqrt{y^6 \cdot y^1} = \sqrt{y^6} \cdot \sqrt{y}$

$= y^3\sqrt{y}$

c) $\sqrt{a^{79}} = \sqrt{a^{78} \cdot a} = \sqrt{a^{78}} \cdot \sqrt{a}$

$= a^{39}\sqrt{a}$

More complex radicals can be simplified using the product rule for radicals and the principles discussed in this section.

EXAMPLE 8 Simplify. **a)** $\sqrt{25x^3}$ **b)** $\sqrt{50x^2}$ **c)** $\sqrt{50x^3}$

Solution Write each expression as the product of square roots, one of which has a radicand that is a perfect square.

a) $\sqrt{25x^3} = \sqrt{25x^2} \cdot \sqrt{x} = 5x\sqrt{x}$

b) $\sqrt{50x^2} = \sqrt{25x^2} \cdot \sqrt{2} = 5x\sqrt{2}$

c) $\sqrt{50x^3} = \sqrt{25x^2} \cdot \sqrt{2x} = 5x\sqrt{2x}$

EXAMPLE 9 Simplify.

a) $\sqrt{50x^2y}$ **b)** $\sqrt{27x^3y^2}$ **c)** $\sqrt{98a^9b^7}$

Solution **a)** $\sqrt{50x^2y} = \sqrt{25x^2} \cdot \sqrt{2y} = 5x\sqrt{2y}$

b) $\sqrt{27x^3y^2} = \sqrt{9x^2y^2} \cdot \sqrt{3x} = 3xy\sqrt{3x}$

c) $\sqrt{98a^9b^7} = \sqrt{49a^8b^6} \cdot \sqrt{2ab} = 7a^4b^3\sqrt{2ab}$

NOW TRY EXERCISE 45

The radicand of your simplified answer should not contain any perfect square factors or any variables with an exponent greater than 1.

Now let's look at an example where we use the product rule to multiply two radicals before simplifying.

EXAMPLE 10 Multiply and then simplify.

a) $\sqrt{2} \cdot \sqrt{8}$ **b)** $\sqrt{2x} \cdot \sqrt{8}$ **c)** $(\sqrt{3x})^2$

Solution **a)** $\sqrt{2} \cdot \sqrt{8} = \sqrt{2 \cdot 8} = \sqrt{16} = 4$

b) $\sqrt{2x} \cdot \sqrt{8} = \sqrt{16x} = \sqrt{16} \cdot \sqrt{x} = 4\sqrt{x}$

c) $(\sqrt{3x})^2 = \sqrt{3x} \cdot \sqrt{3x} = \sqrt{9x^2} = 3x$

EXAMPLE 11 Multiply and then simplify.

a) $\sqrt{8x^3y} \cdot \sqrt{4xy^5}$ **b)** $\sqrt{5ab^8} \cdot \sqrt{6a^5b}$

Solution **a)** $\sqrt{8x^3y} \cdot \sqrt{4xy^5} = \sqrt{32x^4y^6} = \sqrt{16x^4y^6} \cdot \sqrt{2}$

$= 4x^2y^3\sqrt{2}$

b) $\sqrt{5ab^8} \cdot \sqrt{6a^5b} = \sqrt{30a^6b^9} = \sqrt{a^6b^8} \cdot \sqrt{30b}$

$= a^3b^4\sqrt{30b}$

In part **b)**, 30 can be factored in many ways. However, none of the factors are perfect squares, so we leave the answer as given.

NOW TRY EXERCISE 63

Exercise Set 9.2

Concept/Writing Exercises

1. In your own words, state the product rule for square roots and explain what it means.
2. a) In your own words, explain how to simplify a square root containing only a constant.
   b) Simplify $\sqrt{20}$ using the procedure you gave in part **a)**.
3. Explain why the product rule cannot be used to simplify the problem $\sqrt{-4} \cdot \sqrt{-9}$.
4. We learned that for $a \geq 0$, $\sqrt{a^{2 \cdot n}} = a^n$. Explain in your own words what this means.
5. a) Explain how to simplify the square root of a radical containing a variable raised to an odd power.
   b) Simplify $\sqrt{x^{13}}$ using the procedure you gave in part **a)**.
6. a) Explain why $\sqrt{32x^3}$ is not a simplified expression.
   b) Simplify $\sqrt{32x^3}$.
7. a) Explain why $\sqrt{75x^5}$ is not a simplified expression.
   b) Simplify $\sqrt{75x^5}$.

Determine whether the square root on the right-hand side of the equal sign is the simplified form of the square root on the left-hand side of the equal sign. If not, simplify it properly.

8. $\sqrt{75} = 5\sqrt{3}$
9. $\sqrt{32} = 2\sqrt{8}$
10. $\sqrt{x^5} = x^2\sqrt{x}$
11. $\sqrt{x^9} = x\sqrt{x^7}$
12. Use the product rule to write $\sqrt{40}$ as four different products of factors.

Practice the Skills

Simplify.

13. $\sqrt{36}$
14. $\sqrt{16}$
15. $\sqrt{8}$
16. $\sqrt{45}$
17. $\sqrt{96}$
18. $\sqrt{125}$
19. $\sqrt{32}$
20. $\sqrt{52}$
21. $\sqrt{90}$
22. $\sqrt{44}$
23. $\sqrt{80}$
24. $\sqrt{27}$
25. $\sqrt{72}$
26. $\sqrt{128}$
27. $\sqrt{156}$
28. $\sqrt{180}$
29. $\sqrt{256}$
30. $\sqrt{625}$
31. $\sqrt{1600}$
32. $\sqrt{x^4}$
33. $\sqrt{x^8}$
34. $\sqrt{y^{13}}$
35. $\sqrt{x^2y^4}$
36. $\sqrt{xy^2}$
37. $\sqrt{a^{12}b^9}$
38. $\sqrt{x^4y^5z^6}$
39. $\sqrt{a^2b^4c}$
40. $\sqrt{a^3b^9c^{11}}$
41. $\sqrt{2x^3}$
42. $\sqrt{12x^4y^2}$
43. $\sqrt{75a^3b^2}$
44. $\sqrt{125x^3y^5}$
45. $\sqrt{300a^5b^{11}}$
46. $\sqrt{64xyz^5}$
47. $\sqrt{243x^3y^4}$
48. $\sqrt{500ab^4c^3}$
49. $\sqrt{192a^2b^7c}$
50. $\sqrt{112x^6y^8}$
51. $\sqrt{250x^4yz}$
52. $\sqrt{98x^4y^4z}$

Simplify.

53. $\sqrt{7} \cdot \sqrt{7}$
54. $\sqrt{8} \cdot \sqrt{8}$
55. $\sqrt{18} \cdot \sqrt{3}$
56. $\sqrt{60} \cdot \sqrt{5}$
57. $\sqrt{48} \cdot \sqrt{15}$
58. $\sqrt{30} \cdot \sqrt{5}$
59. $\sqrt{3x}\sqrt{7x}$
60. $\sqrt{4x^3}\sqrt{4x}$
61. $\sqrt{4a^2} \cdot \sqrt{12ab^2}$
62. $\sqrt{15x^2}\sqrt{6x^5}$
63. $\sqrt{6xy^3} \cdot \sqrt{12x^2y}$
64. $\sqrt{20xy^4}\sqrt{6x^5}$
65. $\sqrt{21xy}\sqrt{3x^3y^4}$
66. $\sqrt{20x^3y}\sqrt{6x^3y^5}$
67. $\sqrt{15xy^6}\sqrt{6xyz}$
68. $\sqrt{14xyz^5}\sqrt{3xy^2z^6}$
69. $\sqrt{4a^2b^4}\sqrt{9a^4b^6}$
70. $\sqrt{6a^4b^5c^6} \cdot \sqrt{3a^3bc^6}$
71. $(\sqrt{2x})^2$
72. $(\sqrt{6x^2})^2$
73. $(\sqrt{13x^4y^6})^2$
74. $\sqrt{36x^2y^7}\sqrt{2x^4y}$
75. $(\sqrt{5a})^2(\sqrt{3a})^2$
76. $(\sqrt{4ab})^2(\sqrt{3ab})^2$

Problem Solving

Which coefficients and exponents should be placed in the shaded areas to make a true statement? Explain how you obtained your answer.

77. $\sqrt{16x^{\square}y^6} = 4x^2y^3$

78. $\sqrt{\square x^4y^{\square}} = 4x^2y^4$

79. $\sqrt{4x^{\square}y^{\square}} = 2x^3y^2\sqrt{y}$

80. $\sqrt{3x^4y^{\square}} \cdot \sqrt{3x^{\square}y^5} = 3x^5y^7\sqrt{xy}$

81. $\sqrt{2x^{\square}y^5} \cdot \sqrt{\square x^3y^{\square}} = 4x^7y^6\sqrt{x}$

82. $\sqrt{32x^4z^{\square}} \cdot \sqrt{\square x^{\square}z^{12}} = 8x^5z^9\sqrt{z}$

83. **a)** Showing all steps, simplify $(\sqrt{13x^3})^2$.
b) Showing all steps, simplify $\sqrt{(13x^3)^2}$.
c) Compare your results in part **a)** and part **b)**. Are they the same?

84. **a)** Showing all steps, simplify $(\sqrt{7x^4})^2$.
b) Showing all steps, simplify $\sqrt{(7x^4)^2}$.
c) Compare your results in part **a)** and part **b)**. Are they the same?

Simplify. Treat the Δ and ∇ as if they were variables.

85. $\sqrt{200\Delta^{11}}$

86. $\sqrt{180\Delta^7\nabla^{16}}$

87. $\sqrt{5\Delta^{100}} \cdot \sqrt{5\nabla^{36}}$

88. $\sqrt{7\Delta^{10}} \cdot \sqrt{343\nabla^{10}}$

Challenge Problems

Following we illustrate two simplifications involving square roots.

$$\sqrt{x^4} = (x^4)^{1/2} = x^{4(1/2)} = x^2$$

$$\sqrt{x^{2/4}} = (x^{2/4})^{1/2} = x^{(2/4)(1/2)} = x^{1/4}$$

In Section 9.1 we indicated that the square root of an expression may be written as the expression to the $\frac{1}{2}$ power. The rules for exponents that were discussed in Section 6.1 also apply when the exponents are rational numbers. Use the two examples illustrated and the rules for exponents to simplify each of the following. We will discuss rational exponents further in Section 9.7.

89. $\sqrt{x^{2/6}}$

90. $\sqrt{y^{10/12}}$

91. $\sqrt{4x^{4/5}}$

92. $\sqrt{25y^{8/3}}$

93. Is $\sqrt{6.25}$ a rational or an irrational number? Explain how you determined your answer.

94. **a)** In Section 9.4 we will be multiplying expressions like $(\sqrt{a} + \sqrt{b})(\sqrt{a} - \sqrt{b})$ using the FOIL method. Can you find this product now?
b) Multiply $(\sqrt{6} + \sqrt{3})(\sqrt{6} - \sqrt{3})$.

95. The area of a square is found by the formula $A = s^2$. We will learn later that we can rewrite this formula as $s = \sqrt{A}$.
a) If the area is 16 square feet, what is the length of a side?
b) If the area is doubled, is the length of a side doubled? Explain.
c) To double the length of a side of a square, how much must the area be increased? Explain.

96. We know that $\sqrt{a} \cdot \sqrt{b} = \sqrt{a \cdot b}$ if $a \geq 0$ and $b \geq 0$. Does $\sqrt{\frac{a}{b}} = \frac{\sqrt{a}}{\sqrt{b}}$ if $a \geq 0$ and $b > 0$? Try several pairs of values for a and b and see.

97. **a)** Will the product of two rational numbers always be a rational number? Explain and give an example to support your answer.
b) Will the product of two irrational numbers always be an irrational number? Explain and give an example to support your answer.

Group Activity

We learned earlier that $\sqrt{\square} = \square^{1/2}$. For example, $\sqrt{x^6} = (x^6)^{1/2} = x^3$. We can simplify $\sqrt{x^{2n}}$ by writing the expression in exponential form, $\sqrt{x^{2n}} = (x^{2n})^{1/2} = x^{(2n)(1/2)} = x^n$. As a group, simplify the following square roots by writing the expression in exponential form. Show all the steps in the simplification process.

98. $\sqrt{x^{10a}}$

99. $\sqrt{x^{8b}}$

100. $\sqrt{x^{4a}y^{12b}}$

101. $\sqrt{x^{8b}y^{6c}}$

Cumulative Review Exercises

[4.4] **102.** Write the equation $3x + 6y = 9$ in slope–intercept form and indicate the slope and the y-intercept.

[4.5] **103.** Graph $6x - 5y \geq 30$.

[5.3] **104.** Solve the following system of equations.

$$3x - 4y = 6$$
$$5x - 3y = 5$$

[8.2] **105.** Divide $\frac{3x^2 - 16x - 12}{3x^2 - 10x - 8} \div \frac{x^2 - 7x + 6}{3x^2 - 11x - 4}$.

9.3 DIVIDING AND SIMPLIFYING SQUARE ROOTS

1) **Understand what it means for a square root to be simplified.**
2) **Use the quotient rule to simplify square roots.**
3) **Rationalize denominators.**

1) Understand What It Means for a Square Root to be Simplified

In this section we will use a new rule, the quotient rule, to simplify square roots containing fractions. However, before we do that, we need to discuss what it means for a square root to be simplified.

A Square Root Is Simplified When

1. No radicand has a factor that is a perfect square.
2. No radicand contains a fraction.
3. No denominator contains a square root.

All three criteria must be met for an expression to be simplified. Let's look at some radical expressions that *are not simplified.*

Radical	Reason Not Simplified	
$\sqrt{8}$	Contains perfect square factor, 4. $(\sqrt{8} = \sqrt{4} \cdot \sqrt{2} = 2\sqrt{2})$	(Criteria 1)
$\sqrt{x^3}$	Contains perfect square factor, x^2. $(\sqrt{x^3} = \sqrt{x^2} \cdot \sqrt{x} = x\sqrt{x})$	(Criteria 1)
$\sqrt{\frac{1}{2}}$	Radicand contains a fraction.	(Criteria 2)
$\frac{1}{\sqrt{2}}$	Square root in the denominator.	(Criteria 3)

2) Use the Quotient Rule to Simplify Square Roots

The quotient rule for square roots states that the quotient of two square roots is equal to the square root of the quotient.

Quotient Rule for Square Roots

$$\frac{\sqrt{a}}{\sqrt{b}} = \sqrt{\frac{a}{b}}, \quad \text{provided} \quad a \geq 0,\ b > 0 \qquad \text{Rule 3}$$

Examples 1 through 4 illustrate how the quotient rule is used to simplify square roots.

EXAMPLE 1 Simplify.

a) $\sqrt{\frac{27}{3}}$ **b)** $\sqrt{\frac{25}{5}}$ **c)** $\sqrt{\frac{4}{9}}$

Solution When the square root contains a fraction, divide out any factor common to both the numerator and denominator. If the square root still contains a fraction, use the quotient rule to simplify.

a) $\sqrt{\frac{27}{3}} = \sqrt{9} = 3$ **b)** $\sqrt{\frac{25}{5}} = \sqrt{5}$ **c)** $\sqrt{\frac{4}{9}} = \frac{\sqrt{4}}{\sqrt{9}} = \frac{2}{3}$

NOW TRY EXERCISE 17

EXAMPLE 2 Simplify. **a)** $\sqrt{\frac{8x^2}{4}}$ **b)** $\sqrt{\frac{150a^2b}{3b}}$ **c)** $\sqrt{\frac{3x^2y^4}{27x^4}}$ **d)** $\sqrt{\frac{15ab^5c^2}{3a^5bc}}$

Solution First divide out any factors common to both the numerator and denominator; then simplify.

a) $\sqrt{\frac{8x^2}{4}} = \sqrt{2x^2}$ *Simplify the radicand*

$= \sqrt{x^2}\sqrt{2}$ *Product rule*

$= x\sqrt{2}$ *Simplify.*

b) $\sqrt{\frac{150a^2b}{3b}} = \sqrt{50a^2} = \sqrt{25a^2}\sqrt{2} = 5a\sqrt{2}$

c) $\sqrt{\frac{3x^2y^4}{27x^4}} = \sqrt{\frac{y^4}{9x^2}}$ *Simplify the radicand*

$= \frac{\sqrt{y^4}}{\sqrt{9x^2}}$ *Quotient rule*

$= \frac{y^2}{3x}$ *Simplify.*

d) $\sqrt{\frac{15ab^5c^2}{3a^5bc}} = \sqrt{\frac{5b^4c}{a^4}} = \frac{\sqrt{5b^4c}}{\sqrt{a^4}} = \frac{\sqrt{b^4}\sqrt{5c}}{\sqrt{a^4}} = \frac{b^2\sqrt{5c}}{a^2}$

NOW TRY EXERCISE 31

When you are given a fraction containing a radical expression in both the numerator and the denominator, use the quotient rule to simplify, as in Examples 3 and 4.

EXAMPLE 3 Simplify. **a)** $\frac{\sqrt{2}}{\sqrt{8}}$ **b)** $\frac{\sqrt{48}}{\sqrt{3}}$

Solution **a)** $\frac{\sqrt{2}}{\sqrt{8}} = \sqrt{\frac{2}{8}} = \sqrt{\frac{1}{4}} = \frac{\sqrt{1}}{\sqrt{4}} = \frac{1}{2}$

b) $\frac{\sqrt{48}}{\sqrt{3}} = \sqrt{\frac{48}{3}} = \sqrt{16} = 4$

EXAMPLE 4 Simplify. **a)** $\frac{\sqrt{32x^4y^3}}{\sqrt{8xy}}$ **b)** $\frac{\sqrt{75x^8y^4}}{\sqrt{3x^5y^8}}$

Solution **a)** $\frac{\sqrt{32x^4y^3}}{\sqrt{8xy}} = \sqrt{\frac{32x^4y^3}{8xy}} = \sqrt{4x^3y^2} = \sqrt{4x^2y^2} \cdot \sqrt{x} = 2xy\sqrt{x}$

b) $\frac{\sqrt{75x^8y^4}}{\sqrt{3x^5y^8}} = \sqrt{\frac{75x^8y^4}{3x^5y^8}} = \sqrt{\frac{25x^3}{y^4}} = \frac{\sqrt{25x^3}}{\sqrt{y^4}} = \frac{\sqrt{25x^2}\sqrt{x}}{\sqrt{y^4}} = \frac{5x\sqrt{x}}{y^2}$

NOW TRY EXERCISE 35

3) Rationalize Denominators

When the denominator of a fraction contains the square root of a number that is not a perfect square, we generally simplify the expression by **rationalizing the denominator. To rationalize a denominator means to remove all radicals from the denominator.** We rationalize the denominator because it is easier (without a calculator) to obtain the approximate value of a number like $\sqrt{2}/2$ than a number like $1/\sqrt{2}$.

> **To rationalize a denominator with a square root**, multiply *both* the numerator and the denominator of the fraction by the square root that appears in the denominator or by the square root of a number that makes the denominator a perfect square.

EXAMPLE 5 Simplify $\dfrac{1}{\sqrt{3}}$.

Solution Since $\sqrt{3}\cdot\sqrt{3} = \sqrt{9} = 3$, we multiply both the numerator and denominator by $\sqrt{3}$.

$$\frac{1}{\sqrt{3}} = \frac{1}{\sqrt{3}}\cdot\frac{\sqrt{3}}{\sqrt{3}} = \frac{\sqrt{3}}{\sqrt{9}} = \frac{\sqrt{3}}{3}$$

NOW TRY EXERCISE 43

The answer $\dfrac{\sqrt{3}}{3}$ is simplified because it satisfies the three requirements stated earlier.

In Example 5, multiplying both the numerator and denominator by $\sqrt{3}$ is equivalent to multiplying the fraction by 1, which does not change its value.

Avoiding Common Errors

An expression under a square root *cannot* be divided by an expression not under a square root.

CORRECT

$\dfrac{\sqrt{2}}{2}$ cannot be simplified any further.

$\dfrac{\sqrt{6}}{3}$ cannot be simplified any further.

$\dfrac{\sqrt{x^3}}{x} = \dfrac{\sqrt{x^2}\sqrt{x}}{x} = \dfrac{\cancel{x}\sqrt{x}}{\cancel{x}} = \sqrt{x}$

INCORRECT

$\dfrac{\sqrt{\cancel{2}^{1}}}{\cancel{2}_{1}} = \sqrt{1} = \cancel{1}$

$\dfrac{\sqrt{\cancel{6}^{2}}}{\cancel{3}_{1}} = \cancel{\sqrt{2}}$

$\dfrac{\sqrt{\cancel{x^3}^{2}}}{\cancel{x}_{1}} = \sqrt{x^2} \cancel{=} x$

Each of the following simplifications is correct because the constant or variable divided out are not under square roots.

CORRECT

$\dfrac{\overset{2}{\cancel{6}}\sqrt{2}}{\underset{1}{\cancel{3}}} = 2\sqrt{2}$

$\dfrac{\overset{1}{\cancel{4}}\sqrt{3}}{\underset{2}{\cancel{8}}} = \dfrac{\sqrt{3}}{2}$

CORRECT

$\dfrac{\cancel{x}\sqrt{2}}{\cancel{x}} = \sqrt{2}$

$\dfrac{3\overset{x}{\cancel{x^2}}\sqrt{5}}{\cancel{x}} = 3x\sqrt{5}$

EXAMPLE 6 Simplify. **a)** $\sqrt{\frac{2}{3}}$ **b)** $\sqrt{\frac{x^2}{18}}$

Solution **a)**

$$\sqrt{\frac{2}{3}} = \frac{\sqrt{2}}{\sqrt{3}} \qquad \textit{Quotient rule}$$

$$= \frac{\sqrt{2}}{\sqrt{3}} \cdot \frac{\sqrt{3}}{\sqrt{3}} \qquad \textit{Multiply numerator and denominator by } \sqrt{3}.$$

$$= \frac{\sqrt{6}}{\sqrt{9}} \qquad \textit{Product rule}$$

$$= \frac{\sqrt{6}}{3} \qquad \textit{Simplify.}$$

b) $\sqrt{\frac{x^2}{18}} = \frac{\sqrt{x^2}}{\sqrt{18}} = \frac{x}{\sqrt{9} \cdot \sqrt{2}} = \frac{x}{3\sqrt{2}}$

Now rationalize the denominator.

$$\frac{x}{3\sqrt{2}} \cdot \frac{\sqrt{2}}{\sqrt{2}} = \frac{x\sqrt{2}}{3\sqrt{4}} = \frac{x\sqrt{2}}{3 \cdot 2} = \frac{x\sqrt{2}}{6}$$

Part **b)** can also be rationalized as follows:

$$\sqrt{\frac{x^2}{18}} = \frac{\sqrt{x^2}}{\sqrt{18}} = \frac{x}{\sqrt{18}} \cdot \frac{\sqrt{2}}{\sqrt{2}} = \frac{x\sqrt{2}}{\sqrt{36}} = \frac{x\sqrt{2}}{6}$$

Note that $\frac{x}{\sqrt{18}} \cdot \frac{\sqrt{18}}{\sqrt{18}}$ will also give us the same result when simplified.

NOW TRY EXERCISE 57

Exercise Set 9.3

Concept/Writing Exercises

1. What are the three requirements for a square root to be considered simplified?

Explain why each expression is not simplified. Then simplify the expression.

2. $\sqrt{32}$

3. $\sqrt{\frac{1}{3}}$

4. $\frac{3}{\sqrt{5}}$

Explain why each expression can or cannot be simplified. Then simplify the expression if possible.

5. $\frac{\sqrt{3}}{3}$

6. $\frac{4\sqrt{3}}{2}$

7. $\frac{x^2\sqrt{2}}{x}$

8. $\frac{\sqrt{10}}{5}$

9. $\frac{\sqrt{x}}{x}$

10. $\frac{\sqrt{6}}{2}$

11. In your own words, state the quotient rule for square roots and explain what it means.

12. What does it mean to rationalize a denominator?

13. **a)** In your own words, explain how to rationalize the denominator of a fraction of the form $\frac{a}{\sqrt{b}}$.

 b) Rationalize $\frac{a}{\sqrt{b}}$.

14. Explain why the quotient rule cannot be used to simplify the problem $\frac{\sqrt{-10}}{\sqrt{-2}}$.

Practice the Skills

Simplify each expression.

15. $\sqrt{\dfrac{20}{5}}$
16. $\sqrt{\dfrac{24}{6}}$
17. $\sqrt{\dfrac{63}{7}}$
18. $\sqrt{\dfrac{16}{4}}$
19. $\dfrac{\sqrt{18}}{\sqrt{2}}$
20. $\dfrac{\sqrt{3}}{\sqrt{27}}$
21. $\sqrt{\dfrac{1}{36}}$
22. $\sqrt{\dfrac{16}{25}}$
23. $\sqrt{\dfrac{81}{144}}$
24. $\sqrt{\dfrac{4}{121}}$
25. $\dfrac{\sqrt{10}}{\sqrt{1000}}$
26. $\sqrt{\dfrac{20}{80}}$
27. $\sqrt{\dfrac{40x^3}{2x}}$
28. $\sqrt{\dfrac{9ab^4}{3b^3}}$
29. $\sqrt{\dfrac{45x^2}{16x^2y^4}}$
30. $\sqrt{\dfrac{50a^3b^6}{10a^3b^8}}$
31. $\sqrt{\dfrac{16x^5y^3}{100x^7y}}$
32. $\sqrt{\dfrac{14xyz^5}{56x^3y^3z^4}}$
33. $\sqrt{\dfrac{24ab}{24a^5b^3}}$
34. $\sqrt{\dfrac{32x^3y}{32x^7y^3}}$
35. $\dfrac{\sqrt{32x^5}}{\sqrt{8x}}$
36. $\dfrac{\sqrt{24x^2y^2}}{\sqrt{6x^2y^4}}$
37. $\dfrac{\sqrt{81x^5y}}{\sqrt{100xy^3}}$
38. $\dfrac{\sqrt{72}}{\sqrt{36x^2y^6}}$
39. $\dfrac{\sqrt{45ab^6}}{\sqrt{9ab^4c^2}}$
40. $\dfrac{\sqrt{24x^2y^6}}{\sqrt{8x^4z^4}}$
41. $\dfrac{\sqrt{125a^6b^8}}{\sqrt{5a^2b^2}}$
42. $\dfrac{\sqrt{144x^{60}y^{32}}}{\sqrt{12x^{40}y^{18}}}$

Simplify each expression.

43. $\dfrac{1}{\sqrt{5}}$
44. $\dfrac{1}{\sqrt{11}}$
45. $\dfrac{4}{\sqrt{8}}$
46. $\dfrac{3}{\sqrt{3}}$
47. $\dfrac{6}{\sqrt{12}}$
48. $\dfrac{9}{\sqrt{50}}$
49. $\sqrt{\dfrac{2}{3}}$
50. $\sqrt{\dfrac{7}{12}}$
51. $\sqrt{\dfrac{7}{21}}$
52. $\sqrt{\dfrac{5}{7}}$
53. $\sqrt{\dfrac{x^2}{2}}$
54. $\sqrt{\dfrac{x^2}{7}}$
55. $\sqrt{\dfrac{a^2}{8}}$
56. $\sqrt{\dfrac{a^3}{18}}$
57. $\sqrt{\dfrac{x^4}{5}}$
58. $\sqrt{\dfrac{x^3}{11}}$
59. $\sqrt{\dfrac{a^8}{14b}}$
60. $\sqrt{\dfrac{a^7b}{24b^2}}$
61. $\sqrt{\dfrac{8x^4y^2}{32x^2y^3}}$
62. $\sqrt{\dfrac{27xz^4}{6y^4}}$
63. $\sqrt{\dfrac{18yz}{75x^4y^5z^3}}$
64. $\dfrac{\sqrt{25x^5}}{\sqrt{100xy^5}}$
65. $\dfrac{\sqrt{90x^4y}}{\sqrt{2x^5y^5}}$
66. $\dfrac{\sqrt{120xyz^2}}{\sqrt{9xy^2}}$

Problem Solving

67. Will the quotient of two rational numbers (denominator not equal to 0) always be a rational number? Explain and give an example to support your answer.

68. Will the quotient of two irrational numbers (denominator not equal to 0) always be an irrational number? Explain and give an example to support your answer.

If $\sqrt{5} \approx 2.236$ and $\sqrt{10} \approx 3.162$, find to the nearest hundredth:

69. $\sqrt{5} + \sqrt{10}$
70. $\sqrt{5} \cdot \sqrt{10}$
71. $\sqrt{5}/\sqrt{10}$
72. $\sqrt{10}/\sqrt{5}$

If $\sqrt{7} \approx 2.646$ and $\sqrt{21} \approx 4.583$, find to the nearest hundredth:

73. $\sqrt{7} + \sqrt{21}$
74. $\sqrt{7} \cdot \sqrt{21}$
75. $\sqrt{7}/\sqrt{21}$
76. $\sqrt{21}/\sqrt{7}$

77. **a)** Is $\sqrt{10}$ twice as large as $\sqrt{5}$?
b) What number is twice as large as $\sqrt{5}$? Explain how you determined your answer.

78. **a)** Is $\sqrt{21}$ three times as large as $\sqrt{7}$?
b) What number is three times as large as $\sqrt{7}$? Explain how you determined your answer.

In Section 7.6, we solved quadratic equations of the form $ax^2 + bx + c = 0$, by factoring. Not all quadratic equations can be solved by factoring. In Chapter 10, we will be introducing another way to solve quadratic equations, called the **quadratic formula**. *The quadratic formula contains the expression $\sqrt{b^2 - 4ac}$. Evaluate this expression for the following quadratic equations. Note, for example, in the quadratic equation $2x^2 - 4x - 5 = 0$ that $a = 2$, $b = -4$ and $c = -5$.*

79. $x^2 + 3x + 2 = 0$
80. $x^2 + 9x + 20 = 0$
81. $x^2 - 14x - 5 = 0$
82. $x^2 + 4x - 1 = 0$
83. $-2x^2 + 4x + 7 = 0$
84. $-6x^2 - 2x + 1 = 0$

Challenge Problems

Fill in the shaded area to make the expression true. Explain how you determined your answer.

85. $\sqrt{\dfrac{\square}{4x^2}} = 4x^4$
86. $\dfrac{\sqrt{32x^5}}{\sqrt{\square}} = 2x^2$
87. $\dfrac{1}{\sqrt{\square}} = \dfrac{\sqrt{2}}{2}$
88. $\dfrac{3x}{\sqrt{\square}} = \dfrac{3\sqrt{2x}}{2}$

Cumulative Review Exercises

[5.1] **89.** Solve the following system of equations graphically.

$$y = 2x - 2$$
$$2x + 3y = 10$$

[7.4] **90.** Factor $3x^2 - 12x - 96$.

[8.1] **91.** Simplify $\dfrac{x - 1}{x^2 - 1}$.

[8.6] **92.** Solve the equation $x + \dfrac{24}{x} = 10$.

9.4 ADDING AND SUBTRACTING SQUARE ROOTS

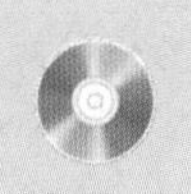

SSM VIDEO 9.4 CD Rom

1) Add and subtract square roots.
2) Rationalize a denominator that contains a binomial.
3) Simplify a fraction containing two numerical terms in the numerator where one is a square root.

1) Add and Subtract Square Roots

Like square roots are square roots having the same radicands. Like square roots are added in much the same manner that like terms are added. **To add like square roots**, add their coefficients and then multiply that sum by the like square root.

Examples of Adding Like Terms

$$2x + 3x = (2 + 3)x = 5x$$
$$4x + x = 4x + 1x = (4 + 1)x = 5x$$

Examples of Adding Like Square Roots

$$2\sqrt{7} + 3\sqrt{7} = (2 + 3)\sqrt{7} = 5\sqrt{7}$$
$$4\sqrt{x} + \sqrt{x} = 4\sqrt{x} + 1\sqrt{x} = (4 + 1)\sqrt{x} = 5\sqrt{x}$$

Note that adding like square roots is an application of the distributive property.

$$2\sqrt{7} + 3\sqrt{7} = (2 + 3)\sqrt{7}$$
$$= 5\sqrt{7}$$

Other Examples of Adding and Subtracting Like Square Roots

$$2\sqrt{5} - 3\sqrt{5} = (2 - 3)\sqrt{5} = -1\sqrt{5} = -\sqrt{5}$$

$$\sqrt{x} + \sqrt{x} = 1\sqrt{x} + 1\sqrt{x} = (1 + 1)\sqrt{x} = 2\sqrt{x}$$

$$6\sqrt{2} + 3\sqrt{2} - \sqrt{2} = (6 + 3 - 1)\sqrt{2} = 8\sqrt{2}$$

$$\frac{2\sqrt{3}}{5} + \frac{1\sqrt{3}}{5} = \left(\frac{2}{5} + \frac{1}{5}\right)\sqrt{3} = \frac{3}{5}\sqrt{3} \text{ or } \frac{3\sqrt{3}}{5}$$

EXAMPLE 1 Simplify if possible.

a) $4\sqrt{3} + 2\sqrt{3} - 2$ b) $\sqrt{5} - 4\sqrt{5} + 5$

c) $5 + 3\sqrt{2} - \sqrt{2} + 3$ d) $2\sqrt{3} + 5\sqrt{2}$

Solution

a) $4\sqrt{3} + 2\sqrt{3} - 2 = (4 + 2)\sqrt{3} - 2 = 6\sqrt{3} - 2$

b) $\sqrt{5} - 4\sqrt{5} + 5 = (1 - 4)\sqrt{5} + 5 = -3\sqrt{5} + 5$

c) $5 + 3\sqrt{2} - \sqrt{2} + 3 = (3 - 1)\sqrt{2} + 5 + 3 = 2\sqrt{2} + 8$

d) Cannot be simplified since the radicands are different.

EXAMPLE 2 Simplify.

a) $2\sqrt{x} - 3\sqrt{x} + 4\sqrt{x}$ b) $3\sqrt{x} + x + 4\sqrt{x}$

c) $x + \sqrt{x} + 2\sqrt{x} + 3$ d) $x\sqrt{x} + 3\sqrt{x} + x$

e) $\sqrt{xy} + 2\sqrt{xy} - \sqrt{x}$

Solution

a) $2\sqrt{x} - 3\sqrt{x} + 4\sqrt{x} = 3\sqrt{x}$

b) $3\sqrt{x} + x + 4\sqrt{x} = x + 7\sqrt{x}$ *Only $3\sqrt{x}$ and $4\sqrt{x}$ can be combined.*

c) $x + \sqrt{x} + 2\sqrt{x} + 3 = x + 3\sqrt{x} + 3$ *Only $\sqrt{x}$ and $2\sqrt{x}$ can be combined.*

d) $x\sqrt{x} + 3\sqrt{x} + x = (x + 3)\sqrt{x} + x$ *Only $x\sqrt{x}$ and $3\sqrt{x}$ can be combined.*

e) $\sqrt{xy} + 2\sqrt{xy} - \sqrt{x} = 3\sqrt{xy} - \sqrt{x}$ *Only $\sqrt{xy}$ and $2\sqrt{xy}$ can be combined.*

NOW TRY EXERCISE 23

Unlike square roots are square roots having different radicands. It is sometimes possible to change unlike square roots into like square roots by simplifying the radicals in an expression. After simplifying, if the terms contain the same radicand they can be combined. Otherwise, they cannot. Examples 3, 4, and 5 illustrate this.

EXAMPLE 3 Simplify $\sqrt{3} + \sqrt{12}$.

Solution Since 12 has a perfect square factor, 4, we write 12 as a product of the perfect square factor and another factor.

$$\begin{aligned}\sqrt{3} + \sqrt{12} &= \sqrt{3} + \sqrt{4 \cdot 3} \\ &= \sqrt{3} + \sqrt{4} \cdot \sqrt{3} && \text{Product rule} \\ &= \sqrt{3} + 2\sqrt{3} \\ &= 3\sqrt{3} && \text{Add like square roots.}\end{aligned}$$

EXAMPLE 4 Simplify $\sqrt{63} - \sqrt{28}$.

Solution Write each radicand as a product of a perfect square factor and another factor.

$$\begin{aligned}\sqrt{63} - \sqrt{28} &= \sqrt{9 \cdot 7} - \sqrt{4 \cdot 7} \\ &= \sqrt{9} \cdot \sqrt{7} - \sqrt{4} \cdot \sqrt{7} && \text{Product rule} \\ &= 3\sqrt{7} - 2\sqrt{7} \\ &= \sqrt{7} && \text{Subtract like square roots.}\end{aligned}$$

EXAMPLE 5 Simplify.

a) $2\sqrt{8} - \sqrt{32}$ **b)** $3\sqrt{12} + 5\sqrt{27} + 2$ **c)** $\sqrt{120} - \sqrt{75}$

Solution

a)
$$\begin{aligned}2\sqrt{8} - \sqrt{32} &= 2\sqrt{4 \cdot 2} - \sqrt{16 \cdot 2} \\ &= 2\sqrt{4}\sqrt{2} - \sqrt{16}\sqrt{2} \\ &= 2 \cdot 2\sqrt{2} - 4\sqrt{2} \\ &= 4\sqrt{2} - 4\sqrt{2} \\ &= 0\end{aligned}$$

b)
$$\begin{aligned}3\sqrt{12} + 5\sqrt{27} + 2 &= 3\sqrt{4 \cdot 3} + 5\sqrt{9 \cdot 3} + 2 \\ &= 3\sqrt{4}\sqrt{3} + 5\sqrt{9}\sqrt{3} + 2 \\ &= 3 \cdot 2\sqrt{3} + 5 \cdot 3\sqrt{3} + 2 \\ &= 6\sqrt{3} + 15\sqrt{3} + 2 \\ &= 21\sqrt{3} + 2\end{aligned}$$

c)
$$\begin{aligned}\sqrt{120} - \sqrt{75} &= \sqrt{4 \cdot 30} - \sqrt{25 \cdot 3} \\ &= \sqrt{4}\sqrt{30} - \sqrt{25}\sqrt{3} \\ &= 2\sqrt{30} - 5\sqrt{3}\end{aligned}$$

Since 30 has no perfect square factors and since the radicands are different, the expression $2\sqrt{30} - 5\sqrt{3}$ cannot be simplified any further.

NOW TRY EXERCISE 31

Avoiding Common Errors

The product rule presented in Section 9.2 was $\sqrt{a} \cdot \sqrt{b} = \sqrt{a \cdot b}$. The same principle **does not apply to** addition.

INCORRECT

~~$\sqrt{a} + \sqrt{b} = \sqrt{a + b}$~~

For example, to evaluate $\sqrt{9} + \sqrt{16}$,

CORRECT	INCORRECT
$\sqrt{9} + \sqrt{16} = 3 + 4$	~~$\sqrt{9} + \sqrt{16} = \sqrt{9 + 16}$~~
$= 7$	~~$= \sqrt{25}$~~
	~~$= 5$~~

2) Rationalize a Denominator That Contains a Binomial

When the denominator of a rational expression is a binomial with a square root term, we again **rationalize the denominator**. We do this by multiplying both the numerator and the denominator of the fraction by the conjugate of the denominator. The **conjugate of a binomial** is a binomial having the same two terms with the sign of the second term changed.

Binomial	Its Conjugate
$3 + \sqrt{2}$	$3 - \sqrt{2}$
$\sqrt{5} - 3$	$\sqrt{5} + 3$
$2\sqrt{3} - \sqrt{5}$	$2\sqrt{3} + \sqrt{5}$
$-x + \sqrt{3}$	$-x - \sqrt{3}$

When a binomial is multiplied by its conjugate using the FOIL method, the sum of the outer and inner terms will be zero.

EXAMPLE 6 Multiply $(3 + \sqrt{5})(3 - \sqrt{5})$ using the FOIL method.

Solution

$$(3 + \sqrt{5})(3 - \sqrt{5}) = \overset{F}{3(3)} + \overset{O}{3(-\sqrt{5})} + \overset{I}{3\sqrt{5}} + \overset{L}{\sqrt{5}(-\sqrt{5})}$$
$$= 9 - 3\sqrt{5} + 3\sqrt{5} - \sqrt{25}$$
$$= 9 - \sqrt{25}$$
$$= 9 - 5 = 4$$

EXAMPLE 7 Multiply $(\sqrt{2} - \sqrt{7})(\sqrt{2} + \sqrt{7})$ using the FOIL method.

Solution

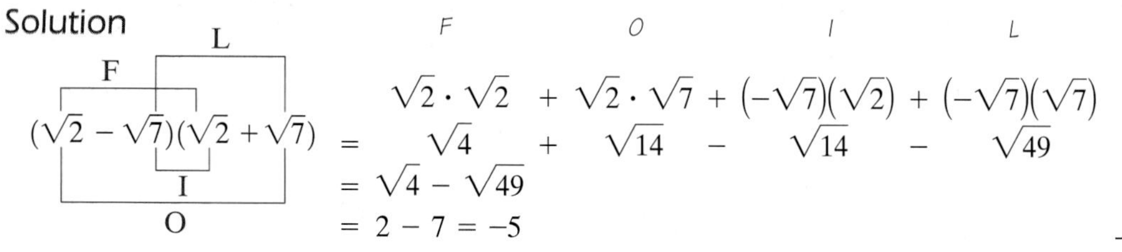

NOW TRY EXERCISE 43

Now let's try some examples where we rationalize the denominator when the denominator is a binomial with one or more radical terms.

EXAMPLE 8 Simplify $\dfrac{5}{2 + \sqrt{3}}$.

Solution To rationalize the denominator, multiply both the numerator and the denominator by $2 - \sqrt{3}$, which is the conjugate of $2 + \sqrt{3}$.

$$\frac{5}{2 + \sqrt{3}} \cdot \frac{2 - \sqrt{3}}{2 - \sqrt{3}} = \frac{5(2 - \sqrt{3})}{(2 + \sqrt{3})(2 - \sqrt{3})} \qquad \textit{Multiply numerator and denominator by } 2 - \sqrt{3}$$

$$= \frac{5(2 - \sqrt{3})}{4 - 3} \qquad \textit{Multiply factors in denominator.}$$

$$= \frac{5(2 - \sqrt{3})}{1} \qquad \textit{Simplify.}$$

$$= 5(2 - \sqrt{3}) = 10 - 5\sqrt{3}$$

Note that $-5\sqrt{3} + 10$ is also an acceptable answer.

EXAMPLE 9 Simplify $\dfrac{4}{\sqrt{5} - \sqrt{3}}$.

Solution Multiply both the numerator and the denominator of the fraction by $\sqrt{5} + \sqrt{3}$, the conjugate of $\sqrt{5} - \sqrt{3}$.

$$\frac{4}{\sqrt{5} - \sqrt{3}} \cdot \frac{\sqrt{5} + \sqrt{3}}{\sqrt{5} + \sqrt{3}} = \frac{4(\sqrt{5} + \sqrt{3})}{(\sqrt{5} - \sqrt{3})(\sqrt{5} + \sqrt{3})}$$ *Multiply numerator and denominator by* $\sqrt{5} + \sqrt{3}$.

$$= \frac{4(\sqrt{5} + \sqrt{3})}{5 - 3}$$ *Multiply factors in denominator.*

$$= \frac{\overset{2}{\cancel{4}}(\sqrt{5} + \sqrt{3})}{\underset{1}{\cancel{2}}}$$ *Simplify.*

$$= 2(\sqrt{5} + \sqrt{3}) = 2\sqrt{5} + 2\sqrt{3}$$

NOW TRY EXERCISE 57

EXAMPLE 10 Simplify $\dfrac{\sqrt{3}}{2 - \sqrt{6}}$.

Solution Multiply both the numerator and the denominator of the fraction by $2 + \sqrt{6}$, the conjugate of $2 - \sqrt{6}$.

$$\frac{\sqrt{3}}{2 - \sqrt{6}} \cdot \frac{2 + \sqrt{6}}{2 + \sqrt{6}} = \frac{\sqrt{3}(2 + \sqrt{6})}{4 - 6}$$

$$= \frac{2\sqrt{3} + \sqrt{3} \cdot \sqrt{6}}{-2}$$ *Distributive Property*

$$= \frac{2\sqrt{3} + \sqrt{18}}{-2}$$ *Product rule*

$$= \frac{2\sqrt{3} + \sqrt{9} \cdot \sqrt{2}}{-2}$$ *Product rule*

$$= \frac{2\sqrt{3} + 3\sqrt{2}}{-2}$$ *Simplify.*

$$= \frac{-2\sqrt{3} - 3\sqrt{2}}{2}$$ *Multiply numerator and denominator by −1.*

EXAMPLE 11 Simplify $\dfrac{x}{x - \sqrt{y}}$.

Solution Multiply both the numerator and the denominator of the fraction by the conjugate of the denominator, $x + \sqrt{y}$.

$$\frac{x}{x - \sqrt{y}} \cdot \frac{x + \sqrt{y}}{x + \sqrt{y}} = \frac{x(x + \sqrt{y})}{x^2 - y} = \frac{x^2 + x\sqrt{y}}{x^2 - y}$$

NOW TRY EXERCISE 63

Remember, you cannot divide out the x^2 terms because they are not factors.

3) Simplify a Fraction Containing Two Numerical Terms in the Numerator Where One Is a Square Root

In Chapter 10, when using the quadratic formula, you will need to simplify fractional expressions that contain a square root in the numerator. For example, you might have to simplify the following expressions.

$$\frac{3 - 12\sqrt{2}}{3}, \quad \frac{5 + 5\sqrt{7}}{10}$$

To Simplify a Fraction Whose Numerator Contains a Constant Term and a Square Root Term

1. Simplify the radical expression as far as possible.
2. If possible, factor out the GCF from each term in the numerator.
3. If possible, divide out a common factor from the GCF that was factored out of the numerator in step 2 and the denominator.

Examples 12 and 13 will illustrate this process.

EXAMPLE 12 Simplify. **a)** $\frac{3 - 12\sqrt{2}}{3}$ **b)** $\frac{-5 + 5\sqrt{7}}{10}$

Solution **a)** Factor 3 from both terms in the numerator. Then divide out the common factor, 3, from the numerator and denominator.

$$\frac{3 - 12\sqrt{2}}{3} = \frac{\cancel{3}(1 - 4\sqrt{2})}{\cancel{3}} = 1 - 4\sqrt{2}$$

b) Factor 5 from both terms in the numerator. Then divide out the common factor, 5, from the numerator and denominator.

$$\frac{-5 + 5\sqrt{7}}{10} = \frac{\cancel{5}(-1 + \sqrt{7})}{\underset{2}{\cancel{10}}} = \frac{-1 + \sqrt{7}}{2}$$

Note that the factoring of the numerators in parts **a)** and **b)** of Example 12 may be checked by multiplying using the distributive property.

a) $3(1 - 4\sqrt{2}) = 3 - 12\sqrt{2}$ **b)** $5(-1 + \sqrt{7}) = -5 + 5\sqrt{7}$

You should always check your factoring by mentally multiplying the factors.

Avoiding Common Errors

Remember, only common *factors* can be divided out. Do not make the mistakes on the right.

CANNOT BE SIMPLIFIED	INCORRECT
$\frac{3 + 2\sqrt{5}}{2}$	$\frac{3 + 2\sqrt{5}}{2} = \frac{3 + \overset{1}{\cancel{2}}\sqrt{5}}{\underset{1}{\cancel{2}}} = 3 + \sqrt{5}$
$\frac{4 + 3\sqrt{5}}{2}$	$\frac{\overset{2}{\cancel{4}} + 3\sqrt{5}}{\underset{1}{\cancel{2}}} = 2 + 3\sqrt{5}$
$\frac{3 + \sqrt{6}}{2}$	$\frac{3 + \sqrt{\cancel{6}^{3}}}{\underset{1}{\cancel{2}}} = 3 + \sqrt{3}$

EXAMPLE 13 Simplify if possible.

a) $\dfrac{12 - \sqrt{27}}{3}$ b) $\dfrac{10 + \sqrt{20}}{5}$ c) $\dfrac{5 + 3\sqrt{10}}{5}$

Solution **a)** First simplify $\sqrt{27}$.

$$\frac{12 - \sqrt{27}}{3} = \frac{12 - \sqrt{9}\sqrt{3}}{3} = \frac{12 - 3\sqrt{3}}{3} = \frac{\cancel{3}(4 - \sqrt{3})}{\cancel{3}} = 4 - \sqrt{3}$$

b) First simplify $\sqrt{20}$.

$$\frac{10 + \sqrt{20}}{5} = \frac{10 + \sqrt{4}\sqrt{5}}{5} = \frac{10 + 2\sqrt{5}}{5} = \frac{2(5 + \sqrt{5})}{5}$$

Since the 2 in the numerator and the 5 in the denominator have no common factors, the expression $\dfrac{10 + \sqrt{20}}{5}$ can be simplified to $\dfrac{10 + 2\sqrt{5}}{5}$ or $\dfrac{2(5 + \sqrt{5})}{5}$, but it cannot be simplified further.

c) First notice that $\sqrt{10}$ cannot be simplified. Since the two terms in the numerator have no common factor, this fraction cannot be simplified.

NOW TRY EXERCISE 77

Exercise Set 9.4

Concept/Writing Exercises

1. What are like square roots? Give an example.
2. What are unlike square roots? Give an example.
3. Under what conditions can two square roots be added or subtracted?
4. In your own words, explain how to add like square roots.
5. Find the conjugate of the following binomials
   a) $4 - \sqrt{7}$
   b) $5 + \sqrt{11}$
   c) $-3 - 2\sqrt{13}$
   d) $\sqrt{5} - 1$
6. Under what conditions do you rationalize the denominator of a rational expression using the conjugate of a binomial?
7. **a)** Explain how to rationalize the denominator in a fraction of the form $\dfrac{a}{b + \sqrt{c}}$.
   b) Using the procedure you gave in part **a)**, rationalize $\dfrac{a}{b + \sqrt{c}}$.
8. **a)** Explain how to rationalize the denominator in a fraction of the form $\dfrac{a}{b - \sqrt{c}}$.
   b) Using the procedure you gave in part **a)**, rationalize $\dfrac{a}{b - \sqrt{c}}$.

Practice the Skills

Simplify each expression.

9. $6\sqrt{2} - 4\sqrt{2}$
10. $8\sqrt{3} + \sqrt{3}$
11. $7\sqrt{5} - 11\sqrt{5}$
12. $4\sqrt{10} + 6\sqrt{10} - \sqrt{10} + 2$
13. $\sqrt{7} + 4\sqrt{7} - 3\sqrt{7} + 6$
14. $10\sqrt{13} - 6\sqrt{13} + 5\sqrt{13}$
15. $7\sqrt{x} + \sqrt{x}$
16. $4\sqrt{x} - 8\sqrt{x}$
17. $-\sqrt{y} + 3\sqrt{y} - 5\sqrt{y}$
18. $3\sqrt{y} - 6\sqrt{y}$
19. $3\sqrt{y} - \sqrt{y} + 3$
20. $3\sqrt{5} - \sqrt{x} + 4\sqrt{5} + 3\sqrt{x}$
21. $\sqrt{x} + \sqrt{y} + x + 3\sqrt{y}$
22. $3 + 4\sqrt{x} - 6\sqrt{x}$
23. $2 + 3\sqrt{y} - 6\sqrt{y} + 5$
24. $3\sqrt{x} - 4\sqrt{x} + 5\sqrt{x} + 6x$
25. $-3\sqrt{7} + \sqrt{7} - 2\sqrt{x} - 7\sqrt{x}$
26. $4\sqrt{5} - 2\sqrt{5} + 3\sqrt{y} - 5\sqrt{y}$

Simplify each expression.

27. $\sqrt{12} + \sqrt{18}$
28. $\sqrt{8} - \sqrt{12}$
29. $\sqrt{300} - \sqrt{27}$
30. $\sqrt{125} + \sqrt{20}$
31. $\sqrt{75} + \sqrt{108}$
32. $\sqrt{60} - \sqrt{135}$
33. $4\sqrt{50} - \sqrt{72} + \sqrt{8}$
34. $-4\sqrt{99} + 3\sqrt{44} + 2\sqrt{11}$
35. $-3\sqrt{125} + 7\sqrt{75}$
36. $4\sqrt{80} - \sqrt{75}$
37. $2\sqrt{360} + 4\sqrt{80}$
38. $6\sqrt{180} - 7\sqrt{108}$
39. $4\sqrt{16} - \sqrt{48}$
40. $3\sqrt{250} + 5\sqrt{160}$

Multiply.

41. $(1 + \sqrt{5})(1 - \sqrt{5})$
42. $(\sqrt{3} + 6)(\sqrt{3} - 6)$
43. $(4 - \sqrt{2})(4 + \sqrt{2})$
44. $(\sqrt{11} - 4)(\sqrt{11} + 4)$
45. $(\sqrt{x} + 3)(\sqrt{x} - 3)$
46. $(\sqrt{x} + 5)(\sqrt{x} - 5)$
47. $(\sqrt{6} + x)(\sqrt{6} - x)$
48. $(\sqrt{3} - y)(\sqrt{3} + y)$
49. $(\sqrt{5x} + \sqrt{y})(\sqrt{5x} - \sqrt{y})$
50. $(\sqrt{x} + \sqrt{y})(\sqrt{x} - \sqrt{y})$
51. $(2\sqrt{x} + 3\sqrt{y})(2\sqrt{x} - 3\sqrt{y})$
52. $(4\sqrt{2x} + \sqrt{3y})(4\sqrt{2x} - \sqrt{3y})$

Simplify each expression.

53. $\dfrac{3}{\sqrt{5} + 2}$
54. $\dfrac{2}{4 - \sqrt{3}}$
55. $\dfrac{2}{\sqrt{6} - 1}$
56. $\dfrac{4}{\sqrt{2} - 7}$
57. $\dfrac{2}{\sqrt{2} + \sqrt{3}}$
58. $\dfrac{5}{\sqrt{6} + \sqrt{3}}$
59. $\dfrac{8}{\sqrt{5} - \sqrt{8}}$
60. $\dfrac{1}{\sqrt{17} - \sqrt{8}}$
61. $\dfrac{5}{\sqrt{y} + 3}$
62. $\dfrac{2}{4 + \sqrt{x}}$
63. $\dfrac{6}{4 - \sqrt{y}}$
64. $\dfrac{5}{3 + \sqrt{x}}$
65. $\dfrac{16}{\sqrt{y} + x}$
66. $\dfrac{4}{x + \sqrt{y}}$
67. $\dfrac{x}{\sqrt{x} + \sqrt{y}}$
68. $\dfrac{\sqrt{3}}{\sqrt{x} - \sqrt{3}}$
69. $\dfrac{\sqrt{x}}{\sqrt{2} - \sqrt{x}}$
70. $\dfrac{x}{\sqrt{x} - y}$

Simplify each expression if possible.

71. $\dfrac{9 - 3\sqrt{2}}{3}$
72. $\dfrac{8 - 12\sqrt{3}}{4}$
73. $\dfrac{12 + 24\sqrt{7}}{3}$
74. $\dfrac{10 + 2\sqrt{5}}{5}$
75. $\dfrac{4 + 3\sqrt{6}}{2}$
76. $\dfrac{4 - 5\sqrt{10}}{4}$
77. $\dfrac{6 + 2\sqrt{75}}{3}$
78. $\dfrac{5 + 2\sqrt{50}}{5}$
79. $\dfrac{-2 + 4\sqrt{80}}{10}$
80. $\dfrac{12 - 6\sqrt{72}}{3}$
81. $\dfrac{16 - 4\sqrt{32}}{8}$
82. $\dfrac{-50 - \sqrt{200}}{60}$

Find the product of the given binomial and its conjugate.

83. $\sqrt{x} - y$
84. $\sqrt{x} + y$
85. $\sqrt{x} + \sqrt{y}$
86. $\sqrt{x} - \sqrt{y}$

Problem Solving

87. If $x > -2$, what is $\sqrt{x + 2} \cdot \sqrt{x + 2}$ equal to?
88. If $x > -5$, what is $\sqrt{x + 5} \cdot \sqrt{x + 5}$ equal to?
89. Is the sum or difference of two rational expressions always a rational expression? Explain and give an example.
90. Is the sum or difference of two irrational expressions always an irrational expression? Explain and give an example.

Fill in the shaded area to make each statement true. Explain how you determined your answer.

91. $\sqrt{\square} - \sqrt{63} = 4\sqrt{7}$

92. $\sqrt{180} + \sqrt{\square} = 9\sqrt{5}$

Find the perimeter and area of the following figures.

93.

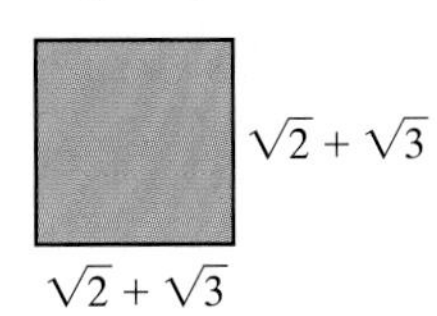

94.

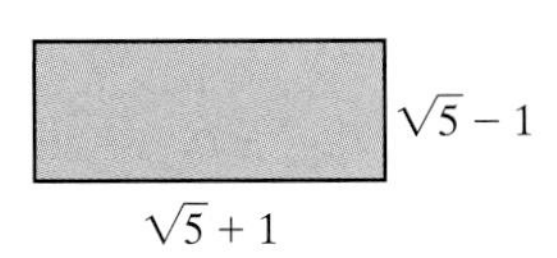

95.

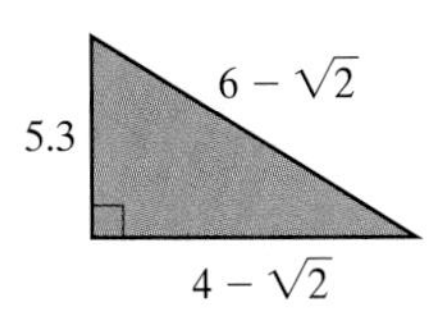

96.

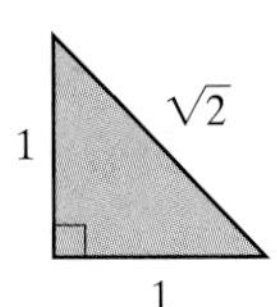

Challenge Problems

Fill in the shaded area to make each statement true. Explain how you determined your answer.

97. $-5\sqrt{\square} + 2\sqrt{3} + 3\sqrt{27} = -9\sqrt{3}$

98. $20\sqrt{2} - 4\sqrt{\square} + 4\sqrt{18} = 12\sqrt{2}$

In Exercises 99–102, assume $x > 0$.

99. What is the conjugate of each expression?
a) $5 + \sqrt{x+2}$ **b)** $6 - \sqrt{x+3}$

100. What is the conjugate of each expression?
a) $2 + \sqrt{x+7}$ **b)** $\sqrt{x+4} + 8$

101. Simplify $\dfrac{\sqrt{x}}{1 - \sqrt{3}}$ by rationalizing the denominator.

102. Simplify $\dfrac{\sqrt{x}}{2 + \sqrt{2}}$ by rationalizing the denominator.

Group Activity

103. Consider $\dfrac{\sqrt{x}}{x + \sqrt{x}}$, where $x > 0$.

a) Individually, simplify $\dfrac{\sqrt{x}}{x + \sqrt{x}}$ by rationalizing the denominator. Compare and correct your answer if necessary.

b) Group Member 1: Substitute $x = 4$ in the original expression and in the results found in part **a)**. Evaluate each expression.

c) Group Member 2: Repeat part **b)** for $x = 6$.

d) Group Member 3: Repeat part **b)** for $x = 9$.

e) As a group, determine what relationship exists between the given expression and the rationalized expression you obtained in part **b)**.

Cumulative Review Exercises

[6.6] **104.** Divide $\dfrac{3x^2 + 4x - 25}{x + 4}$

[7.6] **105.** Solve the equation $2x^2 - x - 36 = 0$.

[8.4] **106.** Subtract $\dfrac{1}{x^2 - 4} - \dfrac{2}{x - 2}$.

[8.7] **107.** Mark DeGroat can stack a cord of wood in 20 minutes. With his wife's help, they can stack the wood in 12 minutes. How long would it take his wife, Terry, to stack the wood by herself?

See Exercise 107

9.5 SOLVING RADICAL EQUATIONS

SSM VIDEO 9.5 CD Rom

1) **Solve radical equations containing only one square root term.**

2) **Solve radical equations containing two square root terms.**

1) Solve Radical Equations Containing Only One Square Root Term

A **radical equation** is an equation that contains a variable in a radicand. Some examples of radical equations are

$$\sqrt{x} = 3 \qquad \sqrt{x+4} = 6 \qquad \sqrt{x-2} = x - 6$$

To Solve a Radical Equation Containing Only One Square Root Term

1. Use the appropriate properties to rewrite the equation with the square root term by itself on one side of the equation. We call this *isolating the square root.*
2. Combine like terms.
3. Square both sides of the equation to eliminate the square root.
4. Solve the equation for the variable.
5. Check the solution in the *original* equation for extraneous roots.

The following examples illustrate this procedure.

EXAMPLE 1 Solve the equation $\sqrt{x} = 8$.

Solution The square root containing the variable is already by itself on one side of the equation. Square both sides of the equation.

$$\sqrt{x} = 8$$

$$(\sqrt{x})^2 = 8^2 \qquad \text{Square both sides.}$$

$$x = 64$$

CHECK:

$$\sqrt{x} = 8$$

$$\sqrt{64} \stackrel{?}{=} 8$$

$$8 = 8 \qquad \text{True}$$

EXAMPLE 2 Solve the equation $\sqrt{x-5} = 4$.

Solution The square root containing the variable is already by itself on one side of the equation. Square both sides of the equation.

$$\sqrt{x-5} = 4$$

$$(\sqrt{x-5})^2 = 4^2 \qquad \text{Square both sides.}$$

$$x - 5 = 16 \qquad \text{Simplify.}$$

$$x - 5 + 5 = 16 + 5 \qquad \text{Add 5 to both sides.}$$

$$x = 21$$

CHECK:
$$\sqrt{x - 5} = 4$$
$$\sqrt{21 - 5} \stackrel{?}{=} 4$$
$$\sqrt{16} \stackrel{?}{=} 4$$
$$4 = 4 \quad \text{True}$$

In the next example, the square root term containing the variable is not by itself on one side of the equation.

EXAMPLE 3 Solve the equation $\sqrt{x} + 5 = 8$.

Solution Since the 5 is outside the square root sign, we first subtract 5 from both sides of the equation to isolate the square root term.

$$\sqrt{x} + 5 = 8$$
$$\sqrt{x} + 5 - 5 = 8 - 5$$
$$\sqrt{x} = 3$$

Now we square both sides of the equation.

$$(\sqrt{x})^2 = 3^2$$
$$x = 9$$

NOW TRY EXERCISE 19

A check will show that 9 is the solution.

HELPFUL HINT

When you square both sides of an equation, you may introduce extraneous roots. An **extraneous root** is a number obtained when solving an equation that is not a solution to the original equation. Equations where both sides are squared in the process of finding their solutions should always be checked for extraneous roots by substituting the numbers found back in the **original** equation.

Consider the equation

$$x = 6$$

Now square both sides.

$$x^2 = 36$$

Note that the original equation $x = 6$ is true only when x is 6. However, the equation $x^2 = 36$ is true for both 6 and -6. When we squared $x = 6$, we introduced the extraneous root -6.

EXAMPLE 4 Solve the equation $\sqrt{x} = -6$.

Solution Begin by squaring both sides of the equation

$$\sqrt{x} = -6$$
$$(\sqrt{x})^2 = (-6)^2$$
$$x = 36$$

CHECK:
$$\sqrt{x} = -6$$
$$\sqrt{36} \stackrel{?}{=} -6$$
$$6 = -6 \quad \text{False}$$

Since the check results in a false statement, the number 36 is an extraneous root and is not a solution to the given equation. Thus, the equation $\sqrt{x} = -6$ has no real solution.

NOW TRY EXERCISE 17

In Example 4, you might have realized without working the problem that there is no solution. In the original equation, the left side is nonnegative and the right side is negative; thus they cannot possibly be equal.

EXAMPLE 5 Solve the equation $\sqrt{2x - 3} = x - 3$.

Solution Square both sides of the equation.

$$(\sqrt{2x - 3})^2 = (x - 3)^2$$

$$2x - 3 = (x - 3)(x - 3) \quad \textit{Simplify on left. Write } (x-3)^2 \textit{ as a product.}$$

$$2x - 3 = x^2 - 6x + 9 \quad \textit{Multiply.}$$

Now solve the quadratic equation as explained in Section 7.6. Make one side of the equation equal to zero by subtracting $2x$ and adding 3 to both sides of the equation.

$$0 = x^2 - 8x + 12 \quad \text{or} \quad x^2 - 8x + 12 = 0$$

Solve for x by factoring.

$$x^2 - 8x + 12 = 0$$

$$(x - 6)(x - 2) = 0 \quad \textit{Factor.}$$

$$x - 6 = 0 \quad \text{or} \quad x - 2 = 0 \quad \textit{Zero-factor property.}$$

$$x = 6 \qquad\qquad x = 2 \quad \textit{Solve for } x.$$

CHECK:

$x = 6$	$x = 2$
$\sqrt{2x - 3} = x - 3$	$\sqrt{2x - 3} = x - 3$
$\sqrt{2(6) - 3} \stackrel{?}{=} 6 - 3$	$\sqrt{2(2) - 3} \stackrel{?}{=} 2 - 3$
$\sqrt{9} \stackrel{?}{=} 3$	$\sqrt{1} \stackrel{?}{=} -1$
$3 = 3$ *True*	$1 = -1$ *False*

The solution is 6. Two is not a solution to the equation.

Remember, when solving a radical equation, begin by isolating the radical, as is shown in Example 6.

EXAMPLE 6 Solve the equation $2x - 5\sqrt{x} - 3 = 0$.

Solution First rewrite the equation so that the square root containing the variable is by itself on one side of the equation.

$$2x - 5\sqrt{x} - 3 = 0$$

$$-5\sqrt{x} = -2x + 3$$

$$\text{or} \quad 5\sqrt{x} = 2x - 3 \quad \textit{Both sides multiplied by } -1.$$

Now square both sides of the equation.

$$(5\sqrt{x})^2 = (2x - 3)^2$$

$$5^2(\sqrt{x})^2 = (2x - 3)^2$$

$$25x = (2x - 3)(2x - 3)$$ *Simplify on left. Write $(2x - 3)^2$ as a product.*

$$25x = 4x^2 - 12x + 9$$ *Multiply.*

$$0 = 4x^2 - 37x + 9$$ *Set one side of equation equal to zero.*

or $$4x^2 - 37x + 9 = 0$$

$$(4x - 1)(x - 9) = 0$$ *Factor.*

$$4x - 1 = 0 \quad \text{or} \quad x - 9 = 0$$ *Zero-factor property*

$$4x = 1 \qquad x = 9$$ *Solve for x.*

$$x = \frac{1}{4}$$

CHECK: $x = \frac{1}{4}$

$$2x - 5\sqrt{x} - 3 = 0$$

$$2\left(\frac{1}{4}\right) - 5\sqrt{\frac{1}{4}} - 3 \stackrel{?}{=} 0$$

$$\frac{1}{2} - 5\left(\frac{1}{2}\right) - 3 \stackrel{?}{=} 0$$

$$\frac{1}{2} - \frac{5}{2} - 3 \stackrel{?}{=} 0$$

$$-\frac{4}{2} - 3 \stackrel{?}{=} 0$$

$$-2 - 3 \stackrel{?}{=} 0$$

$$-5 = 0$$ *False*

$x = 9$

$$2x - 5\sqrt{x} - 3 = 0$$

$$2(9) - 5\sqrt{9} - 3 \stackrel{?}{=} 0$$

$$18 - 5(3) - 3 \stackrel{?}{=} 0$$

$$18 - 15 - 3 \stackrel{?}{=} 0$$

$$0 = 0$$ *True*

NOW TRY EXERCISE 37 The solution is 9; $\frac{1}{4}$ is not a solution.

2) Solve Radical Equations Containing Two Square Root Terms

Consider the radical equations

$$\sqrt{x - 2} = \sqrt{x + 7} \quad \text{and} \quad \sqrt{2x + 6} - \sqrt{3x + 5} = 0$$

These equations are different from those previously discussed because they have two square root terms containing the variable x. To solve equations of this type, rewrite the equation, when necessary, so that there is only one term containing a square root on each side of the equation. Then square both sides of the equation. Examples 7 and 8 illustrate this procedure.

EXAMPLE 7 Solve the equation $\sqrt{4x-4}=\sqrt{3x+6}$.

Solution Since each side of the equation already contains one square root, it is not necessary to rewrite the equation. Square both sides of the equation, then solve for x.

$$(\sqrt{4x-4})^2=(\sqrt{3x+6})^2$$
$$4x-4=3x+6$$
$$x-4=6$$
$$x=10$$

CHECK:
$$\sqrt{4x-4}=\sqrt{3x+6}$$
$$\sqrt{4(10)-4}\stackrel{?}{=}\sqrt{3(10)+6}$$
$$\sqrt{36}\stackrel{?}{=}\sqrt{36}$$
$$6=6 \quad \textit{True}$$

NOW TRY EXERCISE 27 The solution is 10.

EXAMPLE 8 Solve the equation $3\sqrt{x-2}-\sqrt{7x+4}=0$.

Solution Add $\sqrt{7x+4}$ to both sides of the equation to get one square root on each side of the equation. Then square both sides of the equation.

$$3\sqrt{x-2}-\sqrt{7x+4}+\sqrt{7x+4}=0+\sqrt{7x+4}$$
$$3\sqrt{x-2}=\sqrt{7x+4}$$
$$(3\sqrt{x-2})^2=(\sqrt{7x+4})^2 \quad \textit{Square both sides.}$$
$$9(x-2)=7x+4 \quad \textit{Simplify.}$$
$$9x-18=7x+4 \quad \textit{Distributive property}$$
$$2x-18=4$$
$$2x=22$$
$$x=11$$

CHECK:
$$3\sqrt{x-2}-\sqrt{7x+4}=0$$
$$3\sqrt{11-2}-\sqrt{7(11)+4}\stackrel{?}{=}0$$
$$3\sqrt{9}-\sqrt{77+4}\stackrel{?}{=}0$$
$$3(3)-\sqrt{81}\stackrel{?}{=}0$$
$$9-9=0 \quad \textit{True}$$

The solution is 11.

HELPFUL HINT

In Example 6 when we simplified $(5\sqrt{x})^2$ we obtained $25x$ and in Example 8 when we simplified $(3\sqrt{x-2})^2$ we obtained $9(x-2)$. Remember, by the power rule for exponents (discussed in Section 6.1) that when a product of factors is raised to a power, each of the factors is raised to that power. Thus, we see that

$$(5\sqrt{x})^2=5^2(\sqrt{x})^2=25x \quad \text{and} \quad (3\sqrt{x-2})^2=3^2(\sqrt{x-2})^2=9(x-2)$$

Exercise Set 9.5

Concept/Writing Exercises

1. What is a radical equation?
2. What is an extraneous root?
3. Why is it necessary to always check solutions to radical equations?
4. **a)** Write in your own words a step-by-step procedure for solving equations containing a single square root term.
   b) Solve the equation $\sqrt{x+1} - 1 = 1$ using the procedure you gave in part **a)**.

Determine whether the given value is the solution to the equation. If it is not a solution, state why.

5. $\sqrt{x} = 7,\ x = 49$
6. $\sqrt{x} = -7,\ x = 49$
7. $-\sqrt{x} = 8,\ x = 64$
8. $-\sqrt{x} = -8,\ x = 64$
9. $\sqrt{x+4} = 3,\ x = 5$
10. $\sqrt{-x-5} = 3,\ x = 4$

Practice the Skills

Solve each equation. If the equation has no real solution, so state.

11. $\sqrt{x} = 3$
12. $\sqrt{x} = 9$
13. $\sqrt{x} = -5$
14. $\sqrt{x} = -10$
15. $\sqrt{x+5} = 3$
16. $\sqrt{2x-4} = 2$
17. $\sqrt{x} + 5 = -7$
18. $\sqrt{x} - 4 = 8$
19. $\sqrt{x} - 2 = 7$
20. $4 + \sqrt{x} = 9$
21. $11 = 6 + \sqrt{x}$
22. $5 = 7 - \sqrt{x}$
23. $12 + \sqrt{x} = 3$
24. $\sqrt{3x+4} = x - 2$
25. $\sqrt{2x-5} = x - 4$
26. $\sqrt{x^2+8} = x + 2$
27. $\sqrt{3x-7} = \sqrt{x+3}$
28. $5\sqrt{x-4} = 30$
29. $\sqrt{4x+4} = \sqrt{6x-2}$
30. $\sqrt{2x-5} = \sqrt{x+2}$
31. $\sqrt{2x+14} = 3\sqrt{x}$
32. $\sqrt{5x+8} = 2\sqrt{x}$
33. $\sqrt{4x-5} = \sqrt{x+9}$
34. $\sqrt{x^2+3} = x + 1$
35. $3\sqrt{x} = \sqrt{x+8}$
36. $x - 5 = \sqrt{x^2-35}$
37. $4\sqrt{x} = x + 3$
38. $5 + \sqrt{x-5} = x$
39. $\sqrt{2x-3} = 2\sqrt{3x-2}$
40. $12 - 3\sqrt{2x} = 0$
41. $\sqrt{x^2-4} = x + 2$
42. $6\sqrt{3x-2} = 24$
43. $3 + \sqrt{3x-5} = x$
44. $\sqrt{4x+5} + 5 = 2x$
45. $\sqrt{8-7x} = x - 2$
46. $1 + \sqrt{x+1} = x$

Problem Solving

In Section 9.1, we indicated that we could write a square root expression in exponential form. For example, $\sqrt{25}$ can be written $(25)^{1/2}$. Similarly we can write certain exponential expressions as square roots. For example $(25)^{1/2}$ can be written $\sqrt{25}$, and $(x-5)^{1/2}$ can be written $\sqrt{x-5}$. Rewrite the following equations as equations containing square roots. Then solve each equation.

47. $(x+3)^{1/2} = 7$
48. $(x-6)^{1/2} = 5$
49. $(x-2)^{1/2} = (2x-9)^{1/2}$
50. $3(x-9)^{1/2} = 21$

*In Exercises 51–54, **a)** Use the FOIL method to multiply the factors on the left-hand side of the equation. **b)** Solve the equation. Remember to check your answer in the original equation.*

51. $(\sqrt{x}-3)(\sqrt{x}+3) = 40$
52. $(\sqrt{x}+4)(\sqrt{x}-4) = 20$
53. $(7-\sqrt{x})(5+\sqrt{x}) = 35$
54. $(6-\sqrt{x})(2+\sqrt{x}) = 15$
55. The sum of a natural number and its square root is 2. Find the number.
56. The product of 3 and the square root of a number is 18. Find the number.

Challenge Problems

Solve each equation. (Hint: You will need to square both sides of the equation, then isolate the square root containing the variable, then square both sides of the equation again.)

57. $\sqrt{x} + 2 = \sqrt{x+16}$
58. $\sqrt{x+1} = 2 - \sqrt{x}$
59. $\sqrt{x+7} = 5 - \sqrt{x-8}$

Group Activity

Discuss and answer Exercises 60–62 as a group.

60. Radical equations in two variables can be graphed by selecting values for x and finding the corresponding values of y, as was done in Section 4.2 when we graphed linear equations.

a) As a group, discuss what values would be appropriate to select for x if you wanted to graph $y = \sqrt{x}$. Explain your answer.

b) Select four values for x that will result in y being a whole number. List your values for x and y in a table.

c) Individually, plot the points and sketch the graph. Compare your graphs.

d) Is the graph linear? Explain.

e) Is the graph a function? (See Section 4.6.) Explain.

f) Can you determine whether $y = \sqrt{x}$ has an x-intercept and a y-intercept from the graph? If so, give the intercepts.

As a group, repeat parts ***a)–f)*** *of exercise 60 for each the following equations.*

61. $y = \sqrt{x - 2}$

62. $y = \sqrt{x + 4}$

Cumulative Review Exercises

[5.1] **63.** Solve the following system of equations graphically.

$$3x - 2y = 6$$
$$y = 2x - 4$$

[5.2] **64.** Solve the following system of equations by substitution.

$$3x - 2y = 6$$
$$y = 2x - 4$$

[5.3] **65.** Solve the following system of equations using the addition method.

$$3x - 2y = 6$$
$$y = 2x - 4$$

[5.4] **66.** Dan Pellow's boat can travel at a speed of 12 miles per hour with the current and 4 miles per hour against the current in the upper Niagara River by his house in North Tonawonda, New York. Find the speed of the boat in still water and the current.

9.6 RADICALS: APPLICATIONS AND PROBLEM SOLVING

SSM VIDEO 9.6 CD Rom

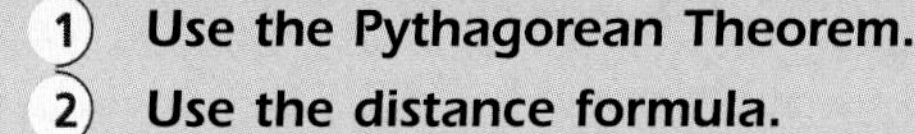

1) Use the Pythagorean Theorem.
2) Use the distance formula.
3) Use radicals to solve application problems.

In this section we will focus on some of the many interesting applications of radicals. We will discuss the Pythagorean Theorem and the distance formula, and then give a few additional applications of radicals.

1) Use the Pythagorean Theorem

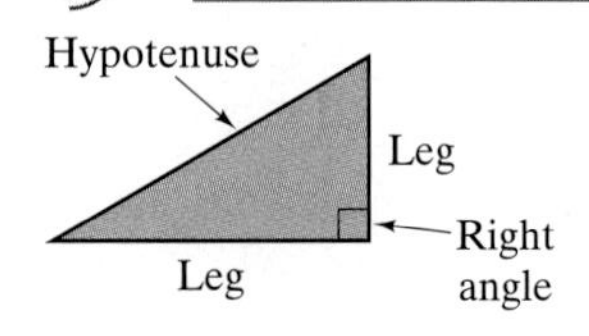

FIGURE 9.3

A **right triangle** is a triangle that contains a right, or 90°, angle (Fig. 9.3). The two shorter sides of a right triangle are called the **legs** and the side opposite the right angle is called the **hypotenuse.** The **Pythagorean Theorem** expresses the relationship between the lengths of the legs of a right triangle and its hypotenuse.

Pythagorean Theorem

The square of the hypotenuse of a right triangle is equal to the sum of the squares of the two legs.

If a and b represent the legs, and c represents the hypotenuse, then

$$a^2 + b^2 = c^2$$

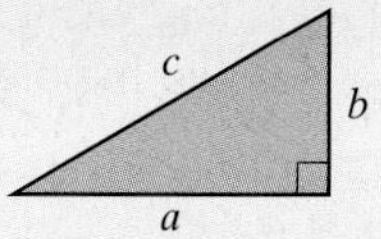

In Section 9.5, when we solved equations containing square roots, we raised both sides of the equation to the second power to eliminate the square roots. When we solve problems using the Pythagorean Theorem, we will raise both sides of the equation to the $\frac{1}{2}$ power to remove the square on one of the variables. We can do this because the rules of exponents presented in Sections 6.1 and 6.2 also apply to fractional exponents. Since lengths are positive, the values of a, b, and c in the Pythagorean Theorem must represent positive values.

EXAMPLE 1 Find the hypotenuse of the right triangle whose legs are 5 feet and 12 feet.

Solution Draw a picture of the problem (Fig. 9.4). When drawing the picture, it makes no difference which leg is called a and which leg is called b.

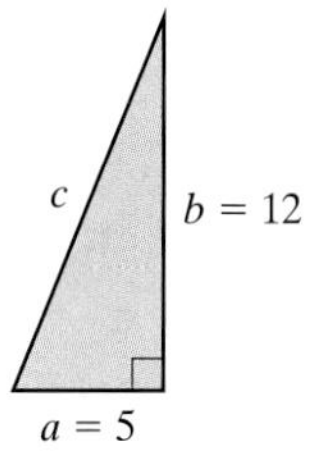

FIGURE 9.4

$$a^2 + b^2 = c^2$$
$$5^2 + (12)^2 = c^2$$
$$25 + 144 = c^2$$
$$169 = c^2$$
$$(169)^{1/2} = (c^2)^{1/2} \quad \text{Raise both sides to the } \tfrac{1}{2} \text{ power.}$$
$$\sqrt{169} = c$$
$$13 = c$$

The hypotenuse is 13 feet.

CHECK:

$$a^2 + b^2 = c^2$$
$$5^2 + (12)^2 \stackrel{?}{=} (13)^2$$
$$25 + 144 \stackrel{?}{=} 169$$
$$169 = 169 \quad \text{True}$$

In Example 1, we could also solve $169 = c^2$ for c by taking the square root of each side of the equation. We will discuss the square root property that allows us to do this in Section 10.1.

EXAMPLE 2 The hypotenuse of a right triangle is 14 inches. Find the second leg if one leg is 9 inches.

Solution First, draw a sketch of the triangle (Fig. 9.5).

$$a^2 + b^2 = c^2$$
$$9^2 + b^2 = (14)^2$$
$$81 + b^2 = 196$$
$$b^2 = 115$$
$$(b^2)^{1/2} = (115)^{1/2}$$
$$b = \sqrt{115} \quad \text{or} \quad \text{approximately 10.72 inches}$$

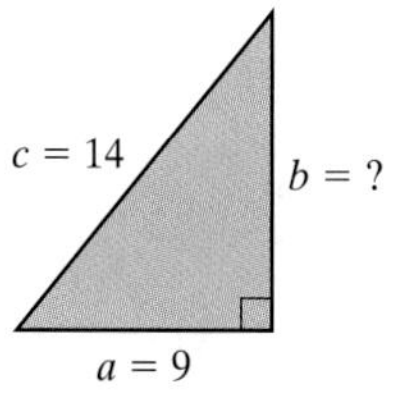

FIGURE 9.5

EXAMPLE 3 Madison Square Garden, which opened in 1926, is located at the corner of 34th St. and 7th Ave. in New York City and is home to the National Basketball Association's New York Knicks. The overall dimension of the basketball floor is 106′ by 57′6″, with a playing surface of 94′ by 50′. Find the length of the diagonal of the playing surface.

Solution **Understand and Translate** First, we draw the playing surface of the basketball court (Fig. 9.6). We are asked to find the length of the diagonal. This length is the hypotenuse, c, of the triangle shown in Figure 9.7

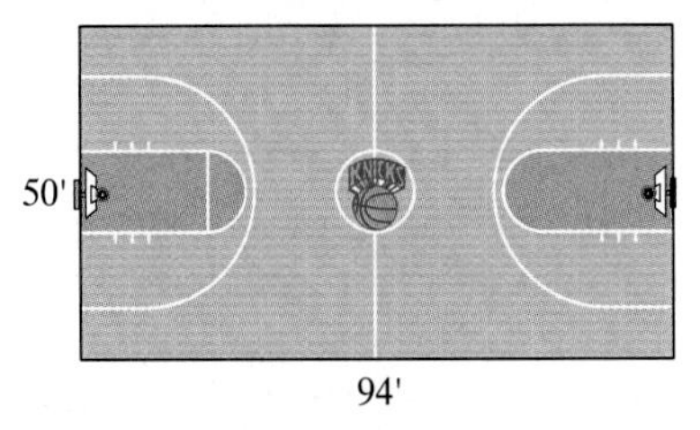

FIGURE 9.6

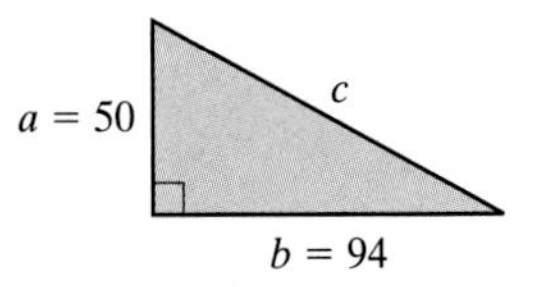

FIGURE 9.7

Carry Out

$$a^2 + b^2 = c^2$$
$$(50)^2 + (94)^2 = c^2$$
$$2500 + 8836 = c^2$$
$$11{,}336 = c^2$$
$$c = \sqrt{11{,}336}, \quad \text{or} \quad \text{approximately } 106.47$$

Answer The length of the diagonal of the playing surface of the basketball court is approximately 106.47 feet.

NOW TRY EXERCISE 25

2) Use the Distance Formula

The **distance formula** can be used to find the distance between two points, (x_1, y_1) and (x_2, y_2), in the Cartesian coordinate system.

Distance Formula

$$d = \sqrt{(x_2 - x_1)^2 + (y_2 - y_1)^2}$$

EXAMPLE 4 Find the length of the line segment between the points $(-1, -4)$ and $(5, -2)$.

Solution The two points are illustrated in Figure 9.8. It makes no difference which point is labeled (x_1, y_1), and which is labeled (x_2, y_2). Let $(5, -2)$ be (x_2, y_2) and $(-1, -4)$ be (x_1, y_1). Thus $x_2 = 5$, $y_2 = -2$ and $x_1 = -1$, $y_1 = -4$.

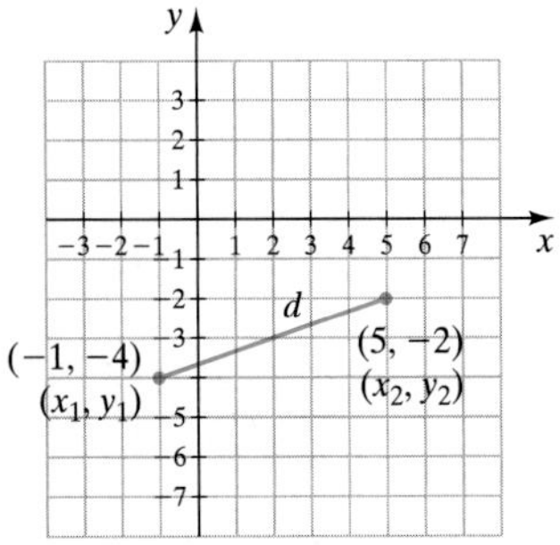

$$d = \sqrt{(x_2 - x_1)^2 + (y_2 - y_1)^2}$$
$$= \sqrt{[5 - (-1)]^2 + [-2 - (-4)]^2}$$
$$= \sqrt{(5 + 1)^2 + (-2 + 4)^2}$$
$$= \sqrt{6^2 + 2^2}$$
$$= \sqrt{36 + 4} = \sqrt{40}, \quad \text{or} \quad \text{approximately } 6.32$$

FIGURE 9.8

NOW TRY EXERCISE 21

Thus, the distance between $(-1, -4)$ and $(5, -2)$ is approximately 6.32 units.

3) Use Radicals to Solve Application Problems

Radicals are often used in science and mathematics courses. Examples 5 through 7 illustrate some scientific applications that use radicals.

EXAMPLE 5

Traffic accident investigators are trained in accident reconstruction. They use, among other things, formulas derived from the laws of physics in their work. One task they face is to try to determine the speed at which a vehicle is traveling before hitting an object. The accident investigators use the formula

$$S = 5.5\sqrt{cl}$$

to estimate the original speed of the vehicle, S, in miles per hour, where c represents the coefficient of friction between the road surface and the tire, and l represents the length of the longest skid mark measured in feet. On dry roads, the value of the coefficient of friction is between 0.69 and 0.75 for most cars. Find the speed of a car that leaves a 40-foot-long skid mark before stopping. Use $c = 0.72$.

Solution **Understand and Translate** Begin by substituting 0.72 for c and 40 for the length in the given formula.

$$S = 5.5\sqrt{cl}$$

$$= 5.5\sqrt{0.72 \times 40}$$

Carry Out

$$= 5.5\sqrt{28.8}$$

$$\approx 29.5$$

Answer A car leaving a 40-foot-long skid mark on a dry surface before stopping is originally traveling at approximately 29.5 miles per hour.

EXAMPLE 6

During the sixteenth and seventeenth centuries, Galileo Galilei, using the (leaning) Tower of Pisa, performed many experiments with objects falling freely under the influence of gravity. He showed, for example, that a rock dropped from, say, 10 feet, hit the ground with a higher velocity than did the same rock dropped from 5 feet. A formula for the velocity of an object in feet per second (neglecting air resistance) after it has fallen a certain distance is

$$v = \sqrt{2gh}$$

where g is the acceleration due to gravity and h is the height the object has fallen in feet. On Earth the acceleration due to gravity, g, is approximately 32 feet per second squared. Find the velocity of a rock after it has fallen 5 feet.

Solution **Understand and Translate** Begin by substituting 32 for g in the given equation.

$$v = \sqrt{2gh} = \sqrt{2(32)(h)} = \sqrt{64h}$$

Carry Out At $h = 5$ feet,

$$v = \sqrt{64(5)} = \sqrt{320} \approx 17.89$$

Answer After a rock has fallen 5 feet, its velocity is approximately 17.9 feet per second.

NOW TRY EXERCISE 43

EXAMPLE 7 Even as far back as the 1600s, people were experimenting with ways to make clocks more accurate. A Dutch astronomer, Christiaan Huygens, is credited with being the first person to suggest using a pendulum mechanism. The *period of a pendulum* (the time required for the pendulum to make one complete swing both back and forth) depends on the length of the pendulum. Finding the length for a pendulum that would cause it to swing back and forth 60 times in one minute was a first step in creating more accurate clocks. Of course, pendulums are not used only in clocks. A detuning pendulum is found on overhead power lines to reduce the sometimes violent motion of conductors (caused by the wind). A pendulum can even be found on the playground—a simple swing, swinging freely back and forth!

The formula for the period, T (in seconds), of a pendulum is

$$T = 2\pi\sqrt{\frac{L}{32}}$$

where L is the length of the pendulum in feet (Fig. 9.9). Find the period of the pendulum if its length is 8 feet. Use 3.14 as an approximation for π.

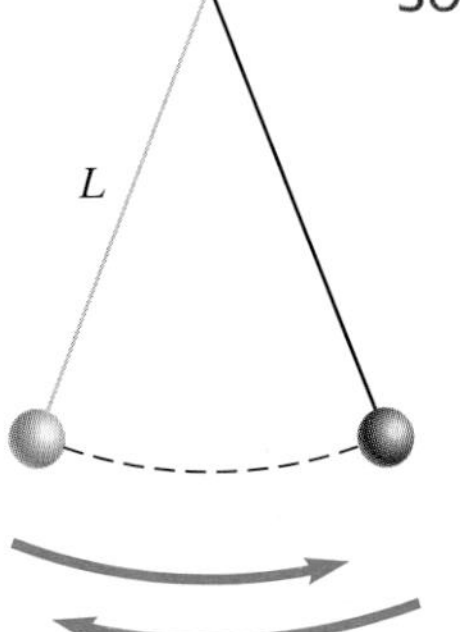

FIGURE 9.9

Solution **Understand and Translate** Substitute 3.14 for π and 8 for L in the formula.

$$T \approx 2(3.14)\sqrt{\frac{8}{32}}$$

Carry Out

$$\approx 6.28\sqrt{\frac{1}{4}}$$

$$\approx 6.28\left(\frac{1}{2}\right) = 3.14 \text{ seconds}$$

Answer A pendulum 8 feet long takes about 3.14 seconds to make one complete swing.

NOW TRY EXERCISE 37

Exercise Set 9.6

Concept/Writing Exercises

1. What is a right triangle?
2. State the Pythagorean Theorem and use your own words to describe what relationship it expresses about a right triangle.
3. Can the Pythagorean Theorem be used with any triangle? Explain.
4. **a)** Write the distance formula.
   b) When do we use the distance formula?
5. What do the following ordered pairs represent in the distance formula: (x_1, y_1) and (x_2, y_2)?
6. Determine the distance between the points $(4, 0)$ and $(-3, 0)$.
7 Determine the distance between the points $(0, -2)$ and $(0, -7)$.
8. What is meant by the period of a pendulum?

Practice the Skills

Use the Pythagorean Theorem to find each indicated quantity. Give the exact value, then round your answer to the nearest hundredth.

9.

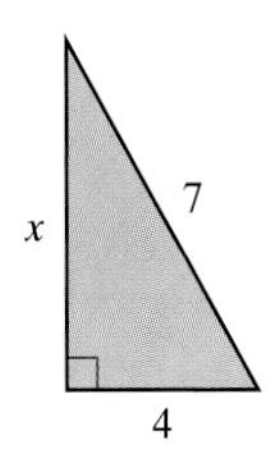

10.

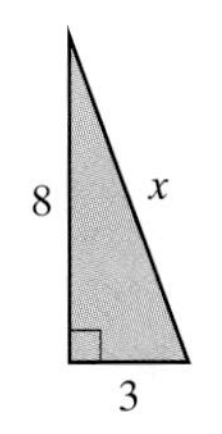

11.

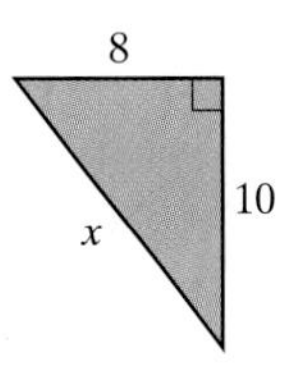

12.

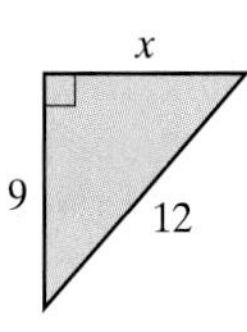

13.

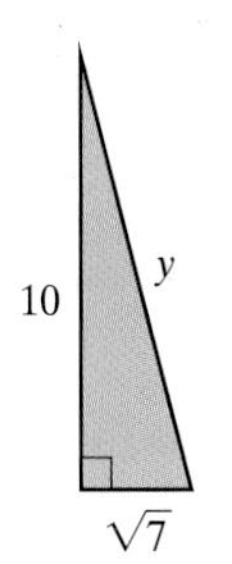

14.

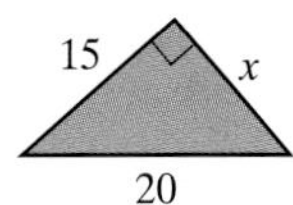

15.

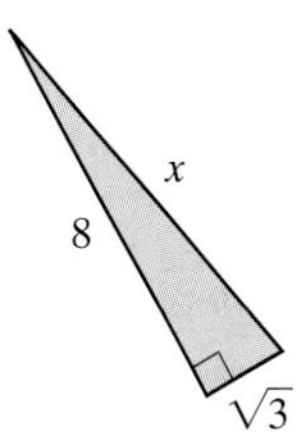

16.

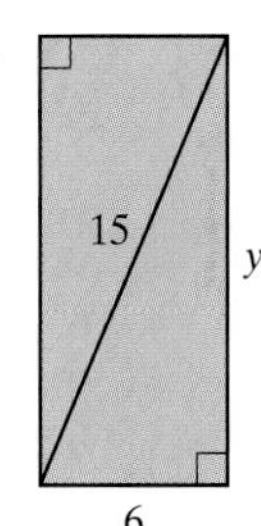

17.

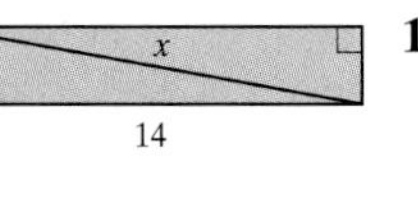

18.

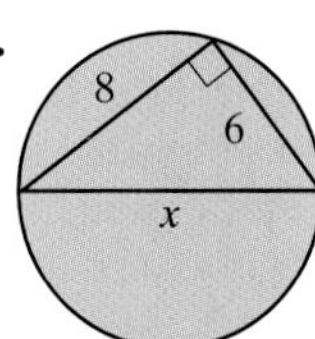

19.

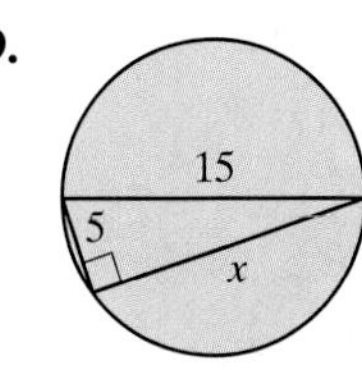

20.

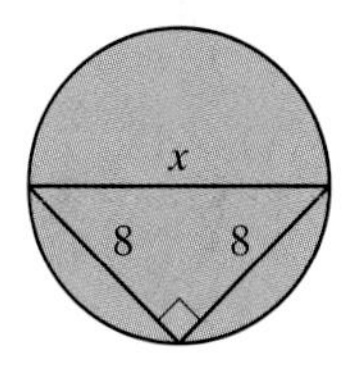

Use the distance formula to find the length of the line segments between each pair of points.

21. $(5, 7)$ and $(2, 9)$
22. $(4, -3)$ and $(-5, 6)$
23. $(-8, 4)$ and $(4, 11)$
24. $(0, 5)$ and $(-6, -4)$

Problem Solving

25. Pro Player Stadium in Miami, Florida is home to both the Florida Marlins and the Miami Dolphins. The football field is 120 yards long from end zone to end zone. Find the length of the diagonal, to the nearest hundredth, from one end zone to the other if the width of the field is 53.3 yards.

26. According to information provided by the Austin Rugby Football Club Web site, the dimensions of a rugby field are 69 meters by 100 meters. What is the length of the diagonal, to the nearest hundredth, from one corner of the field to the opposite corner of the field?

27. Carmen and Rosalie Bongiovanni have a two-story frame house that has not been painted in several years. After buying painting supplies and an 8-meter extension ladder, Carmen is ready to tackle the job. The instructions on the ladder indicate that the base of the ladder should be two meters from the house when in use. If Carmen follows the instructions, how high will the top of the ladder reach on the Bongiovanni house?

28. A local cable company crew is running wire between poles in a suburban neighborhood. At one point, the men working realize they must run a guy wire to provide additional support from the top of a 14-meter pole to a point 6 meters from the base of the pole. How long will the guy wire be?

29. Although the actual size of a boxing ring will vary by country, and by state, in the United States, the rings are squares whose surface area is generally 256 square feet. Peter Matos, manager of the Lakeview Boxing Club, has decided that the ropes along each side of the boxing ring must be replaced before he begins his

membership drive. To determine what to order, Peter must know the length of a side of the boxing ring. Use the formula for the area of a square, $A = s^2$, to determine the length of a side.

30. A formula for the area of a circle is $A = \pi r^2$, where π is approximately 3.14 and r is the radius of the circle. Find the radius of a circle whose area is 45 square inches.

31. Find the radius of a circle whose area is 80 square feet.

32. Each winter, Janice Albert puts a heated circular birdbath in her backyard to provide drinking water for the birds that visit her yard during freezing weather. The birdbath can hold water 2 inches deep, as shown in the figure. If the volume of water that can be held in the birdbath is 402 cubic inches, find the radius of the birdbath. Water volume is calculated by finding the area of the circular bottom of the birdbath and then multiplying the area by the depth of the water. Use the formula $V = \pi r^2 h$.

33. A regulation baseball diamond is a square with 90 feet between bases. How far is second base from home plate?

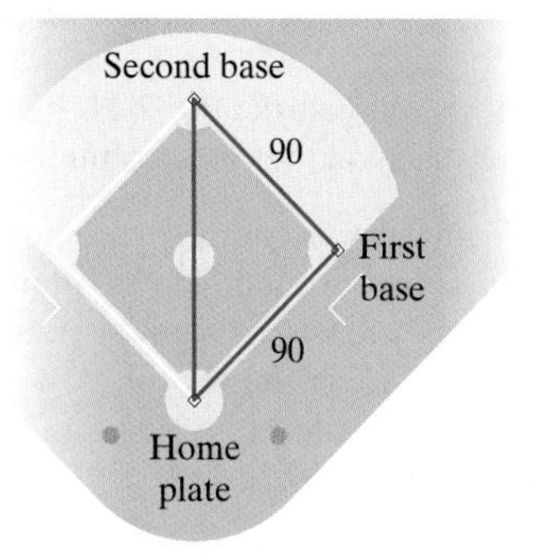

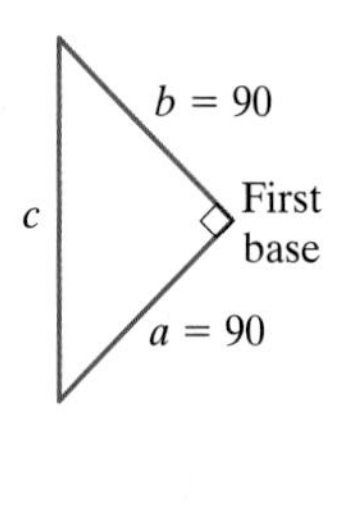

34. The top of a shoe box is a rectangular piece of cardboard with a length of 12 inches and a width of 5 inches. Find the length of the diagonal across the top of the shoe box.

35. A four-wall racquetball court is 20 feet wide, 40 feet long, and 20 feet high. What is the length of the diagonal from corner to corner along the floor?

36. Earning the designation "World's Tallest Building" is quite prestigious. However, there has been some disagreement in the architectural world concerning what is meant by the height of a building. In 1997, the Council on Tall Buildings and Urban Habitat established four categories in defining height: height to the structural top of a building, height to the highest occupied floor, height to the top of the roof, and height to the tip of a spire or antenna. By late 1998, the three buildings in the chart on page 557 shared the honor of being the "World's Tallest Building." If a rock was to be dropped from each of the heights indicated in the last column of the table, determine the velocity of the rock when it strikes the ground. (Use $v = \sqrt{2gh}$; refer to Example 6)

Category	Tower Name	Location	Height in Feet to the Indicated Position
a) Height to structural top	Petronas Towers	Kuala Lumpur, Malaysia	1483
b) Height to the highest occupied floor	Sears Tower	Chicago, USA	1431
c) Height to the tip of the roof	Sears Tower	Chicago, USA	1450
d) Height to the tip of a spire or antenna	World Trade Center	New York City, USA	1728

37. James O'Hara's 7th grade science class is studying simple pendulums. Each member of the class is making a simple pendulum using string cut to various lengths and bobs made of cork, steel, and glass marbles. During their experiments, the class discovered that the period of a simple pendulum is determined by the length of the string, not the material composing the bobs. Mikel Finn made a pendulum with a string measuring 43 inches. What is the period for this pendulum? (Use $T = 2\pi\sqrt{L/32}$; refer to Example 7.)

38. An example of a Foucault Pendulum is on display in the Smithsonian Institute in Washington, D.C. It is used to demonstrate the earth's rotation. It has a cable 52 feet long and a symmetrical, hollow brass bob weighing about 240 pounds. Find the period of the pendulum.

39. The Houston Museum of Natural Science also has a Foucault Pendulum on display. It has a cable 61.6 feet long and a bob weighing 180 pounds. Find the period of the pendulum.

For any planet, its "year" is the time it takes for the planet to revolve once around the sun. The number of "Earth days," N, in a given planet's year, is approximated by the formula

$$N = 0.2(\sqrt{R})^3$$

where R is the mean distance from the sun in millions of kilometers. The mean distance from the sun in millions of kilometers for three planets is illustrated below.

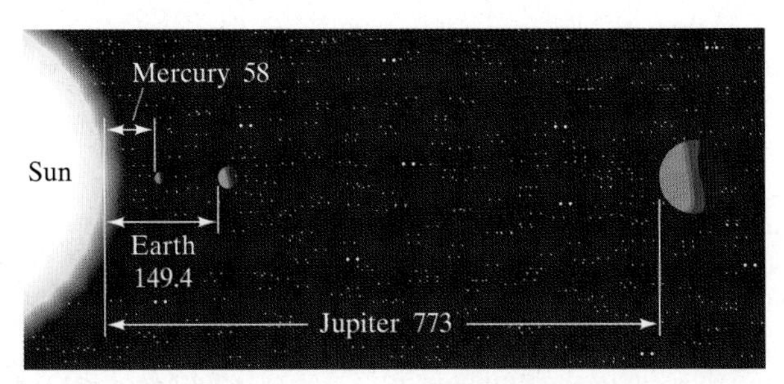

40. Find the number of Earth days in the year of Mercury.

41. Find the number of Earth days in the year of Earth.

42. Find the number of Earth days in the year of Jupiter.

Escape velocity, v_e, is the velocity needed for a spacecraft to escape a planet's gravitational field. This is found by the formula, $v_e = \sqrt{2gR}$, where g is the acceleration due to gravity of the planet and R is the radius of the planet in meters.

43. Find the escape velocity for Earth where $g = 9.81$ meters per second squared and $R = 6{,}370{,}000$ meters.

44. During the Mars Pathfinder mission, NASA dispatched the Sojourner Rover, a robot designed to gather information and transmit pictures and data back to Earth. Future Mars exploration may include sending spacecraft that will land on the surface of the planet and leave after scientific work has been completed. What is the approximate velocity needed for a spacecraft to escape the gravitational field of Mars? The acceleration due to gravity on Mars is approximately 3.6835 meters per second squared. The radius of Mars is approximately 3,393,500 meters.

45. When two forces, F_1 and F_2, pull at right angles to each other as illustrated, the resultant, or the effective force, R, can be found by the formula

$$R = \sqrt{F_1^2 + F_2^2}$$

Two cars at a 90° angle to each other are trying to pull a third car out of the mud, as shown. If car A exerts 600 pounds of force and car B exerts 800 pounds of force, find the resulting force on the car stuck in the mud.

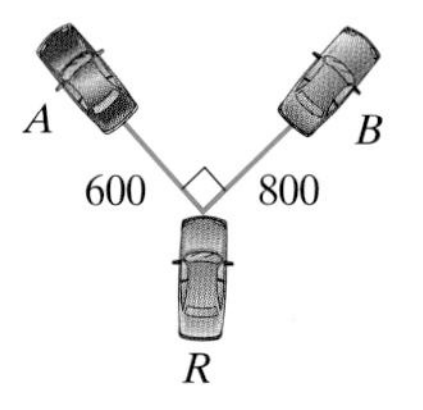

Challenge Problems

46. The length of a rectangle is 3 inches more than its width. If the length of the diagonal is 15 inches, find the dimensions of the rectangle.

47. The acceleration due to gravity on the moon is $\frac{1}{6}$ of that on Earth. If a camera falls from a rocket 100 feet above the surface of the moon, with what velocity will it strike the moon? (Use $v = \sqrt{2gh}$; see Example 6.)

48. Find the length of a pendulum if the period is 2 seconds. (Use $T = 2\pi\sqrt{L/32}$; refer to Example 7.)

49. The length of the diagonal of a rectangular solid (see the figure) is given by

$$d = \sqrt{a^2 + b^2 + c^2}$$

Find the length of the diagonal of a suitcase 37 inches long, 15 inches wide, and 9 inches deep.

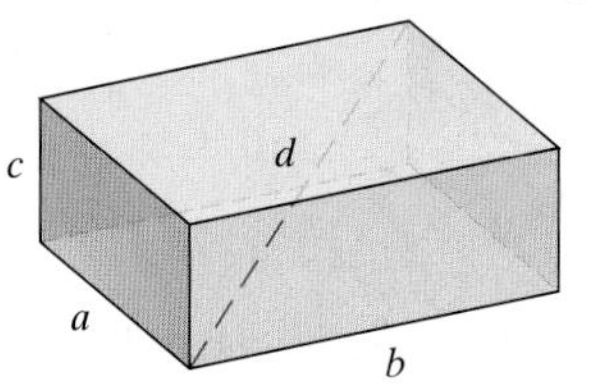

Group Activity

50. A Pythagorean triple is a set of three natural numbers a, b, and c with the property that $a^2 + b^2 = c^2$. Of the three numbers that make up the Pythagorean triple, the largest number is always designated c.

a) As a group, determine whether the numbers 3, 4, and 5 make up a Pythagorean triple.

b) As a group, determine whether the numbers 5, 12, and 13 make up a Pythagorean triple.

c) As a group, determine whether the numbers 8, 15, and 17 make up a Pythagorean triple.

d) Group Member 1: Multiply each of the numbers in part **a)** by 3. Determine whether the new numbers make up a Pythagorean triple.

e) Group Member 2: Multiply each of the numbers in part **b)** by 2. Determine whether the new numbers make up a Pythagorean triple.

f) Group Member 3: Multiply each of the numbers in part **c)** by 4. Determine whether the new numbers make up a Pythagorean triple.

g) As a group, discuss the results from parts **d)**, **e)**, and **f)**.

h) As a group, can you explain how to find additional Pythagorean triples if you know the numbers that make up one Pythagorean triple?

Cumulative Review Exercises

[2.7] **51.** Solve the inequality $2(x + 3) < 4x - 6$.

[5.3] **52.** Solve the following system of equations using the addition method.

$$3x + 4y = 12$$
$$\frac{1}{2}x - 2y = 8$$

[6.2] **53.** Simplify $(4x^{-4}y^3)^{-1}$.

[8.6] **54.** Solve the equation $5 + \frac{6}{x} = \frac{2}{3x}$.

9.7 HIGHER ROOTS AND RATIONAL EXPONENTS

SSM VIDEO 9.7 CD Rom

1) Evaluate cube and fourth roots.
2) Simplify cube and fourth roots.
3) Write radical expressions in exponential form.

1) Evaluate Cube and Fourth Roots

In this section we will use the same basic concepts used in Sections 9.1 through 9.4 to work with radicals with indexes of 3 and 4. Now we introduce **cube** and **fourth roots.**

$\sqrt[3]{a}$ is read "the cube root of a."

$\sqrt[4]{a}$ is read "the fourth root of a."

Note that

$$\sqrt[3]{a} = b \quad \text{if} \quad b^3 = a$$

and

$$\sqrt[4]{a} = b \quad \text{if} \quad b^4 = a, \qquad b > 0$$

Examples

$\sqrt[3]{8} = 2$ since $2^3 = 2 \cdot 2 \cdot 2 = 8$

$\sqrt[3]{-8} = -2$ since $(-2)^3 = (-2)(-2)(-2) = -8$

$\sqrt[3]{27} = 3$ since $3^3 = 3 \cdot 3 \cdot 3 = 27$

$\sqrt[4]{16} = 2$ since $2^4 = 2 \cdot 2 \cdot 2 \cdot 2 = 16$

$\sqrt[4]{81} = 3$ since $3^4 = 3 \cdot 3 \cdot 3 \cdot 3 = 81$

EXAMPLE 1 Evaluate. **a)** $\sqrt[3]{-64}$ **b)** $\sqrt[3]{125}$

Solution **a)** To find $\sqrt[3]{-64}$, we must find the number that when cubed is -64.

$$\sqrt[3]{-64} = -4 \qquad \text{since } (-4)^3 = -64$$

b) To find $\sqrt[3]{125}$, we must find the number that when cubed is 125.

NOW TRY EXERCISE 9

$$\sqrt[3]{125} = 5 \qquad \text{since } 5^3 = 125$$

Note that the cube root of a positive number is a positive number and the cube root of a negative number is a negative number. The radicand of a fourth root (or any even root) must be a nonnegative number for the expression to be a real number. For example, $\sqrt[4]{-16}$ is not a real number because no real number raised to the fourth power can be a negative number.

Using Your Calculator

Evaluating Cube and Higher Roots on a Scientific Calculator

If your scientific calculator contains a y^x or x^y key, then you can find cube and higher roots. To find cube and higher roots, you need to use both the second function key, 2nd, and either the y^x or x^y key. Some calculators use an inverse key INV, in place of the 2nd key. To evaluate $\sqrt[3]{216}$ press the following keys:

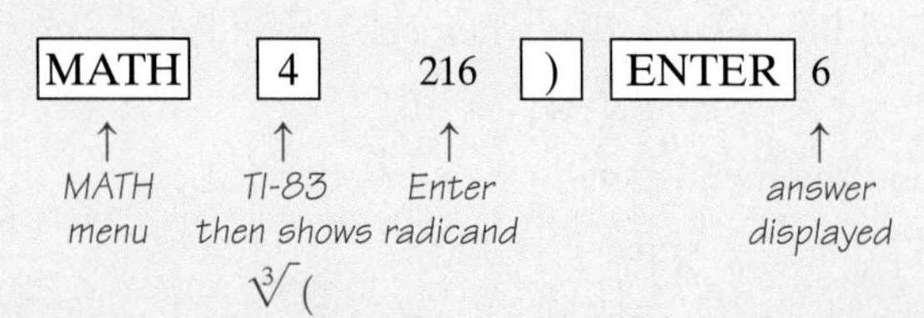

Thus $\sqrt[3]{216} = 6$. To evaluate $\sqrt[4]{158}$, press the following keys:

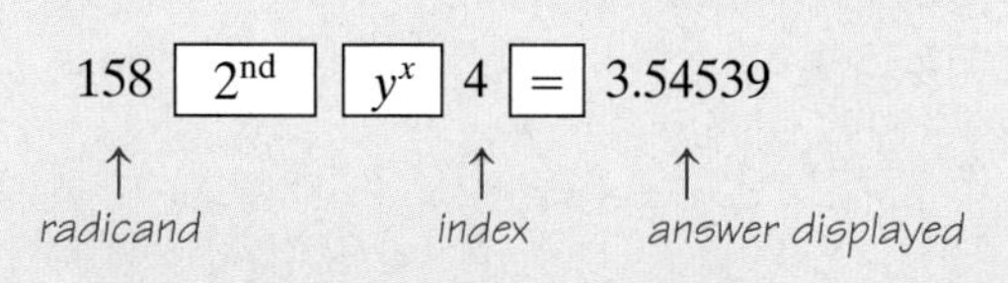

Thus $\sqrt[4]{158} \approx 3.54539$.

To find the odd root of a negative number, find the odd root of that positive number and then place a negative sign before the value. For example, to find $\sqrt[3]{-64}$, find $\sqrt[3]{64}$, which is 4; then place a negative sign before the value to get -4. Thus $\sqrt[3]{-64} = -4$.

Exercises

Evaluate each of the following.

1. $\sqrt[3]{343}$ **2.** $\sqrt[4]{16}$
3. $\sqrt[4]{200}$ **4.** $\sqrt[3]{1000}$
5. $\sqrt[3]{-729}$ **6.** $\sqrt[5]{-1200}$

Using Your Graphing Calculator

Evaluating Cube and Higher Roots on a Graphing Calculator

The procedure to use to find cube and higher roots depends upon your graphing calculator. Read the manual that comes with your graphing calculator for instructions.

On a TI-82 or TI-83 calculator you use the MATH menu to find both cube roots and roots greater than cube roots. To find a cube root you use the MATH menu option: 4. On the TI-83, option 4 is $\sqrt[3]{\ }($ and on the TI-82 option 4 is $\sqrt[3]{\ }$, without the left hand parenthesis.

To evaluate $\sqrt[3]{216}$ on a TI-83 press the following keys:

MATH 4 216) ENTER 6

MATH menu — TI-83 then shows $\sqrt[3]{\ }($ — Enter radicand — answer displayed

To evaluate $\sqrt[3]{-216}$, you press the following keys:

MATH 4 (−) 216) ENTER −6

Thus $\sqrt[3]{-216} = -6$.

To evaluate roots greater than 3 on a TI-82 or TI-83, you use the MATH menu, option $5{:}\sqrt[x]{\ }$. You enter the index before you press the MATH key. To evaluate $\sqrt[4]{158}$ press the following keys.

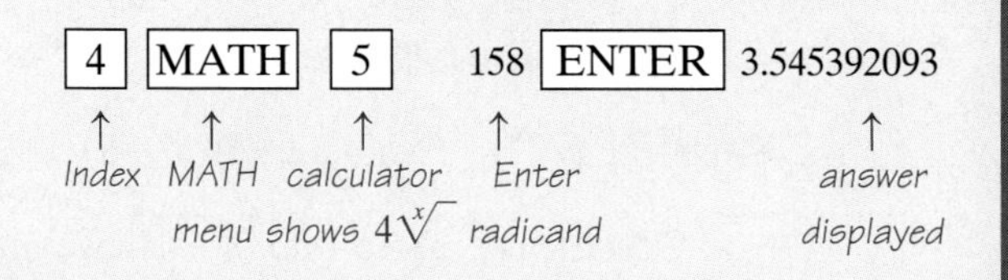

Exercises

Evaluate each of the following.

1. $\sqrt[3]{125}$ **2.** $\sqrt[4]{1296}$
3. $\sqrt[4]{76}$ **4.** $\sqrt[3]{1728}$
5. $\sqrt[3]{-512}$ **6.** $\sqrt[5]{-1000}$

It will be helpful in the explanations that follow if we define perfect cubes. A **perfect cube** is a number that is the cube of a natural number.

$1,\ 2,\ 3,\ 4,\ 5,\ 6,\ 7,\ 8,\ 9,\ 10,\ \ldots$ *Natural numbers*

$1^3,\ 2^3,\ 3^3,\ 4^3,\ 5^3,\ 6^3,\ 7^3,\ 8^3,\ 9^3,\ 10^3,\ \ldots$ *Cubes of natural numbers*

$1,\ 8,\ 27,\ 64,\ 125,\ 216,\ 343,\ 512,\ 729,\ 1000,\ \ldots$ *Perfect cubes*

Note that $\sqrt[3]{1} = 1$, $\sqrt[3]{8} = 2$, $\sqrt[3]{27} = 3$, $\sqrt[3]{64} = 4$, and so on.

Perfect fourth powers can be expressed in a similar manner.

$1,\ 2,\ 3,\ 4,\ 5,\ 6,\ \ldots$ *Natural numbers*

$1^4,\ 2^4,\ 3^4,\ 4^4,\ 5^4,\ 6^4,\ \ldots$ *Fourth powers of natural numbers*

$1,\ 16,\ 81,\ 256,\ 625,\ 1296,\ \ldots$ *Perfect fourth powers*

Note that $\sqrt[4]{1} = 1$, $\sqrt[4]{16} = 2$, $\sqrt[4]{81} = 3$, $\sqrt[4]{256} = 4$, and so on.

You may wish to refer to these numbers when evaluating cube and fourth roots.

2) Simplify Cube and Fourth Roots

The product rule used in simplifying square roots can be expanded to indices greater than 2.

Product Rule for Radicals

$$\sqrt[n]{a}\sqrt[n]{b} = \sqrt[n]{ab}, \quad \text{for } a \geq 0, b \geq 0$$

To simplify a cube root whose radicand is a constant, write the radicand as the product of a perfect cube and another number. Then simplify, using the product rule.

EXAMPLE 2 Simplify.

a) $\sqrt[3]{40}$ **b)** $\sqrt[3]{54}$ **c)** $\sqrt[4]{48}$

Solution **a)** Eight is a perfect cube that is a factor of the radicand, 40. Therefore, we simplify as follows:

$$\sqrt[3]{40} = \sqrt[3]{8 \cdot 5} = \sqrt[3]{8}\sqrt[3]{5} = 2\sqrt[3]{5}$$

b) $\sqrt[3]{54} = \sqrt[3]{27 \cdot 2} = \sqrt[3]{27}\sqrt[3]{2} = 3\sqrt[3]{2}$

c) Write $\sqrt[4]{48}$ as a product of a perfect fourth power and another number, then simplify. From the listing above, we see that 16 is a perfect fourth power. Since 16 is a factor of 48, we simplify as follows:

$$\sqrt[4]{48} = \sqrt[4]{16 \cdot 3} = \sqrt[4]{16}\sqrt[4]{3} = 2\sqrt[4]{3}$$

NOW TRY EXERCISE 25

3) Write Radical Expressions in Exponential Form

A radical expression can be written in **exponential form** by using the following rule.

$$\sqrt[n]{a} = a^{1/n}, \quad a \geq 0 \qquad \text{Rule 4}$$

Examples

$$\sqrt{8} = 8^{1/2} \qquad \sqrt{x} = x^{1/2}$$
$$\sqrt[3]{4} = 4^{1/3} \qquad \sqrt[4]{9} = 9^{1/4}$$
$$\sqrt[3]{x} = x^{1/3} \qquad \sqrt[4]{y} = y^{1/4}$$
$$\sqrt[3]{5x^2} = (5x^2)^{1/3} \qquad \sqrt[4]{3y^2} = (3y^2)^{1/4}$$

Notice $\sqrt{8} = 8^{1/2}$ and $\sqrt{x} = x^{1/2}$, which is consistent with what we learned in Section 9.1. This concept can be expanded as follows.

Power ↘ ↙ Index

$$\sqrt[n]{a^m} = \left(\sqrt[n]{a}\right)^m = a^{m/n}, \text{ for } a \geq 0 \text{ and } m \text{ and } n \text{ integers} \qquad \text{Rule 5}$$

As long as the radicand is nonnegative, we can change from one form to another.

Examples

$$\sqrt[3]{27^4} = \left(\sqrt[3]{27}\right)^4 = 3^4 = 81 \qquad \sqrt[3]{x^3} = x^{3/3} = x^1 = x$$
$$8^{2/3} = \left(\sqrt[3]{8}\right)^2 = 2^2 = 4 \qquad \sqrt[4]{y^8} = y^{8/4} = y^2$$

EXAMPLE 3 Simplify. **a)** $\sqrt[4]{x^{12}}$ **b)** $\sqrt[3]{y^{18}}$

Solution Write each radical expression in exponential form, then simplify.

a) $\sqrt[4]{x^{12}} = x^{12/4} = x^3$

b) $\sqrt[3]{y^{18}} = y^{18/3} = y^6$

NOW TRY EXERCISE 31

The rules of exponents discussed in Sections 6.1 and 6.2 also apply when the exponents are fractions. In Example 4c) we use the negative exponent rule and in Example 6 we use the product and power rules with fractional exponents.

EXAMPLE 4 Evaluate.

a) $8^{4/3}$ **b)** $16^{5/4}$ **c)** $8^{-2/3}$

Solution To evaluate, we write each exponential expression in radical form.

a) $8^{4/3} = \left(\sqrt[3]{8}\right)^4 = 2^4 = 16$

b) $16^{5/4} = \left(\sqrt[4]{16}\right)^5 = 2^5 = 32$

c) Recall from Section 6.2 that $x^{-m} = \dfrac{1}{x^m}$. Thus,

$$8^{-2/3} = \frac{1}{8^{2/3}} = \frac{1}{\left(\sqrt[3]{8}\right)^2} = \frac{1}{2^2} = \frac{1}{4}$$

NOW TRY EXERCISE 45

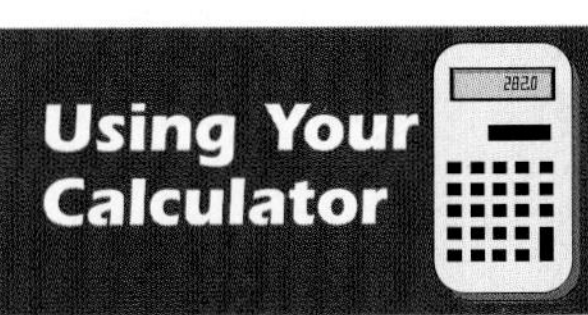

Using Your Calculator

In Chapter 1, page 69, we learned how to use the $\boxed{y^x}$ key or $\boxed{\wedge}$ key to raise a value to an integer exponent greater than 2. You can also use these keys if the exponent is a fraction. We demonstrate how in the following examples.

EXAMPLE 1 Evaluate $27^{2/3}$.

Solution

Scientific calculator:

27 $\boxed{y^x}$ $\boxed{(}$ 2 $\boxed{\div}$ 3 $\boxed{)}$ $\boxed{=}$ 9 ← Answer displayed

Graphing calculator:

27 $\boxed{\wedge}$ $\boxed{(}$ 2 $\boxed{\div}$ 3 $\boxed{)}$ $\boxed{\text{ENTER}}$ 9 ← Answer displayed

The answer is 9. Thus, $27^{2/3} = 9$.

EXAMPLE 2 Evaluate $16^{-3/4}$.

Solution

Scientific Calculator:

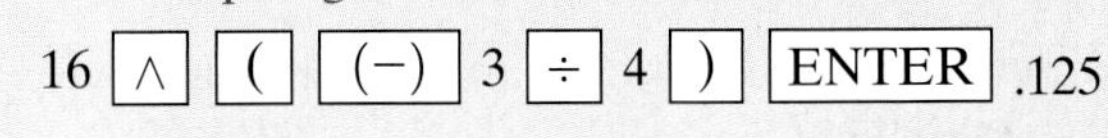

16 $\boxed{y^x}$ $\boxed{(}$ 3 $\boxed{+/-}$ $\boxed{\div}$ 4 $\boxed{)}$ $\boxed{=}$ .125 ← Answer displayed

Graphing calculator:

16 $\boxed{\wedge}$ $\boxed{(}$ $\boxed{(-)}$ 3 $\boxed{\div}$ 4 $\boxed{)}$ $\boxed{\text{ENTER}}$ .125 ← Answer displayed

The answer is 0.125. Thus, $16^{-3/4} = 0.125$.

Exercises

Use your calculator to evaluate each expression.

1. $4^{3/2}$ **2.** $64^{4/3}$

3. $125^{-2/3}$ **4.** $81^{-3/4}$

EXAMPLE 5 Write each radical in exponential form.

a) $\sqrt[3]{a^4}$ **b)** $\sqrt[4]{y^7}$ **c)** $\sqrt[4]{x^{13}}$

Solution **a)** $\sqrt[3]{a^4} = a^{4/3}$ **b)** $\sqrt[4]{y^7} = y^{7/4}$ **c)** $\sqrt[4]{x^{13}} = x^{13/4}$

Avoiding Common Errors

Students may make mistakes simplifying expressions that contain negative exponents. Be careful when working such problems. The following is a common error.

Correct	Incorrect
$27^{-2/3} = \dfrac{1}{27^{2/3}}$	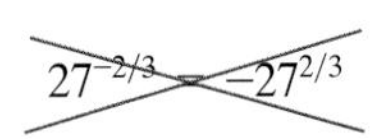 $27^{-2/3} = -27^{2/3}$

The expression $27^{-2/3}$ simplifies to $\frac{1}{9}$. Can you show how?

EXAMPLE 6 Simplify. **a)** $\sqrt{x} \cdot \sqrt[4]{x}$. **b)** $(\sqrt[4]{x^2})^8$

Solution To simplify we change each radical expression to exponential form, then apply the rules of exponents.

a)
$$\begin{aligned}\sqrt{x} \cdot \sqrt[4]{x} &= x^{1/2} \cdot x^{1/4} \\ &= x^{(1/2)+(1/4)} \\ &= x^{(2/4)+(1/4)} \\ &= x^{3/4} \\ &= \sqrt[4]{x^3}\end{aligned}$$

b)
$$\begin{aligned}(\sqrt[4]{x^2})^8 &= (x^{2/4})^8 \\ &= (x^{1/2})^8 \\ &= x^4\end{aligned}$$

NOW TRY EXERCISE 63

This section was meant to give you a brief introduction to roots other than square roots. If you take a course in intermediate algebra, you may study these concepts in more depth.

Exercise Set 9.7

Concept/Writing Exercises

1. Write how the following radicals would be read.
   a) $\sqrt{9}$
   b) $\sqrt[3]{9}$
   c) $\sqrt[4]{9}$
2. In your own words, describe a perfect cube and a perfect fourth power.
3. How do you simplify a cube root whose radicand is a constant?
4. How do you simplify a fourth root whose radicand is a constant?
5. **a)** In your own words, explain how to change an expression written in radical form to exponential form.
   b) Using the procedure you wrote in part **a)**, write $\sqrt[3]{y^7}$ in exponential form.
6. **a)** In your own words, explain how to change an expression written in exponential form to radical form.
   b) Using the procedure you wrote in part **a)**, write $x^{5/4}$ in radical form.

Practice the Skills

Evaluate.

7. $\sqrt[3]{125}$ **8.** $\sqrt[3]{27}$ **9.** $\sqrt[3]{-27}$ **10.** $\sqrt[3]{-125}$

11. $\sqrt[4]{16}$ **12.** $\sqrt[3]{8}$ **13.** $\sqrt[4]{81}$ **14.** $\sqrt[4]{1}$

15. $\sqrt[3]{-1}$ **16.** $\sqrt[3]{1}$ **17.** $\sqrt[3]{-1000}$ **18.** $\sqrt[4]{1296}$

Simplify.

19. $\sqrt[3]{32}$ **20.** $\sqrt[3]{48}$ **21.** $\sqrt[3]{16}$ **22.** $\sqrt[3]{24}$

23. $\sqrt[3]{81}$ **24.** $\sqrt[3]{128}$ **25.** $\sqrt[4]{32}$ **26.** $\sqrt[3]{3000}$

27. $\sqrt[4]{1250}$ **28.** $\sqrt[3]{192}$

Simplify.

29. $\sqrt[4]{x^4}$ **30.** $\sqrt[3]{y^6}$ **31.** $\sqrt[3]{y^{21}}$ **32.** $\sqrt[4]{y^{20}}$

33. $\sqrt[3]{x^{12}}$ **34.** $\sqrt[3]{x^9}$ **35.** $\sqrt[4]{x^{32}}$ **36.** $\sqrt[3]{x^3}$

37. $\sqrt[3]{x^{15}}$ **38.** $\sqrt[3]{x^{18}}$

Evaluate.

39. $16^{3/4}$ **40.** $8^{4/3}$ **41.** $125^{2/3}$ **42.** $64^{3/2}$

43. $1^{2/3}$ **44.** $16^{5/2}$ **45.** $9^{3/2}$ **46.** $64^{2/3}$

47. $27^{4/3}$ **48.** $25^{3/2}$ **49.** $125^{4/3}$ **50.** $256^{2/4}$

51. $8^{-1/3}$ **52.** $16^{-3/4}$ **53.** $27^{-2/3}$ **54.** $64^{-2/3}$

Write each radical in exponential form.

55. $\sqrt[3]{x^5}$ **56.** $\sqrt[4]{x^8}$ **57.** $\sqrt[3]{x^4}$ **58.** $\sqrt[4]{x^{11}}$

59. $\sqrt[4]{y^{15}}$ **60.** $\sqrt[3]{x^9}$ **61.** $\sqrt[3]{y^{21}}$ **62.** $\sqrt[4]{x^5}$

Simplify and write the answer in exponential form.

63. $\sqrt[4]{x}\cdot\sqrt[4]{x^3}$
64. $\sqrt[3]{x}\cdot\sqrt[4]{x}$
65. $\sqrt[4]{x^2}\cdot\sqrt[4]{x^2}$
66. $\sqrt[3]{x^4}\cdot\sqrt[3]{x^5}$
67. $(\sqrt[3]{x^5})^6$
68. $(\sqrt[4]{x^3})^4$
69. $(\sqrt[4]{x^2})^4$
70. $(\sqrt[3]{x^6})^2$
71. Show that $(\sqrt[3]{x})^2=\sqrt[3]{x^2}$ for $x=8$.
72. Show that $(\sqrt[4]{x})^3=\sqrt[4]{x^3}$ for $x=16$.

Problem Solving

The product $\sqrt[3]{2}\cdot\sqrt[3]{2}$ is not an integer, but $\sqrt[3]{2}\cdot\sqrt[3]{2^2}=\sqrt[3]{2^3}=2^{3/3}=2$. Use this information to determine by what radical expression you can multiply each radical by so that the result is an integer.

73. $\sqrt[3]{5}$
74. $\sqrt[3]{7^2}$
75. $\sqrt[3]{6}$
76. $\sqrt[4]{2^3}$
77. $\sqrt[4]{5}$
78. $\sqrt[4]{3^2}$

Fill in the shaded area(s) in each equation to make a true statement.

79. $\sqrt[\blacksquare]{5^2}\cdot\sqrt[\blacksquare]{5}=5$
80. $\sqrt[\blacksquare]{7^2}\cdot\sqrt[\blacksquare]{7^2}=7$
81. $\sqrt[3]{6^{\blacksquare}}\cdot\sqrt[3]{6^2}=6$
82. $\sqrt[5]{2^3}\cdot\sqrt[5]{2^{\blacksquare}}=2$

Challenge Problems

Simplify.

83. $\sqrt[3]{xy}\cdot\sqrt[3]{x^2y^2}$
84. $\sqrt[4]{3x^2y}\cdot\sqrt[4]{27x^6y^3}$
85. $\sqrt[4]{32}-\sqrt[4]{2}$
86. $\sqrt[3]{3x^3y}+\sqrt[3]{24x^3y}$

Group Activity

87. **a)** As a group, discuss why multiplying both the numerator and denominator of $\frac{1}{\sqrt[3]{2}}$ by $\sqrt[3]{2^2}$ will give an integer in the denominator.

b) As a group, rationalize the denominator of $\frac{1}{\sqrt[3]{2}}$ by multiplying both the numerator and denominator by $\sqrt[3]{2^2}$.

c) Explain why we cannot rationalize the denominator by multiplying both the numerator and denominator by $\sqrt[3]{2}$.

d) Group member 1: Determine what you must multiply the numerator and denominator by to rationalize the denominator of $\frac{1}{\sqrt[3]{3}}$.

e) Group member 2: Determine what you must multiply the numerator and denominator by to rationalize the denominator of $\frac{1}{\sqrt[4]{5^3}}$.

f) Group member 3: Determine what you must multiply the numerator and denominator by to rationalize the denominator of $\frac{1}{\sqrt[3]{7^2}}$.

g) Each group member, explain your answer to part **d)**, **e)**, or **f)** to the other members of your group.

h) As a group, rationalize the denominators in parts **d)**, **e)**, and **f)**.

Cumulative Review Exercises

[1.9] **88.** Evaluate $-x^2+4xy-6$ when $x=2$ and $y=-4$.

[4.2] **89.** Graph $y=\frac{2}{3}x-4$.

[7.4] **90.** Factor $3x^2-28x+32$.

[9.3] **91.** Simplify $\sqrt{\frac{64x^3y^7}{2x^4}}$.

SUMMARY

Key Words and Phrases

9.1
Imaginary number
Index of a radical
Irrational number
Negative square root
Perfect square
Perfect square factor
Principal or positive square root
Radical expression
Radical sign
Radicand
Rational number
Writing a radical expression in exponential form
Writing an expression in exponential form as a radical expression

9.2
Perfect square
Perfect square factor
Simplify a square root

9.3
Rationalize a denominator
Simplify a square root

9.4
Conjugate of a binomial
Like square roots
Unlike square roots

9.5
Extraneous root
Isolating the square root
Radical equation

9.6
Distance formula
Hypotenuse of a right triangle
Legs of a right triangle
Pythagorean Theorem
Right triangle

9.7
Cube root
Exponential form
Fourth root
Perfect cube
Perfect fourth power

IMPORTANT FACTS

Numbers that are perfect squares: $1, 4, 9, 16, 25, 36, 49, 64, \ldots$

Numbers that are perfect cubes: $1, 8, 27, 64, 125, 216, 343, 512, \ldots$

Product rule for square roots: $\sqrt{a} \cdot \sqrt{b} = \sqrt{ab}, a \geq 0, b \geq 0$

$\sqrt{a^{2 \cdot n}} = a^n, \ a \geq 0$

$\sqrt{a^2} = a, a \geq 0$

Quotient rule for square roots: $\dfrac{\sqrt{a}}{\sqrt{b}} = \sqrt{\dfrac{a}{b}}, \ a \geq 0, b > 0$

$\sqrt[n]{a} = a^{1/n}, \ a \geq 0$

$\sqrt[n]{a^m} = \left(\sqrt[n]{a}\right)^m = a^{m/n}, \ a \geq 0$

Pythagorean Theorem: $a^2 + b^2 = c^2$

Distance formula: $d = \sqrt{(x_2 - x_1)^2 + (y_2 - y_1)^2}$

Product rule for radicals: $\sqrt[n]{a} \cdot \sqrt[n]{b} = \sqrt[n]{ab}$, for $a \geq 0, \ b \geq 0$

Review Exercises

[9.1] *Evaluate.*

1. $\sqrt{49}$

2. $\sqrt{9}$

3. $-\sqrt{64}$

Write in exponential form.

4. $\sqrt{6}$

5. $\sqrt{26x}$

6. $\sqrt{13x^2y}$

[9.2] *Simplify.*

7. $\sqrt{18}$ **8.** $\sqrt{44}$ **9.** $\sqrt{27x^7y^4}$

10. $\sqrt{125x^4y^6}$ **11.** $\sqrt{48ab^5c^4}$ **12.** $\sqrt{72a^2b^2c^7}$

Simplify.

13. $\sqrt{24} \cdot \sqrt{20}$ **14.** $\sqrt{7y} \cdot \sqrt{7y}$ **15.** $\sqrt{18x} \cdot \sqrt{2xy}$

16. $\sqrt{25x^2y} \cdot \sqrt{3y}$ **17.** $\sqrt{8a^3b} \cdot \sqrt{3b^4}$ **18.** $\sqrt{5ab^3} \cdot \sqrt{20ab^4}$

[9.3] *Simplify.*

19. $\dfrac{\sqrt{50}}{\sqrt{2}}$ **20.** $\sqrt{\dfrac{10}{490}}$ **21.** $\sqrt{\dfrac{7}{28}}$

22. $\dfrac{3}{\sqrt{5}}$ **23.** $\sqrt{\dfrac{a}{6}}$ **24.** $\sqrt{\dfrac{5a}{12}}$

25. $\sqrt{\dfrac{x^2}{3}}$ **26.** $\sqrt{\dfrac{x^5}{8}}$ **27.** $\sqrt{\dfrac{21x^3y^7}{3x^3y^3}}$

28. $\sqrt{\dfrac{30x^4y}{15x^2y^4}}$ **29.** $\dfrac{\sqrt{60}}{\sqrt{27a^3b^2}}$ **30.** $\dfrac{\sqrt{2a^4bc^4}}{\sqrt{7a^5bc^2}}$

[9.4] *Simplify.*

31. $\dfrac{2}{1 - \sqrt{5}}$ **32.** $\dfrac{5}{3 - \sqrt{6}}$ **33.** $\dfrac{\sqrt{2}}{2 + \sqrt{y}}$

34. $\dfrac{2}{\sqrt{x} - 5}$ **35.** $\dfrac{\sqrt{5}}{\sqrt{x} + \sqrt{3}}$

Simplify.

36. $7\sqrt{2} - 4\sqrt{2}$ **37.** $4\sqrt{5} - 7\sqrt{5} - 3\sqrt{5}$ **38.** $3\sqrt{x} - 5\sqrt{x}$

39. $\sqrt{x} + 3\sqrt{x} - 4\sqrt{x}$ **40.** $\sqrt{18} - \sqrt{27}$ **41.** $7\sqrt{40} - 2\sqrt{10}$

42. $2\sqrt{98} - 4\sqrt{72}$ **43.** $7\sqrt{50} + 2\sqrt{18} - 4\sqrt{32}$ **44.** $4\sqrt{27} + 5\sqrt{80} + 2\sqrt{12}$

[9.5] *Solve.*

45. $\sqrt{x} = 4$ **46.** $\sqrt{x} = -5$ **47.** $\sqrt{x - 4} = 3$

48. $\sqrt{3x + 1} = 5$ **49.** $\sqrt{5x + 6} = \sqrt{4x + 8}$ **50.** $4\sqrt{x} - x = 4$

51. $\sqrt{x^2 - 3} = x - 1$ **52.** $\sqrt{4x + 8} - \sqrt{7x - 13} = 0$ **53.** $4\sqrt{5x - 2} = 8$

[9.6] *Find each length indicated by x.*

54.

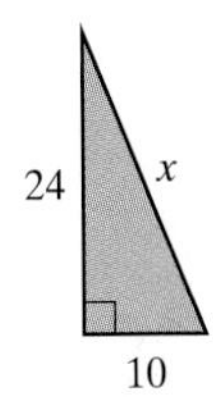

55.

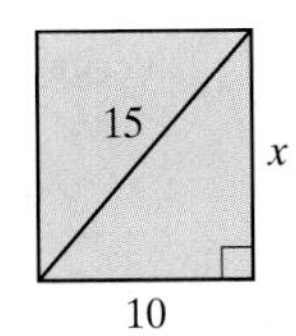

56.

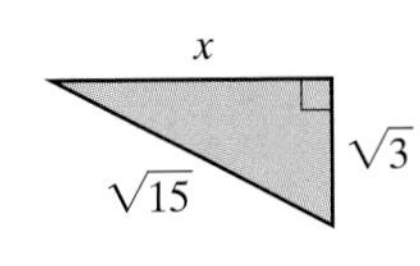

57.

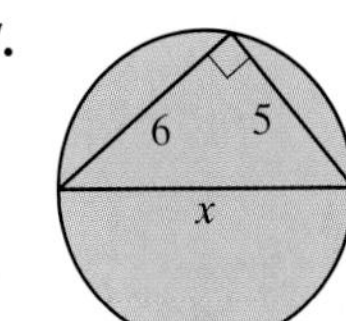

58. Adam Kurtz leans a 12-foot ladder against his grandfather's cottage near Lake Superior. If the base of the ladder is 3 feet from the cottage how high is the ladder on the cottage?

59. Find the diagonal of a rectangle of length 15 inches and width 6 inches.

60. Find the straight-line distance between the points $(4, -3)$ and $(1, 7)$.

61. Find the length of the line segment between the points $(6, 5)$ and $(-6, 8)$.

62. In Section 8.7, Exercise 8, we found that the approximate length of the base of a yield-right-of-way sign in the United States was 36 inches. We calculated this value using an *approximate* area of 558 square inches. These signs are actually in the shape of an **equilateral** triangle with each side, s, measuring 36 inches. Use the formula for the area of an equilateral triangle

$$\text{Area} = \frac{s^2\sqrt{3}}{4}$$

to find a closer approximation of the area of a yield sign. Round the area to the nearest hundredth.

63. Ken Jameston, who owns a farm outside Osceola, Iowa, enjoys climbing the ladder on the side of his silo and looking out toward the horizon. He does this often, and wonders how far he sees. After doing some research, he found that the farthest distance, in miles, he could see can be approximated by the formula $distance = \sqrt{(3/2)h}$, where h represents the height of his vantage point, measured in feet. Approximately how far can Ken see, correct to the nearest hundredth of a mile, if he climbs 40 feet?

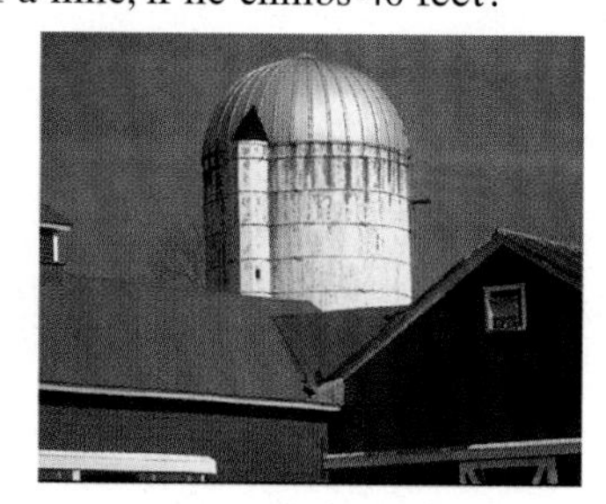

64. One of the Seven Wonders of the World, the Great Pyramid on the Giza Plateau in Egypt, is marveled at by mathematicians, builders, and historians alike because of its precise measurements. The Pyramid has a square base and it rises to a height of 450 feet. It has a volume of 85,503,750 cubic feet. The length of each side of the base of any pyramid with a square base (see figure below) can be found by the formula $s = \sqrt{(3V)/h}$, where V represents the volume, in cubic feet, of the pyramid and h represents the height, in feet, of the pyramid. Find the length of each side, s, of the square base of the Great Pyramid.

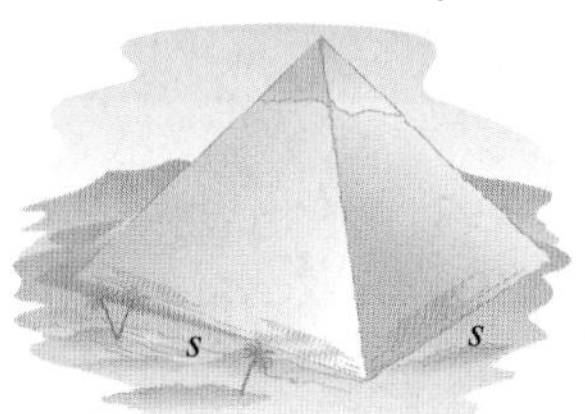

[9.7] *Evaluate.*

65. $\sqrt[3]{64}$ **66.** $\sqrt[3]{-64}$ **67.** $\sqrt[4]{16}$ **68.** $\sqrt[4]{81}$

Simplify.

69. $\sqrt[3]{27}$ **70.** $\sqrt[3]{-8}$ **71.** $\sqrt[4]{32}$ **72.** $\sqrt[3]{48}$

73. $\sqrt[3]{54}$ **74.** $\sqrt[4]{96}$ **75.** $\sqrt[3]{x^{21}}$ **76.** $\sqrt[3]{x^{24}}$

Evaluate.

77. $27^{2/3}$ **78.** $25^{1/2}$ **79.** $27^{-2/3}$

80. $64^{2/3}$ **81.** $8^{-4/3}$ **82.** $49^{3/2}$

Write in exponential form.

83. $\sqrt[3]{x^7}$ **84.** $\sqrt[3]{x^8}$ **85.** $\sqrt[4]{y^9}$

86. $\sqrt{x^5}$ **87.** $\sqrt{y^3}$ **88.** $\sqrt[4]{y^7}$

Simplify.

89. $\sqrt[3]{x} \cdot \sqrt[3]{x^2}$ **90.** $\sqrt[3]{x} \cdot \sqrt[3]{x}$ **91.** $\sqrt[3]{x^2} \cdot \sqrt[3]{x^7}$ **92.** $\sqrt[4]{x^2} \cdot \sqrt[4]{x^6}$

93. $(\sqrt[3]{x^3})^4$ **94.** $(\sqrt[4]{x})^4$ **95.** $(\sqrt[4]{x^8})^3$ **96.** $(\sqrt[4]{x^3})^8$

Practice Test

1. Write $\sqrt{3x}$ exponential form.

2. Write $x^{2/3}$ in radical form.

Simplify.

3. $\sqrt{(x-5)^2}$
4. $\sqrt{80}$
5. $\sqrt{12x^2}$
6. $\sqrt{50x^7y^3}$
7. $\sqrt{8x^2y} \cdot \sqrt{6xy}$
8. $\sqrt{15xy^2} \cdot \sqrt{5x^3y^3}$
9. $\sqrt{\dfrac{5}{125}}$
10. $\dfrac{\sqrt{7x^2y}}{\sqrt{7y^3}}$
11. $\dfrac{1}{\sqrt{6}}$
12. $\sqrt{\dfrac{4x}{5}}$
13. $\sqrt{\dfrac{40x^2y^5}{6x^3y^7}}$
14. $\dfrac{3}{2-\sqrt{7}}$
15. $\dfrac{6}{\sqrt{x}-3}$
16. $\sqrt{48} + \sqrt{75} + 2\sqrt{3}$
17. $5\sqrt{y} + 7\sqrt{y} - \sqrt{y}$

Solve.

18. $\sqrt{x-8} = 4$

19. $2\sqrt{x-4} + 4 = x$

Solve.

20. Find the value of x in the triangle shown.

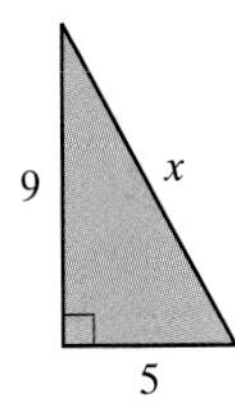

21. Find the length of the line segment between the points $(3, -2)$ and $(-4, -5)$.
22. Evaluate $27^{-4/3}$.
23. Simplify $\sqrt[4]{x^5} \cdot \sqrt[4]{x^7}$.
24. Find the side of a square whose area is 121 square meters.
25. Mary Ellen Baker and her brother, Michael, were in the tree house in their backyard in Rogers, Arkansas. To pass the time, they were dropping raw eggs (much to their mother's dismay) from a ledge in the tree house 10 feet in the air. At what velocity did the eggs hit the ground? Use the formula $v = \sqrt{2gh}$ with $g = 32$ feet per second squared. The velocity will be in feet per second.

Cumulative Review Test

1. Consider the set of numbers

$$\left\{-9, \sqrt{13}, 735, 0.5, 4, \frac{1}{2}\right\}.$$

List those that are **a)** integers; **b)** whole numbers; **c)** rational numbers; **d)** irrational numbers; **e)** real numbers.

2. Evaluate $6x - 5y^2 + xy$, when $x = -5$ and $y = 2$.
3. Solve $-7(3 - x) = 4(x + 2) - 3x$.
4. Solve the inequality $3(x + 2) > 5 - 4(2x - 7)$ and graph the solution on a number line.
5. Find the equation of the line that has a slope of $\frac{2}{5}$ and goes through the point $(-3, 1)$.

6. Write the equation of the graph in the accompanying figure.

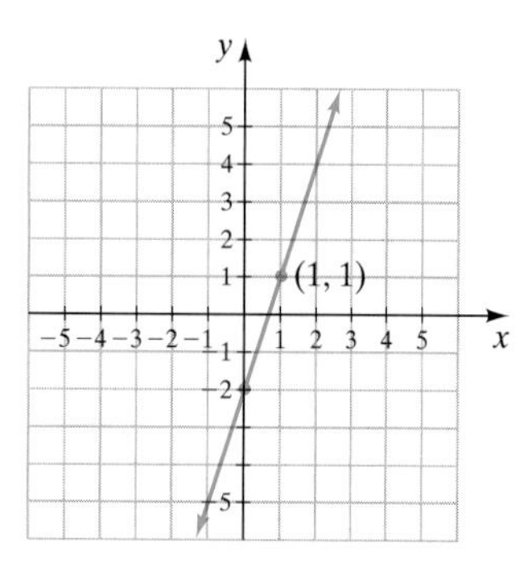

7. Graph $4x - 6y = 24$.

8. Solve the following system of equations by substitution.

$$-2x + 3y = 6$$
$$x - 2y = -5$$

9. Simplify $\frac{4a^3b^{-5}}{28a^8b}$.

10. Factor $3x^3 + x^2 + 6x + 2$.

11. Factor $x^2 + 12x - 28$.

12. Use factoring to solve $x^2 - 3x = 0$.

13. Simplify $\frac{y+5}{8} + \frac{2y-7}{8}$.

14. Solve $\frac{5}{y+2} = \frac{3}{2y-7}$.

15. Simplify $3\sqrt{11} - 4\sqrt{11}$.

16. Simplify $\sqrt{\frac{3z}{28y}}$.

17. Solve $\sqrt{x+5} = 6$.

18. Susan Effing is planning her first holiday meal. She has her grandmother's recipes and has decided to make her grandmother's famous fruitcake. The original recipe calls for 10 cups of flour and will make an 11-pound fruitcake. She would like to make a 3-pound fruitcake. How many cups of flour should Susan use?

19. Between January 1, 1997, and June 30, 1998, campaign disclosure reports showed that approximately $484.3 million had been raised for the House of Representatives and Senate campaigns. This figure represents an 8% increase over the amount raised on campaigns for the House and Senate between January 1, 1995, and June 30, 1996. Approximately how much was raised during this time period?

20. Alan Heard is in the process of getting his pilot's license. As part of his training, he flew his instructor from a small airport near Pasadena, California, to another small airport near San Diego at an average speed of 100 miles per hour. On the return trip, his average speed was 125 miles per hour. How far apart were the two airports if Alan's total flying time was 2 hours?

QUADRATIC EQUATIONS

CHAPTER

10

Use the Angel Web site at www.prenhall.com/angel to be linked to an internet resource that will help you further explore the following application.

Landscaping is important not only because it increases the value of property, but also because of the aesthetic value it contributes to a home, business, or park. Lawns and walkways give people the opportunity to enjoy their surroundings. On page 590 we solve a quadratic equation to determine, given a limited amount of grass seed, how wide of a lawn can be planted around a pottery studio.

Preview and Perspective

Much of this book has dealt with solving linear equations. Quadratic equations are another important category of equations. Quadratic equations were introduced and solved by factoring in Section 7.6. You may wish to review that section now. Recall that quadratic equations are of the form $ax^2 + bx + c = 0$, where $a \neq 0$. Not every quadratic equation can be solved by factoring. In this chapter we present two additional methods to solve quadratic equations: completing the square and the quadratic formula.

The square root property is presented in Section 10.1. In Section 10.2 we solve quadratic equations by completing the square, which uses the square root property. The quadratic formula, discussed in Section 10.3, is most typically used when solving quadratic equations that cannot be factored. We graphed linear equations in Chapter 4. In Section 10.4 we graph quadratic equations. Section 10.4 is a very valuable section because it lays the groundwork for graphing cubic and higher-degree equations in other mathematics courses.

10.1 THE SQUARE ROOT PROPERTY

1) **Know that every positive real number has two square roots.**
2) **Solve quadratic equations using the square root property.**

In Section 7.6 we solved quadratic equations by factoring. Recall that **quadratic equations** are equations of the form

$$ax^2 + bx + c = 0$$

where a, b, and c are real numbers, $a \neq 0$. A quadratic equation in this form is said to be in **standard form**. Solving quadratic equations by factoring is the preferred technique when the factors can be found quickly. To refresh your memory, below we will solve the equation $x^2 - 3x - 10 = 0$ by factoring. If you need further examples, review Section 7.6.

$$x^2 - 3x - 10 = 0$$
$$(x - 5)(x + 2) = 0$$
$$x - 5 = 0 \quad \text{or} \quad x + 2 = 0$$
$$x = 5 \qquad\qquad x = -2$$

The solutions to this equation are 5 and -2.

Not every quadratic equation can be factored easily and many cannot be factored at all. In this chapter we give two techniques, completing the square and the quadratic formula, for solving quadratic equations that cannot be solved by factoring.

1) Know That Every Positive Real Number Has Two Square Roots

In Section 9.1 we stated that every positive number has two square roots. Thus far we have been using only the positive or principal square root. In this section we use both the positive and negative square roots of a number. For example, the positive square root of 49 is 7.

$$\sqrt{49} = 7$$

The negative square root of 49 is −7.

$$-\sqrt{49} = -7$$

Notice that $7 \cdot 7 = 49$ and $(-7)(-7) = 49$. The two square roots of 49 are +7 and −7. A convenient way to indicate the two square roots of a number is to use the plus or minus symbol, ±. For example, the square roots of 49 can be indicated ±7, read "plus or minus 7."

Number	Both Square Roots
36	± 6
100	± 10
3	$\pm\sqrt{3}$

The value of a number like $-\sqrt{5}$ can be found by finding the value of $\sqrt{5}$ on your calculator and then taking its opposite or negative value.

$$\sqrt{5} = 2.24 \quad \text{(rounded to the nearest hundredth)}$$
$$-\sqrt{5} = -2.24$$

Now consider the equation

$$x^2 = 49$$

We can see by substitution that this equation has two solutions, 7 and −7.

CHECK:

$x = 7$		$x = -7$	
$x^2 = 49$		$x^2 = 49$	
$7^2 \stackrel{?}{=} 49$		$(-7)^2 \stackrel{?}{=} 49$	
$49 = 49$	*True*	$49 = 49$	*True*

Therefore, the solutions to the equation $x^2 = 49$ are 7 and −7 (or ±7).

2) Solve Quadratic Equations Using the Square Root Property

In general, for any quadratic equation of the form $x^2 = a$, we can use the **square root property** to obtain the solution.

Square Root Property

If $x^2 = a$, then $x = \sqrt{a}$ or $x = -\sqrt{a}$ (abbreviated $x = \pm\sqrt{a}$).

For example, if $x^2 = 7$, then by the square root property, $x = \sqrt{7}$ or $x = -\sqrt{7}$. We may also write $x = \pm\sqrt{7}$.

EXAMPLE 1 Solve the equation $x^2 - 25 = 0$.

Solution Add 25 to both sides of the equation to get the variable by itself on one side of the equation.

$$x^2 = 25$$

Now use the square root property.

$$x = \pm\sqrt{25}$$
$$x = \pm 5$$

Check in the original equation.

CHECK:

$x = 5$		$x = -5$	
$x^2 - 25 = 0$		$x^2 - 25 = 0$	
$5^2 - 25 \stackrel{?}{=} 0$		$(-5)^2 - 25 \stackrel{?}{=} 0$	
$25 - 25 \stackrel{?}{=} 0$		$25 - 25 \stackrel{?}{=} 0$	
$0 = 0$	True	$0 = 0$	True

NOW TRY EXERCISE 7

EXAMPLE 2 Solve the equation $x^2 + 5 = 86$.

Solution Begin by subtracting 5 from both sides of the equation.

$$x^2 + 5 = 86$$
$$x^2 = 81 \qquad \text{Isolate } x^2.$$
$$x = \pm\sqrt{81} \qquad \text{Square root property}$$
$$x = \pm 9$$

EXAMPLE 3 Solve the equation $x^2 - 6 = 0$.

Solution

$$x^2 - 6 = 0$$
$$x^2 = 6 \qquad \text{Isolate } x^2.$$
$$x = \pm\sqrt{6} \qquad \text{Square root property}$$

NOW TRY EXERCISE 15

EXAMPLE 4 Solve the equation $(x - 3)^2 = 4$.

Solution Begin by using the square root property.

$$(x - 3)^2 = 4$$
$$x - 3 = \pm\sqrt{4} \qquad \text{Square root property}$$
$$x - 3 = \pm 2$$
$$x - 3 + 3 = 3 \pm 2 \qquad \text{Add 3 to both sides}$$
$$x = 3 \pm 2$$
$$x = 3 + 2 \quad \text{or} \quad x = 3 - 2$$
$$x = 5 \qquad\qquad x = 1$$

The solutions are 1 and 5.

EXAMPLE 5 Solve the equation $(5x + 1)^2 = 18$.

Solution

$$(5x + 1)^2 = 18$$
$$5x + 1 = \pm\sqrt{18} \qquad \text{Square root property}$$
$$5x + 1 = \pm\sqrt{9}\sqrt{2} \qquad \text{Simplify } \sqrt{18}.$$
$$5x + 1 = \pm 3\sqrt{2}$$
$$5x + 1 - 1 = -1 \pm 3\sqrt{2} \qquad \text{Subtract 1 from both sides.}$$
$$5x = -1 \pm 3\sqrt{2}$$
$$x = \frac{-1 \pm 3\sqrt{2}}{5} \qquad \text{Divide both sides by 5.}$$

The solutions are $\frac{-1 + 3\sqrt{2}}{5}$ and $\frac{-1 - 3\sqrt{2}}{5}$.

NOW TRY EXERCISE 43

EXAMPLE 6 When creating advertisements, the shape of the ad as well its coloring and print style must be considered. When Antoinette LeMans approached an advertising firm to design a magazine ad for her company, she learned that one of the most appealing shapes is a rectangle whose length is about 1.62 times its width. Any rectangle with these proportions is called a *golden rectangle*. She decided that the ad would have the dimensions of a golden rectangle. Find the dimensions of the ad if it is to have an area of 20 square inches, see Fig. 10.1.

Solution **Understand and Translate:**

Let x = width of rectangle

$1.62x$ = length of rectangle

FIGURE 10.1

$$\text{area} = \text{length} \cdot \text{width}$$

$$20 = (1.62x)x$$

Carry Out:

$$20 = 1.62x^2$$

$$\text{or} \quad 1.62x^2 = 20$$

$$x^2 = \frac{20}{1.62} \approx 12.3$$

$$x \approx \pm\sqrt{12.3} \approx \pm 3.51 \text{ inches}$$

Check and Answer: Since the width cannot be negative, the width, x, is approximately 3.51 inches. The length is about $1.62(3.51) = 5.69$ inches.

CHECK:

$$\text{area} = \text{length} \cdot \text{width}$$

$$20 \stackrel{?}{=} (5.69)(3.51)$$

$$20 \approx 19.97 \qquad \text{True}$$

(There is a slight round-off error due to rounding off decimal answers.)

Note that the answer is not a whole number. In many real-life situations this is the case. You should not feel uncomfortable when this occurs.

NOW TRY EXERCISE 53

Exercise Set 10.1

Concept/Writing Exercises

1. State the square root property.
2. In your own words, explain how to solve the quadratic equation $x^2 = 13$ using the square root property.
3. What is the relationship between the length and width of any golden rectangle?
4. If a golden rectangle has a width of 7 centimeters, what is its length?
5. How many different solutions would you expect for each of the following quadratic equations? Explain.
   a) $x^2 = 6$
   b) $x^2 = 0$
   c) $(x - 2)^2 = 1$
6. Write the standard form of a quadratic equation.

Practice the Skills

Solve.

7. $x^2 = 100$
8. $x^2 = 25$
9. $x^2 = 144$
10. $x^2 = 1$
11. $y^2 = 169$
12. $z^2 = 9$
13. $x^2 = 121$
14. $a^2 = 15$
15. $x^2 - 12 = 0$
16. $w^2 - 24 = 0$
17. $3x^2 = 12$
18. $7x^2 = 7$
19. $2w^2 = 34$
20. $5x^2 = 80$
21. $2x^2 + 1 = 19$
22. $3x^2 - 4 = 8$
23. $9w^2 + 5 = 20$
24. $3y^2 + 8 = 36$
25. $16x^2 - 17 = 56$
26. $2x^2 + 3 = 51$

Solve.

27. $(x - 3)^2 = 25$
28. $(y - 2)^2 = 9$
29. $(x + 3)^2 = 81$
30. $(x + 5)^2 = 25$
31. $(x + 4)^2 = 64$
32. $(x - 4)^2 = 100$
33. $(x + 7)^2 = 32$
34. $(x + 3)^2 = 18$
35. $(x + 6)^2 = 20$
36. $(x - 4)^2 = 50$
37. $(x + 2)^2 = 25$
38. $(x - 11)^2 = 28$
39. $(x - 9)^2 = 100$
40. $(x - 3)^2 = 15$
41. $(2x + 3)^2 = 18$
42. $(3x - 2)^2 = 30$
43. $(4x + 1)^2 = 20$
44. $(5x - 6)^2 = 100$
45. $(2x - 6)^2 = 18$
46. $(5x + 9)^2 = 40$

Problem Solving

47. Write an equation that has the solutions 6 and −6.

48. Write an equation that has the solutions $\sqrt{7}$ and $-\sqrt{7}$.

49. Fill in the shaded area to make a true statement. Explain how you determined your answer. The equation $x^2 - \square = 27$ has the solutions 6 and −6.

50. Fill in the shaded area to make a true statement. Explain how you determined your answer. The equation $x^2 + \square = 45$ has the solutions 8 and −8.

51. **a)** Rewrite $-3x^2 + 9x - 6 = 0$ so that the coefficient of the x^2-term is positive.

    b) Rewrite $-3x^2 + 9x - 6 = 0$ so that the coefficient of the x^2-term is 1.

52. **a)** Rewrite $-\frac{1}{2}x^2 + 3x - 4 = 0$ so that the coefficient of the x^2-term is positive.

    b) Rewrite $-\frac{1}{2}x^2 + 3x - 4 = 0$ so that the coefficient of the x^2-term is 1.

53. Georges Marten is the head groundskeeper at a resort in St. Thomas. After consulting with the resort manager, Georges decided to enlarge the main rectangular garden and alter its dimensions so that it has the length-to-width ratio found in a golden rectangle. Determine the dimensions of the garden if it is to have an area of 2000 square feet.

54. Danielle Delahoussaye has several pages of the rough draft of a manuscript for a novel she is trying to write. She would like to mail it to a friend for his opinion. Since the weight of the pages she is mailing is less than 2 pounds, Danielle decides to use a rectangular Priority Mail envelope available from the United States Post Office. The length of the envelope is about 1.33 times its width. Determine the dimensions of the envelope if its area is approximately 112 square inches.

55. Consider the two squares with sides x and $x + 3$ shown below.

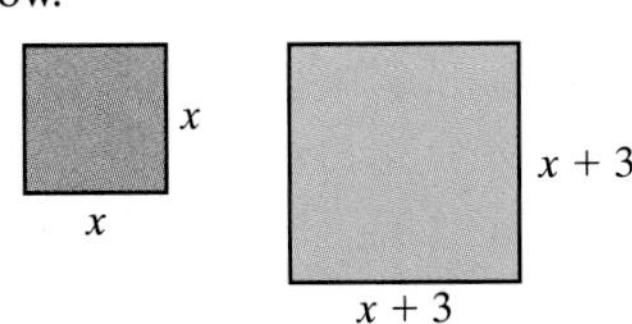

a) Write a quadratic expression for the area of each square.

b) If the area of the square on the left is 36 square inches, what is the length of each side of the square?

c) If the area of the square on the left is 50 square inches, what is the length of each side of the square?

d) If the area of the square on the right is 81 square inches, what is the length of each side of the square?

e) If the area of the square on the right is 92 square inches, what is the length of each side of the square?

Challenge Problems

Use the square root property to solve for the indicated variable. Assume that all variables represent positive numbers. You may wish to review Section 3.1 before working these problems. List only the positive square root.

56. $A = s^2$, for s

57. $I = p^2r$, for p

58. $A = \pi r^2$, for r

59. $a^2 + b^2 = c^2$, for b

60. $I = \dfrac{k}{d^2}$, for d

61. $A = p(1 + r)^2$, for r

Cumulative Review Exercises

[4.4] **62.** Determine the equation of the line illustrated.

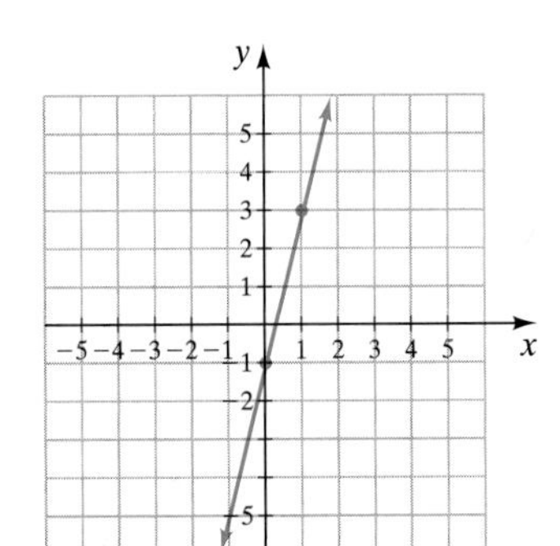

[7.4] **63.** Factor $4x^2 - 10x - 24$.

[8.5] **64.** Simplify $\dfrac{3 - \dfrac{1}{y}}{6 - \dfrac{1}{y}}$.

[9.3] **65.** Simplify $\dfrac{\sqrt{135a^4b}}{\sqrt{3a^5b^5}}$.

10.2 SOLVING QUADRATIC EQUATIONS BY COMPLETING THE SQUARE

SSM | VIDEO 10.2 | CD Rom

1. **Write perfect square trinomials.**
2. **Solve quadratic equations by completing the square.**

Quadratic equations that cannot be solved by factoring can be solved by completing the square or by the quadratic formula. In this section we focus on completing the square.

1 Write Perfect Square Trinomials

A **perfect square trinomial** is a trinomial that can be expressed as the square of a binomial. Some examples follow.

Perfect Square Trinomials	Factors	Square of a Binomial
$x^2 + 6x + 9$	$= (x + 3)(x + 3)$	$= (x + 3)^2$
$x^2 - 6x + 9$	$= (x - 3)(x - 3)$	$= (x - 3)^2$
$x^2 + 10x + 25$	$= (x + 5)(x + 5)$	$= (x + 5)^2$
$x^2 - 10x + 25$	$= (x - 5)(x - 5)$	$= (x - 5)^2$

Notice that each of the squared terms in the preceding perfect square trinomials has a numerical coefficient of 1. When the coefficient of the squared term is 1, there is an important relationship between the coefficient of the x-term and the constant. In every perfect square trinomial of this type, *the constant term is the square of one-half the coefficient of the x-term.*

Consider the perfect square trinomial $x^2 + 6x + 9$. The coefficient of the x-term is 6 and the constant is 9. Note that the constant, 9, is the square of one-half the coefficient of the x-term.

$$x^2 + 6x + 9$$

$$\left[\frac{1}{2}(6)\right]^2 = 3^2 = 9$$

Consider the perfect square trinomial $x^2 - 10x + 25$. The coefficient of the x-term is -10 and the constant is 25. Note that

$$x^2 - 10x + 25$$

$$\left[\frac{1}{2}(-10)\right]^2 = (-5)^2 = 25$$

Consider the expression $x^2 + 8x + \square$. Can you determine what number must be placed in the colored box to make the trinomial a perfect square trinomial? If you answered 16, you answered correctly.

$$x^2 + 8x + \square$$

$$\left[\frac{1}{2}(8)\right]^2 = 4^2 = 16$$

The perfect square trinomial is $x^2 + 8x + 16$. Note that $x^2 + 8x + 16 = (x + 4)^2$.

Let's examine perfect square trinomials a little further.

Perfect Square Trinomial		Square of a Binomial
$x^2 + 6x + 9$	$=$	$(x + 3)^2$
	$\frac{1}{2}(6) = 3$	
$x^2 - 10x + 25$	$=$	$(x - 5)^2$
	$\frac{1}{2}(-10) = -5$	

Note that when a perfect square trinomial is written as the square of a binomial *the constant in the binomial is one-half the value of the coefficient of the x-term in the perfect square trinomial.*

2) Solve Quadratic Equations by Completing the Square

The procedure for solving a quadratic equation by **completing the square** is illustrated in the following example. Some of the equations that you will be solving in the examples could be easily solved by factoring, but they are being solved by completing the square as a means of illustrating the procedure on easy problems before moving on to more difficult problems.

EXAMPLE 1 Solve the equation $x^2 + 6x - 7 = 0$ by completing the square.

Solution First we make sure that the squared term has a coefficient of 1. (In Example 5 we explain what to do if the coefficient is not 1.) Next we wish to get the terms containing a variable by themselves on the left side of the equation. Therefore, we add 7 to both sides of the equation.

$$x^2 + 6x - 7 = 0$$

$$x^2 + 6x = 7$$

Now we determine one-half the numerical coefficient of the x-term. In this example, the x-term is $6x$.

$$\frac{1}{2}(6) = 3$$

We square this number,

$$(3)^2 = (3)(3) = 9$$

and then add this product to both sides of the equation.

$$x^2 + 6x + 9 = 7 + 9$$

or

$$x^2 + 6x + 9 = 16$$

By following this procedure, we produce a perfect square trinomial on the left side of the equation. The expression $x^2 + 6x + 9$ is a perfect square trinomial that can be expressed as $(x + 3)^2$. Therefore,

$$x^2 + 6x + 9 = 16$$

can be written

$$(x + 3)^2 = 16$$

Now we use the square root property,

$$x + 3 = \pm\sqrt{16}$$

$$x + 3 = \pm 4$$

Finally, we solve for x by subtracting 3 from both sides of the equation.

$$x + 3 - 3 = -3 \pm 4$$

$$x = -3 \pm 4$$

$$x = -3 + 4 \quad \text{or} \quad x = -3 - 4$$

$$x = 1 \qquad\qquad x = -7$$

Thus, the solutions are 1 and −7. We check both solutions in the original equation.

CHECK:

$x = 1$		$x = -7$	
$x^2 + 6x - 7 = 0$		$x^2 + 6x - 7 = 0$	
$(1)^2 + 6(1) - 7 \stackrel{?}{=} 0$		$(-7)^2 + 6(-7) - 7 \stackrel{?}{=} 0$	
$1 + 6 - 7 \stackrel{?}{=} 0$		$49 - 42 - 7 \stackrel{?}{=} 0$	
$0 = 0$	*True*	$0 = 0$	*True*

Now let's summarize the procedure.

To Solve a Quadratic Equation by Completing the Square

1. Use the multiplication (or division) property of equality if necessary to make the numerical coefficient of the squared term equal to 1.
2. Rewrite the equation with the constant by itself on the right side of the equation.
3. Take one-half the numerical coefficient of the first-degree term, square it, and add this quantity to both sides of the equation.
4. Replace the trinomial with its equivalent squared binomial.
5. Use the square root property.
6. Solve for the variable.
7. Check your answers in the *original* equation.

EXAMPLE 2 Solve the equation $x^2 - 10x + 21 = 0$ by completing the square.

Solution

$$x^2 - 10x + 21 = 0$$

$$x^2 - 10x = -21 \qquad \textit{Step 2}$$

Take half the numerical coefficient of the x-term, square it, and add this product to both sides of the equation.

$$\frac{1}{2}(-10) = -5, \quad (-5)^2 = 25$$

Now add 25 to both sides of the equation.

$$x^2 - 10x + 25 = -21 + 25 \qquad \textit{Step 3}$$

$$x^2 - 10x + 25 = 4$$

$$\text{or} \qquad (x - 5)^2 = 4 \qquad \textit{Step 4}$$

$$x - 5 = \pm\sqrt{4} \qquad \textit{Step 5}$$

$$x - 5 = \pm 2$$

$$x = 5 \pm 2 \qquad \textit{Step 6}$$

$$x = 5 + 2 \quad \text{or} \quad x = 5 - 2$$

$$x = 7 \qquad\qquad x = 3$$

NOW TRY EXERCISE 13

A check will show that the solutions are 7 and 3.

EXAMPLE 3 Solve the equation $x^2 = 3x + 18$ by completing the square.

Solution Place all terms except the constant on the left side of the equation.

$$x^2 = 3x + 18$$
$$x^2 - 3x = 18 \qquad \text{Step 2}$$

Take half the numerical coefficient of the x-term, square it, and add this product to both sides of the equation.

$$\frac{1}{2}(-3) = -\frac{3}{2}, \qquad \left(-\frac{3}{2}\right)^2 = \frac{9}{4}$$

$$x^2 - 3x + \frac{9}{4} = 18 + \frac{9}{4} \qquad \text{Step 3}$$
$$\left(x - \frac{3}{2}\right)^2 = 18 + \frac{9}{4} \qquad \text{Step 4}$$
$$\left(x - \frac{3}{2}\right)^2 = \frac{72}{4} + \frac{9}{4}$$
$$\left(x - \frac{3}{2}\right)^2 = \frac{81}{4}$$
$$x - \frac{3}{2} = \pm\sqrt{\frac{81}{4}} \qquad \text{Step 5}$$
$$x - \frac{3}{2} = \pm\frac{9}{2}$$
$$x = \frac{3}{2} \pm \frac{9}{2} \qquad \text{Step 6}$$
$$x = \frac{3}{2} + \frac{9}{2} \quad \text{or} \quad x = \frac{3}{2} - \frac{9}{2}$$
$$x = \frac{12}{2} = 6 \qquad x = -\frac{6}{2} = -3$$

NOW TRY EXERCISE 21

The solutions are 6 and -3.

In the following examples we will not display some of the intermediate steps.

EXAMPLE 4 Solve the equation $x^2 - 14x - 1 = 0$ by completing the square.

Solution

$$x^2 - 14x - 1 = 0$$
$$x^2 - 14x = 1 \qquad \text{Step 2}$$
$$x^2 - 14x + 49 = 1 + 49 \qquad \text{Step 3}$$
$$(x - 7)^2 = 50 \qquad \text{Step 4}$$
$$x - 7 = \pm\sqrt{50} \qquad \text{Step 5}$$
$$x - 7 = \pm 5\sqrt{2}$$
$$x = 7 \pm 5\sqrt{2} \qquad \text{Step 6}$$

The solutions are $7 + 5\sqrt{2}$ and $7 - 5\sqrt{2}$.

EXAMPLE 5 Solve the equation $5z^2 - 25z + 10 = 0$ by completing the square.

Solution To solve an equation by completing the square, the numerical coefficient of the squared term must be 1. Since the coefficient of the squared term is 5, we multiply both sides of the equation by $\frac{1}{5}$ to make the coefficient equal to 1.

$$5z^2 - 25z + 10 = 0$$

$$\frac{1}{5}(5z^2 - 25z + 10) = \frac{1}{5}(0) \qquad \textit{Step 1}$$

$$z^2 - 5z + 2 = 0$$

Now we proceed as in earlier examples.

$$z^2 - 5z = -2 \qquad \textit{Step 2}$$

$$z^2 - 5z + \frac{25}{4} = -2 + \frac{25}{4} \qquad \textit{Step 3}$$

$$\left(z - \frac{5}{2}\right)^2 = -\frac{8}{4} + \frac{25}{4} \qquad \textit{Step 4}$$

$$\left(z - \frac{5}{2}\right)^2 = \frac{17}{4}$$

$$z - \frac{5}{2} = \pm\sqrt{\frac{17}{4}} \qquad \textit{Step 5}$$

$$z - \frac{5}{2} = \pm\frac{\sqrt{17}}{2}$$

$$z = \frac{5}{2} \pm \frac{\sqrt{17}}{2} \qquad \textit{Step 6}$$

$$z = \frac{5}{2} + \frac{\sqrt{17}}{2} \quad \text{or} \quad z = \frac{5}{2} - \frac{\sqrt{17}}{2}$$

$$z = \frac{5 + \sqrt{17}}{2} \qquad\qquad z = \frac{5 - \sqrt{17}}{2}$$

The solutions are $\frac{5 + \sqrt{17}}{2}$ and $\frac{5 - \sqrt{17}}{2}$.

NOW TRY EXERCISE 33

Exercise Set 10.2

Concept/Writing Exercises

1. **a)** What is a perfect square trinomial?
   b) Fill in the shaded area to make a perfect square trinomial and explain how you determined your answer.
   $x^2 + 8x$ ▒
2. Fill in the shaded area to make a perfect square trinomial and explain how you determined your answer.
   $x^2 - 10x$ ▒
3. In a perfect square trinomial, what is the relationship between the constant and the coefficient of the x-term?
4. In a perfect square trinomial, if the coefficient of the x-term is 4, what is the constant?
5. In a perfect square trinomial, if the coefficient of the x-term is -12, what is the constant?
6. **a)** In your own words, explain how to solve a quadratic equation by completing the square.
   b) Compare your answer to part **a)** with the procedure given on page 580. Did you omit any steps in your explanation?

Practice the Skills

Solve by completing the square.

7. $x^2 + 3x - 4 = 0$
8. $x^2 + 5x + 6 = 0$
9. $x^2 - 8x + 7 = 0$
10. $x^2 + 8x + 12 = 0$
11. $x^2 + 3x + 2 = 0$
12. $x^2 + 4x - 32 = 0$
13. $z^2 - 2z - 8 = 0$
14. $m^2 - 9m + 14 = 0$
15. $x^2 = -6x - 9$
16. $x^2 = -5x - 4$
17. $x^2 = 2x + 15$
18. $x^2 = -5x - 6$
19. $x^2 + 10x + 24 = 0$
20. $x^2 - 10x + 24 = 0$
21. $x^2 = 15x - 56$
22. $x^2 = 3x + 28$
23. $-4x = -x^2 + 12$
24. $-x^2 - 3x + 40 = 0$
25. $z^2 - 4z = -2$
26. $z^2 + 2z = 6$
27. $8x + 3 = -x^2$
28. $x^2 - x - 3 = 0$
29. $m^2 + 7m + 2 = 0$
30. $x^2 + 3x - 6 = 0$
31. $2x^2 + 4x - 6 = 0$
32. $2x^2 + 2x - 24 = 0$
33. $2x^2 + 18x + 4 = 0$
34. $2x^2 = 8x + 90$
35. $3x^2 - 15x - 18 = 0$
36. $4x^2 = -28x + 32$
37. $3x^2 - 11x - 4 = 0$
38. $3x^2 - 8x + 4 = 0$
39. $3x^2 + 6x = 6$
40. $2x^2 - x = 5$
41. $x^2 - 5x = 0$
42. $2x^2 - 6x = 0$
43. $2x^2 = 12x$
44. $3x^2 = 9x$

Problem Solving

45. **a)** Write a perfect square trinomial that has a term of $20x$.
b) Explain how you constructed your perfect square trinomial.

46. **a)** Write a perfect square trinomial that has a term of $-14x$.
b) Explain how you constructed your perfect square trinomial.

47. When 3 times a number is added to the square of a number, the sum is 4. Find the number(s).

48. When 5 times a number is subtracted from 2 times the square of a number, the difference is 12. Find the number(s).

49. If the square of 3 more than a number is 9, find the number(s).

50. If the square of 2 less than an integer is 16, find the number(s).

51. The product of two positive numbers is 21. Find the two numbers if the larger is 4 greater than the smaller.

Challenge Problems

52. Fill in the shaded area to make a perfect square trinomial and explain how you determined your answer.

$$x^2 \quad \boxed{} \quad + 196$$

53. Fill in the shaded area to make a perfect square trinomial and explain how you determined your answer.

$$x^2 \quad \boxed{} \quad + \frac{9}{100}$$

54. **a)** Solve the equation $x^2 + 3x - 7 = 0$ by completing the square.
b) Check your solution (it will not be a rational number) by substituting the value(s) you obtained in part **a)** for each x in the equation in part **a)**.

55. **a)** Solve the equation $x^2 - 14x - 1 = 0$ by completing the square.
b) Check your solution (it will not be a rational number) by substituting the value(s) you obtained in part **a)** for each x in the equation in part **a)**.

Solve by completing the square.

56. $x^2 + \frac{3}{5}x - \frac{1}{2} = 0.$
57. $x^2 - \frac{2}{3}x - \frac{1}{5} = 0.$
58. $3x^2 + \frac{1}{2}x = 4.$
59. $0.1x^2 + 0.2x - 0.54 = 0$
60. $-5.26x^2 + 7.89x + 15.78 = 0$

Cumulative Review Exercises

[4.4] **61.** Explain how you can determine whether two equations represent parallel lines without graphing the equations.

[5.2, 5.3] **62.** Solve the following system of equations.

$$3x - 4y = 6$$
$$2x + y = 8$$

[8.4] **63.** Simplify $\frac{x^2}{x^2 - x - 6} - \frac{x - 2}{x - 3}.$

[9.5] **64.** Solve the equation $\sqrt{2x + 3} = 2x - 3.$

10.3 SOLVING QUADRATIC EQUATIONS BY THE QUADRATIC FORMULA

1. Solve quadratic equations by the quadratic formula.
2. Determine the number of solutions to a quadratic equation using the discriminant.

1 Solve Quadratic Equations by the Quadratic Formula

Another method that can be used to solve any quadratic equation is the **quadratic formula**. It is the most useful and versatile method of solving quadratic equations.

The standard form of a quadratic equation is $ax^2 + bx + c = 0$, where a is the coefficient of the squared term, b is the coefficient of the first-degree term, and c is the constant.

Quadratic Equation in Standard Form	Values of a, b, and c		
$x^2 - 3x + 4 = 0$	$a = 1,$	$b = -3,$	$c = 4$
$-2x^2 + \frac{1}{2}x - 2 = 0$	$a = -2,$	$b = \frac{1}{2},$	$c = -2$
$3x^2 - 4 = 0$	$a = 3,$	$b = 0,$	$c = -4$
$5x^2 + 3x = 0$	$a = 5,$	$b = 3,$	$c = 0$
$-\frac{1}{2}x^2 + 5 = 0$	$a = -\frac{1}{2},$	$b = 0$	$c = 5$
$-12x^2 + 8x = 0$	$a = -12,$	$b = 8,$	$c = 0$

To Solve a Quadratic Equation by the Quadratic Formula

1. Write the equation in standard form, $ax^2 + bx + c = 0$, and determine the numerical values for a, b, and c.
2. Substitute the values for a, b, and c from step 1 in the quadratic formula below and then evaluate to obtain the solution.

THE QUADRATIC FORMULA

$$x = \frac{-b \pm \sqrt{b^2 - 4ac}}{2a}$$

The quadratic formula can be derived by starting with the equation $ax^2 + bx + c = 0$ and solving the equation for x by completing the square.

EXAMPLE 1 Use the quadratic formula to solve the equation $x^2 + 4x + 3 = 0$.

Solution In this equation $a = 1$, $b = 4$, and $c = 3$. Substitute these values into the quadratic formula and then evaluate.

$$x = \frac{-b \pm \sqrt{b^2 - 4ac}}{2a}$$

$$= \frac{-(4) \pm \sqrt{(4)^2 - 4(1)(3)}}{2(1)}$$

$$= \frac{-4 \pm \sqrt{16 - 12}}{2}$$

$$= \frac{-4 \pm \sqrt{4}}{2}$$

$$= \frac{-4 \pm 2}{2}$$

$$x = \frac{-4 + 2}{2} \quad \text{or} \quad x = \frac{-4 - 2}{2}$$

$$= \frac{-2}{2} = -1 \qquad = \frac{-6}{2} = -3$$

CHECK: $x = -1$

$$x^2 + 4x + 3 = 0$$
$$(-1)^2 + 4(-1) + 3 \stackrel{?}{=} 0$$
$$1 - 4 + 3 \stackrel{?}{=} 0$$
$$0 = 0 \quad \text{True}$$

$x = -3$

$$x^2 + 4x + 3 = 0$$
$$(-3)^2 + 4(-3) + 3 \stackrel{?}{=} 0$$
$$9 - 12 + 3 \stackrel{?}{=} 0$$
$$0 = 0 \quad \text{True}$$

NOW TRY EXERCISE 27

Avoiding Common Errors

The **entire numerator** of the quadratic formula must be divided by $2a$.

CORRECT

$$x = \frac{-b \pm \sqrt{b^2 - 4ac}}{2a}$$

INCORRECT

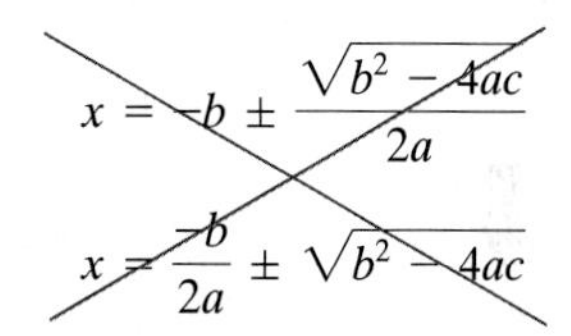

$x = -b \pm \frac{\sqrt{b^2 - 4ac}}{2a}$
$x = \frac{-b}{2a} \pm \sqrt{b^2 - 4ac}$

EXAMPLE 2 Use the quadratic formula to solve the equation $8x^2 + 2x - 1 = 0$.

Solution

$$8x^2 + 2x - 1 = 0$$

$$a = 8, \quad b = 2, \quad c = -1$$

$$x = \frac{-b \pm \sqrt{b^2 - 4ac}}{2a}$$

$$= \frac{-(2) \pm \sqrt{(2)^2 - 4(8)(-1)}}{2(8)}$$

$$= \frac{-2 \pm \sqrt{4 + 32}}{16}$$

$$= \frac{-2 \pm \sqrt{36}}{16}$$

$$= \frac{-2 \pm 6}{16}$$

$$x = \frac{-2 + 6}{16} \quad \text{or} \quad x = \frac{-2 - 6}{16}$$

$$= \frac{4}{16} = \frac{1}{4} \qquad = \frac{-8}{16} = -\frac{1}{2}$$

CHECK: $x = \frac{1}{4}$ $\qquad$ $x = -\frac{1}{2}$

$$8x^2 + 2x - 1 = 0 \qquad 8x^2 + 2x - 1 = 0$$

$$8\left(\frac{1}{4}\right)^2 + 2\left(\frac{1}{4}\right) - 1 \stackrel{?}{=} 0 \qquad 8\left(-\frac{1}{2}\right)^2 + 2\left(-\frac{1}{2}\right) - 1 \stackrel{?}{=} 0$$

$$8\left(\frac{1}{16}\right) + \frac{1}{2} - 1 \stackrel{?}{=} 0 \qquad 8\left(\frac{1}{4}\right) - 1 - 1 \stackrel{?}{=} 0$$

$$\frac{1}{2} + \frac{1}{2} - 1 \stackrel{?}{=} 0 \qquad 2 - 1 - 1 \stackrel{?}{=} 0$$

$$0 = 0 \quad \text{True} \qquad 0 = 0 \quad \text{True}$$

EXAMPLE 3 Use the quadratic formula to solve the equation $2x^2 + 4x - 5 = 0$.

Solution

$$a = 2, \qquad b = 4, \qquad c = -5$$

$$x = \frac{-b \pm \sqrt{b^2 - 4ac}}{2a}$$

$$= \frac{-4 \pm \sqrt{(4)^2 - 4(2)(-5)}}{2(2)}$$

$$= \frac{-4 \pm \sqrt{16 + 40}}{4}$$

$$= \frac{-4 \pm \sqrt{56}}{4}$$

$$= \frac{-4 \pm 2\sqrt{14}}{4}$$

Now factor out 2 from both terms in the numerator; then divide out common factors as explained in Section 9.4.

$$x = \frac{\overset{1}{\cancel{2}}(-2 \pm \sqrt{14})}{\underset{2}{\cancel{4}}}$$

$$x = \frac{-2 \pm \sqrt{14}}{2}$$

Thus, the solutions are

$$x = \frac{-2 + \sqrt{14}}{2} \quad \text{and} \quad x = \frac{-2 - \sqrt{14}}{2}$$

Now let's try two examples where the equation is not in standard form.

EXAMPLE 4 Use the quadratic formula to solve the equation $x^2 = 6x - 4$.

Solution First write the equation in standard form.

$$x^2 - 6x + 4 = 0$$

Set one side of the equation equal to zero.

$$a = 1, \quad b = -6, \quad c = 4$$

$$x = \frac{-b \pm \sqrt{b^2 - 4ac}}{2a}$$

$$= \frac{-(-6) \pm \sqrt{(-6)^2 - 4(1)(4)}}{2(1)}$$ *Substitute.*

$$= \frac{6 \pm \sqrt{36 - 16}}{2}$$ *Simplify.*

$$= \frac{6 \pm \sqrt{20}}{2}$$

$$= \frac{6 \pm \sqrt{4} \cdot \sqrt{5}}{2}$$ *Product rule*

$$= \frac{6 \pm 2\sqrt{5}}{2}$$

$$= \frac{\overset{1}{\cancel{2}}(3 \pm \sqrt{5})}{\underset{1}{\cancel{2}}}$$ *Factor out 2.*

$$= 3 \pm \sqrt{5}$$

NOW TRY EXERCISE 39 $x = 3 + \sqrt{5}$ or $x = 3 - \sqrt{5}$

Avoiding Common Errors

Many students solve quadratic equations correctly until the last step, where they make an error. Do not make the mistake of trying to simplify an answer that cannot be simplified any further. The following are answers that cannot be simplified, along with some common errors.

ANSWERS THAT CANNOT BE SIMPLIFIED	INCORRECT
$\frac{3 + 2\sqrt{5}}{2}$	$\frac{3 + 2\sqrt{5}}{2} = \frac{3 + \overset{1}{\cancel{2}}\sqrt{5}}{\underset{1}{\cancel{2}}} = 3 + \sqrt{5}$
$\frac{4 + 3\sqrt{5}}{2}$	$\frac{\overset{2}{\cancel{4}} + 3\sqrt{5}}{\underset{1}{\cancel{2}}} = 2 + 3\sqrt{5}$

Other examples of common errors of this type are illustrated in the Avoiding Common Errors box on page 540.

EXAMPLE 5 Use the quadratic formula to solve the equation $x^2 = 9$.

Solution First write the equation in standard form.

$$x^2 - 9 = 0$$ *Set one side of the equation equal to 0.*

$$a = 1, \quad b = 0, \quad c = -9$$

$$x = \frac{-b \pm \sqrt{b^2 - 4ac}}{2a}$$

$$= \frac{-0 \pm \sqrt{0^2 - 4(1)(-9)}}{2(1)}$$ *Substitute.*

$$= \frac{\pm\sqrt{36}}{2} = \frac{\pm 6}{2} = \pm 3$$ *Simplify, and solve for x.*

Thus, the solutions are 3 and -3.

The next example illustrates a quadratic equation that has no real number solution.

EXAMPLE 6 Solve the quadratic equation $3x^2 = x - 1$.

Solution

$$3x^2 - x + 1 = 0$$

$$a = 3, \quad b = -1, \quad c = 1$$

$$x = \frac{-b \pm \sqrt{b^2 - 4ac}}{2a}$$

$$= \frac{-(-1) \pm \sqrt{(-1)^2 - 4(3)(1)}}{2(3)}$$

$$= \frac{1 \pm \sqrt{1 - 12}}{6}$$

$$= \frac{1 \pm \sqrt{-11}}{6}$$

Since $\sqrt{-11}$ is not a real number, we can go no further. This equation has no real number solution. *When given a problem of this type, your answer should be "no real number solution." Do not leave the answer blank, and do not write 0 for the answer.*

NOW TRY EXERCISE 45

2) Determine the Number of Solutions to a Quadratic Equation Using the Discriminant

The expression under the square root sign in the quadratic formula is called the **discriminant**.

$$\underbrace{b^2 - 4ac}_{\text{Discriminant}}$$

The discriminant can be used to determine the number of solutions to a quadratic equation.

When the Discriminant Is:

1. **Greater than zero,** $b^2 - 4ac > 0$, the quadratic equation has **two distinct real number solutions.**
2. **Equal to zero,** $b^2 - 4ac = 0$, the quadratic equation has **one real number solution.**
3. **Less than zero,** $b^2 - 4ac < 0$, the quadratic equation has **no real number solution.**

$b^2 - 4ac$	Number of Solutions
Positive	Two distinct real number solutions
0	One real number solution
Negative	No real number solution

EXAMPLE 7 **a)** Find the discriminant of the equation $x^2 - 12x + 36 = 0$.

b) Use the discriminant to determine the number of solutions to the equation.

c) Use the quadratic formula to find the solutions, if any exist.

Solution **a)** $a = 1, \quad b = -12, \quad c = 36$

$$b^2 - 4ac = (-12)^2 - 4(1)(36) = 144 - 144 = 0$$

b) Since the discriminant is equal to zero, there is one real number solution.

c)
$$\begin{aligned} x &= \frac{-b \pm \sqrt{b^2 - 4ac}}{2a} \\ &= \frac{-(-12) \pm \sqrt{0}}{2(1)} \\ &= \frac{12 \pm 0}{2} = \frac{12}{2} = 6 \end{aligned}$$

NOW TRY EXERCISE 9

The only solution is 6.

EXAMPLE 8 Without actually finding the solutions, determine whether the following equations have two distinct real number solutions, one real number solution, or no real number solution.

a) $4x^2 - 4x + 1 = 0$ **b)** $2x^2 + 13x = -15$ **c)** $7x^2 = 3x - 1$

Solution We use the discriminant of the quadratic formula to answer these equations.

a) $b^2 - 4ac = (-4)^2 - 4(4)(1) = 16 - 16 = 0$

Since the discriminant is equal to zero, this equation has one real number solution.

b) First, rewrite $2x^2 + 13x = -15$ as $2x^2 + 13x + 15 = 0$.

$$b^2 - 4ac = (13)^2 - 4(2)(15) = 169 - 120 = 49$$

Since the discriminant is positive, this equation has two distinct real number solutions.

c) First rewrite $7x^2 = 3x - 1$ as $7x^2 - 3x + 1 = 0$.

$$b^2 - 4ac = (-3)^2 - 4(7)(1) = 9 - 28 = -19$$

Since the discriminant is negative this equation has no real number solution.

Now let's look at one of several applications that may be solved using the quadratic formula.

EXAMPLE 9 Laquitia Johnson is planning to plant a grass lawn of uniform width around her rectangular pottery studio, which measures 48 feet by 36 feet. How far will the lawn extend from the studio if Laquitia has only enough seed to plant 4000 square feet of grass?

Solution Let's make a diagram of the pottery studio and lawn area (Fig. 10.2) Let x = the uniform width of the grass area. Then the total length of the larger rectangular area is $2x + 48$. The total width of the larger rectangular area is $2x + 36$.

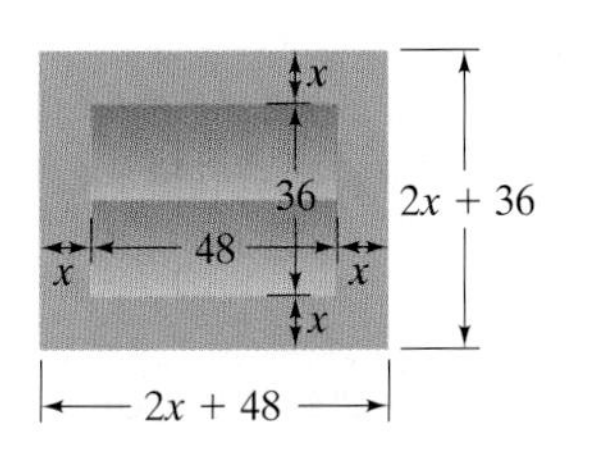

FIGURE 10.2

The lawn area can be found by subtracting the area of the studio from the larger rectangular area.

$$\text{area of studio} = l \cdot w = (48)(36) = 1728 \text{ square feet}$$

$$\begin{aligned}\text{area of large rectangle} &= l \cdot w = (2x + 48)(2x + 36)\\ &= 4x^2 + 168x + 1728 \text{ (studio plus lawn area)}\end{aligned}$$

$$\begin{aligned}\text{lawn area} &= \text{area of large rectangle} - \text{area of studio}\\ &= (4x^2 + 168x + 1728) - (1728)\\ &= 4x^2 + 168x\end{aligned}$$

The total lawn area must be 4000 square feet.

$$\begin{aligned}\text{lawn area} &= 4x^2 + 168x\\ 4000 &= 4x^2 + 168x\\ \text{or}\qquad 4x^2 + 168x - 4000 &= 0\\ 4(x^2 + 42x - 1000) &= 0\\ \frac{1}{\cancel{4}} \cdot \cancel{4}(x^2 + 42x - 1000) &= \frac{1}{4} \cdot 0\\ x^2 + 42x - 1000 &= 0\end{aligned}$$

By the quadratic formula,

$$a = 1, \quad b = 42, \quad c = -1000$$

$$\begin{aligned}x &= \frac{-b \pm \sqrt{b^2 - 4ac}}{2a}\\ &= \frac{-42 \pm \sqrt{(42)^2 - 4(1)(-1000)}}{2(1)}\\ &= \frac{-42 \pm \sqrt{1764 + 4000}}{2}\\ &= \frac{-42 \pm \sqrt{5764}}{2}\\ &\approx \frac{-42 \pm 75.92}{2}\end{aligned}$$

$$x \approx \frac{-42 + 75.92}{2} \quad \text{or} \quad x \approx \frac{-42 - 75.92}{2}$$

$$\approx \frac{33.92}{2} \qquad \approx \frac{-117.92}{2}$$

$$\approx 16.96 \qquad \approx -58.96$$

Since lengths are positive, the only possible answer is $x \approx 16.96$. Thus, there will be a grass lawn about 16.96 feet wide all around the pottery studio.

NOW TRY EXERCISE 57

HELPFUL HINT

Notice in Example 9 that when we had the quadratic equation $4x^2 + 168x - 4000 = 0$ we factored out the common factor 4 to get

$$4x^2 + 168x - 4000 = 0$$
$$4(x^2 + 42x - 1000) = 0$$

We then multiplied both sides of the equation by $\frac{1}{4}$ and used the quadratic equation $x^2 + 42x - 1000 = 0$, where $a = 1$, $b = 42$, and $c = -1000$, in the quadratic formula.

If all the terms in a quadratic equation have a common factor, it will be easier to factor it out first so that you will have smaller numbers when you use the quadratic formula. Consider the quadratic equation $4x^2 + 8x - 12 = 0$.

In this equation $a = 4$, $b = 8$, and $c = -12$. If you solve this equation with the quadratic formula, after simplification you will get the solutions -3 and 1. Try this and see. If you factor out 4 to get

$$4x^2 + 8x - 12 = 0$$
$$4(x^2 + 2x - 3) = 0$$

and then use the quadratic formula with the equation $x^2 + 2x - 3 = 0$, where $a = 1$, $b = 2$, and $c = -3$, you get the same solution. Try this and see.

Exercise Set 10.3

Concept/Writing Exercises

1. **a)** What is the discriminant? **b)** Explain how the discriminant can be used to determine the number of real number solutions a quadratic equation has.
2. How many real number solutions does a quadratic equation have if the discriminant equals

   a) -4? **b)** 0? **c)** $\frac{1}{2}$?
3. Without looking at your notes, write down the quadratic formula. You must memorize this formula.
4. Explain in your own words why a quadratic equation will have two real number solutions when the discriminant is greater than 0, one real number solution when the discriminant is equal to 0, and no real number solution when the discriminant is less than 0. Use the quadratic formula in explaining your answer.
5. What is the first step in solving a quadratic equation using the quadratic formula?
6. A student using the quadratic formula to solve $2x^2 + 5x + 1 = 0$ wrote down the following steps. What is wrong with the student's work?

$$x = \frac{-5 \pm \sqrt{5^2 - 4(2)(1)}}{2(2)}$$

$$= -5 \pm \frac{\sqrt{25 - 8}}{4}$$

$$= -5 \pm \frac{\sqrt{17}}{4}$$

7. A student using the quadratic formula to solve $x^2 = 4x + 7$ wrote down the following steps. What is wrong with the student's work?

$$x = \frac{-4 \pm \sqrt{4^2 - 4(1)(7)}}{2(1)}$$
$$= \frac{-4 \pm \sqrt{16 - 28}}{2}$$
$$= \frac{-4 \pm \sqrt{-12}}{2}$$

8. A student using the quadratic formula to solve $x^2 + 4 = 0$ listed the following steps. What is wrong with the student's work?

$$x = \frac{0 \pm \sqrt{0^2 - 4(1)(4)}}{2(1)}$$
$$= \frac{\pm\sqrt{-16}}{2}$$
$$= \frac{\pm 4}{2}$$
$$= \pm 2$$

Practice the Skills

Determine whether each equation has two distinct real number solutions, one real number solution, or no real number solution.

9. $x^2 + 4x - 6 = 0$
10. $x^2 + 2x - 5 = 0$
11. $2x^2 + x + 1 = 0$
12. $x^2 + x + 18 = 0$
13. $5x^2 + 3x - 7 = 0$
14. $4x^2 - 24x = -36$
15. $2x^2 = 16x - 32$
16. $5x^2 - 4x = 7$
17. $x^2 - 8x + 5 = 0$
18. $x^2 - 5x - 9 = 0$
19. $4x = 8 + x^2$
20. $5x - 8 = 3x^2$
21. $x^2 + 7x - 3 = 0$
22. $2x^2 - 6x + 9 = 0$
23. $4x^2 - 9 = 0$
24. $6x^2 - 5x = 0$

Use the quadratic formula to solve each equation. If the equation has no real number solution, so state.

25. $x^2 - x - 6 = 0$
26. $x^2 + 11x + 10 = 0$
27. $x^2 + 9x + 18 = 0$
28. $x^2 - 3x - 10 = 0$
29. $x^2 - 6x = -5$
30. $x^2 + 5x - 24 = 0$
31. $x^2 = 10x - 24$
32. $x^2 - 49 = 0$
33. $x^2 = 100$
34. $x^2 - 6x = 0$
35. $x^2 - 3x = 0$
36. $z^2 - 17z + 72 = 0$
37. $n^2 - 7n + 10 = 0$
38. $2x^2 - 3x + 2 = 0$
39. $2y^2 - 7y + 4 = 0$
40. $15x^2 - 7x = 2$
41. $6x^2 = -x + 1$
42. $4r^2 + r - 3 = 0$
43. $2x^2 = 5x + 7$
44. $3w^2 - 4w + 5 = 0$
45. $2s^2 - 4s + 3 = 0$
46. $x^2 - 7x + 3 = 0$
47. $4x^2 = x + 5$
48. $x^2 - 3x - 1 = 0$
49. $2x^2 - 7x = 9$
50. $-x^2 + 2x + 15 = 0$
51. $-2x^2 + 11x - 15 = 0$
52. $6y^2 + 9 = -5y$

Problem Solving

53. The product of two consecutive positive integers is 42. Find the two consecutive integers.

54. The length of a rectangle is 3 feet longer than its width. Find the dimensions of the rectangle if its area is 28 square feet.

55. The length of a rectangle is 3 feet smaller than twice its width. Find the length and width of the rectangle if its area is 20 square feet.

56. Sean McDonald picked out a brick pattern at the Weckesser Brick Company to be used for a uniform width brick walkway around his rectangular garden. His garden measures 20 feet by 30 feet. A representative from the brick company called and told him that the pattern he chose was being discontinued, but that they did have enough brick in stock to cover 336 square feet. If Sean uses the brick he originally selected, what will be the width of the walkway?

57. Harold Goldstein and his wife Elaine recently installed a built-in rectangular swimming pool measuring 30 feet by 40 feet. They want to add a decorative tile border of uniform width around all sides of the pool. How wide can they make the tile border if they purchased enough tile to cover 296 square feet?

Challenge Problems

*Find all the values of c that will result in each equation having **a)** two real number solutions, **b)** one real number solution, and **c)** no real number solution.*

58. $x^2 + 6x + c = 0$

59. $2x^2 + 3x + c = 0$

60. $-3x^2 + 6x + c = 0$

61. Farmer Justina Wells wishes to form a rectangular region along a river bank by constructing fencing on three sides, as illustrated in the diagram. If she has only 400 feet of fencing and wishes to enclose an area of 15,000 square feet, find the dimensions of the rectangular region.

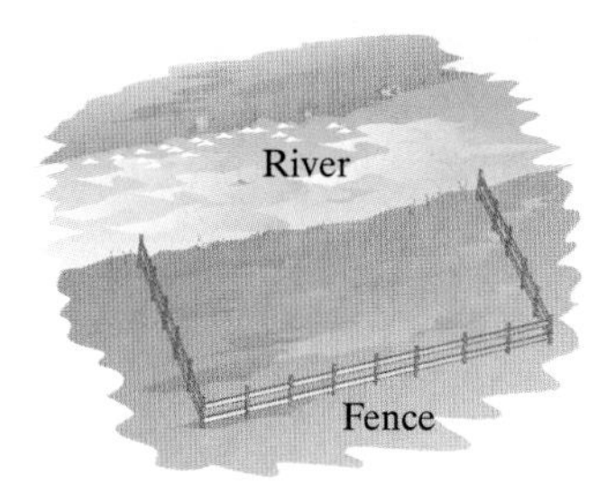

Group Activity

62. In Section 10.4 we will graph quadratic equations. We will learn that the graphs of quadratic equations are *parabolas*. The graph of the quadratic equation $y = x^2 - 2x - 8$ is illustrated below.

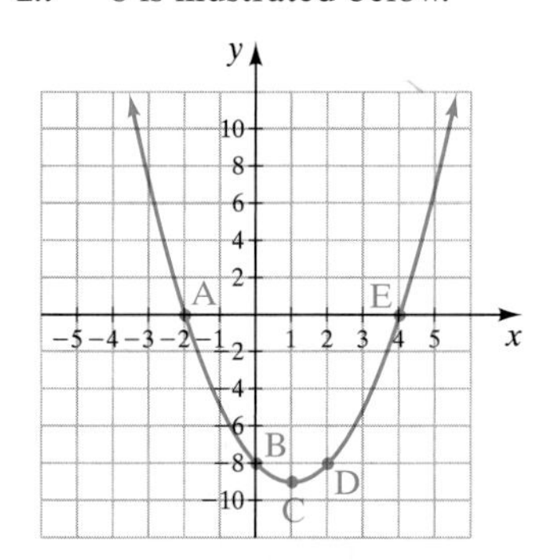

a) Each member of the group, copy the graph in your notebook.

b) Group Member 1: List the ordered pairs corresponding to points A and B. Verify that each ordered pair is a solution to the equation $y = x^2 - 2x - 8$.

c) Group Member 2: List the ordered pairs corresponding to points C and D. Verify that each ordered pair is a solution to the equation $y = x^2 - 2x - 8$.

d) Group Member 3: List the ordered pair corresponding to point E. Verify that the ordered pair is a solution to the equation $y = x^2 - 2x - 8$.

e) Individually, graph the equation $y = 2x - 3$ on the same axes that you used in part **a)**. Compare your graphs with the other members of your group.

f) The two graphs represent the system of equations

$$y = x^2 - 2x - 8$$
$$y = 2x - 3$$

As a group, estimate the points of intersection of the graphs.

g) If we set the two equations equal to each other, we obtain the following quadratic equation in only the variable x.

$$x^2 - 2x - 8 = 2x - 3$$

As a group, solve this quadratic equation. Does your answer agree with the x-coordinates of the points of intersection from part **f)**?

h) As a group, use the values of x found in part **g)** to find the values of y in $y = x^2 - 2x - 8$ and $y = 2x - 3$. Does your answer agree with the y-coordinates of the points of intersection from part **f)**?

Cumulative Review Exercises

[7.6, 10.2, 10.3] *Solve the following quadratic equations by **a)** factoring, **b)** completing the square, and **c)** the quadratic formula. If the equation cannot be solved by factoring, so state.*

63. $x^2 - 13x + 42 = 0$

64. $6x^2 + 11x - 35 = 0$

65. $2x^2 + 3x - 4 = 0$

66. $6x^2 = 54$

10.4 GRAPHING QUADRATIC EQUATIONS

1) Graph quadratic equations in two variables.
2) Find the coordinates of the vertex of a parabola.
3) Use symmetry to graph quadratic equations.
4) Find the x-intercepts of the graph of a quadratic equation.

1) Graph Quadratic Equations in Two Variables

In Section 4.2 we learned how to graph linear equations. In this section we graph quadratic equations of the form

$$y = ax^2 + bx + c, \quad a \neq 0$$

The graph of every quadratic equation of this form will be a **parabola**. The graph of $y = ax^2 + bx + c$ will have one of the shapes indicated in Figure 10.3.

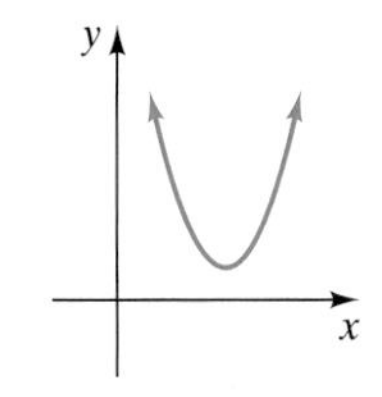

Parabola opens upward
(a)

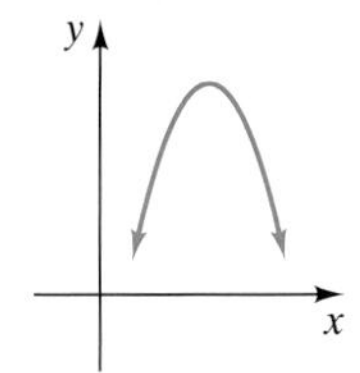

Parabola opens dwonward
(b)

FIGURE 10.3

When a quadratic equation is in the form $y = ax^2 + bx + c$, the sign of a, the numerical coefficient of the squared term, will determine whether the parabola will open upward (Fig. 10.3a) or downward (Fig. 10.3b). When a is positive, the parabola will open upward, and when a is negative, the parabola will open downward. The **vertex** is the lowest point on a parabola that opens upward or the highest point on a parabola that opens downward (Fig. 10.4).

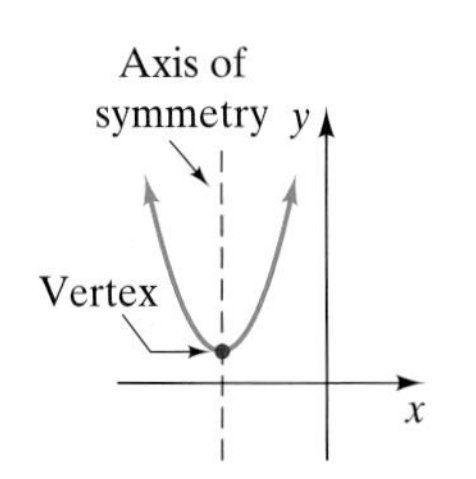

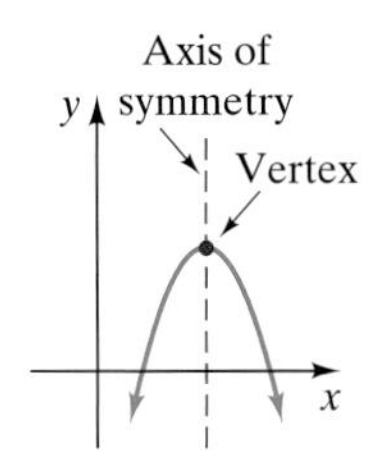

FIGURE 10.4

Graphs of quadratic equations of the form $y = ax^2 + bx + c$ have **symmetry** about a line through the vertex. This means that if we fold the paper along this imaginary line, called the **axis of symmetry**, the right and left sides of the graph will coincide.

One method you can use to graph a quadratic equation is to plot it point by point. When determining points to plot, select values for x and determine the corresponding values for y.

EXAMPLE 1 Graph the equation $y = x^2$.

Solution Since $a = 1$, which is positive, this parabola opens upward.

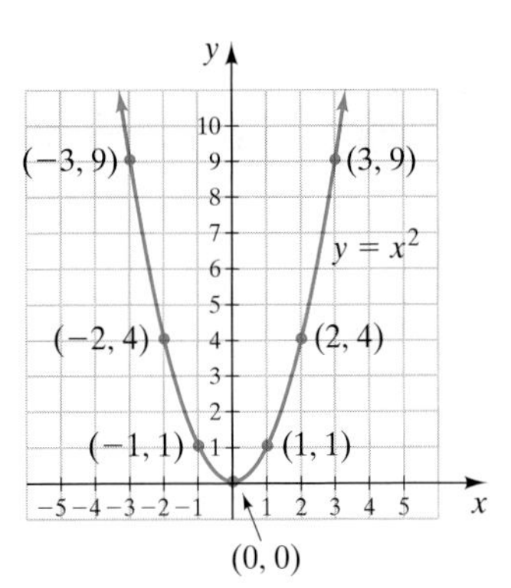

FIGURE 10.5

	$y = x^2$
Let $x = 3$,	$y = (3)^2 = 9$
Let $x = 2$,	$y = (2)^2 = 4$
Let $x = 1$,	$y = (1)^2 = 1$
Let $x = 0$,	$y = (0)^2 = 0$
Let $x = -1$,	$y = (-1)^2 = 1$
Let $x = -2$,	$y = (-2)^2 = 4$
Let $x = -3$,	$y = (-3)^2 = 9$

x	y
3	9
2	4
1	1
0	0
−1	1
−2	4
−3	9

Connect the points with a smooth curve (Fig. 10.5). Note how the graph is symmetric about the line $x = 0$ (or the y-axis).

EXAMPLE 2 Graph the equation $y = -2x^2 + 4x + 6$.

Solution Since $a = -2$, which is negative, this parabola opens downward.

	$y = -2x^2 + 4x + 6$
Let $x = 5$,	$y = -2(5)^2 + 4(5) + 6 = -24$
Let $x = 4$,	$y = -2(4)^2 + 4(4) + 6 = -10$
Let $x = 3$,	$y = -2(3)^2 + 4(3) + 6 = 0$
Let $x = 2$,	$y = -2(2)^2 + 4(2) + 6 = 6$
Let $x = 1$,	$y = -2(1)^2 + 4(1) + 6 = 8$
Let $x = 0$,	$y = -2(0)^2 + 4(0) + 6 = 6$
Let $x = -1$,	$y = -2(-1)^2 + 4(-1) + 6 = 0$
Let $x = -2$,	$y = -2(-2)^2 + 4(-2) + 6 = -10$
Let $x = -3$,	$y = -2(-3)^2 + 4(-3) + 6 = -24$

x	y
5	−24
4	−10
3	0
2	6
1	8
0	6
−1	0
−2	−10
−3	−24

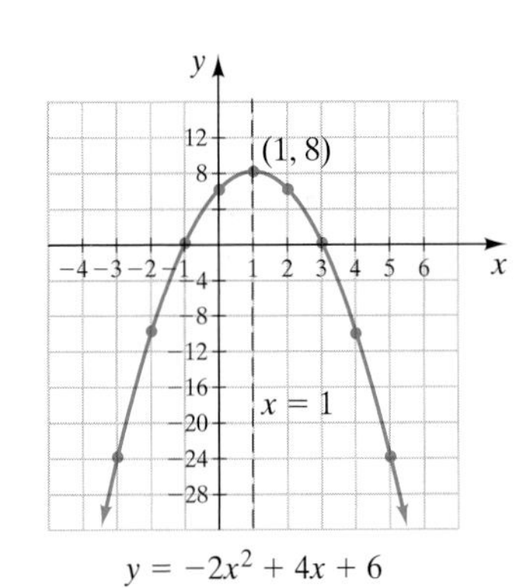

FIGURE 10.6

Note how the graph (Fig. 10.6) is symmetric about the line $x = 1$, which is dashed because it is not part of the graph. The vertex of this parabola is the point $(1, 8)$. Since the y-values are large, the y-axis has been marked with 4-unit intervals to allow us to graph the points $(-3, -24)$ and $(5, -24)$. The arrows on the ends of the graph indicate that the parabola continues indefinitely.

NOW TRY EXERCISE 23

2) Find the Coordinates of the Vertex of a Parabola

When graphing quadratic equations, how do we decide what values to use for x? When the location of the vertex is unknown, this is a difficult question to answer. When the location of the vertex is known, it becomes more obvious which values to use.

In Example 2, the axis of symmetry is $x = 1$, and the x-coordinate of the vertex is also 1. For a quadratic equation in the form $y = ax^2 + bx + c$, both the axis of symmetry and the x-coordinate of the vertex can be found by using the following formula.

Axis of Symmetry and x-Coordinate of the Vertex

$$x = -\frac{b}{2a}$$

In the quadratic equation in Example 2, $a = -2$, $b = 4$, and $c = 6$. Substituting these values in the formula for the axis of symmetry gives

$$x = -\frac{b}{2a} = -\frac{4}{2(-2)} = -\frac{4}{-4} = 1$$

Thus, the graph is symmetric about the line $x = 1$, and the x-coordinate of the vertex is 1.

The y-coordinate of the vertex can be found by substituting the value of the x-coordinate of the vertex into the quadratic equation and solving for y.

$$\begin{aligned} y &= -2x^2 + 4x + 6 \\ &= -2(1)^2 + 4(1) + 6 \\ &= -2(1) + 4 + 6 \\ &= -2 + 4 + 6 \\ &= 8 \end{aligned}$$

Thus, the vertex is at the point (1, 8).

For a quadratic equation of the form $y = ax^2 + bx + c$, the y-coordinate of the vertex can also be found by the following formula.

y-Coordinate of the Vertex

$$y = \frac{4ac - b^2}{4a}$$

For Example 2,

$$\begin{aligned} y &= \frac{4ac - b^2}{4a} \\ &= \frac{4(-2)(6) - 4^2}{4(-2)} \\ &= \frac{-48 - 16}{-8} = \frac{-64}{-8} = 8 \end{aligned}$$

You may use the method of your choice to find the y-coordinate of the vertex. Both methods result in the same value of y.

3) Use Symmetry to Graph Quadratic Equations

One method to use in selecting points to plot when graphing parabolas is to determine the axis of symmetry and the vertex of the graph. Then select nearby values of x on either side of the axis of symmetry. When graphing the equation, make use of the symmetry of the graph.

EXAMPLE 3 **a)** Find the axis of symmetry of the graph of the equation $y = x^2 + 6x + 5$.

b) Find the vertex of the graph.

c) Graph the equation.

Solution **a)** $a = 1, \quad b = 6, \quad c = 5.$

$$x = -\frac{b}{2a} = -\frac{6}{2(1)} = -3$$

The parabola is symmetric about the line $x = -3$. The x-coordinate of the vertex is -3.

b) Now find the y-coordinate of the vertex. Substitute -3 for x in the quadratic equation.

$$y = x^2 + 6x + 5$$
$$y = (-3)^2 + 6(-3) + 5 = 9 - 18 + 5 = -4$$

The vertex is at the point $(-3, -4)$.

c) Since the axis of symmetry is $x = -3$, we will select values for x that are greater than or equal to -3. It is often helpful to plot each point as it is determined. If a point does not appear to lie on the parabola, check it.

$$y = x^2 + 6x + 5$$

Let $x = -2$, $\quad y = (-2)^2 + 6(-2) + 5 = -3$

Let $x = -1$, $\quad y = (-1)^2 + 6(-1) + 5 = 0$

Let $x = 0$, $\quad y = (0)^2 + 6(0) + 5 = 5$

x	y
-2	-3
-1	0
0	5

These points are plotted in Figure 10.7a. The entire graph of the equation is illustrated in Figure 10.7b. Note how we use symmetry to complete the graph. The points $(-2, -3)$ and $(-4, -3)$ are each 1 horizontal unit from the axis of symmetry, $x = -3$. The points $(-1, 0)$ and $(-5, 0)$ are each 2 horizontal units from the axis of symmetry, and the points $(0, 5)$ and $(-6, 5)$ are each 3 horizontal units from the axis of symmetry.

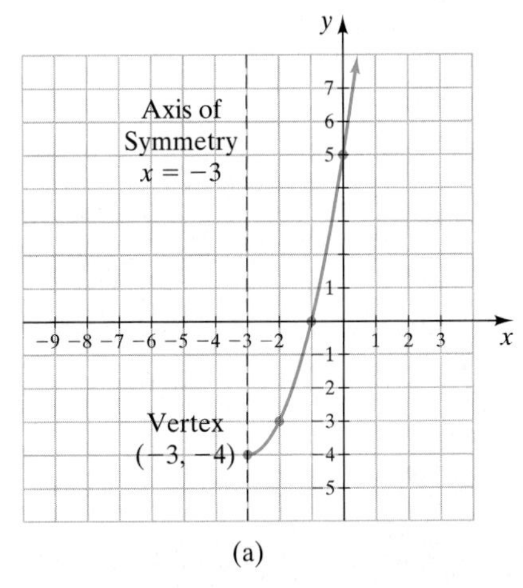

(a)

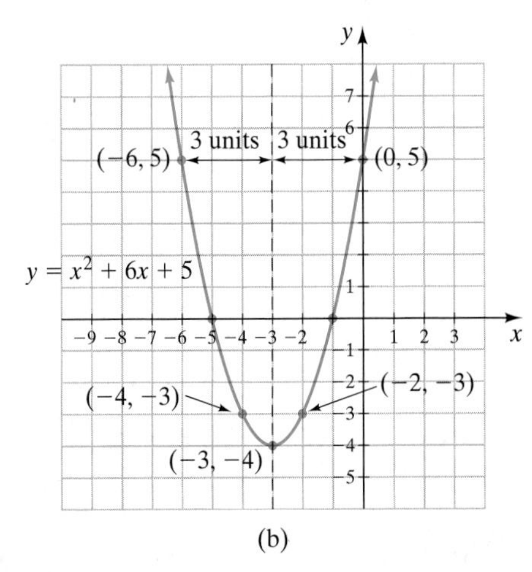

(b)

FIGURE 10.7

EXAMPLE 4 Graph the equation $y = -2x^2 + 3x - 4$.

Solution $a = -2, \quad b = 3, \quad c = -4.$

Since $a < 0$, this parabola will open downward.

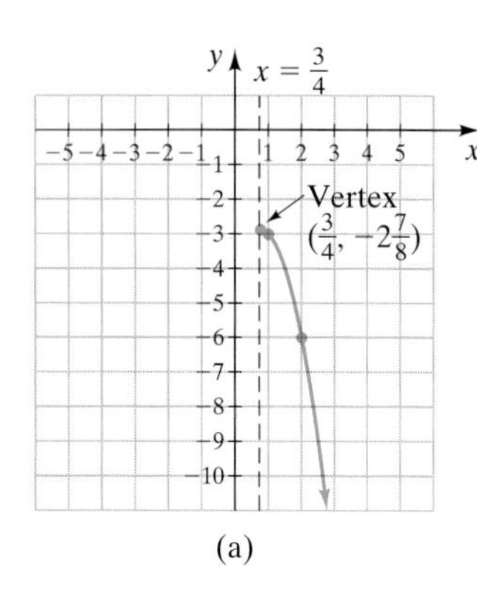

(a)

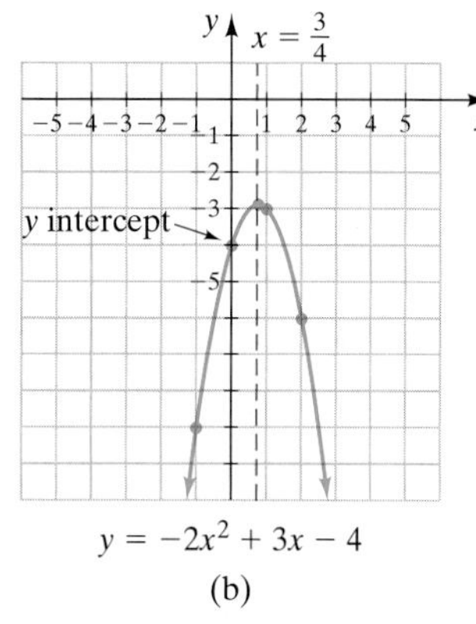

(b)

FIGURE 10.8

NOW TRY EXERCISE 19

$$\text{Axis of symmetry: } x = -\frac{b}{2a}$$

$$= -\frac{3}{2(-2)} = \frac{3}{-4} = \frac{3}{4}$$

Since the x-value of the vertex is a fraction, we will use the formula to find the y-coordinate of the vertex.

$$y = \frac{4ac - b^2}{4a}$$

$$= \frac{4(-2)(-4) - 3^2}{4(-2)} = \frac{32 - 9}{-8} = \frac{23}{-8} = -\frac{23}{8} \left(\text{or } -2\frac{7}{8}\right)$$

The vertex of this graph is at the point $\left(\frac{3}{4}, -\frac{23}{8}\right)$. Since the axis of symmetry is $x = \frac{3}{4}$, we will begin by selecting values of x that are greater than $\frac{3}{4}$.

$$y = -2x^2 + 3x - 4$$

Let $x = 1$, $\quad y = -2(1)^2 + 3(1) - 4 = -3$

Let $x = 2$, $\quad y = -2(2)^2 + 3(2) - 4 = -6$

Let $x = 3$, $\quad y = -2(3)^2 + 3(3) - 4 = -13$

x	y
1	−3
2	−6
3	−13

When the axis of symmetry is a fractional value, be very careful when constructing the graph. You should plot as many additional points as needed. In this example, when $x = 0$, $y = -4$, and when $x = -1$, $y = -9$. Figure 10.8a shows the points plotted on the right side of the axis of symmetry. Figure 10.8b shows the completed graph.

When graphing, we will often have to evaluate quadratic expressions to obtain values for y. You may wish to review *Evaluating Expressions* as explained in the Using Your Calculator box in Section 1.9, page 75.

(a)

(b)

FIGURE 10.9 Not every shape that resembles a parabola is a parabola. For example, the St. Louis Arch (a) resembles a parabola, but it is not a parabola. However, the bridge over the Mississippi near Jefferson Barracks Missouri, which connects Missouri and Illinois (b), is a parabola.

4) Find the x-Intercepts of the Graph of a Quadratic Equation

In Example 4 the graph crossed the y-axis at $y = -4$. Recall from earlier sections that to find the y-intercept we let $x = 0$ and solve for y. The location of the y-intercept is often helpful when graphing quadratic equations.

To find the x-intercept when graphing straight lines in Section 4.2, we set $y = 0$ and found the corresponding value of x. We did this because the value of y where a graph crosses the x-axis is 0. We use the same procedure here when finding the x-intercepts of a quadratic equation of the form $y = ax^2 + bx + c$. To find the x-intercepts of the graph of $y = ax^2 + bx + c$, we set y equal to 0 and solve the resulting equation, $ax^2 + bx + c = 0$. The real number solutions of the equation $ax^2 + bx + c = 0$ will be the x-coordinates of the x-intercepts of the graph of $y = ax^2 + bx + c$. To solve equations of the form $ax^2 + bx + c = 0$, we can use factoring as explained in Section 7.6; completing the square as explained in Section 10.2; or the quadratic formula as explained in Section 10.3.

In Section 10.3 we mentioned that when solving quadratic equations of the form $ax^2 + bx + c = 0$, when the discriminant, $b^2 - 4ac$ is greater than zero, there are two distinct real number solutions; when it is equal to zero, there is one real number solution; and when it is less than zero, there is no real number solution.

A quadratic equation of the form $y = ax^2 + bx + c$ will have either two distinct x-intercepts (Fig. 10.10a), one x-intercept (Fig. 10.10b), or no x-intercepts (Fig. 10.10c). The number of x-intercepts may be determined by the discriminant.

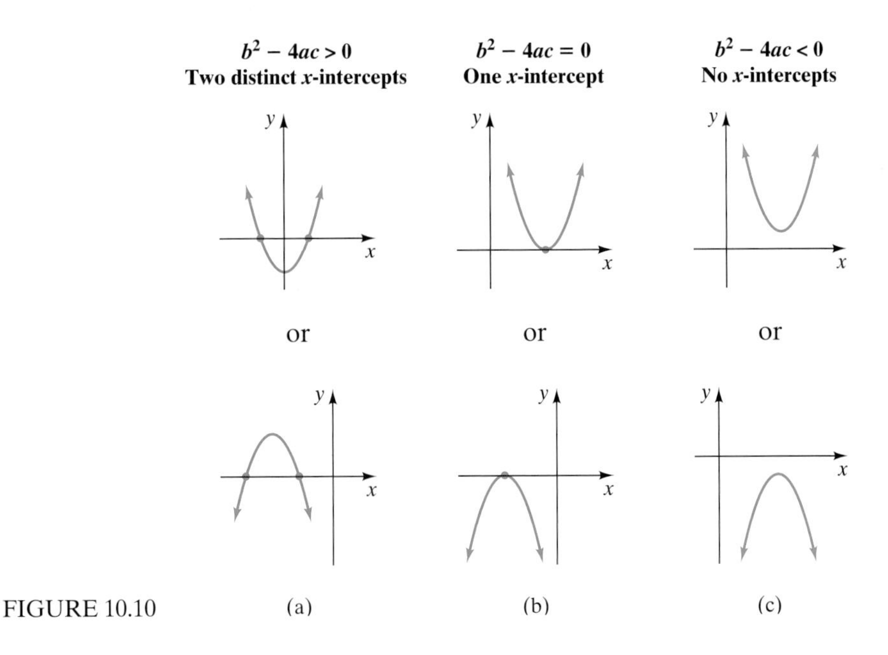

FIGURE 10.10

The x-intercepts can be found graphically. They may also be found algebraically by setting y equal to 0 and solving the resulting equation, as in Example 5.

EXAMPLE 5 **a)** Find the x-intercepts of the graph of the equation $y = x^2 - 6x - 7$ by factoring, by completing the square, and by the quadratic formula.

b) Graph the equation.

Solution **a)** To find the x-intercepts algebraically we set y equal to 0, and solve the resulting equation, $x^2 - 6x - 7 = 0$. We will solve this equation by all three algebraic methods.

METHOD 1: Factoring.

$$x^2 - 6x - 7 = 0$$
$$(x - 7)(x + 1) = 0$$
$$x - 7 = 0 \quad \text{or} \quad x + 1 = 0$$
$$x = 7 \qquad\qquad x = -1$$

METHOD 2: Completing the square.

$$x^2 - 6x - 7 = 0$$
$$x^2 - 6x = 7$$
$$x^2 - 6x + 9 = 7 + 9$$
$$(x - 3)^2 = 16$$
$$x - 3 = \pm 4$$
$$x = 3 \pm 4$$
$$x = 3 + 4 \quad \text{or} \quad x = 3 - 4$$
$$x = 7 \qquad\qquad x = -1$$

METHOD 3: Quadratic formula.

$$x^2 - 6x - 7 = 0$$
$$a = 1, \quad b = -6, \quad c = -7$$
$$x = \frac{-b \pm \sqrt{b^2 - 4ac}}{2a}$$
$$= \frac{-(-6) \pm \sqrt{(-6)^2 - 4(1)(-7)}}{2(1)}$$
$$= \frac{6 \pm \sqrt{36 + 28}}{2}$$
$$= \frac{6 \pm \sqrt{64}}{2}$$
$$= \frac{6 \pm 8}{2}$$
$$x = \frac{6 + 8}{2} \quad \text{or} \quad x = \frac{6 - 8}{2}$$
$$x = \frac{14}{2} = 7 \qquad x = \frac{-2}{2} = -1$$

Note that the same solutions, 7 and -1, were obtained by all three methods. The graph of the equation $y = x^2 - 6x - 7$ will have two distinct x-intercepts. The graph will cross the x-axis at 7 and -1. The x-intercepts are $(7, 0)$ and $(-1, 0)$.

b) Since $a > 0$, this parabola opens upward.

$$\text{axis of symmetry:} \quad x = -\frac{b}{2a} = -\frac{-6}{2(1)} = \frac{6}{2} = 3$$

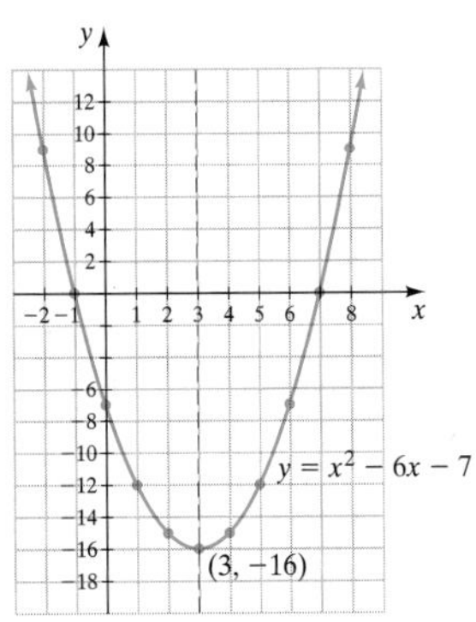

FIGURE 10.11

$$y = x^2 - 6x - 7$$

Let $x = 3$, $y = 3^2 - 6(3) - 7 = -16$

Let $x = 4$, $y = 4^2 - 6(4) - 7 = -15$

Let $x = 5$, $y = 5^2 - 6(5) - 7 = -12$

Let $x = 6$, $y = 6^2 - 6(6) - 7 = -7$

Let $x = 7$, $y = 7^2 - 6(7) - 7 = 0$

Let $x = 8$, $y = 8^2 - 6(8) - 7 = 9$

x	y
3	−16
4	−15
5	−12
6	−7
7	0
8	9

NOW TRY EXERCISE 31

The vertex is at $(3, -16)$. Again we use symmetry to complete the graph (Fig. 10.11). The x-intercepts $(7, 0)$ and $(-1, 0)$ agree with the answer obtained in part **a)**.

Using Your Graphing Calculator

In the Using Your Graphing Calculator box on page 442, we explained how to find the zeros (or roots) of a quadratic equation. Please reread that material now and notice how the graphing calculator can be used to find the x-intercepts of a quadratic equation in two variables. The roots (or zeros) are the x-coordinates of the x-intercepts.

Now let's discuss how to find the vertex of a parabola. In Example 3 we graphed $y = x^2 + 6x + 5$. This graph is shown on the window of a TI-83 in Figure 10.12 using the standard window settings.

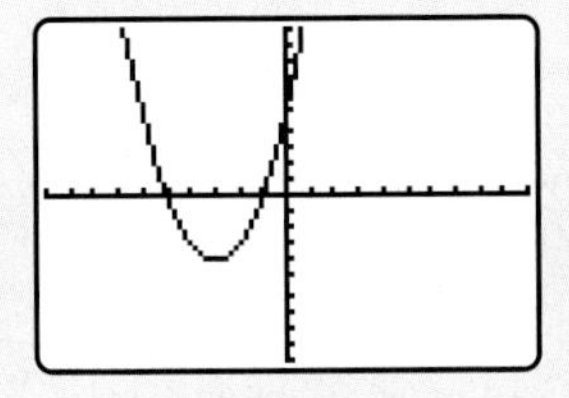

FIGURE 10.12

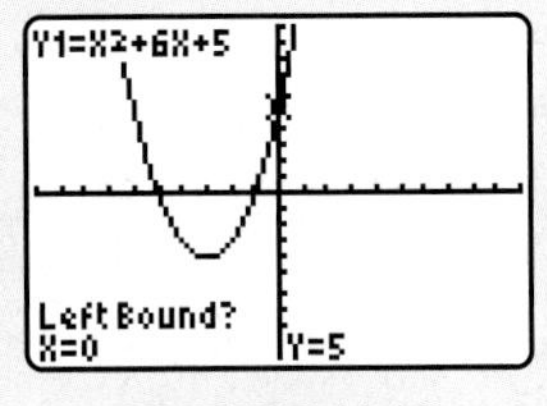

FIGURE 10.13

We can use the CALC (calculate) menu to find the vertex. The vertex of a parabola is also called the minimum point (if the parabola opens upward) or the maximum point (if the parabola opens downward). To obtain the CALC menu on the TI-82 or TI-83, press [2nd] [TRACE]. The CALC menu is shown in Figure 7.7 on page 442.

Since this graph opens upward, it has a minimum point. Scroll down in the CALC menu to item 3, minimum, and press [ENTER], which gives you the graph shown in Figure 10.13. Notice the cursor is at the y-intercept, $(0, 5)$. Use the left-arrow key to move the cursor to the left. Notice the values of X and Y changing as you move the cursor. The value of Y will decrease until the cursor reaches the vertex. When the cursor moves past the vertex, the value of Y will begin to increase. Move the cursor just to the left of the vertex. Press [ENTER]. Notice the words *Left Bound?* change to *Right Bound?*. Now move the cursor just to the right of the vertex and press [ENTER]. The screen now shows *Guess?*. Press [ENTER] again and you get the screen shown in Figure 10.14.

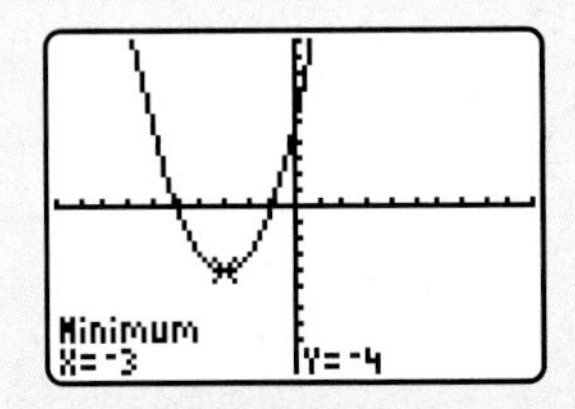

FIGURE 10.14

Figure 10.14 shows that the minimum point of the graph occurs when $x = -3$ and $y = -4$. Thus, the vertex is at $(-3, -4)$.

If the parabola opened downward, the graph would have a maximum point. To obtain the vertex in this case we would use the CALC menu, option 4, maximum, to obtain the vertex. You then would follow the same basic procedure we used here to find the maximum point.

Exercises

*Use your graphing calculator to find the **a)** x-intercepts, **b)** y-intercept, and **c)** vertex of the graph of each equation. You may need to adjust your window settings to show the vertex or y-intercept.*

1. $y = x^2 - 2x - 3$ **2.** $y = x^2 - 7x + 10$

3. $y = -2x^2 - 2x + 12$ **4.** $y = 2x^2 - 7x - 15$

Exercise Set 10.4

Concept/Writing Exercises

1. What do we call the graph of a quadratic equation of the form $y = ax^2 + bx + c, a \neq 0$?
2. What is the vertex of a parabola?
3. Use your own words to explain how to find the coordinates of the vertex of a parabola.
4. What determines whether the graph of a quadratic equation of the form $y = ax^2 + bx + c, a \neq 0$, is a parabola that opens upward or downward? Explain your answer.
5. **a)** What are the x-intercepts of a graph? **b)** How can you find the x-intercepts of a graph algebraically?
6. What does it mean when we say that graphs of quadratic equations of the form $y = ax^2 + bx + c$ have symmetry about an imaginary vertical line through the vertex?
7. **a)** When graphing a quadratic equation of the form $y = ax^2 + bx + c$, what is the equation of the vertical line about which the parabola will be symmetric?

   b) What is this vertical line called?
8. How many x-intercepts will the graph of a quadratic equation have if the discriminant has a value of **a)** 25 **b)** -2 **c)** 0

Practice the Skills

Indicate the axis of symmetry, the coordinates of the vertex, and whether the parabola opens upward or downward.

9. $y = x^2 + 6x - 3$
10. $y = x^2 + 2x - 11$
11. $y = -x^2 + 3x - 4$
12. $y = 3x^2 + 6x - 9$
13. $y = -3x^2 + 5x + 8$
14. $y = x^2 + 3x - 6$
15. $y = 4x^2 + 8x + 3$
16. $y = 3x^2 - 2x + 2$
17. $y = 2x^2 + 3x + 8$
18. $y = -x^2 + x + 8$
19. $y = -5x^2 + 6x - 1$
20. $y = -2x^2 - 6x - 5$

Graph each quadratic equation and determine the x-intercepts, if they exist.

21. $y = x^2 - 4$
22. $y = x^2 + 3$
23. $y = -x^2 + 5$
24. $y = -x^2 - 1$
25. $y = x^2 + 4x + 3$
26. $y = 3x^2 + 3$
27. $y = x^2 + 4x + 4$
28. $y = x^2 + 2x - 15$
29. $y = -x^2 - 2x + 3$
30. $y = x^2 - 4x - 5$
31. $y = x^2 + 5x - 6$
32. $y = -x^2 - 5x + 6$
33. $y = x^2 + 5x - 14$
34. $y = 2x^2 + 4x + 2$
35. $y = x^2 - 6x + 9$
36. $y = x^2 - 4x + 4$
37. $y = x^2 - 6x$
38. $y = -x^2 + 5x$
39. $y = 4x^2 + 12x + 9$
40. $y = x^2 - 2x + 1$
41. $y = -x^2 + 11x - 28$
42. $y = x^2 + x + 1$
43. $y = x^2 - 2x - 16$
44. $y = 2x^2 + 3x - 2$
45. $y = -2x^2 + 3x - 2$
46. $y = -4x^2 - 6x + 4$
47. $y = 2x^2 - x - 15$
48. $y = 6x^2 + 10x - 4$

Using the discriminant, determine the number of x-intercepts the graph of each equation will have. Do not graph the equation.

49. $y = 3x^2 - 2x - 16$
50. $y = -x^2 - 5$
51. $y = 4x^2 - 6x - 7$
52. $y = x^2 - 6x + 9$
53. $y = x^2 - 20x + 100$
54. $y = -4.3x^2 + 5.7x$
55. $y = 1.6x^2 + 2x - 3.9$
56. $y = 5x^2 - 13.2x + 9.3$

Problem Solving

The graph of a quadratic equation of the form $y = ax^2 + bx + c$ is a parabola. The value of a (the coefficient of the squared term in the equation) and the vertex of the parabola, are given. Determine the number of x-intercepts the parabola will have. Explain how you determined your answer.

57. $a = -2$, vertex at $(0, -3)$
58. $a = 5$, vertex at $(4, -3)$
59. $a = -3$, vertex at $(-4, 0)$
60. $a = -1$, vertex at $(2, -4)$
61. Will the equations below have the same x-intercepts when graphed? Explain how you determined your answer.

$$y = x^2 - 2x - 15 \text{ and } y = -x^2 + 2x + 15$$

62. **a)** How will the graphs of the following equations compare? Explain how you determined your answer.

$$y = x^2 - 2x - 8 \text{ and } y = -x^2 + 2x + 8$$

   b) Graph $y = x^2 - 2x - 8$ and $y = -x^2 + 2x + 8$ on the same axes.

Challenge Problems

63. a) Graph the quadratic equation $y = -x^2 + 6x$.
   b) On the same axes, graph the quadratic equation $y = x^2 - 2x$.
   c) Estimate the points of intersection of the graphs. The points represent the solution to the system of equations.

64. a) Graph the quadratic equation $y = x^2 + 2x - 3$.
   b) On the same axes, graph the quadratic equation $y = -x^2 + 1$.
   c) Estimate the points of intersection of the graphs. The points represent the solution to the system of equations.

Cumulative Review Exercises

[4.2] 65. Graph $4x - 6y = 20$.

[4.5] 66. Graph $y < 2$.

[8.4] 67. Subtract $\frac{3}{x+3} - \frac{x-2}{x-4}$.

[8.6] 68. Solve the equation $\frac{1}{3}(x + 6) = 3 - \frac{1}{4}(x - 5)$.

SUMMARY

Key Words and Phrases

10.1
Quadratic equation
Square root property
Standard form of a quadratic equation

10.2
Completing the square
Perfect square trinomial

10.3
Discriminant
Quadratic formula

10.4
Axis of symmetry
Parabola
Symmetry
Vertex of a parabola

IMPORTANT FACTS

Square root property: If $x^2 = a$, then $x = \sqrt{a}$ or $x = -\sqrt{a}$ (or $x = \pm\sqrt{a}$).

Quadratic formula: $x = \dfrac{-b \pm \sqrt{b^2 - 4ac}}{2a}$

Discriminant: $b^2 - 4ac$

If $b^2 - 4ac > 0$ the quadratic equation has two distinct real number solutions.
If $b^2 - 4ac = 0$ the quadratic equation has one real number solution.
If $b^2 - 4ac < 0$ the quadratic equation has no real number solution.

Coordinates of the vertex of a parabola:

$$\left(-\frac{b}{2a}, \frac{4ac - b^2}{4a}\right).$$

Review Exercises

[10.1] *Solve using the square root property.*

1. $x^2 = 100$
2. $x^2 = 12$
3. $2x^2 = 12$
4. $x^2 + 3 = 9$
5. $x^2 - 4 = 16$
6. $2x^2 - 4 = 10$
7. $4x^2 - 30 = 2$
8. $(x - 3)^2 = 12$
9. $(3x - 5)^2 = 50$
10. $(2x + 4)^2 = 30$

[10.2] *Solve by completing the square.*

11. $x^2 - 9x + 18 = 0$
12. $x^2 - 11x + 28 = 0$
13. $x^2 - 18x + 17 = 0$
14. $x^2 + x - 6 = 0$
15. $x^2 - 3x - 54 = 0$
16. $x^2 = -5x + 6$
17. $x^2 - 3x - 8 = 0$
18. $x^2 + 2x - 5 = 0$
19. $2x^2 - 8x = 64$
20. $2x^2 - 4x = 30$
21. $3x^2 + 2x - 8 = 0$
22. $6x^2 - 19x + 15 = 0$

[10.3] *Determine whether each equation has two distinct real number solutions, one real number solution, or no real number solution.*

23. $-2x^2 + 4x + 6 = 0$
24. $-3x^2 + 4x = 9$
25. $x^2 - 10x + 25 = 0$
26. $x^2 - x + 8 = 0$
27. $2x^2 - 3x = 6$
28. $3x^2 - 4x + 5 = 0$
29. $-3x^2 - 4x + 8 = 0$
30. $x^2 - 9x + 6 = 0$

Use the quadratic formula to solve. If an equation has no real number solution, so state.

31. $x^2 - 11x + 18 = 0$
32. $x^2 - 7x - 44 = 0$
33. $x^2 = 10x - 9$
34. $5x^2 - 7x = 6$
35. $x^2 - 18 = 7x$
36. $x^2 - x - 30 = 0$
37. $6x^2 + x - 15 = 0$
38. $-2x^2 + 3x + 6 = 0$
39. $2x^2 + 4x - 3 = 0$
40. $x^2 - 6x + 7 = 0$
41. $3x^2 - 4x + 6 = 0$
42. $3x^2 - 6x - 8 = 0$
43. $7x^2 - 3x = 0$
44. $2x^2 - 5x = 0$

[10.1–10.3] *Solve each quadratic equation using the method of your choice.*

45. $x^2 - 13x + 42 = 0$
46. $x^2 + 15x + 56 = 0$
47. $x^2 - 3x - 70 = 0$
48. $x^2 + 6x = 27$
49. $x^2 - 4x - 60 = 0$
50. $x^2 - x - 42 = 0$
51. $x^2 + 11x - 12 = 0$
52. $x^2 + 6x = 0$
53. $x^2 = 81$
54. $2x^2 + 5x = 3$
55. $2x^2 = 9x - 10$
56. $6x^2 + 5x = 6$
57. $-2x^2 + 6x + 9 = 0$
58. $3x^2 - 11x + 10 = 0$
59. $-3x^2 - 5x + 8 = 0$
60. $x^2 + 3x = 6$
61. $4x^2 - 9x = 0$
62. $3x^2 + 5x = 0$

[10.4] *Indicate the axis of symmetry, the coordinates of the vertex, and whether the parabola opens upward or downward.*

63. $y = x^2 - 4x - 5$
64. $y = x^2 - 12x + 6$
65. $y = x^2 - 3x + 7$
66. $y = -x^2 - 2x + 15$
67. $y = 2x^2 + 7x + 3$
68. $y = -x^2 - 5x$
69. $y = -x^2 - 8$
70. $y = -x^2 - x + 20$
71. $y = -4x^2 + 8x + 5$
72. $y = 3x^2 + 5x - 8$

Graph each quadratic equation and determine the x-intercepts, if they exist. If they do not exist, so state.

73. $y = x^2 - 2x$
74. $y = -3x^2 + 6$
75. $y = x^2 - 2x - 15$
76. $y = -x^2 + 5x - 6$
77. $y = x^2 - x + 1$
78. $y = x^2 + 5x + 4$
79. $y = -x^2 - 4x$
80. $y = x^2 + 4x + 3$
81. $y = -x^2 + x - 3$
82. $y = 3x^2 - 4x - 8$
83. $y = -2x^2 + 7x - 3$
84. $y = x^2 - 5x + 4$

Solve.

85. The product of two positive consecutive even integers is 48. Find the integers.

86. The product of two positive integers is 88. Find the two integers if the larger one is 3 greater than the smaller.

87. Samuel Jones is making a table with a rectangular top for the kitchen. His wife, Jenna, wants the top of the table to be large enough to hold the microwave oven as well as a breadmaker and toaster. After measuring their appliances, Jenna and Samuel determine that the length of the table top should be 6 inches more than twice its width. What are the dimensions of the table top if its area is 920 square inches?

88. Jordan and Patricia Wells recently purchased an oak desk for their son's room. To protect the rectangular writing surface of the desk, they decided to cover the top with a piece of glass. The length of the desktop is 20 inches greater than its width. Find the dimensions of the glass piece Jordan and Patricia should order if the area of the desktop is 1344 square inches.

Practice Test

1. Solve $x^2 - 2 = 26$ using the square root property.

2. Solve $(3x - 4)^2 = 17$ using the square root property.

3. Solve $x^2 - 6x = 40$ by completing the square.

4. Solve $x^2 + 4x = 60$ by completing the square.

5. Solve $x^2 - 8x - 20 = 0$ by the quadratic formula.

6. Solve $2x^2 + 5 = -8x$ by the quadratic formula.

7. Solve $16x^2 = 49$ by the method of your choice.

8. Solve $2x^2 + 9x = 5$ by the method of your choice.

9. Write the quadratic formula.

10. Give an example of a perfect square trinomial.

11. Determine whether $3x^2 - 4x + 2 = 0$ has two distinct real solutions, one real solution, or no real solution. Explain your answer.

12. Determine whether $x^2 + 8x - 3 = 0$ has two distinct real solutions, one real solution, or no real solution. Explain your answer.

13. Find the equation of the axis of symmetry of the graph of $y = -x^2 - 6x + 7$.

14. Find the equation of the axis of symmetry of the graph of $y = 4x^2 - 8x + 9$.

15. Determine whether the graph of $y = -x^2 - 6x + 7$ opens upward or downward. Explain your answer.

16. Determine whether the graph of $y = 4x^2 - 8x + 9$ opens upward or downward. Explain your answer.

17. What is the vertex of the graph of a parabola?

18. Find the vertex of the graph of $y = -x^2 - 10x - 16$.

19. Find the vertex of the graph of $y = 3x^2 - 8x + 9$.

20. Graph the equation $y = x^2 + 2x - 8$ and determine the x-intercepts, if they exist.

21. Graph the equation $y = -x^2 + 6x - 9$ and determine the x-intercepts, if they exist.

22. Graph the equation $y = 2x^2 - 6x$ and determine the x-intercepts, if they exist.

23. The length of a rectangular wall mural is 1 foot greater than 3 times its width. Find the length and width of the mural if its area is 30 square feet.

24. The product of two positive consecutive odd integers is 99. Find the larger of two integers.

25. Shawn Goodwin is 4 years older than his cousin, Aaron. The product of their ages is 45. How old is Shawn?

Cumulative Review Test

1. Evaluate $-3x^2y + 2y^2 - xy$ when $x = 2$ and $y = -3$.
2. Solve the equation $\frac{1}{2}z - \frac{2}{7}z = \frac{1}{5}(3z - 1)$.
3. Find the length of side x.

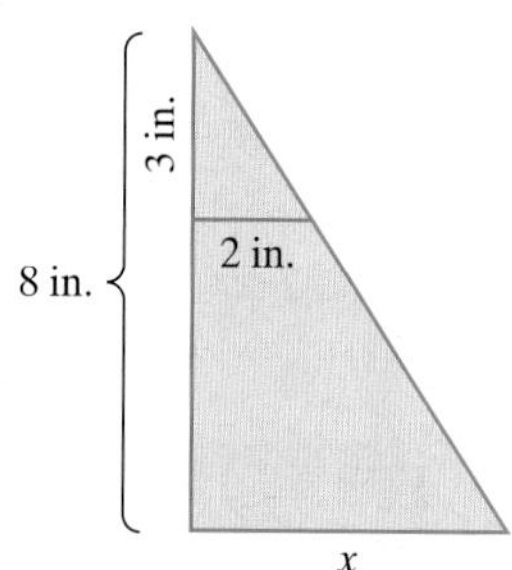

4. Solve the inequality $2(x - 3) \leq 6x - 5$ and graph the solution on a number line.
5. Solve the formula $A = \dfrac{m + n + P}{3}$ for P.
6. Simplify $(2a^4b^3)^3(3a^2b^5)^2$.
7. Divide $\dfrac{x^2 + 6x + 5}{x + 2}$.
8. Factor by grouping $2x^2 - 3xy - 4xy + 6y^2$.
9. Factor $9x^2 - 48x + 15$.
10. Add $\dfrac{4}{a^2 - 16} + \dfrac{2}{(a - 4)^2}$.
11. Solve the equation $x + \dfrac{48}{x} = 14$.
12. Graph the equation $4x - y = 8$.
13. Solve the system of equations by the addition method.

$$3x - 4y = 12$$
$$4x - 3y = 6$$

14. Simplify $\sqrt{\dfrac{3x^2y^3}{54x}}$.
15. Add $2\sqrt{28} - 3\sqrt{7} + \sqrt{63}$.
16. Solve the equation $x - 2 = \sqrt{x^2 - 12}$.
17. Solve the equation $2x^2 + x - 8 = 0$ using the quadratic formula.
18. If 4 pounds of fertilizer can fertilize 500 square feet of lawn, how many pounds of fertilizer are needed to fertilize 3200 square feet of lawn?
19. The length of a rectangular vegetable garden is 3 feet less than three times its width. Find the width and length of the garden if its perimeter is 74 feet.
20. Robert McCloud jogs 3 miles per hour faster than he walks. He jogs for 2 miles and then walks for 2 miles. If the total time of his outing is 1 hour, find the rate at which he walks and jogs.

Appendices

APPENDIX A REVIEW OF DECIMALS AND PERCENT

Decimals

To Add or Subtract Numbers Containing Decimal Points

1. Align the numbers by the decimal points.
2. Add or subtract the numbers as if they were whole numbers.
3. Place the decimal point in the sum or difference directly below the decimal points in the numbers being added or subtracted.

EXAMPLE 1 Add 4.6 + 13.813 + 9.02.

Solution

$$\begin{array}{r} 4.600 \\ 13.813 \\ +\ 9.020 \\ \hline 27.433 \end{array}$$

EXAMPLE 2 Subtract 3.062 from 25.9.

Solution

$$\begin{array}{r} 25.900 \\ -\ 3.062 \\ \hline 22.838 \end{array}$$

To Multiply Numbers Containing Decimal Points

1. Multiply as if the factors were whole numbers.
2. Determine the total number of digits to the right of the decimal points in the factors.
3. Place the decimal point in the product so that the product contains the same number of digits to the right of the decimal as the total found in step 2. For example, if there are a total of three digits to the right of the decimal points in the factors, there must be three digits to the right of the decimal point in the product.

EXAMPLE 3 Multiply 2.34×1.9.

Solution

$$\begin{array}{rl} 2.34 & \longleftarrow \text{ two digits to the right of the decimal point} \\ \times\ 1.9 & \longleftarrow \text{ one digit to the right of the decimal point} \\ \hline 2106 & \\ 234 & \\ \hline 4.446 & \longleftarrow \text{ three digits to the right of the decimal point in the product} \end{array}$$

EXAMPLE 4 Multiply 2.13×0.02.

Solution

$$\begin{array}{rl} 2.13 & \longleftarrow \text{ two digits to the right of the decimal point} \\ \times\ 0.02 & \longleftarrow \text{ two digits to the right of the decimal point} \\ \hline 0.0426 & \longleftarrow \text{ four digits to the right of the decimal point in the product} \end{array}$$

Note that it was necessary to add a zero preceding the digit 4 in the answer in order to have four digits to the right of the decimal point.

To Divide Numbers Containing Decimal Points

1. Multiply both the dividend and divisor by a power of 10 that will make the divisor a whole number.
2. Divide as if working with whole numbers.
3. Place the decimal point in the quotient directly above the decimal point in the dividend.

To make the divisor a whole number, multiply *both* the dividend and divisor by 10 if the divisor is given in tenths, by 100 if the divisor is given in hundredths, by 1000 if the divisor is given in thousandths, and so on. Multiplying both the numerator and denominator by the same nonzero number is the same as multiplying the fraction by 1. Therefore, the value of the fraction is unchanged.

EXAMPLE 5 Divide $\dfrac{1.956}{0.12}$.

Solution Since the divisor, 0.12, is twelve-hundredths, we multiply both the divisor and dividend by 100.

$$\frac{1.956}{0.12} \times \frac{100}{100} = \frac{195.6}{12.}$$

Now we divide.

$$\begin{array}{r} 16.3 \\ 12.\overline{)195.6} \\ \underline{12} \\ 75 \\ \underline{72} \\ 36 \\ \underline{36} \\ 0 \end{array}$$

The decimal point in the answer is placed directly above the decimal point in the dividend. Thus, $1.956/0.12 = 16.3$.

EXAMPLE 6 Divide 0.26 by 10.4.

Solution First, multiply both the dividend and divisor by 10.

$$\frac{0.26}{10.4} \times \frac{10}{10} = \frac{2.6}{104.}$$

Now divide.

$$\begin{array}{r} 0.025 \\ 104\overline{)2.600} \\ \underline{2\ 08} \\ 520 \\ \underline{520} \\ 0 \end{array}$$

Note that a zero had to be placed before the digit 2 in the quotient.

$$\frac{0.26}{10.4} = 0.025$$

Percent

The word *percent* means "per hundred." The symbol % means percent. One percent means "one per hundred," or

$$1\% = \frac{1}{100} \quad \text{or} \quad 1\% = 0.01$$

EXAMPLE 7 Convert 16% to a decimal.

Solution Since 1% = 0.01,

$$16\% = 16(0.01) = 0.16$$

EXAMPLE 8 Convert 2.3% to a decimal.

Solution 2.3% = 2.3(0.01) = 0.023.

EXAMPLE 9 Convert 1.14 to a percent.

Solution To change a decimal number to a percent, we multiply the number by 100%.

$$1.14 = 1.14 \times 100\% = 114\%$$

Often, you will need to find an amount that is a certain percent of a number. For example, when you purchase an item in a state or county that has a sales tax you must often pay a percent of the item's price as the sales tax. Examples 10 and 11 show how to find a certain percent of a number.

EXAMPLE 10 Find 12% of 200.

Solution To find a percent of a number, use multiplication. Change 12% to a decimal number, then multiply by 200.

$$(0.12)(200) = 24$$

Thus, 12% of 200 is 24.

EXAMPLE 11 Monroe County in New York State charges an 8% sales tax.

a) Find the sales tax on a stereo system that cost \$580.

b) Find the total cost of the system, including tax.

Solution **a)** The sales tax is 8% of 580.

$$(0.08)(580) = 46.40$$

The sales tax is \$46.40

b) The total cost is the purchase price plus the sales tax:

$$\text{total cost} = \$580 + \$46.40 = \$626.40$$

APPENDIX B FINDING THE GREATEST COMMON FACTOR AND LEAST COMMON DENOMINATOR

Prime Factorization

In Section 1.2 we mentioned that to simplify fractions you can divide both the numerator and denominator by the *greatest common factor* (GCF). One method to find the GCF is to use *prime factorization*. Prime factorization is the process of writing a given number as a product of prime numbers. *Prime numbers* are natural numbers, excluding 1, that can be divided by only themselves and 1. The first ten prime numbers are 2, 3, 5, 7, 11, 13, 17, 19, 23, and 29. Can you find the next prime number? If you answered 31, you answered correctly.

To write a number as a product of primes, we can use a *tree diagram*. Begin by selecting any two numbers whose product is the given number. Then continue factoring each of these numbers into prime numbers, as shown in Example 1.

EXAMPLE 1 Determine the prime factorization of the number 120.

Solution We will use three different tree diagrams to illustrate the prime factorization of 120.

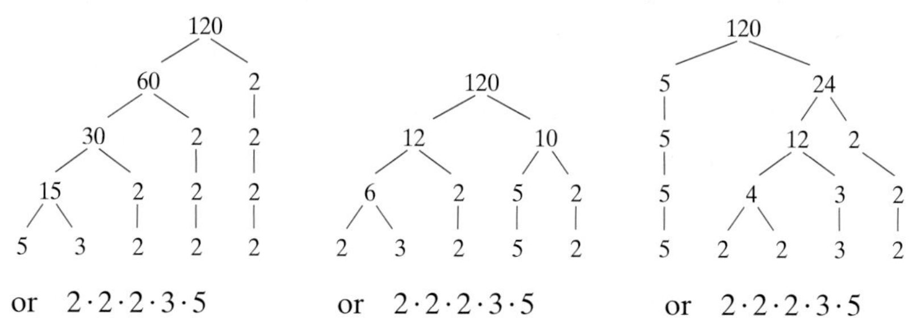

Note that no matter how you start, if you do not make a mistake, you find that the prime factorization of 120 is $2 \cdot 2 \cdot 2 \cdot 3 \cdot 5$. There are other ways 120 can be factored but all will lead to the prime factorization $2 \cdot 2 \cdot 2 \cdot 3 \cdot 5$.

Greatest Common Factor

The greatest common factor (GCF) of two natural numbers is the greatest integer that is a factor of both numbers. We use the GCF when simplifying fractions.

To Find the Greatest Common Factor of a Given Numerator and Denominator

1. Write both the numerator and the denominator as a product of primes.
2. Determine all the prime factors that are common to both prime factorizations.
3. Multiply the prime factors found in step 2 to obtain the GCF.

EXAMPLE 2 Consider the fraction $\frac{108}{156}$.

a) Find the GCF of 108 and 156.

b) Simplify $\frac{108}{156}$.

Solution **a)** First determine the prime factorizations of both 108 and 156.

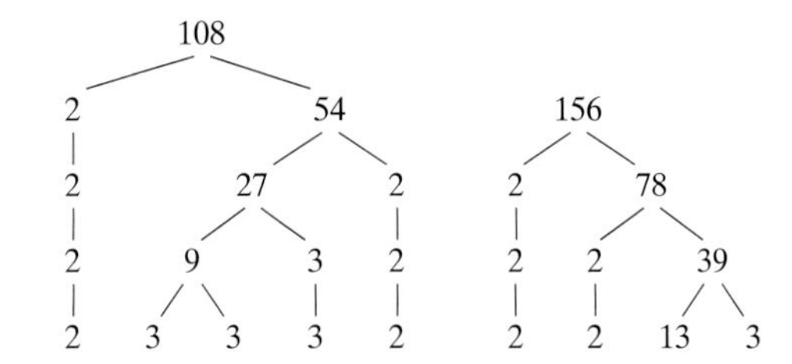

There are two 2's and one 3 common to both prime factorizations; thus

$$\text{GCF} = 2 \cdot 2 \cdot 3 = 12$$

The greatest common factor of 108 and 156 is 12. Twelve is the greatest integer that divides into both 108 and 156.

b) To simplify $\frac{108}{156}$, we divide both the numerator and denominator by the GCF, 12.

$$\frac{108 \div 12}{156 \div 12} = \frac{9}{13}$$

Thus, $\frac{108}{156}$ simplifies to $\frac{9}{13}$.

Least Common Denominator

When adding two or more fractions, you must write each fraction with a common denominator. The best denominator to use is the *least common denominator* (LCD). The LCD is the smallest number that each denominator divides into. Sometimes the least common denominator is referred to as the *least common multiple* of the denominators.

To Find the Least Common Denominator of Two or More Fractions

1. Write each denominator as a product of prime numbers.
2. For each prime number, determine the maximum number of times that prime number appears in any of the prime factorizations.
3. Multiply all the prime numbers, found in step 2. Include each prime number the maximum number of times it appears in any of the prime factorizations. The product of all these prime numbers will be the LCD.

Example 3 illustrates the procedure to determine the LCD.

EXAMPLE 3 Consider $\frac{7}{108} + \frac{5}{156}$.

a) Determine the least common denominator.

b) Add the fractions.

Solution **a)** We found in Example 2 that

$$108 = 2 \cdot 2 \cdot 3 \cdot 3 \cdot 3 \quad \text{and} \quad 156 = 2 \cdot 2 \cdot 3 \cdot 13$$

We can see that the maximum number of 2's that appear in either prime factorization is two (there are two 2's in both factorizations), the maximum number of 3's is three, and the maximum number of 13's is one. Multiply as follows:

$$2 \cdot 2 \cdot 3 \cdot 3 \cdot 3 \cdot 13 = 1404$$

Thus, the least common denominator is 1404. This is the smallest number that both 108 and 156 divide into.

b) To add the fractions, we need to write both fractions with a common denominator. The best common denominator to use is the LCD. Since $1404 \div 108 = 13$, we will multiply $\frac{7}{108}$ by $\frac{13}{13}$. Since $1404 \div 156 = 9$, we will multiply $\frac{5}{156}$ by $\frac{9}{9}$.

$$\frac{7}{108} \cdot \frac{13}{13} + \frac{5}{156} \cdot \frac{9}{9} = \frac{91}{1404} + \frac{45}{1404} = \frac{136}{1404} = \frac{34}{351}$$

Thus, $\frac{7}{108} + \frac{5}{156} = \frac{34}{351}$.

APPENDIX C GEOMETRY

This appendix introduces or reviews important geometric concepts. Table C.1 gives the names and descriptions of various types of angles.

Angles

TABLE C.1

Angle	Sketch of Angle
An **acute angle** is an angle whose measure is between 0° and 90°.	
A **right angle** is an angle whose measure is 90°.	
An **obtuse angle** is an angle whose measure is between 90° and 180°.	
A **straight angle** is an angle whose measure is 180°.	
Two angles are **complementary angles** when the sum of their measures is 90°. Each angle is the complement of the other. Angles A and B are complementary angles.	60°, A, B, 30°
Two angles are **supplementary angles** when the sum of their measures is 180°. Each angle is the supplement of the other. Angles A and B are supplementary angles.	130°, A, B, 50°

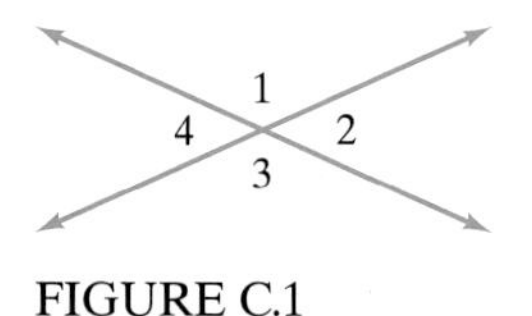

FIGURE C.1

When two lines intersect, four angles are formed as shown in Figure C.1. The pair of opposite angles formed by the intersecting lines are called **vertical angles**.

Angles 1 and 3 are vertical angles. Angles 2 and 4 are also vertical angles. *Vertical angles have equal measures*. Thus, angle 1, symbolized by $\angle 1$, is equal to angle 3, symbolized by $\angle 3$. We can write $\angle 1 = \angle 3$. Similarly, $\angle 2 = \angle 4$.

Parallel and Perpendicular Lines

Parallel lines are two lines in the same plane that do not intersect (Fig. C2). **Perpendicular lines** are lines that intersect at right angles (Fig. C3).

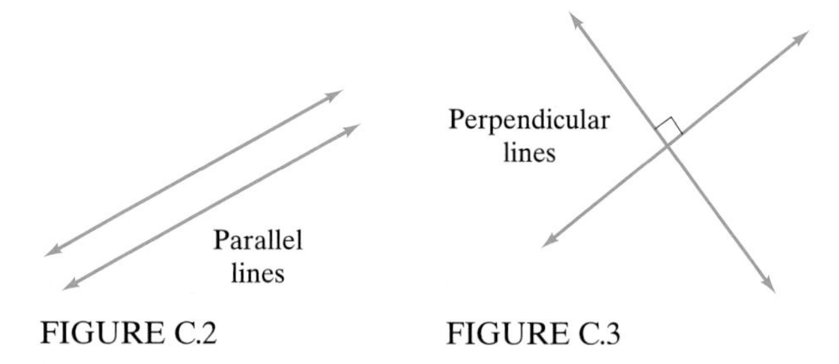

FIGURE C.2

FIGURE C.3

A **transversal** is a line that intersects two or more lines at different points. When a transversal line intersects two other lines, eight angles are formed, as illustrated in Figure C.4. Some of these angles are given special names.

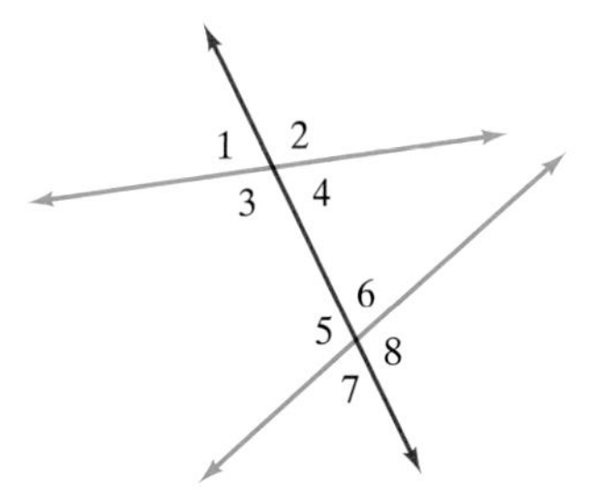

FIGURE C.4

Interior angles: 3, 4, 5, 6

Exterior angles: 1, 2, 7, 8

Pairs of corresponding angles: 1 and 5; 2 and 6; 3 and 7; 4 and 8

Pairs of alternate interior angles: 3 and 6; 4 and 5

Pairs of alternate exterior angles: 1 and 8; 2 and 7

Parallel Lines Cut by a Transversal

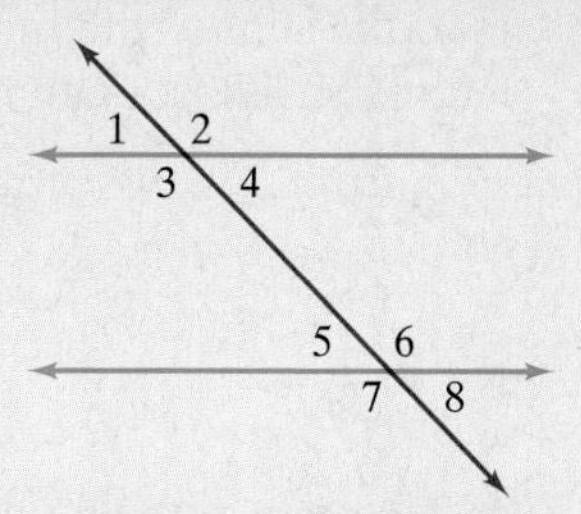

When two parallel lines are cut by a transversal:

1. Corresponding angles are equal ($\angle 1 = \angle 5$, $\angle 2 = \angle 6$, $\angle 3 = \angle 7$, $\angle 4 = \angle 8$).
2. Alternate interior angles are equal ($\angle 3 = \angle 6$, $\angle 4 = \angle 5$).
3. Alternate exterior angles are equal ($\angle 1 = \angle 8$, $\angle 2 = \angle 7$).

EXAMPLE 1 If line 1 and line 2 are parallel lines and $\angle 1 = 112°$, find the measure of angles 2 through 8.

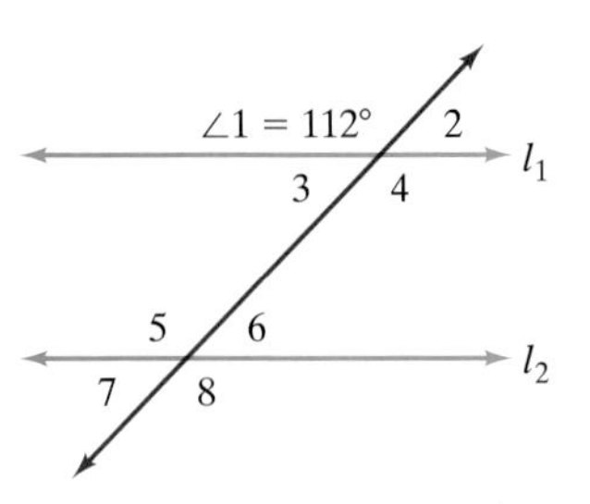

Solution Angles 1 and 2 are supplementary. So $\angle 2$ is $180° - 112° = 68°$. Angles 1 and 4 are equal since they are vertical angles. Thus, $\angle 4 = 112°$. Angles 1 and 5 are corresponding angles. Thus, $\angle 5 = 112°$. It is equal to its verticle angle, $\angle 8$, so, $\angle 8 = 112°$. Angles 2, 3, 6, and 7 are all equal and measure 68°.

Polygons

A **polygon** is a closed figure in a plane determined by three or more line segments. Some polygons are illustrated in Figure C.5.

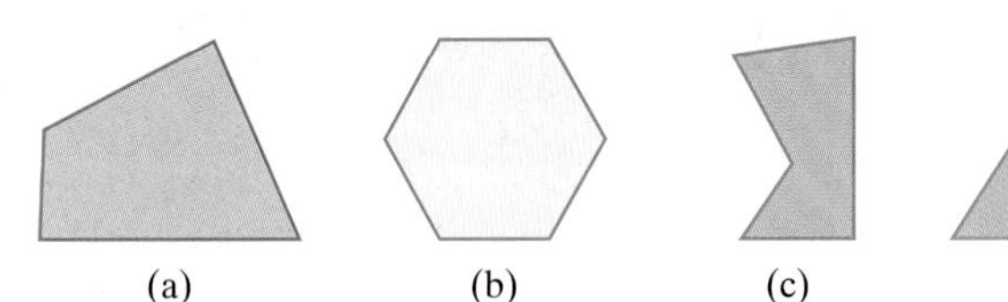

FIGURE C.5 (a) (b) (c) (d)

A **regular polygon** has sides that are all the same length, and interior angles that all have the same measure. Figures C.5(b) and (d) are regular polygons.

Sum of the Interior Angles of a Polygon

The sum of the interior angles of a polygon can be found by the formula

$$\text{Sum} = (n - 2)180°$$

where n is the number of sides of the polygon.

EXAMPLE 2 Find the sum of the measures of the interior angles of **a)** a triangle; **b)** a quadrilateral (4 sides); **c)** an octagon (8 sides).

Solution **a)** Since $n = 3$, we write

$$\begin{aligned}\text{Sum} &= (n - 2)180° \\ &= (3 - 2)180° = 1(180°) = 180°\end{aligned}$$

The sum of the measures of the interior angles in a triangle is 180°.

b) $\text{Sum} = (n - 2)180°$

$= (4 - 2)180° = 2(180°) = 360°$

The sum of the measures of the interior angles in a quadrilateral is 360°.

c) $\text{Sum} = (n - 2)(180°) = (8 - 2)180° = 6(180°) = 1080°$

The sum of the measures of the interior angles in an octagon is 1080°.

Now we will briefly define several types of triangles in Table C.2.

Triangles

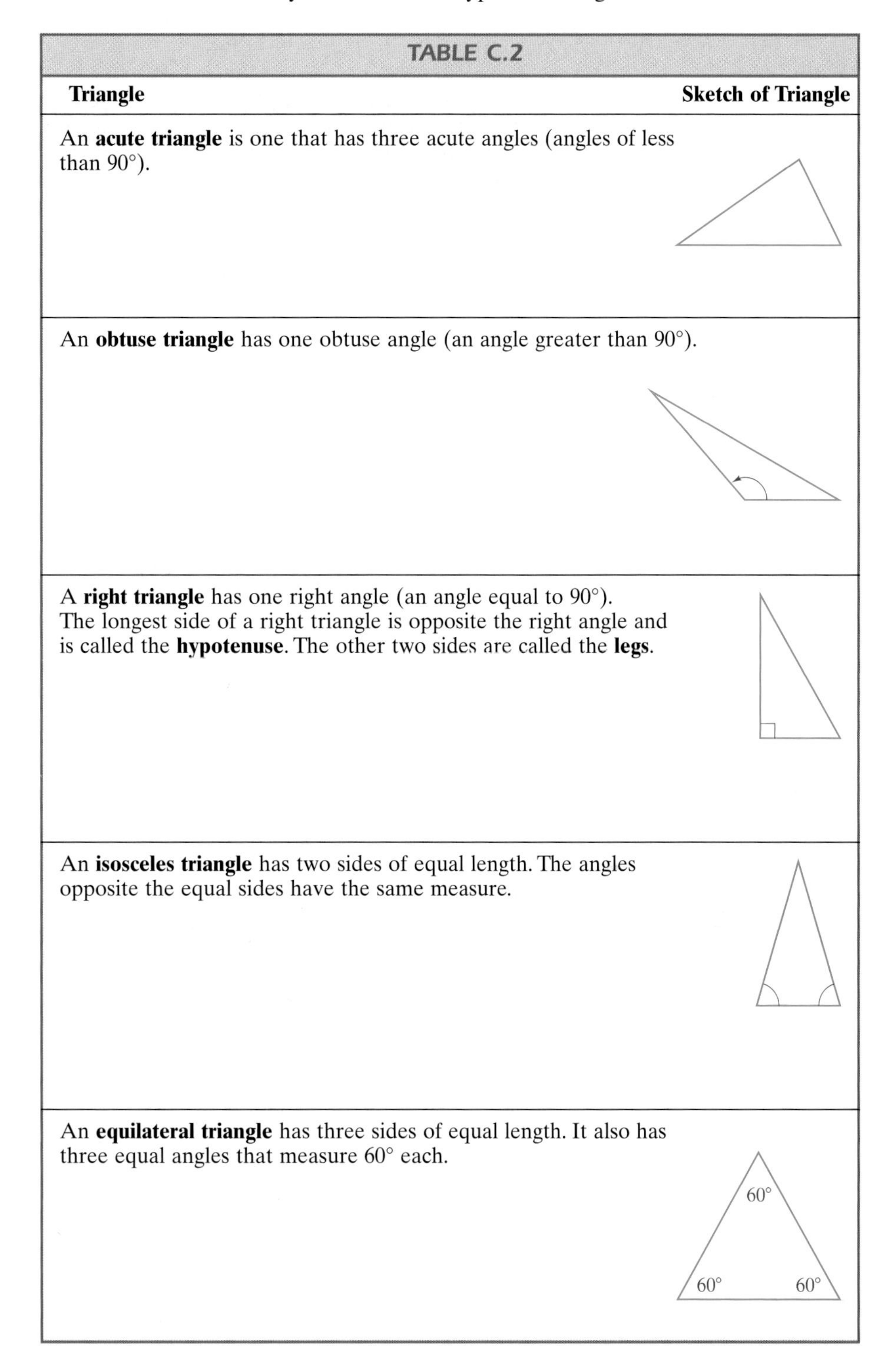

TABLE C.2

Triangle	Sketch of Triangle
An **acute triangle** is one that has three acute angles (angles of less than 90°).	
An **obtuse triangle** has one obtuse angle (an angle greater than 90°).	
A **right triangle** has one right angle (an angle equal to 90°). The longest side of a right triangle is opposite the right angle and is called the **hypotenuse**. The other two sides are called the **legs**.	
An **isosceles triangle** has two sides of equal length. The angles opposite the equal sides have the same measure.	
An **equilateral triangle** has three sides of equal length. It also has three equal angles that measure 60° each.	

When two sides of a *right triangle* are known, the third side can be found using the **Pythagorean Theorem**, $a^2 + b^2 = c^2$, where a and b are the legs and c is the hypotenuse of the triangle. (See Chapter 9 for examples.)

Congruent and Similar Figures

If two triangles are **congruent**, it means that the two triangles are identical in size and shape. Two congruent triangles could be placed one on top of the other if we were able to move and rearrange them.

Two Triangles Are Congruent If Any One of the Following Statements Is True.

1. Two angles of one triangle are equal to two corresponding angles of the other triangle, and the lengths of the sides between each pair of angles are equal. This method of showing that triangles are congruent is called the *angle, side, angle* method.

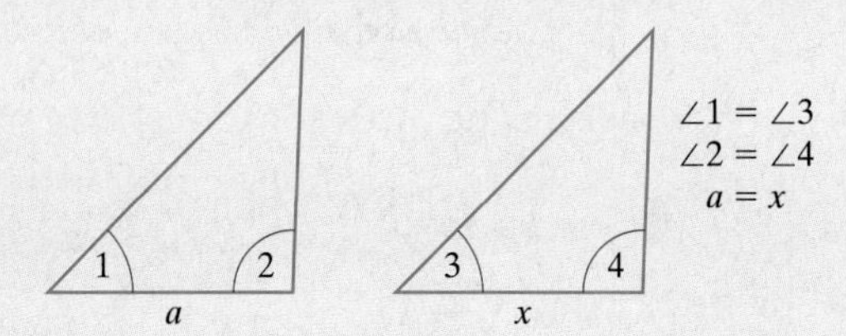

2. Corresponding sides of both triangles are equal. This is called the *side, side, side* method.

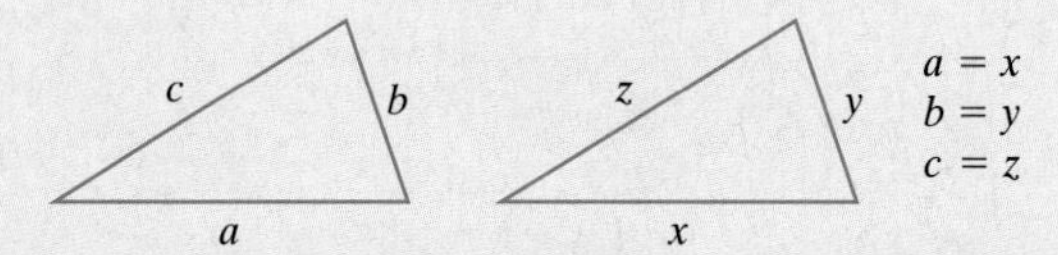

3. Two corresponding pairs of sides are equal, and the angle between them is equal. This is referred to as the *side, angle, side* method.

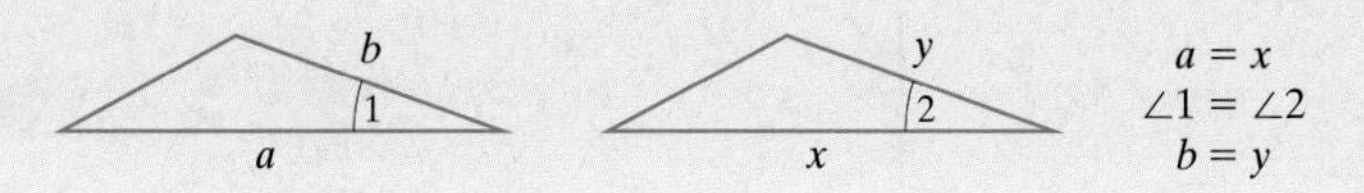

EXAMPLE 3 Determine whether the two triangles are congruent.

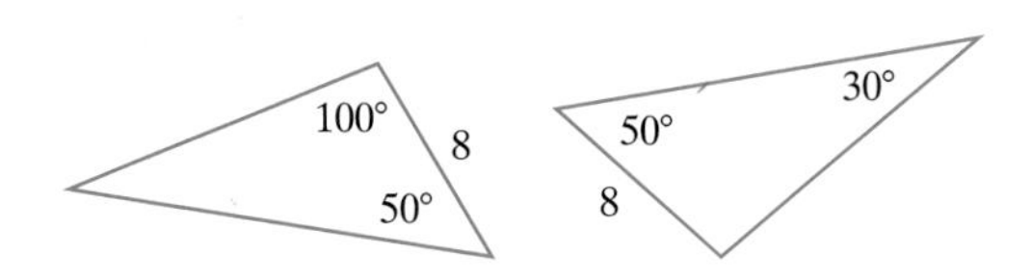

Solution The unknown angle in the figure on the right must measure 100° since the sum of the angles of a triangle is 180°. Both triangles have the same two angles (100° and 50°), with the same length side between them, 8 units. Thus, these two triangles are congruent by the angle, side, angle method.

Two triangles are **similar** if all three pairs of corresponding angles are equal and corresponding sides are in proportion. Similar figures do not have to be the same size but must have the same general shape.

Two Triangles Are Similar If Any One of the Following Statements Is True.

1. Two angles of one triangle equal two angles of the other triangle.

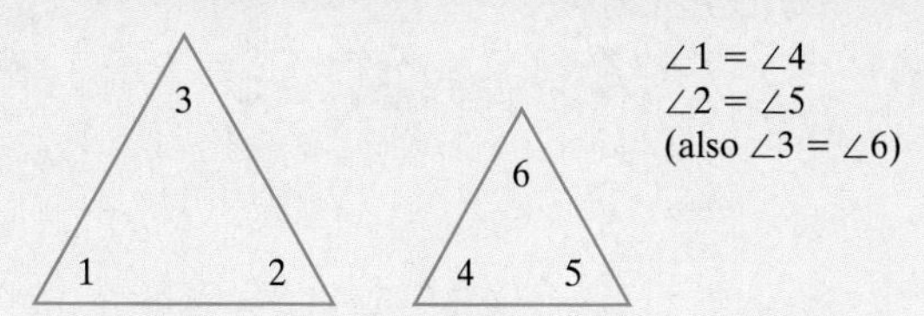

2. Corresponding sides of the two triangles are proportional.

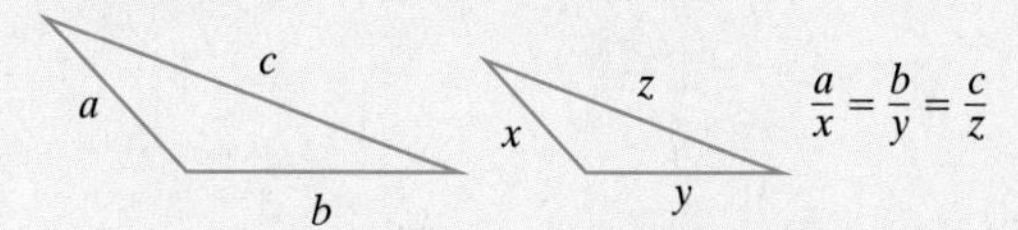

3. Two pairs of corresponding sides are proportional, and the angles between them are equal.

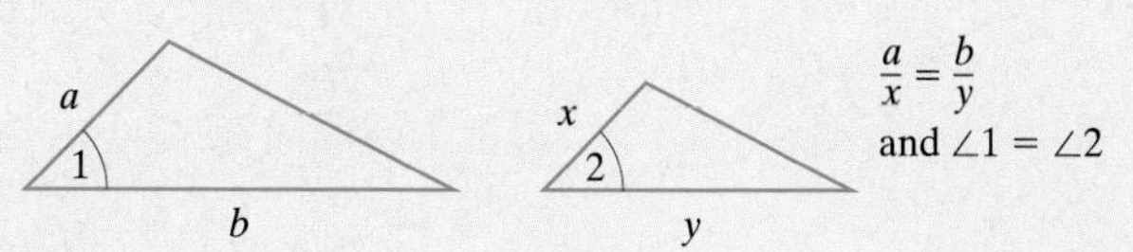

EXAMPLE 4 Are the triangles ABC and $AB'C'$ similar?

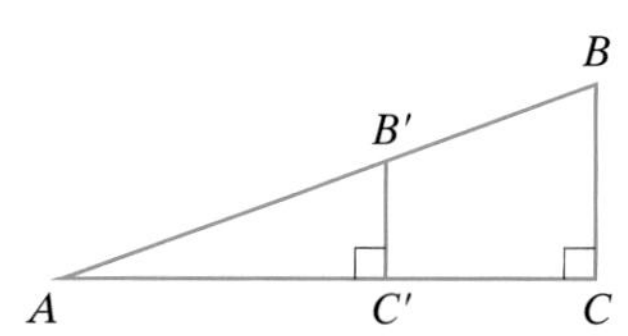

Solution Angle A is common to both triangles. Since angle C and angle C' are equal (both 90°), then $\angle B$ and $\angle B'$ must be equal. Since the three angles of triangle ABC equal the three angles of triangle $AB'C'$, the two triangles are similar.

Answers

CHAPTER 1

Exercise Set 1.1 **11.** Do all the homework and preview the new material that is to be covered in class. **13.** At least 2 hours of study and homework time for each hour of class time is generally recommended. **15. a)** You need to do the homework in order to practice what was presented in class. **b)** When you miss class, you miss important information; therefore, it is important that you attend class regularly. **17.** Answers will vary.

Exercise Set 1.2 **1.** Understand, translate, calculate, check, state answer **3.** Substitute other numbers. **5.** Rank the data. The median is the value in the middle. **7.** Mean **9.** He actually missed a B by 10 points. **11. a)** 76.4 **b)** 74 **13. a)** ≈ \$90.50 **b)** \$83.79 **15. a)** ≈ \$26,132.67 **b)** \$26,272 **17.** \$470 **19.** \$227.05 **21. a)** \$60 **b)** \$25 **23. a)** 19.375 minutes **b)** 88 minutes **c)** 10 minutes **25.** ≈ 18.49 miles per gallon **27.** \$153 **29. a)** \$3963 **b)** \$8453.80 **31. a)** 4106.25 gallons **b)** ≈ \$21.35 **33. a)** \$441 **b)** \$371 **35. a)** \$0.47 **b)** ≈ 3.04 **37. a)** The number of telephones in Hong Kong, ≈ 3.67 million; the number of telephones in the United States, ≈ 202.62 million **b)** The population of the U.S. is greater than the population of Hong Kong. **39.** 23 points per game **41. a)** 48 **b)** He cannot get a C.

Exercise Set 1.3 **1. a)** Variables are letters that represent numbers. **b)** x, y, and z **3.** $5(x), (5)x, (5)(x), 5x, 5 \cdot x$ **5.** Divide out common factors. **7.** b) **9.** c) **11.** Invert the divisor, then multiply. **13.** Multiply the denominator by the whole number, then add the numerator. This number is the numerator of the fraction. The denominator is the denominator in the mixed number. **15.** $\frac{1}{3}$ **17.** $\frac{2}{3}$ **19.** 1 **21.** $\frac{9}{19}$ **23.** $\frac{5}{33}$ **25.** Simplified **27.** $\frac{15}{8}$ **29.** $\frac{43}{15}$ **31.** $\frac{15}{4}$ **33.** $\frac{89}{19}$ **35.** $1\frac{1}{3}$ **37.** $3\frac{3}{4}$ **39.** $7\frac{1}{2}$ **41.** $4\frac{4}{7}$ **43.** $\frac{3}{8}$ **45.** $\frac{1}{9}$ **47.** $\frac{3}{2}$ or $1\frac{1}{2}$ **49.** $\frac{16}{5}$ or $3\frac{1}{5}$ **51.** $\frac{5}{16}$ **53.** 6 **55.** $\frac{77}{40}$ or $1\frac{37}{40}$ **57.** $\frac{43}{10}$ or $4\frac{3}{10}$ **59.** 1 **61.** $\frac{1}{3}$ **63.** $\frac{9}{17}$ **65.** $\frac{6}{5}$ or $1\frac{1}{5}$ **67.** $\frac{1}{18}$ **69.** $\frac{7}{24}$ **71.** $\frac{13}{36}$ **73.** $\frac{97}{120}$ **75.** $\frac{10}{9}$ or $1\frac{1}{9}$ **77.** $\frac{65}{12}$ or $5\frac{5}{12}$ **79.** $19\frac{9}{16}$ inches **81.** $\frac{1}{10}$ **83.** $\frac{7}{10}$ **85.** $7\frac{5}{12}$ tons **87.** $1\frac{7}{8}$ dollars **89.** $1\frac{5}{8}$ inches **91.** $1\frac{3}{8}$ cups **93.** 40 times **95.** $1\frac{1}{2}$ inches **97.** 13 vials **99.** $6\frac{23}{24}$ inches **101. a)** $\frac{* + ?}{a}$ **b)** $\frac{\odot - \square}{?}$ **c)** $\frac{\triangle + 4}{\square}$ **d)** $\frac{x - 2}{3}$ **e)** $\frac{8}{x}$ **103.** 270 pills **105.** Answers will vary. **106.** 15.6 **107.** 16 **108.** Variables are letters used to represent numbers.

Exercise Set 1.4 **1.** A set is a collection of elements. **3.** Answers will vary. **5.** The set of natural numbers is also the set of counting numbers and the set of positive integers. **7. a)** All natural numbers are whole numbers. **b)** All natural numbers are rational numbers. **c)** All natural numbers are real numbers. **9.** $\{\ldots, -3, -2, -1, 0, 1, 2, 3, \ldots\}$ **11.** $\{1, 2, 3, 4, \ldots\}$ **13.** $\{0, 1, 2, 3, \ldots\}$ **15.** True **17.** True **19.** False **21.** False **23.** True **25.** True **27.** False **29.** True **31.** True **33.** True **35.** True **37.** True **39.** True **41.** True **43.** True **45.** True **47. a)** 3, 77 **b)** 0, 3, 77 **c)** 0, −2, 3, 77 **d)** $-\frac{4}{3}, 0, -2, 3, 5\frac{1}{2}, 1.63, 77$ **e)** $\sqrt{8}, -\sqrt{3}$ **f)** $-\frac{4}{3}, 0, -2, 3, 5\frac{1}{2}, \sqrt{8}, -\sqrt{3}, 1.63, 77$ **49.** Answers will vary; three examples are $\frac{1}{2}, -\frac{1}{2}$, and 0.6. **51.** Answers will vary; three examples are $-\frac{2}{3}, \frac{1}{2}$, and 6.3.

53. Answers will vary; three examples are $-\sqrt{7}$, $\sqrt{3}$, and $\sqrt{6}$.
55. Answers will vary; three examples are −13, −5, and −1.
57. Answers will vary; three examples are $-\sqrt{2}$, $-\sqrt{3}$, and $-\sqrt{5}$.
59. Answers will vary; three examples are −7, 1, and 5. **61.** 83
63. a) {1, 3, 4, 5, 8} **b)** {2, 5, 6, 7, 8} **c)** {5, 8} **d)** {1, 2, 3, 4, 5, 6, 7, 8}
65. a) Set B continues beyond 4. **b)** 4 **c)** An infinite number of elements **d)** Infinite set
67. a) An infinite number **b)** An infinite number **69.** $\frac{14}{3}$ **70.** $5\frac{1}{3}$ **71.** $\frac{49}{40}$ or $1\frac{9}{40}$ **72.** $\frac{70}{27}$ or $2\frac{16}{27}$

Exercise Set 1.5

1. a) −6 −5 −4 −3 −2 −1 0 1 2 3 4 5 6 **b)** −6 −5 −4 −3 −2 −1 0 1 2 3 4 5 6
c) Greater than **d)** $-4 < -2$ **e)** $-2 > -4$ **3. a)** 4 is 4 units from 0 on the number line.
b) −4 is 4 units from 0 on the number line. **c)** 0 is 0 units from 0 on the number line. **5.** Yes **7.** 7 **9.** 15
11. 0 **13.** −3 **15.** −15 **17.** $<$ **19.** $<$ **21.** $>$ **23.** $>$ **25.** $>$ **27.** $>$ **29.** $<$ **31.** $>$ **33.** $<$
35. $>$ **37.** $>$ **39.** $>$ **41.** $>$ **43.** $<$ **45.** $>$ **47.** $>$ **49.** $>$ **51.** $<$ **53.** $<$ **55.** $>$ **57.** $=$
59. $=$ **61.** $<$ **63.** $<$ **65.** 4, −4 **67.** Answers will vary; one example is $4\frac{1}{2}$, 5, and 5.5.
69. Answers will vary; one example is −3, −4, and −5. **71.** Answers will vary; one example is 4, 5, and 6.
73. Answers will vary; one example is 3, 4, and 5. **75. a)** Does not include endpoints
b) Answers will vary; one example is 4.1, 5, and $5\frac{1}{2}$. **c)** No **d)** Yes **e)** True **77. a)** 9.15%; January 1995
b) 7.03%; January 1996 **c)** January, February, March, April, August, October, November, December
d) May, June, July, September **e)** \$255.60 **f)** \$348.00 **79.** Greater than **81.** Yes, 0 **84.** $\frac{31}{24}$ or $1\frac{7}{24}$
85. {0, 1, 2, 3, ...} **86.** {1, 2, 3, 4, ...} **87. a)** 5 **b)** 5, 0 **c)** 5, −2, 0 **d)** $5, -2, 0, \frac{1}{3}, -\frac{5}{9}, 2.3$ **e)** $\sqrt{3}$
f) $5, -2, 0, \frac{1}{3}, \sqrt{3}, -\frac{5}{9}, 2.3$

Exercise Set 1.6

1. Addition, subtraction, multiplication, and division **3. a)** No **b)** $\frac{2}{3}$ **5.** Either
7. Answers will vary. **9. a)** Answers will vary. **b)** −82 **c)** Answers will vary. **11.** −5 **13.** 28 **15.** 0
17. $-\frac{5}{3}$ **19.** $-2\frac{3}{5}$ **21.** −0.47 **23.** 11 **25.** 1 **27.** −6 **29.** 0 **31.** 0 **33.** −10 **35.** 0 **37.** −9 **39.** 0
41. −6 **43.** 9 **45.** −54 **47.** −44 **49.** −26 **51.** 5 **53.** −20 **55.** −39 **57.** 91 **59.** −140 **61.** −105
63. a) Positive **b)** 266 **c)** Yes **65. a)** Negative **b)** −373 **c)** Yes **67. a)** Negative **b)** −452 **c)** Yes
69. a) Negative **b)** −1300 **c)** Yes **71. a)** Negative **b)** −22 **c)** Yes **73. a)** Negative **b)** −3880
c) Yes **75. a)** Positive **b)** 1111 **c)** Yes **77. a)** Negative **b)** −2050 **c)** Yes **79.** True **81.** True
83. False **85.** \$277 **87.** 21 yards **89.** 61 feet **91.** 13,796 feet **93. a)** Loss of \$7310 **b)** Gain of \$5230
c) Gain of \$2440 **95.** −22 **97.** 20 **99.** 0 **101.** 55 **103.** 1 **104.** $\frac{43}{16}$ or $2\frac{11}{16}$ **105.** $>$ **106.** $>$

Exercise Set 1.7

1. $5 - 8$ **3.** $\square - *$ **5. b)** $9 + (-12)$ **c)** −3 **7. a)** $a - b$ **b)** $7 - 9$ **c)** −2
9. a) $3 + 6 - 5$ **b)** 4 **11.** 4 **13.** −1 **15.** −6 **17.** −1 **19.** −8 **21.** −6 **23.** 6 **25.** −4 **27.** 2
29. 9 **31.** 0 **33.** −20 **35.** 0 **37.** 4 **39.** −4 **41.** −11 **43.** −16 **45.** −180 **47.** −82 **49.** 140
51. −22 **53.** −52 **55.** −18 **57.** −16 **59.** −2 **61.** 13 **63.** −11 **65.** 11 **67. a)** Positive **b)** 99
c) Yes **69. a)** Negative **b)** −567 **c)** Yes **71. a)** Positive **b)** 1588 **c)** Yes **73. a)** Positive **b)** 196
c) Yes **75. a)** Negative **b)** −448 **c)** Yes **77. a)** Positive **b)** 116.1 **c)** Yes **79. a)** Negative **b)** −69
c) Yes **81. a)** Negative **b)** −1670 **c)** Yes **83. a)** Zero **b)** 0 **c)** Yes **85.** 7 **87.** −2 **89.** −15
91. −2 **93.** 13 **95.** 7 **97.** −32 **99.** −4 **101.** 9 **103.** −9 **105.** 12 **107.** −18 **109. a)** 43 **b)** 143
111. 16 feet below sea level **113.** Dropped 100°F **115. a)** 835,000 **b)** 685,000 **117.** −5 **119. a)** 7 units
b) $5 - (-2)$ **121. a)** 9 feet **b)** −3 feet **122.** {..., −3, −2, −1, 0, 1, 2, 3, ...} **123.** The set of rational numbers together with the set of irrational numbers forms the set of real numbers. **124.** $>$ **125.** $<$

Exercise Set 1.8 **1.** Like signs: product is positive. Unlike signs: product is negative. **3.** 0 **5.** Positive **7.** Positive **9.** Negative **11.** $-\frac{a}{b}$ or $\frac{-a}{b}$ **13. a)** With $3 - 5$ you subtract, but with $3(-5)$ you multiply. **b)** $-2, -15$ **15. a)** With $x - y$ you subtract, but with $x(-y)$ you multiply. **b)** 7 **c)** 10 **d)** -3 **17.** 12 **19.** -9 **21.** -8 **23.** 0 **25.** 42 **27.** 40 **29.** 30 **31.** 0 **33.** -84 **35.** -72 **37.** 84 **39.** 0 **41.** $-\frac{3}{10}$ **43.** $\frac{14}{27}$ **45.** 4 **47.** $-\frac{5}{16}$ **49.** 3 **51.** 4 **53.** 4 **55.** -18 **57.** 12 **59.** -5 **61.** -6 **63.** -9 **65.** 5 **67.** 0 **69.** -3 **71.** -6 **73.** $-\frac{2}{5}$ **75.** $\frac{7}{24}$ **77.** 1 **79.** $-\frac{144}{5}$ **81.** -32 **83.** -20 **85.** -14 **87.** -9 **89.** -20 **91.** -1 **93.** 0 **95.** Undefined **97.** 0 **99.** Undefined **101. a)** Positive **b)** 18,444 **c)** Yes **103. a)** Negative **b)** -16 **c)** Yes **105. a)** Negative **b)** -9 **c)** Yes **107. a)** Negative **b)** $-15{,}219$ **c)** Yes **109. a)** Zero **b)** 0 **c)** Yes **111. a)** Undefined **b)** Undefined **c)** Yes **113. a)** Positive **b)** 3.2 **c)** Yes **115. a)** Positive **b)** 226.8 **c)** Yes **117.** True **119.** True **121.** True **123.** False **125.** False **127.** True **129.** True **131. a)** \$150 **b)** $-\$300$ **133.** The total loss is $4\frac{1}{2}$ points. **135. a)** ≈ 3.13 **b)** ≈ 6.68 **137. a)** 102 to 128 beats per minutes **b)** Answers will vary. **139.** -8 **141.** 1 **143.** Positive **146.** $\frac{25}{7}$ or $3\frac{4}{7}$ **147.** -2 **148.** -3 **149.** 3

Exercise Set 1.9 **1.** Base; exponent **3. a)** 1 **b)** 1, 3, 2, 1 **5. a)** $5x$ **b)** x^5 **7.** Parentheses, exponents, multiplication or division, then addition or subtraction **9.** No **11. a)** 1 **b)** 5 **c)** b **13. a)** Answers will vary. **b)** -191 **15. a)** Answers will vary. **b)** -91 **17.** 16 **19.** 1 **21.** -25 **23.** 9 **25.** -1 **27.** 81 **29.** 36 **31.** 4 **33.** 256 **35.** -16 **37.** 225 **39.** 72 **41. a)** Positive **b)** 343 **c)** Yes **43. a)** Positive **b)** 625 **c)** Yes **45. a)** Negative **b)** -243 **c)** Yes **47. a)** Positive **b)** 1296 **c)** Yes **49. a)** Positive **b)** 789.0481 **c)** Yes **51. a)** Negative **b)** -0.765625 **c)** Yes **53.** 32 **55.** 8 **57.** 13 **59.** 0 **61.** 16 **63.** 29 **65.** -2 **67.** 10 **69.** 121 **71.** $\frac{1}{2}$ **73.** 169 **75.** 90.74 **77.** 36.75 **79.** $\frac{71}{112}$ **81.** $\frac{1}{4}$ **83.** $\frac{1}{17}$ **85. a)** 9 **b)** -9 **c)** 9 **87. a)** 16 **b)** -16 **c)** 16 **89. a)** 36 **b)** -36 **c)** 36 **91. a)** $\frac{1}{4}$ **b)** $-\frac{1}{4}$ **c)** $\frac{1}{4}$ **93.** 2 **95.** 19 **97.** 3 **99.** -1 **101.** -4 **103.** -3 **105.** 114 **107.** 0 **109.** -5 **111.** 15 **113.** -3 **115.** $[(6 \cdot 3) - 4] - 2; 12$ **117.** $9[[(20 \div 5) + 12] - 8]; 72$ **119.** $\left(\frac{4}{5} + \frac{3}{7}\right) \cdot \frac{2}{3}; \frac{86}{105}$ **121.** All real numbers **123.** \$1.12 **125.** \$16,050 **127. a)** .08 **b)** .16 **129.** 1.71 inches **131.** $12 - (4 - 6) + 10$ **137. a)** 4 **b)**

Occupants	Number of Houses
1	3
2	5
3	4
4	6
5	2

c) 59 **d)** 2.95 people per house **138.** \$6.40 **139.** 144 **140.** $\frac{10}{3}$ or $3\frac{1}{3}$

Exercise Set 1.10 **1.** The commutative property of addition states that the sum of two numbers is the same regardless of the order in which they are added; $3 + 4 = 4 + 3$ **3.** The associative property of addition states that the sum of 3 numbers is the same regardless of the way the numbers are grouped; $(2 + 3) + 4 = 2 + (3 + 4)$ **5.** The distributive property states that the product of a number with a sum is the same as the sum of the products of the number with each number in the sum; $2(3 + 4) = 2(3) + 2(4)$ **7.** The associative property involves changing parentheses, and uses only one operation whereas the distributive property uses two operations, multiplication and addition. **9.** Distributive property **11.** Commutative property of multiplication **13.** Distributive property **15.** Associative property of multiplication **17.** Distributive property **19.** $4 + 3$ **21.** $(-6 \cdot 4) \cdot 2$ **23.** $(y)(6)$ **25.** $1 \cdot x + 1 \cdot y$ or $x + y$ **27.** $3y + 4x$ **29.** $3 + (x + y)$ **31.** $3x + (4 + 6)$ **33.** $(x + y)3$ **35.** $4x + 4y + 12$ **37.** Commutative property of addition **39.** Distributive property **41.** Commutative property of addition **43.** Distributive property **45.** Yes **47.** No **49.** No **51.** Yes **53.** No **55.** No **57.** The $(3 + 4)$ is treated as one value. **59.** Commutative property of addition **61.** No; associative property of addition **63.** $\frac{49}{15}$ or $3\frac{4}{15}$ **64.** $\frac{23}{16}$ or $1\frac{7}{16}$ **65.** 45 **66.** −25

Review Exercises

1. 17.2 lb **2.** \$551.25 **3.** \$30 **4.** Less than **5. a)** 78.4 **b)** 79 **6. a)** 80°F **b)** 79°F **7. a)** 1991 **b)** 10%; 80% **c)** 70% **d)** 8 **8. a)** 3¢ **b)** 1970s **c)** None **d)** 11 **9.** $\frac{1}{2}$ **10.** $\frac{9}{25}$ **11.** $\frac{25}{36}$ **12.** $\frac{7}{6}$ or $1\frac{1}{6}$ **13.** $\frac{19}{72}$ **14.** $\frac{17}{15}$ or $1\frac{2}{15}$ **15.** $\{1, 2, 3, \ldots\}$ **16.** $\{0, 1, 2, 3, \ldots\}$ **17.** $\{\ldots, -3, -2, -1, 0, 1, 2, 3, \ldots\}$ **18.** The set of all numbers which can be expressed as the quotient of two integers, denominator not zero **19.** All numbers that can be represented on a real number line **20. a)** 3, 426 **b)** 3, 0, 426 **c)** 3, −5, −12, 0, 426 **d)** $3, -5, -12, 0, \frac{1}{2}, -0.62, 426, -3\frac{1}{4}$ **e)** $\sqrt{7}$ **f)** $3, -5, -12, 0, \frac{1}{2}, -0.62, \sqrt{7}, 426, -3\frac{1}{4}$ **21. a)** 1 **b)** 1 **c)** −8, −9 **d)** −8, −9, 1 **e)** $-2.3, -8, -9, 1\frac{1}{2}, 1, -\frac{3}{17}$ **f)** $-2.3, -8, -9, 1\frac{1}{2}, \sqrt{2}, -\sqrt{2}, 1, -\frac{3}{17}$ **22.** > **23.** > **24.** < **25.** > **26.** < **27.** > **28.** < **29.** = **30.** −9 **31.** 0 **32.** −3 **33.** −6 **34.** −6 **35.** −5 **36.** 8 **37.** −10 **38.** 5 **39.** −5 **40.** 4 **41.** −12 **42.** 5 **43.** −4 **44.** −12 **45.** −7 **46.** 6 **47.** 9 **48.** −28 **49.** 27 **50.** −120 **51.** $-\frac{6}{35}$ **52.** $-\frac{6}{11}$ **53.** $\frac{15}{56}$ **54.** 0 **55.** 48 **56.** 144 **57.** −5 **58.** −3 **59.** −4 **60.** 0 **61.** −8 **62.** 9 **63.** $\frac{56}{27}$ **64.** $-\frac{35}{9}$ **65.** 1 **66.** 0 **67.** 0 **68.** Undefined **69.** Undefined **70.** Undefined **71.** 0 **72.** 24 **73.** −8 **74.** 1 **75.** 3 **76.** −8 **77.** 18 **78.** −2 **79.** −4 **80.** 10 **81.** 1 **82.** 15 **83.** −4 **84.** 36 **85.** 729 **86.** 81 **87.** −27 **88.** −1 **89.** −32 **90.** $\frac{9}{25}$ **91.** $\frac{8}{125}$ **92.** x^2y **93.** $2^2 \cdot 3^3xy^2$ **94.** $5 \cdot 7^2x^2y$ **95.** x^2y^2z **96.** xxy **97.** $xzzz$ **98.** $yyyz$ **99.** $2xxxyy$ **100.** 23 **101.** 23 **102.** 22 **103.** −19 **104.** −39 **105.** −4 **106.** −60 **107.** 10 **108.** 20 **109.** 20 **110.** 114 **111.** 9 **112.** 14 **113.** 26 **114.** 9 **115.** 0 **116.** −3 **117.** −11 **118.** −3 **119.** 21 **120.** 39 **121.** −335 **122.** 353.6 **123.** −2.88 **124.** 117.8 **125.** 729 **126.** −74.088 **127.** Associative property of addition **128.** Commutative property of multiplication **129.** Distributive property **130.** Commutative property of multiplication **131.** Commutative property of addition **132.** Associative property of addition

Practice Test

1. a) \$10.65 **b)** \$0.23 **c)** \$10.88 **d)** \$39.12 **2. a)** 31%; 18% **b)** 24.5%; 24% **c)** 66.5% **3. a)** 58%; 19% **b)** 39% **c)** ≈ 3 times greater **4. a)** 42 **b)** 42, 0 **c)** −6, 42, 0, −7, −1 **d)** $-6, 42, -3\frac{1}{2}, 0, 6.52, \frac{5}{9}, -7, -1$ **e)** $\sqrt{5}$ **f)** $-6, 42, -3\frac{1}{2}, 0, 6.52, \sqrt{5}, \frac{5}{9}, -7, -1$ **5.** < **6.** > **7.** −12 **8.** −11 **9.** −14 **10.** 8 **11.** −24 **12.** $\frac{16}{63}$ **13.** −2 **14.** −69 **15.** 12 **16.** $\frac{27}{125}$ **17.** $2^25^2y^2z^3$ **18.** $2 \cdot 2 \cdot 3 \cdot 3 \cdot 3xxxxxyy$ **19.** 26 **20.** 10 **21.** 11 **22.** 1 **23.** Commutative property of addition **24.** Distributive property **25.** Associative property of addition

CHAPTER 2

Exercise Set 2.1

1. a) The parts that are added **b)** $3x, -4y$, and -5 **c)** $6xy, 3x, -y$, and -9 **3. a)** The parts that are multiplied **b)** They are multiplied together. **c)** They are multiplied together. **5. a)** Numerical coefficient or coefficient **b)** 4 **c)** 1 **d)** −1 **e)** $\frac{3}{5}$ **f)** $\frac{2}{7}$ **7. a)** The parentheses may be removed without having to change the expression within the parentheses. **b)** $x - 8$ **9.** $8x$ **11.** $-x$ **13.** $5y + 3$

15. $3x$ **17.** $-8x + 8$ **19.** $-5x - 5$ **21.** $-2x$ **23.** $-3x + 3$ **25.** $-2x + 11$ **27.** $3x - 6$ **29.** $-5x + y + 2$
31. $2x - 3$ **33.** $x + \frac{23}{5}$ **35.** $0.8x + 6.42$ **37.** $\frac{1}{2}x + 3y + 1$ **39.** $6x + 8y$ **41.** $x + y$ **43.** $-3x - 5y$
45. $-5x + 15$ **47.** $21.72x - 7.11$ **49.** $-\frac{23}{20}x - 5$ **51.** $2x + 12$ **53.** $5x + 20$ **55.** $-2x + 8$ **57.** $-x + 2$
59. $x - 4$ **61.** $\frac{2}{5}x - 2$ **63.** $-0.9x - 1.5$ **65.** $-x + 3$ **67.** $1.4x + 0.35$ **69.** $x - y$ **71.** $-2x - 4y + 8$
73. $3.41x - 5.72y + 3.08$ **75.** $6x - 4y + \frac{1}{2}$ **77.** $x + 3y - 9$ **79.** $3x - 12 - 6y$ **81.** $3x - 8$ **83.** $2x + 1$
85. $14x + 18$ **87.** $4x - 2y + 3$ **89.** $4x - y + 5$ **91.** $7x + 3$ **93.** $x - 9$ **95.** $2x - 2$ **97.** $4x + 5$
99. $2x + 3$ **101.** $2x + 5$ **103.** $y + 4$ **105.** $3x - 5$ **107.** $x + 15$ **109.** $0.2x - 4y - 2.8$
111. $-6x + 7y$ **113.** $\frac{3}{2}x + \frac{7}{2}$ **115.** 2□ + 3⊖ **117.** $2x + 3y + 2$△ **119.** 1, 2, 3, 4, 6, 12 **121.** 2△ + 2□
123. $18x - 25y + 3$ **125.** $6x^2 + 5y^2 + 3x + 7y$ **127.** 7 **128.** −16 **129.** Answers will vary. **130.** −12

Exercise Set 2.2

1. An equation is a statement that shows two algebraic expressions are equal. **3.** Substitute the value in the equation and then determine if it results in a true statement. **5.** Equivalent equations are two or more equations with the same solution. **7.** Add 4 to both sides of the equation to get the variable by itself. **9.** One example is $x + 2 = 1$. **11.** Subtraction is defined in terms of addition. **13.** Yes **15.** No **17.** Yes **19.** Yes **21.** No **23.** Yes **25. a)** $x + 5 = 8$ **b)** $x = 3$ **27. a)** $12 = x + 3$ **b)** $x = 9$ **29. a)** $10 = x + 7$ **b)** $x = 3$ **31. a)** $x + 6 = 4 + 11$ **b)** $x = 9$ **33.** 4 **35.** −7 **37.** −9 **39.** 43 **41.** 22 **43.** 11 **45.** −4 **47.** −5 **49.** −13 **51.** −23 **53.** 2 **55.** −1 **57.** −4 **59.** −20 **61.** 0 **63.** 17 **65.** −26 **67.** −36 **69.** −46.1 **71.** 46.5 **73.** −8.23 **75.** 5.57 **77.** No, the equation is equivalent to $1 = 2$, a false statement. **79. a)** $2x = 8$ **b)** $x = 4$ **81. a)** $20 = 4x$ **b)** $5 = x$ (or $x = 5$) **83.** $x =$ □ + △ **85.** □ = ☺ − △ **88.** 18 **89.** −8 **90.** $2x - 13$ **91.** $10x - 32$

Exercise Set 2.3

1. Answers will vary. **3. a)** $x = -a$ **b)** $x = -5$ **c)** $x = 5$ **5.** Divide by −2 to isolate the variable. **7.** Multiply both sides by 3. **9. a)** $2x = 10$ **b)** $x = 5$ **11. a)** $6 = 3x$ **b)** $x = 2$ **13. a)** $2x = 5$ **b)** $x = \frac{5}{2}$ **15. a)** $4 = 3x$ **b)** $x = \frac{4}{3}$ **17.** 3 **19.** 8 **21.** −3 **23.** −8 **25.** 5 **27.** 3 **29.** $-\frac{3}{2}$ **31.** 11 **33.** −10 **35.** 49 **37.** $-\frac{1}{3}$ **39.** 6 **41.** $\frac{10}{13}$ **43.** 2 **45.** −1 **47.** $-\frac{3}{40}$ **49.** −60 **51.** 150 **53.** −35 **55.** 20 **57.** −50 **59.** 12 **61.** 0 **63.** 22.5 **65.** 6 **67.** −20.2 **69.** $\frac{1}{12}$ **71.** 9 **73. a)** In $5 + x = 10$, 5 is added to the variable, whereas in $5x = 10$, 5 is multiplied by the variable. **b)** $x = 5$ **c)** $x = 2$ **75. a)** 4 **b)** $2x + 6 = 14$ **c)** $x = 4$ **77. a)** 1 **b)** $6 = 2x + 4$ **c)** $x = 1$ **79.** Multiply by $\frac{3}{2}$; 6 **81.** Multiply by $\frac{7}{3}$; $\frac{28}{15}$ **83. a)** ⊡ **b)** Divide both sides of the equation by △. **c)** ⊡ = $\frac{☺}{△}$ **85.** −4 **86.** 0 **87.** $-11x + 38$ **88.** −57

Exercise Set 2.4

1. No, there is an x on both sides of the equation. **3.** $x = \frac{5}{8}$ **5.** $x = -\frac{1}{2}$ **7.** $x = \frac{3}{5}$ **9.** Solve **11. a)** Answers will vary. **b)** Answers will vary. **13. a)** Answers will vary. **b)** $x = -2$ **15. a)** $2x + 4 = 16$ **b)** $x = 6$ **17. a)** $30 = 2x + 12$ **b)** $x = 9$ **19. a)** $3x + 10 = 4$ **b)** $x = -2$ **21. a)** $5 + 3x = 12$ **b)** $x = \frac{7}{3}$ **23.** 3 **25.** −6 **27.** 5 **29.** $\frac{12}{5}$ **31.** −11 **33.** 3 **35.** $\frac{11}{3}$ **37.** $-\frac{19}{16}$ **39.** −10 **41.** 6 **43.** −9 **45.** 3 **47.** 6.8 **49.** 15 **51.** 12 **53.** 0 **55.** 0 **57.** −1 **59.** 6 **61.** −6 **63.** 1 **65.** −4 **67.** 4 **69.** 5 **71.** 0.6 **73.** −1 **75.** $\frac{2}{7}$ **77.** −2.6 **79. a)** You will not have to work with fractions **b)** $x = 3$ **81. a)** $2x = x + 3$ **b)** $x = 3$ **83. a)** $2x + 3 = 4x + 2$ **b)** $x = \frac{1}{2}$ **85.** $\frac{35}{6}$ **87.** −4 **91.** $\frac{49}{40}$ or $1\frac{9}{40}$ **92.** 64 **93.** Isolate the variable on one side of the equation. **94.** Divide both sides of the equation by −4 to isolate the variable.

Exercise Set 2.5 **1.** Answers will vary. **3. a)** An identity is an equation that is true for all real numbers. **b)** All real numbers **5.** Both sides of the equation are identical. **7.** You will obtain a false statement. **9. a)** Answers will vary. **b)** $x = 21$ **11. a)** $2x = x + 6$ **b)** $x = 6$ **13. a)** $5 + 2x = x + 19$ **b)** $x = 14$ **15. a)** $5 + x = 2x + 5$ **b)** $x = 0$ **17. a)** $2x + 8 = x + 4$ **b)** $x = -4$ **19.** 15 **21.** 1 **23.** $\frac{3}{5}$ **25.** $\frac{5}{2}$ **27.** 2 **29.** 1 **31.** 0.1 **33.** 3 **35.** $-\frac{17}{7}$ **37.** No solution **39.** $\frac{34}{5}$ **41.** $-\frac{4}{5}$ **43.** 25 **45.** All real numbers **47.** -1 **49.** 0 **51.** All real numbers **53.** $\frac{21}{20}$ **55.** 14 **57.** $-\frac{5}{3}$ **59.** 0 **61.** 16 **63.** $-\frac{10}{3}$ **65. a)** One example is $x + x + 1 = x + 2$. **b)** It has a single solution. **c)** Answers will vary. **67. a)** One example is $x + x + 1 = 2x + 1$. **b)** Both sides simplify to the same expression. **c)** All real numbers **69. a)** One example is $x + x + 1 = 2x + 2$. **b)** It simplifies to a false statement. **c)** No solution **71.** $* = 6$ **73.** All real numbers **75.** $x = -4$ **79.** ≈ 0.131687243 **80.** Factors are expressions that are multiplied; terms are expressions that are added. **81.** $7x - 10$ **82.** $\frac{10}{7}$ **83.** -3

Exercise Set 2.6 **1.** A ratio is a quotient of two quantities. **3.** c to d, $c:d$, $\frac{c}{d}$ **5.** Need a given ratio and one of the two parts of a second ratio **7.** Yes **9.** No **11.** No, their corresponding angles must be equal but their corresponding sides only need to be in proportion. **13.** 2 : 3 **15.** 3 : 2 **17.** 8 : 1 **19.** 7 : 4 **21.** 2 : 3 **23.** 6 : 1 **25.** 13 : 32 **27.** 8 : 1 **29. a)** 430 : 320 or 43 : 32 **b)** $\approx$ 1.34 : 1 **31. a)** 1,001,000 : 798,000 or 143 : 114 **b)** $\approx$ 1.25 : 1 **33. a)** 40 : 9 **b)** 33 : 53 **35. a)** 63 : 13 **b)** 57 : 13 **37.** 16 **39.** 45 **41.** -100 **43.** -2 **45.** -54 **47.** 6 **49.** 32 inches **51.** 0.72 feet **53.** 19.5 inches **55.** 25 loads **57.** 361.1 miles **59.** $\frac{3}{4}$ feet or 9 inches **61.** 24 teaspoons **63.** $\approx$ 228 inches or $\approx$ 19 feet **65.** 340 trees **67.** $\approx$ 8.75 feet **69.** 0.55 milliliter **71.** 96 seconds or 1 minute 36 seconds **73.** About 260 children **75.** 3.5 feet **77.** 2.9 square yards **79.** 10.5 inches **81. a)** \$10.50 **b)** \$1.36 **83.** \$0.85 **85.** 4 points **87.** Yes, her ratio is 2.12:1. **89.** It must increase. **91.** $\approx$ 41.667 miles **93.** Viper: 117.25 mm; Cadillac: 141 mm **98.** Commutative property of addition **99.** Associative property of multiplication **100.** Distributive property **101.** $x = \frac{3}{4}$

Exercise Set 2.7 **1.** $>$: is greater than; $\geq$: is greater than or equal to; $<$: is less than; $\leq$: is less than or equal to **3. a)** False **b)** True **5.** When multiplying or dividing by a negative number **7.** All real numbers **9.** No solution **11.** $x > 4$; (number line, open circle at 4) **13.** $x \geq -3$; (number line, closed circle at -3) **15.** $x > -5$; (number line, open circle at -5) **17.** $x < 10$; (number line, open circle at 10) **19.** $x \leq -7$; (number line, closed circle at -7) **21.** $x > -\frac{3}{2}$; (number line, open circle at $-\frac{3}{2}$) **23.** $x \leq 1$; (number line, closed circle at 1)  **25.** $x < 0$; (number line, open circle at 0) **27.** $x < \frac{3}{2}$; (number line, open circle at $\frac{3}{2}$) **29.** $x > \frac{35}{9}$; (number line, open circle at $\frac{35}{9}$) **31.** $x > -\frac{8}{3}$; (number line, open circle at $-\frac{8}{3}$)  **33.** $x \leq -\frac{11}{3}$; (number line, closed circle at $-\frac{11}{3}$) **35.** $x \geq -6$; (number line, closed circle at -6)  **37.** $x < 1$; (number line, open circle at 1) **39.** $x < 2$; (number line, open circle at 2) **41.** All real numbers; (number line, 0) **43.** $x > \frac{3}{4}$; (number line, open circle at $\frac{3}{4}$)  **45.** $x > \frac{23}{10}$; (number line, open circle at $\frac{23}{10}$) **47.** No solution; (number line, 0) **49.** $x \geq -\frac{7}{11}$; (number line, closed circle at $-\frac{7}{11}$) 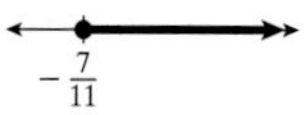 **51. a)** May, September, June, August, and July **b)** January, February, December, March, November, and April **c)** January, February, and December **d)** June, August, and July **53.** $\neq$ **55.** We do not know that y is positive. If y is negative, we must reverse the sign of the inequality. **57.** $x > 4$ **58.** -9 **59.** -25 **60.** $\frac{14}{5}$ or $2\frac{4}{5}$ **61.** 500 kilowatt-hours

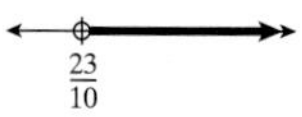

Review Exercises **1.** $3x + 12$ **2.** $3x - 6$ **3.** $-2x - 8$ **4.** $-x - 2$ **5.** $-x + 2$ **6.** $-16 + 4x$ **7.** $18 - 6x$ **8.** $24x - 30$ **9.** $-25x + 25$ **10.** $-4x + 12$ **11.** $-6x + 4$ **12.** $-3 - 2y$ **13.** $-x - 2y + z$ **14.** $-4x + 6y - 14$ **15.** $8x$ **16.** $5y + 2$ **17.** $-3y + 8$ **18.** $5x + 1$ **19.** $-3x + 3y$ **20.** $6x + 8y$ **21.** $9x + 3y + 2$ **22.** $4x + 3y + 6$ **23.** 3 **24.** $-12x + 3$ **25.** $-2x$ **26.** $5x + 7$ **27.** $-10x + 12$ **28.** 0 **29.** $4x - 4$ **30.** $22x - 42$ **31.** $3x - 3y + 6$ **32.** $3x + 2y + 16$ **33.** 3 **34.** $-x - 2y + 4$ **35.** 1 **36.** -13 **37.** 11 **38.** -27 **39.** 2 **40.** $\frac{11}{2}$ **41.** -2 **42.** -3 **43.** 12 **44.** 4 **45.** 6 **46.** -3 **47.** $-\frac{21}{5}$ **48.** -5 **49.** -19 **50.** -1 **51.** 0 **52.** $\frac{1}{5}$ **53.** 3 **54.** $\frac{10}{7}$ **55.** 0 **56.** -1 **57.** No solution **58.** $-\frac{23}{5}$ **59.** All real numbers **60.** $-\frac{4}{3}$ **61.** No solution **62.** All real numbers **63.** $\frac{17}{3}$ **64.** $-\frac{20}{7}$ **65.** 3 : 5 **66.** 5 : 12 **67.** 1 : 1 **68.** 2 **69.** 20 **70.** 9 **71.** $\frac{135}{4}$ **72.** -10 **73.** -24 **74.** 36 **75.** 90 **76.** 40 inches **77.** 1 foot **78.** $x \geq 2$; 2 **79.** $x < 2$; 2 **80.** $x \geq -\frac{12}{5}$; $-\frac{12}{5}$ **81.** No solution; 0 **82.** All real numbers; 0 **83.** $x < -3$; -3 **84.** $x \leq \frac{9}{5}$; $\frac{9}{5}$ **85.** $x > \frac{8}{5}$; $\frac{8}{5}$ **86.** No solution; 0 **87.** All real numbers; 0 **88.** 240 calories **89.** 440 pages **90.** $6\frac{1}{3}$ inches **91.** 9.45 feet **92.** $\approx$ \$0.1171 **93.** 57.3° **94.** 192 bottles

Practice Test **1.** $12x - 24$ **2.** $-x - 3y + 4$ **3.** $-3x + 4$ **4.** $-x + 10$ **5.** $-5x - y - 6$ **6.** $7x - 5y + 3$ **7.** $8x - 1$ **8.** $x = 4$ **9.** $x = -2$ **10.** $x = -3$ **11.** $x = -1$ **12.** $x = -\frac{1}{7}$ **13.** No solution **14.** All real numbers **15.** $x = -45$ **16.** $x = 0$ **17. a)** Conditional equation **b)** Contradiction **c)** Identity **18.** $x > -7$; -7 **19.** $x \leq 12$; 12 **20.** No solution; 0 **21.** All real numbers; 0 **22.** $x = \frac{32}{3}$ feet or $10\frac{2}{3}$ feet **23.** 150 gallons **24.** 50,000 gallons **25.** 175 minutes or 2 hours 55 minutes

Cumulative Review Test **1.** $\frac{16}{25}$ **2.** $\frac{1}{2}$ **3.** > **4.** -6 **5.** -1 **6.** 16 **7.** 3 **8.** 1 **9.** Associative property of addition **10.** $12x + y$ **11.** $3x + 16$ **12.** 3 **13.** -40 **14.** -2 **15.** $-\frac{5}{4}$ **16.** 6 **17.** $x > 10$; 10 **18.** $x \geq -12$; -12 **19.** 158.4 pounds **20.** \$42

CHAPTER 3

Exercise Set 3.1 **1.** A formula is an equation used to express a relationship mathematically. **3.** $i = prt$; i is the amount of interest earned or owed, p is the amount invested or borrowed, r is the rate in decimal form, and t is the length of time invested. **5.** The diameter of a circle is 2 times its radius. **7.** When you multiply a unit by the same unit, you get a square unit. **9.** 16 **11.** 96 **13.** 30.48 **15.** 50.27 **17.** 2 **19.** 15 **21.** 56 **23.** 0.05 **25.** 904.78 **27.** 127.03 **29.** 179.20 **31. a)** $y = -3x + 5$ **b)** -1 **33. a)** $y = \frac{2x + 4}{3}$ **b)** 8 **35. a)** $y = \frac{-3x + 6}{2}$ **b)** 0 **37. a)** $y = \frac{3x - 10}{5}$ **b)** $\frac{2}{5}$ **39. a)** $y = \frac{x + 6}{2}$ **b)** 3 **41. a)** $y = \frac{-x + 8}{2}$ **b)** 6 **43.** $s = \frac{P}{4}$ **45.** $r = \frac{d}{t}$ **47.** $d = \frac{C}{\pi}$ **49.** $h = \frac{2A}{b}$ **51.** $w = \frac{P - 2l}{2}$ **53.** $n = \frac{m - 3}{6}$ **55.** $b = y - mx$ **57.** $r = \frac{A - P}{Pt}$ **59.** $d = \frac{3A - m}{2}$ **61.** $b = d - a - c$ **63.** $y = \frac{-ax + c}{b}$ **65.** $h = \frac{V}{\pi r^2}$ **67.** 35 **69.** $C = 10°$ **71.** $F = 95°$ **73.** $P = 10$

75. $K = 4$ **77.** The area is 4 times as large. **79.** 30 **81.** \$1440 **83.** \$5000 **85.** 25 feet **87.** 558 square inches **89.** $\approx$ 2123.72 square centimeters **91.** 7 square feet **93.** 18 square feet **95.** 14,000 square feet **97.** $\approx$ 11.78 cubic inches **99. a)** $B = \frac{703w}{h^2}$ **b)** $\approx$ 23.91 **101. a)** $\pi = \frac{C}{2r}$ or $\pi = \frac{C}{d}$ **b)** π or about 3.14. **c)** About 3.14 **103. a)** $V = 18x^3 - 3x^2$ **b)** 6027 cubic centimeters **c)** $S = 54x^2 - 8x$ **d)** 2590 square centimeters **106.** 0 **107.** 3 : 2 **108.** $\frac{3}{25} = \frac{x}{13{,}500}$; 1620 minutes or 27 hours **109.** $x \le -17$

Exercise Set 3.2

1. Added to, more than, increased by, and sum indicate addition. **3.** Multiplied by, product of, twice, three times, etc., indicate multiplication. **5.** The cost is increased by 25% of the cost, so the expression should be $c + 0.25c$. **7.** $n + 5$ **9.** $4x$ **11.** $\frac{x}{2}$ **13.** $s + 1.2$ **15.** $0.16P$ **17.** $10P - 9$ **19.** $\frac{8}{9}m + 16{,}000$ **21.** $45 + 0.40x$ **23.** $25x$ **25.** $16x + y$ **27.** $n + 0.04n$ **29.** $t - 0.18t$ **31.** $220 + 80x$ **33.** $275x + 25y$ **35.** Six less than a number **37.** One more than 4 times a number **39.** Seven less than 5 times a number **41.** Four times a number, decreased by 2 **43.** Three times a number subtracted from 2 **45.** Twice the difference between a number and 1 **47. a)** c = Chuck's age **b)** $c + 3$ = Dana's age **49. a)** l = Lois' age **b)** $\frac{1}{3}l$ = Lois' son's age **51. a)** x = the first even integer **b)** $x + 2$ = the next consecutive even integer. **53. a)** c = cost of Camaro **b)** $1.1c$ = cost of Firebird **55. a)** t = percent of profits Tabitha receives **b)** $100 - t$ = percent of profits Shari receives **57. a)** l = average monthly rental rate in Los Angeles **b)** $2l - 177$ = average monthly rental rate in San Jose **59. a)** x = number of coupons redeemed in 1997 (billions) **b)** $2x - 4.4$ = number of coupons redeemed in 1999 **61. a)** n = number of subscribers in 1985 **b)** $125n + 1{,}500{,}000$ = number of subscribers in 1996 **63. a)** s = Dianne's 1998 sales **b)** $s + 0.60s$ = Dianne's 1999 sales **65. a)** s = Kevin's salary last year **b)** $s + 0.15s$ = Kevin's salary this year **67. a)** c = cost of car **b)** $c + 0.07c$ = cost of car with tax **69. a)** p = original pollution level **b)** $p - 0.50p$ = pollution level after decrease **71. a)** x = first number **b)** $x + 5x = 18$ **73. a)** x = smaller integer **b)** $x + (x + 1) = 47$ **75. a)** x = the number **b)** $2x - 8 = 12$ **77. a)** x = the number **b)** $\frac{1}{5}(x + 10) = 150$ **79. a)** s = distance traveled by Southern Pacific train **b)** $s + (2s - 4) = 890$ **81. a)** c = cost of car **b)** $c + 0.07c = 26{,}200$ **83. a)** c = cost of the meal **b)** $c + 0.15c = 32.50$ **85. a)** x = average player's salary **b)** $7.69x - x = 87{,}000$ **87. a)** s = average teacher's salary in South Dakota **b)** $s + (3s - 28{,}784) = 76{,}600$ **89.** Three more than a number is 6. **91.** Three times a number, decreased by 1, is 4 more than twice the number. **93.** Four times the difference between a number and 1 is 6. **95.** Six more than 5 times a number is the difference between 6 times the number and 1. **97.** The sum of a number and the number increased by 4 is 8. **99.** The sum of twice a number and the number increased by 3 is 5. **101.** Answers will vary. **103. a)** $86{,}400d + 3600h + 60m + s$ **b)** 368,125 seconds **109.** $\frac{\frac{1}{2}}{1} = \frac{x}{6.7}$; 3.35 teaspoons **110.** $\frac{1}{3} = \frac{x}{\frac{1}{2}}$; $\frac{1}{6}$ cup **111.** 15 **112.** $y = \frac{3x - 6}{2}$ or $y = \frac{3}{2}x - 3$; 6

Exercise Set 3.3

1. Understand, Translate, Carry out, Check, Answer **3.** Volume · percent = amount **5. a)** $x + (x + 1) = 85$ **b)** 42, 43 **7. a)** $x + (2x + 3) = 27$ **b)** 8, 19 **9. a)** $x + (x + 1) + (x + 2) = 39$ **b)** 12, 13, 14 **11. a)** $(2x - 8) - x = 17$ **b)** 25, 42 **13. a)** $x + (x + 40) = 200$ **b)** Vacuum, 80 decibels; concert, 120 decibels **15. a)** $x + 26x = 121.5$ **b)** Decaf, 4.5 mg; regular, 117 mg **17. a)** $x + (2x - 0.44) = 1.27$ **b)** Germany, 57¢; Japan, 70¢ **19. a)** $20x = 15$ **b)** 75% **21. a)** $6000 + 450x = 10{,}050$ **b)** 9 years **23. a)** $x + (2x - 60) = 153$ **b)** 71, 82 **25. a)** $3000 + 0.03x = 3750$ **b)** \$25,000 **27. a)** $x - 0.20x = 25.99$ **b)** \$32.49 **29. a)** $x + 0.04x = 36{,}400$ **b)** \$35,000 **31. a)** $x + 0.07x = 22{,}800$ **b)** \$21,308.41 **33. a)** $(2x + 5) - x = 65$ **b)** Budget, \$60 million; projected gross earnings, \$125 million **35. a)** $x - (2x - 8.6) = 1.5$ **b)** Boxster, 7.1 seconds; Corvette, 5.6 seconds **37. a)** $40{,}000 + 2400x = 49{,}600 + 800x$ **b)** 6 years **39. a)** $x + x + 2x + 3x = 91$ **b)** Younger workers, 13 hours; third, 26 hours; fourth, 39 hours **41. a)** $x - 0.60x = 24$ **b)** 60 gallons **43. a)** $22 + 22x = 23.76$ **b)** 8% **45. a)** $x - 0.10x - 20 = 250$ **b)** \$300 **47. a)** $64.3 = x + 0.63x$

b) ≈ 39.45 million pounds **49. a)** $x + (x + 1) + (5x + 2) = 24$ **b)** Personal, 3; bills & statements, 4; advertisements, 17 **51. a)** 99 minutes **b)** *One Rate Plus Plan*; save $5.05 **53. a)** $466.50 **b)** $418.50 **c)** $2000 **d)** ≈ 42 months or 3.5 years **e)** Citibank **55. a)** 18.75 months or about 1.56 years **b)** Countrywide **57. a)** ≈ 9.2 months **b)** $325.50 per month **59. a)** $\frac{74 + 88 + 76 + x}{4} = 80$ **b)** 82 **61. a)** 0.75 year, or 9 months **b)** $375 **63.** $\frac{17}{12}$ **64.** Associative property of addition **65.** Commutative property of multiplication **66.** Distributive property **67.** 56 pounds **68.** $b = 2M - a$

Exercise Set 3.4 **1.** The area remains the same. **3.** The volume is eight times as great. **5.** An isosceles triangle is a triangle with 2 equal sides. **7.** 180° **9.** 9.5 inches **11.** $A = 47°; B = 133°$ **13.** 50°; 60°; 70° **15.** 4 meters; 4 meters; 2 meters **17.** Width is 36 feet; length is 78 feet **19.** Smaller angles are 50°; larger angles are 130° **21.** 63°, 73°, 140°, 84° **23.** Width is 4 feet; height is 7 feet **25.** Width is 2.5 feet; height is 5 feet **27.** Width is 10 feet; length is 14 feet **29. a)** $A = S^2 - s^2$ **b)** 45 square inches **32.** $<$ **33.** $>$ **34.** −8 **35.** $-2x - 4y + 6$ **36.** $y = \frac{-2x + 9}{3}$ or $y = -\frac{2}{3}x + 3; 1$

Exercise Set 3.5 **1.** 50 mph **3.** 2.4 cm **5.** 2736 mph **7.** 250 cubic cm per hour **9.** ≈ 64.4 days **11.** ≈ 189.75 mph **13.** 2.4 hours **15.** 70 mph **17.** 2 hours **19. a)** 35 mph, 40 mph **b)** ≈ 30.43 knots; ≈ 34.78 knots **21. a)** 2.4 miles **b)** 112.0 miles **c)** 26.2 miles **d)** 140.6 miles **e)** 9.51 hours **23.** *Apollo*: 5 mph; *Pythagoras*: 9 mph **25. a)** 1.25 hours **b)** 43.75 miles **27. a)** 2 seconds **b)** 100 feet **29. a)** 275 feet **b)** 120 feet per minute **31.** 570 mph; 600 mph **33.** Bridge: 1.25 feet per day; road: 2.45 feet per day **35.** $2400 at 5%; $7000 at 7% **37.** $2400 at 6%; $3600 at 4% **39.** November **41.** 8 hours at Home Depot; 10 hours at veterinary clinic **43.** $6.07 per pound **45.** ≈ 11.1% **47.** $1\frac{2}{3}$ liters **49.** 4 ounces **51.** 1.9 gallons of Clorox; 2.1 gallons of the shock treatment **53.** 3.25% **55.** 40% pure juice. **57. a)** 32 shares of GE; 160 shares of PepsiCo **b)** $32 **59.** ≈ 5.74 hours **63. a)** $\frac{22}{13}$ or $1\frac{9}{13}$ **b)** $\frac{35}{8}$ or $4\frac{3}{8}$ **64.** All real numbers **65.** $\frac{3}{4}$ or 0.75 **66.** $x \le \frac{1}{4}$

Review Exercises **1.** 25.13 **2.** 18 **3.** 48 **4.** 25 **5.** 21 **6.** 5000 **7. a)** $y = x - 2$ **b)** 8 **8. a)** $y = -2x - 3$ **b)** −27 **9. a)** $y = \frac{4}{3}x - 5$ **b)** −1 **10. a)** $y = \frac{2}{3}x - 4$ **b)** −8 **11.** $m = \frac{F}{a}$ **12.** $h = \frac{2A}{b}$ **13.** $t = \frac{i}{pr}$ **14.** $w = \frac{P - 2l}{2}$ **15.** $h = \frac{V}{\pi r^2}$ **16.** $B = 2A - C$ **17.** $108 **18.** 6 inches **19.** The sum of a number and the number increased by 5 is 9. **20.** The sum of a number and twice the number decreased by 1 is 10. **21.** 33 and 41 **22.** 118 and 119 **23.** 38 and 7 **24.** $18,000 **25.** $2000 **26.** $650 **27. a)** ≈ 166.7 months or 13.9 years **b)** Mellon Bank **28. a)** ≈ 27.4 months or 2.3 years **b)** Yes **29.** 45°, 55°, 80° **30.** 30°, 40°, 150°, 140° **31.** Width: 15.5 feet; length: 19.5 feet **32.** Width: 50 feet; length: 80 feet **33.** 30 gallons per hour **34.** 6.5 mph **35.** 2 hours **36.** 4 hours **37.** $4000 at 8%; $8000 at $7\frac{1}{4}$% **38.** 1.2 liters of 10%; 0.8 liters of 5% **39.** 103 and 105 **40.** $450 **41.** $12,000 **42.** 42°, 50°, 88° **43.** 8 years **44.** 70°, 70°, 110°, 110° **45. a)** 500 copies **b)** King Kopie by $5 **46. a)** 10 minutes **b)** 600 feet **47.** 60 pounds of $3.50; 20 pounds of $4.10 **48.** Older brother: 55 mph; younger brother: 60 mph **49.** 0.4 liters

Practice Test **1.** 9% **2.** 18 feet **3.** 145 **4.** 85 **5.** ≈ 7.96 **6. a)** $y = \frac{4}{3}x - 3$ **b)** 13 **7.** $R = \frac{P}{I}$ **8.** $a = 3A - b$ **9.** $c = \frac{D - Ra}{R}$ or $c = \frac{D}{R} - a$ **10.** 3106.86 square feet **11. a)** x = the amount of money Matthew receives **b)** $300 - x$ = the amount that Karen receives **12. a)** z = the projected gross income from *Mask of Zorro* **b)** $2z - 45$ = the projected gross income from *Deep Impact* **13.** The sum of a number and the number increased by 3 is 7. **14. a)** x = smaller integer **b)** $x + (2x - 10) = 158$ **c)** 56 and 102 **15. a)** x = smallest integer **b)** $x + (x + 1) + (x + 2) = 42$ **c)** 13, 14, and 15

16. a) c = cost of furniture before tax **b)** $c + 0.06c = 2650$ **c)** \$2500
17. a) x = price of the most expensive meal he can order **b)** $x + 0.15x + 0.07x = 40$ **c)** \$32.79
18. a) x = amount of profit Kathleen and Corrina each receive **b)** $x + x + 2x = 120{,}000$
c) Kathleen and Corrina: \$30,000; Kristen: \$60,000
19. a) x = number of times the plow is needed for the costs to be equal **b)** $80 + 5x = 50 + 10x$ **c)** 6 times
20. a) x = number of months for the total cost of both mortgages to be the same.
b) $451.20x + 600 = 430.20x + 2400$ **c)** $\approx$ 85.7 months or 7.1 years
21. a) x = length of the smallest side **b)** $x + (x + 15) + 2x = 75$ **c)** 15, 30, and 30 inches
22. a) x = measure of each smaller angle **b)** $x + x + (2x + 30) + (2x + 30) = 360$ **c)** 50°, 50°, 130°, 130°
23. a) x = width of the driveway **b)** $2x + 2(4x + 12) = 144$ **c)** Width = 12 feet, length = 60 feet
24. a) x = rate Carrie digs **b)** $84x + 84(x + 0.2) = 67.2$ **c)** Carrie, 0.3 foot per minute; Don, 0.5 foot per minute
25. a) x = amount of 20% salt solution to be added **b)** $0.20x + (0.40)(60) = 0.35(x + 60)$ **c)** 20 liters

Cumulative Review Test

1. 35.36 million tons **2. a)** 129,262 **b)** 255,908 **c)** 3.04 times greater **3. a)** 7.2 parts per million **b)** 6 parts per million **4.** $\frac{5}{9}$ **5.** $\frac{7}{24}$ inch **6. a)** $\{1, 2, 3, 4, \ldots\}$ **b)** $\{0, 1, 2, 3, \ldots\}$ **c)** A rational number is a quotient of two integers, denominator not 0. **7. a)** 3 **b)** $|-5|$ **8.** 6 **9.** $2x - 28$ **10.** 9 **11.** $\frac{1}{5}$ **12.** No solution **13.** 9 gallons **14.** $x \leq 1$; (number line, closed dot at 1, shaded left) **15.** $\approx$ 113.10 **16. a)** $y = -\frac{1}{2}x + 2$ **b)** 4 **17.** $l = \frac{P - 2w}{2}$ **18.** 6, 23 **19.** 40 minutes **20.** 40°, 45°, 90°, 185°

CHAPTER 4

Using Your Graphing Calculator, 4.1

1.

−20, 40, 5, −10, 60, 10

2.

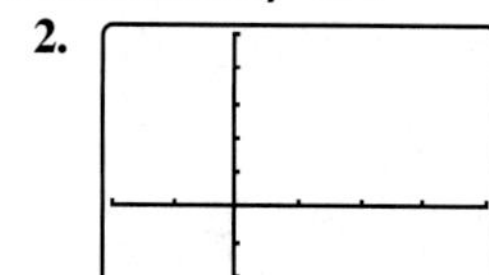

−200, 400, 100, −500, 1000, 200

3. 50 **4.** 100

Exercise Set 4.1

1. The x-coordinate **3. a)** x-axis **b)** y-axis **5.** Axis is singular; axes is plural. **7.** An illustration of the set of points whose coordinates satisfy the equation **9.** A straight line **11.** $ax + by = c$

13.

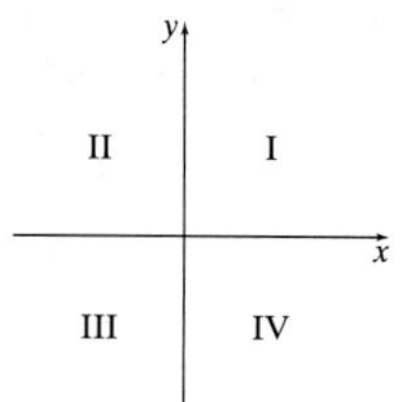

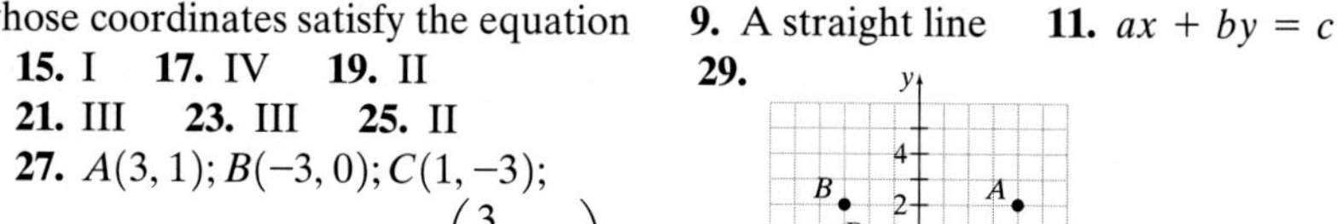

15. I **17.** IV **19.** II **21.** III **23.** III **25.** II

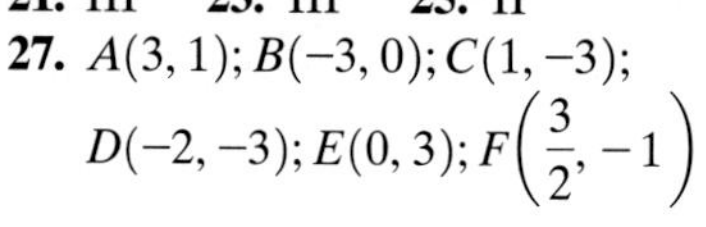

27. $A(3, 1)$; $B(-3, 0)$; $C(1, -3)$; $D(-2, -3)$; $E(0, 3)$; $F\left(\frac{3}{2}, -1\right)$

29.

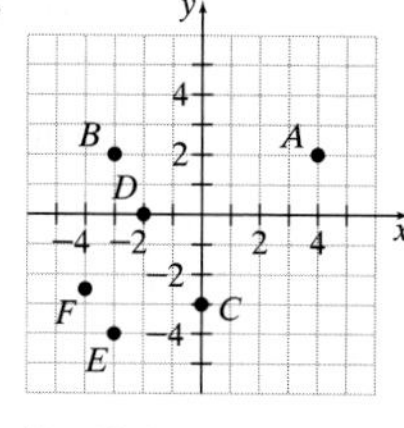

31.

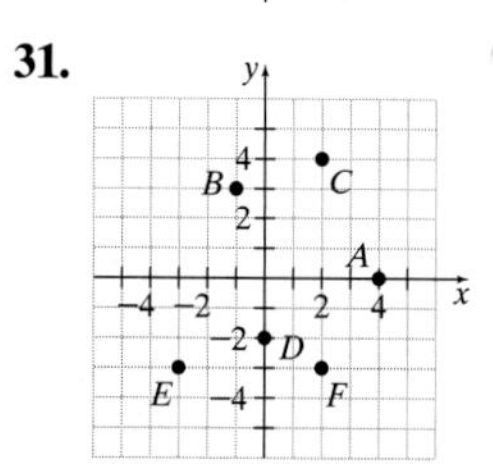

33. The points are collinear.

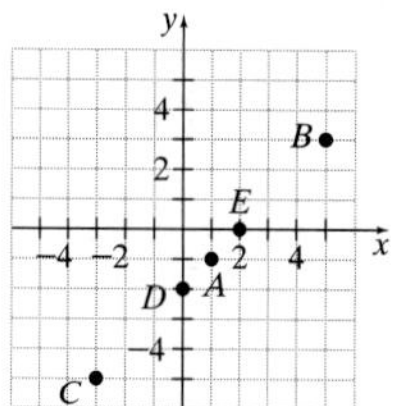

35. (3, 0) is not on the line.

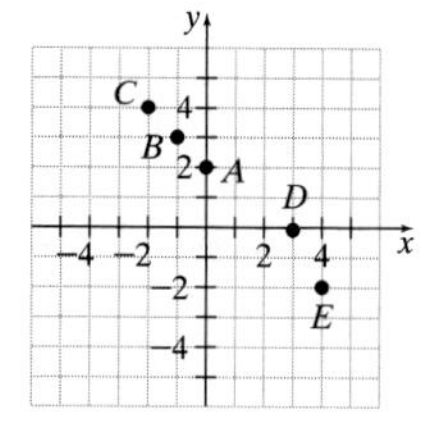

37. a) Point d) does not satisfy the equation.
b)

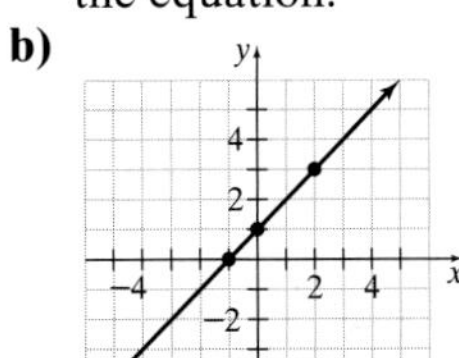

39. a) Point a) does not satisfy the equation.
b)

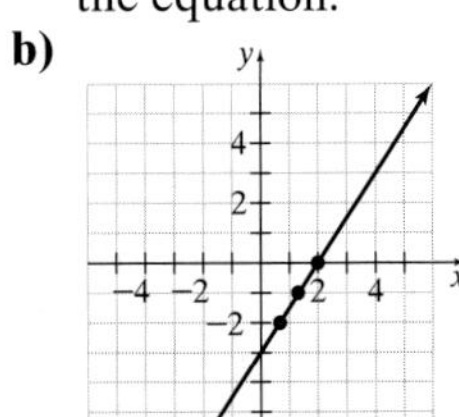

41. a) Point a) does not satisfy the equation.
b)

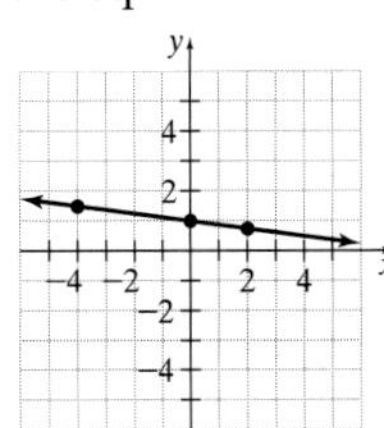

43. 4 **45.** −2 **47.** 2 **49.** $\frac{11}{3}$ **51.** 0 **58.** An equation of the form $ax + b = c$
59. A linear equation that has only one solution **60.** An equation that is true for all real numbers
61. $C \approx 18.84$ inches; $A \approx 28.27$ square inches **62.** $y = \frac{2x-6}{5} = \frac{2}{5}x - \frac{6}{5}$

Using Your Graphing Calculator, 4.2

1.

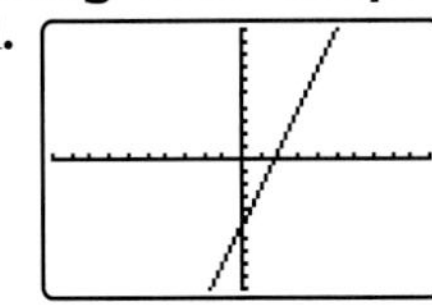

−10, 10, 1, −10, 10, 1

2.

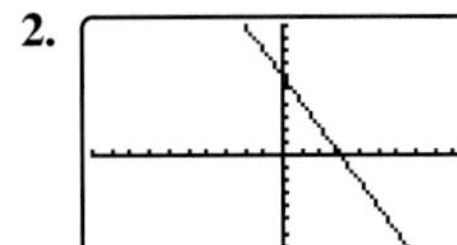

−10, 10, 1, −10, 10, 1

3.

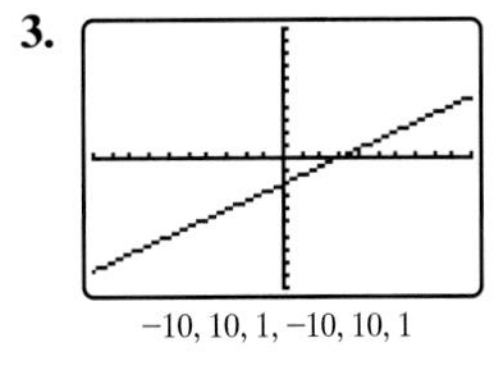

−10, 10, 1, −10, 10, 1

4.

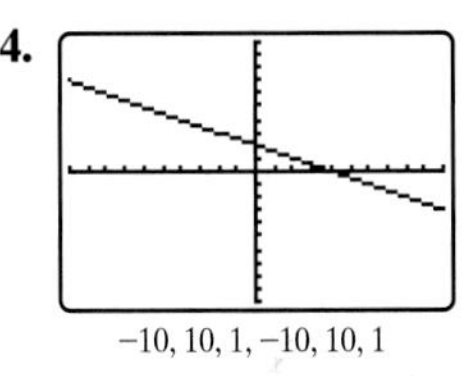

−10, 10, 1, −10, 10, 1

Using Your Graphing Calculator, 4.2

1. (2, 0), (0, −8) **2.** (−3, 0), (0, −6) **3.** (2.5, 0), (0, −2) **4.** (1.5, 0), (0, −4.5)

Exercise Set 4.2

1. x-intercept: substitute 0 for y and find the corresponding value of x; y-intercept: substitute 0 for x and find the corresponding value of y **3.** A horizontal line
5. You may not be able to read exact answers from a graph. **7.** Yes **9.** 0 **11.** $\frac{11}{2}$ **13.** 3 **15.** 2 **17.** $\frac{8}{3}$ **19.** $-\frac{17}{2}$

21.

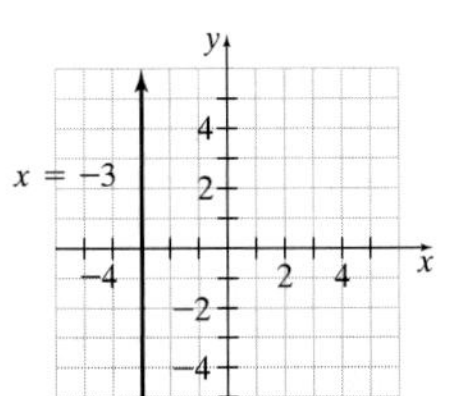

23.

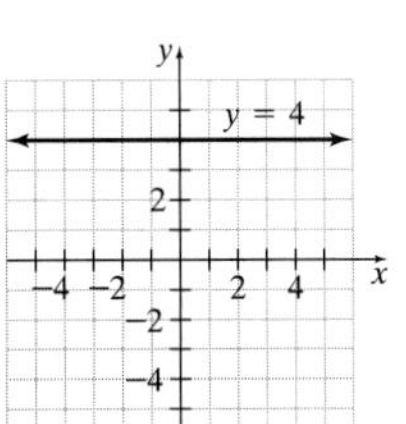

25.

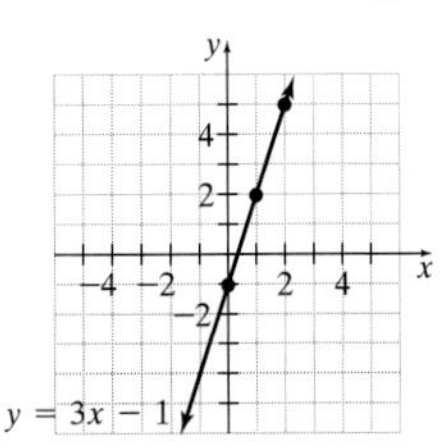

27.

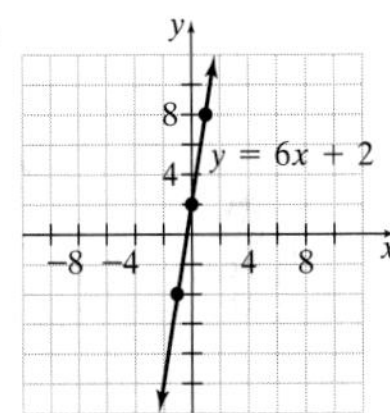

29.

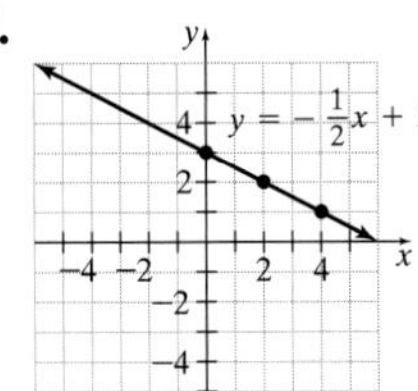

31.

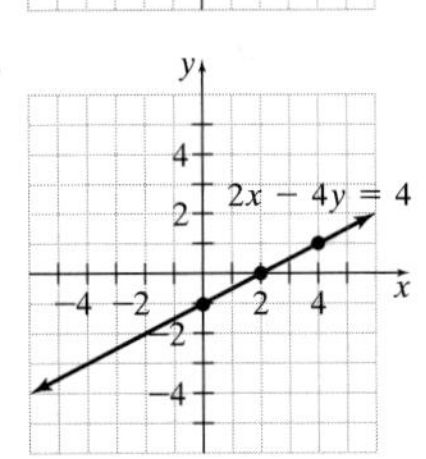

33.

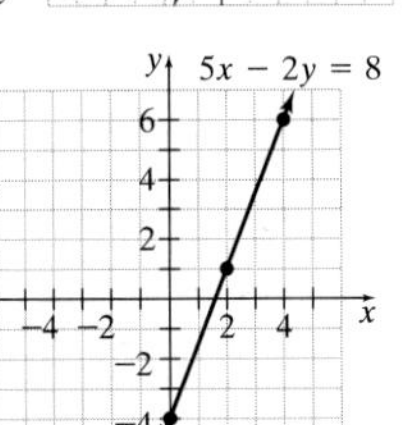

35.

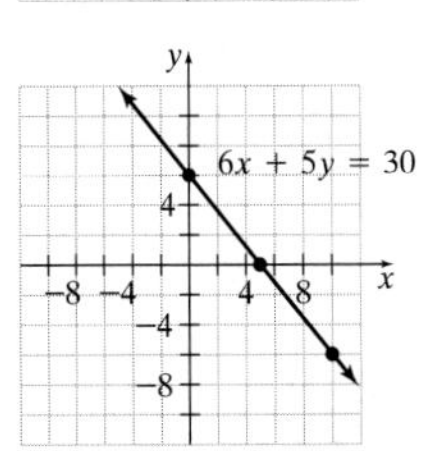

37.

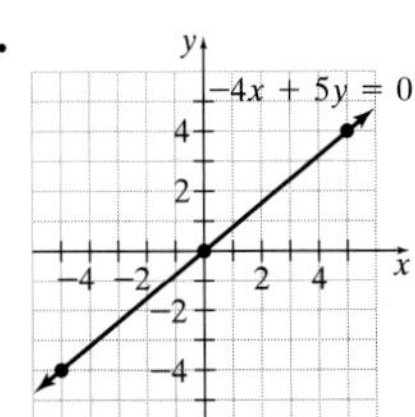

39.

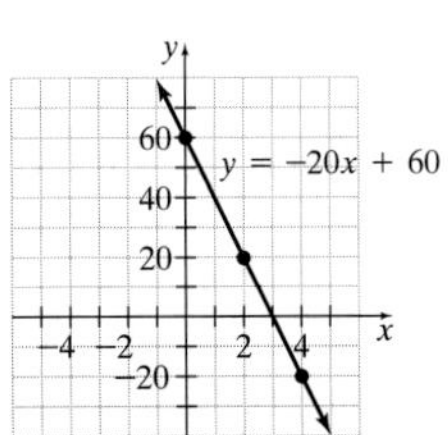

41.

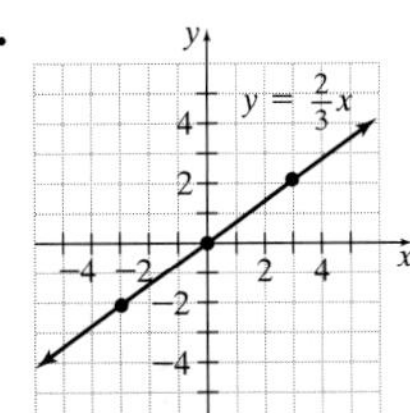

43.

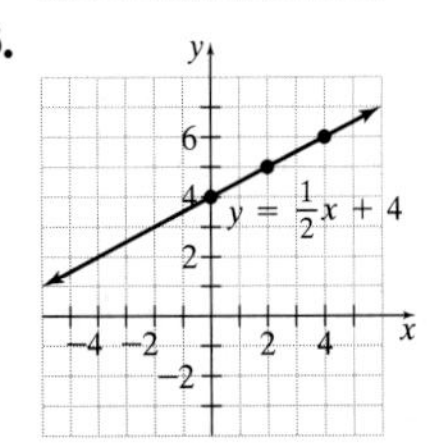

45.

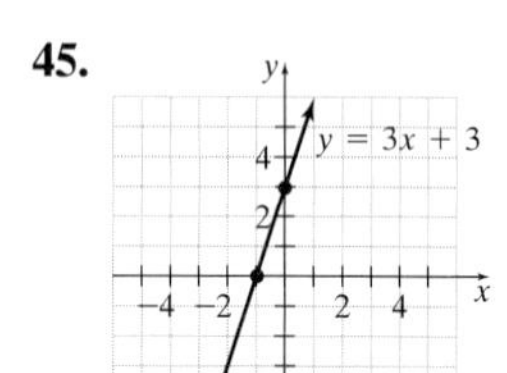

47. $y = 2x - 3$

49. $y = -6x + 5$

51.

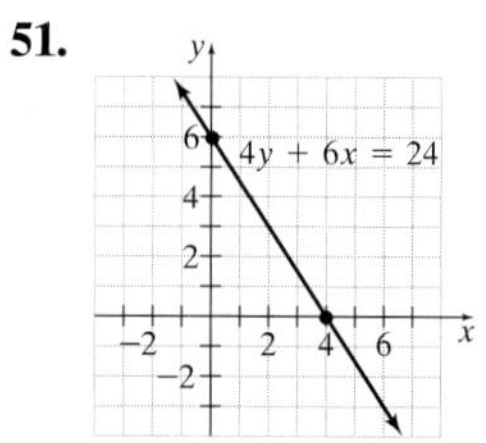

53.

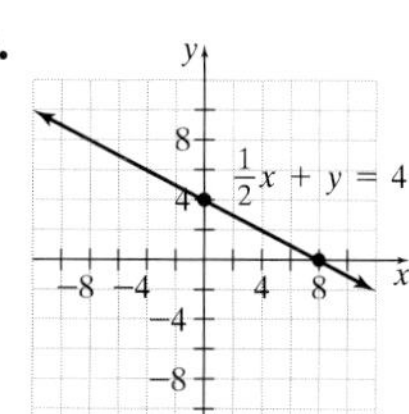

55.

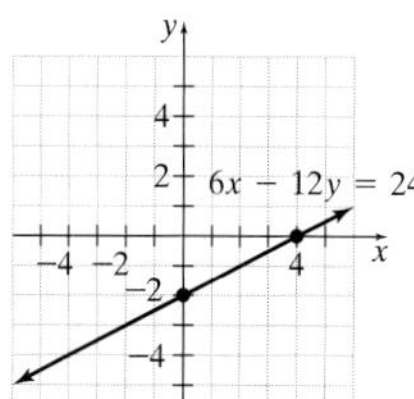

57.

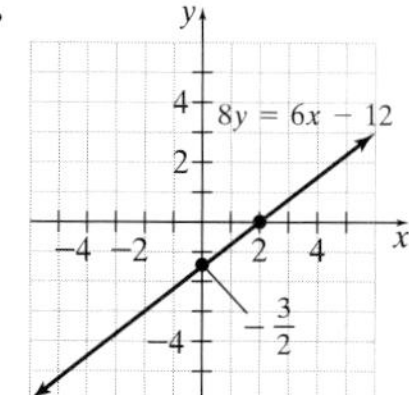

59. $30y + 10x = 45$

61.

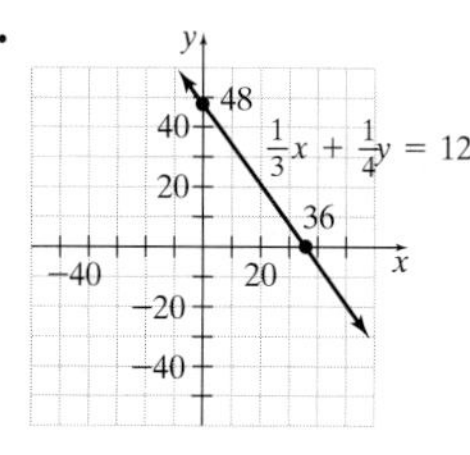

63.

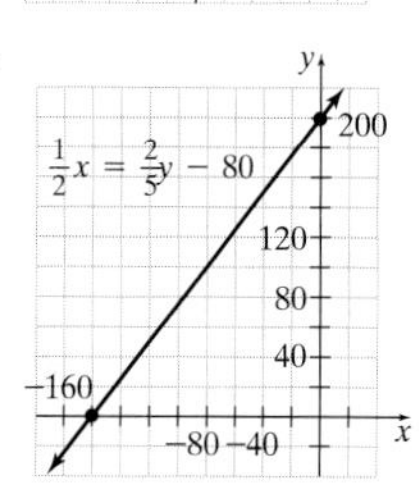

65. $x = -3$ **67.** $y = 6$ **69.** 4 **71.** 2 **73.** Yes

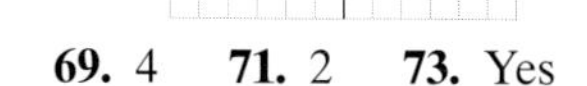

75. a) $C = s + 50$
b)

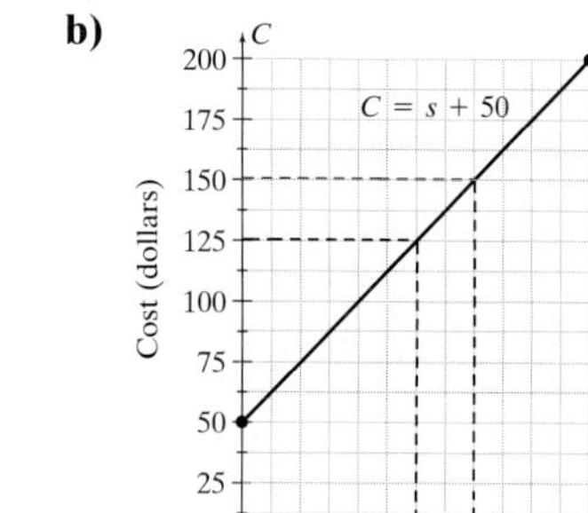

c) \$150 **d)** 75 square feet

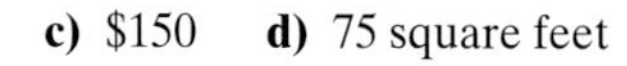

77. a) $C = m + 40$
b)

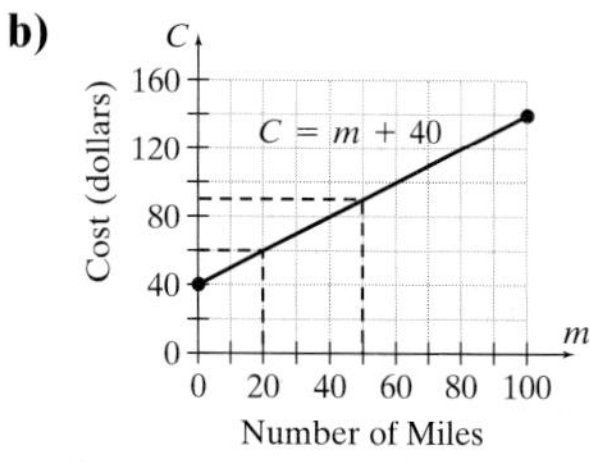

c) \$90 **d)** 20 miles

79. a) $P = 1.5n - 200$ (1000, 1300)
b) \$550 **c)** 800 tapes

81. 5; 4 **83.** 6; 4 **85.**

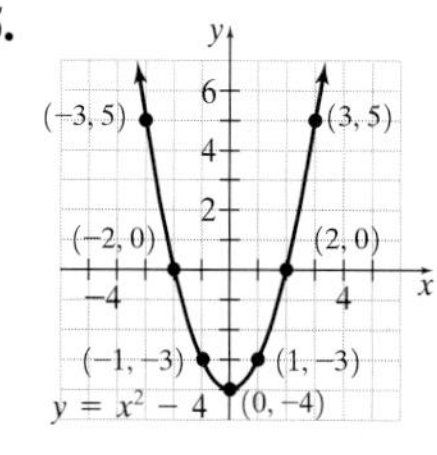

88. −18 **89.** 18 **90.** 6.67 ounces **91.** 9, 28

Exercise Set 4.3

1. The slope of a line is the ratio of the vertical change to the horizontal change between any two points on the line. **3.** Rises from left to right **5.** Lines that rise from the left to right have a positive slope; lines that fall from left to right have a negative slope. **7.** No, since we cannot divide by 0, the slope is undefined. **9.** 5 **11.** $\frac{1}{2}$ **13.** 0 **15.** 0 **17.** Undefined **19.** $-\frac{3}{8}$ **21.** $\frac{2}{3}$ **23.** $m = 2$ **25.** $m = -\frac{3}{2}$ **27.** $m = -\frac{3}{2}$ **29.** $m = \frac{7}{4}$ **31.** $m = 0$ **33.** $m = -\frac{1}{3}$ **35.** $m = 0$ **37.** First **39. a)** 26 **b)** 14.6 **41.** (4, 6) (Answers will vary.) **43.** (3, 3) (Answers will vary.) **45.** $\frac{225}{68}$

47. a)

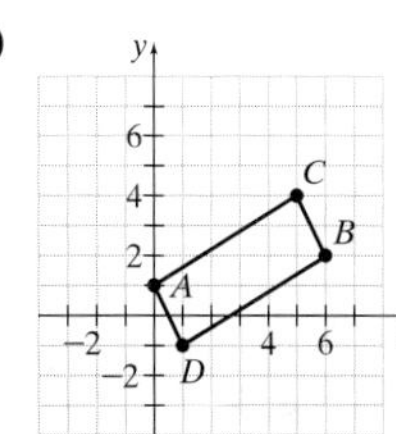

b) AC, $m = \frac{3}{5}$; CB, $m = -2$; DB, $m = \frac{3}{5}$; AD, $m = -2$ **c)** Yes, opposite sides are parallel. **51.** 0 **52. a)** $\frac{3}{2}$ **b)** 0 **53.** 2 **54.** x-intercept: (3, 0); y-intercept: (0, −5)

Using Your Graphing Calculator, 4.4

1. a)

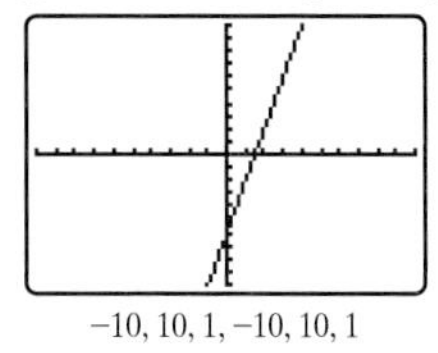

−10, 10, 1, −10, 10, 1

b)

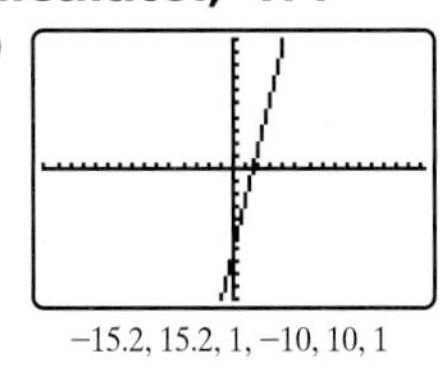

−15.2, 15.2, 1, −10, 10, 1

c)

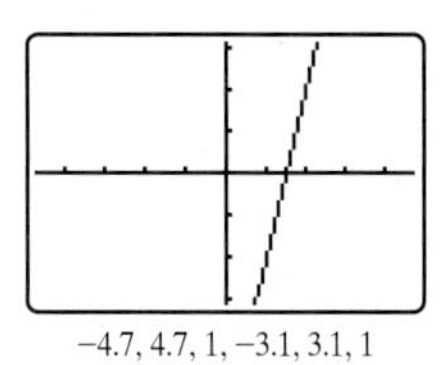

−4.7, 4.7, 1, −3.1, 3.1, 1

2. a)

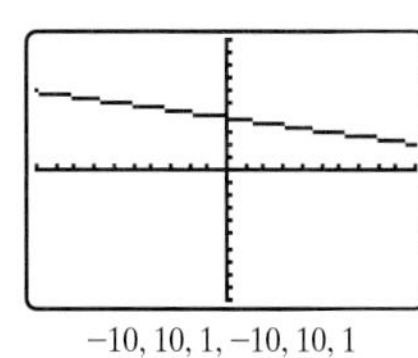

−10, 10, 1, −10, 10, 1

b)

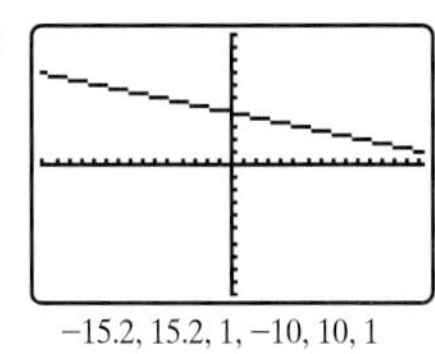

−15.2, 15.2, 1, −10, 10, 1

c)

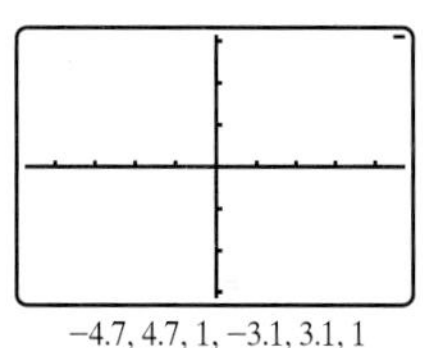

−4.7, 4.7, 1, −3.1, 3.1, 1

Exercise Set 4.4

1. $y = mx + b$ **3.** $y = 4x - 2$ **5.** If slopes are the same and the y-intercepts are different, the lines are parallel. **7.** $y - y_1 = m(x - x_1)$ **9.** 3; (0, −7) **11.** $\frac{4}{3}$; (0, −5)

13. 1; (0, −1)

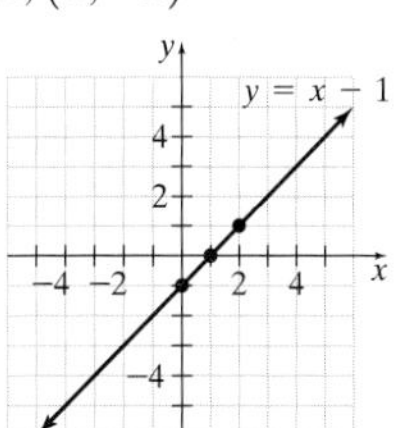

15. 3; (0, 2)

$y = 3x + 2$

17. −4; (0, 0)

$y = -4x$

19. 2; (0, −3)

$-2x + y = -3$

21. $\frac{5}{2}$; (0, −5)

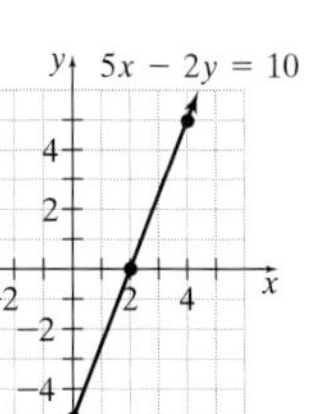

23. $-\frac{1}{2}$; $\left(0, \frac{3}{2}\right)$

$5x + 10y = 15$

25. 3; (0, 4)

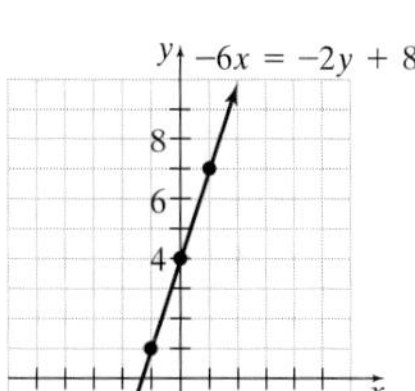

27. $\frac{3}{2}$; (0, 2)

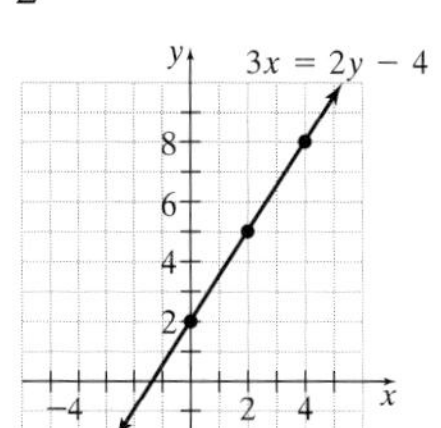

29. $y = x + 2$ **31.** $y = -\frac{1}{3}x + 2$ **33.** $y = \frac{1}{3}x + 5$ **35.** $y = 2x - 1$ **37.** Yes **39.** No **41.** Yes **43.** No **45.** $y = 3x + 2$ **47.** $y = -2x - 3$ **49.** $y = \frac{1}{2}x - \frac{9}{2}$ **51.** $y = \frac{2}{5}x + 6$ **53.** $y = 3x + 10$ **55.** $y = -\frac{3}{2}x$ **57.** $y = \frac{1}{2}x - 2$ **59.** $y = 5.2x - 1.6$ **61. a)** $y = 5x + 60$ **b)** \$210 **63. a)** Slope-intercept form **b)** Point-slope form **c)** Point-slope form **65. a)** No **b)** $y + 4 = 2(x + 5)$ **c)** $y - 10 = 2(x - 2)$ **d)** $y = 2x + 6$ **e)** $y = 2x + 6$ **f)** Yes **67.** $y = -2x + 4$ **69.** $y = \frac{3}{4}x + 2$ **74.** $<$ **75.** False **76.** True **77.** False **78.** False **79.** 125

Exercise Set 4.5 **1.** Points on the line satisfy the = part of the inequality. **3.** The shading is on opposite sides of the line.

5.
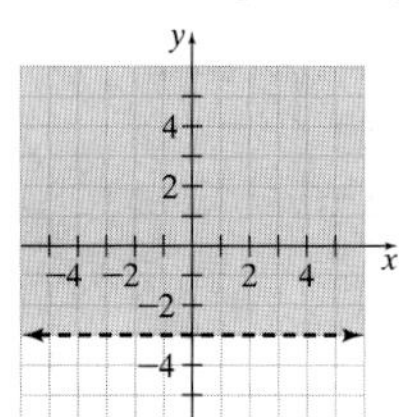
7.
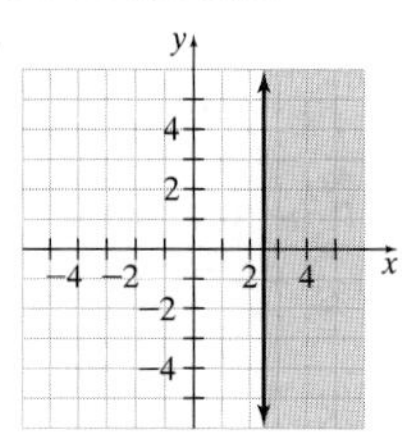
9.
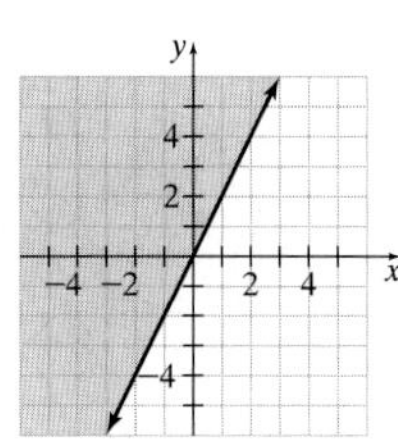
11.
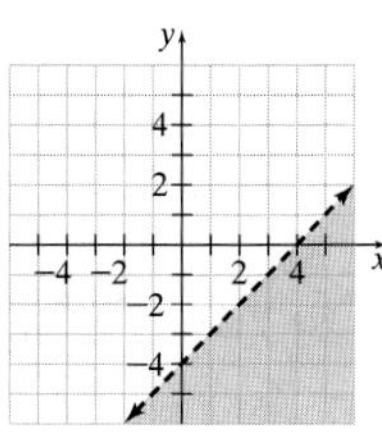
13.
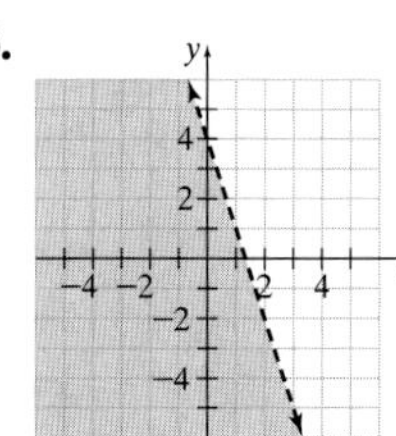
15.
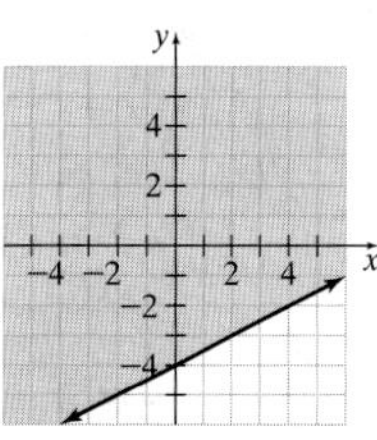
17.
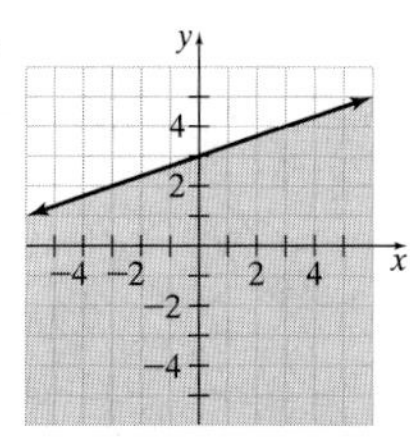
19.
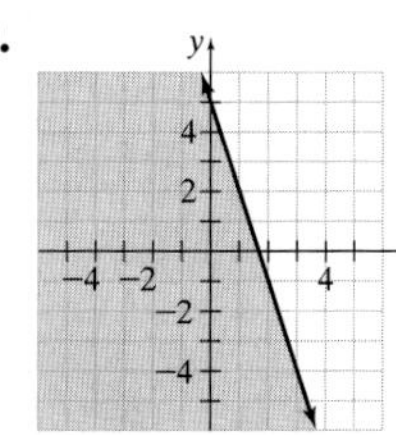
21.
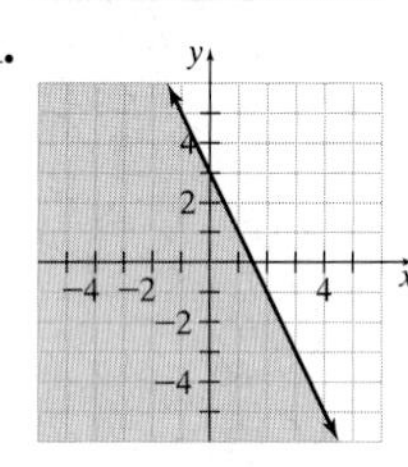
23.
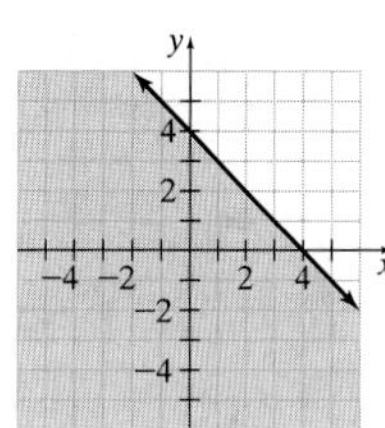

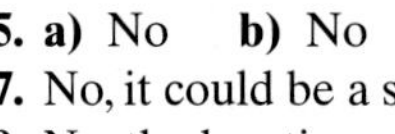
25. a) No **b)** No **c)** Yes **d)** Yes **27.** No, it could be a solution to $ax + by = c$. **29.** No, the location of an ordered pair which satisfies the first inequality lies on one side of the line while an ordered pair which satisfies the other inequality lies either on the line or on the other side of the line.

31. a) $x + y \geq 100$

b)
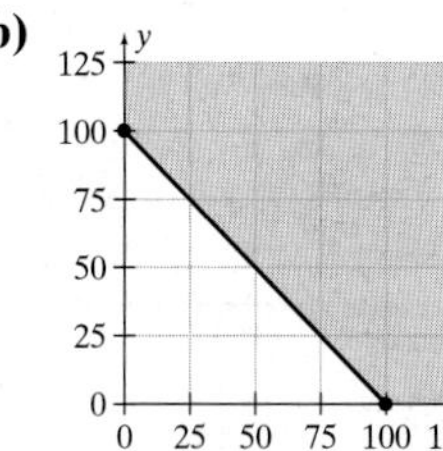
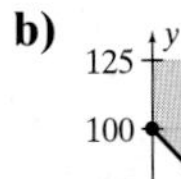

35. a) 2 **b)** 2, 0 **c)** $2, -5, 0, \frac{2}{5}, -6.3, -\frac{23}{34}$ **d)** $\sqrt{7}, \sqrt{3}$ **e)** $2, -5, 0, \sqrt{7}, \frac{2}{5}, -6.3, \sqrt{3}, -\frac{23}{34}$ **36. a)** 0 **b)** Undefined **37.** When evaluating an expression, the order of operations is parentheses, exponents, multiplication or division (left to right), and addition or subtraction (left to right). **38.** $-\frac{2}{3}$

Exercise Set 4.6 **1.** A relation is any set of ordered pairs. **3.** A function is a set of ordered pairs in which each first component corresponds to exactly one second component **5. a)** The set of first components in the set of ordered pairs. **b)** The set of second components in the set of ordered pairs. **7.** No, each x must have a unique y for it to be a function. **9.** Function; Domain: $\{1, 2, 3, 4, 5\}$; Range: $\{1, 2, 3, 4, 5\}$ **11.** Relation; Domain: $\{1, 2, 3, 6, 7\}$; Range: $\{-2, 0, 2, 4, 5\}$ **13.** Relation; Domain: $\{0, 1, 3, 5\}$; Range: $\{-4, -1, 0, 1, 2\}$ **15.** Relation; Domain: $\{0, 1, 3\}$; Range: $\{-3, 0, 2, 5\}$ **17.** Function; Domain: $\{0, 1, 2, 3, 4\}$; Range: $\{3\}$ **19. a)** $\{(1, 4), (2, 5), (3, 5), (4, 7)\}$ **b)** Function **21. a)** $\{(-5, 4), (0, 7), (6, 9), 6, 3)\}$ **b)** Not a function **23.** Function **25.** Function **27.** Not a function **29.** Function **31.** Not a function **33.** Function **35. a)** 14 **b)** −2 **37. a)** 6 **b)** 1 **39. a)** 5 **b)** 11 **41. a)** 3 **b)** 5

43.
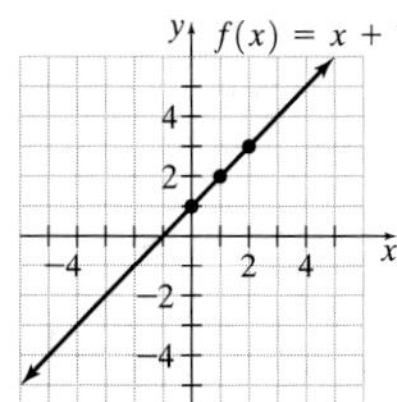

45.
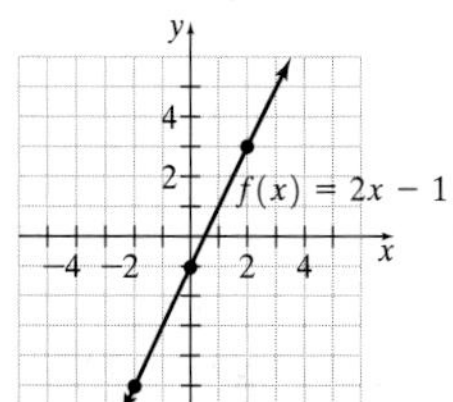

47.
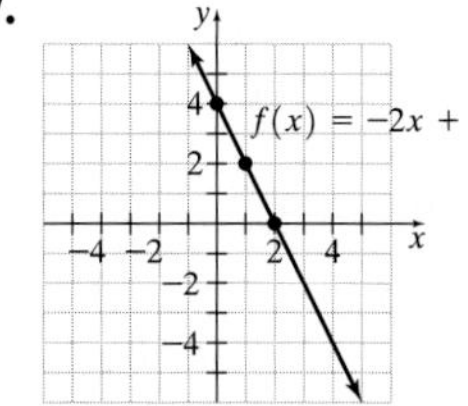

49.
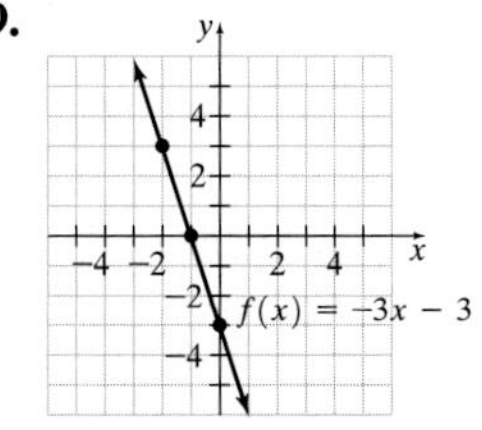

51. No, some value of x must correspond to two values of y. **53.** Yes, it passes the vertical line test.

55. a)
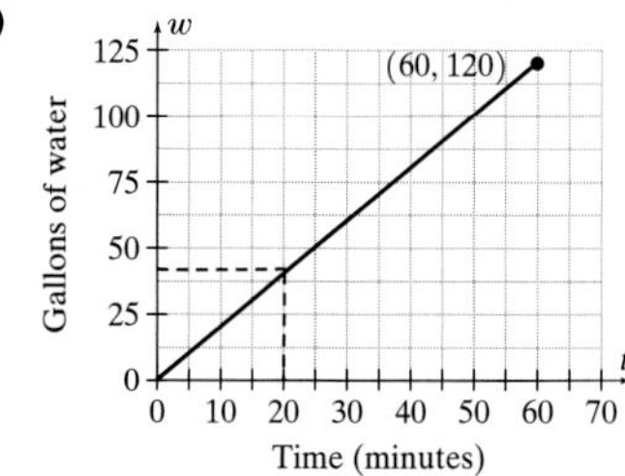

b) 40 gallons

57. a)
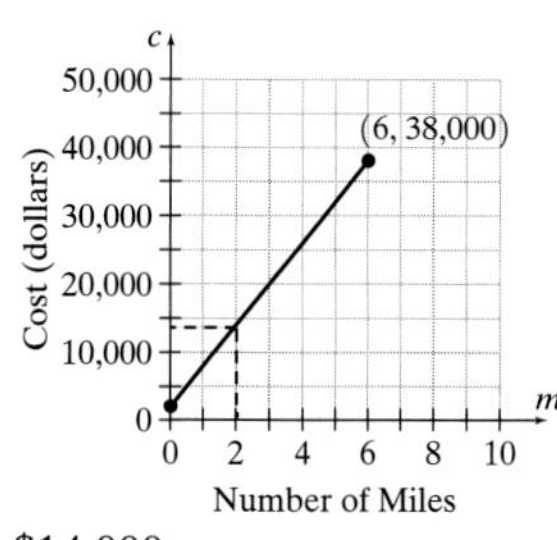

b) \$14,000

59. a)
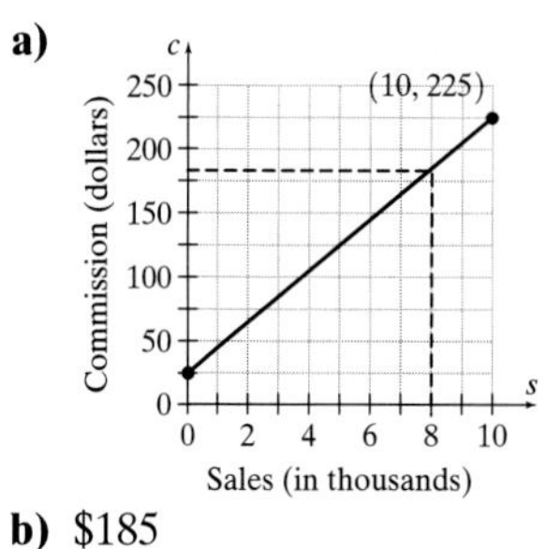

b) \$185

61. a)
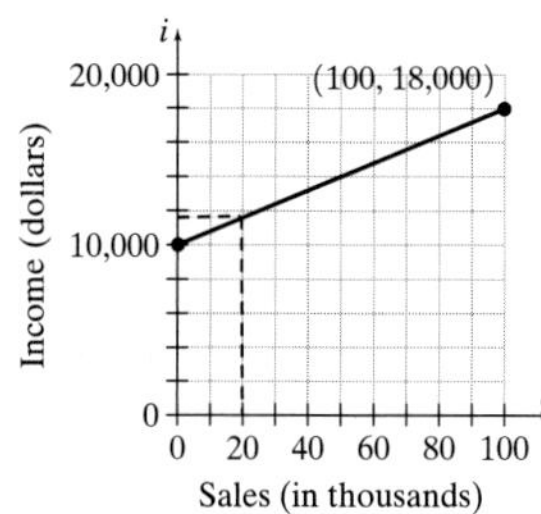

b) \$11,600

63. Yes, it passes the vertical line test.
65. No, the vertical line $x = 1$ intersects the graph at more than one point.
67. a) $\frac{29}{8}$ **b)** $\frac{29}{9}$ **c)** 4.42 **71.** $\frac{8}{63}$
72. a) Commutative property of multiplication
b) Associative property of addition **c)** Distributive property
73. -14 **74.** 13 miles

Review Exercises

1.
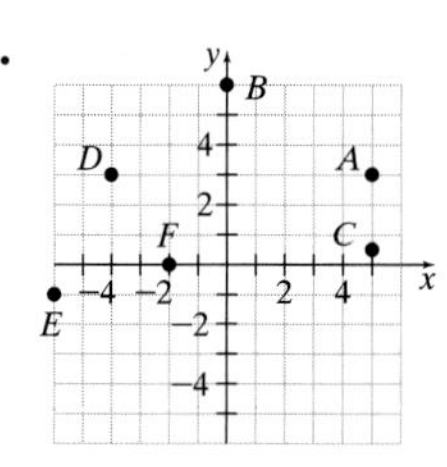

2. Not collinear **3.** b) and d)
4. a) 2 **b)** -4 **c)** $\frac{16}{3}$ **d)** $\frac{8}{3}$

5.
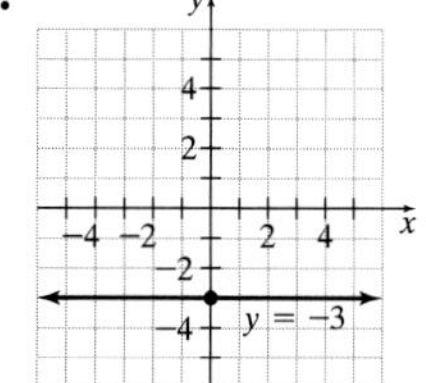

6.
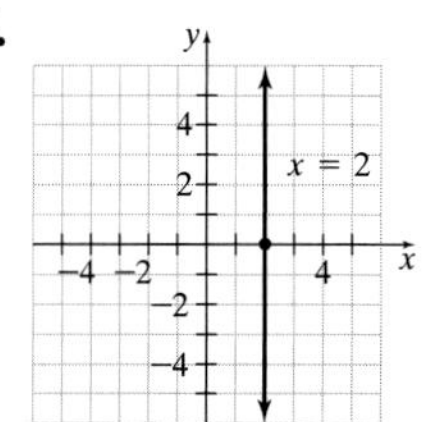

7.
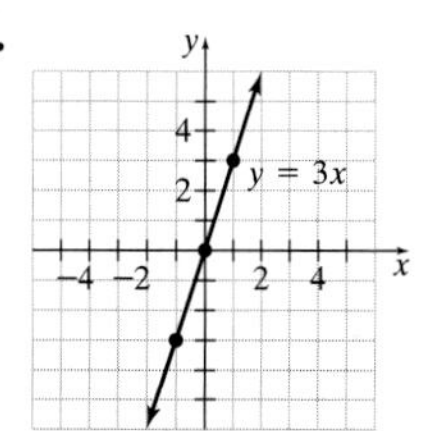

8.
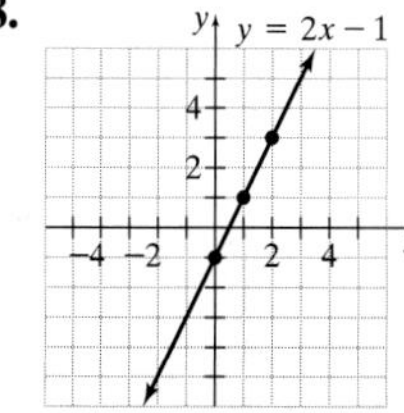

9.
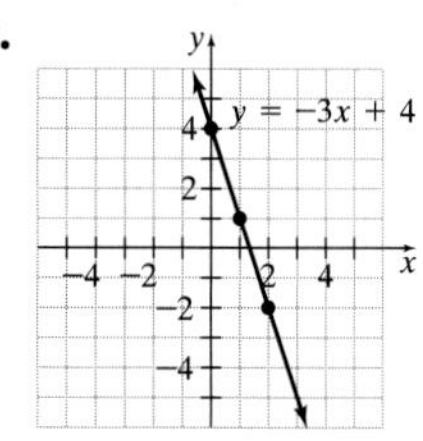

10.
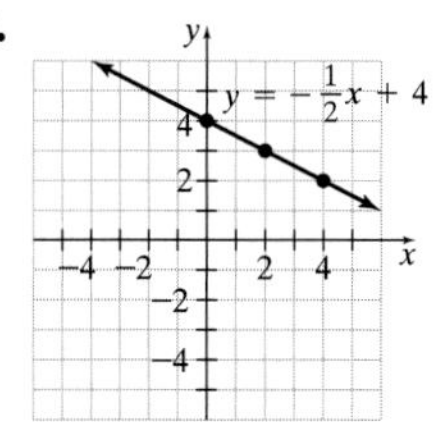

11.
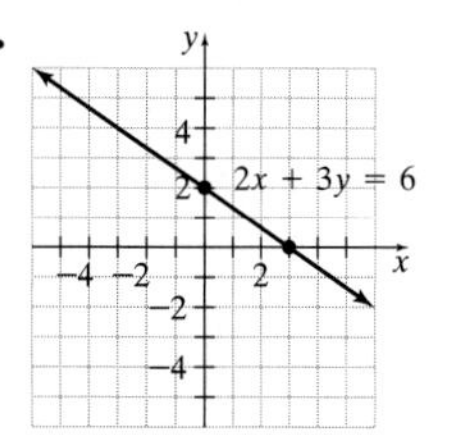

12.
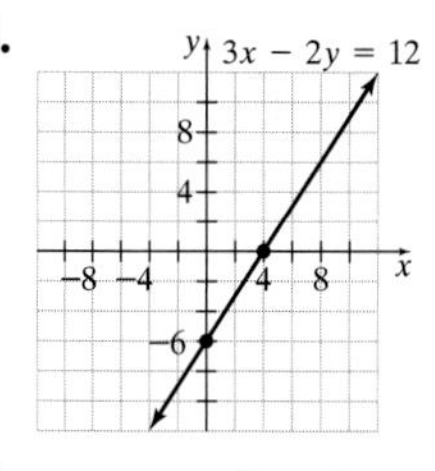

13.
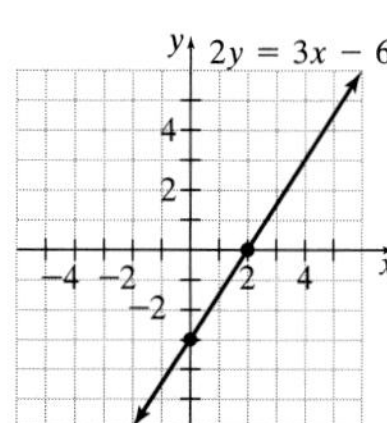

14. $-5x - 2y = 10$
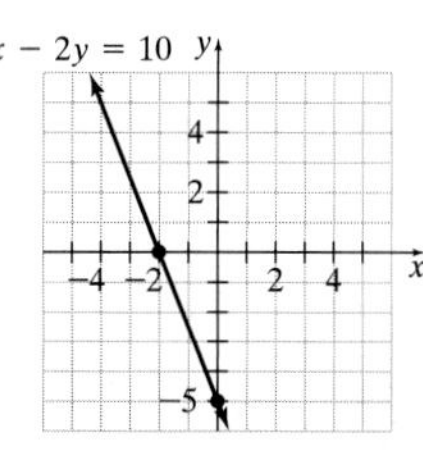

15.
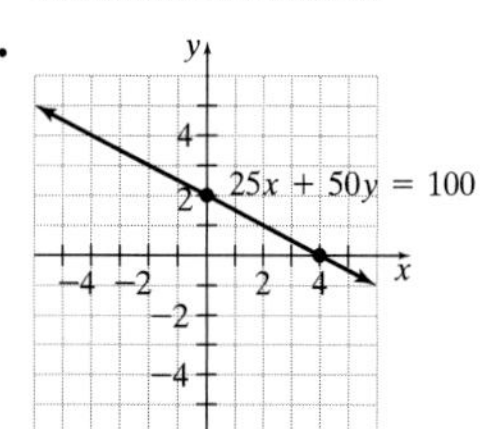

16.
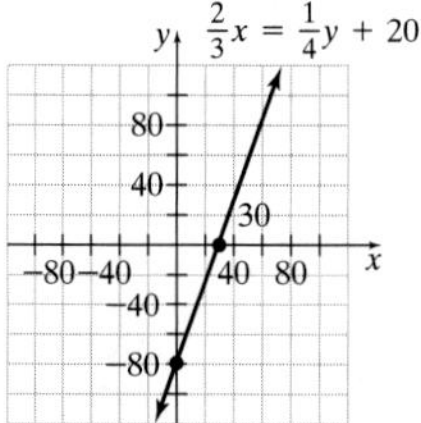

17. $-\frac{6}{5}$ **18.** $-\frac{1}{12}$ **19.** -2 **20.** 0 **21.** Undefined

22. The slope of a line is the ratio of the vertical change to the horizontal change between any two points on the line.

23. $-\frac{5}{7}$ **24.** $\frac{1}{4}$ **25. a)** 5250 **b)** $-21{,}000$ **26.** $-\frac{9}{7}; \left(0, \frac{15}{7}\right)$ **27.** Slope is undefined; no y-intercept

28. $m = 0; (0, -3)$ **29.** $y = 2x + 2$ **30.** $y = -\frac{1}{2}x + 2$ **31.** Parallel **32.** Not parallel **33.** $y = 2x - 2$

34. $y = -\frac{2}{3}x + 4$ **35.** $y = 2$ **36.** $x = 4$ **37.** $y = -\frac{7}{2}x - 4$ **38.** $x = -4$

39.

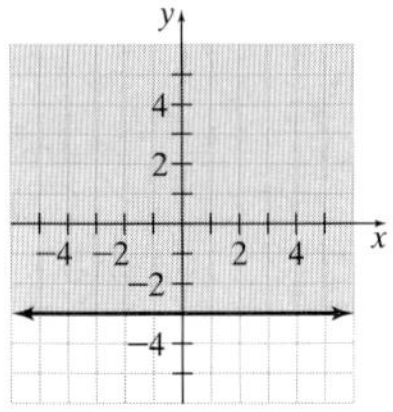

40.

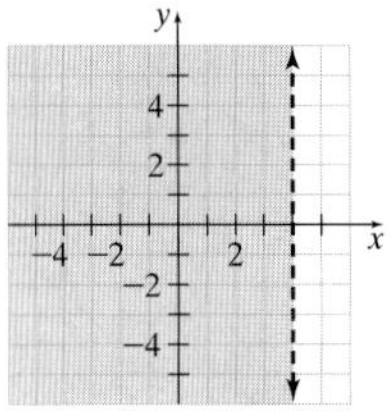

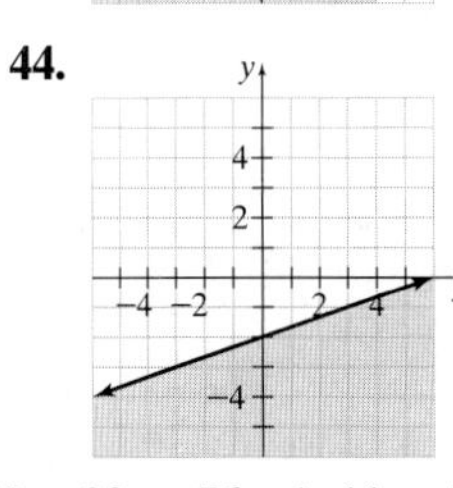

41.

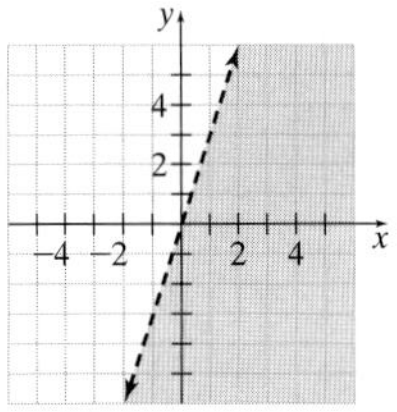

42.

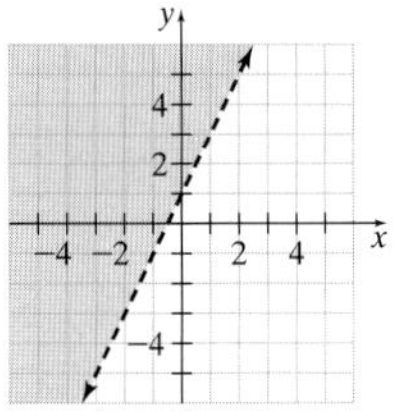

43.

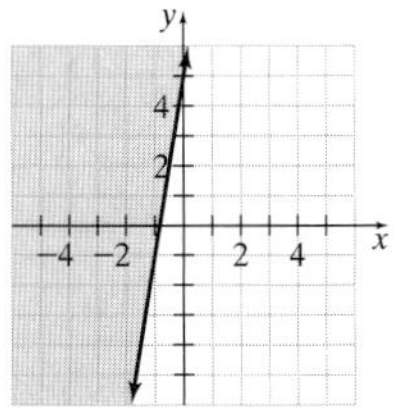

44.

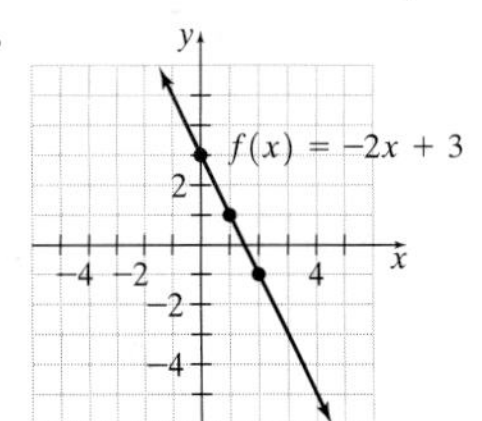

45. Function; Domain: $\{0, 1, 2, 4, 6\}$; Range: $\{-3, -1, 2, 4, 5\}$
46. Not a function; Domain: $\{3, 4, 6, 7\}$; Range: $\{0, 1, 2, 5\}$
47. Not a function; Domain: $\{3, 4, 5, 6\}$; Range: $\{-3, 1, 2\}$
48. Function; Domain: $\{-2, 3, 4, 5, 9\}$; Range: $\{-2\}$
49. a) $\{(1, 3), (4, 5), (7, 2), (9, 2)\}$ **b)** Function
50. a) $\{(4, 1), (6, 3), (6, 5), (8, 7)\}$ **b)** Not a function
51. Function **52.** Not a function **53.** Function

54. Function **55. a)** 7 **b)** −28 **56. a)** 11 **b)** −37 **57. a)** −4 **b)** −8 **58. a)** 12 **b)** 76
59. Yes, it passes the vertical line test. **60.** Yes, each year has a unique number of bankruptcies.

61.

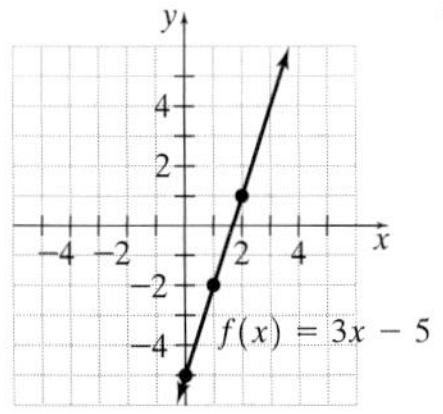

62.

63. a)

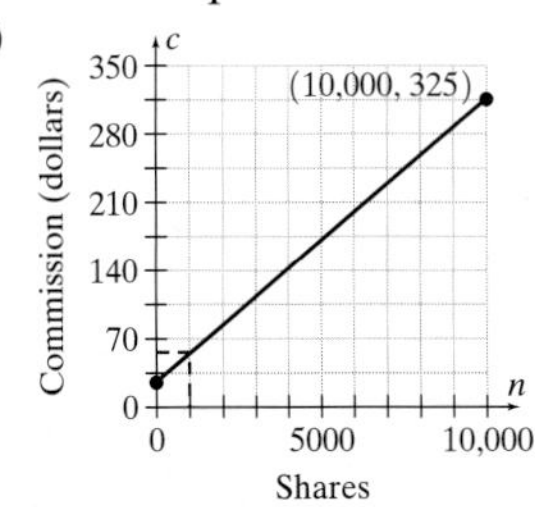

b) \$55

64. a)

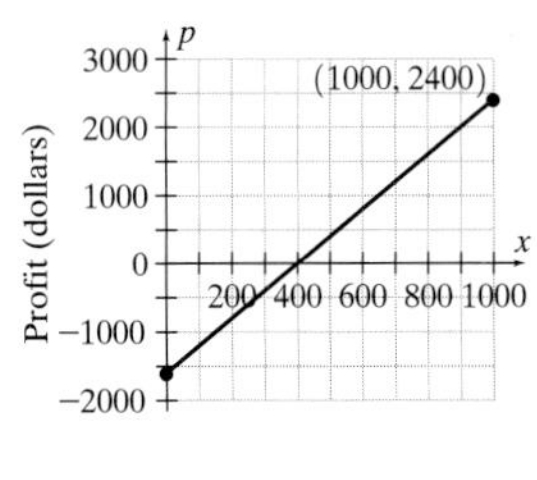

b) \$0

Practice Test **1.** A graph is an illustration of the set of points that satisfy an equation. **2. a)** III **b)** IV
3. a) $ax + by = c$ **b)** $y = mx + b$ **c)** $y - y_1 = m(x - x_1)$ **4.** a) and b) **5.** $-\frac{4}{3}$ **6.** $\frac{4}{9}; \left(0, -\frac{5}{3}\right)$
7. $y = -x - 1$

8.

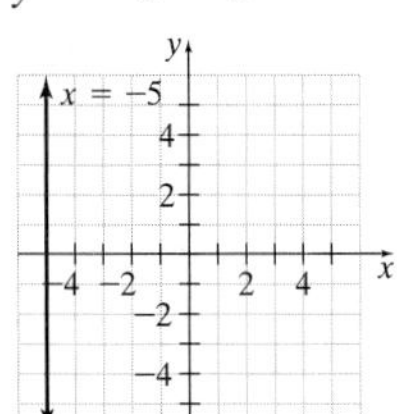

9.

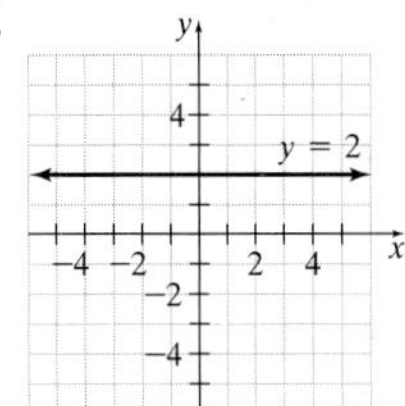

10.

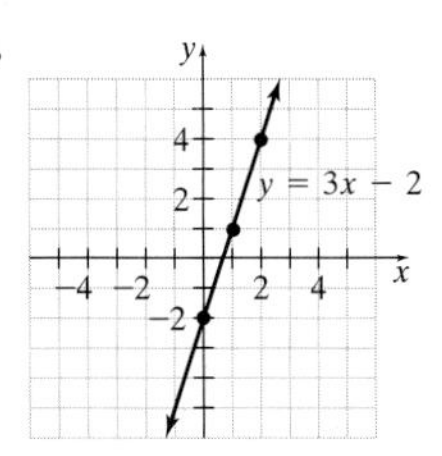

11. a) $y = \frac{1}{2}x - 2$

b)

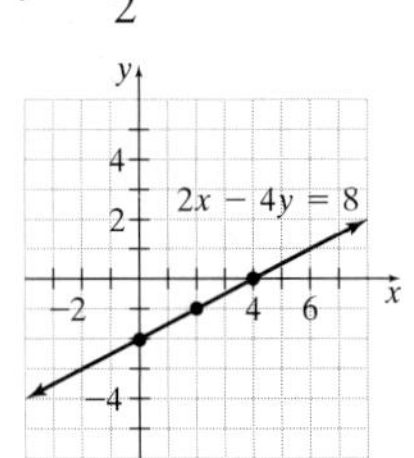

12.

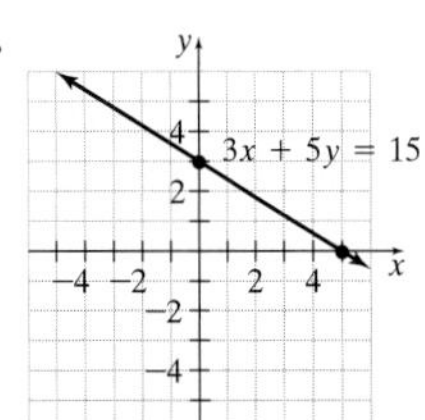

13. $y = 3x$

14. $y = -\frac{3}{7}x + \frac{2}{7}$

15. The lines are parallel since they have the same slope but different y-intercepts.

16. 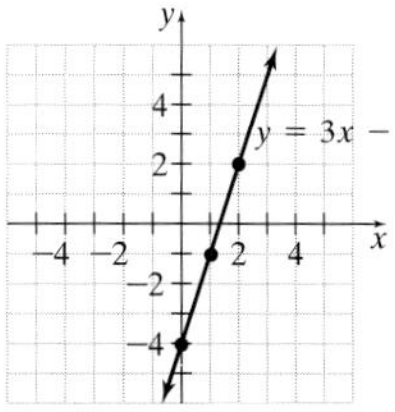

17. 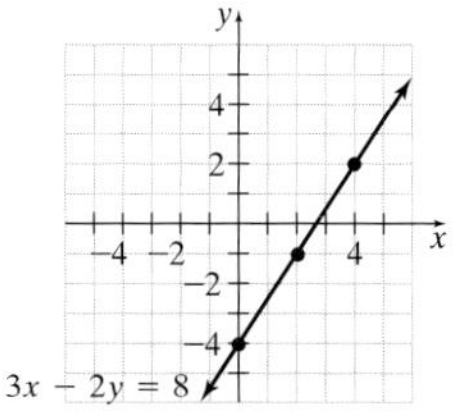

18. A set of ordered pairs in which each first component corresponds to exactly one second component. **19. a)** Not a function; 1, a first component, is paired with more than 1 value **b)** Domain: $\{1, 3, 5, 6\}$; Range: $\{-4, 0, 2, 3, 5\}$

20. a) Function; it passes the vertical line test
b) Not a function; a vertical line can be drawn that intersects the graph at more than one point **21. a)** 14 **b)** 9

22. 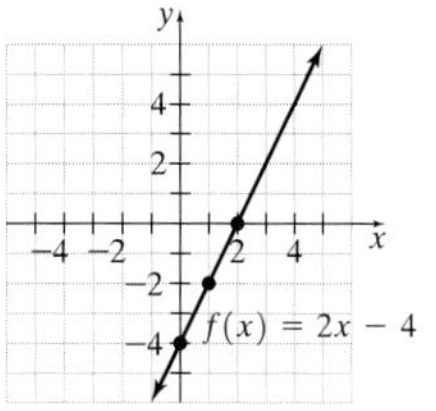

23. 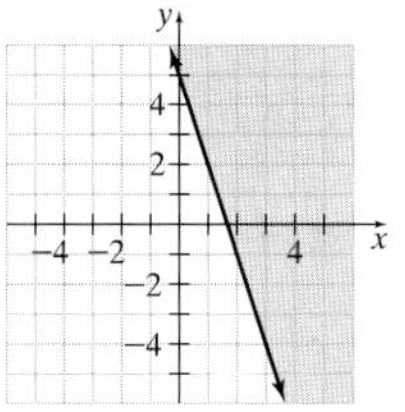

24. 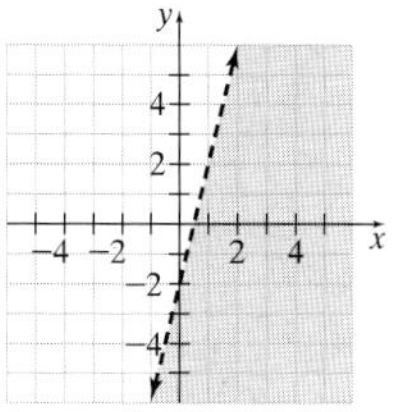

25. a) 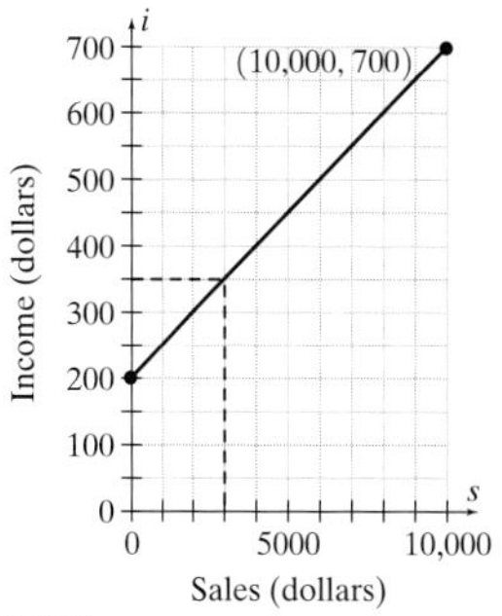

b) \$350

Cumulative Review Test

1. a) $\{1, 2, 3, \dots\}$ **b)** $\{0, 1, 2, 3, \dots\}$ **2. a)** Distributive property **b)** Commutative property of addition **3.** -1 **4.** 12 **5.** -40 **6.** 20 **7.** All real numbers

8. $x < -5$;  **9.** $w = \dfrac{v}{lh}$ **10.** \$3.33 **11.** 4 **12.** Width is 3 feet; length is 10 feet **13.** 2 hours

14. Answers will vary.

15. 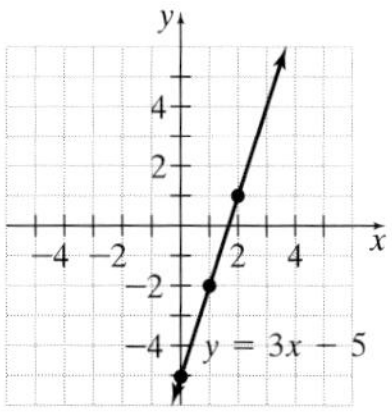

16. 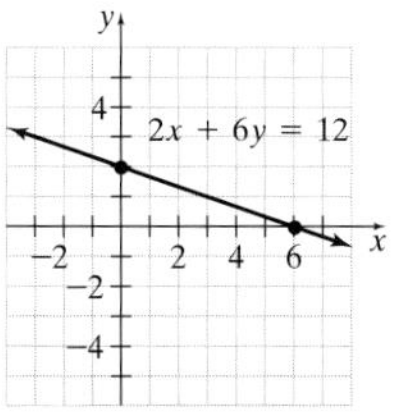

17. $-\dfrac{3}{5}$; $\left(0, \dfrac{12}{5}\right)$

18. 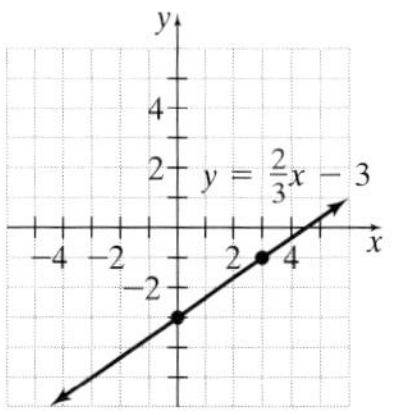

19. $y - 2 = 3(x - 5)$

20. a) Not a function; a vertical line can be drawn that intersects the graph at more than one point.
b) Yes; each x-value is paired with exactly one y-value.

CHAPTER 5

Using Your Graphing Calculator, 5.1

1. $(-3, -4)$ **2.** $(-4, 3)$ **3.** $(3.1, -1.5)$ **4.** $(-0.6, 4.8)$

Exercise Set 5.1

1. The solution to a system of equations represents the ordered pairs that satisfy all the equations in the system. **3.** Write the equations in slope–intercept form and compare their slopes and y-intercepts. If the slopes are the same and the y-intercepts are different, the system has no solution. If the slopes and y-intercepts are the same, the system has an infinite number of solutions. **5.** The point of intersection can only be estimated. **7.** b) **9.** a) **11.** none **13.** a), c) **15.** a) **17.** Consistent; one solution **19.** Dependent; infinite number of solutions **21.** Consistent; one solution **23.** Inconsistent; no solution **25.** One solution **27.** No solution **29.** One solution **31.** Infinite number of solutions **33.** No solution **35.** No solution

37.
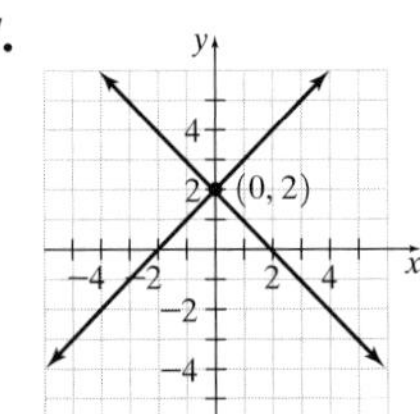

39.
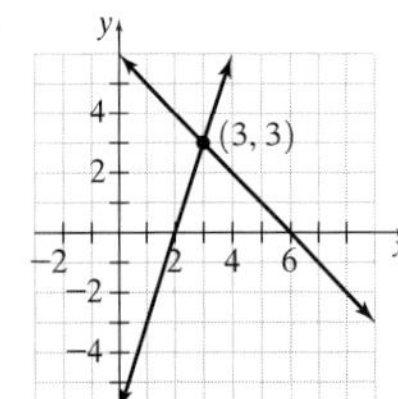

41.
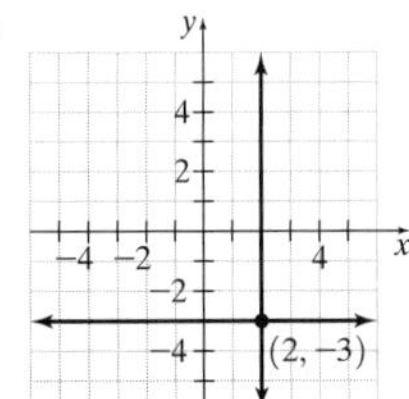

43.
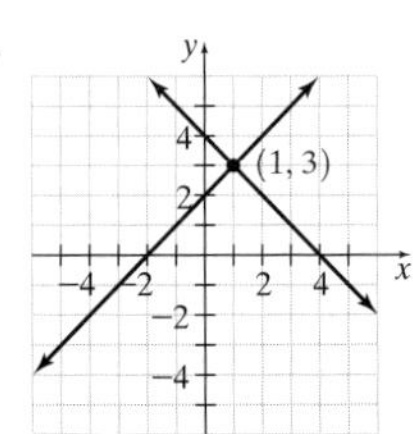

45.
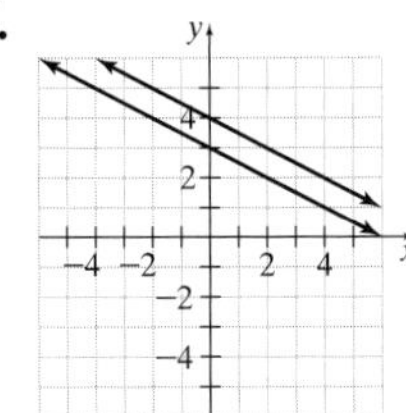
Inconsistent

47.
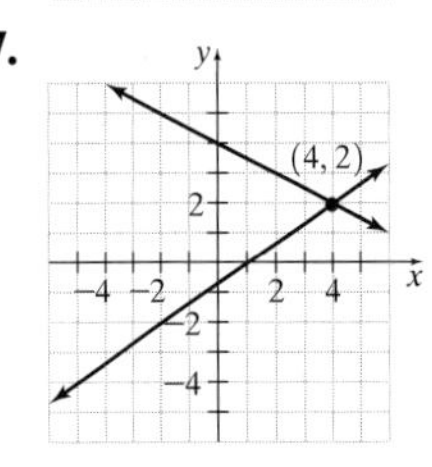

49.
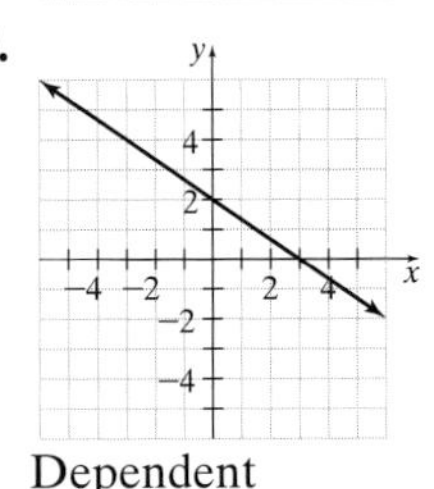
Dependent

51.
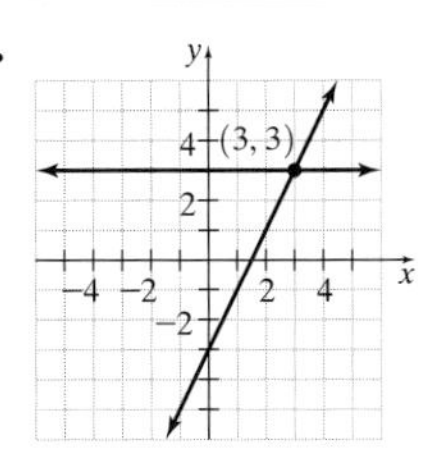

53.
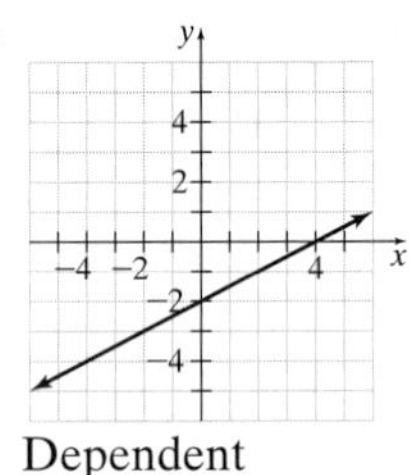
Dependent

55.
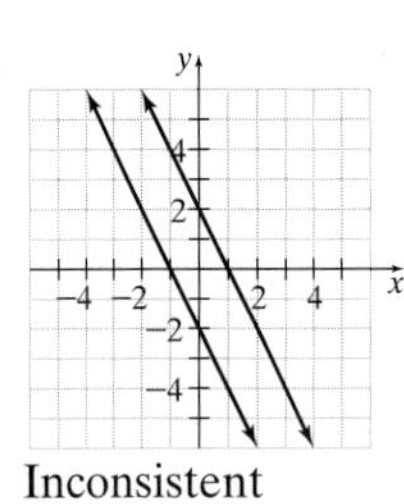
Inconsistent

57.
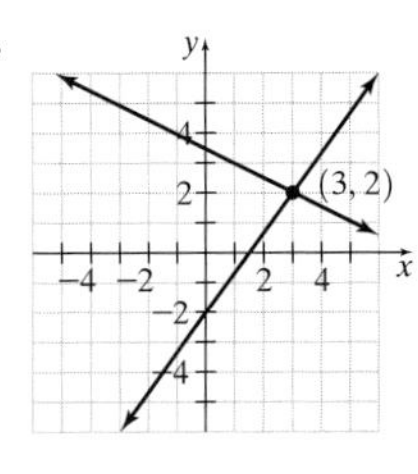

59.
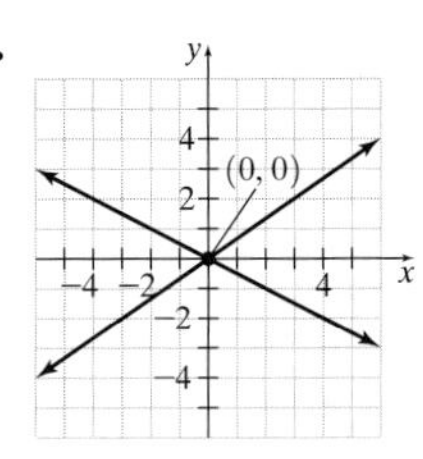

61. Lines are parallel; they have the same slope and different y-intercepts. **63.** The system has an infinite number of solutions. If the two lines have two points in common then they must be the same line. **65.** The system has no solutions. Distinct parallel lines do not intersect. **67.** One; (4, 3) **69.** 6 years **71.** 250 shares **79.** $-2x + 18$ **80.** 1 **81.** $\frac{1}{2}$ **82.** $h = \frac{2A}{b}$

Exercise Set 5.2 **1.** The x in the first equation, since both 6 and 9 are divisible by 3. **3.** You will obtain a false statement, such as $3 = 0$. **5.** (3, 1) **7.** (−1, −1) **9.** No solution **11.** (3, −8) **13.** Infinite number of solutions **15.** (2, 5) **17.** Infinite number of solutions **19.** (2, 1) **21.** No solution **23.** (−1, 0) **25.** $\left(-\frac{59}{37}, \frac{9}{37}\right)$ **27. a)** 15 months **b)** Yes **29.** 100 tapes **33.** 7.14 years **34.** 21 days or more

35.
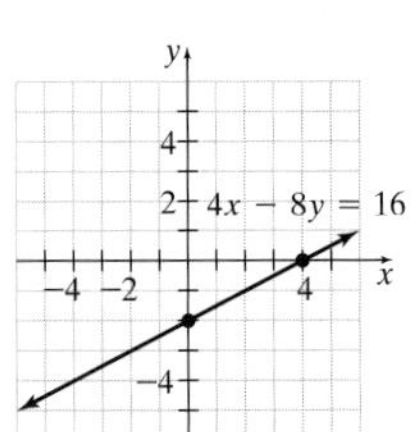

36. Slope: $\frac{3}{5}$; y-intercept: $\left(0, -\frac{8}{5}\right)$

Exercise Set 5.3 **1.** Multiply the top equation by 2. **3.** You will obtain a false statement, such as $0 = 6$. **5.** (5, 1) **7.** (−2, 3) **9.** $\left(4, \frac{11}{2}\right)$ **11.** No solution **13.** (−8, −26) **15.** (4, −2) **17.** (5, 4) **19.** (3, −1) **21.** Infinite number of solutions **23.** $\left(\frac{2}{5}, \frac{8}{5}\right)$ **25.** No solution **27.** (0, 0) **29.** $\left(10, \frac{15}{2}\right)$ **31.** (−5, −13) **33.** $\left(\frac{20}{39}, -\frac{16}{39}\right)$ **35.** $\left(\frac{14}{5}, -\frac{12}{5}\right)$ **37.** Answers will vary. **39. a)** (200, 100) **b)** Same solution; dividing both sides of an equation by a nonzero number does not change the solution. **41.** (8, −1) **45.** 125 **46.** 7 **47.** $x > -1$; [number line: open circle at −1, shaded to the right] **48.** 14

Exercise Set 5.4

1. $x + y = 29, y = x + 3$; 13, 16 **3.** $A + B = 90, B = A + 18$; $A = 36°, B = 54°$ **5.** $A + B = 180, A = B + 48$; $A = 114°, B = 66°$ **7.** $2w + 2h = 124, w = h + 8$; $h = 27$ inches, $w = 35$ inches **9.** $c + w = 100, 450c + 430w = 44{,}400$; 70 acres corn, 30 acres wheat **11.** $d = 5m, 26d + 85m = 10{,}750$; 250 Disney, 50 Microsoft **13.** $k + c = 4.7, k - c = 3.4$; kayak, 4.05 miles per hour; current, 0.65 miles per hour **15.** $c = 50 + 0.85m, c = 100 + 0.80m$; $1000 **17. a)** $c = 18 + 0.02n, c = 25 + 0.015n$; 1400 **b)** Office Copier Depot **19. a)** $c = 200 + 20y, c = 260 + 16y$; 15 square yards **b)** Tom Taylor's **21.** $x + y = 8000, 0.10x + 0.08y = 750$; $2500 at 8%; $5500 at 10% **23.** $10{,}000x + 6000y = 740, x = y + 0.01$; CD, 5%; money market, 4% **25.** $e = m + 4, 3e = 3.2m$; Melissa, 60 miles per hour; Elizabeth, 64 miles per hour **27.** $50j + 50l = 30, j = l + 0.1$; John's, 0.35 miles per hour; Leigh's, 0.25 miles per hour **29.** $5a = 8d, a = d + 0.3$; 0.5 hour **31.** $x + y = 10, 0.25x + 0.50y = 0.40(10)$; 4 liters of 25%; 6 liters of 50% **33.** $x + y = 380$, $3x + 5y = 1500$; 180 **35.** $x + y = 100, 0.05x + 0.00y = 0.035(100)$; 70 gallons 5%; 30 gallons skim

37. $x + y = 8, 12x + 6y = 10(8)$; $2\frac{2}{3}$ ounces drink, $5\frac{1}{3}$ ounces juice **39.** $9t = d, 5t = d - 0.5$; 1.125 miles

43. a) Commutative property of addition **b)** Associative property of multiplication **c)** Distributive property

44. $\frac{1}{2}$ **45.** $w = 3$ feet, $l = 8$ feet **46.** A graph is an illustration of the set of points that satisfies an equation.

Exercise Set 5.5

1. Yes, the solution to a system of linear inequalities contains all of the ordered pairs which satisfy both inequalities. **3.** Yes, when the lines are parallel. One possible system is $x + y > 2, x + y < 1$.

5. 

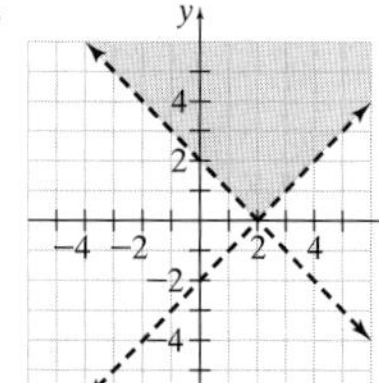

7.

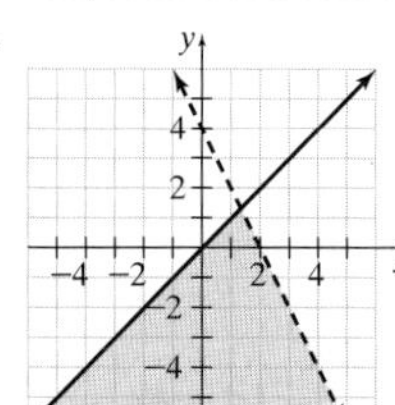

9.

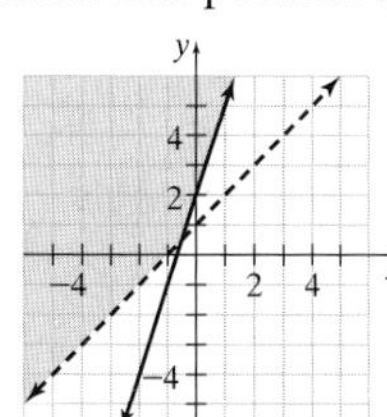

11.

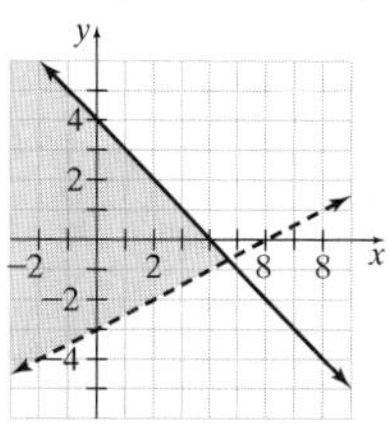

13.

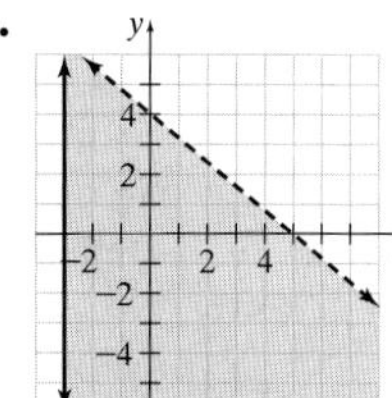

15.

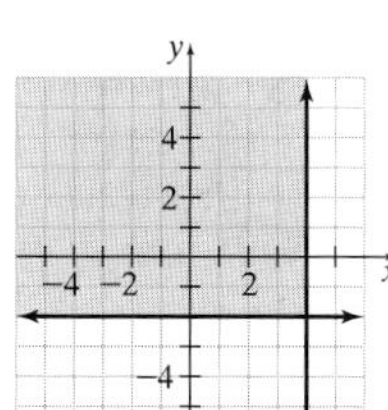

17.

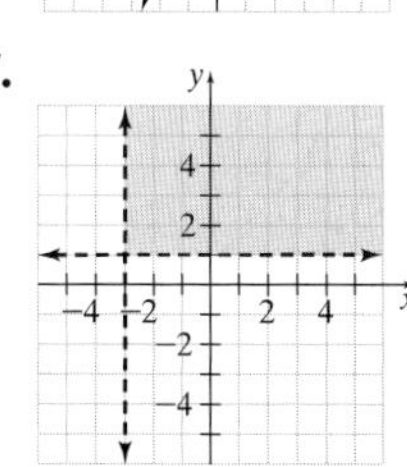

19.

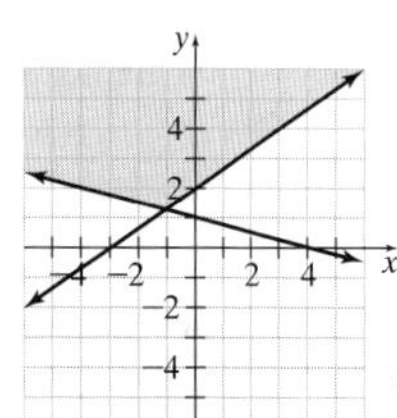

21. No, the system can have no solution or infinitely many solutions.

24. All real numbers; ⟵—+—⟶ 0 **25.** $y = \frac{2}{5}x - \frac{6}{5}$ **26.**

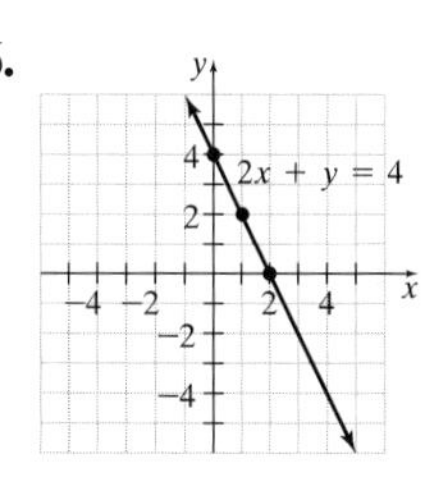

27. $-\frac{8}{7}$

Review Exercises

1. c) **2.** a) **3.** Consistent, one solution **4.** Inconsistent, no solution **5.** Dependent, infinite number of solutions **6.** Consistent, one solution **7.** No solution **8.** One solution **9.** Infinite number of solutions **10.** One solution

11.

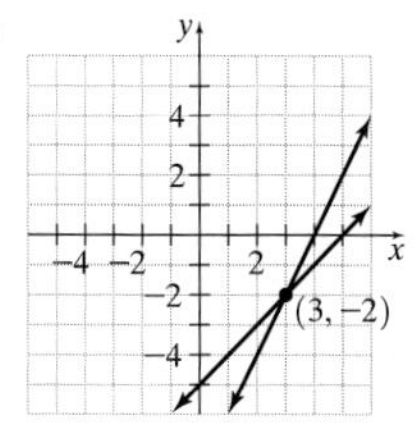

12.

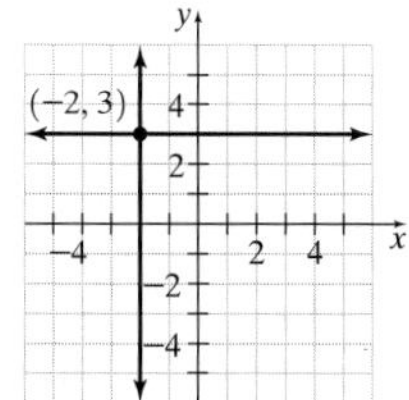

13.

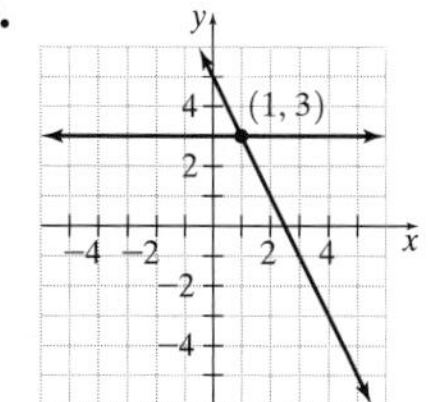

14.

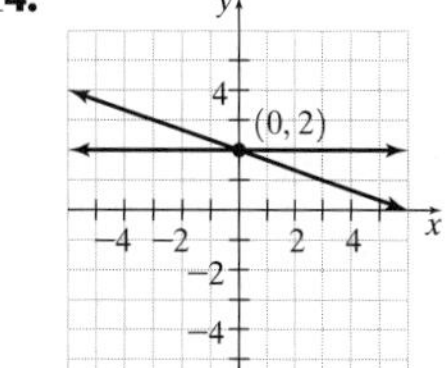

15. 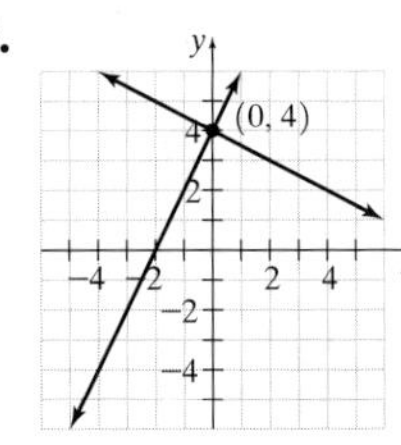

16. 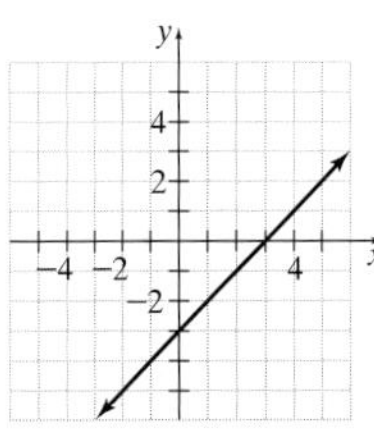
Infinite number of solutions

17. 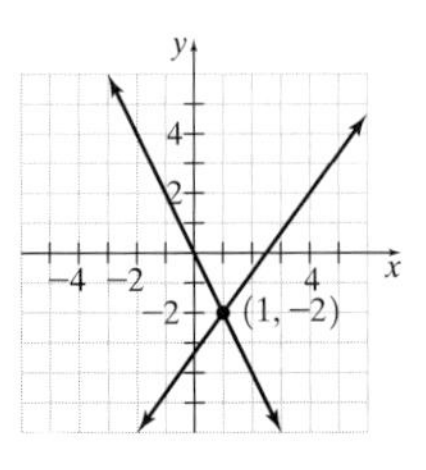

18. 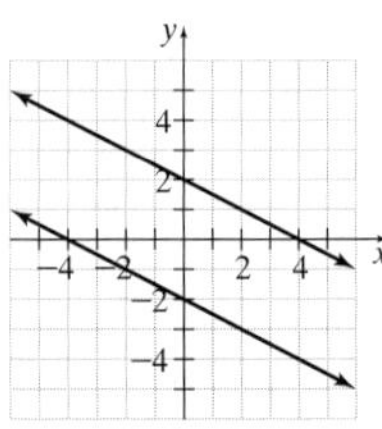
No solution

19. (5, 2) **20.** (−3, 2) **21.** (5, 4) **22.** (−18, 6) **23.** No solution **24.** Infinite number of solutions
25. $\left(\frac{5}{2}, -1\right)$ **26.** $\left(\frac{20}{9}, \frac{26}{9}\right)$ **27.** (8, −2) **28.** (1, −2) **29.** (−7, 19) **30.** $\left(\frac{32}{13}, \frac{8}{13}\right)$ **31.** $\left(-1, \frac{13}{3}\right)$
32. No solution **33.** Infinite number of solutions **34.** $\left(-\frac{78}{7}, -\frac{48}{7}\right)$ **35.** $x + y = 48$, $y = 2x - 3$; 17 and 31
36. $x + y = 600$, $x - y = 530$; plane: 565 miles per hour; wind: 35 miles per hour
37. $c = 20 + 0.5x$, $c = 35 + 0.4x$; 150 miles **38.** $x + y = 16{,}000$, $0.04x + 0.06y = 760$; \$10,000 at 4%, \$6000 at 6%
39. $m = l + 6$, $5l + 5m = 600$; Liz's: 57 miles per hour; Mary's: 63 miles per hour
40. $g + a = 40$, $0.6g + 0.45a = 20.25$; 15 pounds of Green Turf's, 25 pounds of Agway's
41. $x + y = 6$, $0.3x + 0.5y = 0.4(6)$; 3 liters of each

42. 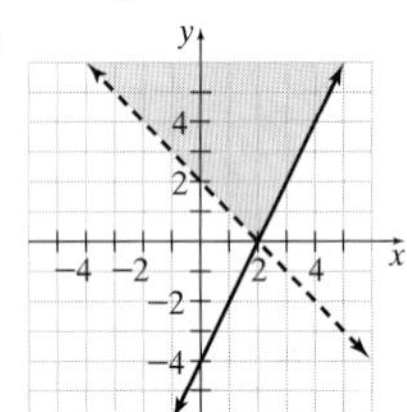

43. 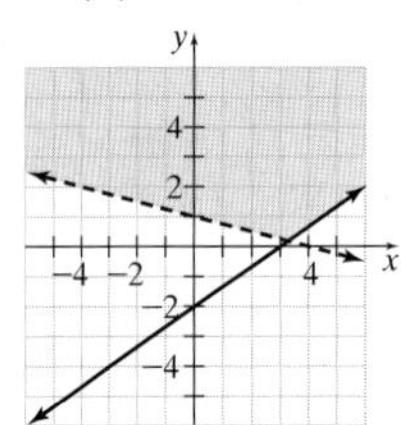

44. 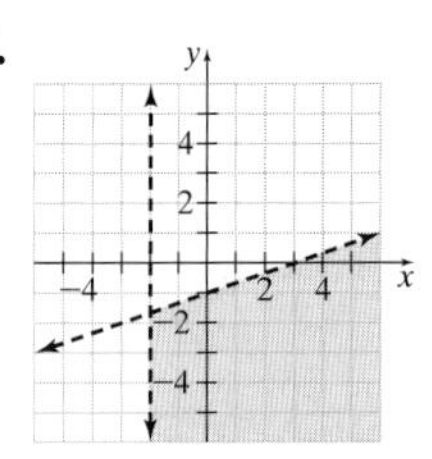

45. 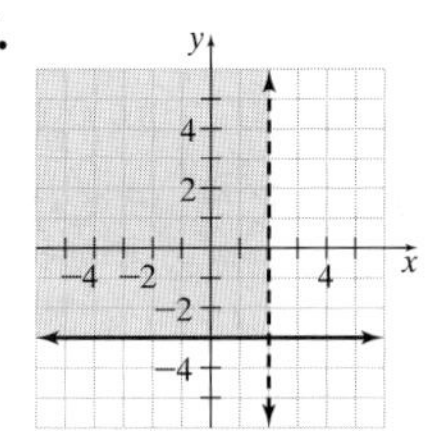

Practice Test **1.** b) **2.** Inconsistent, no solution **3.** Consistent, one solution
4. Dependent, infinite number of solutions **5.** No solution **6.** One solution **7.** Infinite number of solutions
8. a) You will obtain a false statement, such as $6 = 0$. **b)** You will obtain a true statement, such as $0 = 0$.

9. 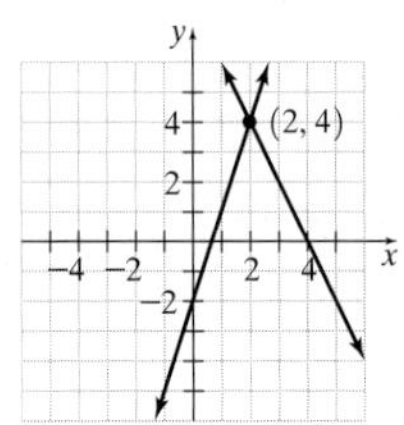

10. 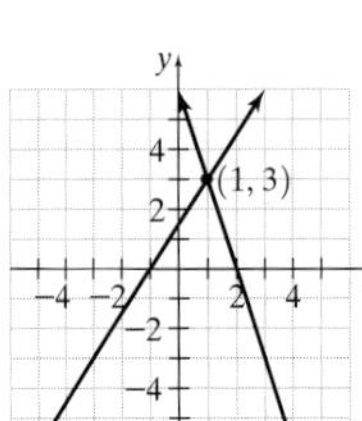

11. 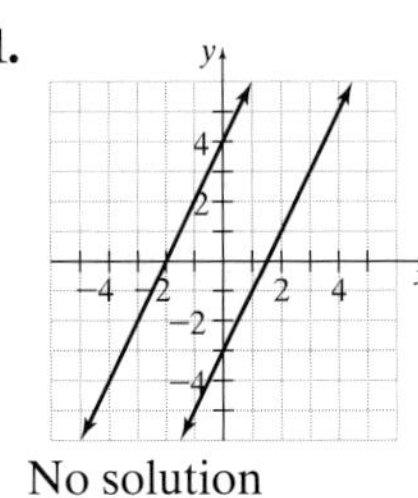
No solution

12. $\left(\frac{7}{2}, -\frac{5}{2}\right)$ **13.** (3, 1)
14. (6, 23) **15.** (5, −5)
16. $\left(\frac{44}{19}, \frac{48}{19}\right)$
17. Infinite number of solutions
18. (2, 2) **19.** $\left(-\frac{40}{7}, \frac{52}{7}\right)$

20. $\left(\frac{50}{19}, \frac{8}{19}\right)$ **21.** $c = 40 + 0.08m$, $c = 45 + 0.03m$; 100 miles
22. $l + b = 20$, $6l + 4.5b = 5(20)$; butterscotch: $13\frac{1}{3}$ pounds; lemon: $6\frac{2}{3}$ pounds
23. $h = r + 4$, $3h = 3.2r$; Deja's boat, 60 miles per hour; Dante's boat, 64 miles per hour

24. 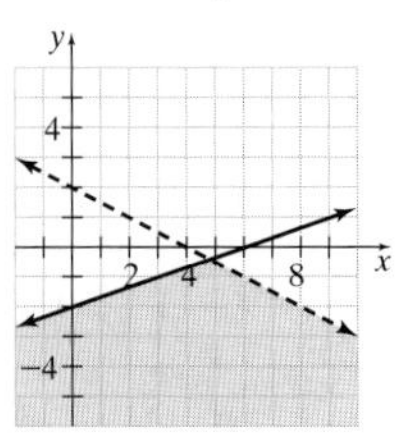

25. 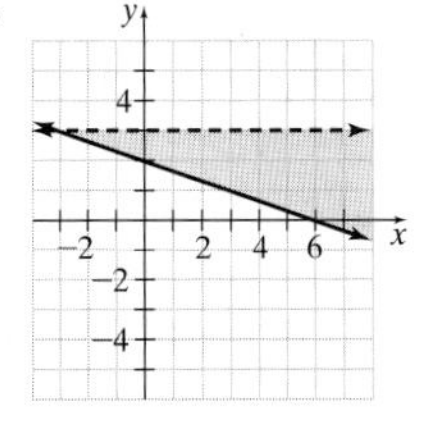

Cumulative Review Test

1. a) 1900–1909 **b)** 1990–1997 **c)** 1990–1997 **2.** 83 **3.** 84.4 **4.** $\frac{51}{40}$

5. a) 7 **b)** $-6, -0.2, \frac{3}{5}, 7, 0, -\frac{5}{9}, 1.34$ **c)** $\sqrt{7}, -\sqrt{2}$ **d)** $-6, -0.2, \frac{3}{5}, \sqrt{7}, -\sqrt{2}, 7, 0, -\frac{5}{9}, 1.34$ **6.** $|-4|$

7. -182 **8.** -34 **9.** $-x + 12$ **10.** -2 **11.** 62.5 pounds **12.** $x \le 5$;  **13.** $w = \frac{P - 2l}{2}$

14. \$5833.33 **15.** 20°, 40°, 120°

16.

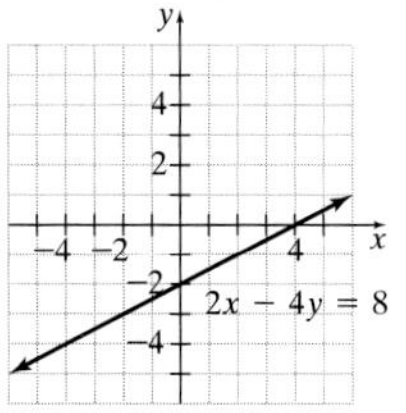

17.

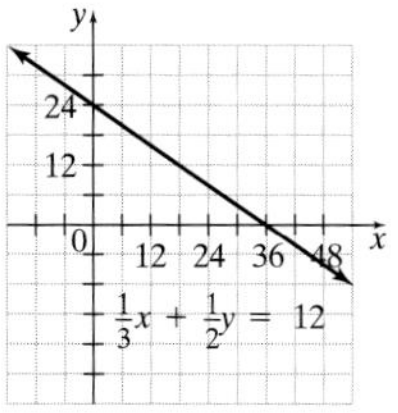

18. Infinite number of solutions

19.

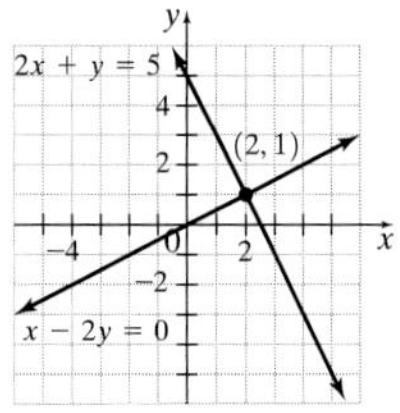

20. $\left(-\frac{28}{3}, -18\right)$

CHAPTER 6

Exercise Set 6.1

1. In the expression c^r, c is the base, r is the exponent. **3. a)** $\frac{x^m}{x^n} = x^{m-n}, x \ne 0$ **b)** Answers will vary. **5. a)** $(x^m)^n = x^{m \cdot n}$ **b)** Answers will vary. **7.** $x^0 \ne 1$ when $x = 0$ **9.** x^6 **11.** y^3 **13.** $3^5 = 243$ **15.** y^5 **17.** z^8 **19.** y^7 **21.** 2 **23.** x^7 **25.** $5^2 = 25$ **27.** y **29.** 1 **31.** y^3 **33.** 1 **35.** 3 **37.** 1 **39.** 1 **41.** x^{10} **43.** x^{25} **45.** x^3 **47.** x^{12} **49.** x^{15} **51.** $1.69x^2$ **53.** $-27x^9$ **55.** $8x^6y^3$ **57.** $\frac{x^3}{64}$ **59.** $\frac{y^4}{x^4}$ **61.** $\frac{216}{x^3}$ **63.** $\frac{27x^3}{y^3}$ **65.** $\frac{9x^2}{25}$ **67.** $\frac{16y^{12}}{x^4}$ **69.** $\frac{x^4}{y^2}$ **71.** $\frac{5x^2}{y^2}$ **73.** $\frac{1}{8x^2y^2}$ **75.** $\frac{7}{3x^5y^3}$ **77.** $-\frac{3y^2}{x^3}$ **79.** $-\frac{2}{x^3y^2z^5}$ **81.** $\frac{8}{x^6}$ **83.** $27y^9$ **85.** 1 **87.** $\frac{x^4}{y^4}$ **89.** $\frac{z^{24}}{16y^{28}}$ **91.** $64x^3y^{12}$ **93.** $9x^2y^8$ **95.** $18x^3y^9$ **97.** $10x^6y^3$ **99.** $10x^2y^7$ **101.** $8x^3y^3$ **103.** $-\frac{x^6}{y^6}$ **105.** x^2 **107.** $6.25x^6$ **109.** $\frac{x^6}{y^4}$ **111.** $-\frac{x^{15}}{y^6}$ **113.** $-216x^9y^6$ **115.** $-8x^{12}y^6z^3$ **117.** $49x^4y^8$ **119.** $108x^5y^7$ **121.** $73.96x^4y^{10}$ **123.** $x^{11}y^{13}$ **125.** x^8z^8 **127.** Cannot be simplified **129.** Cannot be simplified **131.** $5z^2$ **133.** Cannot be simplified **135.** 18 **137.** 1

139. The sign will be negative because a negative number with an odd number for an exponent will be negative. This is because $(-1)^m = -1$ when m is odd.

141. $7x^2$ **143.** $3x^2 + 4xy$ **145.** $\frac{9x^2}{8y^3}$ **148.** All real numbers **149. a)** 4 inches, 9 inches **b)** $w = \frac{P - 2l}{2}$

150.

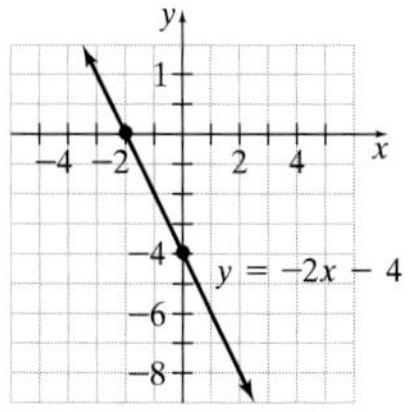

151. $\frac{2}{3}$

Exercise Set 6.2

1. Answers will vary. **3.** No, it is not simplified because of the negative exponent; $\frac{x^5}{y^3}$

5. The given simplification is not correct since $(y^4)^{-3} = \frac{1}{y^{12}}$. **7. a)** The numerator has one term, x^5y^2. **b)** The factors of the numerator are x^5 and y^2. **9.** The sign of the exponent changes when a factor is moved from the numerator to the denominator of a fraction. **11.** $\frac{1}{x^4}$ **13.** $\frac{1}{3}$ **15.** x^3 **17.** x **19.** 16 **21.** $\frac{1}{x^6}$ **23.** $\frac{1}{y^{20}}$

25. $\frac{1}{x^8}$ **27.** 64 **29.** y^2 **31.** x^2 **33.** 9 **35.** $\frac{1}{r^2}$ **37.** p^3 **39.** $\frac{1}{x^4}$ **41.** 27 **43.** $\frac{1}{27}$ **45.** z^9 **47.** $\frac{1}{x^{10}}$
49. y^6 **51.** $\frac{1}{x^4}$ **53.** $\frac{1}{x^{15}}$ **55.** $\frac{1}{x^8}$ **57.** y^{10} **59.** 1 **61.** 1 **63.** 1 **65.** x^8 **67.** 1 **69.** $\frac{1}{4}$ **71.** $\frac{1}{36}$ **73.** x^3
75. $\frac{1}{9}$ **77.** 125 **79.** $\frac{1}{9}$ **81.** 1 **83.** $\frac{1}{36x^4}$ **85.** $\frac{3y^2}{x^2}$ **87.** $\frac{5}{x^5y^2}$ **89.** $\frac{y^9}{x^{15}}$ **91.** $\frac{20}{y^5}$ **93.** $-12x^2$ **95.** $12x^5$
97. $\frac{8z^5}{y}$ **99.** $2x^5$ **101.** $\frac{4}{x^2}$ **103.** $\frac{x^4}{2y^5}$ **105.** $\frac{8x^6}{y^2}$ **107. a)** Yes **b)** No **109.** $16\frac{1}{16}$ **111.** $25\frac{1}{25}$ **113.** -2
115. $\frac{2}{3}$ **117.** $\frac{11}{6}$ **119.** $2, -3$ **121.** $\frac{5}{2}$ **125.** -18 **126.** 18 **127.** ≈ 6.67 ounces **128.** 9, 28

Exercise Set 6.3 **1.** A number in scientific notation is written as a number greater than or equal to 1 and less than 10 that is multiplied by some power of 10. **3. a)** Answers will vary. **b)** 4.69×10^{-3} **5.** 6 places to the left **7.** The exponent is negative when the number is less than 1. **9.** Negative since $0.00734 < 1$. **11.** 1.0×10^{-6} **13.** 4.2×10^4 **15.** 4.50×10^2 **17.** 5.3×10^{-2} **19.** 1.9×10^4 **21.** 1.86×10^{-6} **23.** 9.14×10^{-6} **25.** 1.101×10^5 **27.** 8.87×10^{-1} **29.** 7400 **31.** 40,000,000 **33.** 0.0000213 **35.** 625,000 **37.** 9,000,000 **39.** 535 **41.** 2991 **43.** 10,000 **45.** 120,000,000 **47.** 243,000 **49.** 0.02262 **51.** 320 **53.** 2500 **55.** 20 **57.** 4.2×10^{12} **59.** 4.5×10^{-7} **61.** 2.0×10^3 **63.** 1.75×10^2 **65.** $8.3 \times 10^{-4}, 3.2 \times 10^{-1}, 4.6, 4.8 \times 10^5$ **67. a)** $\$6.85 \times 10^8$ **b)** $\$1.37 \times 10^8$ **69. a)** $\$2.149 \times 10^9$ **b)** $\$2.71 \times 10^8$ **71.** 8,640,000,000 ft^3 **73. a)** 1.43×10^4 kg **b)** 8.30×10^3 kg **c)** 6.0×10^3 or 6000 kg **75. a)** 16,000,000 **b)** 500,000,000 **c)** 16 million **d)** 500 million **77. a)** 2.504 **b)** 2.504 **c)** Yes **d)** 0.0906 Newtons **79.** $\$1.612 \times 10^{10}$ **81. a)** 1.184×10^{10} **b)** $\approx 3.06 \times 10^5$ or about 306,000 people per day **83.** 11 **86.** 0 **87. a)** $\frac{3}{2}$ **b)** 0 **88.** 2 **89.** $-\frac{y^{12}}{64x^9}$

Exercise Set 6.4 **1.** A polynomial is an expression containing the sum of a finite number of terms of the form ax^n, where a is a real number and n is a whole number. **3. a)** The exponent on the variable is the degree of the term. **b)** The degree of the polynomial is the same as the degree of the highest degree term in the polynomial. **5.** $(3x + 2) - (4x - 6) = 3x + 2 - 4x + 6$ **7.** Because the exponent on the variable in a constant term is 0. **9. a)** Answers will vary. **b)** $4x^3 + 0x^2 + 5x - 7$ **11.** No, it contains a fractional exponent. **13.** Binomial **15.** Monomial **17.** Binomial **19.** Monomial **21.** Binomial **23.** Polynomial **25.** Trinomial **27.** $2x + 3$, first **29.** $x^2 - 2x - 4$, second **31.** $3x^2 + x - 8$, second **33.** Already in descending order, first **35.** Already in descending order, third **37.** $4x^3 - 3x^2 + x - 4$, third **39.** $-2x^3 + 3x^2 + 5x - 6$, third **41.** $3x - 2$ **43.** $-2x + 11$ **45.** $-5x - 1$ **47.** $x^2 + 6.6x + 0.8$ **49.** $2x^2 + 8x + 5$ **51.** $x^2 + 3x - 3$ **53.** $-3x^2 + x + \frac{17}{2}$ **55.** $5.4x^2 - 5x + 4$ **57.** $-2x^3 - 3x^2 + 4x - 3$ **59.** $2x^2 - 3xy + 5$ **61.** $5x^2y - 3x + 2$ **63.** $7x - 1$ **65.** $5x^2 + x + 5$ **67.** $4x^2 + 2x - 4$ **69.** $2x^3 - x^2 + 6x - 2$ **71.** $5x^3 - 7x^2 - 2$ **73.** $x - 6$ **75.** $3x + 4$ **77.** $-4x + 3$ **79.** $6x^2 + 7x - 8.5$ **81.** $8x^2 + x + 4$ **83.** $-7x^2 + 5x - 9$ **85.** $8x^3 - 7x^2 - 4x - 3$ **87.** $2x^3 - \frac{23}{5}x^2 + 2x - 2$ **89.** $3x - 2$ **91.** $2x^2 - 9x + 14$ **93.** $-x^3 + 11x^2 + 9x - 7$ **95.** $3x + 8$ **97.** $4x + 7$ **99.** $x^2 - 3x - 3$ **101.** $4x^2 - 5x - 6$ **103.** $4x^3 - 7x^2 + x - 2$ **105.** Answers will vary. **107.** Answers will vary. **109.** Sometimes **111.** Sometimes **113.** Answers will vary; one example is: $x^5 + x^4 + x$ **115.** No, all three terms would have to be degree 5 or 0. Therefore at least two of the terms could be combined. **117.** $a^2 + 2ab + b^2$ **119.** $4x^2 + 3xy$ **121.** $-12x + 18$ **123.** $8x^2 + 28x - 24$ **125.** < **126.** False **127.** True **128.** False **129.** False **130.** $\frac{y^3}{8x^9}$

Exercise Set 6.5 **1.** The distributive property is used when multiplying a monomial by a polynomial. **3.** First, Outer, Inner, Last **5.** Yes **7.** $(a + b)^2 = a^2 + 2ab + b^2$; $(a - b)^2 = a^2 - 2ab + b^2$ **9.** No, $(x + 5)^2 = x^2 + 10x + 25$ **11.** Answers will vary. **13.** Answers will vary. **15.** $3x^3y$ **17.** $20x^5y^6$ **19.** $-28x^6y^{15}$ **21.** $54x^6y^{14}$ **23.** $3x^6y$ **25.** $5.94x^8y^3$ **27.** $3x + 12$ **29.** $-6x^2 + 6x$ **31.** $-16y - 10$ **33.** $-2x^3 + 4x^2 - 10x$ **35.** $-20x^3 + 30x^2 - 20x$ **37.** $0.5x^5 - 3x^4 - 0.5x^2$ **39.** $0.6x^2y + 1.5x^2 - 1.8xy$ **41.** $xy - y^2 - 3y$ **43.** $x^2 + 7x + 12$ **45.** $6x^2 + 3x - 30$ **47.** $4x^2 - 16$ **49.** $-6x^2 - 8x + 30$ **51.** $-12x^2 + 32x - 5$ **53.** $4x^2 - 10x + 4$ **55.** $12x^2 - 30x + 12$ **57.** $x^2 - 1$ **59.** $4x^2 - 12x + 9$

61. $-4x^2 + 32x - 28$ **63.** $-4x^2 + 2x + 12$ **65.** $x^2 - y^2$ **67.** $6x^2 - 5xy - 6y^2$ **69.** $6x^2 + 2xy + 6x + 2y$ **71.** $x^2 + 0.9x + 0.18$ **73.** $xy - 2x - 2y + 4$ **75.** $x^2 - 25$ **77.** $9x^2 - 9$ **79.** $x^2 + 2xy + y^2$ **81.** $x^2 - 0.4x + 0.04$ **83.** $16x^2 + 40x + 25$ **85.** $0.16x^2 + 0.8xy + y^2$ **87.** $2x^3 + 10x^2 + 11x - 3$ **89.** $12x^3 + 5x^2 + 13x + 10$ **91.** $-14x^3 - 22x^2 + 19x - 3$ **93.** $-4x^3 + 6x^2 + 26x - 24$ **95.** $6x^4 + 5x^3 + 5x^2 + 10x + 4$ **97.** $x^4 - 3x^3 + 5x^2 - 6x$ **99.** $a^3 + b^3$ **101.** Yes **103.** No **105.** 6, 3, 1 **107. a)** $(x + 2)(2x + 1)$ or $2x^2 + 5x + 2$ **b)** 54 square feet **c)** 1 foot **109. a)** $a + b$ **b)** $a + b$ **c)** Yes **d)** $(a + b)^2$ **e)** $a^2 + ab + ab + b^2 = a^2 + 2ab + b^2$ **f)** $a^2 + 2ab + b^2$ **111.** $6x^6 - 18x^5 + 3x^4 + 35x^3 - 54x^2 + 38x - 12$ **113.** 13 miles **114.** $\frac{1}{16y^4}$ **115. a)** -216 **b)** $\frac{1}{216}$ **116.** $-6x^2 - 2x + 8$

Exercise Set 6.6

1. To divide a polynomial by a monomial, divide each term in the polynomial by the monomial. **3.** $1 + \frac{5}{y}$ **5.** Terms should be listed in descending order. **7.** $\frac{x^3 + 0x^2 + 5x - 1}{x + 2}$ **9.** $(x + 5)(x - 3) - 2 = x^2 + 2x - 17$ **11.** $\frac{x^2 - 2x - 15}{x + 3} = x - 5$ or $\frac{x^2 - 2x - 15}{x - 5} = x + 3$ **13.** $\frac{2x^2 + 5x + 3}{2x + 3} = x + 1$ or $\frac{2x^2 + 5x + 3}{x + 1} = 2x + 3$ **15.** $\frac{4x^2 - 9}{2x + 3} = 2x - 3$ or $\frac{4x^2 - 9}{2x - 3} = 2x + 3$ **17.** $x + 2$ **19.** $x + 3$ **21.** $\frac{3}{2}x + 4$ **23.** $-3x + 2$ **25.** $3x + 1$ **27.** $\frac{1}{2}x + 4$ **29.** $-\frac{4}{3} + 4x$ **31.** $1 + \frac{2}{x} - \frac{3}{x^2}$ **33.** $-2x^3 + \frac{3}{x} + \frac{4}{x^2}$ **35.** $x^3 + 4x - \frac{3}{x^3}$ **37.** $3x^2 - 2x + 6 - \frac{5}{2x}$ **39.** $-x^2 - \frac{3}{2}x + \frac{2}{x}$ **41.** $3x^4 + x^2 - \frac{10}{3} - \frac{3}{x^2}$ **43.** $x + 3$ **45.** $2x + 3$ **47.** $2x + 4$ **49.** $x + 3$ **51.** $x + 5 - \frac{3}{2x - 3}$ **53.** $2x + 3$ **55.** $2x - 3 + \frac{2}{4x + 9}$ **57.** $4x - 3 - \frac{3}{2x + 3}$ **59.** $3x^2 - 5$ **61.** $2x^2 + \frac{12}{x - 2}$ **63.** $x^2 + 3x + 9 + \frac{19}{x - 3}$ **65.** $x^2 + 3x + 9$ **67.** $2x^2 + x - 2 - \frac{2}{2x - 1}$ **69.** $-x^2 - 7x - 5 - \frac{8}{x - 1}$ **71.** No; for example $\frac{x + 2}{x} = 1 + \frac{2}{x}$ which is not a binomial. **73.** $2x^2 + 11x + 16$ **75.** First degree **77.** $4x$ **79.** Since the shaded areas minus 2 must equal 3, 1, 0, and −1, respectively, the shaded areas are 5, 3, 2, and 1, respectively. **81.** $2x^2 - 3x + \frac{5}{2} - \frac{3}{2(2x + 3)}$ **83.** $-3x + 3 + \frac{1}{x + 3}$ **86. a)** 2 **b)** 2, 0 **c)** $2, -5, 0, \frac{2}{5}, -6.3, -\frac{23}{34}$ **d)** $\sqrt{7}, \sqrt{3}$ **e)** $2, -5, 0, \sqrt{7}, \frac{2}{5}, -6.3, \sqrt{3}, -\frac{23}{34}$ **87. a)** 0 **b)** Undefined **88.** Parentheses, exponents, multiplication or division from left to right, addition or subtraction from left to right **89.** $-\frac{2}{3}$ **90.**

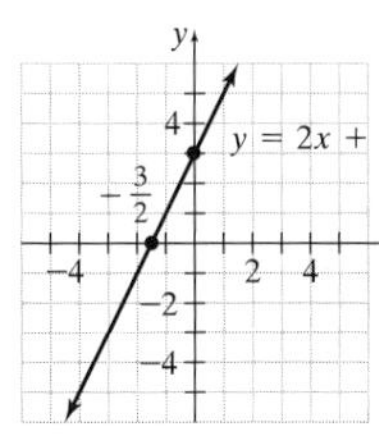

91.

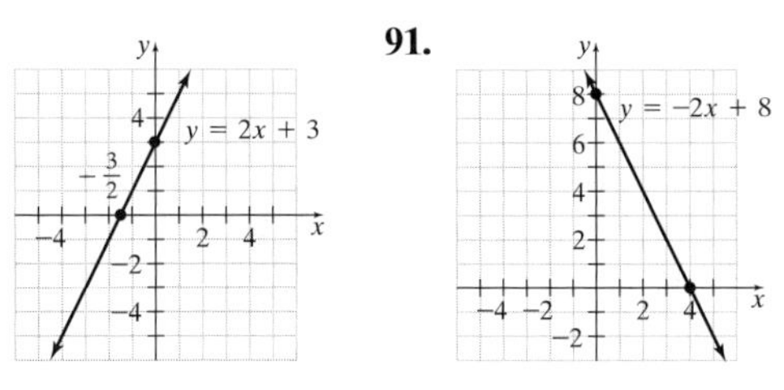

Review Exercises

1. x^6 **2.** x^6 **3.** 243 **4.** 32 **5.** x^3 **6.** 1 **7.** 25 **8.** 16 **9.** $\frac{1}{x^2}$ **10.** y^3 **11.** 1 **12.** 4 **13.** 1 **14.** 1 **15.** $25x^2$ **16.** $27x^3$ **17.** $16y^2$ **18.** $-27x^3$ **19.** $16x^8$ **20.** $-x^{12}$ **21.** x^{12} **22.** $\frac{4x^6}{y^2}$ **23.** $\frac{16y^4}{x^2}$ **24.** $24x^5$ **25.** $\frac{4x}{y}$ **26.** $18x^3y^6$ **27.** $9x^2$ **28.** $24x^7y^7$ **29.** $16x^8y^{11}$ **30.** $16x^4y^8$ **31.** $\frac{16x^6}{y^4}$ **32.** $27x^{12}y^3$ **33.** $\frac{1}{x^3}$ **34.** $\frac{1}{27}$ **35.** $\frac{1}{25}$ **36.** x^3 **37.** x^7 **38.** 9 **39.** $\frac{1}{y^3}$ **40.** $\frac{1}{x^5}$ **41.** $\frac{1}{x^3}$ **42.** $\frac{1}{x^4}$ **43.** x^5

44. x^7 **45.** $\frac{1}{x^6}$ **46.** $\frac{1}{9x^8}$ **47.** $\frac{x^9}{64y^3}$ **48.** $\frac{4y^2}{x^4}$ **49.** $10y^2$ **50.** $\frac{27y^3}{x^6}$ **51.** $\frac{x^4}{16y^6}$ **52.** $\frac{6}{x}$ **53.** $10x^2y^2$
54. $\frac{24y^2}{x^2}$ **55.** $\frac{12x^2}{y}$ **56.** $3y^5$ **57.** $\frac{3y}{x^3}$ **58.** $\frac{7x^5}{y^4}$ **59.** $\frac{4y^{10}}{x}$ **60.** $\frac{1}{2x^2y^5}$ **61.** 3.64×10^5 **62.** 1.53×10^{-1}
63. 7.63×10^{-3} **64.** 4.7×10^4 **65.** 1.37×10^6 **66.** 3.14×10^{-4} **67.** 0.0042 **68.** 0.000652 **69.** 970,000
70. 0.00000438 **71.** 0.914 **72.** 11,030,000 **73.** 57.8 **74.** 1260 **75.** 245 **76.** 340,000 **77.** 0.0325
78. 0.00003 **79.** 1.2×10^9 **80.** 4.8×10^9 **81.** 9.2×10^0 **82.** 5.0×10^{-4} **83.** 3.4×10^{-3} **84.** 3.4×10^7
85. a) $\$9.72 \times 10^{10}$ **b)** $\$9.21 \times 10^{10}$ or \$92,100,000,000 **86.** 3.0524×10^{15} miles **87.** Not a polynomial
88. Monomial, zero **89.** $x^2 + 3x - 4$, trinomial, second **90.** $4x^2 - x - 3$, trinomial, second
91. Binomial, third **92.** Not a polynomial **93.** $-4x^2 + x$, binomial, second **94.** Not a polynomial
95. $2x^3 + 4x^2 - 3x - 7$, polynomial, third **96.** $3x + 7$ **97.** $7x + 9$ **98.** $-3x - 5$ **99.** $-3x^2 + 10x - 15$
100. $11x^2 - 2x - 3$ **101.** $x - 6.7$ **102.** $-2x + 2$ **103.** $4x^2 - 12x - 15$ **104.** $6x^2 - 18x - 4$
105. $-5x^2 + 8x - 19$ **106.** $4x + 2$ **107.** $7x^2 + 10x$ **108.** $-15x^2 - 12x$ **109.** $6x^3 - 12x^2 + 21x$
110. $-3x^3 + 6x^2 + x$ **111.** $24x^3 - 16x^2 + 8x$ **112.** $x^2 + 9x + 20$ **113.** $-12x^2 - 21x + 6$
114. $4x^2 - 24x + 36$ **115.** $-6x^2 + 14x + 12$ **116.** $x^2 - 16$ **117.** $3x^3 + 7x^2 + 14x + 4$
118. $3x^3 + x^2 - 10x + 6$ **119.** $-12x^3 + 10x^2 - 30x + 14$ **120.** $x + 2$ **121.** $5x + 6$ **122.** $8x + 4$
123. $2x^2 + 3x - \frac{4}{3}$ **124.** $2x - 4 - \frac{3}{x}$ **125.** $4x^4 - 2x^3 + \frac{3}{2}x - \frac{1}{x}$ **126.** $-8x + 2$ **127.** $\frac{5}{3}x - 2 + \frac{5}{x}$
128. $\frac{5}{2}x + \frac{5}{x} + \frac{1}{x^2}$ **129.** $x + 4$ **130.** $4x + 2$ **131.** $5x - 2 + \frac{2}{x + 6}$ **132.** $2x^2 + 3x - 4$ **133.** $2x - 3$

Practice Test **1.** $12x^5$ **2.** $27x^3y^6$ **3.** $4x^3$ **4.** $\frac{x^3}{8y^6}$ **5.** $\frac{y^4}{4x^6}$ **6.** $\frac{x^5y}{3}$ **7.** 4 **8.** 1.26×10^9
9. 2.0×10^{-7} **10.** Trinomial **11.** Monomial **12.** Not a polynomial **13.** $6x^3 - 2x^2 + 5x - 5$, third degree
14. $3x^2 - 3x + 1$ **15.** $-2x^2 + 4x$ **16.** $3x^2 - x + 3$ **17.** $-8x^2 - 16x$ **18.** $8x^2 + 2x - 21$
19. $-12x^2 - 2x + 30$ **20.** $6x^3 + 2x^2 - 35x + 25$ **21.** $4x^2 + 2x - 1$ **22.** $-x + 2 - \frac{5}{3x}$ **23.** $4x + 5$
24. $3x - 2 - \frac{2}{4x + 5}$ **25. a)** 5.73×10^3 **b)** $\approx 7.78 \times 10^5$

Cumulative Review Test **1.** 11 **2.** 13 **3.** −3 **4.** $x > -\frac{9}{2}$;  **5.** $y = 5x + 5$ **6.** $-\frac{1}{2}$
7. Function; Domain: {−4, 1, 2, 5}; Range: {−5, 5, 7, 13} **8.** (4, −1) **9.** (5, 3) **10.** $135x^8y^{13}$
11. $-7x^2 - 5x + 2$, second **12.** $3x^2 + 9x - 2$ **13.** $4x^2 - 1$ **14.** $6y^2 - y - 15$ **15.** $6x^3 - 13x^2 + 9x - 2$
16. $2x + 4 - \frac{3}{x}$ **17.** $2x + 5$ **18.** \$3.33 **19.** Length = 10 feet; width = 4 feet **20.** Bob, 56.5 mph; Nick, 63.5 mph

CHAPTER 7

Exercise Set 7.1 **1.** A prime number is an integer greater than 1 that has exactly two factors, itself and 1.
3. The number 1 is not a prime number; it is called a unit. **5.** The greatest common factor is the greatest number that divides into all the numbers. **7.** A factoring problem may be checked by multiplying the factors.
9. $2^4 \cdot 3$ **11.** $2 \cdot 3^2 \cdot 5$ **13.** $2^3 \cdot 5^2$ **15.** 4 **17.** 14 **19.** 18 **21.** x **23.** $3x$ **25.** 1 **27.** xy **29.** x^3y^5
31. 5 **33.** x^2y^2 **35.** x **37.** $x + 3$ **39.** $2x - 3$ **41.** $3x - 4$ **43.** $x - 4$ **45.** $5(x + 2)$ **47.** $5(3x - 1)$
49. Cannot be factored **51.** $3x(3x - 4)$ **53.** $2p(13p - 4)$ **55.** $2x(2x^2 - 5)$ **57.** $12x^8(3x^4 + 2)$
59. $3y^3(8y^{12} - 3)$ **61.** $x(1 + 3y^2)$ **63.** Cannot be factored **65.** $4xy(4yz + x^2)$ **67.** $2xy^2(17x + 8y^2)$
69. $25x^2yz(z^2 + x)$ **71.** $y^2z^3(13y^3 - 11xz^2)$ **73.** $3(x^2 + 2x + 3)$ **75.** $3(3x^2 + 6x + 1)$ **77.** $4x(x^2 - 2x + 3)$
79. Cannot be factored **81.** $3(5p^2 - 2p + 3)$ **83.** $4x^3(6x^3 + 2x - 1)$ **85.** $xy(8x + 12y + 9)$
87. $(x + 3)(x + 4)$ **89.** Cannot be factored **91.** $(4x + 1)(2x + 1)$ **93.** $(4x + 1)(2x + 1)$ **95.** $3(\bigstar + 2)$
97. $7\Delta(5\Delta^2 - \Delta + 2)$ **99.** $2(x - 3)[2x^2(x - 3)^2 - 3x(x - 3) + 2]$ **101.** $x^{1/3}(x^2 + 5x + 6)$
103. $(x + 3)(x + 2)$ **105.** $-3x + 17$ **106.** 2 **107.** $y = \frac{4x - 20}{5}$ or $y = \frac{4}{5}x - 4$

108. a) $(6, 0), (0, 4)$ **b)**

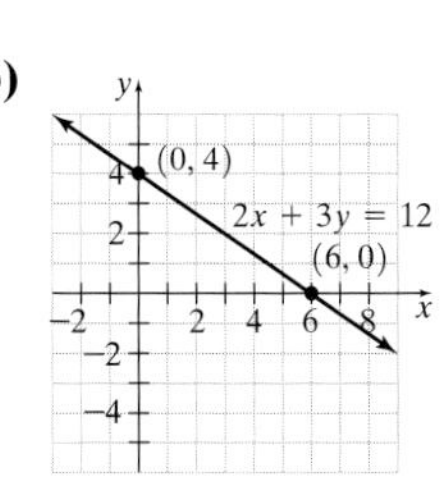

109. $\frac{9y^2}{4x^6}$

Exercise Set 7.2

1. The first step is to factor out a common factor, if one exists. **3.** $x^2 + 4x - 2x - 8$; found by multiplying the factors. **5.** Answers will vary. **7.** $(x + 2)(x + 3)$ **9.** $(x + 4)(x + 3)$ **11.** $(x + 5)(x + 2)$ **13.** $(x - 5)(x + 3)$ **15.** $(2x + 7)(2x - 7)$ **17.** $(3x + 1)(x + 3)$ **19.** $(3x - 1)(2x + 1)$ **21.** $(8x + 1)(x + 4)$ **23.** $(x + 1)(3x - 2)$ **25.** $(2x - 3)(x - 2)$ **27.** $(3x + 5)(x - 3)$ **29.** $(x - 3y)(x + 2y)$ **31.** $(x - 3y)(3x + 2y)$ **33.** $(2x - 5y)(5x - 6y)$ **35.** $(x + a)(x + b)$ **37.** $(x - 2)(y + 4)$ **39.** $(a + b)(a + 3)$ **41.** $(x + 5)(y - 1)$ **43.** $(4 - x)(3 + 2y)$ **45.** $(z^2 + 1)(z + 3)$ **47.** $(x^2 - 3)(x + 4)$ **49.** $2(x + 4)(x - 6)$ **51.** $4(x + 2)(x + 2) = 4(x + 2)^2$ **53.** $x(3x - 1)(2x + 3)$ **55.** $x(x + 2y)(x - 3y)$ **57.** $(x + 4)(y + 2)$ **59.** $(x + 5)(y + 6)$ **61.** $(a + b)(x + y)$ **63.** $(d + 3)(c - 4)$ **65.** $(a + b)(c - d)$ **67.** No; $xy + 2x + 5y + 10$ is factorable; $xy + 10 + 2x + 5y$ is not factorable in this arrangement. **69.** $(\Delta - 5)(\Delta + 3)$ **71. a)** $3x^2 + 6x + 4x + 8$ **b)** $(3x + 4)(x + 2)$ **73. a)** $2x^2 - 6x - 5x + 15$ **b)** $(2x - 5)(x - 3)$ **75. a)** $4x^2 - 20x + 3x - 15$ **b)** $(4x + 3)(x - 5)$ **77.** $(\bigstar + 2)(\odot + 3)$ **79.** $\frac{6}{5}$ **80.** 30 pounds of chocolate wafers, 20 pounds of peppermints. **81.** $5x^2 - 2x - 3 + \frac{5}{3x}$ **82.** $x + 3$

Exercise Set 7.3

1. Since 8000 is positive, both signs will be the same. Since 180 is positive, both signs will be positive. **3.** Since −8000 is negative, one sign will be positive, the other will be negative. **5.** Since 8000 is positive, both signs will be the same. Since −240 is negative, both signs will be negative. **7.** The trinomial $x^2 + 3xy - 18y^2$ is obtained by multiplying the factors using the FOIL method. **9.** The trinomial $3x^2 - 3y^2$ is obtained by multiplying all the factors and combining like terms. **11.** Find two numbers whose product is c, and whose sum is b. The factors are (x + first number) and (x + second number). **13.** $(x - 3)(x - 4)$ **15.** $(x + 2)(x + 4)$ **17.** $(x + 4)(x + 3)$ **19.** Cannot be factored **21.** $(y - 15)(y - 1)$ **23.** $(x + 3)(x - 2)$ **25.** $(r - 5)(r + 3)$ **27.** $(b - 9)(b - 2)$ **29.** Cannot be factored **31.** $(a + 11)(a + 1)$ **33.** $(x + 15)(x - 2)$ **35.** $(x + 2)^2$ **37.** $(k + 3)^2$ **39.** $(x - 5)^2$ **41.** $(w - 15)(w - 3)$ **43.** $(x + 13)(x - 3)$ **45.** $(x - 5)(x + 4)$ **47.** $(y + 7)(y + 2)$ **49.** $(x + 16)(x - 4)$ **51.** Cannot be factored **53.** $(x + 8)(x - 10)$ **55.** $(x - 5)(x - 13)$ **57.** $(x - 3y)(x - 5y)$ **59.** $(x - 2y)^2$ **61.** $(x + 5y)(x + 3y)$ **63.** $7(x - 5)(x - 1)$ **65.** $5(x + 3)(x + 1)$ **67.** $2(x - 4)(x - 3)$ **69.** $x(x + 3)(x - 6)$ **71.** $2x(x + 7)(x - 4)$ **73.** $x(x + 4)^2$ **75.** Both negative; one positive and one negative; one positive and one negative; both positive **77.** $x^2 + 5x + 4 = (x + 1)(x + 4)$ **79.** $x^2 + 11x + 28 = (x + 7)(x + 4)$ **81.** $(x + 0.4)(x + 0.2)$ **83.** $\left(x + \frac{1}{5}\right)\left(x + \frac{1}{5}\right) = \left(x + \frac{1}{5}\right)^2$ **85.** $-(x + 2)(x + 4)$ **87.** $(x + 20)(x - 15)$ **89.** 6 **90.** 19.6%

91.

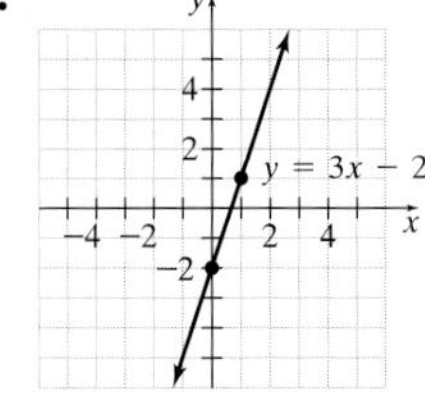

92. $2x^3 + x^2 - 16x + 12$ **93.** $3x + 2 - \frac{2}{x - 4}$ **94.** $(x - 2)(3x + 5)$

Using Your Graphing Calculator, 7.4

1. $(3x + 4)(2x - 7)$ **2.** $(5x - 8)(2x + 3)$ **3.** $(6x - 7)(2x + 5)$ **4.** $(9x + 4)(3x + 5)$

Exercise Set 7.4

1. Factoring trinomials is the reverse process of multiplying binomials. **3.** The constant, c, of the trinomial **5.** $(2x + 1)(x + 5)$ **7.** $(3x + 2)(x + 4)$ **9.** $(3x + 1)(x + 1)$ **11.** $(2x + 5)(x + 4)$ **13.** $(2x - 1)(2x + 3)$ **15.** $(5y - 3)(y - 1)$ **17.** Cannot be factored

19. $(6x + 1)(x + 3)$ **21.** Cannot be factored **23.** $(5y - 1)(y - 3)$ **25.** $(7x + 1)(x + 6)$
27. $(7x - 1)(x - 1)$ **29.** $(x - 1)(3x - 7)$ **31.** $(5z + 4)(z - 2)$ **33.** $(4y - 3)(y + 2)$ **35.** $(5x - 1)(2x - 5)$
37. $(4x + 5)(2x - 3)$ **39.** $3(2x + 1)(x + 5)$ **41.** $2(3x + 5)(x + 1)$ **43.** $x(2x + 1)(3x - 4)$
45. $4x(3x + 1)(x + 2)$ **47.** $2x(3x + 5)(x - 1)$ **49.** $5(6x + 1)(2x + 1)$ **51.** $(2x + y)(x + 2y)$
53. $(2x - y)(x - 3y)$ **55.** $2(2x - y)(3x + 4y)$ **57.** $3(2x + 3y)(x - 4y)$
59. $4x^2 - 27x - 7$; obtained by multiplying the factors. **61.** $10x^2 + 35x + 15$; obtained by multiplying the factors.
63. $2x^4 - x^3 - 3x^2$; obtained by multiplying the factors.
65. a) Dividing the trinomial by the binomial gives the second factor. **b)** $6x + 11$ **67.** $15x - 32$
69. $(8x - 3)(x - 12)$ **71.** $(8x + 5)(2x - 9)$ **73.** $3(8x - 7)(3x + 20)$
75. $x + 4$; the product of the three first terms must equal $2x^3$, and the product of the constants must equal -36.
76. 49 **77.** $\approx$ 133.33 mph **78. a)** 11 **b)** 17 **79.** $12xy^2(3x^3y - 1 + 2x^4y^4)$ **80.** $(x - 9)(x - 6)$

Exercise Set 7.5

1. a) $a^2 - b^2 = (a + b)(a - b)$ **b)** Answers will vary.
3. a) $a^3 - b^3 = (a - b)(a^2 + ab + b^2)$ **b)** Answers will vary. **5.** No **7.** $(y + 4)(y - 4)$
9. $(y + 10)(y - 10)$ **11.** $(x + 7)(x - 7)$ **13.** $(x + y)(x - y)$ **15.** $(3y + 5z)(3y - 5z)$
17. $4(4a + 3b)(4a - 3b)$ **19.** $(5x + 6)(5x - 6)$ **21.** $(z^2 + 9x)(z^2 - 9x)$ **23.** $9(x^2 + 3y)(x^2 - 3y)$
25. $(6m^2 + 7n)(6m^2 - 7n)$ **27.** $20(x + 3)(x - 3)$ **29.** $4(2x + 5y^2)(2x - 5y^2)$ **31.** $(x + y)(x^2 - xy + y^2)$
33. $(a - b)(a^2 + ab + b^2)$ **35.** $(x + 2)(x^2 - 2x + 4)$ **37.** $(x - 3)(x^2 + 3x + 9)$ **39.** $(a + 1)(a^2 - a + 1)$
41. $(3x - 1)(9x^2 + 3x + 1)$ **43.** $(3a - 5)(9a^2 + 15a + 25)$ **45.** $(3 - 2y)(9 + 6y + 4y^2)$
47. $(3x + 4y)(9x^2 - 12xy + 16y^2)$ **49.** $8(x + 1)^2$ **51.** $b(a + 3)(a - 3)$ **53.** $3(x + 2)^2$
55. $5(x - 3)(x + 1)$ **57.** $3(x + 3)(y - 2)$ **59.** $2(x + 5)(x - 5)$ **61.** $3y(x + 3)(x - 3)$
63. $3y^2(x + 1)(x^2 - x + 1)$ **65.** $2(x - 2)(x^2 + 2x + 4)$ **67.** $2(x + 4)(3x - 2)$ **69.** $x(3x + 2)(x - 4)$
71. $(x + 2)(4x - 3)$ **73.** $25(b + 2)(b - 2)$ **75.** $a^3b^2(a + 2b)(a - 2b)$ **77.** $3x^2(x - 3)^2$ **79.** $x(x^2 + 25)$
81. $(y^2 + 4)(y + 2)(y - 2)$ **83.** $5(3a - 2b)(4a + b)$ **85.** $(2a - 3)(b + 2)$ **87.** $9(1 + y^2)(1 + y)(1 - y)$
89. You cannot divide both sides of the equation by $(a - b)$ because it equals 0. **91.** $2◆^4(◆^2 + 2✱^2)$
93. $(x^2 + 1)(x^4 - x^2 + 1)$ **95.** $(x - 3 + 2y)(x - 3 - 2y)$ **97.** $(x + y)(x - y)(x^2 - xy + y^2)(x^2 + xy + y^2)$
98. $x \le 1$;  **99.** $d = \dfrac{2A - hb}{h}$ **100.** $-\dfrac{3}{2}$ **101.** $\dfrac{8x^9}{27y^{12}}$ **102.** $\dfrac{1}{x^5}$

Using Your Graphing Calculator, 7.6

1. 4, −2 **2.** 2, 5 **3.** 2, 1 **4.** −1, 5

Exercise Set 7.6

1. Answers will vary. **3.** $ax^2 + bx + c = 0$ **5. a)** The zero-factor property may only be used when one side of the equation is equal to 0 **b)** −2, 3 **7.** 0, −1 **9.** 0, 8 **11.** $-\dfrac{5}{2}, 3$ **13.** 4, −4 **15.** 0, 12
17. 0, −3 **19.** −5, 4 **21.** −2, −10 **23.** 3, −6 **25.** −4, 2 **27.** 2, −21 **29.** −5, −6 **31.** 5, −3 **33.** 30, −1
35. 6, −3 **37.** $\dfrac{1}{3}, 7$ **39.** $\dfrac{2}{3}, -1$ **41.** −4, 3 **43.** $-\dfrac{3}{4}, \dfrac{1}{2}$ **45.** 2, 3 **47.** 0, 16 **49.** 10, −10 **51.** 3, −3
53. 9, −9 **55.** 30 **57. a)** 4 **b)** 9 **59.** $x^2 - 12x + 36 = 0$ **61.** 3 meters **63.** 4 seconds **65.** 9 and 13
67. Length: 12 feet; width: 3 feet **69.** Length: 15 feet; width: 10 feet **71.** 8 cm; 6 cm
73. $x^3 - 8x^2 + 15x = 0$; the factors are x, $x - 3$, and $x - 5$. **75.** 3 and 6 **79.** You cannot divide by zero
80. $-x^2 + 7x - 4$ **81.** $6x^3 + x^2 - 10x + 4$ **82.** $2x - 3$ **83.** $2x - 3$

Review Exercises

1. y^2 **2.** $3p$ **3.** $5a^2$ **4.** $4x^2y^2$ **5.** 1 **6.** 1 **7.** $x - 5$ **8.** $x + 5$ **9.** $3(x - 3)$
10. $6(2x + 1)$ **11.** $4y(6y - 1)$ **12.** $5p^2(11p - 4)$ **13.** $12ab(4a - 3b)$ **14.** $6xy(1 - 2x)$
15. $6x^4y^2(10 + x^5y - 3x)$ **16.** Cannot be factored **17.** Cannot be factored **18.** $(x - 2)(5x + 3)$
19. $(3x - 2)(x - 1)$ **20.** $(2x + 1)(4x - 3)$ **21.** $(x + 2)(x + 5)$ **22.** $(x + 4)(x - 3)$
23. $(y - 7)(y - 7) = (y - 7)^2$ **24.** $(3a - 1)(a - b)$ **25.** $(3x + 2)(y + 1)$ **26.** $(x - 2y)(x + 3)$
27. $(5x - 1)(x + 4)$ **28.** $(x + 4y)(5x - y)$ **29.** $(4x - 5y)(x + 3y)$ **30.** $(2a - b)(3a - 5b)$
31. $(a + 1)(b - 1)$ **32.** $(3x + 2y)(x - 3y)$ **33.** $(7a - b)(a + 2b)$ **34.** $(3x - 1)(2x + 3)$
35. $(x + 1)(x - 8)$ **36.** Cannot be factored **37.** $(x - 6)(x - 7)$ **38.** $(x + 5)(x - 4)$ **39.** $(x + 6)(x + 5)$
40. $(x - 8)(x - 7)$ **41.** Cannot be factored **42.** Cannot be factored **43.** $x(x - 9)(x - 8)$
44. $x(x - 8)(x + 5)$ **45.** $(x + 3y)(x - 5y)$ **46.** $4x(x + 5y)(x + 3y)$ **47.** $(2x + 3)(x + 4)$
48. $(3x - 1)(x - 4)$ **49.** $(4x - 5)(x - 1)$ **50.** $(5x + 2)(x - 3)$ **51.** $(3x - 1)(3x + 2)$
52. $(5x - 2)(x - 6)$ **53.** Cannot be factored **54.** $(6x + 1)(x + 5)$ **55.** $(5x - 3)(x + 8)$
56. $(3x - 2)(2x + 5)$ **57.** $(4x + 5)(2x - 7)$ **58.** $(3x - 1)^2$ **59.** $x(3x - 2)^2$ **60.** $2x(3x + 1)(3x - 5)$
61. $(8a + b)(2a - 3b)$ **62.** $(2a - 3b)(2a - 5b)$ **63.** $(x + 5)(x - 5)$ **64.** $(x + 10)(x - 10)$

65. $4(x+2)(x-2)$ **66.** $9(3x+y)(3x-y)$ **67.** $(7+x)(7-x)$ **68.** $(8+x)(8-x)$
69. $(4x^2+7y)(4x^2-7y)$ **70.** $(10x^2+11y^2)(10x^2-11y^2)$ **71.** $(x-y)(x^2+xy+y^2)$
72. $(x+y)(x^2-xy+y^2)$ **73.** $(x-1)(x^2+x+1)$ **74.** $(x+2)(x^2-2x+4)$ **75.** $(a+3)(a^2-3a+9)$
76. $(x-2)(x^2+2x+4)$ **77.** $(5a+b)(25a^2-5ab+b^2)$ **78.** $(3-2y)(9+6y+4y^2)$
79. $3(3x^2+5y)(3x^2-5y)$ **80.** $2(x-4y)(x^2+4xy+16y^2)$ **81.** $(x-6)(x-10)$ **82.** $3(x-3)^2$
83. $4(a+4)(a-4)$ **84.** $4(y+3)(y-3)$ **85.** $8(x+3)(x-1)$ **86.** $(x-9)(x+3)$
87. $(3x-1)(3x+2)$ **88.** $(4x-1)(x+2)$ **89.** $8(x-1)(x^2+x+1)$ **90.** $y(x-3)(x^2+3x+9)$
91. $b(a+3)(a-5)$ **92.** $3x(2x+3)(x+5)$ **93.** $(x+3y)(x+2y)$ **94.** $(2x-5y)(x+2y)$
95. $(2x-5y)^2$ **96.** $(5a+7b)(5a-7b)$ **97.** $(x+2)(y-7)$ **98.** $y^5(4+5y)(4-5y)$
99. $2x(x+4y)(x+2y)$ **100.** $(2x-3y)(3x+7y)$ **101.** $x^2(4x+1)(4x-3)$ **102.** $(a^2+1)(a+1)(a-1)$

103. $0, -6$ **104.** $2, -8$ **105.** $-5, \frac{3}{4}$ **106.** $0, 3$ **107.** $0, -4$ **108.** $0, -2$ **109.** $-2, -5$ **110.** $-4, 3$ **111.** $1, 2$

112. $-1, -4$ **113.** $2, 4$ **114.** $1, -2$ **115.** $\frac{1}{4}, -\frac{3}{2}$ **116.** $\frac{1}{2}, -8$ **117.** $2, -2$ **118.** $\frac{10}{7}, -\frac{10}{7}$ **119.** $\frac{3}{2}, \frac{1}{4}$

120. $\frac{3}{2}, \frac{5}{2}$ **121.** 12, 13 **122.** 6, 8 **123.** 4, 14 **124.** Width: 7 feet; length: 9 feet **125.** 9 inches **126.** 80 dozen

Practice Test **1.** $3y^2$ **2.** $3xy^2$ **3.** $2xy(5x^2y-4)$ **4.** $4a^2b(2a-3b+7)$ **5.** $(x+2)(x-8)$
6. $(5x+2)(x-3)$ **7.** $(a-5b)(a-4b)$ **8.** $(x+4)(x+8)$ **9.** $(x+8)(x-3)$ **10.** $(5a-3b)(5a+2b)$
11. $4(x+2)(x-6)$ **12.** $x(2x-1)(x-1)$ **13.** $(3x+2y)(4x-3y)$ **14.** $(x+3y)(x-3y)$

15. $(x+3)(x^2-3x+9)$ **16.** $\frac{3}{5}, 1$ **17.** $0, 6$ **18.** $-8, 8$ **19.** 7 **20.** $-2, -3$ **21.** $3, 4$

22. 4, 9 **23.** 9, 11 **24.** Length: 6 meters; width: 4 meters **25.** 10 seconds

Cumulative Review Test **1.** -515 **2.** 9 **3.** $\approx 298{,}545{,}455$ **4.** $\frac{31}{4}$ **5.** $x \geq 5$;

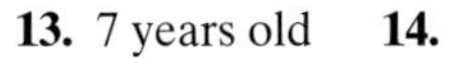

6. $y = \frac{2}{3}x - 2$ **7.** $\frac{1}{4}$ hour **8.** Longer than 2 days **9.** $(6, 0), (0, 4)$ **10.** $-\frac{13}{9}$ **11.** $y = \frac{1}{2}x - 5$

12.

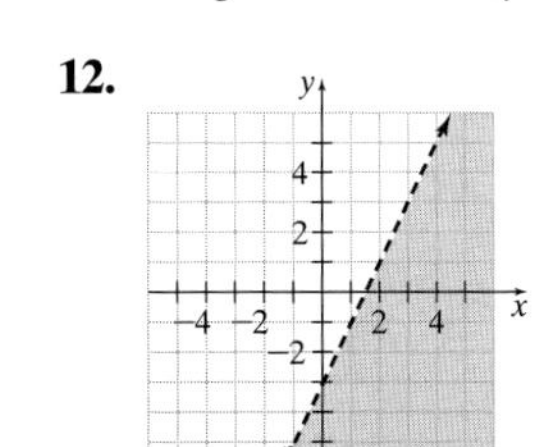

13. 7 years old **14.**

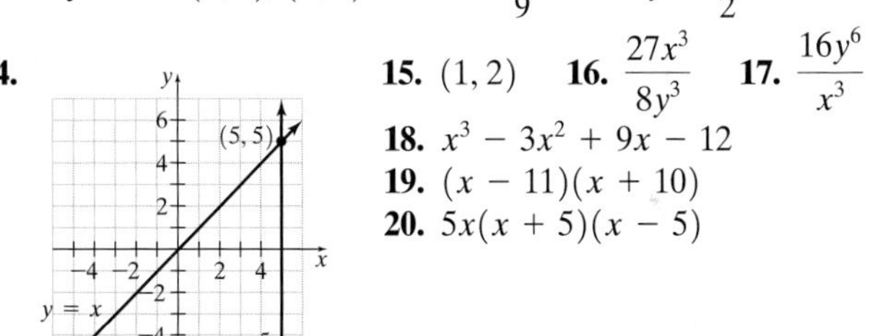

15. $(1, 2)$ **16.** $\frac{27x^3}{8y^3}$ **17.** $\frac{16y^6}{x^3}$
18. $x^3 - 3x^2 + 9x - 12$
19. $(x-11)(x+10)$
20. $5x(x+5)(x-5)$

CHAPTER 8

Using Your Graphing Calculator, 8.1 **1.** Yes **2.** Yes **3.** No **4.** Yes

Exercise Set 8.1 **1. a–b)** Answers will vary. **3.** The value of the variable does not make the denominator equal to 0. **5.** There is no factor common to both terms in the numerator. **7.** The denominator cannot be 0. **9.** $x \neq 3$ **11.** No **13.** All real numbers except $x = 0$. **15.** All real numbers except $x = 3$.

17. All real numbers except $x = 2$ and $x = -2$. **19.** All real numbers except $x = 2$ and $x = -8$. **21.** $\frac{1}{1+y}$

23. 5 **25.** $\frac{x^2+6x+3}{2}$ **27.** $x+1$ **29.** $\frac{x}{x+2}$ **31.** $\frac{x-3}{x-2}$ **33.** $\frac{x+1}{x+2}$ **35.** -1 **37.** $-(x+2)$

39. $-\frac{x+6}{2x}$ **41.** $-(x+3)$ **43.** $3x+2$ **45.** $3x-4$ **47.** $2x-3$ **49.** $x+4$ **51.** $\frac{x-4}{x+4}$

53. x^2+2x+4 **55.** $\frac{\Delta}{3}$ **57.** $\frac{\Delta}{2\Delta+3}$ **59.** -1 **61.** $x+2$; $(x+2)(x-3) = x^2-x+6$

63. $x^2+7x+12$; $(x+3)(x+4) = x^2+7x+12$ **65. a)** $x \neq -3, x \neq 2$ **b)** $\frac{1}{x-2}$

67. a) $x \neq 0, x \neq -5, x \neq \frac{3}{2}$ **b)** $\frac{1}{x(2x-3)}$ **69.** 1, numerator and denominator are identical. **72.** $y = x - 2z$
73. 28°, 58°, and 94° **74.** (3, 2) **75.** $\frac{x^4}{9y^2}$ **76.** $9x^2 - 10x - 17$

Exercise Set 8.2

1. Answers will vary. **3.** $x^2 - 2x - 8$; numerator must be $(x-4)(x+2)$
5. $x^2 - x - 12$; denominator must be $(x-4)(x+3)$ **7.** $\frac{xy}{6}$ **9.** $\frac{80x^4}{y^6}$ **11.** $\frac{36x^9y^2}{25z^7}$ **13.** $\frac{-3x+2}{3x+2}$ **15.** 1
17. $\frac{1}{a^2-b^2}$ **19.** $\frac{x+3}{2(x+2)}$ **21.** $x+3$ **23.** $3x^2y$ **25.** $\frac{10z}{x}$ **27.** $\frac{1}{6a^2b}$ **29.** $\frac{3x^2}{2}$ **31.** $\frac{x+6}{x}$ **33.** $\frac{x-8}{x+2}$
35. $\frac{x+3}{x-1}$ **37.** -1 **39.** $4x^2y^2$ **41.** $\frac{5}{3ab^2c^2}$ **43.** $\frac{3y^2}{a^2}$ **45.** $\frac{32x^7}{35my^2}$ **47.** $\frac{1}{2}$ **49.** $3x$ **51.** $\frac{1}{6\Delta^3}$ **53.** $\frac{\Delta + ☺}{9(\Delta - ☺)}$
55. $x^2 + 3x + 2$ **57.** $x^2 - 3x - 10$ **59.** $x^2 - 3x + 2$ **61.** $\frac{x-3}{x+3}$ **63.** 1 **68.** $20x^4y^5z^{11}$
69. $2x^2 + x - 2 - \frac{2}{2x-1}$ **70.** $3(x-5)(x+2)$ **71.** 5, −2

Exercise Set 8.3

1. Answers will vary. **3.** Answers will vary. **5.** $5(x+5)$ **7.** $3x(x-3)$
9. a) The negative sign in $-(3x^2 - 4x + 5)$ was not distributed. **b)** $\frac{6x - 2 - 3x^2 + 4x - 5}{x^2 - 4x + 3}$ **11.** $\frac{3x-1}{8}$
13. $\frac{4x+3}{7}$ **15.** $\frac{x+6}{x}$ **17.** 2 **19.** $\frac{5x+7}{x-1}$ **21.** $\frac{2x+1}{2x^2}$ **23.** $\frac{1}{x-4}$ **25.** 0 **27.** $\frac{-3x-10}{x+7}$ **29.** $x-3$
31. $-\frac{1}{x-2}$ **33.** 1 **35.** $\frac{2x+9}{2x-9}$ **37.** $\frac{x+5}{x+6}$ **39.** $\frac{3}{4}$ **41.** $\frac{x-5}{x+2}$ **43.** $\frac{3x+2}{x-4}$ **45.** $\frac{6x+1}{x-8}$ **47.** 7
49. $3x$ **51.** $20x$ **53.** x^3 **55.** $6x+5$ **57.** $x^2(x+1)$ **59.** $36x^3y$ **61.** $72x(x+5)$ **63.** $x(x+1)$
65. $120x^2y^3$ **67.** $6(x+4)(x+2)$ **69.** $(x+6)(x+5)$ **71.** $(x-8)(x+3)(x+8)$
73. $(x+4)(x-4)(x+1)$ **75.** $(x+5)(x+1)(x+3)$ **77.** $(x-3)^2$ **79.** $(x-6)(x-1)$
81. $(3x-2)(x+6)(3x-1)$ **83.** $(2x+1)^2(4x+3)$ **85.** $x^2 + x - 9$; sum of numerators must be $2x^2 - 5x - 6$
87. $x^2 + 9x - 10$; sum of numerators must be $5x - 7$ **89.** $(x-5)(x+4)$ **91.** $40\Delta^4☺^5$
93. $(\Delta+3)(\Delta-3)(\Delta-1)$ **95.** $\frac{-20x-4}{x+7}$ **97.** $(x-3)(x+3)$ **99.** $(x+2)(x-2)(3x-1)$
101. $\frac{92}{45}$ or $2\frac{2}{45}$ **102.** $-\frac{1}{5}$ **103.** 2.25 oz **104. a)** 70 hours **b)** Plan 2
105. **106.** 1

Exercise Set 8.4

1. For each fraction, divide the LCD by the denominator. **3. a)** Answers will vary.
b) $\frac{x^2 + x - 9}{(x+2)(x-3)(x-2)}$ **5. a)** $6y^2$ **b)** $\frac{2xy+5}{6y^2}$ **c)** Yes **7.** $\frac{16}{3x}$ **9.** $\frac{3x+8}{2x^2}$ **11.** $\frac{3x+5}{x}$
13. $\frac{3x+10}{5x^2}$ **15.** $\frac{35y+12x}{20x^2y^2}$ **17.** $\frac{y^2+x}{y}$ **19.** $\frac{9a+1}{6a}$ **21.** $\frac{4x^2+2y}{xy}$ **23.** $\frac{30a^2-4b}{5a^2b}$ **25.** $\frac{10x-12}{x(x-4)}$
27. $\frac{11p+6}{p(p+3)}$ **29.** $\frac{-4x^2+9x+15}{4x(3x+5)}$ **31.** $\frac{2}{p-3}$ **33.** $\frac{14}{x+7}$ **35.** $\frac{a+6}{2(a-2)}$ **37.** $\frac{20x}{(x-5)(x+5)}$
39. $\frac{5x-12}{(x+3)(x-3)}$ **41.** $\frac{-x+6}{(x+2)(x-2)}$ **43.** $\frac{11}{(x-3)(x-4)}$ **45.** $\frac{6x-8}{(x+4)(x-2)}$ **47.** $\frac{x^2+3x-18}{(x+5)^2}$

49. $\dfrac{2a + 13}{(a - 8)(a - 1)(a + 2)}$ **51.** $\dfrac{4x - 3}{(x + 3)(x + 2)(x - 2)}$ **53.** $\dfrac{3x^2 - 8x - 3}{(2x + 1)(3x - 2)(x + 3)}$
55. $\dfrac{2x^2 - 3x - 4}{(4x + 3)(x + 2)(2x - 1)}$ **57.** $\dfrac{1}{x - 3}$ **59.** All real numbers except $x = 0$.
61. All real numbers except $x = 3$ and $x = -6$. **63.** $\dfrac{4}{\Delta - 2}$ **65.** All real numbers except $a = -b$ and $a = 0$.
67. 0 **69.** $\dfrac{2x - 3}{2 - x}$ **71.** $\dfrac{6x + 5}{(x + 2)(x - 3)(x + 1)}$ **74.** 1215 **75.** $x > -8$, (number line: open circle at -8, arrow to the right)
76. $4x - 3 - \dfrac{4}{2x + 3}$ **77.** -1

Exercise Set 8.5 **1.** A complex fraction is a fraction whose numerator or denominator (or both) contains a fraction. **3. a)** Numerator, $\dfrac{x - 1}{5}$; denominator, $\dfrac{2}{x^2 + 5x + 6}$ **b)** Numerator, $\dfrac{1}{2y} + x$; denominator, $\dfrac{3}{y} + x$
5. $\dfrac{11}{4}$ **7.** $\dfrac{57}{32}$ **9.** $\dfrac{3}{448}$ **11.** $\dfrac{xy^3}{18}$ **13.** $6xz^2$ **15.** $\dfrac{ab - a}{1 + a}$ **17.** $\dfrac{3}{x}$ **19.** $\dfrac{5x - 1}{4x - 1}$ **21.** $\dfrac{y^2 + x^2}{y(x - y)}$ **23.** $-\dfrac{a}{b}$
25. -1 **27.** $a^2(a - b)$ **29.** $b - a$ **31.** $\dfrac{a^2 + b}{b(b + 1)}$ **33.** $\dfrac{y + x}{y - x}$ **35.** $\dfrac{ab^2 + b^2}{a^2(b + 1)}$ **37. b)–c)** $-\dfrac{342}{245}$
39. b)–c) $\dfrac{x - y + 3}{2x + 2y - 7}$ **41. a)** $\dfrac{\frac{5}{12x}}{\frac{8}{x^2} - \frac{4}{3x}}$ **b)** $\dfrac{5x}{96 - 16x}$ **43.** $\dfrac{y + x}{2xy}$ **45.** $x + y$ **47. a)** $\dfrac{2}{7}$ **b)** $\dfrac{4}{13}$
49. $\dfrac{a^3b + a^2b^3 - ab^2}{a^3 - ab^3 + b^2}$ **51.** $\dfrac{17}{2}$ **52.** $(5, -3)$ **53.** A polynomial is an expression containing a finite number of terms of the form ax^n where a is a real number and n is a whole number. **54.** $\dfrac{x^2 - 9x + 2}{(3x - 1)(x + 6)(x - 3)}$

Exercise Set 8.6 **1. a)** Answers will vary. **b)** $\dfrac{2}{3}$ **3. a)** The problem on the left is an expression to be simplified while the problem on the right is an equation to be solved. **b)** Left: Write the fractions with the LCD, $12(x - 1)$, then combine numerators; Right: Multiply both sides of the equation by the LCD, $12(x - 1)$, then solve. **c)** Left: $\dfrac{x^2 - x + 12}{12(x - 1)}$; Right: 4, -3 **5.** You need to check for extraneous solutions when there is a variable in a denominator. **7.** No, because there is no variable in any denominator. **9.** 12 **11.** 52 **13.** 30 **15.** -1 **17.** -1
19. 4 **21.** 16 **23.** $-\dfrac{1}{5}$ **25.** $\dfrac{5}{8}$ **27.** $\dfrac{14}{3}$ **29.** No solution **31.** -1 **33.** $-\dfrac{12}{7}$ **35.** No solution **37.** 38
39. -14 **41.** $-4, -5$ **43.** 4 **45.** $-\dfrac{5}{2}$ **47.** 30 **49.** 24 **51.** No solution **53.** -3 **55.** -4 **57.** 5, since 3 must equal $x - 2$. **59.** 0, since the terms are the same, they cannot be opposites, therefore they must be 0.
61. x can be any real number since the sum on the left is also $\dfrac{2x - 4}{3}$. **63.** 15 cm
65. No, it is impossible for both sides of the equation to be equal. **67.** More than $6\frac{1}{3}$ hours **68.** 40°, 140°
69. 75 minutes **70.** A linear equation has the form $ax + b = c, a \neq 0$, while a quadratic equation has the form $ax^2 + bx + c = 0, a \neq 0$. Linear equation example: $3x + 5 = 12$; quadratic equation example: $4x^2 + 3x - 6 = 0$.

Exercise Set 8.7 **1.** Some examples are $A = \frac{1}{2}bh$, $A = \frac{1}{2}h(b_1 + b_2)$, $V = \frac{1}{3}\pi r^2h$, and $V = \frac{4}{3}\pi r^3$
3. It represents 1 complete task. **5.** $90 = w\left(\frac{2}{3}w + 4\right)$; length = 10 inches, width = 9 inches
7. $42 = \frac{1}{2}x(x + 5)$; base = 12 cm, height = 7 cm **9.** $\dfrac{1}{x} - \dfrac{1}{10x} = 3$; $\dfrac{3}{10}$, 3 **11.** $\dfrac{3 + x}{4} = \dfrac{5}{2}$; 7
13. $\dfrac{4}{r - 2} = \dfrac{6}{r + 2}$; 10 mph **15.** $\dfrac{d}{30} + \dfrac{d}{30} = \dfrac{1}{2}$; 7.5 miles **17.** $\dfrac{1600}{4r} + \dfrac{500}{r} = 6$; 150 mph, 600 mph

19. $\frac{d}{25} + \frac{70 - d}{50} = 1.6$; 10 miles driving, 60 miles on the train

21. $\frac{d}{600} + \frac{2800 - d}{500} = 5$; 3 hours at 600 mph and 2 hours at 500 mph **23.** $\frac{d}{30} = \frac{d}{40} + 20$; 2400 meters

25. $\frac{t}{6} + \frac{t}{8} = 1; 3\frac{3}{7}$ hours **27.** $\frac{t}{2} - \frac{t}{3} = 1$; 6 hours **29.** $\frac{3}{8} + \frac{3}{t} = 1; 4\frac{4}{5}$ hours **31.** $\frac{5}{12} + \frac{t}{15} = 1; 8\frac{3}{4}$ days

33. $\frac{t}{6} + \frac{t+3}{4} = 1; \frac{3}{5}$ hour or 36 minutes **35.** $\frac{t}{60} + \frac{t}{50} - \frac{t}{30} = 1$; 300 hours **37.** $\frac{4}{x} + 5x = 12$; 2 or $\frac{2}{5}$

39. $\frac{p}{4} - 1 = \frac{p}{8}$; 8 pints **41.** $-\frac{3}{2}x - \frac{9}{2}$ **42.**

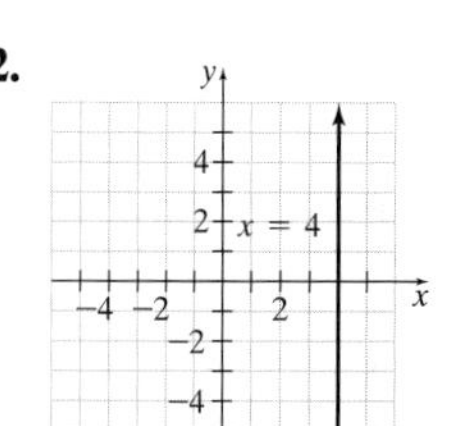

43. 1 **44.** $\frac{3x^2 - 9x - 15}{(2x + 3)(3x - 5)(3x + 1)}$

Review Exercises

1. All real numbers except $x = 5$. **2.** All real numbers except $x = 3$ and $x = 5$. **3.** All real numbers except $x = \frac{1}{5}$ and $x = -1$. **4.** $\frac{1}{x - 1}$ **5.** $x^2 + 4x + 12$ **6.** $3x + 2y$ **7.** $x + 4$ **8.** $x + 5$ **9.** $-(2x - 1)$ **10.** $\frac{x - 6}{x + 2}$ **11.** $\frac{x - 3}{x - 2}$ **12.** $\frac{x - 8}{2x + 3}$ **13.** $\frac{3}{8ab^2}$ **14.** $6xz^2$ **15.** $\frac{16b^3c^2}{a^2}$ **16.** $-\frac{1}{3}$ **17.** $-2x$ **18.** 1 **19.** 64 **20.** $\frac{32z}{x^3}$ **21.** $\frac{5}{a - b}$ **22.** $\frac{1}{3(a + 3)}$ **23.** 1 **24.** $3y(x - y)$ **25.** $\frac{x - 3}{x + 3}$ **26.** 3 **27.** 9 **28.** $\frac{4}{x + 10}$ **29.** $4x - 3$ **30.** $3x + 4$ **31.** 21 **32.** $x + 3$ **33.** $20x^2y^3$ **34.** $x(x + 1)$ **35.** $(x + 2)(x - 3)$ **36.** $x(x + 1)$ **37.** $(x + y)(x - y)$ **38.** $x - 7$ **39.** $(x + 7)(x - 5)(x + 2)$ **40.** $\frac{5}{2y}$ **41.** $\frac{10x + y}{5xy}$ **42.** $\frac{5x^2 - 12y}{3x^2y}$ **43.** $\frac{6x + 10}{x + 2}$ **44.** $\frac{x^2 - 2xy - y^2}{xy}$ **45.** $\frac{11x + 16}{x(x + 4)}$ **46.** $\frac{-x - 4}{3x(x - 2)}$ **47.** $\frac{3x + 13}{(x + 2)^2}$ **48.** $\frac{2x - 8}{(x - 3)(x - 5)}$ **49.** $\frac{4x + 29}{(x + 5)(x + 2)}$ **50.** $\frac{4x - 12}{x - 4}$ **51.** $\frac{4ab + 8b}{a - 2}$ **52.** $\frac{3x - 3}{(x + 3)(x - 3)}$ **53.** $\frac{(x + y)y^2}{2x^3}$ **54.** $\frac{4}{(x + 2)(x - 3)(x - 2)}$ **55.** $\frac{8x - 29}{(x + 2)(x - 7)(x + 7)}$ **56.** $\frac{x}{x + y}$ **57.** $\frac{2x + 6y}{5(x - 3y)}$ **58.** $a - 3$ **59.** $\frac{3a^2 - 7a + 2}{(a + 1)(a - 1)(3a - 5)}$ **60.** $\frac{8x - 16}{x}$ **61.** 4 **62.** $\frac{26}{55}$ **63.** $\frac{bc}{3}$ **64.** $\frac{16x^3z^2}{y^3}$ **65.** $\frac{ab - a}{a + 1}$ **66.** $\frac{xy + 1}{y^3}$ **67.** $\frac{4x + 2}{x(6x - 1)}$ **68.** $\frac{2}{x}$ **69.** x **70.** $\frac{2a + 1}{2}$ **71.** $\frac{x + 1}{-x + 1}$ **72.** $\frac{3x^2 - x^2y}{y(y - x)}$ **73.** 25 **74.** 1 **75.** 6 **76.** 6 **77.** 52 **78.** −20 **79.** No solution **80.** 12 **81.** $\frac{1}{2}$ **82.** −6 **83.** 28 **84.** $2\frac{8}{11}$ hours **85.** $16\frac{4}{5}$ hours **86.** $\frac{1}{5}$, 1 **87.** Robert: 2.1 mph; Tran: 5.6 mph

Practice Test

1. −1 **2.** $\frac{x^2 + x + 1}{x + 1}$ **3.** $\frac{2x^2z}{3y^3}$ **4.** $a + 3$ **5.** $\frac{x^2 - 6x + 9}{(x + 3)(x + 2)}$ **6.** −1 **7.** $\frac{x - 3y}{3}$ **8.** $\frac{5}{y + 5}$ **9.** $-\frac{x + 5}{x - 2}$ **10.** $\frac{3x - 1}{y}$ **11.** $\frac{7x^2 - 6x - 11}{x + 3}$ **12.** $\frac{10x + 3}{2x^2}$ **13.** $\frac{4y^2 - 3}{xy^3}$ **14.** $\frac{-3x + 12}{x + 4}$ **15.** $\frac{-1}{(x + 4)(x - 4)}$ **16.** $\frac{45}{28}$ **17.** $\frac{x^2 + x^2y}{y}$ **18.** $\frac{2x + 3}{2 - 5x}$ **19.** 2 **20.** $-\frac{12}{7}$ **21.** 12 **22.** $3\frac{1}{13}$ hours **23.** 1 **24.** Base: 6 inches; height: 9 inches **25.** 2 miles

Cumulative Review Test **1.** 123 **2.** $\frac{17}{8}$ **3.** $\frac{27y^6}{x^9}$ **4.** $R = \frac{P - 2E}{3}$ **5.** $8x^2 + 5x + 4$
6. $12x^3 - 38x^2 + 39x - 15$ **7.** $(6a - 5)(a - 1)$ **8.** $13(x + 3)(x - 1)$ **9.** $y = \frac{4}{3}x - 4$ **10.** $(1, -4)$
11.

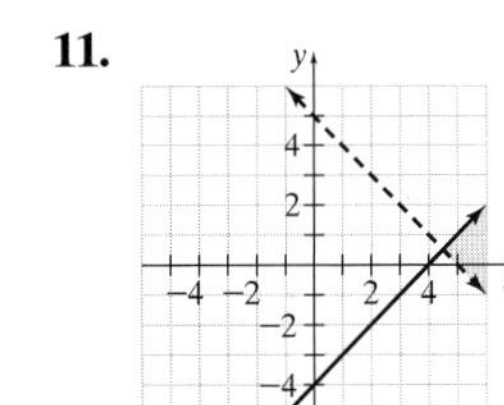

12. $4, \frac{3}{2}$ **13.** $\frac{x + 3}{2x + 1}$ **14.** $\frac{x^2 - 8x - 12}{(x + 4)(x - 5)}$ **15.** $\frac{6x + 2}{(x - 5)(x + 2)(x + 3)}$
16. $-\frac{3}{2}$ **17.** No solution **18.** \$2000
19. 20 pounds sunflower seed; 30 pounds premixed assorted seed mix
20. First leg, 3.25 miles; second leg, 9.5 miles

CHAPTER 9

Using Your Calculator, 9.1 **1.** 3.31662479 **2.** 12.28820573 **3.** Error **4.** 5.196152423

Exercise Set 9.1 **1.** The principal square root of a positive real number, x, is the positive number whose square equals x. **3.** Answers will vary. **5.** Answers will vary. **7.** Yes, since $4^2 = 16$. **9.** No, because the square root of a negative number is not a real number. **11.** Yes, since $\sqrt{\frac{9}{25}} = \frac{3}{5}$ which is a rational number. **13.** 0 **15.** 1 **17.** -9 **19.** 20 **21.** -5 **23.** 12 **25.** 13 **27.** -1 **29.** 9 **31.** -11 **33.** $\frac{1}{2}$ **35.** $\frac{4}{3}$ **37.** $-\frac{5}{6}$ **39.** $\frac{6}{7}$ **41.** 3.1622777 **43.** 3.8729833 **45.** 8.9442719 **47.** 9 **49.** 9.8488578 **51.** 1.7320508 **53.** True **55.** True **57.** False **59.** True **61.** True **63.** True **65.** $3^{1/2}$ **67.** $(17)^{1/2}$ **69.** $(6y)^{1/2}$ **71.** $(12x^2)^{1/2}$ **73.** $(15ab^2)^{1/2}$ **75.** $(50a^3)^{1/2}$ **77.** Rational: $7.24, \frac{5}{7}, 0.666\ldots, 5, \sqrt{\frac{4}{49}}, \frac{3}{7}, -\sqrt{9}$; Irrational: $\sqrt{\frac{5}{16}}$; Imaginary: $\sqrt{-9}, -\sqrt{-16}$ **79.** 6 and 7 **81. a)** Square 4.6 and compare the result to 20. **b)** 4.6 **83.** $-\sqrt{7}, -\sqrt{4}, -\frac{1}{2}, 2.5, 3, \sqrt{16}, 4.01, 12$ **85.** $\sqrt{4} = 2; 6^{1/2} \approx 2.45; -\sqrt{9} = -3; -(25)^{1/2} = -5; (30)^{1/2} \approx 5.48; (-4)^{1/2}$, imaginary number **87. a)** Yes **b)** No **c)** No **d)** Yes **e)** No **89. a)** Yes **b)** Yes **c)** a **d)** Answers will vary. **91.** $x^{3/2}$ **93.** x^3 **95.** $\frac{4}{11}$ **96.** 16 **97.** -6 **98.** -1

Using Your Graphing Calculator, 9.2 **1.** $6\sqrt{5}$ **2.** $5\sqrt{7}$ **3.** $8\sqrt{3}$ **4.** $19\sqrt{3}$

Exercise Set 9.2 **1.** Answers will vary. **3.** The product rule cannot be used when radicands are negative. **5. a)** Answers will vary **b)** $x^6\sqrt{x}$ **7. a)** There can be no perfect square factors nor any exponents greater than 1 in the radicand of a simplified expression. **b)** $5x^2\sqrt{3x}$ **9.** No; $4\sqrt{2}$ **11.** No; $x^4\sqrt{x}$ **13.** 6 **15.** $2\sqrt{2}$ **17.** $4\sqrt{6}$ **19.** $4\sqrt{2}$ **21.** $3\sqrt{10}$ **23.** $4\sqrt{5}$ **25.** $6\sqrt{2}$ **27.** $2\sqrt{39}$ **29.** 16 **31.** 40 **33.** x^4 **35.** xy^2 **37.** $a^6b^4\sqrt{b}$ **39.** $ab^2\sqrt{c}$ **41.** $x\sqrt{2x}$ **43.** $5ab\sqrt{3a}$ **45.** $10a^2b^5\sqrt{3ab}$ **47.** $9xy^2\sqrt{3x}$ **49.** $8ab^3\sqrt{3bc}$ **51.** $5x^2\sqrt{10yz}$ **53.** 7 **55.** $3\sqrt{6}$ **57.** $12\sqrt{5}$ **59.** $x\sqrt{21}$ **61.** $4ab\sqrt{3a}$ **63.** $6xy^2\sqrt{2x}$ **65.** $3x^2y^2\sqrt{7y}$ **67.** $3xy^3\sqrt{10yz}$ **69.** $6a^3b^5$ **71.** $2x$ **73.** $13x^4y^6$ **75.** $15a^2$ **77.** 4 **79.** Exponent on x, 6; on y, 5 **81.** Coefficient, 8; exponent on x, 12; exponent on y, 7 **83. a)** $13x^3$ **b)** $13x^3$ **c)** Yes **85.** $10\Delta^5\sqrt{2\Delta}$ **87.** $5\Delta^{50}\nabla^{18}$ **89.** $x^{1/6}$ **91.** $2x^{2/5}$ **93.** Rational since $\sqrt{6.25} = 2.5$ and 2.5 is a terminating decimal number. **95. a)** 4 feet **b)** No; the area increased $\sqrt{2}$ or ≈ 1.414 times **c)** 4 times **97. a)** Yes **b)** No, for example $\sqrt{2} \cdot \sqrt{2} = 2$.

102. $y = -\frac{1}{2}x + \frac{3}{2}; m = -\frac{1}{2}; \left(0, \frac{3}{2}\right)$ **103.**

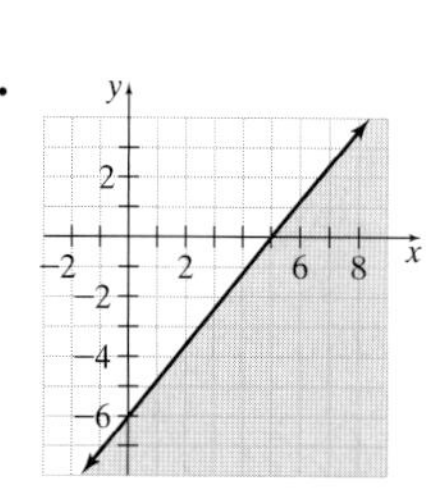

104. $\left(\frac{2}{11}, -\frac{15}{11}\right)$ **105.** $\frac{3x + 1}{x - 1}$

Exercise Set 9.3

1. No perfect square factors in any radicand, no radicand contains a fraction, no square roots in any denominator **3.** The radicand contains a fraction, $\frac{\sqrt{3}}{3}$. **5.** Cannot be simplified **7.** Simplifies to $x\sqrt{2}$ **9.** Cannot be simplified **11.** Answers will vary. **13. a)** Answers will vary. **b)** $\frac{a}{\sqrt{b}} = \frac{a}{\sqrt{b}} \cdot \frac{\sqrt{b}}{\sqrt{b}} = \frac{a\sqrt{b}}{b}$ **15.** 2 **17.** 3 **19.** 3 **21.** $\frac{1}{6}$ **23.** $\frac{9}{12} = \frac{3}{4}$ **25.** $\frac{1}{10}$ **27.** $2x\sqrt{5}$ **29.** $\frac{3\sqrt{5}}{4y^2}$ **31.** $\frac{2y}{5x}$ **33.** $\frac{1}{a^2b}$ **35.** $2x^2$ **37.** $\frac{9x^2}{10y}$ **39.** $\frac{b\sqrt{5}}{c}$ **41.** $5a^2b^3$ **43.** $\frac{\sqrt{5}}{5}$ **45.** $\sqrt{2}$ **47.** $\sqrt{3}$ **49.** $\frac{\sqrt{6}}{3}$ **51.** $\frac{\sqrt{3}}{3}$ **53.** $\frac{x\sqrt{2}}{2}$ **55.** $\frac{a\sqrt{2}}{4}$ **57.** $\frac{x^2\sqrt{5}}{5}$ **59.** $\frac{a^4\sqrt{14b}}{14b}$ **61.** $\frac{x\sqrt{y}}{2y}$ **63.** $\frac{\sqrt{6}}{5x^2y^2z}$ **65.** $\frac{3\sqrt{5x}}{xy^2}$ **67.** Yes **69.** 5.40 **71.** 0.71 **73.** 7.23 **75.** 0.58 **77. a)** No **b)** $2\sqrt{5}$ **79.** 1 **81.** $\sqrt{216} = 6\sqrt{6}$ **83.** $\sqrt{72} = 6\sqrt{2}$ **85.** $64x^{10}$ **87.** 2 **89.**

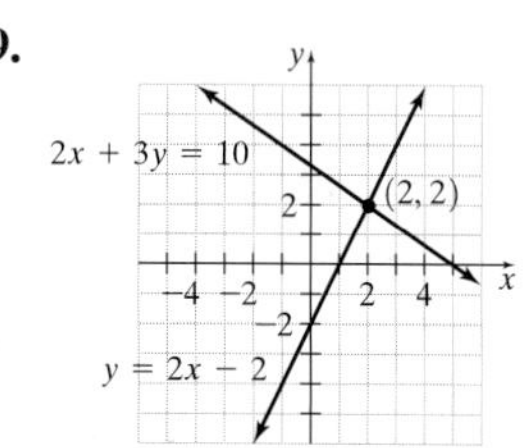

90. $3(x + 4)(x - 8)$ **91.** $\frac{1}{x + 1}$ **92.** 4, 6

Exercise Set 9.4

1. Like square roots are square roots having the same radicand. One example is $\sqrt{3}$ and $5\sqrt{3}$. **3.** Only like square roots can be added or subtracted. **5. a)** $4 + \sqrt{7}$ **b)** $5 - \sqrt{11}$ **c)** $-3 + 2\sqrt{13}$ **d)** $\sqrt{5} + 1$ **7. a)** Multiply the numerator and denominator by $b - \sqrt{c}$. **b)** $\frac{ab - a\sqrt{c}}{b^2 - c}$ **9.** $2\sqrt{2}$ **11.** $-4\sqrt{5}$ **13.** $2\sqrt{7} + 6$ **15.** $8\sqrt{x}$ **17.** $-3\sqrt{y}$ **19.** $2\sqrt{y} + 3$ **21.** $\sqrt{x} + 4\sqrt{y} + x$ **23.** $7 - 3\sqrt{y}$ **25.** $-2\sqrt{7} - 9\sqrt{x}$ **27.** $2\sqrt{3} + 3\sqrt{2}$ **29.** $7\sqrt{3}$ **31.** $11\sqrt{3}$ **33.** $16\sqrt{2}$ **35.** $-15\sqrt{5} + 35\sqrt{3}$ **37.** $12\sqrt{10} + 16\sqrt{5}$ **39.** $16 - 4\sqrt{3}$ **41.** -4 **43.** 14 **45.** $x - 9$ **47.** $6 - x^2$ **49.** $5x - y$ **51.** $4x - 9y$ **53.** $3\sqrt{5} - 6$ **55.** $\frac{2\sqrt{6} + 2}{5}$ **57.** $-2\sqrt{2} + 2\sqrt{3}$ **59.** $\frac{-8\sqrt{5} - 16\sqrt{2}}{3}$ **61.** $\frac{5\sqrt{y} - 15}{y - 9}$ **63.** $\frac{24 + 6\sqrt{y}}{16 - y}$ **65.** $\frac{16\sqrt{y} - 16x}{y - x^2}$ **67.** $\frac{x\sqrt{x} - x\sqrt{y}}{x - y}$ **69.** $\frac{\sqrt{2x} + x}{2 - x}$ **71.** $3 - \sqrt{2}$ **73.** $4(1 + 2\sqrt{7})$ **75.** Cannot be simplified **77.** $\frac{2(3 + 5\sqrt{3})}{3}$ **79.** $\frac{-1 + 8\sqrt{5}}{5}$ **81.** $2 - 2\sqrt{2}$ **83.** $x - y^2$ **85.** $x - y$ **87.** $x + 2$ **89.** Yes **91.** 343 **93.** Perimeter: $4(\sqrt{2} + \sqrt{3})$ units; area: $5 + 2\sqrt{6}$ square units **95.** Perimeter: $15.3 - 2\sqrt{2}$ units; area: $10.6 - 2.65\sqrt{2}$ square units **97.** 48 **99. a)** $5 - \sqrt{x + 2}$ **b)** $6 + \sqrt{x + 3}$ **101.** $\frac{-\sqrt{x} - \sqrt{3x}}{2}$ **104.** $3x - 8 + \frac{7}{x + 4}$ **105.** $\frac{9}{2}, -4$ **106.** $\frac{-2x - 3}{(x + 2)(x - 2)}$ **107.** 30 minutes

Exercise Set 9.5

1. A radical equation is an equation that contains a variable in a radicand. **3.** It is necessary to check solutions because they may be extraneous. **5.** Yes **7.** No; $-\sqrt{64} = -8$ **9.** Yes **11.** 9 **13.** No solution **15.** 4 **17.** No solution **19.** 81 **21.** 25 **23.** No solution **25.** 7 **27.** 5 **29.** 3

31. 2 **33.** $\frac{14}{3}$ **35.** 1 **37.** 1, 9 **39.** No solution **41.** -2 **43.** 7 **45.** No solution **47.** $\sqrt{x+3} = 7$; 46
49. $\sqrt{x-2} = \sqrt{2x-9}$; 7 **51. a)** $x - 9 = 40$ **b)** 49 **53. a)** $35 + 2\sqrt{x} - x = 35$ **b)** 0, 4 **55.** 1
57. 9 **59.** 9 **63.**

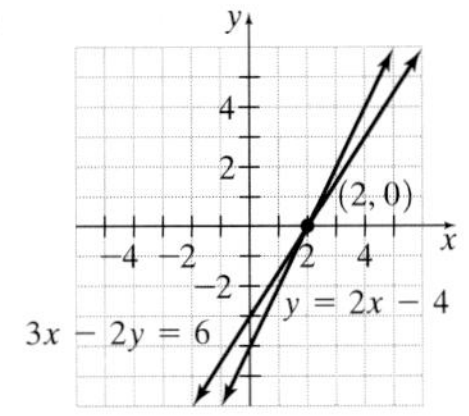

64. (2, 0) **65.** (2, 0) **66.** Boat, 8 mph; current, 4 mph

Exercise Set 9.6 **1.** A right triangle is a triangle that contains a 90° angle. **3.** No; only with right triangles.
5. They represent the two points in the coordinate plane that you are trying to find the distance between. **7.** 5
9. $\sqrt{33} \approx 5.74$ **11.** $\sqrt{164} \approx 12.81$ **13.** $\sqrt{107} \approx 10.34$ **15.** $\sqrt{67} \approx 8.19$ **17.** $\sqrt{202} \approx 14.21$ **19.** $\sqrt{200} \approx 14.14$
21. $\sqrt{13} \approx 3.61$ **23.** $\sqrt{193} \approx 13.89$ **25.** $\sqrt{17{,}240.89} \approx 131.30$ yards **27.** $\sqrt{60} \approx 7.75$ meters **29.** 16 feet
31. $\approx \sqrt{25.48} \approx 5.05$ feet **33.** $\sqrt{16{,}200} \approx 127.28$ feet **35.** $\sqrt{2000} \approx 44.72$ feet **37.** $6.28\sqrt{0.11} \approx 2.08$ seconds
39. $6.28\sqrt{1.925} \approx 8.71$ seconds **41.** About 365 Earth days **43.** $\sqrt{19.62(6{,}370{,}000)} \approx 11{,}179.42$ meters per second
45. $\sqrt{1{,}000{,}000} = 1000$ pounds **47.** ≈ 32.66 feet per second **49.** ≈ 40.93 inches
51. $x > 6$ **52.** $\left(7, -\frac{9}{4}\right)$ **53.** $\frac{x^4}{4y^3}$ **54.** $-\frac{16}{15}$

Using Your Calculator, 9.7 (page 560) **1.** 7 **2.** 2 **3.** ≈ 3.76060 **4.** 10 **5.** -9 **6.** ≈ -4.12891

Using Your Graphing Calculator, 9.7 (page 560) **1.** 5 **2.** 6 **3.** ≈ 2.952591724 **4.** 12 **5.** -8
6. ≈ -3.981071706

Using Your Calculator, 9.7 (page 563) **1.** 8 **2.** 256 **3.** .04 **4.** $.\overline{037}$

Exercise Set 9.7 **1. a)** The square root of 9 **b)** The cube root of 9 **c)** The fourth root of 9
3. Write the radicand as a product of a perfect cube and another number. **5. a)** Answers will vary.
b) $y^{7/3}$ **7.** 5 **9.** -3 **11.** 2 **13.** 3 **15.** -1 **17.** -10 **19.** $2\sqrt[3]{4}$ **21.** $2\sqrt[3]{2}$ **23.** $3\sqrt[3]{3}$ **25.** $2\sqrt[4]{2}$
27. $5\sqrt[4]{2}$ **29.** x **31.** y^7 **33.** x^4 **35.** x^8 **37.** x^5 **39.** 8 **41.** 25 **43.** 1 **45.** 27 **47.** 81 **49.** 625
51. $\frac{1}{2}$ **53.** $\frac{1}{9}$ **55.** $x^{5/3}$ **57.** $x^{4/3}$ **59.** $y^{15/4}$ **61.** y^7 **63.** x **65.** x **67.** x^{10} **69.** x^2 **71.** Both equal 4
73. $\sqrt[3]{5^2}$ **75.** $\sqrt[3]{6^2}$ **77.** $\sqrt[4]{5^3}$ **79.** 3 **81.** 1 **83.** xy **85.** $\sqrt[4]{2}$ **88.** -42
89.

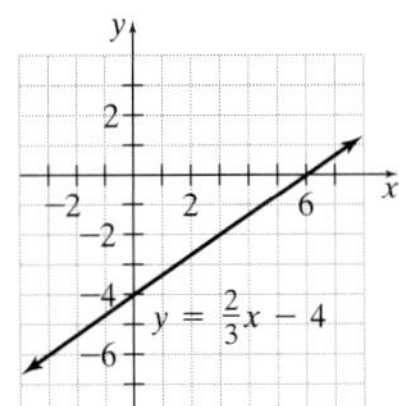

90. $(3x - 4)(x - 8)$ **91.** $\frac{4y^3\sqrt{2xy}}{x}$

Review Exercises **1.** 7 **2.** 3 **3.** -8 **4.** $6^{1/2}$ **5.** $(26x)^{1/2}$ **6.** $(13x^2y)^{1/2}$ **7.** $3\sqrt{2}$ **8.** $2\sqrt{11}$
9. $3x^3y^2\sqrt{3x}$ **10.** $5x^2y^3\sqrt{5}$ **11.** $4b^2c^2\sqrt{3ab}$ **12.** $6abc^3\sqrt{2c}$ **13.** $4\sqrt{30}$ **14.** $7y$ **15.** $6x\sqrt{y}$ **16.** $5xy\sqrt{3}$
17. $2ab^2\sqrt{6ab}$ **18.** $10ab^3\sqrt{b}$ **19.** 5 **20.** $\frac{1}{7}$ **21.** $\frac{1}{2}$ **22.** $\frac{3\sqrt{5}}{5}$ **23.** $\frac{\sqrt{6a}}{6}$ **24.** $\frac{\sqrt{15a}}{6}$ **25.** $\frac{x\sqrt{3}}{3}$
26. $\frac{x^2\sqrt{2x}}{4}$ **27.** $y^2\sqrt{7}$ **28.** $\frac{x\sqrt{2y}}{y^2}$ **29.** $\frac{2\sqrt{5a}}{3a^2b}$ **30.** $\frac{c\sqrt{14a}}{7a}$ **31.** $\frac{-1-\sqrt{5}}{2}$ **32.** $\frac{15+5\sqrt{6}}{3}$ **33.** $\frac{2\sqrt{2}-\sqrt{2y}}{4-y}$
34. $\frac{2\sqrt{x}+10}{x-25}$ **35.** $\frac{\sqrt{5x}-\sqrt{15}}{x-3}$ **36.** $3\sqrt{2}$ **37.** $-6\sqrt{5}$ **38.** $-2\sqrt{x}$ **39.** 0 **40.** $3\sqrt{2} - 3\sqrt{3}$ **41.** $12\sqrt{10}$
42. $-10\sqrt{2}$ **43.** $25\sqrt{2}$ **44.** $16\sqrt{3} + 20\sqrt{5}$ **45.** 16 **46.** No solution **47.** 13 **48.** 8 **49.** 2 **50.** 4
51. 2 **52.** 7 **53.** $\frac{6}{5}$ **54.** 26 **55.** $\sqrt{125} \approx 11.18$ **56.** $\sqrt{12} \approx 3.46$ **57.** $\sqrt{61} \approx 7.81$

58. $\sqrt{135} \approx 11.62$ feet **59.** $\sqrt{261} \approx 16.16$ inches **60.** $\sqrt{109} \approx 10.44$ **61.** $\sqrt{153} \approx 12.37$
62. ≈ 561.18 square inches **63.** $\sqrt{60} \approx 7.75$ miles **64.** 755 feet **65.** 4 **66.** -4 **67.** 2 **68.** 3 **69.** 3
70. -2 **71.** $2\sqrt[4]{2}$ **72.** $2\sqrt[3]{6}$ **73.** $3\sqrt[3]{2}$ **74.** $2\sqrt[4]{6}$ **75.** x^7 **76.** x^8 **77.** 9 **78.** 5 **79.** $\frac{1}{9}$ **80.** 16
81. $\frac{1}{16}$ **82.** 343 **83.** $x^{7/3}$ **84.** $x^{8/3}$ **85.** $y^{9/4}$ **86.** $x^{5/2}$ **87.** $y^{3/2}$ **88.** $y^{7/4}$ **89.** x **90.** $\sqrt[3]{x^2}$ **91.** x^3
92. x^2 **93.** x^4 **94.** x **95.** x^6 **96.** x^6

Practice Test **1.** $(3x)^{1/2}$ **2.** $\sqrt[3]{x^2}$ **3.** $x - 5$ **4.** $4\sqrt{5}$ **5.** $2x\sqrt{3}$ **6.** $5x^3y\sqrt{2xy}$ **7.** $4xy\sqrt{3x}$
8. $5x^2y^2\sqrt{3y}$ **9.** $\frac{1}{5}$ **10.** $\frac{x}{y}$ **11.** $\frac{\sqrt{6}}{6}$ **12.** $\frac{2\sqrt{5x}}{5}$ **13.** $\frac{2\sqrt{15x}}{3xy}$ **14.** $-2 - \sqrt{7}$ **15.** $\frac{6\sqrt{x} + 18}{x - 9}$ **16.** $11\sqrt{3}$
17. $11\sqrt{y}$ **18.** 24 **19.** 4, 8 **20.** $\sqrt{106} \approx 10.30$ **21.** $\sqrt{58} \approx 7.62$ **22.** $\frac{1}{81}$ **23.** x^3
24. 11 meters **25.** $\sqrt{640} \approx 25.30$ feet per second

Cumulative Review Test **1. a)** $-9, 735, 4$ **b)** $735, 4$ **c)** $-9, 735, 0.5, 4, \frac{1}{2}$ **d)** $\sqrt{13}$
e) $-9, \sqrt{13}, 735, 0.5, 4, \frac{1}{2}$ **2.** -60 **3.** $\frac{29}{6}$ **4.** $x > \frac{27}{11}$;

5. $y = \frac{2}{5}x + \frac{11}{5}$ **6.** $y = 3x - 2$

7.

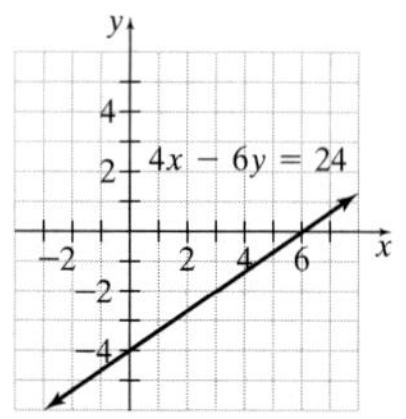

8. (3, 4) **9.** $\frac{1}{7a^5b^6}$ **10.** $(x^2 + 2)(3x + 1)$ **11.** $(x + 14)(x - 2)$ **12.** 0, 3
13. $\frac{3y - 2}{8}$ **14.** $\frac{41}{7}$ **15.** $-\sqrt{11}$ **16.** $\frac{\sqrt{21yz}}{14y}$ **17.** 31 **18.** $2\frac{8}{11}$ cups
19. \$448.4 million **20.** ≈ 111.1 miles

CHAPTER 10

Exercise Set 10.1 **1.** If $x^2 = a$, then $x = \sqrt{a}$ or $x = -\sqrt{a}$ **3.** In any golden rectangle, the length is about 1.62 times its width. **5. a)** 2 **b)** 1 **c)** 2 **7.** 10, -10 **9.** 12, -12 **11.** 13, -13 **13.** 11, -11
15. $2\sqrt{3}, -2\sqrt{3}$ **17.** 2, -2 **19.** $\sqrt{17}, -\sqrt{17}$ **21.** 3, -3 **23.** $\frac{\sqrt{15}}{3}, -\frac{\sqrt{15}}{3}$ **25.** $\frac{\sqrt{73}}{4}, -\frac{\sqrt{73}}{4}$ **27.** 8, -2
29. 6, -12 **31.** 4, -12 **33.** $-7 + 4\sqrt{2}, -7 - 4\sqrt{2}$ **35.** $-6 + 2\sqrt{5}, -6 - 2\sqrt{5}$ **37.** 3, -7 **39.** 19, -1
41. $\frac{-3 + 3\sqrt{2}}{2}, \frac{-3 - 3\sqrt{2}}{2}$ **43.** $\frac{-1 + 2\sqrt{5}}{4}, \frac{-1 - 2\sqrt{5}}{4}$ **45.** $\frac{6 + 3\sqrt{2}}{2}, \frac{6 - 3\sqrt{2}}{2}$ **47.** $x^2 = 36$ **49.** 9
51. a) $3x^2 - 9x + 6 = 0$ **b)** $x^2 - 3x + 2 = 0$ **53.** Width ≈ 35.14 feet, length ≈ 56.93 feet
55. a) Left, x^2; right, $(x + 3)^2$ **b)** 6 inches **c)** $\sqrt{50} \approx 7.07$ inches **d)** 9 inches **e)** $\sqrt{92} \approx 9.59$ inches
57. $p = \sqrt{\frac{I}{r}}$ **59.** $b = \sqrt{c^2 - a^2}$ **61.** $r = \sqrt{\frac{A}{p}} - 1$ **62.** $y = 4x - 1$ **63.** $2(2x + 3)(x - 4)$ **64.** $\frac{3y - 1}{6y - 1}$
65. $\frac{3\sqrt{5a}}{ab^2}$

Exercise Set 10.2 **1. a)** A perfect square trinomial is a trinomial that can be expressed as the square of a binomial. **b)** +16; The constant term is the square of half the coefficient of the x-term.
3. The constant is the square of half the coefficient of the x-term. **5.** 36 **7.** 1, -4 **9.** 7, 1 **11.** $-2, -1$ **13.** 4, -2
15. -3 **17.** 5, -3 **19.** $-4, -6$ **21.** 8, 7 **23.** 6, -2 **25.** $2 + \sqrt{2}, 2 - \sqrt{2}$ **27.** $-4 + \sqrt{13}, -4 - \sqrt{13}$
29. $\frac{-7 + \sqrt{41}}{2}, \frac{-7 - \sqrt{41}}{2}$ **31.** 1, -3 **33.** $\frac{-9 + \sqrt{73}}{2}, \frac{-9 - \sqrt{73}}{2}$ **35.** 6, -1 **37.** 4, $-\frac{1}{3}$ **39.** $-1 + \sqrt{3}, -1 - \sqrt{3}$
41. 5, 0 **43.** 6, 0 **45. a)** $x^2 + 20x + 100$ **b)** Answers will vary. **47.** 1, -4 **49.** 0, -6 **51.** 3, 7
53. $+\frac{3}{5}x$ or $-\frac{3}{5}x$ **55. a)** $7 + 5\sqrt{2}, 7 - 5\sqrt{2}$ **57.** $\frac{5 + \sqrt{70}}{15}, \frac{5 - \sqrt{70}}{15}$ **59.** $-1 + \sqrt{6.4}, -1 - \sqrt{6.4}$

61. If the slopes are the same and the y-intercepts are different, the equations represent parallel lines.
62. $\left(\frac{38}{11}, \frac{12}{11}\right)$ **63.** $\frac{4}{(x+2)(x-3)}$ **64.** 3

Exercise Set 10.3 **1. a)** $b^2 - 4ac$ **b)** Discriminant: greater than 0, two solutions; equal to 0, one solution; less than 0, no real solution **3.** $x = \frac{-b \pm \sqrt{b^2 - 4ac}}{2a}$ **5.** The first step is to write the equation in standard form.
7. The values used for b and c are incorrect because the equation was not first put in standard form.
9. Two distinct real number solutions **11.** No real number solution **13.** Two distinct real number solutions
15. One real number solution **17.** Two distinct real number solutions **19.** No real number solution
21. Two distinct real number solutions **23.** Two distinct real number solutions **25.** 3, −2 **27.** −3, −6 **29.** 5, 1
31. 6, 4 **33.** 10, −10 **35.** 3, 0 **37.** 5, 2 **39.** $\frac{7+\sqrt{17}}{4}, \frac{7-\sqrt{17}}{4}$ **41.** $\frac{1}{3}, -\frac{1}{2}$ **43.** $\frac{7}{2}, -1$ **45.** No solution
47. $\frac{5}{4}, -1$ **49.** $\frac{9}{2}, -1$ **51.** $\frac{5}{2}, 3$ **53.** 6, 7 **55.** Width = 4 feet; length = 5 feet **57.** 2 feet
59. a) $c < \frac{9}{8}$ **b)** $c = \frac{9}{8}$ **c)** $c > \frac{9}{8}$ **61.** 300 feet long by 50 feet wide, or 100 feet long by 150 feet wide **63.** 7, 6
64. $\frac{5}{3}, -\frac{7}{2}$ **65.** Cannot be solved by factoring; $\frac{-3+\sqrt{41}}{4}, \frac{-3-\sqrt{41}}{4}$ **66.** 3, −3

Using Your Graphing Calculator, 10.4 **1. a)** (3, 0), (−1, 0) **b)** (0, −3) **c)** (1, −4)
2. a) (5, 0), (2, 0) **b)** (0, 10) **c)** (3.5, −2.25) **3. a)** (−3, 0), (2, 0) **b)** (0, −12) **c)** (−0.5, 12.5)
4. a) $\left(-\frac{3}{2}, 0\right)$, (5, 0) **b)** (0, −15) **c)** (1.75, −21.125)

Exercise Set 10.4 **1.** A parabola **3.** Answers will vary. **5. a)** The x-intercepts are where the graph crosses the x-axis. **b)** The x-intercepts are found by setting $y = 0$ and solving for x. **7. a)** $x = -\frac{b}{2a}$
b) This line is called the axis of symmetry. **9.** $x = -3$, $(-3, -12)$, upward
11. $x = \frac{3}{2}$, $\left(\frac{3}{2}, -\frac{7}{4}\right)$, downward **13.** $x = \frac{5}{6}$, $\left(\frac{5}{6}, \frac{121}{12}\right)$, downward **15.** $x = -1$, $(-1, -1)$, upward
17. $x = -\frac{3}{4}$, $\left(-\frac{3}{4}, \frac{55}{8}\right)$, upward **19.** $x = \frac{3}{5}$, $\left(\frac{3}{5}, \frac{3}{5}\right)$, downward

21.
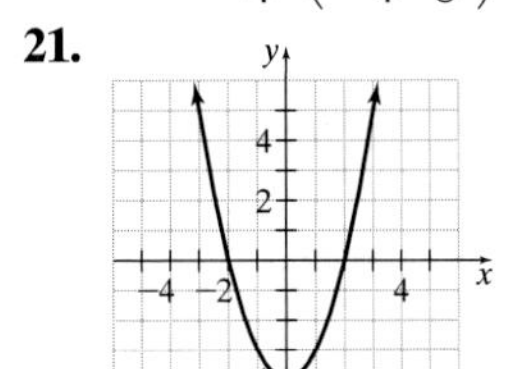

23.
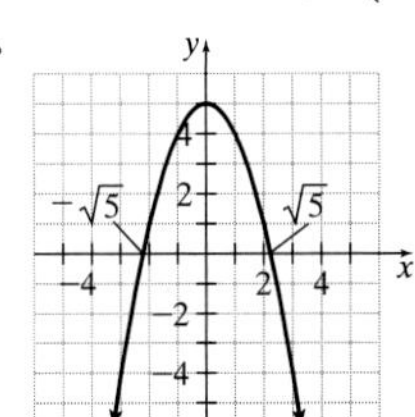

25.
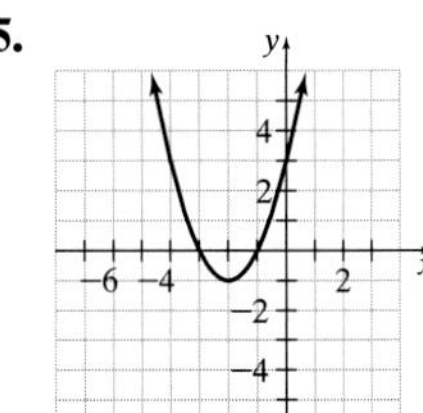

27.
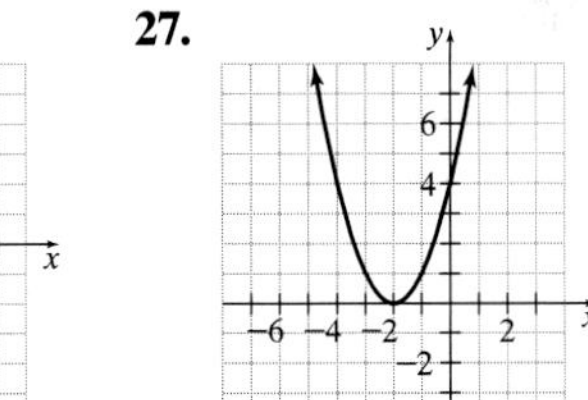

29.
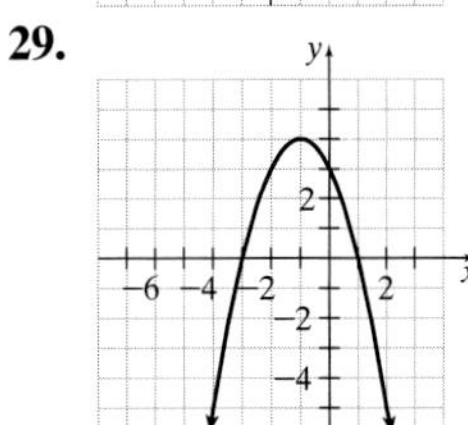

31.
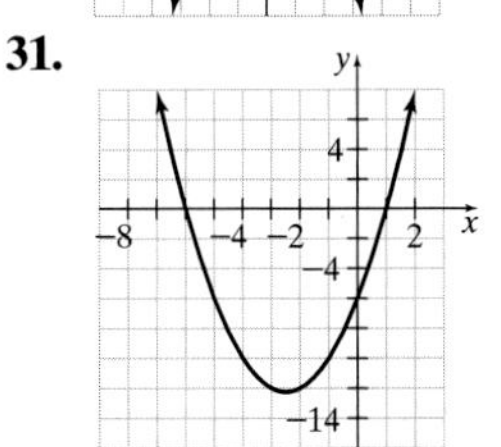

33.
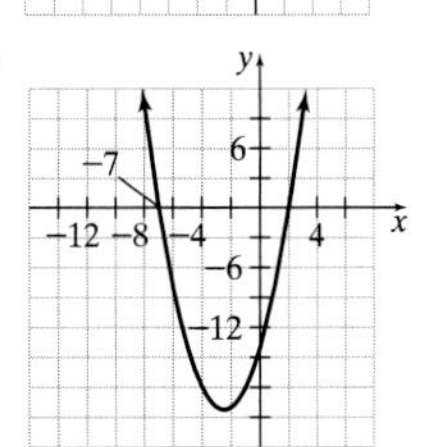

35.
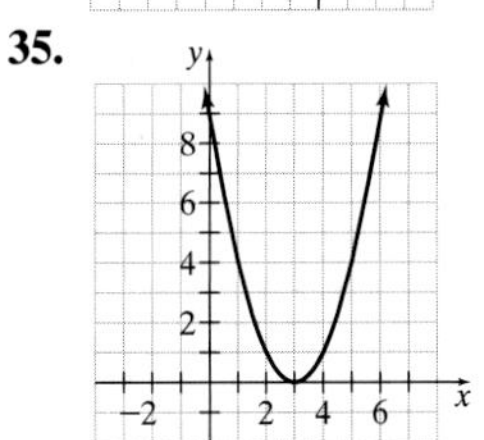

37.
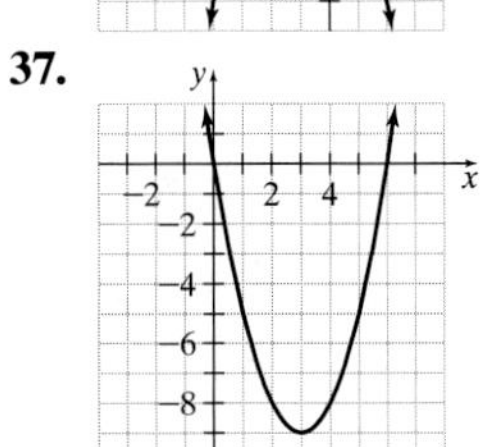

39.
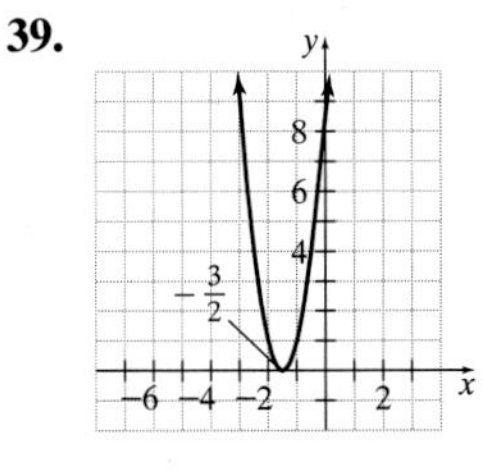

41.
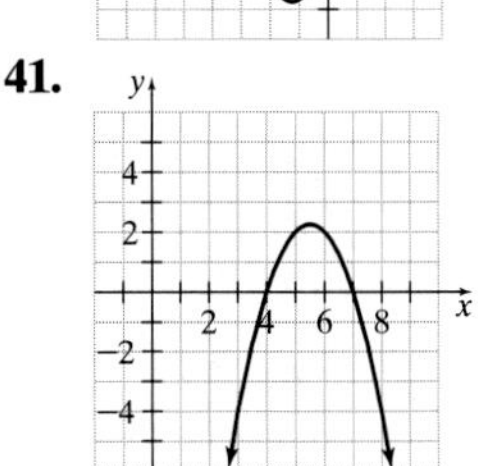

43.
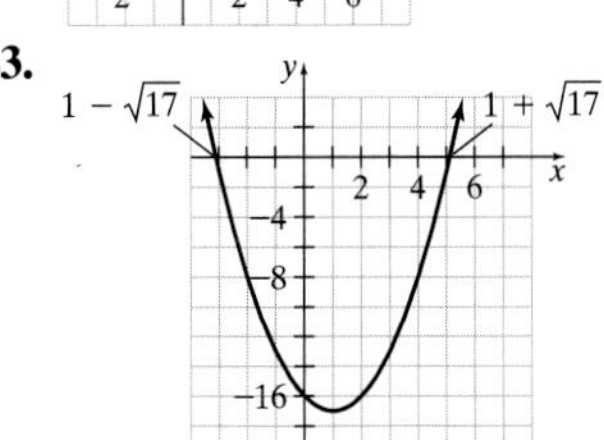

45.

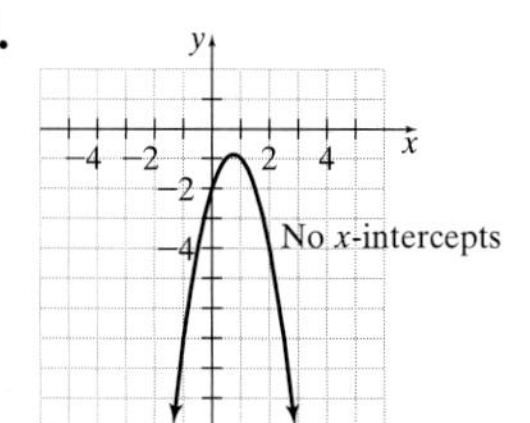

47.

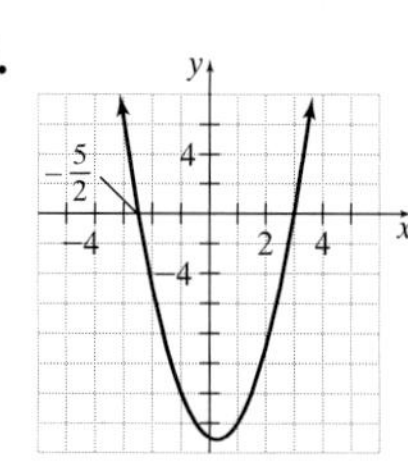

49. Two **51.** Two **53.** One **55.** Two

57. None; the vertex is below the x-axis and the parabola opens downward

59. One; the vertex of the parabola is on the x-axis.

61. Yes; if y is set to 0, both equations have the same solutions, 4 and -2.

63. a)

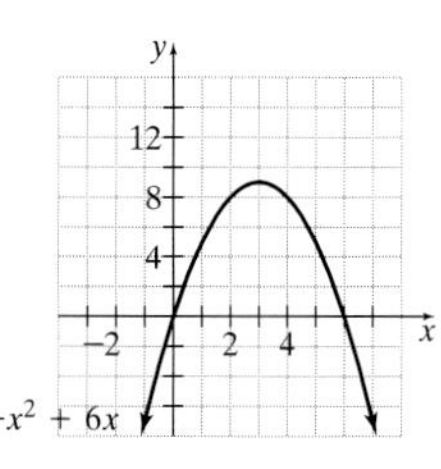

b)

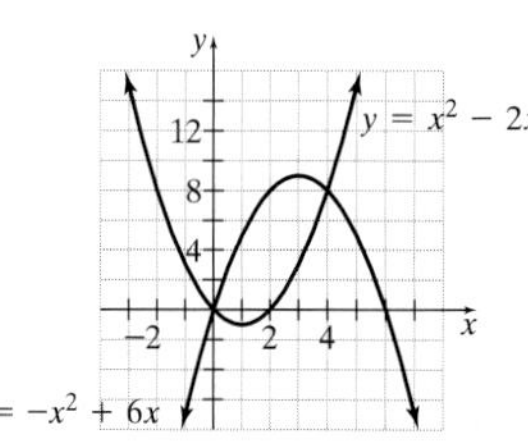

c) $(0, 0), (4, 8)$

65.

66.

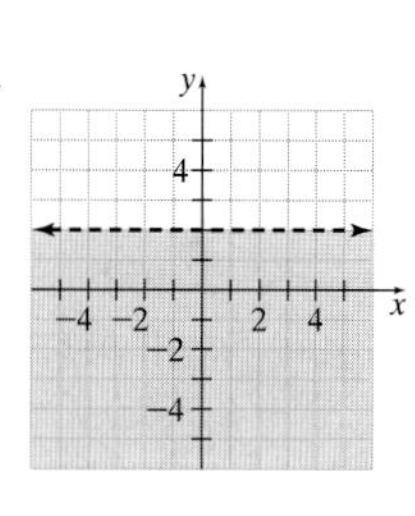

67. $\dfrac{-x^2 + 2x - 6}{(x + 3)(x - 4)}$ **68.** $\dfrac{27}{7}$

Review Exercises

1. $10, -10$ **2.** $2\sqrt{3}, -2\sqrt{3}$ **3.** $\sqrt{6}, -\sqrt{6}$ **4.** $\sqrt{6}, -\sqrt{6}$ **5.** $2\sqrt{5}, -2\sqrt{5}$ **6.** $\sqrt{7}, -\sqrt{7}$ **7.** $2\sqrt{2}, -2\sqrt{2}$ **8.** $3 + 2\sqrt{3}, 3 - 2\sqrt{3}$ **9.** $\dfrac{5 + 5\sqrt{2}}{3}, \dfrac{5 - 5\sqrt{2}}{3}$ **10.** $\dfrac{-4 + \sqrt{30}}{2}, \dfrac{-4 - \sqrt{30}}{2}$

11. $6, 3$ **12.** $7, 4$ **13.** $17, 1$ **14.** $2, -3$ **15.** $9, -6$ **16.** $1, -6$ **17.** $\dfrac{3 + \sqrt{41}}{2}, \dfrac{3 - \sqrt{41}}{2}$

18. $-1 + \sqrt{6}, 1 - \sqrt{6}$ **19.** $8, -4$ **20.** $5, -3$ **21.** $\dfrac{4}{3}, -2$ **22.** $\dfrac{5}{3}, \dfrac{3}{2}$ **23.** Two solutions **24.** No real solution

25. One solution **26.** No real solution **27.** Two solutions **28.** No real solution **29.** Two solutions

30. Two solutions **31.** $9, 2$ **32.** $11, -4$ **33.** $9, 1$ **34.** $2, -\dfrac{3}{5}$ **35.** $9, -2$ **36.** $6, -5$ **37.** $\dfrac{3}{2}, -\dfrac{5}{3}$

38. $\dfrac{3 + \sqrt{57}}{4}, \dfrac{3 - \sqrt{57}}{4}$ **39.** $\dfrac{-2 + \sqrt{10}}{2}, \dfrac{-2 - \sqrt{10}}{2}$ **40.** $3 + \sqrt{2}, 3 - \sqrt{2}$ **41.** No real solution

42. $\dfrac{3 + \sqrt{33}}{3}, \dfrac{3 - \sqrt{33}}{3}$ **43.** $\dfrac{3}{7}, 0$ **44.** $0, \dfrac{5}{2}$ **45.** $6, 7$ **46.** $-7, -8$ **47.** $10, -7$ **48.** $-9, 3$ **49.** $-6, 10$

50. $7, -6$ **51.** $1, -12$ **52.** $0, -6$ **53.** $9, -9$ **54.** $\dfrac{1}{2}, -3$ **55.** $\dfrac{5}{2}, 2$ **56.** $\dfrac{2}{3}, -\dfrac{3}{2}$ **57.** $\dfrac{3 + 3\sqrt{3}}{2}, \dfrac{3 - 3\sqrt{3}}{2}$

58. $2, \dfrac{5}{3}$ **59.** $1, -\dfrac{8}{3}$ **60.** $\dfrac{-3 + \sqrt{33}}{2}, \dfrac{-3 - \sqrt{33}}{2}$ **61.** $0, \dfrac{9}{4}$ **62.** $0, -\dfrac{5}{3}$ **63.** $x = 2, (2, -9)$, upward

64. $x = 6, (6, -30)$, upward **65.** $x = \dfrac{3}{2}, \left(\dfrac{3}{2}, \dfrac{19}{4}\right)$, upward **66.** $x = -1, (-1, 16)$, downward

67. $x = -\dfrac{7}{4}, \left(-\dfrac{7}{4}, -\dfrac{25}{8}\right)$, upward **68.** $x = -\dfrac{5}{2}, \left(-\dfrac{5}{2}, \dfrac{25}{4}\right)$, downward **69.** $x = 0, (0, -8)$, downward

70. $x = -\dfrac{1}{2}, \left(-\dfrac{1}{2}, \dfrac{81}{4}\right)$, downward **71.** $x = 1, (1, 9)$, downward **72.** $x = -\dfrac{5}{6}, \left(-\dfrac{5}{6}, -\dfrac{121}{12}\right)$, upward

73. 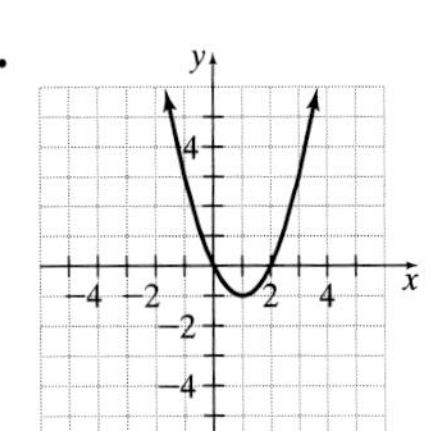

74. 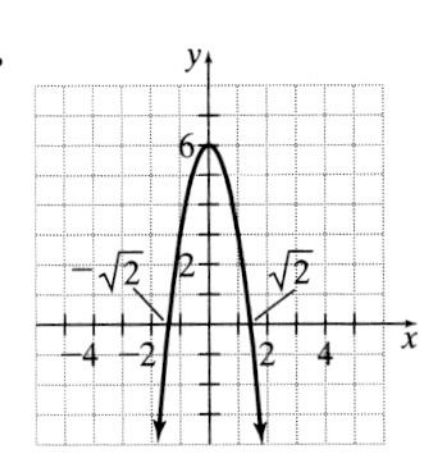

75. 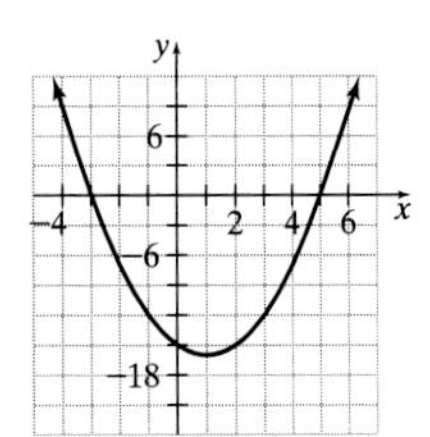

76. 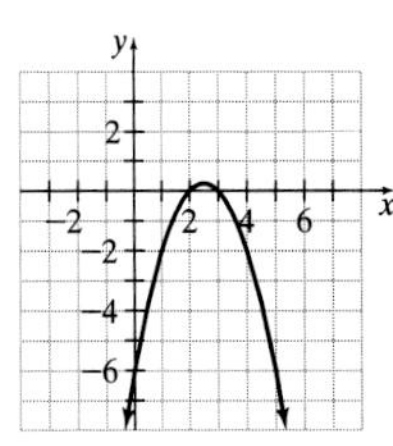

77. 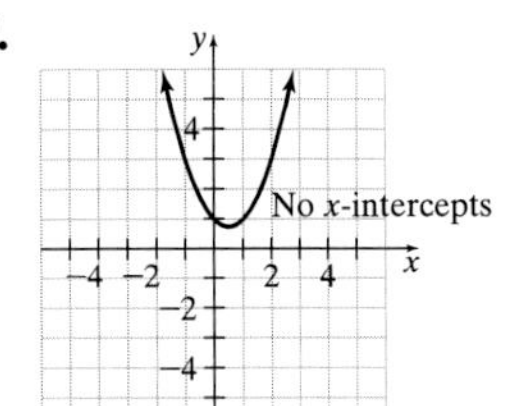

78. 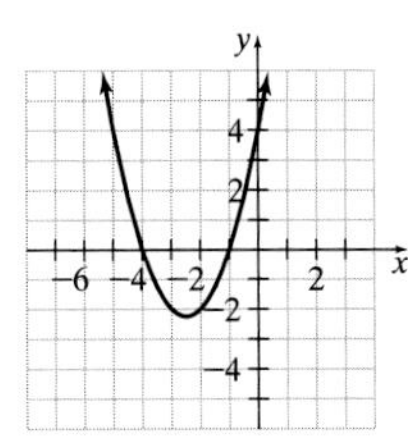

79. 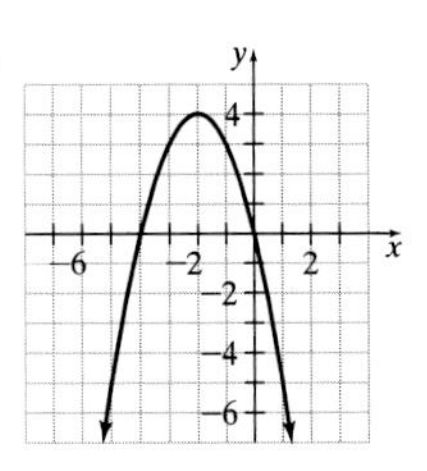

80. 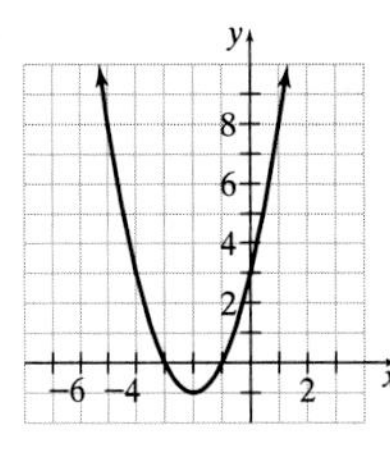

81. 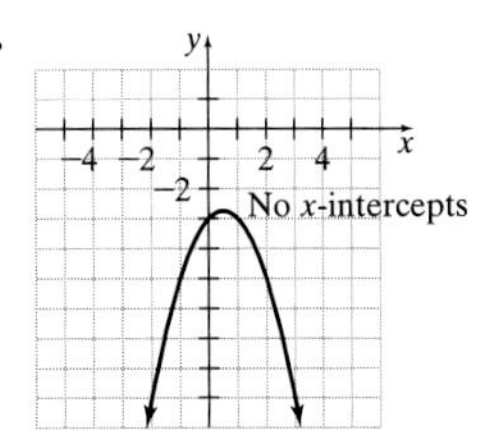

82. 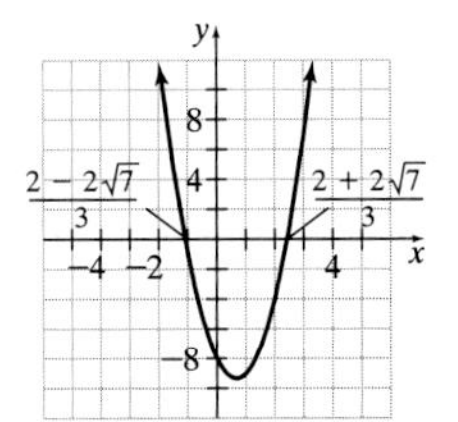

83. 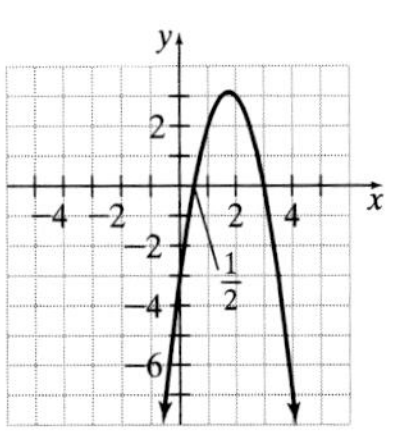

84. 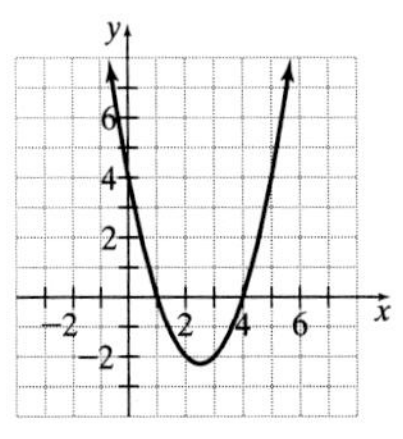

85. 6, 8 **86.** 8, 11 **87.** Width = 20 inches, length = 46 inches **88.** Width = 28 inches, length = 48 inches

Practice Test

1. $2\sqrt{7}, -2\sqrt{7}$ **2.** $\frac{4+\sqrt{17}}{3}, \frac{4-\sqrt{17}}{3}$ **3.** 10, −4 **4.** 6, −10 **5.** 10, −2 **6.** $\frac{-4+\sqrt{6}}{2}, \frac{-4-\sqrt{6}}{2}$ **7.** $\frac{7}{4}, -\frac{7}{4}$ **8.** $-5, \frac{1}{2}$ **9.** $x = \frac{-b \pm \sqrt{b^2-4ac}}{2a}$ **10.** Answers will vary. **11.** No real solution **12.** Two distinct real solutions **13.** $x = -3$ **14.** $x = 1$ **15.** Downward; $a < 0$ **16.** Upward; $a > 0$ **17.** The vertex is the lowest point on a parabola that opens upward; highest point on a parabola that opens downward. **18.** $(-5, 9)$ **19.** $\left(\frac{4}{3}, \frac{11}{3}\right)$

20. 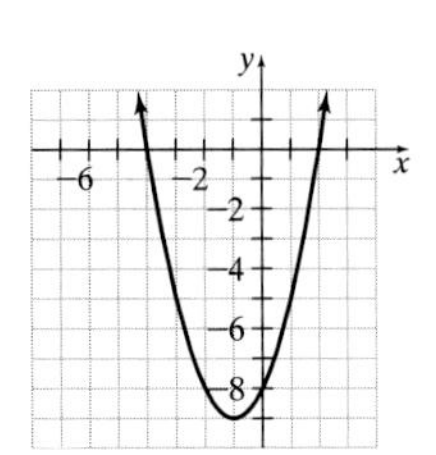

21. 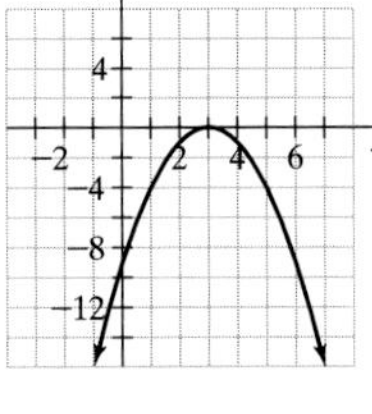

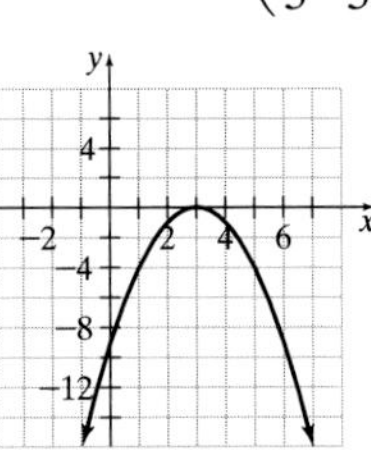

22. 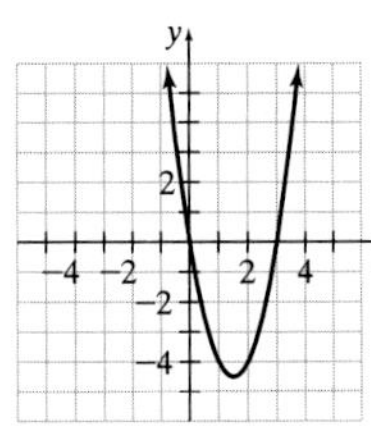

23. Width = 3 feet, length = 10 feet **24.** 11 **25.** 9

Cumulative Review Test

1. 60 **2.** $\frac{14}{27}$ **3.** $5\frac{1}{3}$ inches **4.** $x \geq -\frac{1}{4}$; 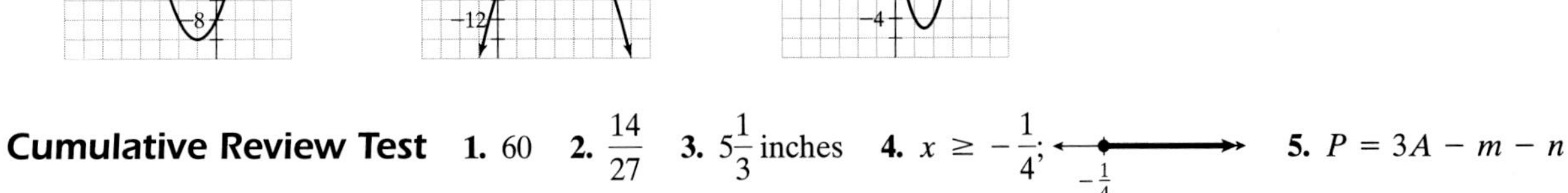 **5.** $P = 3A - m - n$ **6.** $72a^{16}b^{19}$ **7.** $x + 4 - \frac{3}{x+2}$ **8.** $(x-2y)(2x-3y)$ **9.** $3(3x-1)(x-5)$ **10.** $\frac{6a-8}{(a+4)(a-4)^2}$ **11.** 6, 8

12. 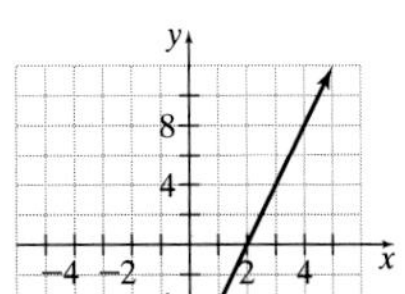

13. $\left(-\frac{12}{7}, -\frac{30}{7}\right)$ **14.** $\frac{y\sqrt{2xy}}{6}$ **15.** $4\sqrt{7}$ **16.** 4 **17.** $\frac{-1+\sqrt{65}}{4}, \frac{-1-\sqrt{65}}{4}$ **18.** 25.6 pounds **19.** Width = 10 feet, length = 27 feet **20.** Walks 3 mph, jogs 6 mph

INDEX

Photo credits:

1. Myrleen Ferguson/PhotoEdit. **6.** Gary Conner/PhotoEdit. **9.** AP/Wide World Photos. **10.** Allen R. Angel. **19.** Uniphoto Stock Agency New York. **57.** James Blank/Stock Boston. **78.** Allen R. Angel. **91.** Andy Lyonc/Allsport Photography (USA), Inc. **140.** Jaime Squire/Allsport Photography (USA), Inc. **146.** Tony Freeman/PhotoEdit. **146.** Y. Gladu/Photo Researchers, Inc. **147.** Allen R. Angel. **147.** Allen R. Angel. **161.** SuperStock, Inc. **165.** Allen R. Angel. **173.** SuperStock, Inc. **174.** Allen R. Angel. **183.** Zefa (UK)/The Stock Market. **184.** Uniphoto Stock Agency New York. **186.** Allen R. Angel. **198.** Allen R. Angel. **199.** Michael Newman/PhotoEdit. **217.** Allen R. Angel. **219.** Uniphoto Stock Agency New York. **221.** Allen R. Angel. **229.** Spencer Grant/PhotoEdit. **230.** Sheila Terry/Science Photo Library/Photo Researchers, Inc. **247.** Jon Feingersh/The Stock Market. **258.** Sylvie Chappaz/Photo Researchers, Inc. **282.** Uniphoto Stock Agency New York. **293.** National Hurricane Center, National Oceanic and Atmospheric Administration. **304.** D. Young-Wolff/PhotoEdit. **310.** Adam Jones/Photo Researchers, Inc. **320.** Mike Mazzaschi/Stock Boston. **320.** National Hurricane Center, National Oceanic and Atmospheric Administration. **325.** Jeff Greenberg/PhotoEdit. **330.** Peter Skinner/Photo Researchers, Inc. **332.** Tom Prettyman/PhotoEdit. **341.** Lawrence Livermore National Laboratory/U.S. Dept. of Energy/IBM/LLNL/DOE. **362.** Fabian Falcon/Stock Boston. **364.** Allen R. Angel. **365.** John Sanford/Science Photo Library/Photo Researchers, Inc. **393.** MCMXCI/Uniphoto Stock Agency New York. **395.** Bruce Plotkin/Liaison Agency, Inc. **409.** Eric Futran/Liaison Agency, Inc. **417.** Keith Wood/Tony Stone Images. **431.** Robert LaBerge/Allsport Photography (USA), Inc. **444.** Stock Boston. **446.** SuperStock, Inc. **447.** Gary Cralle/The Image Bank. **450.** Allen R. Angel. **453.** SuperStock, Inc. **475.** SuperStock, Inc. **498.** Jeremy Horner/Tony Stone Images. **501.** Kerrick James/Tony Stone Images. **502.** Tami Dawson/Photo Resource Hawaii. **508.** Allen R. Angel. **509.** Tom Stewart/The Stock Market. **513.** Allen R. Angel. **514.** Allen R. Angel. **515.** David R. Frazier/David R. Frazier Photolibrary, Inc. **543.** SuperStock, Inc. **553.** David R. Frazier/David R. Frazier Photolibrary, Inc. **553.** Charles Phelps/National Geographic Society. **556.** David Lassman/The Image Works. **556.** Allen R. Angel. **557.** Corbis. **558.** NASA/JPL/Caltech/The Image Works. **568.** J. Marshall/The Image Works. **571.** Alan Becker/The Image Bank. **576.** Reuters/Rafael Perez/Archive Photos. **598.** Srenco Photography. **598.** Allen R. Angel.

Chapter 9 Roots and Radicals

Rules of Radicals

$\sqrt{a} \cdot \sqrt{b} = \sqrt{ab},\ a \ge 0, b \ge 0$

$\sqrt{a^2} = a,\ a \ge 0$

$\sqrt{a^{2n}} = a^n,\ a \ge 0$

$\sqrt[n]{a} = a^{1/n},\ a \ge 0$

$\sqrt[n]{a^m} = \left(\sqrt[n]{a}\right)^m = a^{m/n},\ a \ge 0$

$\dfrac{\sqrt{a}}{\sqrt{b}} = \sqrt{\dfrac{a}{b}},\ a \ge 0, b \ge 0$

A Square Root is Simplified When

1. No radicand has a factor that is a perfect square.
2. No radicand contains a fraction.
3. No denominator contains a square root.

To Solve a Radical Equation

1. Use the appropriate properties to rewrite the equation with the square root term by itself on one side of the equation.
2. Combine like terms.
3. Square both sides of the equation to remove the square root.
4. Solve the equation for the variable.
5. Check the solution in the original equation for extraneous roots.

Pythagorean theorem: $a^2 + b^2 = c^2$

Distance formula: $d = \sqrt{(x_2 - x_1)^2 + (y_2 - y_1)^2}$

Chapter 10 Quadratic Equations

Standard form of quadratic equation: $ax^2 + bx + c = 0,\ a \ne 0$
A quadratic equation can be solved by factoring, completing the square, or by using the quadratic formula.

Square root property: If $x^2 = a$, then $x = \sqrt{a}$ or $x = -\sqrt{a}$ $\left(\text{or } x = \pm\sqrt{a}\right)$

Quadratic formula: $x = \dfrac{-b \pm \sqrt{b^2 - 4ac}}{2a}$

Discriminant: $b^2 - 4ac$

If $b^2 - 4ac > 0$, then the quadratic equation has two distinct real number solutions.
If $b^2 - 4ac = 0$, then the quadratic equation has one real number solution.
If $b^2 - 4ac < 0$, then the quadratic equation has no real number solution.

The graph of $y = ax^2 + bx + c$ will be a parabola with vertex at $\left(-\dfrac{b}{2a}, \dfrac{4ac - b^2}{4a}\right)$, that opens upwards when $a > 0$ and downwards when $a < 0$. The **axis of symmetry** of a parabola is $x = -\dfrac{b}{2a}$

$y = ax^2 + bx + c$

Index of Applications*

* Partial Listing